THE WILEY BICENTENNIAL–KNOWLEDGE FOR GENERATIONS

Each generation has its unique needs and aspirations. When Charles Wiley first opened his small printing shop in lower Manhattan in 1807, it was a generation of boundless potential searching for an identity. And we were there, helping to define a new American literary tradition. Over half a century later, in the midst of the Second Industrial Revolution, it was a generation focused on building the future. Once again, we were there, supplying the critical scientific, technical, and engineering knowledge that helped frame the world. Throughout the 20th Century, and into the new millennium, nations began to reach out beyond their own borders and a new international community was born. Wiley was there, expanding its operations around the world to enable a global exchange of ideas, opinions, and know-how.

For 200 years, Wiley has been an integral part of each generation's journey, enabling the flow of information and understanding necessary to meet their needs and fulfill their aspirations. Today, bold new technologies are changing the way we live and learn. Wiley will be there, providing you the must-have knowledge you need to imagine new worlds, new possibilities, and new opportunities.

Generations come and go, but you can always count on Wiley to provide you the knowledge you need, when and where you need it!

WILLIAM J. PESCE
PRESIDENT AND CHIEF EXECUTIVE OFFICER

PETER BOOTH WILEY
CHAIRMAN OF THE BOARD

Visualizing HUMAN BIOLOGY

Kathleen Anne Ireland

David J. Tenenbaum

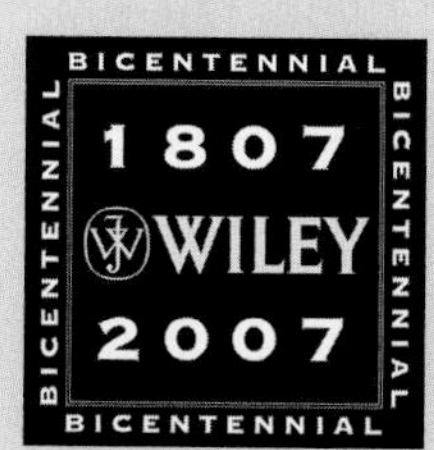

In collaboration with

THE NATIONAL GEOGRAPHIC SOCIETY

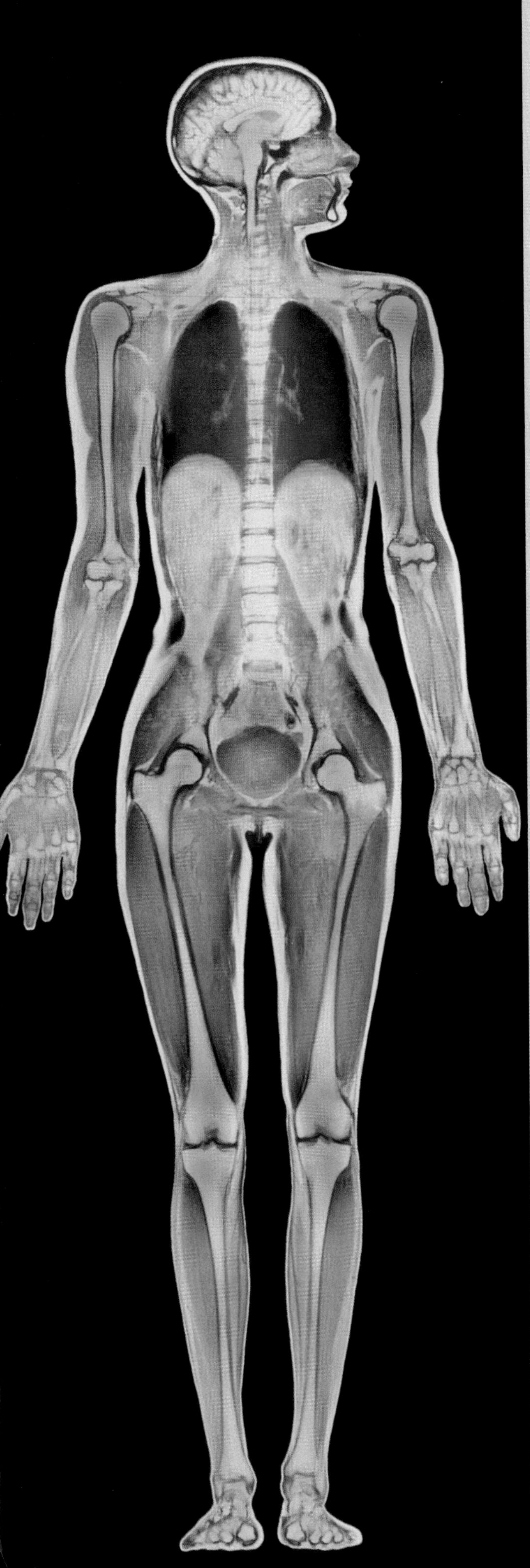

CREDITS

PUBLISHER Kaye Pace
MANAGING DIRECTOR Helen McInnis
EXECUTIVE EDITOR Bonnie Roesch
DIRECTOR OF DEVELOPMENT Barbara Heaney
DEVELOPMENT EDITOR Karen Trost
PROJECT EDITOR Lorraina Raccuia
EDITORIAL ASSISTANT Lauren Morris
EXECUTIVE MARKETING MANAGER Jeffrey Rucker
MARKETING MANAGER Ashaki Charles
SENIOR MEDIA EDITOR Linda Muriello
PRODUCTION MANAGER Kelly Tavares; Full Service Production provided by Camelot Editorial Services, LLC
PRODUCTION ASSISTANT Courtney Leshko
CREATIVE DIRECTOR Harry Nolan
COVER DESIGNER Harry Nolan
INTERIOR DESIGN Vertigo Design
PAGE LAYOUT ARTIST Karin Kincheloe
PHOTO RESEARCHERS Tara Sanford/Mary Ann Price/ Stacy Gold, National Geographic Society
ILLUSTRATION COORDINATOR Sandra Rigby
COVER CREDITS
Main image: Joel Sartore/NG Image Collection
Small images (left to right): Dennis Kunkel/Phototake
Digital Vision/Getty Images
M. Kulyk/Photo Researchers, Inc.
James Cavallini/NG Image Collection
Alfred Pasieka/Photo Researchers, Inc.

This book was set in Times New Roman by Preparé, Inc., and printed and bound by Quebecor World. The cover was printed by Phoenix Color.

Wiley 200th Anniversary logo designed by Richard J. Pacifico.

To order books or for customer service, please call 1-800-CALL WILEY (225-5945).

ISBN 978-047168932-4

Printed in the United States of America

10 9 8 7 6 5 4 3 2 1

FOREWORD *From the Publisher*

Visualizing Human Biology is designed to help your students learn effectively. Created in collaboration with the National Geographic Society and our Wiley Visualizing Consulting Editor, Professor Jan Plass of New York University, *Visualizing Human Biology* integrates rich visuals and media with text to direct students' attention to important information. This approach represents complex processes, organizes related pieces of information, and integrates information into clear representations. Beautifully illustrated, *Visualizing Human Biology* shows your students what the discipline is all about, its main concepts and applications, while also instilling an appreciation and excitement about studying the human body.

Visuals, as used throughout this text, are instructional components that display facts, concepts, processes, or principles. They create the foundation for the text and do more than simply support the written or spoken word. The visuals include diagrams, graphs, photographs, illustrations, schematics, animations, and videos.

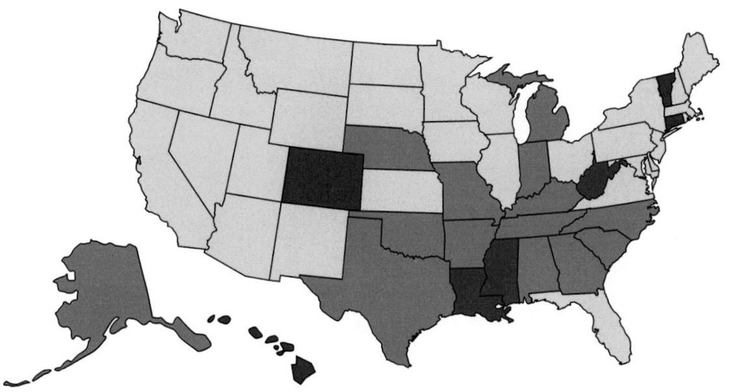

Why should a textbook based on visuals be effective? Research shows that we learn better from integrated text and visuals than from either medium separately. Beginners in a subject benefit most from reading about the topic, attending class, and studying well-designed and integrated visuals. A visual, with good accompanying discussion, really can be worth a thousand words!

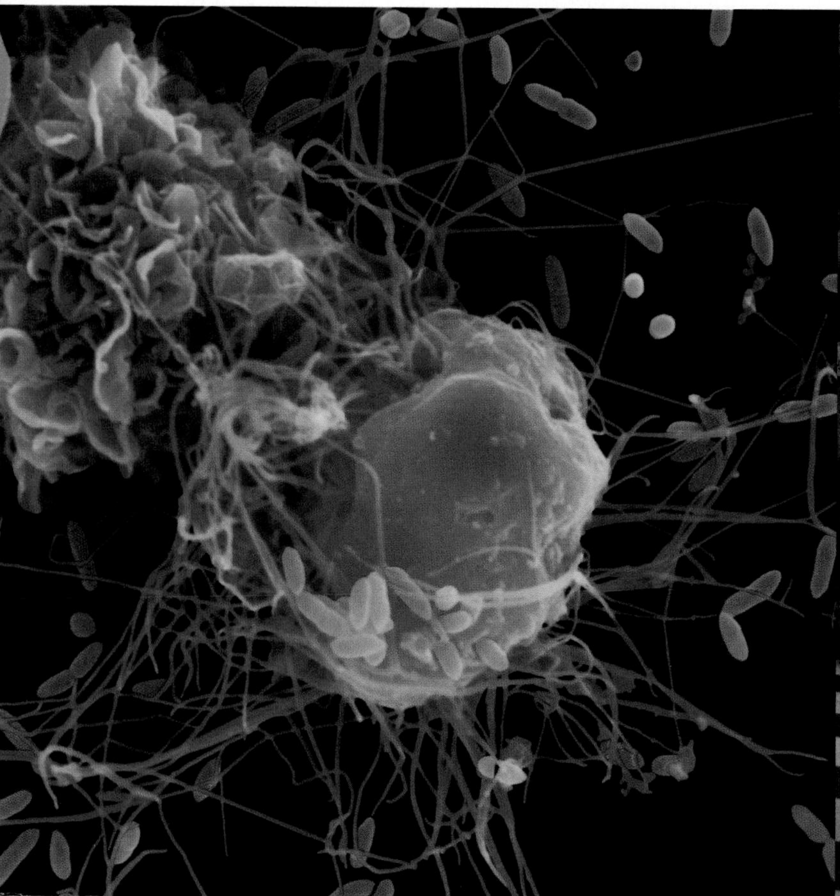

Well-designed visuals can also improve the efficiency with which information is processed by a learner. The more effectively we process information, the more

likely it is that we will learn. This processing of information takes place in our working memory. As we learn we integrate new information in our working memory with existing knowledge in our long-term memory.

Have you ever read a paragraph or a page in a book, stopped, and said to yourself: "I don't remember one thing I just read?" This may happen when your working memory has been overloaded, and the text you read was not successfully integrated into long-term memory. Visuals don't automatically solve the problem of overload, but well-designed visuals can reduce the number of elements that working memory must process, thus aiding learning.

You, as the instructor, facilitate your student's learning. Well-designed visuals, used in class, can help you in that effort. Here are six methods for using the visuals in *Visualizing Human Biology* in classroom instruction.

1. **Assign students to study visuals in addition to reading the text.**
   Instead of assigning only one medium of presentation, it is important to make sure your students know that the visuals are just as essential as the text.

2. **Use visuals during class discussions or presentations.**
   By pointing out important information as the students look at the visuals during class discussions, you can help focus students' attention on key elements of the visuals and help them begin to organize the information and develop an integrated model of understanding. The verbal explanation of important information combined with the visual representation can be highly effective.

3. **Use visuals to review content knowledge.**
   Students can review key concepts, principles, processes, vocabulary, and relationships displayed visually. Better understanding results when new information in working memory is linked to prior knowledge.

4. **Use visuals for assignments or when assessing learning.**
   Visuals can be used for comprehension activities or assessments. For example, students could be asked to identify examples of concepts portrayed in visuals. Higher-level thinking activities that require critical thinking, deductive and inductive reasoning, and

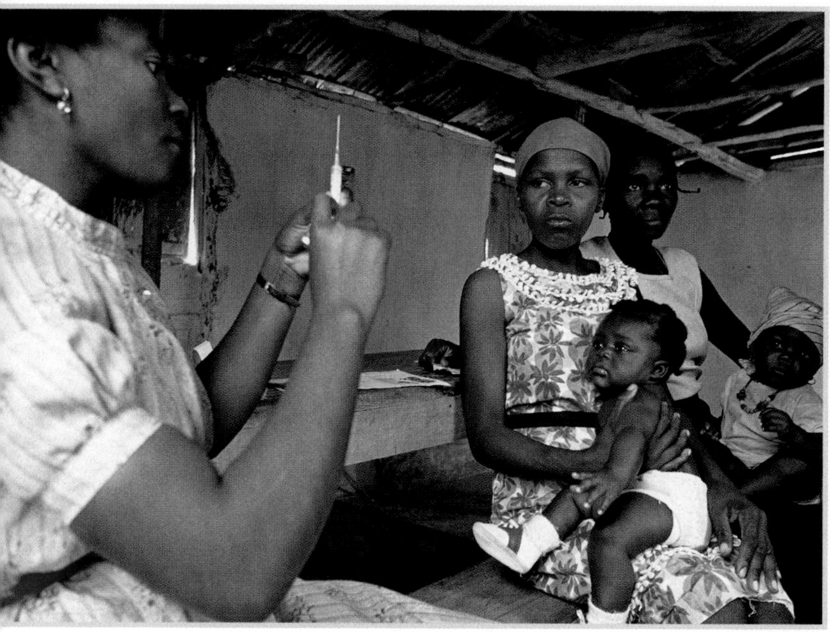

prediction can also be based on visuals. Visuals can be very useful for drawing inferences, for predicting, and for problem solving.

5. **Use visuals to situate learning in authentic contexts.** Learning is made more meaningful when a learner can apply facts, concepts, and principles to realistic situations or examples. Visuals can provide that realistic context.

6. **Use visuals to encourage collaboration.** Collaborative groups often are required to practice interactive processes such as giving explanations, asking questions, clarifying ideas, and argumentation. These interactive, face-to-face processes provide the information needed to build a verbal mental model. Learners also benefit from collaboration in many instances such as decision making or problem solving.

Visualizing Human Biology not only aids student learning with extraordinary use of visuals, but it also offers an array of remarkable photos, media, and film from the National Geographic Society collections. Students using *Visualizing Human Biology* also benefit from the long history and rich, fascinating resources of National Geographic.

National Geographic has also performed an invaluable service in fact-checking *Visualizing Human Biology:* they have verified every fact in the book with two outside sources, ensuring the accuracy and currency of the text.

Given all of its strengths and resources, *Visualizing Human Biology* will immerse your students in the discipline, its main concepts and applications, while also instilling an appreciation and excitement about the study of the human body.

Additional information on learning and instructional design is provided in a special guide to using this book, *Learning from Visuals: How and Why Visuals Can Help Students Learn*, prepared by Matthew Leavitt of Arizona State University. This article is available at the Wiley Web site: www.wiley.com/college/visualizing. The online *Instructor's Manual* also provides guidelines and suggestions on using the text and visuals most effectively.

Visualizing Human Biology also offers a rich selection of visuals in the supplementary materials that accompany the book. To complete this robust package the following materials are available: Test Bank with visuals used in assessment, PowerPoints, Image Gallery to provide you with the same visuals used in the text, interactive animations on the Web, and videos from National Geographic.

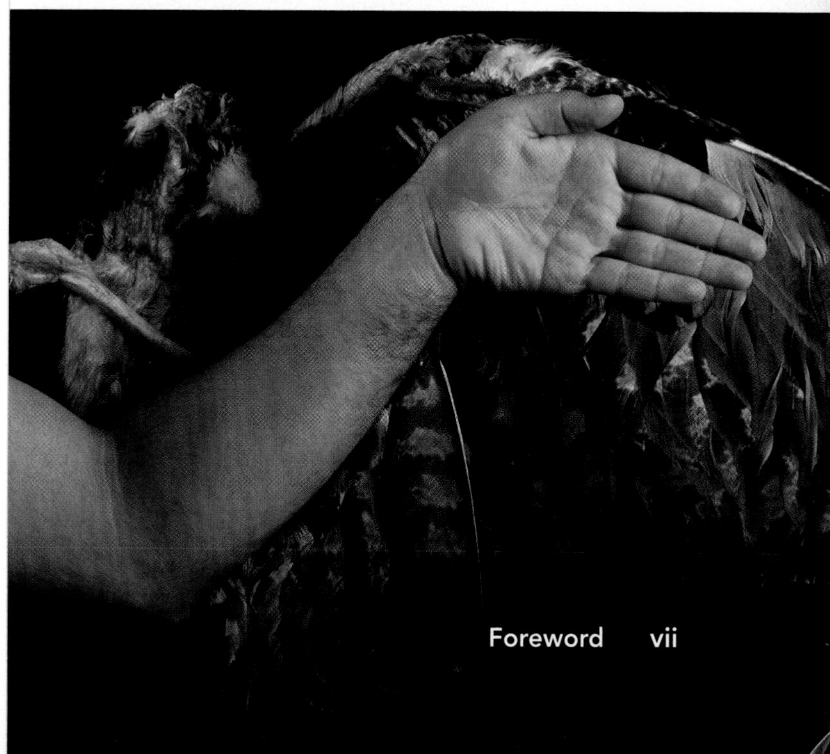

PREFACE

V*isualizing Human Biology* immerses students in the world around them by demonstrating the interaction between humans and the environment, while focusing on the human organism. Both structures and functions of the human body are covered in an engaging, personable style, following the natural curiosity of students. Topics such as evolution, ecology, and chemistry are introduced in a nonthreatening and logical fashion. Designed for introductory courses in human biology, the book answers authentic student questions in each chapter. Intriguing photographs, many from the National Geographic Society's collection, sprinkled throughout the book remind the student that human biology is actually a study of humans in the world. Topics that regularly appear in the media are introduced and explained in the context of human biology.

ORGANIZATION

Any course in human biology must introduce the student to science through a focus on human beings; the authors achieve this by stressing the role of the human in the environment. This theme cements the broad-ranging information in human biology courses together, providing an organizing principle that simultaneously relates human biology to the student's daily experience.

Organized by this theme of "humans in the environment," the chapters are filled with stimulating images and photographs that create natural breaks in the information, allowing the reader to set and achieve attainable study goals. Each chapter begins with an engaging vignette designed to stimulate a desire for more information. The chapters are grouped in larger, cohesive units that ensure the logical flow of information in the entire course.

This text begins the study of life by focusing on the definitive characteristics of life. A general understanding of chemistry, cell biology, and tissues rounds out **Unit 1: Introduction to the Study of Life**. **Unit 2: Movement through the Environment** discusses the human systems involved in movement: the skeletal, muscular, and nervous systems. **Unit 3: Protection from the Environment** describes how the integmentary and lymphatic systems protect the body against injury and invasion by pathogens. **Unit 4: Thriving within the Environment** describes how the cardiovascular and respiratory systems transport nutrients and oxygen to the tissues, and how food is digested and wastes are eliminated. **Unit 5: Populating the Environment** covers the action of the endocrine system, which brings us to sexual maturity and the reproductive system. Fertilization and development complete this unit. Finally, **Unit 6: Adapting to and Affecting the Environment** relates this information to inheritance and DNA, human evolution, and the ecological balance of the biosphere, which serve to tie the entire book together.

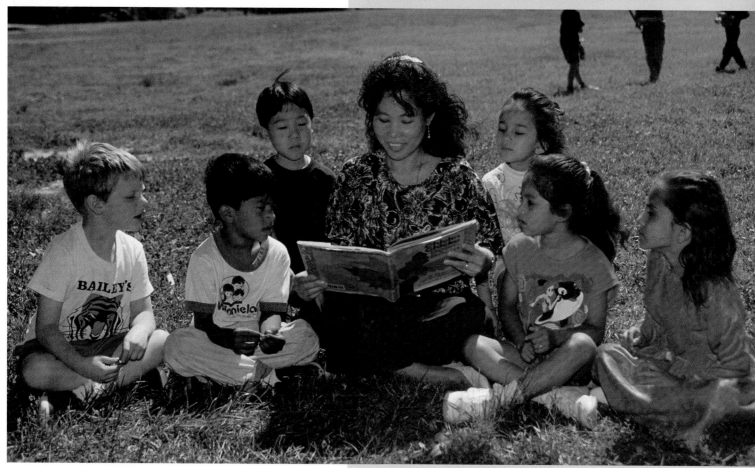

ILLUSTRATED

A number of pedagogical features using visuals have been developed specifically for *Visualizing Human Biology*. This Illustrated Book Tour provides a guide to the pedagogical features of the book.

CHAPTER INTRODUCTIONS Unique chapter openers visually grab the students' attention and connect them with the material that follows, and relate the chapter topic to current events and students' daily lives. For example, Chapter 4: Tissues discusses the production of artificial blood for victims of accidents or wars.

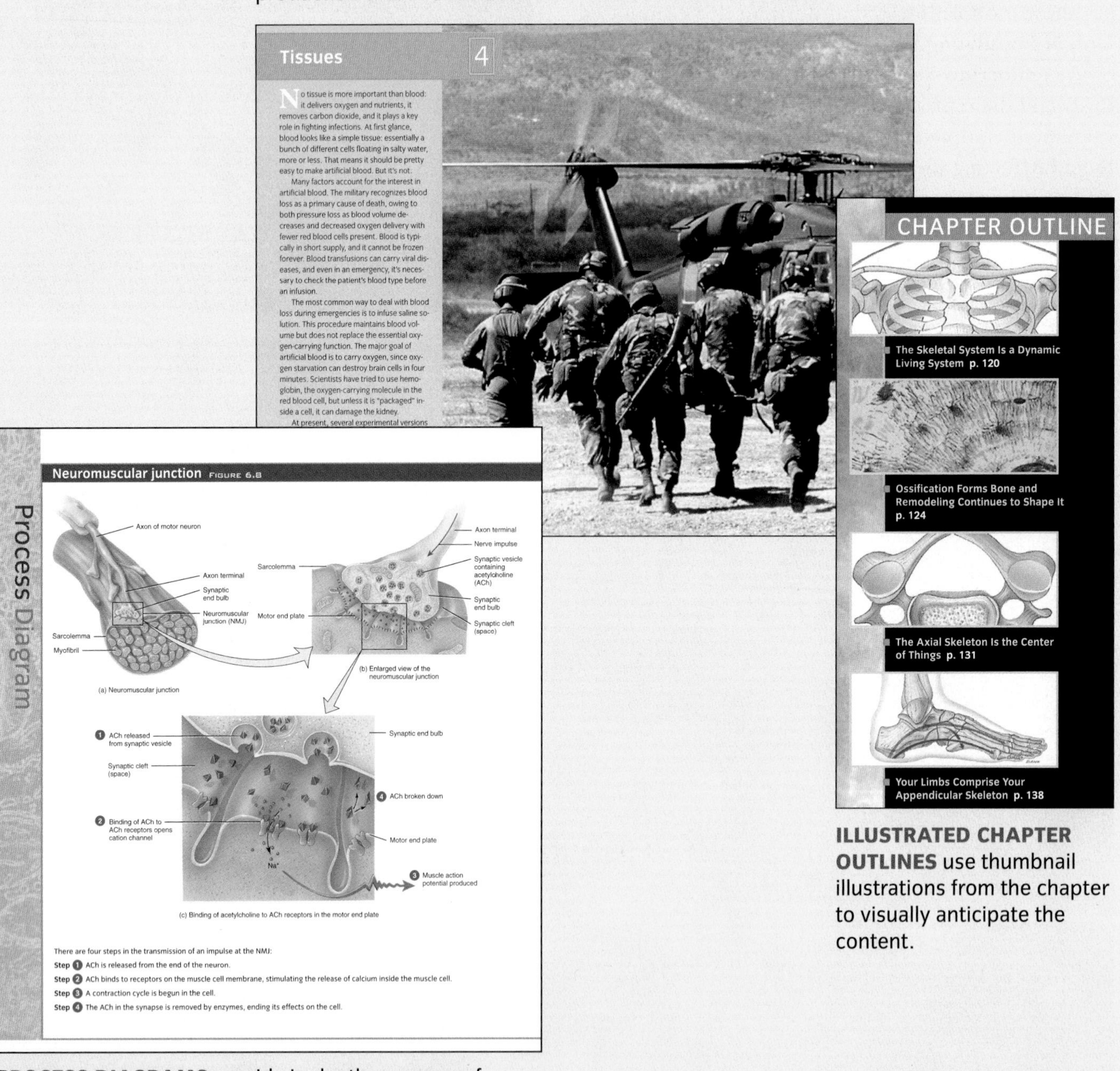

Tissues 4

No tissue is more important than blood: it delivers oxygen and nutrients, it removes carbon dioxide, and it plays a key role in fighting infections. At first glance, blood looks like a simple tissue: essentially a bunch of different cells floating in salty water, more or less. That means it should be pretty easy to make artificial blood. But it's not.

Many factors account for the interest in artificial blood. The military recognizes blood loss as a primary cause of death, owing to both pressure loss as blood volume decreases and decreased oxygen delivery with fewer red blood cells present. Blood is typically in short supply, and it cannot be frozen forever. Blood transfusions can carry viral diseases, and even in an emergency, it's necessary to check the patient's blood type before an infusion.

The most common way to deal with blood loss during emergencies is to infuse saline solution. This procedure maintains blood volume but does not replace the essential oxygen-carrying function. The major goal of artificial blood is to carry oxygen, since oxygen starvation can destroy brain cells in four minutes. Scientists have tried to use hemoglobin, the oxygen-carrying molecule in the red blood cell, but unless it is "packaged" inside a cell, it can damage the kidney.

At present, several experimental versions

CHAPTER OUTLINE

Process Diagram

Neuromuscular junction FIGURE 6.8

There are four steps in the transmission of an impulse at the NMJ:

Step 1 ACh is released from the end of the neuron.

Step 2 ACh binds to receptors on the muscle cell membrane, stimulating the release of calcium inside the muscle cell.

Step 3 A contraction cycle is begun in the cell.

Step 4 The ACh in the synapse is removed by enzymes, ending its effects on the cell.

ILLUSTRATED CHAPTER OUTLINES use thumbnail illustrations from the chapter to visually anticipate the content.

PROCESS DIAGRAMS provide in-depth coverage of physiological processes correlated with clear, step-by-step narrative, enabling students to grasp important concepts with less effort.

BOOK TOUR

"I WONDER . . ." BOXES ask students' everyday questions about life science and the human body. These questions come directly from the students in Kathleen Ireland's classroom. Some examples include: "I Wonder. . . Why do trained athletes have heart attacks?"; "I Wonder. . . What happens when we learn?"; and "I Wonder. . . Why are asthma rates going up?"

I WONDER . . .

Why are asthma rates going up?

Asthma, a constriction of the bronchi that causes wheezing and shortness of breath, has become an epidemic. An estimated 14 to 30 million Americans have asthma, including at least 6 million children, many in inner cities. The disease sends about 500,000 people to the hospital each year. The rate of asthma diagnoses has doubled since the 1980s.

Part of the increase may be due to better diagnosis, but could something else be increasing this disease? The answer would lie in the environmental causes of asthma and/or in the human response to those causes. Asthma can result from exposure to irritants and allergens, including pollen, cockroaches, mold, cigarette smoke, air pollutants, respiratory pathogens, exercise, cold air, and some medicines. Researchers have examined these exposures and found important clues to the asthma epidemic:

- Asthma hospitalizations peak just after school starts in the fall. In a Canadian study, schoolchildren aged 5 to 7 were going to emergency rooms a few days before preschoolers and adults. Of the wheezing schoolchildren, 80 to 85 percent had active rhinovirus (common cold) infections, as did 50 percent of adults. The research suggests that the common cold is spread by children (partly because immune systems are still developing) and that rhinovirus infections may trigger many asthma attacks.
- Poverty and environmental pollution both help to explain why inner-city Americans have such high rates of asthma. One potent asthma allergen is the cuticle (shell) of a cockroach, an insect often found in crowded inner cities. Compounding this problem are the chaotic home lives characteristic of inner-city families, which can also interfere with timely administration of medicines to control asthma symptoms.
- Children who lived on a farm before age 5 have significantly lower rates of asthma, wheezing, and use of asthma medicine, compared to children who live in town. Although allergies play a key role in asthma, the farm children did not have lower rates of hay fever, an allergic reaction to pollen.

None of these studies exactly explains the surge in asthma diagnoses, but the last one does offer a clue. Some scientists suspect that early exposure to dirt and/or infectious disease somehow "tunes" the immune system to reduce the hyperactive reaction that contributes to the inflammation of asthma. Early exposure to endotoxin, a component of the cell wall of gram-negative bacteria, has been associated with low rates of asthma. But the picture is complicated: Endotoxin also inflames lung tissue in healthy people, and some studies have linked it to more wheezing, not less.

With the causes of the asthma epidemic still uncertain, the best take-home message is this: Most cases of asthma are controllable. If you suffer from it, know what triggers your symptoms and take action to reduce your exposure to them. Take your preventative medications as prescribed and get the suggested immunizations to prevent viral infections from triggering attacks.

Ethics and Issues

Attention deficit hyperactivity disorder: does drug treatment make sense?

Attention deficit hyperactivity disorder (ADHD) is one of the most common mental disorders among children. Characteristically, ADHD causes difficulties in concentration, taking directions, sitting still, and cooperating, all of which can lead to learning and social difficulties.

In terms of brain physiology, it is not clear what causes ADHD. Some think ADHD may even be related to sleep deprivation: researchers have found abnormal levels of **sleep apnea** (the periodic cessation of breathing during sleep: *a*− = without, *pnea* = breath) among ADHD children. This breathing problem causes repeated awakenings at night, interfering with deep sleep. If this observation is correct, stimulants could merely be masking a condition of sleepiness that might better be treated more specifically.

Whatever the cause, the diagnosis of ADHD is growing more common. Widely varying statistics show that it affects 1 to 6 percent of American youths. ADHD is also being diagnosed among adults, with an estimated 1 percent of Americans aged 20 to 64 taking stimulants for the condition. Among adults, ADHD is less likely to cause hyperactivity than restlessness, difficulty paying attention, impulsive behavior, and frustration with failing to reach goals.

What can be done to treat ADHD? One approach is behavioral; parents try to shape behavior by rewarding desirable activity and imposing consequences for actions they want to discourage. The behavioral approach can be combined with, or replaced by, treatment with stimulant drugs, especially forms of amphetamine. Curiously, although amphetamines stimulate most people, they calm people with ADHD. This unexpected effect is actually a hallmark of the disease.

Still, the widespread use of prescription medication for ADHD is making some people nervous, especially those who suspect that an ADHD diagnosis is mainly a tactic to make business for psychiatrists and the pharmaceutical industry. These are reasons for concern:

1. Among 12- to 17-year-olds, abuse of prescription drugs is rising faster than abuse of illegal drugs, and amphetamines are addictive in some people.
2. Some college students with ADHD prescriptions say the amphetamines give them extra focus and energy during tests.
3. Stimulants have been linked to the death of 19 children and 6 adults (among an estimated 4 million people taking stimulants for ADHD) due to heart problems that may be related to the stimulants. The U.S. Food and Drug Administration is considering stronger warning labels on the packages. Although some unexplained deaths are inevitable among any group of 4 million people, the news should prompt doctors to evaluate heart health before prescribing stimulants for ADHD.
4. Shouldn't we just "let boys be boys?" According to this logic, boys typically have more of the "ADHD personality characteristics," like impulsivity, excess energy, and difficulty with planning. Should being male be considered a mental illness, especially in a society plagued by drug abuse?

Like other challenges of parenting, ADHD forces parents to persist, improvise, and decide. Behavioral therapy can be wearing, and it may require assistance from teachers and others who are important to the child. Stimulant drugs can send a message that psychological problems can be fixed with a pill. But if the consequences of failing to treat ADHD are negative enough, parents must choose a treatment strategy and philosophy, and carry it through.

Although scientists are improving their understanding of brain function, much remains to be understood, including the integration of different portions of the brain, and the function of various nuclei and neurotransmitters. As neuroscientists probe deeper into the brain's structure and function, we may learn to treat or even prevent some of the severe mental disorders that afflict our fellow humans.

Health, Wellness, and Disease

Lessons of the AIDS epidemic

The AIDS worldwide epidemic, which sprang seemingly from nowhere in the early 1980s, continues to spread around the world. Despite limited progress in some areas, the toll of death and disease remains high. About 40 million people were infected with HIV at the end of 2005, a year that saw 5 million new infections. More than 25 million have died since the epidemic was identified in 1981, and AIDS has slashed life expectancy for those living in southern Africa.

Africa is believed to be where HIV originated. The virus is structurally similar to simian immunodeficiency virus (SIV), which infects our closest relatives, chimpanzees. The virus may have "jumped species" when a hunter ate a chimp infected with a mutated form of SIV that could infect humans, or was bitten by an infected chimp while hunting. A similar danger exists today. Scientists say a deadly avian flu virus may spread from poultry to people, starting a global epidemic of a virus that spreads through the air (see discussion at the beginning of this chapter).

Although East Africa has been considered a focus of AIDS, scientists now concede that previous estimates of infection were overstated because they were based on women who visited prenatal clinics. Because these women were more sexually active than average, their high rates of HIV were not fully representative of all women. Despite the over-reporting, the AIDS epidemic is thriving in southern Africa, where at least 20 percent of pregnant women in six countries carry the virus.

AIDS is spreading in Eastern Europe and the Russian Federation, especially among drug injectors and prisoners. In Asia, 8 million people are infected, including 1 million infected during 2005. In the United States, about 1 million people are living with HIV, and about 42,000 people have AIDS.

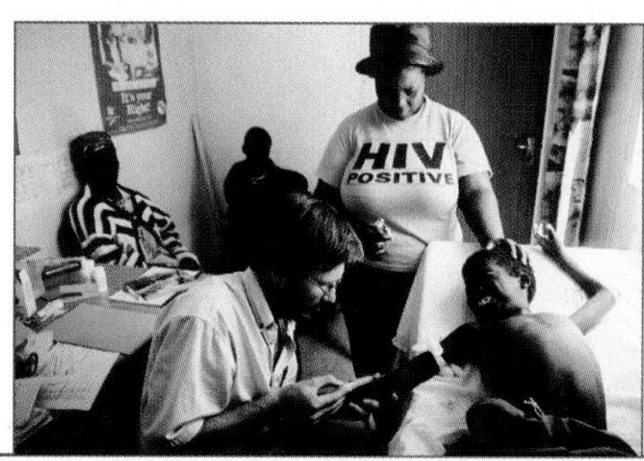

ETHICS AND ISSUES BOXES in each chapter highlight controversial topics in human biology to encourage critical thinking about such current issues as stem cell research, use of prescription drugs to treat youth behavioral problems, the human role in global warming, the use of antibiotics in livestock, eating disorders, and the hazard of bird flu.

HEALTH, WELLNESS, AND DISEASE BOXES provide information and insight on personal and community health. Topics include cancer, HIV/AIDS, fad diets, STDs, fetal alcohol syndrome, genetic engineering, and the human impact on the globe.

OTHER PEDAGOGICAL FEATURES

LEARNING OBJECTIVES located at each section head indicate in behavioral terms what the student must be able to do to demonstrate understanding of chapter concepts.

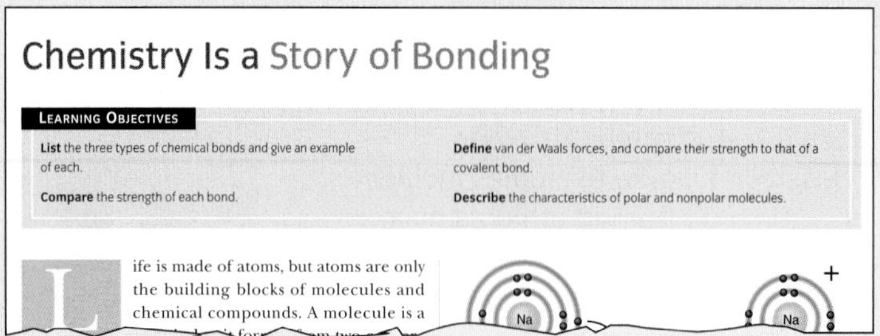

Chemistry Is a Story of Bonding
Learning Objectives
List the three types of chemical bonds and give an example of each.
Compare the strength of each bond.
Define van der Waals forces, and compare their strength to that of a covalent bond.
Describe the characteristics of polar and nonpolar molecules.
Life is made of atoms, but atoms are only the building blocks of molecules and chemical compounds. A molecule is a
Na
Na

CONCEPT CHECK questions at the end of each section allow students to test their comprehension of the learning objectives.

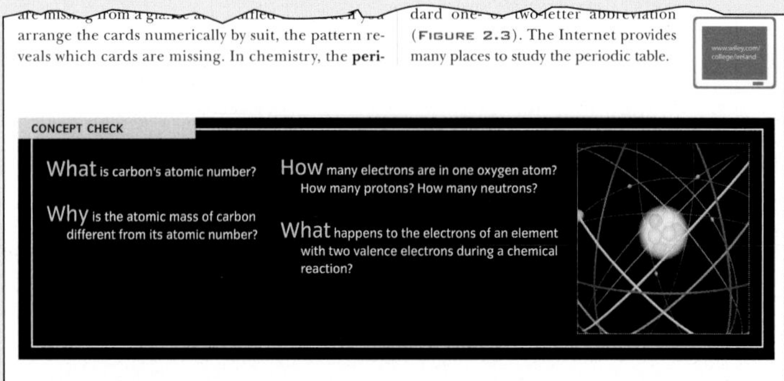

arrange the cards numerically by suit, the pattern reveals which cards are missing. In chemistry, the peri-
dard one- or two-letter abbreviation (Figure 2.3). The Internet provides many places to study the periodic table.
CONCEPT CHECK
What is carbon's atomic number?
Why is the atomic mass of carbon different from its atomic number?
How many electrons are in one oxygen atom? How many protons? How many neutrons?
What happens to the electrons of an element with two valence electrons during a chemical reaction?

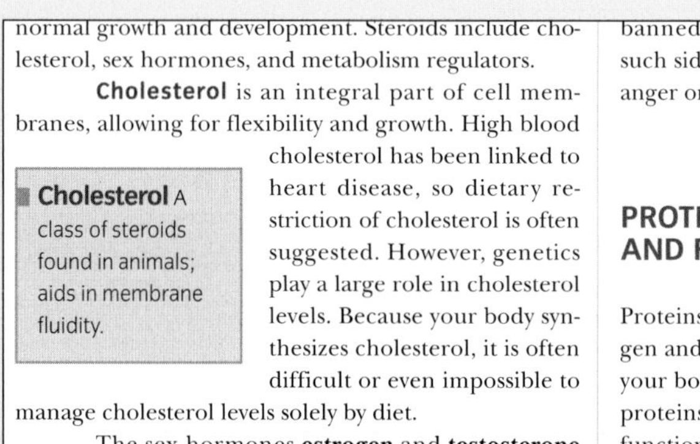

normal growth and development. Steroids include cholesterol, sex hormones, and metabolism regulators.
Cholesterol is an integral part of cell membranes, allowing for flexibility and growth. High blood cholesterol has been linked to heart disease, so dietary restriction of cholesterol is often suggested. However, genetics play a large role in cholesterol levels. Because your body synthesizes cholesterol, it is often difficult or even impossible to manage cholesterol levels solely by diet.
The sex hormones estrogen and testosterone
Cholesterol A class of steroids found in animals; aids in membrane fluidity.

MARGINAL GLOSSARY TERMS (IN GREEN) introduce each chapter's important terms. Other important terms appear in black boldface and are defined in the text.

IMAGES, many originating from National Geographic's rich archives, are used liberally, with captions that are integral to the information presented.

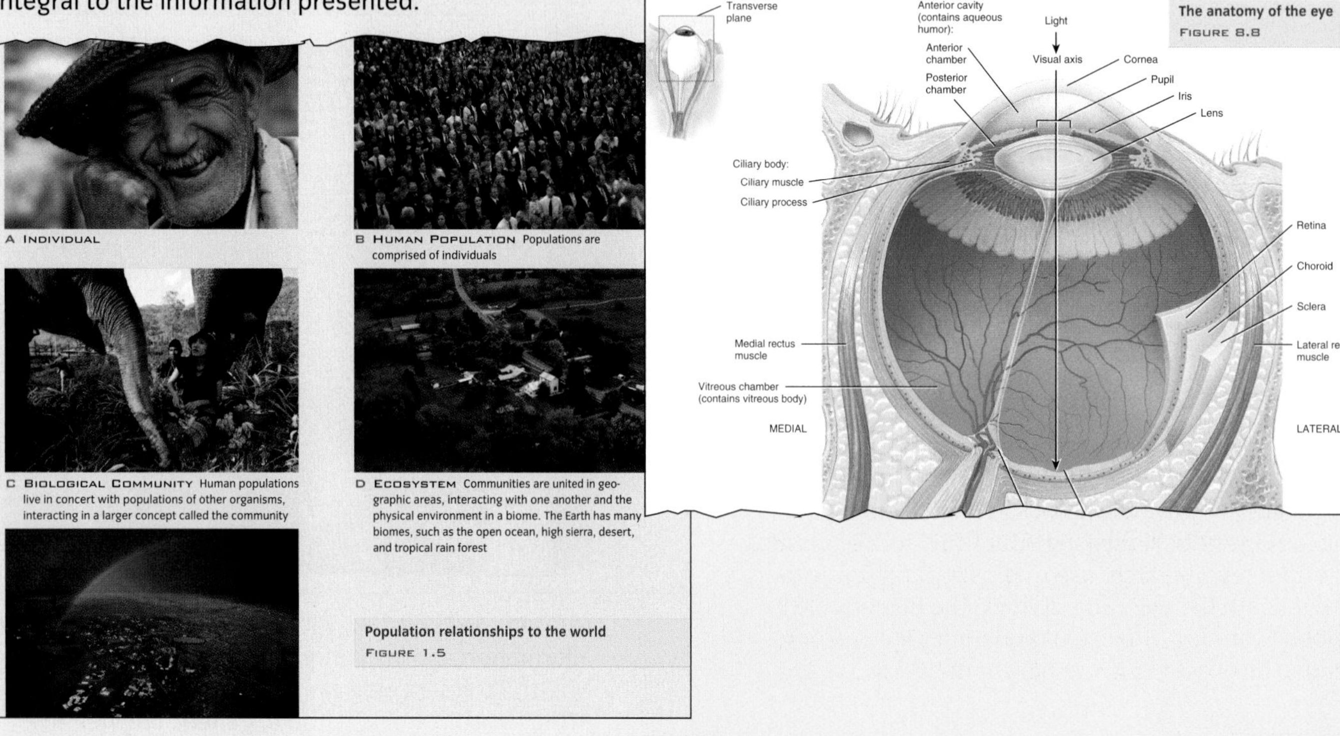

A Individual
B Human Population Populations are comprised of individuals
C Biological Community Human populations live in concert with populations of other organisms, interacting in a larger concept called the community
D Ecosystem Communities are united in geographic areas, interacting with one another and the physical environment in a biome. The Earth has many biomes, such as the open ocean, high sierra, desert, and tropical rain forest
Population relationships to the world
Figure 1.5
The anatomy of the eye
Figure 8.8
Transverse plane
Anterior cavity (contains aqueous humor):
Anterior chamber
Posterior chamber
Light
Visual axis
Cornea
Pupil
Iris
Lens
Ciliary body:
Ciliary muscle
Ciliary process
Retina
Choroid
Sclera
Medial rectus muscle
Vitreous chamber (contains vitreous body)
MEDIAL
LATERAL

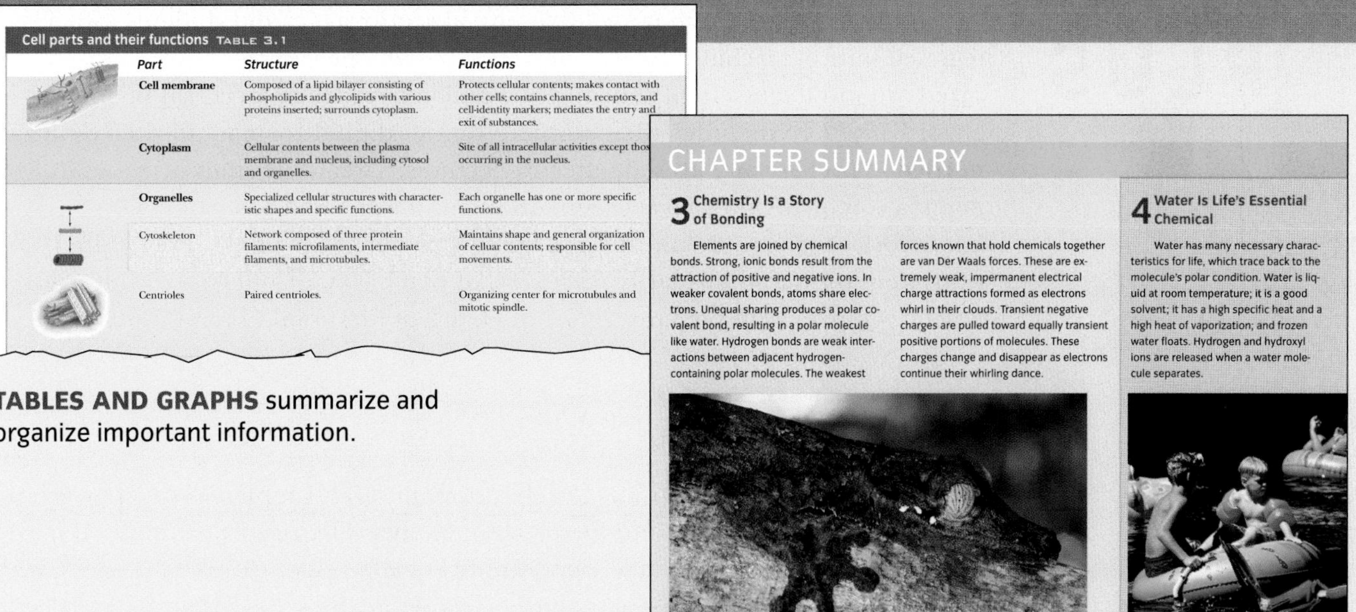

Cell parts and their functions TABLE 3.1

	Part	Structure	Functions
	Cell membrane	Composed of a lipid bilayer consisting of phospholipids and glycolipids with various proteins inserted; surrounds cytoplasm.	Protects cellular contents; makes contact with other cells; contains channels, receptors, and cell-identity markers; mediates the entry and exit of substances.
	Cytoplasm	Cellular contents between the plasma membrane and nucleus, including cytosol and organelles.	Site of all intracellular activities except thos occurring in the nucleus.
	Organelles	Specialized cellular structures with characteristic shapes and specific functions.	Each organelle has one or more specific functions.
	Cytoskeleton	Network composed of three protein filaments: microfilaments, intermediate filaments, and microtubules.	Maintains shape and general organization of celluar contents; responsible for cell movements.
	Centrioles	Paired centrioles.	Organizing center for microtubules and mitotic spindle.

CHAPTER SUMMARY

3 Chemistry Is a Story of Bonding

Elements are joined by chemical bonds. Strong, ionic bonds result from the attraction of positive and negative ions. In weaker covalent bonds, atoms share electrons. Unequal sharing produces a polar covalent bond, resulting in a polar molecule like water. Hydrogen bonds are weak interactions between adjacent hydrogen-containing polar molecules. The weakest forces known that hold chemicals together are van Der Waals forces. These are extremely weak, impermanent electrical charge attractions formed as electrons whirl in their clouds. Transient negative charges are pulled toward equally transient positive portions of molecules. These charges change and disappear as electrons continue their whirling dance.

4 Water Is Life's Essential Chemical

Water has many necessary characteristics for life, which trace back to the molecule's polar condition. Water is liquid at room temperature; it is a good solvent; it has a high specific heat and a high heat of vaporization; and frozen water floats. Hydrogen and hydroxyl ions are released when a water molecule separates.

TABLES AND GRAPHS summarize and organize important information.

The **CHAPTER SUMMARY** revisits each learning objective. Each portion of the Chapter Summary is illustrated with a relevant photo or diagram from its section of the chapter.

KEY TERMS listed at the end of each chapter provide a reference for students.

CRITICAL THINKING QUESTIONS encourage a creative approach to the topic, challenging students to think more broadly about chapter concepts.

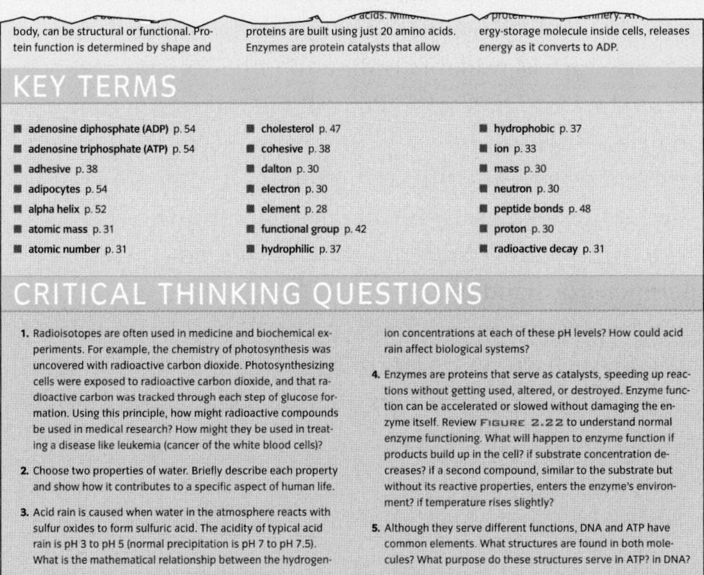

body, can be structural or functional. Protein function is determined by shape and proteins are built using just 20 amino acids. Enzymes are protein catalysts that allow ergy-storage molecule inside cells, releases energy as it converts to ADP.

KEY TERMS

- adenosine diphosphate (ADP) p. 54
- adenosine triphosphate (ATP) p. 54
- adhesive p. 38
- adipocytes p. 54
- alpha helix p. 52
- atomic mass p. 31
- atomic number p. 31
- cholesterol p. 47
- cohesive p. 38
- dalton p. 30
- electron p. 30
- element p. 28
- functional group p. 42
- hydrophilic p. 37
- hydrophobic p. 37
- ion p. 33
- mass p. 30
- neutron p. 30
- peptide bonds p. 48
- proton p. 30
- radioactive decay p. 31

CRITICAL THINKING QUESTIONS

1. Radioisotopes are often used in medicine and biochemical experiments. For example, the chemistry of photosynthesis was uncovered with radioactive carbon dioxide. Photosynthesizing cells were exposed to radioactive carbon dioxide, and that radioactive carbon was tracked through each step of glucose formation. Using this principle, how might radioactive compounds be used in medical research? How might they be used in treating a disease like leukemia (cancer of the white blood cells)?
2. Choose two properties of water. Briefly describe each property and show how it contributes to a specific aspect of human life.
3. Acid rain is caused when water in the atmosphere reacts with sulfur oxides to form sulfuric acid. The acidity of typical acid rain is pH 3 to pH 5 (normal precipitation is pH 7 to pH 7.5). What is the mathematical relationship between the hydrogen-ion concentrations at each of these pH levels? How could acid rain affect biological systems?
4. Enzymes are proteins that serve as catalysts, speeding up reactions without getting used, altered, or destroyed. Enzyme function can be accelerated or slowed without damaging the enzyme itself. Review FIGURE 2.22 to understand normal enzyme functioning. What will happen to enzyme function if products build up in the cell? if substrate concentration decreases? if a second compound, similar to the substrate but without its reactive properties, enters the enzyme's environment? if temperature rises slightly?
5. Although they serve different functions, DNA and ATP have common elements. What structures are found in both molecules? What purpose do these structures serve in ATP? in DNA?

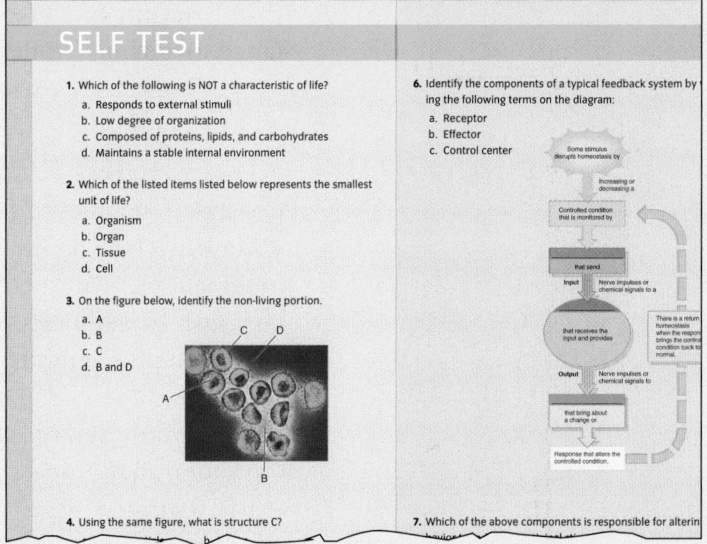

SELF TEST

1. Which of the following is NOT a characteristic of life?
   a. Responds to external stimuli
   b. Low degree of organization
   c. Composed of proteins, lipids, and carbohydrates
   d. Maintains a stable internal environment
2. Which of the listed items listed below represents the smallest unit of life?
   a. Organism
   b. Organ
   c. Tissue
   d. Cell
3. On the figure below, identify the non-living portion.
   a. A
   b. B
   c. C
   d. B and D
4. Using the same figure, what is structure C?
6. Identify the components of a typical feedback system by ing the following terms on the diagram:
   a. Receptor
   b. Effector
   c. Control center
7. Which of the above components is responsible for alterin

A **SELF TEST** at the end of each chapter, which incorporates modified versions of several of the illustrations from the chapter, challenges the students to test their knowledge of the material they have just learned. The answers for the tests are found at the end of the book.

WILEYPLUS

WileyPLUS, a powerful online tool that provides instructors and students with an integrated suite of teaching and learning resources in one easy-to-use Web site, is organized around the essential activities you and your students perform in and out of class. *WileyPLUS* includes the complete text in digital format; all the resources needed to prepare and present effective lectures; a wealth of study activities for students; a complete suite of assessment questions for pre- and post-quizzing, exams, or homework assignments; and student and instructor grade books. Prelecture Questions were prepared by Marilynn Bartels of Black Hawk College and the Student Practice Quizzes were prepared by Keith Hench at Kirkwood Community College.

ANCILLARY SUPPORT

www.wiley.com/college/Ireland

To support the printed text, *Visualizing Human Biology* is supplemented by a rich resource of additional visual and media components. The concepts discussed in the text are supported by National Geographic Society video lecture-launchers, with content from the National Geographic Society, animations, PowerPoints, and a complete image bank.

VIDEOS

A rich collection of videos from the award-winning National Geographic Film Collection have been selected to accompany and enrich the text. Each unit includes at least one video clip, available online as digitized streaming video, to illustrate and expand on a concept or topic to aid student understanding.

Accompanying each video are appropriate commentary and questions designed to enhance student understanding.

The videos are available on the main Web site: www.wiley.com/college/ireland

POWERPOINT PRESENTATIONS AND IMAGE GALLERY

A complete set of dynamic, colorful Lecture PowerPoints prepared by Aimee T. Lee of the University of Southern Mississippi is available online to enhance classroom presentations. With specially created lecture notes, the PowerPoints allow you to use the visuals from the book during class discussions.

All photographs, figures, and other visuals from the text are online for your use in the classroom. These digital files can easily be incorporated into your own PowerPoint or SmartBoard (and Symposium) presentations, or can be converted into your own overhead transparencies or handouts.

As noted, using visuals in class discussions allows you to focus attention on key elements of the visuals so that students can organize the information and develop greater understanding. The classroom explanation of important information combined with the visual representation is highly effective.

TEST BANK (AVAILABLE IN WILEYPLUS AND ELECTRONIC FORMAT)

The visuals from the textbook are also carried through in the Test Bank, which was prepared by Corey Paulin of Allegany College of Maryland.

The test generation program has approximately 50 test items per chapter, with at least 20 percent incorporating visuals from the book. The test items include multiple choice, true/false, and short-answer questions.

INSTRUCTOR'S RESOURCES

The Instructor's Resources begins with a special introduction on *Using Visuals in the Classroom*, prepared by Matthew Leavitt of Arizona State University, which provides guidelines and suggestions on using visuals to teach the course. Kathleen Ireland has carefully prepared Instructor's Resources that will further help instructors use visuals and prepare interesting lectures. For each chapter, she has prepared:

- Guide to Using Video Lecture Launchers
- Chapter Objectives
- Teaching Tips
- Additional Activities
- Answers to Critical Thinking Questions

ANIMATIONS www.wiley.com/college/Ireland

A robust suite of multimedia learning resources have been designed for *Visualizing Human Biology*. Red Web icons throughout the text direct students to animations or additional Web content. Animations visually support the learning of a difficult concept or process, many of them built around a specific feature such as a Process Diagram, or key visual in the chapter. The animations go beyond the content and visual presented in the book, providing students with the opportunity to interact with the animation by completing activities.

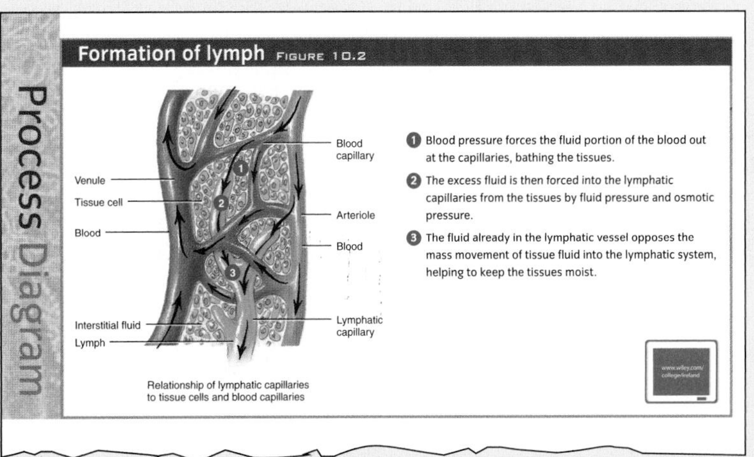

ACKNOWLEDGMENTS

This book would not have come about without the tremendous work, creative energy, and tireless support of the Wiley team. Kathleen especially thanks Executive Editor Bonnie Roesch, who took a chance on a new author, giving her the opportunity to share her enthusiasm for the subject and her unconventional teaching style by putting it to paper. Kathleen and David sincerely appreciate Bonnie's bravery and strong support of this project from start to finish. Special thanks to Kaye Pace, Executive Publisher, who oversaw the project and thanks to Barbara Heaney, Director of Development for all her guidance. Jeffrey Rucker, Marketing Manager for the Visualizing Series, and Ashaki Charles, Marketing Manager for Biology, also helped make this book a success. Thanks to Mary O'Sullivan, Development Editor, who helped us start this project. Perhaps most responsible for shaping this book and handling the recurrent crises with calm professionalism and a touch of humor was Project Editor Lorraina Raccuia. Production Manager Kelly Tavares, Christine Cervoni of Camelot Editorial Services, and Development Editor Karen Trost all spent countless hours on the project, ensuring that the tone and attitude of the text mirrored our original idea. The amazing images and illustrations reflect the patient efforts of photo editors Mary Ann Price and Tara Sanford, as well as Sandra Rigby, Illustration Coordinator. We love the look of the book—thank you, Hope Miller, for designing such a delightful cover. The interior is equally engaging, due to the design of Harry Nolan, Creative Director. We appreciate the talents of our page layout artist, Karim Kinchelor, who laid out the pages to be visually attractive and pedagogically effective, creating a student-friendly style. Media is a large part of this text, and Linda Muriello, Senior Media Editor, did a fantastic job knitting it together. Mahalo nui loa. Thanks to all!

CLASS TESTING AND STUDENT FEEDBACK

In order to make certain that *Visualizing Human Biology* met the needs of current students, we asked several instructors to class test a chapter. The feedback that we received from students and instructors confirmed our belief that the visualizing approach taken in this book is highly effective in helping students to learn. We wish to thank the following instructors and their students who provided us with helpful feedback and suggestions:

Christine Barrow, Prince George's Community College
Robert Chesney, William Paterson University
William Cushwa, Clark College
Alison Elgart, Florida Gulf Coast University
Sheldon R. Gordon, Oakland University
Nancy Mann, Cuesta College
Lisa Maranto, Prince George's Community College
Erin Morrey, Georgia Perimeter College
Polly Phillips, Florida International University
Melody Ricci, Victor Valley College
Beverly A. Schieltz, Wright State University
Alicia Steinhardt, Hartnell Community College

PROFESSIONAL FEEDBACK

Throughout the process of developing the concept of visual pedagogy for Wiley's Visualizing textbooks, including *Visualizing Human Biology* and others, we benefited from the comments and constructive criticism provided by the instructors and colleagues listed below. We offer our sincere appreciation to these individuals for their helpful reviews:

Loren Ammerman, Angelo State University
Curt Anderson, Idaho State University
Bert Atsma, Union County College
Thomas Bahl, Aquinas College
Jamie M. Chapman, Central Community College
Elizabeth A. Cowles, Eastern Connecticut State University
Gerard Cronin, Salem Community College
Michael Davis, Central Connecticut State
Ellie Dias, Westfield State College
James Dunn, Laredo Community College
Linda Celia Ellis, University of Toronto at Scarborough
Steve Fields, Winthrop University
Heidi L. Forman, Buffalo State University
Mary Louise Greeley, Salve Regina University
Robert Greene, Niagara University
Gretel Guest, Alamance Community College
Martin E. Hahn, William Paterson University
Richard R. Jurin, University of Northern Colorado
Martin A. Kapper, Central Connecticut State University
Jonathan Karp, Rider University
Pushkar N. Kaul, Clark Atlanta University
Johanna Kruckeburg, Kirkwood Community College
M. Leal, Sacred Heart University
Marty Lowe, Bergen Community College
Carol Mack, Erie Community College
Nancy Jean Mann, Cuesta College
James C. Marker, University of Wisconsin
Kelly Murray, University of Wisconsin
Corey Paulin, Allegany College of Maryland
Polly K. Phillips, Florida International University
Mary C. Reese, Mississippi State University
Melody L. Ricci, Victor Valley College
Alexandra P. Robins, Ivy Tech Community College
April Rottman, Rock Valley College
Nahed Salama, Rockland Community College
D. Mike Satterwhite, Lewis-Clark State College
Sr. Donna Schroeder, College of St. Scholastica
Harry A. Schutte, Ohio University
Charlie Shaeff, Central Virginia Community College
Barkur S. Shastry, Oakland University
Beverly Shieltz, Wright State University
Patricia Singer, Simpson College
Gregory Smutzer, Temple University
Ruth Sporer, Rutgers University
David Tauck, Santa Clara University
James Thompson, Austin Peay State University
Judith Valeras, University of Northern Colorado
Suzanne M. Walsh, Westfield State College
Murray Weinstein, Erie Community College
Emily Williamson, Mississippi State University
Richard Worthington, University of Texas

FOCUS GROUP AND TELESESSION PARTICIPANTS

A number of professors and students participated in focus groups and telesessions, providing feedback on the text, visuals, and pedagogy in *Visualizing Human Biology* and textbooks for other academic disciplines. Our thanks to the following participants for their helpful comments and suggestions.

Sylvester Allred, *Northern Arizona University;* David Bastedo, *San Bernardino Valley College;* Ann Brandt-Williams, *Glendale Community College;* Natalie Bursztyn, *Bakersfield College;* Stan Celestian, *Glendale Community College;* O. Pauline Chow, *Harrisburg Area Community College;* Diane Clemens-Knott, *California State University, Fullerton;* Mitchell Colgan, *College of Charleston;* Linda Crow, *Montgomery College;* Smruti Desai, *Cy-Fair College;* Charles Dick, *Pasco-Hernando Community College;* Donald Glassman, *Des Moines Area Community College;* Mark Grobner, *California State University, Stanislaus;* Michael Hackett, W*estchester Community College;* Gale Haigh, *McNeese State University;* Roger Hangarter, *Indiana University;* Michael Harman, *North Harris College;* Terry Harrison, *Arapahoe Community College;* Javier Hasbun, *University of West Georgia;* Stephen Hasiotis, *University of Kansas;* Adam Hayashi, *Central Florida Community College;* Laura Hubbard, *University of California, Berkeley;* James Hutcheon, *Georgia Southern University;* Scott Jeffrey, *Community College of Baltimore County, Catonsville Campus;* Matther Kapell, *Wayne State University;* Arnold Karpoff, *University of Louisville;* Dale Lambert, *Tarrant County College NE;* Arthur Lee, *Roane State Community College;* Harvey Liftin, *Broward Community College;* Walter Little, *University at Albany, SUNY;* Mary Meiners, *San Diego Miramar College;* Scott Miller, *Penn State University;* Jane Murphy, *Virginia College Online;* Bethany Myers, *Wichita State University;* Terri Oltman, *Westwood College;* Keith Prufer, *Wichita State University;* Ann Somers, *University of North Carolina, Greensboro;* Donald Thieme, *Georgia Perimeter College;* Kip Thompson, *Ozarks Technical Community College;* Judy Voelker, *Northern Kentucky University;* Arthur Washington, *Florida A&M University;* Stephen Williams, *Glendale Community College;* Feranda Williamson, *Capella University.*

For my heart and soul: "the G-bud" and "baby M," Gregory and Marcus Tatum.

Kathleen Anne Ireland

For Frances Tanenbaum: Always a writer, always an editor.

David J. Tenenbaum

CONTENTS *in Brief*

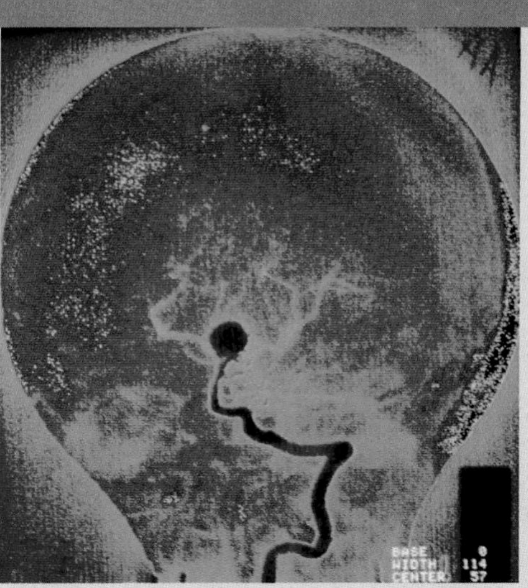

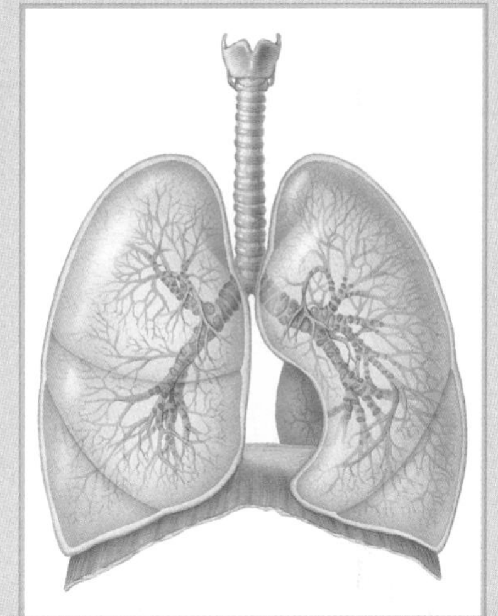

CONTENTS

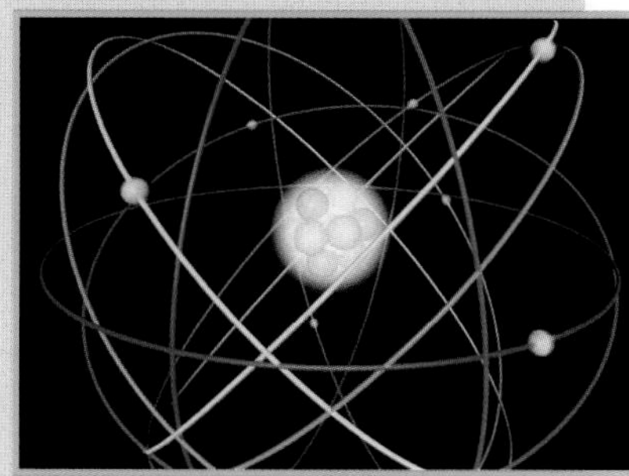

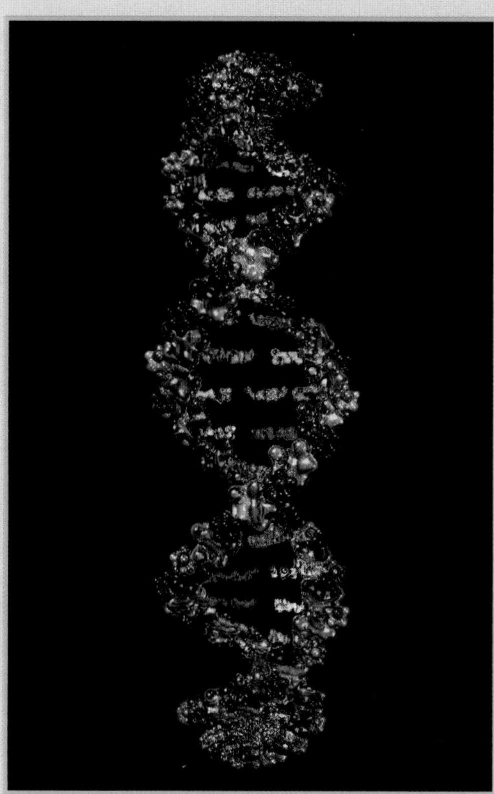

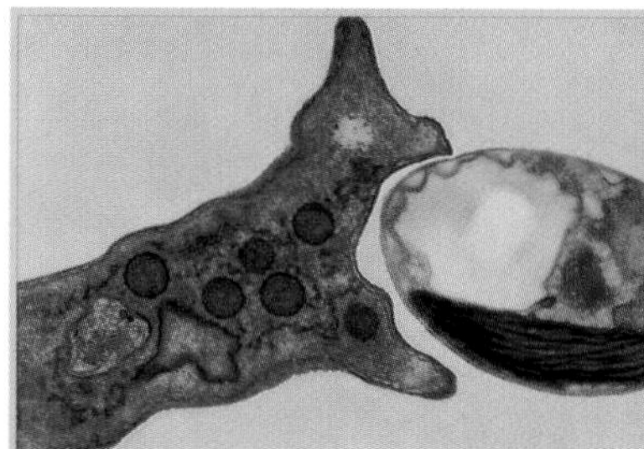

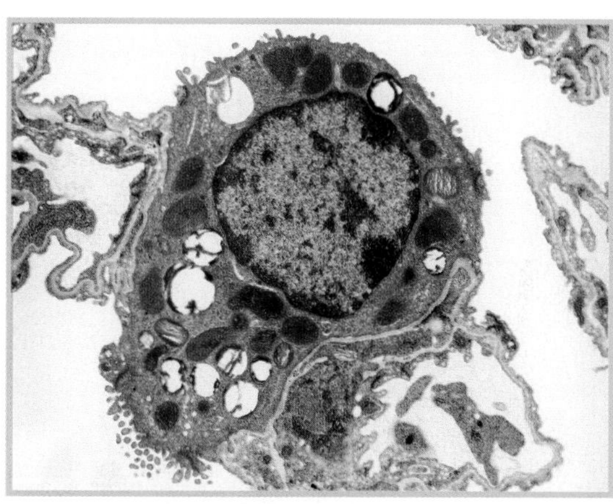

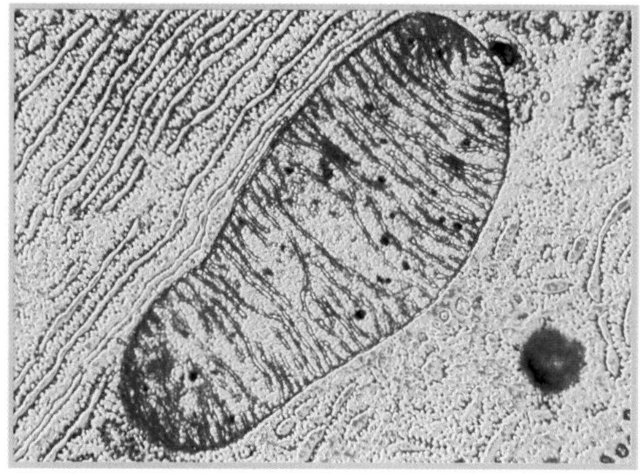

NATIONAL GEOGRAPHIC

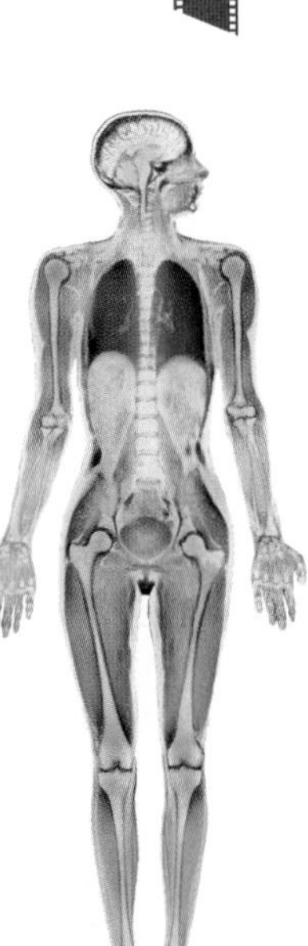

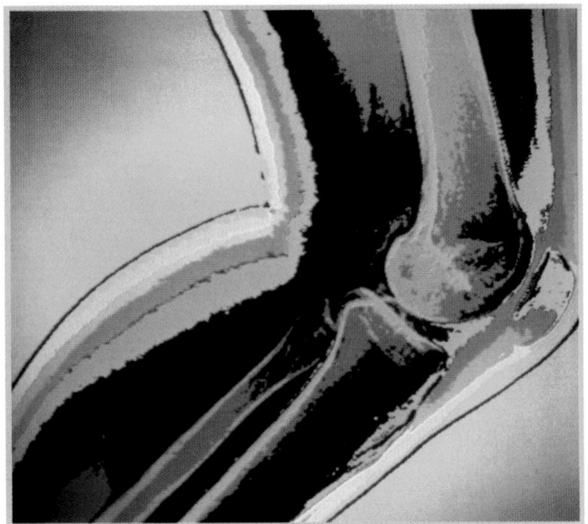

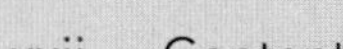

6 The Muscular System 156

7 The Nervous System 186

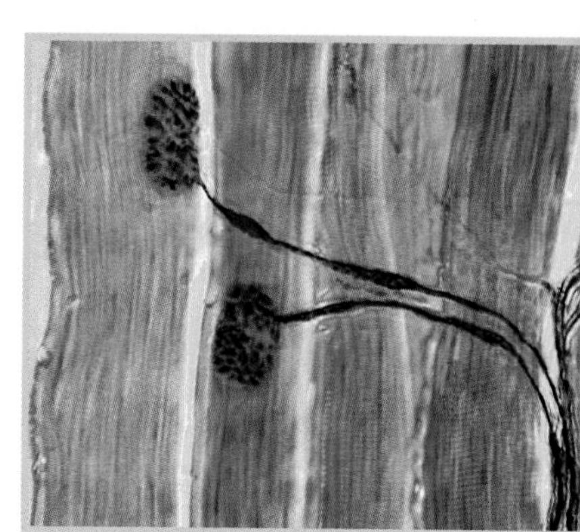

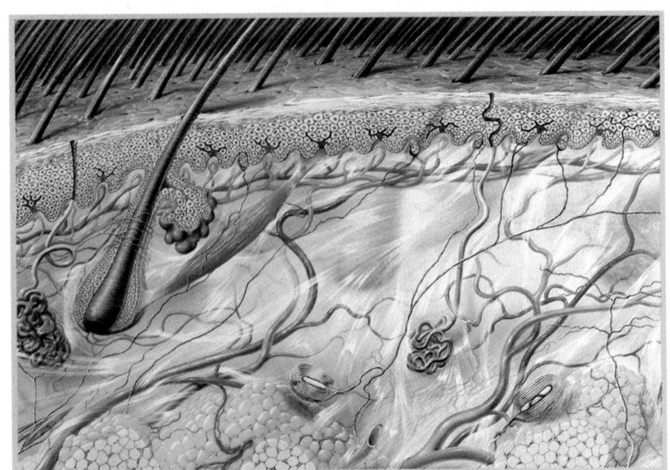

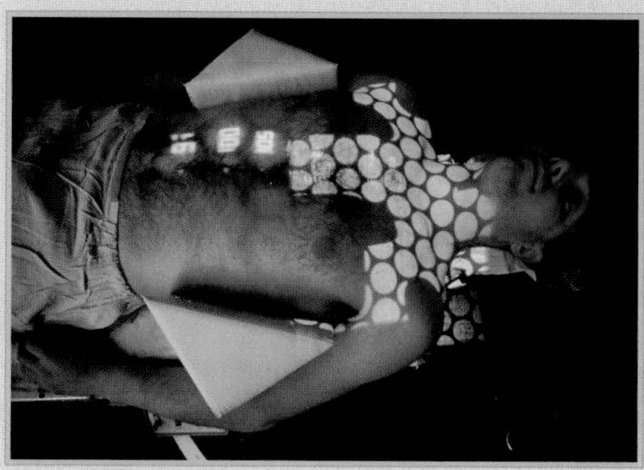

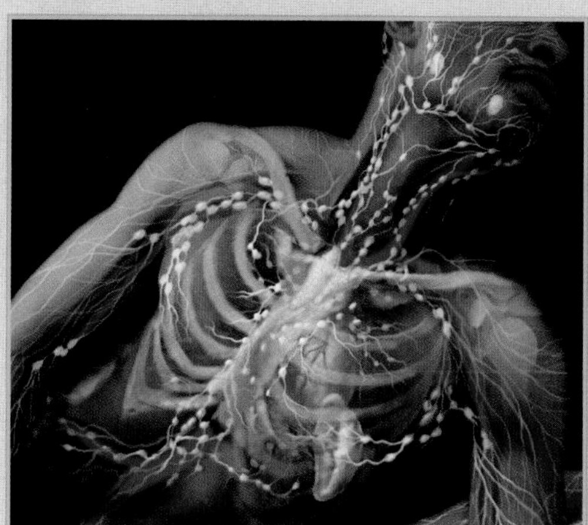

NATIONAL GEOGRAPHIC

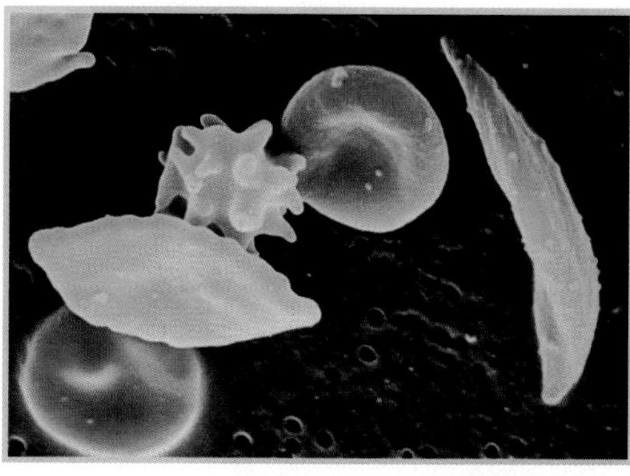

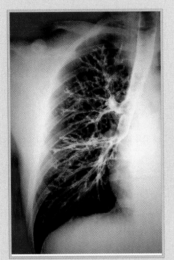

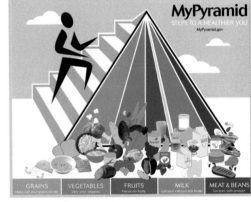

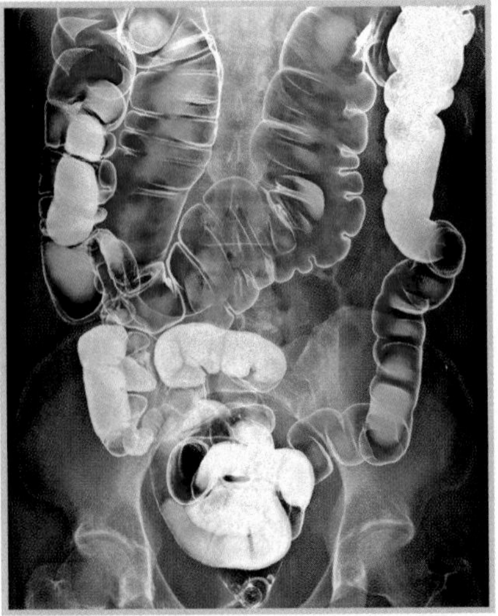

GEORGE'S

STOP GMOs
STOP
PROTOCOLE SUR LA BIOSÉCURITÉ
MIEUX VAUT PRÉVENIR QUE GUÉRIR
PRINCIPE PRECAUTION
DDT C'ETAIT
OGM
YES

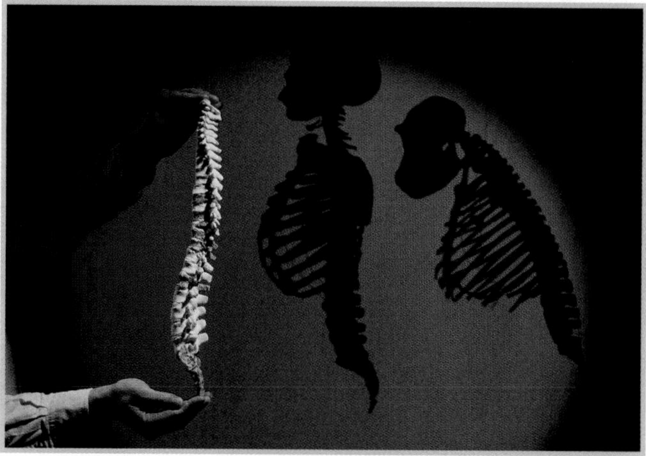

New York's Only
5-Star Bank
GUY & GALLARD

FOR INSTRUCTORS

Wiley**PLUS** is built around the activities you perform in your class each day. With Wiley**PLUS** you can:

Prepare & Present

Create outstanding class presentations using a wealth of resources such as PowerPoint™ slides, image galleries, interactive simulations videos, and more. You can even add materials you have created yourself.

Create Assignments

Automate the assigning and grading of homework or quizzes by using the provided question banks, or by writing your own.

Track Student Progress

Keep track of your students' progress and analyze individual and overall class results.

Now Available with WebCT and Blackboard!

"It has been a great help, and I believe it has helped me to achieve a better grade."

Michael Morris,
Columbia Basin College

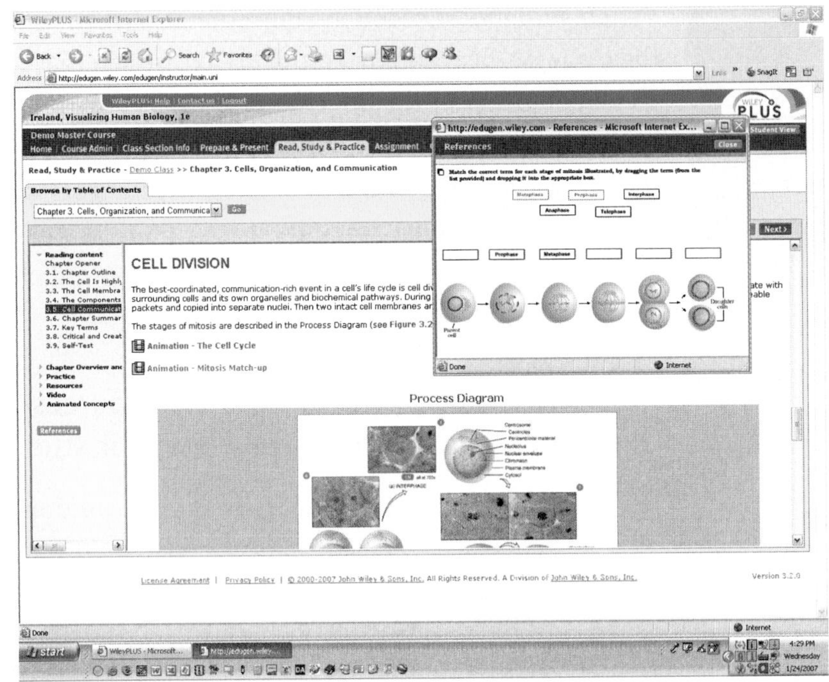

FOR STUDENTS

You have the potential to make a difference!

WileyPLUS is a powerful online system packed with features to help you make the most of your potential and get the best grade you can!

With Wiley**PLUS** you get:

- A complete online version of your text and other study resources.
- Problem-solving help, instant grading, and feedback on your homework and quizzes.
- The ability to track your progress and grades throughout the term.

For more information on what *WileyPLUS* can do to help you and your students reach their potential, please visit www.wiley.com/college/*wileyplus*.

76% of students surveyed said it made them better prepared for tests. *

*Based on a survey of 972 student users of *WileyPLUS*

Visualizing
HUMAN BIOLOGY

What Is Life?

1

The day begins with the startling blare of the clock radio. Instinctively your hand fumbles for the snooze button, seeking a few extra moments of peace in the warm bed. You eventually arise, don some clothes, and stumble toward the kitchen to get some food. If you are like many older students returning to college, you stop to wake your children for school. Watching your children haphazardly prepare for their day, you start thinking about your own. Oh, it is a big day! Today your Human Biology class begins. What will the class cover? Will the material be fun to learn? The feeling was subtle, but was that a twinge of anxiety?

What is human biology? Human biology is the study of life, starting from the human experience. The field covers the anatomy and physiology of the human body, as well as the back-and-forth interplay between humans and the environment. If you, like your classmates, are indeed human, this commonality will be used to create a foundation for building an understanding of the broader issues of biology.

As you delve into this area of study, you'll look at what makes something alive, how scientists see the world, and how the scientific approach can help you become a better citizen. You will also look at how the body moves through the environment and how organ systems can fail and cause disease. Finally, you will look at how an individual organism protects itself within the environment and at how populations of organisms cooperate or compete in their ecosystems. The beauty of studying biology from a human perspective is that it guides you through the entire living realm starting from your own experience.

CHAPTER OUT

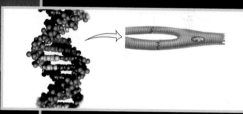

Living Organisms Display Nine Specific Characteristics

LEARNING OBJECTIVES

List the characteristics of life.

Define homeostasis and relate it to the study of life.

Reflect again on the start of your day. It has just demonstrated many of the characteristics of life (TABLE 1.1). Several appeared during your first minutes of awakening. Life is defined by the ability to **respond to external stimuli** (remember waking to the alarm?). Objects that are alive can **alter their environment**, as you did by silencing the dreadful noise. You **sensed your environment** when you felt the chill of the morning, then you **adapted to your environment** by covering yourself with clothes to maintain your internal temperature. Living things **require energy**, which plants get by synthesizing compounds using solar power and which animals get by ingesting nutrients, aka breakfast. All of us are proof that living organisms **reproduce**. On the average foggy-headed morning, you doubtless failed to notice three other characteristics of life: (1) life is composed of **materials found only in living objects** (your body contains proteins, lipids, carbohydrates, and nucleic acids: DNA and RNA); (2) living organisms maintain a stable internal environment, a property called **homeostasis**; and (3) life exhibits a **high degree of organization**, which extends from microscopic units, called **cells**, into increasingly complex tissues, **organs**, **organ systems**, and individual **organisms**. Not all organisms display these nine characteristics, leading to confusion as to their classification. (See I Wonder. . . Are viruses considered living organisms?)

Cell The smallest unit of life, contained in a membrane or cell wall.

Organ A structure composed of more than one tissue having one or more specific functions.

Organ system A group of organs that perform a broad biological function, such as respiration or reproduction.

Organism One living individual.

Characteristics of life TABLE 1.1

Characteristics of life
Respond to external stimuli
Alter the environment
Sense the environment
Adapt to the environment
Use energy
Reproduce
Contain materials found only in living organisms
Maintain a constant internal environment (homeostasis)
Have a high degree of organization

I WONDER . . .

Are viruses considered living organisms?

Viruses are among the smallest agents that can cause disease, and they cause some of the worst diseases around. Scientists think that smallpox, caused by the *variola* virus, killed more people in the past few centuries than all wars combined. HIV, the human immunodeficiency virus, causes AIDS, whose death toll continues to mount year after year.

Because viruses are less than 1 micron (millionth of a meter) across, they were not discovered until early in the nineteenth century. Viruses are much smaller than bacteria, which are single-celled organisms that are truly alive.

We know viruses can kill. To determine whether they are alive, we refer to the required characteristics of life, and we observe that viruses lack many of them, such as:

- cells (viruses are basically a protein coat surrounding a few genes, made of either DNA or RNA);
- the ability to reproduce;
- the ability to metabolize or respire; and
- a mechanism to store or process energy.

Viruses can reproduce but only if they can slip inside a host cell and seize control of its internal machinery. Viruses are more complex than prions, the distorted proteins that cause bovine spongiform encephalopathy—mad cow disease. However, viruses are far simpler than even a bacterial cell. So although viruses are not alive, they are the ultimate parasite.

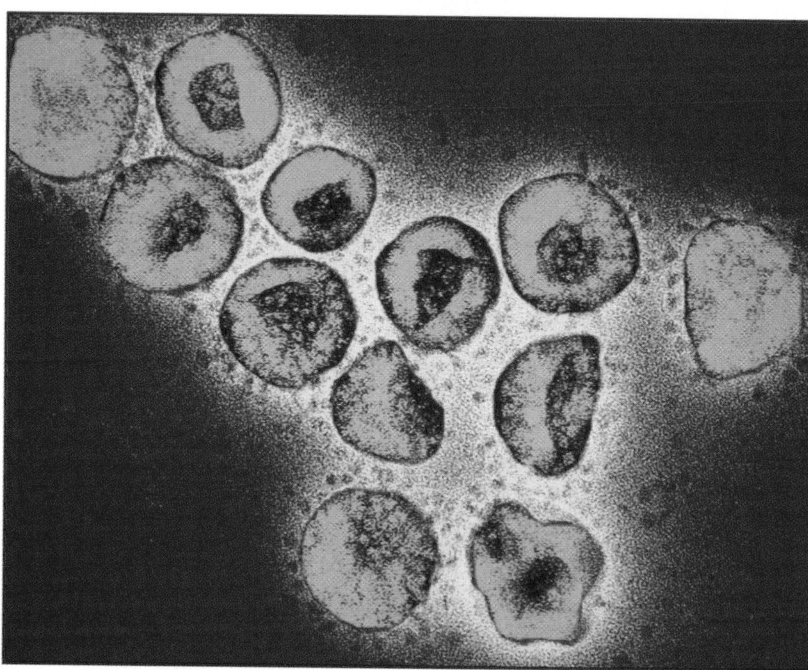

The colorized blue cells in this photograph are surrounded by very small, circular viral particles. The tremendous size difference between typical cells and viruses is evident here. The picture shows the Corona virus, the cause of the common cold, and the magnification is TEM X409,500.

CONCEPT CHECK

Define cell biology.

List six of the nine characteristics of life.

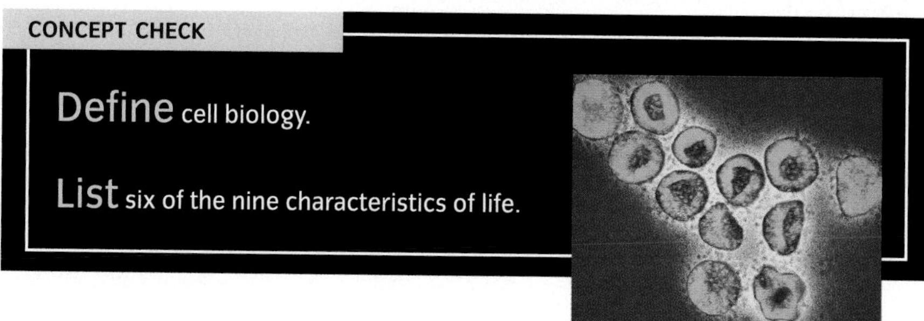

Living Things Must Maintain Homeostasis

Learning Objectives

Define homeostasis and describe how it plays a role in everyday activities.

Contrast negative and positive feedback systems.

One key element of life is **homeostasis**, a word that means "staying the same" (*homeo* = unchanging; *stasis* = standing). Humans, along with other organisms, can function properly only if they stay within narrow ranges of temperature and chemistry. Homeostasis allows you to respond to changes in your internal environment by modifying some aspect of your behavior, either consciously or unconsciously. When you are chilled, you consciously look for ways to warm yourself. This morning, you clothed yourself in an attempt to remain warm. If your clothing was not enough, your body would begin to shiver to generate internal heat through chemical reactions. Blood vessels near the surface of your skin would constrict and carry less blood, thereby reducing heat loss through **radiation**. These changes are attempts to maintain homeostasis. (See Health, Wellness, and Disease: Homeostasis and blood chemistry, on p. 8.)

Radiation The transfer of heat from a warm body to the surrounding atmosphere.

Homeostasis helps an organism stay alive, often through the use of **feedback systems**, or loops (Figure 1.1). The most common type of feedback system in the human is **negative feedback**. Negative feedback systems operate to reduce or eliminate the changes detected by the stimulus receptor (Figures 1.2 and 1.3). Negative feedback prevents you from breathing fast enough to pass out or from drinking so much water that your blood chemistry becomes dangerously unbalanced. Feedback is so important that we will return to it when we discuss each organ system.

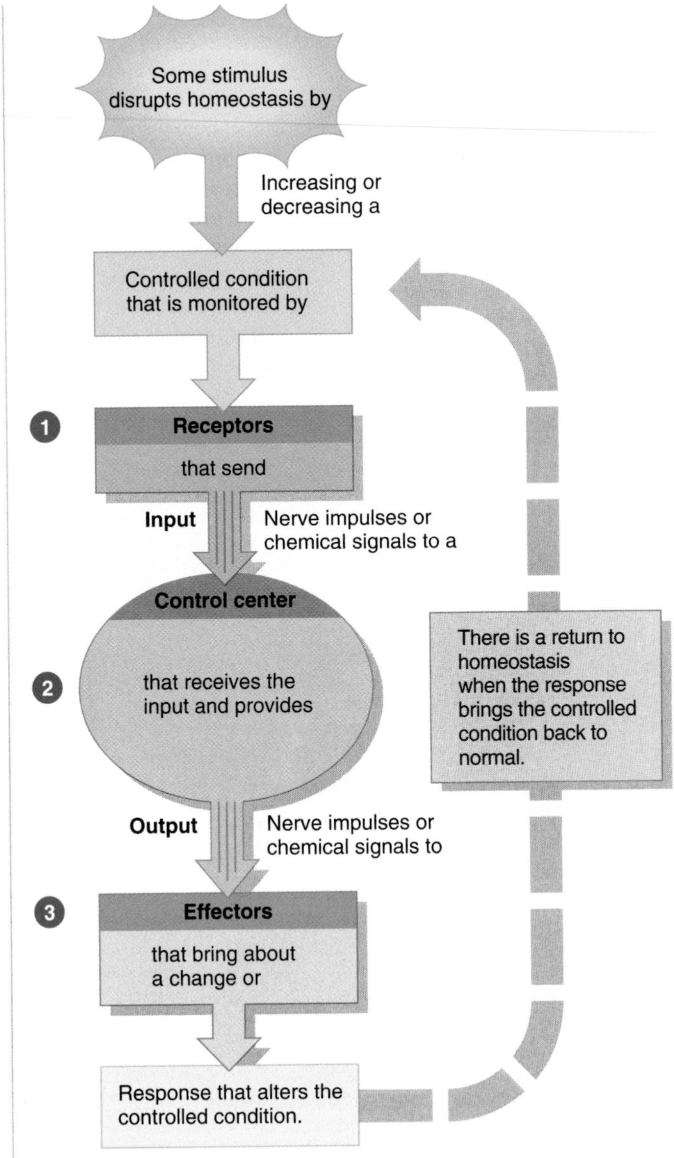

Feedback loop Figure 1.1

A feedback system requires three components: 1 a stimulus **receptor** that monitors the environment and reports any changes, 2 a **control center** that registers the signal from the stimulus receptor and formulates a response, and 3 an **effector** or effectors to carry out that response.

Process Diagram

Thermostat regulation coupled to body temperature regulation

FIGURE 1.2

Thermoregulation is a good example of negative feedback.

1. Drop in temperature disrupts homeostasis.
2. Body temperature decreases.
3. Temperature-sensitive receptors in the skin detect falling body temperature and signal thermoregulatory cells in the brain, which make you feel chilled.
4. The brain starts a response that raises your internal temperature by signaling effector organs to generate heat and reduce heat loss.
5. The large muscles of your legs and chest begin to move, causing shivering that generates internal heat. The blood vessels of your skin constrict, slowing blood flow near the skin, reducing heat loss. Normally, these responses restore homeostasis and keep your body within its range of normal temperature.

1 Some stimulus disrupts homeostasis by

Decreasing

2 Body temperature

3 **Receptors**

Thermoreceptors in skin

Input

Nerve impulses

4 **Control center**

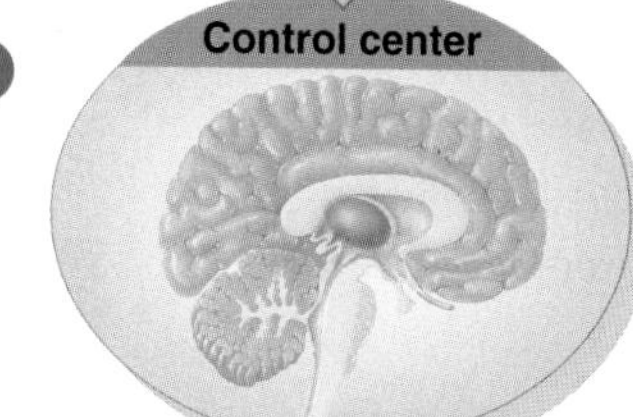

Output

Nerve impulses or chemical signals

5 **Effectors**

Skin constricts to decrease heat loss through the skin

Skeletal muscles contract in a repetitive cycle called shivering

Increase in body temperature

Return to homeostasis when response brings body temperature back to normal

Snowshoe hare FIGURE 1.3

A thermoregulatory feedback system is found in humans and other **endothermic** animals. The snowshoe hare thermoregulates using negative feedback, as do humans. Many endothermic animals can restrict blood flow to reduce heat loss through their extremities. We find our fingers and toes turning white in cold climates due to a restriction in the blood flow through these extremities. Many cold-climate mammals also thermoregulate through behavioral adaptions, such as migrating, building burrows, clustering for warmth, or hibernating.

Endothermic Organisms that maintain an internal temperature within a narrow range despite environmental conditions.

Health, Wellness, and Disease

Homeostasis and blood chemistry

Blood is a tissue that unites most of the body. Because the body's countless chemical reactions require specific environmental conditions, maintaining blood chemistry through homeostasis is critical to health. Two key aspects of blood chemistry are water concentration and levels of various ions, charged particles called electrolytes. Key electrolytes include calcium, sodium, and potassium. If homeostasis fails, and concentrations of water or electrolytes stray outside normal ranges, disease or death can result.

Water. The body has several mechanisms to regulate water balance in blood and the fluid around cells. The brain triggers thirst when it detects a low blood volume and changes in blood electrolyte levels. Homeostatic mechanisms also direct excretion: the kidney responds to a hormone that can speed or slow urine production to restore water concentrations to the normal range.

Calcium is essential for blood clotting and for muscle and nerve function. Excess calcium can cause painful kidney stones; too little calcium can lead to tremors and other neurological symptoms.

Potassium is a key electrolyte in the blood and fluid around cells. Potassium ions affect the cell membrane and are needed for proper nerve and muscle function. If potassium is exhausted due to prolonged exercise or other causes, muscles may fail to respond and the heart rhythm can deteriorate. As many athletes know, bananas are a good source of potassium.

Sugars. Sugar in the blood, especially glucose, is a key source of energy. Glucose must enter a cell before it can be used, and this entry requires the hormone insulin. In diabetes, the body either fails to make functional insulin (type I diabetes) or it cannot use the insulin it has (type II diabetes). Diabetes causes a massive departure from homeostasis, and high levels of sugar in the blood damage small blood vessels in the eyes, legs, and kidneys. Diabetes is growing rapidly as the population gets fatter and less active and is a leading cause of blindness and amputation.

When blood sugar gets too low, we feel hungry and try to eat to restore blood sugar levels. The immediate effects of low blood sugar can include light-headedness and weakness. Low blood sugar starts a complicated homeostatic process that we can summarize like this: "Where am I going to get my next meal?" If blood sugar stays low, the body may break down stored fat, using it as an alternative source of fuel. If starvation continues, enzymes (specific biological proteins that cause chemical reactions) are activated that break down the protein in muscle. This releases amino acids (the building blocks of protein) into the blood. The amino acids are then taken up by the liver, which converts some of them into sugar that the body can use. If this goes on too long, it results in muscle wasting.

Sports drinks replenish water, potassium, and blood sugar, helping restore homeostasis after vigorous exercise.

CONCEPT CHECK

What negative feedback system is at work right now in your body (other than thermoregulation)?

Identify the stimulus, receptor, effector, and control center using your example.

Biological Organization Is Based on Structure

LEARNING OBJECTIVES

Understand the organizational pattern of all biology.

Explain how atoms, and therefore the entire field of chemistry, relate to the study of life.

One of the oldest techniques for dealing with our world is to categorize it and divide it into manageable chunks. Imagine trying to understand this paragraph if the sentences were not chunked into words through the use of spaces. Similarly, the natural world seems overwhelming and chaotic until we organize it. Biology is organized in steps, from micro to macro: Small units make up larger units, which in turn form still larger units. We see this in both artificial and natural organization in biology. In artificial classification (taxonomy), a system of names is used to identify organisms and show their relationship. Natural organization, in contrast, emerges from the structure of organisms. Both natural and artificial organization help us make sense of the living world. Natural organization appears in the human body as it does in the rest of the living realm.

Natural organization is based on a system of increasing complexity. Each level in the hierarchy is composed of groups of simpler units from the previous level, arranged to perform a specific function. The smallest particles that usually matter in biology are atoms (FIGURE 1.4, p. 10). **Atoms** are defined as the smallest unit of an element that has the properties of an element. Atoms combine to form molecules—larger units that can have entirely different properties than the atoms they contain. You already know some of the molecules we will discuss, like water, glucose, and DNA. Molecules then combine to form **cells**, which are the smallest unit of life. We will take a closer look at the cell in Chapter 3. Groups of similar cells with similar function combine to form **tissues**. The human body has four major tissue types: muscular, nervous, epithelial, and connective. Tissues working together form organs, such as the kidney, stomach, liver, and heart. **Organs** with the same general function combine to form **organ systems**. For example, the respiratory system includes organs that work together to exchange gas between cells and the atmosphere; organs in the skeletal system support the body and protect the soft internal organs. A suite of organ systems combine to form the human **organism**. Notice that each layer of complexity involves a group of related units from the preceding layer. This type of hierarchy is found throughout biology and the natural world.

Taking a global view of the organization found in the natural world, we see that the concept of hierarchy does not stop at the individual. The individual human organism lives in groups of humans called **populations** (FIGURE 1.5, p. 11).

Population All representatives of a specific organism found in a defined area.

A human being illustrates the largest level of structural organization: the organism.

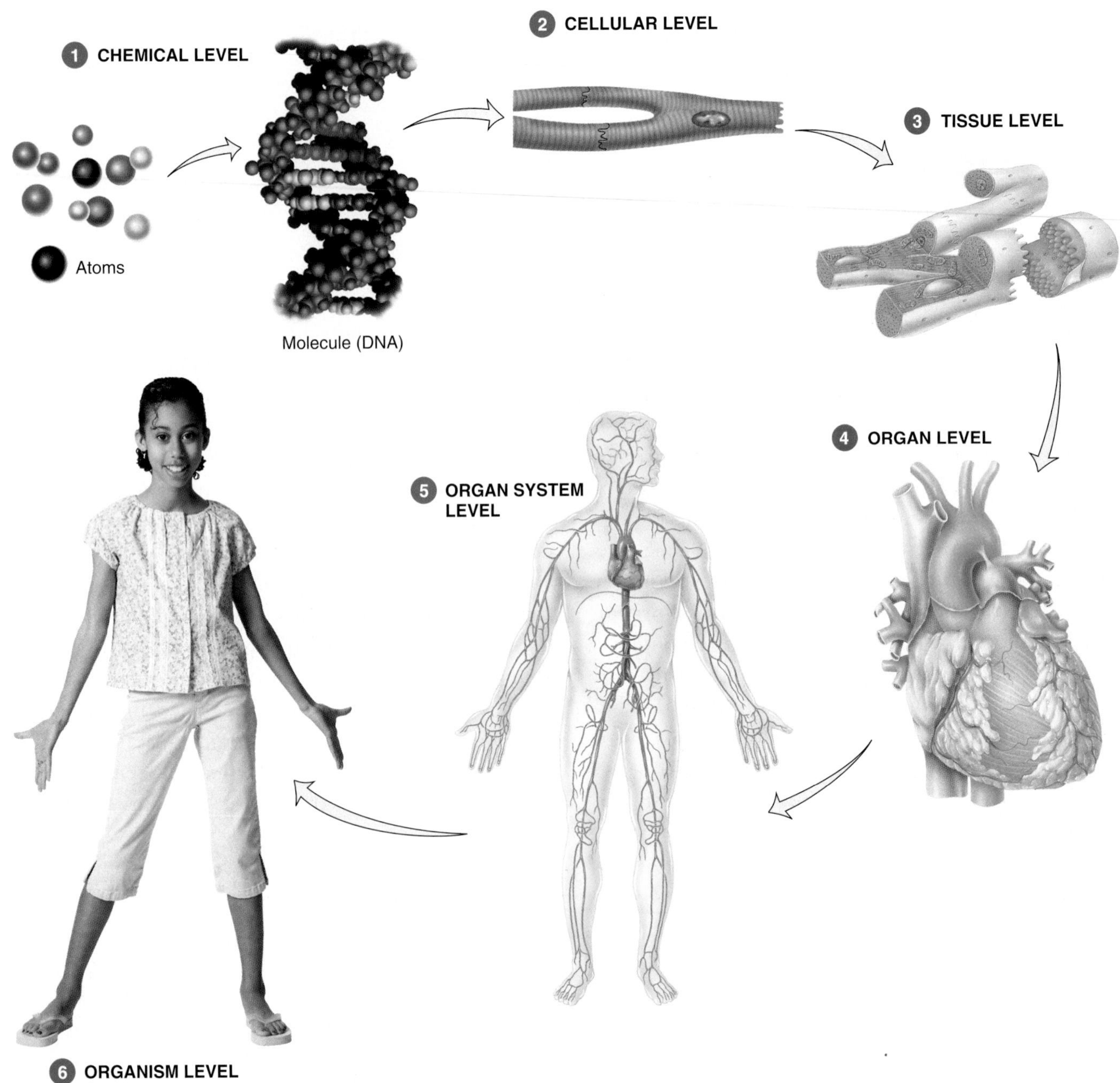

Hierarchy of organization of life FIGURE 1.4

Natural organization: from atom to organism.

1. Chemical level: the chemical "components" that are arranged into cells (for example, protein).
2. Cellular level: the smallest unit of life; a component bounded by a membrane or cell wall. In multicellular organisms, cells are usually specialized to perform specific functions (muscle cell).
3. Tissue level: an assemblage of similar cells (muscle).
4. Organ level: an assemblage of tissues, which often have several functions (heart).
5. Organ system level: the group of organs that carries out a more generalized set of functions (cardiovascular system).
6. Organism level: *Homo sapiens*.

Can wolves and dogs produce offspring?

FIGURE 1.7

Taxonomy is not always straightforward. Organisms can get different names based on different criteria and depending on whether you prefer to "lump" or "split" close relatives. For example, the Smithsonian Institution and the American Society of Mammalogists classify the domestic dog and the gray wolf as one species, *Canus lupus*. Within this species they include the domestic dog (*C. lupus familiaris*), the gray wolf (*C. lupus lupus*), and even the dingo (*C. lupus dingo*). These three canines can interbreed and produce viable offspring. Other taxonomists consider the dingo and the domestic dog separate species (*C. dingo* and *C. familiaris*, respectively) because they do not interbreed in the wild. Thus "lumpers" and "splitters" can both find justification in nature. But those tabloid-news headlines about fearsome wolf-dogs may be true; they are members of the same species, and they can interbreed.

Kingdom A high-level taxonomic classification.

Species A precise taxonomic classification, consisting of organisms that can breed and produce offspring capable of breeding.

Viable Capable of remaining alive.

genus, and **species**. Each category defines the organisms more tightly, resulting in a hierarchy of similarity. The final category, species, implies reproductive isolation, meaning (with a few exceptions) that members of a particular species can produce **viable** offspring only if they breed with each other (FIGURE 1.7).

Tax-Talk Taxonomists capitalize the first letter of all classification terms except species (*Homo sapiens*). The species name is always preceded by the entire genus name, unless you have just used the genus; then you can abbreviate it: "In regard to *Homo sapiens*, we must note that *H. sapiens* . . .". Genus and species names are either underlined or written in italics (FIGURE 1.8, p. 14).

Each successive category refines the characteristics of "human" to the point where only humans are classified in the final category, *Homo sapiens*. And despite the amazingly complex and pervasive cultural differences that exist between populations of humans, we are all members of the same species.

www.wiley.com/college/ireland

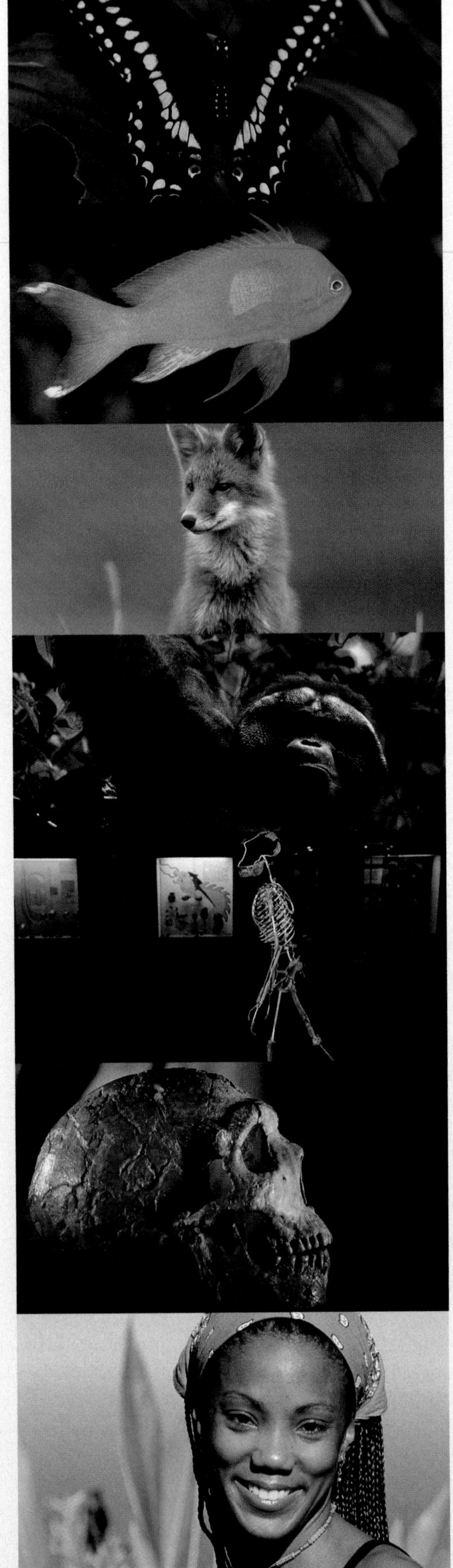

Meet your human taxonomy:

A Kingdom = Animalia (all multicellular organisms that ingest nutrients rather than synthesize them)

B Phylum = Vertebrata (all animals with a vertebral column or **dorsal hollow notocord** (a structure along the top of animals) protecting their central nervous system

C Class = Mammalia (all vertebrates with placental development, mammary glands, hair or fur, and a tail located behind the anus)

D Order = Primates (mammals adapted to life in trees)

E Family = Hominidae (primates that move primarily with bipedal—two-footed—locomotion)

F Genus = **Homo** (hominids with large brain cases, or skulls)

G Species = *H. sapiens* (the largest brain case of the genus *Homo*; *"sapiens"* loosely translates as "knowing")

Human taxonomy FIGURE 1.8

CONCEPT CHECK

List the taxonomic categories from most inclusive to least inclusive.

What can you discover about an organism by comparing its full taxonomic classification to that of a human?

Scientists Approach Questions Using the Scientific Method

LEARNING OBJECTIVES

List the steps in the scientific method in order.

Define hypothesis and theory.

Science is a field with specific goals and rules. The overall goals are to provide sound theories regarding the phenomena we observe, using rules embodied by the **scientific method**. When a question arises about the natural world, the scientific method provides the accepted, logical path to the answer (FIGURE 1.9).

A scientific experiment is an exercise in vandalism: Our goal is to prove our hypothesis wrong. Say our hypothesis is that the rooster's crow causes the sun to rise within the next 20 minutes. How could we test this hypothesis? Could we force the rooster to crow at midnight, and wait 20 minutes for a glow on the eastern horizon? Could we prevent the rooster from crowing in

Scientific method FIGURE 1.9

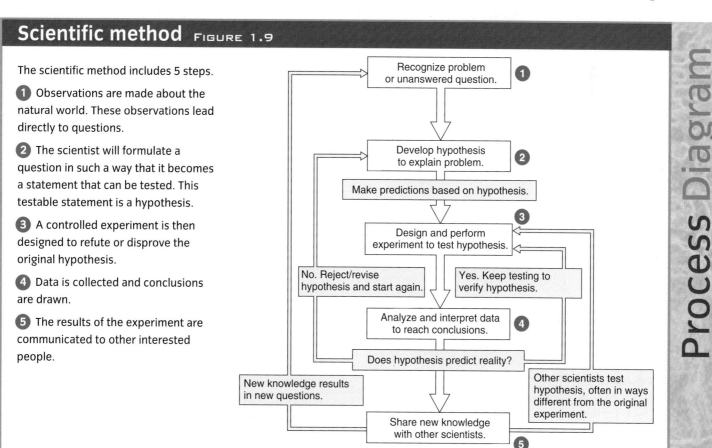

Process Diagram

The scientific method includes 5 steps.

1. Observations are made about the natural world. These observations lead directly to questions.
2. The scientist will formulate a question in such a way that it becomes a statement that can be tested. This testable statement is a hypothesis.
3. A controlled experiment is then designed to refute or disprove the original hypothesis.
4. Data is collected and conclusions are drawn.
5. The results of the experiment are communicated to other interested people.

Recognize problem or unanswered question. 1

Develop hypothesis to explain problem. 2

Make predictions based on hypothesis.

3 Design and perform experiment to test hypothesis.

No. Reject/revise hypothesis and start again.

Yes. Keep testing to verify hypothesis.

Analyze and interpret data to reach conclusions. 4

Does hypothesis predict reality?

New knowledge results in new questions.

Other scientists test hypothesis, often in ways different from the original experiment.

Share new knowledge with other scientists. 5

the morning? In either case, if the sun rose as usual, our hypothesis would be disproved, and we would need to find a better hypothesis (FIGURE 1.10).

This silly example shows how scientists may manipulate factors that (according to the hypothesis) seem related to the observation, all in an attempt to disprove the hypothesis. We develop a hypothesis using **inductive reasoning**—creating a general statement from our observations. We design the experiment, however, with **deductive reasoning**, moving from the general hypothesis to a specific situation. An "if, then" statement is an ideal basis for a scientific experiment: "If situation A (rooster crows) occurs, then result B (sunrise) will follow." In our experiment, we changed situation A and monitored any changes in result B.

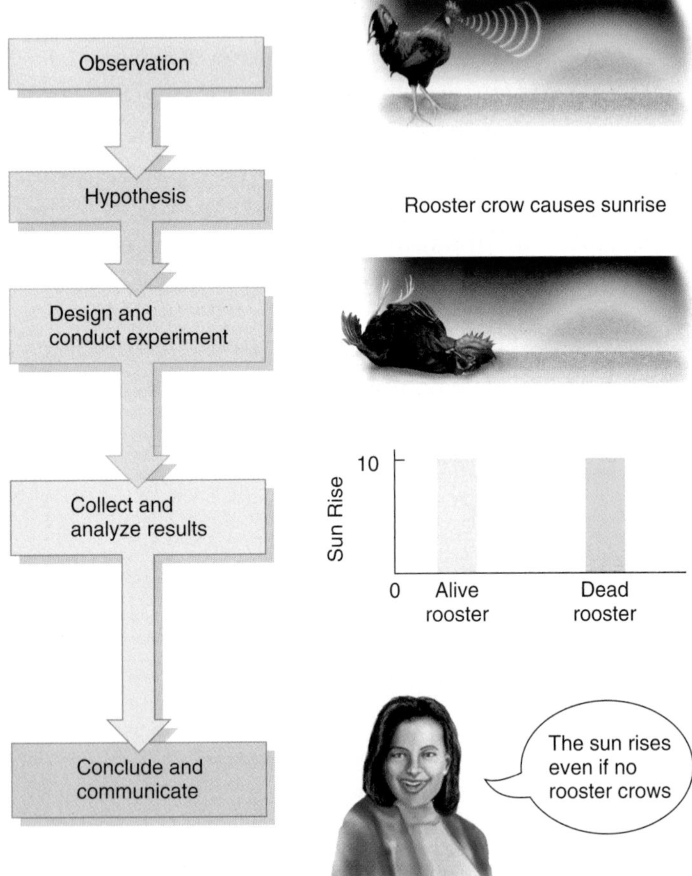

Experimental design FIGURE 1.10

The scientific method is rooted in logic. If we can prove that our hypothesis does not apply to even one situation, then it is wrong. After we analyze this data and draw conclusions from it, we may have to junk our hypothesis, or conclude that it applies to a more limited range of circumstances.

When designing and running the experiment, we must control all potential variables. Otherwise, we cannot draw any valid conclusions. In the rooster example, it would be a good idea to muzzle all nearby roosters. Otherwise, how would we know if a bird in the next chicken coop had caused the sunrise? Similarly, in testing new medicines, scientists use a "double-blind" experiment: nobody knows whether each research subject is getting real medicine or a fake, called a "placebo." This prevents expectations that the drug will work from actually causing a change in the subject's health. The "placebo effect" can be powerful, but the goal is to test the drug, not the research subject's expectations.

Finally, our hypothesis must be testable. If we cannot think of a situation where we could disprove it, there is no experiment to devise. Learning to assess situations with the scientific method takes some practice, but it's a skill that can be useful throughout life.

Let's take an example from human biology to show the process of testing a hypothesis. Have you seen those hand lotions that claim to be "skin firming"? Sounds great, but how would we test this claim? Under the scientific method, we consider the marketing claim to be the observation, so we must develop a testable hypothesis from the observation: "Using this hand cream for one month will cause measurable tightening of the skin on the back of the hand." Now we restate the hypothesis as an "if, then" statement: "If the cream does firm the skin, then using the cream on the back of the hand for one month will reduce the skinfold measurement." This is a testable statement that lends itself to controlled experimentation. First, we will assess each person's skin tautness by measuring the skin fold that can be pulled up on the back of the hand. Then we will randomly divide the participants into two groups: a control group and an experimental group. We will treat each group in an identical manner, except that the control group will use Brand X hand cream without the firming agent and the experimental group will get

Statistical significance An experimental result that would occur by chance in less than 1 experiment in 20; the accepted level in modern science.

Variable A factor that can be changed in an experiment to test whether and how it affects the outcome.

Brand X with the firming agent. After using the cream for one month, we will repeat the skinfold measurements and analyze our data, looking for changes in skin tautness between the two groups as evidence for either accepting or refuting the hypothesis. If the experimental group displays a change in tautness that would occur by chance in less than 1 experiment in 20, the change is said to have **statistical significance**, and the hypothesis is supported: the cream does tighten the skin.

Because biologists cannot always control all factors, or **variables**, that might affect the outcome, they often use observation as a form of experimentation. If you were interested in the effects of mercury on the human brain, it would not be ethical to dose people with mercury, but you could perform an observational study. You could measure blood levels of mercury, or you could ask your subjects about past diet (food, especially fish, is the major source of mercury exposure). Then you would use statistical tests to look for a relationship between mercury exposure and intelligence. Finally, you could try to confirm or refute your results with controlled experiments in lab animals. Does mercury make rats faster or slower at negotiating a maze (a standard test for rat intelligence)? Observational studies are also a mainstay of field biology (FIGURE 1.11).

Theory A general uniting principle of science, upheld by observation and many experiments.

Observation, experimentation, and analysis are the basis for scientific reasoning. Once a hypothesis has survived rigorous testing without being disproved, it is accepted as a **theory**. Theories are not facts, but rather extremely well-supported explanations of the natural world that nobody has disproved. To a scientist, a theory is much more than a hypothesis or a belief—it's our best effort to date to explain nature. Many fields of science may be involved in supporting a theory. The theory of evolution through natural selection, for example, is supported by taxonomists, geologists, paleontologists, geneticists, and even embryologists. Many scientists have tried, but none has refuted the basic hypothesis first described by Charles Darwin in 1859. Another key theory, the cell theory, will be discussed in Chapter 3.

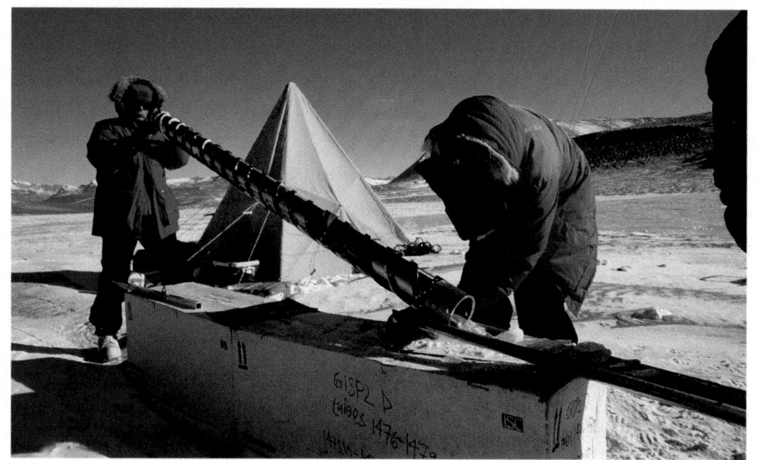

Scientists in field collecting experimental data

FIGURE 1.11

Field ecologists, like many other biologists, must rely on observational studies rather than controlled experiments.

Global warming—A human solution to a human problem?

After years of debate and intense study, scientists around the globe are convinced: Burning fossil fuels and other activities that release "greenhouse" gases like carbon dioxide into the atmosphere are changing our planet. Greenhouse gases make the atmosphere act much like a greenhouse—trapping heat and warming the climate. Over the past century, largely due to greenhouse gases, average temperatures have risen 0.5° to 1.0°C. Over the next century, scientists predict a rise of another 2° to 5°C. This heat-up sounds minor, but it has already had global effects.

Glaciers are melting on almost every continent. The ice cap on Mount Kilimanjaro in Africa could disappear in a few decades. If snowcaps melt in the mountains of Colorado and California, water supplies in adjacent parched regions could quickly decline (melting snow is a major source of water in both regions).

The Arctic Ocean could become ice-free during the summer after 50 years or so. Polar bears, which live much of their lives on the ice, are particularly endangered, as are indigenous people of the North.

Animal migrations are already affected. Birds are leaving their wintering grounds earlier in the year. Caribou in North America have drowned trying to migrate across rivers that used to be ice-covered during the migration.

Coral reefs are centers of biological diversity in the oceans, but rising ocean temperatures have helped cause massive decline of coral.

As temperatures warm, mosquitoes are moving to higher elevations, spreading malaria and other diseases.

Sea level is already rising, in part because water expands as it warms. Some low-lying island nations in the Pacific have started making plans to leave their ancestral homes. Coastal cities around the globe could be forced to relocate or build flood walls, and the sea-level rise may accelerate as ice melts in Greenland and elsewhere.

Climate scientists, climatologists, also worry about feedback mechanisms. If the Arctic Ocean melts, the water may capture solar energy affecting global temperatures. Normally, tundra in the north stores vast amounts of methane, which traps even more heat than carbon dioxide. If the tundra warms, this methane could be released, causing yet more warming.

A hotter Earth is likely to be a drier Earth, and that means more forest fires. Change in the weather could hurt agricultural productivity, and it is likely to cause more species to go extinct.

The bad news is that global warming may pose the biggest environmental threat in human history. The good news is that since humans caused the problem, humans can also solve it. But finding solutions will be a huge task. We must change how we work, live, and move about the planet, and we must do so before the effects of warming get even more severe.

Solutions can come in two areas: (1) energy efficiency and social changes to reduce the demand for energy, and (2) alternative energy sources that release no greenhouse gases. Without solutions the alternative is grim: In just a century or two, the Earth could warm up as much as it has warmed in the 10,000 years since the last ice age. Back then, towering glaciers covered the American Midwest. That's a lot of change to absorb in 100 to 200 years.

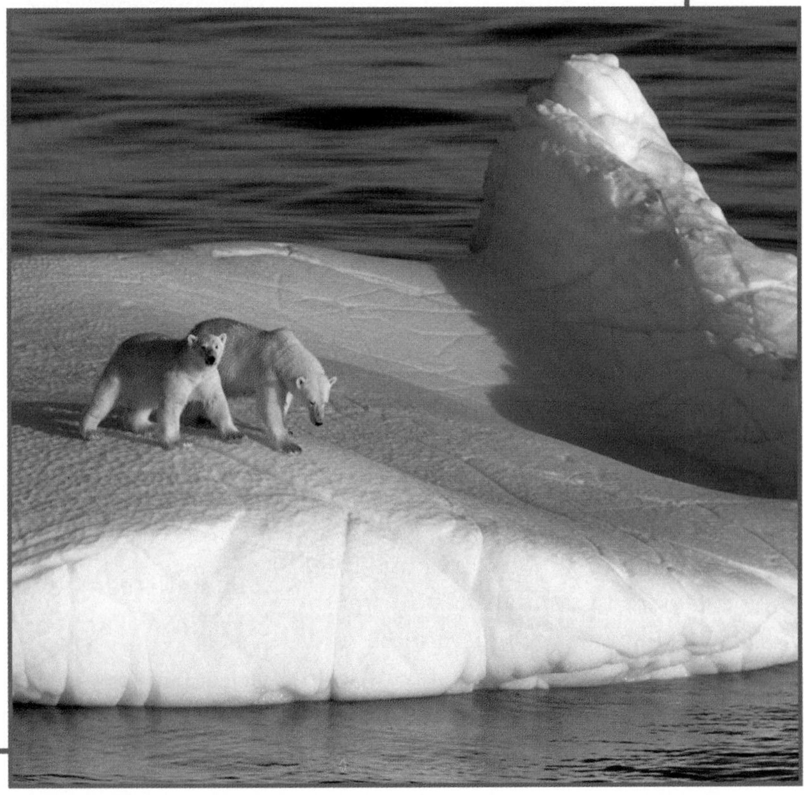

Science is not a perfect, set-in-stone answer to questions about the natural world, but rather a dynamic, ever-changing collection of ideas. New information can change or destroy accepted explanations for the natural world. For example, doctors once blamed contagious disease on ill humors, miasmas, and evil spirits. Through the work of nineteenth-century biologist Louis Pasteur, it became clear that many diseases were caused by microscopic organisms. In his breakthrough experiment, Pasteur sterilized some grape juice and showed that it did not ferment into wine. Then he added yeast, and the juice fermented. When Pasteur showed through experiment that invisible organisms can also cause disease, he helped establish the germ theory of disease. Although it's called a theory, the germ theory is the universally accepted scientific explanation for infectious disease.

More recently, the accepted role of the cell nucleus has come into question. Based on experiments, biologists used to consider the nucleus the cell's control center, but new evidence suggests it actually functions more like a library for genetic data. The actual control of gene expression and cellular activity seems to reside outside the nucleus, in specific RNA molecules. The theory of nuclear control in the cell is under serious scrutiny, and further experiments could alter it.

Scientific studies are part of the daily news. As technology advances, humans confront scientific hypotheses and experimental results almost every day. We see advertisements for new drugs. We hear that fossil fuels are warming the globe (see Ethics and Issues: Global warming). We see countless new technologies in the field of consumer electronics. In medicine, we hear about a steady stream of new surgeries and wonder drugs. About the only way to wade through the morass of information in the media is to understand and use the scientific process. Responsible citizens living in technological cultures sometimes must make decisions about contested scientific issues they read about in the media.

Some reports have linked the radiation from cell phones to brain tumors, but other reports find no connection. A few concerned citizens have demanded that manufacturers produce "safer" cell phones, with lower radiation emissions. Can you think of an experiment that would resolve this issue, at least in principle? As you read about the scientific studies on this issue, ask yourself, what types of controlled and observational experiments underlie the claims about cell phones and cancer? Are they convincing?

The ability to question and criticize is useful in many aspects of human biology, whether it is our constantly changing understanding of obesity or the danger of food additives or environmental chemicals. Critically analyze the data, experiments, and claims before you accept what you read. There are plenty of opinions out there; don't accept any until you consider the evidence and reach an informed decision.

CONCEPT CHECK

What type of reasoning combines many observations into one general statement to be tested?

What is the difference between a hypothesis and a theory?

When might the scientific method be helpful in your daily life?

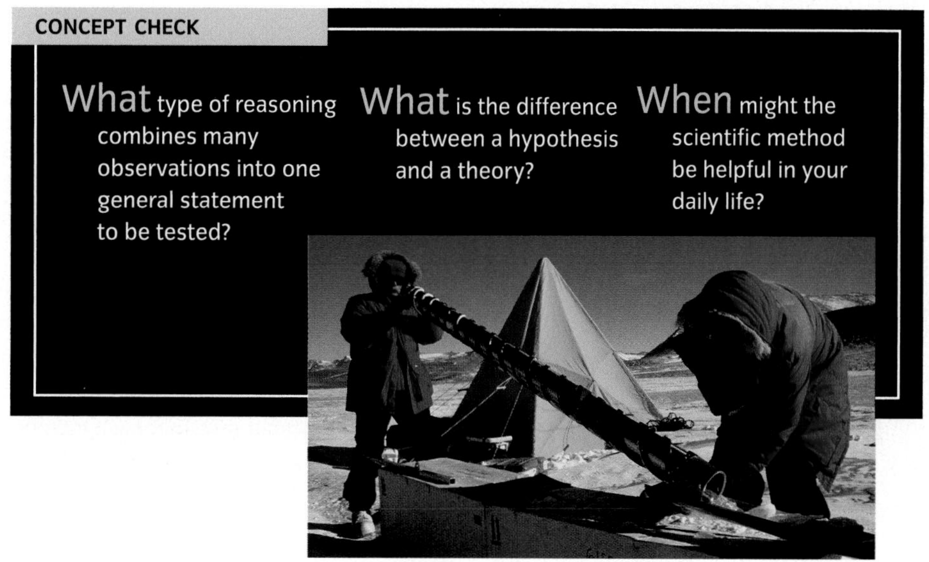

Scientific Findings Often Lead to Ethical Dilemmas

LEARNING OBJECTIVES

Define altruistic behavior.

Why is a basic understanding of science essential to being a productive citizen?

Humans have evolved as social animals, following the rules and expectations that make life possible in groups. This cultural structure that overlies the biological structure of human life certainly adds interest to our study of human biology. Culture generally requires that people accept responsibility for other individuals within the population, rather than merely surviving and protecting their young. Although **altruistic** behavior does appear among some primates, it helps distinguish humans from other life forms and creates one basis for the governments and laws people have established.

Altruistic Putting the needs of others ahead of, or equal to, personal needs.

When individuals must make judgments and act for the good of the group rather than the individual, they must make **ethical decisions**, and ethical decisions should be informed decisions.

Ethical decision A decision based on the principles of right and wrong, rather than on financial, personal, or political gain.

Where does that information come from? Scientific research provides our basic understanding of the natural world. Although humans can and do add their interpretations and values to the results of science, science itself is judgment free. Scientific results are neither good nor bad; they are just the best current idea of how the material world operates. When Pasteur and his peers discovered that germs cause many diseases, that was neither good nor bad—it was just true.

The ability to analyze scientific issues is essential in an informed society and turns out to be more important as scientifically based issues become even more

Nuclear power plant FIGURE 1.12

Nuclear power poses an interesting mix of scientific and political issues. Atomic fission can provide a large amount of electricity, and it does not create greenhouse gases, which are warming the globe and threatening harm to the biosphere. However, radioactive waste is dangerous, and nuclear plants can melt down and spew vast amounts of radiation, as one did at Chernobyl in the Soviet Union in 1986. The decision to use nuclear power is a political decision, not a scientific one, and therefore it is imperative that each member of society understand the scientific data on nuclear reactors, as well as the social ramifications of that information.

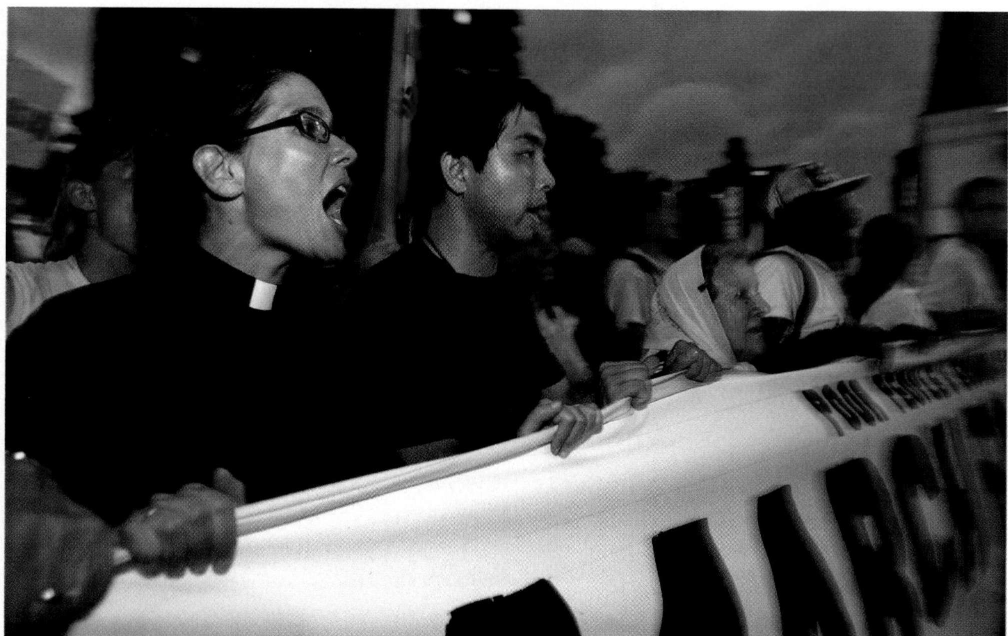

Antinuclear power rally
FIGURE 1.13

Nuclear power has its pluses and minuses. To take a position, you should know about global warming, radioactive waste, and the costs and benefits of other technologies for making electricity. All these issues are scientific issues.

common and complex. Science seeks to explain the natural world, but the uses of science, both beneficial and harmful, grow from human choices. Sometimes people choose to use scientific discoveries to improve the environment and the human condition, and sometimes to carry out evil designs (FIGURES 1.12 and 1.13). The germ theory of disease can be used to help cure disease—or to invent biological warfare.

Many ethically charged scientific issues, such as stem-cell research, environmental conservation, or genetically modified food, have both personal and political ramifications. Each of these requires an understanding of the science and the societal issues. An informed voting public requires that each individual draw logical and defensible conclusions from scientific information (FIGURE 1.14).

Speak no evil, see no evil, hear no evil
FIGURE 1.14

Is this any way to run an informed citizenry?

CONCEPT CHECK

What is altruistic behavior?

Why is it important to understand scientific information?

CHAPTER SUMMARY

1 Living Organisms Display Nine Specific Characteristics

Cell biology is the study of life. One characteristic of life is organization. Living things are organized from microscopic to macroscopic. All life is also composed of cells and is responsive to the environment. Life adapts, uses energy, and reproduces. Living organisms are composed of carbohydrates, lipids, proteins, and nucleic acids. In order to maintain life, these organisms must maintain a relatively constant internal environment.

2 Living Things Must Maintain Homeostasis

Homeostasis (*homeo* = unchanging; *stasis* = standing) is maintaining a constant internal environment. Homeostasis is accomplished via negative feedback systems, which include a receptor, a control center, and an effector. Negative feedback counters the original stimulus. A second type of feedback, positive feedback, is rare in the body and usually destructive.

3 Biological Organization Is Based on Structure

The natural organization of life on Earth is based on a system of increasing complexity. The base of this hierarchy is atoms, meaning that the basis of biology is actually chemistry. Atoms combine to form molecules. Molecules join together to form cells. Similar cells form tissues; tissues with a common function form organs; organs with similar functions form organ systems, and a group of organ systems all functioning together form an organism.

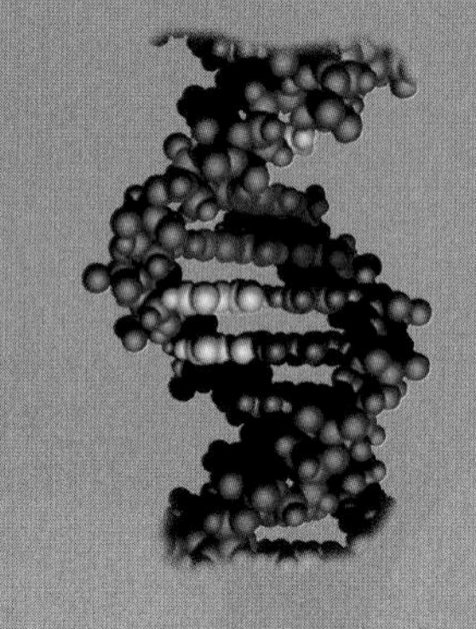

4 Biological Classification Is Logical

Taxonomy is the study of classification. Organisms are classified based on shared characteristics. Each successive level of classification gets more restrictive, until only one interbreeding species is described by a binomial name indicating its genus and species.

5 Scientists Approach Questions Using the Scientific Method

Science is more a way of thinking than a body of knowledge. The steps of the scientific method include:

- Observation: witnessing an unusual or unexpected phenomenon
- Hypothesis: formulating an educated guess as to why the phenomenon occurs
- Experiment: designing and running a controlled experiment to test the validity of the hypothesis
- Collecting results and analysis: recording the results of the experimental procedure and determining the meaning of the results obtained from the experiment
- Communicating the findings: preparing a paper, presenting a poster, or speaking about the results of the experiment

6 Scientific Findings Often Lead to Ethical Dilemmas

Science in and of itself is neither inherently good nor bad. It is in the use of scientific principles that value judgments are made. Science can be used for either the betterment of society or its destruction. Individuals who understand the ramifications of the science are the ones who should make this choice. In democratic nations, however, these ethical decisions are placed in the hands of the voting populace. In order to make the right choices, we must all understand at least a little bit about the functioning of the biological world in which we live.

KEY TERMS

- **altruistic** p. 20
- **cell** p. 4
- **endothermic** p. 7
- **ethical decision** p. 20
- **kingdom** p. 13
- **organ** p. 4
- **organ system** p. 4
- **organism** p. 4
- **population** p. 9
- **radiation** p. 6
- **species** p. 13
- **statistical significance** p. 17
- **taxonomy** p. 12
- **theory** p. 17
- **variable** p. 17
- **viable** p. 13

CRITICAL THINKING QUESTIONS

1. Gerald proudly displays his pet rock, complete with its cardboard cage, in his bedroom. His sister, Marianne, has a Chia Pet® in her bedroom. The Chia Pet ® is a planter shaped like a puppy, with sprouts simulating fur growing on the puppy planter's back and head. Using the characteristics of life listed in the beginning of this chapter, argue that either pet is alive. Explain why the other pet is NOT alive.

2. When considering the increasing complexity of atoms, molecules, cells, and tissues, you may notice that each step has characteristics that were absent in the previous level. These characteristics, called emergent properties, demonstrate that the whole organism is more than the sum of its individual parts. Consider the heart, an organ with a variety of tissues. In what way is the heart more than the sum of the tissues it comprises?

3. One of many negative feedback systems in the body is the regulation of blood sugar via the production of insulin and glucagon. When this system functions, insulin is produced as blood sugar rises, enabling cells to capture glucose. As blood sugar decreases, insulin production stops and glucagon production begins. Glucagon stimulates liver and muscle cells to release stored glucose to maintain blood glucose levels. What would happen to this negative feedback system if insulin production could not stop?

4. Taxonomy places organisms in smaller and smaller categories, each with more restrictive criteria, until a particular organism is defined so tightly that no other can share that classification. Look at the classification for humans. Where would an organism diverge from the human lineage if it had not developed during an arboreal (tree-dwelling) existence? Where on the taxonomic tree would a bipedal placental mammal with a tiny brain case diverge?

5. Dr. Pamela Sullivan claims that her new toothpaste whitens teeth five times faster than other toothpastes. How would you design a controlled experiment to test Dr. Sullivan's hypothesis?

SELF TEST

1. Which of the following is NOT a characteristic of life?
   a. Responds to external stimuli
   b. Low degree of organization
   c. Composed of proteins, lipids, and carbohydrates
   d. Maintains a stable internal environment

2. Which of the listed items listed below represents the smallest unit of life?
   a. Organism
   b. Organ
   c. Tissue
   d. Cell

3. On the figure below, identify the non-living portion.
   a. A
   b. B
   c. C
   d. B and D

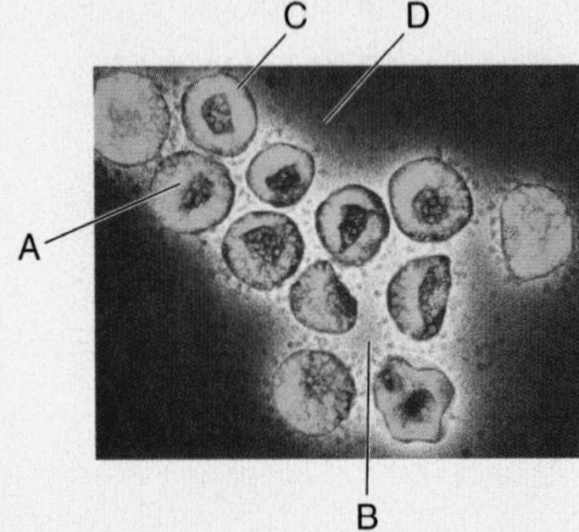

4. Using the same figure, what is structure C?
   a. A viral particle
   b. Tissue
   c. The cell
   d. An organ

5. Homeostasis is maintained most often by
   a. positive feedback systems.
   b. negative feedback systems.
   c. endothermic animals.
   d. viruses.
   e. radiation.

6. Identify the components of a typical feedback system by writing the following terms on the diagram:
   a. Receptor
   b. Effector
   c. Control center

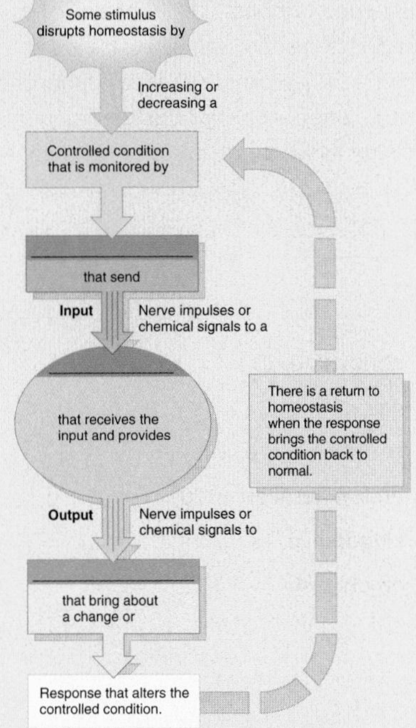

7. Which of the above components is responsible for altering behavior to reduce the original stimulus?
   a. Receptor
   b. Effector
   c. Control center

8. This organism is demonstrating what type of homeostatic mechanism?
   a. Negative feedback
   b. Positive feedback
   c. Ion control
   d. Water balance

9. What level of organization is indicated by the figure below?

   a. Cellular level
   b. Organ level
   c. Organ system level
   d. Chemical level

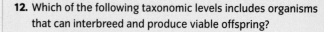

10. Of the levels listed, which is most complicated?

   a. Organism level
   b. Atomic level
   c. Organ level
   d. Organ system level
   e. Chemical level

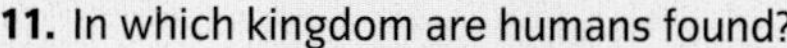

11. In which kingdom are humans found?

   a. A
   b. B
   c. C
   d. D
   e. E
   f. F

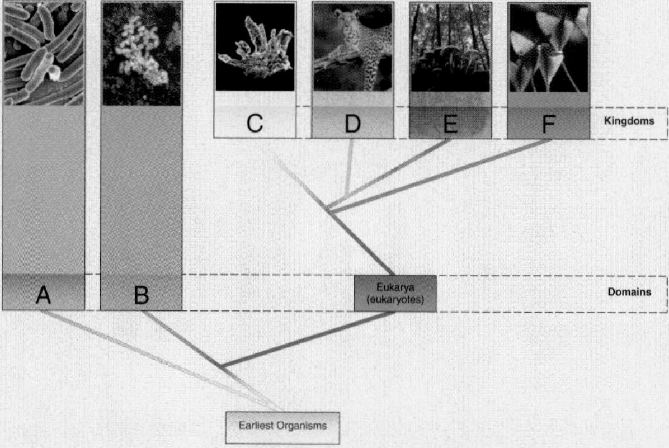

12. Which of the following taxonomic levels includes organisms that can interbreed and produce viable offspring?

   a. Genus
   b. Species
   c. Family
   d. Phylum

13. True or False? Anyone can employ the scientific method to answer questions they have about the world around them.

14. True or false? One of the criteria of the scientific method is the sharing of information.

15. Creating a general statement from an observation is referred to as

   a. theorizing.
   b. deductive reasoning.
   c. inductive reasoning.
   d. hypothesizing.

16. In the figure below, what step of the scientific method is most likely being practiced?

   a. Hypothesizing
   b. Observing
   c. Communicating
   d. Experimenting

17. If a scientific discovery has both personal and political ramifications, it would be best to

   a. rely on the media to inform you of the best use of the discovery.
   b. read one small article in your local paper to stay informed.
   c. read and evaluate every article that you can find on the subject.
   d. ask your neighbors what they think, and go along with their opinion.

Everyday Chemistry of Life

2

The entire science of biology rests on a foundation of chemistry, and the exact interactions among chemicals determine everything from birth to death. In fact, everything in our universe is made of chemicals, from the planets to the oceans to you. Understanding the chemistry of human life—and of life in general—must rest on an understanding of chemicals and chemical interactions. Yet when you mention "chemistry," most people exhibit a classic "fight or flight" reaction. This is sad, because chemistry is actually a vibrant and wholly understandable subject. And once you know a few basic concepts about biological chemistry, you will suddenly find yourself understanding more about the health-related information we hear every day:

"Cut your cholesterol." Why? What's cholesterol, and can I live without it?

"Stop eating so much saturated fat. Unsaturated fats are better for you." Why? What's the difference?

We use chemistry in our daily lives, without realizing it. When we bathe, cook, clean, or even water our plants, we do so because we know certain chemical reactions will occur—even if we can't name them. We use water to remove soil from our hands, adding detergent to loosen greasy or stubborn soil. We turn to more powerful organic solvents when we need to thin some enamel paint or to remove an oil spill. We water our plants when they wilt, and we add nutrient chemicals when their growth slows. We eat when we feel hungry, expecting our bodies to convert food to the energy we need to function.

Life: It's all about chemicals.

CHAPTER OUTLINE

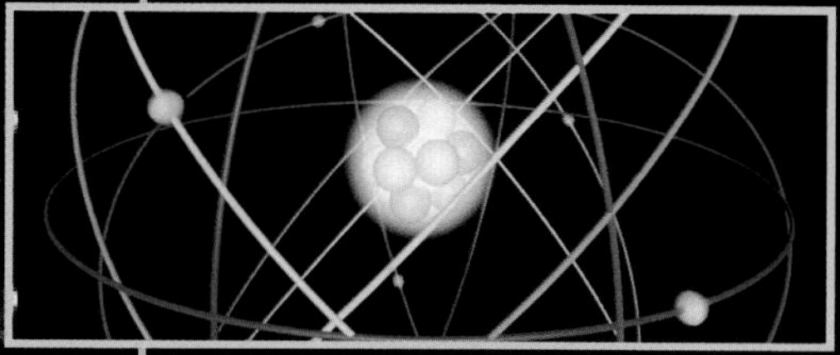

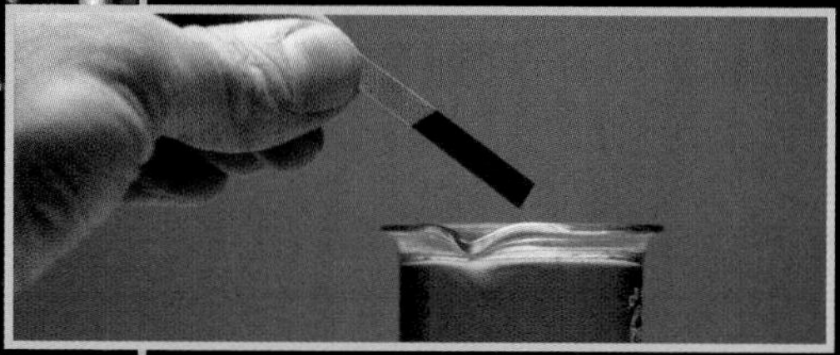

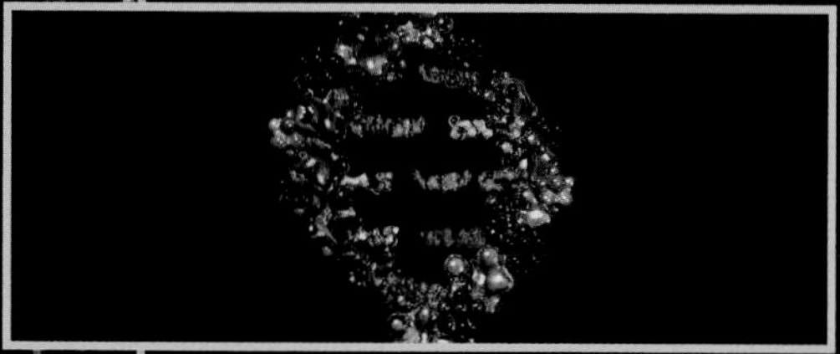

Life Has a Unique Chemistry

Learning Objectives

Identify the four most common chemicals in living organisms.

Define trace elements and give two examples.

Humans and the rest of the living realm are made of multiple chemicals, but four **elements** predominate: oxygen, carbon, hydrogen, and nitrogen. We also contain relatively high quantities of calcium, phosphorus and sulfur, sodium, chlorine, and magnesium (FIGURE 2.1). Although trace elements are less abundant in the body, some of them are necessary for life, such as iron, iodine, and selenium. Most of these trace elements are for sale at your local pharmacy, next to the multivitamins. Sometimes our diets do not provide enough of these trace elements. But they are needed in extremely small doses, minute traces in fact, so taking supplemental minerals will enhance bodily performance ONLY if your diet lacks them to begin with. (See I Wonder . . . Do I need to take dietary supplements . . . ?)

Element
A substance made entirely of one type of atom; cannot be broken down via chemical processes.

Chemistry is a story of bonds made and bonds broken. Bonds between atoms determine how chemical compounds form, fall apart, and re-form. When we metabolize sugar, for example, we are essentially combining its carbon and hydrogen atoms with oxygen, forming carbon dioxide and water. These reactions produce heat and energy that the body uses for just about every purpose. If we don't use sugar and related compounds right away, some of them are converted to fat—larger molecules that store even more energy in their chemical bonds.

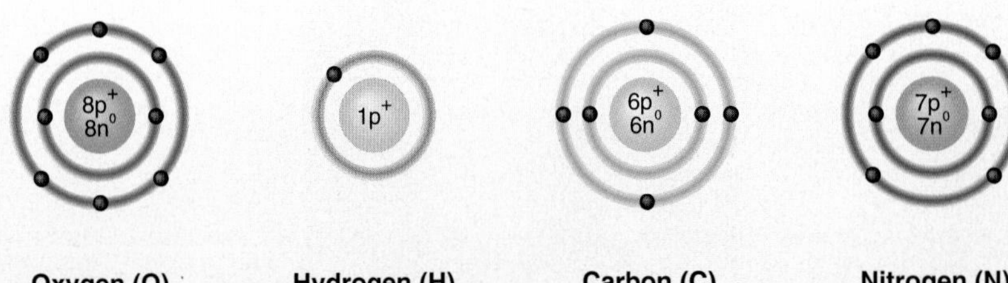

The atomic structures of oxygen, hydrogen, carbon, and nitrogen FIGURE 2.1

These four elements are most prevalent in living organisms.

I WONDER . . .

Do I need to take dietary supplements to stay healthy? Will they help me lose weight?

To answer these questions, we must understand the roles of minerals and vitamins in the body. Your body needs many micronutrients in small, steady amounts. Vitamins, for example, aid in growth, vision, digestion, blood clotting, and mental alertness. They catalyze reactions, and they are essential to the use of carbohydrates, fats, and proteins. Vitamins supply no calories, but without them you could not obtain energy from your food. Minerals are the main components of teeth and bone and help control nerve impulses and maintain fluid balance. You need at least 250 milligrams (mg) of calcium, phosphorus, magnesium, sodium, potassium, and chloride each day. Chromium, copper, fluoride, iodine, iron, manganese, molybdenum, selenium, and zinc are considered trace minerals; the daily requirement of trace minerals is less than 20 mg.

Micronutrients must be obtained from your diet because you have no mechanism for manufacturing them. If a change in diet causes a micronutrient deficiency, supplements will improve functioning and health. Complaints of fatigue, poor vision, digestive difficulties, even malaise, can sometimes be traced to the lack of a specific, essential mineral or vitamin. But dietary supplements will help ONLY if your diet lacks the micronutrient you are supplementing. Taking extra micronutrients will do nothing at best and at worst could cause other health problems. Too much vitamin B-3 can injure the liver, and excess vitamin A can weaken bones. And chromium? Well, we know that it helps your body use sugar and that a lack of chromium leads to nerve problems and decreased sugar usage. But there is no evidence that excess chromium will improve sugar use.

Many people take vitamins A, C, and E, in the hope that these antioxidants will slow the oxidative damage that accompanies aging. But vegetables contain a huge mix of antioxidants, and they seem to be far more effective at sustaining health than anything you can buy in a bottle.

When it comes to micronutrients and antioxidants, your best bet is to eat a healthy diet, just like mother said!

We cannot leave the subject without mentioning the mushrooming supplement business. For many reasons, these compounds are not regulated by the federal government, and the result is a mishmash of untested products on the shelves. Aunt Mary says one helps her rheumatism. Mr. Lopez says another helps him sleep at night. Ibrahim thinks a third one helps with that trick knee he injured in college.

Maybe. But maybe not. In fact, there is absolutely no incentive for supplement makers to test their products against placebos (inert look-alikes) or against medicines that do work. So they are marketed with vague claims like "Improves joint health" or "Take as recommended by your provider." The truth is that, although the plant kingdom has provided medical ingredients for thousands of years, nobody knows the effect of most nutritional supplements. Good? We don't know. Bad? Perhaps. Helpful in some cases, harmful in others? Could be.

Science is about data, and where there is no data, there is no way to answer the question. Sorry, but forewarned is forearmed.

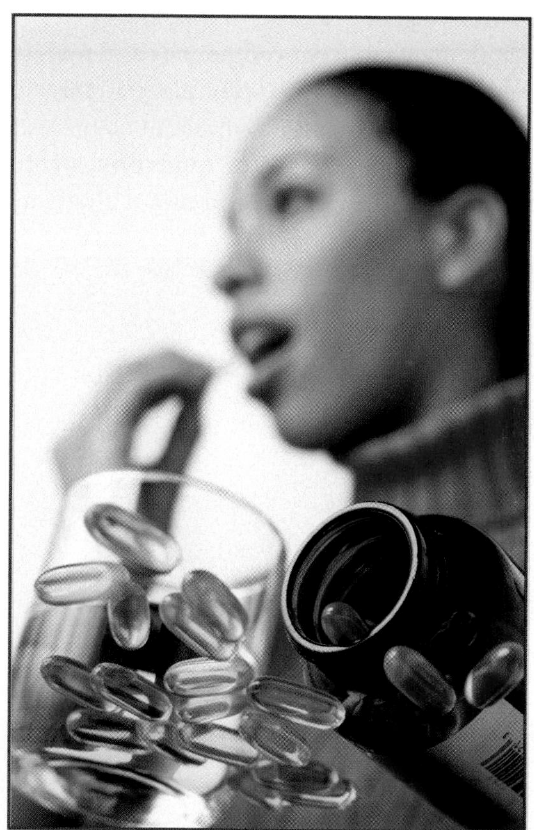

CONCEPT CHECK

List the most common chemicals in the human body.

Why do multivitamins and mineral supplements often make no difference in overall health?

Atomic Structure Is the Foundation of Life

LEARNING OBJECTIVES

Recognize the structure of an atom.

Understand the structure of the periodic table and the meaning of the numbers on it.

Define the relationship between valence electrons and atomic reactivity.

toms are mostly empty space. A cloud of electrons orbits the nucleus. The electrons stay in orbit through electrical attraction to the proton.

Atoms are the basis for the chemical world, and each atom is the smallest possible sample of a particular element (FIGURE 2.2A). Most atoms can react with other elements to form **compounds** and **molecules**. (A molecule can contain any two or more atoms; a compound must contain at least two different elements.)

Atoms are composed of **neutrons**, **protons**, and **electrons**. Neutrons and protons are always in the nucleus, at the center of the atom, and the electrons move rapidly around the nucleus. Elements are defined by the number of protons; all atoms of a particular element have the same number of protons. Protons and neutrons each have a **mass** of approximately 1 **dalton**; electrons are far less massive.

Neutron The neutral particle in the atomic nucleus.

Proton The positive particle in the atomic nucleus.

Electron The negative particle, found in orbitals surrounding the nucleus.

Mass The amount of "substance" in an object ("weight" is the mass under a particular amount of gravity).

Dalton A unit of mass, equal to the mass of one proton.

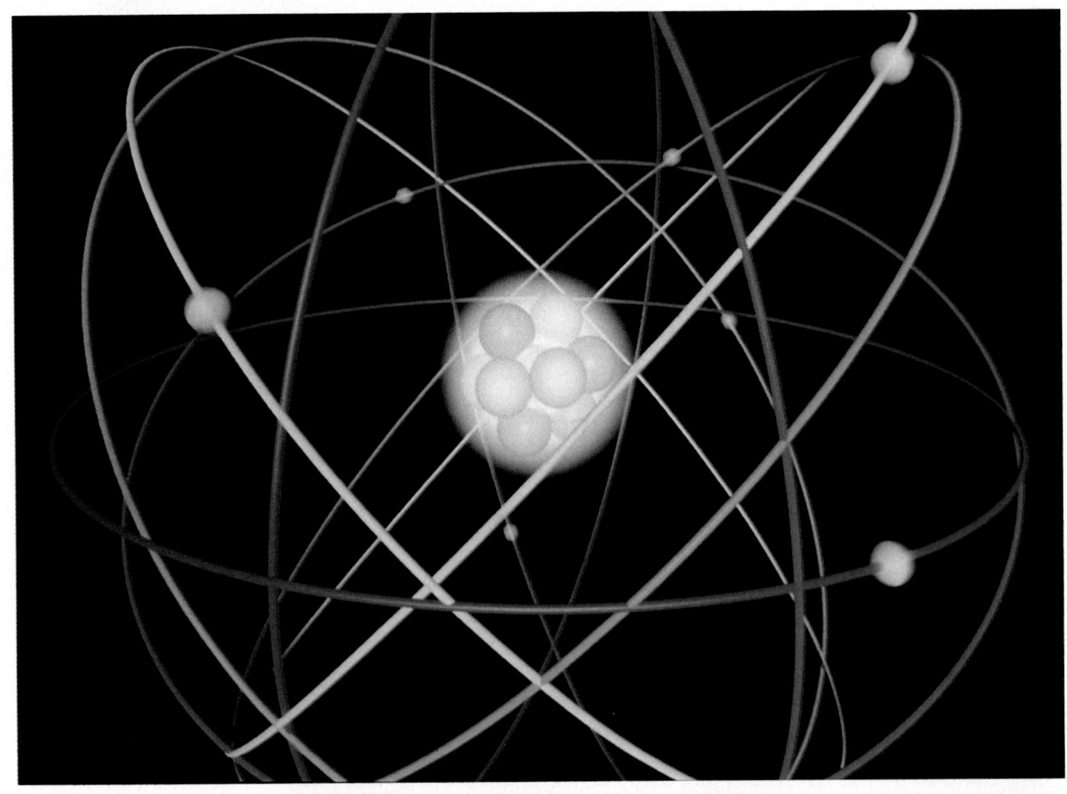

Representative diagram of an atom FIGURE 2.2A

Many elements have **isotopes**, with the same number of protons but a different number of neutrons. All isotopes of a particular element are chemically identical but have different masses, owing to the change in neutrons. The number of neutrons equals the atomic mass minus the **atomic number**. The **atomic mass** in the periodic table is an average mass for the element.

Adding or subtracting protons from a nucleus creates a new element through a nuclear reaction. New elements form inside stars, nuclear reactors, and nuclear bombs. They also form through **radioactive decay**, which breaks down **radioisotopes**. When these unstable isotopes break apart, they release energy and form less massive atoms, which may break again into other elements. The emitted energy can be helpful or harmful, as you can see in the Health, Wellness, and Disease box on page 32.

Atomic number The number of protons in the nucleus of an atom.

Atomic mass The mass of the atom; different isotopes have different atomic masses.

Radioactive decay Change of an atom into another element through division of the nucleus and the release of energy.

The number of electrons always equals the number of protons in a neutral (uncharged) atom. Protons have a positive charge and a mass, whereas electrons carry a negative charge but no appreciable mass. The electromagnetic attraction between protons and electrons prevents the electrons from leaving the atom (FIGURE 2.2B). The positive–negative attraction between proton and electron resembles the north–south attraction between refrigerator magnets and steel refrigerator doors.

But what prevents electrons from slamming into the protons? Magnets, after all, tend to stick to the fridge door. The answer comes from a branch of physics called quantum mechanics, which treats electrons as waves as well as particles. These waves must orbit the nucleus in complete waves; fractional waves are not allowed. An electron wave cannot drop down half a wave—so it stays in a specific orbit, or jumps up or down a full orbit.

Electrons repel each other because they all carry negative charge. This repulsion is much like what happens when you try to force the north poles of two magnets together. This repulsion, combined with the wave behavior just mentioned, channels electrons into specific energy levels, called **orbitals**, which define the most likely location of electrons at any given moment. Orbitals are best imagined as clouds of electrons surrounding the nucleus.

The outermost energy level of electrons, or **valence shell**, is most important in chemistry and biology, because that is where atoms bond together. The Roman numeral above each column in the periodic table

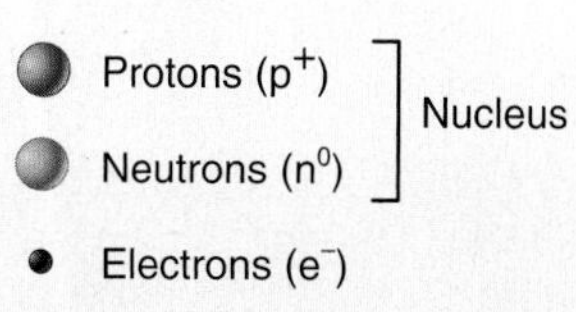

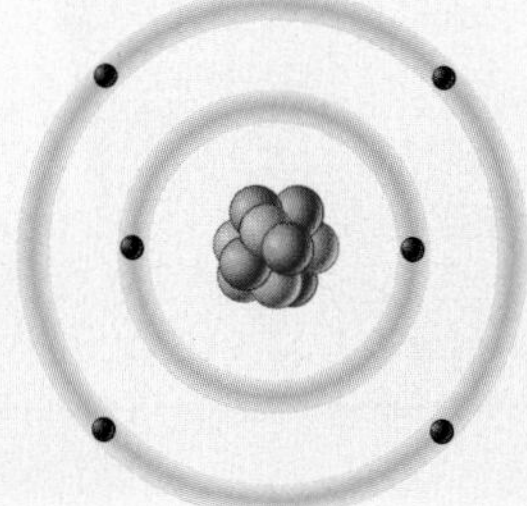

The electron shell model

FIGURE 2.2B

The outer shell of electrons determines whether and how an atom bonds with other atoms.

Radioisotopes in science and medicine

Radioactive elements are atoms that decay spontaneously, releasing energy as they break down into other, smaller atoms. Many people fear radiation, but these elements, called radioisotopes, do have roles in science and medicine. Take the natural radioisotope carbon 14. Whereas most (C_{12}) carbon has 6 neutrons, a small percentage has 8 neutrons (C_{14}). After 5,730 years, half of the C_{14} in any sample converts through radioactive decay to the stable element N_{14}. This time span is called the half-life of C_{14}, and it is essential to carbon dating. Because living things respire, organic materials start out with the same C_{12}/C_{14} ratio as the atmosphere. After an organism dies, it stops taking in C_{14}, which starts decaying to N_{14}. We can measure how much C_{14} is present and figure out when it started decaying based on the fact that every 5,730 years half the C_{14} in a sample turns into nitrogen.

The decay of other radioisotopes, including uranium, helps date even older objects. Radioactive dating has helped estimate the age of the Earth and of fossils and has been used to test whether certain religious artifacts were actually present during the time of Christ.

Radioactive isotopes are produced in nuclear reactors for two basic medical purposes: (1) to serve as *diagnostic* tools, or radiotracers, and (2) to *treat* cancer with radiation therapy, which takes advantage of the destructive effects of ionizing radiation, radiation that is capable of knocking electrons free.

Various organs and tissues absorb specific chemicals. The thyroid, for instance, absorbs iodine, whereas the brain uses large amounts of glucose. Radiotracers work because all isotopes of a particular element are chemically identical, so the tracers move just like the usual form of the element. Radio tracers usually have short half-lives and emit relatively harmless radiation that is easy to track from outside of the body. Radiotracers are commonly used to detect blocked cardiac arteries by tracking blood flow through the heart.

Some radiation therapy uses isotopes to produce destructive, local irradiation. This intense energy is directed to destroy specific tissues without harming surrounding tissue. Cells that are dividing quickly are particularly susceptible to ionizing radiation. These cells are found not only in cancers, but also in the digestive tract and the hair-forming follicles on the scalp, which explains the nausea and hair loss that may accompany radiation treatment. Radioisotopes may be implanted temporarily in the body to destroy local tissue.

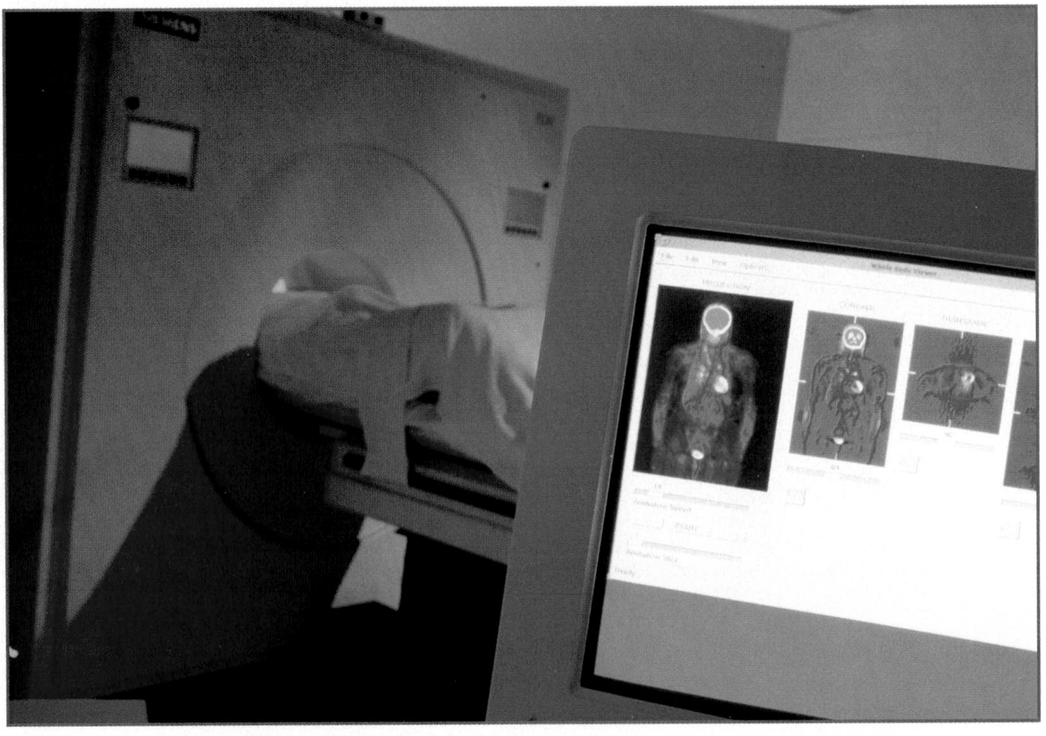

Ion Charged atom.

indicates the number of valence electrons of all the atoms in that column. That number tells us how the atom will react with other atoms:

- An atom with one to three electrons in its valence shell can lose electrons, forming a positive **ion**. The positive charge results when electrons are lost, because the number of protons does not change.
- An atom with five to seven electrons in its outer shell tends to grab electrons to "fill" the valence shell with eight electrons. These atoms become negative ions able to participate in a chemical reaction. Just as protons and electrons attract, so do positive and negative ions.
- An atom with eight electrons in the valence shell will usually not bond, because the valence orbital is full. Elements with eight valence electrons include "noble gases" like neon and argon.

Chemistry encompasses a vast amount of information that can be useful only if it is organized. A card player knows it's almost impossible to tell which cards are missing from a glance at a shuffled deck. But if you arrange the cards numerically by suit, the pattern reveals which cards are missing. In chemistry, the **periodic table** (see Appendix A at the back of the book for a full version of the periodic table) organizes all elements in a logical pattern, according to atomic number. As you now know, the atomic number is the number of **protons** in the nucleus. The table also reveals an element's reactivity—its ability to bond with other elements, as reflected in the valence electrons. Elements in a particular column have the same number of valence electrons, and thus similar reactive properties. If we are familiar with any element in a column, we can predict the reactivity of other elements in that column.

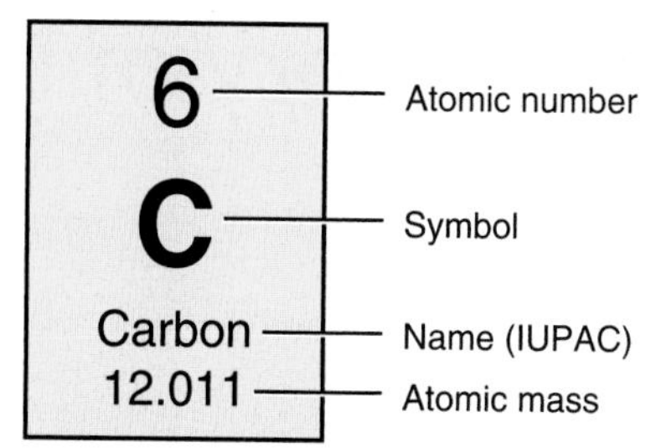

Carbon as it appears on the periodic table

FIGURE 2.3

The periodic table is the fundamental tool of chemistry. It lists the atomic number and atomic mass.

The periodic table lists each element by a standard one- or two-letter abbreviation (FIGURE 2.3). The Internet provides many places to study the periodic table.

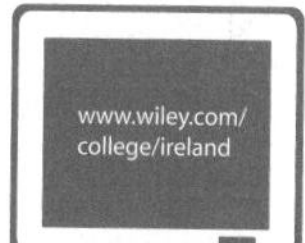

CONCEPT CHECK

What is carbon's atomic number?

Why is the atomic mass of carbon different from its atomic number?

How many electrons are in one oxygen atom? How many protons? How many neutrons?

What happens to the electrons of an element with two valence electrons during a chemical reaction?

Chemistry Is a Story of Bonding

LEARNING OBJECTIVES

List the three types of chemical bonds and give an example of each.

Compare the strength of each bond.

Define van der Waals forces, and compare their strength to that of a covalent bond.

Describe the characteristics of polar and nonpolar molecules.

Life is made of atoms, but atoms are only the building blocks of molecules and chemical compounds. A molecule is a chemical unit formed from two or more atoms. H_2 for example, is a molecule of hydrogen. A **compound** is a molecule with unlike atoms: CO_2, carbon dioxide, is both a molecule and a compound. The chemical properties of a compound have little or nothing to do with the properties that make up the atoms. Sodium, for example, is a soft metal that burns when exposed to air. Chlorine is a toxic gas at room temperature. But sodium chloride is table salt.

Chemical bonds are a matter of electrons. Atoms without a "filled" valence shell adhere to one another by sharing or moving electrons. Atoms can bond in three common ways, ranging in strength from the strong ionic bonds of salts to the weak hydrogen bonds that hold DNA molecules together.

The **ionic bond** holds ions in a compound, based on the strong attraction between positive and negative ions—something like the north–south attraction between two refrigerator magnets discussed previously. The interactions between sodium and chlorine show a typical ionic bond (FIGURE 2.4).

Many ions in the human body, including calcium (Ca^{2+}), sodium (Na^+), potassium (K^+), hydrogen (H^+), phosphate ($PO_4^{\ 3-}$), bicarbonate (HCO_3^-), chloride (Cl^-), and hydroxide (OH^-), can form ionic bonds. All these ions play significant roles in homeostasis. In some people, too much sodium can raise blood pressure. Too little calcium causes soft, weak bones as in rickets, and potassium and calcium imbalances can cause heart irregularities. The other ions are vital to maintaining the blood's acid/base balance. If ion levels do not stay within normal range, cellular functions can cease, leading to the death of tissue, organs, and even the organism.

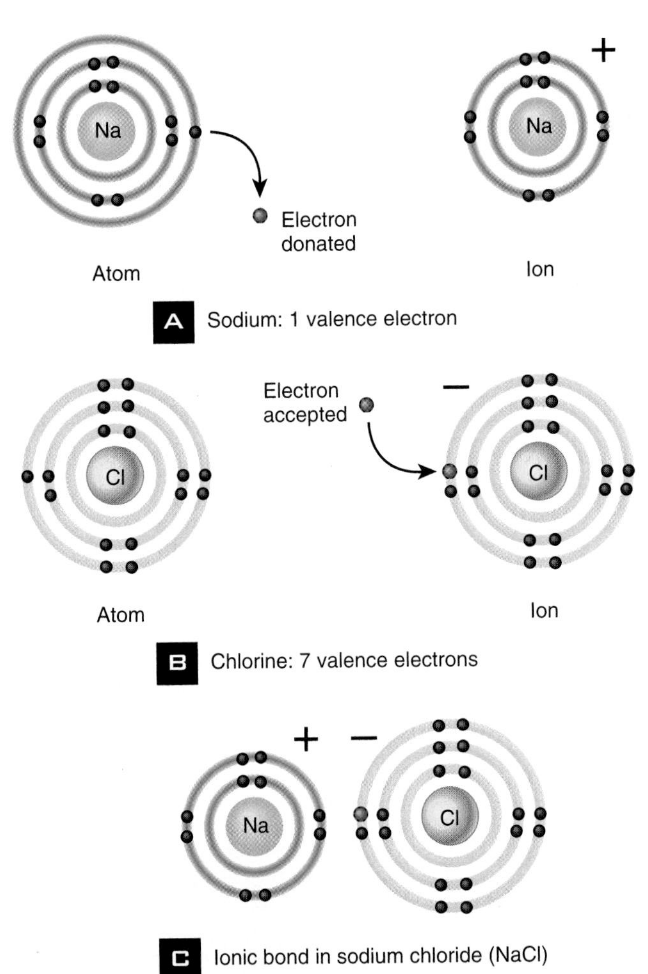

The ionic bond of an NaCl salt molecule
FIGURE 2.4

A typical ionic bond: Sodium atoms have one electron in the outer orbital. If this electron is stripped away, the atom becomes a sodium ion (Na^+). Chlorine atoms have seven valence electrons, so they tend to attract free electrons, forming a chloride ion (Cl^-). The attraction between the two ions is an ionic bond.

Although ions are common in the body, **covalent bonds** are actually more important than ionic bonds to living tissue. In covalent bonds, atoms share electrons; electrons are not donated by one atom and grabbed by another, as in an ionic bond. Covalent bonds commonly involve carbon, oxygen, nitrogen, or hydrogen, the elements predominant in life. In a covalent bond, atoms share electrons so that each gets to complete its valence shell.

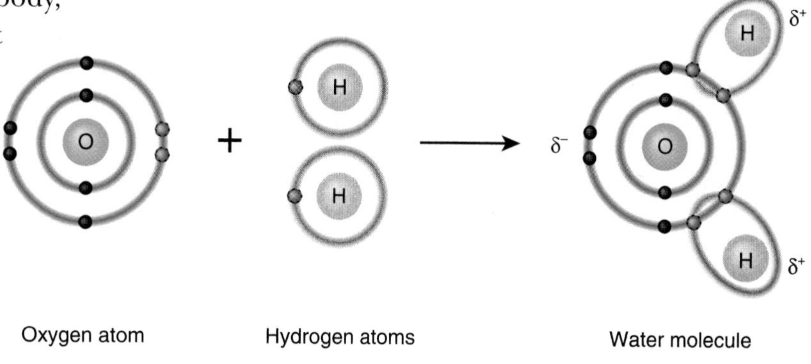

Polar covalent bonds FIGURE 2.5

Two atoms share one pair of electrons in a single covalent bond, as occurs in a hydrogen molecule. Single covalent bonds are shown in chemical diagrams as one line: H—H.

In a double covalent bond, two pairs of electrons are shared. For example, two oxygen atoms form an oxygen molecule (O_2) by sharing four electrons. Each oxygen atom has six electrons in its valence shell; with the addition of two more electrons, it gets that stable shell of eight electrons. Double covalent bonds are shown in chemical diagrams with a double line: O═O.

In a triple covalent bond, three pairs of electrons are shared. This is the way that two atoms of nitrogen form a nitrogen molecule (N_2). Nitrogen has five electrons in its valence shell; by sharing six electrons between the two, each can add three electrons, making eight in the valence shell. Triple covalent bonds are shown as a triple line N≡N.

Carbon has four electrons in the valence shell, so it can complete the valence shell by sharing four electrons. When carbon is in a covalent bond, the electrons are distributed equally between the atoms. Neither atom has a strong enough charge to pull the electrons off the other, but the electromagnetic force of the nuclei does affect the placement of those electrons. Rather than strip electrons from one atom and carry them on the other, the two atoms share the electrons equally. This creates a **nonpolar** molecule (one that is electrically balanced).

Most covalent bonds in the human body are nonpolar. In some cases, however, one atom has a stronger attraction for the shared electrons (it reminds us of trying to share a cell phone with an older sibling). Unequal electron-sharing on the atomic level creates **polar covalent bonds** (FIGURE 2.5).

In a polar covalent bond, shared electrons reside preferentially near one nucleus, forming a polar molecule. Part of the molecule has a slight negative charge, because the electrons are there more often. The other part of the molecule carries a slight positive charge. Water, a compound that is essential to all forms of life, is a polar molecule; the polar bonds account for many of water's life-giving characteristics.

The **hydrogen bond** is weak but is still vital to biology. When a hydrogen atom is part of a polar covalent bond, the hydrogen end of the molecule tends to be more positive, leaving the other end more negative (FIGURE 2.6). This results in a molecule with a charge gradient along its length. Although the hydrogen bond is too weak to bond atoms in the same way as covalent or ionic bonds, it does cause attractions between nearby molecules. Hydrogen bonds join the two

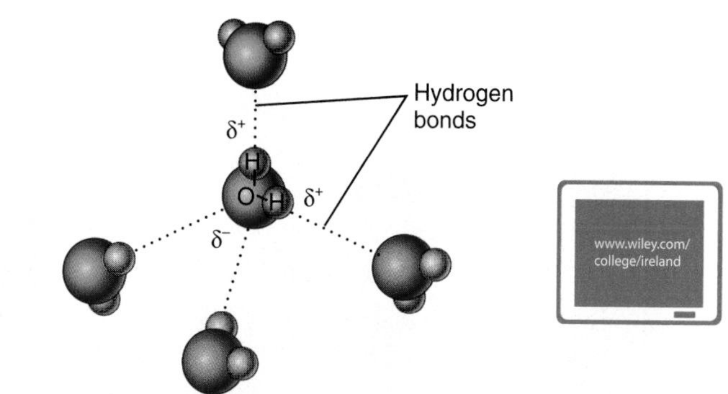

Hydrogen bonds between water molecules FIGURE 2.6

Hydrogen bonds are weak, but they play vital roles in biology.

Van der Waals forces in nature

FIGURE 2.7

Do you recognize a lizard called the gecko? We now know that it uses van der Waals forces to walk up walls and across ceilings. The foot pads of this lizard are designed to enhance the surface area in contact with the wall. Van der Waals forces literally stick the gecko's foot to the tree branch, as shown here.

strands of DNA (your genetic material) in the nucleus of your cells. They also help shape proteins, the building blocks of living bodies.

Hydrogen bonds occur between water molecules because the partially negative oxygen atom in one molecule is attracted to the partially positive hydrogen atoms of another molecule.

A fourth category of atomic interaction, **van der Waals force**, has interesting implications for biology. These forces are extremely weak, resulting from intermittent electromagnetic interactions between resonating molecules. As atoms vibrate and electrons whirl in their clouds, various regions briefly become positive or negative. Van der Waals forces occur when these intermittent charges attract adjacent molecules that briefly have opposite charges (FIGURE 2.7).

Bonds do more than hold atoms together in molecules. They also contain energy. Some bonds absorb energy when they form. These **endothermic** reactions include the formation of longer-chain sugars from short-chain, simple sugars. Endothermic reactions are used to store energy in the body for later release.

In an **exothermic** reaction, energy is released when the bond is formed. A common exothermic reaction is simple combustion: $C + O_2 = CO_2$. A second is the burning of hydrogen: $2\ H_2 + O_2 = 2\ H_2O$.

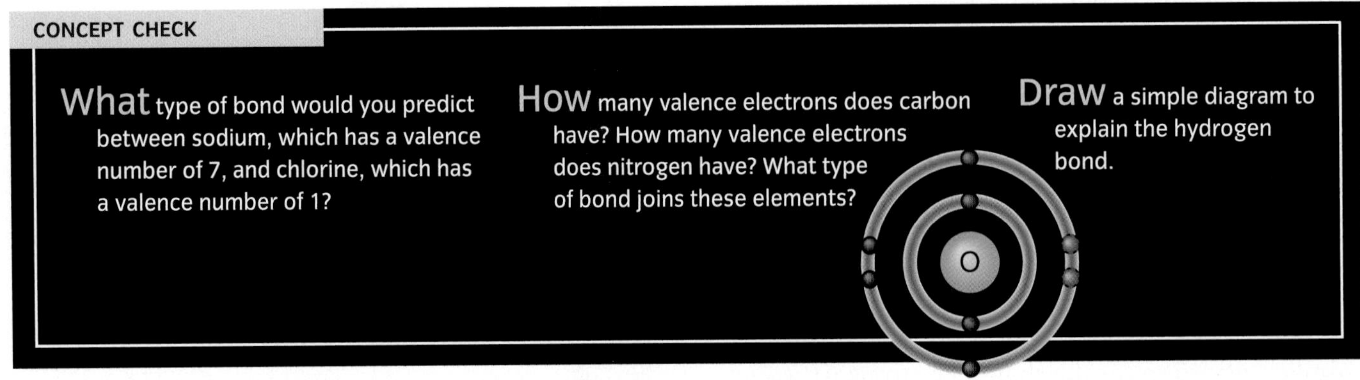

CONCEPT CHECK

What type of bond would you predict between sodium, which has a valence number of 7, and chlorine, which has a valence number of 1?

How many valence electrons does carbon have? How many valence electrons does nitrogen have? What type of bond joins these elements?

Draw a simple diagram to explain the hydrogen bond.

Water Is Life's Essential Chemical

LEARNING OBJECTIVES

Define the six properties of water that are critical to life.

Explain the biological importance of heat of vaporization and latent heat of water.

We all know water (FIGURE 2.8). We drink water; swim in it; surf, ski, and float on it; and use it to maintain our lawns and plants, and even cool our vehicles and heat some of our homes. It is the most abundant molecule in living organisms, making up between 60 and 70 percent of total body weight. Our bodies need water to carry out the basic functions of digestion, excretion, respiration, and circulation. Without adequate water, the body's chemical reactions would fail, and our cells would cease to function—we would die. These six properties of water are critical to life:

1. Water is liquid at room temperature, whereas most compounds with similar molecular weights are gases. At sea level, water becomes a gas (vaporizes) only at or above 100°C. Water remains liquid due to the hydrogen-bond attraction between molecules.

2. Water is able to dissolve many other substances, and therefore, is a good solvent. The two atoms of hydrogen and one of oxygen have polar covalent bonds, making the molecule polar. This polar characteristic sets up a lattice of water molecules in solution. As water molecules move, the hydrogen bonds between them continually make and break. Substances that are surrounded by water are subjected to constant electromagnetic pulls, which separate charged particles—causing the compound to break down, or dissolve. Polar covalent molecules align so that negative ends and positive ends sit on the respective complementary areas of the solute and pull it apart. **Hydrophilic** substances, such as NaCl (salt), carry a charge and are immediately separated in water. **Hydrophobic** substances are not soluble in water. Hydrophobic substances include large, uncharged particles like fats and oils. In the human body, fats and oils separate cells from the surrounding fluids of the body.

Hydrophilic Having an affinity for water.

Hydrophobic Lacking affinity for water.

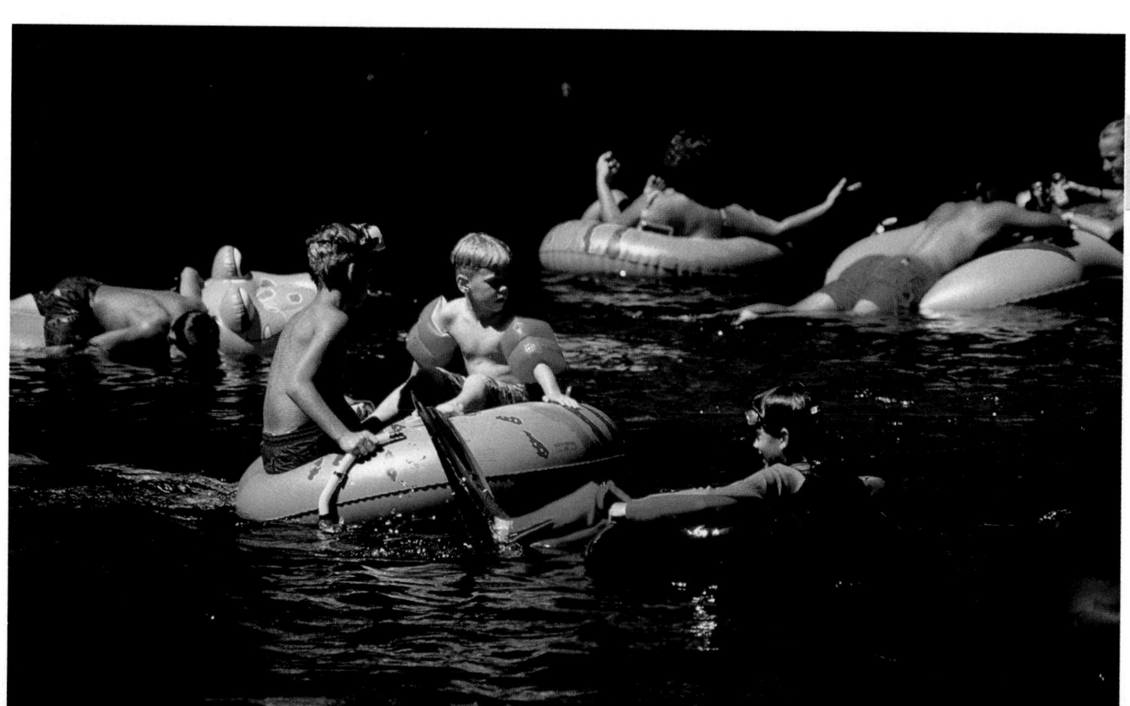

Water
FIGURE 2.8
Water is life—and the chemistry of biology begins with the study of water.

The polar molecules of water, combined with its ability to form hydrogen bonds with other charged particles, allow it to dissolve most substances, which aids in chemical reactions. Even though water cannot dissolve hydrophobic compounds, it is still called the "universal solvent."

3. Water is both **cohesive** and **adhesive**, allowing it to fill vessels and spaces within the body. This property also allows water to line membranes and provide lubrication. Your blood is 92 percent water, which allows it to stick to the sides of the vessels and fill them completely.

4. Water has a high **specific heat**—it takes a lot of energy to raise or lower its temperature. It takes one **calorie** of energy to raise the temperature of one gram of water one degree Celsius. (A different calorie is used in dieting: It is actually a kilocalorie: 1 kcal equals 1,000 calories.) Water therefore serves as a temperature buffer in living systems. Water does the same for the Earth. Look at a weather map and compare the temperature ranges for coastal and inland areas. The temperature range is much smaller near the coast. The highest and lowest temperatures ever recorded both come from inland areas. Vostok, Antarctica, located in the center of that contiment, hit an amazingly frigid −89°C in 1983.

Cohesive
Having the ability to stick to itself.

Adhesive
Having the ability to stick to other surfaces.

5. Water has a high **heat of vaporization**, a measure of the amount of heat needed to vaporize the liquid. A large amount of heat energy, 540 calories, is needed to convert 1 gram of water to vapor. This is important for thermal homeostasis. Your body cannot survive unless it remains in a narrow temperature range, and a great deal of excess heat is generated by cellular activity. Much of this heat is lost through the evaporation of water from your skin. As your core temperature rises, your body responds by increasing sweat production to increase evaporative heat loss. (A second homeostatic regulation to maintain the all-important temperature is increase in blood flow, which transfers heat from the core to the skin.)

6. **Ice floats**. As water cools, the molecules lose energy and move more slowly. The hydrogen bonds that continuously break and re-form in the liquid cease to break, and the water turns solid. The bonds hold a specific distance between the molecules, making solid water slightly less dense than liquid water. Freezing a can of soda shows what happens: As the water inside freezes, the can deforms and may even rip open. Frostbite can occur if tissues freeze. The water within and between the cells expands, bursting and crushing the cells. The tissue dies because its cellular integrity is lost. On the positive side, ice that forms on lakes stays at the surface, allowing fish to survive in the cold (but liquid) water near the bottom.

CONCEPT CHECK

What is meant by "water is the universal solvent"? What property of the molecule allows it to dissolve hydrophilic compounds?

How does water's heat of vaporization help you maintain a steady internal temperature?

Why is it important to biological systems that ice floats?

Hydrogen Ion Concentration Affects Chemical Properties

Learning Objectives

Develop an understanding of the pH scale.

Grasp the biological significance of acids, bases, and buffers.

One of the most important ions is hydrogen, H^+, which is simply a bare proton. In pure water, some of the molecules dissociate, releasing equal numbers of H^+ and hydroxide ions (OH^-). Pure water is neutral. If the concentration of H^+ increases, the solution becomes acidic; if the OH^- concentration increases, it becomes basic, or alkaline.

Acidity matters to the human body, because it affects the rate of many chemical reactions and the concentration of many chemicals. As we'll see shortly, the body has many mechanisms for maintaining proper acidity, through the use of buffer systems.

Lemon juice, orange juice, cranberry juice, vinegar, and coffee are common acids. They taste "sharp" and can cause mouth sores or indigestion if consumed in large quantities. The bite in carbonated beverages results from the formation of carbonic acid in the drink. When these beverages go "flat," the acid content is reduced because the carbonic acid has been converted to carbon dioxide, which leaves the solution as carbonation bubbles. A freshly opened typical carbonated drink is about pH 2.5. After a few hours, it would have increased to about 8 and would taste terrible.

The **pH scale** measures the concentration of H^+ and OH^-. Put another way, it measures acidity or alkalinity (Figures 2.9 and 2.10).

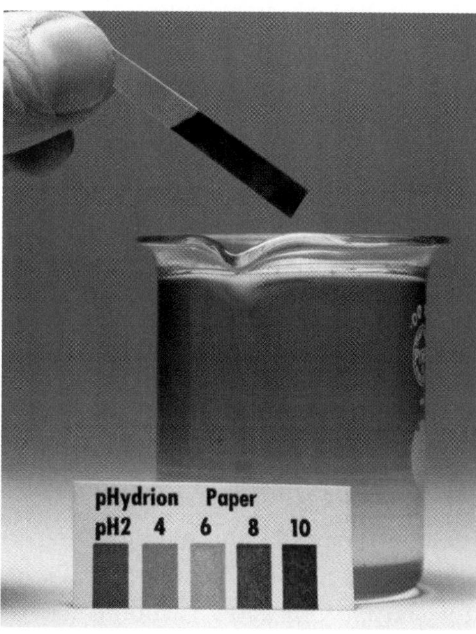

Testing the pH of a solution Figure 2.9

The pH scale is logarithmic: a change of 1 means a 10× change in H^+ concentration.

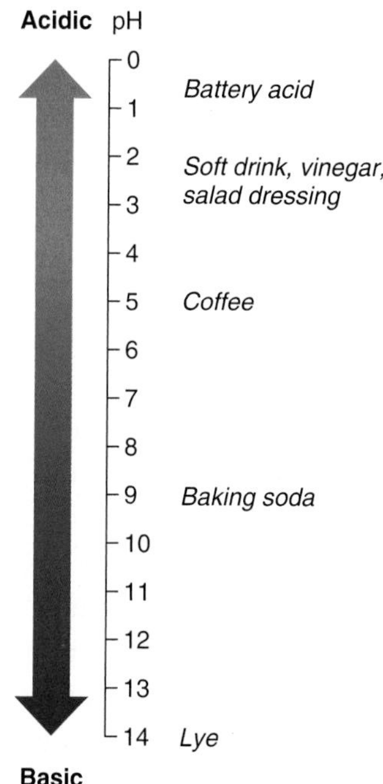

Scale of acidic and basic household items Figure 2.10

Hazardous material FIGURE 2.11

When a household cleanser contains strong acid or base, it should carry a warning like this.

The pH scale ranges from 0 to 14. Pure water registers 7, meaning it has equal numbers of H^+ and OH^- ions. In pure water, 10^{-7} **moles** of molecules dissociate per liter (1 mole = 6.023×10^{23} atoms, molecules, or particles). Lower pH numbers indicate higher H^+ concentrations and greater acidity. The pH scale is logarithmic: Each one unit represents a tenfold change in H^+ concentration. So a change from pH 3 to pH 8 would reduce the H^+ concentration by a factor of 100,000.

Strong acids dissociate (break apart) almost completely in water, adding a great deal of H^+ to the solution. For example, HCl, hydrochloric acid, dissociates into H^+ and Cl^- ions. Weak acids dissociate poorly, adding fewer H^+. Hydrochloric acid, one of the strongest acids used in the laboratory and also found in your stomach, is pH 2. Concentrated hydrochloric acid can injure the skin in minutes or dissolve a steel nail in a few days, which is a bit frightening when you realize soft drinks are very nearly one pH unit (not quite 10 times) less corrosive (FIGURE 2.11)!

A basic solution has more OH^- ions than H^+ ions and a pH of 7.01 to 14. Like acids, bases are classified as strong or weak, depending on the concentration of OH^-. Like strong acids, strong bases are harmful to living organisms because they destroy cell structure. Common bases include soaps such as lye, milk of magnesia, and ammonia. Basic solutions generally taste bitter and feel slippery, a feeling you may have noticed the last time you cleaned with ammonia.

A pH indicator measures a solution's acidity or alkalinity. One of the first pH indicators was litmus, a vegetable dye that changes color in the presence of acid or base. Litmus turns from blue to red in the presence of acids, and from red to blue with bases. This test is simple and so definitive that it has become part of our language. In extreme-sport circles, you might hear, "That jump is the litmus test for fearless motocross riders."

Acids and bases cannot coexist. If both H^+ and OH^- are present, they tend to neutralize each other. When a base dissociates in water, it releases hydroxide ions into the solution. But if a base dissociates in an acidic solution, its OH^- ions bond to H^+ ions, forming water, which tends to neutralize the solution. Similarly, when you add acid to a solution, it may start out by neutralizing any OH^- that is present, reducing the pH.

Your body cannot withstand a shift in acidity any better than it can a shift in temperature. The pH of your blood must stay between 7.4 and 7.5 for your cells to function. If you breathe too slowly, the carbon dioxide level in the blood will rise, and the blood can become more acidic. Because pH is critical to biological systems, various homeostatic mechanisms exist to keep it in the safe range. One of the most important of these are biological **buffers**, compounds that stabilize pH by absorbing excess H^+ or OH^- ions.

One of the most common buffering systems for blood pH consists of carbonic acid, H_2CO_3, and bicarbonate ion, HCO_3^-. In water, carbonic acid dissociates into H^+ and HCO_3^-. The H^+ can bond to OH^-, forming water, whereas the bicarbonate ion can bond to a hydrogen ion, re-forming carbonic acid. The carbonic acid-bicarbonate system works in either direction. When excess H^+ is present (the system is acidic), bicarbonate and hydrogen ion combine, forming carbonic acid.

$$HCO_3^- + H^+ \longrightarrow H_2CO_3$$

When hydrogen ion levels are too low, carbonic acid becomes a source of hydrogen ion:

$$HCO_3^- + H^+ \longleftarrow H_2CO_3$$

Chemists write this as a reversible reaction, with a double-ended arrow in the middle to indicate that it can go in either direction, depending on conditions around the reaction.

$$HCO_3^- + H^+ \rightleftharpoons H_2CO_3$$

A similar buffering system is used in some common anti-acid medicines. Many contain calcium carbonate, $CaCO_3$, which dissociates into calcium ion, Ca^{2+}, and carbonate ion, CO_3^{2-}, which neutralizes acid by accepting one or two hydrogen ions and becoming HCO_3^- or H_2CO_3. Calcium carbonate also provides calcium needed by the bones.

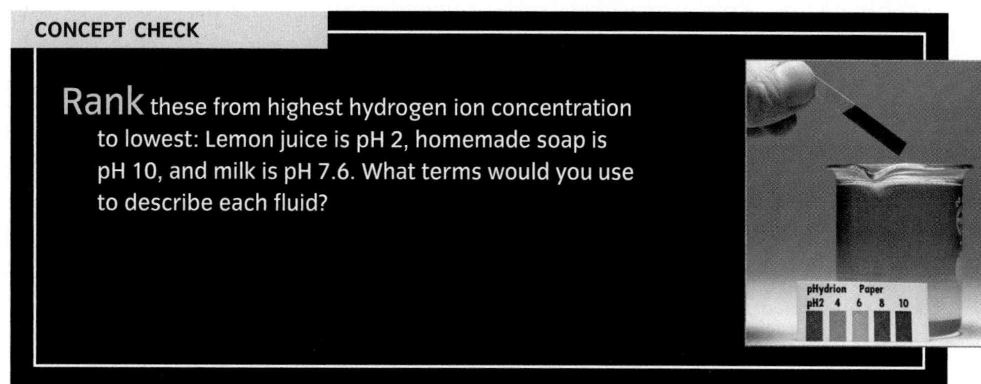

CONCEPT CHECK

Rank these from highest hydrogen ion concentration to lowest: Lemon juice is pH 2, homemade soap is pH 10, and milk is pH 7.6. What terms would you use to describe each fluid?

There Are Four Main Categories of Organic Chemicals

LEARNING OBJECTIVES

Identify the main categories of organic compounds.

Define the roles of carbohydrates, lipids, proteins, and nucleic acids in the human body.

Explain the function of ATP in energy storage and usage.

When we discuss life, we are discussing organic chemistry. Scientists used to think that all organic chemicals were made by organisms. Although that's not true, organic chemicals are usually made by organisms, and they always contain carbon. In terms of bonding, carbon is astonishingly flexible. With four valence electrons, it can bond covalently with four other atoms, leading to an almost infinite set of carbon structures, from simple methane, CH_4, to highly complex chains with many dozen carbon atoms. In organic compounds, carbon often bonds with two carbons and two

Functional group A subunit on an organic molecule that helps determine how it reacts with other chemicals.

hydrogens, thereby filling hydrogen's valence shell as well. The resulting hydrocarbon compounds can be chain or ring structures. Attached to the carbon/hydrogen core are **functional groups** (FIGURE 2.12) that determine the compound's reactivity. Carboxyl (COOH), amino (NH_3^+), and phosphate (PO_4^-) are some important functional groups.

Organic compounds are grouped into four main categories: **carbohydrates**, **lipids**, **proteins**, and **nucleic acids**.

CARBOHYDRATES ARE THE BEST ENERGY SOURCE FOR THE HUMAN BODY

Carbohydrates are the most abundant organic molecules in organisms. A carbohydrate is composed of carbon, hydrogen, and oxygen in a ratio of 1:2:1. Many carbohydrates are **saccharides** (sugars). Glucose (FIGURE 2.13) and fructose are both simple sugars. They are called **monosaccharides** because they have one ring of six carbons, with twelve hydrogens and six oxygens attached. **Oligosaccharides** and **polysaccharides** are longer sugar chains (*oligo* = few, and *poly* = many). **Disaccharides**, such as sucrose and lactose, are common in the human diet. **Glycogen** (FIGURE 2.14) is a polysaccharide sugar molecule stored in animal tissue. It is a long chain of glucose molecules, with a typical branching pattern. Glycogen is stored in muscle and liver, where it is readily broken down when needed.

Unlike glycogen, starch is a fairly long, straight chain of sugars. Plants store energy in **starch**, often in roots, tubers, and grains. **Cellulose** (Figure 2.14), another polysaccharide, has a binding pattern similar to glycogen. Cellulose is often used in structural fibers in plants and is the main component of paper. The difference between cellulose and glycogen depends on which particular carbon on the sugar ring connects the branches to the main chain. This small difference

Name and Structural Formula	Occurrence and Significance
Hydroxyl R–O–H	*Alcohols* contain an —OH group, which is polar and hydrophilic. Molecules with many —OH groups dissolve easily in water.
Sulfhydryl R–S–H	*Thiols* have an —SH group, which is polar and hydrophilic. Certain amino acids, the building blocks of proteins, contain —SH groups, which help stabilize the shape of proteins.
Carbonyl R–C(=O)–R or R–C(=O)–H	*Ketones* contain a carbonyl group within the carbon skeleton. The carbonyl group is polar and hydrophilic.
Carboxyl R–C(=O)–OH	*Carboxylic* acids contain a carboxyl group at the end of the carbon skeleton. All amino acids have a —COOH group at one end.
Ester R–C(=O)–O–R	*Esters* predominate in dietary fats and oils and also occur in our body triglycerides. Aspirin is an ester of salicylic acid, a pain-relieving molecule found in the bark of the willow tree.
Phosphate R–O–P(=O)(–O⁻)–O⁻	*Phosphates* contain a phosphate group (—PO_4^{2-}), which is very hydrophilic due to the dual negative charges. An important example is adenosine triphosphate (ATP), which transfers chemical energy between organic molecules during chemical reactions.
Amino R–N(H)–H or R–N(H)(H)–H	*Amines* have an —NH_2 group. At the pH of body fluids, most amino groups have a charge of 1^+. All amino acids have an amino group at one end.

Functional groups FIGURE 2.12

These functional groups are found on a variety of organic molecules. The "R" seen above stands for the "reactive group" or the "radical" and is usually a long string of carbon molecules.

All atoms written out

Standard shorthand

Glucose FIGURE 2.13

The glucose molecule, $C_6H_{12}O_6$, can be diagrammed in two ways.

makes cellulose indigestible to humans, whereas glycogen is an easily digestible source of quick energy. (Cows and other ruminants digest cellulose with the aid of countless bacteria in their digestive systems.)

Despite the hoopla surrounding the high-protein Atkins diet, carbohydrates are the best energy source for the human body: We are efficient carbohydrate-burning machines. Restricting intake of carbohydrates and increasing intake of other organic compounds puts biochemical stress on the whole body. When digesting proteins, for example, we generate nitrogenous wastes, which can release potentially harmful nitrogen compounds into our blood. These need to be detoxified and removed by our kidneys and liver.

Water is needed to digest carbohydrates. In the process of **hydrolysis**, digestive enzymes insert a water molecule between adjacent monosaccharides in the chain, disrupting a covalent bond between sugars and releasing one sugar molecule. In this manner, digestive enzymes separate glucose molecules from glycogen and starch. Once glucose enters a cell, it can be completely metabolized into carbon dioxide and water, producing energy through the process of cellular respiration described in Chapter 9. Because we lack the enzymes needed to remove sugar molecules from cellulose, all the cellulose we eat travels through our digestive system intact. This "fiber" is not converted into fuel, but it is essential for proper digestion and defecation.

Glycogen and cellulose FIGURE 2.14

Glycogen, the human body's primary polysaccharide, is a long chain of links called glucose monomers.

Cellulose is a polysaccharide found in plant tissue. Animals cannot digest cellulose directly, but cows and some other animals host bacteria that can digest cellulose for them, releasing the stored energy.

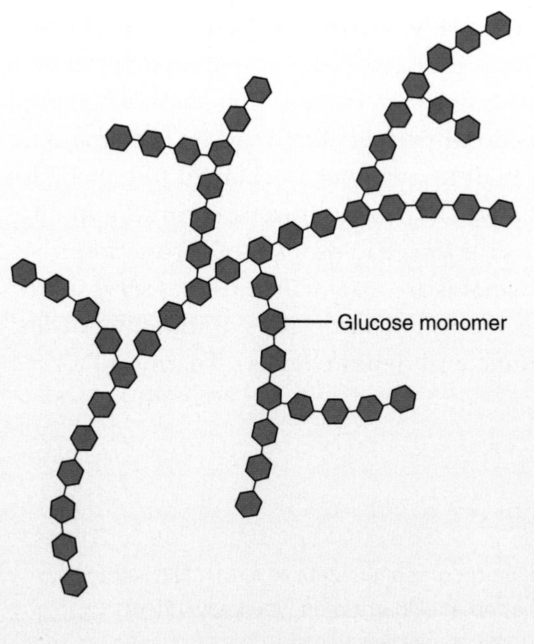

Glycogen

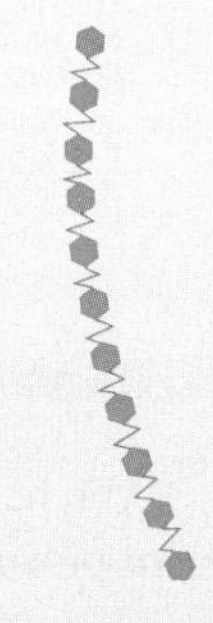

Cellulose

LIPIDS ARE LONG CHAINS OF CARBONS

Lipids, such as oils, waxes, and fats, are long-chain organic compounds that are not soluble in water—they are hydrophobic (FIGURE 2.15). Although most of the human body is aqueous, it must be divided into compartments, typically cells. Because water does not dissolve lipids, they are the perfect barrier between aqueous compartments like the cells. Lipids, like other organic compounds, are composed of carbon, hydrogen, and oxygen, but NOT with the 1:2:1 ratio of carbohydrates. The carbon-hydrogen ratio is often 1:2, but lipids have far fewer oxygens than carbohydrates. Lipids have a high energy content (9 kilocalories per gram), and most people enjoy the "richness" they impart to food.

Humans store excess caloric intake as fats, so reducing lipids is a common dietary tactic. As the proportion of lipids in the body rises, people become overweight or obese, as discussed in Chapter 3. The average human male contains approximately 12 to18 percent fat, and the average female ranges from 18 to 24 percent. This percentage changes as we age, as discussed in the Ethics and Issues box.

Fatty acids are energy-storing lipids. A fatty acid is a long chain of hydrogens and carbon, sometimes with more than 36 carbons. A carboxyl (acid) group is attached to the end carbon, which gives it the name "fatty acid." The other carbons are almost exclusively bonded to carbons or hydrogens. These chains are hydrophobic; the carboxyl group is the only hydrophilic location. Generally, the longer the hydrocarbon chain, the less water-soluble the fatty acid will be.

You have no doubt heard about two types of fatty acid: **saturated** and **unsaturated** fats. Saturated fats have no double bonds between carbons in the fat chains. For this reason, they are completely *saturated* with hydrogens and cannot hold any more. The straight chains of hydrocarbons in a saturated fat allow the individual chains to pack close together. Saturated fats, such as butter and other animal fats, are solid at room temperature. Unsaturated fats have at least one double bond between adjacent carbons. This puts a crimp in the straight carbon chain, causing rigid bends that prevent close packing of the molecules. This makes unsaturated fats liquid at room temperature. More double bonds increase fluidity. Examples of unsaturated fats include vegetable oils and the synthetic fats added to butter substitutes. Some vegetable oils are "hydrogenated" to remain solid at room temperature. Soy oil, for example, is commonly hydrogenated to make a solid vegetable shortening for cooking. Hydrogenating adds hydrogens, removes double bonds, and straightens the molecular arrangement of the fats. This allows the lipid to act like an animal fat and to be solid or semisolid at room temperature.

A **triglyceride** is three fatty acids attached to a glycerol backbone. Triglycerides, the most abundant fat in the body, can store two to three times as much energy per gram as carbohydrates. Triglycerides are manufactured by the body as nonpolar, uncharged storage molecules. In adipose (fat) tissue (discussed in Chapter 4), excess calories are stored in droplets of triglycerides.

Eicosanoids are essential lipids that serve as raw materials for two important classes of inflammatories: **prostaglandins** and **leukotrienes**. To manufacture eicosanoids, we must eat plant foods containing linoleic

Lipids FIGURE 2.15

If you have used hydrocortisone cream as a topical treatment for poison ivy or another skin irritation, you know it feels greasy, indicating that it contains lipids.

Aging: Making the most of a natural biological process

The human body is composed of many materials, including fat, muscle, bone, and water. The simplest way to appreciate how aging changes the amount, location, and nature of these materials is to compare a toddler, teenager, and an older adult. Aging starts when development stops. Humans reach physical maturity around age 30, and thereafter begin to decline into old age. Muscle mass typically starts to decline between age 30 and 40, although vigorous exercise can greatly retard the onset and slow the rate of decline. Endurance athletes typically reach their prime in their 30s, long after athletes whose sports depend on pure speed or agility.

Some age-related changes concern body fat. Babies have a higher proportion of body fat, partly because they need insulation due to their high ratio of surface area to volume. Adults need less of that insulation due to greater activity and a lower ratio of surface area to volume. When older people regain body fat, it's not for its insulating value but as a byproduct of a slowed metabolism and often a lower level of physical activity. The proportion of body fat may increase by 30 percent with age. Many people say "fat replaces shrinking muscle," but the muscle is not actually turning to fat. As muscle mass declines, fat is coincidentally added. Much of this fat is deposited in the abdomen of men and the thighs of women. The increasing percentage of fat can create a positive feedback situation; with less muscle tissue metabolizing fuel, one can gain weight without changing caloric intake or activity.

Other age-related changes include:

- A decrease in subcutaneous fat, causing the skin to thin and wrinkle.
- A decrease in height (roughly a centimeter per decade after age 40).
- A decrease in bone density and increased curvature of the spine at the hip and shoulder.
- Weight loss.
- Neurological changes, including slowed reflexes and difficulties with short-term memory.
- Tissue changes that reduce resilience and flexibility and may lead to joint problems.
- Impairment in the inner ear's balance mechanism, which reduces stability.

But aging isn't all bad news, especially when you consider the alternative. Recent articles show that healthy lifestyle choices made during middle age can greatly improve health later on. Behaviors such as regular physical exercise, engaging in brain-stimulating activities like learning new skills or challenging puzzles, maintaining an optimistic mental outlook, and finding meaning in life can all help keep body and soul together. Think about the means that some people use to try to stop the aging process, such as increased exercise: Will you alter your lifestyle as you mature?

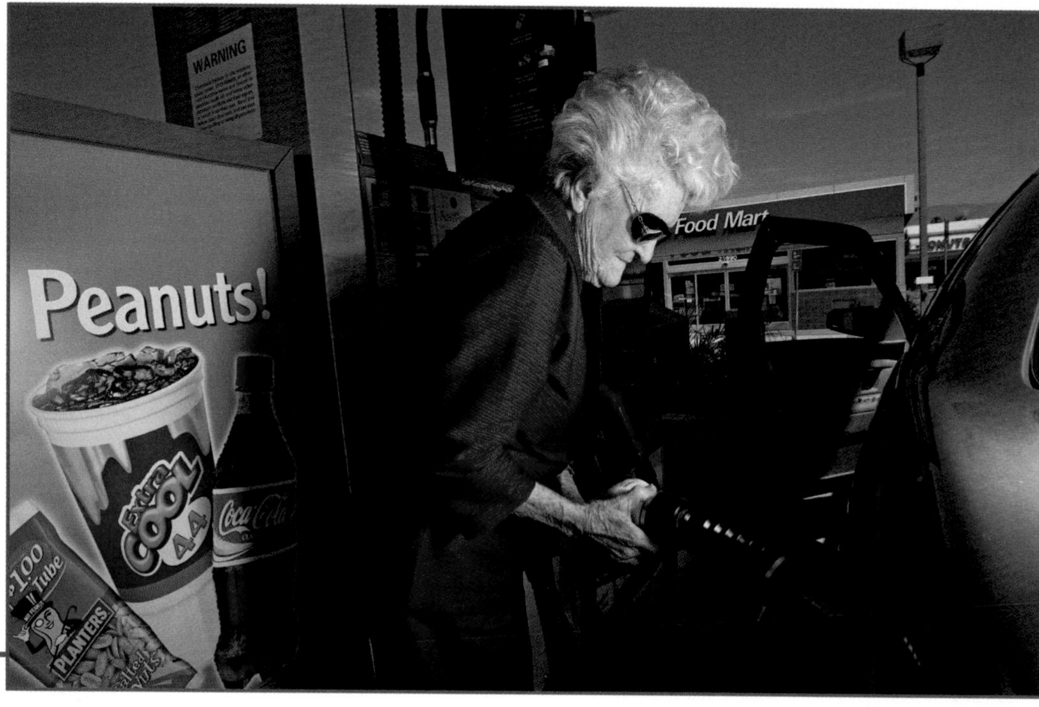

acid, such as corn and safflower oils. Prostaglandins are short-chain fatty acids that regulate local signaling processes. When nearby cells detect some prostaglandins, they respond immediately with the sensation of pain. Other prostaglandins signal the start of uterine contractions during labor. Aspirin blocks prostaglandins from reaching their cellular target, whereas ibuprofen competes only for the site where prostaglandins bind to cells. Ibuprofen acts more like the game of musical chairs, with the pain receptor as the chair and prostaglandin as the other player. When many ibuprofen molecules are around, prostaglandin is less likely to "find the chair" (occupy the receptor and stimulate pain). Because aspirin blocks prostaglandins entirely, it is more effective against some pain. Leukotrienes are inflammatory compounds released by some white blood cells. They cause the typical airway constriction in asthma and are key to long-term hypersensitivity of the airways.

Phospholipids (FIGURE 2.16) are another key group of lipids. These fats have two fatty acids and one phosphate group attached to a glycerol backbone. The fatty acids comprise the hydrophobic tail, whereas the phosphate group serves as a hydrophilic head. This unique structure allows phospholipids to form double layers (bilayers) that attract water on their edges and yet repel water from their center. The cell membrane, explored in the next chapter, is one such bilayer.

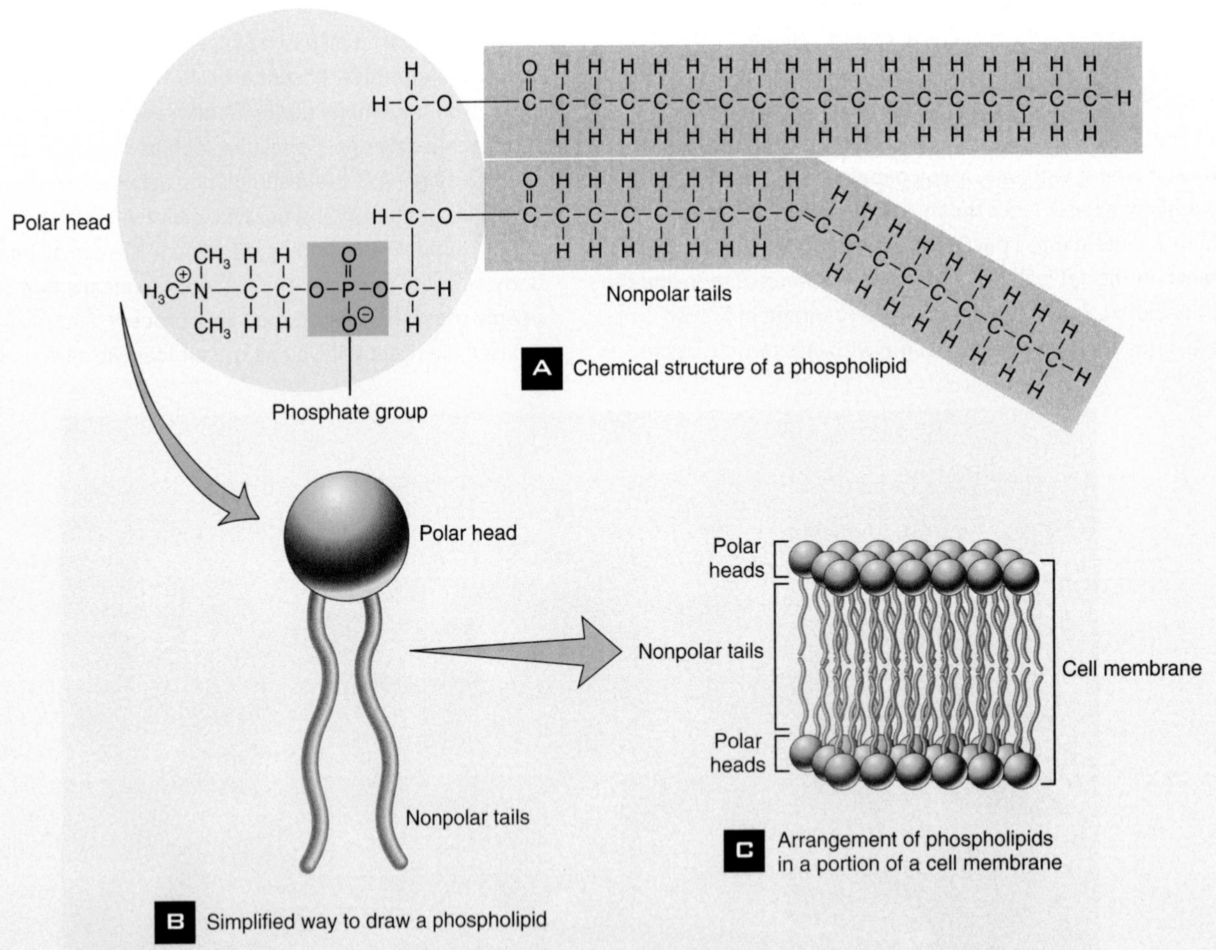

Phospholipids FIGURE 2.16

A phospholipid molecule has a polar head and a nonpolar tail.

Steroids are a final group of lipids that often makes news (FIGURE 2.17). These are large molecules with a common four-ring structure, important to normal growth and development. Steroids include cholesterol, sex hormones, and metabolism regulators.

Cholesterol is an integral part of cell membranes, allowing for flexibility and growth. High blood cholesterol has been linked to heart disease, so dietary restriction of cholesterol is often suggested. However, genetics play a large role in cholesterol levels. Because your body synthesizes cholesterol, it is often difficult or even impossible to manage cholesterol levels solely by diet.

> **Cholesterol** A class of steroids found in animals; aids in membrane fluidity.

The sex hormones **estrogen** and **testosterone** are two steroids that are responsible for the enormous changes of puberty. Anabolic steroids, which are related to testosterone, stimulate growth of the muscles. Anabolic steroids have important medical value as replacement hormones for males and females with low levels of testosterone or human growth hormone. Although many athletes have taken anabolic steroids to increase muscle mass and improve performance, they are banned in most sports. Anabolic steroids can cause such side effects as shrunken testicles and outbursts of anger or other psychological problems.

PROTEINS ARE BOTH STRUCTURAL AND FUNCTIONAL

Proteins contain carbon, hydrogen, oxygen, and nitrogen and are the most abundant organic compounds in your body. You contain more than 2 million different proteins. Some provide structural support, and others function in physiological processes. Proteins provide a framework for organizing cells and a mechanism for moving muscles. They are responsible for transporting substances in the blood, strengthening tissues, regulating metabolism and nervous communications, and even fighting disease.

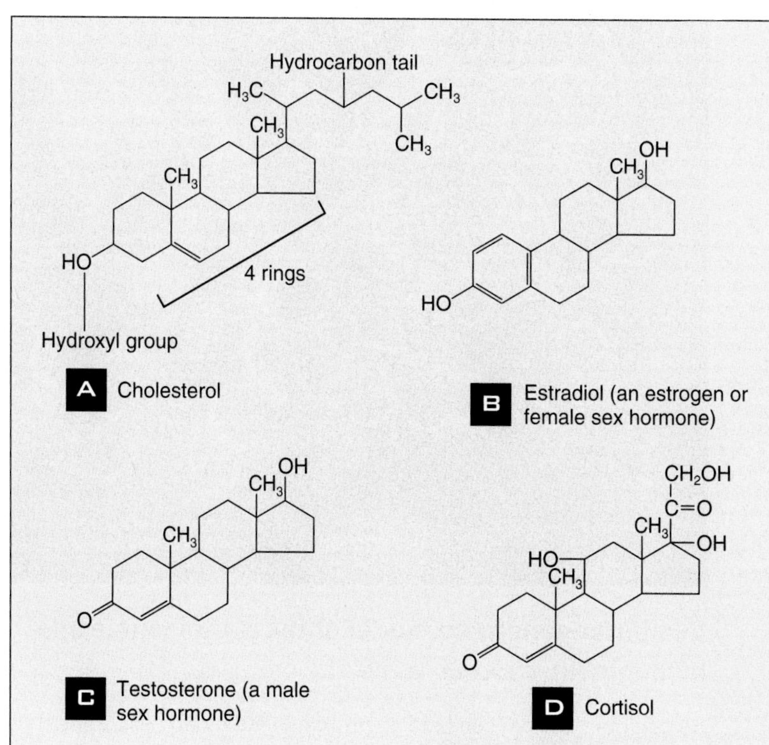

Steroids FIGURE 2.17

The body synthesizes cholesterol into other steroids, which play essential regulatory roles as hormones. Regulatory hormones such as cortisone maintain salt and calcium balance in the fluids of the body.

The millions of different proteins are all formed from just 20 building blocks, called **amino acids**. An amino acid is composed of a central carbon atom with four groups attached to it: (1) a hydrogen atom, (2) an amino group ($-NH_2$), (3) a carboxyl group ($-COOH$), and (4) a radical group or side chain (R). The R group determines the activity of the amino acid (FIGURE 2.18).

> **Peptide bonds**
> The bond between the carboxyl group of one amino acid and the amino group of the adjacent amino acid.

Individual amino acids combine to form proteins, using **peptide bonds** that form between the amino group of one amino and the carboxyl group of the next. The resulting two-amino-acid compound is called a dipeptide. As more amino acids join the growing chain, it becomes a **polypeptide**. As a rule of thumb, when the amino acid count exceeds 100, the compound is called a protein (FIGURE 2.19).

Insulin, the hormone that stimulates the cellular uptake of glucose, was the first polypeptide whose sequence of amino acids was determined. Frederick Sanger and his coworkers determined the sequence in 1955, and Sanger earned the first of two Nobel prizes for chemistry in 1958. (His second Nobel was awarded in 1980 for his work in determining the nucleotide sequence of a virus that attacks bacteria.) Insulin is a short polypeptide, with only 51 amino acids. Titin, the largest protein isolated so far from humans, is found in muscles and contains over 38,000 amino acids.

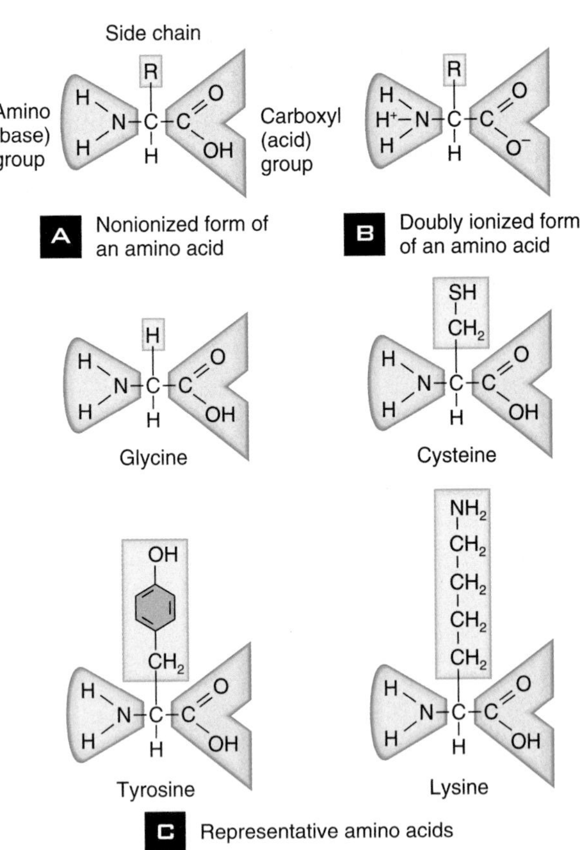

Amino acid structure FIGURE 2.18

Amino acids are the building blocks of proteins. Twenty amino acids combine into millions of proteins.

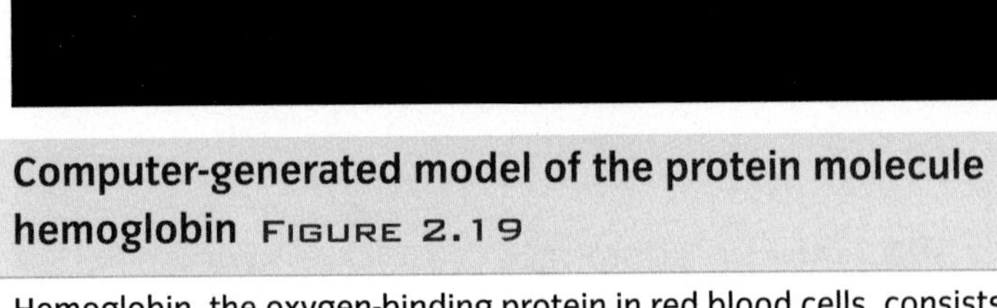

Computer-generated model of the protein molecule hemoglobin FIGURE 2.19

Hemoglobin, the oxygen-binding protein in red blood cells, consists of two chains of 141 amino acids and two chains of 146 amino acids, for a total of 574 amino acids. The iron in hemoglobin gives blood its red color.

The folding and interacting of adjacent amino acids determine the shape of a protein. The folding brings different amino acids together. If they repel one another, the protein bends outward. If they attract via weak hydrogen bonds, they bend inward (FIGURE 2.20).

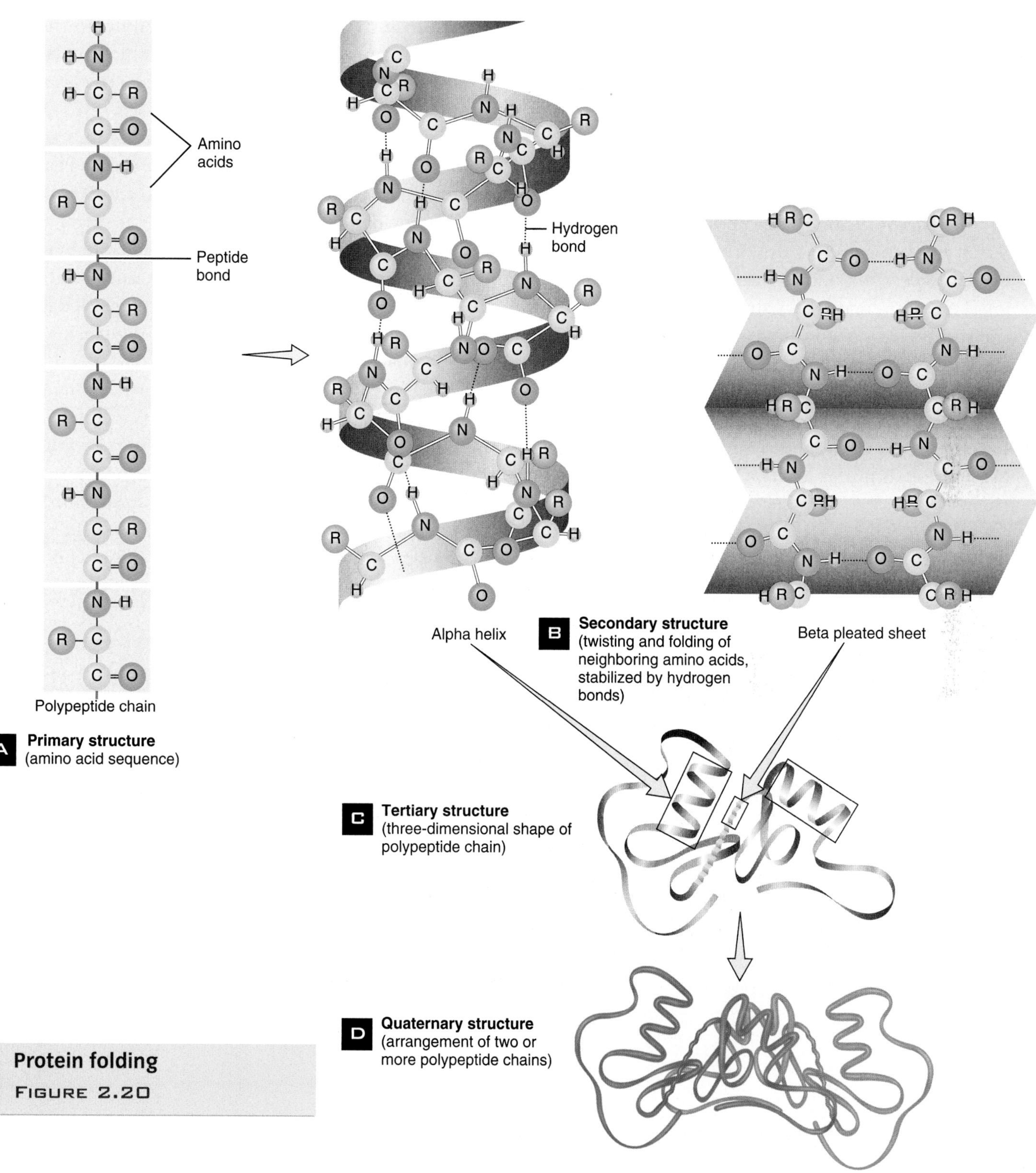

Protein folding
FIGURE 2.20

Proteins have four levels of structural complexity. Their **primary structure** is the unique order of amino acids in the chain (FIGURE 2.20A). Nearby amino acids interact via hydrogen bonds to form either alpha helixes or beta, pleated sheets, which is the **secondary structure** (FIGURE 2.20B). The **tertiary structure** emerges from interactions of the helical or pleated sheets, creating a complex coiling and folding (FIGURE 2.20C). Tertiary structure is a result of the hydrophobic and hydrophilic portions of the molecule twisting to either associate with water or to "hide" from it inside the molecule. One example of tertiary structure is the disulfide bond that occurs between two cysteine amino acids. Each carries sulfur ions. When the secondary structures hold them close, the sulfurs form a covalent, disulfide bond that permanently bends the protein. The **quaternary structure** (FIGURE 2.20D) emerges from the looping of two or more strands around one another. (Some proteins have only one strand, but many, including hemoglobin, are composed of two or more polypeptide chains.)

The final shape of a protein is either **globular** or **fibrous**. Globular proteins are round and usually water-soluble. These are often functional proteins, such as enzymes and contractile proteins. Fibrous proteins are stringy, tough, and usually insoluble. They provide the framework for supporting cells and tissues.

The shape of a protein molecule determines its function, and the final shape is determined by its primary structure. Changing even one amino acid can alter the folding pattern, with devastating effects on the protein's function (FIGURE 2.21).

In sickle cell anemia, a change of one amino acid from the normal hemoglobin protein creates a protein that fails to deliver oxygen correctly. When normal hemoglobin releases its oxygen to a tissue, the protein remains globular. But a "sickled" hemoglobin molecule becomes sharp, deforming the entire red blood cell into the sickle shape. These cells can get lodged in small blood vessels, causing pain and interfering with oxygen flow to the tissues.

Proteins and their bonds are susceptible to minor changes in the environment, such as increased temperature or decreased pH. When a protein unfolds, or radically alters its folding pattern in response to environmental changes, we say it is **denatured**. This happens when we cook. As we heat eggs, proteins in the clear whites unfold, forming a cloudy mass. This reaction is not reversible; denaturing is permanent.

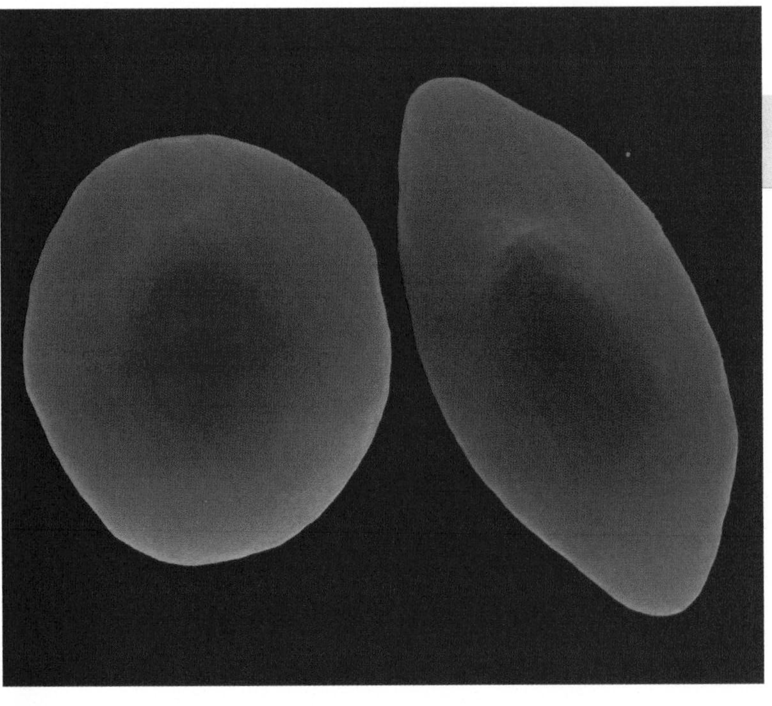

Microscan of normal and sickled red blood cells (sickle cell anemia) FIGURE 2.21

Enzyme activity FIGURE 2.22

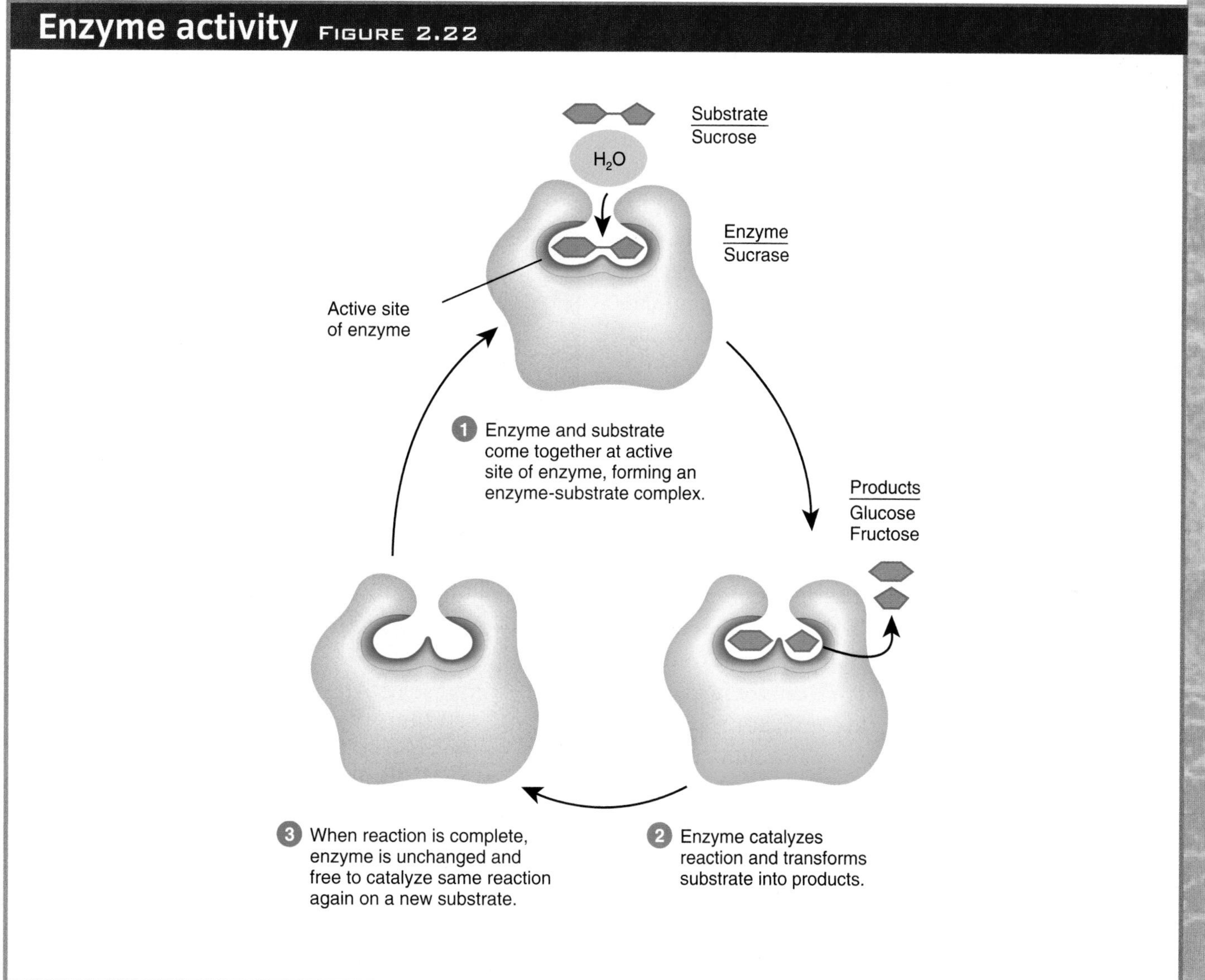

Enzymes are a special class of functional proteins. Enzymes serve as **catalysts** for biochemical reactions—meaning that they facilitate the reaction without being altered during it. Catalysts bring the reactants, or substrates, together, so a reaction can occur much more quickly. Enzymes rely on shape to function properly. The **active site** of the protein is shaped to bind to one specific substrate. After the substrate binds, the enzyme provides an environment for the specific chemical reaction to occur. After the reaction, the enzyme releases the products of the reaction and is ready to bind to another substrate molecule (FIGURE 2.22).

NUCLEIC ACIDS ARE INFORMATION MOLECULES

The final class of organic compounds is the **nucleic acid**. These are large molecules composed of carbon, hydrogen, oxygen, nitrogen, and phosphorus. Nucleic acids store and process an organism's hereditary information. The two types of nucleic acid are **deoxyribonucleic acid (DNA)** and **ribonucleic acid (RNA)**.

DNA exists in the center of our cells, in the nucleus. It contains the hereditary (genetic) information of the cell. DNA encodes the information needed to build proteins, to regulate physiological processes, and to maintain homeostasis. The genes that make each individual and each organism unique are usually made of DNA (FIGURES 2.23 and 2.24).

The sugar in DNA is a deoxyribose, meaning it lacks an oxygen, whereas RNA contains a simple ribose sugar. DNA has four bases: adenine (A), thymine (T), cytosine (C), and guanine (G). RNA has uracil (U) instead of thymine. DNA is a double-stranded molecule. To fit the two DNA strands of one macromolecule together neatly and precisely, the strands lie antiparallel to one another—meaning that although they lie parallel, they run in opposite directions. The phosphate end of one strand opposes the hydroxyl terminus of the other.

During DNA replication, this antiparallel configuration provides a logical explanation for why one strand is replicated with ease, whereas the other one is copied in "fits and starts." The enzyme responsible for duplicating the DNA can read in only one direction. It replicates DNA smoothly from 5′ to 3′, just as you read easily from left to right. The enzyme cannot read in the opposite direction, slowing the replication process. Imagine how much more slowly you would read these words if they made sense only from right to left. In addition, James Watson and Francis Crick, who discovered DNA's structure, could not make their model mathematically fit without the antiparallel configuration. The antiparallel arrangement of DNA strands is paramount to the entire molecule—one strand must be upside down compared to the other.

RNA is a messenger unit, not a storage unit, and it may occur inside or outside the nucleus. RNA serves to regulate cellular metabolism, produce proteins, and govern developmental timing. RNA is usually a single-stranded molecule. However, nucleic acids are more stable when paired. To achieve stability, RNA strands will fold back on themselves, pairing up A:U and C:G, similar to DNA. The shape of the RNA molecule often dictates its function.

In DNA, the phosphate group hangs off one end, technically the **five prime (5′)** terminus, of each base. The opposite end, with its hydroxyl group, is the **three prime (3′)** terminus. The 3 and the 5 indicate the carbon number of the ribose sugar that holds the phosphate or the reactive hydroxyl group. The two chains of DNA nucleotides wrap around one another in an **alpha helix**, held together by hydrogen bonds between bases. In natu-

Alpha helix Spiral chain of amino acids, resembling a twisted ladder.

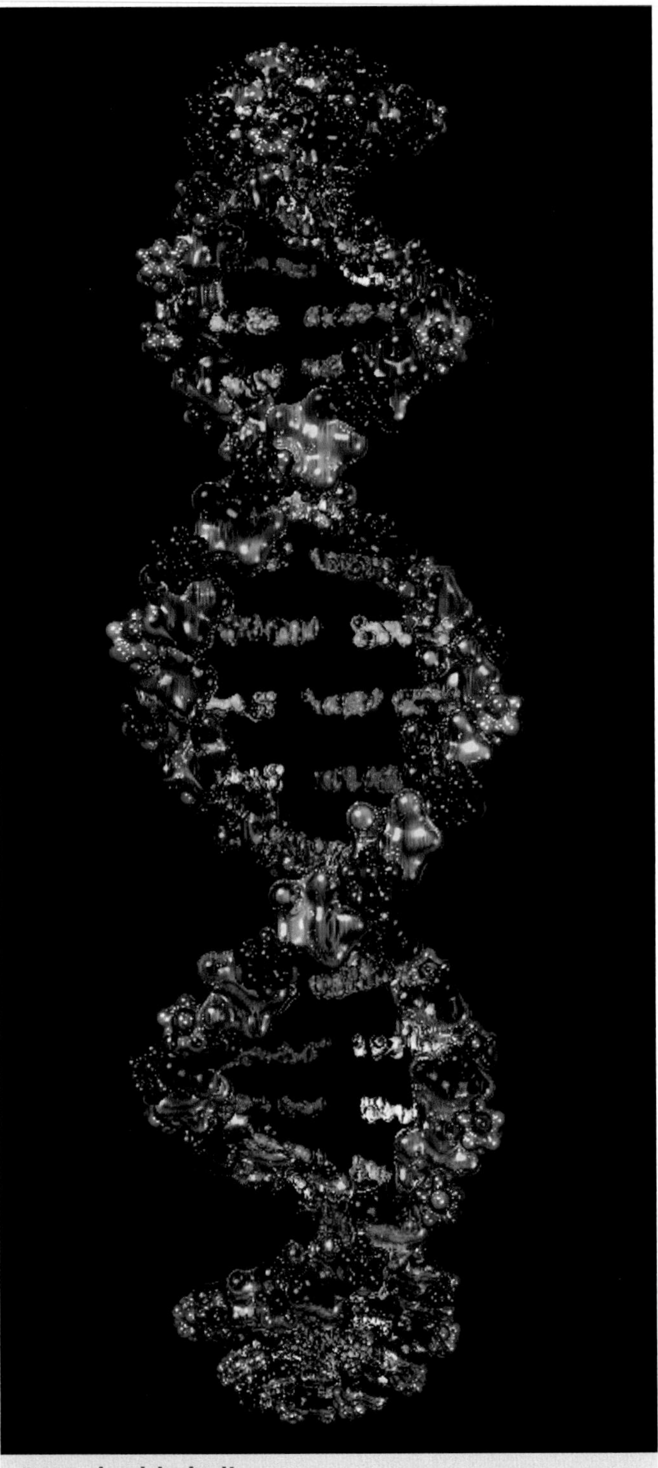

DNA double helix FIGURE 2.23

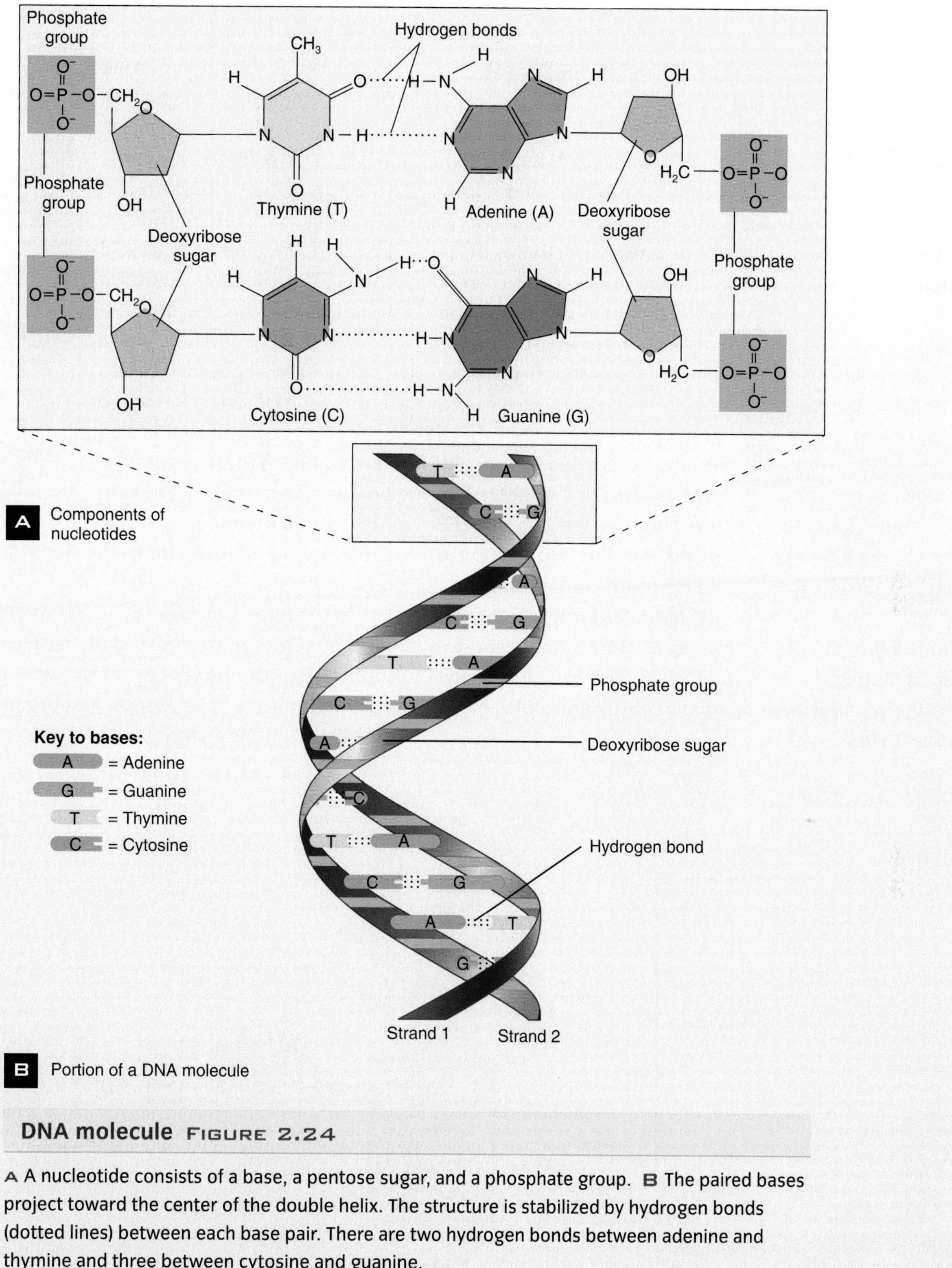

DNA molecule FIGURE 2.24

A A nucleotide consists of a base, a pentose sugar, and a phosphate group. B The paired bases project toward the center of the double helix. The structure is stabilized by hydrogen bonds (dotted lines) between each base pair. There are two hydrogen bonds between adenine and thymine and three between cytosine and guanine.

rally occurring DNA, the ratio of adenine to thymine is always 1:1 and the ratio of cytosine to guanine is again 1:1. This indicates that A bonds to T and C to G. Every time you find an adenine base on one strand of DNA, you will see it linked to a thymine on the complementary strand.

HIGH-ENERGY COMPOUNDS POWER CELLULAR ACTIVITY

Life requires energy. Most often energy is available in spurts, rather than as a continuous stream all day long. We eat food, which our bodies convert to usable energy. Soon after a meal, lots of this energy circulates in the blood, but without a way to store the excess, we would have to eat almost continuously. Our energy storage system provides short- and long-term storage. Short-term energy storage uses a high-energy system that is reversible and instantly available. The most common storage is **ATP**, or **adenosine triphosphate**. ATP powers all cellular activity, from forming proteins to contracting muscles (FIGURE 2.25). Long-term storage includes glycogen in muscles and liver, and triglycerides packed into specialized storage cells called **adipocytes**.

Adenosine triphosphate (ATP) The primary energy molecule that can be used to perform cellular functions.

ATP is an adenine bonded to a ribose sugar with three phosphates attached. When ATP is hydrolyzed, the third phosphate bond breaks, releasing inorganic phosphate (P_i) and the energy that held the ATP molecule together, forming **adenosine diphosphate (ADP)**. This released energy drives cellular activity. The ATP-ADP energy storage system is readily available and renewable. When glucose is broken down, the released energy can be used to recombine the inorganic phosphate to the ADP, generating a new ATP molecule.

Adipocytes Specialized cells (fat cells) that store large quantities of lipid.

Adenosine diphosphate (ADP) The molecule that results when ATP releases one phosphate group.

Without chemistry, there is no life, but how does life emerge from the many molecules we have examined? In the next chapter, we will look further up the hierarchy, to cells, tissues, and organs, to see the basic organization of an organism.

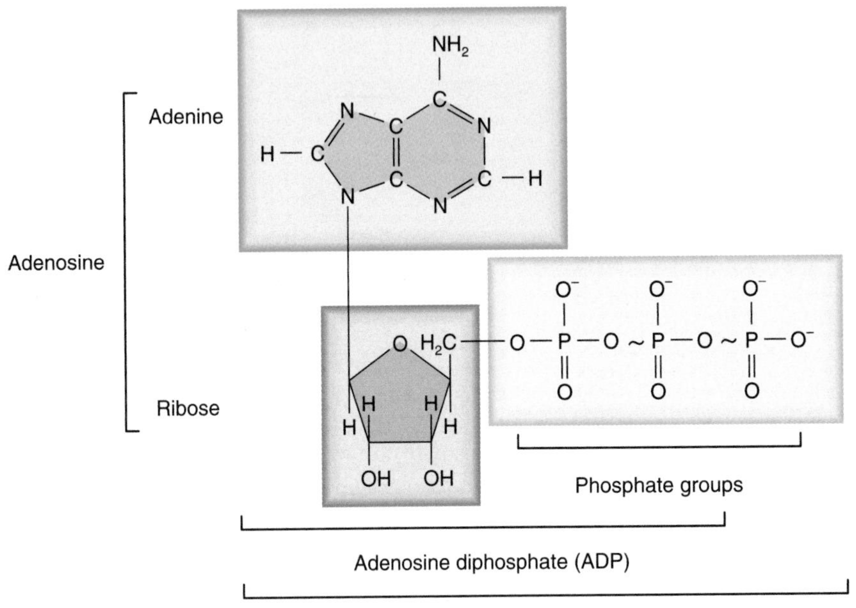

Adenosine triphosphate (ATP) and adenosine diphosphate (ADP)
FIGURE 2.25

CONCEPT CHECK

Describe the four levels of protein folding. How do they relate to one another?

List the classes of lipids and the function of each.

What is the basic component of all carbohydrates?

Compare and contrast the structure of DNA and RNA.

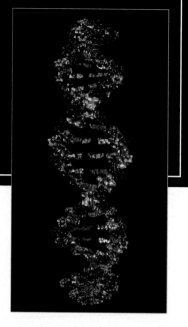

CHAPTER SUMMARY

1 Life Has a Unique Chemistry

All life is based on the chemical elements. The four most common elements in living organisms are carbon, hydrogen, oxygen, and nitrogen. The remainder of the elements that comprise living organisms are considered trace elements because they appear in small, or trace, amounts only.

2 Atomic Structure Is the Foundation of Life

The atoms of any particular element contain a specific number of protons in the nucleus, as well as a cloud of electrons around the nucleus. The outside, or valence, electrons determine the chemical reactivity of an atom.

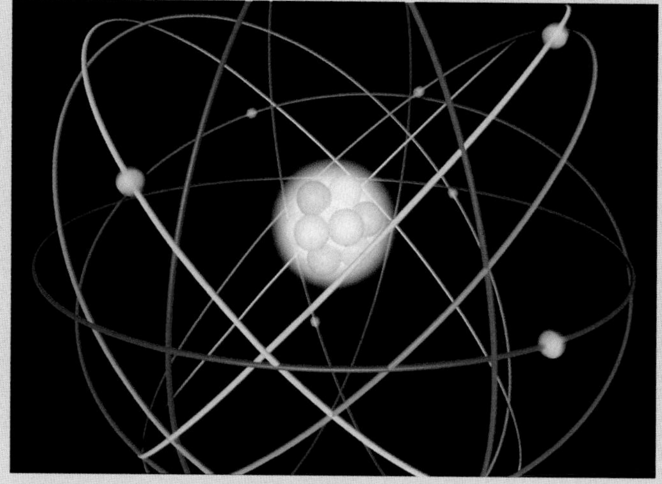

CHAPTER SUMMARY

3 Chemistry Is a Story of Bonding

Elements are joined by chemical bonds. Strong, ionic bonds result from the attraction of positive and negative ions. In weaker covalent bonds, atoms share electrons. Unequal sharing produces a polar covalent bond, resulting in a polar molecule like water. Hydrogen bonds are weak interactions between adjacent hydrogen-containing polar molecules. The weakest forces known that hold chemicals together are van Der Waals forces. These are extremely weak, impermanent electrical charge attractions formed as electrons whirl in their clouds. Transient negative charges are pulled toward equally transient positive portions of molecules. These charges change and disappear as electrons continue their whirling dance.

4 Water Is Life's Essential Chemical

Water has many necessary characteristics for life, which trace back to the molecule's polar condition. Water is liquid at room temperature; it is a good solvent; it has a high specific heat and a high heat of vaporization; and frozen water floats. Hydrogen and hydroxyl ions are released when a water molecule separates.

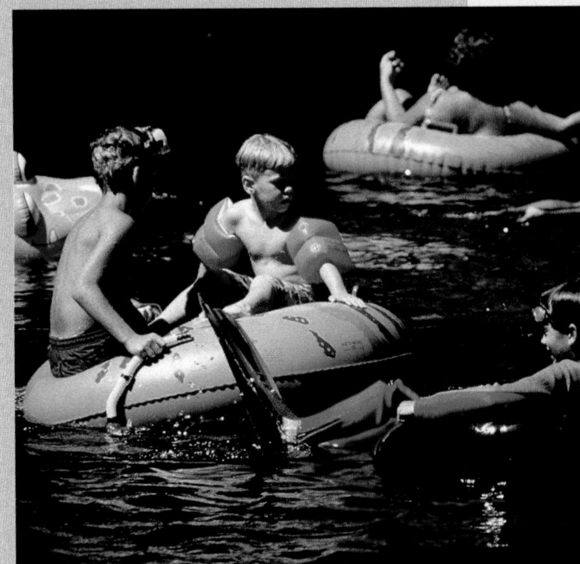

5 Hydrogen Ion Concentration Affects Chemical Properties

The hydrogen ion concentration in any solution is indicated by the pH of that solution. pH 1 is highly acidic; pH 14 is extremely basic. Pure water is pH 7. Acids donate hydrogen ions to solutions, whereas bases add hydroxyl ions. When mixed together, acids and bases usually neutralize and form water. Buffers are weak acids that stabilize the pH of solutions by absorbing excess hydrogen or hydroxyl ions.

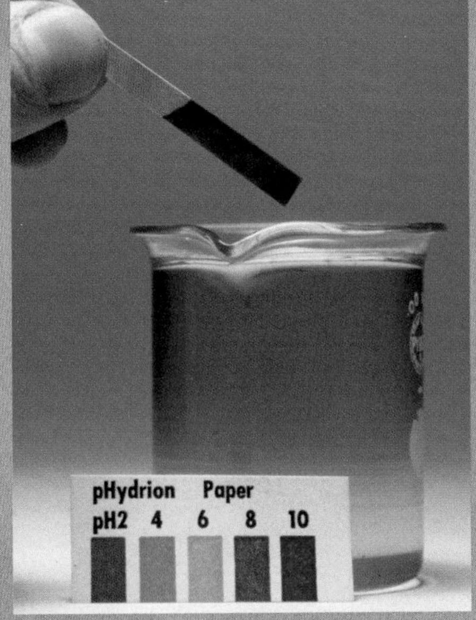

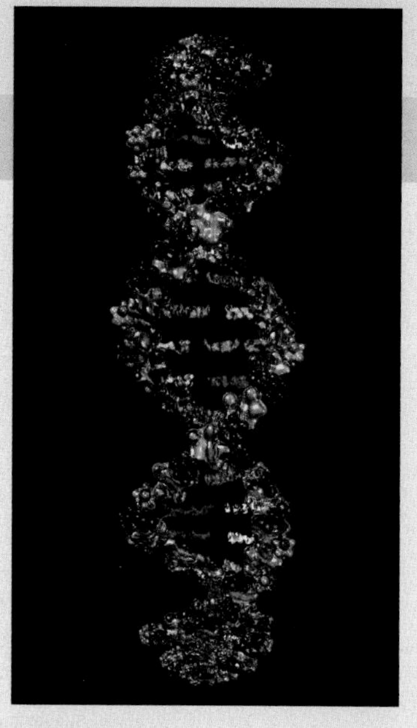

6 There Are Four Main Categories of Organic Chemicals

Biochemistry is the study of biological molecules. The carbohydrate glucose is a key source of ready energy. Lipids store energy, serve in the cell membrane, and are the basis for sex hormones. Phospholipids make up the cell membrane, which is vital to cellular function. Proteins provide structure and chemical processing. Nucleic acids store data in our genes and transfer information.

Proteins, the building blocks of the body, can be structural or functional. Protein function is determined by shape and the sequence of amino acids. Millions of proteins are built using just 20 amino acids. Enzymes are protein catalysts that allow faster chemical reactions. Enzymes have an active site, where substrate molecules bind before the reaction takes place.

Nucleic acids store and carry information in the cell. DNA is a double-stranded helix made of four bases (A, C, T, and G), and occurs in the nucleus. DNA codes for specific proteins, depending on the sequence of bases. The single-stranded molecule RNA serves mainly to carry DNA data to protein-making machinery. ATP, the energy-storage molecule inside cells, releases energy as it converts to ADP.

KEY TERMS

- **adenosine diphosphate (ADP)** p. 54
- **adenosine triphosphate (ATP)** p. 54
- **adhesive** p. 38
- **adipocytes** p. 54
- **alpha helix** p. 52
- **atomic mass** p. 31
- **atomic number** p. 31
- **cholesterol** p. 47
- **cohesive** p. 38
- **dalton** p. 30
- **electron** p. 30
- **element** p. 28
- **functional group** p. 42
- **hydrophilic** p. 37
- **hydrophobic** p. 37
- **ion** p. 33
- **mass** p. 30
- **neutron** p. 30
- **peptide bonds** p. 48
- **proton** p. 30
- **radioactive decay** p. 31

CRITICAL THINKING QUESTIONS

1. Radioisotopes are often used in medicine and biochemical experiments. For example, the chemistry of photosynthesis was uncovered with radioactive carbon dioxide. Photosynthesizing cells were exposed to radioactive carbon dioxide, and that radioactive carbon was tracked through each step of glucose formation. Using this principle, how might radioactive compounds be used in medical research? How might they be used in treating a disease like leukemia (cancer of the white blood cells)?

2. Choose two properties of water. Briefly describe each property and show how it contributes to a specific aspect of human life.

3. Acid rain is caused when water in the atmosphere reacts with sulfur oxides to form sulfuric acid. The acidity of typical acid rain is pH 3 to pH 5 (normal precipitation is pH 7 to pH 7.5). What is the mathematical relationship between the hydrogen-ion concentrations at each of these pH levels? How could acid rain affect biological systems?

4. Enzymes are proteins that serve as catalysts, speeding up reactions without getting used, altered, or destroyed. Enzyme function can be accelerated or slowed without damaging the enzyme itself. Review Figure 2.22 to understand normal enzyme functioning. What will happen to enzyme function if products build up in the cell? if substrate concentration decreases? if a second compound, similar to the substrate but without its reactive properties, enters the enzyme's environment? if temperature rises slightly?

5. Although they serve different functions, DNA and ATP have common elements. What structures are found in both molecules? What purpose do these structures serve in ATP? in DNA?

SELF TEST

1. The four most common elements in the human body include

    a. calcium.
    b. sodium.
    c. carbon.
    d. nitrogen.
    e. both c and d are correct.

2. Identify the particle indicated as A in the figure below.

    a. Proton
    b. Neutron
    c. Electron
    d. Orbital

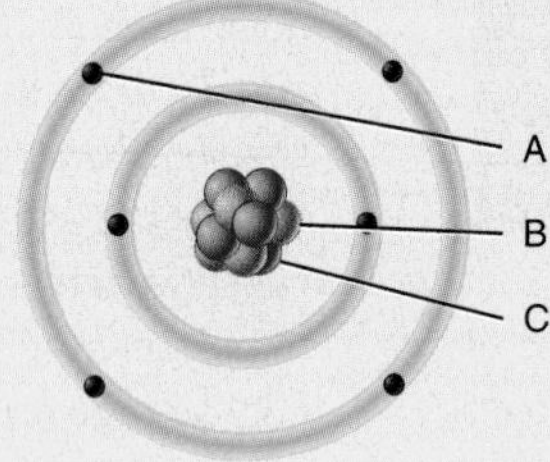

3. The particle indicated as C in the figure above carries a ___________ charge.

    a. positive
    b. negative
    c. neutral

4. Which of the identified particles in the figure above has a mass of less than 1 dalton?

    a. A
    b. B
    c. C
    d. All of the above carry a mass of 1 dalton.

5. Carbon has an atomic mass of 12.01. It has an atomic number of 6. How many neutrons are in a carbon atom nucleus?

    a. 12
    b. 6
    c. 18
    d. 4

6. An atom with 7 electrons in its valence orbital will most likely form what type of chemical bond?

    a. Hydrogen
    b. van der Waals
    c. Ionic
    d. Covalent

7. The type of bond indicated here is a/an

    a. ionic bond.
    b. covalent bond.
    c. polar covalent bond.
    d. hydrogen bond.

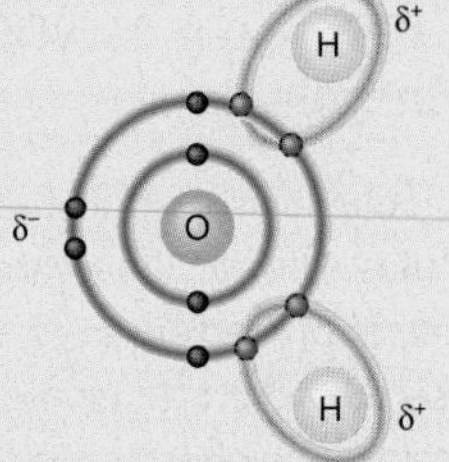

8. Some atoms are held together in compounds by attractive forces of positive and negative charges. Which of the following bond types rely on these attractive forces?

    a. Ionic bond
    b. Covalent bond
    c. Hydrogen bond
    d. All of the above utilize positive/negative attraction.

9. A substance that is attracted to water or dissolves in water is referred to as

    a. hydrophobic.
    b. hydrophilic.
    c. cohesive.
    d. adhesive.

10. Water serves as a temperature buffer because it

    a. is cohesive.
    b. is capable of dissolving many compounds.
    c. has a high specific heat.
    d. has a high heat of vaporization.

11. The key characteristic of the molecule highlighted in this photograph, giving it many of the unique properties it demonstrates, is

    a. its polarity.
    b. its distinct shape.
    c. the two closely held hydrogen ions.
    d. the large oxygen molecule.

12. On this pH scale, what is the hydrogen ion concentration difference between human blood (pH 7) and ammonia (pH 11)?

a. 10 units
b. 100 units
c. 1000 units
d. 10,000 units

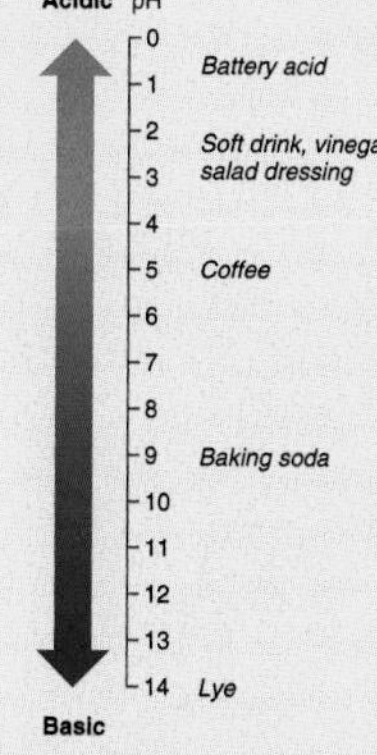

13. This figure illustrates a/an

a. carbohydrate.
b. lipid.
c. protein.
d. nucleic acid.

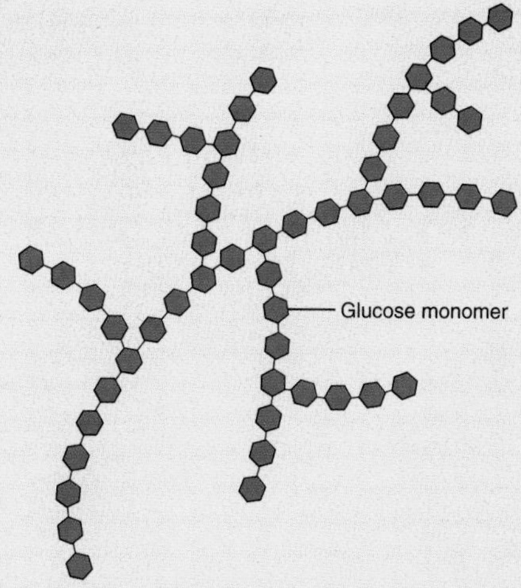

14. Another name for glycogen is

a. oligosaccharide.
b. monosaccharide.
c. polysaccharide.
d. disaccharide.

15. A/an ___________ fat is a solid at room temperature and includes straight long hydrocarbon chains with no double bonds.

a. unsaturated
b. saturated
c. hydrophilic

16. The class of lipid that has both a hydrophilic and a hydrophobic end is

a. steroids.
b. eicosanoids.
c. phospholipids.
d. triglycerides.

17. The level of protein structural complexity that determines the final shape of the protein is the

a. first level, the amino acid sequence.
b. second level, the folding of the amino acid sequence.
c. third level, the interaction of the folded amino acid sequence.
d. fourth level, the interaction of two or more subunits of the finished protein.

18. This figure illustrates that enzymes

a. require substrate.
b. are specific catalysts.
c. have an active site.
d. All of the above options are correct.

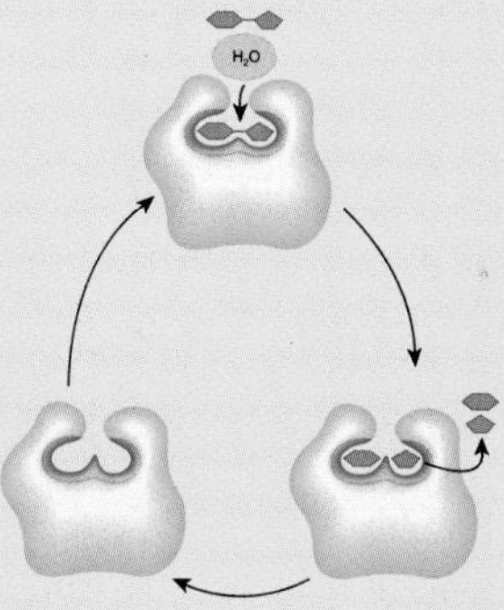

19. In DNA, which base complements adenine?

a. cytosine.
b. guanine.
c. thymine.
d. uracil.

20. The energy in ATP is carried in

a. the first phosphate bond.
b. nitrogenous base.
c. adenine.
d. the last phosphate bond.

Cells, Organization, and Communication

3

What is the largest organism? The answer depends on your definition of "largest." Among animals, the blue whale is the largest animal on Earth, and possibly the largest animal ever. This sea mammal can weigh over 100 metric tons and stretch 35 meters from head to fluke. Blue whales feed on krill, which look like miniature shrimp. By the early 1960s, blue whales had nearly gone extinct due to whaling. They were hunted for their large stores of blubber, a lipid used for lighting and lubrication before the petroleum age. Luckily, most nations outlawed the hunting of blue whales, and they are slowly rebounding.

In terms of area, the largest organism is a newly discovered fungus, *Armillaria ostoyae*. One fungal individual covers 10 square kilometers of Oregon forest floor. By mass, the largest organism is the giant sequoia (*Sequoia sempervirens*), a tree native to California's humid coastal forests. Giant sequoias can reach 110 meters in height, with a mass of about 2,500 metric tons. Like the blue whale, the giant sequoia has been threatened (it makes good lumber), but some reserves have been set aside for protection from the chainsaw.

Ironically, these giants are a stunning example of the success of the smallest unit of life—the cell.

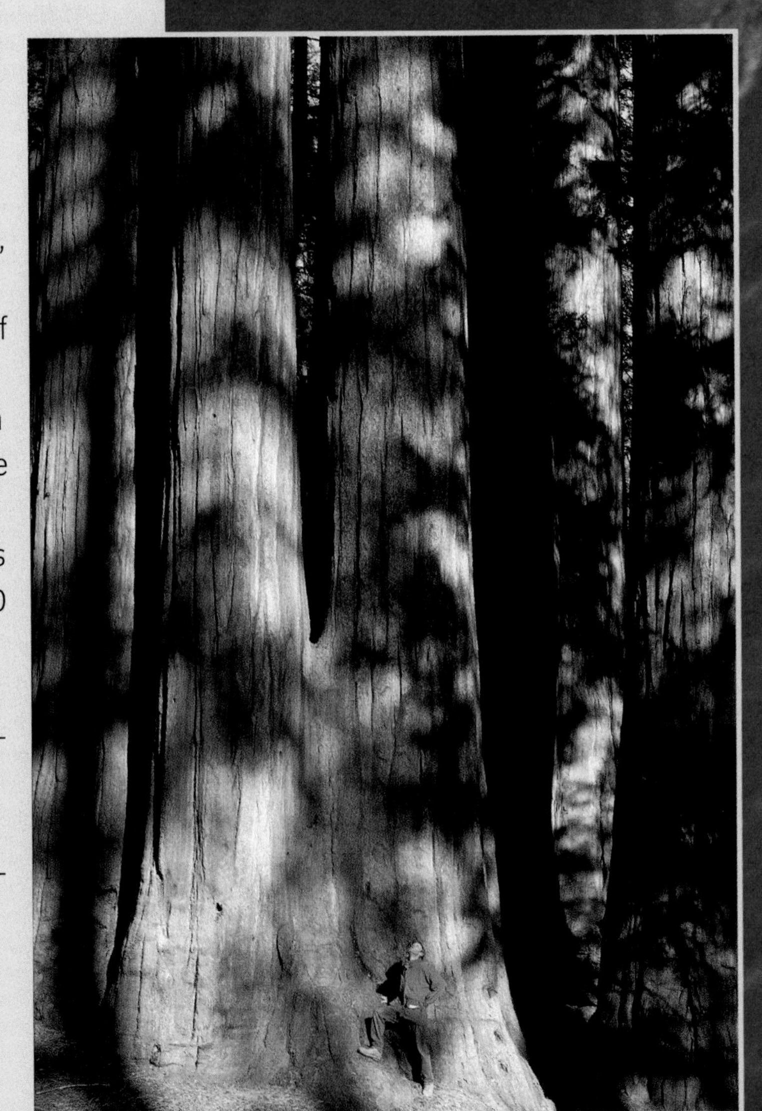

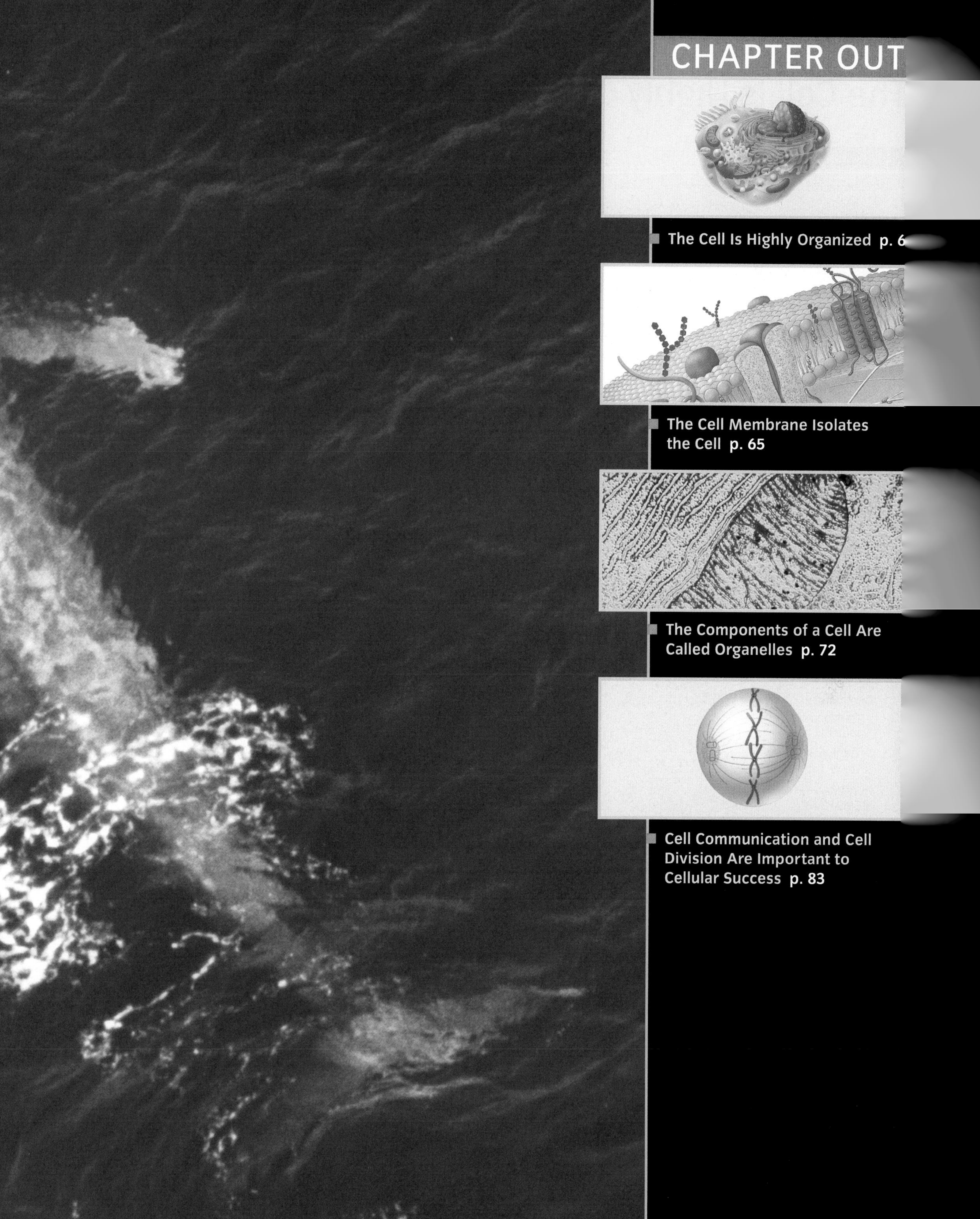

CHAPTER OUT

The Cell Is Highly Organized

Learning Objectives

Outline the cell theory.

Relate the size of cells to the instrument used to view them.

Describe the difference between organelles and cytoplasm.

Cells are the building blocks of life. Every living thing is composed of cells, from the smallest bacterium to the blue whale or the giant sequoia. These giants have vastly more cells than single-celled bacteria, and more organization, both inside and outside those bacterial cells. However, all animals' structure ultimately comes down to cells, as all animals are multicellular.

You can think of cells as packages. Because life requires certain chemical conditions, organisms must concentrate some chemicals and exclude others. The tiny compartments that have the right conditions for the many chemical reactions that sustain life are called cells (FIGURE 3.1).

The study of cells is called **cytology**, and scientists who study cells are called **cytologists**. All cells, regardless of source, have similar characteristics, as defined by **cell theory**. This represents the latest version of our centuries-old understanding about cells:

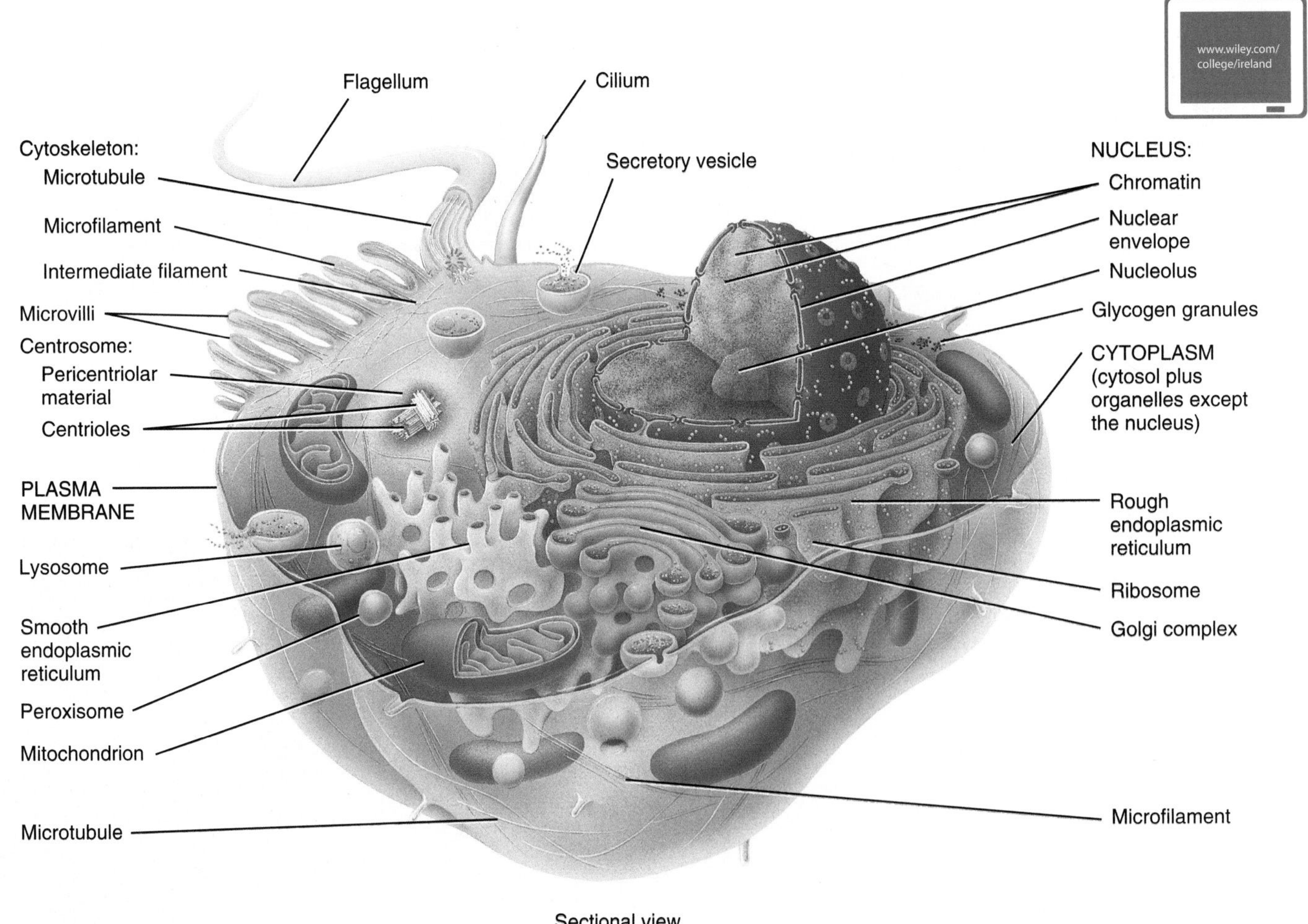

Typical animal cell FIGURE 3.1

1. All living things are composed of cells.
2. All cells arise from preexisting cells through cell division.
3. Cells contain hereditary material, which they pass to another cell during cell division.
4. The chemical composition of all cells is quite similar.
5. The metabolic processes associated with life occur within cells.

While all cells share these characteristics, they can be remarkably different in shape and size. Cells can be as large as an ostrich egg or smaller than a dust speck (a typical liter of blood, for example, contains more than 5.9×10^{12} red blood cells). Because most cells are microscopic, you need millions to make up a typical mammal: the human body contains trillions of cells, and virtually all but one type is invisible without a microscope. (See I Wonder . . . on p. 64.) Our egg, the only human cell visible to the naked eye, is approximately as big as this period: .

The cell is a highly organized structure that is defined by a barrier called the plasma membrane (in animals) or cell wall (in plants and bacteria). Inside the plasma membrane is a fluid called **cytosol,** which supports multiple types of **organelles**, each with a function vital to the life of the cell.

Organelle Typically a membrane-bound structure suspended in the cytosol; hairlike projections from the cell may also be called organelles.

Cellular organization is evident with a quick glance at a magnified cell. Inside the cell, membrane-bound compartments can be seen. These compartments are organelles, small structures whose overall goal is to maintain cellular homeostasis. Some organelles break down nutrients, others are tiny factories that churn out structural and functional proteins, and still others extend through the plasma membrane to the surface of the cell and circulate the surrounding fluid so that waste materials and nutrients can diffuse into or out of the cell.

Cytosol contains water, dissolved compounds, and small molecules called **inclusions**. These molecules vary by the type of cell, and may include **keratin** for waterproofing, **melanin** for absorbing ultraviolet light, and **carotenes,** which are precursors to vitamin A.

Keratin Tough fibrous proteins that form hard structures such as hair and nails.

Melanin A dark brown, UV-light-absorbing pigment produced by specific cells.

Carotene A yellow-orange pigment.

CONCEPT CHECK

What are the five statements that make up the cell theory?

Define "organelle."

Give three examples of inclusions that can be found in the cytoplasm of a typical human cell.

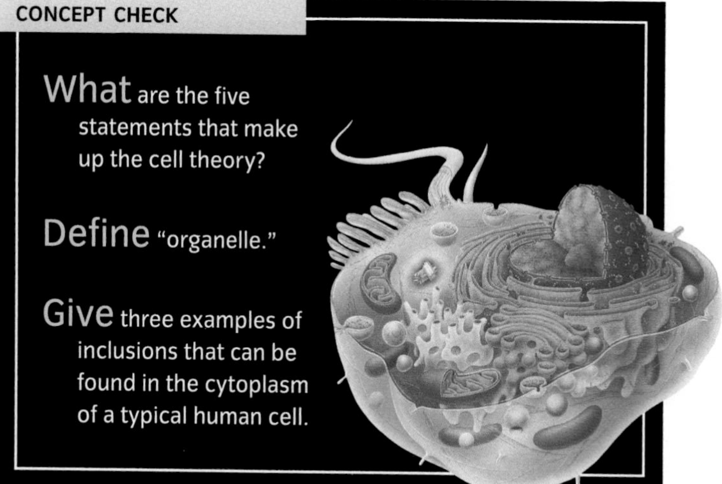

How can you see cells?

Most cells are microscopic—meaning you cannot see them without a microscope. Microscopes make images of "objects" and are measured by their magnification and **resolution**. Magnification is a factor comparing image size to object size. Resolution measures the size of objects that can be distinguished from each other.

Cytologists use several types of microscopes to see cells and their organelles. A **compound light** microscope uses a series of lenses to focus rays of light that pass through an object. The maximum magnification of a light microscope is about 1000 times (written 1000×), enough to see cells and some larger organelles. Images taken with a light microscope are called photomicrographs.

For imaging sub-cellular structures, the cytologist's instrument of choice is the **electron microscope**. Your eye can resolve objects that are approximately 0.2 millimeters apart (approximately one thickness of a human hair). A light microscope can resolve objects 0.2 micrometers apart (one thousand times closer). An electron microscope can distinguish objects just 10 nanometers apart. Electron microscopes can also achieve a magnification of 50,000×.

A **scanning electron microscope (SEM)** provides a three-dimensional look at cellular surfaces. The sample must be coated with a material that conducts electricity, and stabilized to withstand the vacuum conditions inside the microscope. An intense electron beam knocks electrons away from the sample. These electrons are registered by the microscope's detector to create a black-and-white image of extreme detail.

A **transmission electron microscope (TEM)** produces images akin to photomicrographs, although with higher magnification and better resolution. Instead of measuring knocked-away electrons, this microscope measures electrons that pass through the sample. In terms of magnification, compound light microscopes are the least powerful microscopic tool discussed. Electron microscopes are far more powerful, and TEM microscopes have the highest resolution of any tool currently available. A TEM microscope built in 2000 at Hitachi's Advanced Research Laboratory is capable of distinguishing rows of *atoms* only half an angstrom apart!

Microscopes have played a heroic role in the development of biology. The Dutch lens maker Anton van Leeuwenhoek started making simple microscopes around 1660. With no scientific training but an open mind, Leeuwenhoek discovered bacteria, sperm, and other basic cells.

At about the same time, English scientist Robert Hooke built compound microscopes, and made astonishingly detailed drawings of insects, feathers, and other life forms. Hooke made the key discovery of cells as the basis for life.

Later, as biology turned to treating disease, pioneering German biologist Robert Koch used a microscope to view the bacteria that cause tuberculosis and cholera. These images helped cement the germ theory of disease. More recently, electron microscopes have shown that biology's phenomenal degree of organization extends down to the smallest scale imaginable—the molecular scale.

SEM Scanning electron microscope; device that directs a beam of electrons across the object; image is produced from reading the scattered electrons.

TEM Transmission electron microscope; device that makes an image by passing electrons through an object.

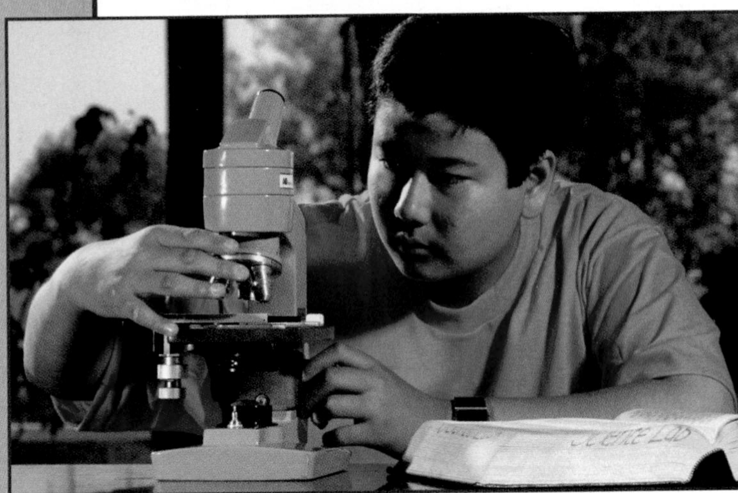

Light microscope

You may have the opportunity to use a compound light microscope such as this one in your laboratory class. The microscope enables you to view human cells, such as your own cheek cells.

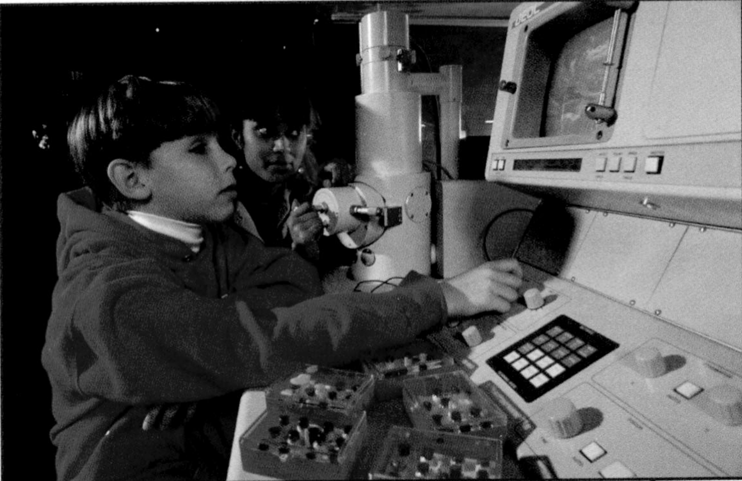

Scanning electron microscope

The scanning electron microscope achieves high resolution, but can only be used for nonliving samples.

The Cell Membrane Isolates the Cell

LEARNING OBJECTIVES

Discuss the structure of the cell membrane.

Explain movement across the membrane, both passive and active.

Define osmosis and relate it to the actions of hypotonic and hypertonic solutions.

Compare the subtle differences in the main categories of active transport.

The obvious place to start studying cellular anatomy is the plasma membrane, the structure that separates the cell from the extracellular fluid. This membrane is composed of two layers of **phospholipids**, interspersed with proteins, fats, and sugars (FIGURE 3.2). The phospholipids are arranged in a double layer, or bilayer, with the **hydrophilic**, water-loving heads (the charged, phosphate ends of the molecule) oriented toward the aqueous environment both inside and outside the cell. The **hydrophobic**, water-fearing, non-polar, lipid portion of the molecules is sandwiched in the center.

Phospholipids Compounds containing phosphoric acid and a fatty acid.

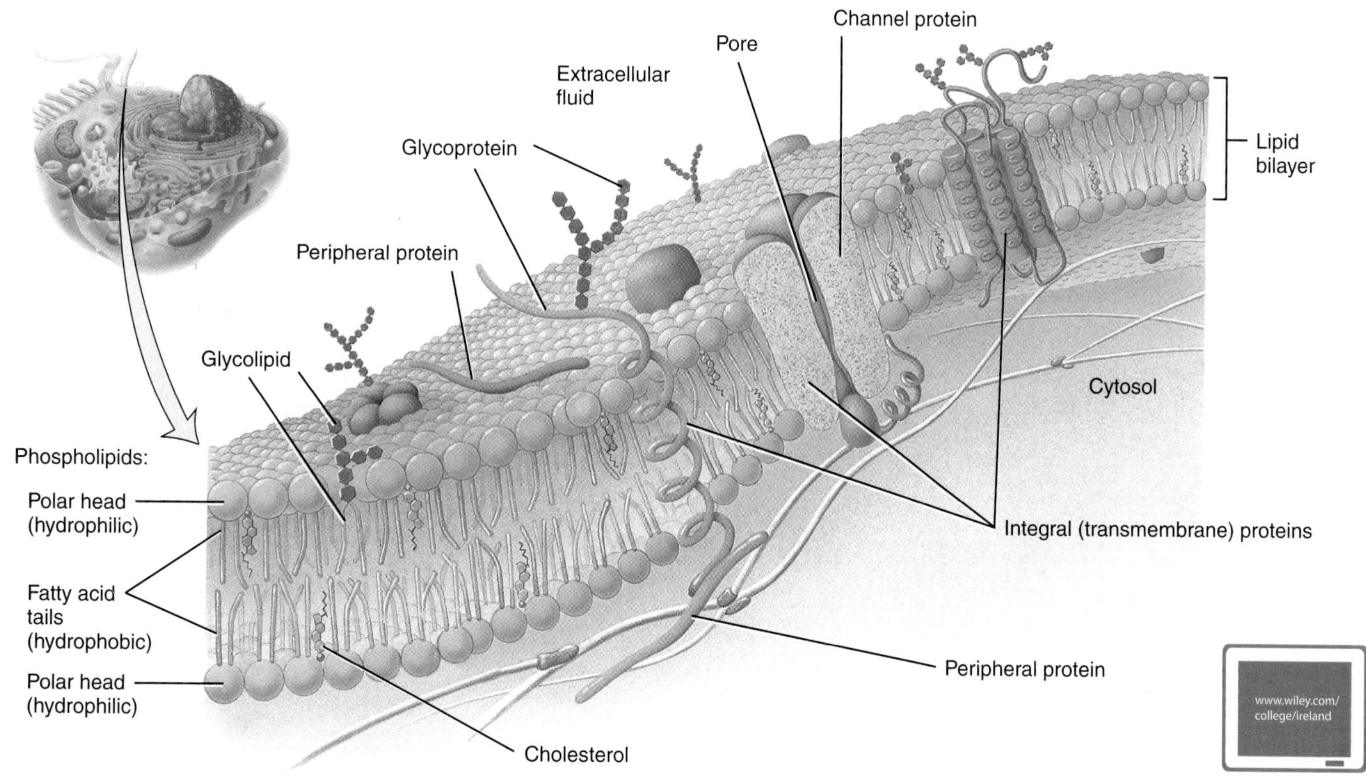

Cell membrane FIGURE 3.2

Some of the proteins and lipids associated with the cell membrane have sugars attached to their external surface and are called **glycoproteins** and **glycolipids**.

Glycoprotein Proteins plus a carbohydrate.

Glycolipid Lipid plus at least one carbohydrate group.

The glycoproteins and glycolipids form a layer called the **glycocalyx**; which is unique enough to define the cell as belonging to a specific organism. Both blood type and tissue type are defined by the specific structures on the glycocalyx. For example, each person's white blood cells carry a group of identifying proteins called the human leukocyte antigens (HLAs) that serve as markers indicating that our cells belong to us. HLA is used to match tissues before organ transplants. Because HLA is inherited, if we need a transplant, we can often find a close tissue match within our immediate family.

The cell membrane is not a static structure. At 37°C, its phospholipids are liquid, not solid, so the basic structure of the membrane is a continually swirling fluid. (Cholesterol, a necessary component of the cell membrane, helps to maintain this fluidity.) The proteins embedded in the membrane are in constant motion, floating around in the fluid phospholipid bilayer. Picture a beach ball covered in Vaseline and rolled in the sand. As the Vaseline warms in the sun, it will begin to flow around the ball (inner cytosol of the cell) causing the embedded sand grains to swirl with it. Similarly, the glycocalyx and embedded proteins in the fluid phospholipid bilayer swirl around the cell membrane.

MOVEMENT ACROSS THE MEMBRANE CAN BE PASSIVE OR ACTIVE

The phospholipid bilayer defines the cell and protects it from the aqueous environment. Without membrane lipids, the cell would literally disintegrate, much like a cracker dropped into a glass of juice. But the plasma membrane cannot maintain cellular homeostasis unless it allows some compounds in and out of the cell. In fact, rather than being a simple plastic bag, the membrane is a **semipermeable** barrier that allows nutrients to enter the cell and waste and secretory products to exit it. Some ions and molecules cross freely; others can be moved across the membrane with the expenditure of some energy, and still others cannot cross at all. Movement across the membrane can be either passive or active.

Passive movement includes **filtration**, **diffusion**, and **facilitated diffusion**. None of these activities requires the cell to expend energy. Filtration is the movement of solutes in response to fluid pressure. Your kidneys separate waste products from the blood via filtration.

DIFFUSION MOVES MOLECULES FROM HIGH CONCENTRATIONS TO LOW CONCENTRATIONS

Diffusion is the movement of a substance toward an area of lower concentration. Open a perfume bottle and set it in the corner of a room. Within a short time, the perfume will diffuse from the bottle and permeate

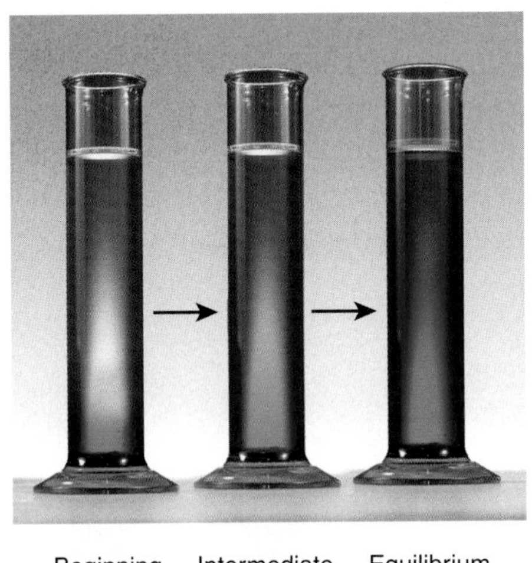

Diffusion FIGURE 3.3

At equilibrium, net diffusion stops, but the random movement of particles continues.

the room. Warm the room, and the diffusion speeds up. Diffusion results from the random movement of the molecules, which eventually tends to balance out the molecule's concentrations (FIGURE 3.3). The same phenomenon occurs continuously in your cells. Lipid-soluble compounds and gases can diffuse across the cell membrane as if it weren't there, traveling right through the phospholipid bilayer. The driving force for the movement of oxygen from the atmosphere into the deepest tissues of the body is merely diffusion.

While lipid-soluble molecules can diffuse freely through it, the phospholipid bilayer blocks the diffusion of **aqueous**, or water-soluble, solutes. This is a potential problem, as many aqueous solutes, such as glucose, are essential compounds that must be able to penetrate the cell membrane. To solve this problem, the lipid bilayer has **integral** and **peripheral proteins** that serve as channels and receptors for aqueous solutes to enter and exit the cell.

Integral protein A protein that spans the plasma membrane.

Peripheral protein A protein that sits on the inside or the outside of the cell membrane.

Solute Salts, ions, and compounds dissolved in a solvent, forming a solution; water is the most common solvent in the human body.

Isotonic A solution with the same concentration as the cell cytoplasm.

The most abundant compound in the body is water. To maintain homeostasis, cells must allow water to move between the intracellular fluid (ICF) and the extracellular fluid (ECF). Diffusion of water across a semipermeable membrane such as the cell membrane is termed **osmosis.** In osmosis, water moves in a direction that tends to equalize **solute** concentration on each side of the membrane. In effect, locations with higher solute concentrations seem to "pull" water toward them.

Water cannot cross the phospholipid bilayer, so it must travel through proteins. Usually, the extracellular fluid is **isotonic** to the cells, and water flows equally into and out of the cell through transport proteins. If you place a cell in a **hypotonic** solution (water with a lower concentration of solutes than the cytosol), the cell will take in water and may even burst. In contrast, a **hypertonic** solution (with a higher concentration of solutes), will remove water from the cell, and cause it to shrivel up (FIGURE 3.4). (See the Health, Wellness, and Disease feature on p. 70 for applications of these principles to medical treatment.)

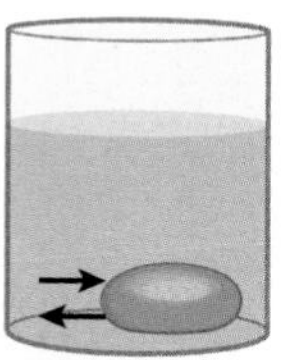

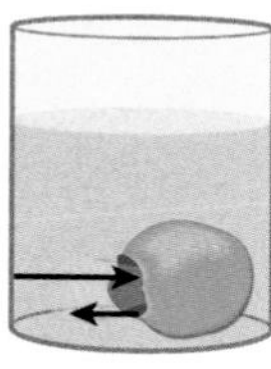

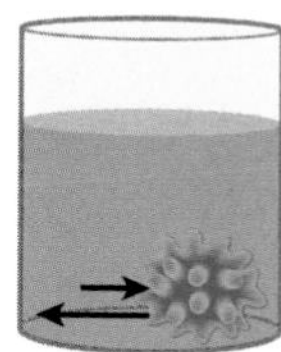

A Illustrations showing direction of water movement

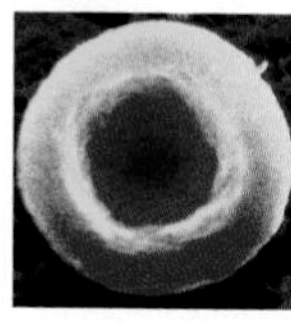

RBC undergoes hemolysis

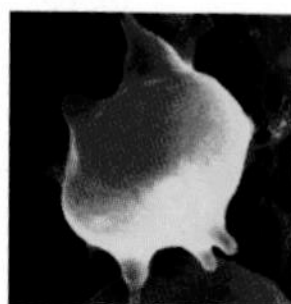

B Scanning electron micrographs (all 800x)

Hypotonic and hypertonic solutions
FIGURE 3.4

Osmosis can occur quite rapidly when cells are placed in hypotonic or hypertonic solutions. Hemolysis is an almost instantaneous process, and crenation (the shriveling of red blood cells) in hypertonic solutions takes less than 2 minutes.

FACILITATED DIFFUSION

When solutes are transported across the membrane down their concentration gradients (from high concentration to low concentration) by **transport proteins**, no energy is expended. This type of movement is called facilitated diffusion and is the main avenue through which glucose is moved into cells. After a meal, blood glucose is higher than cellular glucose. However, in order to diffuse into the cell, glucose needs a "doorway"

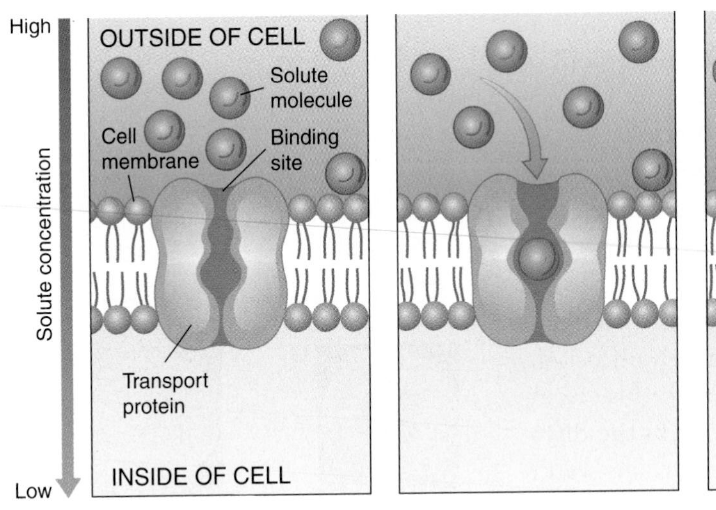

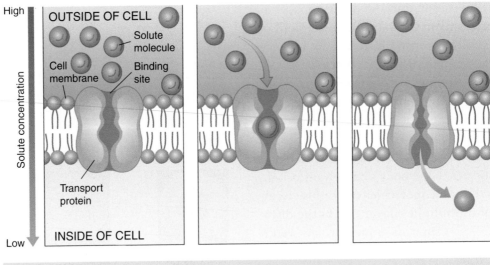

Facilitated diffusion FIGURE 3.5

through the phospholipid bilayer. It would make very little sense to expend energy just to get glucose into the cell to make energy (FIGURE 3.5).

During osmosis, as water diffuses toward areas of lower water concentration (and higher solute concentration) across the semipermeable membrane, it creates osmotic pressure. The Greek letter Ψ (psi) stands for **water potential**. This is a calculation of the osmotic pressure of resting cells in an isotonic solution. The two components of water potential are the pressure exerted on the cell by its environment and the solute concentration of the cells, so:

$$\Psi = \Psi_{\text{atmospheric}} + \Psi_{\text{solute concentration}}$$

Water potential is useful for calculating the concentration of an isotonic solution, and is used most often by botanists to predict water movement in and out of plant cells.

ACTIVE TRANSPORT USES ENERGY TO MOVE MOLECULES ACROSS MEMBRANES

When energy is consumed to move a molecule or ion against the concentration gradient, we call the process **active transport**, or solute pumping. Osmosis and other forms of diffusion move molecules "down" their concentration gradients without additional energy. Active transport is used to concentrate molecules inside cells at levels that exceed the extracellular concentration, using energy derived from the breakdown of ATP into ADP. Active transport accounts for the almost complete uptake of digested nutrients from the lumen of the intestine, the sequestering of iodine in thyroid gland cells, and the return to the blood of the vast majority of sodium ions filtered from the blood by the kidneys.

Active transport can move atoms, ions, or molecules into the cell **(endocytosis)** or out of it (**exocytosis**). In endocytosis, extracellular molecules and particles are taken into the cell via vesicle formation. Just as punching a partially inflated balloon caves in the balloon wall, endocytosis begins with depression of the cell membrane. Particles in the extracellular fluid flow into the new dimple and get trapped within the vesicle that forms when the two sides touch and are pinched off inside the cell. The two forms of endocytosis are **pinocytosis** and **phagocytosis** (FIGURE 3.6).

> **Pinocytosis** Cell drinking, or taking in a small quantity of the extracellular fluid.

> **Phagocytosis** Cell eating, or taking in of large molecules and particles using vacuoles.

Exocytosis is used to remove secretory products or waste products from the cell. Vesicles form within

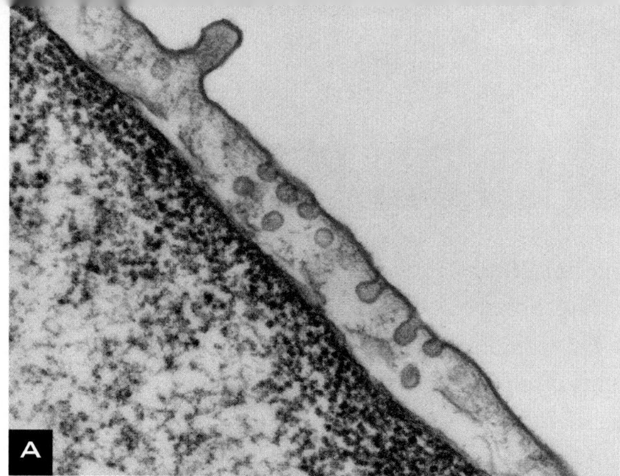

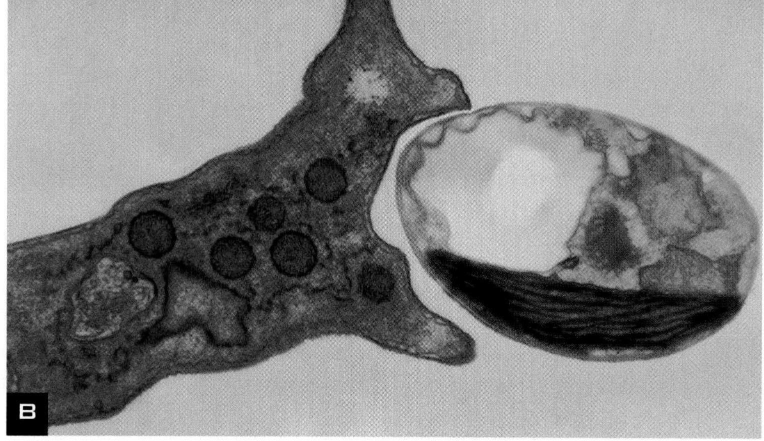

Pinocytosis and phagocytosis FIGURE 3.6

A Pinocytosis produces extremely small vesicles filled with the fluid surrounding the cell. This is an electron micrograph of a capillary epithelial cell membrane showing the process of pinocytosis, magnified 12,880×.

B During phagocytosis the cell surrounds and ingests a large particle. In this image, the cell on the left is in the act of engulfing the round cell on the right.

the cell, usually from one of two organelles, the Golgi apparatus or a lysosome. They travel to the inner wall of the cell membrane and fuse with it (think of two soap bubbles fusing into one larger bubble where they touch). This fusion releases the vesicle's contents into the extracellular fluid (FIGURE 3.7).

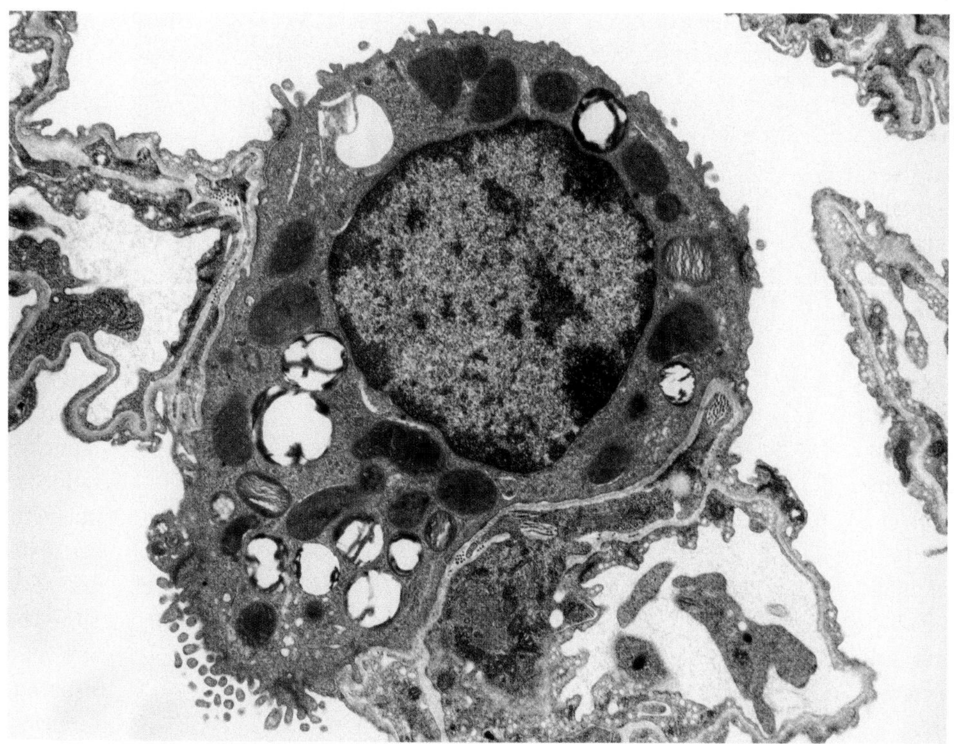

Exocytosis FIGURE 3.7

This alveolar cell, found in the lungs, is secreting materials. The transport vesicles move up to the cell membrane and fuse with the cell membrane. When the two membranes fuse, the contents of the vesicles are dumped into the extracellular fluid.

The science of intravenous fluids

Few visits to the hospital seem complete without some form of intravenous (IV) therapy. Intravenous ("within a vein") treatment allows doctors to place medicine right inside your bloodstream. IV treatment can convey vitamins, glucose, coagulants, anticoagulants, chemotherapy drugs, antibiotics, and electrolytes.

These medicines and agents are commonly added to a carrier called IV solution. Have you ever wondered why so much IV solution is marked "0.9% saline," or "5% dextrose"? The answer is rooted in osmosis: the diffusion of water through a semipermeable membrane. Molecules diffuse from areas of high concentration to areas of lower concentration. Water diffuses, or osmoses, toward areas with higher solute concentration (where the water concentration is lower).

Diffusion explains the focus on solute concentration in IV solution. Blood plasma is normally isotonic, and doctors usually want IV solution to be isotonic as well, which is why they use 0.9% saline or 5% dextrose IV solution. The total solute (salt or sugar) concentration in these IV bags is the same as in human plasma.

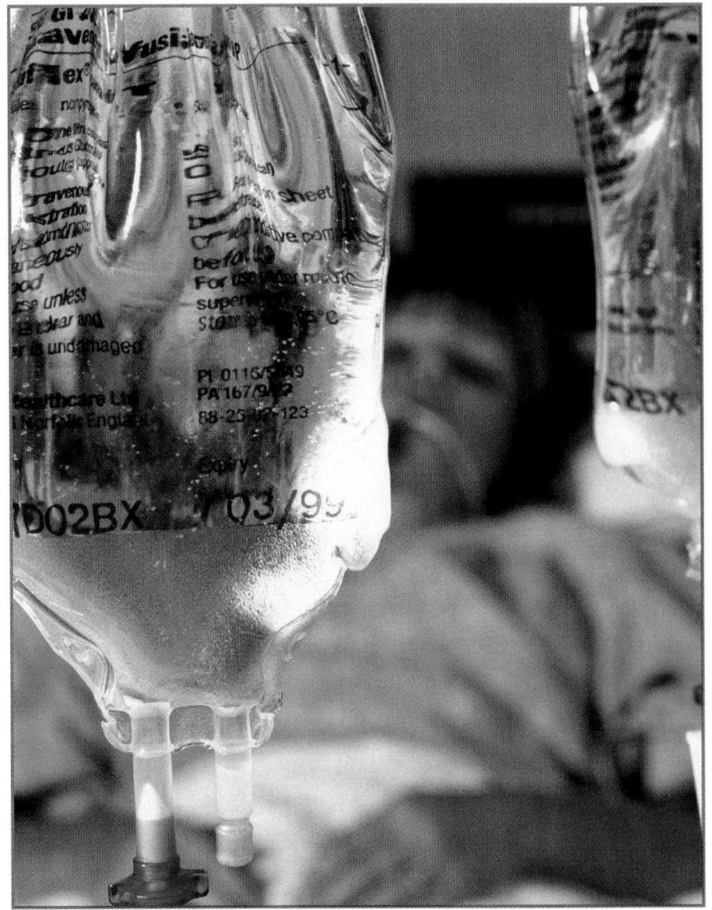

In some cases, other IV solutions are more appropriate. A less-concentrated ("hypotonic") solution would, according to the principles of diffusion, move water across the plasma membrane into cells, helping to reverse dehydration.

A hypertonic IV solution (with a higher solute concentration than the cytosol) will cause osmotic pressure that removes water from the cells. Because this will cause cells to shrivel, a hypertonic IV solution is used to treat swelling. In **cerebral edema**, for example, hypertonic IV treatment is often the first line of treatment. If it does not reduce swelling of the brain, physicians may take drastic steps like surgically removing part of the skull to relieve the pressure.

Cerebral edema Fluid accumulation in the brain or cerebral area.

Osmotic pressure plays a key role in two common conditions: diabetes and hypertension (high blood pressure). Diabetes results from either a shortage of insulin or a failure to respond to insulin. In either case, glucose fails to enter the cells, and the blood becomes hypertonic, which tends to remove water from the cells. Such an alteration in solute concentration can cause some of the widespread changes in cellular metabolism seen in diabetes.

Osmosis also plays a key role in many cases of hypertension, which affects 65 million Americans and is implicated in heart disease and stroke, two of the top three killers. Excess sodium ions in the blood change the osmotic conditions and reduce the kidneys, ability to excrete water, resulting in an increase in blood pressure. The exact role of salt in hypertension is debated, however, as reducing salt intake does not always reduce blood pressure among hypertensive people. Clearly, the homeostatic mechanisms that regulate blood pressure are so important that many other factors are involved.

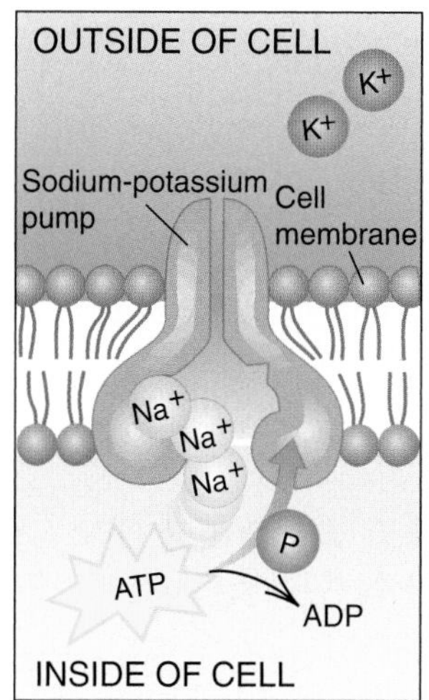

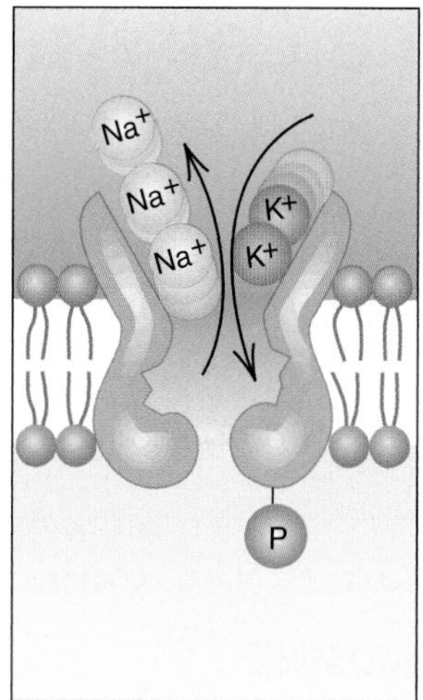

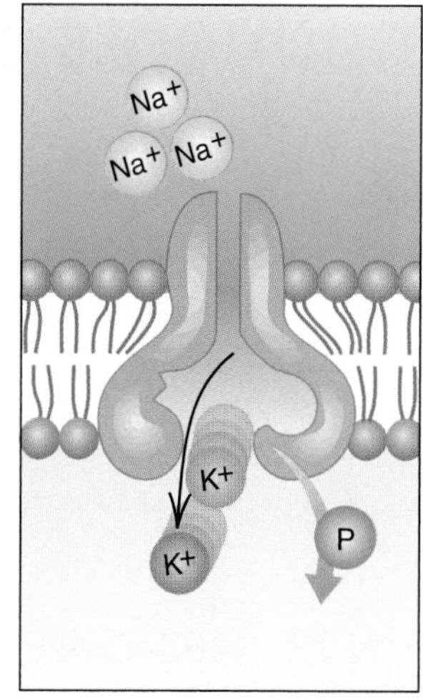

Na^+/K^+ ATPase FIGURE 3.8

The sodium/potassium pump transfers two potassium ions into the cell for every three sodium ions it removes. The movement of ions happens simultaneously.

Often small molecules or ions are moved by intramembrane pumps, as transport proteins are sometimes called. These protein structures may transport ions or small molecules in either direction across the plasma membrane. Pumps often have reciprocal functions—pumping one molecule or ion into the cell while simultaneously removing a second chemical species from the cell. For example, **sodium/potassium ATPase** act as a common reciprocal pump, moving two potassium ions into the cell while pumping three sodium ions out of it (FIGURE 3.8). We will discuss this pump again when we cover neurophysiology.

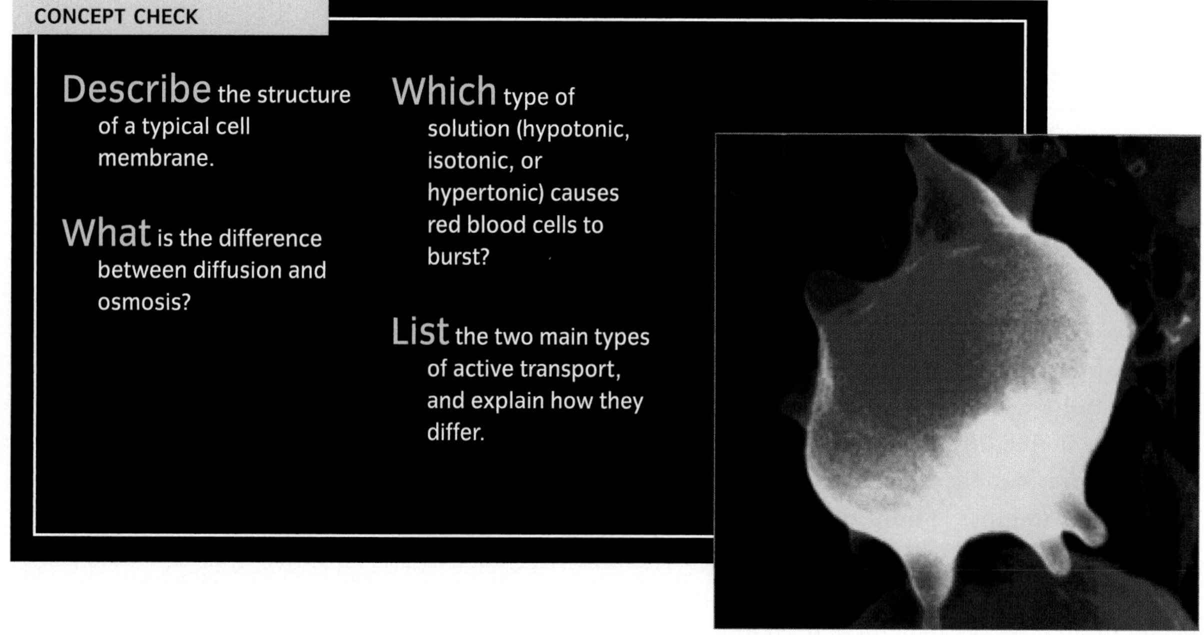

CONCEPT CHECK
Describe the structure of a typical cell membrane.
What is the difference between diffusion and osmosis?
Which type of solution (hypotonic, isotonic, or hypertonic) causes red blood cells to burst?
List the two main types of active transport, and explain how they differ.

The Components of a Cell Are Called Organelles

LEARNING OBJECTIVES

List the main organelles of a typical animal cell and describe their function.

Compare eukaryotic and prokaryotic cells.

Identify the organelles specific to plant cells

CYTOSKELETON IS THE POWER BEHIND THE MEMBRANE

Cytologists used to view the cytosol as a water bath, but it is actually a highly organized chemical soup pervaded by a support structure called the **cytoskeleton**. The cytoskeleton lies directly underneath the plasma membrane, and is attached to it in many places. Composed mainly of three types of filament, the cytoskeleton extends throughout the cytosol, providing shape, support, and a scaffold for suspending and moving organelles. Unlike your bony skeleton, the cytoskeleton is continuously changing shape, forming and breaking down. This gives cells a plasticity, or fluid resiliency, that allows them to change shape or move organelles quickly.

Cytoskeleton The internal framework of a cell.

The cytoskeleton has three types of protein structure: **microfilaments**, **intermediate filaments**, and **microtubules**. Microfilaments, the thinnest elements, are responsible for cellular locomotion, muscle contractions, and movement during cell division. They also establish the basic shape and strength of the cell. Intermediate filaments are much stronger than microfilaments and protect the cell from mechanical stresses. Microtubules are long strings of the globular protein tubulin, coiled tightly into a tube. Microtubules are used as tracks for organelle movement, and are instrumental in chromosome movement during cell division. The proteins of these cytoskeletal elements are what give them their characteristic functions. The microfilaments are composed mostly of actin, a protein that, under the proper conditions, will cause movement in a predictable fashion. We discuss this protein far more extensively when looking at skeletal muscle contraction. Intermediate filaments are composed of extremely tough, supportive proteins found nowhere else in the cell.

FLAGELLA AND CILIA KEEP THINGS MOVING

Many cells have projections from their surface that can move either the entire cell or the extracellular fluid. **Flagella** are single, long whiplike structures that propel the cell forward. The only human cell that moves by flagellum is the sperm.

Cilia are shorter extensions that look like hairs or eyelashes, and they are far more common in the human body (FIGURE 3.9). They beat synchronously in what is referred to as a "power stroke" to move mucus across the surface of the cell, or to circulate the extracellular fluid to increase diffusion. Cilia line the upper respiratory tract, moving mucus upward and sweeping out debris and pathogens. Cilia lining the fallopian tubes move the egg from the ovary to the uterus.

Cilia movement FIGURE 3.9

Cilia are formed from an inner core of microtubules, extending from the cytoskeleton.

ENDOPLASMIC RETICULUM: PROTEIN AND HORMONE MANUFACTURING SITE

Within the cytosol of many cells lie networks of folded membranes, called the **endoplasmic reticulum** or **ER** (literally "within fluid network"). The membranes of the ER are directly connected to the double membrane surrounding the cell nucleus.

Human cells have two types of ER, rough and smooth. **Rough endoplasmic reticulum** (RER) is a processing and sorting area for proteins synthesized by the **ribosomes** that stud its outer membrane (FIGURE 3.10). Ribosomes are small nonmembrane-bound organelles composed of protein and ribosomal RNA. They serve as protein factories, synthesizing proteins that may be included in other organelles or in the plasma membrane itself, or are exocytosed through secretory vesicles.

Smooth endoplasmic reticulum, or **SER**, is responsible for the synthesis of fatty acids and steroid hormones, such as testosterone. SER has no ribosomes. In the liver, enzymes that break down drugs and alcohol are stored in the SER.

In both RER and SER, the end product is a vesicle filled with product ready for the next step in processing. These vesicles form from the ER and usually move substances from the ER to the cell membrane for exocytosis, or to the Golgi complex for further packaging.

GOLGI COMPLEX: COMPLICATED CHEMICAL FACTORY

This organelle is one of the few to retain the name of its discoverer, Camillo Golgi, who discovered it in 1898. The **Golgi complex**, or Golgi apparatus, is usually found

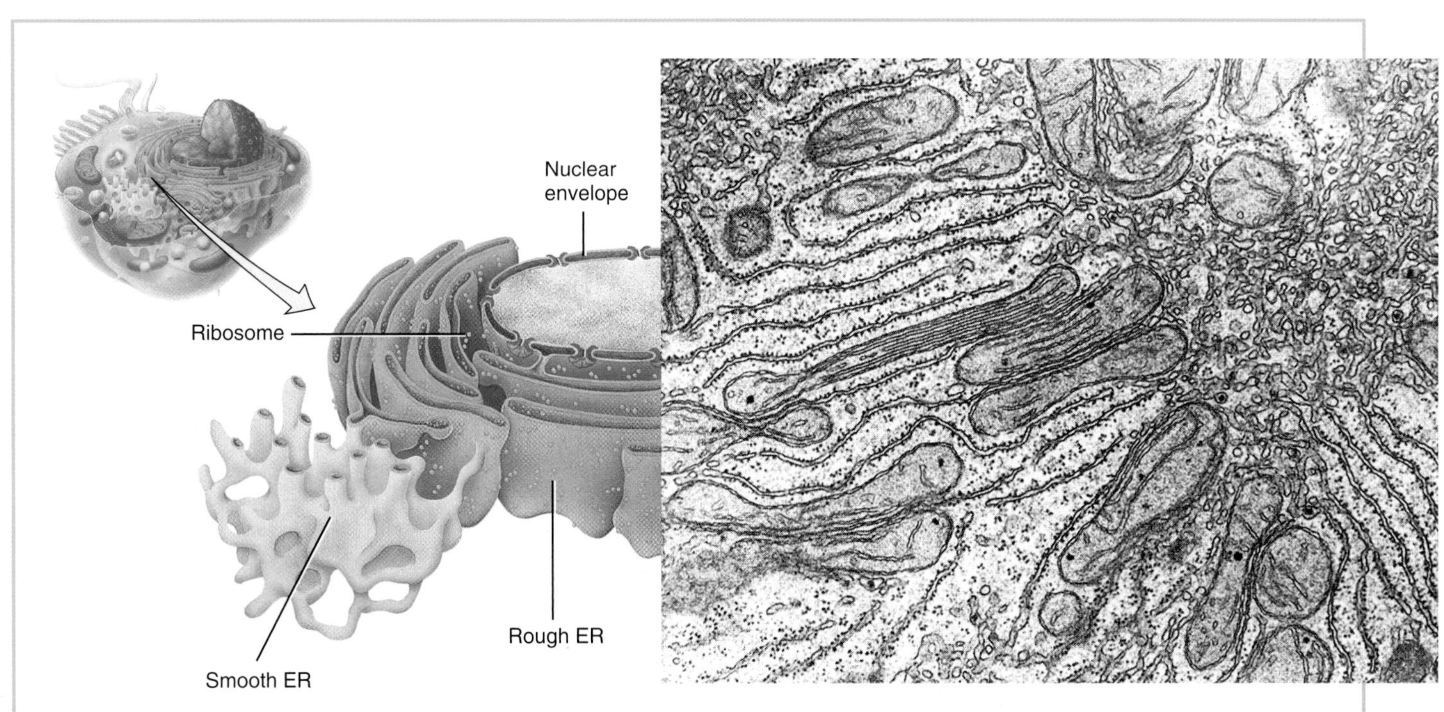

Smooth endoplasmic reticulum and rough endoplasmic reticulum FIGURE 3.10

The cell is packed with ER. The thin tubules without ribosomes studding their surface are the channels of the SER. The RER is concentrated in the lower left of the micrograph. As can be seen in the view of the whole cell at the left, RER is found immediately outside the cell nucleus, while SER is a continuation of the RER tubules.

near the end of the SER and resembles a stack of pancakes called **saccules** (FIGURE 3.11). Saccules are slightly curved, with concave and convex faces. The concave portions usually face the ER, and the convex portions face the plasma membrane. Vesicles are found at the edges of these saccules.

Saccule Small circular vesicle used to transport substances within a cell.

The precise role of the Golgi complex is debated. Clearly it is involved with processing of proteins and fatty acids, but exactly how does it do that? Some scientists believe that vesicles from the ER fuse with the lowest saccule of the Golgi complex, and then the saccules "move up" in ranking toward the upper saccule. From there, the Golgi complex membrane reforms the vesicle, which transports completed proteins to their destination (FIGURE 3.12). Other scientists believe that the original vesicles from the ER fuse with the top saccule of the Golgi complex right from the start. The enzymes within this top saccule complete the processing of the proteins or fatty acids in the vesicle, which are then transported to their functional areas.

In either case, the vesicles that leave the Golgi complex migrate all over the cell. Some fuse with the cell membrane, others fuse with lysosomes, and still others become lysosomes. It seems that the Golgi complex completes the processing of proteins and fatty acids, readying the products for use in other organelles or in the cell membrane.

LYSOSOMES: SAFE CHEMICAL PACKAGES

Lysosomes are chemical packages produced by the Golgi complex that contain **hydrolytic enzymes** powerful enough to digest an entire cell from the inside. The lysosome sequesters these digestive enzymes for use in decomposing macromolecules that have entered the cell via endocytosis (FIGURE 3.13). When a lysosome (*lyse* means to break open or break apart)

Hydrolytic enzymes Proteins that help decompose compounds by splitting bonds with water molecules.

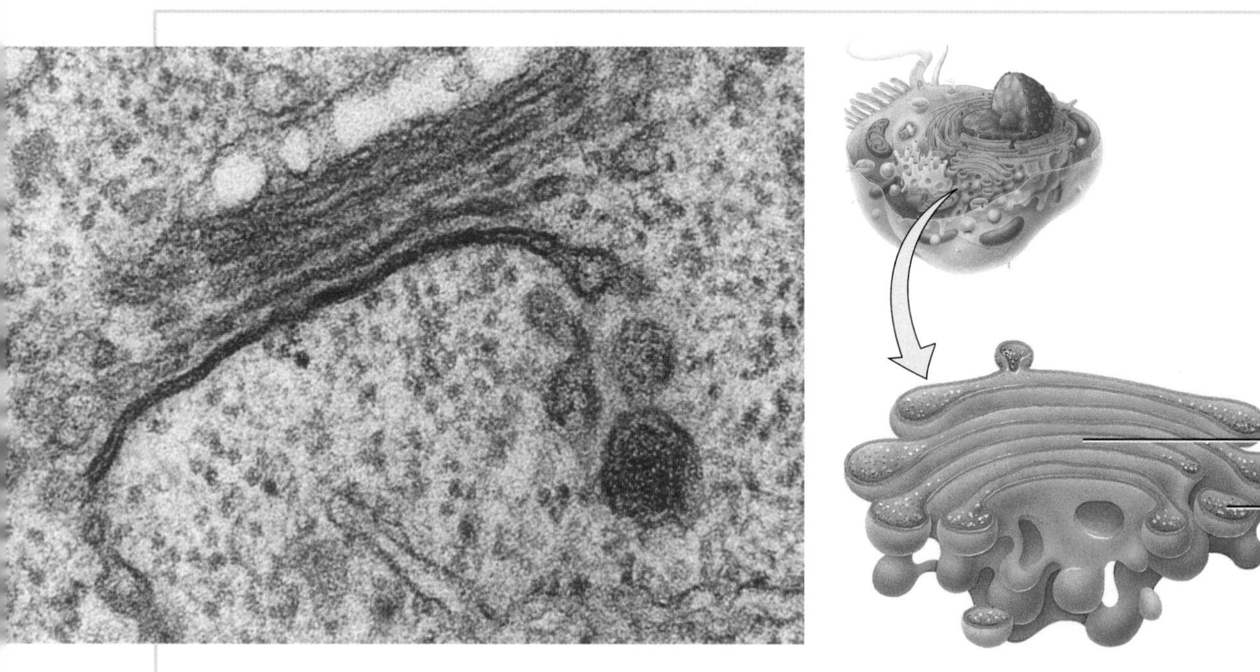

Golgi complex FIGURE 3.11

The color-enhanced blue Golgi complex in this cell clearly shows the "stack of pancakes" appearance of this organelle.

Membrane cycling FIGURE 3.12

Membrane is constantly cycling through the cell. New membrane is being made in the ER, transported through the Golgi complex, and finally fused with the plasma membrane.

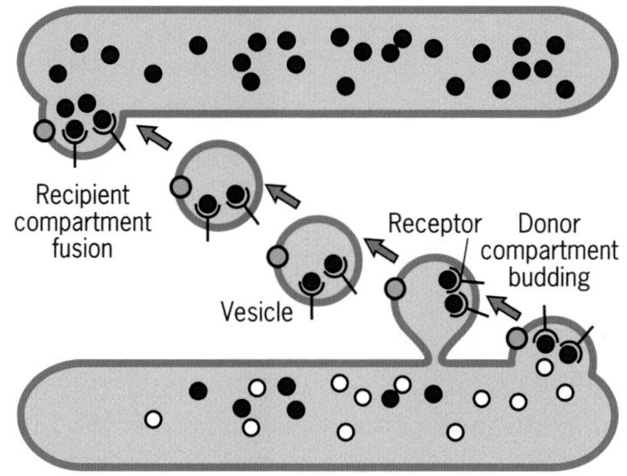

A

In the diagram above (A), you see a portion of membrane synthesized at the RER with an associated membrane protein associated. Following that small green bubble on the membrane through the green stacks of the Golgi complex and out at the green vesicles, you can see that this protein eventually winds up as an integral part of the membrane.

In the illustration to the right (B), you see another vesicle filled with pinkish digestive enzymes coming from the RER. These vesicles also move through the Golgi complex, but the new membrane does not fuse with the plasma membrane. Instead these vesicles become the lysosomes, fusing with endocytosed materials and digesting both the material and the membrane encircling it.

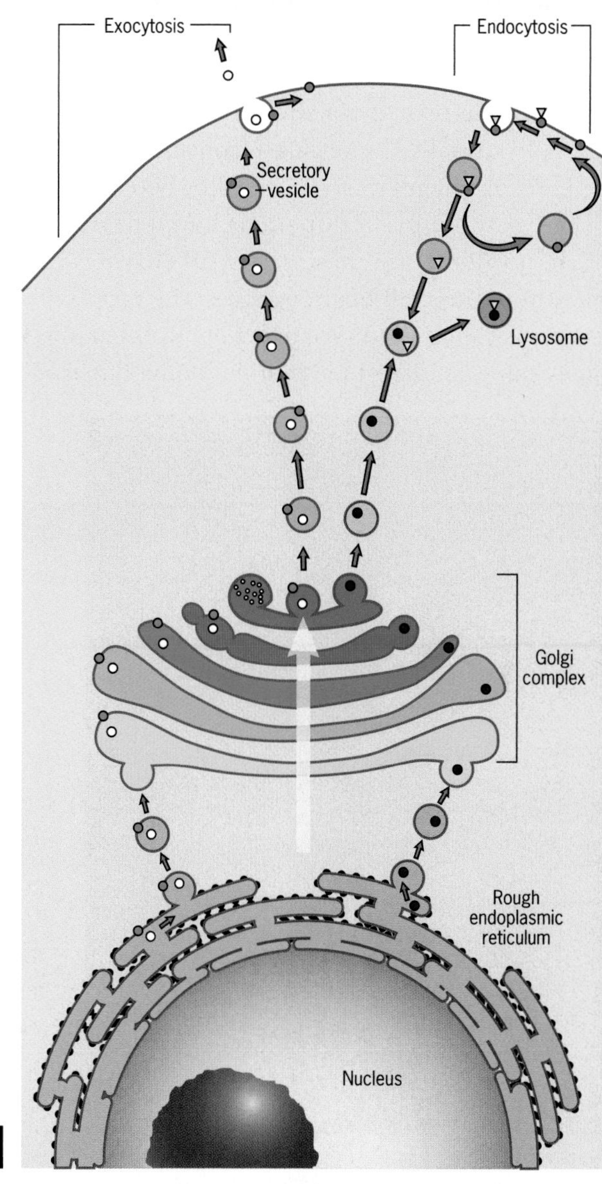

B

Process Diagram

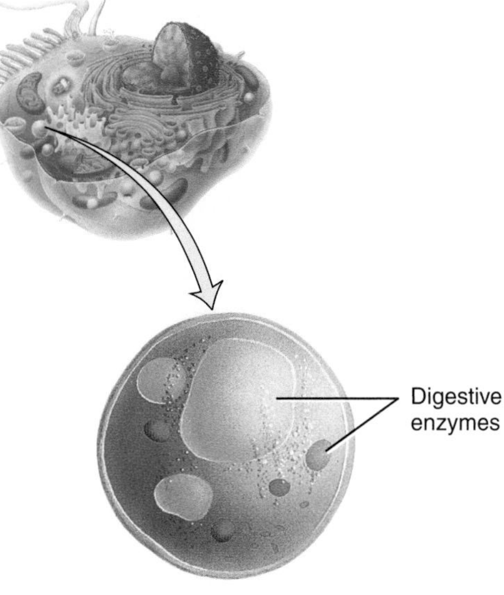

A Lysosome

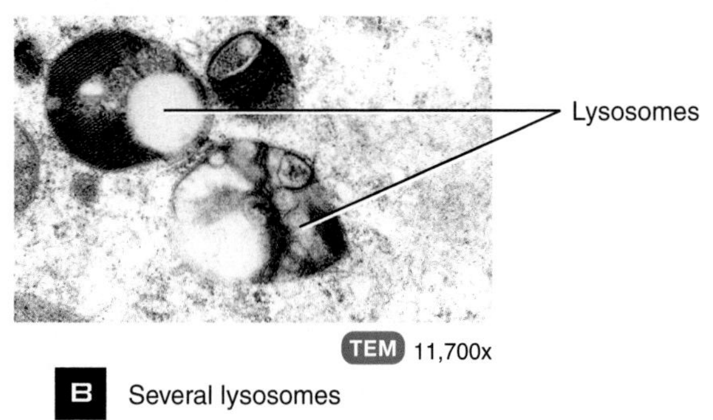

B Several lysosomes

Lysosome FIGURE 3.13

The lysosome sequesters digestive enzymes for use in decomposing macromolecules that have entered the cell via endocytosis, or for autolysis (self-destruction).

fuses with an endocytotic vesicle, it pours its contents into the vesicle. The hydrolytic enzymes immediately begin breaking down the vesicle's contents. In this way, the lysosome provides a site for safe digestion in the cell. Additionally, bacteria are routinely destroyed in the body by phagocytosis followed by lysosomal activity. If the lysosome breaks open, as happens during cell death, it will release these powerful enzymes into the cell, where they will begin to digest the cell itself. This process is called autolysis, literally self-breaking. Lysosomes can even digest parts of the body. The frog's tail is lost not by developmental changes in DNA processing but rather by lysosomes bursting and digesting cells in the tail.

THE CELL'S LIBRARY IS THE NUCLEUS

The **nucleus** contains a cell's genetic library, and is usually the largest organelle in a cell (FIGURE 3.14). (Mature human red blood cells, however, have no nucleus.) This organelle is approximately 5 micrometers in diameter in most human cells. It is covered, like the cell itself, by two layers of membrane, called the **nuclear envelope**. The envelope is punctuated by **nuclear pores,** which allow molecules to enter and exit the nucleus. The DNA in the nucleus is the cell's library, which is "read" by molecules called RNA. After RNA makes a perfect impression of the DNA, it leaves the nucleus and serves as templates for proteins. The process of forming RNA is called **transcription**, which means to "write elsewhere" (FIGURE 3.15A).

Nuclear membrane FIGURE 3.14

In this freeze-fractured electron micrograph of the nuclear membrane, the nuclear pores are clearly visible. These pores are ringed by proteins, seen here as depressions around the central pore. The two layers of the nuclear membrane have separated in the center of the image, providing a clear view of both membranes.

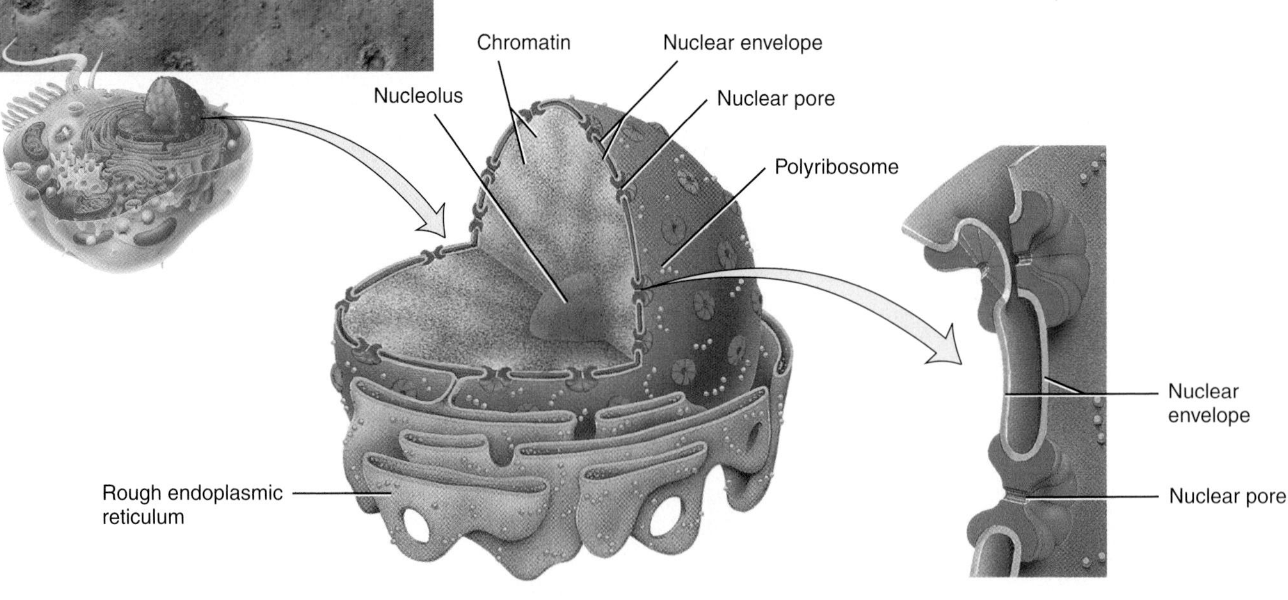

Transcription and translation FIGURE 3.15

During transcription, the genetic information in DNA is copied to RNA.

Step 1 The DNA within the nucleus unwinds at the gene that codes for the protein that is to be produced. RNA polymerase sits on this open portion of the strand and begins to form an RNA copy of that information, shown in red on the diagram. The copy is done via base pair matching, using uracil instead of thymine. There is no thymine in RNA! Making a copy like this is similar to your going into the reserve section of the library and copying notes word-for-word from a reserved text. You need the notes to be exact copies because you cannot take a reserved text from the library.

Step 2 The messenger RNA (the single strand in red) is then modified by snRNP molecules. These molecules recognize "junk DNA" or bits of code that do not code for anything useful in the final product. These redundant or nonessential portions of the mRNA are then snipped out of the final messenger RNA by molecules known as snRNPs. A snRNP is an energy-driven molecule that locates sections of redundant or meaningless DNA and cuts that sequence from the newly formed mRNA.

Step 3 Once the messenger RNA meets approval, it leaves the nucleus through a nuclear pore, shown here as the brown spool-looking structure at the bottom of the image. Once outside the nucleus, this message can be translated into a protein.

The Process of Translation, or decoding the nucleic acid message to form a protein, is fully described in Part B on page 78.

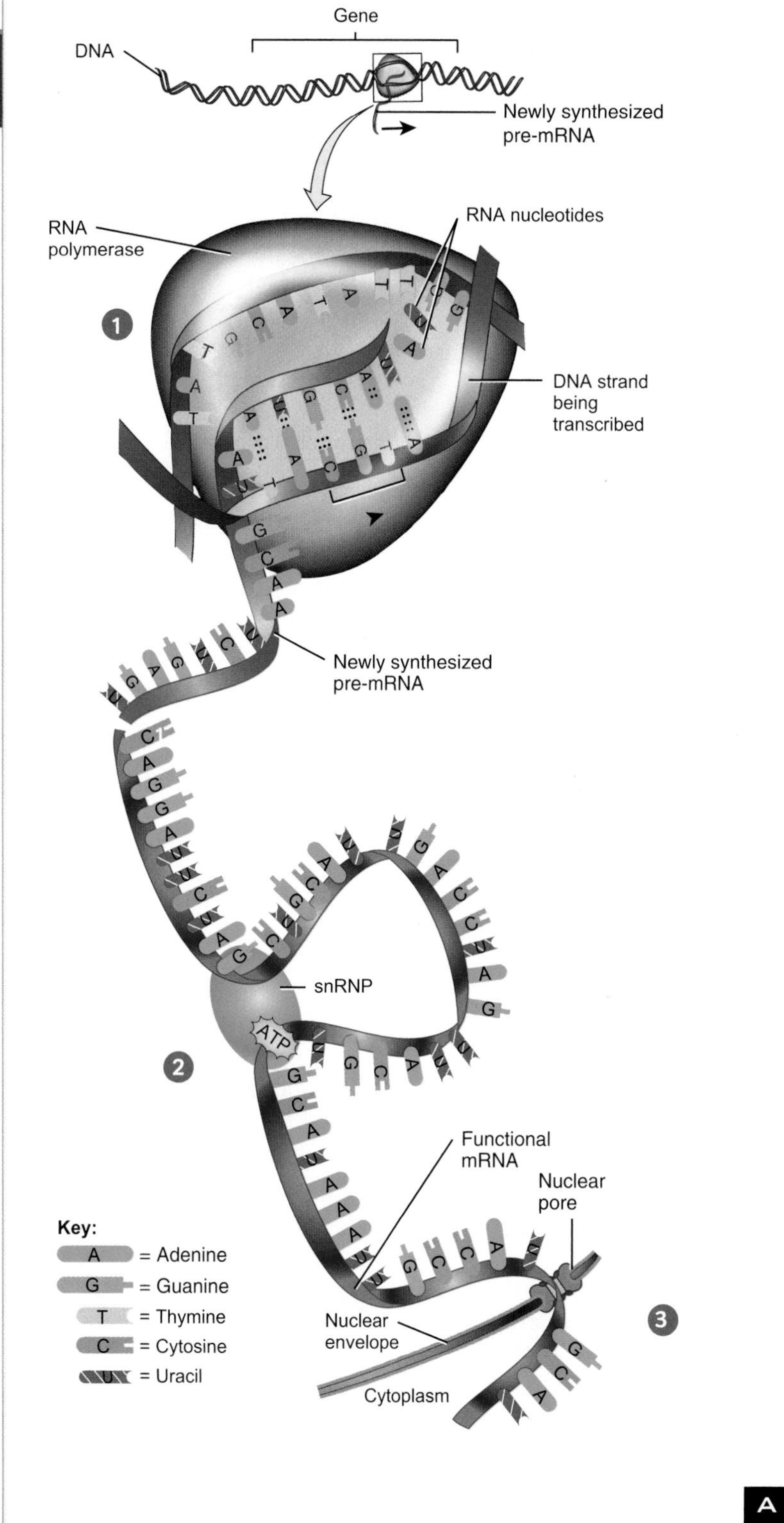

Process Diagram

(continues)

Once RNA is formed within the nucleus, it leaves via the nuclear pores. In the cytoplasm, this message is "read" by ribosomes. The single strand of mRNA is fed into the center of a ribosome, where transfer RNA matches up to it in three-base pair sections. Each tRNA carries an amino acid, specified by those same three-base pairs that are matching up to the mRNA. As the mRNA is passed through the ribosome, the

Process Diagram

Transcription and translation FIGURE 3.15

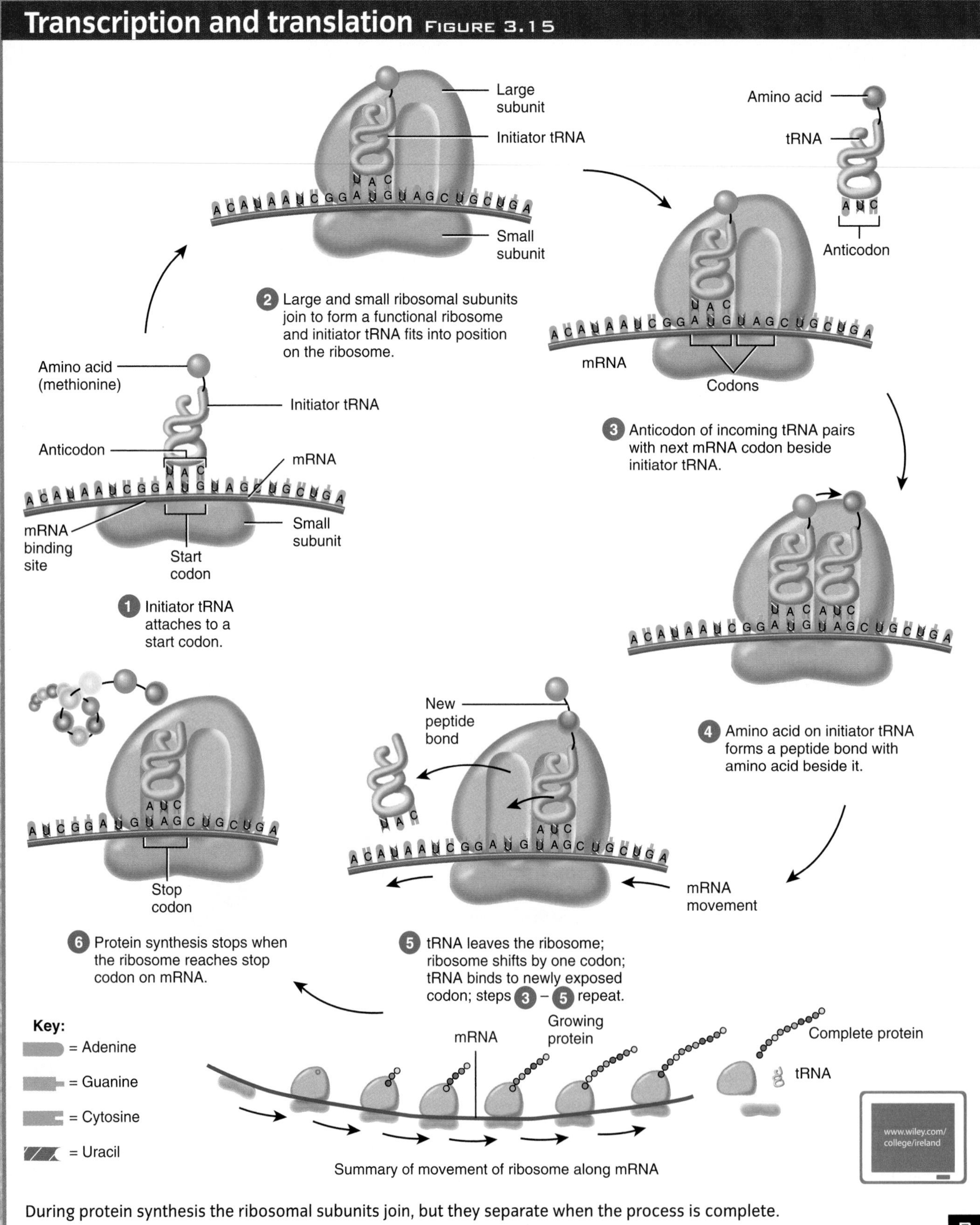

Large subunit
Initiator tRNA
Small subunit
2 Large and small ribosomal subunits join to form a functional ribosome and initiator tRNA fits into position on the ribosome.
Amino acid
tRNA
Anticodon
mRNA
Codons
3 Anticodon of incoming tRNA pairs with next mRNA codon beside initiator tRNA.
Amino acid (methionine)
Initiator tRNA
Anticodon
mRNA
mRNA binding site
Start codon
Small subunit
1 Initiator tRNA attaches to a start codon.
4 Amino acid on initiator tRNA forms a peptide bond with amino acid beside it.
New peptide bond
mRNA movement
Stop codon
6 Protein synthesis stops when the ribosome reaches stop codon on mRNA.
5 tRNA leaves the ribosome; ribosome shifts by one codon; tRNA binds to newly exposed codon; steps 3 – 5 repeat.
Key:
= Adenine
= Guanine
= Cytosine
= Uracil
mRNA
Growing protein
Complete protein
tRNA
Summary of movement of ribosome along mRNA
www.wiley.com/college/ireland

During protein synthesis the ribosomal subunits join, but they separate when the process is complete.

B

message it carries is read in triplets and the correct amino acids are brought to the growing peptide chain. This "decoding" of the nucleic acid message, converting it to an amino acid sequence, is referred to as translation. See FIGURE 3.15 for a full description of the events in transcription and translation.

The DNA within the nucleus of an active cell (neither resting nor dividing) is present as a threadlike molecule called **chromatin** that looks diffuse and grainy under a light microscope. Before cell division, these threads condense and coil into individually visible **chromosomes** (FIGURE 3.16). Imagine trying to sort yarn into two equal piles. It would be impossible until you coil the yarn into balls. The same is true of the chromatin in the nucleus. The process of forming chromosomes facilitates nuclear division by organizing and packaging the DNA.

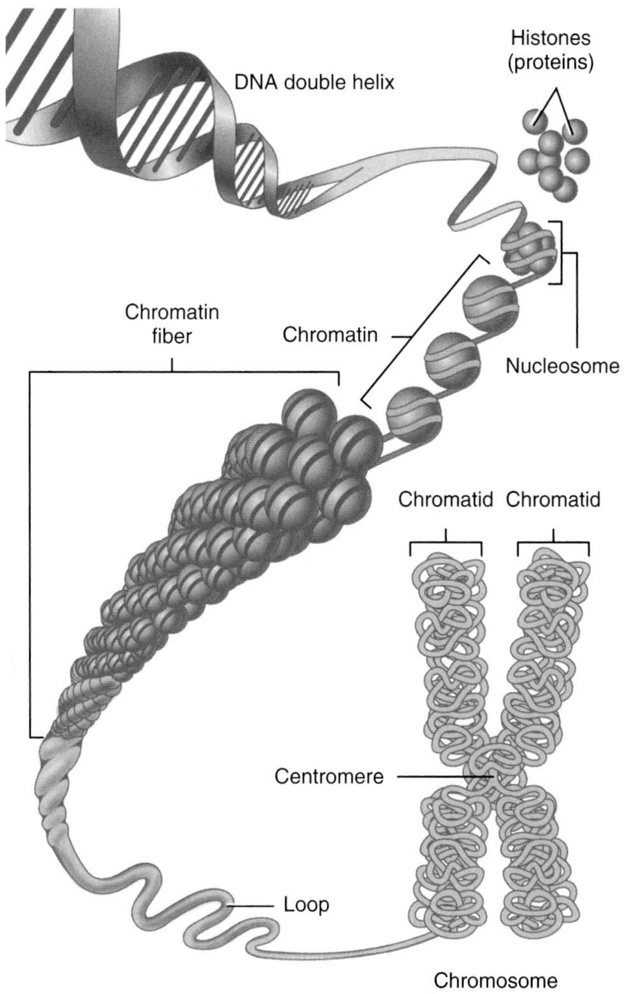

Chromosome FIGURE 3.16

A chromosome is a highly coiled and folded DNA molecule that is combined with proteins. The two arms of the chromosome are identical pieces of DNA.

The nucleus of most active cells contains darker areas of chromatin, called **nucleoli** (singular: nucleolus). Nucleoli produce ribosomal RNA and assemble ribosomes. Completed ribosomes then pass through the nuclear pores into the cytosol where some attach to the RER and others remain as free ribosomes. Because a cell's need for ribosomes changes throughout the cell cycle, nucleoli appear and disappear in the **nucleoplasm**.

Nucleoplasm
Fluid within the nucleus, containing the DNA.

MITOCHONDRIA ARE ENERGY FACTORIES

The last of the major organelles is the **mitochondrion** (plural: mitochondria). This bean-shaped organelle has a smooth outer membrane and a folded inner membrane, with folds called **cristae** (FIGURE 3.17). The mitochondria convert digested nutrients into usable

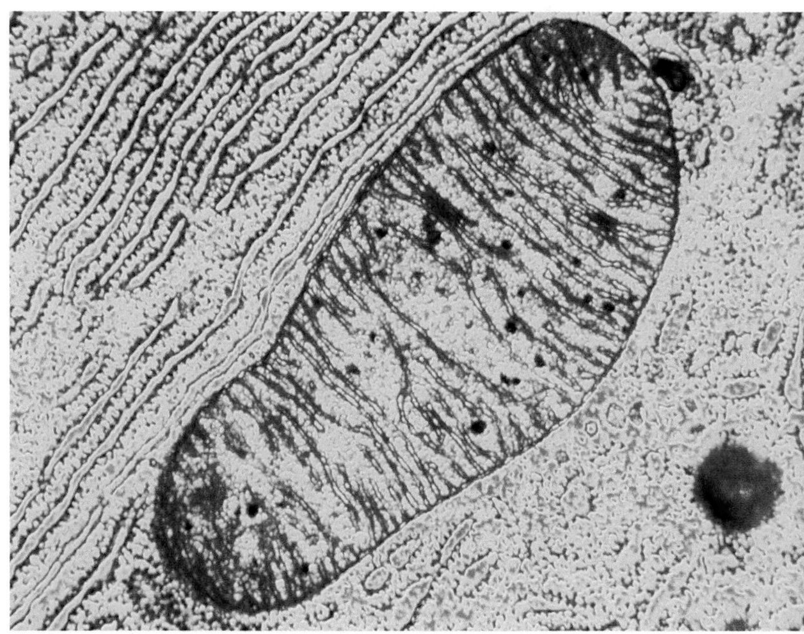

Mitochondrion FIGURE 3.17

Mitochondrial reactions FIGURE 3.18

Mitochondria break down glucose to produce ATP. This process is completed in four steps, the first of which happens outside the mitochondrial walls. The formation of ATP occurs on the inner walls of the mitochondrion. These processes will be discussed in more detail in Chapter 13.

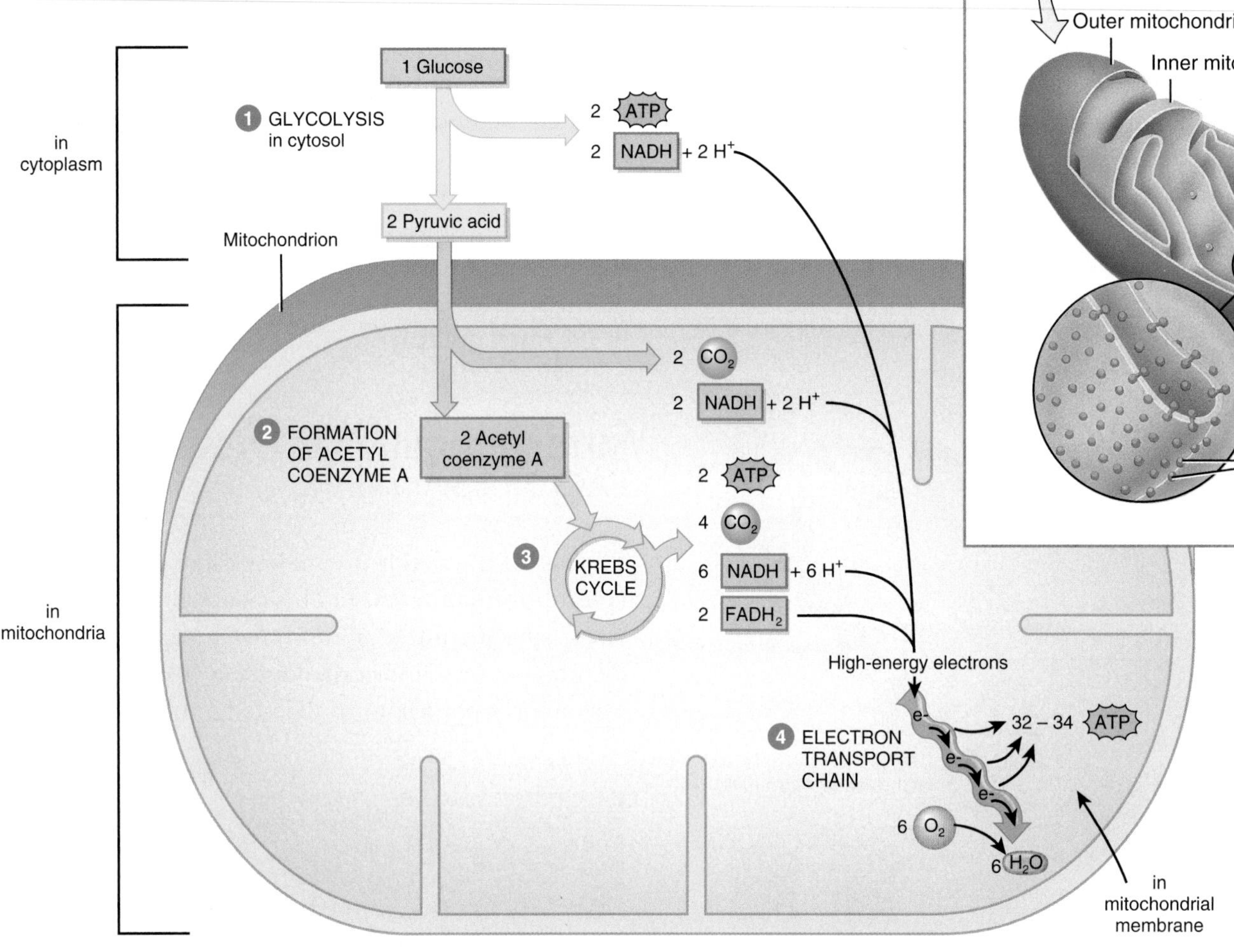

1. Glucose is brought into the cell via facilitated diffusion, where it is broken down in a series of chemical reactions called glycolysis. Glycolysis releases energy in two ATP molecules and two molecules of pyruvic acid.

2. Pyruvic acid then gets taken into the mitochondrion, where it is converted to acetyl co-A.

3. Acetyl co-A feeds into the Krebs cycle, another series of biochemical reactions that release energy from the acetyl co-A in the form of ATP, NADH, and $FADH_2$.

4. In the final step of the mitochondrial reactions, the NADH and $FADH_2$ formed during glycolysis and the Krebs cycle are transported to the inner membrane of the mitochondrion. There they are used to drive a final series of reactions called the electron transport chain. This final series converts the energy stored in the NADH and $FADH_2$ into usable ATP for the cell.

energy for the body, in the form of ATP. Virtually every move you make, every step you take, can be traced to mitochondria. Each cell has many mitochondria, all undergoing cellular respiration producing the ATP your cells need to survive. ATP forms within the inner membrane of the mitochondrion (FIGURE 3.18). Mitochondria require oxygen, and produce carbon dioxide in their endless production of ATP. In the final analysis, we inhale oxygen to serve our mitochondria, and we exhale the carbon dioxide they produce while generating ATP.

Mitochondria can divide, replicating these energy-producing organelles when our cells need more ATP. Cells in active tissues, like skeletal muscle and liver, have more mitochondria than cells in less-active tissue. This ability to reproduce has long intrigued cellular biologists. Mitochondria resemble bacteria in size and chemical composition, and carry their own DNA to pattern their proteins. Some scientists hypothesize that these organelles were once free-living bacteria that evolved from a **symbiotic** relationship into a type of ultimate, intimate symbiosis. Perhaps billions of years ago, a bacterial cell traded a free-living existence for a safe and constant environment in which to carry out its life processes. In this "you scratch my back and I'll scratch yours" arrangement, the sheltering cell receives a supply of ATP in return for protecting the mitochondria, delivering oxygen to it and disposing of its waste carbon dioxide. Also, as mentioned above, mitochondria have their own DNA. Mitochondria are not constantly reshuffled through sexual reproduction and are inherited only through the maternal lineage. The relatively stable DNA in mitochondria means they can help trace human migrations and evolution.

Symbiotic Intimate co-existing of two organisms in a mutually beneficial relationship.

See TABLE 3.1 on page 82 for a summary of cell parts and their functions.

CELLS ARE DIFFERENT IN PLANTS AND BACTERIA

The typical cell described above is a **eukaryotic** cell. All fungi, plants, and animals are composed of eukaryotic cells.

Eukaryotic Cells that contain a distinct membrane-bound nucleus.

Plant cells differ slightly from human cells. Because plants lack the skeleton found in most animals, their structure emerges from cell walls that surround their cells (FIGURE 3.19).

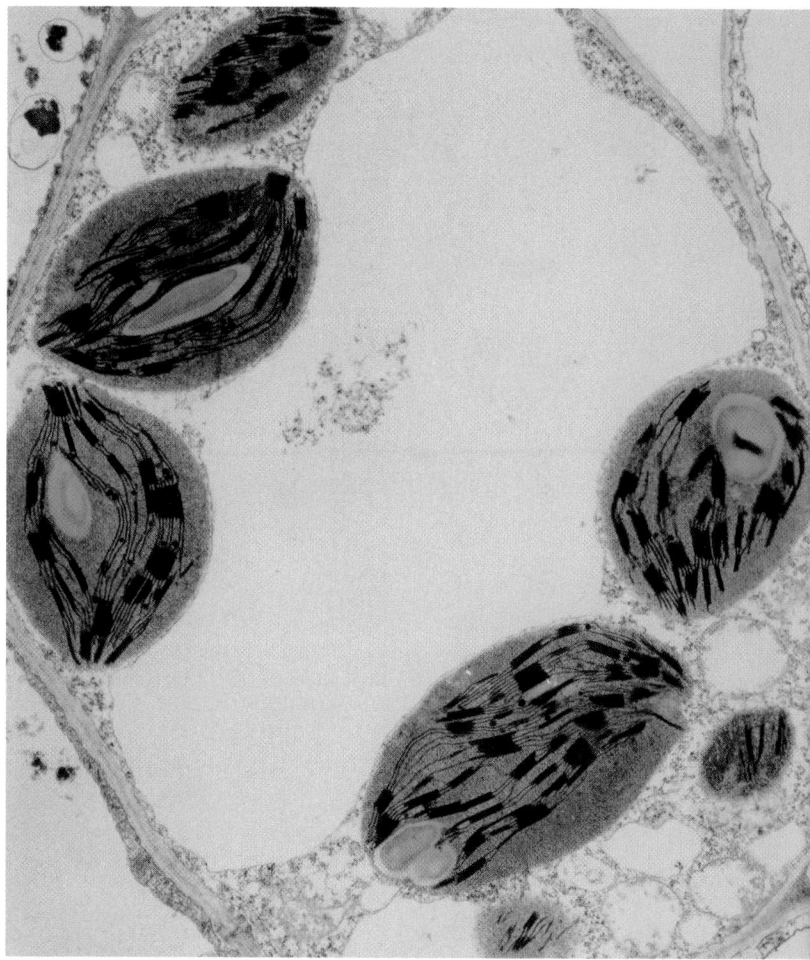

Plant cell FIGURE 3.19

Plant cell walls, composed of tough **cellulose**, provide rigid support immediately superficial to the cell membrane. The large, membrane-bound organelle called the **central vacuole** (not found in animal cells) maintains cell **turgor**. Many plant cells also have chloroplasts—organelles where photosynthesis occurs. These are similar in structure to mitochondria in that they have an inner and an outer membrane, and they are responsible for producing energy and making simple sugars from carbon dioxide. Like mitochondria, chloroplasts may have originated as bacteria that were "adopted" by the plant cell.

Turgor Internal pressure in living cells.

Cell parts and their functions TABLE 3.1

Part	Structure	Functions
Cell membrane	Composed of a lipid bilayer consisting of phospholipids and glycolipids with various proteins inserted; surrounds cytoplasm.	Protects cellular contents; makes contact with other cells; contains channels, receptors, and cell-identity markers; mediates the entry and exit of substances.
Cytoplasm	Cellular contents between the plasma membrane and nucleus, including cytosol and organelles.	Site of all intracellular activities except those occurring in the nucleus.
Organelles	Specialized cellular structures with characteristic shapes and specific functions.	Each organelle has one or more specific functions.
Cytoskeleton	Network composed of three protein filaments: microfilaments, intermediate filaments, and microtubules.	Maintains shape and general organization of celluar contents; responsible for cell movements.
Centrioles	Paired centrioles.	Organizing center for microtubules and mitotic spindle.
Cilia and flagella	Motile cell surface projections with inner core of microtubules.	Cilia move fluids over a cell's surface; a flagellum moves an entire cell.
Ribosome	Composed of two subunits containing ribosomal RNA and proteins; may be free in cytosol or attached to rough ER.	Protein synthesis.
Endoplasmic reticulum (ER)	Membranous network of folded membranes. Rough ER is studded with ribosomes and is attached to the nuclear membrane; smooth ER lack ribosomes.	Rough ER is the site of synthesis of glycoproteins and phospholipids; smooth ER is the site of fatty acid and steroid synthesis.
Golgi complex	A stack of 3 to 20 flattened membranous sacs called cisterns.	Accepts proteins from rough ER; stores, packages, and exports proteins.
Lysosome	Vesicle formed from Golgi complex; contains digestive enzymes.	Fuses with and digests contents of vesicles; digests worn-out organelles, entire cells, and extracellular materials.
Peroxisome	Vesicle containing oxidative enzymes.	Detoxifies harmful substances.
Mitochondrion	Consists of outer and inner membranes, cristae, and matrix.	Site of reactions that produce most of a cell's ATP.
Nucleus	Consists of nuclear envelope with pores, nucleoli, and chromatin (or chromosomes).	Contains genes, which control and direct most cellular activities.

Prokaryotic A cell with no internal membrane-bound compartments, usually having only ribosomes and chromosomes as recognizable organelles.

Not all life forms are eukaryotic. Bacteria, for example, do not have membrane-bound organelles. They have no chloroplasts, no mitochondria, no ER, no Golgi complex, or even a nucleus (FIGURE 3.20). These **prokaryotic** cells do not compartmentalize functions like eukaryotic cells, and their genetic material is loose within the cytoplasm.

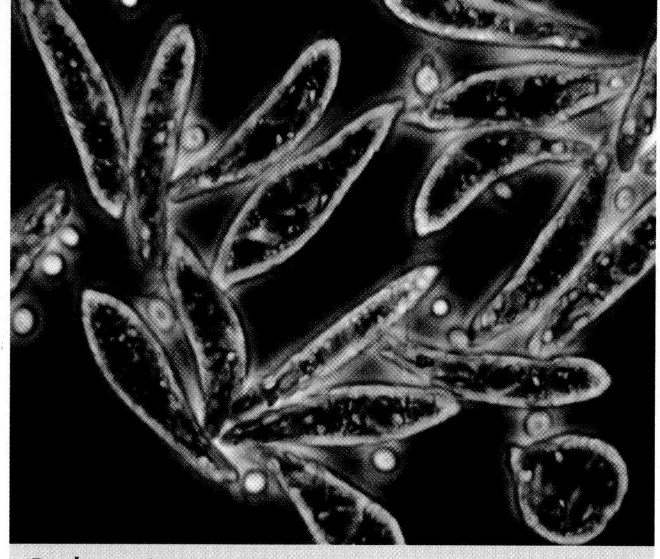

Prokaryote FIGURE 3.20

Note the smaller size of these bacterial cells. No compartmentalization is seen in these cells: no nucleus, no membrane bound organelles of any sort.

CONCEPT CHECK

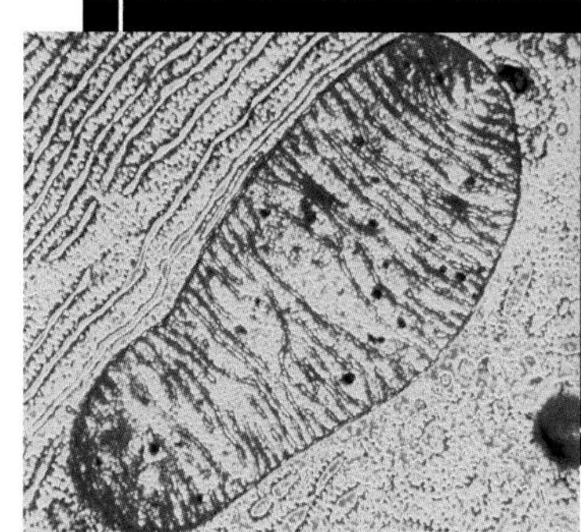

Describe the relationship between the Golgi complex and the lysosome.

What is located in the nucleus?

What evidence supports the idea that mitochondria were originally symbiotic bacteria?

List three differences between plant and animal cells.

Cell Communication and Cell Division Are Important to Cellular Success

LEARNING OBJECTIVES

Explain cellular signaling as it relates to the human body.

Define hormone.

Identify the steps necessary before mitosis can begin.

Trace the steps in mitotic cellular division.

To maintain stability and organization inside the human body, communication is essential. Cells must communicate with one another to function as a tissue. Tissues must send signals throughout the organ for the organ to function properly. Organs in a particular system must communicate to carry out the system's process. This only makes sense. Think how little you could accomplish in your personal life without communication among individuals in your community. How would schooling prepare you for life if no one discussed what it means to be an educated citizen? What would become of government if there weren't any communication among constituents? On a really grand scale, how would your personal life fare without a cell phone or Internet connections? Just as society requires communication for survival, cells of the body require communication in order to maintain homeostasis.

The signals sent from cell to cell include information about the timing of cell divisions, the health of adjacent cells, and the status of the external environment. Cells communicate with one another via chemical messengers or physical contact (FIGURE 3.21). Cell signaling can be accomplished via three routes:

1. Circulating **hormones** can be released into the bloodstream, potentially reaching every cell.
2. Local hormones, called **paracrines,** can be released to affect only cells in the vicinity. Neurons use paracrines to stimulate nearby nerve, muscle, or glandular cells by releasing short-lived chemicals called neurotransmitters.
3. Cells of epithelial and muscular tissues can interact with other cells directly through physical connections at cell-to-cell junctions.

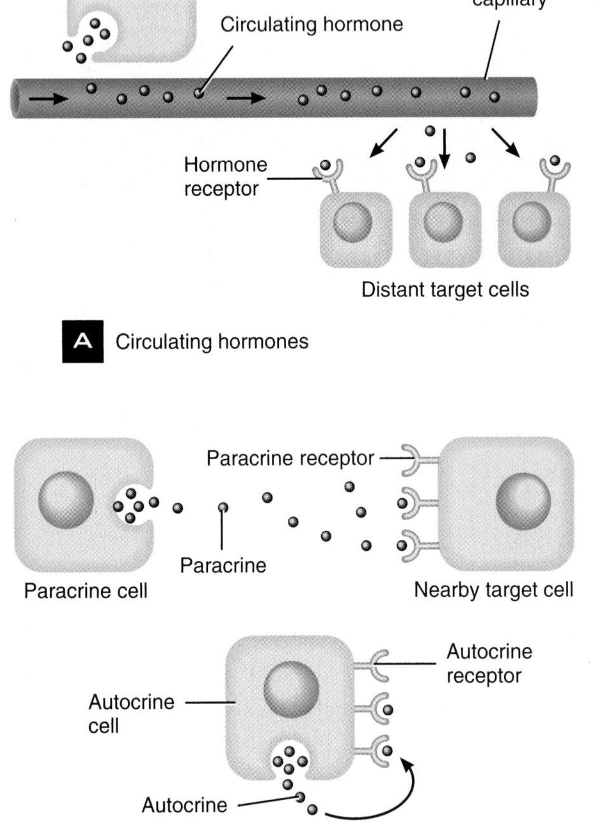

Cell signaling mechanisms FIGURE 3.21

Circulating hormones are carried through the bloodstream to act on distant target cells. Paracrines act on neighboring cells.

Hormone A secretion produced in one area of the body that travels to and alters the physiological activity of remote cells.

The differences in these types of signals lie mostly in the speed and distance of signal transmission, and the selectivity of reaching the target cells. Much hormonal communication is long-distance, carrying information to distant cells that will alter their functioning. The target cells are often remote from the secreting cells. For example, the pituitary gland in the center of the brain secretes a hormone that stimulates reproductive organs in the pelvic cavity.

Paracrine communication is mostly used when quick responses are required. Cells that are infected with a virus, for example, may secrete the paracrine interferon. Interferon alerts the surrounding cells, warning them of the viral invasion, and ideally helping them to fight it. Neurons must respond instantly to information; therefore, they secrete neurotransmitters directly into the space between cells. Sending neurotransmitters into the bloodstream would be too slow for nerve impulses.

Gap junctions, such as those between neurons, are used for instantaneous communications. They occur across very small distances and are extremely specific. Unlike endocrine communication, which has long-lasting effects, gap junction communications are immediate and short-lived.

Cell-to-cell junctions occur in tissues like your skin, where cells are in direct contact with one another. Skin cells are knit tightly together, so any change in one cell will be immediately passed to the next. Muscular tissue has similar cell-to-cell junctions, allowing the entire muscle to contract as one organ.

Should we try to cure deadly diseases with embryonic stem cells?

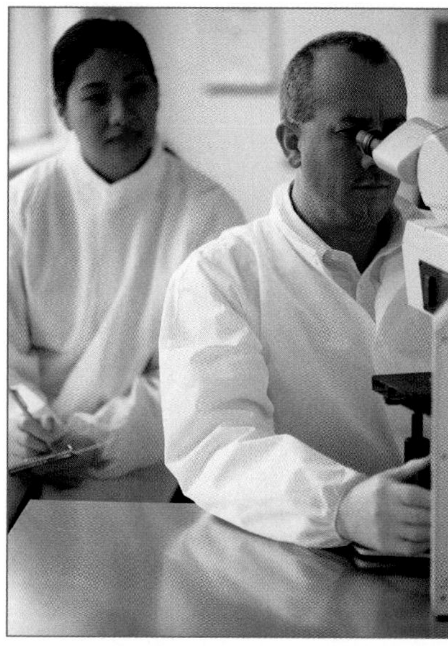

Many of the worst diseases around are actually diseases of the cells. When Parkinson's disease robs someone of the ability to move, the damage is caused by the death of a few million cells in a specific region of the brain. Spinal-cord injury is usually localized to a specific part of that essential nervous tissue. And type 1 diabetes is caused by the death of beta cells in the pancreas.

The ideal treatment is to replace dead or diseased cells with new "spare parts" grown from stem cells, flexible cells that naturally give rise to more specialized cells during development.

The most versatile type of stem cell is the embryonic stem cell (ESC). First grown in the laboratory at the University of Wisconsin in 1998, ESCs appear in the blastocyst a few days after conception. Because ESCs can form any body cell, they theoretically could be used to treat many conditions caused by cell death, such as type 1 diabetes, Alzheimer's disease, spinal cord injury, and Parkinson's disease.

ESC therapy would begin with "therapeutic cloning": inserting the patient's genes into an egg cell. After some cell divisions, ESCs carrying the patient's genetics would be extracted. The technology is still at a primitive stage, but it could sidestep the problems of organ shortage and immune attack.

ESCs are controversial, since the embryo is destroyed when ESCs are removed. To religions that believe that life begins at conception, ESC research is no better than abortion: it is murder. The U.S. allows federal support for research on ESCs only if they were derived before 2001. Some groups maintain that the potential medical benefits of ESC research outweigh the harm. Reform Judaism, for example, "believes strongly in the promise that stem cell research holds for finding a cure or treatment" for many deadly diseases.

Some opponents of ESC research prefer to focus on the more developed adult stem cells, which do not require killing an embryo. Some adult stem cells are already used to restore the immune system (bone-marrow transplants are actually transplants of bone-marrow stem cells, which form blood cells). But many scientists say stem cell research is too new to wall off one entire field—ESCs—and base our hopes solely on adult stem cells.

Advocates of ESC research also note that most ESCs come from surplus embryos donated by parents after successful in-vitro fertilization. Is discarding surplus embryos more ethical than using them to cure horrendous diseases?

What's your take? Would you draw an absolute line against the destruction of embryos, or do you believe adults with grave illnesses have more rights than a tiny ball of cells that may, under the right circumstances, grow into a person? If you oppose the destruction of embryos for stem-cell research, does that ethically oblige you to try to save the many embryos that fail to implant themselves in the uterine wall? (See also the box on reproductive cloning in Chapter 16).

CELL DIVISION

The best-coordinated, communication-rich event in a cell's life cycle is cell division, or **mitosis**. To carry out this complicated process, the cell must communicate with surrounding cells and its own organelles and biochemical pathways. During mitosis, DNA and organelles are duplicated, and DNA is condensed into manageable packets and sorted into separate nuclei. Then two intact cell membranes are formed, each containing all of the organelles and DNA of the parent cell.

The stages of mitosis are described in the Process Diagram (**FIGURE 3.22** on pp. 86–87).

Process Diagram

Mitosis FIGURE 3.22

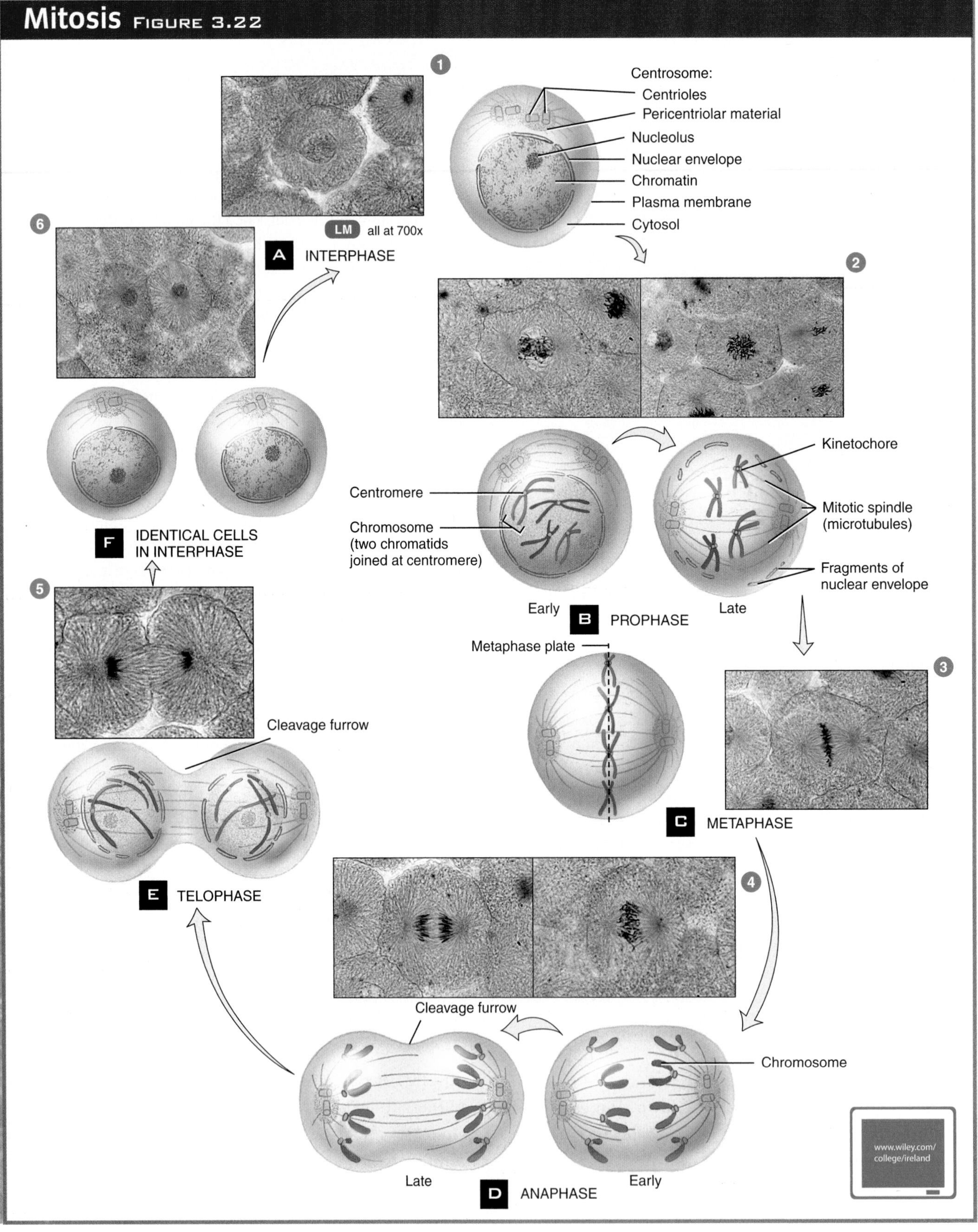

1
Centrosome:
Centrioles
Pericentriolar material
Nucleolus
Nuclear envelope
Chromatin
Plasma membrane
Cytosol
LM all at 700x
A INTERPHASE
2
Kinetochore
Centromere
Chromosome
(two chromatids
joined at centromere)
Mitotic spindle
(microtubules)
Fragments of
nuclear envelope
Early
B PROPHASE
Late
Metaphase plate
3
C METAPHASE
4
Cleavage furrow
Chromosome
Late
D ANAPHASE
Early
Cleavage furrow
E TELOPHASE
5
F IDENTICAL CELLS
IN INTERPHASE
6
www.wiley.com/
college/ireland

Process Diagram

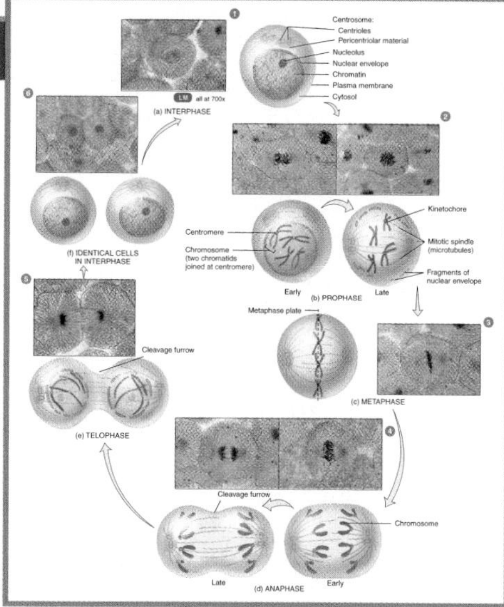

1 **Interphase** is the "resting" phase. The cell is not dividing, but rather carrying out its normal duties. The nuclear membrane is intact, the DNA is loose and unwound in chromatin threads, and nucleoli are present. In a cell that is destined to divide (some, like skeletal muscle and nerve cells, do not divide), the DNA doubles during interphase, but interphase is not considered part of mitosis.

2 In **prophase**, the nuclear membrane disappears; the chromatin condenses and becomes visible in the cell as chromosomes; the centrioles (which also doubled during interphase) separate and migrate to opposite ends of the cell. As the centrioles migrate, the spindle apparatus is formed. This is a network of microtubules that attach to the centromeres. Prophase is the longest phase of mitosis.

3 In **metaphase**, the middle phase of mitosis, the chromosomes are lined up on the central axis of the cell. As soon as the chromosomes are aligned, anaphase begins.

4 In **anaphase**, the spindle apparatus shortens, pulling the two arms of the chromosome from the centromeres. As the spindle fibers shorten, the chromosome breaks apart at the centromere, and the two arms are pulled away from each other. Anaphase is very quick, but it is here that the doubled genetic material separates into the DNA needed for each daughter cell.

5 **Telophase** is the final phase of mitosis. The chromosomes, now separated into two equal groups, de-condense into chromatin, and the DNA returns to the thread-like appearance. Nuclear envelopes form around these chromatin groups. The center of the cell pinches to form a cleavage furrow. The furrow deepens, eventually separating the cell into two separate cells, each with a nucleus containing the same amount of DNA as the parent cell.

6 The two daughter cells contain identical genetic material, and are clones of the single parent cell. Once division is completed, the daughter cells are in interphase, meaning they have begun a new growth phase. Eventually they will each reach the size of the original cell. They may undergo mitosis as well, individually moving through the cycle again.

CONCEPT CHECK

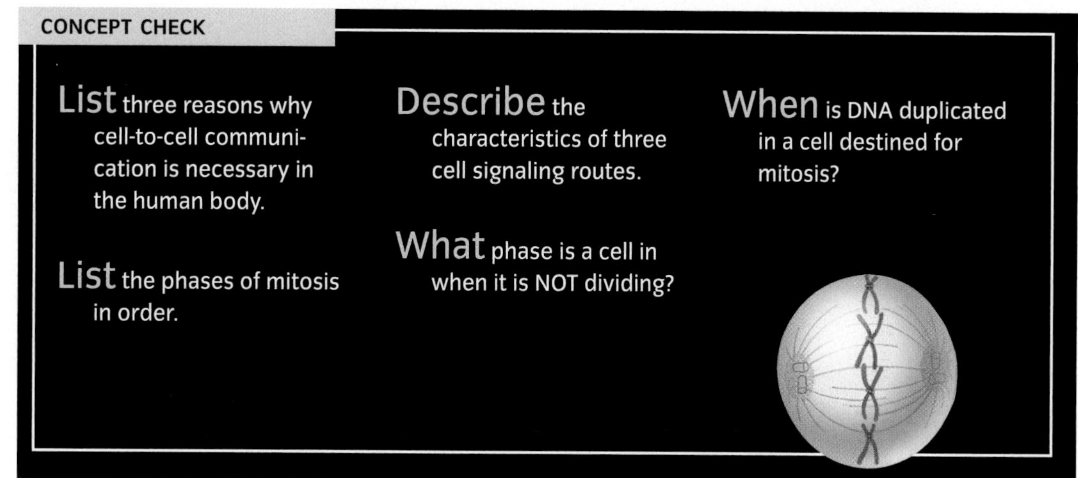

List three reasons why cell-to-cell communication is necessary in the human body.

List the phases of mitosis in order.

Describe the characteristics of three cell signaling routes.

What phase is a cell in when it is NOT dividing?

When is DNA duplicated in a cell destined for mitosis?

CHAPTER SUMMARY

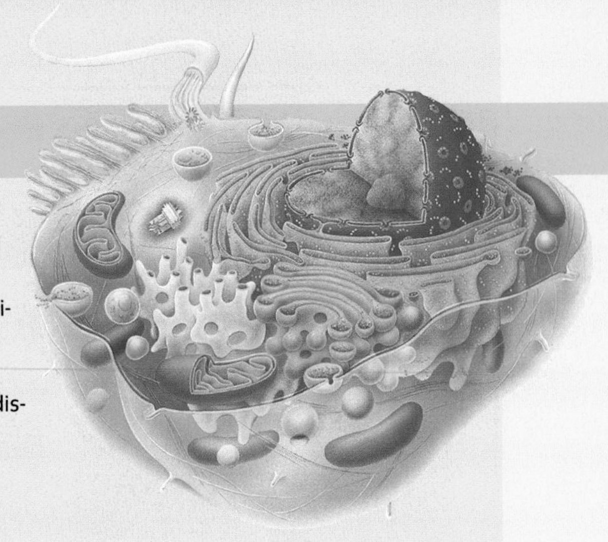

1 The Cell Is Highly Organized

According to the cell theory, all life is composed of cells. Cells come from preexisting cells, they contain hereditary material, and they are composed of similar chemical compounds. These cells have a membrane that separates them from the environment, as well as internal compartments designed to carry out specific functions. The cell membrane (or plasma membrane) is composed of a phospholipid bilayer. It is vital to cellular function, allowing some ions to pass freely and requiring energy expenditure to transport others. The membrane has embedded proteins and surface proteins that help distinguish a particular organism's immune identity. Osmosis and diffusion both transport molecules across the cell membrane, as do the active transport processes of endocytosis and exocytosis.

2 The Cell Membrane Isolates the Cell

The cell membrane is composed of a phospholipid bilayer, studded with proteins and covered on the surface with the glycocalyx. This liquid membrane is selectively permeable,allowing some substances free access to the cell while others are excluded. Passive transport across the membrane requires no energy, and includes filtration, diffusion, and facilitated diffusion. Osmosis describes the movement of water across the cell membrane. Solutions can be defined as isotonic, hypotonic, or hypertonic, depending on the concentration of water relative to that in the cell. Active transport requires ATP, and includes moving substances into the cell (endocytosis) and out of the cell (exocytosis) against their concentration gradients.

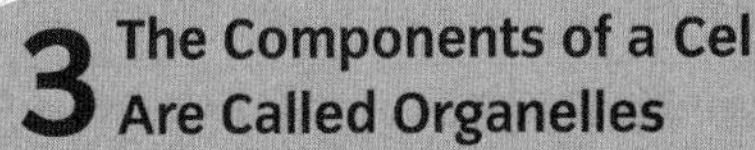

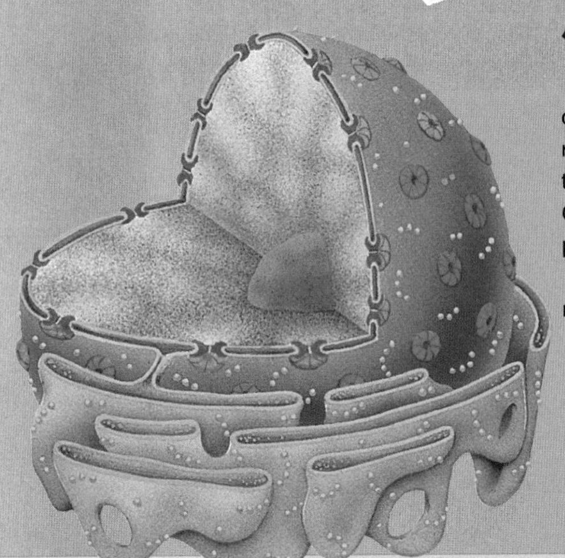

3 The Components of a Cell Are Called Organelles

A typical animal cell has the following organelles: nucleus, nucleolus, RER, SER, ribosomes, Golgi complex, lysosomes, centrioles, cytoskeleton, and mitochondria. Cilia are found on cells that must move fluid past them, and sperm carry a flagellum.

The cell is a dynamic place, where membrane is constantly being created and used. New membrane made at the RER is processed while moving to the Golgi apparatus and then to a transport vesicle destined to leave the cell. When the vesicle fuses with the cell membrane, the new phospholipid bilayer is spliced into place.

4 Cell Communication and Cell Division Are Important to Cellular Success

Cells communicate with one another through chemicals. Hormones carry information long distances in the body, while paracrine hormones convey information locally. Some cells, such as those of the skin or muscles, interact through direct physical contact as well.

Cells divide through a process called mitosis. In a cell destined for mitosis, the DNA is replicated during interphase. The steps in mitosis are:

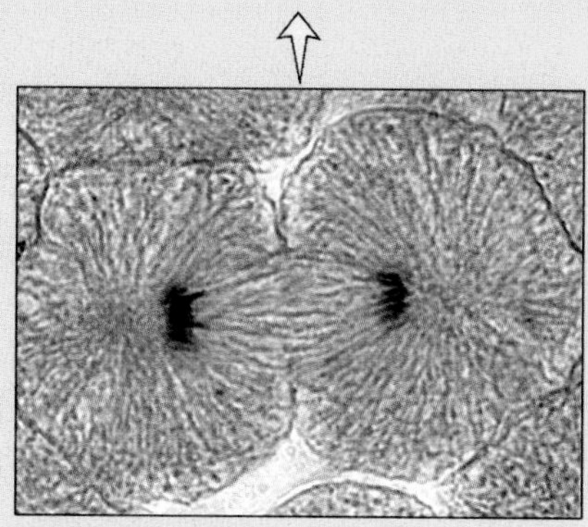

1. Prophase: when the nuclear membrane breaks down, the chromatin condenses into visible chromosomes, the centrioles separate, the spindle apparatus forms, and the spindle fibers attach to the centers of the chromosomes.
2. Metaphase: during which the chromosomes are aligned on the central plane of the cell.
3. Anaphase: which is the fastest phase of mitosis. During anaphase the sister chromosomes are pulled apart and separated, one to each pole of the cell.
4. Telophase: The final stage of mitosis. The chromosomes de-condense; two nuclear membranes now appear, one surrounding each group of separated chromosomes; the spindle apparatus dissolves; and a cleavage furrow appears.

Cytokinesis (cell division) marks the end of this process, resulting in two identical daughter cells both in interphase.

KEY TERMS

- **carotene** p. 53
- **cerebral edema** p. 60
- **cytoskeleton** p. 62
- **eukaryotic** p. 71
- **glycolipid** p. 56
- **glycoprotein** p. 56
- **hormone** p. 74
- **hydrolytic enzymes** p. 64
- **integral protein** p. 57
- **isotonic** p. 57
- **keratin** p. 53
- **melanin** p. 53
- **nucleoplasm** p. 69
- **organelle** p. 53
- **peripheral protein** p. 57
- **phagocytosis** p. 58
- **phospholipids** p. 55
- **pinocytosis** p. 58
- **prokaryotic** p. 73
- **saccule** p. 64
- **SEM** p. 54
- **solute** p. 57
- **symbiotic** p. 71
- **TEM** p. 54
- **turgor** p. 71

CRITICAL THINKING QUESTIONS

1. As a research assistant in a cytology lab, you are handed a stack of photographs from an electron microscope. Each represents a different type of cell. You are asked to identify photos of animal cells that secrete large amounts of protein, do not divide, and include a mechanism for moving their secretions along their surfaces. What organelles would be required by this cell? Which organelles would you not expect to see?

2. Assume you are now a lead scientist in a cytology lab, studying membrane proteins. You have placed a radioactive marker on an embedded membrane protein immediately after the DNA was translated. Trace the pathway this membrane protein would likely take while moving from its formation to its destination in the cell membrane. What organelles will it pass through? Where will it be located within these organelles?

3. Mutations are permanent changes in the nucleotide sequence in DNA. A point mutation is the loss or gain of a single nucleotide. If a base was lost from the DNA sequence, how would this affect subsequent transcription and translation of that gene?

SELF TEST

1. Which of the following is NOT a part of the cell theory?

   a. All living things are composed of cells.
   b. Cells cannot arise from preexisting cells.
   c. Chemically all cells are quite similar.
   d. Metabolism occurs within cells.

2. An organelle can be defined as

   a. dissolved compounds in the cytosol.
   b. a structure within the cytosol that performs at least one vital cellular function.
   c. a phospholipids bilayer.
   d. the smallest unit of life.

3. Within a human cell, it is common to find

   a. cytosol.
   b. melanin.
   c. ribosomes.
   d. all of the above.

4. True or False? The cell membrane is made up of phospholipids, which have a hydrophilic phosphate head and a hydrophobic lipid tail.

5. Movement across the cell membrane can be passive or active. Which of the following is an example of active transport?

   a. Diffusion
   b. Filtration
   c. Osmosis
   d. Sodium/potassium ATPase

6. On the figure below, identify the glycocalyx.

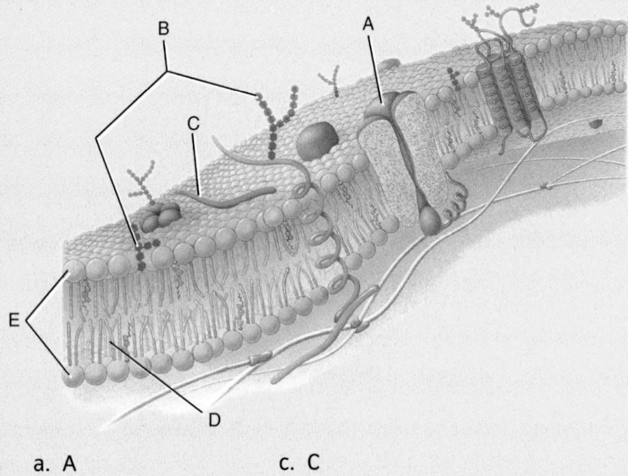

   a. A
   b. B
   c. C
   d. D

7. Using the same figure, what is the function of structure E?

   a. Identifying the cell as self
   b. Preventing water entry into the cell (hydrophobic end of the lipid)
   c. Allowing proteins to enter the cell
   d. Allowing cellular interaction with the aqueous environment of the body

8. Again looking at the same figure, indicate which label is identifying the integral proteins.

   a. A
   b. B
   c. C
   d. D
   e. E

9. Putting a cell in a hypotonic solution will result in that cell

   a. shrinking as water passes out of the cell membrane.
   b. expanding as water moves into the cell.
   c. remaining static, with no net water movement across the membrane.
   d. expanding as proteins move into the cell.
   e. shrinking as proteins move out of the cell.

10. The process of ______________ removes secretory products or wastes from a cell.

    a. endocytosis
    b. pinocytosis
    c. exocytosis
    d. phagocytosis

11. Identify the organelles indicated on this figure by matching the structure on the figure with the names below:

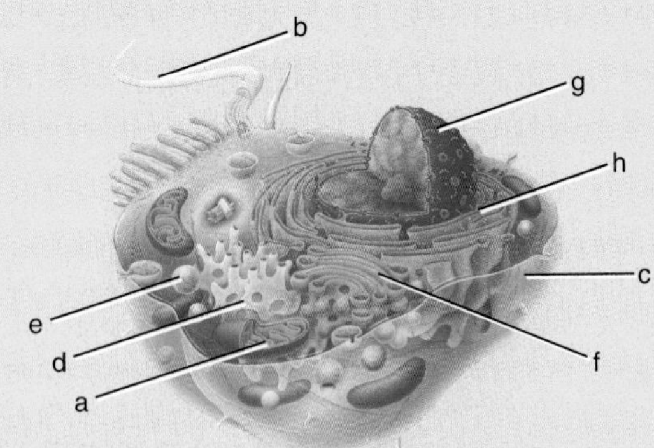

    a. Mitochondrion
    b. Flagella
    c. RER
    d. SER
    e. Lysosome
    f. Golgi complex
    g. Nucleus
    h. Nucleolus

12. What is the function of lysosomes?

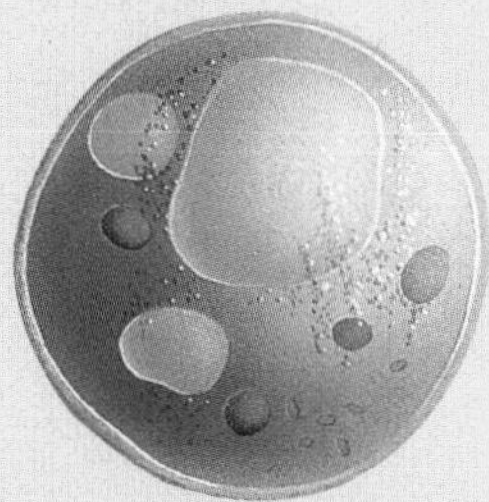

a. ATP production
b. Protein packaging and processing
c. Housing the DNA
d. Digesting worn-out organelles

13. The functions of the cytoskeleton include

a. providing cellular shape.
b. supporting organelles within the cytoplasm.
c. cellular locomotion.
d. all of the above are correct.

14. Which organelle is thought to have been a bacterial symbiont that is now permanently incorporated into eukaryotic cells?

a. Mitochondrion
b. Golgi complex
c. Ribosomes
d. Nucleus

15. The organelles responsible for moving fluid past the surface of a cell are

a. microvilli.
b. flagella.
c. cilia.
d. RER.

16. When a protein is formed, it moves from the ribosome to the RER and then on to the ______________, where it is processed for use either in the cell or in the extracellular matrix.

a. SER
b. Golgi complex
c. lysosome
d. nucleus

17. True or False? The largest organelle in most human cells is the organelle that houses the DNA of the cell.

18. Some cells communicate with one another through paracrines, which can be defined as

a. cell-to-cell contact.
b. long range hormones.
c. local hormones.
d. gap junctions.

19. What stage of mitosis is indicated by the image below?

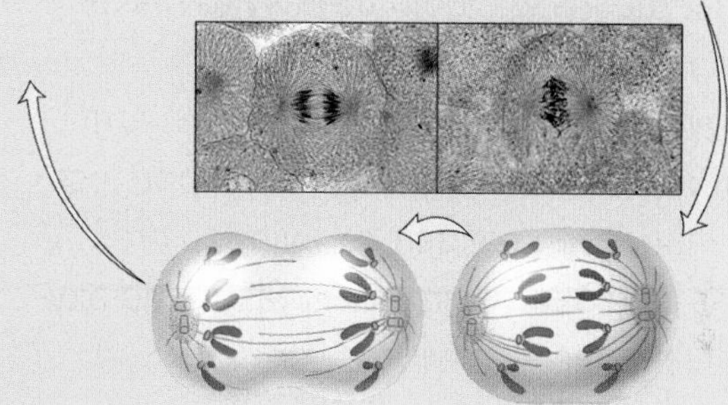

a. Prophase
b. Metaphase
c. Anaphase
d. Telophase
e. Interphase

20. In which cellular phase stage does DNA duplicate?

a. Prophase
b. Metaphase
c. Anaphase
d. Telophase
e. Interphase

Tissues

4

Tissues are groups of cells with similar function, and no tissue is more important than blood: it delivers oxygen and nutrients, it removes carbon dioxide, and it plays a key role in fighting infections. At first glance, blood looks simple: essentially a bunch of cells floating in salty water, more or less. That means it should be easy to make artificial blood. But it's not.

Many factors explain the interest in artificial blood. The military recognizes blood loss as a primary cause of death, owing to pressure loss as blood volume decreases and oxygen delivery declines with fewer red blood cells present. Blood is typically in short supply, and it cannot be frozen forever. Blood transfusions can carry viral diseases, and even in an emergency, it's necessary to check the patient's blood type before an infusion.

The most common way to deal with blood loss during emergencies is to infuse saline solution. This procedure maintains blood volume but does not replace the essential oxygen-carrying function. The major goal of artificial blood is to carry oxygen, since oxygen starvation can destroy brain cells in four minutes. Scientists have tried to use hemoglobin, the oxygen-carrying molecule in the red blood cell, but unless it is "packaged" inside a cell, it can damage the kidney.

At present, several experimental versions of oxygen-carrying artificial blood are being tested. But blood is a complex and subtle tissue, and so artificial blood is likely to serve as a supplement, not a replacement, for the natural stuff.

CHAPTER OUTLINE

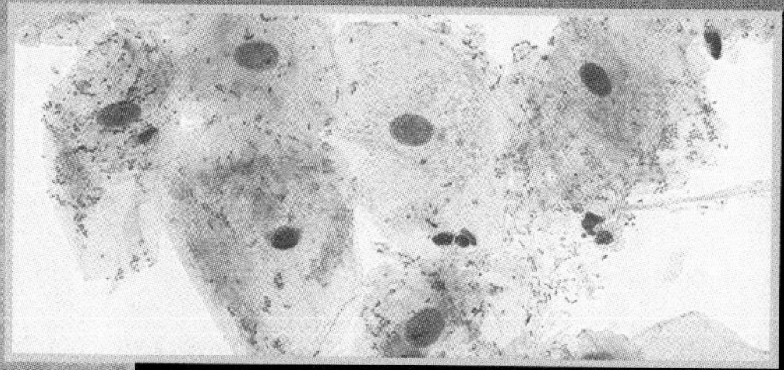

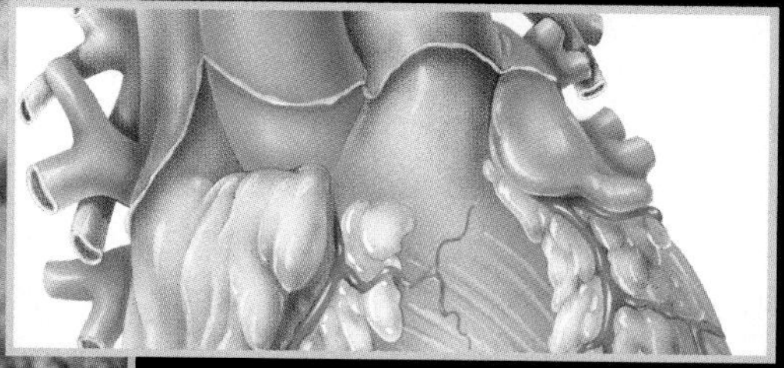

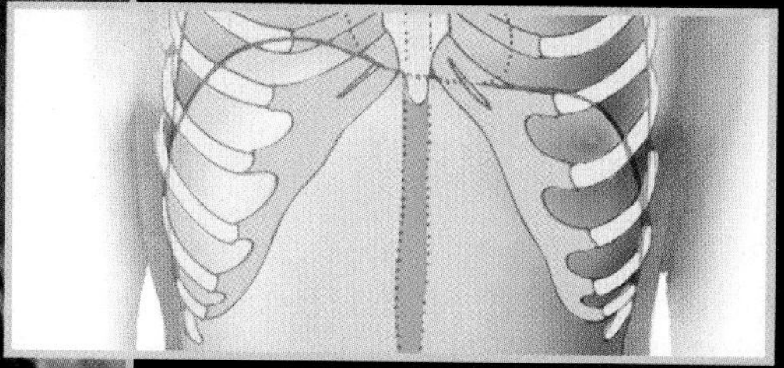

Cells Are the Building Blocks of Tissues

LEARNING OBJECTIVES

List the four tissue types in the human body.

Describe the function of each tissue type, explaining its unique characteristics.

Outline the various types of epithelial, connective, and muscular tissue.

A tissue is a group of similar cells and extracellular substances that have combined to perform a single function. The human body has four tissue types. **Epithelial** tissue covers the body, lines all cavities, and composes the glands. **Connective** tissue connects the structures of the body, providing structural support and holding organs together. Stretchy and strong, connective tissue maintains the body's integrity. **Muscular** tissue provides movement and heat. **Nervous** tissue responds to the environment by detecting, processing, and coordinating information.

EPITHELIAL TISSUE IS AT THE SURFACE

Epithelial tissue (or simply epithelium) is composed of cells laid together in sheets (FIGURE 4.1). One side of these cells is oriented toward the body cavity or external environment and may have cilia or **microvilli**. The other surface is joined to deeper connective tissue at the basement membrane. This basement layer, an acellular membrane, is composed of a collection of polysaccharides and proteins that help to cement the epithelial tissue to the underlying structures.

Microvilli Folded parts of the cell membrane that increase the cell's surface area.

Epithelium is little more than cells tightly connected, one to the next. It has neither blood vessels nor any extracellular substances between the cells. The classes of epithelium are identified by both the number of cell layers and the shape of the cells in the upper layer.

Simple epithelium has one layer of cells and usually functions as a diffusion or absorption membrane. The lining of your blood vessels and the respiratory membranes of your lungs are simple epithelium. **Stratified** epithelia have many layers of cells and are designed for protection. Examples are found in the outer layer of your skin and the ducts of your salivary glands.

Epithelial cells can be flattened, cube-like, or columnar. Each shape mirrors the function of the tissue. Flattened cells, reminiscent of fried eggs, are called **squamous** cells. Squamous epithelium is thin enough to form a membrane through which compounds can move via diffusion. **Cuboidal** and **columnar** epithelia are plumper and usually compose mucous membranes, in which the epithelial cells secrete mucus and other compounds (FIGURE 4.2).

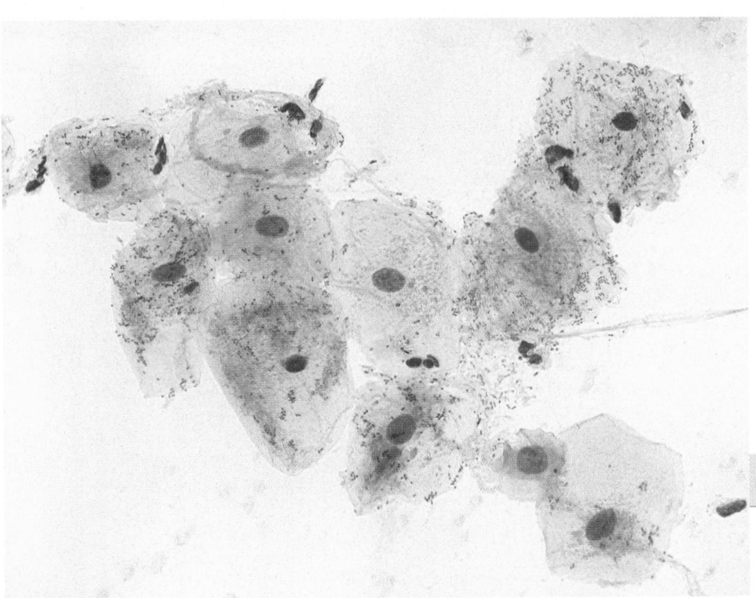

Epithelial cells FIGURE 4.1

These epithelial cells have been removed from the body; however, their strong cell-to-cell attachments still hold them together.

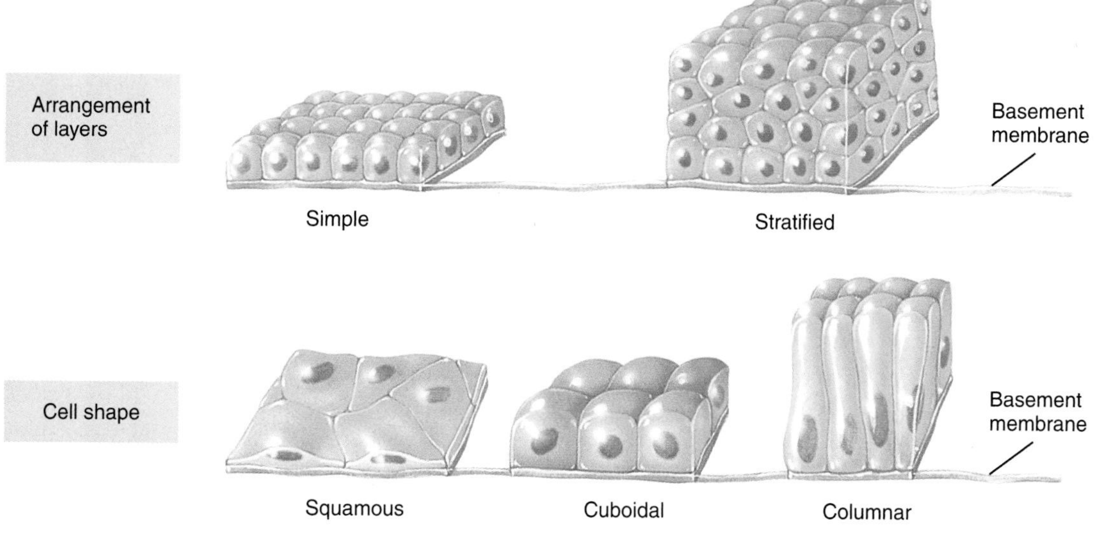

Cell shapes and arrangements FIGURE 4.2

Cell shapes and the arrangement of layers are the basis for classifying epithelial tissues. The shape of the topmost layer of cells determines the name of stratified epithelium because this layer is not deformed by those above it. Layers of flattened cells that look like "piles of tiles" would be classified as stratified squamous epithelium, whereas the image here shows layers of plump cells, classified as stratified cuboidal epithelium.

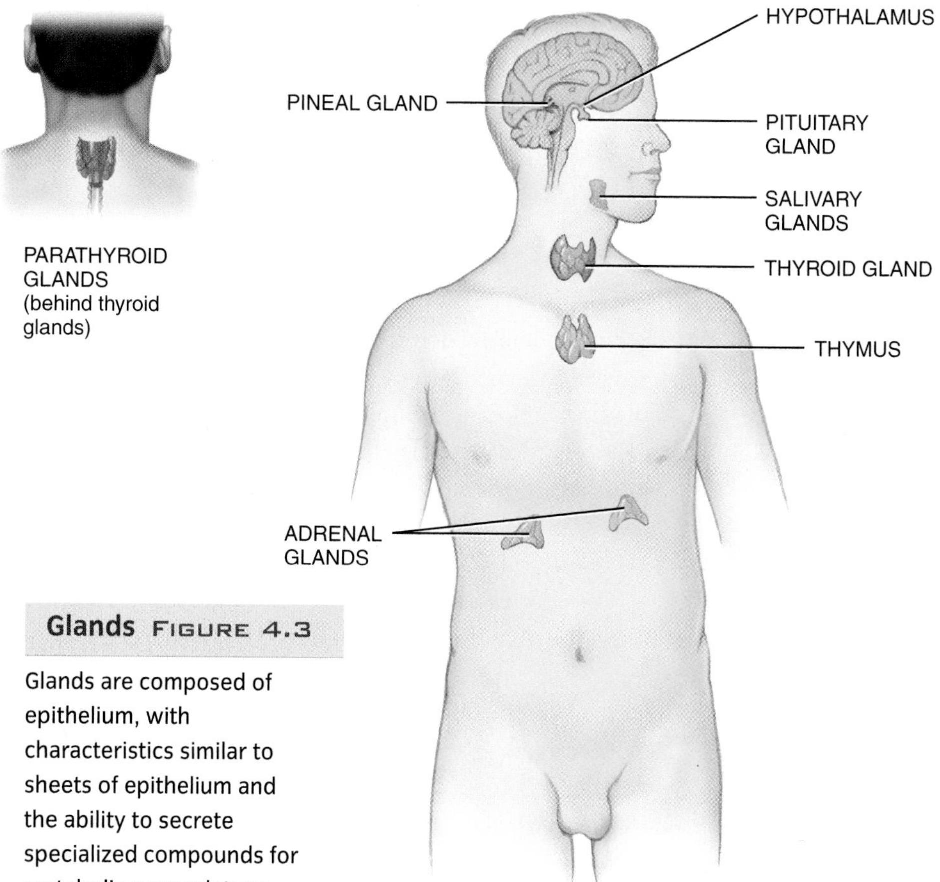

Glands FIGURE 4.3

Glands are composed of epithelium, with characteristics similar to sheets of epithelium and the ability to secrete specialized compounds for metabolic or regulatory purposes.

Glands are composed of epithelial tissue (FIGURE 4.3) and classified by how their secretions are released. Glands that secrete into ducts are **exocrine** glands. Salivary glands and sweat glands are exocrine glands. Each one secretes its products into a duct that directs the secretion to the surface of the gland.

Endocrine glands have no ducts. Instead, they secrete directly into the extracellular fluid surrounding the gland. Endocrine glands secrete hormones that are then picked up by the bloodstream and carried throughout the body. The adrenal, thyroid, and pituitary glands are all endocrine glands.

CONNECTIVE TISSUE KEEPS IT TOGETHER

As the name implies, connective tissue connects bodily structures. It binds, supports, and anchors the body and is the most abundant type of tissue in the body. Connective tissue is composed of cells suspended in a noncellular **matrix**. The matrix, or "ground substance," is secreted by the connective tissue cells, and it determines the characteristics of the connective tissue. The matrix can be liquid, gel-like, or solid, depending on the cells. The ground substance of all connective tissue contains fibers of **collagen** (for strength) and **elastin** (for flexibility, stretch, and recoil). Collagen is one of the main components of all connective tissue and consequently is the most abundant protein in the animal kingdom.

The nature of the ground substance leads us to classify connective tissue as **soft connective tissue** or as **specialized connective tissue** (FIGURE 4.4). Cartilage, bone, blood, and lymph are types of connective tissues.

CARTILAGE JOINS AND CUSHIONS

Avascular
Without blood vessels.

Cartilage is a unique connective tissue because it is **avascular** (other types of connective tissue all have rich blood supplies). (FIGURES 4.5 and 4.6 p. 98) **Chondrocytes**, the cartilaginous cells, secrete a gel-like matrix that eventually surrounds and imprisons them, segregating them from direct contact with one another or any nutrient supply. Cartilage heals slowly because nutrients must diffuse through the matrix to the chondrocytes; nutrients cannot reach the cells directly via the bloodstream. Each chondrocyte resides in a small "lake" within the matrix called a lacuna. The fluid bathing the cell in this lacuna diffuses through the matrix to and from the blood supply. This indirect route is far slower than bringing the fluid directly to the cells and is the reason cartilage is so slow to repair itself. Osteoarthritis is a serious disease of the joints, targeting the cartilage found within them. It is difficult to treat, in part because the cartilage is avascular and therefore does not respond quickly to medications (see the I Wonder . . . feature on arthritis on p. 102).

BONE IS SIMILAR TO STEEL-REINFORCED CONCRETE

Bone is a hard mineralized tissue found in the skeleton, which is a defining characteristic of vertebrates (FIGURE 4.7, p. 99). Bone cells secrete an **osteoid** substance that eventually hardens and surrounds the cells in an **ossified** matrix. This "osteoid ground substance" includes proteins, water, calcium, and phosphorous salts. Once the matrix ossifies, the cells remain in contact with one another through small channels called **canaliculi**. Like other connective tissues, bone has collagen fibers in the matrix for flexible support. Young bone has a higher percentage of collagen fibers than older bone, accounting for the greater flexibility of bones in infants and young people. Where an adult's bone would snap under excessive force, a young child's bone will bend. The convex surface may fray, like a bent green stick, but the bone does not break.

Osteoid
Bone matrix before it is calcified.

BLOOD AND LYMPH COMMUNICATE WITH THE ENTIRE BODY

Blood and lymph are considered fluid connective tissues because their matrix is not a solid. Blood is composed of specialized cells that are carried in the fluid matrix, or **plasma**. The main function of blood is to transport nutrients, gases, hormones, and wastes. Chapter 11 devotes an entire section to blood.

Lymph is another fluid connective tissue. It is derived from the **interstitial fluid** that bathes the cells and is collected in the lymphatic vessels. Like blood, lymph includes cells as well as proteins and other compounds in its fluid matrix. Chapter 10 deals with lymph in greater detail.

Plasma
The clear, yellowish fluid portion of blood.

Interstitial fluid
Fluid that fills the spaces between cells of tissues.

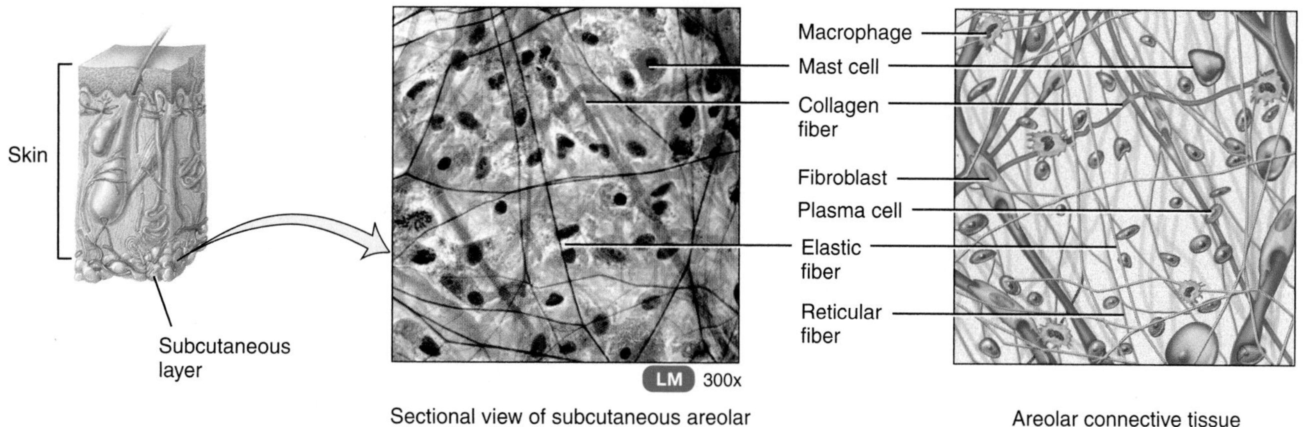

Sectional view of subcutaneous areolar connective tissue

Areolar connective tissue

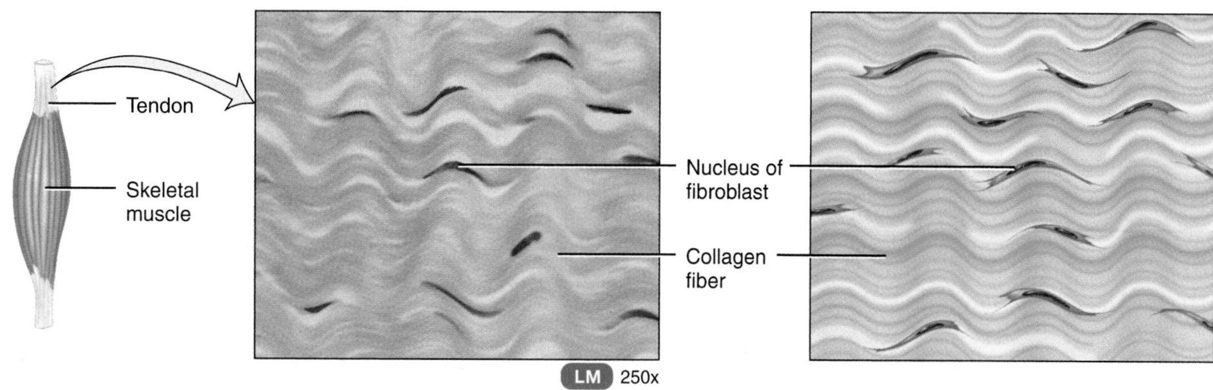

Sectional view of dense regular connective tissue of a tendon

Dense regular connective tissue

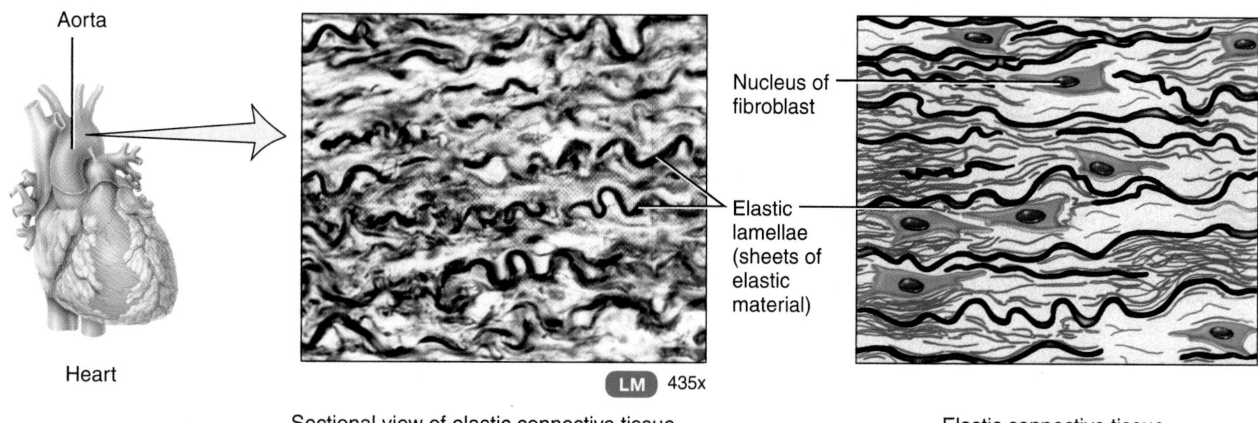

Sectional view of elastic connective tissue of aorta

Elastic connective tissue

Soft connective tissues: loose and dense FIGURE 4.4

Soft connective tissue has a matrix composed of a semifluid ground substance, *fibroblasts* that secrete fibers, and white blood cells that fight infection. The fibers of the matrix can be either loosely arranged or densely packed together. Loose connective tissue is sometimes called *areolar* connective tissue. Dense connective tissue includes the *dense irregular* tissue of the dermis of the skin, where the collagen fibers are arranged in a network, and the *dense regular* tissue of tendons, where the collagen fibers are aligned to resist tearing. Elastic connective tissue is made up of freely branching elastic fibers, with fibroblasts in the spaces between fibers.

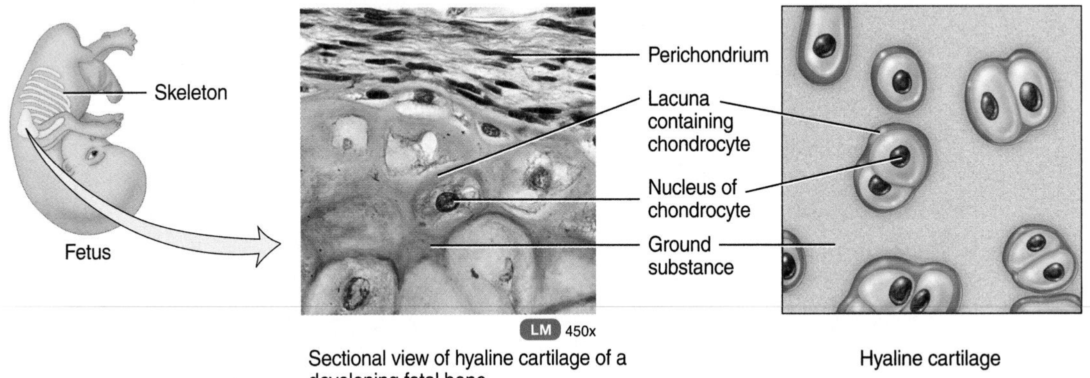

Sectional view of hyaline cartilage of a developing fetal bone

Hyaline cartilage

Hyaline cartilage FIGURE 4.5

The most common type of cartilage is *hyaline cartilage*. The matrix of hyaline cartilage contains many collagen fibers and looks crystal blue in living tissue. Hyaline cartilage covers the ends of bones, allowing them to slide against one another without damage. It is also found in your nose and **trachea**. During development, your entire skeleton was modeled in hyaline cartilage, which then ossified—that is, turned to bone.

Trachea The main trunk of the respiratory tree.

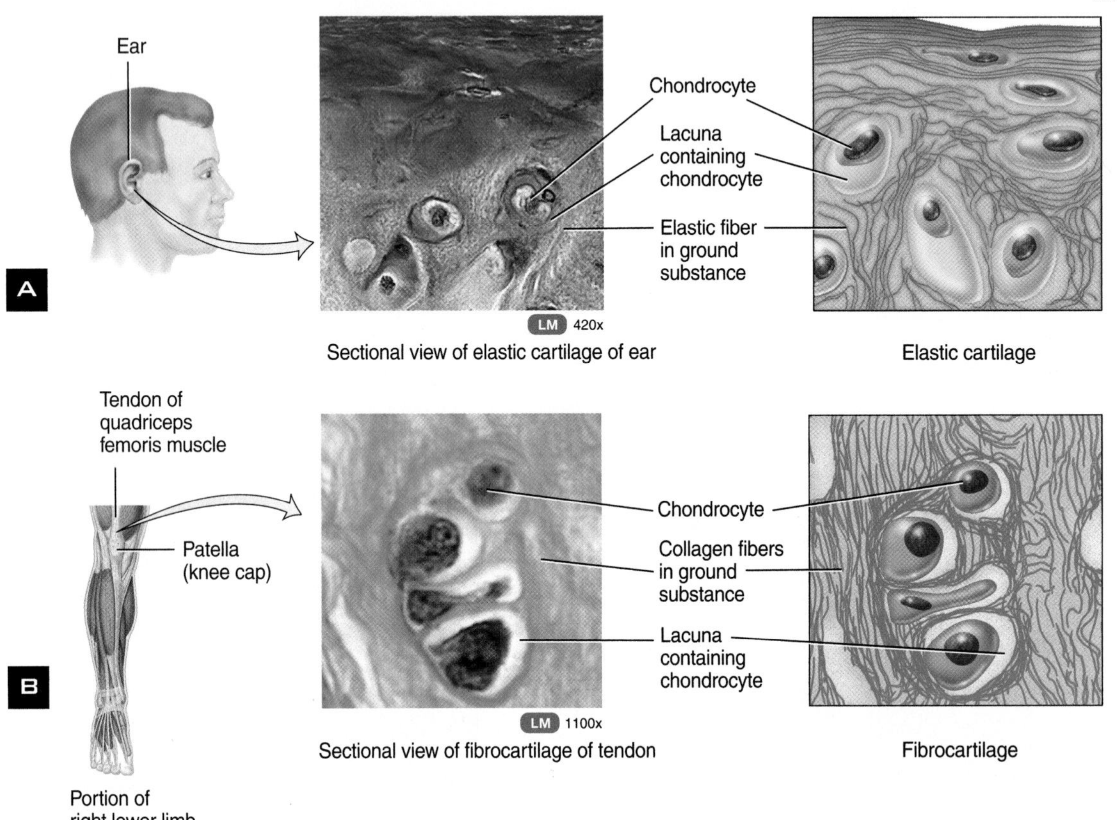

Sectional view of elastic cartilage of ear

Elastic cartilage

Sectional view of fibrocartilage of tendon

Fibrocartilage

Elastic cartilage and fibrocartilage FIGURE 4.6

A Elastic cartilage contains many elastic fibers in the matrix. Elastic cartilage allows the outer ear to bend, then return to its original shape. The **epiglottis** that prevents food and liquid from entering your respiratory tract also contains elastic cartilage. When you swallow, the epiglottis bends to cover the opening of the trachea. Afterward, the epiglottis snaps back to its original position, allowing air to flow through the windpipe.

B The matrix of fibrocartilage is packed with collagen fibers, so it is found where extra strength is needed. Cushions in your knee joints and the disks between the vertebrae are made of fibrocartilage.

Epiglottis Large, leaf-shaped piece of cartilage lying over top of the larynx.

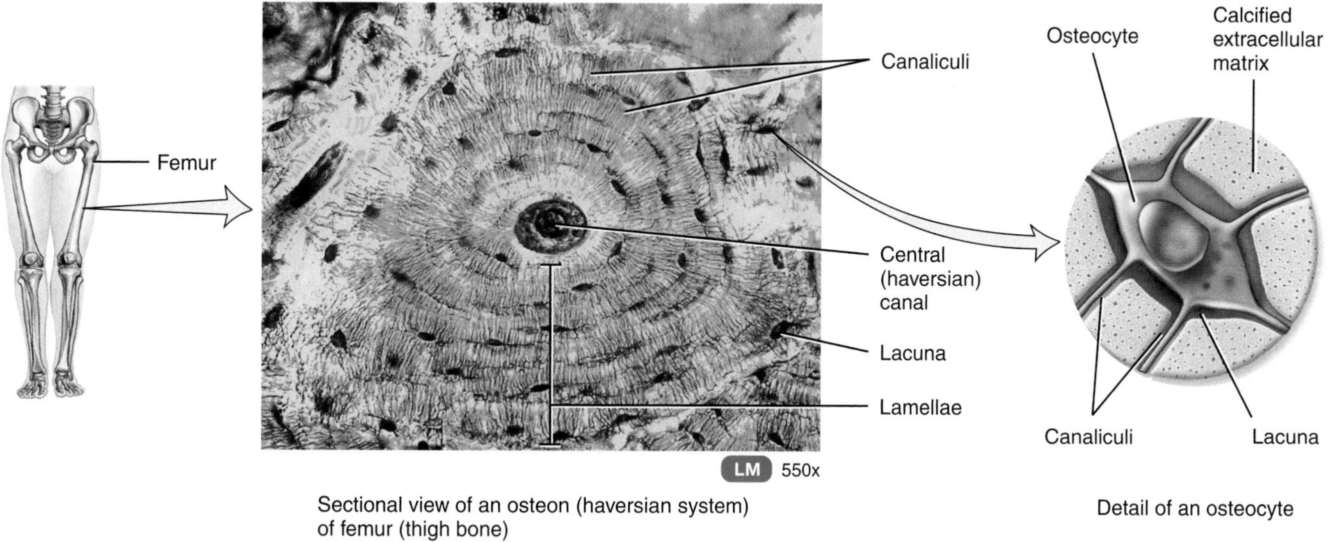

Sectional view of an osteon (haversian system) of femur (thigh bone)

Detail of an osteocyte

Compact bone FIGURE 4.7

Bone consists of a hard matrix surrounding living cells. Bone has both a blood supply and a nerve supply running through it. The matrix of compact bone is found in long cylinders called osteons or Haversian systems. Lighter spongy bone has less structure, formed in struts and supports rather than a solid mass.

EVEN FAT HAS A JOB TO DO

Adipose tissue contains fat cells—cells that are specialized for lipid storage. Unlike other connective tissues, adipose tissue does not have an extensive extracellular matrix (FIGURE 4.8). Its matrix is a soft network of fibers holding the cell together and binding it to surrounding tissues. **Cellulite** "bumps" on the skin indicate where the adipose matrix is connected to the skin. The

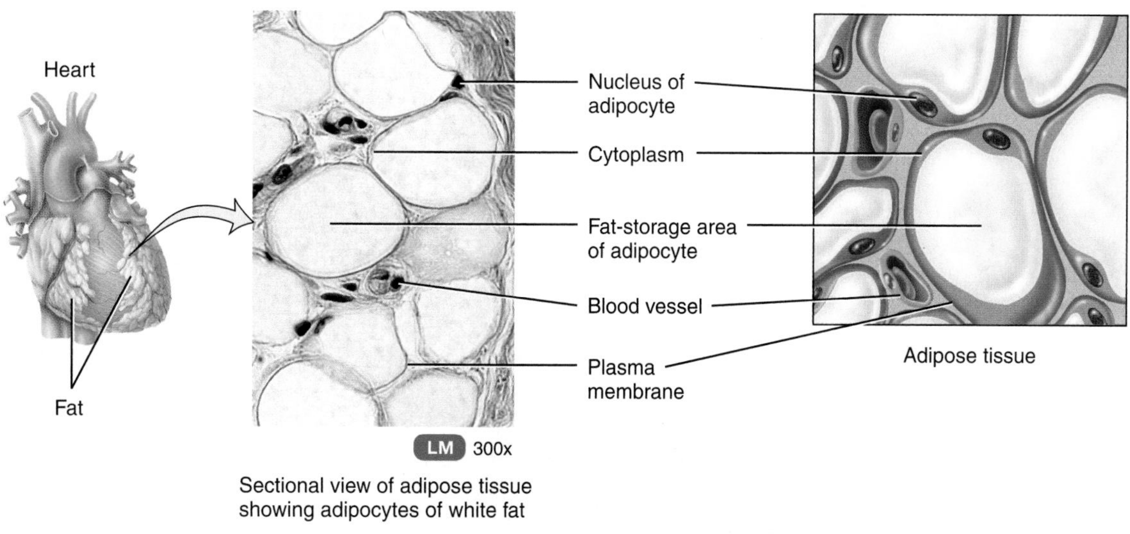

Sectional view of adipose tissue showing adipocytes of white fat

Adipose tissue

Adipose tissue FIGURE 4.8

The nucleus and cytoplasm in adipocytes play second fiddle to the main action: the huge droplet of stored lipid.

adipose cells within the fibrous matrix can expand with the swelling of the fat droplets they contain, while the matrix fibers cannot stretch as far. The different stretching capacities of these two components of adipose tissue form dimples on the skin. Cellulite is a normal function of fat deposition and storage. It is not an inherently evil tissue that must be removed from the body, despite what you may have read in the supermarket tabloids. Even newborns have cellulite!

MUSCULAR TISSUE MOVES US

The function of muscular tissue is to contract. The cells get shorter, generating force and often movement. The three types of muscular tissue are: **skeletal muscle, smooth muscle**, and **cardiac muscle**. Skeletal muscle tissue is highly organized, with the cells lying parallel to each other, much like a cable. When stimulated, groups of muscle cells contract in unison (FIGURE 4.9).

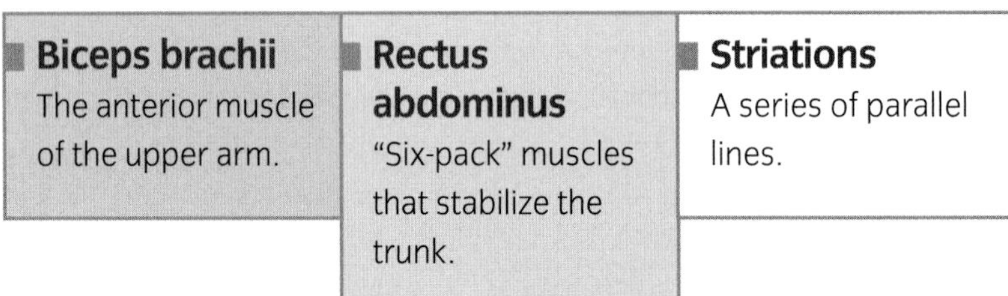

Biceps brachii
The anterior muscle of the upper arm.
Rectus abdominus
"Six-pack" muscles that stabilize the trunk.
Striations
A series of parallel lines.

Comparison of three types of muscle tissue FIGURE 4.9

Skeletal muscle is the tissue that makes up the muscles that move your limbs and stabilize your trunk, including your **biceps brachii** and **rectus abdominus**. This tissue is composed of long, multinucleate cells with visible **striations**. The cells of skeletal muscle extend the length of the muscle and are arranged in parallel groups called *fascicles*. Skeletal muscle is described in full detail in Chapter 5. Because you consciously control muscle contractions, skeletal muscle is called voluntary muscle.

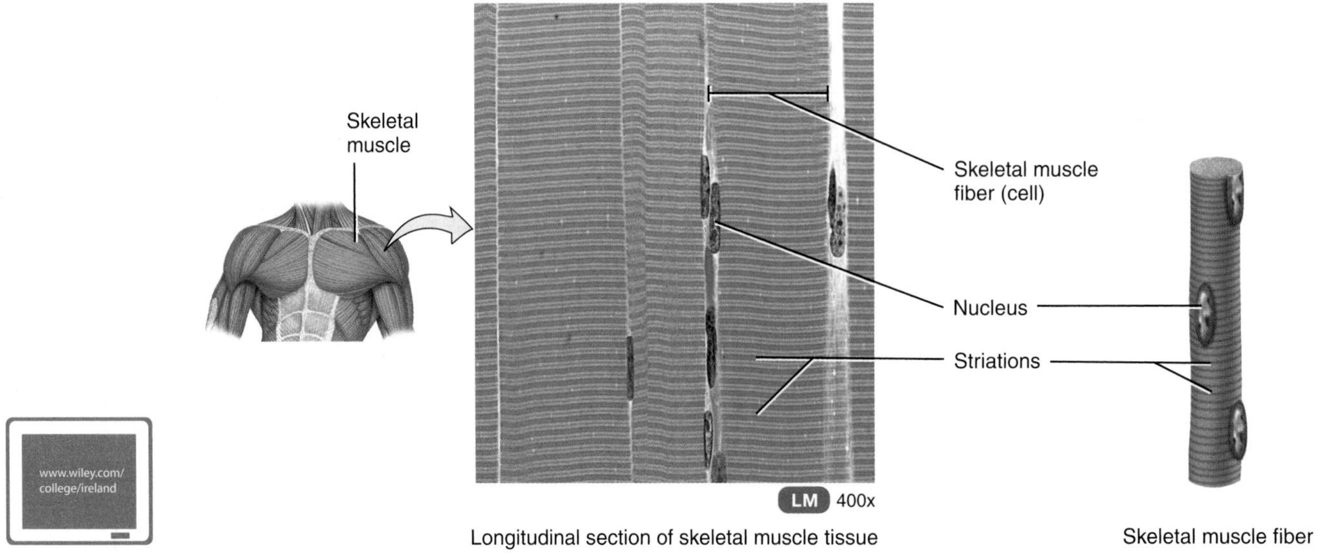

Longitudinal section of skeletal muscle tissue

Skeletal muscle fiber

Smooth muscle lines hollow organs such as the blood vessels and the digestive tract. Smooth muscle cells are short, cylindrical cells that taper at both ends and have only one nucleus. They are not striated, and are not under voluntary control. This is helpful. Wouldn't it be nerve-wracking to consciously manage the diameter of your blood vessels to maintain blood pressure, or to consciously create the rhythmic constrictions that the digestive tract uses to move food during digestion?

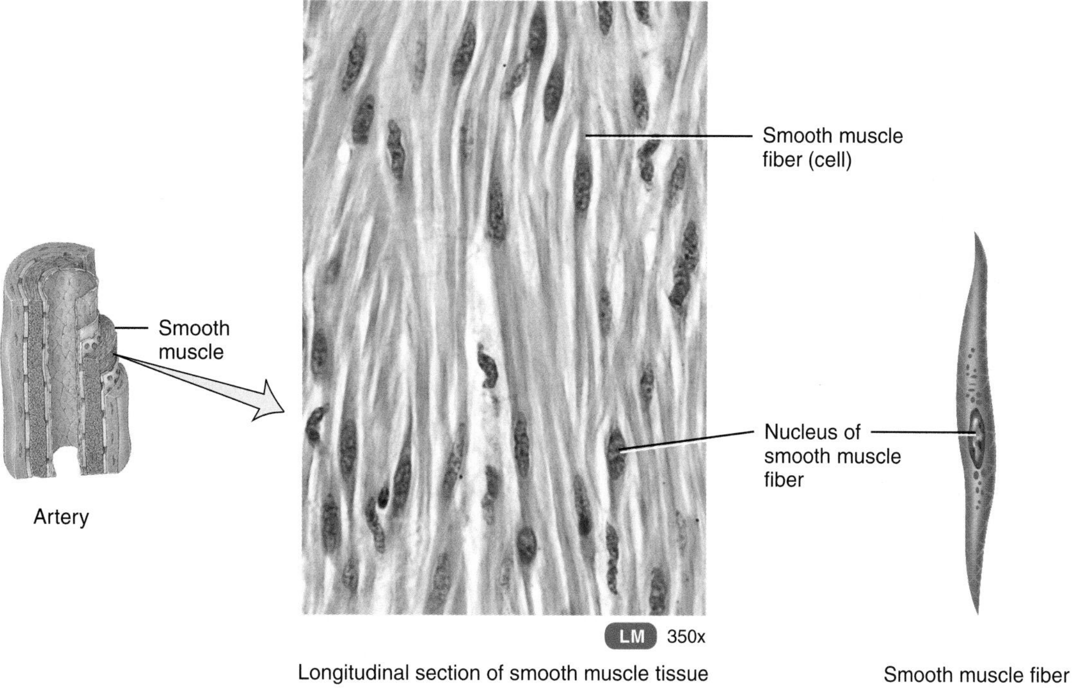

Longitudinal section of smooth muscle tissue

Smooth muscle fiber

Cardiac (heart) muscle has short, branched, striated cells, with one nucleus at the center of each cell. Specialized communication junctions called intercalated discs facilitate the heartbeat by transmitting the signal to contract. Intercalated discs are gap junctions where the closely knit cell membranes help to spread the contraction impulse while also binding the cells together. Cardiac muscle will be described in more detail in Chapter 11.

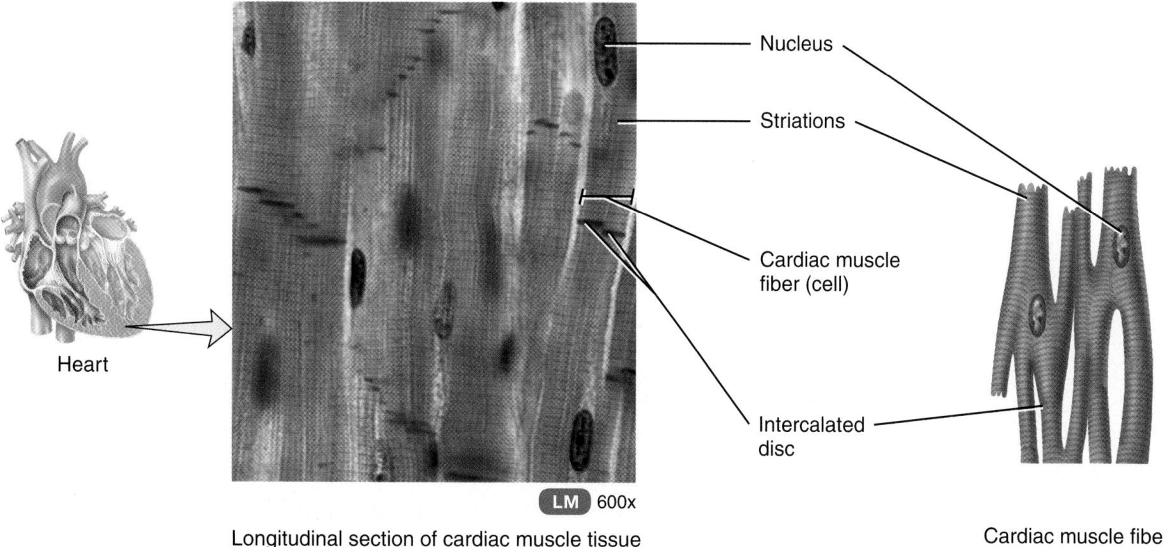

Longitudinal section of cardiac muscle tissue

Cardiac muscle fibers

I WONDER . . .

What is arthritis?

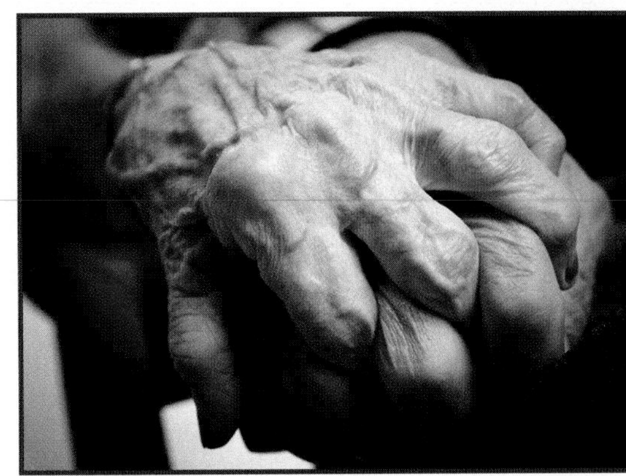

Arthritis is a general term for degradation of the joints, with a variety of causes. Rheumatoid arthritis results from an autoimmune attack on the joint. Osteoarthritis results from various sorts of wear and tear. Only one type of arthritis, the one that results from an infection, can be cured (using antibiotics). Other forms must be managed to reduce pain and improve quality of life.

Osteoarthritis affects 21 million Americans. The disease damages cartilage that normally forms a smooth bearing surface inside joints. Pain begins as bones rub on bones. Osteoarthritis tends to be worst when it affects the spine, hips, and knees.

Rheumatoid arthritis is an inflammation of the synovium, which lines the joints. The disease can seriously deform the hands, but it often affects joints throughout the body. Rheumatoid arthritis is two or three times as common among women, indicating that females are genetically more susceptible to this type of autoimmune attack. About 2 million Americans have the disease.

Common symptoms of arthritis include stiffness, especially in the morning, swelling, warmth or redness in the joint, and constant or recurring pain. Many patients reduce their movement to avoid pain, but inactivity harms muscles, ligaments, and the cardiovascular system. A proper diagnosis must precede treatment, as doctors want to rule out other diseases that can affect the joints, such as lupus and fibromyalgia. After a good patient history, diagnostic tools may include

- X-rays, to detect abnormal growth or erosion in the joints.
- CT (computed tomography) or MRI (magnetic resonance imaging) scans, to view soft tissue, such as cartilage.
- An arthroscopic (tube-mounted) camera, to determine the extent of damage inside a joint.
- Blood analysis, to test for proteins and antibodies that accompany an immune attack, a hallmark of rheumatic arthritis.

Treatment options include various antiinflammatory drugs, rehabilitative therapy, pain management, and, in extreme cases, joint replacement. Both heat and cold can improve symptoms by reducing muscle spasms and pain. Exercises can improve flexibility, endurance, and strength, helping restore mobility and quality of life. A variety of new medicines called biological response modifiers may limit inflammation in rheumatoid arthritis by interfering with an immune protein called tumor necrosis factor. Surgeons may fuse bones to prevent movement at the affected joint, or may replace the joint with a metal joint.

Arthritis research continues. Scientists want to understand the role of genetics or a prior infection in triggering joint damage. What exactly is going wrong with the immune system and cells in the joint? Because joint damage can be permanent, researchers hope to intervene to stop the damage at an early stage. That explains the interest in "biological markers"—unique compounds or proteins that are associated with arthritic processes. As the population ages, we can expect more interest in arthritis, and as we learn more about the disease process, we can expect more therapeutic progress as well.

NERVOUS TISSUE IS THE BODY'S PHONE AND COMPUTER SYSTEM

Nervous tissue, the final type of tissue in the human body, is "irritable," which means it responds to changes in the environment. Nervous tissue contains two categories of cells: **neurons** (FIGURE 4.10) and **neuroglia**.

Neuroglia are the supporting cells of nervous tissue ("glia" means "glue"). It was once thought that these cells merely held the neurons together. Now we know that the various neuroglial cells have specific supporting roles. Neuroglia do not send or receive electrical impulses. Instead, they improve nutrient flow to the neurons, provide physical support, remove debris, and provide electrical insulation.

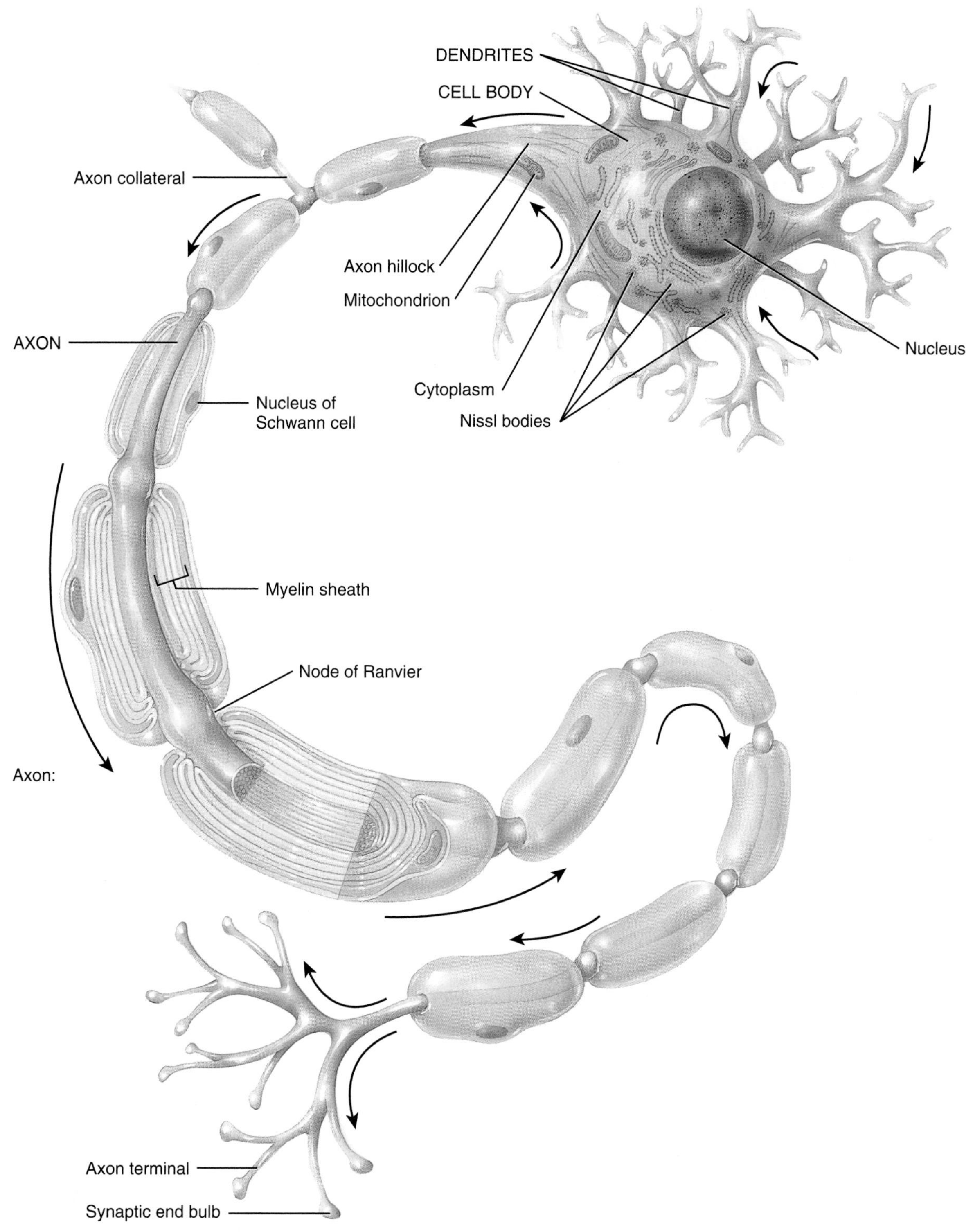

Neuron FIGURE 4.10

Neurons are the cells that carry electrical impulses. They can be extremely short, like those found within the spinal cord and brain, or they can be longer than 125 centimeters, like those that extend from the spinal cord to the end of the great toe. The cell body of a neuron has long, slender projections. One group of projections, called *dendrites*, receives impulses from other neurons, bringing the information to the cell body. The other projection, called the *axon*, transmits impulses from the cell body to other cells. Individual neurons may have many dendrites, but each can have only one axon.

Health, Wellness, and Disease

Are we making progress in the struggle against cancer?

Cancer is one of the scariest of all diagnoses. In 2004, cancer passed heart disease to become the largest cause of death in the United States, claiming about 560,000 lives. Lung cancer will kill an estimated 160,000 Americans in 2007; other major killers include colorectal (55,000), breast (41,000), pancreatic (32,000), and prostate (27,000) cancer.

After decades of research effort, the five-year cancer survival rate has been inching up, from 50 percent in the mid-1970s to 65 percent in 2000. The most famous survivor is cyclist Lance Armstrong. After nearly dying of testicular cancer, Armstrong went on to win the Tour de France an unprecedented seven times. Testicular cancer is one of the most curable cancers, with a five-year survival rate of 96 percent. For breast cancer, five-year survival is up to 88 percent. To some extent, cancer survival is affected by the nature of the organ where it originates. Organs with large blood supply, like the lung (five-year survival: 15 percent) and liver, tend to have lower survival rates.

Cancer is tenacious: tumor cells are hard to kill. In essence, cancer is uncontrolled cell replication. Normally, all cells are kept under tight regulation, dividing only when necessary. Cancer results from some sort of "insult" to this regulatory apparatus. The insult may be caused by a chemical or by radiation, or it may arise when a mistake is made in DNA while a cell is replicating. This genetic defect can promote cell growth by producing hormones or other signaling molecules that encourage cells to divide. Or it may produce receptors that become hypersensitive to a signal to divide. Defects can add up if other mutations impair genes that normally restrain cell division or the natural DNA repair process. When a cancer cell divides, the two new cells carry the same defects, and cancer is under way.

A large number of carcinogens can cause cancer. The best known is tobacco smoke, which contains compounds that cause lung, bronchial, and throat cancer. The ultraviolet portion of sunlight can cause skin cancer. A variety of organic chemicals can cause the kind of mutation that will eventually become cancer. Alcohol and estrogen can both promote cancer by stimulating cell division, which in turn raises the risk of defective copying of DNA. Viruses are another major causes of cancers, probably because they must take command of a cell's genetic machinery in order to copy themselves.

At some point, cancer cells may leave their origin (the "primary" site) and move through the blood or lymph to a distant site. These metastatic tumors may continue to grow even if the primary tumor is removed or killed. Breast cancer often metastasizes to the bones or lungs, and lung cancer to the brain.

Cancer treatment has long focused on killing or removing the primary tumor and then attacking any metastatic tumors. Surgical removal of cancerous tumors began even before the discovery of anesthetics. After the discovery of X-rays in 1895, radiation began to be employed, as it is particularly deadly to cells that are dividing. The third leg of the cancer-fighting triad is chemotherapy: compounds that kill fast-growing cells.

Though helpful, each kind of therapy has drawbacks. Surgery and radiation do not effectively treat metastatic tumors. Because radiation and chemotherapy kill dividing cells, they are particularly toxic to cells that normally divide rapidly. These occur in the scalp (causing the typical hair loss of chemotherapy) and, more importantly, in the gastrointestinal tract, causing bleeding and nausea.

Advanced diagnostic techniques are often used to monitor the success of cancer treatments. For example, PET (positron emission tomography) scans can measure glucose usage from outside the body. Cancer cells show up because they are heavy users of this simple sugar.

Newer forms of cancer therapy reflect the fact that cancer, at its root, is many different diseases, caused by many problems with cells and the intercellular signaling system in tissues. Gradually, the traditional approach to treatment—killing tumor cells—is giving way to more targeted therapies that can either identify cancer cells more precisely or block the signals that cause their growth.

A good example of this approach involves human epidermal growth factor. This natural compound attaches to receptors on cell membranes and signals them to grow. In about 25 percent of breast cancers, a genetic defect causes extra copies of this receptor to form on the cancer cells, causing them to overreact to this normal growth signal and grow uncontrollably. In 1998, a genetic engineering firm began selling a protein called Herceptin that blocks this pathway. The compound was an early sign of how better science can work to defeat cancer, but it is only effective in cancers with one specific defect.

A second "intelligent" approach is even more experimental: using the body's immune system to fight cancer with a vaccine. Cancer vaccines may be designed to fight emerging cancers or to kill existing tumors. The vaccine might consist of a portion of the tumor cell membrane that is not found on normal cells. Vaccination would prime the immune system to make antibodies that would attach only to tumor cells and then direct an immune attack on the cell. The concept is promising, but progress has been slow.

Inevitably, however, the many specific diseases we group together as "cancer" will succumb to some form of this more targeted attack. In medicine, as in so many fields, knowledge is power.

Nerves are clusters of neurons and their projections, sheathed in connective tissue. Because nerves exist in the body's periphery, they are part of the **peripheral nervous system**. Sensory nerves conduct sensory messages from the body's **sensory organs** to the spinal cord, which routes the information to the brain. Motor nerves carry impulses that cause muscular movement or glandular secretion from the spinal cord to the muscles and glands. The brain and spinal cord contain neurons that receive and integrate information and stimulate motor neurons to fire. These information-processing neurons occur in the central axis of the body, so they comprise the **central nervous system**. The breakdown of the nervous system and the histology of nervous tissue are covered extensively in Chapter 7.

As you have seen, tissues are composed of cells working together to perform a single function. In most cases, the cells divide and reproduce only enough to perform the function of the tissue. Sometimes though, the integrity of the tissue can be damaged through uncontrolled cellular growth. When cancer strikes a tissue, not only can it cause malfunctioning of that tissue, but it can also spread to other areas of the body. Cancer is a disease of the tissues that is capable of destroying the entire body. (See the Health Wellness and Disease feature "Are we making progress in the struggle against cancer?")

CONCEPT CHECK

What is the difference between simple and stratified epithelium?

List and briefly describe three types of cartilage.

Which types of muscular tissue are involuntary?

What are the two main types of cells in nervous tissue?

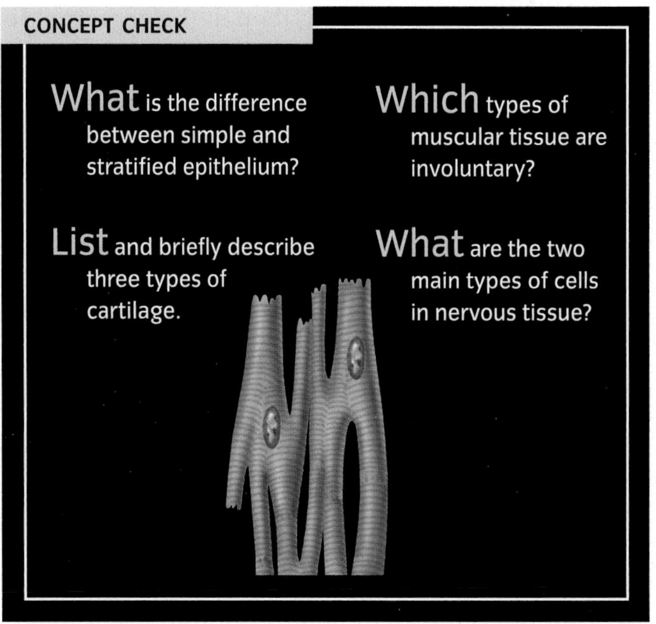

Organization Increases with Organs, Organ Systems, and the Organism

LEARNING OBJECTIVES

Explore how organisms display the hierarchy of life.

Outline the role organ systems play in maintaining homeostasis.

ecall that one characteristic of life is a high degree of organization (FIGURE 4.11). A layered of organization, or hierarchy, is visible in all life forms:

Atom

Molecule

Organelle

Cell

Tissue

Organ

Organ system

Organism

While each organ system has one specialized function, the continuation of life requires that these systems be integrated into a whole, cohesive unit. The ultimate level of organization, then, is the **organism**. In human biology, the human being is the pinnacle of organization, although humans are also part of a larger social and ecological framework, as discussed later in this text. You are composed of cells cooperating in tissues, which are in turn positioned together to efficiently carry out an organ's processes. Organs then work together to perform a larger function, such as cleansing the blood, comprising an organ system. There are 11 organ systems in the human body, as follows:

Integumentary (protecting and covering), skeletal (support), muscular (movement and heat production), nervous (sensing and responding), cardiovascular (transporting fluids and oxygen), respiratory (gas exchange), urinary (fluid balance), endocrine (regulating sequential growth and development), digestive (obtaining nutrients), lymphatic (immunity), and reproductive systems (continuation of species). All 11 organ systems, integrated and working together, maintain life as you know it. When something goes wrong with an organ, the system as well as the entire organism suffers. Replacement organs are usually in short supply, necessitating the creation of new medical solutions. (See the Ethics and Issues feature "Should we, could we, grow new organs?" on p. 108.)

Levels of structural organization

FIGURE 4.11

The four main types of tissues—epithelial, connective, muscular, and nervous—join together in specific proportions and patterns to form organs, such as the heart, brain, and stomach. Each organ has a specific and vital function. Organs that interact to perform a specific task comprise an organ system. For example, the heart and the blood vessels together make up the cardiovascular system. All 11 organ systems in the human will be discussed in the remainder of this text. Ten of these systems help maintain individual homeostasis, while the reproductive system maintains the human population.

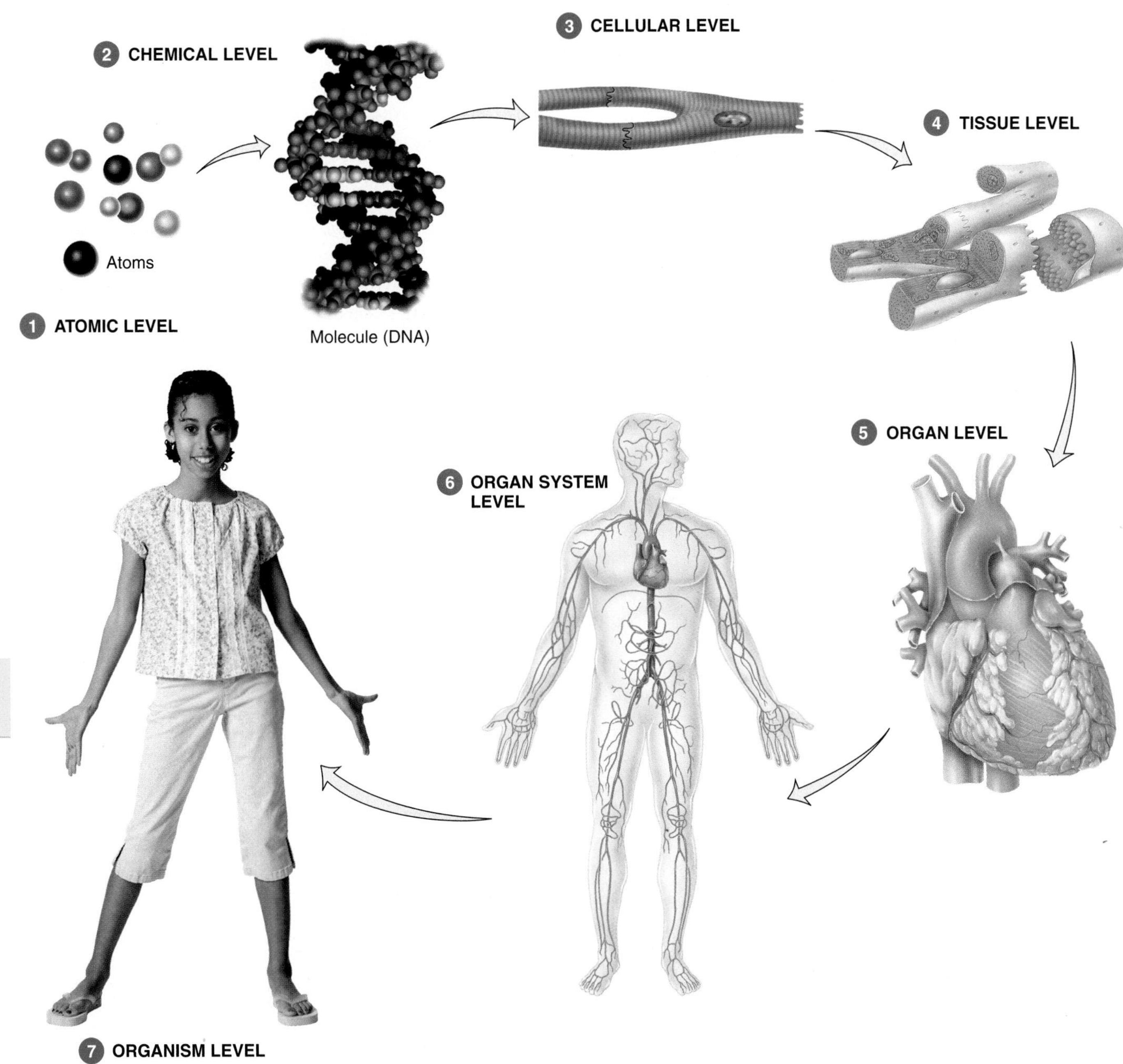
2 CHEMICAL LEVEL
3 CELLULAR LEVEL
4 TISSUE LEVEL
Atoms
1 ATOMIC LEVEL
Molecule (DNA)
5 ORGAN LEVEL
6 ORGAN SYSTEM LEVEL
7 ORGANISM LEVEL

Ethics and Issues

Should we, could we, grow new organs?

It sounds like a Doctor Seuss rhyme, and growing human organs in the lab sounds like the plot of the next-generation Frankenstein story. But there is a basis of truth to it. Of course, the need is there. We have many more organ requests than donated organs every year. According to the National Women's Health Center, every day 63 people in America receive organ transplants, but another 16 die for lack of appropriate organs. Growing the needed organ from just a few of the patient's own cells would solve this problem, giving everyone a seemingly limitless supply of replacement organs. It might be the perfect solution to our current transplant woes—but is it feasible?

Amazingly enough, researchers at Wake Forest University have recently been able to do just that. Urinary bladders are not on the list of organs that are usually donated, but there is a growing need for replacement bladders. Patients with bladder cancer, spina bifida, and other debilitating nervous and urinary diseases often need replacement bladders. In Wake Forest University laboratories, Dr. Anthony Atala and his team of researchers have been able to grow new, functional urinary bladders for seven young patients ranging from toddlers to adolescents. How did they do this? The process seems deceptively simple on the surface. The doctors take samples of healthy bladder cells and muscle tissue from the patient and grow these in an incubator under sterile conditions. When these tissues are growing well and appear to be healthy, the doctors move them from the original sample dishes to a sterile mold of a bladder. By layering these new tissues over the mold, they give the tissues a "pattern" to follow. Within a few weeks, the patient's own tissues have grown over the bladder mold, creating a brand new bladder. The newly created bladder is smaller than the patient's original bladder, but once placed inside the patient's pelvic cavity, the new bladder grows to reach adult size. Even more amazingly, when these newly created bladders were properly positioned in the patients' bodies, all seven worked just as they should.

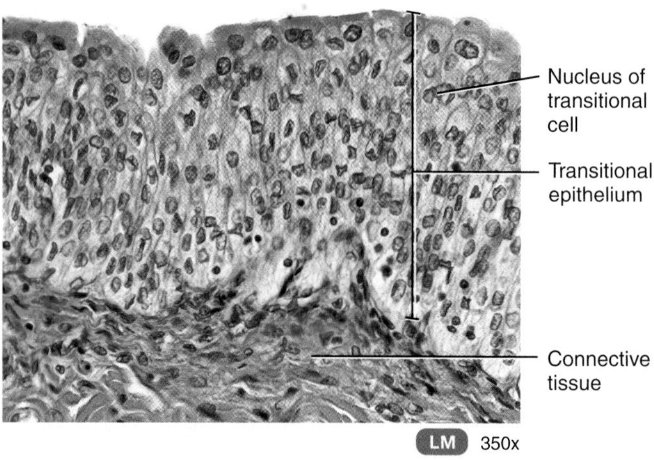

Sectional view of transitional epithelium of urinary bladder in relaxed state

The transitional epithelium that lines the bladder is relatively easy to grow in the laboratory. After all, when stretched to line a full bladder, it is essentially a single sheet of cells similar to the epidermis of the skin. Here it is seen in the relaxed state, as it would appear when the bladder is empty.

Although this seems very promising, there are concerns. The urinary bladder is a simple organ, basically composed of a layer of transitional epithelium covering smooth muscle and connective tissue. Its main function is to store urine, without absorbing the contents back into the bloodstream, and then to empty. More complicated organs, such as the heart and lungs, will be more difficult to culture. News of preliminary successes like these seven bladders may be causing unfounded hope in those still on organ waiting lists.

Another concern that arises from this exciting news is the possibility of "organ farms," where healthy clients provide cells and tissues, perhaps for monetary reward, and organs are created before the need arises. Can you imagine a competitive market for human organs? There is currently a small black market for kidneys and hearts. With the success of technologies such as this, would we see a new "biological market" filled with organs grown to specifications? In many ways this is a brave new world. Let's hope we handle it wisely.

The goal of the organism is to maintain homeostasis, keeping the internal environment stable despite constant internal and external changes. You put food into the digestive tract, requiring water and energy to digest it into nutrients, which are consumed during movement and metabolic activity. You lose fluids through sweating, breathing, and urinating. You alter your dissolved gas concentrations with every breath. Every muscular contraction changes your blood chemistry and internal temperature. Each subtle change in body chemistry must be corrected in order to maintain homeostasis. Alterations in one system affect the functioning of all other systems; metabolism in the muscles requires oxygen, which is delivered through the respiratory and cardiovascular systems. You are a finely balanced machine, and every mechanical action, every chemical reaction, requires that balance be restored. Negative feedback loops keep your vital statistics in acceptable ranges despite the myriad changes you put your body through every day.

CONCEPT CHECK

Order the following terms by increasing complexity: cell, molecule, organism, organelle, organ, tissue, organ system.

Define organ system.

Give three examples of natural processes in your body that are balanced via homeostatic mechanisms.

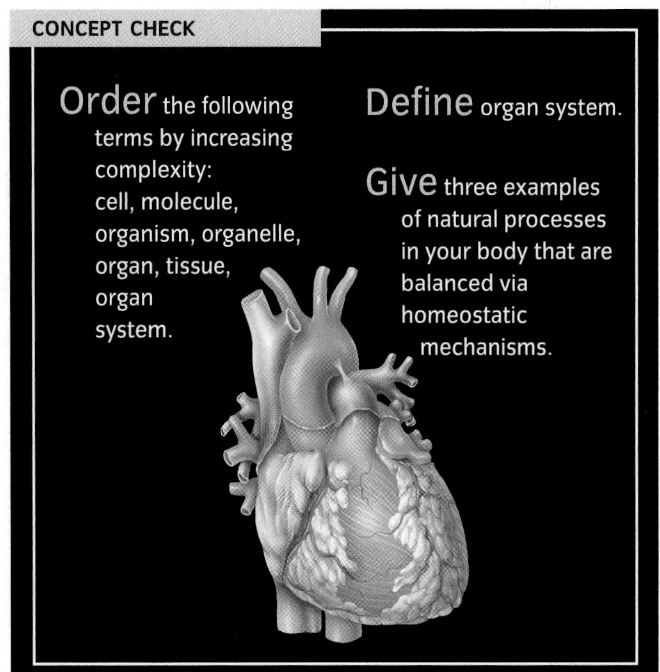

Scientists Use a Road Map to the Human Body

LEARNING OBJECTIVES

Learn to use anatomical directional terms.

Identify the body cavities and the organs that each contains.

Studying human biology—human anatomy and physiology—is a daunting task because we are concerned not only with the location of organs and organ systems, but also their interconnection. To discuss these complicated matters clearly, we need a system to precisely name the structures of the body. Whenever we talk about an organ's placement, or the appearance of a por-

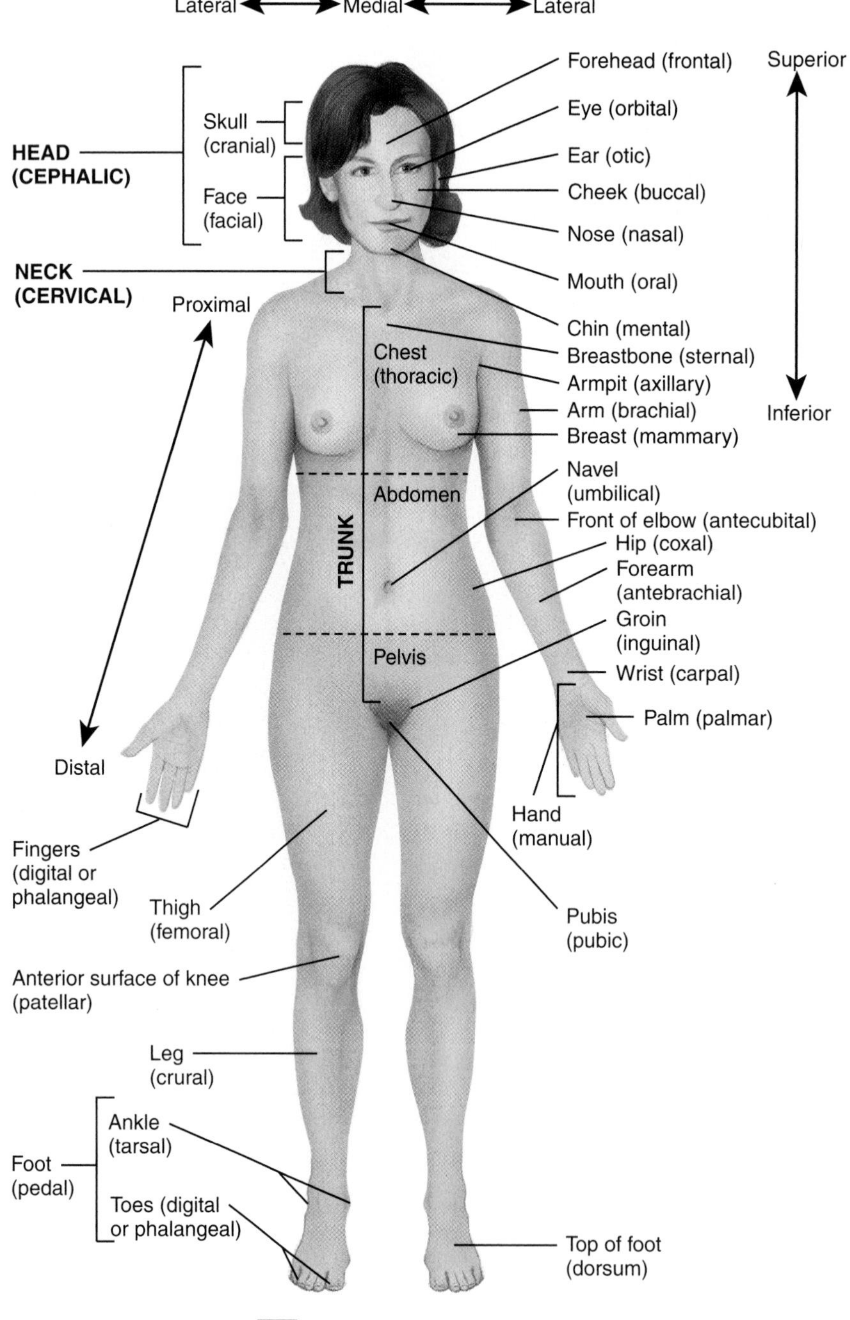

A Anterior view

Proximal/distal Opposite terms meaning near the core of the body versus farther from the core.

Superior/inferior Opposite terms meaning above and below.

tion of the body, we assume we have placed the body in the **anatomical position** (FIGURE 4.12). Using this position as a standard allows us to make sense of directional terms such as **proximal** and **distal**, **superior** and **inferior**.

The body has natural boundaries that we exploit for describing position in human biology. The body has two large cavities, the ventral and dorsal cavities (FIGURE 4.13 on p. 112). The **ventral cavity** comprises the entire ventral (or belly) aspect of your torso. The ventral portion of the body contains distinct sections. The **thoracic cavity** includes the chest area and houses the heart, lungs, vessels, and lymphatic system of the **mediastinum**. The "guts" are found within the **abdominal cavity**, which is lined with **peritoneum**. The organs of the urinary system and the reproductive system are located in the **pelvic cavity**.

Mediastinum The broad area between the lungs.

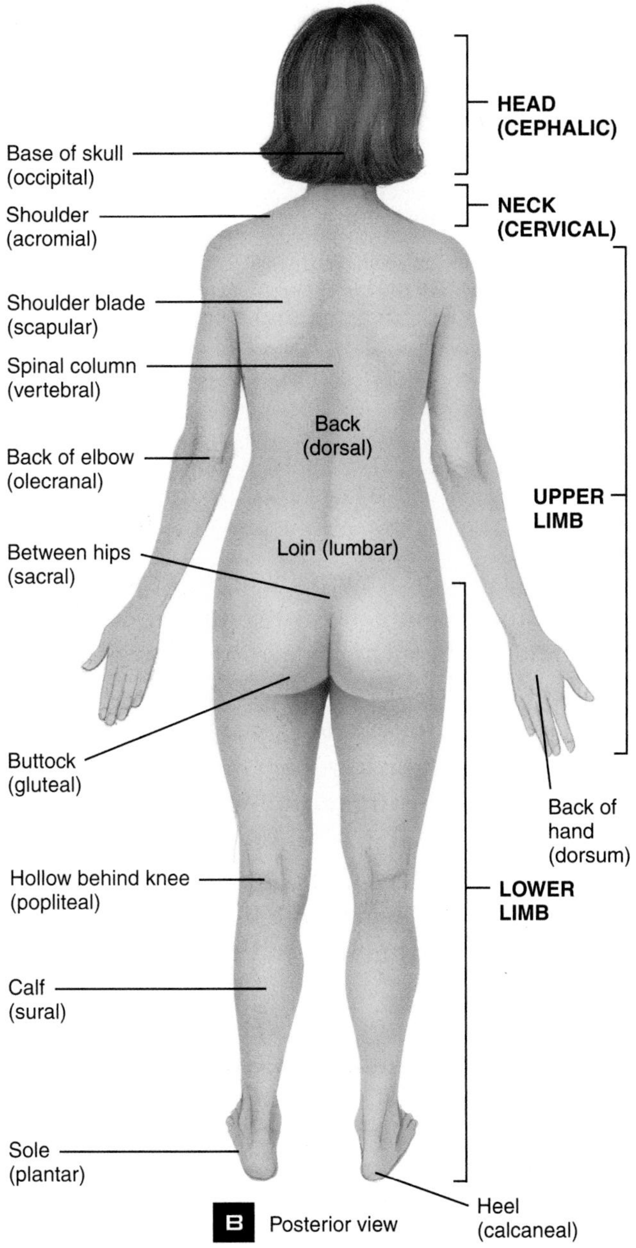

Anatomical position with directional terms

FIGURE 4.12

In the "anatomical position," the bones of the forearm lie straight instead of crossing over one another as they do when our hands rest by our sides.

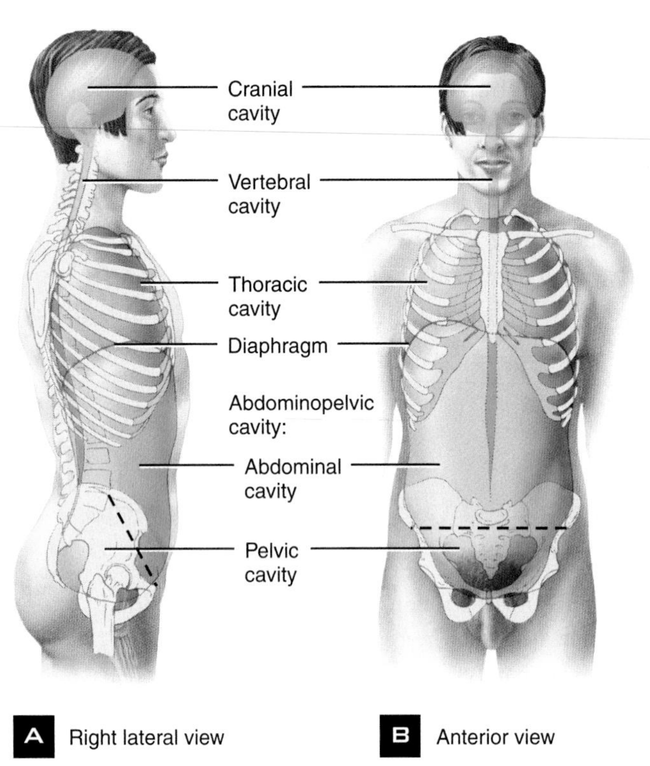

CAVITY	COMMENTS
Cranial cavity	Formed by cranial bones and contains brain.
Vertebral cavity	Formed by vertebral column and contains spinal cord and the beginnings of spinal nerves.
Thoracic cavity	Chest cavity; contains pleural and pericardial cavities and mediastinum.
Pleural cavity	Each surrounds a lung; the serous membrane of the pleural cavities is called the pleura.
Pericardial cavity	Surrounds the heart; the serous membrane of the pericardial cavity is called the pericardium.
Mediastinum	Central portion of thoracic cavity between the lungs; extends from sternum to vertebral column and from neck to diaphragm; contains heart, thymus, esophagus, trachea, and several large blood vessels.
Abdominopelvic cavity	Subdivided into abdominal and pelvic cavities.
Abdominal cavity	Contains stomach, spleen, liver, gallbladder, small intestine, and most of large intestine; the serous membrane of the abdominal cavity is called the peritoneum.
Pelvic cavity	Contains urinary bladder, portions of large intestine, and internal organs of reproduction.

Body cavities FIGURE 4.13

Meninges
Three protective membranes covering the brain and spinal cord.

The **dorsal body cavity** includes the **cranial** cavity housing the brain, and the **vertebral** cavity which contains the spinal cord. The **meninges** line these two continuous cavities.

Medical specialists often refer to the nine **abdominopelvic** regions and four **quadrants** of the body when diagnosing pain. Use of this terminology allows us to describe a particular area housing just a few abdominal organs (FIGURE 4.14).

As we study human biology, we will refer to these regions, quadrants, and cavities as landmarks for identifying the position of organs and the relationships between them. They also provide a common language to facilitate communication about location or organ function. In the coming age of computer-controlled surgery and online medical diagnoses, having a common language becomes even more important. Digital clinical assistance, or even distance education in this field, would be impossible without these conventions.

Knowing the organization of the chemicals, cells, and tissues that make up the human body is a prerequisite for understanding how humans function in the environment. Armed with this basic knowledge, an in-depth look at humans and their environment becomes much more interesting. Ultimately, the goal of the text is to explore the relationship between human physiology and the environment in which humans live.

Body regions and cavities FIGURE 4.14

Clavicles
Midclavicular lines
Right
Left
RIGHT HYPOCHONDRIAC REGION
EPIGASTRIC REGION
LEFT HYPOCHONDRIAC REGION
Subcostal line
RIGHT LUMBAR REGION
UMBILICAL REGION
LEFT LUMBAR REGION
Transtubercular line
RIGHT INGUINAL (ILIAC) REGION
HYPOGAS-TRIC (PUBIC) REGION
LEFT INGUINAL (ILIAC) REGION

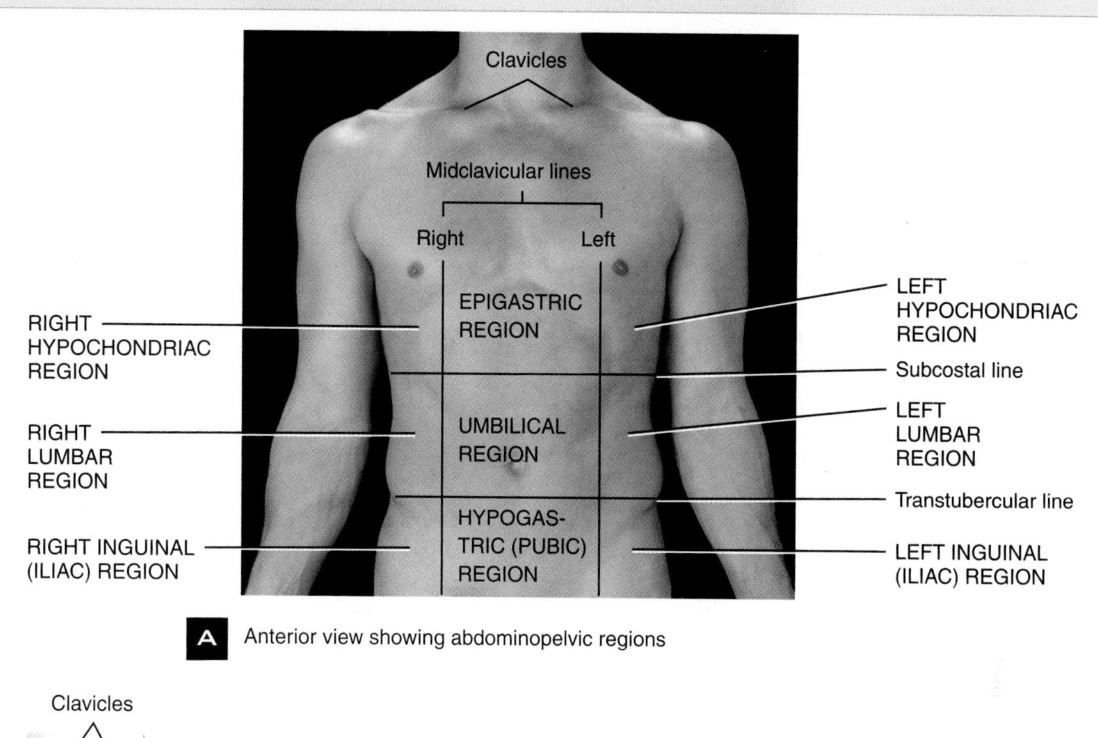

A Anterior view showing abdominopelvic regions

Clavicles
Midclavicular lines
Right
Left
RIGHT HYPOCHONDRIAC REGION
EPIGASTRIC REGION
LEFT HYPOCHONDRIAC REGION
RIGHT LUMBAR REGION
UMBILICAL REGION
LEFT LUMBAR REGION
RIGHT INGUINAL (ILIAC) REGION
HYPOGAS-TRIC (PUBIC) REGION
LEFT INGUINAL (ILIAC) REGION

B Anterior view showing location of abdominopelvic regions

RIGHT UPPER QUADRANT (RUQ)
LEFT UPPER QUADRANT (LUQ)
RIGHT LOWER QUADRANT (RLQ)
LEFT LOWER QUADRANT (LLQ)

C Anterior view showing location of abdominopelvic quadrants

CONCEPT CHECK

Identify a body part that is proximal to the hand.

What are the divisions of the ventral cavity?

What organs are found in the dorsal cavity?

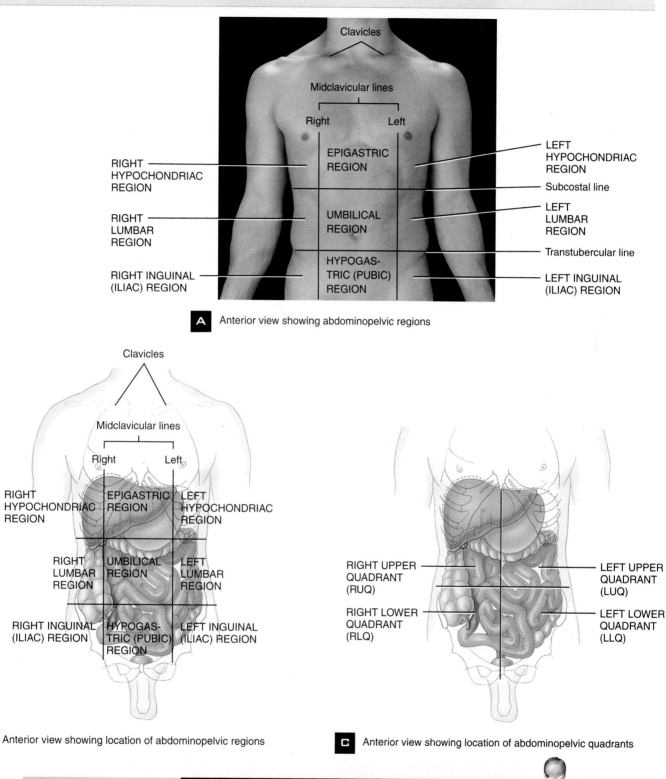

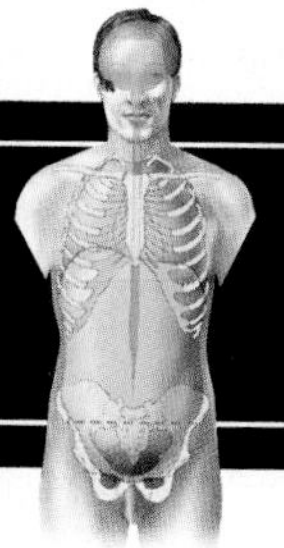

CHAPTER SUMMARY

1 Cells Are the Building Blocks of Tissues

The human body has four tissue types: epithelial, connective, nervous, and muscular. Epithelium covers and lines all body cavities and is classified based on cell shape (squamous, cuboidal, or columnar) and number of cell layers (simple or stratified). Connective tissue can be soft and loose or dense. Cartilage, bone, blood, and lymph are all examples of connective tissue. Muscular tissue can contract, and it comes in three varieties: smooth, skeletal, and cardiac muscle. Nervous tissue includes the impulse-carrying neurons and the neuroglia, which provide support for neurons.

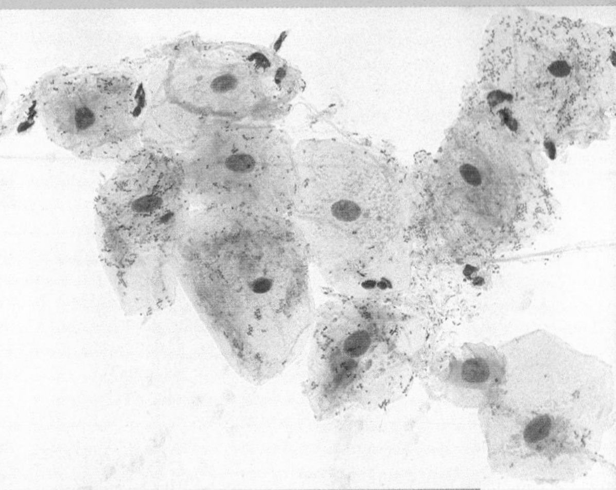

2 Organization Increases with Organs, Organ Systems, and the Organism

Tissues are grouped together in organs. Organs performing a similar function come together in organ systems. A group of organ systems comprises an organism. The 11 organ systems of the human are the skeletal (bones providing support and protection), muscular (muscles for movement and heat generation), nervous (sensing and responding to the environment), cutaneous (skin serving as a protective and sensitive layer), lymphatic (providing specific immunity), cardiovascular (transporting oxygen and nutrients to cells), respiratory (obtaining oxygen and removing carbon dioxide), digestive (obtaining nutrients), urinary (maintaining fluid balance), reproductive (producing new individuals), and endocrine (regulating sequential growth and development).

3 Scientists Use a Road Map to the Human Body

When discussing the placement of human anatomical structures, we assume the body is in the anatomical position. This is a face-forward position, with the palms of the hands forward. The two main body cavities are the dorsal cavity and the ventral cavity. The dorsal cavity includes the cranial cavity, holding the brain, and the spinal cavity, surrounding the spinal cord. The ventral cavity includes the thoracic cavity, the abdominal cavity, and the pelvic cavity. The ventral cavity can be subdivided into quadrants for specifically pinpointing the location of an organ, a structure, or a physiological event in the body.

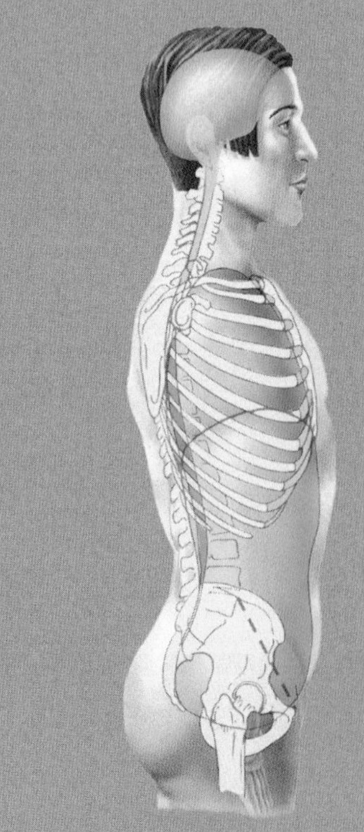

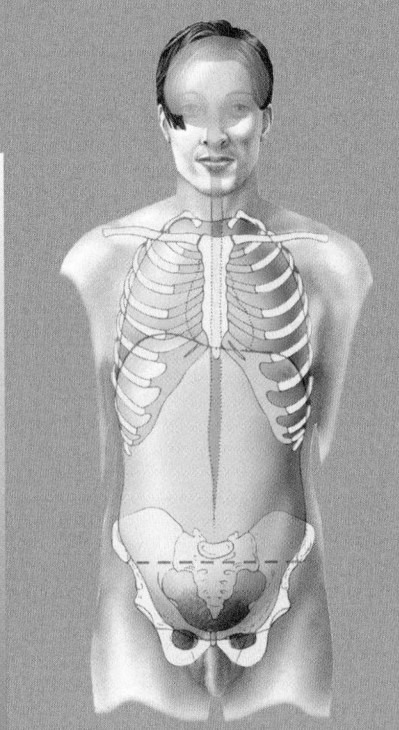

KEY TERMS

- **avascular** p. 96
- **biceps brachii** p. 100
- **epiglottis** p. 98
- **interstitial fluid** p. 96
- **mediastinum** p. 111
- **meninges** p. 112
- **microvilli** p. 94
- **osteoid** p. 96
- **plasma** p. 96
- **proximal/distal** p. 111
- **rectus abdominus** p. 100
- **striations** p. 100
- **superior/inferior** p. 111
- **trachea** p. 98

CRITICAL THINKING QUESTIONS

1. As a research assistant in a cytology lab, you are handed a stack of photographs from an electron microscope. Each represents a different type of animal cell. You are asked to identify photos of cells that secrete large amounts of protein and include a mechanism for moving their secretions along their surface. What organelles would be required by this cell? In which tissues might these cells be found?

2. The digestive tract has two surfaces; an inner surface that lines the gut and allows food to pass, and an outer surface that separates the gut from the rest of the abdominal organs. What specific tissue would you expect to find on each of these surfaces? Would the inner surface have the same lining as the outer? Why or why not?

3. There are many types of connective tissue in the body, from adipose to bone to blood. What is it that makes these tissues different? More importantly, what are the unifying characteristics found in all connective tissues?

4. You are given the opportunity to create artificial skin in a laboratory to help burn patients. Remember that the skin must be protective, relatively water-tight, and yet have some sensory function. What tissues will you need for this organ? Which type of epithelium will you use for the outer layer? What tissue will you need to house the blood vessels and the nerves? Will you need muscular tissue? Nervous tissue?

5. Physicians often use the quadrants and regions of the body to diagnose pathologies. If a patient complained of stabbing pain in the abdominal cavity, which organs might be involved? Look at FIGURE 4.13 to help with your diagnosis. How would you describe the location of the urinary bladder using the nine abdominopelvic regions given in FIGURE 4.14?

SELF TEST

1. The four major tissue types that comprise the human body include all of the following EXCEPT
   a. epithelial tissue.
   b. muscular tissue.
   c. areolar tissue.
   d. nervous tissue.
   e. connective tissue.

2. The tissue that can be found covering and lining openings in the body is
   a. epithelial tissue.
   b. muscular tissue.
   c. areolar tissue.
   d. nervous tissue.
   e. connective tissue.

3. The tissues that do not have a blood supply include
   a. epithelial tissue only.
   b. epithelial and connective tissue.
   c. some types of connective tissue only.
   d. epithelial and some types of connective tissue.

4. Identify the tissue type pictured
   a. stratified epithelium.
   b. cuboidal epithelium.
   c. simple epithelium.
   d. columnar epithelium.

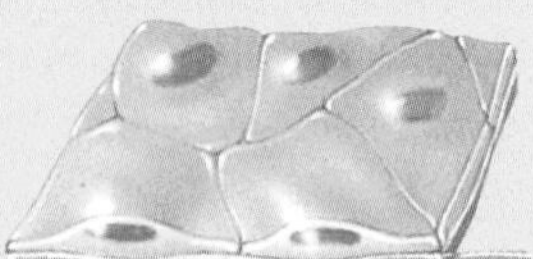

5. The function of the tissue pictured is most likely
   a. a diffusion membrane.
   b. a protective membrane.
   c. a contractile organ.
   d. a connective support.

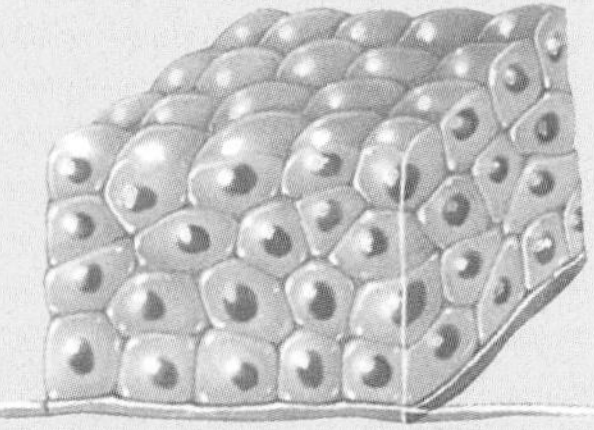

6. The specific type of cell that comprises most diffusion membranes is a
   a. squamous epithelial cell.
   b. cuboidal epithelial cell.
   c. columnar epithelial cell.
   d. exocrine cell.

7. The structure labeled A in this diagram is
   a. fibroblast.
   b. collagen fiber.
   c. matrix.
   d. white blood cell.

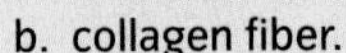

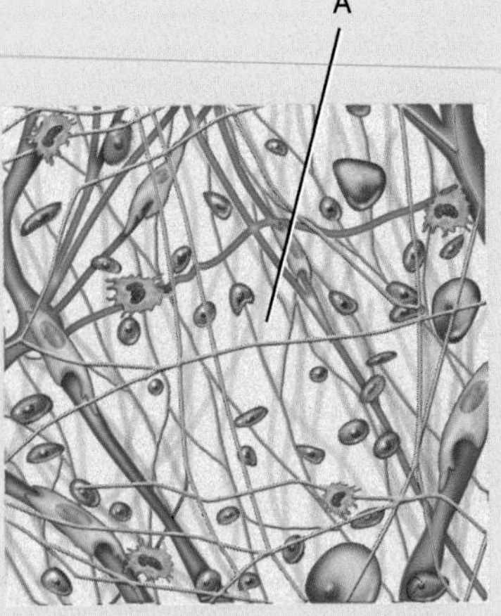

8. The connective tissue that is composed of regular, linear arrangements of collagen fibers packed tightly into the matrix is called
   a. areolar connective tissue.
   b. loose connective tissue.
   c. dense irregular connective tissue.
   d. dense regular connective tissue.

9. Identify the type of connective tissue illustrated below.
   a. Bone
   b. Hyaline cartilage
   c. Elastic cartilage
   d. Lymph
   e. Fibrocartilage

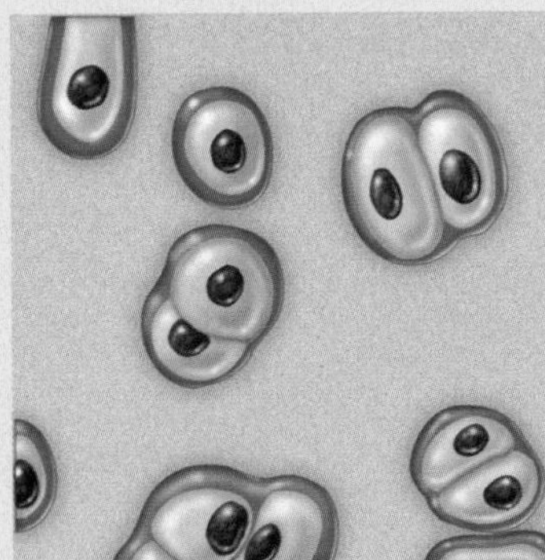

10. The connective tissue whose functions include lipid storage and organ protection is
    a. blood.
    b. lymph.
    c. adipose.
    d. bone.

11. Identify the tissue shown.

   a. Hyaline cartilage
   b. Skeletal muscle
   c. Cardiac muscle
   d. Smooth muscle

12. Which type of muscle tissue can be described as involuntary, striated, and connected via intercalated discs?

   a. Skeletal muscle
   b. Cardiac muscle
   c. Smooth muscle
   d. Two of these have the listed characteristics.

13. Identify the structure labeled as A on this image

   a. Neuroglia
   b. Dendrites
   c. Axon
   d. Neuron body

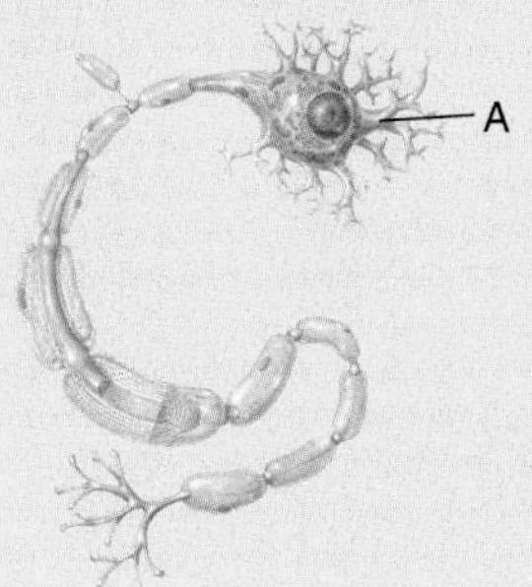

14. The functions of neuroglia include

   a. improving nutrient flow to neurons.
   b. supporting neurons.
   c. sending and receiving electrical impulses.
   d. two of the above are correct.
   e. all of the above are correct.

15. The correct order from least to most complex is:

   a. organ, organ system, organelle, organism.
   b. cell, tissue, organism, organ system.
   c. tissue, organ, organ system, organism.
   d. cell, organelle, tissue, organ.

16. Which term correctly describes the relationship indicated as "A" on this figure?

   a. Superior
   b. Inferior
   c. Proximal
   d. Distal

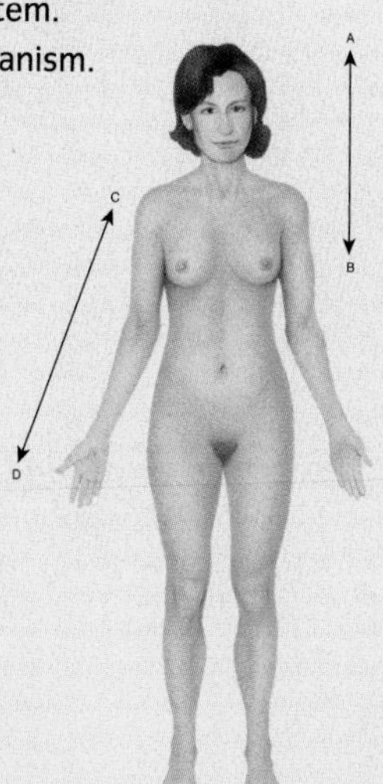

17. Which term correctly describes the relationship indicated as C in the figure accompanying question 16?

   a. Superior
   b. Inferior
   c. Proximal
   d. Distal

18. The ___________ houses the heart, lungs, vessels, and lymphatics of the mediastinum.

   a. ventral cavity
   b. abdominal cavity
   c. cranial cavity
   d. thoracic cavity

19. What membrane lines the indicated cavity?

   a. pericardial membrane
   b. meninges
   c. peritoneum
   d. mediastinum

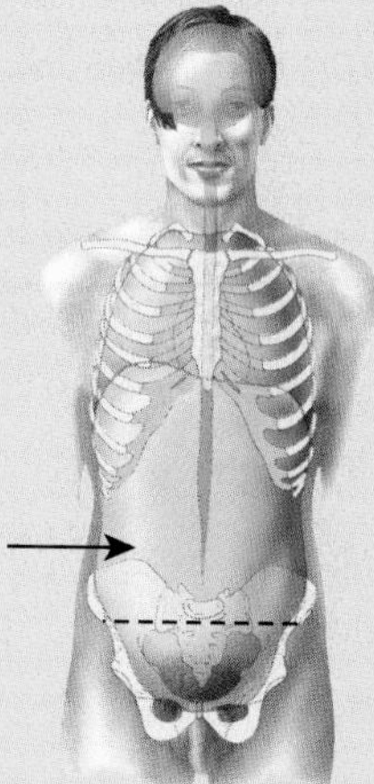

20. Which label indicates the quadrant in which the majority of the liver lies?

   a. A
   b. B
   c. C
   d. D

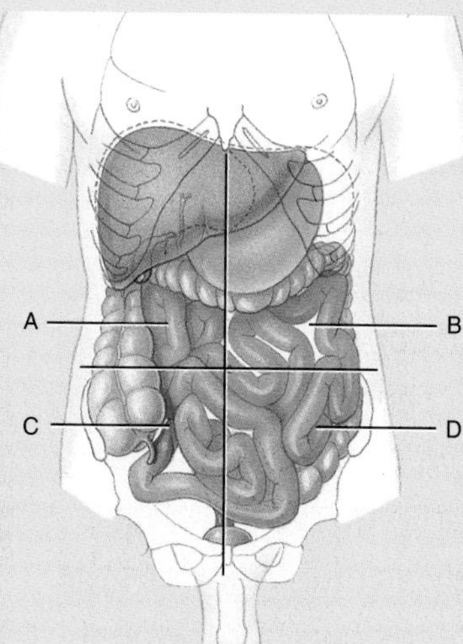

The Skeletal System

5

It was a surfer's dream: sunny, calm air, with perfect sets. . . . The waves were coming in fairly quickly; each set seeming to build on the last. Kainoa paddled furiously to catch the largest wave. The wave lifted the back of the board, and he was on. Up he popped, standing as his board skimmed along the wave face. As he inched up to walk the nose, the tip grabbed the wave, pearling into the depths. His board went down, then abruptly upward, and he tumbled into the angry water. Turning and rolling in the wave, Kainoa felt a stabbing pain in his right shoulder. As he struggled to the surface, the pain shot down his right arm and he couldn't move his arm. Reaching the surface, Kainoa turned to face the next wave just as another surfer came barreling through. His board caught Kainoa square in the face, cracking his chin and right cheek. At the hospital, doctors delivered the bad news: Kainoa had broken his clavicle, most likely when his board jumped from the wave and hit him. The second board's impact fractured his zygomatic bone and compressed his temporomandibular joint (the joint between his jaw and skull). Both injuries made movement extremely painful. What was the source of the pain? How would these injuries be treated? How could he speed the healing process?

CHAPTER OUT

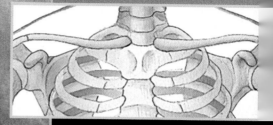

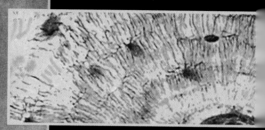

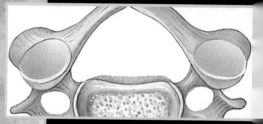

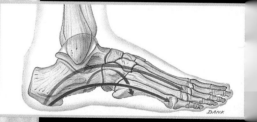

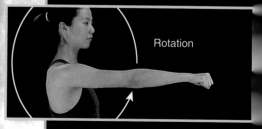

The Skeletal System Is a Dynamic Living System

LEARNING OBJECTIVES

List the functions of the skeletal system.

Define the divisions of the skeletal system.

Classify the bones in the human body by structure.

Chances are, when you think about your bones, you envision a rigid, unchanging support, something like a scaffold. True, the bones are rigid, but they are not a simple, inflexible framework. The skeletal system, like your other body systems, is a dynamic, living entity, with many essential functions (FIGURE 5.1). Although support is the most obvious function, the skeletal system also anchors the skeletal muscles, protects soft tissues and organs, produces blood cells (a process called **hematopoiesis**), stores and releases minerals such as calcium and phosphorus, and moves gracefully courtesy of the many joints. Bones look stable and static, but they change throughout life. The entire calcium content of your femur (the large thigh bone) is replaced every five to seven years in a sedentary person—and more often in those who are physically active.

Two hundred and six named bones comprise the skeleton that underlies the adult human form. This number varies a bit from person to person because small bones can exist within some tendons. The skeleton is divided into the **axial** skeleton (the central axis of the body) and the **appendicular** skeleton (the appendages [arms, legs, hands, and feet] and girdles holding them to the central axis) (FIGURE 5.2).

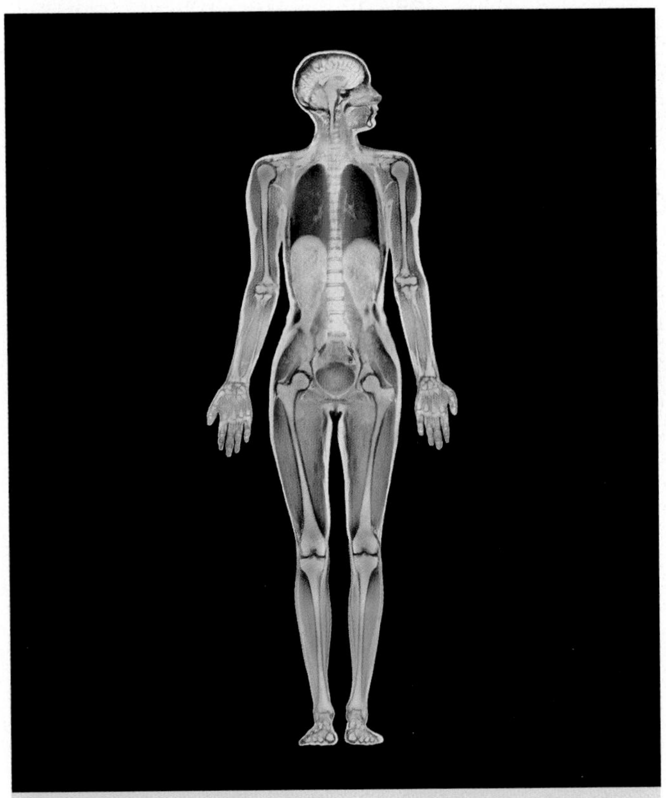

MRI of female body in frontal section FIGURE 5.1

Because bones are denser than surrounding body tissue, the skeleton is relatively easy to visualize using X-rays or MRI. This is an MRI of a female in frontal section, clearly showing many of her 206 bones.

The skeleton FIGURE 5.2

The bones of the skull and thorax make up the axial skeleton. The arms, hands, legs, and feet, along with the bones that secure these limbs to the body, make up the appendicular skeleton.

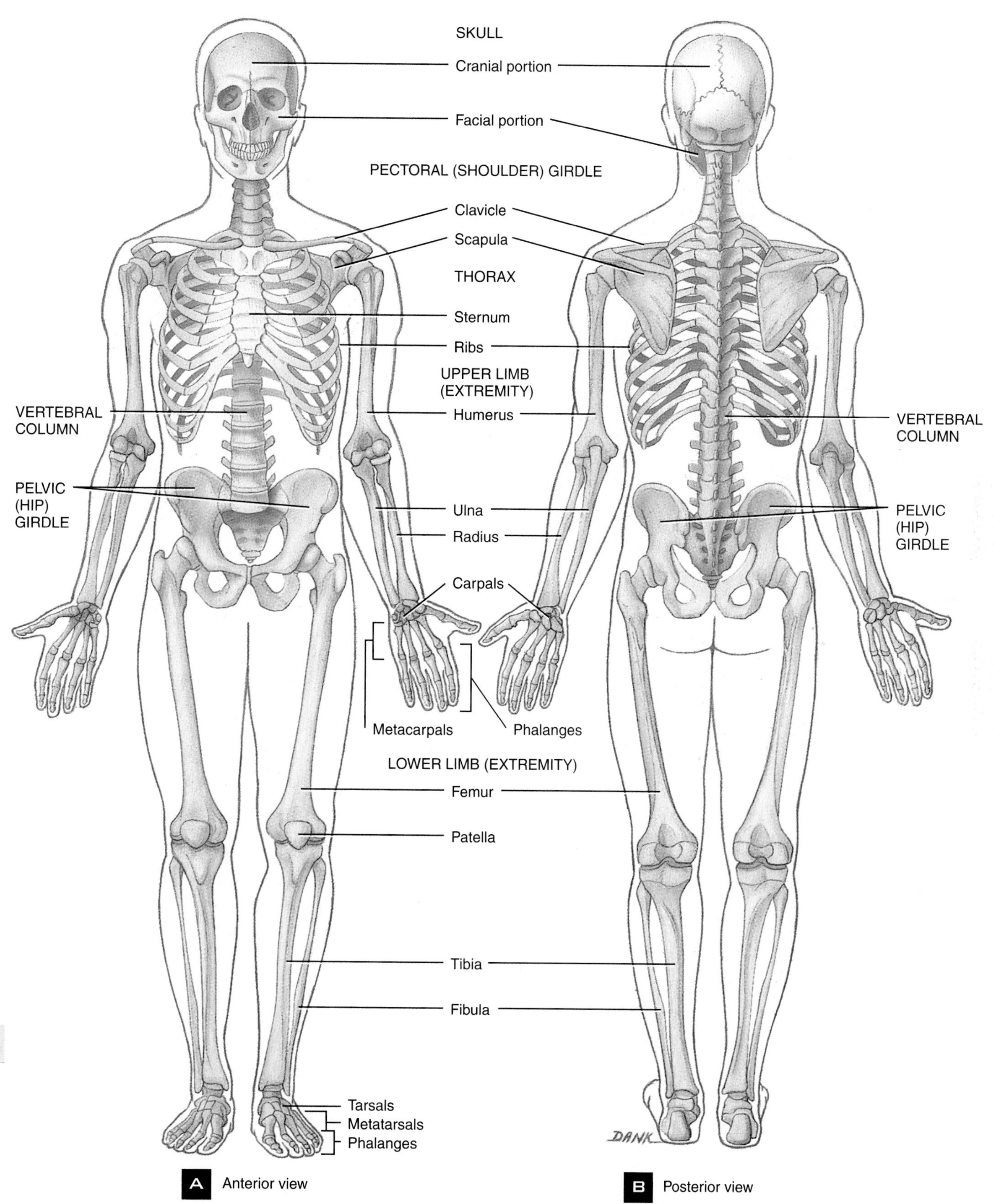
SKULL
Cranial portion
Facial portion
PECTORAL (SHOULDER) GIRDLE
Clavicle
Scapula
THORAX
Sternum
Ribs
UPPER LIMB
(EXTREMITY)
Humerus
VERTEBRAL
COLUMN
PELVIC
(HIP)
GIRDLE
Ulna
Radius
Carpals
Metacarpals
Phalanges
LOWER LIMB (EXTREMITY)
Femur
Patella
Tibia
Fibula
Tarsals
Metatarsals
Phalanges
DANK
A Anterior view
B Posterior view

In the body, "form follows function," and this is nowhere more true than in the skeletal system (FIGURE 5.3). Every bone in your body is designed to perform a specific task. For example, the long bone known as the femur must be strong and have a slight anterior curve to bear the weight of the upper torso. The bones of

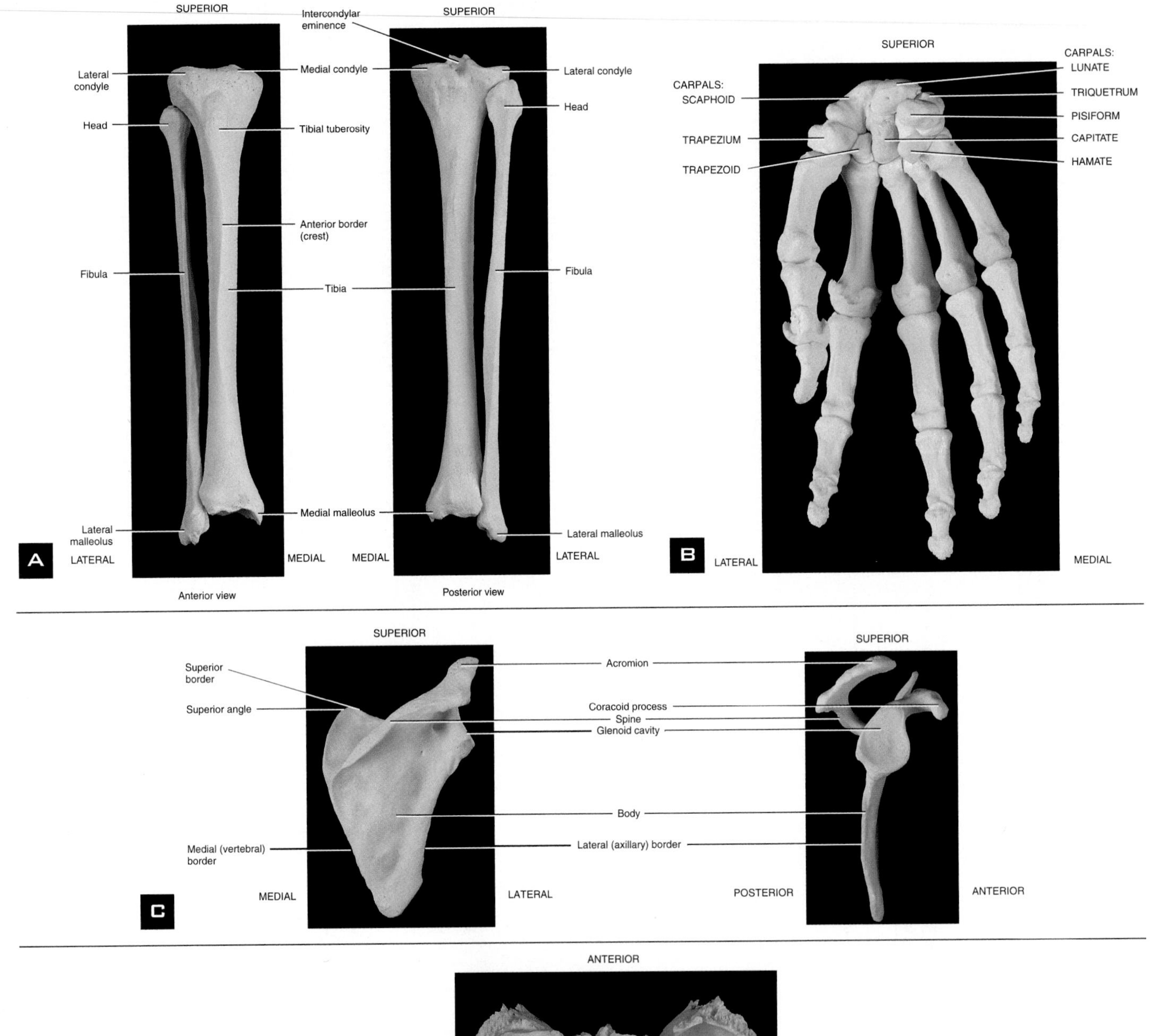

Classification of bone FIGURE 5.3

We classify the bones according to shape. A Long bones **are longer than they are wide;** B short bones like those of the wrist **are akin to small cubes;** C flat bones **are very thin in one dimension;** D irregular bones **have odd shapes;** E sesamoid bones **form inside tendons;** F Wormian bones **are embedded in the sutures between the main skull bones.**

the skull must curve into a "bowl" that houses and protects the brain. Later we'll look at the knee and shoulder—two supreme examples of form following function.

Bones are composed of compact or spongy mineralized tissue that is structured for maximum strength and minimum weight. If you have ever broken a bone, you learned that the tissue has a blood supply and nervous connections—factors that help to explain the large, painful bruise that develops when most bones break.

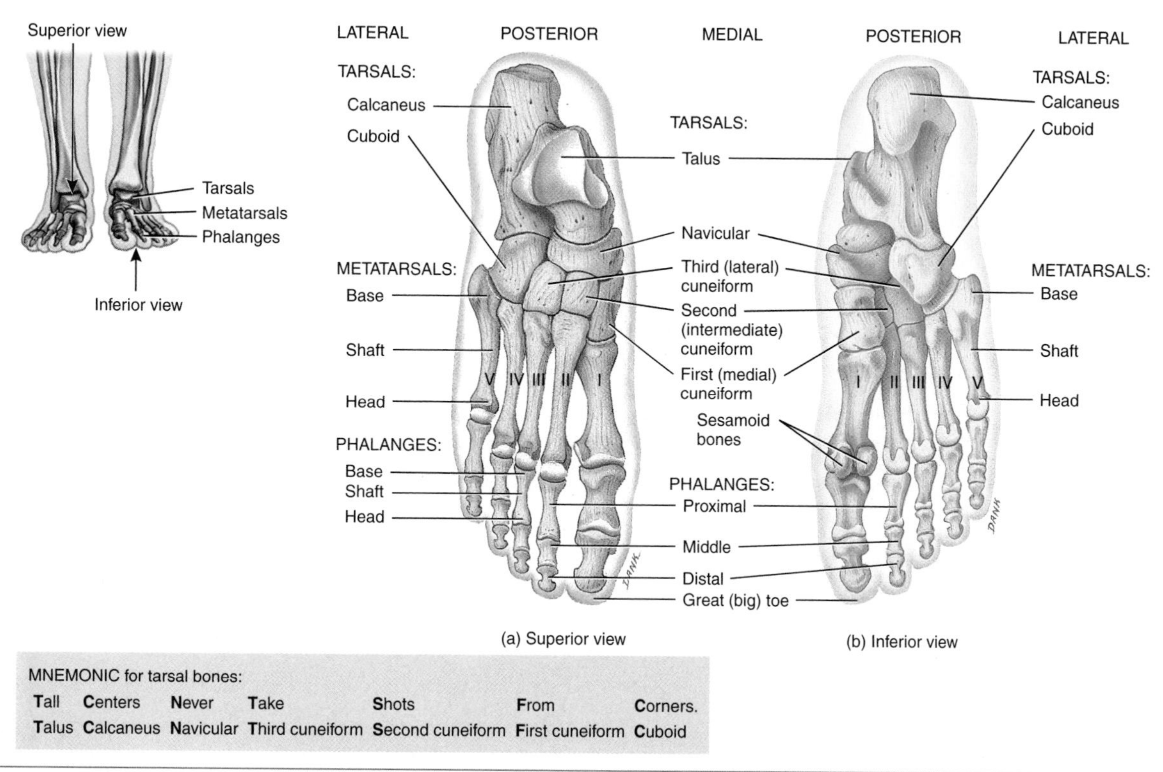

E

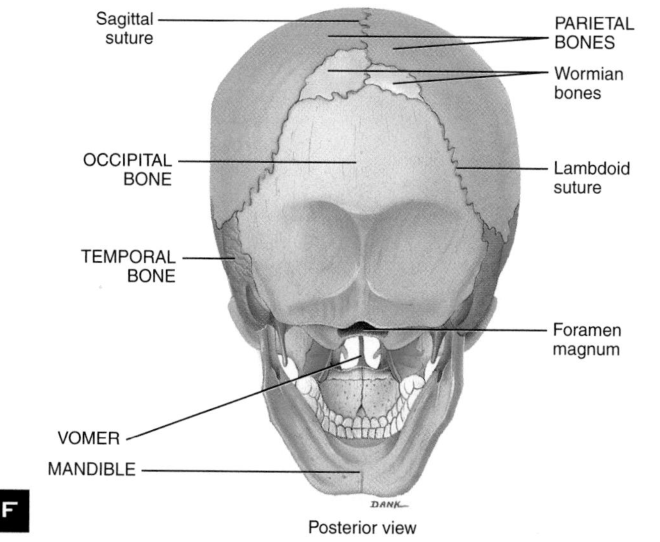

F

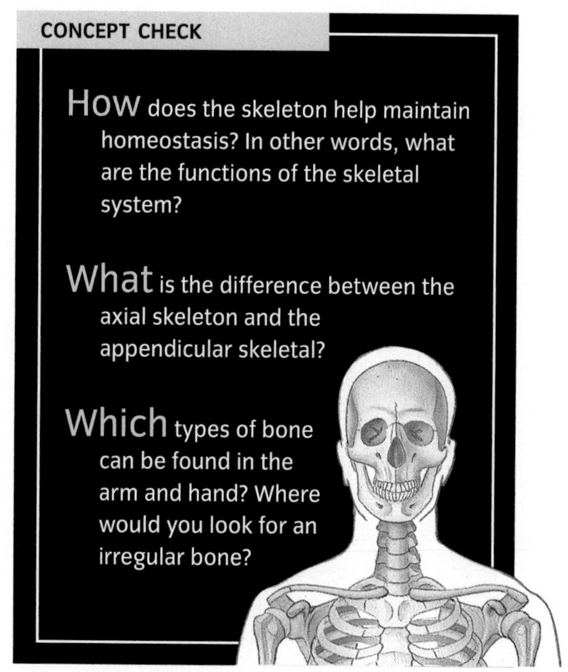

CONCEPT CHECK
How does the skeleton help maintain homeostasis? In other words, what are the functions of the skeletal system?
What is the difference between the axial skeleton and the appendicular skeletal?
Which types of bone can be found in the arm and hand? Where would you look for an irregular bone?

Ossification Forms Bone, and Remodeling Continues to Shape It

LEARNING OBJECTIVES

Differentiate compact from spongy bone.

Identify the parts of a typical long bone.

Describe the formation of a typical long bone and a flat bone.

Explain the steps in bone remodeling and repair.

Discuss, briefly, how different lifestyles affect the appearance of the skeletal system.

Bones are a form of connective tissue produced by immature bone cells called **osteoblasts**. Ossification—bone formation—can be **endochondral** (FIGURE 5.4) or **intramembranous**. Most of your bones are "endochondral," meaning that they were formed within cartilage.

Not only do long bones grow longer, they also grow thicker. This so-called **appositional** growth occurs at the outer surface of the bone. The innermost cellu-

Process Diagram

Endochondral ossification FIGURE 5.4

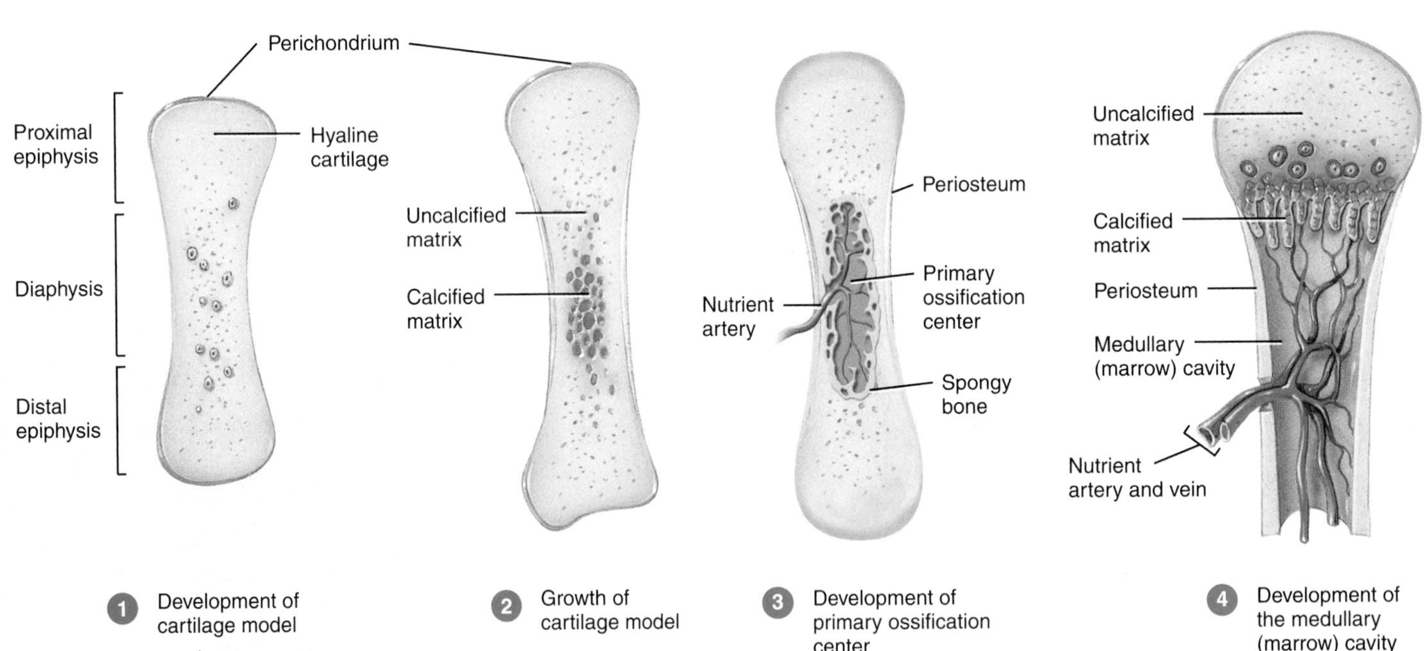

1 In endochondral ossification, a hyaline cartilage model of each bone forms in the embryo. 2 The hyaline cartilage model expands into the space the final bone will occupy. 3 A blood vessel invades the central portion of the model, stimulating osteoblasts to begin producing bone. 4 The marrow cavity forms.

Osteoblasts Immature bone cells not yet surrounded by bony matrix.

Endochondral Within cartilage.

Intramembranous Between membranes.

lar layer of the **periosteum**, the membrane that covers the bone, differentiates into osteoblasts and begins to add matrix to the exterior. Accumulating matrix entraps these osteoblasts, which mature into **osteocytes**, creating new bone tissue around the exterior of the bone.

Intramembranous ossification (FIGURE 5.5, p. 126) forms the flat bones of the skull, clavicle, and mandible. Again, the name suggests how the process occurs. Bone is laid down within embryonic connective tissue, surrounded by the developing periosteum. These bones form deep in the dermis of the skin and thus are often called dermal bones. Dermal bones may also form in the connective tissues of joints, in the kidneys, or in skeletal muscles when subjected to excessive stress (bones that form outside the usual areas are called **heterotopic bones**).

Osteocytes Mature bone cells surrounded by bone matrix.

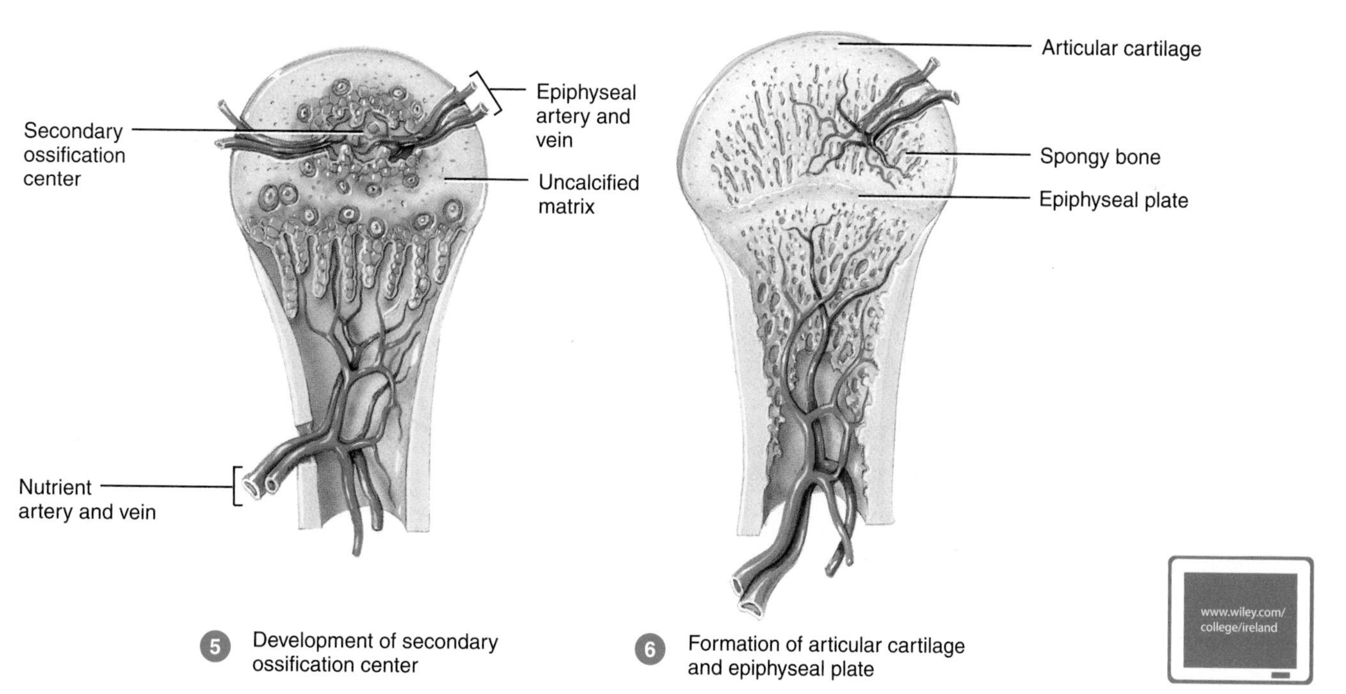

5 After birth, a second blood vessel invades each end of the developing bone, again stimulating osteoblast activity.

6 The epiphyses (long bone ends) are ossified, leaving a central area of cartilage called the epiphyseal plate, which continues growing through adolescence. Cartilage on the surface of the epiphyses also remains, forming articular cartilage.

At maturity, the epiphyseal plate closes as the ossification rate exceeds cartilage growth, and the bone's length is essentially static.

Intramembranous ossification FIGURE 5.5

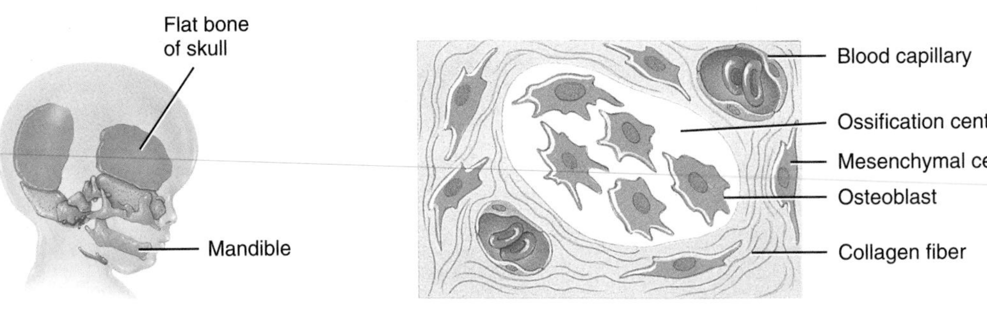

1 Development of ossification center

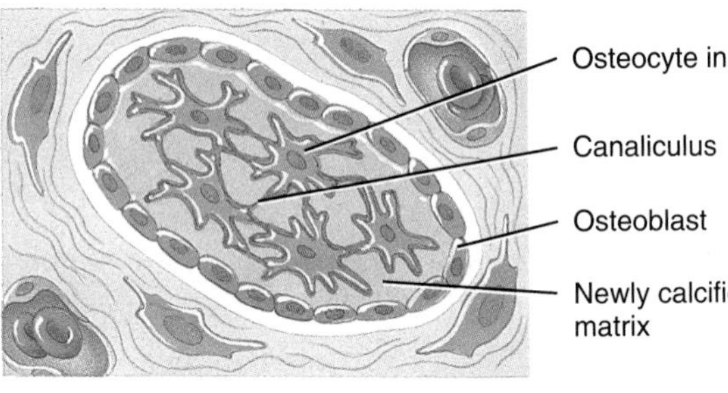

2 Calcification

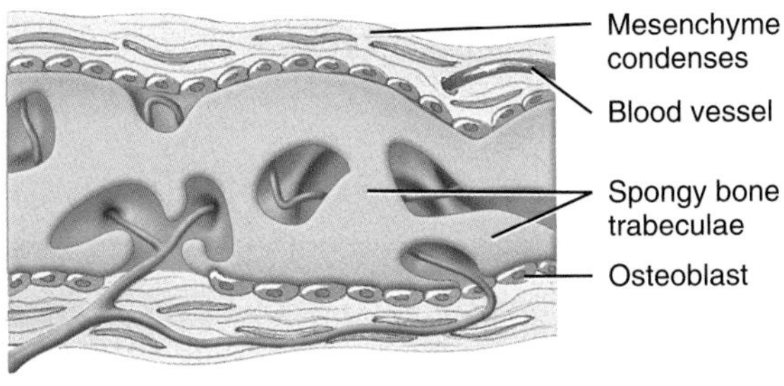

3 Formation of trabeculae

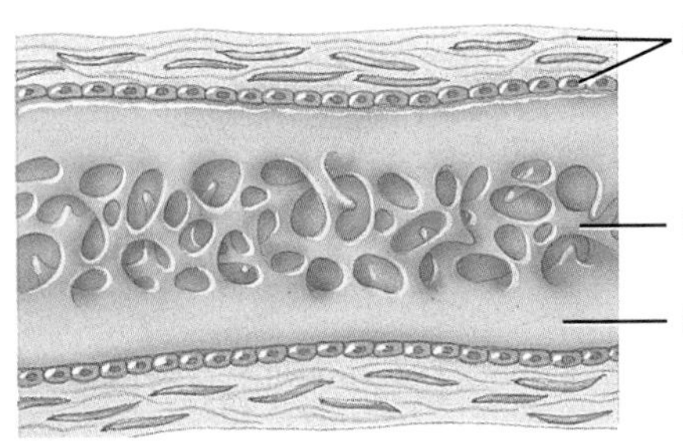

4 Development of the periosteum

Intramembranous ossification follows these steps:

1. Embryonic cells come together and begin to secrete organic material called osteoid. This matrix becomes mineralized by the enzyme alkaline phosphatase. The embryonic cells trapped in the matrix transform into osteoblasts.
2. These new osteoblasts form an ossification center, from which the developing bone grows outward in small spicules. Osteoblasts trapped in the newly formed bone mature into osteocytes, as discussed above.
3. Blood vessels grow into the area to provide nutrients and oxygen to the developing bone and osteocytes.
4. Eventually ossification slows, and the connective tissue surrounding the new bone becomes organized into the periosteum.

BONY TISSUE COMES IN TWO FORMS

Bone structure may be compact (dense) or spongy. Compact bone material usually occurs at the edges of the bone and is composed of many individual **Haversian systems** (FIGURE 5.6). These are concentric rings of matrix laid by osteocytes and formed surrounding a central canal. One complete Haversian system is called an **osteon**. Despite its strength and inflexible structure, bone is living tissue, and as such the cells

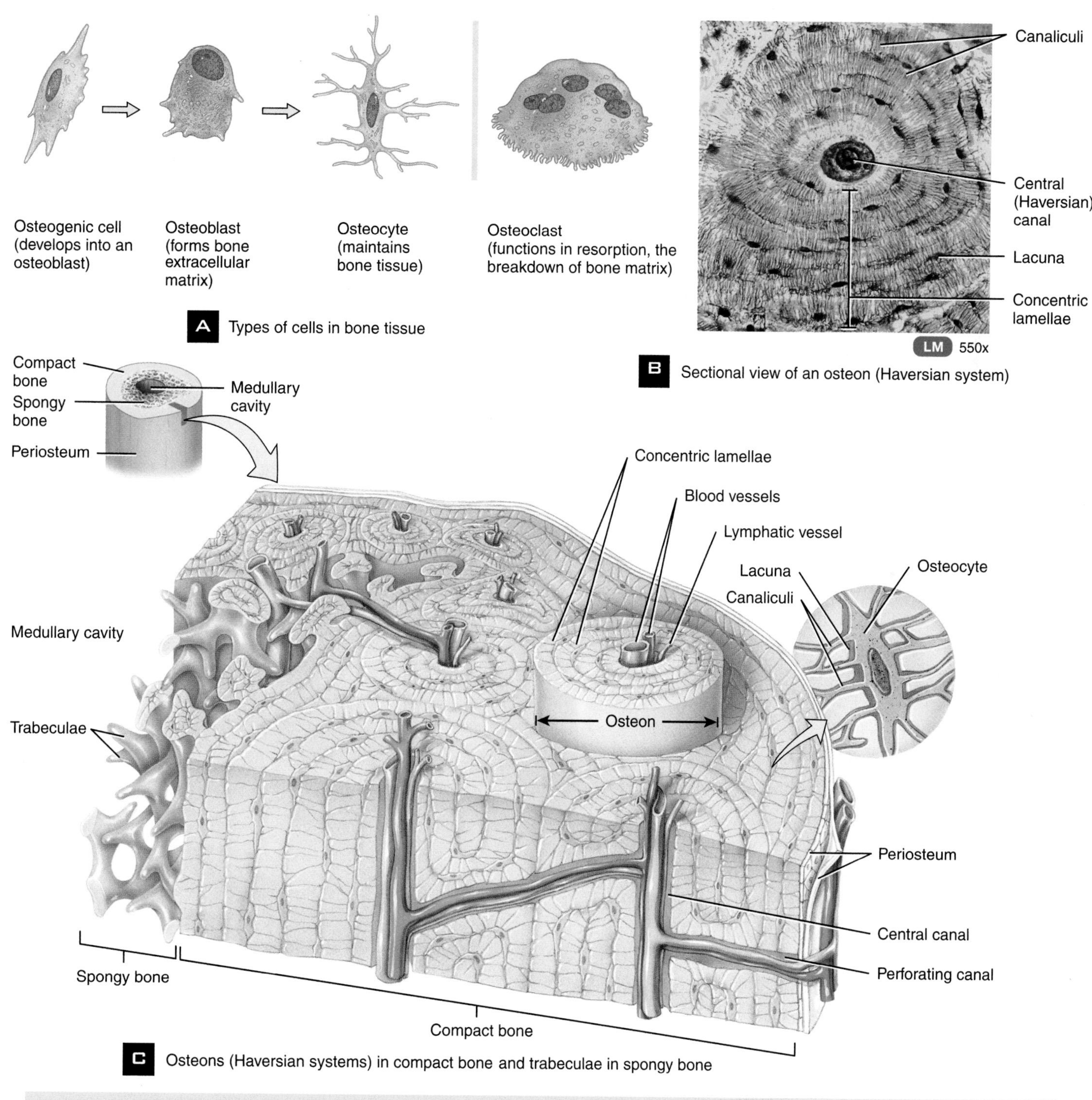

Composition of bone FIGURE 5.6

Note the similar appearance of the osteocytes in both types of bone, despite the different patterns of matrix deposition.

within the bone must receive a constant nutrient supply and be able to dispose of wastes. They require a blood supply just like all other living cells. The central canal of the osteon houses the blood and nerve supply for the bone tissue. Individual cells lie within small holes in the matrix called **lacunae**. Because tissue cells must contact one another, bone cells communicate via small canals cut into the matrix. These canals, or **canaliculi**, allow fluid carrying vital nutrients and signaling chemicals to pass between cells. Osteons communicate via larger perforating (Volkmann's) canals that run perpendicular to the long axis of the Haversian systems and connect one central canal with the next.

In a typical bone, dense compact bone surrounds the organ and spongy bone comprises the inner support. Spongy bone is less organized than compact bone and lacks Haversian systems. Instead, spongy bone has **trabeculae**, or struts, that form in response to stress. These struts are composed of osteocytes surrounded by matrix similar to the osteon of compact bone. Instead of being laid in concentric rings, the matrix looks like short, interconnecting support rods.

Both forms of bone are held inside a protective covering called the periosteum. This covering houses the nerve and blood supply for the bone. The periosteum also provides a constant supply of fresh osteoblasts to lay new matrix. As osteoblasts produce bony matrix, they mature into osteocytes. Mature osteocytes are caught in the calcified matrix they secrete and rest in the lacunae described earlier.

The shaft of a long bone, or **diaphysis**, is composed of dense bone surrounding a central canal (FIGURE 5.7). (In flat bones, as in the skull or sternum, a layer of spongy bone is sandwiched between two thin layers of compact bone.) The central canal of the long bone houses the marrow; blood cells form in red marrow, and energy is stored in yellow marrow. The ends of the bones, or **epiphyses**, include the **epiphyseal plate**, an area of cartilage where long bones continue to grow during childhood and adolescence. When bones cease growing, this cartilage is replaced by bone, leaving the **epiphyseal line**. Wherever two bones meet, you will find a layer of hyaline cartilage. This **articulating cartilage** prevents bone from grinding against bone at a joint.

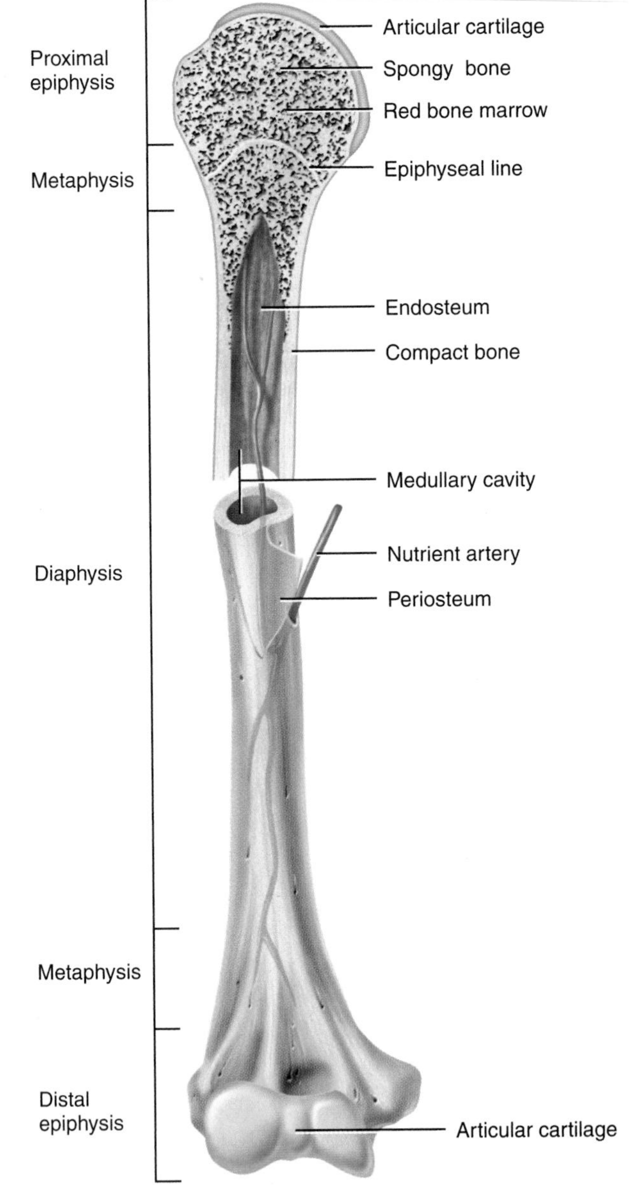

Long bone with parts identified FIGURE 5.7

BONE CONSTANTLY UNDERGOES REMODELING AND REPAIR

Bones are dynamic structures, constantly being remodeled and perfected to suit the needs of the body, and continuously making subtle changes in shape and density to accommodate your lifestyle. Although long bones cease growing in length at maturity, they do change shape throughout life. The calcium within each bone is being removed and new calcium is added in response to blood calcium levels and the amount of stress placed on the bones.

Remodeling of existing bone occurs via a different process than original ossification because it takes advantage of the interplay between **osteoclasts** and osteoblasts. Osteoclasts are large cells that adhere to the surface of bony tissue and release acids and enzymes. The end result of the activity of these cells is the breakdown of the bony matrix and the addition of calcium and other minerals to the bloodstream. Osteoblasts build the mineral structure back up, pulling calcium and minerals from the bloodstream. The osteoblasts first secrete an organic matrix called osteoid. They then cause an increase in local calcium concentration around the osteoid, converting the osteoid to bone. This process takes upwards of three months to complete. As usual, rebuilding takes much longer than destruction, but the overall outcome of osteoclast and osteoblast activity is a cyclic process that tears down and rebuilds the bony matrix.

The bones are a storehouse for calcium needed in physiological processes such as nerve impulse transmission and muscle contraction. When the blood calcium level drops, osteoclasts go to work to release stored calcium to the blood. Conversely, when the blood calcium level rises, the osteoblasts create new matrix, removing excess calcium from the blood.

When you begin a new exercise regime, you may be thinking about the muscular or cardiovascular benefits, but athletic stress can also remodel your bones. Extra support is added where muscles exert a stronger pull, so skeletal strength matches muscular development. High-impact exercises are particularly good at stressing the skeleton to activate osteoblasts to increase bone density. Removing skeletal stress by reducing activity reduces osteoblast stimulation and thus reduces bone density.

When a bone breaks, the repair process is something like a drastic version of the remodeling process, as you can see in the Health, Wellness, and Disease box on page 130.

For bone to heal, the ends of the fracture must be aligned and immobilized. When alignment is possible without disturbing the skin, the process is called "closed reduction." In "open reduction," the skin must be cut, and often metal screws, plates, or pins are used to fix the bones in place. Open reduction is more likely to be needed in "compound fractures," which have more than one break and often include a tear or opening in the skin with the original injury. After either type of reduction, a cast, splint, or other external paraphernalia is generally needed to immobilize the fracture.

Still, complete immobilization may not be ideal for healing bone. Limited movement, stress, or partial weight-bearing activities can actually help the bones grow, because those stresses on the bone matrix stimulate bone deposition.

Health, Wellness, and Disease

Healing broken bones

Broken bones are pretty common, especially among active children and young adults. How does a broken bone heal? Healing involves the formation of a **fracture hematoma** around the break, bringing in blood and nutrients to begin repairs. The fractured area slowly fills with spongy bone, creating a large, bony **callus**. The callus will remain part of the bone for a long period as it is slowly remodeled to return to normal size. Typically, broken bones are held fixed in position until the healing process can be completed.

Bone repair occurs in four stages:

1. **Fracture hematoma forms.** Blood leaks from broken vessels near the fracture, and a clot forms within a few hours of the break. Dead blood cells accumulate, and other blood cells start to remove them.
2. **Fibrocartilaginous callus forms.** Actual repair begins as fibroblasts are produced by the periosteum and start making collagen fibers. **Chondroblasts**, also derived from the periosteum, start to make **fibrocartilage**. Within about three weeks of the injury, a fibrocartilaginous callus forms from these two types of connective tissue.
3. **Bony callus forms.** Osteoblasts start to produce spongy bone tissue at the ends of the broken bone, beginning in areas with healthy bone and good vascularization. Fibrocartilage also converts into spongy bone tissue.
4. **Bone remodels.** Osteoclasts gradually resorb dead bone tissue from the damage site. Spongy bone is converted into compact bone. The healed bone is often thicker and stronger than the original bone. The callus remains as a visible thickened bump on the bone for many years after the break is healed.

Fracture hematoma A bruise that develops over the site of a fractured bone.

Chondroblasts Immature cartilage cells, not yet completely surrounded by the cartilage matrix.

Fibrocartilage Cartilage with strengthening fibers in the matrix.

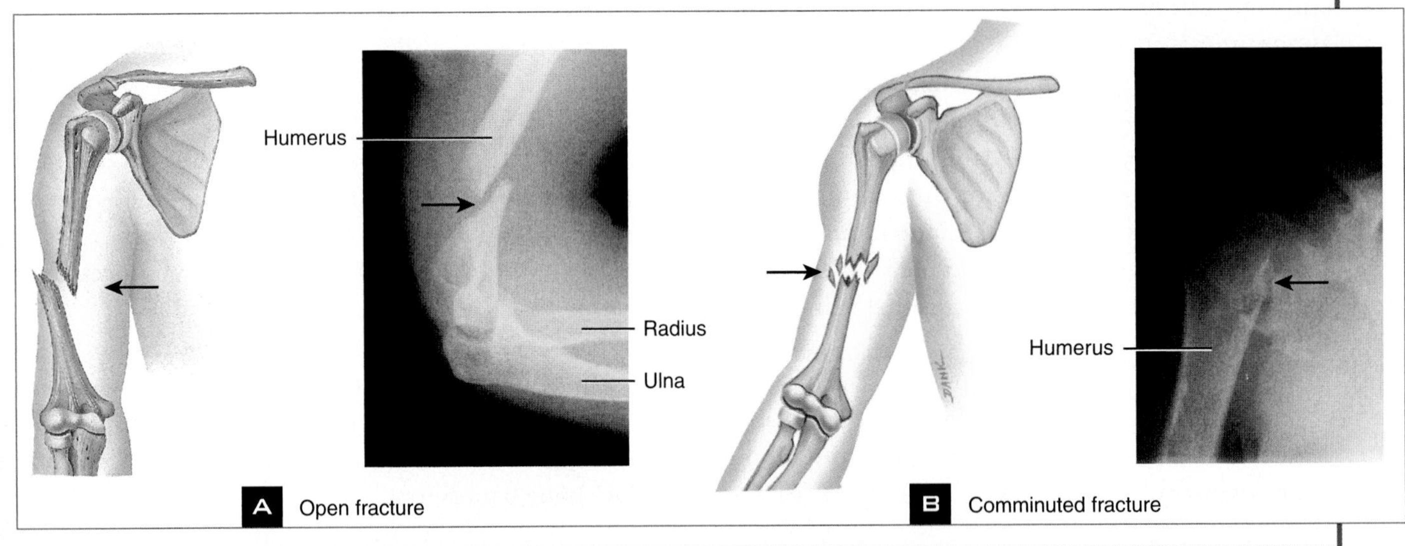

A Open fracture
B Comminuted fracture

CONCEPT CHECK

Describe the formation of the femur. How does this differ from the formation of the sternum?

List the cellular steps that occur during appositional growth. Which cells are active?

Describe an osteon and its relationship to compact bone.

Explain the functions of the periosteum.

What does inactivity do to the skeletal system?

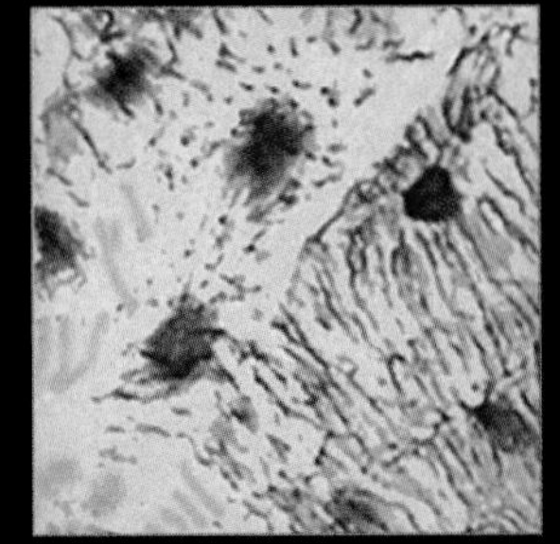

The Axial Skeleton Is the Center of Things

Learning Objectives

List the major bones of the axial division of the skeletal system.

Explain how each bone relates to the anatomy or physiology of the area where it's located.

he axial skeleton includes the 8 cranial and 14 facial bones as well as the hyoid bone, ribs, and vertebrae.

PURE PROTECTION: BONES OF THE FACE AND SKULL

Cranial bones, collectively known as the skull, surround and protect the brain (FIGURE 5.8). Of these, the parietal and temporal bones are paired, whereas the frontal bone, occipital bone, ethmoid, and sphenoid are single bones. All eight cranial bones are held together by fixed joints called **sutures**.

The **frontal** bone at the forehead protects the frontal lobe of the brain. The frontal bone originates as two frontal bones that start fusing in early development. This fusion continues so that by age 8 or so, the suture is difficult to locate. The frontal bone can be the source of misery: when the lining of the large sinuses in the frontal

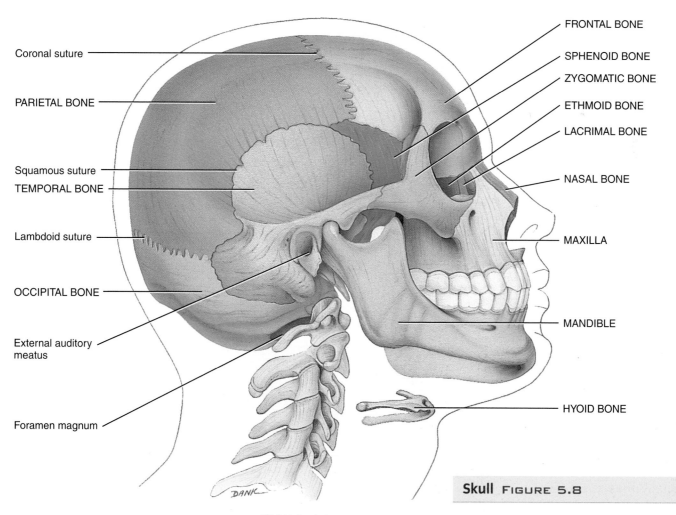

Skull FIGURE 5.8

Parietal
Walls, as in the walls of the cranial cavity.

Articulates
Joins; an articulation is a joint holding two bones together.

bone becomes inflamed, you get a sinus headache.

The **parietal** bones protect the upper sides of the head, whereas the **temporal** bones protect the middle sides of the head and support the ears. This bone underlies the area commonly referred to as the temples. The lower jaw (**mandible**) **articulates** with the temporal bones. The mandible is the only bone of the skull that is not fused to the rest. The **auditory ossicles** (bones used for hearing) are found within the temporal bone; these 6 bones are the smallest of the 206 in the human body.

Cribriform plate
A fragile, porous area of the ethmoid bone at the superior portion of the nasal cavity.

The entire back of the skull is a single bone, called the **occipital** bone. An opening in this bone, called the **foramen magnum** (big hole), allows the spinal cord to extend from its protective cranium into the vertebral foramen.

Two cranial bones comprise the floor of the brain bucket, or cranial cavity. The **ethmoid** forms the floor of the front portion of the cranial cavity. It articulates with the frontal bone and a few bones of the face. The **cribriform plate** lies within the ethmoid. This unique sieve-like structure allows olfactory nerves to extend from the olfactory bulb of the brain into the mucous membrane of the nasal passageway.

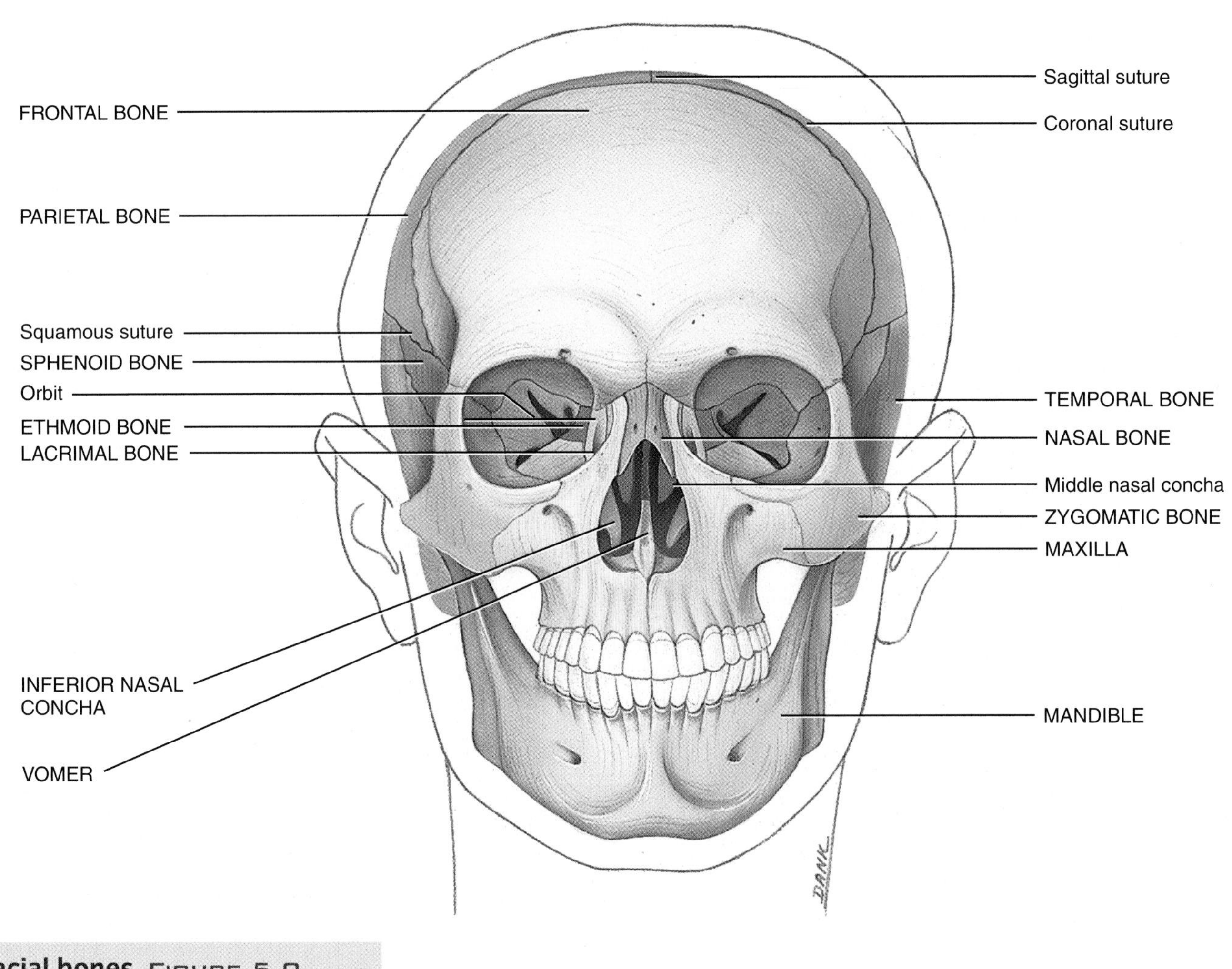

Facial bones FIGURE 5.9

Sella turcica (literally "Turk's saddle"), an oddly shaped depression in the sphenoid bone that holds the pituitary gland.

The final bone of the cranium, the **sphenoid**, articulates with all other cranial bones. The sphenoid provides the base for the cranium, supporting the brain. It is shaped somewhat like a bat, and includes the **sella turcica**, which completely encases the pituitary gland.

The 14 facial bones support the distinctive facial features we so closely associate with our identity. Anatomically, the facial bones protect the entrances to the respiratory and digestive systems, and the sensory organs (FIGURE 5.9).

Two facial bones are single, and 12 occur in pairs. The paired **maxillae** and **palatine** bones make up the front (maxillae) and roof of the mouth (the palatine bones). When these bones do not form properly, a cleft palate may result (FIGURE 5.10).

The small, thin, paired **nasal** bones form the bridge of the nose. In a self-defense course, you may have learned that it is possible to shove these bones up into the brain, through the cribriform plate of the ethmoid bone, by pounding upward on an attacker's nose—not recommended unless your life is in danger!

On either side of the nose are the small, paired **lacrimal** bones. The root of this word, *lacrima*, means tears. A small passage in these bones allows the tears to collect and pass through the skull into the nasal cavity.

Your cheekbones are one of the most memorable facial features, since they create the relief and depth of your face. These bones, the paired **zygomatic** bones, bulge outward and help protect the eyes. A blow to this bone can cause a black eye.

Within the nasal cavity lies the final pair of facial bones, the **inferior nasal conchae** (a conch is a snail with a helical shell). These bones form the swirling surface of the nasal cavity, helping to warm and moisten the air we inhale.

Fossa A pit, groove, or depression.

The **mandible**, the only bone of the skull attached by a movable joint, articulates with the mandibular **fossae** of the temporal bone at the temporomandibular joint (TMJ). Two **mental foramina**, small holes in the front of the mandible, connect with the **mandibular canal**, which provides passage for the blood vessels and nerves that supply the lower teeth. You may already be familiar with this canal, because it is the structure dentists seek when anesthetizing the sensory nerves of the lower teeth.

The single **hyoid** bone, which lies below the tongue, is the only bone of the skeleton that is not directly attached to any other bony structure. The hyoid bone is instead suspended by the throat muscles. This bone is of forensic interest because it can reveal death by strangulation; it is crushed only by pressure applied to the throat.

The **vomer** is the bony separation between nasal passages. A deviated septum, or broken nose, occurs when the cartilage that this bone supports shifts from its central location to block one passageway. Most deviated septa are caused by a blow to the soft tissues of the nose, as may occur during a boxing match. Surgery is required to reposition the cartilage, alleviating the breathing difficulty caused by the misaligned cartilage.

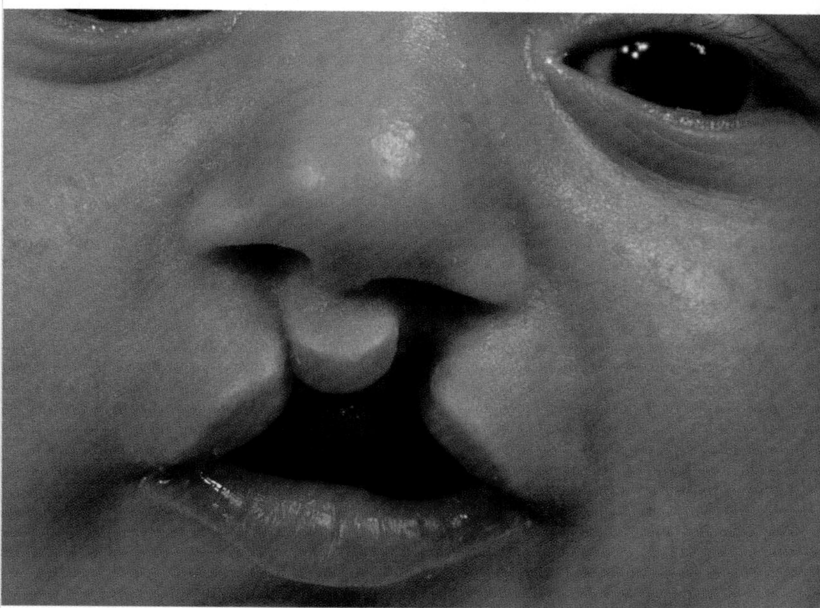

Cleft palate FIGURE 5.10

A cleft palate results if the maxillae and palatine bones do not completely fuse during development.

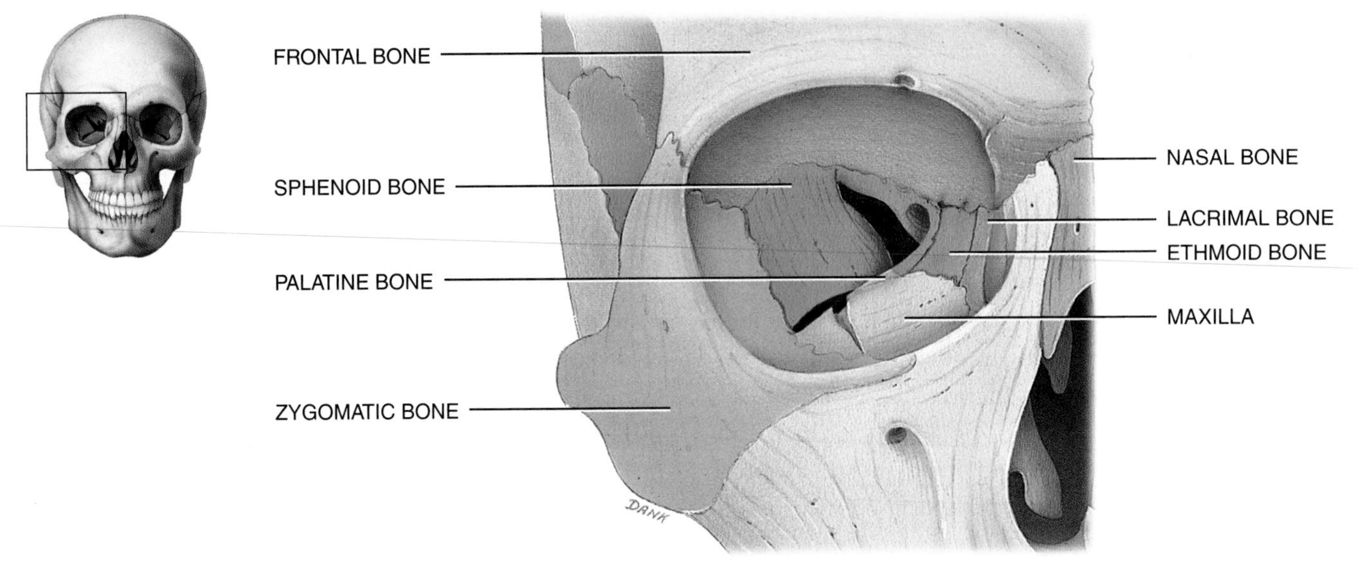

Anterior view showing the bones of the right orbit

Orbital complex FIGURE 5.11

The bony **orbital complex** (FIGURE 5.11) of the eye is composed of seven bones, three cranial and four facial bones, which together provide the bony support for the eye. These bones house the eye and provide attachment sites for the muscles that move the eye. The orbital opening is much larger than the eyeball. The excess space is filled with fat, protecting and cushioning the eye. When you run or otherwise jar your body, the eye literally bounces around in this protective fat!

The **nasal complex** (FIGURE 5.12) includes bones that encase the nasal cavities and the **paranasal sinuses**. These sinuses are air-filled chambers connected to the nasal cavity, lined with a membrane continuous with that of the nasal cavity. The purpose of this membrane is to warm, moisten, and filter the air we breathe. Small irritants will cause the membrane to produce excess mucus to wash the irritants away. Pepper is a fine example of this response. The sneeze is an attempt to literally blow the offending particles from the nasal cavity. If that fails, the membrane will secrete copious amounts of mucus. A bacterial infection, on the other hand, causes the membrane itself to swell. If the membrane of the nasal cavity becomes infected,

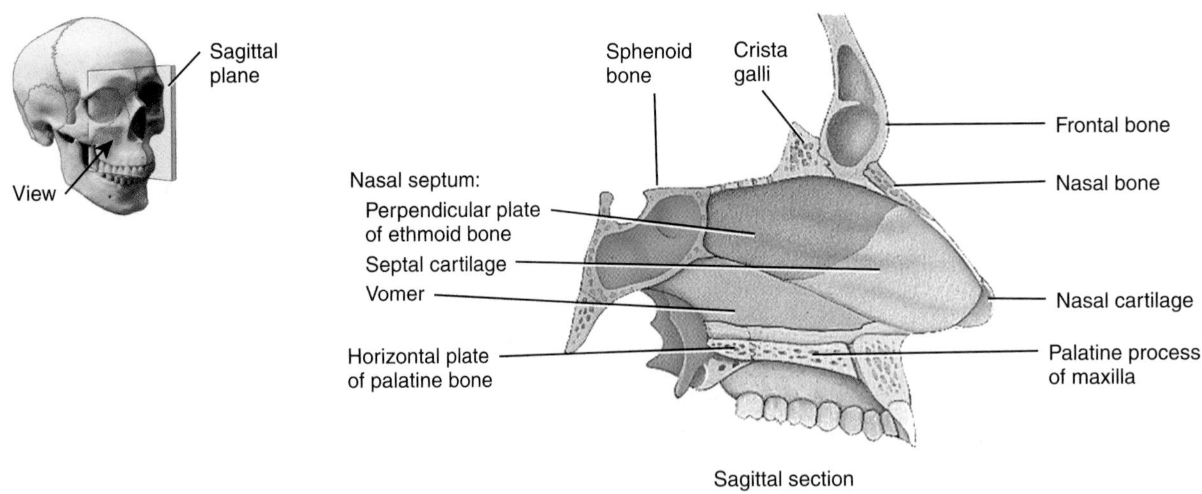

Sagittal section

Nasal complex FIGURE 5.12

the swelling can extend into the paranasal sinuses, closing the drainage opening of the nasal cavity. Pressure builds in the sealed cavity, causing pain and inducing us to run to the nearest drug store for some pharmaceutical help. Antihistamines such as pseudoephedrine, diphenhydramine chloride, and chlorpheniramine can all reduce swelling in these mucous membranes.

VERTEBRAE, RIBS, AND STERNUM FORM THE BALANCE OF THE AXIAL SKELETON

The remainder of the axial skeleton is composed of the **vertebrae** (FIGURE 5.13), **ribs**, and **sternum**. These bones allow upright posture and protect vital organs of the thoracic cavity.

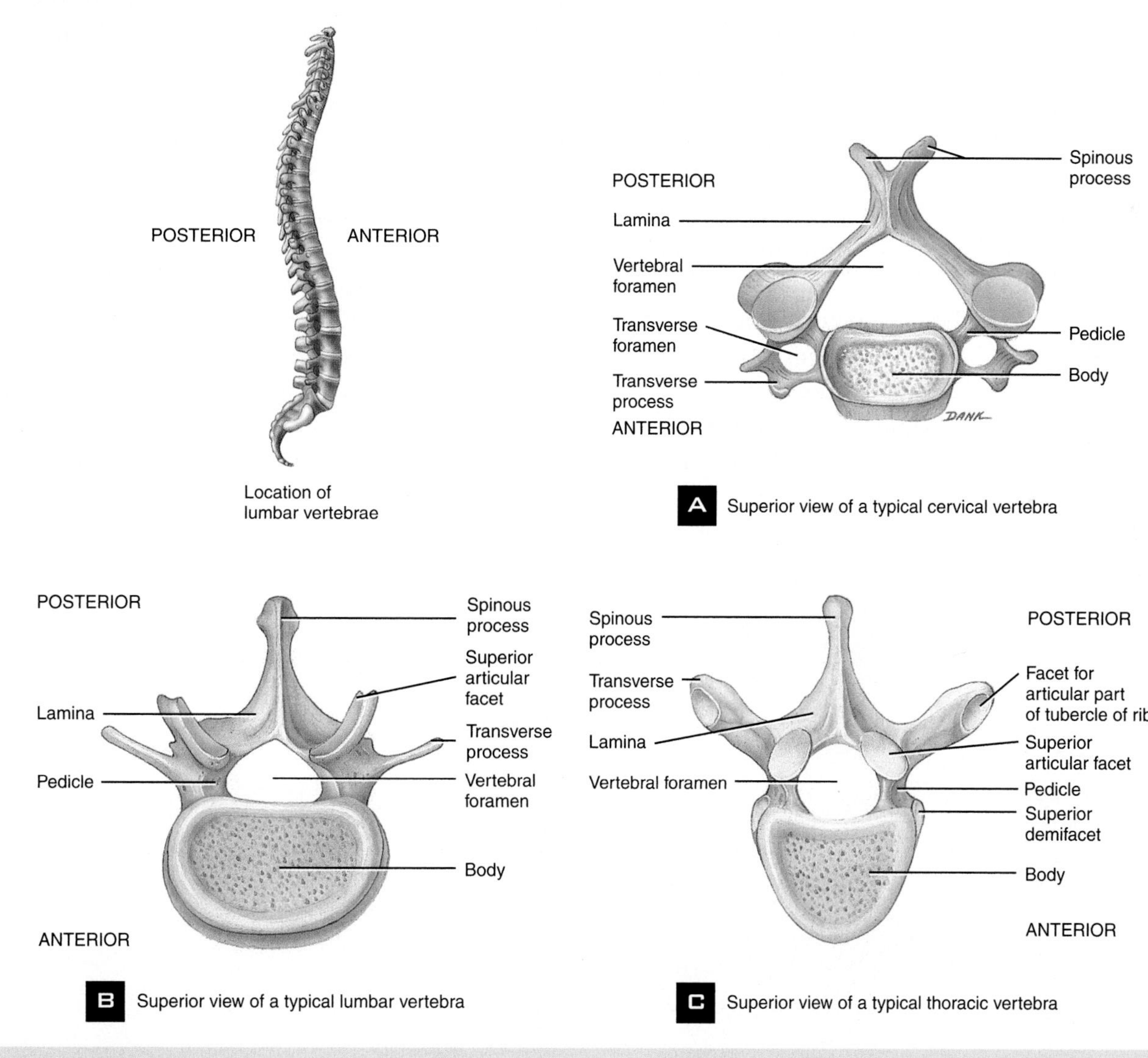

Vertebrae FIGURE 5.13

The parts of a vertebra include the vertebral body, the spinous process (the bumps that run down the middle of your back), the articulating surfaces that connect one vertebra to the next in your spinal column, and the vertebral foramen where the spinal cord lies. Cervical vertebrae (A) are thinner and more delicate than the rest of the vertebrae. Thoracic vertebrae (C) each articulate with a rib. Lumbar vertebrae (B) have heavy bodies capable of supporting the weight of the torso.

There are 24 vertebrae, one **sacrum**, and three to five **coccyx** bones in the adult vertebral column. A typical vertebra is composed of three parts: the **vertebral body**, the **vertebral arch**, and the **vertebral articular processes**. The articular processes serve as points of attachment between adjacent vertebra and sites for muscle attachment. The column is divided into the cervical region (vertebrae C1–C7), the thoracic region (T1–T12), and the lumbar region (L1–L5). Moving down the column, the bodies of the vertebrae grow larger, because they must support more weight. Between each vertebra is a pad of fibrocartilage called the intervertebral disc. The disc serves as a shock absorber, preventing vertebrae from rubbing against one another and crushing under the body's weight. These discs also allow limited motion between vertebrae.

The sacrum is actually five fused vertebrae that form a solid base for the **pelvic girdle**, with openings along their length for the exit of sacral nerves. The tailbone, or coccyx, is our post-anal tail. (As mammals, we must have a tail, although it is hardly obvious!) Our tail is made of three to five small bones that extend off the sacrum, completing the inner curve of the pelvis. In females, these bones are tilted further outward than in males, so they do not interfere during childbirth. Even so, some infants break their mother's coccyx during childbirth.

> **Pelvic girdle**
> The bones that connect the leg to the axial skeleton; the hip bones.

Osteoporosis, a disease that causes progressive bone weakening, often attacks the axial skeleton. The disease results from an imbalance in bone homeostasis, making bones fragile and less able to support weight, and increasing the chance of fracture. Painful vertebral fractures can cause a "dowager's hump" that can reduce height by several inches. Read the Ethics and Issues box on page 137 for more information on this disease.

Ribs attach to the thoracic vertebrae to form a structure sometimes referred to as the thoracic cage (FIGURE 5.14). We have seven pairs of **true ribs** and five pairs of **false ribs**. The true ribs attach directly to the sternum or make a direct connection with the costal (rib) cartilage, which in turn is directly associated with the sternum. False ribs either attach to the costal cartilage (ribs 8, 9, and 10), which then joins the sternum, or their lateral ends are free (sometimes called floating ribs 11 and 12). Despite what you may have heard, males and females have the same number of ribs.

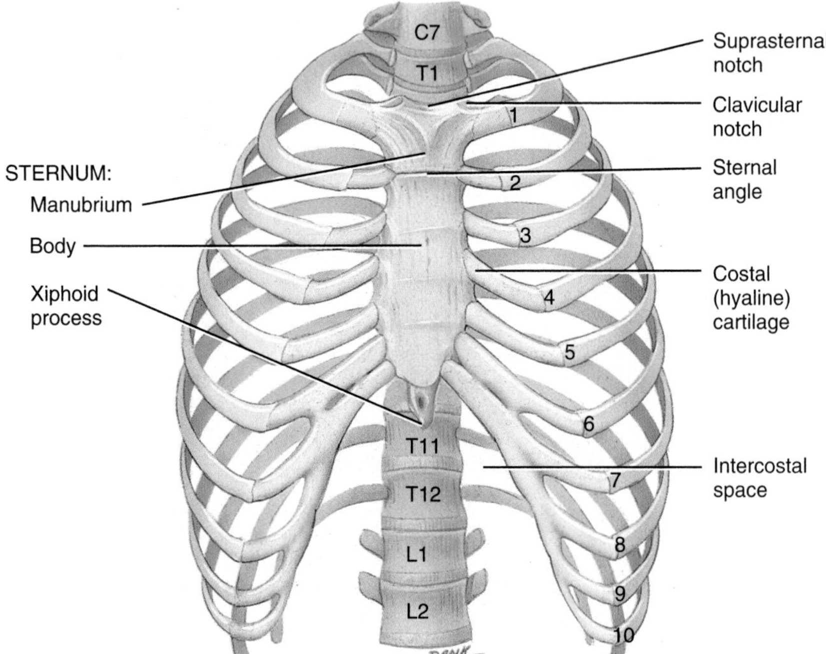

Anterior view of skeleton of thorax

Thoracic cage FIGURE 5.14

Ethics and Issues

How to confront the osteoporosis crisis

Although we tend to think of bones as solid and unchanging, the disease called osteoporosis proves otherwise. Bone normally exists in a state of balance, with the calcium-rich extracellular matrix being built up by osteoblasts and torn down by osteoclasts. Osteoporosis is a thinning of the bone structure caused by an imbalance between the processes of bone deposition and resorption.

Osteoporosis increases the risk for fractures in the hip, back, and wrists. Advanced osteoporosis can cause kyphosis (a humpback), severe pain, and shortened stature as the long bones and especially vertebrae are crushed under the body's weight.

Osteoporosis causes an estimated 1.5 million fractures a year in the United States, and the number is predicted to grow along with the number of older people in the population. Hip fractures, largely due to osteoporosis, are a major cause of death, disability, and loss of independence among the elderly. Within a year of a hip fracture, one-third of these patients spend time in a nursing home. Discovering how to slow, prevent, or even reverse this bone thinning ought to be a top priority.

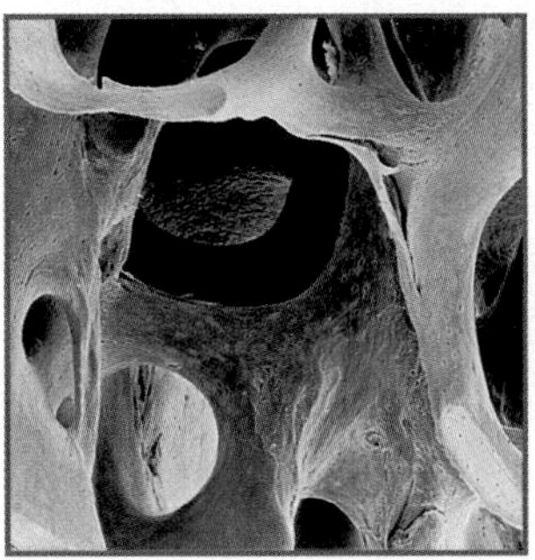

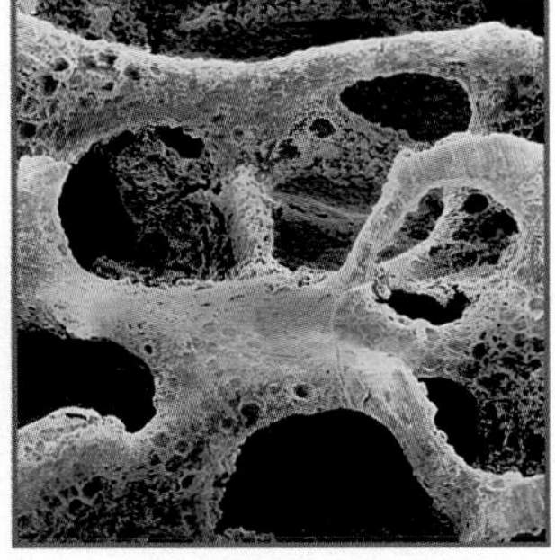

According to a recent scientific report, low bone density (osteopenia, an early sign of possible osteoporosis) affects 30 million women and 14 million men above age 50 in the United States alone. The risk is higher among aging women owing to the hormonal changes of menopause. Estrogen and testosterone both stimulate osteoblasts, but testosterone remains higher in aging men than estrogen does in aging women. Additionally, women also live longer on average, giving their bones more years to thin.

Other risk factors for osteoporosis are

- Asian or European ancestry
- Light stature ("small-boned")
- Smoking cigarettes
- Inactivity
- Lacking calcium and vitamin D in the diet
- Drinking caffeinated beverages, which speeds calcium release into the urine

Bone density can be measured with X-ray devices, and many doctors recommend baseline measurements, especially in women who are approaching menopause. Although osteoporosis cannot be cured, medicine, diet, and exercise can all help shift the balance away from resorption and toward bone mineralization. The drug bisphosphonate and the hormone calcitonin both inhibit osteoclasts. Although estrogen supplements assist bone mineralization, they also raise the risk of breast cancer and do not benefit the heart, which was a major goal of estrogen supplements after menopause. Both factors have reduced the popularity of hormone replacement for treating osteoporosis.

Increasing calcium intake seems like an obvious way to increase bone mineralization, but a large study recently found that calcium supplements were surprisingly ineffective. Vitamin D is essential to utilizing dietary calcium, but the effectiveness of vitamin D supplements is uncertain.

Perhaps the best recommendation is exercise. Impact and stress on the bones, as occurs with weight-lifting and running, triggers activity of the osteoblasts, helping mineralize the bones.

The sternum, or breastbone, protects the anterior of the chest. The three parts of the sternum are the **manubrium**, which articulates with the appendicular skeleton; the **body**; and a small tab of cartilage at the end of the body, the **xyphoid process**. The diaphragm and rectus abdominus muscles (the six-pack muscles so dramatically featured in body-building magazines) attach to the xyphoid process.

If you take a CPR course, you will be trained to locate the xyphoid process and avoid it as you depress the chest wall. Force can easily break the xyphoid process from the sternum, piercing the liver and causing life-threatening internal bleeding. This is NOT ideal if you want to save that life!

CONCEPT CHECK

Which cranial bone serves as the keystone of the cranium?

What is the overall function of the facial bones?

Why is the hyoid bone often studied in criminal investigations?

Differentiate between true ribs and false ribs.

List and describe the five types of vertebrae in the vertebral column.

Your Limbs Comprise Your Appendicular Skeleton

LEARNING OBJECTIVES

List and Identify the bones of the appendicular skeleton.

Understand the characteristics of the pectoral and pelvic girdles.

The appendicular skeleton (FIGURE 5.15) includes all the bones that are attached, or appended, to the axial skeleton. Specifically, it includes the **pectoral girdle**, the upper appendages (arms and hands), the pelvic girdle, and the lower appendages (legs and feet).

Pectoral girdle The bones that attach the arm to the axial skeleton; the shoulder bones.

The clavicle is the bone that most commonly breaks in car or bicycle accidents. To stop their fall, most people naturally respond by using their hands for protection, which transfers the shock

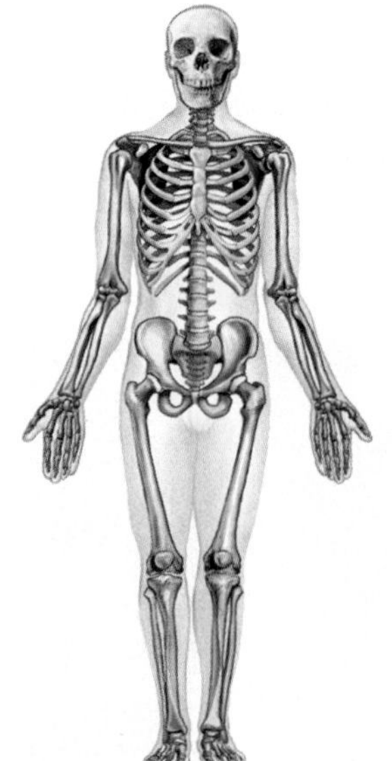

Appendicular skeleton FIGURE 5.15

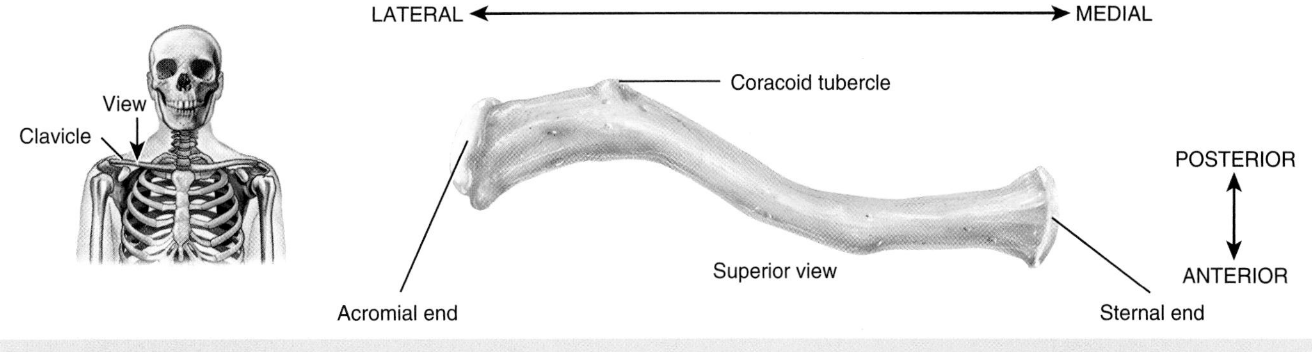

Clavicle FIGURE 5.16

of landing up the strong arm bones, concentrating it on the **clavicle** (FIGURE 5.16). This pressure is generally opposite the strong axis of the clavicle, which breaks the bone.

The **scapulae** (the singular form is *scapula*) are the "chicken wings" on your back. These bones connect to the strong back muscles and articulate only with the clavicles, which gives each shoulder joint greater range of motion (FIGURE 5.17).

The **humerus** is the longest and strongest bone in the upper appendicular skeleton. The anatomical neck of the humerus is actually quite thick and strong. The surgical neck is the thinner area of the humerus distal to the neck, where the musculature of the arm

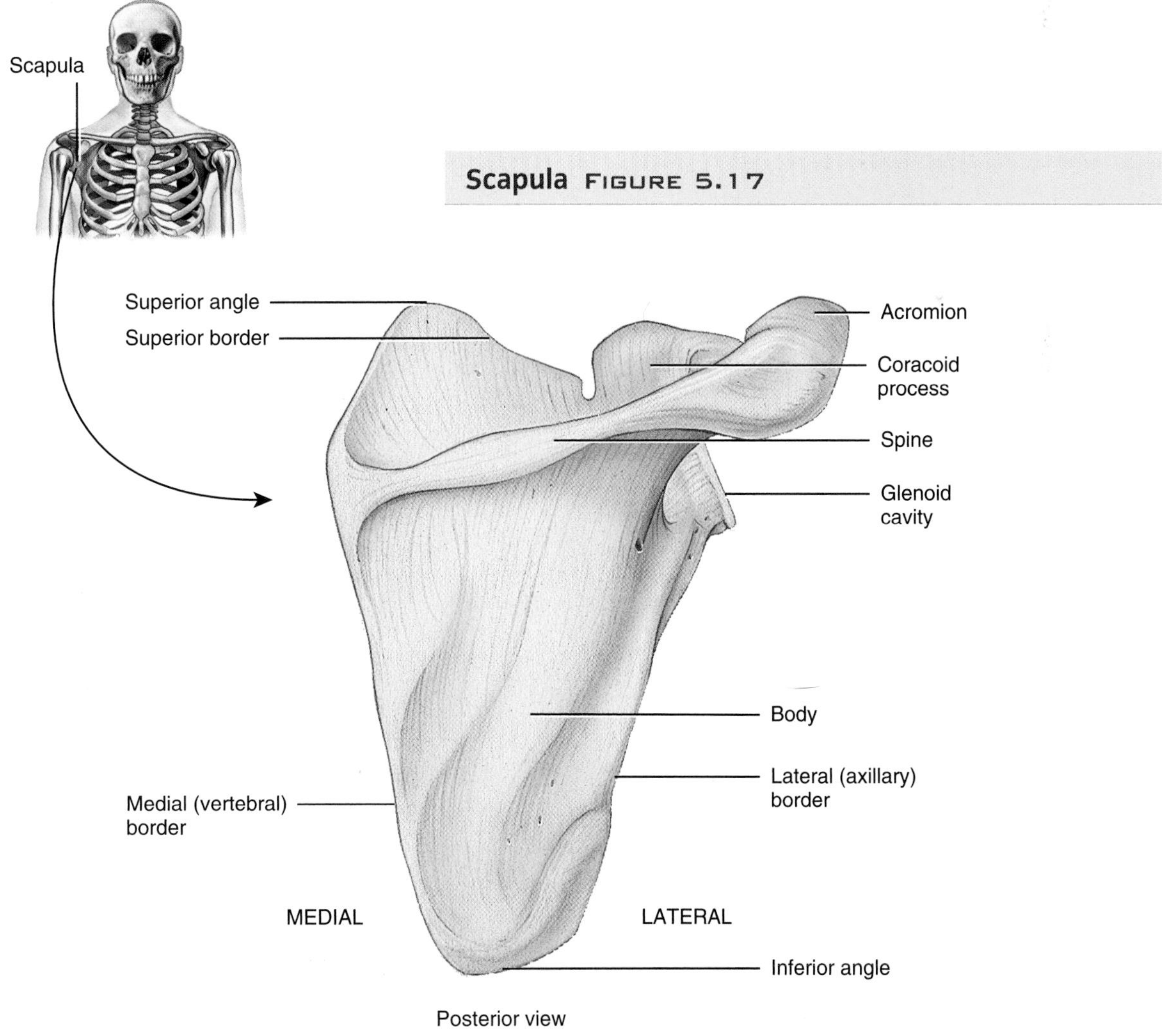

Scapula FIGURE 5.17

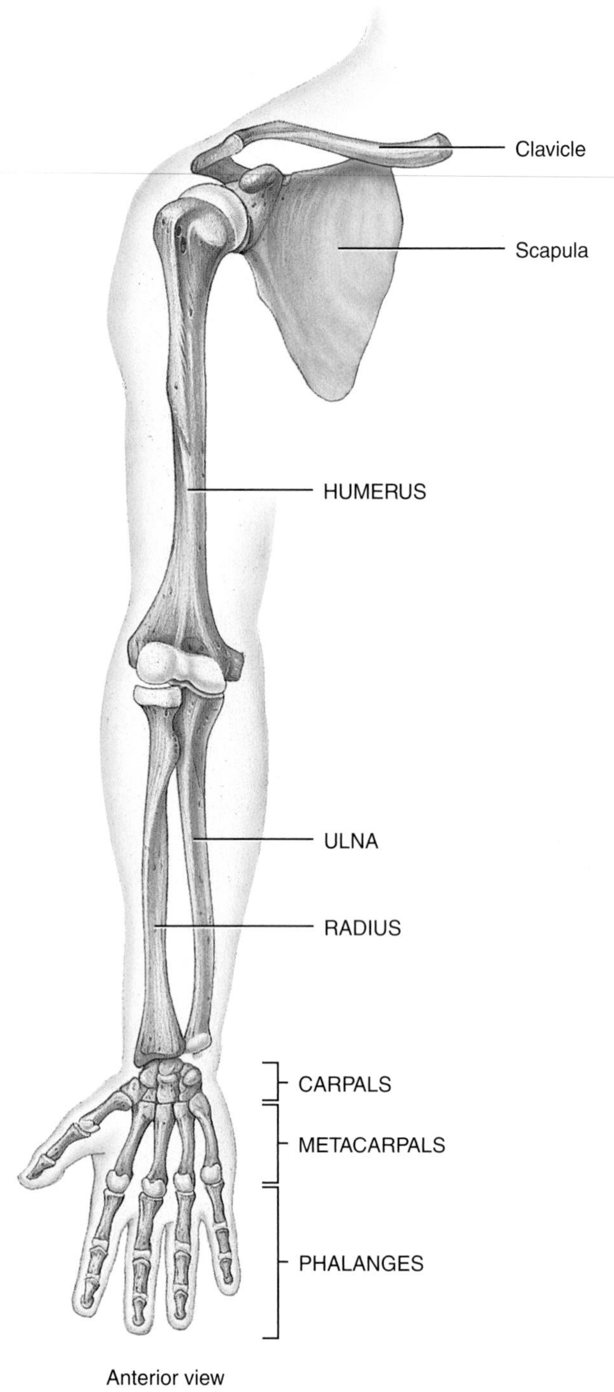

Right upper limb FIGURE 5.18

Each upper limb includes a humerus, ulna, radius, carpals, metacarpals, and phalanges.

does not cover the humerus well. You can feel this area by running your hand approximately one third of the way down the arm, until you feel your unprotected bone. Most breaks to the humerus occur at the surgical neck rather than the anatomical neck (FIGURE 5.18). Distal to the humerus is a pair of bones in an area commonly known as the forearm. The **ulna** is on the medial side of the forearm, the same side as your little finger, and is the longer of the two bones. The **radius** is on the thumb side of the forearm. One way to learn this arrangement is to memorize the mnemonic "p.u." (the **p**inky is on the **u**lna side). The *elbow* is the joint formed by the distal end of the humerus and the proximal ends of the radius and ulna; a large projection of the ulna called the *olecranon* forms the point of the elbow. At the other end of the forearm, the radius is in more direct contact with the next set of upper limb bones, the carpals.

The wrist bones **(carpals)** are in two rows of four short bones. The **scaphoid** bone, **lunate** bone, **triquetrum**, and **pisiform** bone make up the proximal row. The distal row of carpal bones includes the **trapezium**, **trapezoid** bone, **capitate** bone, and **hamate** bone. All eight of these bones are bound by a large tendon on the palmar side of the hand. With heavy use, the tendons passing between the carpals and this tendon can swell. Because there is no extra room, this compresses the nerves. This "carpal tunnel syndrome" causes pain and numbness in the hand.

The **metacarpals** make up the structure of the hand. If you make a fist, the distal tips of the metacarpals are those protruding knuckles. A "boxer's fracture" is a shearing of the distal end of a metacarpal, which makes the knuckle recede (FIGURE 5.19).

The **phalanges**—finger bones—are considered long bones. Each finger has three bones: the proximal, middle, and distal phalanx. The thumb (pollex) has only two phalanges. With excessive writing a small sesamoid bone can develop in the tendon of the thumb because the tendon rubs over the joint between the proximal phalanx and metacarpal.

In the anatomical position, the phalanges of the hand reach below the beginning of the lower limb. The lower limb, or leg, originates at the pelvic girdle. This girdle, composed of the hip bones and lower vertebrae, is much denser, stronger, and less flexible than

A Anterior view

B Posterior view

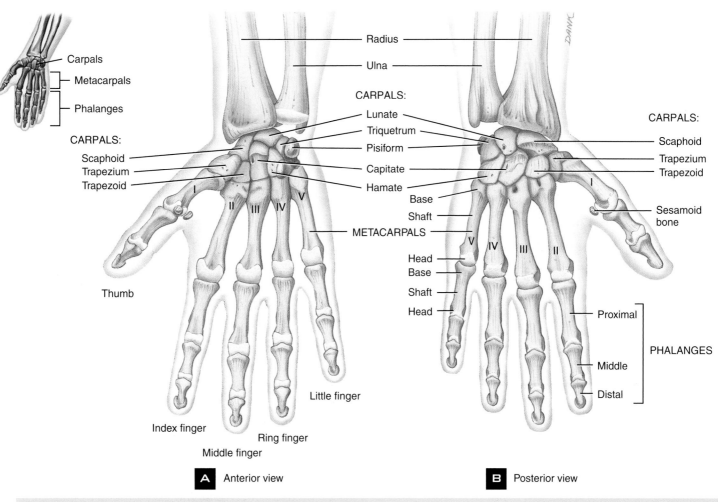

Carpals, metacarpals, and phalanges FIGURE 5.19

the appendicular girdle. The **os coxa** (hip bone) emerges from three bones that fuse in early puberty: the ilium, ischium, and pubic bone. The femur articulates at the junction of these three bones. The **acetabulum** is the curved recess that serves as a socket for the head of the femur (FIGURE 5.20).

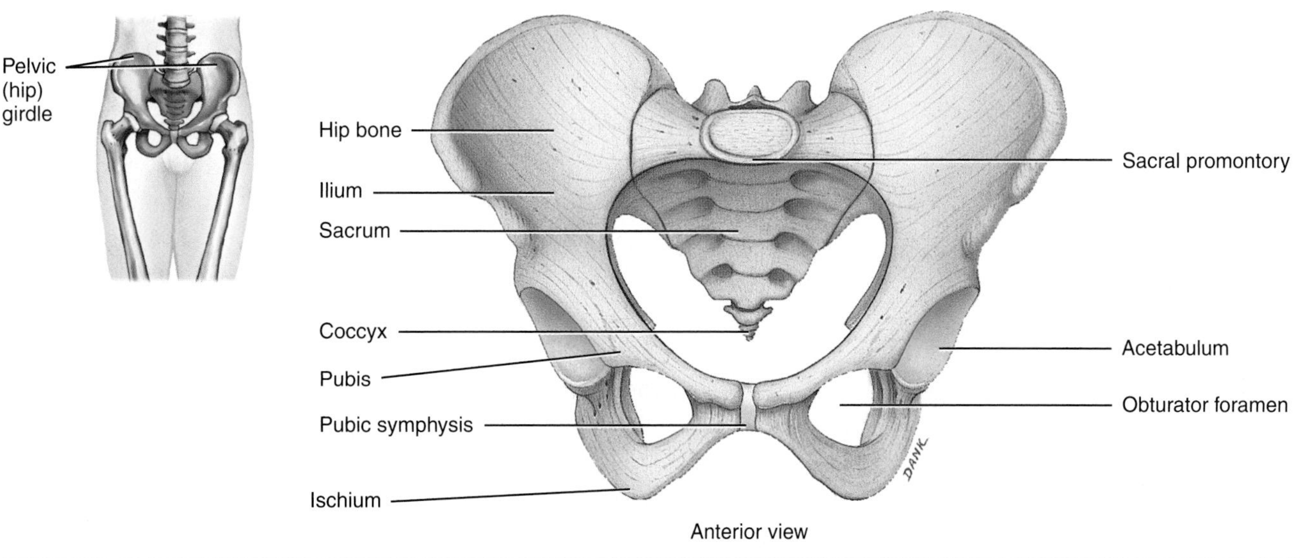

Pelvis FIGURE 5.20

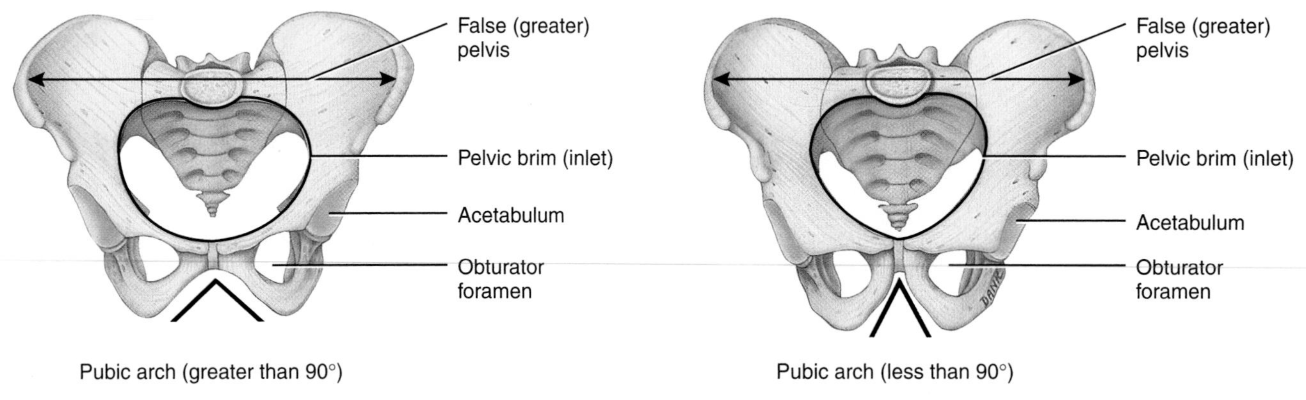

Female versus male pelvis FIGURE 5.21

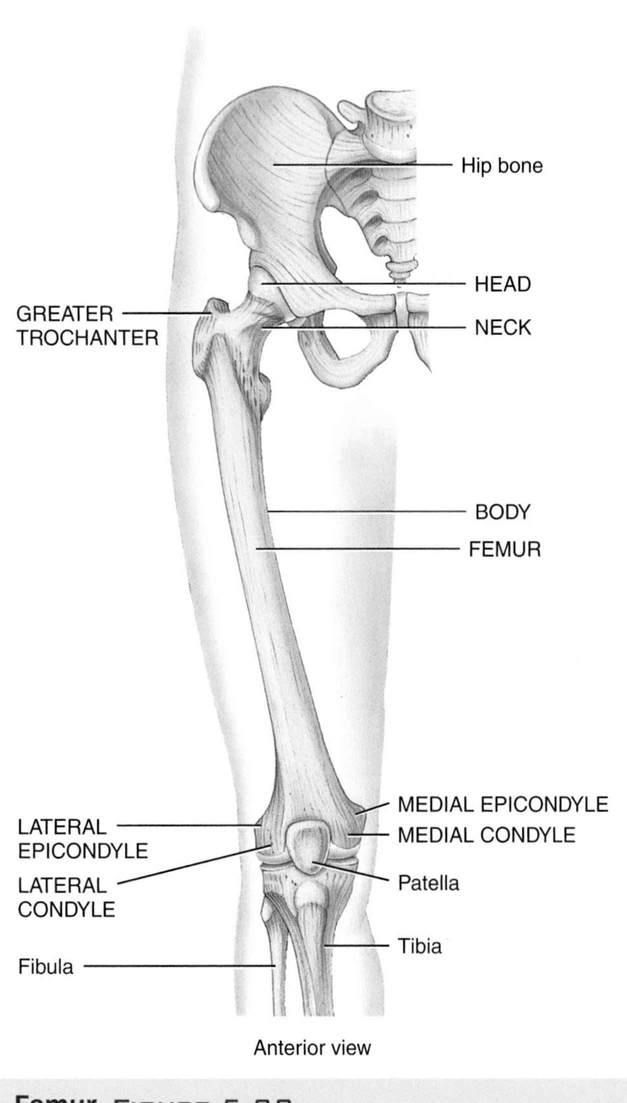

Femur FIGURE 5.22

The pelvis is technically made of two os coxae, plus the sacrum and the coccyx. Between each of the two os coxae is a pad of fibrocartilage called the symphysis pubis, which serves the same purpose as the intervertebral discs. Each os coxa articulates with the sacrum posteriorly. The sacroiliac joint, made famous by a comedy team from vaudeville and early television called The Three Stooges ("Oh my aching sacroiliac!"), lies between the sacrum and the ilium.

Male and female hip bones are visibly different (FIGURE 5.21). Female hip bones are shallower, broader, and more dished, and have an enlarged pelvic outlet, a wider, more circular pelvic inlet, and a broader pubic angle. Each of these modifications eases childbirth by enlarging or smoothing the portion of the birth canal in the pelvis. Sadly, these modifications also change the angle of attachment of the female femur. This slight shift alters the position of the knee joint, leading to a knock-kneed appearance and increasing the chance of knee and ankle injuries among women athletes. (See the I Wonder . . . feature "How do forensic scientists learn about someone's movement from the bones?" on p. 144 for more on what can be learned from analyzing a skeleton.)

The femur is the longest and heaviest bone of the body (FIGURE 5.22). The fovea capitus, a ligament, lies inside the hip joint capsule and connects the head of the femur to the acetabulum. This is the only ligament that lies completely within a joint, probably to improve stability. The neck of the femur joins the shaft at a 125° angle, putting huge stress on the neck. This makes it susceptible to breaking as bones thin and weaken with

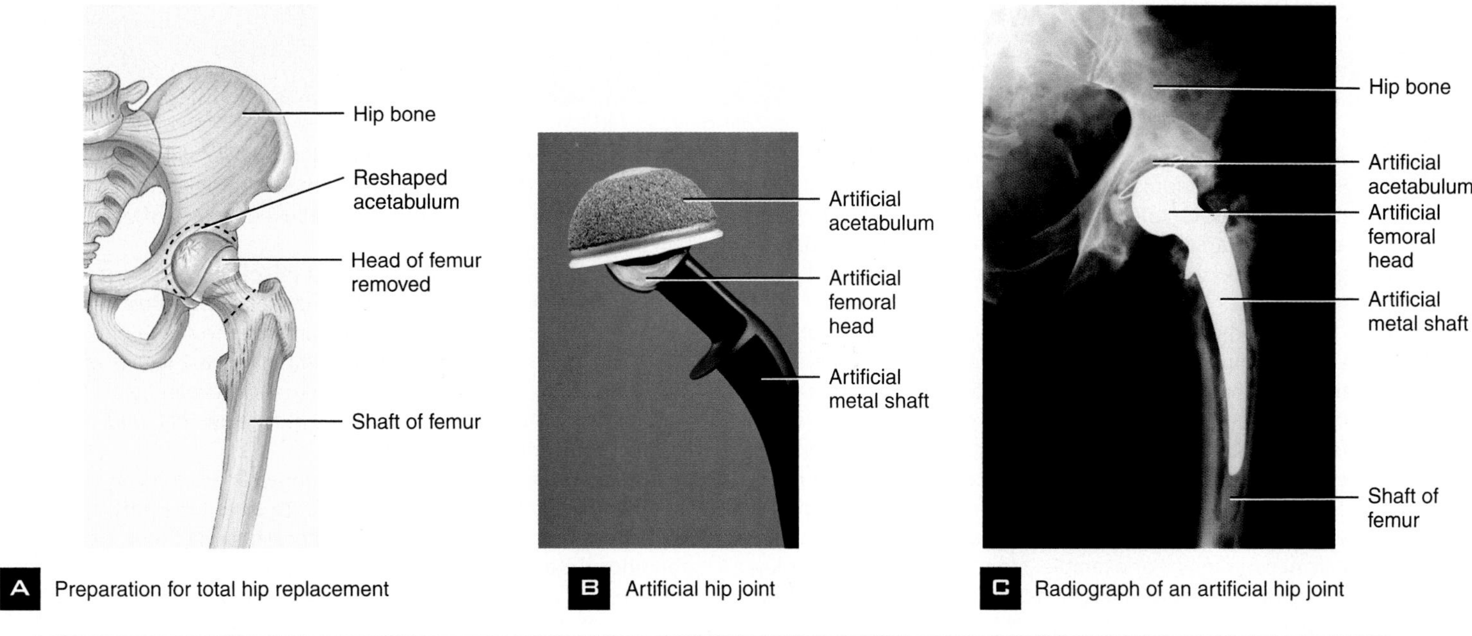

A Preparation for total hip replacement

B Artificial hip joint

C Radiograph of an artificial hip joint

Total hip replacement FIGURE 5.23

age. A total hip replacement is a surgical procedure that replaces the head of the femur, the femoral neck, and a portion of the femoral shaft with metal parts. Hip replacements are common in elderly women suffering from osteoporosis. In younger patients, hip replacements have grown rarer, now that resurfacing of the hip joint can improve hip movement without the incapacitating hip replacement surgery (FIGURE 5.23).

The patella, or kneecap, forms within the tendon of the quadriceps femoris, the powerful muscle that straightens the knee. Interestingly, although the patella is counted among the 206 bones, humans are born with a cartilaginous blob for a kneecap. The bone develops as the large tendon associated with the anterior thigh rubs across the distal end of the femur and the proximal end of the tibia. Ossification centers in the patella respond to this friction by laying down a bony matrix. The patella shows up by age 2 in most females and at age 3 to 5 in males (FIGURE 5.24).

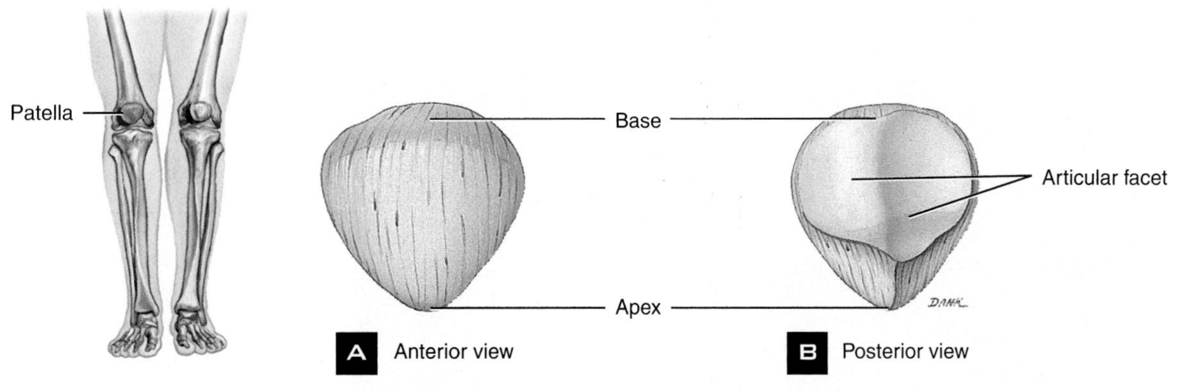

A Anterior view

B Posterior view

Patella FIGURE 5.24

I WONDER . . .

How do forensic scientists learn about someone's movement from the bones?

Almost every episode of the television series *CSI* (Crime Scene Investigation) seems to mention skeletal identification. Scientific technologies now available allow us to glean an amazing amount of information from bones, including gender, age, physical conditioning, and even diet and homeland.

You may have heard people described as "big boned." Whereas most bones vary little in shape from one person to the next, they do differ in size and density. One femur looks like another, but there are individual differences between femurs, just as there are between faces. Weight training, exercise, and hard work increase bone mass because bone responds to stress by growing stronger. Extremely heavy wear in certain locations on the skeleton is evidence of specific occupations: For example, stone masons and brick layers used to develop enlarged acromion processes on their scapulae as a result of continually carrying loads on the shoulder.

Anthropologists and forensic scientists also look at the size and ruggedness of muscle attachment scars, which form where muscles pull on bones. Bone placed under severe muscular stress becomes stronger and thicker at the attachment site, so larger muscle scars suggest larger, stronger muscles. Usually, males have larger muscles and larger bones, so heavier bones with pronounced muscle scars suggest a male skeleton. A second indication of sex is the pelvic girdle. Because of the demands of childbirth, women's pelvic bones tend to be wider than men's.

Skeletal remains also suggest the age at death, based on maturity of the bones (skull bones, for example, do not fuse until age 1 or 2). And of course skeletons reveal height and suggest body weight, helping to indirectly indicate general health.

Teeth, a form of bone, can also suggest both overall health and diet. Heavy wear on the molars can indicate a diet high in fiber-rich, uncooked food. Massive cavities indicate a diet rich in sugar or simple carbohydrates. Deformities in the legs may indicate rickets, caused by lack of vitamin D in the diet.

Bones can carry evidence of past infections, accident, or foul play. Healed breaks can show evidence of a violent past, whereas unhealed breaks can suggest the cause of death and the timing of the break—within the last weeks of life. Knife marks on the long bones may show evidence of cannibalism or foul play; tooth marks may show that wild animals consumed the remains.

Isotopic analysis of elements in bones and teeth can identify the origin of the specimen. Teeth mineralize early in life, so the isotopes they contain represent childhood conditions. Bones continually remineralize, so bone isotopes tell us about the more recent environment. Isotopes in the teeth indicated that the Iceman, a mummified man found in 1991 in a glacier in the Alps who died about 5,200 years ago, had been born in Central or Southern Europe. Isotopes in the bones suggested that he had probably spent part of his summers at high altitude. He was probably herding animals, which is how people have made a living in the summer in the Alps for 6,000 years.

Forensic scientists have derived DNA from bone marrow to identify victims of mass murders in Bosnia and South America. This "DNA fingerprint" identification, combined with other evidence from skeletal remains, can help relatives achieve some closure after massacres or political "disappearances." It can also help prosecutors gain convictions of the guilty parties.

In cases where bodies cannot be identified, computer reconstruction of facial bones can produce a "mask," an informed estimate of the person's appearance. In this way, forensic scientists can literally "give a face to the dead."

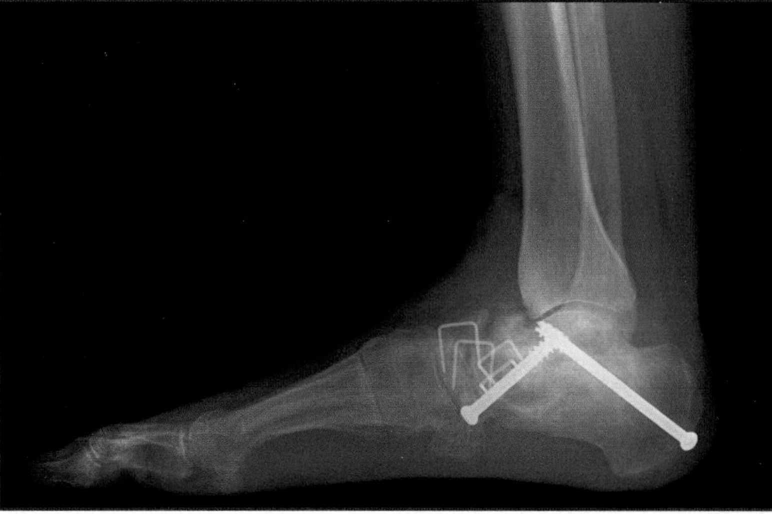

The ankle bones of this person will be easy to identify for many years to come. The scars left from the pins and staples will remain visible to the forensic scientist long after the injury is forgotten by the victim.

The tibia lies closest of any bone to the exterior, covered only with skin and periosteum at the tibial crest. The tibial tuberosity is the bump that you can feel on the proximal anterior surface of the tibia. Here the patellar tendon attaches to the leg. The medial ankle bump is actually part of the tibia, called the medial malleolus (FIGURE 5.25).

The fibula is a small, nonweight-bearing bone in the lower leg that is an important site for muscular attachment. The fibula is bound tightly to the tibia by the **interosseus** membrane. The lateral ankle bump is part of the fibula. This bone is often used as a source of bone tissue for grafting because part can be removed with little loss of function (FIGURE 5.26).

Interosseus
Between bones.

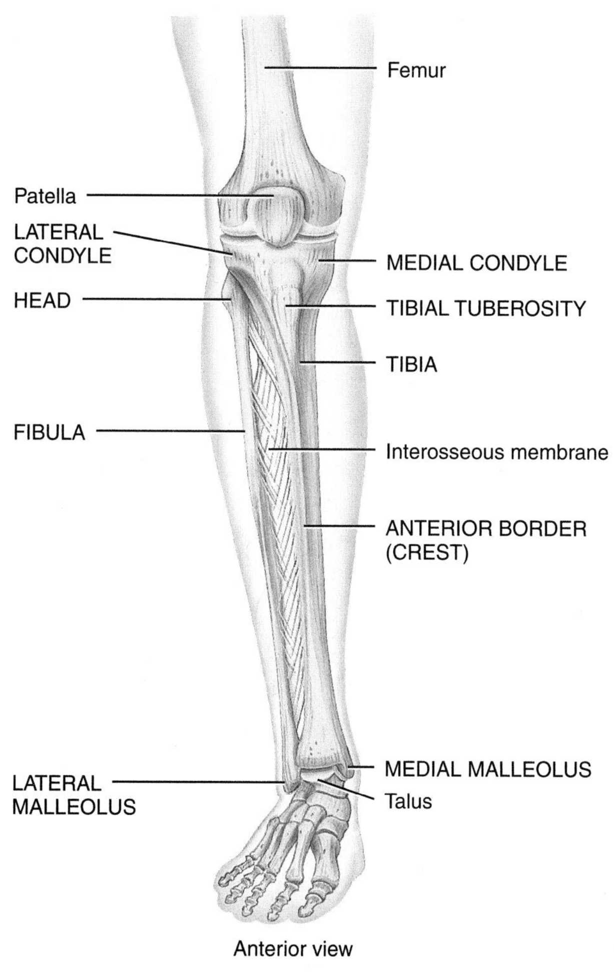

Tibia FIGURE 5.25

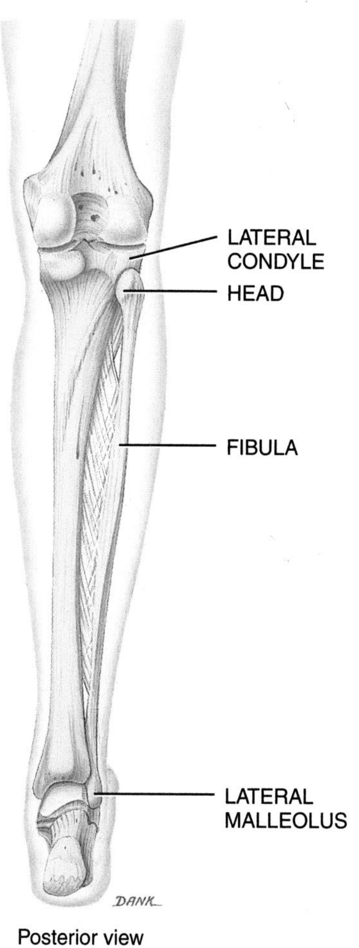

Fibula FIGURE 5.26

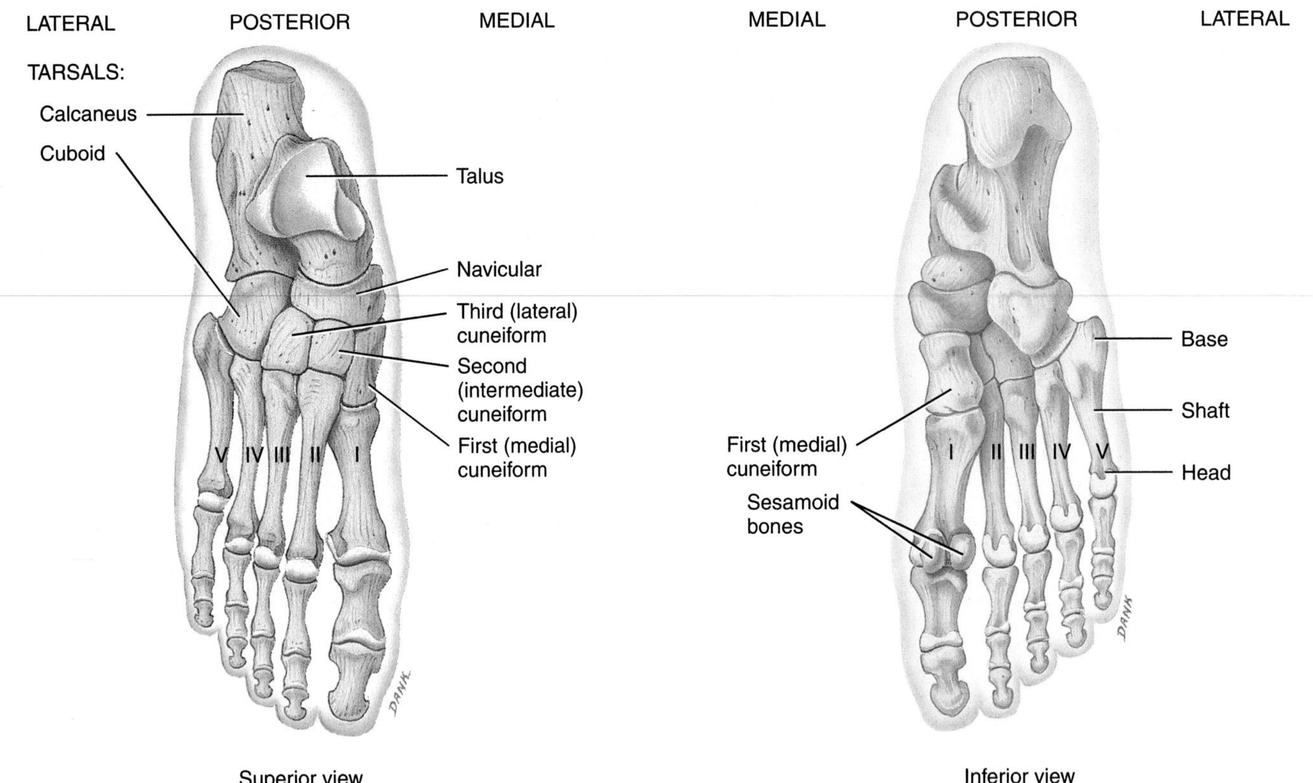

Tarsals FIGURE 5.27

Metatarsals FIGURE 5.28

Seven tarsal bones in the ankle transfer weight from the leg to the foot. The **talus** articulates with the tibia and takes the entire force of each step before transferring it to the rest of the ankle bones. The **calcaneus** (heel bone) is the largest of the tarsal bones. During walking, weight transfer progresses from the tibia to the talus to the calcaneus, then to the ground. The **calcaneus (Achilles) tendon** runs from the muscles of the leg to the projection on the back of the calcaneus. The remaining bones of the ankle include the **navicular** bone, the **cuboid** bone (so named because it is quite cube-like!) and the third **(lateral)**, second **(intermediate)**, and first **(medial) cuneiform** bones (FIGURE 5.27).

The metatarsal bones, which comprise the body of the foot, are comparable to the metacarpals in the hand (FIGURE 5.28). The phalanges in the toes are

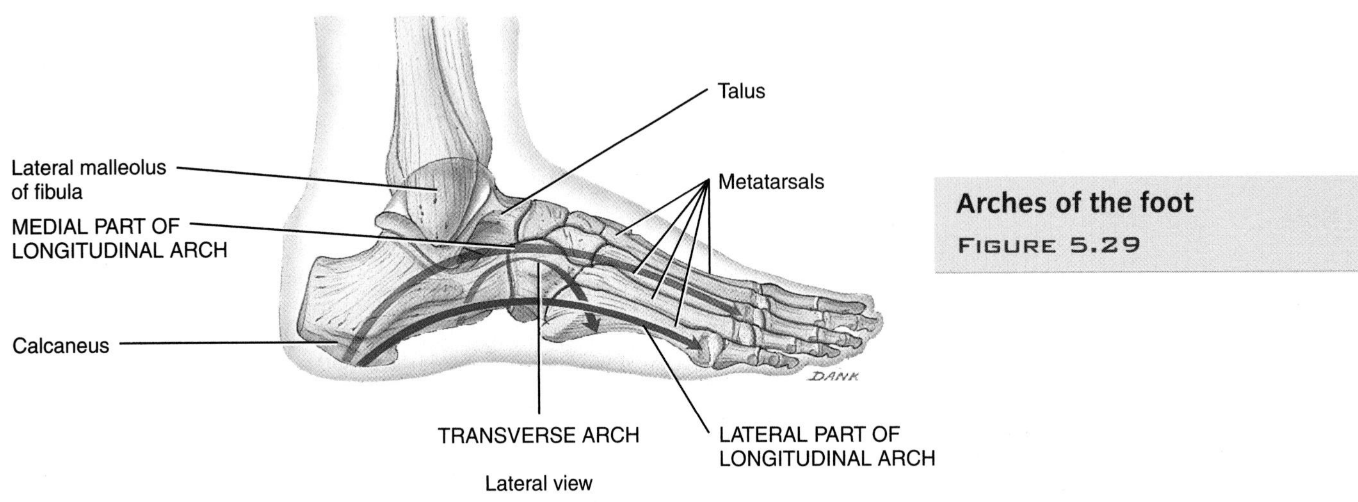

Arches of the foot
FIGURE 5.29

Ligament Dense regular connective tissue connecting bone to bone.

Tendon Dense regular connective tissue connecting muscle and bone.

similar to those of the fingers, but shorter. The great toe is called the **hallux**.

The **arches of the foot** (FIGURE 5.29) are maintained by **ligaments** and **tendons** surrounding the metatarsals and the tarsals. The longitudinal arch transfers weight from heel to toe along the metatarsals. The transverse arch is the portion of the foot that shoe manufacturers call the arch. The greater the curvature of this arch, the greater the flexibility it affords the foot.

CONCEPT CHECK

Which bones make up the pectoral girdle? Which ones combine to form the pelvic girdle?

List three differences between male and female pelvic girdles.

Which bones of the leg, ankle, and foot are responsible for transferring your weight to the ground with every step?

Joints Link the Skeletal System Together

LEARNING OBJECTIVES

Discuss the different types of joints, classified by structure and their particular movement.

Define the typical movements of diarthrotic joints.

The skeletal system provides internal scaffolding from which the skin, muscles, and organs are suspended. The skeleton, however, must not only support and protect, but also flex and move. This is accomplished by the joints of the body, which exist wherever two bones meet. These joints can be classified by function or by structure. Functionally, joints are immovable (**synarthrotic**) (FIGURE 5.30), semimovable (**amphiarthrotic**), or freely movable (**diarthrotic,** or **synovial**). Structurally, a joint is considered a bony fusion,

Synarthrotic joint FIGURE 5.30

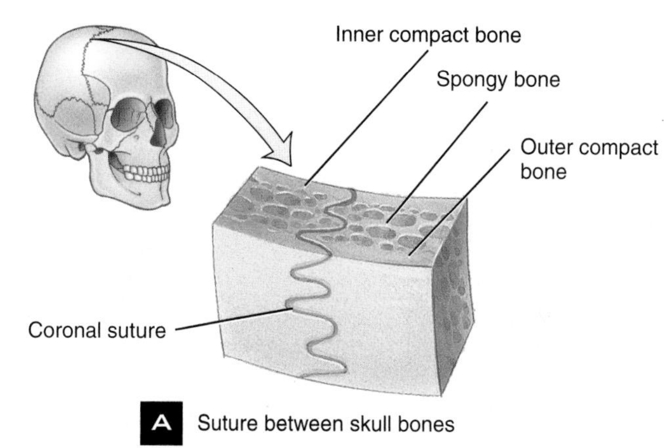

A Suture between skull bones

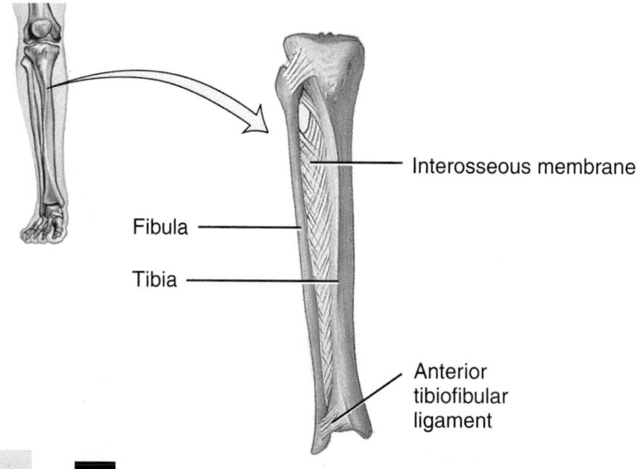

B Syndesmoses between tibia and fibula

or a fibrous, cartilaginous, or synovial joint. The term *synovial* is confusing, however, because it describes both the function and structure of a movable joint. "Synovial" describes the fluid in the joint (structure) and any structure that secretes synovial fluid (function).

Bones in synarthrotic joints are fused together, so movement is impossible. The fibrous connections holding the teeth in the jaw are synarthrotic joints. Others are found on the skull: The **coronal** suture borders the frontal and parietal bones (where a crown would be worn). The **sagittal** suture lies along the midsagittal plane. The **lambdoidal** suture outlines the occipital bone, and the **squamous** suture follows the fusion between the temporal and parietal bones on each side of the head.

Amphiarthrotic joints allow some movement. In intervertebral joints, a fibrocartilage pad (the intervertebral disc) lies between adjacent vertebrae, allowing limited flexing movements (FIGURE 5.31). The disc is composed of an outer fibrous ring, the **annulus fibrosus**, and an inner pulp, the **nucleus pulposus**. Many people suffer back trouble caused by a herniated disc, often called a slipped or ruptured disc. A rupture in the annulus fibrosus allows the nucleus pulposus to escape and press on a nerve, causing extreme pain in the entire region that the nerve innervates. Another type of amphiarthrotic joint occurs between the tibia and the fibula, which are connected by tight ligaments of the interosseus membrane. These ligaments bind the two bones so the tibia can move only slightly with respect to the fibula.

Diarthrotic, or synovial, joints are the most common joints. These joints serve as the fulcrum of a lever, so the force generated by contracting muscle can move a load. Synovial joints allow free movement between two bones. These joints are characterized by a complex joint structure bounded by a joint capsule containing **synovial fluid**. Tendons, ligaments, **bursae**, and **menisci** are often associated with synovial joints (FIGURE 5.32). Accessory ligaments outside the joint help to stabilize and reinforce

Synovial fluid Fluid secreted by the inner membrane of a synovial joint, similar in viscosity to egg white.

Bursae Fluid-filled sac between the bones or tendons of a joint and the skin positioned to reduce friction.

Menisci (singular: *meniscus*) Fat pads within joints that cushion bones and assist in "fit."

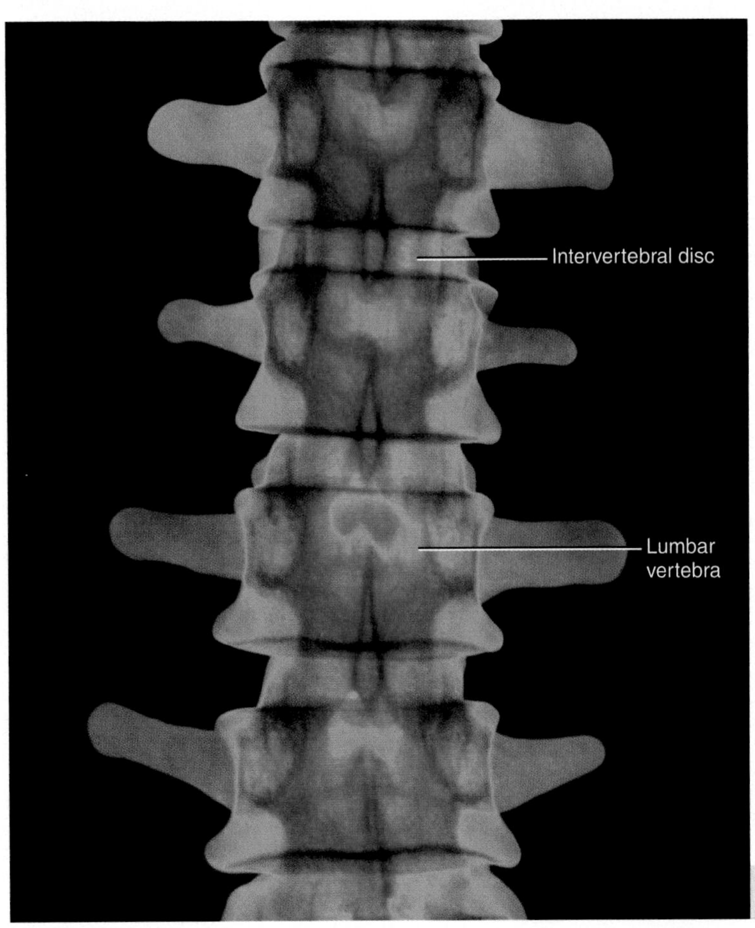

Spine FIGURE 5.31

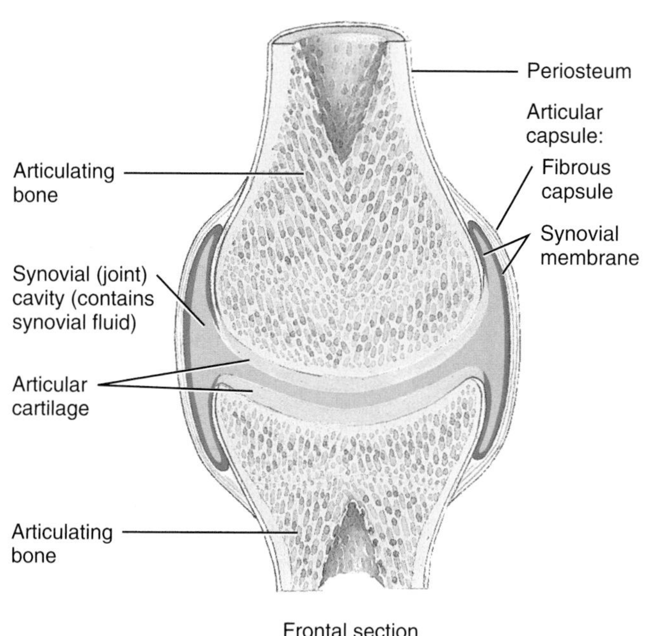

Typical synovial joint FIGURE 5.32

the joint capsule. Some joints, like the hip and shoulder, even have ligaments inside the joint capsule. In the knee, the **anterior** and **posterior cruciate ligaments** are inside the joint capsule.

When a joint moves, so do the overlying tissues. To reduce friction and absorb shock from this movement, fluid-filled sacs called bursae are found in the connective tissue surrounding many diarthrotic joints. These sacs can be damaged, resulting in inflammation of the bursae. Bursitis, as this is called, is usually attributed to severe, repetitive motion at a joint. Throwing a baseball, swinging a golf club, or playing tennis are all common causes of bursitis of the shoulder. Kneeling for long periods can cause "water on the knee," an inflammation of the bursa around the patella.

Another supportive structure associated with synovial joints is a meniscus, or fat pad. This structure can subdivide the synovial cavity, channel the flow of synovial fluid, or improve the "fit" between the bones and the joint capsule. The medial and lateral menisci of the knee help stabilize the knee and provide lateral support (FIGURE 5.33). These are commonly injured in side impacts in games such as football and rugby.

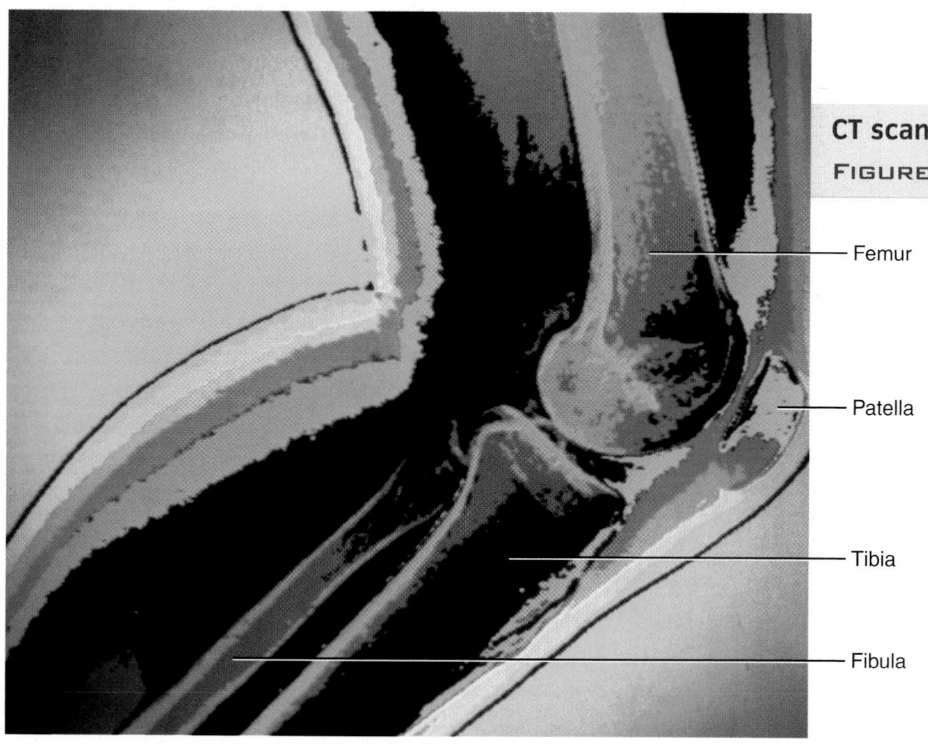

CT scan of knee in sagittal section
FIGURE 5.33

Movement Figure 5.34

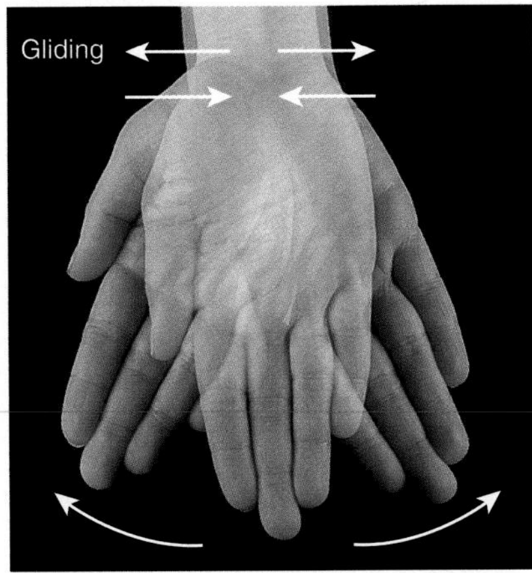

A Gliding motion

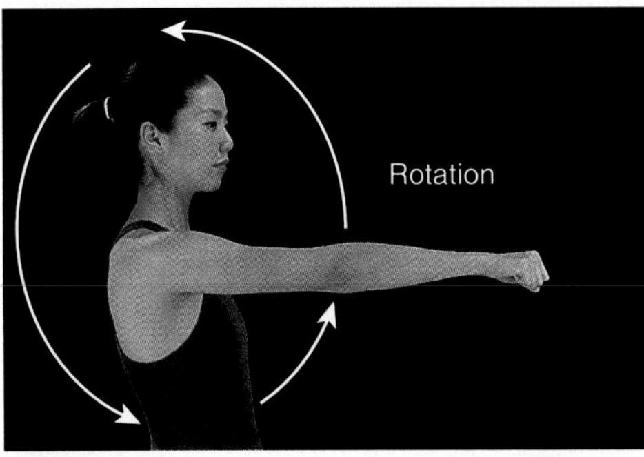

B Rotation

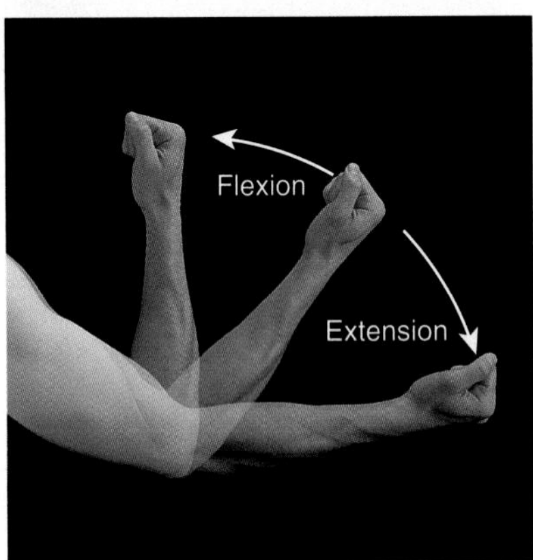

C Flexion and extension

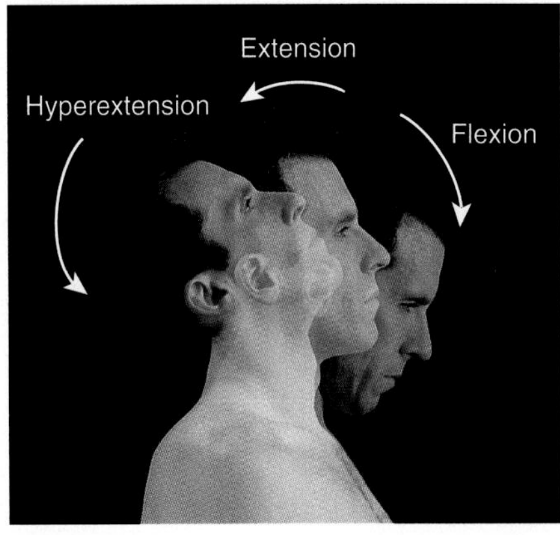

D Flexion and hyperextension

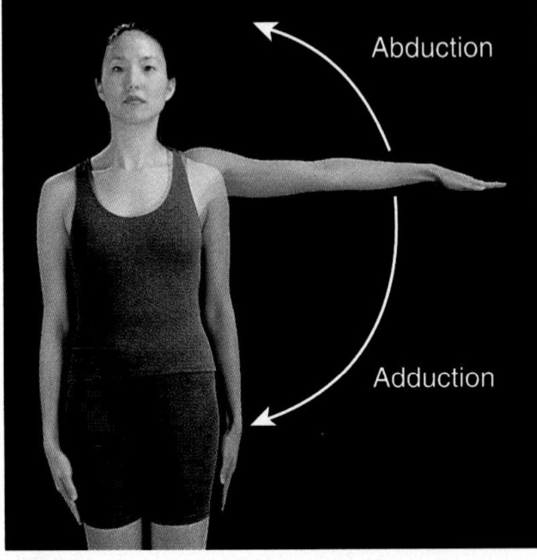

E Adduction and abduction

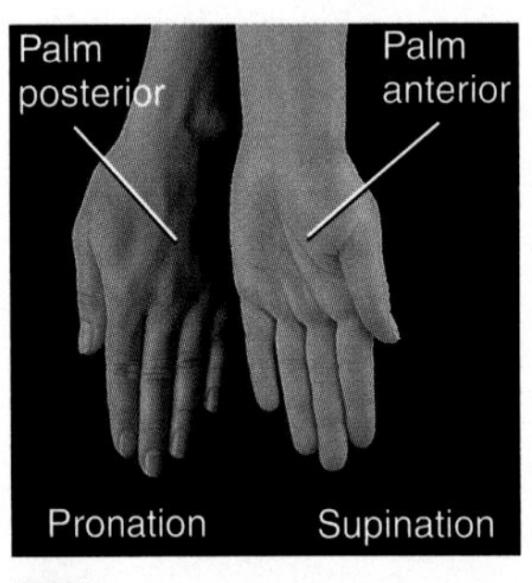

F Pronation and supination

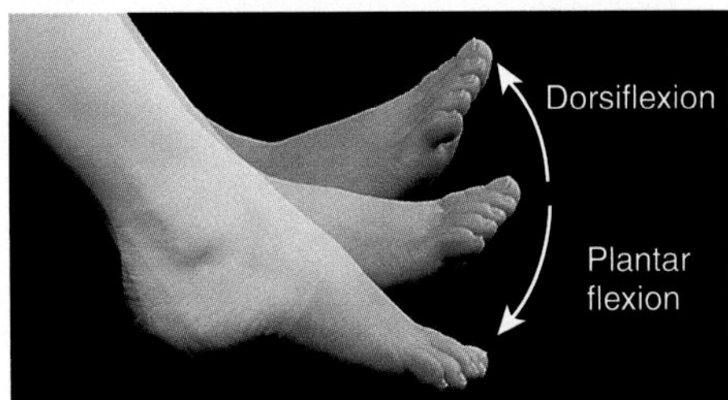

G Plantar flexion and dorsiflexion

A Gliding motion occurs when one surface slides past the other, as in the ankle or wrist.

B Rotation is the movement of a bone in all directions and all planes, as you may have done when making arm circles in gym class. This is possible only at the pectoral girdle; you can move your humerus in literally any direction, although the joint has more flexibility in some planes than others.

C Flexion and extension are opposite motions of hinge joints. Flexion decreases the angle at the joint; extension increases it. When standing in the anatomical position, all the joints are at full extension.

D Joints such as the neck and wrist can be hyperextended (moved beyond full extension). Tipping your head back to look at the ceiling hyperextends the neck.

E Adduction and abduction are opposite movements that bring appendages closer to, or farther from, the body midline, respectively. Clasping your hands in front of your body, as in prayer, shows adduction, whereas flinging your arms outward as if to say “the fish that got away was THIS big . . .” shows abduction.

F Supination and pronation are movements at the wrist, involving the radius and the ulna. In supination, the two bones are parallel; in pronation, the radius is twisted over the ulna. Holding your hands palms upward to cup a bowl of soup shows supination, whereas holding your hands palms downward shows pronation. Interestingly, the radius and ulna are the only two bones that you can twist one over the other without damage.

G The ankle joint has its own specific movements. In plantar flexion, the foot is flexed so that the toes point downward, as if you were planting a seed with your toe. Dorsiflexion is the opposite, with the dorsal surface of the foot flexed so the toes point upward.

Monaxial
One plane of movement.

Diaxial
Two planes of movement.

Triaxial
All three planes of movement.

Diarthrotic joints can permit movement in one, two, or three planes. **Monaxial joints** include the hinge joint of the elbow and ankle. The wrist is a typical **diaxial joint**. The hips and shoulder are **triaxial joints**. The type of movement at synovial joints is also classified as gliding, extension, flexion, hyperextension, adduction, abduction, rotation, pronation, and supination (FIGURE 5.34).

The skeleton, together with all of its sophisticated joints, provides a framework for movement, but it does not generate movement on its own. That is the task of the muscular system, which we will investigate next.

CONCEPT CHECK

How do the joint classifications based on structure relate to the joint classifications based on function? Can you see any similarities between these two systems?

What are the possible actions of Your knee joint? Your spinal joints? Your shoulder joint?

Describe the movements your ankle and wrist make that other joints cannot.

CHAPTER SUMMARY

1 The Skeletal System Is a Dynamic Living System

The skeletal system is composed of the axial and appendicular skeletons. It includes 206 bones, each properly named and thoroughly described. Functions of the skeletal system include support, movement, protection, calcium storage, and hematopoiesis. All bones are composed of bony tissue arranged to maximize strength and minimize weight.

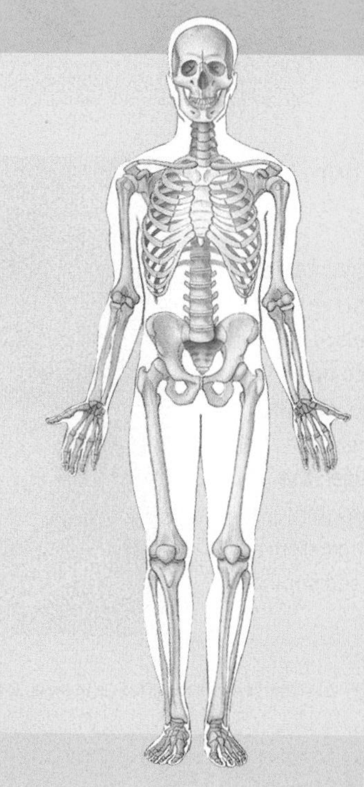

CHAPTER SUMMARY

2 Ossification Forms Bone, and Remodeling Continues to Shape It

The formation of bone is called ossification. Most of your bones were formed from the ossification of a cartilage model, starting with a primary ossification center and the hollowing of the medullary canal. Eventually, a secondary ossification center developed at the ends of the bone, leaving the cartilage growth plate, or epiphyseal plate. Compact bone is composed of Haversian systems, or osteons, stacked for strength. Spongy bone is less dense, formed of trabeculae that are easily deconstructed and moved should stresses develop in new areas.

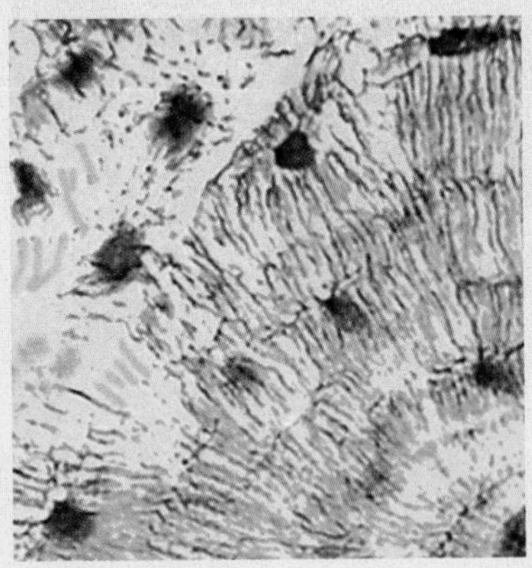

3 The Axial Skeleton Is the Center of Things

The axial skeleton includes 8 cranial bones, 14 facial bones, the thoracic cage, and the vertebrae, which protect the spinal cord. The skull includes many fused (synarthrotic) joints.

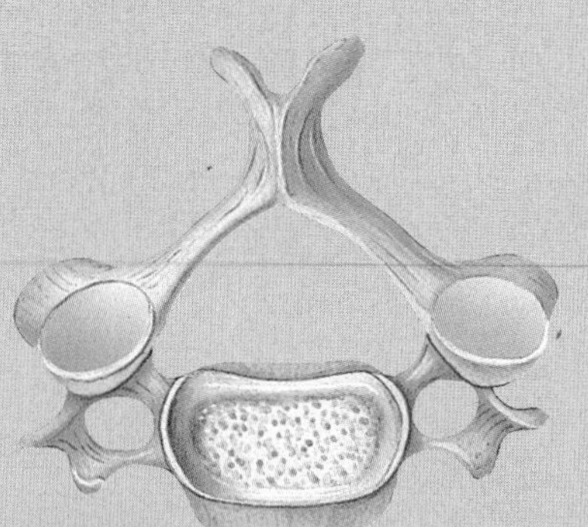

4 Your Limbs Comprise Your Appendicular Skeleton

The pectoral girdle, arms, hands, pelvic girdle, legs, and feet comprise the appendicular skeleton, which features a wide variety of joints and bone forms.

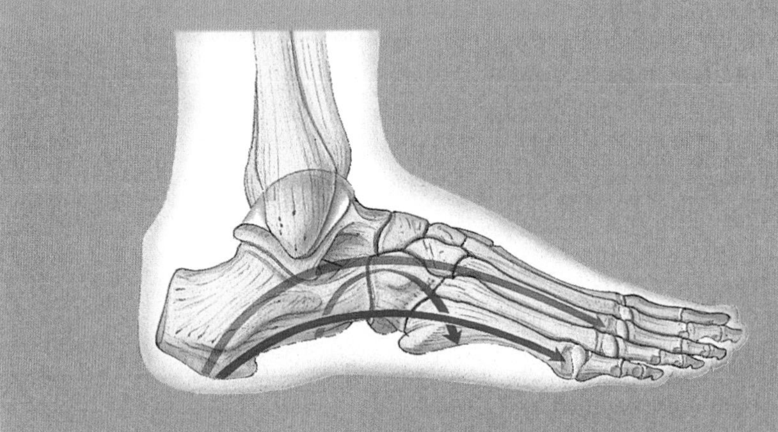

5 Joints Link the Skeletal System Together

The skeletal system moves only at the joints. The most maneuverable and common joint of the skeletal system is the synovial joint, which has menisci, ligaments, tendons, bursae, and a capsule to complement the bony structure. Synarthrotic joints provide great strength but allow no movement. Amphiarthrotic joints, like the one found between the two pubic bones, are slightly movable. Many joint movements are paired, including flexion and extension, adduction and abduction, pronation and supination, and plantar flexion and dorsiflexion.

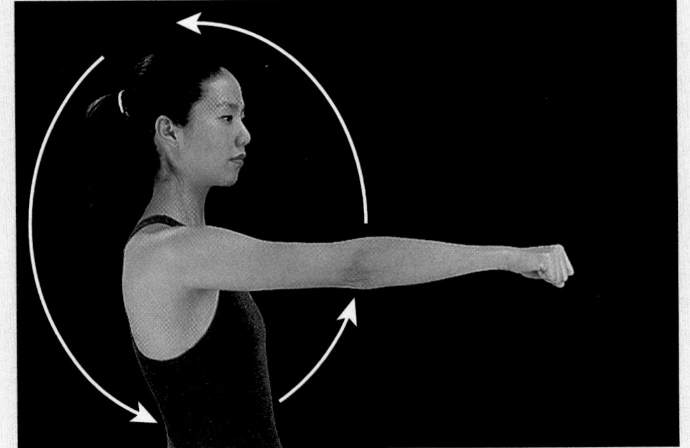

KEY TERMS

- **articulates** p. 132
- **bursae** p. 148
- **chondroblasts** p. 130
- **cribriform plate** p. 132
- **diaxial** p. 151
- **endochondral** p. 125
- **fibrocartilage** p. 130
- **fossa** p. 133
- **fracture hematoma** p. 130
- **interosseus** p. 145
- **intramembranous** p. 125
- **ligament** p. 147
- **menisci** p. 148
- **monaxial** p. 151
- **osteoblasts** p. 125
- **osteocytes** p. 125
- **parietal** p. 132
- **pectoral girdle** p. 138
- **pelvic girdle** p. 136
- **sella turcica** p. 133
- **synovial fluid** p. 148
- **tendon** p. 147
- **triaxial** p. 151

CRITICAL THINKING QUESTIONS

1. The bones of the hand, wrist, and arm are similar in appearance to those of the foot, ankle, and leg. Compare these two appendages, listing the similar bones and noting any significant structural or functional differences.

2. When hiking in the backwoods, you come across a human skeleton. What clues can you use to determine the identity of the deceased? How would you determine gender? Can you determine age, dietary preferences, general health, and occupation? What markings or other signs would you consider valuable clues?

3. We know bone formation can occur in areas that usually do not support bony tissue. For example, masons who regularly carry bricks on their shoulder often develop dermal bones in the skin on the anterior aspect of the shoulder. Describe the process of ossification that would occur here. Is this type of bone formation more like endochondral ossification or intermembranous ossification?

4. Bone remodeling is achieved through hormonal regulation of the osteoclasts and osteoblasts. Calcitonin is a hormone that prevents osteoclast activity, resulting in the overall deposition of bone, whereas parathyroid hormone stimulates removal of calcium from bones. Which type of bone cell does calcitonin stimulate? What about parathyroid hormone? Explain what role these hormones could play in osteoporosis.

5. Draw a typical long bone, labeling at least five structures. Indicate the likely location of compact bone and of spongy bone. Relate the structure of compact bone to its function in the long bone.

SELF TEST

1. The functions of the skeletal system include all of these EXCEPT
   a. hematopoiesis.
   b. support.
   c. protection.
   d. calcium storage.
   e. movement.

2. The axial skeleton includes
   a. the carpals.
   b. the phalanges.
   c. the ribs.
   d. the clavicle.

3. Identify the type of bone indicated as A on this diagram.
   a. Long
   b. Short
   c. Flat
   d. Irregular
   e. Sesamoid

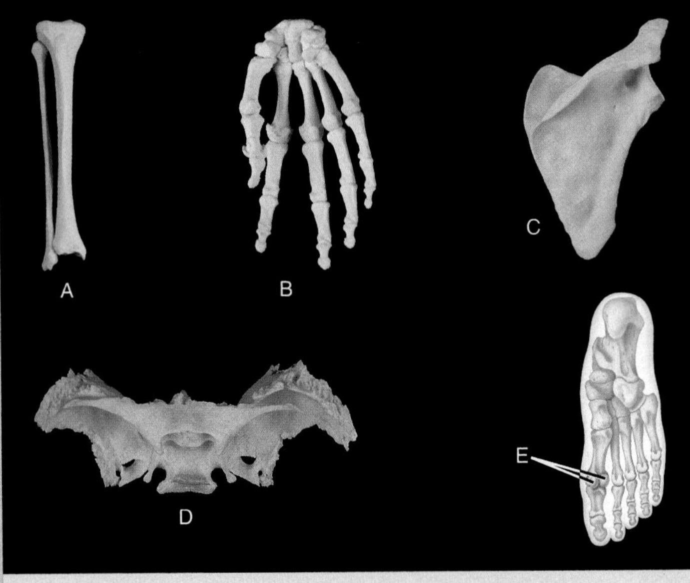

4. Which of the bone types indicated on the above figure is formed inside tendons?
   a. A
   b. B
   c. C
   d. D
   e. E

5. True or False? Endochondral ossification involves ossification of a cartilage skeletal model.

6. The cells responsible for the formation of the osteoid indicated in this figure are called
   a. osteoclasts.
   b. osteocytes.
   c. osteoblasts.
   d. osteons.

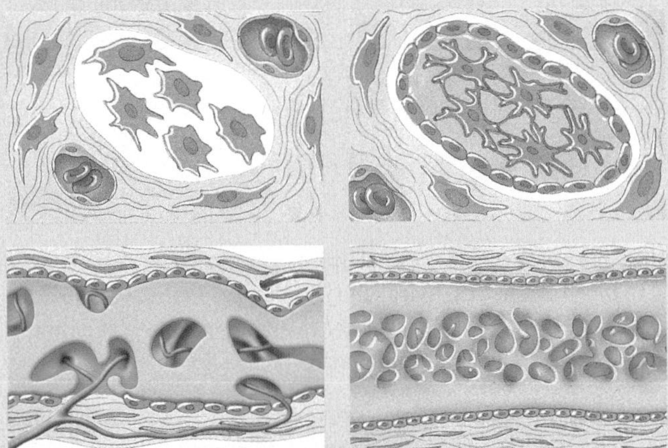

7. Concentric rings of matrix laid down by osteocytes surrounding a central canal are referred to as
   a. trabeculae.
   b. Haversian systems.
   c. lacunae.
   d. canaliculi.

8. Identify the portion of a long bone indicated as B in this figure.
   a. Diaphysis
   b. Epiphysis
   c. Medullary canal
   d. Articulating cartilage

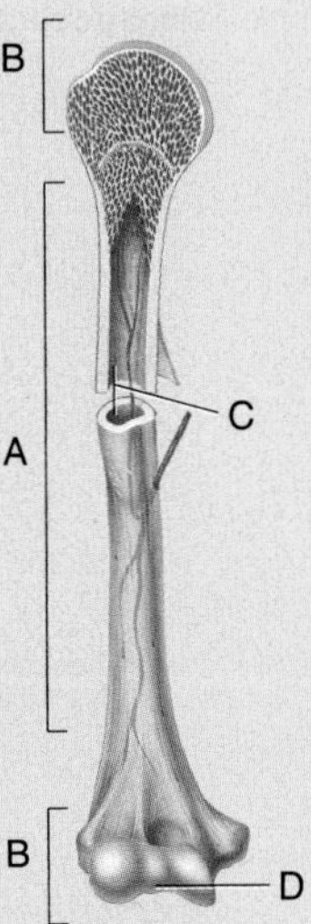

9. During bone remodeling, the cell responsible for breaking down bone and releasing the stored calcium is the

a. osteoblast.
b. osteoclast.
c. osteocyte.
d. osteon.

10. The first step in bone fracture healing is the formation of a

a. fracture hematoma.
b. fibrous callus.
c. bony callus.
d. periosteum.

11. The bone that forms the base of the cranial cavity, touching all other cranial bones, is the

a. ethmoid.
b. parietal bone.
c. frontal bone.
d. sphenoid.
e. temporal bone.

12. The function of the orbital complex is to

a. support the eye.
b. protect the eye.
c. provide a site for muscle attachment.
d. All of the above.

13. The thinnest and most delicate of the vertebrae are the

a. thoracic vertebrae.
b. lumbar vertebrae.
c. cervical vertebrae.
d. sacral vertebrae.
e. coccyx.

14. Risk factors for osteoporosis include all of the following EXCEPT

a. inactivity.
b. light stature.
c. vitamin D surplus.
d. Asian or European ancestry.

15. Identify the bone in this figure.

a. Clavicle
b. Scapula
c. Humerus
d. Carpal

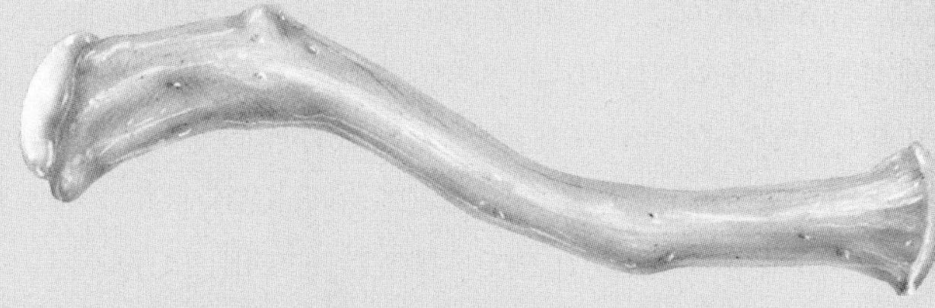

16. "Boxer's fractures" affect the ___________ bones.

a. carpal
b. tarsal
c. metacarpal
d. metatarsal

17. ___________ hip bones are thicker, narrower, and have a narrow pubic angle.

a. Male
b. Female
c. Infant
d. Adult

18. Identify the bone seen in this figure.

a. Femur
b. Fibula
c. Patella
d. Talus

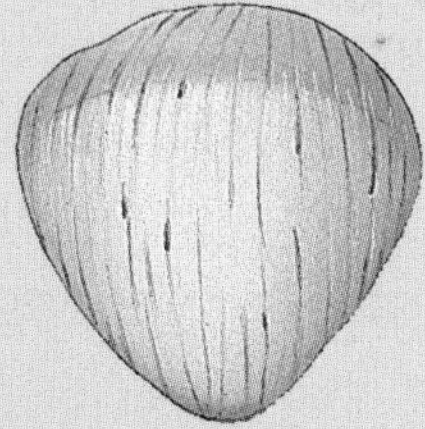

19. The type of joint typically found in the skull is a

a. synarthrotic joint.
b. synovial joint.
c. amphiarthrotic joint.
d. diarthrotic joint.

20. The movement permitted at the knee joint can best be described as

a. gliding.
b. rotation.
c. flexion and extension.
d. adduction and abduction.

The Muscular System

6

"Join our Gym!" scream the ads. "Live longer, live healthier. Change your body by working out with us!" Who are they kidding? Can 40 minutes of exercise a few times a week make me look better, feel better, have more energy, even live longer? Can these benefits emerge from pushing my muscular system to do more than walk from the couch to the fridge? Sounds fishy. No matter how many people are skeptical of these ads, fitness centers are always on the lookout for new members.

Once you understand the workings of the muscular system, you might give these pitches a bit more credence. Skeletal muscles, the ones you use to work out at the gym, run from danger, or even sit upright, are built like a set of nested cables that amplify the tiny force of molecular "machines" to produce your every motion, from the fine motor actions of signing your name to the huge force needed to shove a car uphill. When you begin a long-term exercise program of weight lifting or cycling, your muscles respond by changing in ways that improve their ability to lift weights or cycle tomorrow. In a "toned" muscle, individual fibers fire at random, and that causes them to hold their shape—and to use extra fuel. Kilogram for kilogram, a toned individual can eat more calories without gaining weight. But keeping in shape has important psychological benefits as well. Exercise liberates compounds that make you feel better. And there is mounting evidence that exercise is good for the immune system and can even help you live longer.

CHAPTER OUT
The Muscular System Has Many Functions p. 158
Skeletal Muscles Are Contractile Organs p. 159
Muscle Contraction Occurs as Filaments Slide Past One Anothe p. 167
Whole-Muscle Contractions Eme from Tiny Impulses p. 172
Muscles Require Energy to Work Smoothly and Powerfully p. 175

The Muscular System Has Many Functions

LEARNING OBJECTIVES

List the functions of the muscular system.

Understand how skeletal muscles promote blood flow.

Movement through the environment is a defining characteristic of animal life. Humans move by applying tension to the bones and joints of the skeletal system, thereby propelling us through the world to find food, shelter, and clothing and to satisfy various social and emotional needs. This movement is generated by the muscular system, in close interaction with the nervous and skeletal systems.

Muscular tissue is contractile tissue. Studies of the muscular system usually focus on **skeletal muscle** and its connective-tissue covering. The human body has two other types of muscle tissue—cardiac muscle (Chapter 11) and smooth muscle—which are not found in the skeletal muscles.

Skeletal muscle Contractile tissue composed of protein filaments arranged to move the skeletal system.

Beyond manipulating our environment and moving through it, skeletal muscles have other functions. They help stabilize movement at joints. When you lift a heavy object, muscles in your forearms stabilize the wrists to prevent **flexion**—or **extension**—of the hand. You may have seen a weight lifter lose a lift when the bar tilted sideways. The muscles' stabilizing ability was overtaxed, allowing the joints to unwillingly flex.

The contraction of muscles in the appendages aids in the flow of lymph and blood through the body. When they contract, skeletal muscles squeeze blood vessels and convert them into pumps. Pregnant women are reminded to walk to help push the additional blood volume in their legs toward the heart. Muscles also protect internal organs, as exemplified by the "six-pack" muscles lying in front of the digestive organs.

Skeletal muscle has yet another function. Think how your body responds to cold. Shivering is the random contraction of muscles designed not to produce movement but to maintain thermal homeostasis by generating heat. Muscles are **heat-producing organs**. In fact, they are the largest producers of internal heat in the human body, making heat whenever they are used and even (to a limited degree) while at rest.

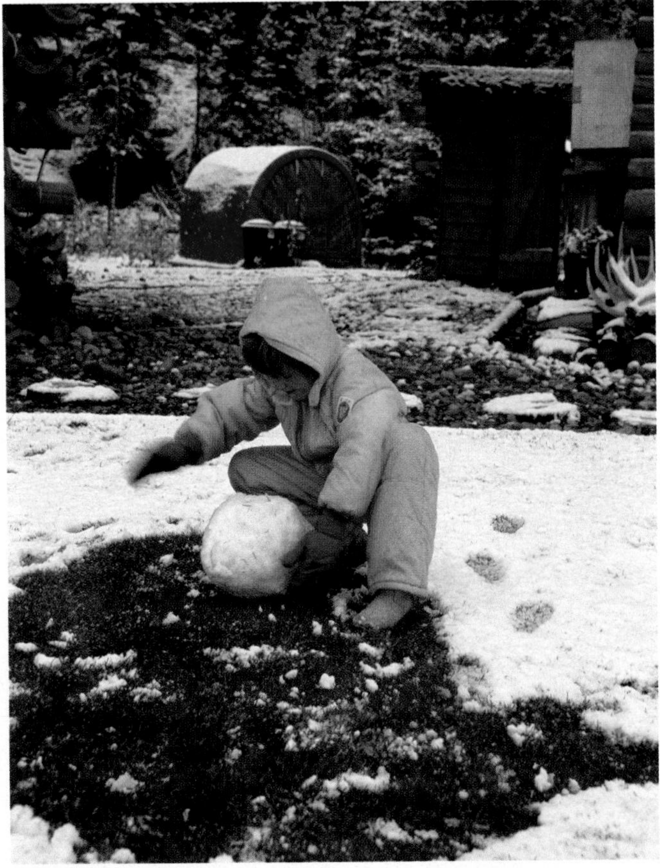

The muscular system is at work here, allowing this child to move in precisely defined patterns as she creates her snow sculpture. She may also be shivering slightly in the freezing air.

CONCEPT CHECK

What is the primary function of the muscular system?

Explain two other functions of the muscular system.

Skeletal Muscles Are Contractile Organs

LEARNING OBJECTIVES

Explain the difference between origin and insertion.

Define the relationship between muscle agonists and antagonists.

Describe the anatomy of a skeletal muscle.

Diagram the arrangement of proteins in the sarcomere.

Most people do not consider muscles to be organs, but they fit the definition: A muscle is made of tissues that are combined to perform a specific job in the organism. All human skeletal muscles have a similar function and structure. They **contract**, or get shorter, to produce movement. Muscles can relax to their original ("resting") length or even elongate beyond that. In general, each skeletal muscle has an **origin**, an end that remains stationary when the organ shortens, and an **insertion**, an end that moves during contraction (FIGURE 6.1). Knowing the origin and insertion of any skeletal muscle offers clues about its function. If you mentally pull the insertion toward the origin, you can visualize the effect of contraction.

To coordinate and control body movements, most human skeletal muscles function as a member of an **antagonistic** or **synergistic pair**. One or more muscles provide movement (the **prime mover** or **agonist**) while a second muscle or group opposes that movement (the **antagonist**). Moving your hand to your shoulder requires the simultaneous contraction of the prime movers, the

Antagonistic (synergistic) pair Muscles with opposing actions working together to provide smooth and controlled movements.

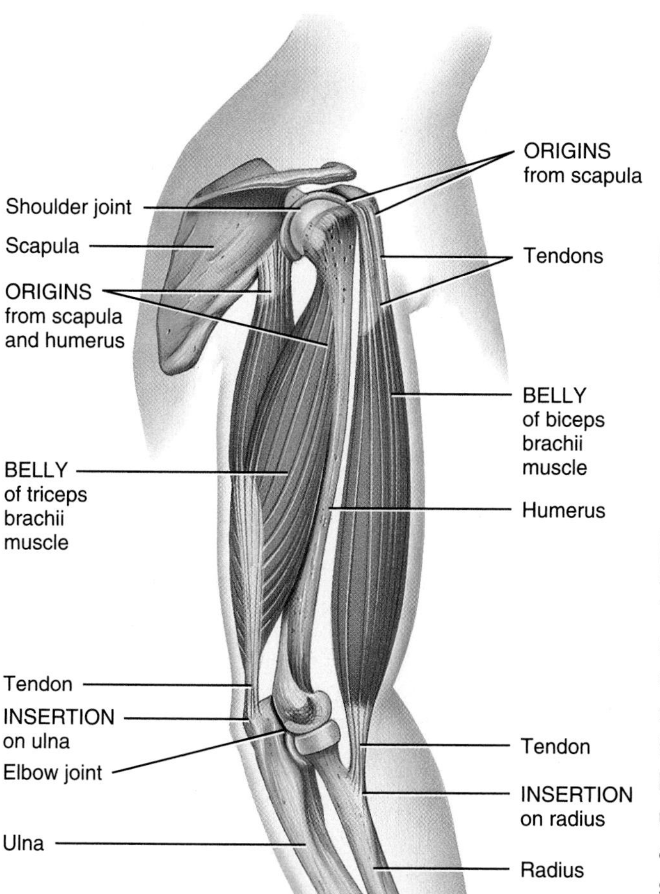

Origin and insertion of a skeletal muscle

Muscle origin and insertion FIGURE 6.1

To raise your hand toward your shoulder, the humerus must remain stable while the radius and ulna pivot upward. This movement is accomplished by contraction of the *brachialis* and *biceps brachii* muscles. The origin for the brachialis is at the upper end of the humerus, and the insertion is the proximal end of the ulna. When the brachialis muscle contracts, the humerus remains stationary and the ulna moves toward it. The origin for the biceps brachii is on the scapula, and its insertion is on the radius. When this muscle contracts, the scapula above the humerus remains in place and the radius moves upward to meet it.

brachialis and biceps brachii muscles, and relaxation of the antagonist, the triceps brachii. These muscle pairs can often be identified by simply looking carefully at the superficial muscles. Occasionally the prime mover will be on the anterior surface and the antagonist will be on the posterior surface (FIGURES 6.2 and 6.3).

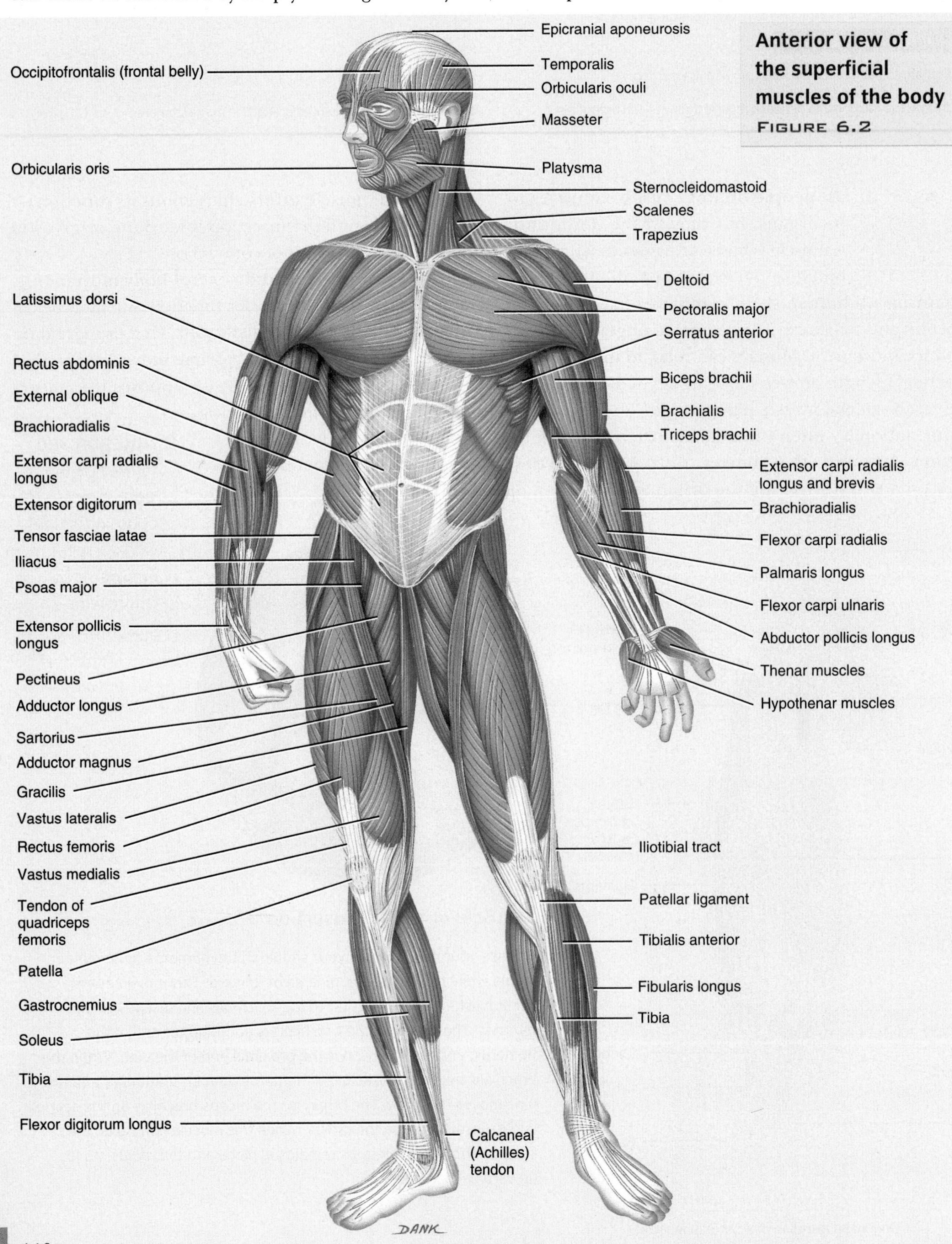

Anterior view of the superficial muscles of the body
FIGURE 6.2

Posterior view of the superficial muscles of the body

FIGURE 6.3

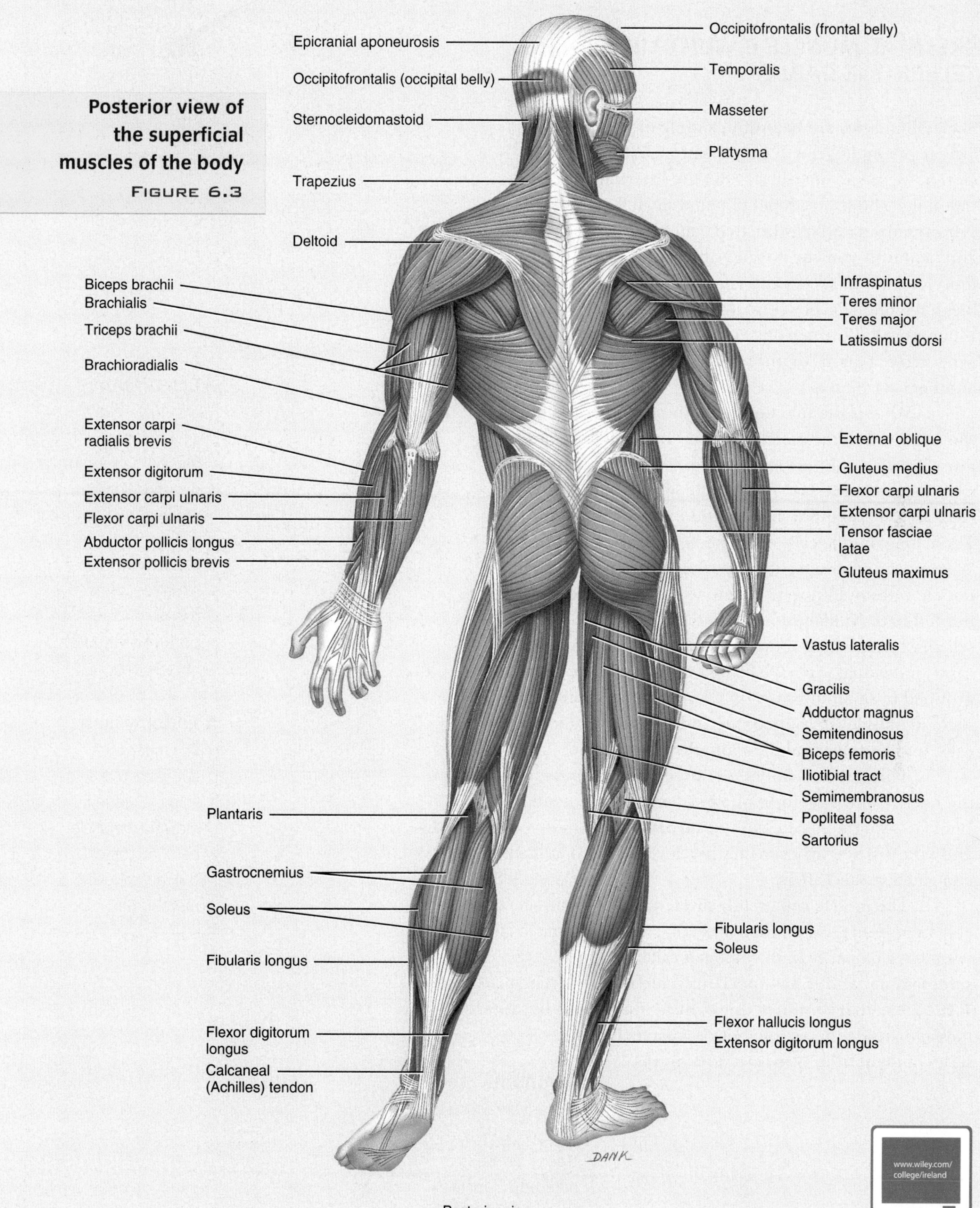

Epicranial aponeurosis
Occipitofrontalis (occipital belly)
Sternocleidomastoid
Trapezius
Deltoid
Occipitofrontalis (frontal belly)
Temporalis
Masseter
Platysma
Biceps brachii
Brachialis
Triceps brachii
Brachioradialis
Extensor carpi radialis brevis
Extensor digitorum
Extensor carpi ulnaris
Flexor carpi ulnaris
Abductor pollicis longus
Extensor pollicis brevis
Infraspinatus
Teres minor
Teres major
Latissimus dorsi
External oblique
Gluteus medius
Flexor carpi ulnaris
Extensor carpi ulnaris
Tensor fasciae latae
Gluteus maximus
Vastus lateralis
Gracilis
Adductor magnus
Semitendinosus
Biceps femoris
Iliotibial tract
Semimembranosus
Popliteal fossa
Sartorius
Plantaris
Gastrocnemius
Soleus
Fibularis longus
Flexor digitorum longus
Calcaneal (Achilles) tendon
Fibularis longus
Soleus
Flexor hallucis longus
Extensor digitorum longus
DANK
www.wiley.com/college/ireland

Posterior view

SKELETAL MUSCLE IS BUILT LIKE TELEPHONE CABLE

Skeletal muscles are beautiful, simple organs, with an awe-inspiring degree of organization. When we look closely, we see an amazingly effective internal configuration that shows how repetition and small forces, properly organized and coordinated, can produce strength and beauty. If you cut through the center of a skeletal muscle, you will see an internal structure that resembles a telephone cable (FIGURE 6.4). Skeletal muscle is composed of numerous elongated structures, running from origin to insertion, one nested inside another.

What is the function of all these connective tissue layers within the skeletal muscle? Individual skeletal muscle cells are long—sometimes 30 centimeters (or even longer in the *sartorius* muscle of the thigh). Muscle cells are also quite slender and exceedingly fragile. These long, fragile cells must shorten, creating tension. Without connective tissue support, the soft tissue of the muscle cell would not be able to withstand the tension needed to provide movement, and the cell would rip itself apart rather than shorten the organ. In a telephone cable, individual wires are coated with insulation, then grouped in small units within a larger cable. Similarly, skeletal muscle is grouped into individually protected cells, into fascicles, and then into the entire organ.

This "nested fibers" arrangement extends to the microscopic organization of skeletal muscle tissue. Look at a single muscle cell, or **myofiber**, and you will see an even smaller level of elongated, nested fibers.

The muscle cell itself is covered in a cell membrane very much like that discussed in Chapter 3. In this case it is called a **sarcolemma**, and it has specialized areas, **T tubules**, that conduct the contraction message. Inside the sarcolemma is a parallel series of **myofibrils** (FIGURE 6.5, p. 164).

T tubules Tubes formed in the sarcolemma cross through the muscle cell, carrying contractile impulses to the opposite side of the cell.

Myofibrils Linearly arranged groups of the contractile proteins actin and myosin.

Epimysium The outermost covering on a muscle, separating one muscle from the next.

Perimysium An inner connective tissue covering supporting a group of muscle cells.

Endomysium The innermost connective tissue lining, covering individual muscle cells, on top of the muscle cell membrane.

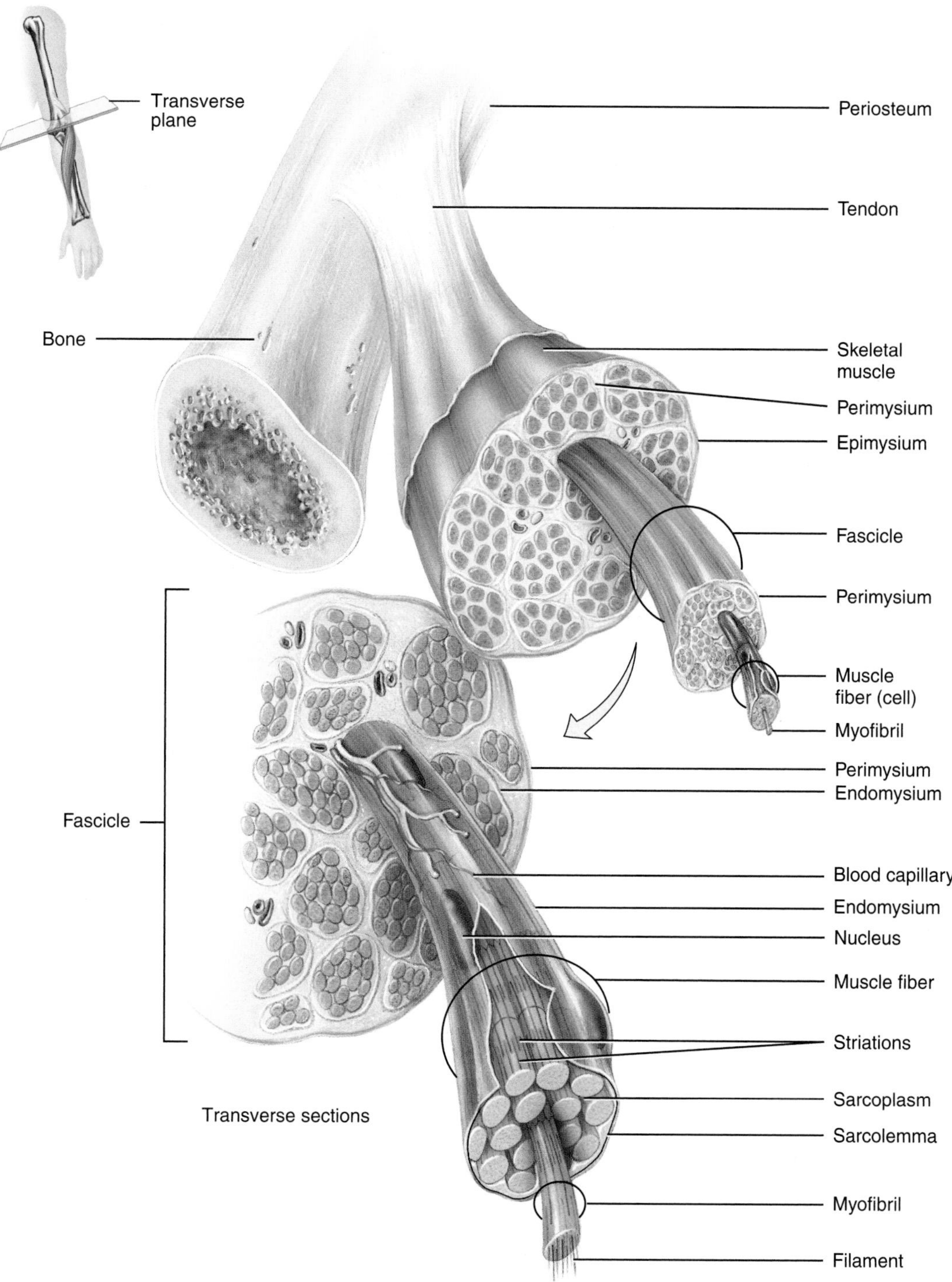

Anatomy of a muscle FIGURE 6.4

The outermost lining of skeletal muscle is the *deep fascia* or **epimysium**. Within this lining, blood vessels, nerves, and bundles of muscle cells are surrounded by a second lining, the **perimysium**. Each group of covered muscle cells is called a *fascicle*. (If you drag a fork across the top of a raw T-bone steak, those little tabs you see are the fascicles.) Within the perimysium is yet another lining, the **endomysium**, which surrounds individual muscle cells. (*Epi* = on top of; *peri* = around, like the perimeter of a circle; *endo* = within; and *my* is the root for "muscle.")

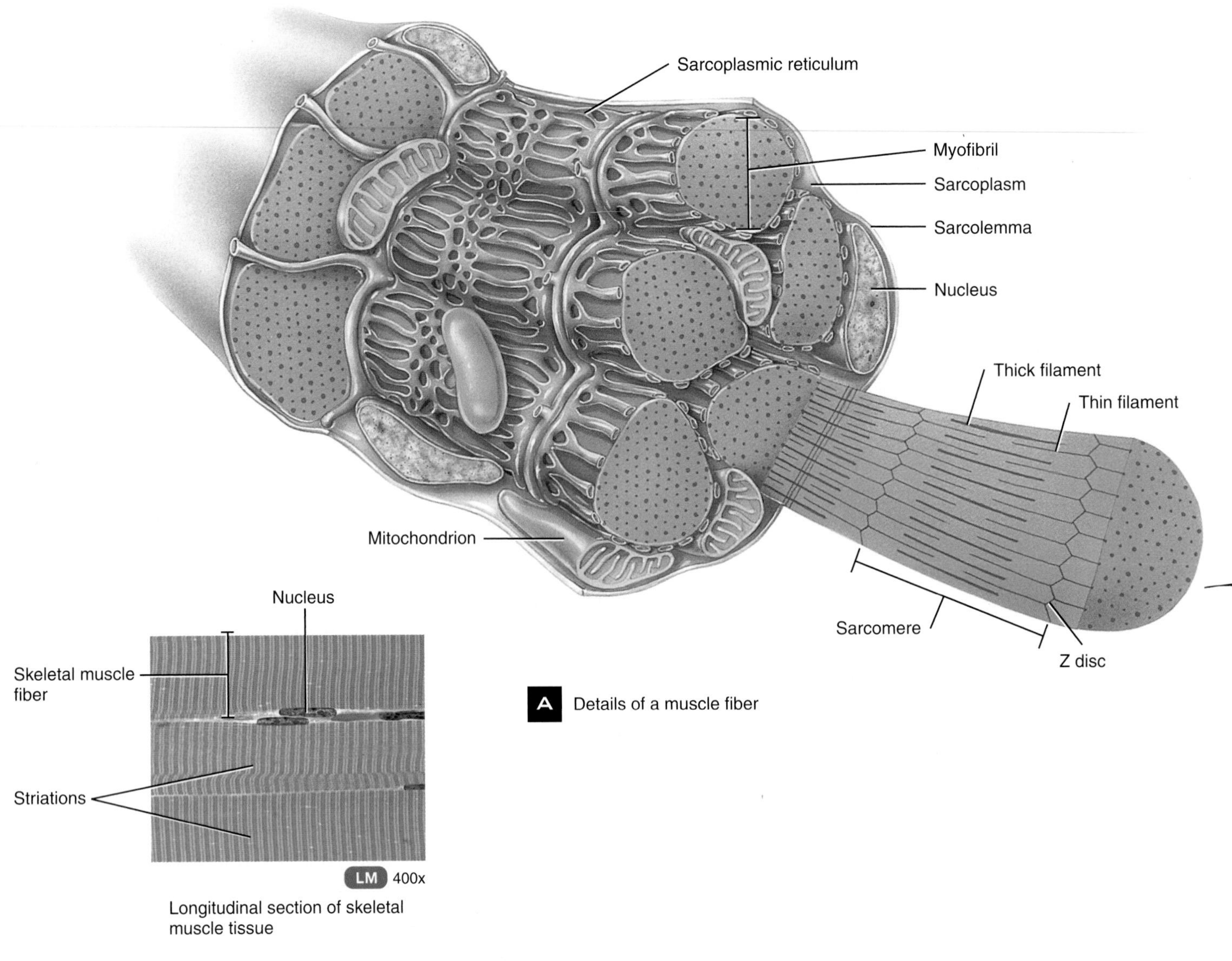

Longitudinal section of skeletal muscle tissue

A Details of a muscle fiber

Organization of skeletal muscle from the gross to the molecular level FIGURE 6.5

PROTEINS DRIVE MUSCLES

Inside these myofibrils, we find one final level of nested, elongated structures, microfilaments composed of the proteins **actin** and **myosin**. These two microscopic proteins interact in a way that causes the entire muscle tissue to shorten and therefore produce movement.

If you interweave your fingers and slide them together, you can approximate the interaction of actin and myosin. These proteins are held in regular arrangements in contractile units, or **sarcomeres**, which are stacked end to end in the myofibrils. Although each sarcomere is quite small, when they all contract at once, the force generated is large enough to tap your toe or leap tall buildings in a single bound. Every one of our body movements originates in the interaction of these tiny proteins within the highly organized skeletal muscle: blinking, shoveling snow, playing the piano, or bench-pressing 200 kilograms.

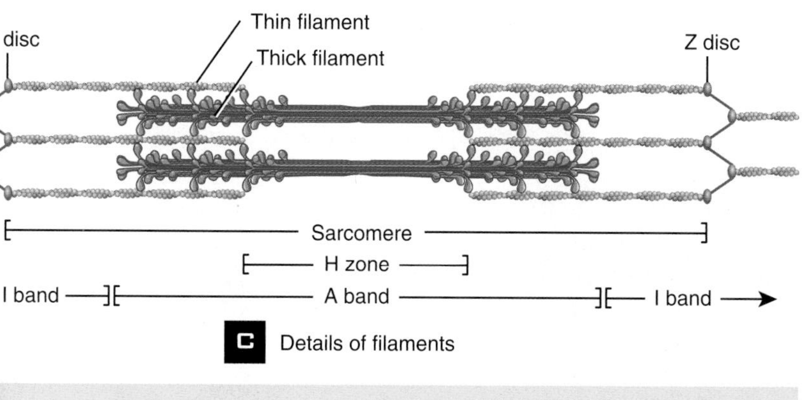

THE SARCOMERE IS BUILT FOR CONTRACTION

If you examine a sarcomere, you'll get clues to the nature of muscular contraction. When relaxed, bands are visible in individual sarcomeres. All of these sarcomeres, and consequently their bands, line up with in the muscle cell, visible as continuous dark and light areas on the cell. This alignment of sarcomeres and banded appearance produce striations in the muscle cell as a whole. We refer to skeletal muscle as striated tissue. The ends of the sarcomere make thin dark lines, called Z discs, that run transverse to the length of the muscle cell (think, "Z is the end of the alphabet and Z is the end of the sarcomere"). Attached to the Z discs, and extending to the middle of the sarcomere on each side, are thin actin filaments. Thick myosin filaments are suspended in the center of the sarcomere between the actin filaments.

Passing light through a sarcomere reveals patterns of light and shadow due to the relative thickness of these structures (FIGURE 6.5). The bands in a sarcomere are named for their ability to block light. The **I bands** are between the Z disks and the myosin thick

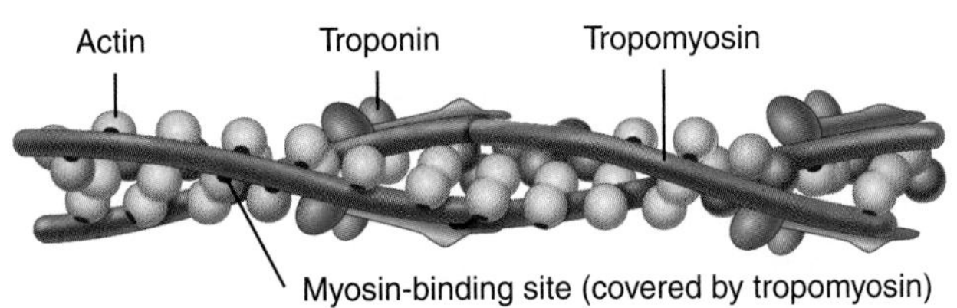

Portion of a thin filament

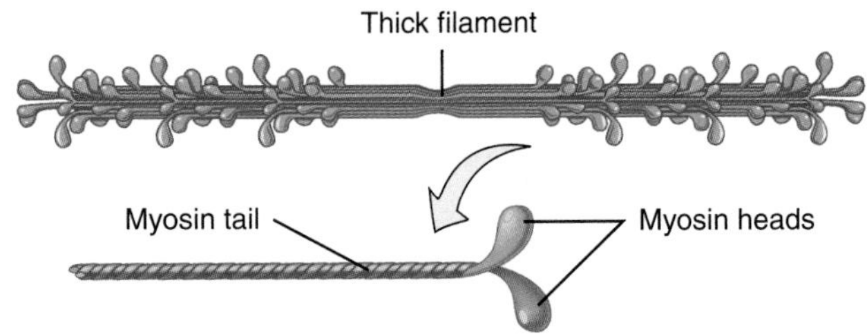

One thick filament (above) and a myosin molecule (below)

The arrangement of actin, troponin, and tropomyosin in a thin filament FIGURE 6.6

Actin is a **globular** protein, looking much like a string of pearls in the thin filaments of the sarcomere. Each thin filament is composed of two strands of actin, twisted about one another. Within the grooves of this double strand lie two additional small globular proteins that respond to calcium ion concentration. These accessory proteins, *troponin* and *tropomyosin*, cover the active site on each actin molecule, where they regulate contraction.

Globular
Spherical or round.

Myosin filament
FIGURE 6.7

The thick filament is composed of a grouping of myosin proteins oriented with their golf-club heads toward the Z lines in both directions and their shafts bundled together in the H zone. This arrangement leaves the thick filaments with a central area at the H zone where there are no heads. Heads extend off the filament in both directions, toward both Z discs. Many myosin heads extend from the thick filaments, arranged 360° around the filament. These heads are positioned so they do not overlap one another, but provide a continuous swirl of extended heads throughout the A bands.

filaments, where only actin is found. These bands are light-colored because only the thin actin filaments are blocking the light ("I" stands for "isotropic," meaning light is not altered as it shines through). The portion of the sarcomere where myosin resides is thicker, so it blocks light, and is called the **A band**, which stands for anisotropic (an = without or against). In the center of the sarcomere, the **H zone** is a light portion where the thinner central portions of the myosin filaments are grouped and overlapping actin is absent. The H zone is important in contraction because it is the zone into which actin is pulled as the sarcomere contracts. The T-tubules necessary for contraction are at the junction of the I bands and A bands in human skeletal muscle. Other animals, such as frogs, have only one T-tubule per sarcomere at the Z line.

The biochemistry of muscle contraction emerges from the structure of actin and myosin (FIGURE 6.6 and FIGURE 6.7). Actin is a thin, globluar protein with an area that will interact with myosin. In comparison, myosin is a larger, heavier protein, shaped like a golf club with a double head. The myosin head includes an area that will interact with the actin molecule under the correct circumstances.

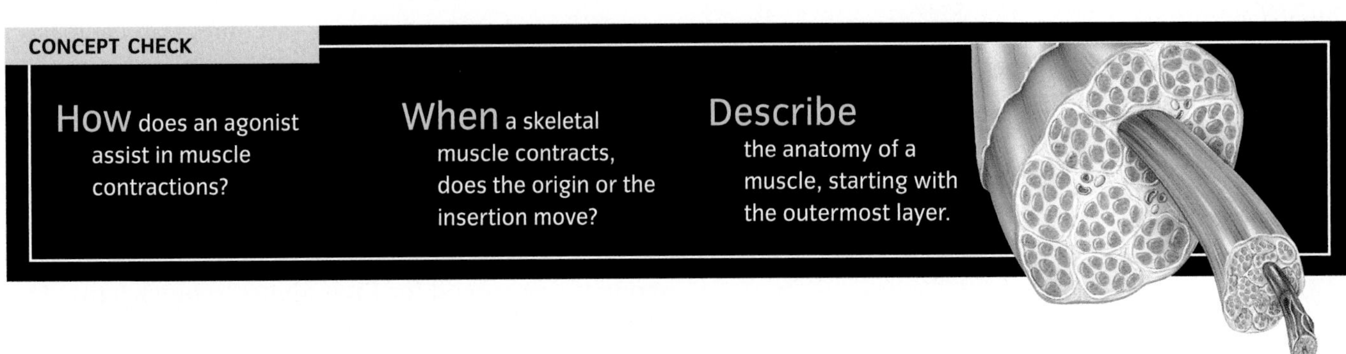

CONCEPT CHECK

How does an agonist assist in muscle contractions?

When a skeletal muscle contracts, does the origin or the insertion move?

Describe the anatomy of a muscle, starting with the outermost layer.

Muscle Contraction Occurs as Filaments Slide Past One Another

LEARNING OBJECTIVES

Describe the appearance of the neuromuscular junction.

Outline the steps in the sliding filament model.

The contraction of skeletal muscle stems from the movement of actin and myosin, as described in the **sliding filament model**, proposed in 1969. The use of the word "model" indicates that although we know quite a bit about the mechanics of sarcomere contraction, the picture emerging from research laboratories is continually refining that understanding.

The contraction of a skeletal muscle starts when an impulse from a motor neuron (nerve cell carrying information to a muscle) reaches an area called the **neuromuscular junction.** Here the motor neuron ends very close to a group of muscle cells, separated only by a small fluid-filled space called the **synapse,** or synaptic cleft.

Nerves send a contraction impulse across the synapse via chemical messengers, called neurotransmitters. The most common of these messengers is **acetylcholine**, abbreviated **ACh**. When acetylcholine is released from the axon terminal, it diffuses across the synaptic cleft and binds to receptors on the surface of the muscle cell membrane, delivering the chemical signal to contract. This impulse to contract is then passed through the entire muscle cell via the specialized membrane structures called T tubules.

Inside the muscle cell is a particular organelle called the **sarcoplasmic reticulum (SR)**, which looks much like the endoplasmic reticulum discussed in Chapter 3. The sarcoplasmic reticulum stores calcium ions and releases them when acetylcholine binds to the surface of the cell. Calcium is held within the SR by a protein called calcium sequestrin. The storage and release of calcium from the SR is accomplished by an enzyme on the surface of the sarcoplasmic reticulum, calcium-magnesium ATPase, which removes calcium from the cytoplasm and moves it into the SR. Calcium-magnesium ATPase works by converting ATP to ADP, powering a calcium "pump." It may surprise you to learn that free calcium inside the cell is toxic. Calcium-magnesium ATPase removes excess calcium from the muscle cell cytosol and adds it to the inner chamber of the SR, thereby ensuring cell survival (FIGURE 6.8 on p. 168).

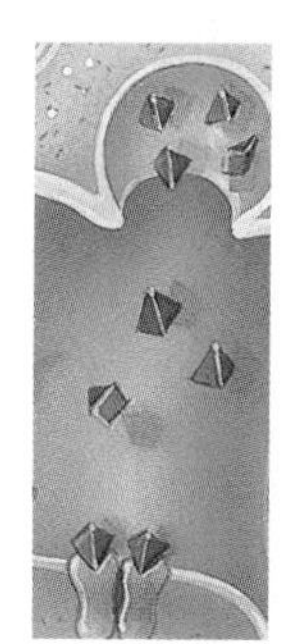

What happens next is a series of chemical reactions that proceed like a line of falling dominoes (FIGURE 6.9 on p. 169). The sliding filament model explains our best understanding of how muscle cells shorten.

Note that neither actin nor myosin undergoes any kind of chemical transformation, nor do they intertwine as the muscle cell contracts. Actin merely slides over the myosin filament, pulling the Z discs with it, hence the name "sliding filament model." This cycle of myosin grabbing exposed actin sites and ratcheting inward continues until (1) the removal of acetylcholine from the sarcolemma stimulates calcium-magnesium ATPase to pull calcium back into the sarcoplasmic reticulum, or (2) the supply of ATP is exhausted. Without a fresh supply of ATP, the myosin heads cannot release the actin molecule.

This is exactly what happens during **rigor mortis**. The death of the muscle cells causes the sarcoplasmic reticulum, and specifically calcium sequestrin, to lose its ability to hold calcium. This triggers a release of the stored calcium, which causes the stored ATP in the myosin heads to begin a contraction cycle. Lacking oxygen supply and blood flow, the ATP used during cross-bridge formation cannot be replaced. Myosin heads cannot release the actin without fresh ATP, so the sarcomere is stuck in the cross-bridge condition until the actin and myosin proteins begin to decompose. The extent of decomposition of these proteins is one clue that coroners use to determine time of death.

Neuromuscular junction FIGURE 6.8

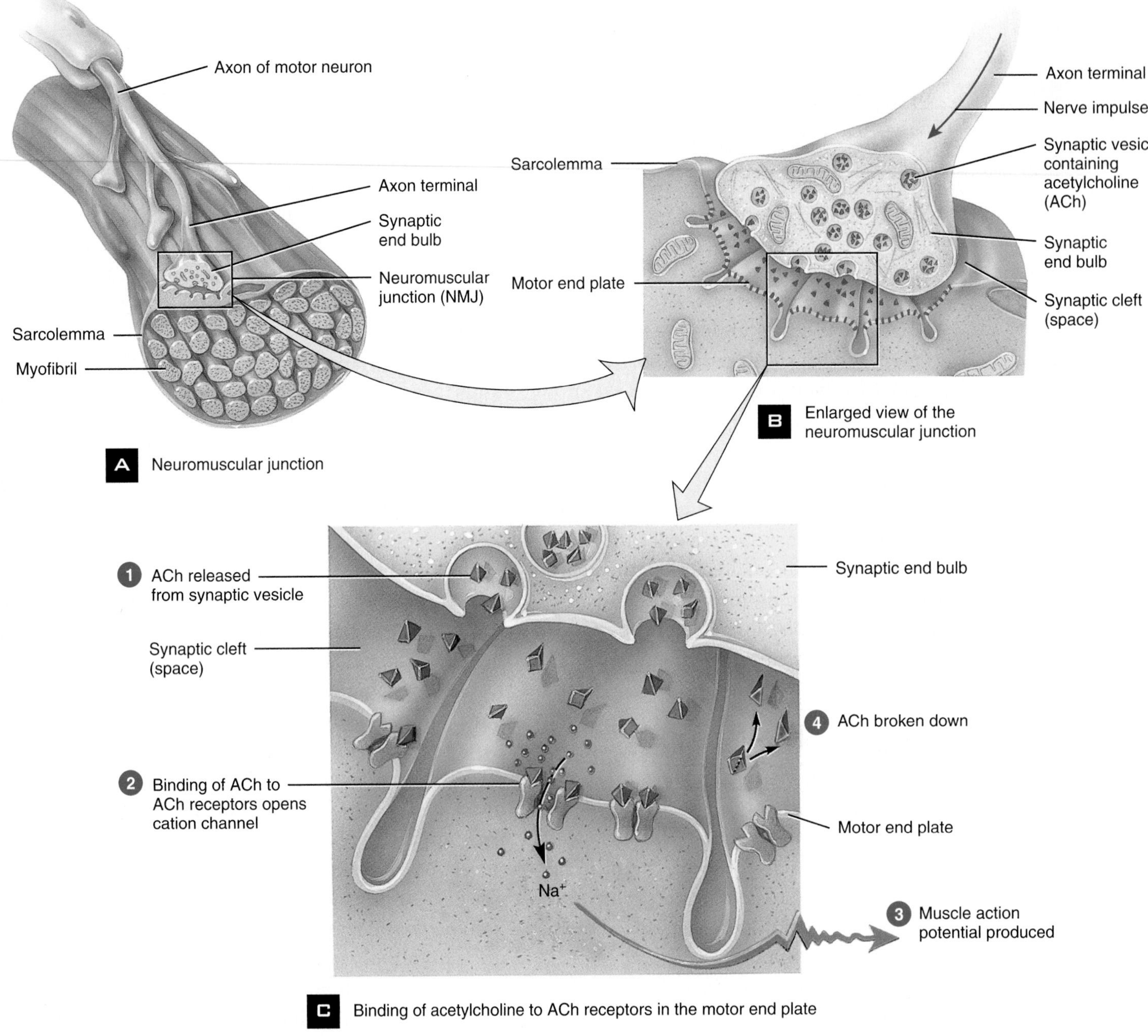

C Binding of acetylcholine to ACh receptors in the motor end plate

There are four steps in the transmission of an impulse at the NMJ:

Step 1 ACh is released from the end of the neuron.

Step 2 ACh binds to receptors on the muscle cell membrane, stimulating the release of calcium inside the muscle cell.

Step 3 A contraction cycle is begun in the cell.

Step 4 The ACh in the synapse is removed by enzymes, ending its effects on the cell.

Contraction cycle FIGURE 6.9

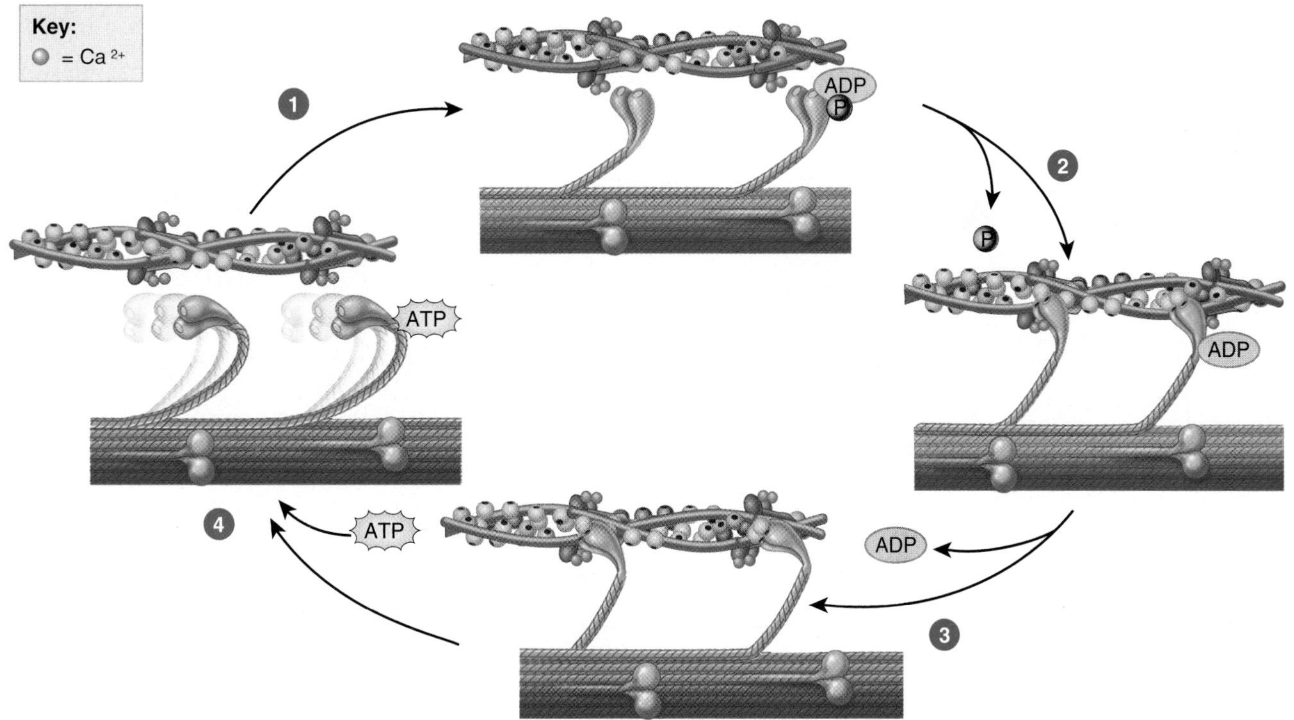

Step 1 Calcium binds to troponin, one of the two accessory proteins on the actin thin filament. The troponin molecule in this diagram is shown in blue, and the bound calcium ions are the smaller purple structures attached. This binding shifts the position of the troponin, which in turn shifts the position of the second accessory protein on the thin filament, tropomyosin, exposing the binding site on the actin filament. Meanwhile, the myosin heads in the thick filament are sitting at the ready, with the energy to contract stored right in the split golf-club head (remember that ATP is the molecule of energy in the body). The myosin heads obtain a molecule of ATP and immediately split it into ADP and a phosphate group (P), releasing energy from the bond, which is then ready for muscle contraction.

Step 2 With actin-binding sites exposed, the myosin head is able to reach toward the actin binding site and react, using the energy from the split ATP in the myosin head. Linking the myosin head to the actin binding site creates a cross-bridge between the thick and thin filaments of the sarcomere.

Step 3 As the myosin head releases energy, it bends toward the center of the sarcomere in the **power stroke** of the cycle. This slight bend pulls the actin filament across the myosin filament toward the H zone.

Step 4 With the addition of fresh ATP, the myosin head will drop the actin, return to the ready position, and immediately grab a new actin-binding site. This process will continue until the calcium is re-sequestered and the troponin and tropomyosin are returned to their precontraction state.

Process Diagram

Summary of events in contraction and relaxation of skeletal muscle

FIGURE 6.10

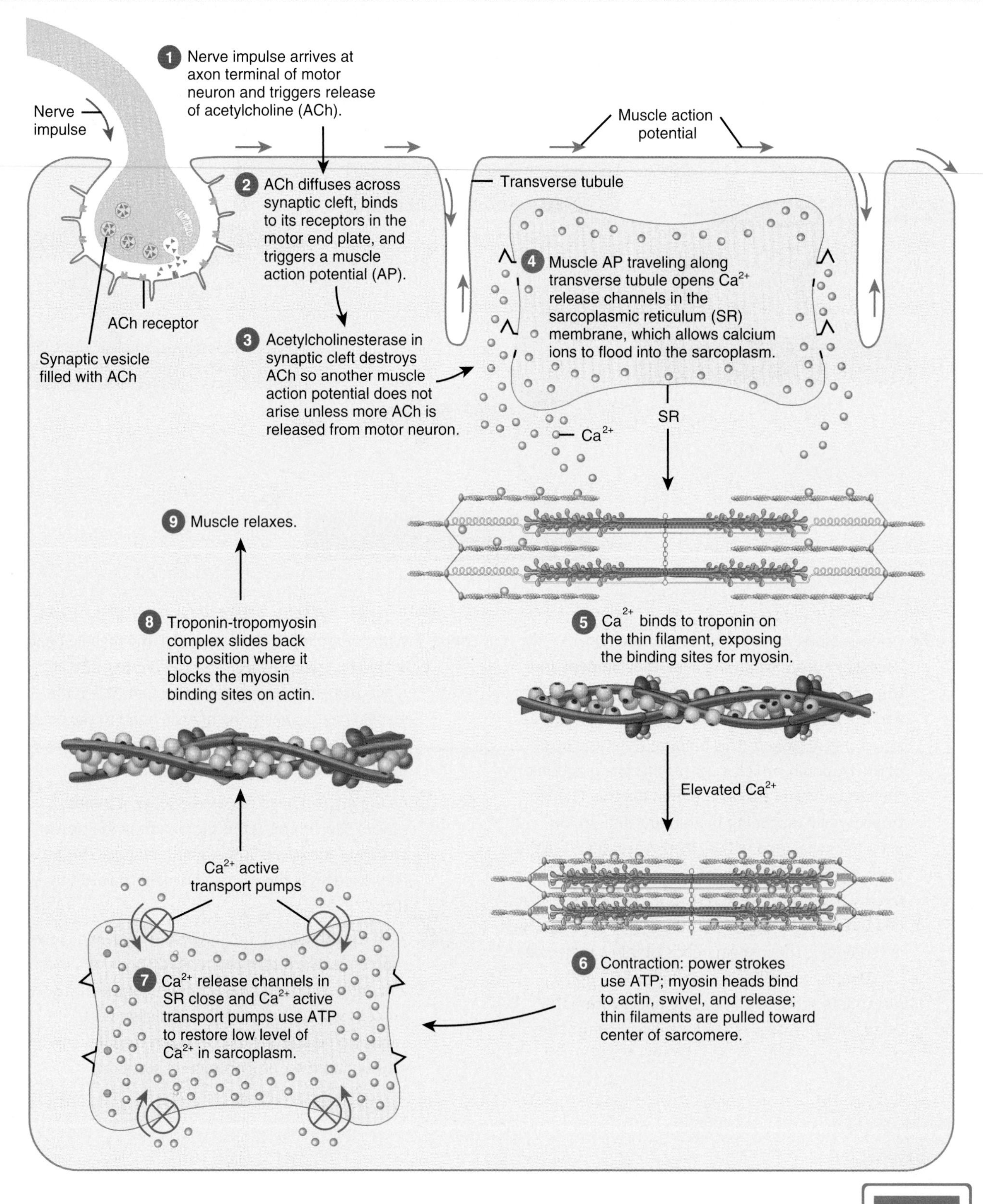

Nerve impulse
1 Nerve impulse arrives at axon terminal of motor neuron and triggers release of acetylcholine (ACh).
Muscle action potential
Transverse tubule
2 ACh diffuses across synaptic cleft, binds to its receptors in the motor end plate, and triggers a muscle action potential (AP).
ACh receptor
Synaptic vesicle filled with ACh
3 Acetylcholinesterase in synaptic cleft destroys ACh so another muscle action potential does not arise unless more ACh is released from motor neuron.
4 Muscle AP traveling along transverse tubule opens Ca^{2+} release channels in the sarcoplasmic reticulum (SR) membrane, which allows calcium ions to flood into the sarcoplasm.
Ca^{2+}
SR
9 Muscle relaxes.
8 Troponin-tropomyosin complex slides back into position where it blocks the myosin binding sites on actin.
5 Ca^{2+} binds to troponin on the thin filament, exposing the binding sites for myosin.
Elevated Ca^{2+}
Ca^{2+} active transport pumps
7 Ca^{2+} release channels in SR close and Ca^{2+} active transport pumps use ATP to restore low level of Ca^{2+} in sarcoplasm.
6 Contraction: power strokes use ATP; myosin heads bind to actin, swivel, and release; thin filaments are pulled toward center of sarcomere.

Process Diagram

www.wiley.com/college/ireland

If we zoom out from the microscopic scale, hundreds of simultaneous, asynchronous ratchet-like movements pull the thin filaments of each individual sarcomere into the H zone. Because the thin filaments are attached to the Z discs, this pulls the Z discs along with the actin, shortening the sarcomere. With millions of sarcomeres lined up in each muscle cell, and many muscle cells innervated by one motor neuron, these tiny chemical reactions shorten the entire muscle. See FIGURE 6.10 for a summary of these events.

The strength of an individual contraction event is also dictated by the sliding filament model. There is a strong correlation between the degree of overlap of the thick and thin filaments in the sarcomere and the amount of tension produced during contraction. Sarcomeres generate maximum tension with optimal overlap. When the Z discs of the sarcomeres are pushed too closely together (the muscle is understretched), or when the thick and thin filaments barely overlap (overstretched), the muscle fiber cannot generate much tension (FIGURE 6.11).

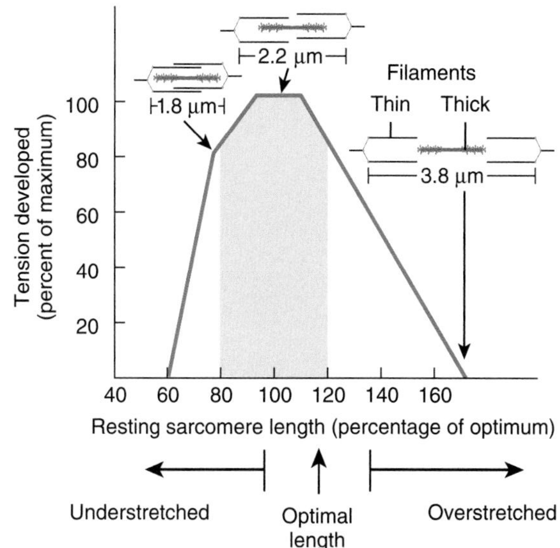

Length-tension relationship in a skeletal muscle fiber FIGURE 6.11

This graph illustrates the strength of contraction at various-length sarcomeres. As you can see, the sarcomeres that are squished too much cannot contract. As the thick and thin filaments slide apart, they reach an optimal length where contraction is quite forceful. As the sarcomere gets even longer, the power of the contraction lessens. Eventually the sarcomere will be too long for any overlap to exist between thick and thin filaments. With no overlap, there can be no contraction.

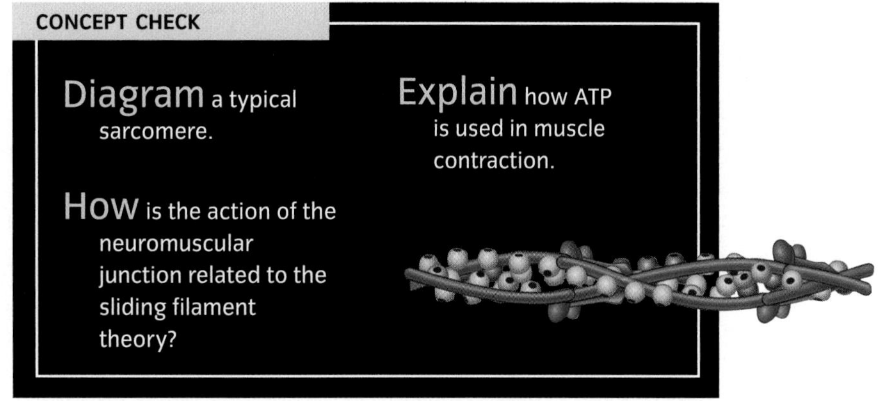

CONCEPT CHECK
Diagram a typical sarcomere.
How is the action of the neuromuscular junction related to the sliding filament theory?
Explain how ATP is used in muscle contraction.

Whole-Muscle Contractions Emerge from Tiny Impulses

Learning Objectives

Define the all-or-none basis of muscle contraction.

Identify the parts of a single muscle twitch.

Explain summation, treppe, and tetanus.

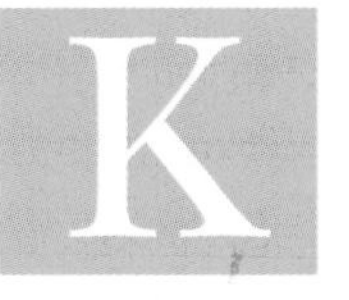

Knowing the biochemistry of contraction and muscle anatomy, we now have a good foundation for discussing whole-muscle contraction. How does an entire large muscle like that of your thigh contract and generate movement?

Muscle cells are grouped in **motor units**, composed of one motor neuron and the set of muscle cells it controls (FIGURE 6.12). The entire motor unit contracts when it receives a signal from the motor neuron, which causes the release of the calcium ions that triggers that sliding action just discussed. Muscle cells contract on an **all-or-nothing** basis. Nothing happens when the nerve stimulus is too weak to cause the release of calcium from the sarcoplasmic reticulum. In muscle cells, when the **threshold stimulus** is reached, calcium is released and the entire muscle cell contracts. **Graded contraction** is not possible at the cellular level. The all-or-nothing nature is similar to a mouse trap baited with cheese. A mouse can nibble the cheese and remove small amounts without consequence. But as soon as the mouse removes enough cheese, the trap snaps shut, trapping the hungry rodent.

Threshold stimulus The minimal amount of stimulation needed to cause a response.

Graded contraction A smooth transition from a small, weak contraction to a forceful contraction.

A myogram records a single contraction of one motor unit, called a single twitch (FIGURE 6.13).

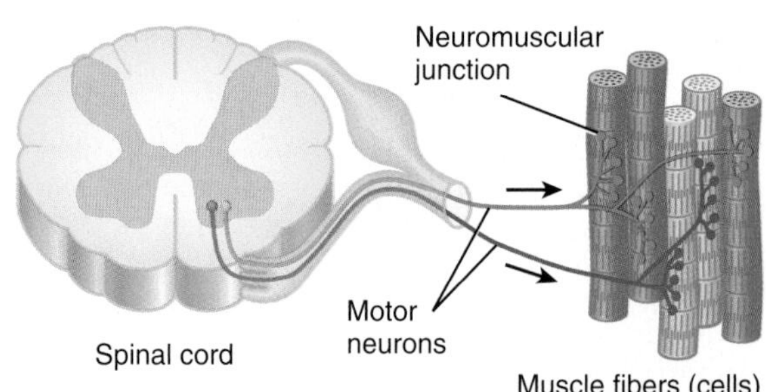

Motor unit FIGURE 6.12

Each motor unit is individually controlled. Contraction strength depends on how many motor units are stimulated, not on how powerfully the individual cells contract. Few motor units are stimulated during a weak contraction, but feats of strength require many motor units.

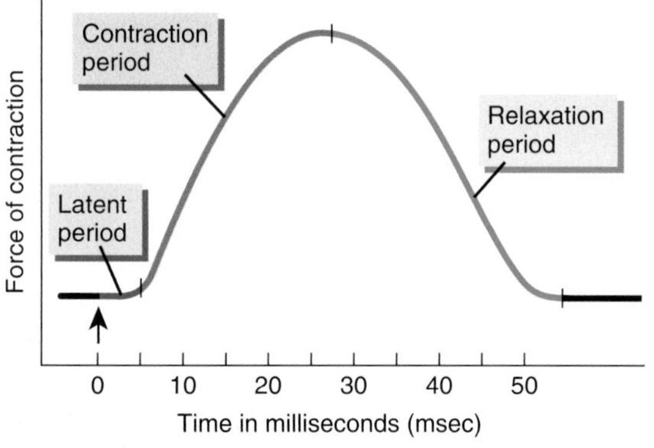

Myogram FIGURE 6.13

During the *latent period*, calcium ions are moving, actin active sites are being exposed and myosin heads are taking up slack in the myofibers, but contraction is not visible from outside the cell. Once the slack is taken up, the cell suddenly and visibly shortens, causing the sharp rise in the myogram at the *contraction* period. As the calcium is re-sequestered and the actin filaments with associated Z discs are released from the myosin cross bridges, the sarcomeres slide back to their original location. On the myogram, the return to baseline is called the *relaxation period*.

Treppe
The increased strength of contraction after successive identical stimuli.

Single twitches are not effective in producing body movement, because they last only a fraction of a second. To produce a meaningful amount of contraction, the motor unit requires multiple stimuli, reaching the muscle cell in such quick succession that it has no time to relax. Each contraction builds on the heels of the last, until the muscle cell is continuously contracted. This buildup of contractions is called **summation,** and the phenomenon of increased strength after successive identical stimuli is known as **treppe**. Warming up prior to an athletic event takes advantage of treppe. As the athlete continues to perform the motions required in competition, the muscles undergo treppe. Each successive contraction gets a little stronger. By the time the event starts, the muscles are ready to perform at peak strength. Once continuous contraction is achieved, the muscle is said to be in **tetanus**. (This continuous, and normal, contraction of the muscle is not the same as the bacterial infection also called tetanus.) The neck muscles of an adult are in tetanus most of the day. It is unusual to see adults' heads bobbing like a newborn's (FIGURE 6.14)—unless they are trapped in a boring lecture!

Summation explains how single twitches can provide sustained movement, but how is the strength of contraction monitored and regulated? You know you are capable of graded contractions—you can pick up a pencil with ease, using the same muscles that you would use to pick up a big stack of weighty textbooks. The answer is that contractions are graded by recruiting more motor units, under the brain's control. Before lifting, your brain makes an assumption, based on your experience, about the weight of the object, and begins the contraction by stimulating the appropriate number of motor units. If the original number of recruited motor units is incorrect, the brain will adjust by either recruiting more motor units or releasing some extra ones. We have all been fooled at some time. A small bar of silver is far heavier than it looks and can make us feel foolish

Muscle tension FIGURE 6.14

As adults, our postural muscles remain in tetanus throughout the day. Newborns, however, are not yet able to do this. As his neuromuscular junctions develop, this infant will be able to keep his head up for short periods of time. The "head bobbing" stage will last only a few days, after which tetanus is achieved in the neck muscles, and the baby will be able to observe and interact with the surroundings for long periods.

Scene from the film *Twister* FIGURE 6.15

on our first attempt to lift it. Conversely, lifting a piece of movie-set Styrofoam requires far less force than the brain may rally. On the set of the 1996 disaster flick *Twister* (FIGURE 6.15), the semi-trailer that is blown into the air was made of large chunks of Styrofoam. The stagehands threw these Styrofoam chunks into the air after unintentionally using too many motor units to lift them.

Even during tetanus, a small number of muscle cells are relaxing. The pattern of contraction and relaxation is asynchronous. If all the cells functioned in unison, the muscle would bounce between completely contracted to totally relaxed and back to completely contracted! That's a recipe for jittery, stuttering motion.

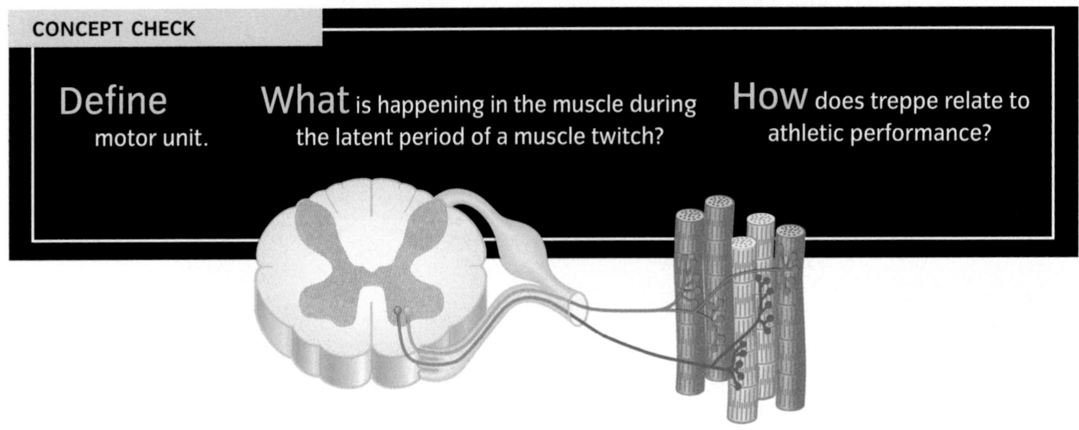

CONCEPT CHECK
Define motor unit.
What is happening in the muscle during the latent period of a muscle twitch?
How does treppe relate to athletic performance?

Muscles Require Energy to Work Smoothly and Powerfully

Learning Objectives

Describe the health benefits of toned muscles.

Compare aerobic and anaerobic energy pathways.

Explain the roles of ATP and creatine phosphate in muscle contraction.

Describe the different types of muscle fiber.

Now that we have examined the anatomy and physiology of skeletal muscles, it's time to look at how they work together to produce smooth, powerful movement. Let's start by looking at ATP, the general-purpose source of readily available energy inside cells.

Aerobic pathway Metabolic pathway that requires oxygen to burn glucose completely.

Anaerobic pathways Metabolic pathways that occur in the cytoplasm and burn glucose to lactic acid, releasing some energy.

CONTRACTION ENERGY CAN BE PRODUCED AEROBICALLY OR ANAEROBICALLY

The body can make ATP for muscular contraction through either the aerobic or anaerobic pathways. The highly efficient **aerobic pathway** burns (oxidizes) glucose, forming water, carbon dioxide, and ATP in the mitochondria. This pathway produces the largest amount of ATP and is the dominant method of energy production.

During heavy muscle activity, oxygen supply cannot keep up with the energy demands. ATP production then shifts to the **anaerobic pathways**. Anaerobic pathways are less efficient, producing far fewer ATP molecules per glucose molecule. Anaerobic pathways produce lactic acid, which is detrimental to sarcomere functioning. Lactic acid is eventually removed from the tissue by conversion to pyruvic acid, which gets shunted into the **TCA (Krebs) cycle** and the **electron transport chain** (Figure 6.16).

TCA (Krebs) cycle The citric acid cycle, step two in the production of ATP from glucose, carried out in the mitochondrial cristae.

Electron transport chain Step three in aerobic respiration wherein electrons are passed along in a series of chemical reactions, eventually producing ATP.

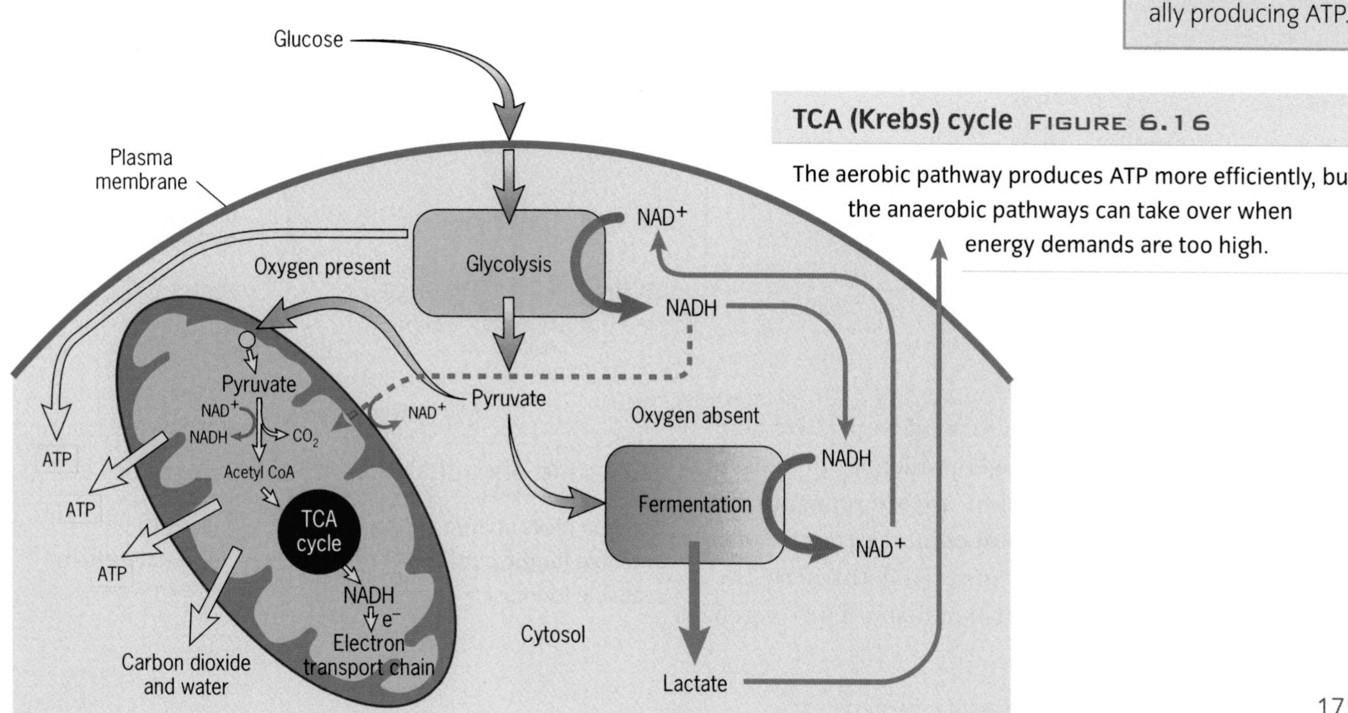

TCA (Krebs) cycle Figure 6.16

The aerobic pathway produces ATP more efficiently, but the anaerobic pathways can take over when energy demands are too high.

Process Diagram

The conversion of lactic acid to pyruvic acid and then on to ATP FIGURE 6.17

The conversion of lactic acid to pyruvic acid (FIGURE 6.17) requires oxygen, which is one reason we breathe heavily after exertion. We are repaying the **oxygen debt** incurred as a result of increased muscular activity. The added oxygen is carried through the bloodstream to the lactic acid-laden tissue. The oxygen reacts with the lactic acid, converting it to pyruvic acid and then to coenzyme A, which the mitochondria can use.

> **Oxygen debt**
> The amount of oxygen needed to convert the lactic acid produced by anaerobic respiration into pyruvic acid and burn it entirely to CO_2, H_2O, and energy.

Creatine phosphate is important in the anaerobic phase of muscle energy production because it stores energy much as ATP does, in a phosphate bond. Creatine is a highly reactive compound that picks up the phosphates released when the myosin heads drop the actin active site. Recall that the ATP stored in the myosin head is broken into ADP and a free phosphate ion prior to myosin grabbing the actin active site. This free phosphate is released when the myosin head bends toward the center, sliding the actin filament. This freed phosphate ion reacts with creatine to form creatine phosphate. Creatine phosphate then provides a reserve of phosphate for the formation of ATP from ADP. As long as there is a fresh supply of creatine, this cycle will prolong the contracting ability of the tissue (FIGURE 6.18). Even the most fit person will eventually experience muscle fatigue. The Health, Wellness, and Disease box, "Muscle fatigue, muscle woes" (p. 178), explains what happens in these instances.

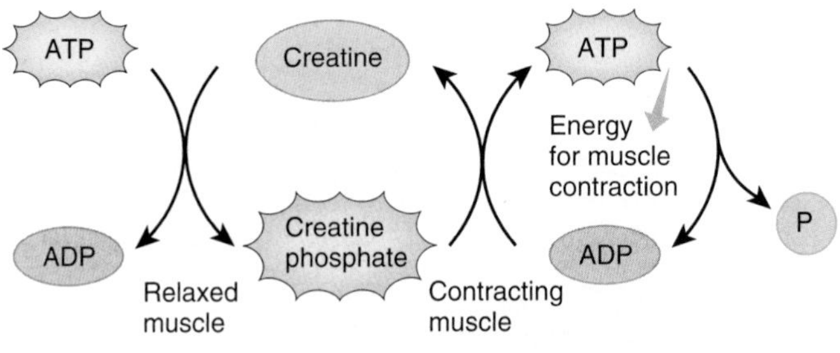

Creatine phosphate reaction FIGURE 6.18

Creatine picks up free phosphate groups, making them available for the conversion of ADP to ATP, increasing energy available for muscle contraction.

MUSCLE TWITCHES CAN BE FAST, INTERMEDIATE, OR SLOW

What causes some muscles to enlarge with exercise, whereas others seem to get stronger without any outward or visible changes? There are three types of muscle cells—**fast twitch** (or fast glycolytic), **intermediate** (or fast oxidative-glycolytic), and **slow twitch** (or slow oxidative) (FIGURE 6.19). Slow twitch muscle cells appear red, have a large blood supply, have many mitochondria within their sarcolemma, and store an oxygen-carrying protein called **myoglobin**. These cells are sometimes called **nonfatiguing** or **aerobic** cells. Everything about these muscle cells is designed to provide oxygen to the mitochondria to sustain the supply of ATP within the sarcomeres. Distance running and other aerobic sports stimulate these cells. In these muscle cells, efficiency and strength come not from increasing mass but from using oxygen more efficiently.

Fast twitch, or **anaerobic,** muscle cells are almost total opposites. Fast twitch cells provide a short burst of extreme energy and contraction power, but they fatigue quickly. Fast twitch cells are thicker, contain fewer mitochondria, usually contain larger **glycogen** reserves, and have a less developed blood supply. These are the cells that are responsible for hypertrophy. Because short bursts of power come from these fibers, exercises that continuously require bursts of power will enlarge them. Weight training puts demands on fast twitch fibers, resulting in the hypertrophy (muscle enlargement) we associate with body-building.

> **Glycogen**
> A large polysaccharide easily broken down to release individual glucose molecules.

Although training can alter the functioning of both red (slow twitch) and white (fast twitch) fibers, it does not change their proportions. Training can cause fast twitch fibers to function more like slow twitch fibers, providing more endurance with increased exercise. Despite this, your percentage of fast and slow twitch fibers is genetically determined. The ratio can, however, differ for each muscle group. You may have a preponderance of fast twitch fibers in your shoulder and back muscles, whereas your quadriceps muscle group may contain more slow twitch fibers. Olympic-caliber athletes are often those blessed with higher percentages of red or white fibers than the average person. Sprinters, obviously, benefit from a high proportion of fast twitch muscles, while long-distance skiers need more aerobic muscle cells.

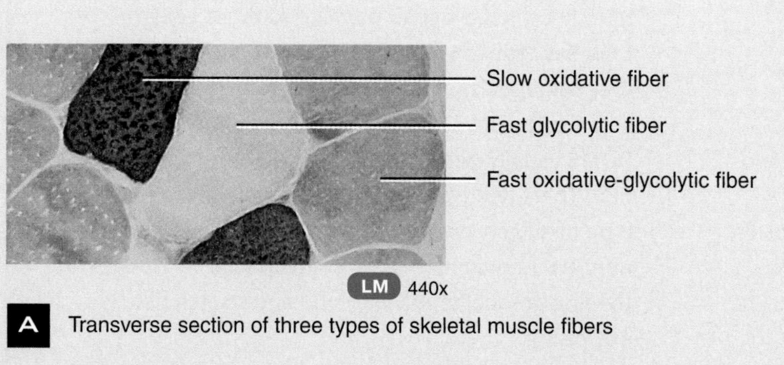

A Transverse section of three types of skeletal muscle fibers

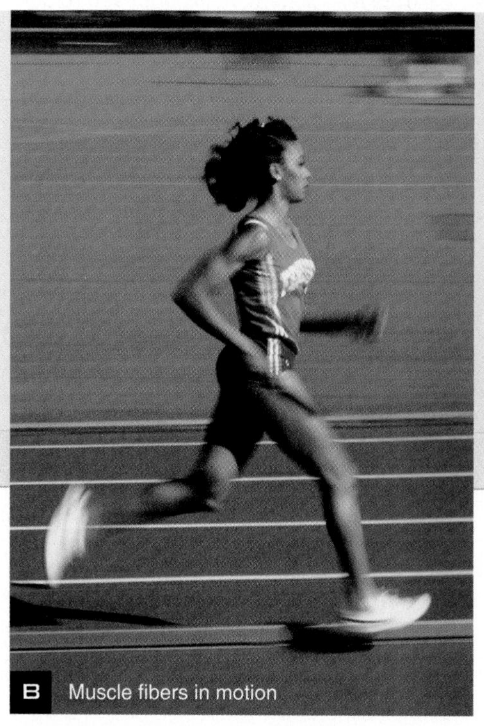

B Muscle fibers in motion

Three types of muscle fibers FIGURE 6.19

Slow oxidative fibers are synonymous with slow twitch fibers—note the thinner diameter and greater capillary flow through these cells as compared to the fast glycolytic fibers. Fast glycolytic fibers are the white fibers, carrying lots of glycogen and plenty of immediately available energy.

Health, Wellness, and Disease

Muscle fatigue, muscle woes

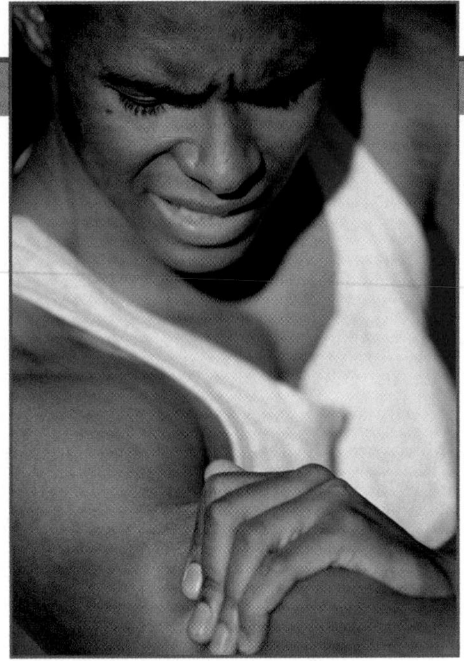

Ever watched the finish of the Iron Man endurance race? Hours after the front-runners pass the finish line come the also-rans. Some cross the line with a wobble-legged stagger and possibly without any control of bowel or bladder. These athletes are suffering from muscle fatigue—their muscles have virtually stopped responding to the nervous impulse to contract.

The physical inability to contract seems to result from deficiencies in the chemicals needed for contraction. Some scientists think it is due to a brief shortage of oxygen in the contracting muscles. Others blame fatigue on a shortage of ATP, the do-it-all energy source. Still others blame fatigue on a buildup of lactic acid as a result of anaerobic respiration. As lactic acid is released into the bloodstream, it can lead to **acidosis** (lower blood pH). Acidosis affects the central nervous system and can lead to decreased contractile and metabolic functioning.

The tiredness you sometimes experience during exercise is not physical fatigue but a response to alarm signals from stressed-out muscle tissue, which may help maintain homeostasis by maintaining pH at safe levels. The exact threshold of fatigue varies widely among humans. Scientists have recently learned that women apparently use aerobic respiration more effectively than men, which could explain why women's muscles are more resistant to fatigue. Fortunately, it's fairly easy to recover from muscle fatigue (as opposed to dehydration). Rest, adequate oxygen intake, fluids, and electrolytes usually do the trick.

Exercise may play a role in other common muscle problems:

- Muscle spasms (tics or twitches) are involuntary, and often painful, muscle contractions.
- Muscle cramps are persistent spasms that may result if overuse causes microscopic tears in muscle fibers, or from dehydration, often combined with electrolyte imbalance. Cramps often afflict muscles used for postural control—in the back and neck, for example. Fluids, electrolytes, stretching, and reasonable exercising can all help prevent and treat muscle cramps.
- Delayed-onset muscle soreness (DOMS) can result from an unusual increase in the severity or quantity of resistance training. DOMS is marked by temporary soreness, tenderness, and stiffness, and is associated with elevated levels of myoglobin and the enzyme creatine phosphokinase (CPK) in the blood. CPK facilitates the transfer of phosphate from creatine phosphate to ADP, and its appearance outside muscle cells is a sign of damage to those cells. DOMS is probably caused by tears in muscle tissue that damage the cells, by muscle spasms, or by tears in the connective tissue framework of muscles. DOMS usually peaks about two days after the exertion. There seems to be no correlation between the type of activity performed and the level of pain experienced. To prevent DOMS, build up your resistance training gradually, and warm up and stretch before exercising.

Endurance training can gradually transform fast twitch fibers to **fast oxidative-glycolytic fibers** (FOG, or intermediate fibers). These fibers are slightly larger in diameter, have more mitochondria, a greater blood supply, and more endurance than typical white fibers. The vast majority of human muscles are composed of these intermediate fibers. Regardless of which type of muscle fiber you wish to enhance, there are many benefits to an exercise regime. The I Wonder . . . feature, "What are the holistic benefits of physical exercise?" outlines a few of these benefits.

What are the holistic benefits of physical exercise?

I WONDER . . .

Finally, we have enough knowledge to judge the claims of the fitness centers, and yes, exercise has health benefits. Resistance training concentrates the enzymes that help create ATP and increases the efficiency of the enzymes in the myosin heads that break down ATP, both of which help increase muscle tone. Toned muscle uses ATP continuously, and faster. The body's shape can change as it burns extra calories to sustain the muscle mass and the white muscle fibers that have enlarged through hypertrophy.

Exercise has many other specific benefits:

- **The heart:** The heart is a muscle, and muscles get stronger with use. A stronger heart is better able to resist setbacks like clogged arteries or rhythm problems that can cause major trouble in weaker hearts. Aerobic exercise can reduce blood pressure, which reduces the exposure to strokes. As the American Heart Association says, "Physical inactivity is a major risk factor for heart disease and stroke and is linked to cardiovascular mortality."
- **The blood:** Exercise changes the lipid profile, increasing high-density lipoprotein, which seems to protect the arteries, and reducing low-density lipoprotein and triglycerides, which are associated with hardening and narrowing of the arteries, especially in the heart.
- **The immune system:** Many studies show a link between physical exercise and resistance to colds and other viral infections.
- **The endocrine system:** Exercise is a key recommendation for slowing the course of type II diabetes. Exercise can also reduce the symptoms, probably because it increases the body's use of insulin and reduces the level of glucose in the blood.
- **The psyche:** Exercise is great for reducing stress. The body makes compounds called endorphins, which are related to opiates and may have a calming effect. Curiously, recent research shows that exercise may be more reliable at reducing negative feelings like anxiety or depression than at inducing positive feelings, and that the benefits appear at exercise levels that do not cause endorphin formation.
- **The skeletal system:** Impact exercises—especially running—help keep the bones dense, warding off osteoporosis. Increasing the muscular tug on the bones has also been shown to increase bone density.
- **The individual:** Exercise has an antiaging component. Although it does not exactly reverse aging, it does help older people retain or improve their ability to do what gerontologists call the activities of daily life. Repeated studies show that exercise prevents injury and increases balance, endurance, walking speed, and rates of spontaneous physical activity throughout life.

All in all, both aerobic exercise and resistance training have a place in maintaining and restoring health. Exercise is better for you than any pill and cheaper than most healthcare regimes!

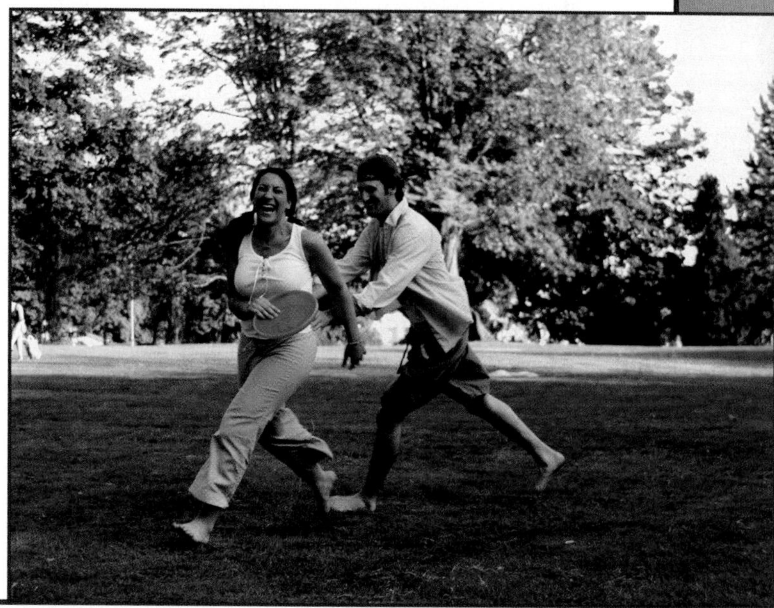

TONED MUSCLES WORK BETTER, LOOK BETTER

Muscle tone
Constant partial contraction of muscle when the body is "in shape."

When muscles are used often, we say they have "good **muscle tone**." What we are really saying is that even at rest, some muscle cells are always contracted. In a toned muscle, individual cells sporadically contract and relax, causing no movement but keeping the muscle taut. We can see muscle definition through the skin, due to this partial contraction. Increased tone is an important benefit of regular exercise, and not just for the "buff" look. Toned muscles are more effective at burning energy, meaning they use more ATP per gram than less-toned muscle tissue. People who are in shape can eat more without gaining weight because that continual, low-level contraction burns ATP.

Here is our first indication about the truth of those fitness club claims that exercise will help control weight. Making ATP takes energy in the form of calories.

Ethics and Issues

The dangers of steroid hormones

For those who want a shortcut to big, powerful muscles, testosterone and related steroid hormones have long been the drugs of choice. Testosterone and estrogen are the steroid hormones that cause sex-linked traits to emerge during puberty. In males, testosterone causes the voice to deepen and (of interest here) the skeletal muscles to grow; in females, estrogen causes growth of the breasts and plays a key role in regulating fertility. Steroid hormones that cause muscle growth are called **anabolic steroids**. Both males and females produce testosterone, which enlarges body mass by increasing the production of proteins and red blood cells.

Steroid hormones are based on cholesterol, and their lipid structure gives them the ability to diffuse through the plasma membrane. Once inside muscle cells, anabolic steroids stimulate the formation of proteins such as actin, myosin, and dystrophin, which bulk up existing muscle cells. Skeletal muscles seldom divide, so after puberty most muscle growth comes from enlargement of individual cells, called hypertrophy. Resistance training can also cause hypertrophy, and adding anabolic steroids greatly speeds the process.

Since the 1950s, athletes have injected testosterone and related compounds to sprint faster, jump higher, and lift heavier. (The hormones must be injected because they cannot withstand stomach acid.) But anabolic steroids cause more than just muscular hypertrophy. Severe and dangerous side effects include acne, masculinization of females (the former East Germany used to be famous for its bearded women swimmers), the much publicized uncontrollable anger called "roid rage," liver dysfunction, testicular cancer, kidney disease, and kidney failure. Even diabetes and hypertension have been linked to steroid abuse. Former users have commented that steroids can be addictive: The hormones produce an energetic high that disappears when use stops.

Anabolic steroids Lipid-soluble cholesterol-based compounds that stimulate increased muscle development, among other effects.

The side effects of anabolic steroids are so severe that they are regulated by the same laws covering morphine. Anabolic steroids, along with a long list of other performance-enhancing drugs, are now banned by a growing list of amateur and professional sports organizations. The bans seem to be working: After professional baseball banned steroids, some sluggers rapidly lost weight—and stopped hitting so many home runs.

Maintaining a toned muscle mass requires more ATP and therefore more calories than maintaining a less athletic body. Bottom line, a well-exercised body burns more calories in a day than an inactive body.

Exercise or chemical compounds can also change the size of a muscle. See the Ethics and Issues box, "The dangers of steroid hormones," for a discussion of steroid abuse. The muscular system is the organ system that can be altered most greatly by lifestyle choices. Scientists think the total number of muscle fibers is essentially set at birth, so how do we alter the appearance of this system? Through muscle enlargement or **hypertrophy** (hyper = above; trophy = to grow). Scientists believe hypertrophy is caused by the addition of new myofibrils within the endomysium of individual muscle cells, which thickens individual myofibers. This means hypertrophic muscles should have thicker muscle cells, packed with more sarcomeres than non-hypertrophic muscles cells. Exercise that requires muscle to contract to at least 75 percent of maximum

Evolution of humans

FIGURE 6.20

Only in the twentieth century did humans gain the luxury of choosing to work their muscular systems for better fitness. It is interesting to contemplate how different your muscular system, and therefore your entire body, would be if you depended on it for survival.

tension will cause hypertrophy. Body builders use this knowledge to create their sculpted figures. Interestingly, aerobic exercises like cycling and dancing will not cause hypertrophy, but they still provide the cardiovascular and metabolic effects of increased muscle tone.

THE MUSCULAR SYSTEM HOLDS ONE OF OUR KEYS TO SURVIVAL

To see the muscular system in a different light, consider how the human lifestyle has changed in the past 20,000 years or so. We no longer live like the other animals, where our muscular system must function at peak performance to provide nutrition and keep us safe (FIGURE 6.20). Yet our muscles were designed to provide movement, to manipulate the environment, and to help maintain homeostasis by generating internal heat. The muscular system protects the organs in our viscera and maintains our upright posture. Today, although we still need to satisfy these functions to stay alive, our technologies fulfill many of those needs. We heat our homes, wear clothing, and even add protective garb such as athletic pads to defend our internal organs from damaging blows. Our muscular system can grow flaccid without substantially endangering our survival, at least in the short term.

Throughout this unit we have explored how the skeletal and muscular systems work together not only to provide us with a structural framework but also to provide a means of locomotion. In the next chapter, we will explore the nervous system, which you have already seen has an intimate connection to muscles and movement.

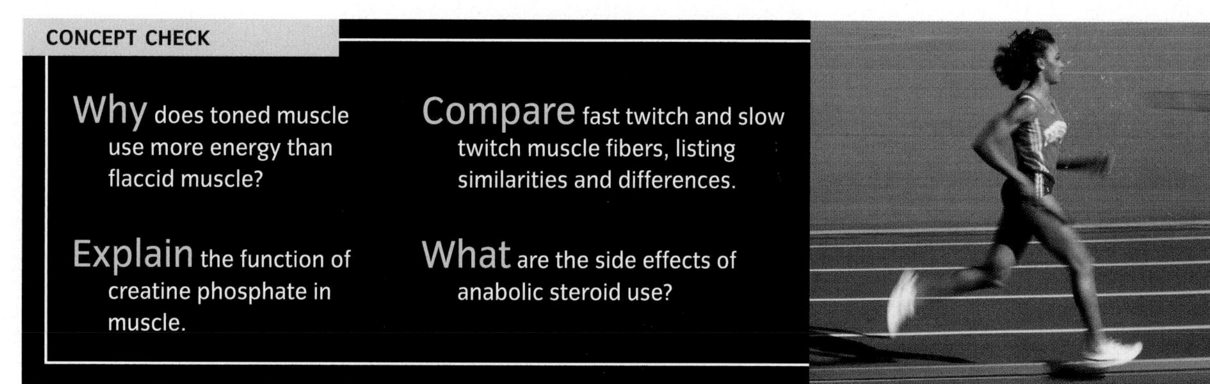

CONCEPT CHECK

Why does toned muscle use more energy than flaccid muscle?

Explain the function of creatine phosphate in muscle.

Compare fast twitch and slow twitch muscle fibers, listing similarities and differences.

What are the side effects of anabolic steroid use?

CHAPTER SUMMARY

1 The Muscular System Has Many Functions

The muscular system is composed of skeletal muscle tissue. Its most obvious function is to generate movement; however, it also generates heat, stabilizes joints, helps move lymphatic fluid through the body, and supports and protects soft internal organs.

2 Skeletal Muscles Are Contractile Organs

Muscles are highly organized organs, protected by layers of connective tissue. The epimysium covers the entire organ, with fascicles of muscle cells covered in perimysium and individual myofibers surrounded by endomysium. Skeletal muscles extend from the immovable origin to the movable insertion and often work in antagonistic pairs. Within the muscle itself, the proteins actin and myosin are arranged in patterned bundles called sarcomeres. Thick myosin filaments are surrounded by thinner actin filaments, attached directly to the Z discs. The sarcomere is the contractile unit of skeletal muscle, extending from one Z disc to another.

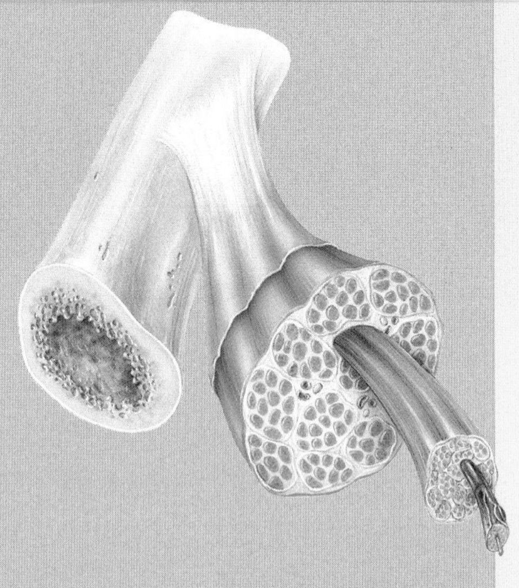

3 Muscle Contraction Occurs as Filaments Slide Past One Another

The contraction of muscle is controlled by nervous impulses passed to the muscle cell at the neuromuscular junction. The nerve cell dumps acetylcholine into the synapse between it and the muscle cell, beginning muscular contraction. A series of chemical interactions causes the filaments of myosin and actin to slide past each other, shortening the sarcomere and the muscle. Initially, calcium is released into the muscle cell, binding to the troponin and tropomyosin of the actin filament. The actin active sites are exposed so they can react with the energy-rich myosin heads. ATP is used to relax the contracted muscle.

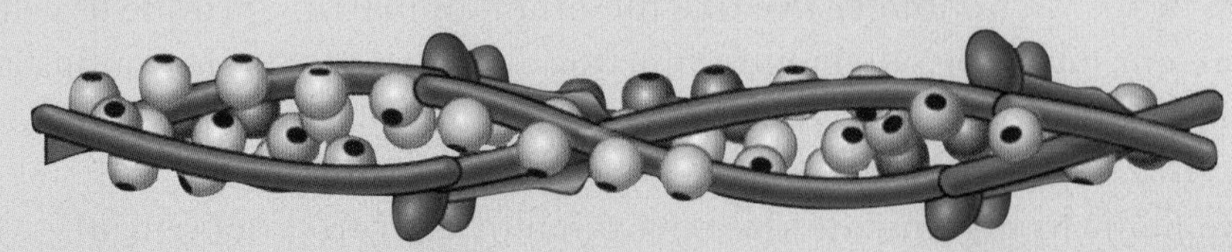

4 Whole-Muscle Contractions Emerge from Tiny Impulses

Whole-muscle contractions produce powerful movement from the combined simultaneous contraction of millions of sarcomeres. Muscle cells contract in an all or nothing fashion, with identifiable phases. Each motor unit contracts in unison when the motor neuron fires with enough force to reach threshold. Initially, the muscle cells show no outward signs of shortening; this is the latent period. Ions are moving, and slack in the muscle is being taken up. During the contraction phase, the muscle visibly shortens. During the relaxation phase, the muscle returns to its original length. Tetanus is a combination of many twitches, overlapping to produce constant tension in the muscle, a common occurrence in the muscles of your back and neck.

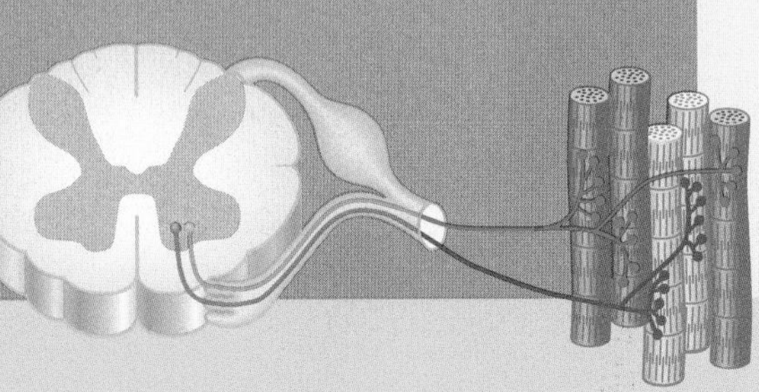

5 Muscles Require Energy to Work Smoothly and Powerfully

Muscle cells can produce different types of movement. All movement requires ATP, either stored in the cell, or produced via metabolic pathways. Cellular respiration cannot keep pace with the energy demands of strenuously used muscles, so creatine phosphate is employed to store inorganic phosphate for conversion of ADP to ATP. Individual muscle cells respond differently to twitch impulses. Muscle fibers can be fast and easily exhausted, slow and nonfatiguing, or somewhere in between. Fast twitch fibers have a large supply of ready energy, with limited ability to remove waste or create ATP. Slow twitch fibers have less immediate energy but more ability to create ATP and remove wastes. Most skeletal muscle in the human body is composed of intermediate fibers.

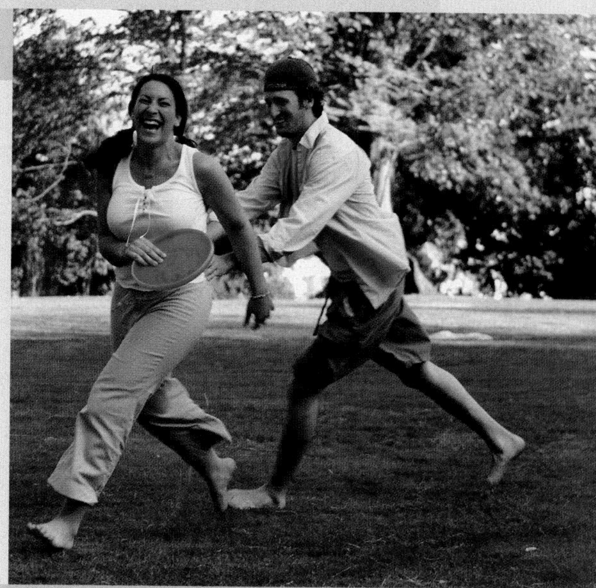

KEY TERMS

- **aerobic pathway** p. 175
- **anabolic steroids** p. 180
- **anaerobic pathways** p. 175
- **antagonistic (synergistic) pair** p. 159
- **electron transport chain** p. 175
- **endomysium** p. 162
- **epimysium** p. 162
- **globular** p. 166
- **glycogen** p. 177
- **graded contraction** p. 172
- **muscle tone** p. 179
- **myofibrils** p. 162
- **oxygen debt** p. 176
- **perimysium** p. 162
- **skeletal muscle** p. 158
- **T tubules** p. 162
- **TCA (Krebs) cycle** p. 175
- **threshold stimulus** p. 172
- **treppe** p. 173

CRITICAL THINKING QUESTIONS

1. In Greek mythology Achilles was a heroic warrior, undefeated in many battles. His undoing was an arrow to the tendon of the gastrocnemius muscle (see FIGURE 6.3 on p. 161 for the exact position). Using the terms *origin, insertion,* and *belly,* explain the location of his wound. In common language, why did the arrow end Achilles' fighting career? Anatomically speaking, what destroyed his fighting ability?

2. Briefly describe the structure of a muscle cell, starting with the sarcolemma and ending with the structure of the sarcomere. Why do you suppose muscle cells are set up this way? Where is their strength? Are any weaknesses created by this arrangement? *Hint*: Envision the cut end of a rope or cable. What happens to the arrangement of the fibers? What happens to a rope if you apply tension from the side rather than the end?

3. When a muscle is stimulated to contract by an external electrical source instead of the motor neurons, there is a period before external movement appears. We know ATP is being used immediately after the current is applied. What is happening during this latent period? How would you expect the latent period to compare between a toned, "in shape" individual and someone without good muscle tone?

4. List the sources of energy that are readily available for muscle contraction. What happens in endurance events? Where do the muscles of the leg get their steady energy supply during a grueling athletic event like a marathon? Does it make sense for endurance athletes to take in nutrients during events?

5. We know training affects muscle fibers by making them more efficient. Specifically how does this occur? Assume you have begun endurance training for the Tour de France. What will this training do for your red muscle fibers? For your white and intermediate muscle fibers? Can training alter the proportion of these fibers?

SELF TEST

1. The functions of the muscular system include all of the following EXCEPT

a. manipulating and moving through our environment.
b. stabilizing joints during motion.
c. generating heat.
d. aiding in blood flow through the body.
e. All of the above are functions of the muscular system

2. Looking at your own biceps brachii (the muscle that allows you to flex your arm), locate its insertion.

a. The humerus
b. The elbow
c. The radius
d. The carpals

3. The muscle that is primarily responsible for any action of the body is referred to as the

a. antagonist.
b. agonist.
c. synergist.
d. fixator.

4. Identify the outermost layer of connective tissue surrounding a muscle (identified as A in the figure).

a. Epimysium
b. Endomysium
c. Perimysium
d. Myofiber

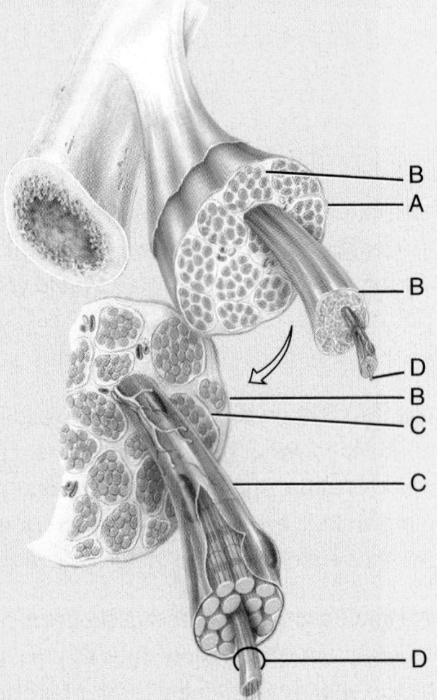

5. The structure indicated as A in this figure serves to

a. sequester calcium.
b. house actin and myosin.
c. protect the muscle cell.
d. carry the impulse to contract quickly through the entire cell.

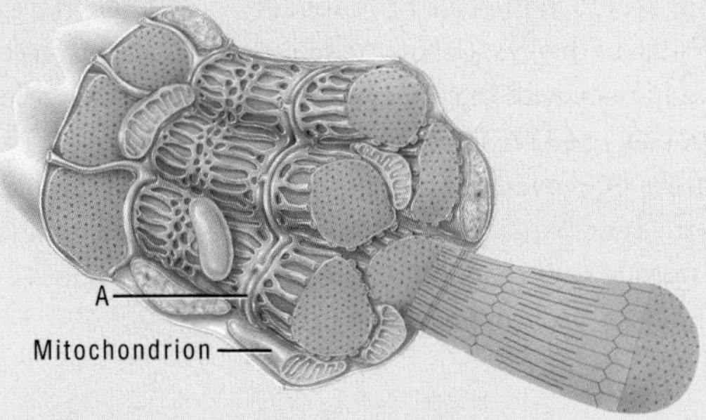

6. The contractile unit of skeletal muscle is the

a. sarcomere.
b. sarcolemma.
c. epimysium.
d. actin.

7. The Z discs are represented in this image by structure

a. A.
b. B.
c. C.
d. D.

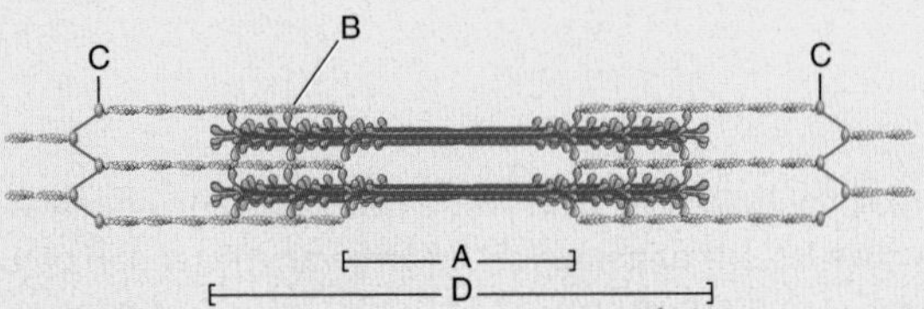

8. The globular protein instrumental in muscle contraction is found in the ____________ of the sarcomere.

a. A band
b. I band
c. Z disc
d. middle
e. Both a and b are correct.

9. The events at the neuromuscular junction begin with

a. ACh binding to receptors on the muscle cell.
b. neurotransmitter being dumped into the neuromuscular synapse.
c. calcium being released from the SR.
d. sliding filaments.

10. The protein to which calcium binds in step one of this image is

a. actin.
b. myosin.
c. troponin.
d. tropomyosin.

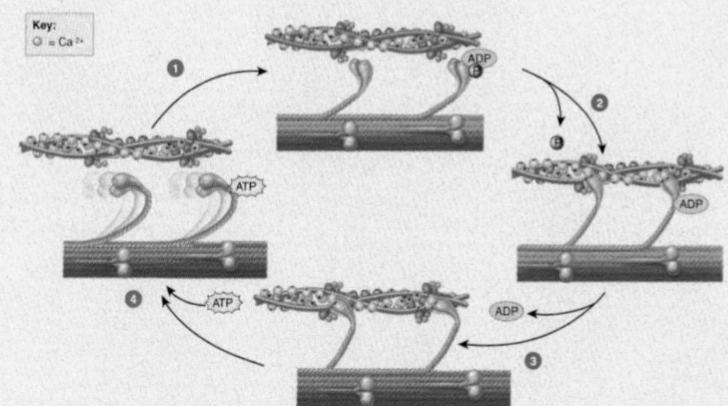

11. True or False? Once calcium binds to the proper protein, moving it off the active site, myosin heads bend toward the center of the sarcomere.

12. This graph indicates that

a. muscle tension is independent of sarcomere length.
b. muscle tension increases steadily with increasing sarcomere length.
c. powerful contractions can only be generated in a very narrow range of sarcomere lengths.
d. sarcomeres with Z discs nearly touching generate more power than those with Z discs spread far apart.

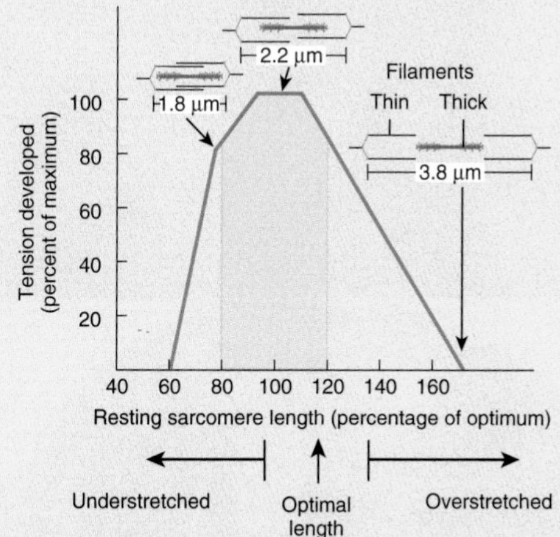

13. The portion of the myogram indicated as B corresponds to what action?

a. Relaxation
b. Latent period
c. Contraction
d. Treppe

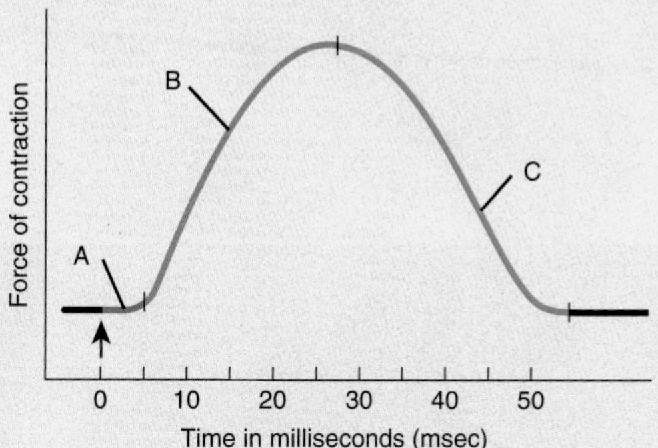

14. The most efficient production of energy for muscular contraction is

a. aerobic pathways.
b. anaerobic pathways.
c. lactic acid metabolism.
d. creatine phosphate.

15. The muscle fiber that is quick to contract and quick to fatigue is the

a. fast glycolytic fiber.
b. slow oxidative fiber.
c. non-fatiguing fiber.
d. aerobic fiber.

16. Identify the type of movement produced by the muscle indicated as B on the figure.

a. Supination
b. Extension
c. Adduction
d. Flexion
e. Abduction

17. Identify the muscle indicated as A on the diagram at right.

a. Rectus abdominus
b. Triceps brachii
c. Quadriceps group
d. Pectoralis major

18. The antagonist for the muscle labeled B on the same figure is the

a. hamstrings.
b. gluteus maximus.
c. deltoid.
d. triceps brachii.

19. Identify the type of movement produced by the muscle indicated as B on the figure.

a. Extension
b. Plantar flexion
c. Rotation
d. Dorsiflexion

20. The muscle identified as A on the same diagram moves the

a. spinal column.
b. base of the skull.
c. shoulder.
d. forearm.

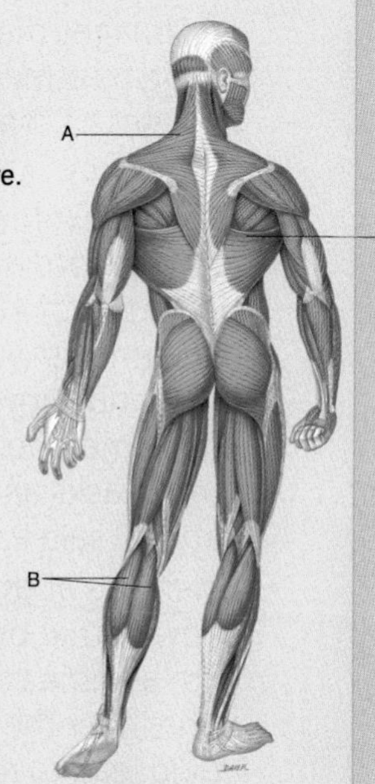

The Nervous System

7

What part of your brain is involved in language? In memory? In emotions? Neuroscientists used to answer these questions by looking at different types of brain damage and relating them to specific neurological problems. Now, highly sophisticated machines are peeking inside living human brains—and showing an astonishing level of detail about learning, emotions, and memory. Chief among these harmless techniques is functional magnetic resonance imaging, or fMRI. Regular MRI shows the location of soft tissue; fMRI tracks the movement of glucose through the brain. Because glucose is the basic fuel for the brain, fMRI shows which areas are active at any moment.

Results from just the past couple of years show how much fMRI can reveal about brain function:

- Before surgery to correct epilepsy, fMRI can locate speech centers, which are often damaged by this surgery. By identifying where in the brain the patient forms words, surgeons can avoid damaging the ability to speak.
- Brain images show differences between the brains of dyslexic children and normal readers. Images made after intensive language treatment show how the brain changes as the children gain language proficiency.
- Men and women use their brains differently, according to fMRI studies from the University of Alberta. "Sometimes males and females would perform the same tasks and show different brain activation, and sometimes they would perform different tasks and show the same brain activation," said PhD student Emily Bell.
- Scientists at the University of Wisconsin showed that brain regions associated with asthma can be activated when patients hear the word "wheeze." The study could lead to new drugs and/or a better appreciation of the brain's role in asthma.

CHAPTER OUT
The Nervous System Makes Sens
of Everything p. 188
The Nervous System Is Categoriz
by Function and Structure p. 19
Nerve Tissue Is Made of Neurons
and Glial Cells p. 193
Neurons Work through Action
Potentials p. 194
The Brain and Spinal Cord Are Ce
to the Nervous System p. 202
The Peripheral Nervous System

The Nervous System Makes Sense of Everything

LEARNING OBJECTIVES

List the functions of the nervous system.

Describe the main difference between the endocrine system and the nervous system.

Lift this book. Turn the page. Scan the words with your eyes and understand them with your brain. All of these conscious movements are directed by the nervous system. Brush a bothersome hair off your face. Listen to tires crunch the pavement as a car drives past the open window. Smell the flowers outside. All of these sensations are brought to you compliments of the nervous system. Every conscious action that occurs in your body is governed by the nervous system. So are most of the "unconscious" or automatic actions that maintain homeostasis.

When skeletal muscles contract, they do so in response to stimuli from the nervous system. We plan our movement in the brain, and the nervous system transmits that plan to the muscles. At the muscles, the nervous system stimulates contraction but stimulates only those motor units needed for that particular task. In Chapter 6 you learned about neuromuscular junctions. Review FIGURE 6.8 on page 168 for a quick reminder of this structure.

Although this type of nervous system activity is familiar, the nervous system has numerous other functions, some better understood than others. The nervous system is used to communicate from one end of the body to another. The nervous system receives and integrates stimuli and formulates an appropriate response. The stimulus can be an external change, such as a shift in temperature or sound, or it can be an internal change, such as a localized decrease in blood pressure or a general increase in carbon dioxide levels in the tissues. Whatever the change, the nervous system's job is to immediately detect it and adapt in order to maintain homeostasis.

Often that change will involve the endocrine system, which produces hormones that work in concert with the nervous system. The nervous system usually initiates immediate short-term responses, using **neurons** (FIGURE 7.1) and **neurotransmitters** to produce near-instant results. In contrast, the endocrine system relies on chemical interactions of hormones and target cells, which take longer to initiate a response than neural responses but tend to last longer. Your development from infancy to adulthood is driven by hormones, whereas your startled jump at the sound of a car's backfire is caused by the nervous system.

Neuron A nerve cell that sends and receives electrical signals.

Neurotransmitter A chemical used to transmit a nervous impulse from one cell to the next.

Neuron FIGURE 7.1

The neuron is the functional unit of the nervous system. These remarkable cells are responsible for carrying sensory information into the brain, formulating a response, and sending that response out to the proper organs. The arrows indicate the direction of impulse propagation.

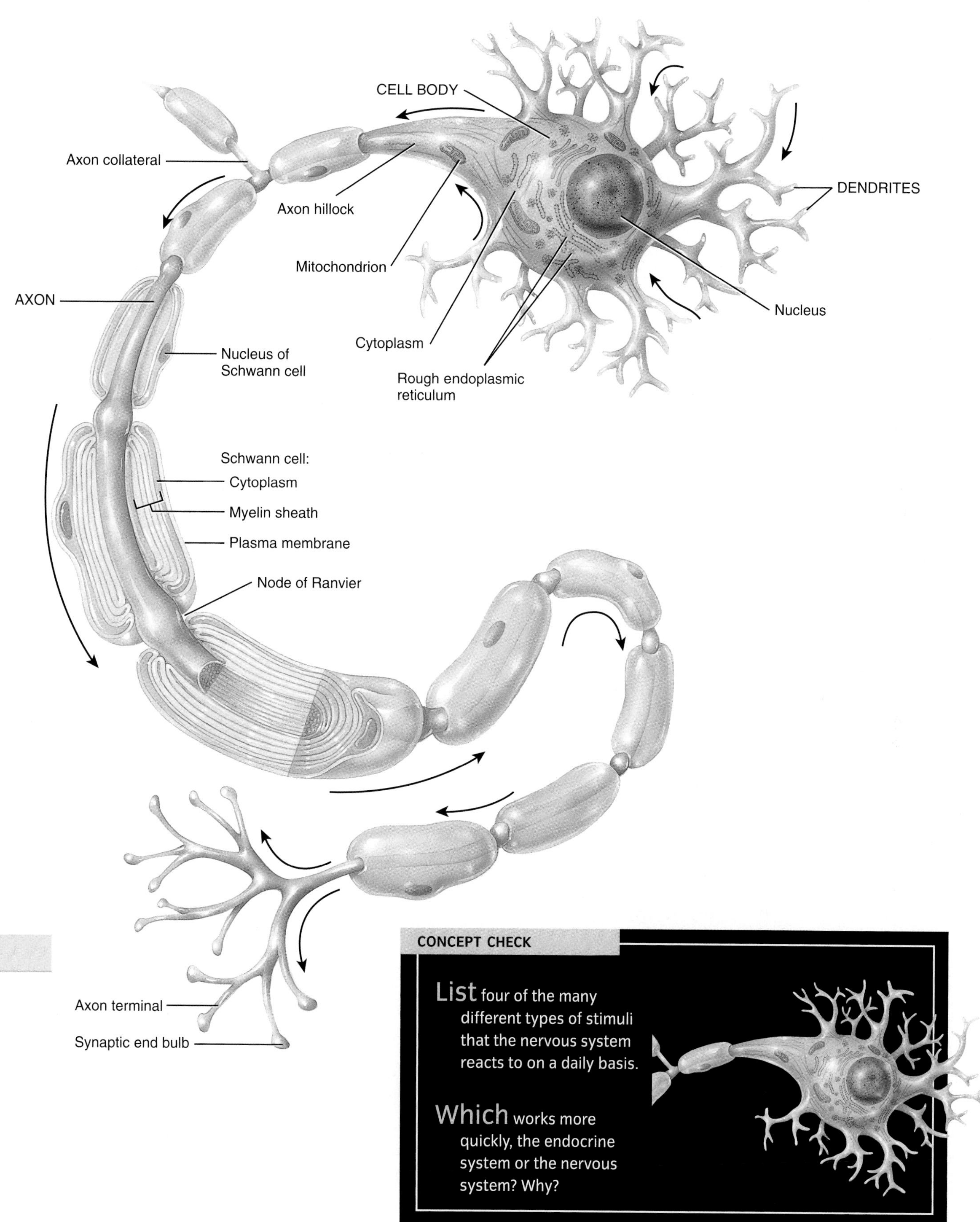

CONCEPT CHECK

List four of the many different types of stimuli that the nervous system reacts to on a daily basis.

Which works more quickly, the endocrine system or the nervous system? Why?

The Nervous System Is Categorized by Function and Structure

Learning Objectives

Outline the major divisions of the nervous system.

List the three types of receptors in the afferent nervous system.

Explain the differences between the somatic and autonomic divisions of the efferent peripheral nervous system.

Describe the functions of the sympathetic and parasympathetic divisions of the autonomic nervous system.

The nervous system has two components: the **central nervous system** (CNS) and the **peripheral nervous system** (PNS) (FIGURE 7.2). This distinction is based mainly on location. The CNS includes the **brain** and **spinal cord**. It lies encased in the axial skeleton and is covered by the meninges.The CNS is the main integration center of the body. Sensory information comes in to the CNS, where it is analyzed and an appropriate motor response is generated. The motor response is usually directed toward muscular or glandular tissue.

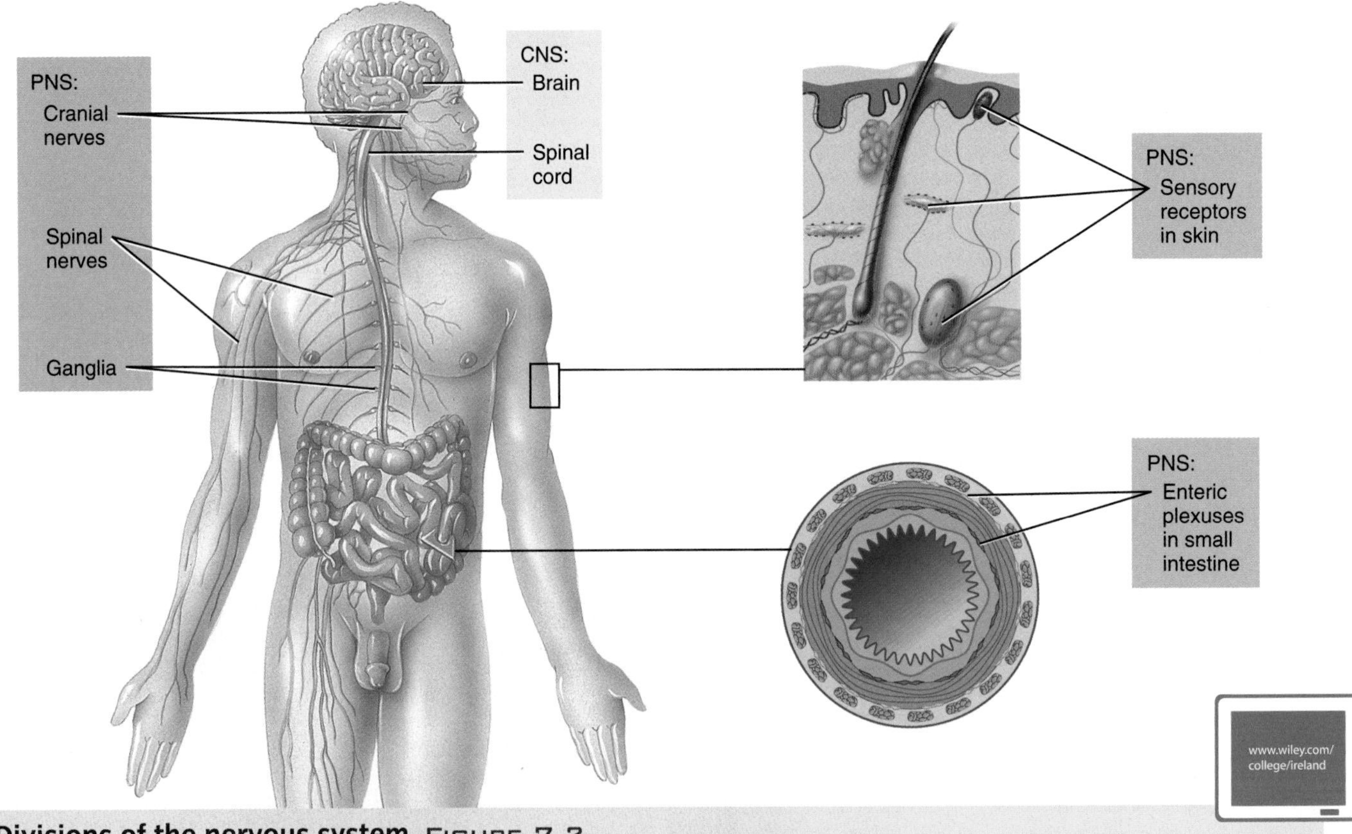

Divisions of the nervous system FIGURE 7.2

The two main divisions of the nervous system are the central nervous system (CNS), consisting of the brain and spinal cord and the peripheral nervous system (PNS), consisting of all nervous tissue outside the CNS. Subdivisions of the peripheral nervous system are the somatic nervous system, the autonomic nervous system, and the enteric nervous system.

Afferent
Sensory impulses moving toward the brain.

Efferent
Motor impulses moving from the brain to the effector organ.

The PNS is composed of all the **afferent** and **efferent** neurons that extend from the CNS. The neurons of the PNS are arranged in bundles called **nerves** (FIGURE 7.3). Nerves can be motor, sensory, or mixed, depending on what type of neurons they contain.

Most information going to and from the central nervous system travels through the peripheral nervous system. Information reaches the CNS from the afferent division of the peripheral nervous system. The PNS picks up this information with one of three types of receptors: special senses, general sensory receptors, or visceral receptors. These receptors allow us to experience many different sensations. Our **special senses** enable us to see, hear, taste, and smell the external world. Our skin has **general sensory receptors** that inform us about external temperature as well as light touch, pressure, and pain. Within our bodies, visceral receptors monitor **proprioception** and organ functioning. Stomach aches and sore throats are examples of visceral sensory input.

Motor responses are formulated in the CNS and taken to the muscles or glands by the efferent division of the PNS. Here again, the impulses can travel on different pathways. To consciously move skeletal muscle, we plan an activity in the CNS and then direct the muscles to carry it out through motor commands sent by the **somatic division** of the PNS. This division is sometimes called the voluntary division, because the motor commands are consciously, and therefore voluntarily, controlled. However, the involuntary movement of reflexes is also part of this division. The same motor neurons that stimulate reflexive movements are used for conscious movements.

Proprioception
Stimuli from within the body giving information on body position and posture.

Somatic division
Division of the nervous system involved in conscious movement.

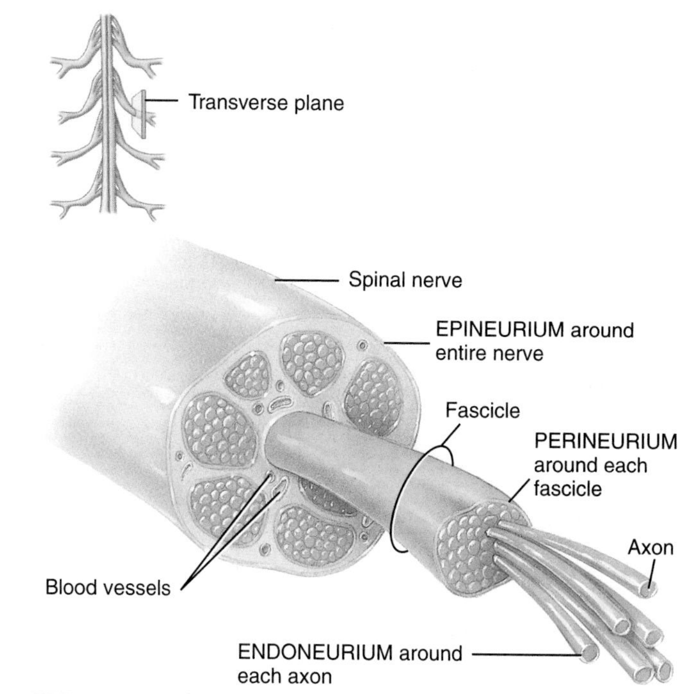

A Transverse section showing the coverings of a spinal nerve

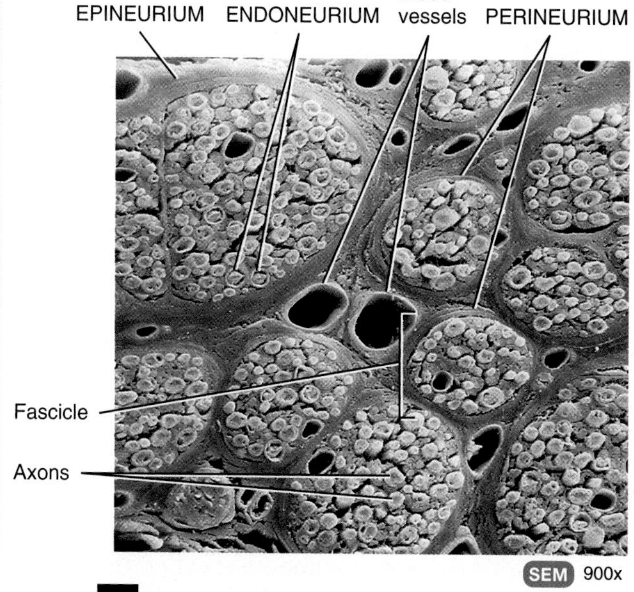

B Transverse section of 12 nerve fascicles

Spinal nerve
FIGURE 7.3

Nerves are bundles of individual neurons, held together with connective tissue wrappings. Outside of the nerve is the protective epineurium. Within the nerve, the perineurium sheathes groups of neurons; individual neurons are covered with an endoneurium.

Outline of the nervous system, comparing the sympathetic and parasympathetic characteristics TABLE 7.1

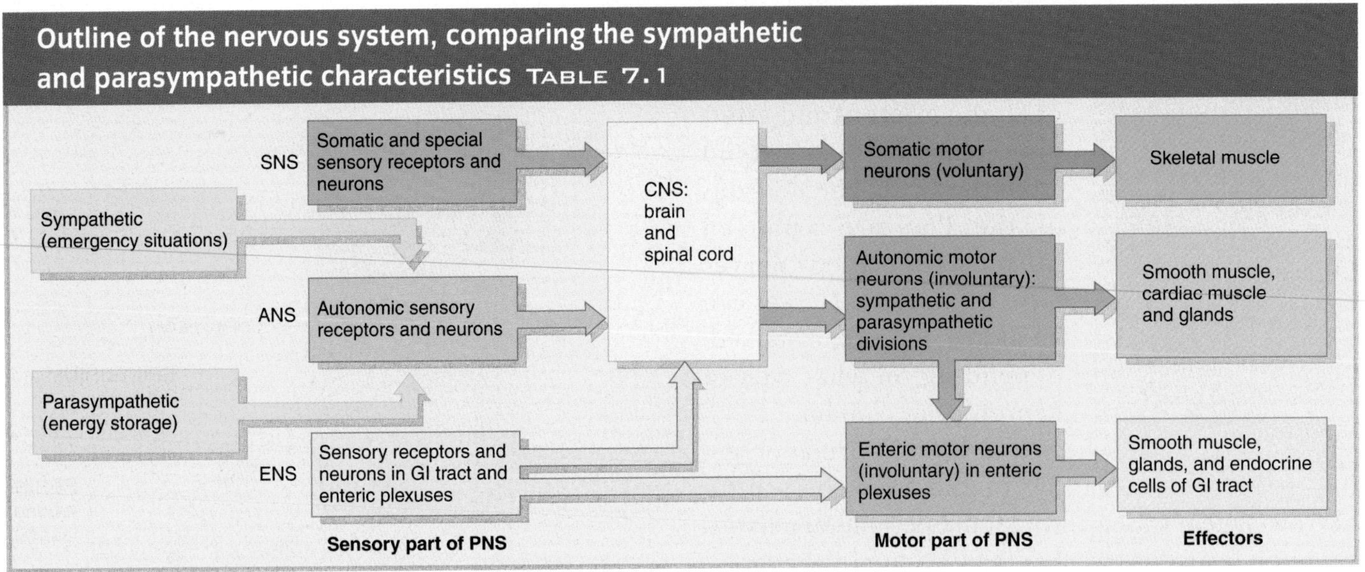

THE AUTONOMIC NERVOUS SYSTEM WORKS WHILE YOU SLEEP

Autonomic division (ANS) Division of the nervous system regulating functions such as blood vessel diameter and stomach activity.

The **autonomic division** of the PNS, also known as the ANS, is a control system that governs your body's responses to subtle changes in homeostasis with involuntary, unconscious reactions. For example, the CNS continually generates responses to sensory input concerning blood pressure, blood gases, and visceral functioning. You are not aware of these inputs, nor do you control the motor responses that travel through the autonomic nervous system.

The autonomic nervous system has two subdivision (TABLE 7.1). The first subdivision, the **sympathetic division,** includes those nerves that control the body when it is actively moving and burning energy. The sympathetic division is sometimes called the "fight or flight" division, because it is triggered when we feel threatened and must choose to remove ourselves from the danger (flight) or stay and "fight." The **parasympathetic division** is responsible for digestion, energy storage, and relaxation.

These divisions are nicely separated by the contradictory demands of human life. Sometimes we must conserve energy and rest; other times we must move quickly and expend energy. When competing in an athletic event, or running from an alligator, we must use energy. Conversely, after a feast (alligator steak, anyone?), we must digest the meal and store the energy.

Almost every organ of your body has dual innervation, meaning that it is stimulated and controlled by both the sympathetic and the parasympathetic divisions. The two systems work antagonistically to maintain homeostasis, with only one system stimulating the organ at any given time. You can easily determine which system is in control by looking at the organ's activity. If the organ is burning energy, releasing oxygen or glucose into the bloodstream, or otherwise aiding in sharp mental capacity and quick responses, the sympathetic division is working. If the organ's function is conducive to rest and relaxation, you can bet the parasympathetic division is in control.

The functions of these two divisions are easy to remember. The sympathetic division is sympathetic to your plight. It is active when you need quick energy and rapid movement. The parasympathetic division starts with "P," like potato. When this system is active you are relaxing—acting like a "couch potato."

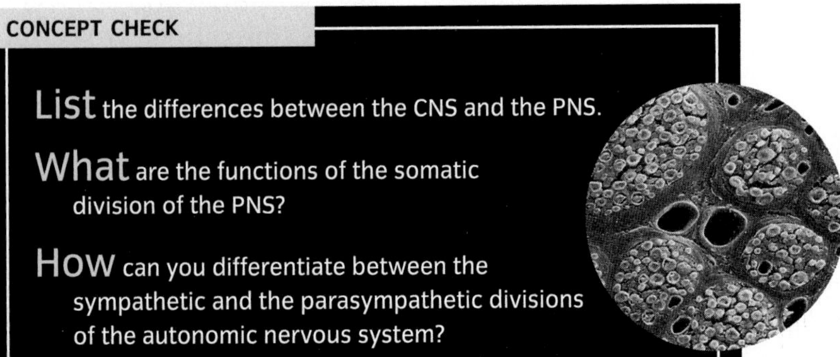

CONCEPT CHECK

List the differences between the CNS and the PNS.

What are the functions of the somatic division of the PNS?

How can you differentiate between the sympathetic and the parasympathetic divisions of the autonomic nervous system?

Nerve Tissue Is Made of Neurons and Glial Cells

LEARNING OBJECTIVES

List the functional categories of neurons.

Describe the function of the supporting neuroglia.

Nervous tissue, one of the four main tissue types, is composed of neurons and supporting cells called **neuroglia** or simply glia (singular *neuroglion*). The types and functions of the neuroglia are listed in TABLE 7.2. The three classes of neurons are based on function: **sensory neurons, motor neurons,** and **interneurons** (FIGURE 7.4). Each type has a distinctive shape, allowing ready identification. Despite their anatomical differences, all neurons

Neuroglia
Supporting and protecting cells within the nervous system, including cells that provide nutrients, remove debris, and speed impulse transmission.

Neuroglia size and shape, location, and function TABLE 7.2

Name	*Location*	*Function*
Satellite cells	PNS	Regulate oxygen, carbon dioxide, nutrient, and neurotransmitter levels around ganglia
Schwann cells	PNS	Surround axons in PNS, causing myelination of axons and faster impulse transmission, aid in repair after injury
Oligodendrocytes	CNS	Myelinate CNS neurons, provide structural support
Astrocytes	CNS	Maintain blood-brain barrier, regulate nutrient, ion and dissolved gas concentrations, absorb and recycle neurotransmitters, form scar tissue after injury
Microglia	CNS	Clean up cellular debris and pathogens via phagocytosis
Ependymal cells	CNS	Line ventricles and central canal of cord, assist in CSF production

www.wiley.com/college/ireland

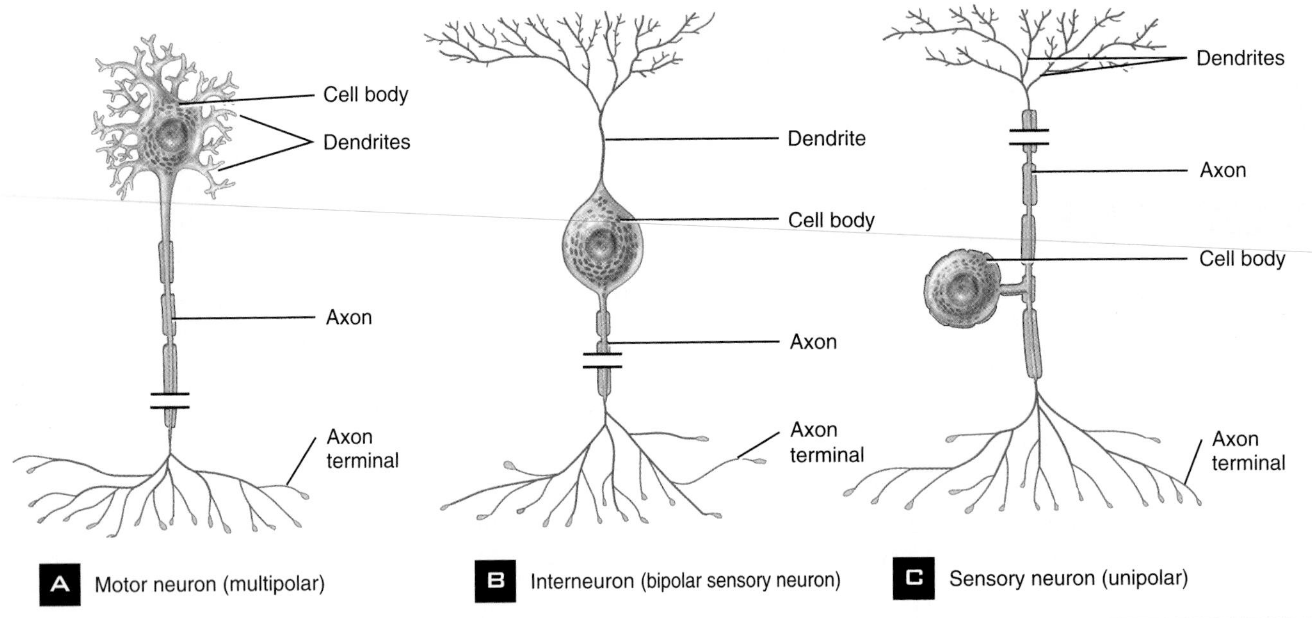

Motor neurons, interneurons, and sensory neurons FIGURE 7.4

have a cell body, one axon, and at least one dendrite. The dendrite(s) bring information to the cell body. There can be many dendrites, with the branches providing many avenues for incoming impulses. The single axon routes the nerve impulse from the cell body to another neuron or an effector organ. The axon can have many terminal branches, so each time the nerve fires, it can stimulate more than one cell.

CONCEPT CHECK

List three types of neuroglia and give their functions.

What are the anatomical differences between sensory and motor neurons?

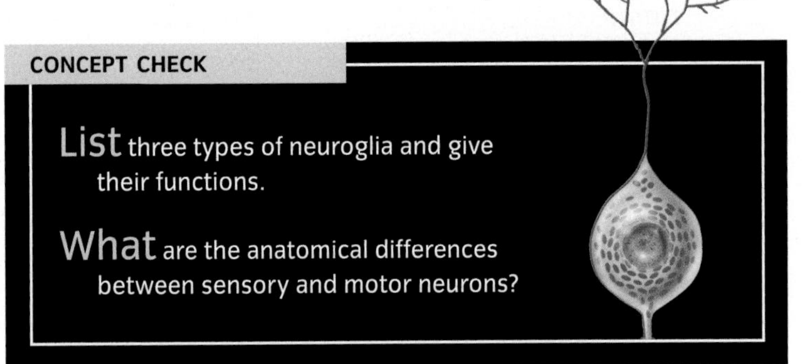

Neurons Work through Action Potentials

LEARNING OBJECTIVES

Differentiate action potential from membrane potential.

Describe the types of channels found in neuron membranes.

List the events in an action potential.

An action potential is a brief change in electrical conditions at a neuron's membrane that occurs when a neural signal arrives; it is what happens when we say a neuron "fires." What, at the molecular level, allows neurons to carry electrical impulses? How do these oddly shaped cells receive, integrate, and respond to information? The answer begins with the electrical conditions surrounding the neuronal membrane, called the neurolemma. These electrical conditions create a **membrane potential** across the neurolemma that is exploited when the nerve fires (FIGURE 7.5).

A resting neuron, as seen in FIGURE 7.6, has a membrane potential of −70 mV. The levels of positive sodium ions and negative chloride ions are higher outside the

Membrane potential
The difference in electrical charge between two sides of a membrane.

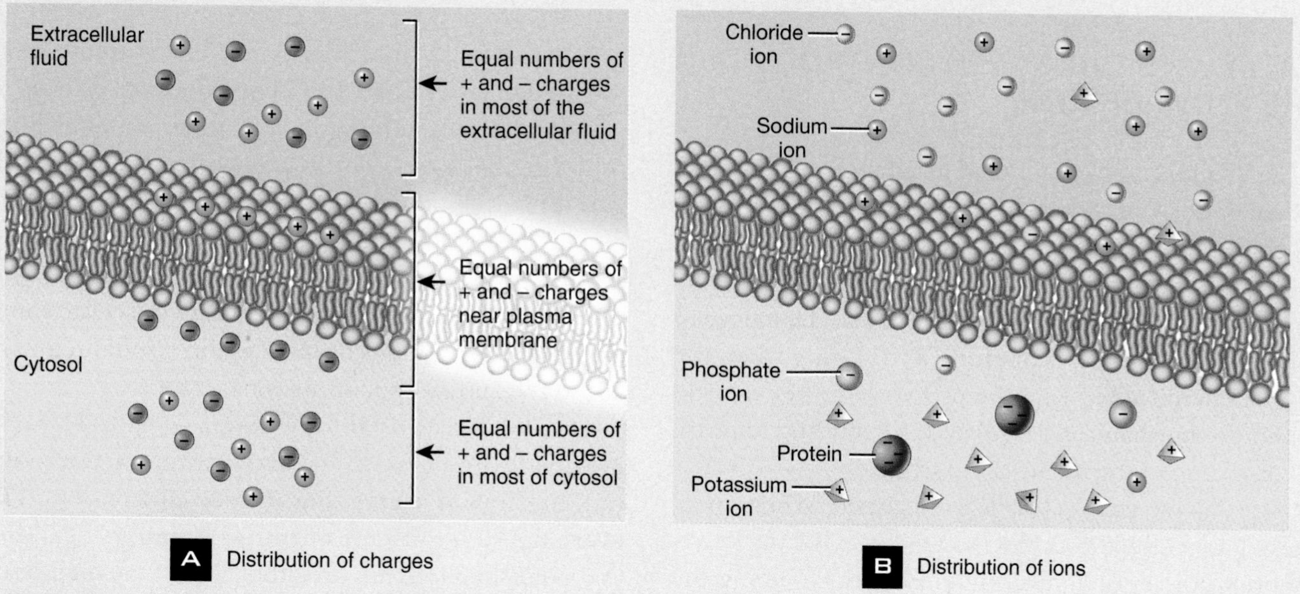

Membrane potential FIGURE 7.5

Unlike most body cells, neurons can significantly alter their *membrane potential*. The charge difference across the neurolemma (the membrane potential) alternates between −70 mV and +30 mV during a typical nerve impulse. This cyclic change of charge across the neurolemma from −70mV to +30 mV and back to −70 mV is called the nerve impulse, or *action potential*. Charge differences are controlled by the movement of sodium and potassium ions entering and leaving the neuron.

neuron than inside. Conversely, positive potassium ions are more concentrated inside the neuron than outside. Large, negatively charged proteins trapped in the neuron help to maintain the negative charge across the membrane. In the absence of a selectively permeable membrane, these differences would rapidly disappear as the ions diffused down their respective concentration gradients. Sodium would diffuse into the cell, potassium would diffuse out, and the negative charges would balance.

This diffusion does not happen, however, because ions cannot simply diffuse through the lipid bilayer of the cell membrane. Instead, they must travel through channel proteins that serve as portals for ion diffusion. Channel proteins can be either passive or active. Passive channels are "leaky" and allow a constant trickle of ions. Active channel proteins allow no ion movement unless stimulated. This means the rate of ion movement across the nerve cell membrane depends on the physical state of the channel proteins, which can vary greatly from moment to moment. This variation in ion concentration across the cell membrane allows neurons to generate action potentials.

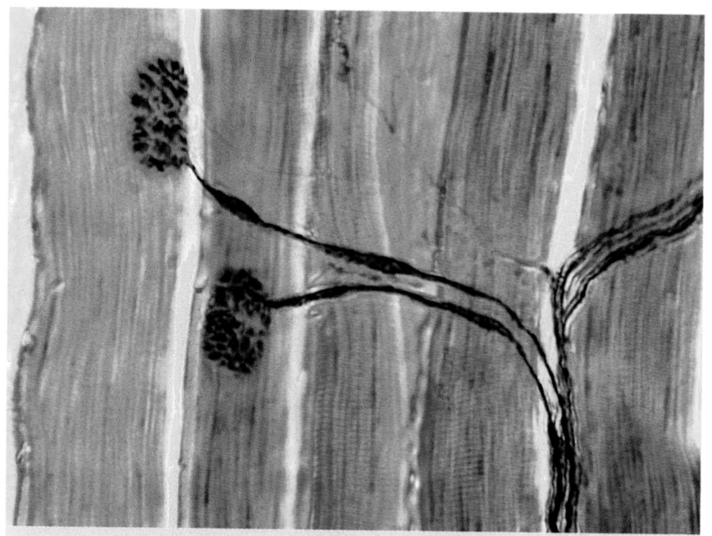

Neuron and neuromuscular junction
FIGURE 7.6

Resting neurons are visually no different from neurons undergoing an action potential. One way to determine the physiological state of a neuron is to measure the resting potential, and another is to look for the release of neurotransmitter. Resting neurons have a membrane potential near −70 mV and are not actively releasing neurotransmitters.

GATES AND CHANNELS CONTROL THE FLOW OF IONS

Active channels are often called **gated** channels, because they allow ion transport only under specific environmental conditions. Some gated channels are **voltage-gated**, opening and closing in response to transmembrane voltage changes (FIGURE 7.7A). Others are **ligand-gated** (chemically regulated) opening and closing when the proper chemical binds to them (FIGURE 7.7B). Still others are **mechanically regulated**, responding to physical distortion of the membrane surface.

At rest, gated channels are closed. When open, these gates allow ions to cross the membrane in response to their concentration gradients, changing the transmembrane potential and generating a nerve impulse. The steps of an action potential are outlined in FIGURE 7.8.

At the end of the action potential, the transmembrane potential is −90 mV. From the moment the sodium channels open until they reclose, the neuron cannot respond to another action potential. There are two phases to this inactive period. The **absolute refractory period** lasts from 0.4 to 1.0 milliseconds. During this period, sodium and potassium channels are returning to their original states. The **relative refractory period** begins when the sodium channels are again in resting condition, and continues until the transmembrane potential stabilizes at −70 mV. The **sodium potassium exchange pump** (Na^+/K^+ATPase) helps stabilize the cell at the initial ion concentrations by moving three sodium ions out of the cell and two potassium ions into it.

Scientists used to believe Na^+/K^+ATPase was needed for the neuron to carry another action potential, but now it seems that it need not operate after every nerve impulse. Enormous numbers of sodium and potassium ions are on either side of the membrane, and the subtle concentration changes of one action potential do not block impulse transmission. It would take literally thousands of consecutive action potentials to alter the ion concentrations enough to destroy the overall mechanism. The Na^+/K^+ATPase merely helps return the local membrane potentials quickly so a second action potential can be generated.

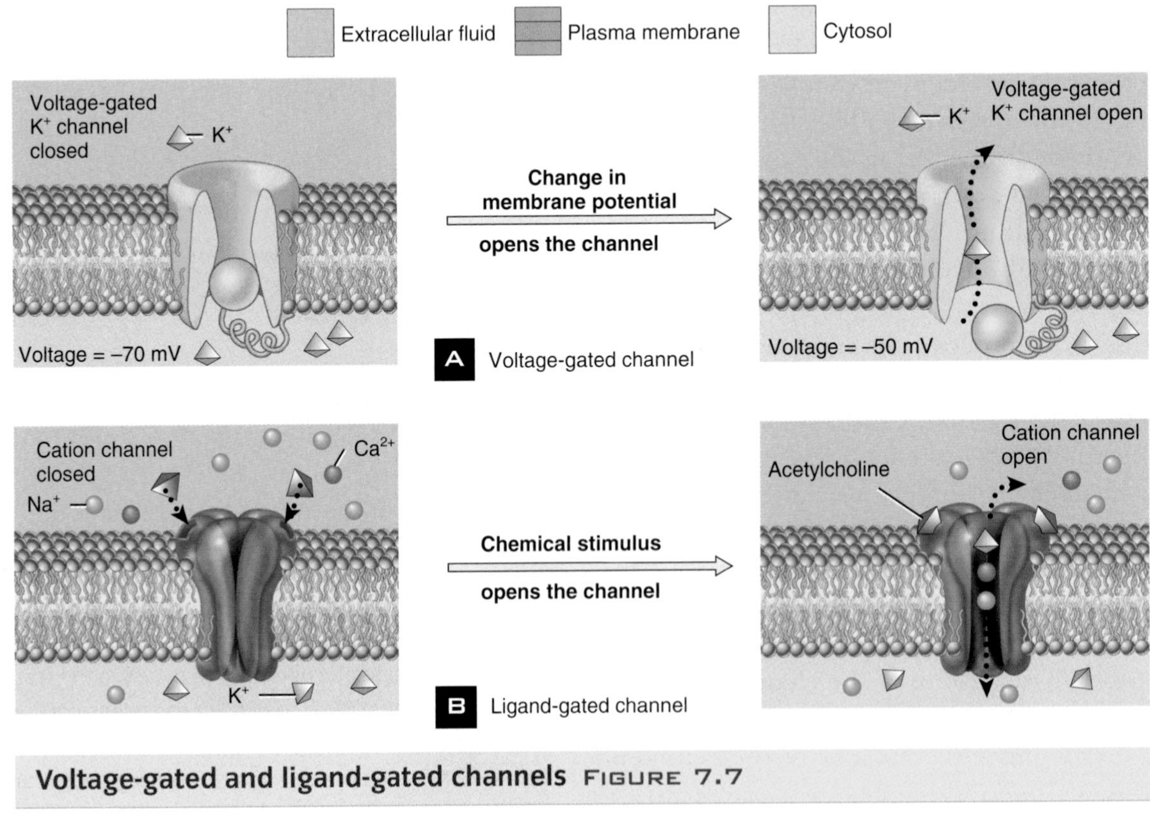

Voltage-gated and ligand-gated channels FIGURE 7.7

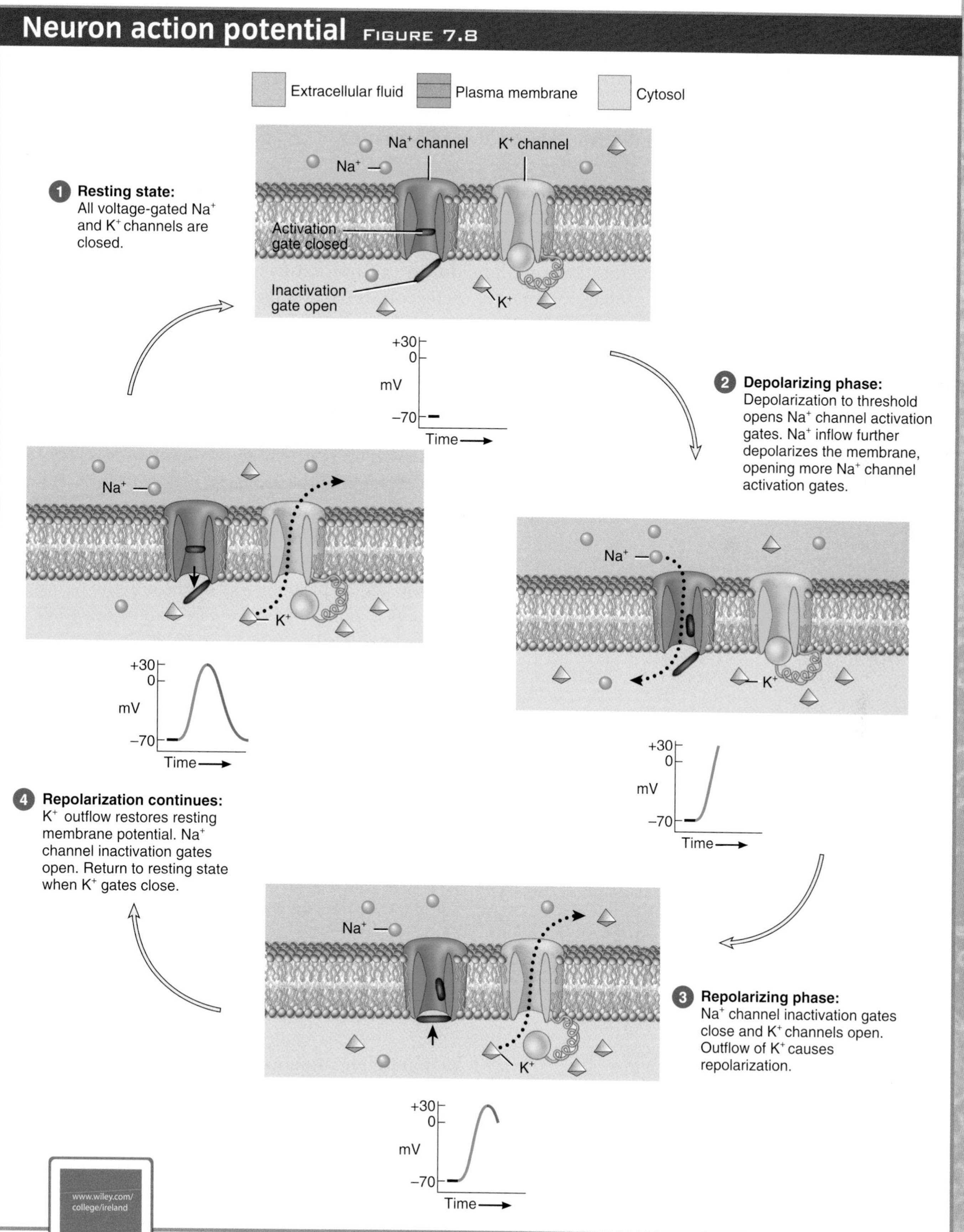
Neuron action potential FIGURE 7.8
Extracellular fluid
Plasma membrane
Cytosol
Na+ channel
K+ channel
Na+
Activation gate closed
Inactivation gate open
K+
1 Resting state:
All voltage-gated Na+ and K+ channels are closed.
+30
0
mV
–70
Time
2 Depolarizing phase:
Depolarization to threshold opens Na+ channel activation gates. Na+ inflow further depolarizes the membrane, opening more Na+ channel activation gates.
3 Repolarizing phase:
Na+ channel inactivation gates close and K+ channels open. Outflow of K+ causes repolarization.
4 Repolarization continues:
K+ outflow restores resting membrane potential. Na+ channel inactivation gates open. Return to resting state when K+ gates close.
www.wiley.com/college/ireland
Process Diagram

ACTION POTENTIALS WORK AT DIFFERENT SPEEDS

Nerves can propagate action potentials at different speeds. Nerve impulses are sent along the axon in wave-like fashion. Impulses always begin at the swollen base of the axon, the axon hillock. These impulses travel along the membrane to the axon terminus, where they stimulate the release of neurotransmitters. Propagation speed can be influenced by the diameter of the axon (thin axons propagate faster) and by the amount of **myelin** on the axon (FIGURE 7.9). When the axon is wrapped in a myelin sheath, action potentials travel in a jumping pattern. The actual movement of sodium and potassium ions occurs only at the nodes, those stretches of naked axon visible between the cells that create the myelin sheath. This allows the action potential to travel much faster, jumping from one node to the next rather than moving steadily down the length of the axon.

> **Myelin**
> White lipids and phospholipids wrapped around neural processes that aids in faster transmission.

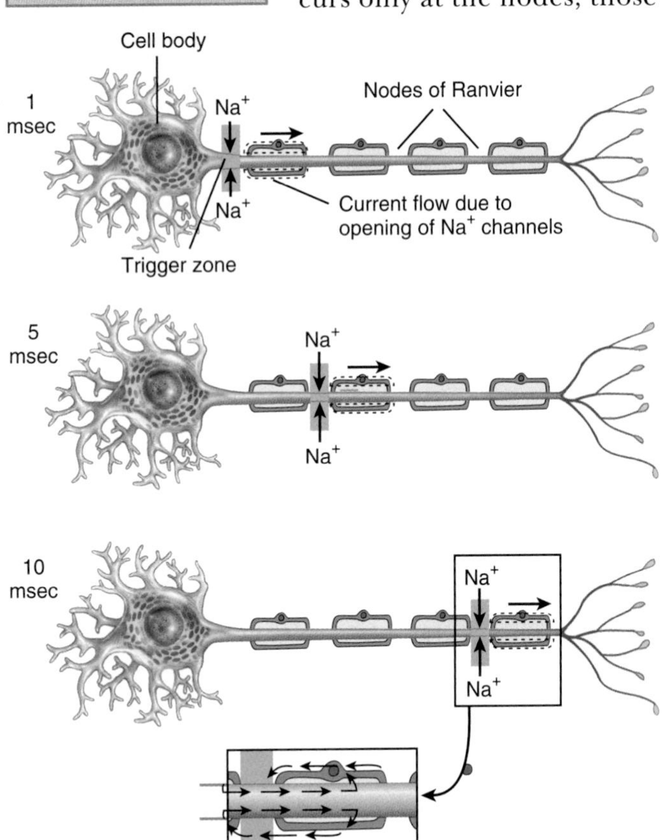

Impulse conduction in a myelinated neuron
FIGURE 7.9

In the PNS, the neuroglial cells responsible for myelination are called **Schwann cells** (FIGURE 7.10). These cells wrap around the axon, providing a covering of phospholipids. Schwann cells also aid in regeneration of neural axons. If the axon is damaged, the Schwann cells remain in place, providing a tube through which the regenerating axon can grow. In this way, the axon terminus remains in association with the same muscular or glandular cells when it regenerates after being severed.

Schwann cells are not present in the CNS, where myelin is provided by **oligodendrocytes** (FIGURE 7.11). These are large cells with branching appendages that touch and protect many axons. If an axon is damaged in the CNS, the oligodendrocyte retreats, leaving no tube or pathway to aid in axonal regrowth. This is partially why damage to the neurons in the CNS is generally not repaired and why spinal-cord injuries are usually permanent.

Although PNS neurons can recover from some damage, neurons in neither the PNS nor the CNS can regenerate if the cell body is damaged. Axons will regenerate only if they are damaged beyond the axon hillock. As far as we know, new neurons do not form in adult CNS tissue with the exception of one small area of the brain called the hippocampus. Interestingly, depression seems to be linked to the inability to generate new neurons in this area. For the most part, however, when a CNS neuron is damaged beyond repair, it is lost.

SYNAPSES SEPARATE ONE NEURON FROM ANOTHER

Action potentials move along the neural membrane as a local change in voltage. Ions flow back and forth across the membrane as gated channels open and close, causing the alteration in voltage associated with the action potential. At the **terminal bulb**, however, the impulse must be transferred to the next neuron in line; there is no membrane to carry it. Neurons do not phys-

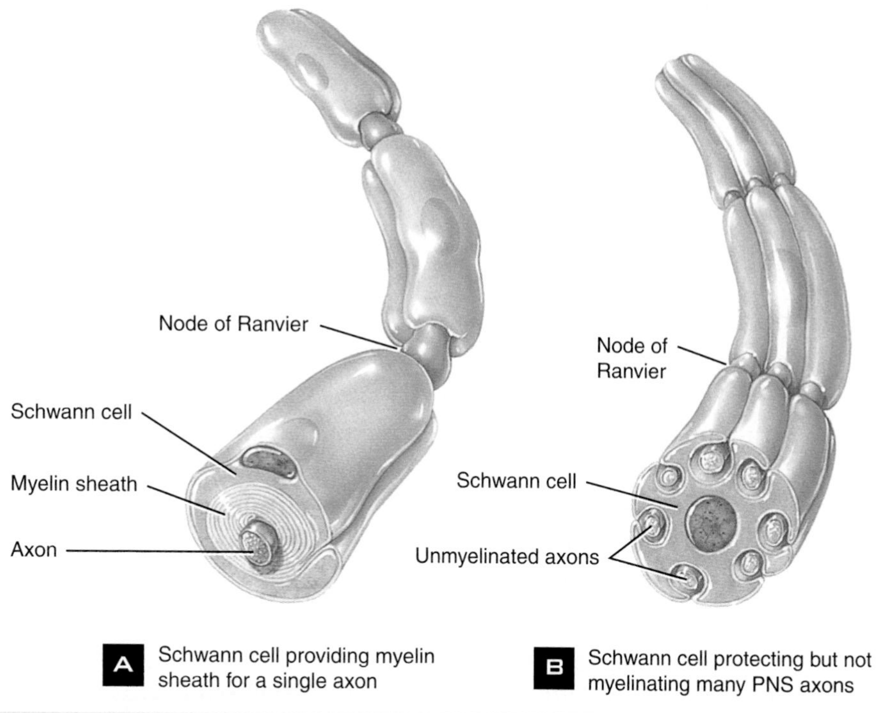

Schwann cell FIGURE 7.10

These cells individually wrap and protect the delicate and often extremely long axons of PNS neurons. They secrete compounds that aid in the regeneration of severed neuronal processes, as sometimes happens when we receive a deep wound.

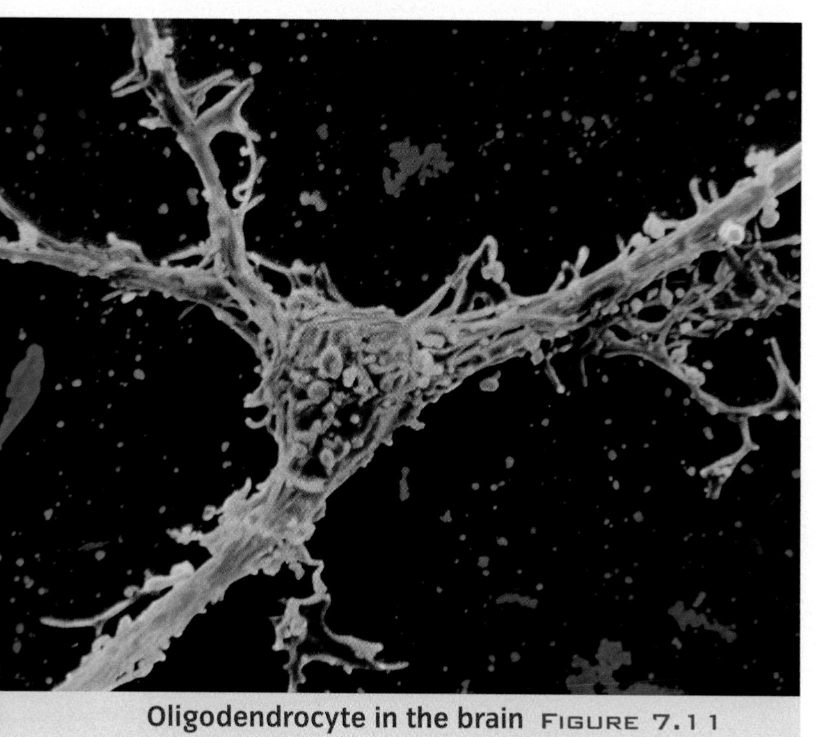

Oligodendrocyte in the brain FIGURE 7.11

ically touch one another; instead they are separated by a gap called a synapse. Neurotransmitters released from the terminal bulb diffuse into the synapse, just as they do at the neuromuscular junction. They traverse this space, called the synaptic cleft, by simple diffusion. Neurotransmitters leave the **presynaptic neuron** and diffuse toward the **postsynaptic neuron**, where they settle on receptors and initiate a reaction.

Terminal bulb
The swollen terminal end of the axon that releases neurotransmitters into the synapse.

Presynaptic neuron
The neuron that lies before the synapse, whose axon leads to the synapse.

Postsynaptic neuron
The neuron that begins after passing the synapse, whose dendrites pick up diffusing neurotransmitters.

NEUROTRANSMITTERS CARRY THE MESSAGE ACROSS THE SYNAPSE

Neurotransmitters are specific chemicals that carry an impulse across a synaptic cleft. We currently have identified and studied more than 45 neurotransmitters, each with a slightly different effect on the postsynaptic neuron (TABLE 7.3). The most common neurotransmitters are **acetylcholine** (ACh) and **norepinephrine** (NE). As described in Chapter 6, ACh stimulates muscle contractions when picked up by receptors on the muscle cell membrane. Once released, it is broken down quite rapidly by the enzyme **acetylcholinesterase**. ACh is present on the muscle cell and in the synapse for approximately 20 milliseconds.

Norepinephrine (NE) is responsible for the adrenaline rush we experience during tense situations. NE, unlike ACh, is mostly reabsorbed by the presynaptic neuron instead of being broken down. Reabsorption takes longer, so NE can remain effective for 1 to 2 seconds at a time.

GRADED RESPONSES CREATE FINE NEURAL CONTROL

Action potentials are "all or nothing" events, meaning that once the threshold is reached, the nerve will fire completely. Because a single neuron cannot create a partial action potential, we vary the strength of nervous stimulation by changing the number of neurons that are firing.

Graded responses can be obtained by **hyperpolarizing** or **depolarizing** individual neural membranes. The hyperpolarized neuron requires a larger stimulus to reach threshold and begin an action potential. The depolarized neuron is the opposite: It requires less of a "kick" to begin an action potential, because its resting potential is closer to the action potential threshold. But once threshold is reached, the neuron generates an action potential that is indistinguishable from any other action potential.

These hyperpolarized and depolarized neurons result from alterations in the resting membrane potential of postsynaptic neurons. Two types of postsynaptic

Neurotransmitters TABLE 7.3

Class	Name	Location	Effects
Acetylcholine	Acetylcholine	Throughout CNS and PNS, neuromuscular junctions, parasympathetic division	Contracts muscle, causes glandular secretions, general parasympathetic functions
Biogenic amine	Norepinephrine	Hypothalamus, brain stem, cerebellum, spinal cord, cerebral cortex, and most sympathetic division junctions	Attention, consciousness, control of body temperature
Biogenic amine	Epinephrine	Thalamus, hypothalamus, midbrain, spinal cord	Uncertain, but thought to be similar to norepinephrine
Biogenic amine	Dopamine	Hypothalamus, midbrain, limbic system, cerebral cortex, retina	Regulates subconscious motor functions, emotional responses, addictive behaviors, and pleasurable experiences
Biogenic amine	Serotonin	Hypothalamus, limbic system, cerebellum, spinal cord	Maintains emotional states, moods, and body temperature
Biogenic amine	Histamine	Hypothalamus	Sexual arousal, pain threshold, thirst, and blood pressure control
Amino acid	Glutamate	Cerebral cortex and brain stem	Excitatory, aids in memory and learning
Amino acid	GABA	Cerebral cortex	Inhibitory, shows potential as an anti-anxiety drug
Neuropeptide	Substance P	Spinal cord, hypothalamus, digestive tract	Pain sensation, controls digestive functions
Neuropeptide	Neuropeptide Y	Hypothalamus	Stimulates appetite and food intake
Opioids	Endorphins and enkephalins	Thalamus, hypothalamus, brain stem	Pain control, behavioral effects

Psychoactive drugs: getting a good reception at the synapse

Compounds that affect the nervous system are called psychoactive drugs. Psychoactive drugs, whether legal like alcohol or illegal like marijuana, share one key feature: They affect the synapses where neurons communicate with one another. Psychoactive drugs most often act through one of two essential mechanisms: They may block or activate neural receptors in the synapses, or they may change the concentration of neurotransmitters that naturally occur in the synapses.

In either case, the result is a change in the nature and/or intensity of the message that is passed across the synapse. Psychoactive drugs can subtly change these messages, prevent them entirely, or create false messages where none was intended.

Many psychoactive drugs affect the neurotransmitter dopamine, which is involved in movement and emotion. Normally, after a neurotransmitter is released in the synapse, it affects a change in the postsynaptic cell and is quickly taken up (removed) from the synapse. Cocaine and amphetamine both inhibit the removal, or "reuptake," of dopamine, so dopamine remains in the synapse, continuing to activate the receptors for an abnormal period. The excess stimulation of dopamine receptors explains the "high" of cocaine and amphetamine.

Neurotransmitter reuptake is also impaired by the "selective serotonin reuptake inhibitors," the SSRI drugs, including Zoloft and Prozac. Many patients with anxiety and depression apparently lack an adequate supply of the neurotransmitter serotonin. SSRI drugs change conditions at the synapse by inhibiting serotonin reuptake, which has the effect of stretching the supply of serotonin and making it more likely that a serotonin signal will transit the synapse.

Surprisingly, the nervous system contains many receptors specifically tuned to associate with compounds found in opiate drugs. The well-studied "mu opiate receptors" can occur on excitatory or inhibitory neurons. Excitatory neurons increase the activity of other neurons; inhibitory neurons slow or stop nerve signals. Opiate drugs like heroin or morphine can activate these receptors. The body produces related compounds called "endogenous opioids" that also activate these receptors. Endogenous opioids may account for the feeling of well-being that follows physical exercise and could explain the ability to ignore pain during an emergency.

The nervous system also contains a great number of receptors for chemicals found in marijuana. These receptors exist in high concentration "around the hippocampus, cortex, olfactory areas, basal ganglia, cerebellum, and spinal cord," according to Roger Pertwee, a cannabis expert at the University of Aberdeen (United Kingdom). "This pattern accounts for the effects of cannabinoids on memory, emotion, cognition, and movement." Cannabinoid receptors are also found in the male and female reproductive systems, and on neurons that cause nausea and vomiting.

The prevalence of receptors for psychoactive drugs helps explain the power of these drugs and also demonstrates how evolution finds new uses for old molecules. And it might even help treat disease: The discovery of cannabinoid receptors in a part of the eye that controls fluid pressure could eventually lead to drugs to control glaucoma, a blinding disease caused by excess pressure inside the eye.

Psychoactive drugs affect behavior and judgment, as they alter the functioning of cerebral synapses. Many of these compounds are illegally sold on the black market. Here a police officer guards confiscated drugs obtained during a successful "drug bust."

potential can be developed. **Excitatory postsynaptic potentials (EPSPs)** cause slight depolarization of the neuron. The membrane potential is already closer to threshold, so a smaller stimulus is needed to begin the action potential. Think of being in a frustrating situation: Maybe you are trying to study for a human biology test while your roommates are listening to music with a driving beat. The longer this goes on, the more frustrated you become. When your roommate asks if you want something to eat, you snap at her. Normally, having to answer this question would not elicit such a reaction, but when you are already angry, it does. This quick reaction to a smaller stimulus mimics an EPSP.

Inhibitory postsynaptic potentials (IPSPs) cause the opposite reaction in the postsynaptic neuron. IPSPs hyperpolarize the neuron, meaning the membrane potential is further from that needed to generate an action potential, so a larger stimulus is required to begin an action potential. Using the above example, if you were wearing headphones with relaxing music, you could block out the noise, and your roommate would need to tap your shoulder to get your attention. She would need to raise the input level to receive the normal response. Many prescription and recreational drugs affect the events of the synapse, as discussed in the Health, Wellness, and Disease box on page 201. They can alter the potential of the pre- and postsynaptic neurons, affect the diffusion of neurotransmitters, or even mimic the effect of the neurotransmitters on the postsynaptic neuron.

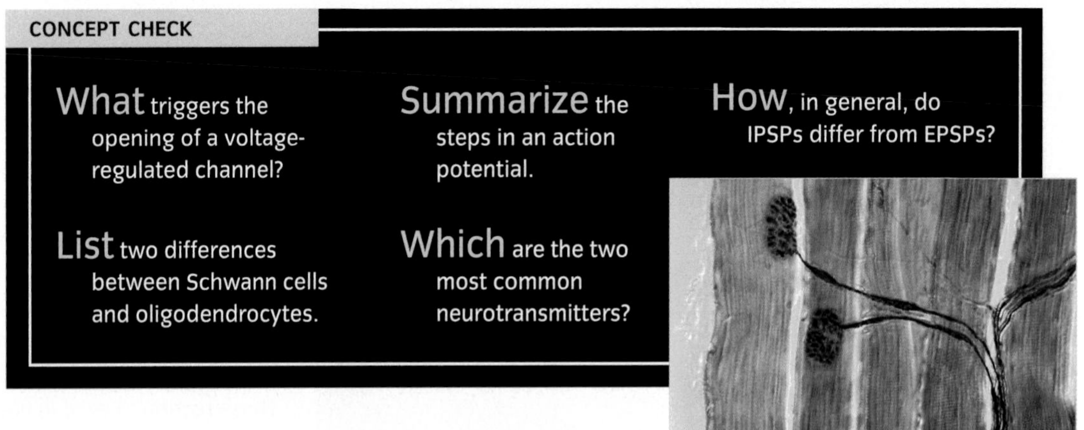

CONCEPT CHECK
What triggers the opening of a voltage-regulated channel?
List two differences between Schwann cells and oligodendrocytes.
Summarize the steps in an action potential.
Which are the two most common neurotransmitters?
How, in general, do IPSPs differ from EPSPs?

The Brain and Spinal Cord Are Central to the Nervous System

LEARNING OBJECTIVES

Describe the anatomy and coverings of the brain.

Explain the functions of the various parts of the brain.

Explore the anatomy of the spinal cord.

List the steps in a typical reflex.

The human brain occupies approximately 1,165 cubic centimeters and weighs about 1,400 grams. In terms of complexity, nothing that we know of in the universe is even close. Although brains look pretty unexciting from the outside, they conceal an amazing level of detail, all of which emerges from just a few types of cells, specifically and purposefully connected. We'll

start our examination of the brain by looking at how it is protected from injury.

THE MENINGES AND CEREBROSPINAL FLUID PROTECT AND NOURISH THE CENTRAL NERVOUS SYSTEM

The axial skeleton provides bony protection for the CNS. The **meninges** and **cerebrospinal fluid (CSF)**, in turn, protect the CNS from the axial skeleton, providing a soft lining and cushion that nourishes and protects the delicate neural structures. The meninges are a series of three connective tissue coverings between the nervous tissue and the bone, surrounding and protecting the brain and spinal cord (FIGURE 7.12). The cerebrospinal fluid within the meninges nourishes the neurons and absorbs shock.

Cerebrospinal fluid (CSF)
A liquid similar to plasma, but with less dissolved material that maintains uniform pressure within the brain and spinal cord.

The outer covering of the meninges, called **dura mater,** is a tough connective tissue layer immediately beneath the skull. Below the dura mater is the **arachnoid**. This layer is thin and fragile and looks like a spider web. Cerebrospinal fluid flows between the strands of the arachnoid. The inner layer of the meninges is called **pia mater**. This extremely thin layer is attached to the neurons and cannot be peeled off without damaging them.

Meningitis, an inflammation of these three layers of connective tissue, is extremely difficult to treat because the environment of the brain is isolated and controlled, so medications cannot be easily introduced. Meningitis can be life-threatening because the swollen membranes compress the neurons of the brain and spinal cord. Meningitis can be viral or bacterial. Although a new vaccine shows promise in controlling viral outbreaks, at present viral meningitis has no cure. Physicians merely treat the symptoms and hope that the patient is strong enough to recover after the virus runs its course. Bacterial meningitis causes other concerns. Normal doses of antibiotic are ineffective because they seldom if ever get from the blood to the cerebrospinal fluid of the brain and on to the meninges. It is difficult to prescribe the proper amount of antibiotic—too little will not reach the infection, and too much can kill the patient.

Cerebrospinal fluid provides a constant environment for the central nervous system, as well as a cushion in which the organs float. Every time you move your head, your brain floats within the cranium. When you lift your head from your pillow in the morning, the brain sloshes toward the occipital bone. Because fluid is noncompressible, the CSF around the brain prevents the fragile surface of the brain from striking the cranium. Otherwise, the delicate outer portion of the brain would bang against the bones every time you moved your head, destroying neural connections and ultimately the tissue itself.

Meninges FIGURE 7.12

The meninges lie directly under the skull, between the bone and the brain. Here you can see the skin on the left side of the head. The sequential layers visible from left to right are the periosteum of the skull bones, the bone itself, the dura mater, the arachnoid, and the pia mater lying directly on top of the gyri (bumps) and sulci (grooves) of the brain.

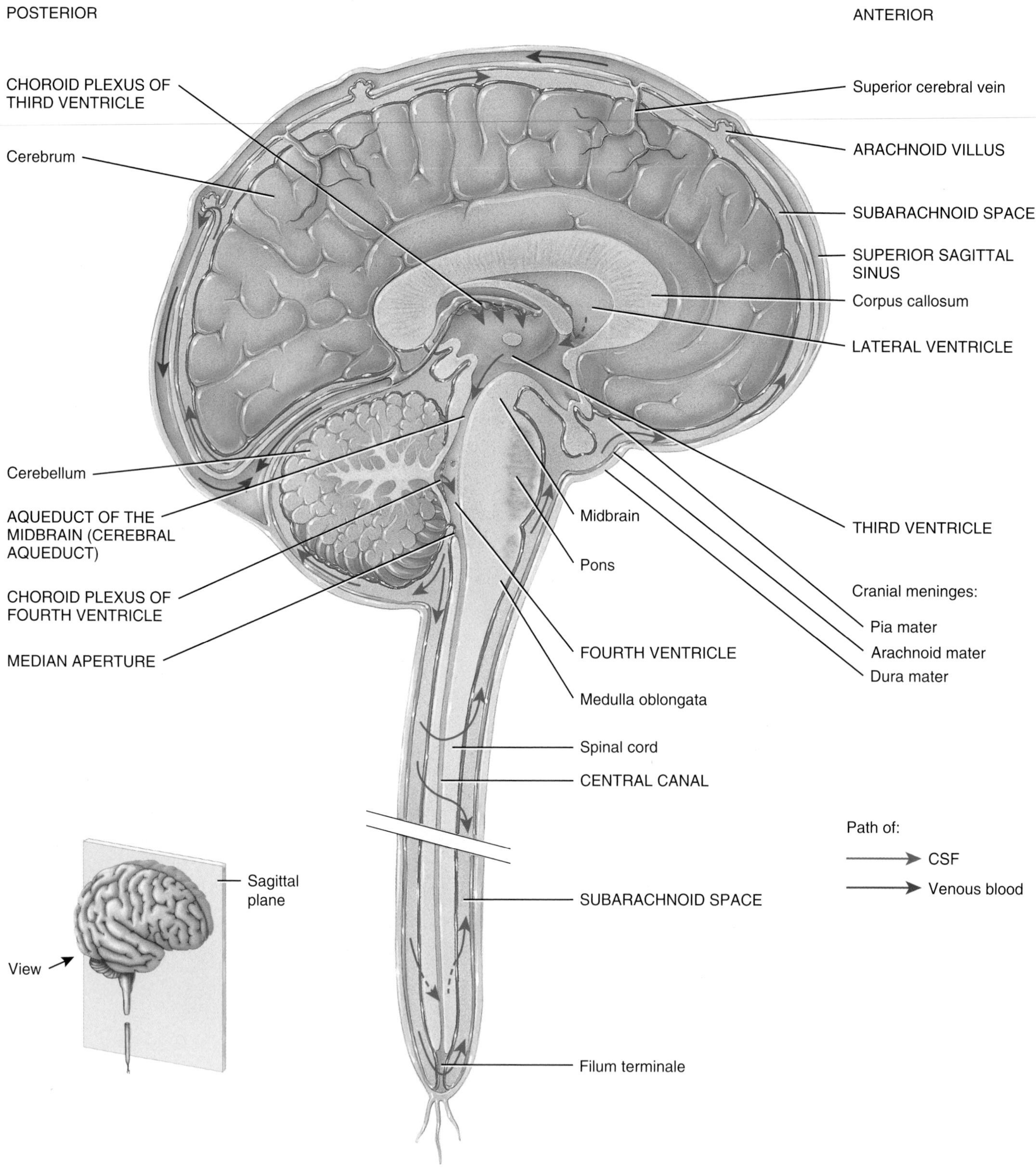
POSTERIOR
ANTERIOR
CHOROID PLEXUS OF THIRD VENTRICLE
Cerebrum
Superior cerebral vein
ARACHNOID VILLUS
SUBARACHNOID SPACE
SUPERIOR SAGITTAL SINUS
Corpus callosum
LATERAL VENTRICLE
Cerebellum
AQUEDUCT OF THE MIDBRAIN (CEREBRAL AQUEDUCT)
CHOROID PLEXUS OF FOURTH VENTRICLE
MEDIAN APERTURE
Midbrain
Pons
THIRD VENTRICLE
Cranial meninges:
Pia mater
Arachnoid mater
Dura mater
FOURTH VENTRICLE
Medulla oblongata
Spinal cord
CENTRAL CANAL
Path of:
CSF
Venous blood
Sagittal plane
View
SUBARACHNOID SPACE
Filum terminale

Sagittal section of brain and spinal cord

VENTRICLES MAKE CEREBROSPINAL FLUID

The brain may look like a solid mass of nervous tissue, but nothing could be further from the truth. Four rather large cavities in the brain are filled with CSF. These cavities (FIGURE 7.13) are literally holes in your head, but we call them **ventricles.**

CSF is continuously produced and absorbed, creating a constant flow. If drainage back to the blood and the heart gets blocked, CSF builds up within the brain, adding a watery fluid under the skull that is rightly named hydrocephaly ("water head"). In infants whose skull bones have not yet fused, hydrocephaly forces the entire cranial cavity to expand at the fontanels. Once the skull has ossified, there are no fontanels, and hydrocephaly compresses the neurons of the cortex, effectively shutting down parts of the brain. This can be corrected by surgically implanting a shunt to drain the excess fluid.

CSF formation helps maintain the **blood-brain barrier**, which permits only certain ions and nutrients to cross the vessels of the choroid plexus, resulting in a controlled environment for CNS neurons. Bacteria and viruses thus have difficulty entering the brain. Unfortunately, when bacteria do enter, they are difficult to treat, because the blood-brain barrier also keeps most antibiotics out.

Each ventricle contains a choroid plexus, which forms CSF. CSF flows throughout the central nervous system, starting in the ventricles and flowing down toward the spinal cord. It flows down the central canal of the spinal cord, then up the outside of the cord, and around the outside of the brain. CSF is absorbed into the bloodstream in the subarachnoid space.

CSF formation and flow FIGURE 7.13

THE BRAIN HAS FOUR MAIN PARTS

A first glance at the brain shows four major parts: the brain stem, the diencephalon and midbrain, the cerebellum, and the cerebrum. Although the entire brain is basically involved in the integration of sensory input and motor responses, each section has different roles.

THE BRAIN STEM IS AN ANCIENT ROOT OF LIFE

The brain stem contains vital centers that regulate heart rate, breathing, and blood pressure (FIGURE 7.14 on p. 206). The brain stem is the portion of the brain closest, anatomically and physiologically, to the spinal cord. The **medulla oblongata** and the **pons** make up the brain stem.

The medulla oblongata contains the vital centers of the brain stem associated with heart rate, respiratory function, and blood pressure. These centers, found in many animals, indicate that the medulla oblongata evolved in ancient times. Here also are reflex centers for sneezing, coughing, hiccupping, and swallowing. Motor impulses generated in the higher centers of the brain travel through the medulla oblongata on their way to the PNS.

You may have heard that the right side of the brain controls the left side of the body and vice versa. This is basically true, because 80 percent of the motor information from the right side of the brain enters the medulla oblongata and crosses to the left side before leaving the CNS. The crossing of these tracts is visible on the anterior surface of the medulla oblongata. The structures that can be seen crossing over one another are the pyramids (descending motor tracts), and the technical term for crossing is decussing,

Medulla oblongata Portion of the brain stem immediately adjacent to the spinal cord, associated with heart rate, breathing controls, and blood pressure.

Pons The area superior to the medulla oblongata involved in transfer of information and respiratory reflexes.

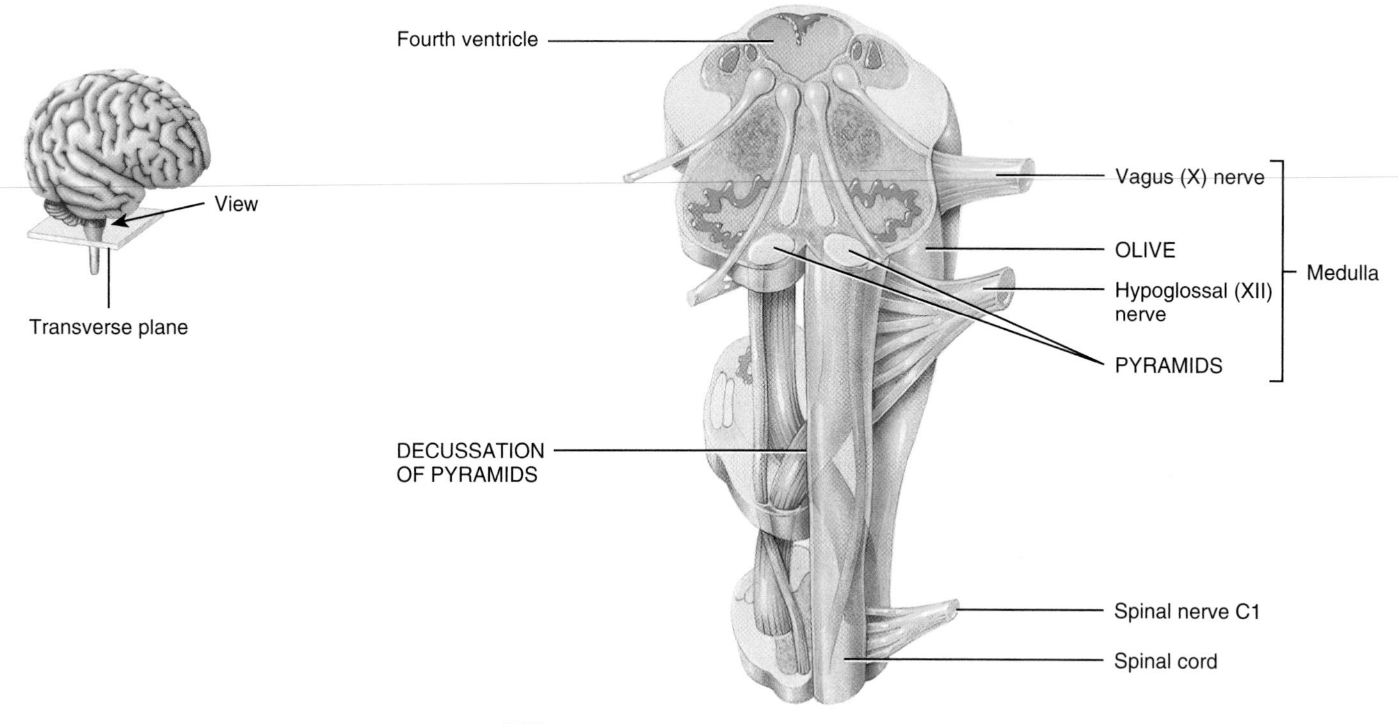

A Transverse section and anterior surface of medulla oblongata

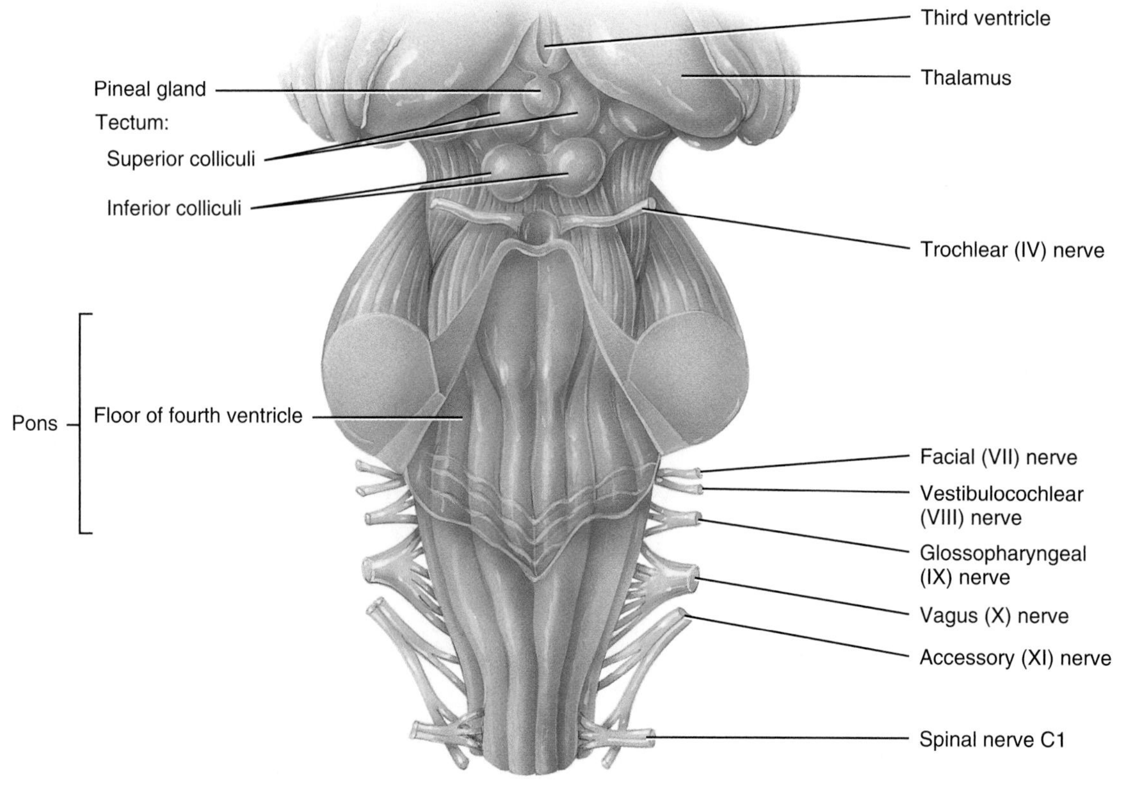

B Posterior view of midbrain in relation to brain stem

Brain stem (medulla oblongata and pons) FIGURE 7.14

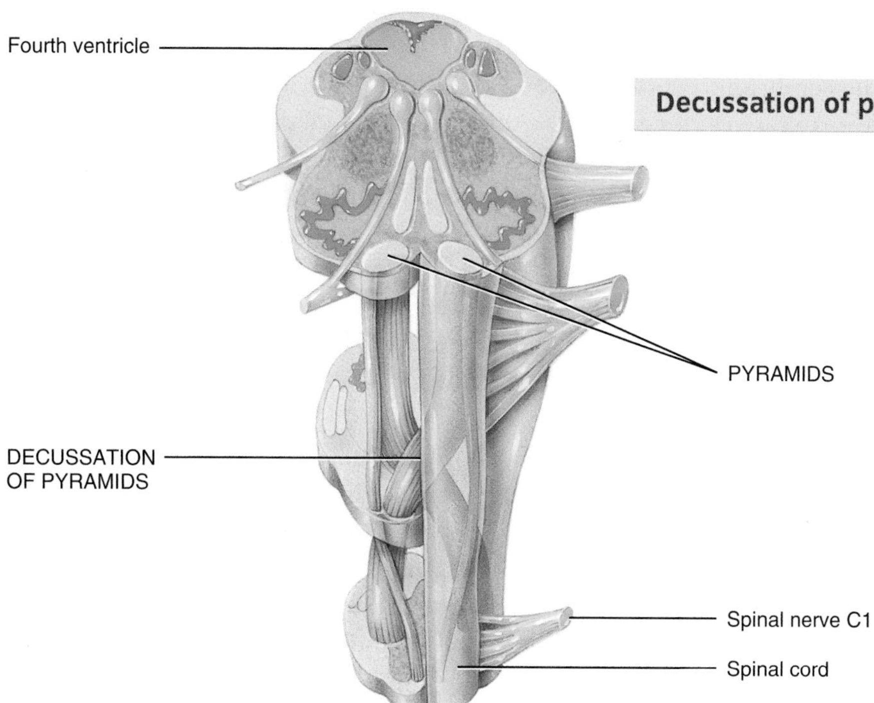

Decussation of pyramids in the brain stem FIGURE 7.15

therefore the entire structure is referred to as the **decussation of pyramids** (FIGURE 7.15).

The pons focuses on respiration. Most of the pons is composed of **tracts** that carry information up to the brain, down from the brain to the spinal cord, or laterally from the pons to the cerebellum. The only vital center found in the pons is related to respiratory reflex. The **apneustic** and **pneumotaxic** reflexes begin in the pons. The apneustic center triggers breathing even when we consciously hold the diaphragm still. Despite the threats of countless children, you cannot hold your breath until you die. If you tried your hardest, you would eventually pass out, and the apneustic center would immediately restart your breathing. The pneumotaxic center works oppositely, because it is charged with preventing overinflation of the lungs. When stretch receptors in the lungs are stimulated, the pneumotaxic center sends a motor response, causing you to exhale.

Tracts
Axons and/or dendrites with a common origin, destination, and function.

THE CEREBELLUM FOCUSES ON MUSCLES AND MOVEMENT

Posterior to the brain stem, we see something that looks like a smaller brain hanging off the back of the brain. This small, round structure is the cerebellum (FIGURE 7.16). It has two main functions: maintaining muscle

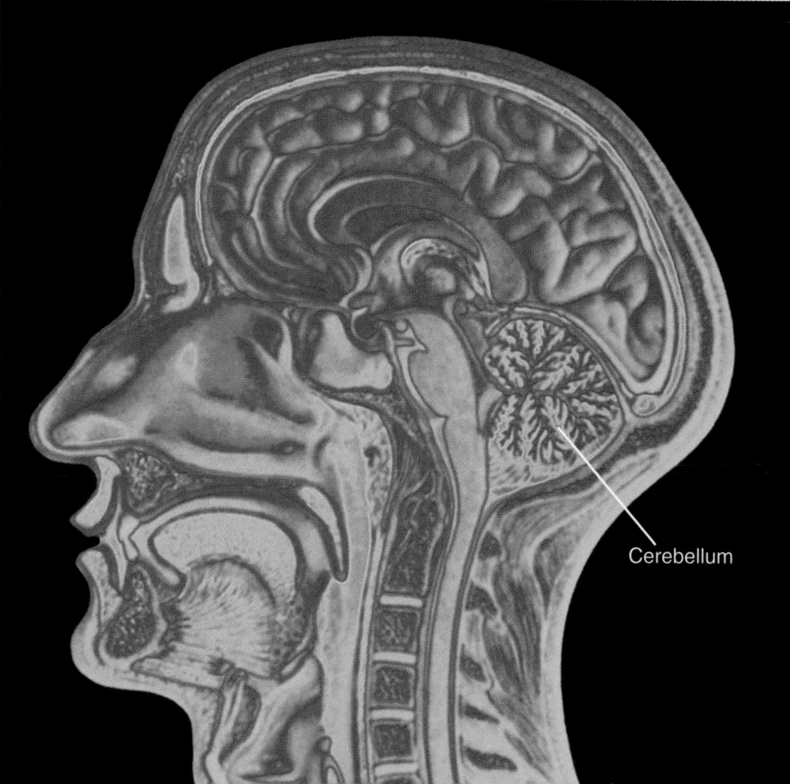

The cerebellum FIGURE 7.16

In this colorized scan, the cerebellum can be seen below the cerebrum.

tone, posture, and balance; and fine-tuning conscious and unconscious movements directed by the cerebrum. Although we walk without thinking, the process requires exact coordination. That smooth gait, with its leg lifts and counterbalancing arm swings, is directed by the cerebellum.

One job of the cerebellum is to understand where the limbs are located, using proprioception. This sensory capability allows you to lift your legs and move them forward without glancing at them, because your brain knows where your feet are at all times. The nervous pathways associated with proprioception run from the muscles and joints to the cerebellum.

The cerebellum is also important in learning motor skills. Riding a bike, learning to swim, or even learning new information through repeatedly writing notes are all examples of cerebellar learning. New research indicates that the cerebellum may also play a role in sensory integration by receiving input from sensory neurons and directing it to inner portions of the cerebrum. Abnormal cerebellar anatomy has been detected in autistic children, suggesting a link between cerebellar function and autism.

THE DIENCEPHALON IS A RELAY CENTER

The diencephalon includes the central portion of the brain and functions mainly as a relay center for sensory information from the body and motor responses from the cerebrum (FIGURE 7.17). Within this portion of the brain, conscious and unconscious sensory information and motor commands are integrated. Centers for visual and auditory startle reflexes are located here.

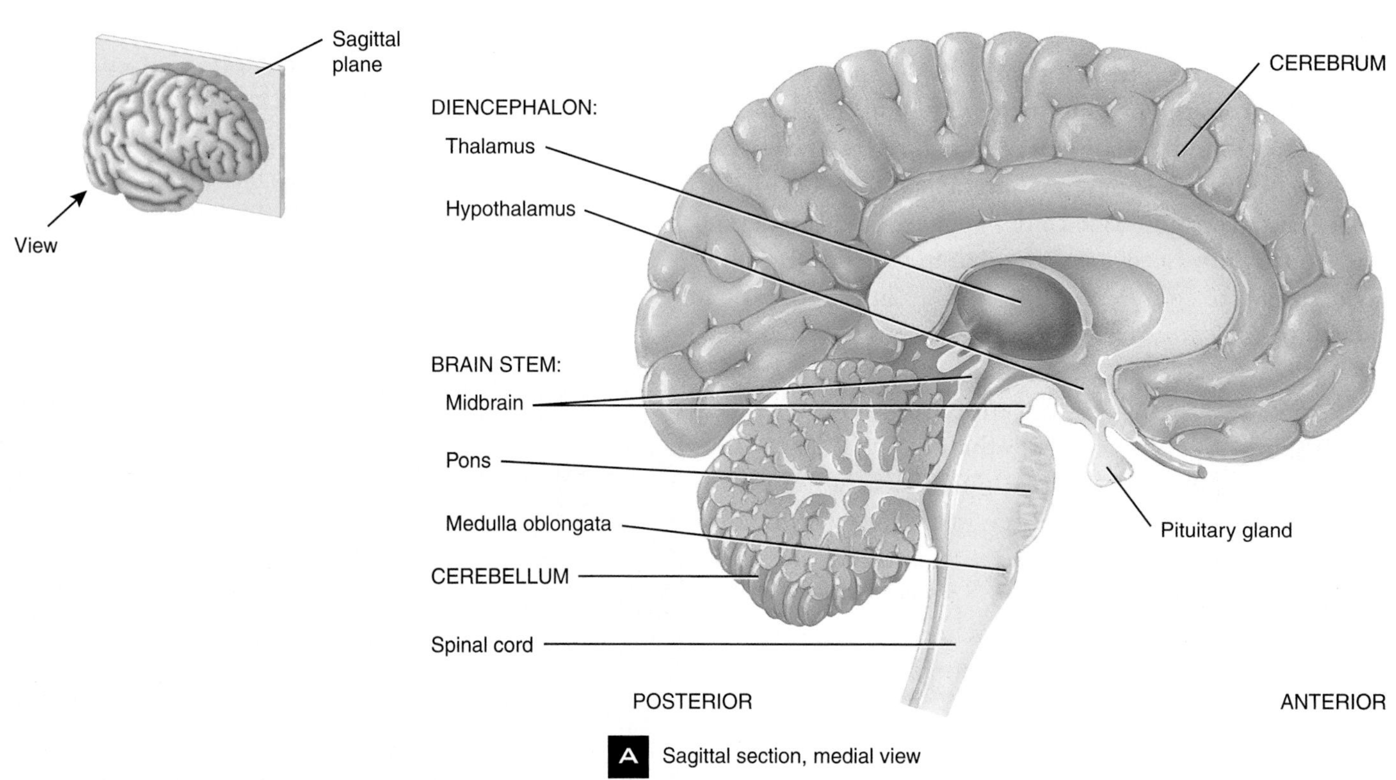

A Sagittal section, medial view

The diencephalon FIGURE 7.17

The auditory reflex causes you to "jump" when you hear a car backfire. The visual reflex can also cause you to jump when you are focused on reading or studying and something flits by your peripheral vision. If you jump and rapidly turn your head to catch that fleeting vision, you've had a visual reflex.

The **thalamus** and **hypothalamus** are also located in the diencephalon. The thalamus is a relay station for most incoming sensory information. Stimuli are sent from the thalamus to the appropriate portions of the cerebrum. The **limbic system**, which is responsible for our emotions, communicates with the anterior portion of the thalamus. This communication forms a physical link between incoming sensory information and emotions.

The hypothalamus is, as the name implies, below the thalamus. It secretes hormones that control the anterior pituitary gland, monitor water balance, and stimulate smooth muscle contraction. The hypothalamus also regulates our circadian rhythm, body temperature, heart rate, and blood pressure.

THE CEREBRUM IS A CENTRAL PROCESSING CENTER

The **cerebrum** is the largest portion of the brain (FIGURE 7.18, p. 211). It is here that information is processed and integrated, and appropriate responses are generated. The cerebrum contacts all other parts of the brain and is our center for higher thought processes. It is here that we learn, remember, and plan activities. Learning is the subject of many research studies, and we are only beginning to understand how the brain learns and remembers facts. (See I Wonder. . . What happens when we learn? on p. 210.)

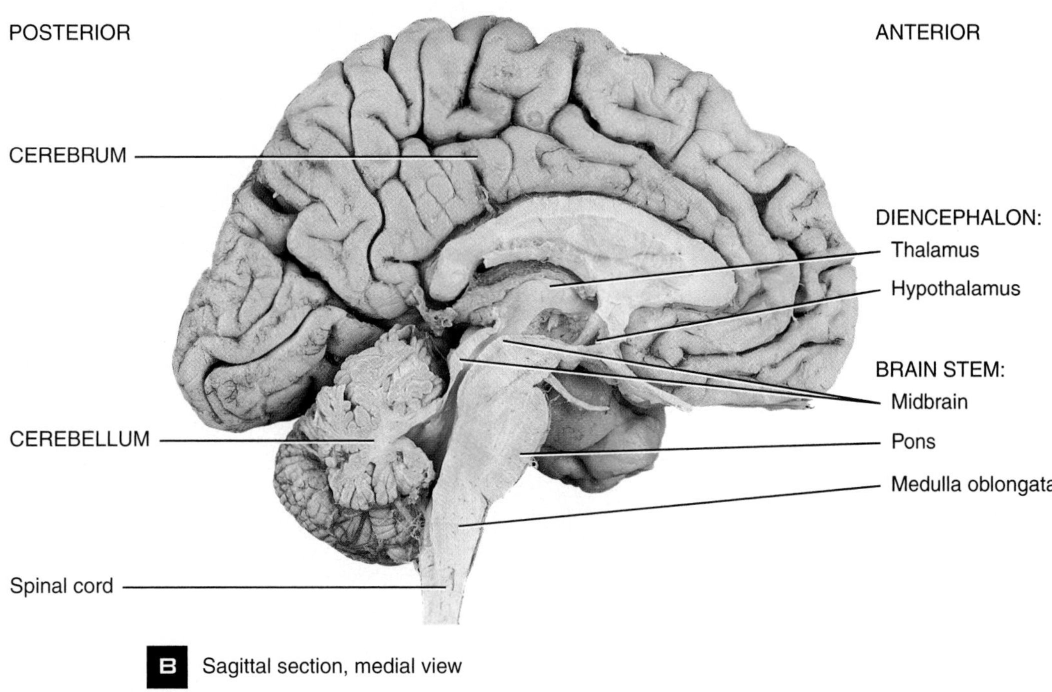

B Sagittal section, medial view

I WONDER . . .

What happens when we learn?

Understanding learning is one of the toughest challenges in neuroscience. Brains are sometimes compared to computers, but whereas it's easy to point to where a hard drive stores certain information, that is seldom possible in the brain. The brain stores information here and there, in complex, threadlike networks of neurons. Our learned ability to speak, for example, is stored separately from our memory of last year's birthday party. And both are stored separately from our ability to paddle a canoe or whistle a song.

Learning is a type of memory, and memory occurs in three phases. Immediate memory prevents us from being bewildered by maintaining information in our consciousness so that we know, for example, where we are. Short-term memory helps us carry out tasks—keeping a conversation going, say, or remembering why we are writing a letter. Although much of our short-term memory is quickly erased, some of it gets adopted in long-term memory. This memory can survive for life, or it can fade, but it is what many people mean when they say "memory." Scientists believe these three types of memory may exist in different parts of the brain.

Several types of change occur when the brain remembers something, but we call them all "neural plasticity," meaning changes in the brain that alter its ability to do something. The neural plasticity associated with learning has several components. For example, during learning, specific proteins are synthesized in the brain (we know this is true because when we block protein synthesis, we block learning). Synapses change in neural pathways so that impulses can travel through them faster and more easily, a change we call potentiation. When we learn to ride a bike, for example, the neural pathways that tell us to steer to avoid falling are potentiated. The next time we ride, these reactions happen faster, and take less conscious effort , until they eventually are triggered automatically whenever we ride a bike.

Neural plasticity also changes the dendrites, the neural processes that bring impulses to the cell body. Recent studies teaching skills to rats looked specifically at the rat hippocampus and found that certain ion channels in the membrane at the dendrites become more numerous after just 10 minutes of training.

Learning does not exist in a vacuum; the brain's ability to learn is related to what else is going on. Lab studies show that fight-or-flight conditions drastically reduce the ability to learn. People with post-traumatic stress disorder have difficulty learning, probably because of high levels of stress hormones. Emotional stress may even cause amnesia, which can destroy our memory of who we are, without harming the skill of tying a shoe.

Memory and learning play a critical role at both ends of life. Learning to swim, play guitar, or distinguish the peripheral from the central nervous system may all occur while we are young. In our final years, diseases like Alzheimer's can undo the learning of a lifetime, leaving us bewildered and frustrated over simple tasks we used to accomplish with ease. One final point in our "scratch-the-surface" overview of learning: The topic remains a black hole of neuroscience. Expect to learn a lot more about learning in the years to come.

Association areas help us identify objects we have already seen, recognize familiar faces and voices, and remember winter holidays from a whiff of fresh-baked cookies.

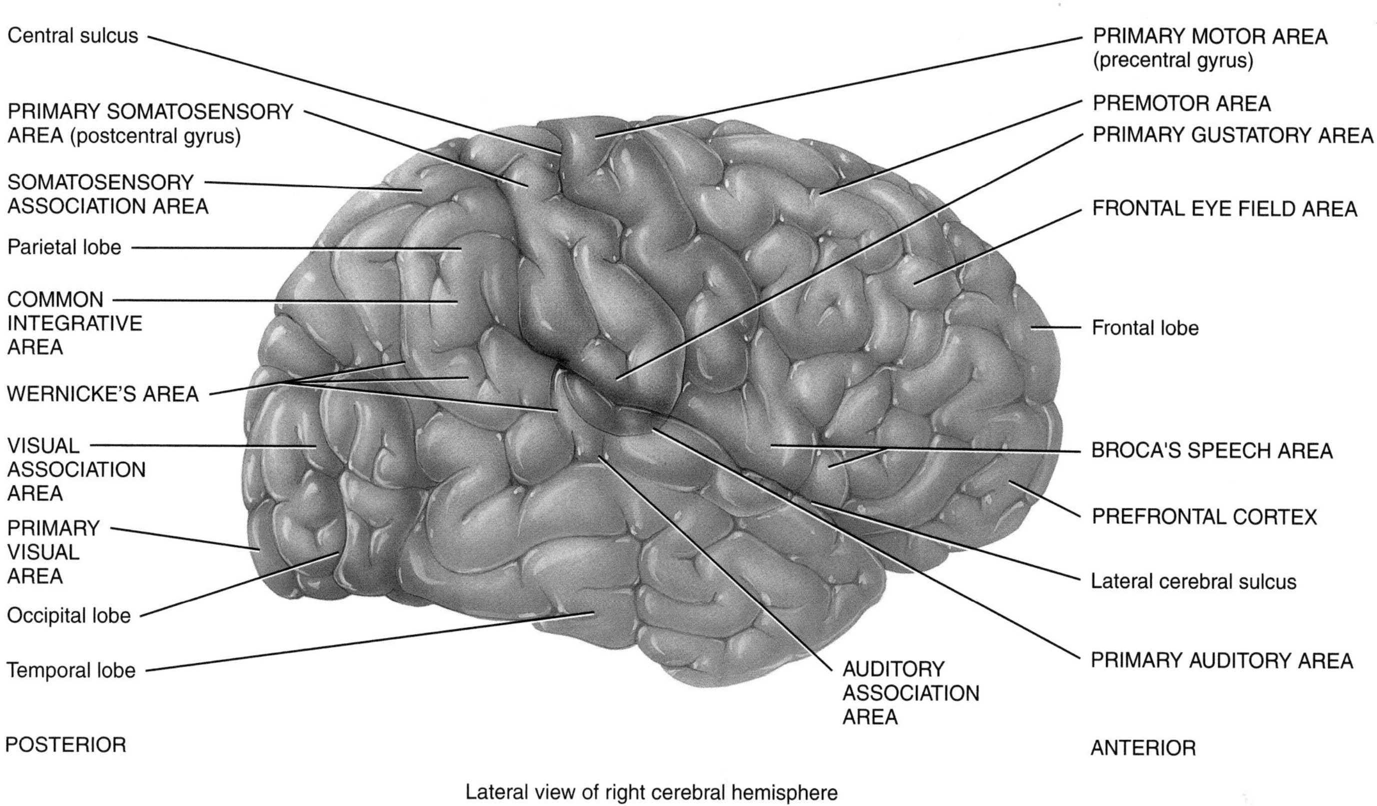

Cerebrum with lobes and their general functions indicated FIGURE 7.18

The surface of the cerebrum has creases or **sulci** that separate individual raised portions called **gyri**. The surface of the cerebrum is composed of **gray matter**, whereas the interior is white. Gray matter is mainly cell bodies and nonmyelinated neural processes—in other words, naked axons and dendrites. In the gray matter, connections are made as axons meet dendrites. The cerebral **cortex** is entirely gray matter, folded to provide a larger surface area for these neural connections. It contains billions of cell bodies responsible for sensations, voluntary movements, and thought.

Sulci (sulcus) Shallow grooves on the surface of the brain.

Gyri (gyrus) Elevations separating individual sulci; the bumps on the brain.

Cortex Thin outer layer of any organ.

The white matter inside the cerebrum contains myelinated axons that carry information to the spinal cord or other areas of the brain. Myelinated axons are covered in lipids, giving this tissue its characteristic white appearance and allowing for faster impulse transmission. Information is passed from one area of the brain to another via tracts of white matter.

THE CEREBRAL HEMISPHERES ARE HOMES OF LOGIC AND ARTISTRY

The cerebrum has two hemispheres that are quite similar anatomically. Both hemispheres are divided into lobes with general functions assigned to each. For example, the occipital lobe is where vision is interpreted, and the frontal lobe is involved in conscious thought processes. The cortex of each lobe has motor areas, sensory areas, and **association** areas that integrate new information with stored memories. The **primary motor** area, in the frontal lobe just in front of the central sulcus, formulates voluntary motor commands. Each portion of the body is represented in the primary motor area. The more control we have over movements of a particular body part, the larger the section of the primary motor area devoted to it, as seen in the **homunculus** diagram (FIGURE 7.19).

Sensory information from the skin and skeletal muscles is received in the primary somatosensory area of the cortex, just behind the primary motor area. As with the primary motor area, sensations from each body part go to a specific segment of this gyrus. The larger the segment of primary somatosensory area devoted to the body part, the more sensory receptors are found in that part. Interestingly, when any of the nerves along these sensory pathways are stimulated, the brain interprets the sensation as coming from the organ at the distal end of the pathway, regardless of the source of the stimulation. This causes **referred pain**, which also occurs when we interpret a painful stimulus from an internal organ as pain in our skin or surface organs. This may happen because the visceral sensory pathways often join with or cross cutaneous sensory pathways in the spinal cord. When the pain stimulus reaches the brain, it is interpreted as coming from the skin, which is the usual site of injury. A typical example is the pain of appendicitis. Although the appendix lies in the lower right abdomen, appendicitis pain is usually described as right behind the umbilicus, or belly button.

A few specialized motor actions are governed by areas outside the primary motor area. The formation of words, for example, is organized in **Broca's area**, on the right frontal lobe.

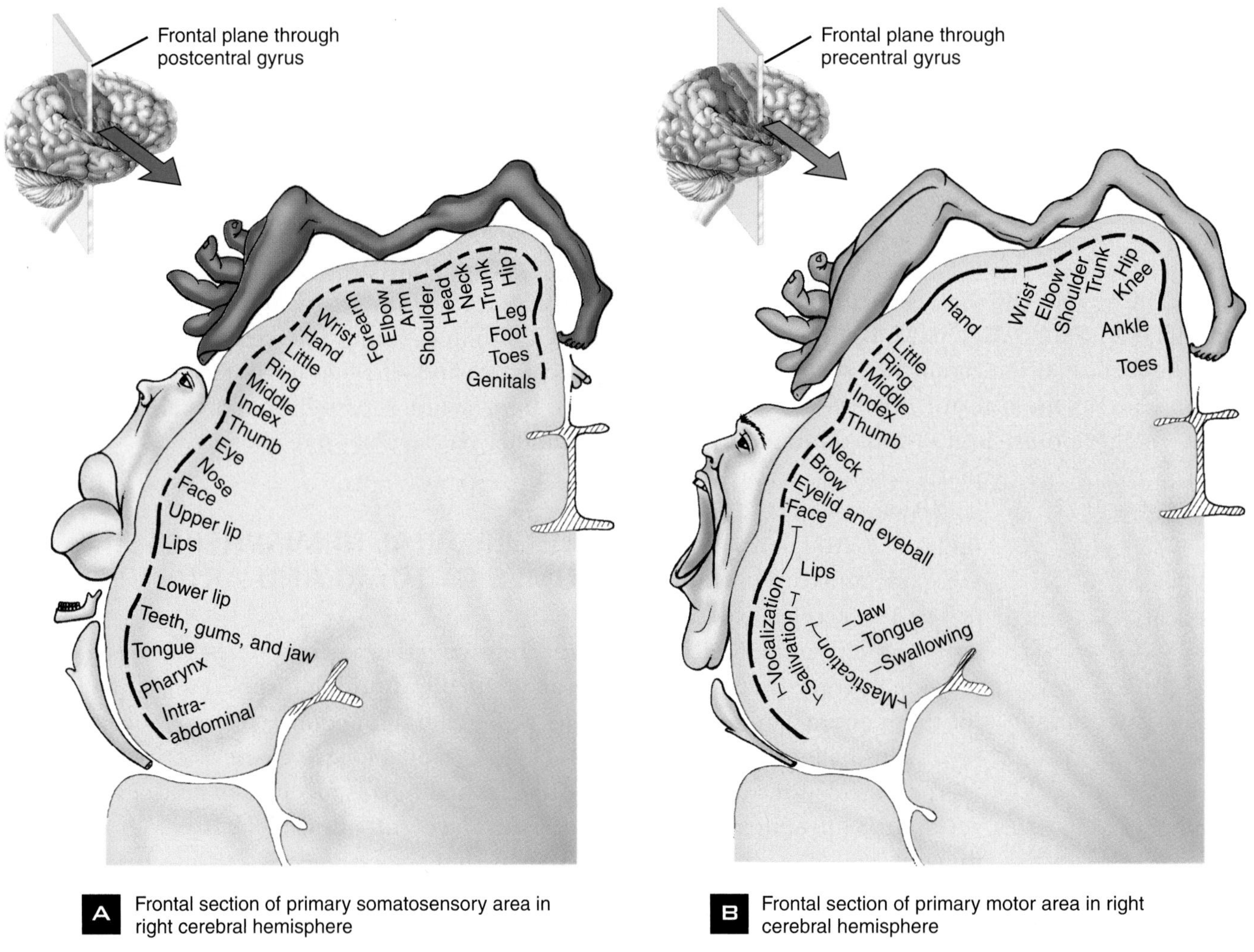

A Frontal section of primary somatosensory area in right cerebral hemisphere

B Frontal section of primary motor area in right cerebral hemisphere

Sensory homunculus and motor homunculus FIGURE 7.19

The left and right cerebral hemispheres are distinct in some important ways. In most people, the right hemisphere analyzes sensory input, recognizes faces, and functions in spatial relationships. Emotional interpretation of conversation is a function of the right hemisphere. When you hear someone say "that's just great," your right hemisphere determines whether the speaker was actually impressed or speaking sarcastically. The left hemisphere usually includes the general language interpretation and speech centers, and it controls writing and speaking. The left hemisphere is more active during mathematical calculations, categorizing items, and making logical decisions, leading some to call it the "dominant" or "categorical" hemisphere.

Special senses
The five senses of the body: hearing, vision, taste, smell, and touch.

Special senses (see Chapter 8) are integrated in specific areas of the cerebral cortex. For example, the entire occipital lobe is devoted to visual interpretation. Auditory interpretation occurs in the primary auditory area of the temporal lobe. We even have a primary taste area in the parietal lobes that permits us to differentiate the taste of chocolate from coffee. No word yet on how that works with mocha java . . .

ASSOCIATION AREAS LINK THINGS TOGETHER

Association areas of the cerebral cortex integrate and coordinate information from many sources. For example, the somatosensory association area processes sensory information from the skin and muscles. The visual association area associates new visual information with stored visual images. The auditory association area does the same thing with new auditory information.

Although we can assign functions to each part of the brain, the various parts do not function alone. The brain is a network of incomprehensible complexity. Stimuli are integrated throughout the cortex, and responses are generated from many areas. The left and right sides of the brain connect through the transverse tracts of the corpus callosum, sharing information and generating different responses (FIGURE 7.18, p. 211). In this way, despite **hemispheric lateralization** of some tasks, the entire cerebrum is aware of incoming sensory information as well as outgoing motor responses.

Hemispheric lateralization
The isolation of a task to either the left or right hemisphere of the cerebrum.

THE RETICULAR ACTIVATING SYSTEM IS THE BRAIN'S ALARM CLOCK

The **reticular formation** serves as an important connection between various parts of the brain. This series of **nuclei** and tracts extends throughout the brain, receiving sensory information, parceling it to the higher centers, and directing motor responses to the appropriate body areas. The **reticular activating system** (RAS) is a portion of the reticular formation, important in maintaining alertness. Look around the next time you are trapped listening to a long-winded lecture. If your reticular activating system is doing its job, you will remain alert and attentive. But you might see some people whose RAS is not working so well. Their heads will be drooping; they might even be napping.

Nuclei
Areas of concentrated neuronal cell bodies in the brain.

The RAS may also be important in our ability to learn. One symptom of **Attention Deficit Hyperactivity Disorder** (ADHD) is the inability to filter out extraneous noises and focus on what is important. The RAS is responsible for this filtering, allowing you to study while the radio is on. It is possible that ADHD is partly due to poor function of the RAS. You can read more about ADHD in the Ethics and Issues box on page 216.

Common mental disorders TABLE 7.4

Class of disorder	*Common types*	*Symptoms*	*Treatment*
Anxiety disorders	Phobias	Extreme fear or dread	Medications, cognitive and behavioral therapy
	Panic disorders	Sudden intense feelings of terror for no apparent reason	Medications, cognitive and behavioral therapy
	Obsessive compulsive disorder	Anxiety coping strategies that include repetitive actions or words or ritualistic behaviors	Medications, cognitive and behavioral therapy
Mood disorders	Depression and bipolar disorders	Depression: extreme sadness, sleeping or eating disturbances, changes in activity or energy levels. Bipolar disorder: violent mood swings	Psychotherapy, antidepressants, lithium
Schizophrenia	Schizophrenia	Chemical imbalances in the brain that lead to hallucinations, delusions, withdrawal, poor speech and reasoning	Prescription antipsychotic medications such as Haloperidol (Haldol) and Loxitane
Dementias	Alzheimer's	Loss of mental function and memory, decline in physical abilities	Increased nursing care
Eating disorders	Anorexia nervosa	Preoccupation with food and unnatural fear of becoming fat, self-starvation or over-exercising	Psychotherapy, lifestyle changes
	Bulimia	Bingeing and purging, cycles of huge caloric intake, with self-induced vomiting	Psychotherapy, lifestyle change

Humans suffer from many other mental disorders. TABLE 7.4 gives some information on the most common of these ailments.

THE SPINAL CORD CONNECTS TO ALMOST EVERYWHERE

The spinal cord, which extends from the brain into the vertebral column, is the second organ of the CNS (FIGURE 7.20). The spinal cord is composed of white tracts surrounding gray matter (opposite the arrangement in the brain). This means that the exterior of the spinal cord is composed of communication tracts running up and down the spinal cord, while the interior is composed of connections between spinal nerves. The spinal cord is the main route of communication between the brain and the body. Sensory information enters the spinal cord via the **dorsal root** and is transferred to an upward tract heading toward the brain.

Motor impulses generated in the brain are passed through the downward tracts of the spinal cord to the nerves of the body. These tracts are often called pyramids. The pyramids are continuations of the tracts in the medulla oblongata that cross to carry information generated in one hemisphere over to the opposite side of the body.

REFLEXES BYPASS THE BRAIN

Sensory information that demands immediate attention may initiate a **reflex**. Reflexes are extremely quick responses to sensory stimuli, running through the spinal cord from the dorsal root immediately to the ventral root and bypassing the brain. Evolution honed this brilliant system to keep our vertebrate ancestors safe from danger. Incoming sensory information is transferred to an association neuron in the innermost portion of the spinal cord and then directly to a motor

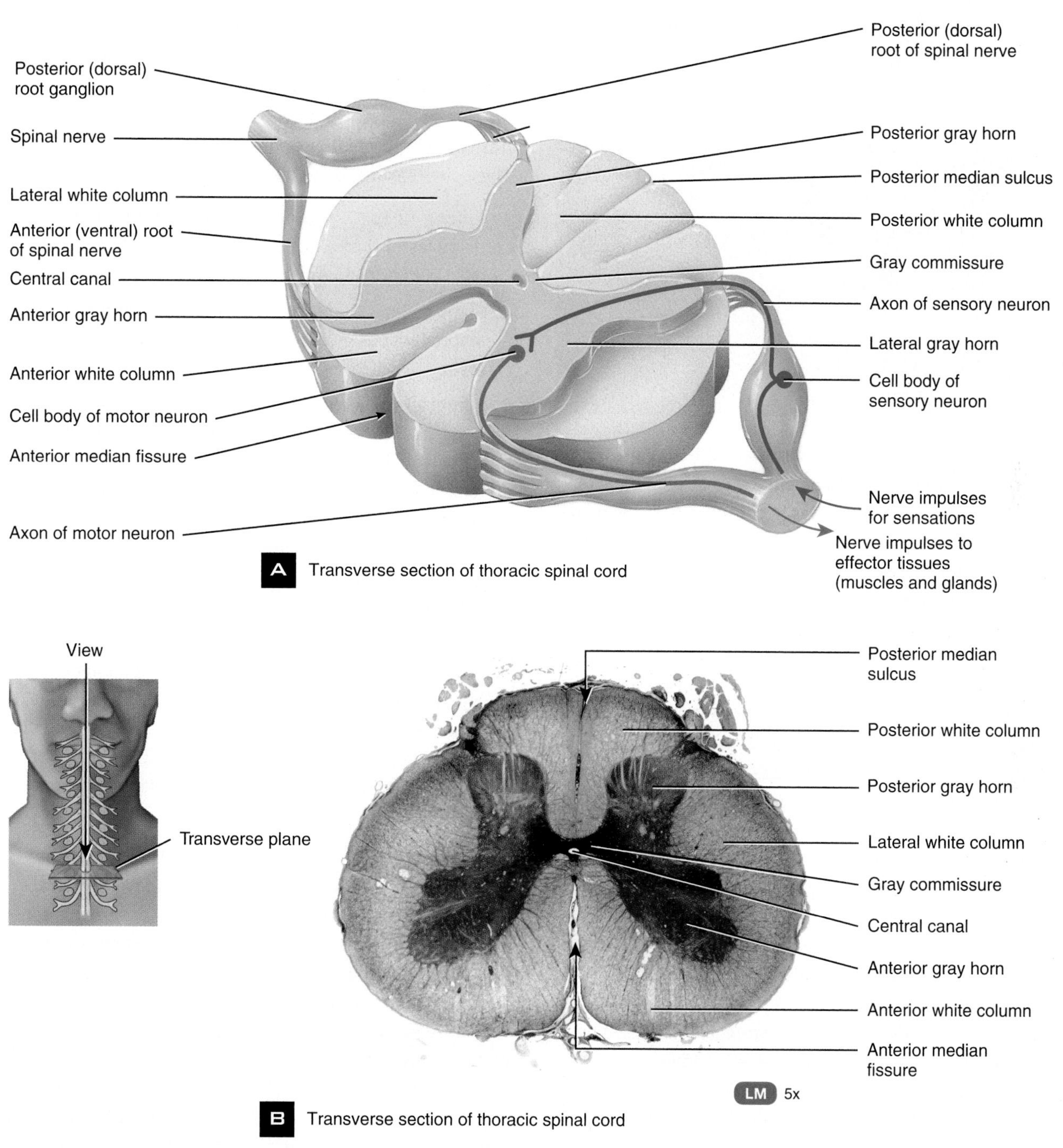

A Transverse section of thoracic spinal cord

B Transverse section of thoracic spinal cord

Spinal cord FIGURE 7.20

Ethics and Issues

Attention deficit hyperactivity disorder: does drug treatment make sense?

Attention deficit hyperactivity disorder (ADHD) is one of the most common mental disorders among children. Characteristically, ADHD causes difficulties in concentration, taking directions, sitting still, and cooperating, all of which can lead to learning and social difficulties.

In terms of brain physiology, it is not clear what causes ADHD. Some think ADHD may even be related to sleep deprivation: researchers have found abnormal levels of **sleep apnea** (the periodic cessation of breathing during sleep: *a*– = without, *pnea* = breath) among ADHD children. This breathing problem causes repeated awakenings at night, interfering with deep sleep. If this observation is correct, stimulants could merely be masking a condition of sleepiness that might better be treated more specifically.

Whatever the cause, the diagnosis of ADHD is growing more common. Widely varying statistics show that it affects 1 to 6 percent of American youths. ADHD is also being diagnosed among adults, with an estimated 1 percent of Americans aged 20 to 64 taking stimulants for the condition. Among adults, ADHD is less likely to cause hyperactivity than restlessness, difficulty paying attention, impulsive behavior, and frustration with failing to reach goals.

What can be done to treat ADHD? One approach is behavioral; parents try to shape behavior by rewarding desirable activity and imposing consequences for actions they want to discourage. The behavioral approach can be combined with, or replaced by, treatment with stimulant drugs, especially forms of amphetamine. Curiously, although amphetamines stimulate most people, they calm people with ADHD. This unexpected effect is actually a hallmark of the disease.

Still, the widespread use of prescription medication for ADHD is making some people nervous, especially those who suspect that an ADHD diagnosis is mainly a tactic to make business for psychiatrists and the pharmaceutical industry. These are reasons for concern:

1. Among 12- to 17-year-olds, abuse of prescription drugs is rising faster than abuse of illegal drugs, and amphetamines are addictive in some people.
2. Some college students with ADHD prescriptions say the amphetamines give them extra focus and energy during tests.
3. Stimulants have been linked to the death of 19 children and 6 adults (among an estimated 4 million people taking stimulants for ADHD) due to heart problems that may be related to the stimulants. The U.S. Food and Drug Administration is considering stronger warning labels on the packages. Although some unexplained deaths are inevitable among any group of 4 million people, the news should prompt doctors to evaluate heart health before prescribing stimulants for ADHD.
4. Shouldn't we just "let boys be boys?" According to this logic, boys typically have more of the "ADHD personality characteristics," like impulsivity, excess energy, and difficulty with planning. Should being male be considered a mental illness, especially in a society plagued by drug abuse?

Like other challenges of parenting, ADHD forces parents to persist, improvise, and decide. Behavioral therapy can be wearing, and it may require assistance from teachers and others who are important to the child. Stimulant drugs can send a message that psychological problems can be fixed with a pill. But if the consequences of failing to treat ADHD are negative enough, parents must choose a treatment strategy and philosophy, and carry it through.

Although scientists are improving their understanding of brain function, much remains to be understood, including the integration of different portions of the brain, and the function of various nuclei and neurotransmitters. As neuroscientists probe deeper into the brain's structure and function, we may learn to treat or even prevent some of the severe mental disorders that afflict our fellow humans.

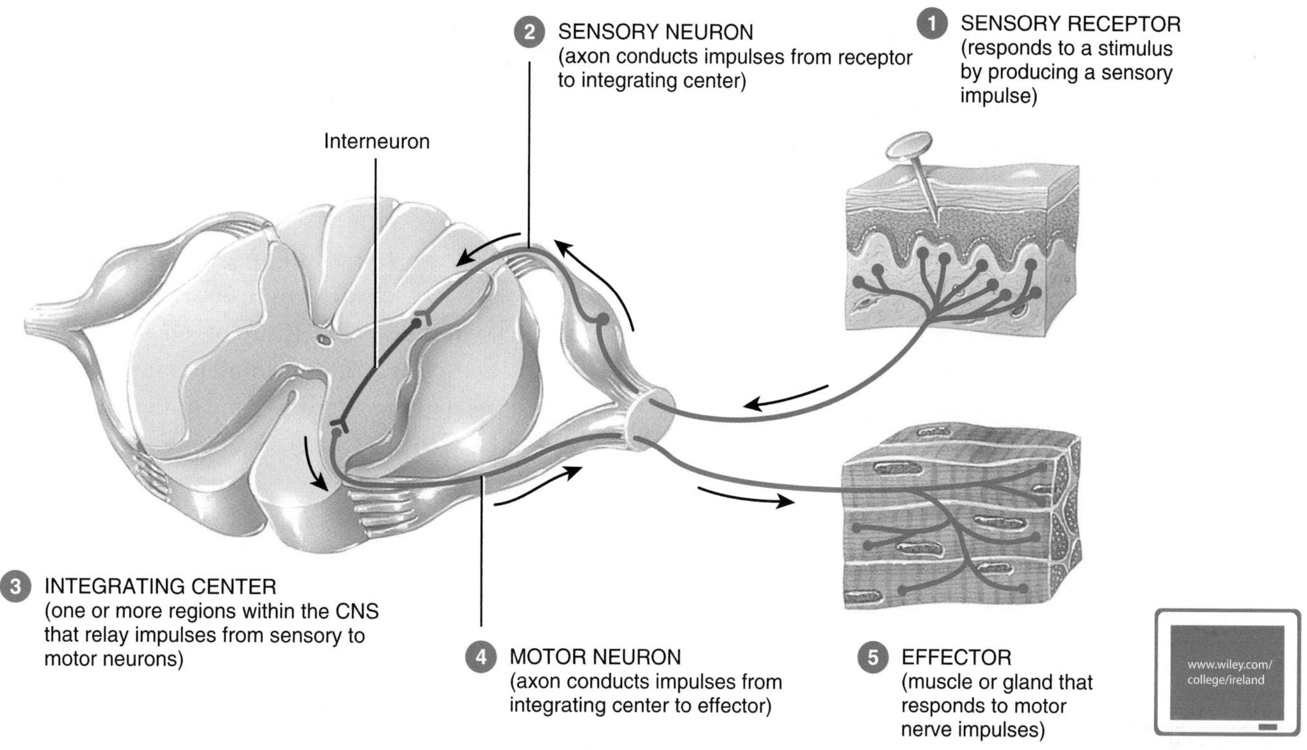

Reflex arc FIGURE 7.21

neuron. The motor neuron transmits an immediate response through the ventral root to the effector organ.

Reflexes generate an immediate, life-saving motor response. You pull your hand from an open flame even before you consciously recognize the heat. As you pull your hand away, the "that's hot!" information is still traveling to your brain. There, a series of motor responses begins, causing you to rub your hand, inspect it for burns, and exclaim in surprise or pain. Fortunately, before all these brain-initiated motor responses can occur, the reflex has already removed your hand from danger. (See FIGURE 7.21.)

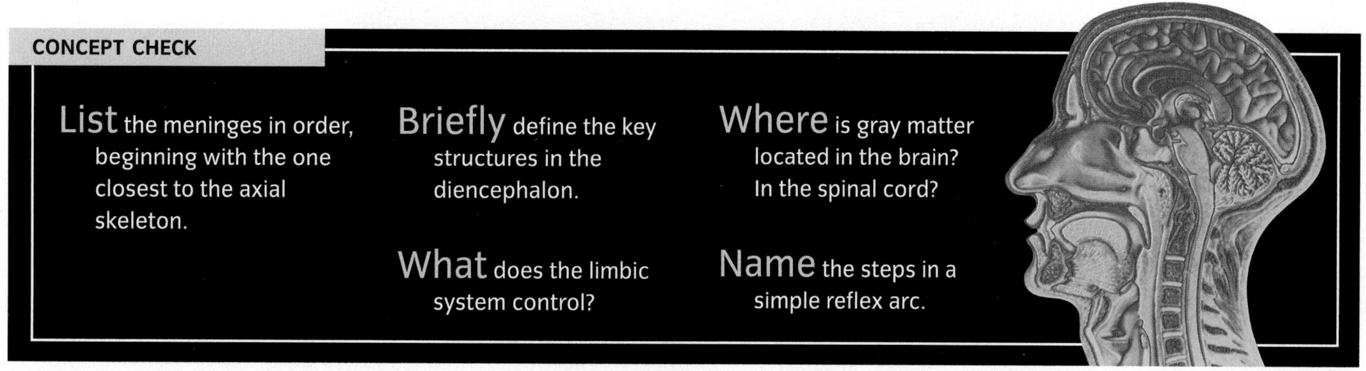

CONCEPT CHECK

List the meninges in order, beginning with the one closest to the axial skeleton.

Briefly define the key structures in the diencephalon.

What does the limbic system control?

Where is gray matter located in the brain? In the spinal cord?

Name the steps in a simple reflex arc.

The Peripheral Nervous System Operates Beyond the Central Nervous System

LEARNING OBJECTIVES

Describe the difference between spinal and cranial nerves.

Compare the sympathetic and parasympathetic aspects of the PNS.

Compare the anatomy and the functions of the cranial and spinal nerves.

The peripheral nervous system (PNS) is composed of all neural tissue other than the brain and spinal cord. The PNS includes the nerves that protrude from these structures. The 12 nerves that extend from the brain are called the **cranial nerves** (TABLE 7.5). These nerves are identified by name and a Roman numeral (FIGURE 7.22). Some are sensory only, others

Brain and cranial nerves FIGURE 7.22

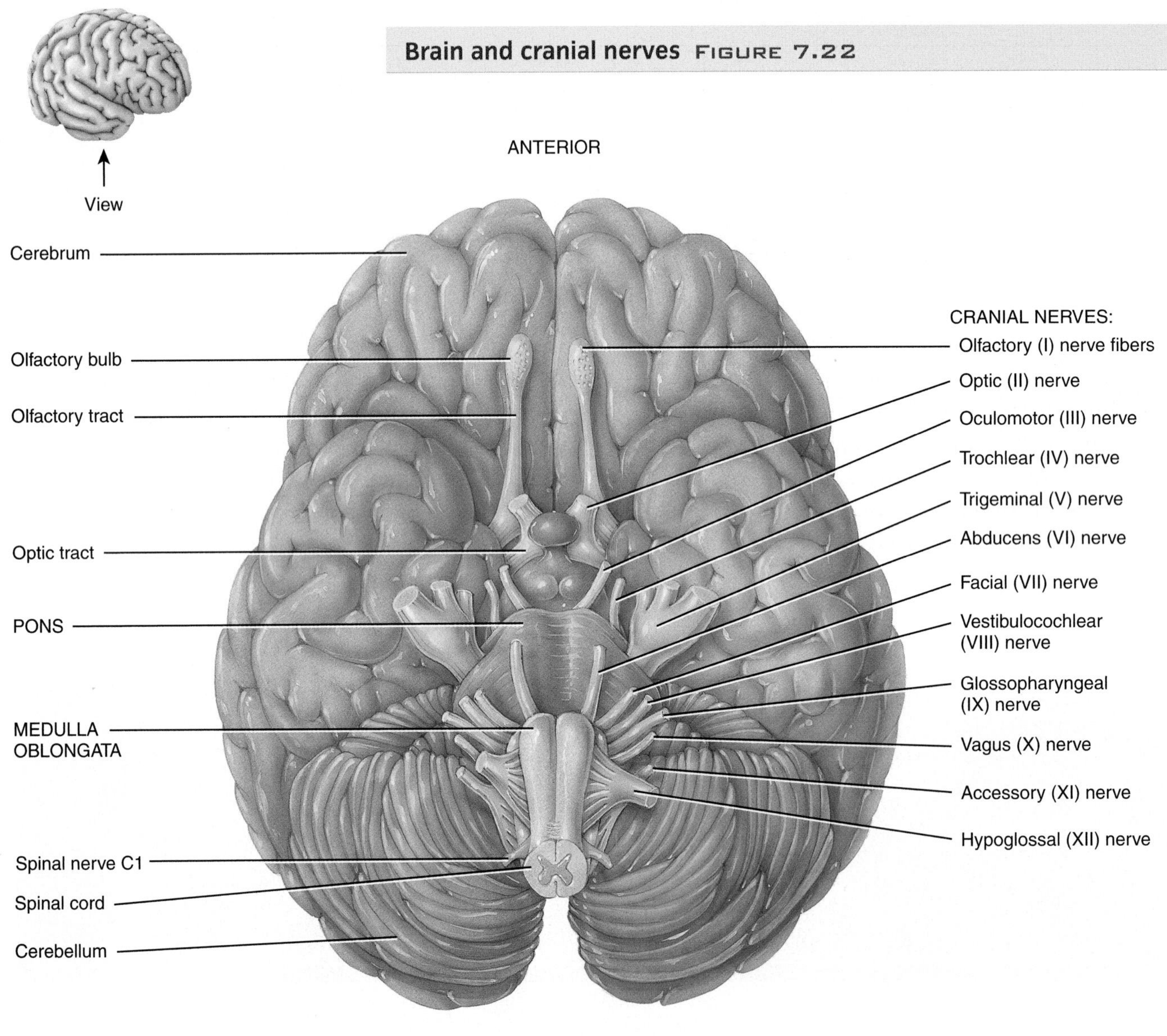

Inferior aspect of brain

Cranial nerves TABLE 7.5

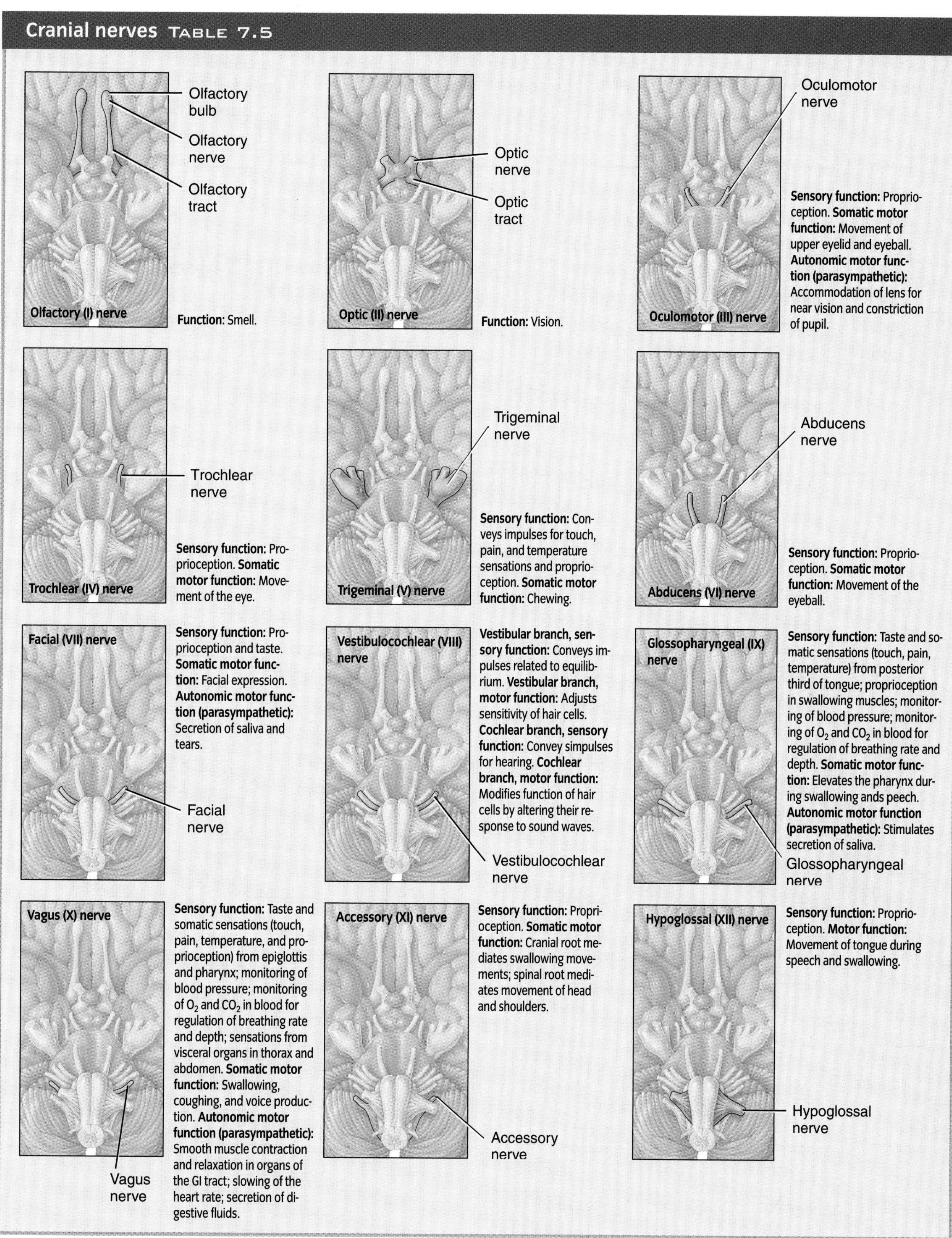

Olfactory bulb

Olfactory nerve

Olfactory tract

Olfactory (I) nerve

Function: Smell.

Optic nerve

Optic tract

Optic (II) nerve

Function: Vision.

Oculomotor nerve

Oculomotor (III) nerve

Sensory function: Proprioception. **Somatic motor function:** Movement of upper eyelid and eyeball. **Autonomic motor function (parasympathetic):** Accommodation of lens for near vision and constriction of pupil.

Trochlear nerve

Trochlear (IV) nerve

Sensory function: Proprioception. **Somatic motor function:** Movement of the eye.

Trigeminal nerve

Trigeminal (V) nerve

Sensory function: Conveys impulses for touch, pain, and temperature sensations and proprioception. **Somatic motor function:** Chewing.

Abducens nerve

Abducens (VI) nerve

Sensory function: Proprioception. **Somatic motor function:** Movement of the eyeball.

Facial (VII) nerve

Sensory function: Proprioception and taste. **Somatic motor function:** Facial expression. **Autonomic motor function (parasympathetic):** Secretion of saliva and tears.

Facial nerve

Vestibulocochlear (VIII) nerve

Vestibular branch, sensory function: Conveys impulses related to equilibrium. **Vestibular branch, motor function:** Adjusts sensitivity of hair cells. **Cochlear branch, sensory function:** Convey simpulses for hearing. **Cochlear branch, motor function:** Modifies function of hair cells by altering their response to sound waves.

Vestibulocochlear nerve

Glossopharyngeal (IX) nerve

Sensory function: Taste and somatic sensations (touch, pain, temperature) from posterior third of tongue; proprioception in swallowing muscles; monitoring of blood pressure; monitoring of O_2 and CO_2 in blood for regulation of breathing rate and depth. **Somatic motor function:** Elevates the pharynx during swallowing ands peech. **Autonomic motor function (parasympathetic):** Stimulates secretion of saliva.

Glossopharyngeal nerve

Vagus (X) nerve

Sensory function: Taste and somatic sensations (touch, pain, temperature, and proprioception) from epiglottis and pharynx; monitoring of blood pressure; monitoring of O_2 and CO_2 in blood for regulation of breathing rate and depth; sensations from visceral organs in thorax and abdomen. **Somatic motor function:** Swallowing, coughing, and voice production. **Autonomic motor function (parasympathetic):** Smooth muscle contraction and relaxation in organs of the GI tract; slowing of the heart rate; secretion of digestive fluids.

Vagus nerve

Accessory (XI) nerve

Sensory function: Proprioception. **Somatic motor function:** Cranial root mediates swallowing movements; spinal root mediates movement of head and shoulders.

Accessory nerve

Hypoglossal (XII) nerve

Sensory function: Proprioception. **Motor function:** Movement of tongue during speech and swallowing.

Hypoglossal nerve

are motor only, and the remainder serve both functions. Most cranial nerves carry impulses that deal with the head, neck, and facial regions. However, the **vagus nerve** (X) branches to the throat, voice box, and abdominal organs.

Thirty-one pairs of **spinal nerves** extend from the spinal cord. These are all mixed nerves, carrying both sensory and motor information. Each spinal nerve connects with body structures near the region where it originates (FIGURE 7.23).

Sensory neurons carry information to the CNS. They join other motor and sensory neural processes to form a spinal nerve. These sensory neurons separate from the motor neurons after they enter the spinal cord. Sensory neurons enter the spinal cord at the back, through the dorsal root of the spinal nerve. Their cell bodies are located just outside the CNS, in the dorsal root ganglia. Motor neurons exit the spinal cord at the front. Their cell bodies are within the CNS, and their axons extend out through the ventral root of the spinal cord. These processes can be long. The axon of the motor neuron that moves the great toe reaches from the sacral area of the spinal cord down the entire length of the leg, a distance of up to 1 meter! (See FIGURE 7.20.)

THE PNS ALSO CONTAINS SYMPATHETIC AND PARASYMPATHETIC NERVES

Autonomic nerves—the ones you do not consciously control—are also part of the PNS. Along with the physiological differences in sympathetic and parasympathetic divisions discussed previously, these nerves display anatomical differences (FIGURE 7.24).

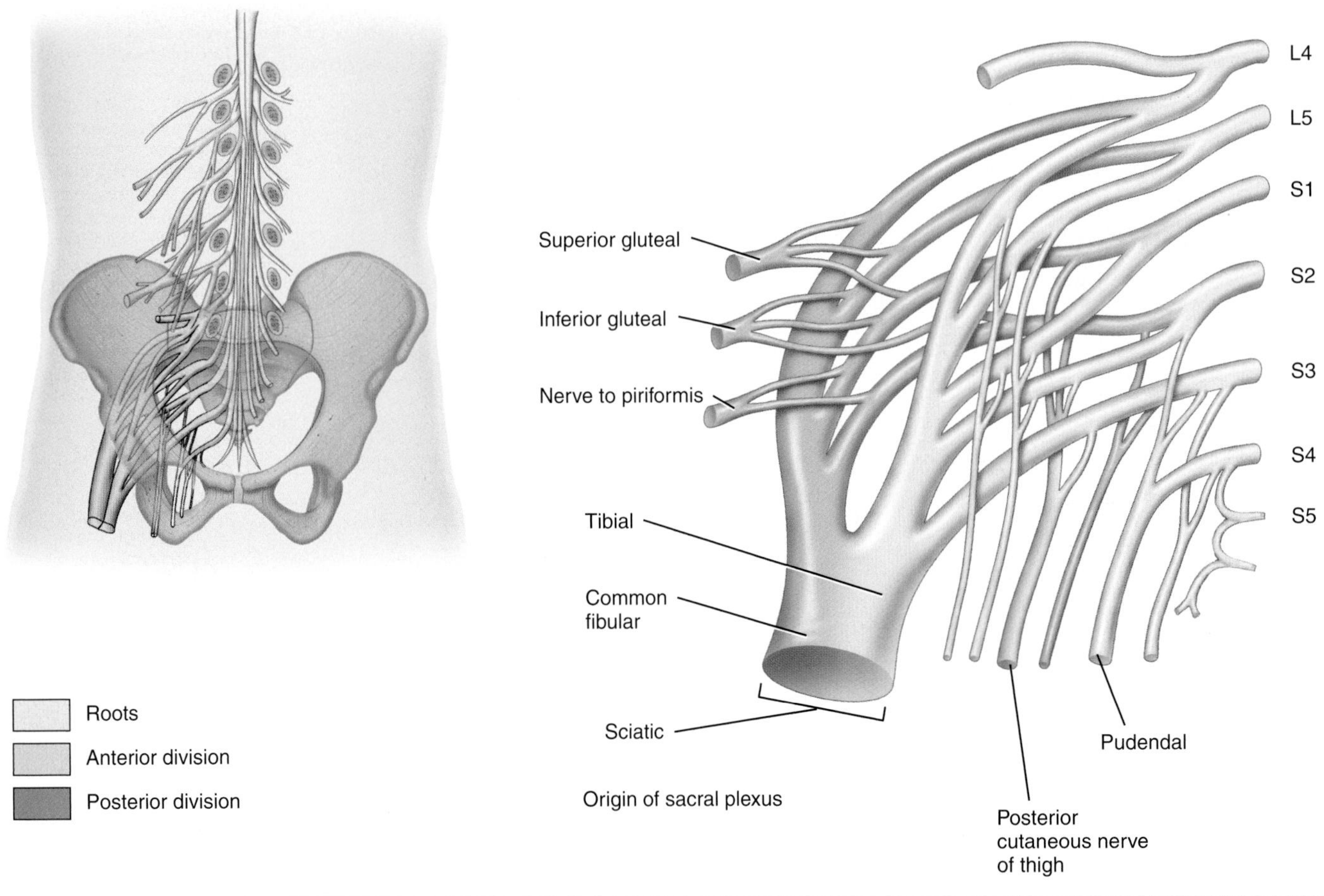

Sacral nerve FIGURE 7.23

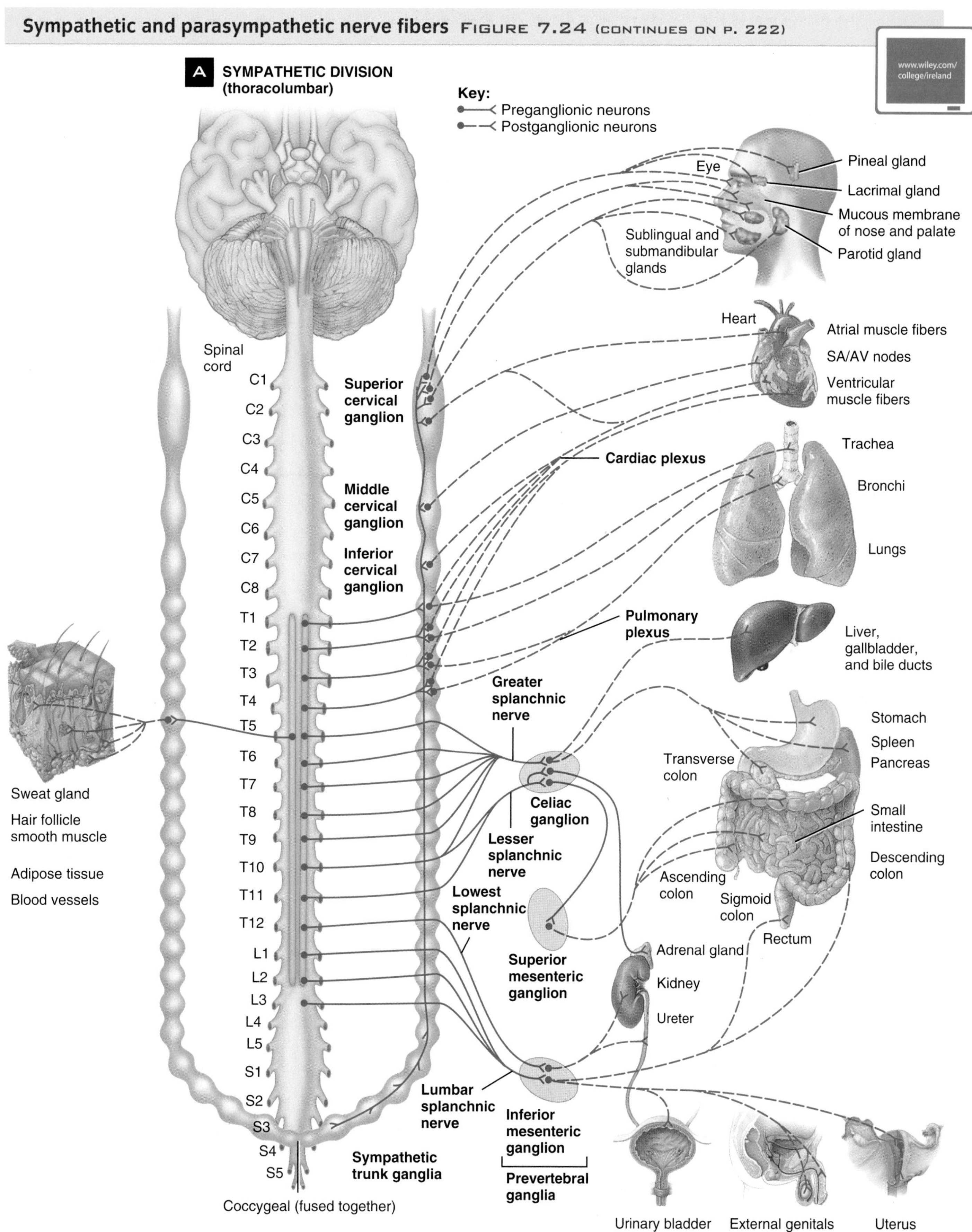
www.wiley.com/college/ireland
A SYMPATHETIC DIVISION (thoracolumbar)
Key:
Preganglionic neurons
Postganglionic neurons
Eye
Pineal gland
Lacrimal gland
Mucous membrane of nose and palate
Parotid gland
Sublingual and submandibular glands
Heart
Atrial muscle fibers
SA/AV nodes
Ventricular muscle fibers
Cardiac plexus
Trachea
Bronchi
Lungs
Pulmonary plexus
Liver, gallbladder, and bile ducts
Greater splanchnic nerve
Stomach
Spleen
Pancreas
Transverse colon
Celiac ganglion
Small intestine
Lesser splanchnic nerve
Descending colon
Ascending colon
Sigmoid colon
Lowest splanchnic nerve
Rectum
Superior mesenteric ganglion
Adrenal gland
Kidney
Ureter
Spinal cord
C1
C2
C3
C4
C5
C6
C7
C8
T1
T2
T3
T4
T5
T6
T7
T8
T9
T10
T11
T12
L1
L2
L3
L4
L5
S1
S2
S3
S4
S5
Superior cervical ganglion
Middle cervical ganglion
Inferior cervical ganglion
Sweat gland
Hair follicle smooth muscle
Adipose tissue
Blood vessels
Lumbar splanchnic nerve
Inferior mesenteric ganglion
Prevertebral ganglia
Sympathetic trunk ganglia
Coccygeal (fused together)
Urinary bladder
External genitals
Uterus

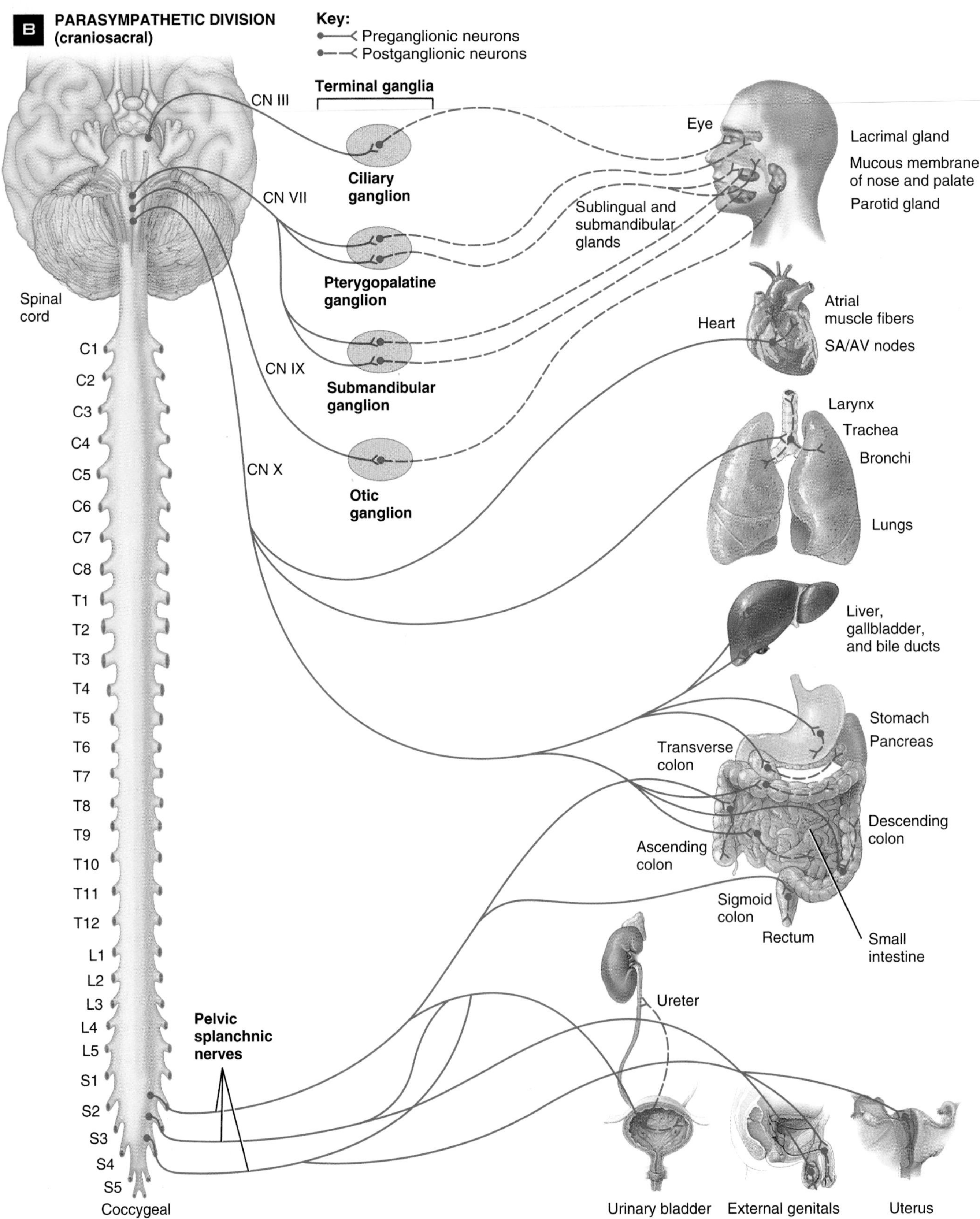
B
PARASYMPATHETIC DIVISION (craniosacral)
Key:
Preganglionic neurons
Postganglionic neurons
Terminal ganglia
CN III
Ciliary ganglion
CN VII
Pterygopalatine ganglion
CN IX
Submandibular ganglion
CN X
Otic ganglion
Spinal cord
C1
C2
C3
C4
C5
C6
C7
C8
T1
T2
T3
T4
T5
T6
T7
T8
T9
T10
T11
T12
L1
L2
L3
L4
L5
S1
S2
S3
S4
S5
Coccygeal
Pelvic splanchnic nerves
Eye
Lacrimal gland
Mucous membrane of nose and palate
Parotid gland
Sublingual and submandibular glands
Heart
Atrial muscle fibers
SA/AV nodes
Larynx
Trachea
Bronchi
Lungs
Liver, gallbladder, and bile ducts
Stomach
Pancreas
Transverse colon
Descending colon
Ascending colon
Sigmoid colon
Rectum
Small intestine
Ureter
Urinary bladder
External genitals
Uterus

The sympathetic nervous system includes nerves in the thoracic and lumbar region of the spinal cord only. Sympathetic fibers extend from the spinal cord to a series of **ganglia** (group of cell bodies) called the **sympathetic chain**, on either side of the spinal cord. At these ganglia, neurons from the CNS synapse with a second neuron that extends to the effector organ. Thus sympathetic neurons leaving the spinal cord are shorter than those leaving the sympathetic chain. We call the neurons leaving the spinal cord and synapsing in the ganglia preganglionic. Those that leave the ganglion and synapse with the effector organ are called postganglionic.

Parasympathetic fibers are found only in the cranial, cervical, and sacral region of the cord. These neurons leave the spinal or cranial nerve and join a ganglion near or in the effector organ. The parasympathetic preganglionic fibers are long, and the postganglionic neurons are extremely short.

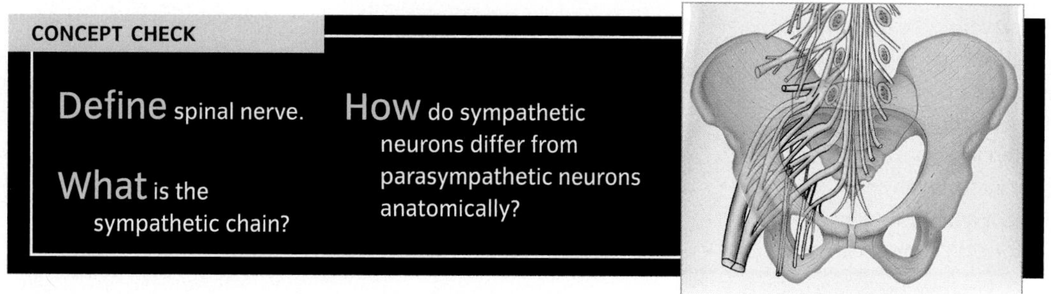

CONCEPT CHECK
Define spinal nerve.
What is the sympathetic chain?
How do sympathetic neurons differ from parasympathetic neurons anatomically?

CHAPTER SUMMARY

1 The Nervous System Makes Sense of Everything

The nervous system is responsible for maintaining homeostasis by reacting almost instantaneously to stimuli. It works in concert with the endocrine system to maintain homeostasis. The work of the system is performed by neurons, supported by neuroglial cells.

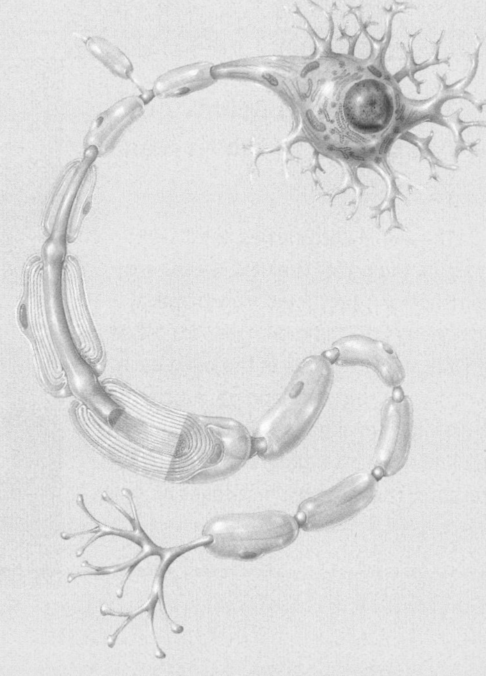

CHAPTER SUMMARY

2 The Nervous System Is Categorized by Function and Structure

The nervous system is divided into the central and peripheral nervous systems. The CNS includes the brain and spinal cord and is the main integration center of the body. The PNS includes the autonomic, sensory, and somatic nerves of the body. The autonomic division is further subdivided into the sympathetic and parasympathetic divisions. A nerve is composed of a bundle of neurons, protected by layers of connective tissue. Sensory information enters the CNS, which analyzes it and sends a motor response through the PNS to muscular or glandular tissue.

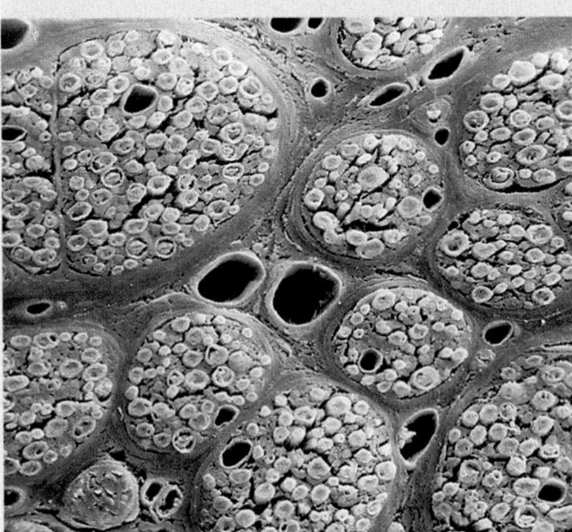

3 Nerve Tissue Is Made of Neurons and Glial Cells

The nervous system contains neurons and neuroglial cells. Neurons carry impulses, whereas glial cells carry out supporting functions. Sensory neurons detect conditions in the environment or body, motor neurons carry instructions to the body, and interneurons connect the two systems. Dendrites bring signals to the cell body, and the long axons deliver signals to other neurons or tissue.

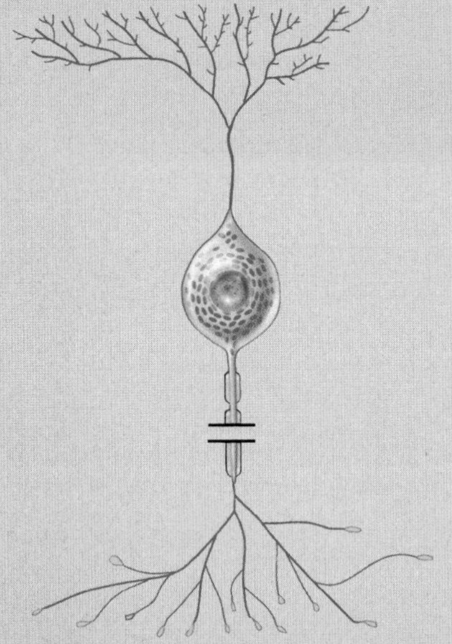

4 Neurons Work through Action Potentials

An action potential is a brief change in electrical conditions at a neuron's membrane that occurs when a neuron "fires." An action potential occurs when the charge differential across the neuron's membrane suddenly reverses polarity, as a result of changing ion concentrations inside and outside the neuron. Impulse speed is determined by axon diameter, degree of myelination, and other factors. Neurotransmitters carry signals from one neuron to the next across a tiny gap called the synapse. IPSPs and EPSPs also influence the generation of action potentials.

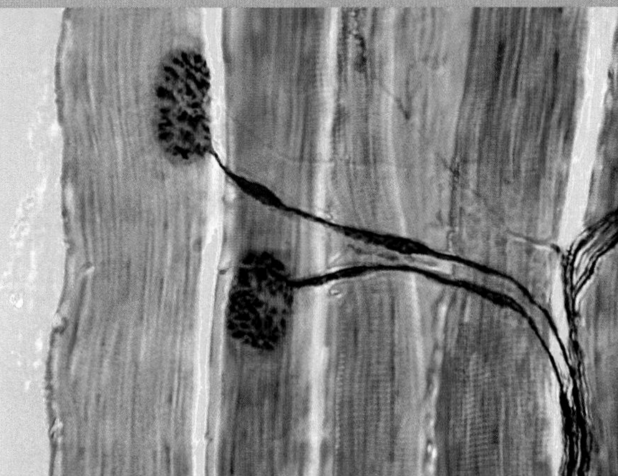

5 The Brain and Spinal Cord Are Central to the Nervous System

The spinal cord carries impulses to and from the brain. The CNS organs are nourished and protected from physical damage by CSF and meninges. The lobes and internal structures of the brain each have distinct, but overlapping, functions. The brain stem contains vital centers that regulate heart rate, breathing, and blood pressure. The cerebellum focuses on muscles and movement. The diencephalon is a relay center between other parts of the brain, whereas the cerebrum is a central processing center, home of logic and skills. The reticular activating system is the brain's alarm clock. Reflexes are two- or three-neuron circuits that bypass the brain to allow fast retreat from injury.

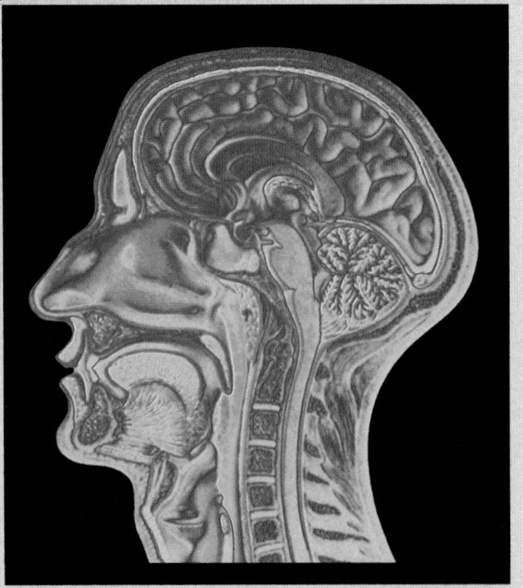

6 The Peripheral Nervous System Operates Beyond the Central Nervous System

The peripheral nervous system includes the nerves that protrude from the brain and spinal cord. The PNS originates with 12 cranial nerves and 31 pairs of spinal nerves. Peripheral nerves may be sensory, motor, or mixed. The autonomic nerves are not under conscious control. Sympathetic autonomic nerves control visceral organs in the thoracic and lumbar regions of the spinal cord. Parasympathetic autonomic nerve fibers emerge from the cranial, cervical, and sacral regions of the spinal cord.

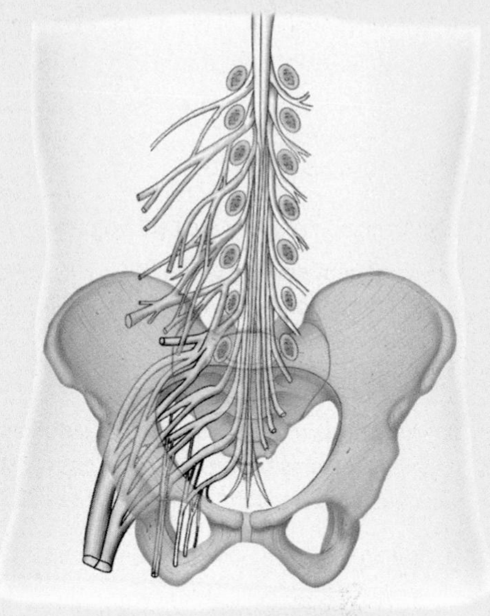

KEY TERMS

- **afferent** p. 191
- **autonomic division** p. 191
- **cerebrospinal fluid (CSF)** p. 203
- **cortex** p. 211
- **efferent** p. 191
- **gyri** p. 211
- **hemispheric lateralization** p. 213
- **medulla oblongata** p. 205
- **membrane potential** p. 194
- **myelin** p. 198
- **neuroglia** p. 193
- **neuron** p. 188
- **neurotransmitter** p. 188
- **nuclei** p. 213
- **pons** p. 205
- **postsynaptic neuron** p. 199
- **presynaptic neuron** p. 199
- **proprioception** p. 191
- **somatic division** p. 191
- **special senses** p. 213
- **sulci** p. 211
- **terminal bulb** p. 199
- **tracts** p. 207

CRITICAL THINKING QUESTIONS

1. Compare the structure of a nerve to the structure of a muscle. What explains the anatomical similarities? What are the main differences?

2. Review the steps in an action potential, as well as the definition of IPSP and EPSP. Using what you know, describe a neuron that is exhibiting an IPSP. How would the ion concentrations across the membrane be different from an EPSP? Can you predict what ion conditions would cause an EPSP?

3. Why are reflexes faster than conscious thought? Why is the response slower when the brain is involved? Why do we even have reflexes?

SELF TEST

1. The functional unit of the nervous system is
   a. the brain.
   b. the brain and spinal cord.
   c. the neuron.
   d. the neuroglia.

2. Information reaches the CNS from the
   a. afferent division of the PNS.
   b. efferent division of the PNS.
   c. motor neurons.
   d. sympathetic division.

3. True or False? The division of the autonomic nervous system that is responsible for digestion, energy storage, and relaxation is the parasympathetic division of the PNS.

4. Identify the type of neuroglion shown.
   a. Astrocyte
   b. Motor neuron
   c. Microglion
   d. Oligodendrocyte

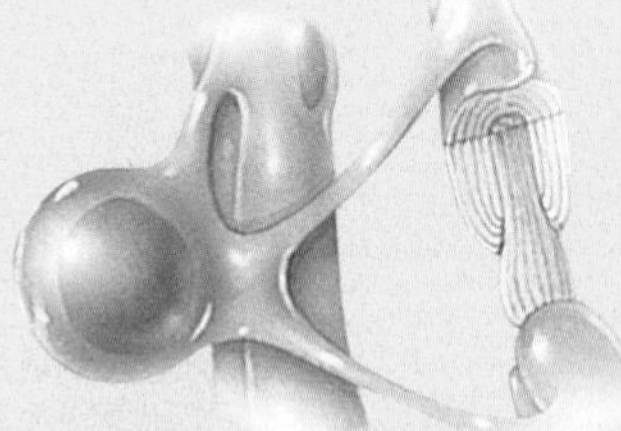

5. The neuron pictured here is responsible for
   a. sending and receiving sensory information.
   b. sending and receiving motor information.
   c. integrating information from sensory and motor neurons.
   d. Neuron function cannot be determined from neuron anatomy.

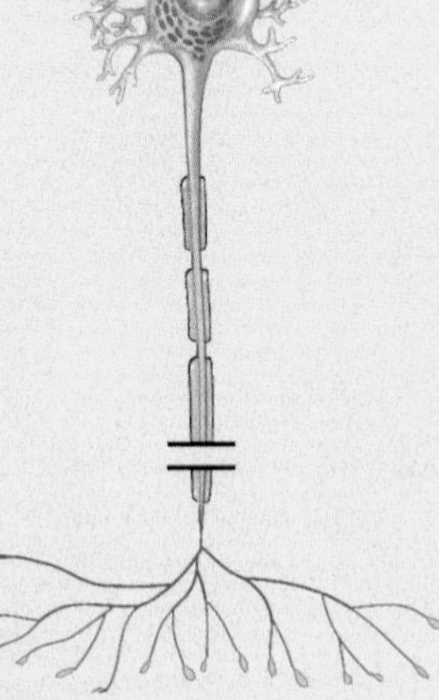

6. The type of membrane protein that allows ions to enter the cell only during a shift in membrane voltage is a
   a. mechanically regulated channel.
   b. ligand-gated channel.
   c. voltage-gated channel.
   d. leaky gated channel.

The next few questions refer to the image below

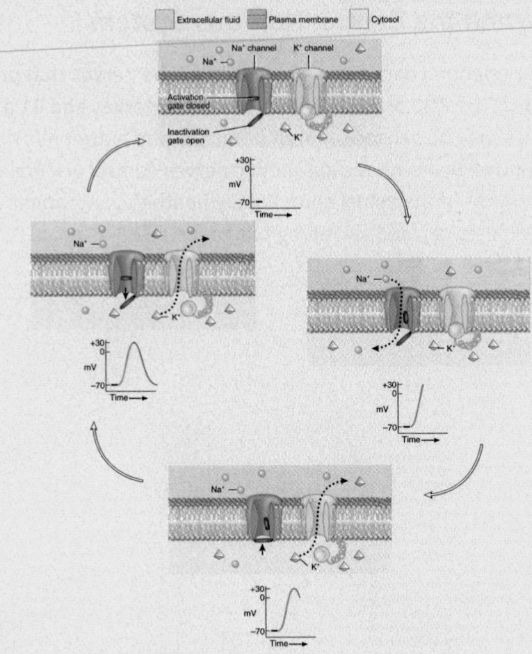

7. The original membrane potential of a resting neuron is
   a. −70 mV.
   b. +90 mV.
   c. 0 mV.
   d. dependent on neuron location.

8. The first ion to enter the neuron at the beginning of an action potential is
   a. calcium.
   b. potassium.
   c. sodium.
   d. ATP.

9. The period of time immediately after an action potential, during which the neuron cannot send a second action potential, is the
   a. relative refractory period.
   b. absolute refractory period.
   c. dead zone.
   d. sodium/potassium ATPase period.

10. The function of the cell shown in the diagram below is to
    a. myelinate PNS neurons.
    b. myelinate CNS neurons.
    c. increase action potential propagation speed.
    d. decrease action potential propagation speed.
    e. Both a and c are correct.

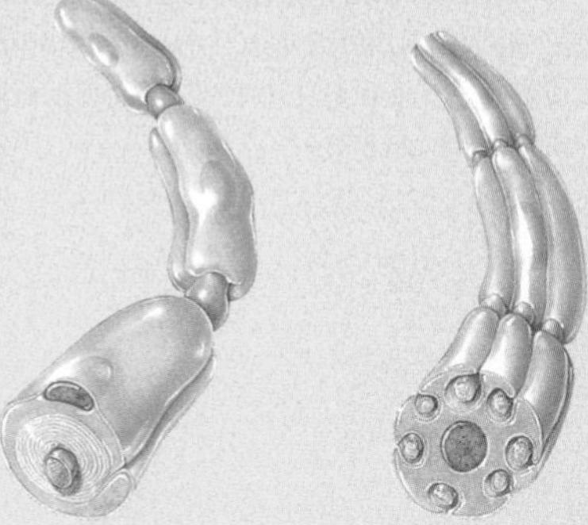

11. What cell provides this same function in the brain?

a. Schwann cell
b. Astrocyte
c. Oligodendrocyte
d. Microglial cell

12. True or False? An EPSP causes a slight hyperpolarization of the neuron cell membrane, making it more difficult to initiate an action potential.

13. Identify the specific layer of the meninges indicated by the letter A on this figure.

a. Dura mater
b. Pia mater
c. Arachnoid

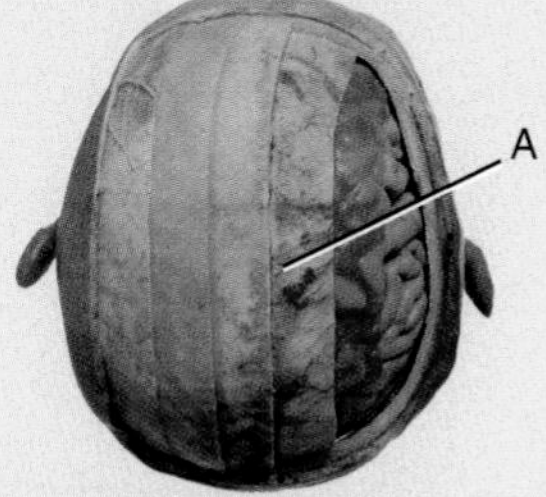

14. The ventricles in your brain are the site of

a. sensory input.
b. CSF formation.
c. memory formation.
d. CSF absorption.

15. Identify the portion of the brain indicated in this figure.

a. Brain stem
b. Cerebrum
c. Cerebellum
d. Diencephalon

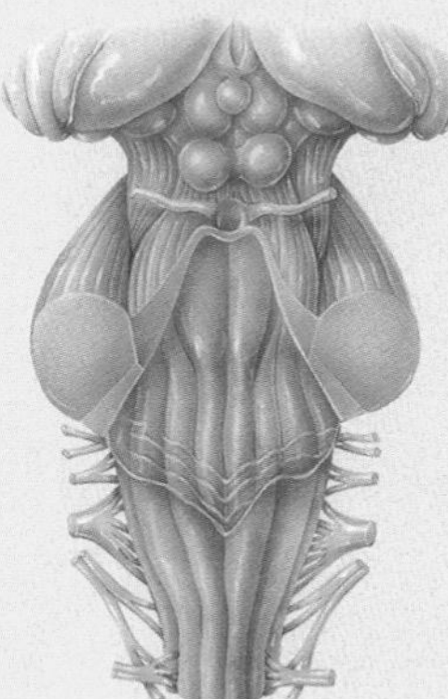

16. The functions of the structure shown below include

a. sensory interpretation.
b. proprioception.
c. learning.
d. heart rate control.

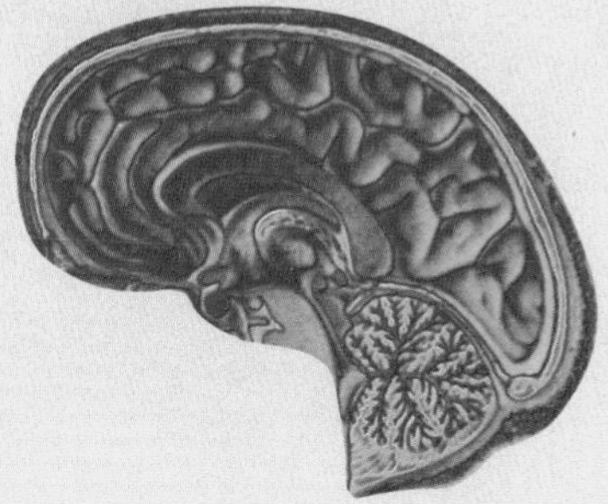

17. The portion of the brain that is responsible for emotions is the

a. hypothalamus.
b. thalamus.
c. reticular formation.
d. limbic system.

18. The surface of the spinal cord is white, indicating that it functions as

a. a highway for information traveling up and down the cord.
b. an integration center, where impulses are connected to one another and then passed to the brain.
c. an insulation layer surrounding the functioning neurons underneath.
d. In nervous tissue, color does not indicate function.

19. The correct sequence of structures in a reflex is

a. sensory receptor → sensory neuron → spinal cord → brain → spinal cord → motor neuron → effector organ.
b. sensory receptor → spinal cord → brain → motor neuron → effector organ.
c. sensory receptor → motor neuron → spinal cord → sensory neuron → effector organ.
d. sensory receptor → sensory neuron → spinal cord → motor neuron → effector organ.
e. sensory receptor → effector organ → brain → motor neuron → spinal cord.

20. Which of the two divisions of the autonomic division of the PNS has the longer postsynaptic neurons?

a. Sympathetic division
b. Parasympathetic division

21. What is the function of the autonomic division of the PNS shown in this figure?

a. Increased digestive activity
b. Increased respiratory and heart rate
c. Increased urinary output
d. Decreased mental alertness

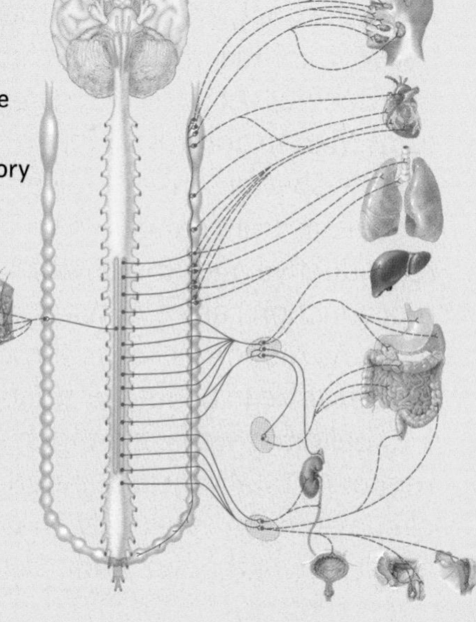

The Special Senses

8

Rollercoasters and tilt-a-whirls are notorious for inducing nausea, but some people get similar problems from the little swerves and dips of a journey by car, boat, or plane. These folks break into a cold sweat and get a headache. They get nauseous and feel listless or uneasy.

The syndrome goes by many names: carsickness, seasickness, airsickness, or, more generically, motion sickness. Many people suffer from it, at least under some conditions, even 70 percent of first-time astronauts.

The problem seems to arise from a war between the senses. In the back seat of a car, most of what you see is stationary in relation to you, so your eyes tell your brain that you are not moving. But other senses say you are. The seat presses against your skin on each bump, your joints flex, and your inner ear registers changes in direction. As your brain struggles with what to believe, the conflicting messages cause inner turmoil, the release of stress hormones, and misery.

The commonsense prescription for motion sickness is to try to reduce this sensory war. Move to the front seat of a car or the deck of a boat, and stare at the horizon. Show your brain that you are indeed moving. Generally, it's easier to prevent symptoms than to control them.

The special senses evolved to protect organisms from danger so they can reach reproductive age, and anything that affects reproduction can have species-wide effects. As motion sickness shows, the special senses can be fooled and delude us into believing we face danger, and can even render us totally incapacitated in certain circumstances. Obviously, these senses can affect us whether we want them to or not.

CHAPTER OUTLINE

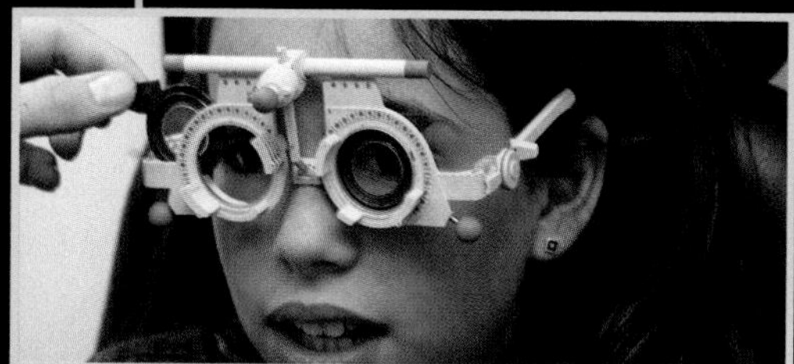

The Special Senses Tell Us about Our Environment

LEARNING OBJECTIVES

Describe the special senses.

Discuss the physiology of balance and hearing.

Relate the structure of the outer, middle, and inner ear to the functions of each.

Explain the physiology of the chemical senses of taste and smell.

The intricate functioning of the nervous system is best appreciated when discussing our special senses. These extremely sensitive receptors supply us with detailed information about the world around us, including the sights, sounds, smells, and tastes present in our surroundings. The wealth of information they provide occupies most of our brain and forms the basis for our logical and rational decisions.

We rely on our senses to get through even the simplest task. To eat an apple, we first locate it visually, and we may scan it for rotten spots or an appealing color. Picking it up, we gain more information from the firmness of the skin and the fruit's density. We may even raise the apple to our nose and smell it before taking the first bite. Consciously or not, we assess that first bite to make sure it tastes right. Each of these small, practically automatic actions supplies information to the brain through the special senses.

Our special senses include **photoreceptors** for vision, **mechanoreceptors** for hearing and balance, and **chemoreceptors** for smell and taste. (There is an in-depth discussion of mechanoreceptors in the skin in Chapter 9.) We are extremely visual creatures, using our eyes to provide most of our clues about the environment. Hearing is our second most acute sense, providing enough information to allow us to move through the environment even when we cannot see. Our sense of balance, or equilibrium, is closely allied with hearing in that both reside in the ear. However, balance is often overruled by our strong visual perceptions. The thrill of amusement park rides and the awful feeling of seasickness both come from our brain trying to reconcile visual information with conflicting balance information from the inner ear. The senses of smell and taste are interwoven to provide us with a subtle palate for food and a wide range for detecting aromas.

SMELLING AND TASTING ARE CHEMICAL SENSES

Both **olfaction** and **gustation** are chemical senses, because these sensory receptors respond to chemicals dissolved in the mucus lying over them.

Olfaction
The sense of smell.

Gustation
The sense of taste.

Olfaction occurs in the upper chambers of the nasal passages, on the roof of the nasal cavity. We take deep breaths when we smell something to flood the upper portion of the nasal cavity with inhaled odor. Olfactory cells extend from the **olfactory bulb** (at the end of cranial nerve I) through the cribriform plate of the ethmoid, and into the mucus lining of the nasal cavity. The sensory receptors themselves are a small yellow patch of olfactory epithelium in the lining of the nasal cavity. Each cell is a modified neuron that ends in approximately six to twelve olfactory cilia, which bear one of many thousands of specific olfactory receptors. When the receptor binds its specific odor molecule, a sensory impulse is sent to the olfactory bulb and on to the brain. Neural connections between the olfactory bulb and the limbic system explain why smells trigger memories and emotions. The perfume industry depends on this neurological connection between odor and emotion. (See "I Wonder . . . What is the role of odor in emotional communication?" on p. 232.)

The sense of taste is closely allied with olfaction. Have you noticed how food loses its appeal when you suffer nasal congestion? In the mouth, food is broken down by the grinding of the teeth and chemically degraded by enzymes in saliva. Taste buds in the lining of the mouth and on the surface of the tongue (FIGURE 8.1) can distinguish only four categories of taste: sweet, sour, salty, and bitter. Like the olfactory epithelium, individual taste buds respond to only one class of chemical compound, although taste buds respond to only four classes of compounds rather than the thousands that olfactory neurons recognize. When stimulated, taste buds send information on to the brain. At the brain, the stimuli are analyzed to determine the overall taste of the food we are eating.

We rarely classify a food as tasting simply sweet or bitter. We describe coffee as "rich" or "full bodied." One major caffeine purveyor even describes its flavors for November as "elegant sweet fruit" and "intense floral notes." The subtle differences in food tastes are actually due to the involvement of olfaction. Food in the mouth is dissolved in the

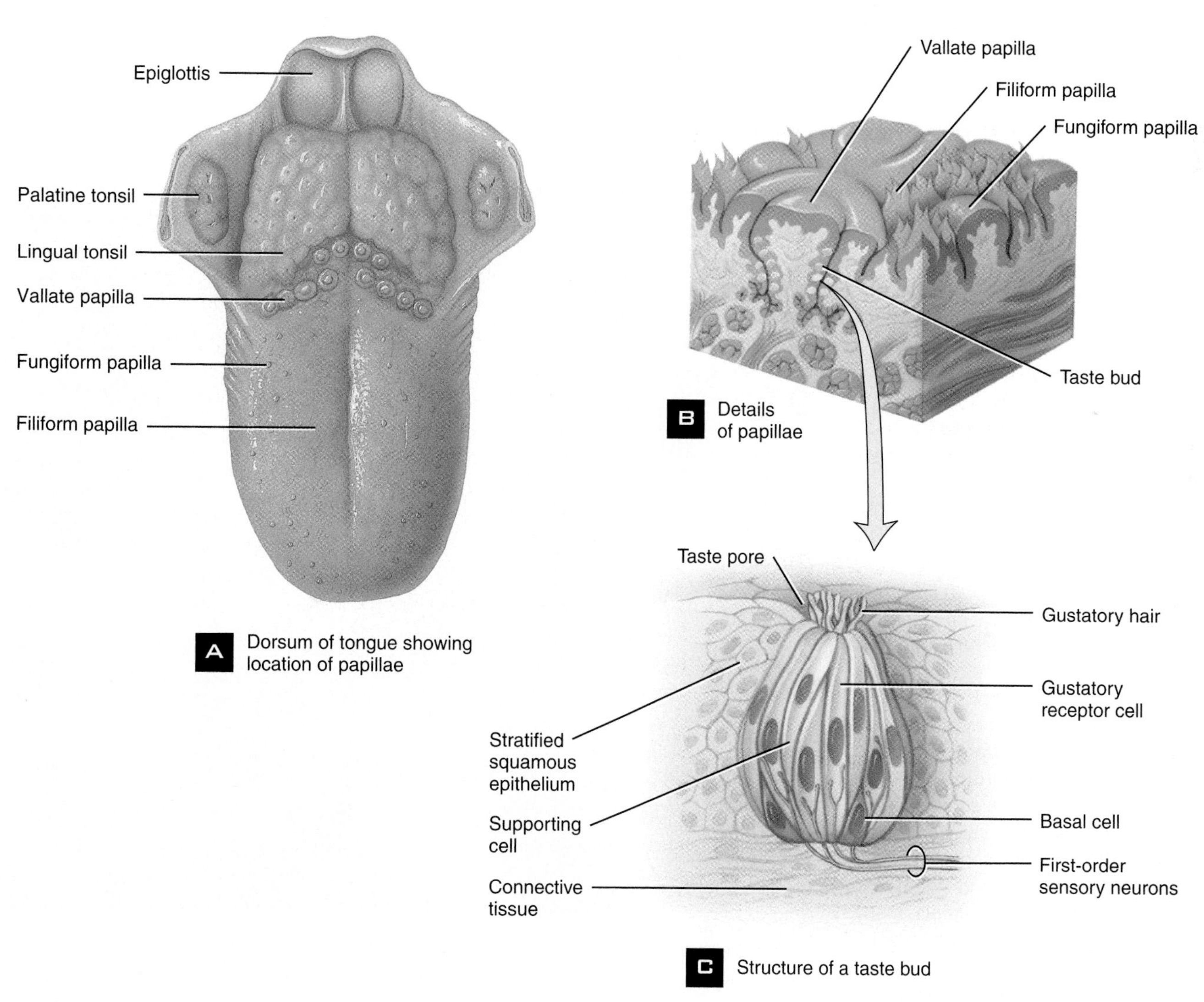

Taste bud anatomy FIGURE 8.1

I WONDER . . .

What is the role of odor in emotional communication?

Animals use chemicals for all sorts of communication, and not surprisingly it turns out that humans do likewise. Communication can occur both with chemicals that we recognize (odors) and with chemicals that we can't name, or even consciously detect. The many examples of chemical communication start with newborns, who use olfaction to find their mother's breast. Scientists have recently learned that newborns cry less when they can smell their mother's amniotic fluid.

Many people have observed that the memories that odors evoke are curiously powerful, and the connection between odor and emotion is well established. Scientists have learned that neural messages about scents go directly to the limbic system, the emotional center of the brain. For example, after exposing subjects to an unpleasant odor, scientists have recorded increased blood flow in the amygdala, a portion of the brain's limbic system that is central to emotion.

Some of the most interesting studies have placed sweat samples in front of subjects' noses. One such study used samples taken from people who were either exercising or under the stress of preparing for an academic exam. When the female subjects were asked to respond emotionally to photos of faces, their response differed depending on which scent they had smelled. The sweat from a stressed individual caused the females to respond more negatively to the facial images.

Odors can also give insights into the emotions of strangers. In a fascinating study, scientists asked women and men to watch a frightening movie and a happy one and after each one, collected smelly underarm pads. Women subjects could distinguish the odors collected from "happy" men and women. Male subjects could also do that—but only with samples taken from women.

Human chemical communication reflects animal use of chemicals to identify individuals, mark social rank and territories, and signal reproductive status. In animals, many of these behavior-affecting chemicals are called pheromones. Moths, for example, release vanishingly small concentrations of pheromones to attract mates. Honey bees use pheromones when sharing information, such as the route to food sources, with the rest of the hive. Many vertebrates use pheromones to signal readiness for mating.

Do humans also respond to pheromones? Some perfume makers are eager to market the idea that pheromones can facilitate *Homo sapiens,* dating and mating, but the claim is still debatable. For years, scientists denied that humans could respond to pheromones because we do not have a vomeronasal organ, the anatomical structure in the nasal passages which other vertebrates use to detect pheromones. Now, it appears that we do have such an organ, although it may deteriorate after birth. The exact role of pheromones and the vomeronasal organ is uncertain in humans. But amid a cascade of bizarre discoveries about how people communicate with chemicals, more olfactory surprises would not be too astonishing.

Not all communication in the animal kingdom is auditory or even visual. Bees have a strong communication system that allows them to share information on the location and type of pollen to be found. Humans also communicate using emotional cues along with, or sometimes rather than, the spoken word.

Uvula
The tab of soft tissue that hangs down in the back of the throat, visible as a pointed tab.

mucus as we chew. At the back of the oral cavity, posterior to the **uvula**, lies a hole connecting to the nasal cavity (FIGURE 8.2). During swallowing, the uvula closes this hole, preventing swallowed items from being propelled out the nose. When the hole is open, food in the mouth can be smelled by the olfactory epithelium. The combination of the food's texture (determined by the tongue), taste (sweet, sour, salty, or bitter), and odor (determined by the olfactory epithelium) are all related to our description of the taste of a food. The food industry has become quite skilled at "engineering" new foods with specific tastes and "mouth feels."

HEARING AND BALANCE ARE DETECTED BY MECHANORECEPTORS

Our sense of hearing gives us the ability to detect the slightest noises. The recent movie *Ray* documented the life story of rhythm-and-blues legend Ray Charles. Born with normal vision, Ray lost his sight in grade school. Few of us use our ears as well as Ray Charles did, even though we were born with the same capability. In a touching scene, 10-year-old Ray learned he could "see" by listening carefully. By following the sound of a cricket's feet on the wood floor, he located and caught the cricket. He turned to his mother and told her he knew she was present, even though she was quiet. He wanted to know why she was crying as she watched him discover his world through sound instead of sight.

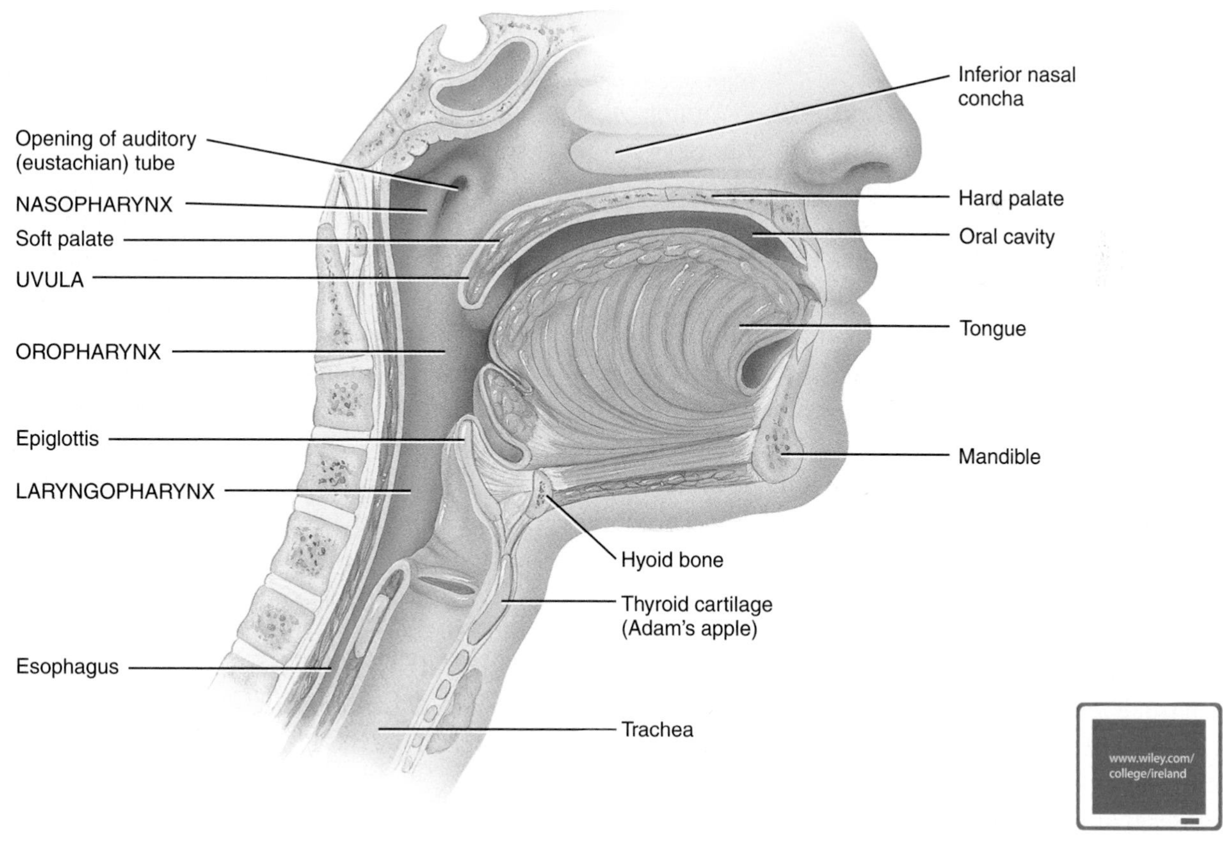

Mouth and uvula FIGURE 8.2

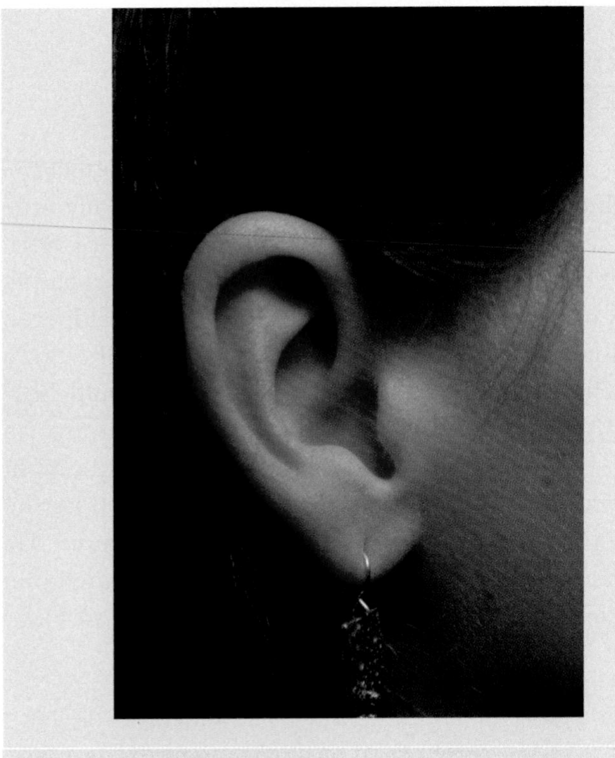

The ear FIGURE 8.3

The ear (FIGURE 8.3), as we all know, houses our sense of hearing. It has three functional parts: the **outer**, **middle**, and **inner** ear (FIGURE 8.4). The outer ear is composed of the **pinna** and external **auditory canal**, both of which capture sound waves and funnel them to the middle ear.

The ear drum, or **tympanic membrane**, marks the beginning of the middle ear. Compression waves in the air (sound) cause the membrane to vibrate, converting sound into mechanical motion. Attached to the inside of the tympanic membrane is the **malleus**, one of the three smallest bones in the human body. The vibrating tympanic membrane moves the malleus, which in turn moves the **incus** through a synovial joint. One more small bone, the **stapes**, is joined to this chain through another tiny synovial joint. The stapes is the final small bone, or ossicle, of the middle ear.

These three bones can dampen or amplify the movement of the tympanic membrane. Extremely loud noises that cause tremendous vibration of the tympanic membrane are dampened in the middle ear when tiny skeletal muscles tighten at these synovial joints. We can hear soft noises more clearly as these muscles relax, allowing the bones to move freely.

Beyond the stapes is the inner ear (FIGURE 8.4). The **stapes** connects to the **oval window**, a membrane that functions like the tympanic membrane. The oval window bounces in response to movement of the stapes, creating fluid waves in the inner ear.

The entire middle and inner ear are actually within a hollow portion of the temporal bone. The middle ear is filled with air and communicates with the external environment through the **eustachian**, or auditory, **tube**. Air pressure must be almost equal on both sides of the tympanic membrane for it to freely vibrate in response to sound waves. When we pop our ears, we are actually opening the auditory tube, allowing air to equilibrate on both sides of the eardrum.

The cochlea is a coiled tube, built like a snail shell (FIGURE 8.5). If we unwound it, the cochlea would be a straight tube, extending from the oval window at the beginning of the inner ear to the round window. The cochlear tube has three compartments. The uppermost compartment, continuous with the oval window, is called the **vestibular canal**. At the tip of the snail shell, this compartment rounds the end of the tube and forms the **tympanic canal** at the bottom of the cochlea. The tympanic canal ends at the round window. These two chambers form a U-shaped fluid-filled passage for the pressure waves generated at the oval window.

Within the center of the cochlea is a third chamber. This chamber houses the organ that converts mechanical vibration into sensory input, the **organ of Corti**. The flattened **tectorial membrane** lies on top of the organ of Corti. The membrane rests on the top of hair cells, with the hairs, or **stereocilia**, just touching the membrane. The hair cells of the organ of Corti are directly linked to the **vestibulocochlear** nerve, cranial nerve VIII.

THE MECHANICS OF HEARING INVOLVES WAVES, HAIRS, AND NEURONS

When the tectorial membrane is deformed and the underlying hairs are bent, as happens in response to sound, a nerve impulse is created in the neuron of

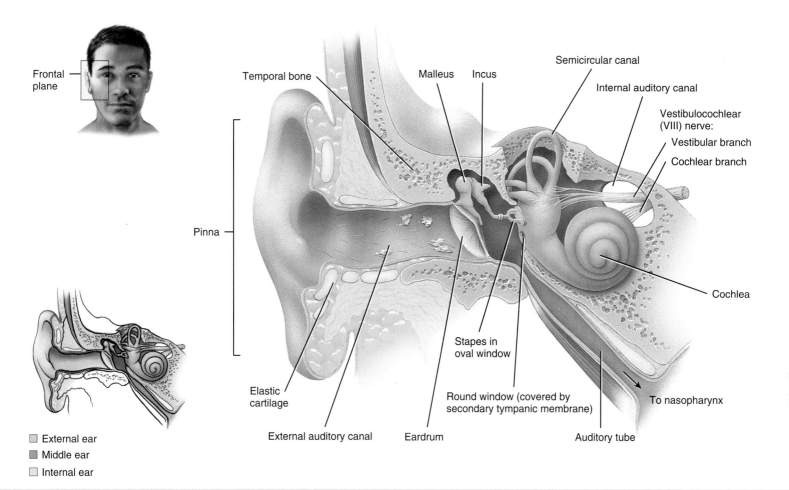

Frontal plane

Temporal bone
Malleus
Incus
Semicircular canal
Internal auditory canal
Vestibulocochlear (VIII) nerve:
Vestibular branch
Cochlear branch
Pinna
Cochlea
Stapes in oval window
Elastic cartilage
Round window (covered by secondary tympanic membrane)
To nasopharynx
External auditory canal
Eardrum
Auditory tube

External ear
Middle ear
Internal ear

The parts of the ear FIGURE 8.4

The parts of the ear include the outer ear, the middle ear, and the inner ear. The inner ear is a completely self-contained, fluid-filled chamber within the temporal bone. In this fluid floats a *membranous labyrinth*. There are two parts to these floating membranes:

1. the *cochlea* contains the organ used in hearing;
2. the *vestibule* and *semicircular canals* contain the organs of balance.

The cochlea FIGURE 8.5

Process Diagram

Hearing FIGURE 8.6

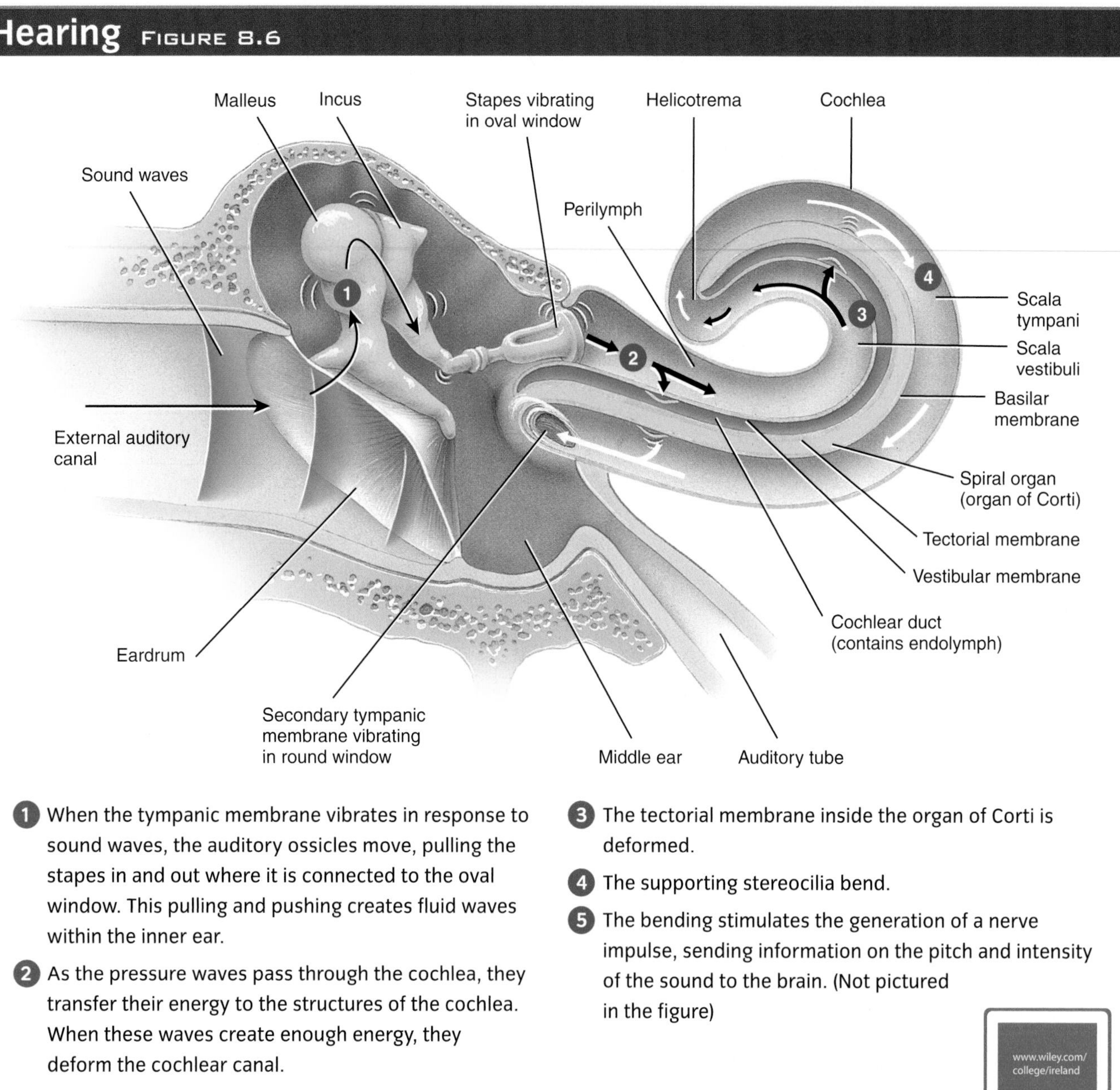

1. When the tympanic membrane vibrates in response to sound waves, the auditory ossicles move, pulling the stapes in and out where it is connected to the oval window. This pulling and pushing creates fluid waves within the inner ear.
2. As the pressure waves pass through the cochlea, they transfer their energy to the structures of the cochlea. When these waves create enough energy, they deform the cochlear canal.
3. The tectorial membrane inside the organ of Corti is deformed.
4. The supporting stereocilia bend.
5. The bending stimulates the generation of a nerve impulse, sending information on the pitch and intensity of the sound to the brain. (Not pictured in the figure)

www.wiley.com/college/ireland

that particular hair cell. This impulse is carried to the brain, where it is interpreted as a particular pitch. Each part of the tectorial membrane is sensitive to a different pitch, allowing us to receive discrete information concerning the sounds we hear. Lower frequency noises vibrate the organ of Corti near the tip of the cochlea, whereas higher frequency noises cause vibrations at the base. The nerves from each portion of the cochlea lead to specific areas of the brain, further enhancing our ability to discriminate sounds (FIGURE 8.6).

EQUILIBRIUM IS ALSO HOUSED IN THE INNER EAR

Many people are surprised to learn that the sense of balance is also housed in the ear. The vestibule and semicircular canals of the inner ear house structures responsible for the two types of equilibrium—static and dynamic (FIGURE 8.7).

Static equilibrium Static equilibrium is the physical response to gravity that tells us which direction is

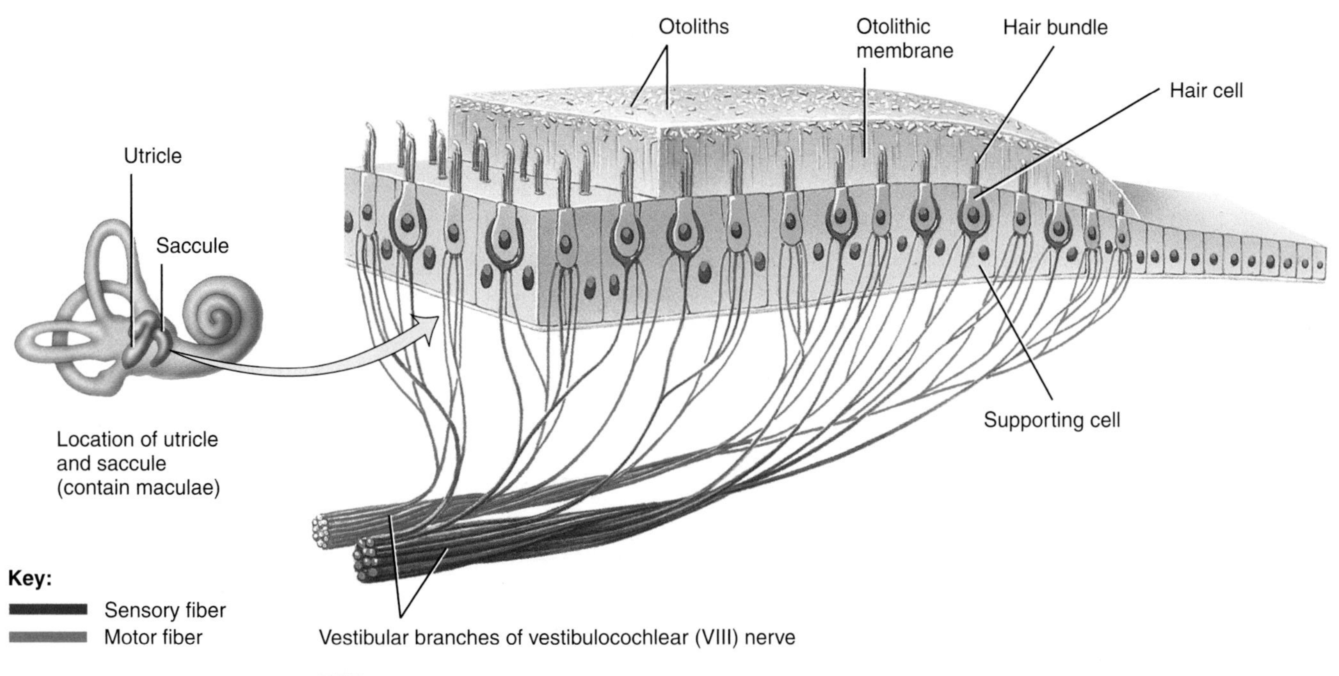

A The location of the utricle and saccule and the structure of the macula.

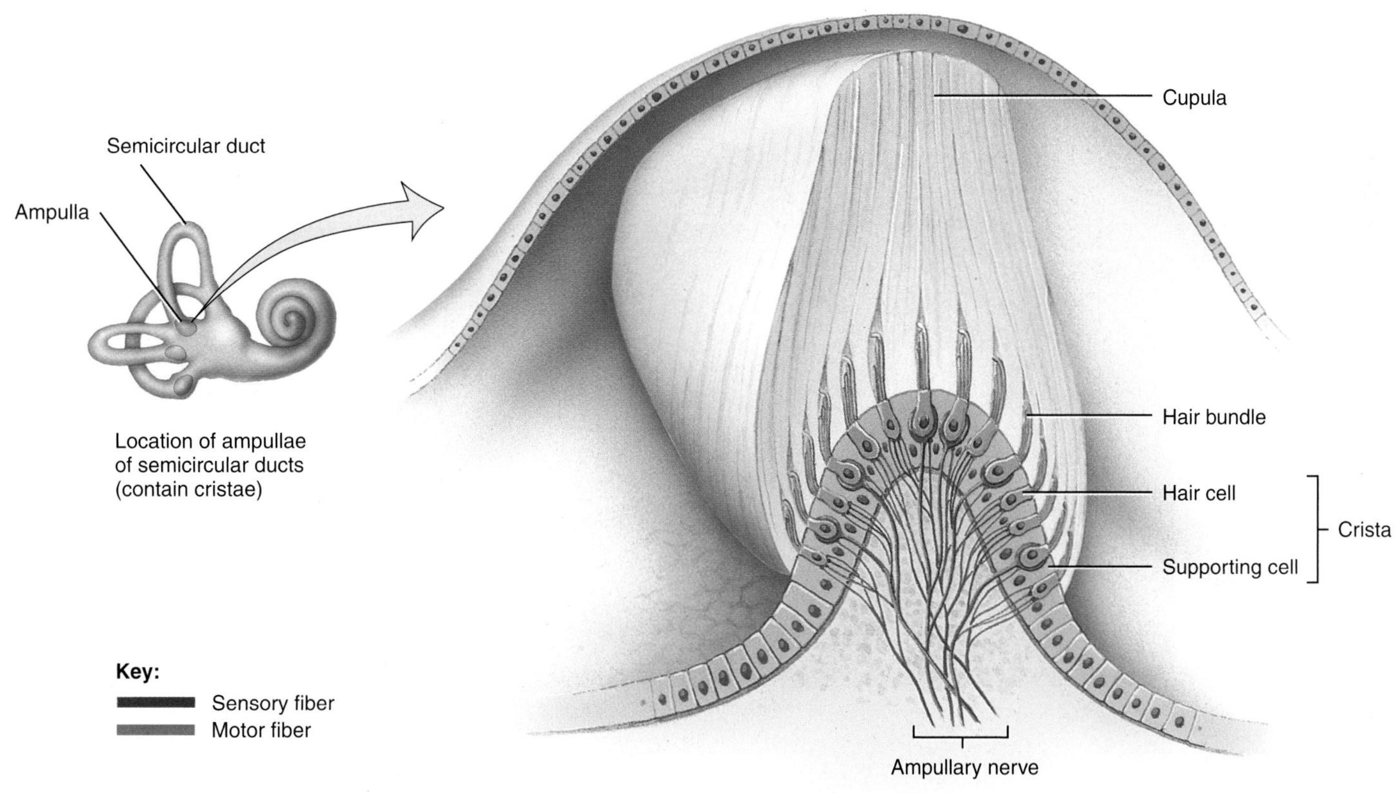

B The location of the semicircular canals, and the structure of the crista

Inner ear structures of balance FIGURE 8.7

down. The **utricle** and **saccule** are structures located in the vestibule of the inner ear. Much as in the sense of hearing, these two structures initiate a nerve impulse when hairs within them bend. The utricle and saccule contain two gelatinous blobs situated at right angles to one another in the vestibule, called the **maculae**. Each of these organs contains tiny pieces of bone that respond to gravity. These organs are held in the vestibule by hair cells. The ends of the hairs are stuck in the gelatin, causing them to respond to movement of the organ.

The utricle and saccule are arranged at right angles to one another, so that when the head is upright one of them is always vertical and the other horizontal. As gravity pulls on the vertical element, the hairs associated with it bend. As before, this bending causes a nerve impulse to be generated, except that this impulse goes to the area of the brain that interprets static equilibrium. As head position changes with respect to gravity, these impulses change in frequency and direction, continually providing information on the up-and-down placement of your head.

Dynamic Equilibrium Your sense of dynamic equilibrium detects acceleration or deceleration of your head. This sense originates in three semicircular canals situated so that each one lies in a separate plane; X (the horizontal plane, or the plane this book lies on when you lay it flat on the table), Y (the vertical plane, or the plane this book lies on when you stand it upright on the table with the spine facing you), and Z (transverse plane, or the plane that this book lies on when you again stand it upright on the table, this time with the cover facing you). The fluid in each tube rocks in response to acceleration in its particular plane. At the base of each semicircular canal is a swelling. This swollen area houses the dynamic equilibrium receptor, a flame-shaped **cupula** of gel with hairs embedded. As the fluid in the semicircular canal rocks through the swollen base of the canal, it pushes on the cupula and bends its hairs, again sending a nerve impulse to the brain. These structures are responsible for the strange feeling you get in an elevator. The fluid in the canals responds to the acceleration of your head, but your eyes perceive no motion, so you get that familiar flipping feeling in your stomach.

CONCEPT CHECK

Give examples of sensations you might receive from a chemoreceptor, a mechanoreceptor, and a photoreceptor.

Briefly describe the anatomy of your olfactory sense.

Why do we consider some foods so tasty, when we can sense only four tastes?

Follow the pathway of sound through the ear.

Compare the functioning of static and dynamic equilibrium. What do these senses have in common?

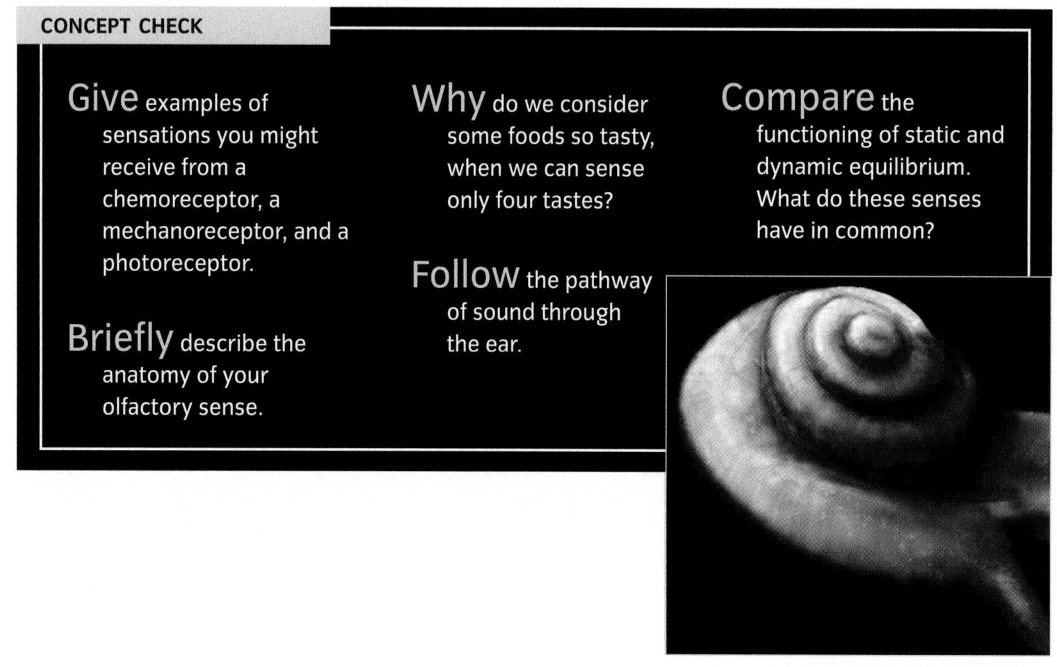

Vision Is Our Most Acute Sense

LEARNING OBJECTIVES

Decribe the anatomy of the eye.

Follow the pathway of light through the eye.

Explain nearsightedness and farsightedness, listing the proper corrective measures.

Discuss the chemical reactions involved in photoreception.

We are visual creatures. We perceive the world primarily through our eyes, devoting a large percentage of our brain to the interpretation of visual images. Despite the enormous importance of our eyes, they are simple structures (FIGURE 8.8). The eye works like a pinhole camera. It regulates the amount of light that enters the photoreceptor area and then captures that

The anatomy of the eye

FIGURE 8.8

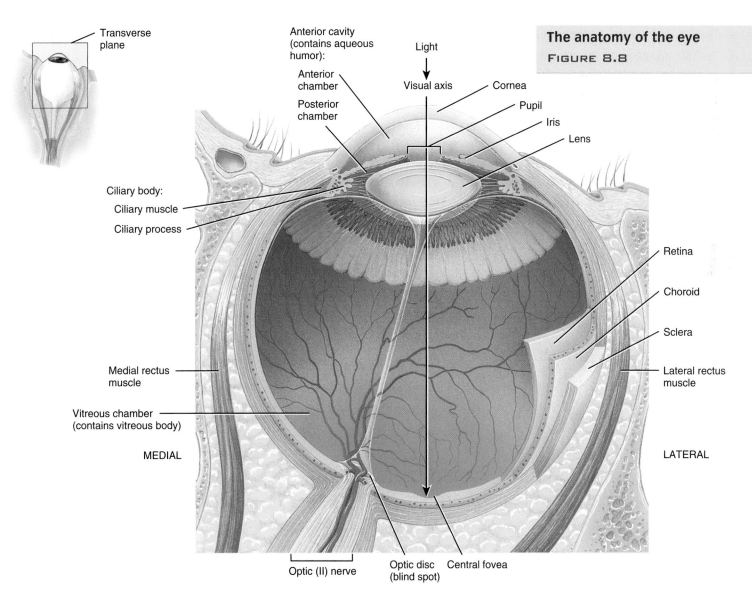

Superior view of transverse section of right eyeball

light as an image. The brain, like film in a camera, interprets that image and makes sense of what is seen.

The eye has three layers, or tunics: the **fibrous tunic** (sclera), the **vascular tunic** (choroid), and the **nervous tunic** (retina). The outermost layer, the fibrous tunic, is composed of dense connective tissue forming the white **sclera** and the clear **cornea**. The fibrous tunic is protected by the eyelids, eyelashes, and eyebrows, which prevent dust and particles from entering the eye.

The fibrous tunic provides a stiff outer covering for attaching the six extrinsic muscles that connect the eyeball to the bony orbit. **Lateral, medial, superior**, and **inferior rectus muscles** roll the eye left and right, up and down, in its socket, whereas the **superior** and **inferior oblique muscles** pull the eye obliquely. For example, when you contract your superior oblique muscle, your eye rolls upward and inward. The oblique muscles also help stabilize the eye as it is pulled by the four rectus muscles.

The anterior sclera and cornea are bathed continuously by **lacrimal gland** secretions, or tears. These glands lie in the superior lateral aspect of the orbit (the upper, outer corner of the eye). The tears wash across the eye and are collected in holes, called **lacrimal punctae**, on either side of the nasal cavity. These holes drain into the nasal epithelium, helping to moisten that as well. When we cry, the lacrimal secretions exceed the carrying capacity of the lacrimal punctae, and the tears spill over the lower eyelid onto the face. The tears collected by the lacrimal punctae can also overflow the nasal epithelium, causing a runny nose (FIGURE 8.9).

Immediately beneath the fibrous tunic is a dark pigmented layer, the **choroid**. This layer houses the blood supply for the eye and contains melanin to absorb light. (Imagine how difficult it would be to interpret visual images if light bounced around inside the eye. With the light not absorbed, we would see repeated images, rather like a house of mirrors.) The choroid ensures that light strikes the **retina** only once.

The choroid is visible as the **iris,** the colored portion in the front of the eye. The iris is a muscular diaphragm that regulates light entering the eye. When contracted, concentric muscles close down the **pupil**, whereas radial muscles dilate, or open, it. The color of the iris is a reflection of the amount of melanin produced by the choroid. Dark eyes have more light-absorbing melanin on both sides of the choroid. Lighter eyes have less melanin on the underside of the choroid, which is what we see through the cornea.

> **Pupil**
> The hole in the center of the iris.

Immediately behind the iris, the choroid thickens and becomes the ciliary body. This structure holds the lens in place, pulling it to change the shape of the lens to accommodate near and far vision.

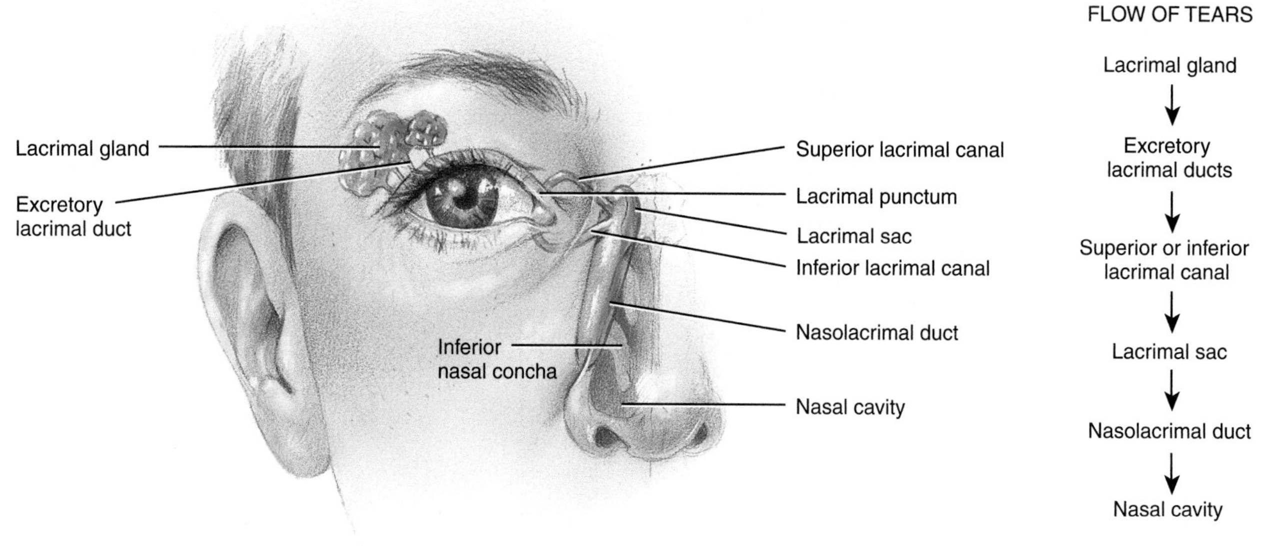

Lacrimal system FIGURE 8.9

Eye structures and their functions TABLE 8.1	
FIBROUS TUNIC 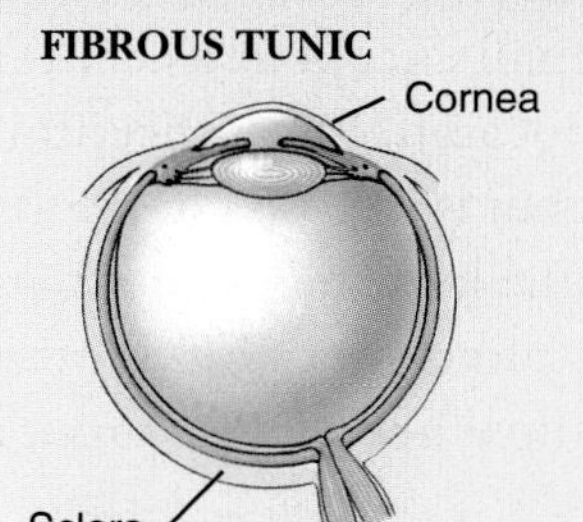	*Cornea:* Admits and refracts (bends) light. *Sclera:* Provides shape and protects inner parts.
 **VASCULAR TUNIC** 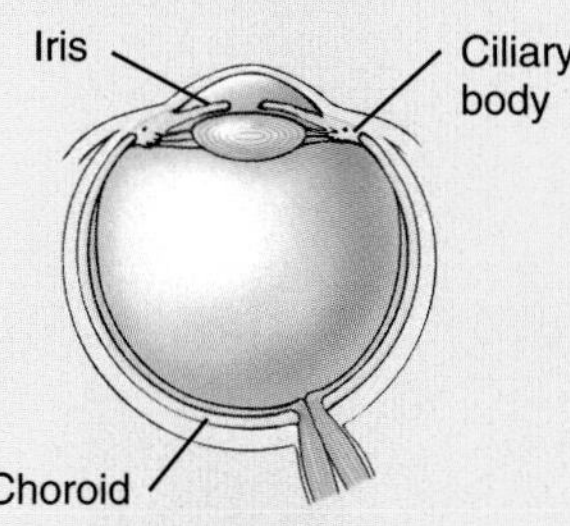	*Iris:* Regulates amount of light that enters eyeball. *Ciliary body:* Secretes aqueous humor and alters shape of lens for near or far vision (accommodation). *Choroid:* Provides blood supply and absorbs scattered light.
NERVOUS TUNIC 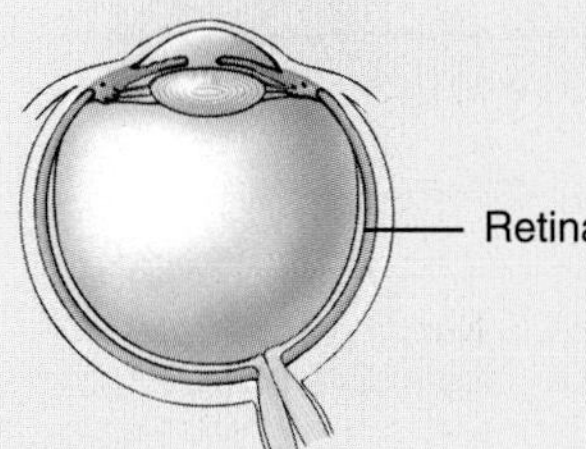	Receives light and converts it into nerve impulses. Provides output to brain via axons of ganglion cells, which form the optic (II) nerve.
LENS 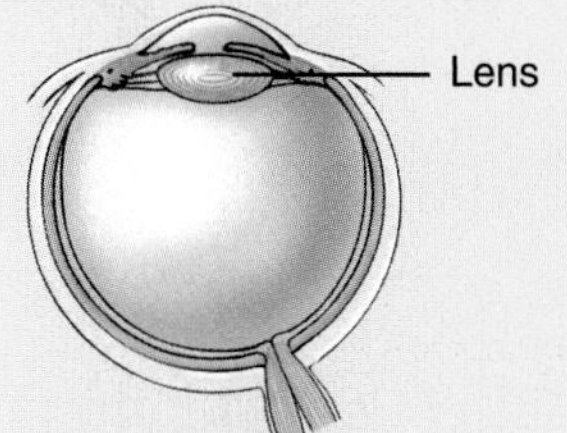	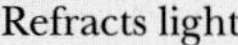 Refracts light.
ANTERIOR CAVITY 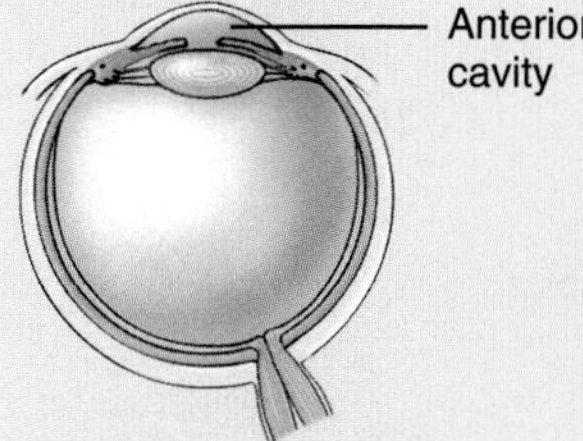	Contains aqueous humor that helps maintain shape of eyeball and supplies oxygen and nutrients to lens and cornea.
 **VITREOUS CHAMBER** 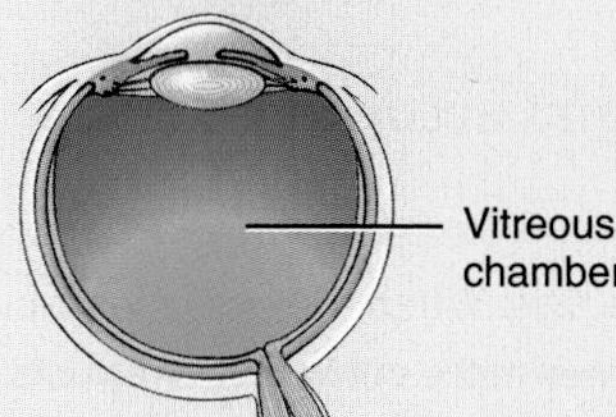	Contains vitreous body that helps maintain shape of eyeball and keeps the retina attached to the choroid.

The lens and cornea are both bathed in **aqueous humor**, a fluid that is constantly filtered from the blood. The aqueous humor is returned to the blood via the canals of Schlemm, at the junction of the cornea and the sclera. These canals get constricted in glaucoma, causing an increase in pressure that can eventually destroy the light-sensitive cells in the retina. See TABLE 8.1 for a complete listing of the structures of the eye and their functions.

THE LENS CHANGES SHAPE TO ACHIEVE OPTIMAL OPTICS

Visual acuity requires the eye to focus entering light onto the nervous tunic at the back of the eyeball. The lens, immediately behind the pupil, is held inside a connective-tissue covering that connects to the ciliary body. When the muscles of the ciliary body contract, the entire ring of the ciliary body gets smaller. This releases pressure on the connective tissue covering the lens, and the lens bulges. When the muscle relaxes, the ring of the ciliary body enlarges, pulling the lens flat and enabling the eye to focus on nearby objects. This changing of lens shape, called **accommodation**, gets more difficult with age. This is because with each passing year, the lens continues to add layers that resemble the layers of an onion. These extra layers make the lens thicker and stiffer, so it resists flattening when the ciliary body relaxes. Starting around age 45 or 50, this flattening becomes so difficult that many people need reading glasses. These glasses enlarge the image before it reaches the pupil, giving the lens a larger image to bring into focus.

> **Visual acuity**
> The resolving power of the eye.

Nearsightedness and farsightedness are both caused by the lens's inability to accommodate light properly. In nearsightedness, the eye is too long for the lens to focus the light rays on the retina. The focal point winds up in the vitreous humor (the fluid in the back chamber of the eye), and the image is spreading out and fuzzy again when it hits the photoreceptors. A concave lens will spread the light rays farther before they enter the eye, correcting this problem. Farsightedness is the opposite of nearsightedness. The lens focuses the image from the pupil behind the retina. A convex lens will begin the process of focusing the light rays before they enter the eye, moving the focal point forward to the retina itself.

Astigmatism is another common abnormality of the eye. In this case, the cornea is imperfectly shaped, resulting in an uneven pattern of light hitting the retina. Some areas of the image are in focus, but not others. A carefully crafted lens that compensates for the uneven flaws of the cornea can correct this problem. (FIGURES 8.10 and 8.11.) See the Health, Wellness, and Disease box, "What are the most common visual impairments, for more information on visual troubles."

Behind the lens lies a large chamber filled with **vitreous humor**, a gel-like fluid that holds the third tunic, the retina, in place. Unlike aqueous humor, which forms continuously through our life, vitreous humor is formed during fetal development and remains in the eye for life. The retina spreads out over the inside rear of the eye somewhat like the cloth of an umbrella, which is spread over the umbrella frame. Unlike the umbrella cloth, however, the retina is not physically attached to the back of the eyeball except at its center, where only the blind spot is located. Photoreceptors line the surface of the retina that is exposed to the vitreous humor, directed toward the lens and pupil. There are no photoreceptors in the blind spot of the retina because it is in this area that retinal nerves dive through

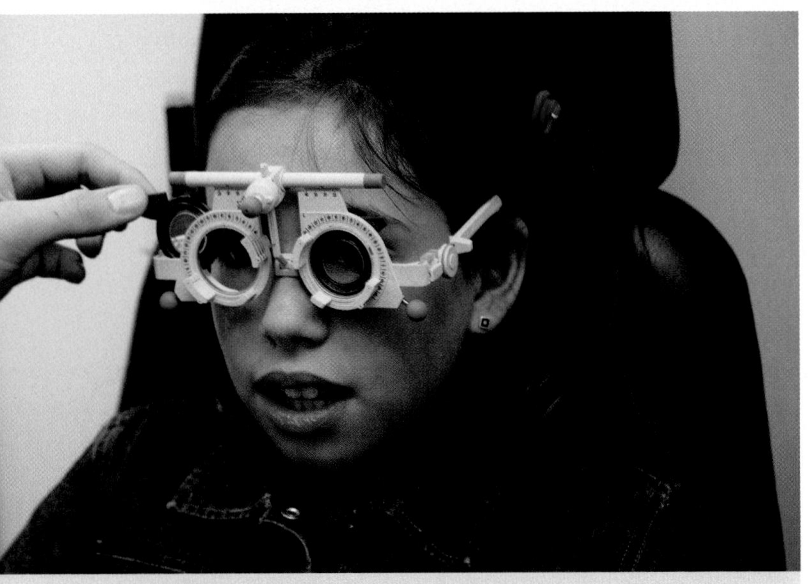

An eye exam FIGURE 8.10

Vision check-ups are an important part of maintaining good health. Visual acuity and astigmatism are routinely monitored during a simple eye exam. Visual acuity is determined by reading successively smaller type until the letters are too blurred to distinguish. Astigmatism can be diagnosed by observing a diagram of a wheel with spokes extending in all directions. If a few of these spokes are not distinct, the eye may be out of round in those areas.

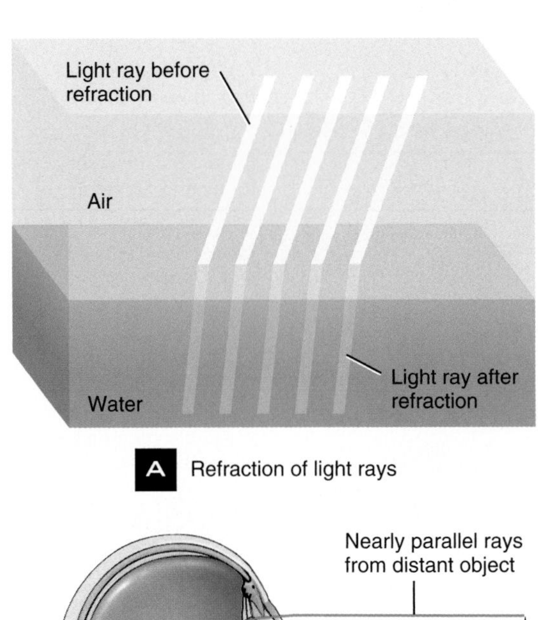

A Refraction of light rays

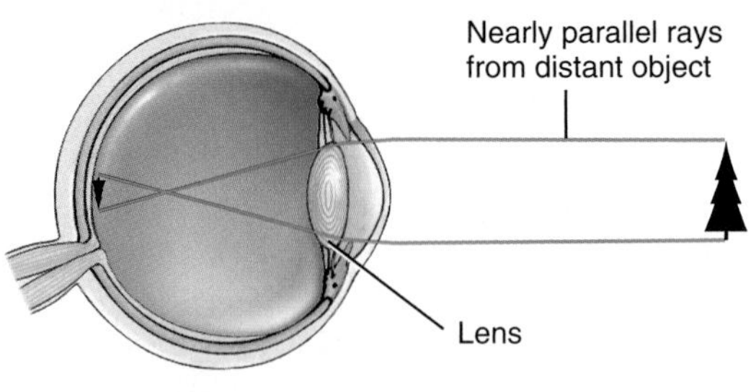

B Viewing distant object

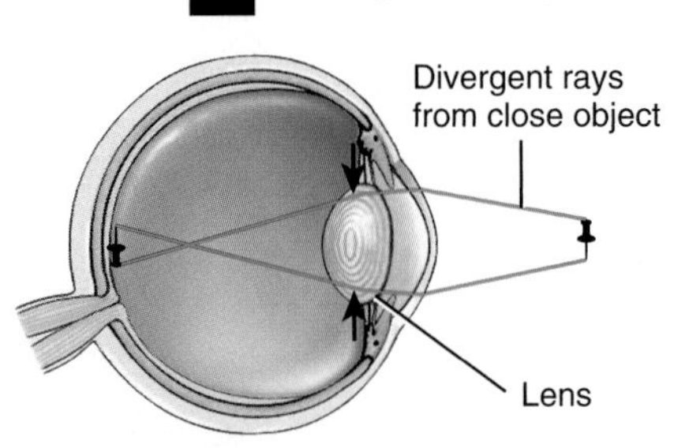

C Accommodation

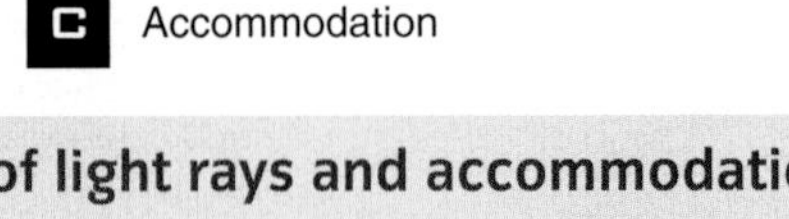

Refraction of light rays and accommodation FIGURE 8.11

Refraction is the bending of light rays. When the eye is focusing on a close object, the lens becomes more convex and refracts the light rays more.

What are the most common visual impairments?

The eye is an astonishing sensory organ, far superior to even a good camera. But things can go wrong with eyes, just as they can with cameras. Many serious vision impairments grow from physiological problems like a shortage of blood in the retina or of a necessary visual chemical. But the most common vision problems emerge from simple anatomy, causing eyes that are nearsighted, farsighted, or astigmatic. In nearsighted, or myopic, eyes, images of distant objects focus in front of the retina, while close-up images do focus on the retina. This explains why nearsighted people can "see" close-up objects. Eyes that are farsighted, or hyperopic, have the opposite problem: close-up images are focused behind the retina, while distant images focus properly.

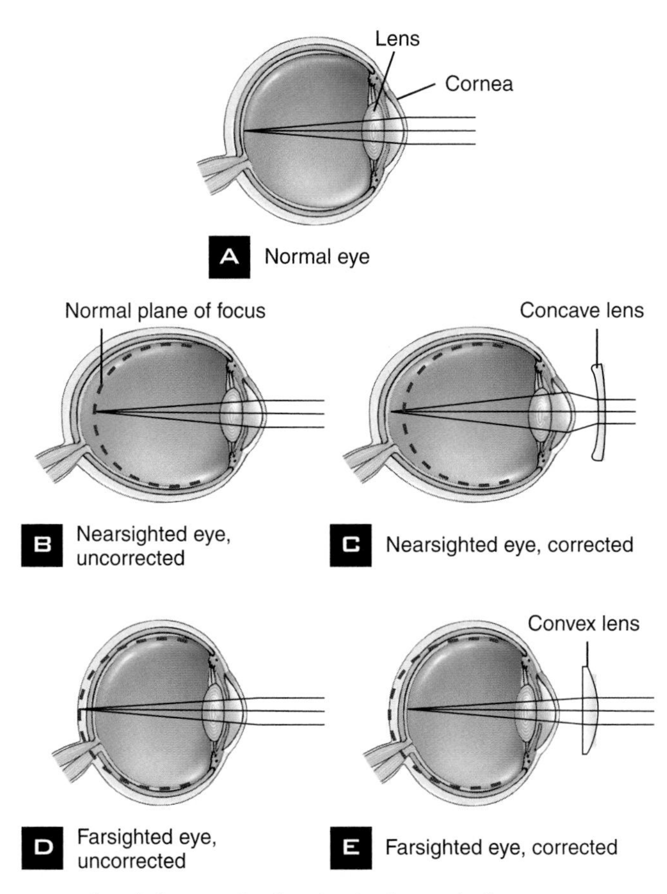

Normal and abnormal refraction in the eyeball.

In the eye, the lens and cornea both focus light rays so that they converge on the retina. In myopia and hyperopia, the eyeball may simply be the wrong size for this focusing. Myopia and hyperopia are quite common, each affecting between 10 and 40 percent of various populations. Astigmatism is caused by a nonsymmetrical cornea or lens. Typically, astigmatism causes a blur in vertical or horizontal images, but not both. Astigmatism may be even more common than myopia or hyperopia, and it can coexist with either of them.

Myopia often causes children to squint, hold a book up close, or sit unusually close to a television or computer monitor. Hyperopia is often less troublesome early in life, because young lenses are often flexible enough to focus on close-up objects. Myopia and hyperopia can cause headaches or eyestrain, owing to the extra muscular effort needed to distort the lens to focus images on the retina.

Both myopia and hyperopia seem to have a genetic component, for they tend to run in families. Environment may also be a factor, as eye function is closely related to health of the individual and clarity of the surrounding air. For example, a diet low in vitamin A will cause visual disturbances, as vitamin A is necessary to form rhodopsin. And of course we have all experienced watery, painful eyes while sitting near a smoky bonfire, or trying to see clearly after swimming in a chlorinated pool. Many studies have associated myopia with doing close work. Some also link myopia with the use of night lights, a previous surgical correction for a retinal problem, or other events. Eye specialists are not in agreement on how to interpret these studies.

Eyeglasses and contact lenses are the traditional technologies used to help the lens and cornea focus in myopia, hyperopia, and astigmatism. Today, corrective surgery is becoming a more viable method to reshape the cornea to achieve visual acuity.

Ophthalmologists are still struggling to treat blinding diseases, such as glaucoma, macular degeneration, and retinitis pigmentosa. Fortunately, however, the most common vision problems are easy to detect and resolve.

the retina toward the brain. The vitreous humor maintains slight pressure on the retina, pressing it flat against the back of the eye. Because it is not attached, the retina can be "detached" if the eye is hit hard enough to slosh the vitreous humor—even a momentary movement may allow the retina to fold. If this happens, light cannot reach the photoreceptors inside the fold, so they detect nothing.

PHOTORECEPTORS DETECT LIGHT IN THE RETINA

The retina (FIGURE 8.12) is composed entirely of neurons in layers containing rods and cones, bipolar cells, and ganglionic cells. The rods and cones are the neurons that detect light—the photoreceptors. The ganglionic cells and bipolar cells are interneurons that carry the action potential generated by the photoreceptors to the brain. The cones respond to bright light, providing color vision and resolution that is high enough to allow us to distinguish tiny individual structures such as human hairs. The rods function in low levels of light, providing only vague images. These two types of cells are unevenly distributed. Cones are concentrated near the center of the retina, where incoming light is strongest. In fact, the area of the retina immediately behind the pupil is slightly yellow owing to the high concentration of cones, and it is called the **macula lutea**. This area provides our highest resolution, allowing us to discriminate subtle differences in objects needed, for example, to read. Rods are spread across the periphery of the retina. They are not terribly good at resolution, but they do respond in extremely low light.

The layers of neurons in the human eye seem backwards, because the photoreceptors are against the back of the eye, oriented toward the brain rather than the source of light. Light rays must pass through the entire retina before they stimulate the photoreceptors at the back. This so-called indirect retina is found in most mammals. Interestingly, the squid and octopus have eyes that are anatomically very similar to our own, except that they do NOT have an indirect retina. Their photoreceptors are directly behind the vitreous humor, so light strikes them first. As a result, they do not have a blind spot, which is doubtless helpful in the deep, dark depths of the ocean.

Rods and cones operate using different chemical mechanisms. When a photon of light hits a rod, a neural response is initiated via the chemical **rhodopsin**. The energy from the photon splits rhodopsin into two compounds (retinal and opsin), releasing energy that starts a series of events ultimately resulting in a closing of ion gates on the photoreceptor membrane. When the ion gates on the photoreceptor close, ion movement ceases, an action potential is generated, and the brain receives a single bit of visual information.

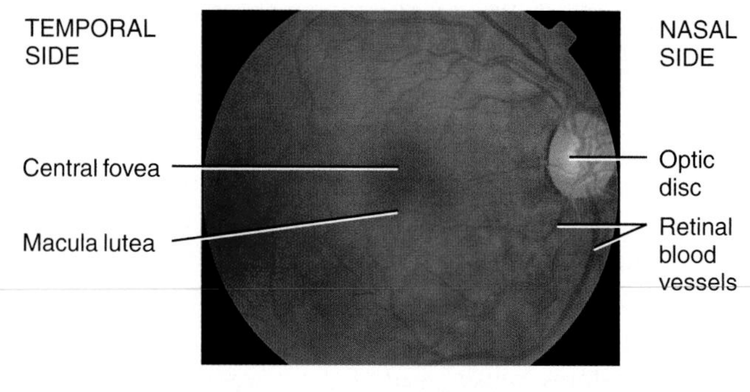

Anatomy of the retina FIGURE 8.12

Rhodopsin is easily **bleached**, meaning a slight increase in light can cause it to fall apart and not be able to recombine. Until the light is reduced, rhodopsin cannot regenerate. This means that the rods cannot detect another photon when in bright light. If rhodopsin is not put back together, there can be no further action potentials.

When you stargaze, you are using rods. You may know that to see an especially dim star, it's better to focus to one side of the star. Why? Because rods are not found directly behind the pupil but rather on the periphery of the retina. The dim starlight is not strong enough to stimulate the cones directly behind the pupil, but strong enough to stimulate the rods. You may also be aware that you see far more stars after 15 to 20 minutes of looking at the heavens. After this period, bleached rhodopsin has entirely re-formed in the rods.

Cones—the source of fine-detailed color vision—operate slightly differently. You have three types of cones, which are sensitive to different wavelengths of light, representing red, green, or blue. Cones also use the visual pigments retinal and opsin but with slight variation. Although the retinal and opsin in rods fall apart and do not regenerate in bright light, these chemicals readily regenerate in the cones.

Macula lutea
Macula = spot; lutea = yellow. The area of the retina immediately behind the pupil.

Rhodopsin
Visual pigment that responds to low levels of white light.

Photoreceptor impulse generation FIGURE 8.13

In a rod or cone, light is absorbed by a photopigment. This initiates an action potential in the photoreceptor, which starts the journey to the brain.

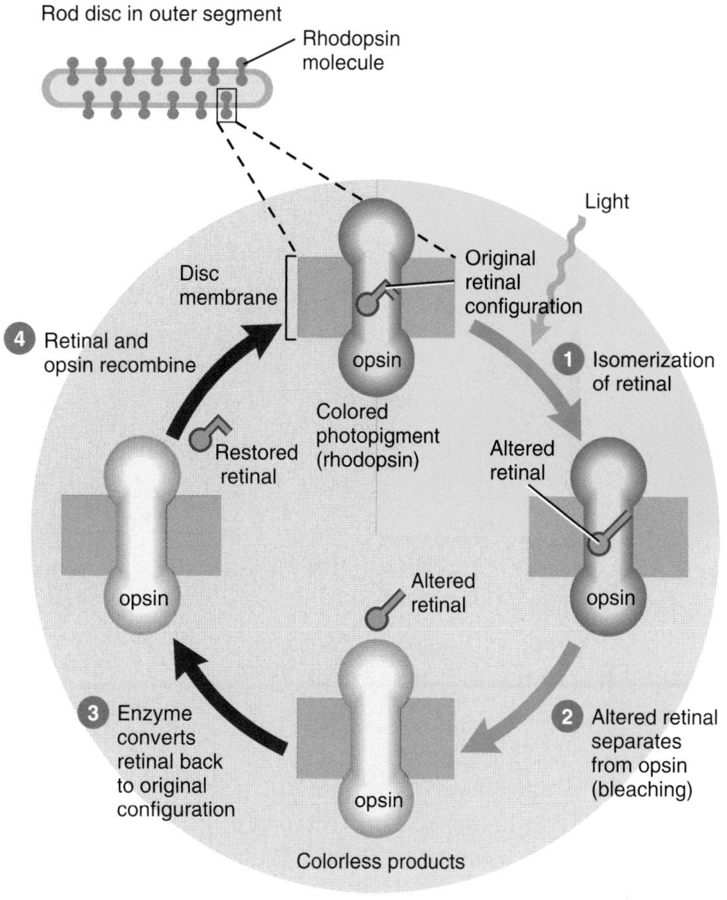

These physiological responses explain how our eyes respond to sudden changes in light. When the lights first go down in a movie theater, they dim slowly to give our eyes time to adjust to the dark. The rhodopsin in the rods, which had bleached in the bright light, gets time to regenerate. After the rods resume working, we can see nearby chairs even in near darkness. Cones respond almost immediately to brightening light. If you leave a theater, you can soon see in the lobby. But if you reenter a dark theater, you may experience momentary panic because the sudden dark effectively blinds you. If you exit a dark theater for the sunlit outdoors, the rhodopsin in your rods, which were providing vision in low light, suddenly bleaches, sending information to your brain that you experience as a "white flash." In the bright light, rhodopsin cannot regenerate, and the rods remain defunct, but cones will quickly start sending impulses to the brain, your pupils will close, and your vision will be restored (FIGURE 8.13).

Regardless of whether the visual nerve impulse came from a rod or a cone, it travels from the retina to the brain in basically the same pathway. The impulse first passes toward the front of the eye, from the photoreceptors to the bipolar neurons. These bipolar neurons transmit the impulse to the ganglionic cells in the anterior of the retina. Ganglionic cells collect impulses from a small cluster of bipolar cells and pass them to the brain via the optic nerve (cranial nerve II) (FIGURE 8.14, p. 246).

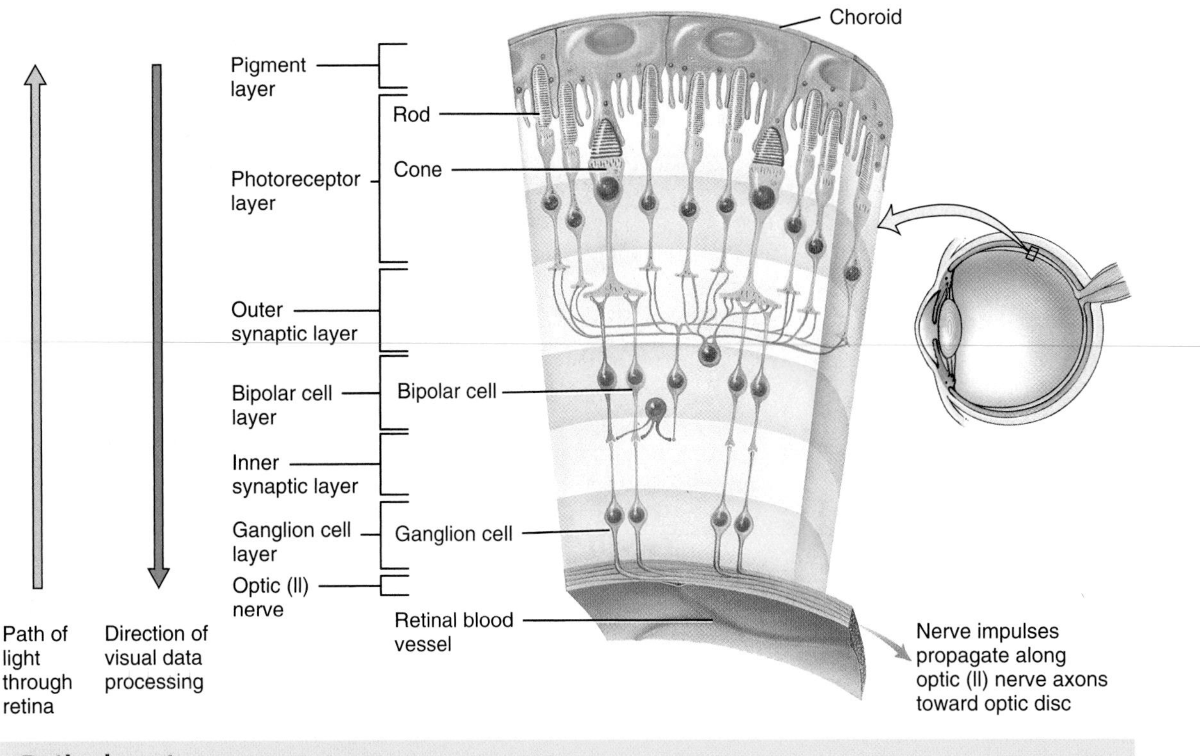

Retinal anatomy FIGURE 8.14

The ganglionic cells are in the front of the retina, and the brain is behind it. To reach the brain, axons of the ganglionic cells must penetrate the retina, which they do by literally diving through the retina. This location can have no photoreceptors, which explains why a blind spot is located just off-center in each eye. We generally do not recognize the blind spot owing to our **stereoscopic** vision. Each eye sees a slightly different view of the world because the eyes are placed slightly apart, angled just a little bit away from one another. Our brains meld these two views into one continuous field of vision. Objects that fall on the blind spot of the right retina are seen by the left retina, and vice versa. The brain fills in the missing details from each view, providing us an unobstructed perception of our environment, and disguising the blind spot.

Stereoscopic Depth perception gained through use of the visual field of both eyes.

Exactly how the brain interprets the flood of information it receives from the eyes is a field of study in and of itself. Vision is so important that it occupies more space in the brain than any other special sense. We know the impulses travel along the optic nerve, through the thalamus to the occipital lobe of the brain. Some impulses cross to the opposite side of the brain at the **optic chiasma**. The view from the right eye is partially projected on the left side of the visual cortex of the cerebrum, and that from the left is partially projected on the right side. Additionally, the image reaching the occipital lobe is upside down and inverted. The brain must flip and invert the image before it makes sense to us. All of this occurs continuously and almost instantaneously, without your even knowing it.

Optic chiasma The physical crossing of the left and right optic nerves.

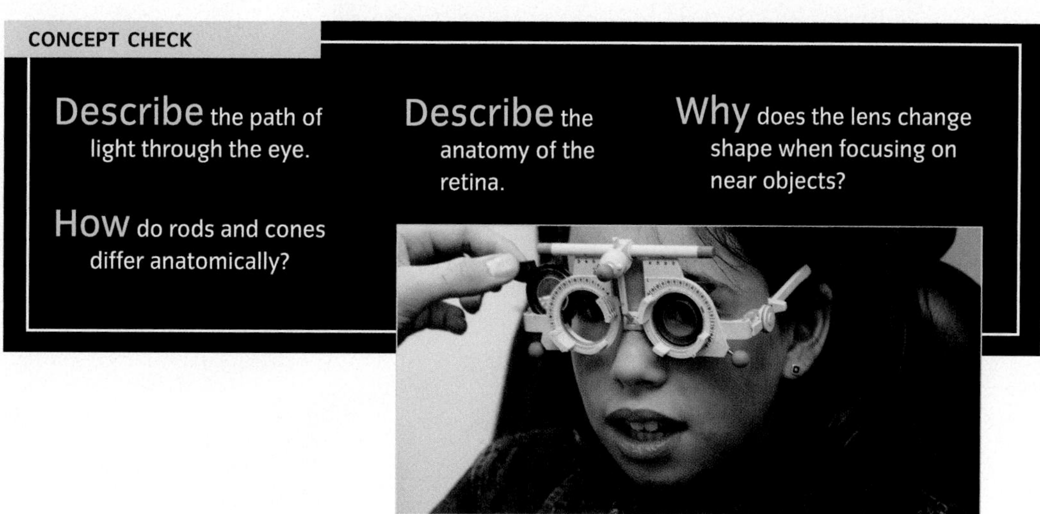

CONCEPT CHECK

Describe the path of light through the eye.

How do rods and cones differ anatomically?

Describe the anatomy of the retina.

Why does the lens change shape when focusing on near objects?

The Special Senses Are Our Connection to the Outside World

LEARNING OBJECTIVES

Discuss how society views sensory loss.

Understand conduction deafness and nerve deafness.

Although aging can impair many of the special senses, most people can still lead productive, active lives even with some decline in their ability to perceive the world. Mild eyesight defects are usually easy to correct with eyeglasses. In fact, an entire market has been created for designer eyewear. And several surgical techniques, such as laser eye surgery, can improve the focusing of light rays, permitting many to see well without corrective lenses (FIGURE 8.15).

Laser eye surgery FIGURE 8.15

LASIK surgery refers to the use of lasers to alter the shape of the cornea. A small flap of cornea is cut and lifted back, exposing the center of the cornea. This middle corneal tissue is then vaporized in small, precisely controlled sections, causing the cornea to lie in a shape conducive to clear vision when the flap is replaced. Radial keratotomy is essentially the same process, except that instead of lifting a flap of cornea, the outer layer is removed completely.

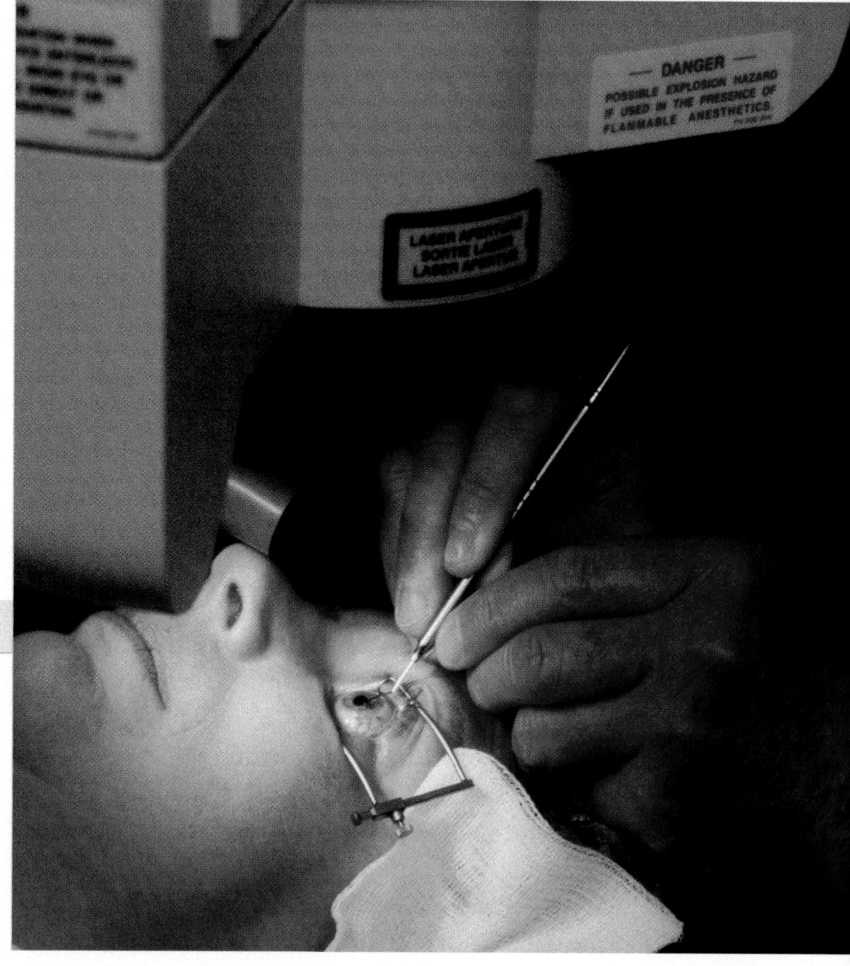

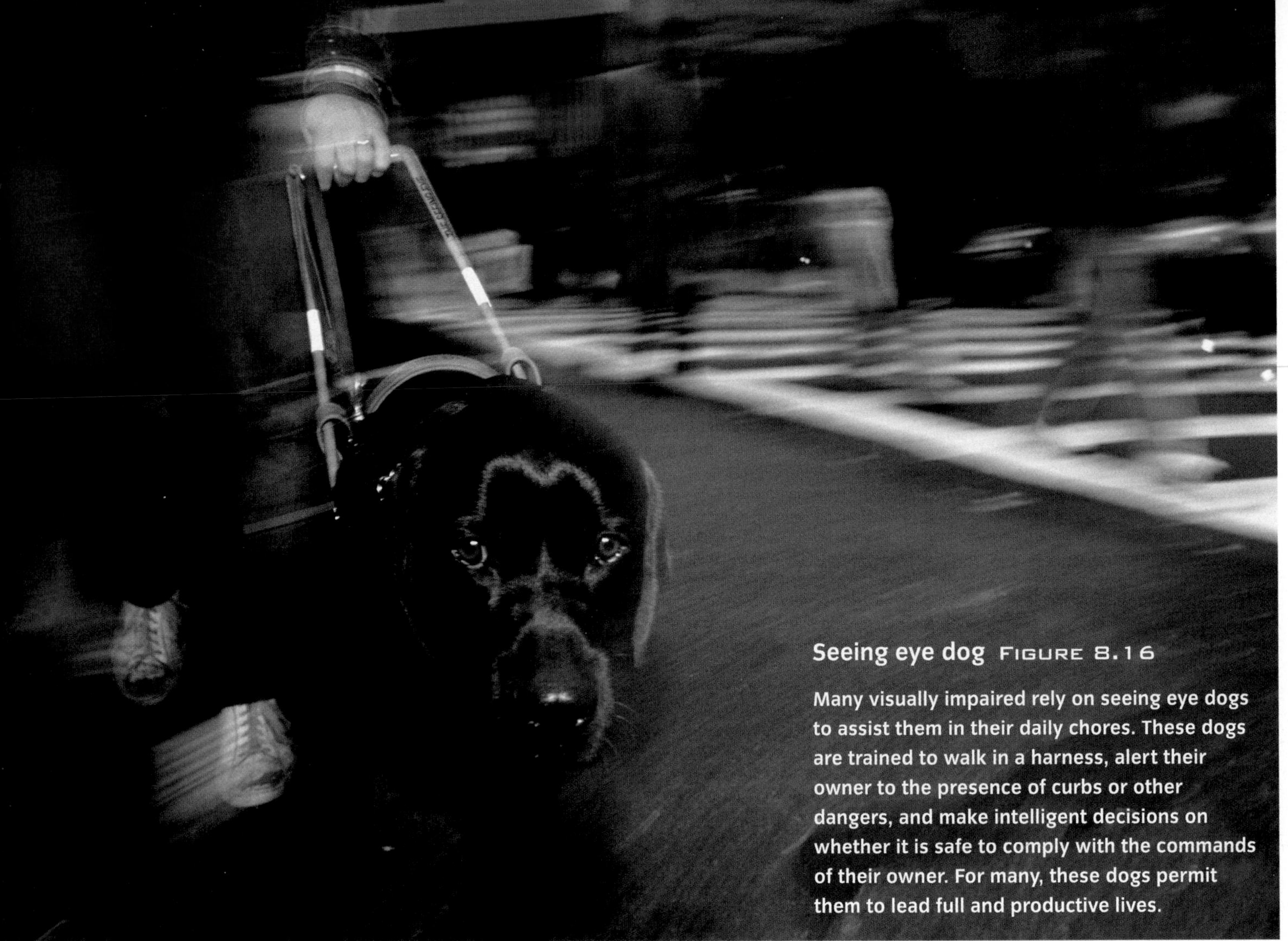

Seeing eye dog FIGURE 8.16

Many visually impaired rely on seeing eye dogs to assist them in their daily chores. These dogs are trained to walk in a harness, alert their owner to the presence of curbs or other dangers, and make intelligent decisions on whether it is safe to comply with the commands of their owner. For many, these dogs permit them to lead full and productive lives.

Complete loss of sight is another story, however. The blind are not easily assimilated into mainstream culture. As mentioned earlier, we humans are extremely visual organisms, relying mainly on sight to get us through the world. Our social and economic systems require us to pick up visual cues, leaving blind people to function in a society designed for the sighted. Despite the use of Braille on elevator buttons and a few restaurant menus, many blind people must obtain aid from a sighted person or a seeing-eye dog to function (FIGURE 8.16). Simply getting around can be challenging.

Like vision, hearing can diminish with age. Some hearing loss is due to mechanical malfunctions. In **conduction deafness,** sound is poorly conducted from the outer ear to the inner ear, as would happen, for example, if the ossicles were prevented from moving easily. Hearing aids can help those suffering from conduction deafness by increasing the amplitude of sound that enters the ear. But deafness is often due to neurological malfunction rather than a conduction problem. If auditory troubles are caused by **nerve deafness**, a hearing aid does not help because the problem is that the sound is either not detected or the nerve impulse is not transmitted to the brain. Cochlear implants convert sound vibrations into electrical impulses and have shown some promise in treating nerve deafness. Just like blindness, deafness can be life-threatening. Sirens, smoke alarms, even the ringing of a phone are all auditory cues that warn us of danger. Visual cues have been added to most fire and hazard alarms in public buildings to assist those with hearing loss. In addition, many phones are available with a visual ring cue. For an alternative viewpoint on deafness, see the Ethics and Issues box. Being born deaf is not always considered a disability.

Being born deaf is not always considered a disability

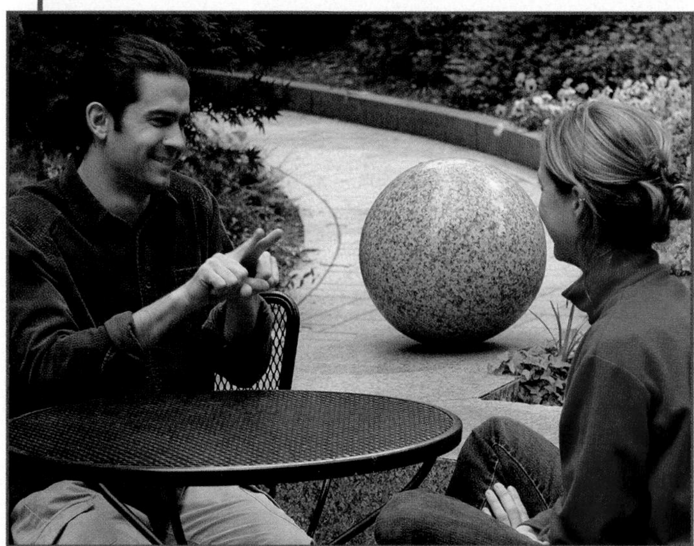

Is deafness a disability? Should it be corrected? To most hearing people, these questions have obvious answers: "yes," and "whenever possible." But many deaf people answer differently. To understand their reason, you need to understand something about the culture of deafness. Deaf people live among us, but many live separate lives because they "speak" with their hands, using American Sign Language (ASL). Sign language uses the words, but not the grammar, of English, and it can be just as subtle as English. According to the Web site aslinfo.com, the use of sign language is a central and desirable aspect of deaf culture: "Far from seeing the use of sign as a handicap, deaf people regard ASL as their natural language, which reflects their cultural values and keeps their traditions and heritage alive."

A massive student protest in 2006 at Gallaudet University in Washington, D.C., long a center of deaf culture, emphasized the central role of ASL. The protests erupted after it was learned that the university's newly named president had not learned sign language until age 23. The appointee had also gone to regular schools rather than the "residential" schools that are a major avenue of socialization for the deaf. These schools assist in the unique ASL socialization of deaf people because everyone in these schools speaks with sign instead of English. The new president's background aroused fears among the protestors that henceforth speech and lip reading might play a greater role than ASL at Gallaudet, thereby excluding deaf people who communicate best with sign.

Deaf culture is changing, and ASL is a major barometer of this change. For many years, "mainstreaming" was the goal of education for the deaf, and so deaf people were forced to try the difficult task of reading lips. This practice continued at most schools for the deaf, such as Gallaudet, for many years. Not until 1988 was a deaf person, fluent in ASL, appointed president of Gallaudet.

Today the cultural battle involving the deaf has entered a new phase, as a rising number of deaf young people are gaining the ability to hear through the use of cochlear implants. These devices detect sound with a microphone and directly stimulate the auditory nerves, bypassing the defective cochlea, the cause of the deafness. The implants have allowed many deaf young people to enter mainstream culture rather than become immersed in a deaf culture reliant on ASL. That prospect constitutes yet another threat to the deaf community for whom ASL is central. As aslinfo.com has noted, although it sounds contrary to usual parental concerns, many deaf parents hope their children will be deaf, to keep deaf culture alive.

Historically, discrimination and isolation have spawned high culture and distinctive art forms: jazz, klezmer, and gypsy music, for example, all flourished under oppressive conditions. What do you think? Should we try to "cure" deafness, or should we support a flourishing deaf culture based on American Sign Language?

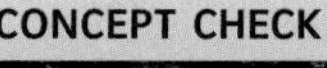

CONCEPT CHECK

What is the difference between conduction deafness and nerve deafness?

Can any type of nerve deafness be corrected?

Can lack of visual acuity be corrected?

CHAPTER SUMMARY

1 The Special Senses Tell Us about Our Environment

Special senses include smell, taste, hearing, balance, and vision. Smell and taste are chemical senses, requiring that compound be dissolved in mucus before being sensed. In the ear, sound waves are converted into mechanical motion, and then nerve impulses travel to the brain. Static equilibrium is monitored by the maculae in the saccule and utricle, and the cristae of the semicircular canals provide our sense of dynamic equilibrium. Taste originates at chemoreceptors on the tongue, and smell (olfaction) originates in chemoreceptors in the nose.

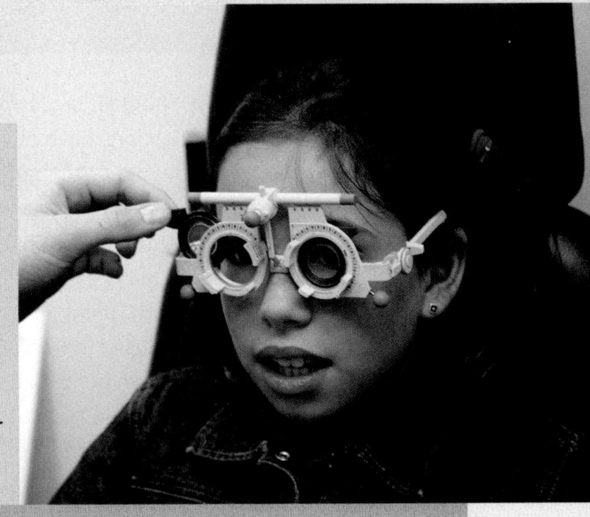

2 Vision Is Our Most Acute Sense

Vision begins with the eye, where light is converted to nerve impulses, and concludes in the occipital lobe of the brain, where these impulses are organized and interpreted. Vision is the best-developed of the special senses in the human, and its interpretation occupies more of the brain than any other special sense. The pathway of light through the eye begins with the cornea and aqueous humor. Light passes through the pupil, is focused by the lens, and strikes the retina. When there are problems focusing the light rays, glasses or laser surgery can help.

3 The Special Senses Are Our Connection to the Outside World

Loss of visual or auditory acuity causes difficulty functioning in our society. We are visual beings—everything from road signs to menus to walking paths are designed for the sighted. Braille is used to present text messages to those who cannot see well enough to read, and trained dogs give the blind a degree of independence in a sighted world. Deafness can be caused by conduction problems or neurological malfunction. Hearing aids and cochlear implants can restore hearing to many patients.

KEY TERMS

- **gustation** p. 230
- **macula lutea** p. 244
- **olfaction** p. 230
- **optic chiasma** p. 246
- **pupil** p. 240
- **rhodopsin** p. 244
- **stereoscopic** p. 246
- **uvula** p. 233
- **visual acuity** p. 241

CRITICAL THINKING QUESTIONS

1. Some people are born with a condition in which the cribriform plate of the ethmoid bone is not formed properly. The tiny perforations that allow the olfactory neurons to extend into the upper nasal passageway are not present, and the cribriform plate is instead a solid bone. How would this affect the sense of smell? The sense of taste?

2. As we age, we lose many of the taste buds of our youth. The function of those remaining does not change, but the sheer number of taste buds declines over time. How do you think this might alter our perception of food? Can you relate this phenomenon to your own life? (Did you always enjoy the foods that you now enjoy?)

3. When you ride an elevator, why does your stomach feel like it is "dropping" when you ascend? Which sensory organ(s) account for this sickening feeling, and what perceptual conflict helps create it?

4. A cataract is a clouded lens, usually associated with age. How would a cataract affect vision? Trace the light entering an eye with a cataract, listing possible effects of the clouded lens. From what you know about the pathway of light through the eye, what might correct these visual disturbances?

5. Why do hearing aids not help a person suffering from nerve deafness? What is the difference between nerve deafness and conduction deafness? Which is easier to correct, and why?

SELF TEST

1. All of the following are special senses EXCEPT

a. vision.
b. equilibrium.
c. olfaction.
d. proprioception.

2. The following sense involves mechanoreceptors:

a. smell.
b. taste.
c. touch.
d. balance.
e. c and d are correct.

3. True or False? We are capable of distinguishing only four different odors.

4. Identify the structure seen in this figure.

a. Olfactory neuron
b. Gustatory neuron
c. Taste bud
d. Retina

5. What is the function of the structure labeled A?

a. Prevent choking
b. Block the passageway from mouth to nose
c. Prevent mucus from entering mouth
d. Allow foodstuffs to enter nasal cavity

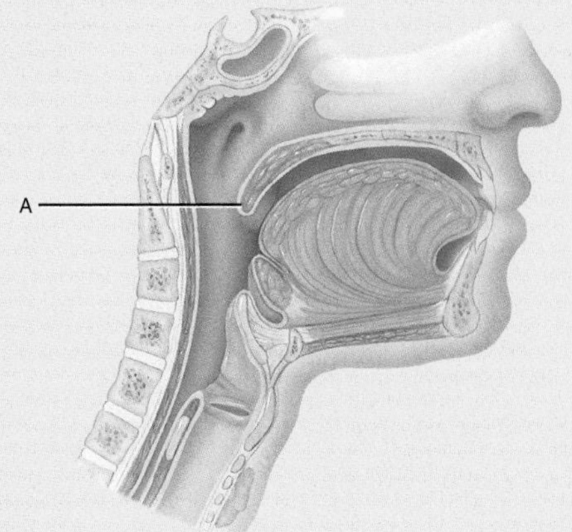

The next five questions all relate to this figure

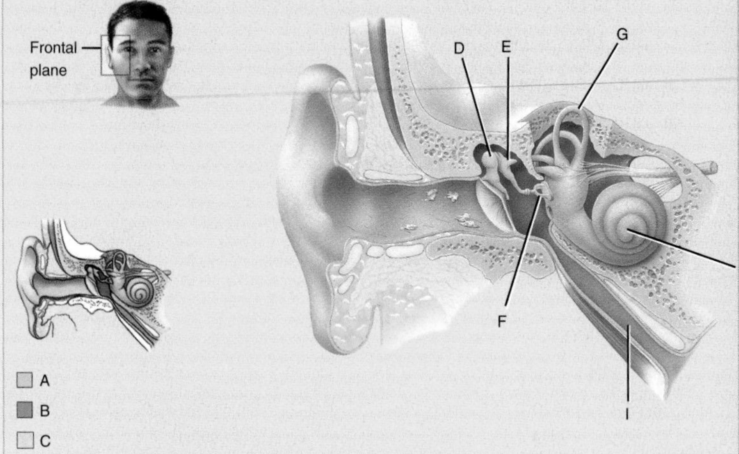

6. What is the function of the area(s) labeled A?

a. Collect and transmit sound
b. Convert sound waves to vibrations
c. Dampen loud sounds
d. Equilibrate pressure on either side of the tympanic membrane

7. What is the function of the structure labeled G?

a. Hearing
b. Static equilibrium
c. Dynamic equilibrium
d. Olfaction

8. Which area is responsible for transmitting sound waves into vibrations?

a. A
b. B
c. C
d. H

9. The label that indicates the ossicle responsible for transmitting vibrations to the inner ear is

a. D.
b. E.
c. F.
d. I.

10. What is the function of the structure labeled H?

a. Hearing
b. Static equilibrium
c. Dynamic equilibrium
d. Olfaction

11. The bending of hairlike projections generates nerve impulses and begins the process of

    a. hearing.
    b. dynamic equilibrium.
    c. static equilibrium.
    d. All of the above.

12. The layer (tunic) of the eye that includes the whites and the cornea is the

    a. vascular tunic.
    b. fibrous tunic.
    c. nervous tunic.
    d. innermost tunic.

13. The function of the structure labeled A on this image is to

    a. produce tears.
    b. collect tears.
    c. moisten the nasal passages.
    d. spill tears onto the cheek bones.

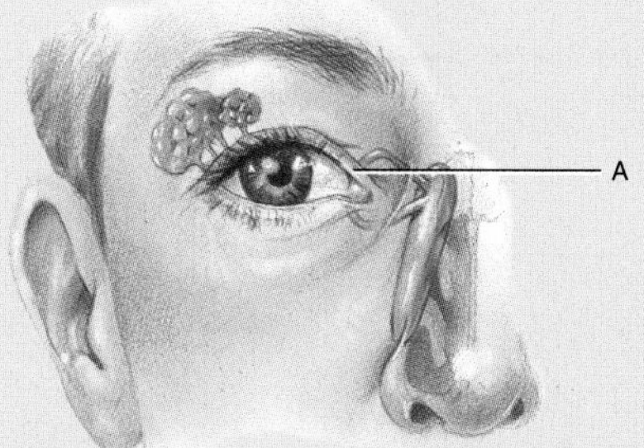

Questions 14–16 require the use of the following diagram:

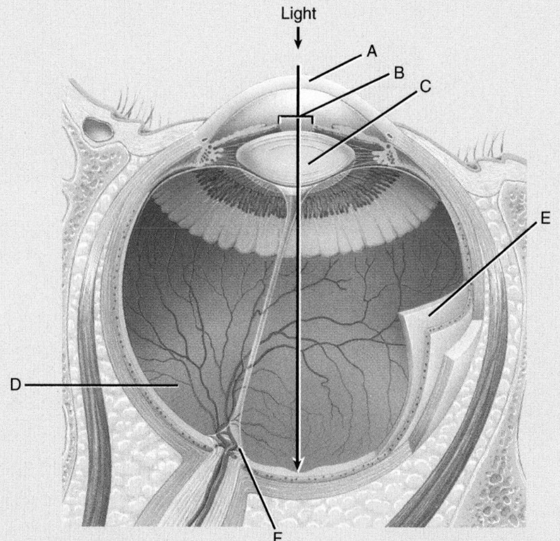

14. What is the name of the structure labeled as E?

    a. Aqueous humor
    b. Choroid
    c. Lens
    d. Vitreous humor
    e. Retina

15. What is the function of the structure labeled B?

    a. Focuses light entering the eye
    b. Directs the amount of light entering the eye
    c. Sends light rays on the retina
    d. Sends visual impulses to the brain

16. Which of these structures continues to grow during your lifetime?

    a. A
    b. B
    c. C
    d. D
    e. E

17. Nearsightedness is caused by the lens focusing images

    a. in front of the retina.
    b. behind the retina.
    c. unevenly on the retina.
    d. directly on the macula lutea of the retina.

18. Correction for farsightedness usually requires

    a. a concave lens.
    b. a convex lens.
    c. laser surgery to reshape and smooth the cornea.
    d. a carefully crafted lens that matches the contours of the cornea.

19. True or False? The cones allow us to see indistinct shapes in low light, but they bleach and are ineffective when light levels increase.

20. The correct layers of neurons in the retina, from anterior to posterior in the eye, is

    a. bipolar neurons → ganglionic neurons → rods and cones → back of eye.
    b. rods and cones → bipolar neurons → ganglionic neurons → back of eye.
    c. ganglionic neurons → rods and cones → bipolar neurons → back of eye.
    d. ganglionic neurons → bipolar neurons → rods and cones → back of eye.

21. Despite being a visual society, people with impaired vision can function by taking advantage of

    a. braille menus and buttons.
    b. seeing-eye animals.
    c. cochlear implants.
    d. ASL.
    e. Both a and b are correct.
    f. Both c and d are correct.

Defense Against Disease: Stress, Nonspecific Immunity, and the Integumentary System

9

Children in war-torn countries are often trapped by violence. You see them carrying weapons and walking in menacing packs on newscasts, and you have to wonder: How old are these children—12? 15? Did they volunteer for combat? Are they more like modern-day slaves? Do they even understand what they are doing? One of the most disturbing aspects of recent civil wars has been the rise of child soldiers. Too vulnerable to say no, they maim and mutilate until, often, they wind up getting a taste of their own medicine.

One thing is sure: Few situations are more stressful than combat. War creates immediate and long-term psychological and physiological stresses, ranging from adrenaline rushes and exposure to wounds, disease, and death in the short term, to post-traumatic stress disorder and fatal diseases in the long term. Money spent on the military siphons funds from public health services, such as immunization or malaria prevention. Conflict spreads infectious disease: In Uganda, HIV infection skyrocketed during the turbulent 1980s, then fell during the relatively peaceful 1990s. As the medical infrastructure dissolves, rape, anarchy, poverty, and unsanitary conditions all increase. Whereas the stress of combat lowers our defenses against disease and facilitates homeostatic imbalances, much milder stresses can also cause physiological and psychological problems. In this chapter, we examine the body's initial defenses against stress and pathogens, and look at helpful and harmful responses to stresses in the environment.

CHAPTER OUTL

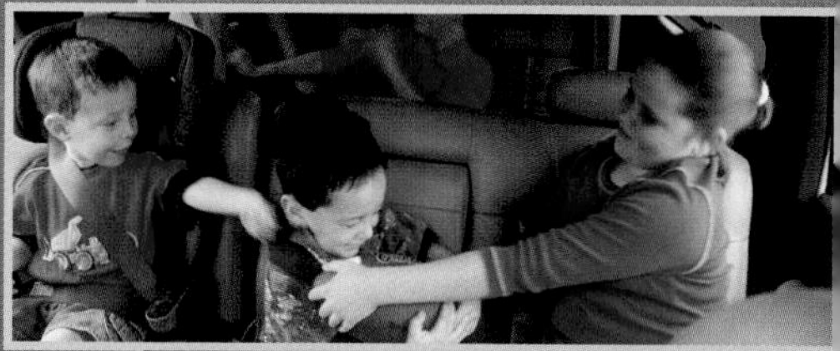

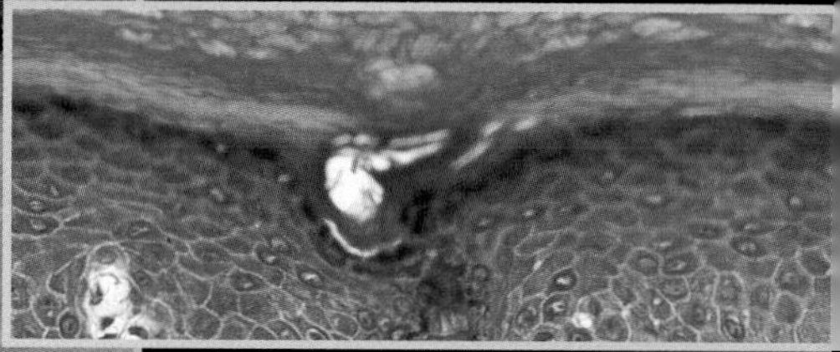

NATIONAL GEOGRAPHIC

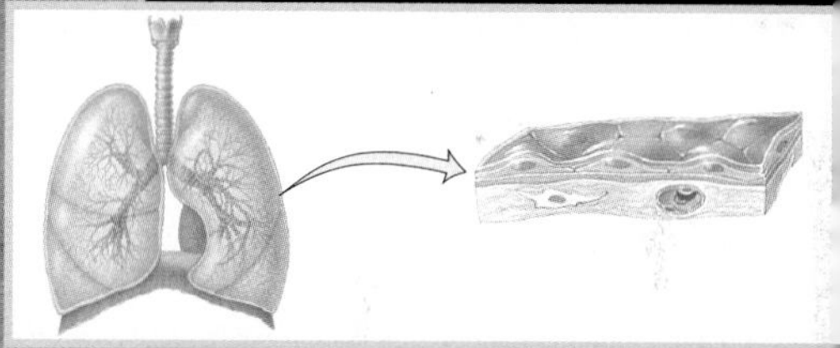

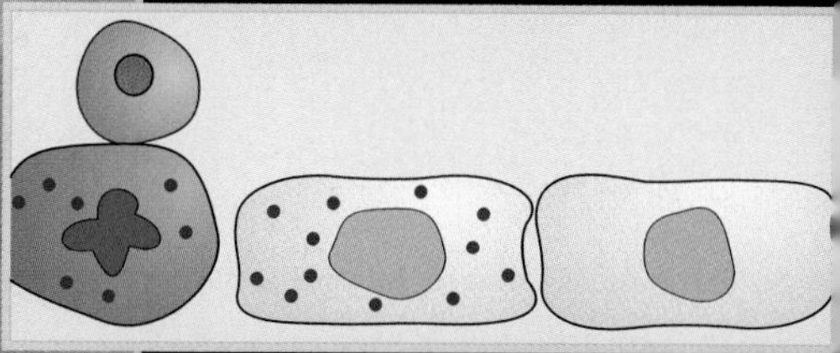

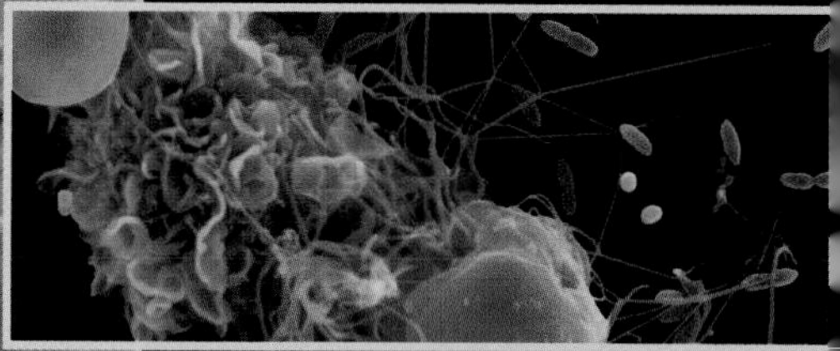

What Is Stress?

LEARNING OBJECTIVES

Define stress and the body's immediate response to it.

List the innate defenses.

Explain specific and nonspecific immunity.

Stress! It comes in many shapes and sizes, but scientifically, we define stress as any force that pushes the body out of optimum homeostatic conditions. Scientifically, stress can arise from many situations that we do not normally consider stressful, such as digesting food, exercising, waking after a long sleep, or even walking outdoors after a few hours indoors. Indeed, if you think about it, the events that take place during daily living affect the body's internal chemistry, causing stress and an imbalance that must be corrected.

Technically, "stressors" are any factor that causes stress. Some stressors are obvious. We've already seen the example of child soldiers, and we know that having an infectious disease, ingesting a toxic chemical, or being exposed to winter storms also stresses the body. If the original stress resulted from moving to a cold area, you might generate heat by shivering. If the

Innate defenses TABLE 9.1

Component	Functions
First Line of Defense: Skin and Mucous Membranes	
Physical Factors	
Epidermis of skin	Forms a physical barrier to the entrance of microbes.
Mucous membranes	Inhibit the entrance of many microbes, but not as well as intact skin.
Mucus	Traps microbes in respiratory and gastrointestinal tracts.
Hairs	Filter out microbes and dust in nose.
Cilia	Together with mucus, trap and remove microbes and dust from upper respiratory tract.
Lacrimal apparatus	Tears dilute and wash away irritating substances and microbes.
Saliva	Washes microbes from surfaces of teeth and mucous membranes of mouth.
Urine	Washes microbes from urethra.
Defecation and vomiting	Expel microbes from body.
Chemical Factors	
Sebum	Forms a protective acidic film over the skin surface that inhibits growth of many microbes.
Lysozyme	Antimicrobial substance in perspiration, tears, saliva, nasal secretions, and tissue fluids.
Gastric juice	Destroys bacteria and most toxins in stomach.
Vaginal secretions	Slight acidity discourages bacterial growth; flush microbes out of vagina.
Second Line of Defense: Internal Defenses	
Antimicrobial Proteins	
Interferons (IFNs)	Protect uninfected host cells from viral infection.
Complement system	Causes bursting of microbes, promotes phagocytosis, and contributes to inflammation.
Natural killer (NK) cells	Kill infected target cells by releasing granules that contain perforin. Phagocytes then kill the released microbes.
Phagocytes	Ingest foreign particulate matter.
Inflammation	Confines and destroys microbes and initiates tissue repair.
Fever	Intensifies the effects of interferons, inhibits growth of some microbes, and speeds up body reactions that aid repair.

stressor is an increase in blood sugar caused by eating an ice cream sundae, the pancreas will secrete insulin to reduce blood sugar levels. Other stressors—conforming to social expectations, for example—are less obvious. Have you felt uneasy while trapped in a painfully slow checkout line? Did you fantasize pushing to the head of the line or loudly urging the cashier to "speed it up"? School tests and grades are another familiar source of stress. How many students show signs of that stress on college campuses during finals week?

Stressors place physiological demands on the body, which can cause cells to halt routine activities and instead respond to the immediate demands of that stressor. The physiological changes associated with stress may alter sleep patterns or even the personality. But regardless of the stressor, the body's response follows a familiar pattern: opposing the stressor, accommodating to it, and finally succumbing to it. This pattern, called the General Adaptation Syndrome, will be described shortly.

Invasions of fungal, bacterial, or viral **pathogens** are a second important category of stressors. Our inborn ability to defend against these daily stresses is called **innate immunity**. The most obvious of these defenses is our outer layer of epithelium—the **cutaneous** membrane or the skin, which is often called our first line of defense. Innate immunity includes physical barriers, such as the skin and mucous membranes, chemical deterrents such as the **complement system** and **interferon**, and general pathogen-fighting measures like **phagocytes**, inflammation, and fever.

We are born with these general protective defenses (TABLE 9.1), which are equally active regardless of whether the threat is a bacterial invasion in the moist environment of your throat or a younger sibling taking over your space in the car (FIGURE 9.1). The mucous membranes continuously secrete mucus to "wash" the membrane surface. Your skin remains as an intact waterproof barrier defending you from chemicals, as well as your brother's elbows. Whatever the stress is, these nonspecific, innate defenses will respond the only way they can, repeating the same defense each time. If these defenses fail to ward off the threat, **specific immunity** attempts to eradicate that specific invader. The mechanisms of specific immunity, including the interactions of white blood cells, antibodies, and macrophages, are discussed in Chapter 10.

Pathogen
Agent that produces disease.

Complement system
A series of plasma proteins that, when activated, associate in a specific order to destroy pathogenic bacteria.

Interferon
A protein produced by virally infected cells that helps other cells respond to viral infection.

Phagocytes
Cells that endocytose (engulf) pathogens.

A typically response to the stresses of having siblings in the car! FIGURE 9.1

CONCEPT CHECK

Define stress, giving three examples.

Describe three physiological responses to stress.

How else can the body protect against pathogens if innate immunity fails?

The General Adaptation Syndrome Helps Overcome Stress

LEARNING OBJECTIVES

Describe briefly the three phases of the General Adaptation Syndrome.

Explain the role of epinephrine in combating stress?

You may have heard that "fight or flight" is a common response to danger. Fight or flight is one of our innate, automatic physiologic responses to stress, and in fact is the first of the three stages of the GAS, or General Adaptation Syndrome. This is a series of predictable responses to stress that are an attempt to adapt and deal with the original stressor. The three stages of this reaction are: (1) alarm, (2) resistance, and (3) exhaustion (FIGURE 9.2). During the alam stage, we feel that sudden rush of adrenaline, that immediate jolt of energy that provides the speed, power, and quickness of wit to remove ourselves from danger. It is initiated by the autonomic nervous system, discussed in more detail in Chapter 7. If this fight-or-flight response fails to overcome the stress, however, the body continues working through with the other stages of GAS: resistance and exhaustion.

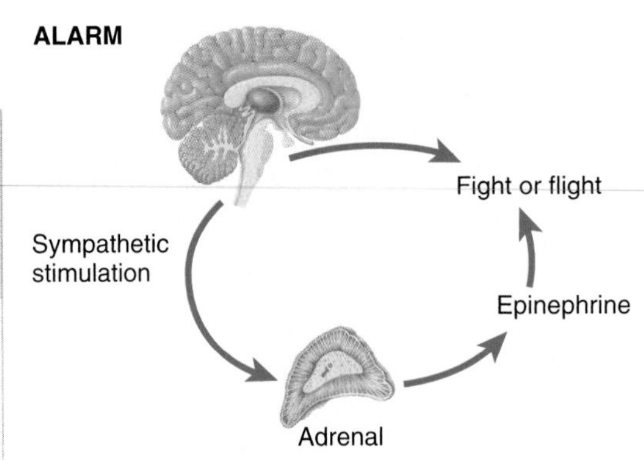

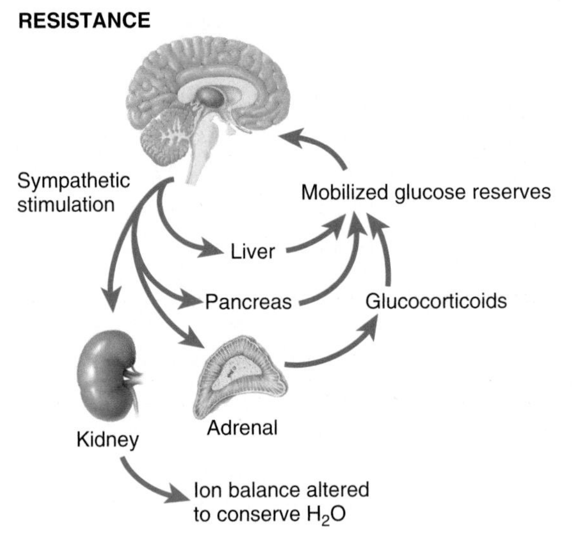

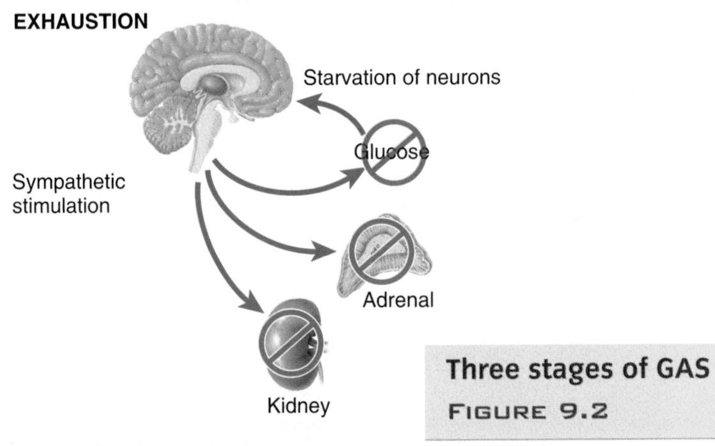

Three stages of GAS
FIGURE 9.2

DURING THE ALARM PHASE, WE MAY FIGHT OR FLEE

The alarm phase occurs when the individual detects danger, and the body first starts to deal with it. Alarm is characterized by immediate, almost frenetic, action. The "fight-or-flight" nervous system (also called the sympathetic division of the autonomic nervous system) takes over, and the body jumps into action. Energy reserves are mobilized, blood sugar increases sharply, and the body prepares to defend itself or flee. The alarm phase is controlled by the release of the hormone **epinephrine**, also known as adrenaline. This is the hormone responsible for our feelings of fear and for "adrenaline rushes" (FIGURE 9.3).

> **Epinephrine**
> A hormone released from the adrenal gland in response to stress.

In sports, the nervous state before competition shows the alarm phase in action: You experience height-

Responses to stressors during the alarm phase of GAS Figure 9.3

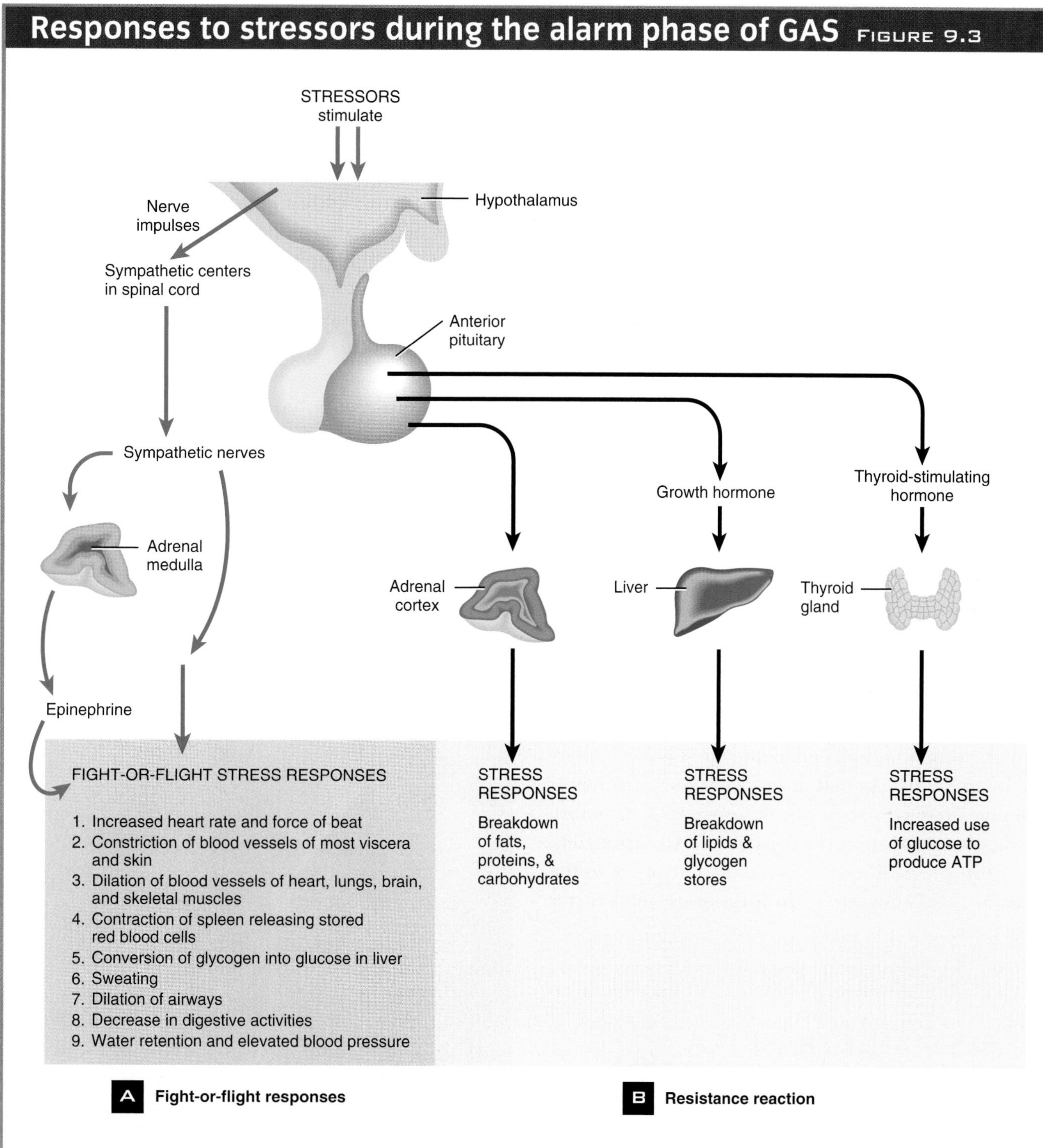

Process Diagram

ened mental alertness and increased energy becomes available to the skeletal muscles, as energy stored in glycogen and lipids is released. The circulatory system shunts blood to the organs, mainly the skeletal muscles, needed for fighting or fleeing, and away from the skin, kidneys, and digestive organs. Your body, after all, is acting as if your life depends on leaving the situation—or fighting your way out of it—with maximum haste. To save your life, is it more important to digest your last meal or to prime your skeletal muscles for action? (After all, if you run too slowly when being chased by a tiger, that last meal may literally be your last meal.)

Shifting the blood flow away from the digestive organs will often produce "butterflies" in the stomach. (An intriguing but poorly understood "enteric nervous system" helps regulate the activity of the digestive system and may also play a role in the nervous stomach.)

Although other hormones may be involved in the alarm phase, especially if the stressor is causing blood loss, epinephrine is the key hormone at this point. Epinephrine boosts blood pressure, heart rate, and respiratory rate, all of which speed the delivery of highly oxygenated blood to the skeletal muscles. Sweat production also increases, resulting in what is often called a "cold sweat."

Although changes effected during the alarm phase will help the body operate at peak performance while confronting or avoiding a stressor, these changes are less appropriate as responses to social stresses. Increasing heart rate and blood glucose will not speed up a checkout line, but they will boost your frustration level. We call a severe and inappropriate triggering of the alarm phase a "panic attack." Occasionally, a person may experience episodes of free-floating panic, with a racing heart, profuse sweating, and an inexplicable feeling of dizziness and nausea. These symptoms are characteristic of panic disorder, a chronic state characterized by panic attacks that often occur during times of prolonged stress, such as during pregnancy, or before marriage or graduation (FIGURE 9.4). Unfortunately, these physiological responses are inappropriate to the situation, and they often do little more than foster more panic.

THE RESISTANCE PHASE IS A RESPONSE TO PROLONGED STRESS

During the resistance phase, the body concentrates on surviving the stress rather than evading it. The individual is likely to feel tired, irritable, and emotionally fragile. He or she may overreact to simple daily irritants or commonplace events. This phase may begin within a few hours of the onset of stress, after the alarm phase has failed to eliminate the stressor. During the resistance phase, the brain consumes immense amounts of glucose that it obtains from the blood. A series of hormones, including glucocorticoids, epinephrine, growth hormones, and thyroid hormones, ensure that lipid and protein reserves are continuously tapped to maintain the high blood sugar level needed by the brain. The skeletal muscles become more concerned with survival than rapid movement, and they begin to break down proteins in response to other hormonal triggers. The breakdown of lipids sustains the high fuel supply even during starvation, as the liver begins converting stored carbohydrates into glucose. In addition, blood volume is conserved by maintaining water and sodium in the body, which can simultaneously raise blood pressure. Potassium and hydrogen ions are lost at abnormally high rates. Some of the glucocorticoid hormones responsible for maintaining the resistance phase inhibit wound healing, so wounds can become infected before they heal, adding to the overall stress on the body.

The resistance phase lasts until the stress is removed, lipid reserves are depleted, or complications

Marriage may often cause a momentary feeling of panic FIGURE 9.4

During a panic attack, your heart pounds, and you may feel sweaty, dizzy, or nauseous. This is often our reaction immediately before taking a life-changing step.

arise from the altered body chemistry. Poor nutrition, physical damage to the heart, liver, or kidneys, or even emotional trauma can abruptly end the resistance phase.

THE EXHAUSTION PHASE CAN BE TERMINAL

Resistance requires us to maintain extreme physiological conditions, and prolonged resistance can lead to the exhaustion phase, which is a polite way of saying, "death through organ failure and system shutdown." During exhaustion, homeostasis breaks down through the depletion of lipid reserves and the loss of normal blood electrolyte balance. Accumulating damage to vital organs may cause the affected organ systems to collapse. Mineral imbalances, due to sodium retention and potassium loss, may cause neurons to fail and thus result in the failure of skeletal and cardiac muscle.

POST-TRAUMATIC STRESS DISORDER IS A STRESS THAT SEEMS NEVER-ENDING

After severe stress, such as witnessing or being victimized by warfare, rape, or violent crime, some people develop post-traumatic stress disorder (PTSD). This disorder is a type of stress reaction that may get worse, not better, with time. Biologically, PTSD looks like a prolonged experience of the resistance phase of GAS, with a similar picture of hormonal activation. In addition, research has shown that victims of PTSD show abnormal brain patterns and changes in the volume of certain areas of the brain, especially in the amygdala, a center associated with emotion and fear, and the hypothalamus, the homeostasis center. These changes help explain the symptoms of PTSD: fear, heightened vigilance, panic reactions, inability to concentrate, and memory disorders. PTSD can usually be treated with psychotherapy or psychoactive drugs.

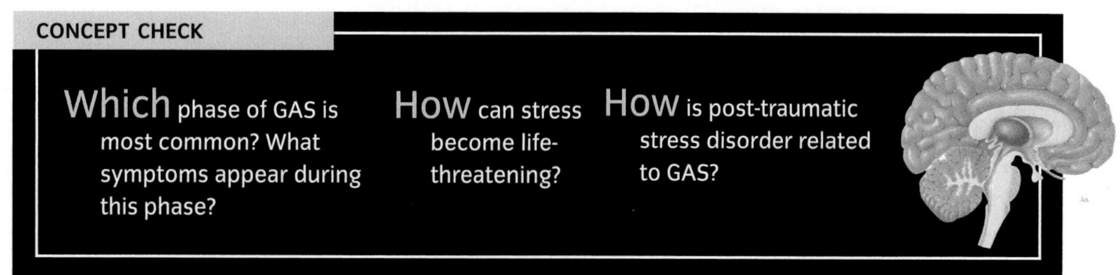

CONCEPT CHECK
Which phase of GAS is most common? What symptoms appear during this phase?
How can stress become life-threatening?
How is post-traumatic stress disorder related to GAS?

The Skin Is the Primary Physical Barrier

LEARNING OBJECTIVES

Describe the structure of the skin.

Explore the role of the skin in innate defense.

e can think of GAS as a set of behavioral defenses—activities that the body undertakes to cope with prolonged stresses. But the body has other innate, or inborn, defenses. The most obvious of these is our skin. This outer layer of epithelium is a **cutaneous** membrane that is often called our first line of defense. Other forms of innate immunity, including physical barriers, chemical deterrents, and general anti-pathogen measures, will be discussed later in this chapter.

The skin is the largest organ of the human body. It encases the body, protecting it from desiccation (drying out) and preventing the entry of disease-causing microbes. Sensory receptors in the skin monitor the immediate environment, noting light touch, heavier pressure, and temperature. The skin also has vital homeostatic functions such as helping the body regulate water content and temperature. Finally, the skin produces vitamin D, which is necessary for bone growth and development.

THE EPIDERMIS IS A DEAD DEFENSIVE LAYER

The skin is composed of a superficial **epidermis** and a deeper **dermis** (FIGURE 9.5). Underlying the dermis is the **hypodermis** (where we receive injections with a "hypodermic" needle). The hypodermis is not technically part of the skin but rather the layer

Epidermis
The outermost, nonvascular layer of the skin.

Dermis
The underlying, vascularized, connective tissue layer of the skin.

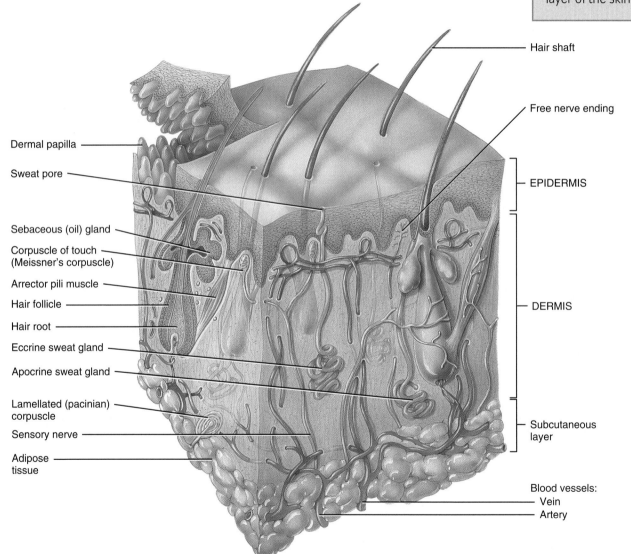

Sectional view of skin and subcutaneous layer

Components of the skin FIGURE 9.5

of connective tissue that holds the skin to the deeper organs. The hypodermis is composed of areolar connective tissue and includes a store of adipose tissue, a large blood supply, and many connective tissue fibers.

We spend a lot of energy and time on the appearance of our epidermis, especially the outermost layer of cells. The Ethics and Issues box "Skin and society: Beauty is only skin deep" on page 264 discusses this in more detail.

The epidermis is composed of stratified squamous epithelium, but most of the cells are dead. These squamous cells are produced deep within this tissue, in a layer immediately above the dermis. As these cells divide, they continually push the daughter cells upward, away from the nutrient source in the dermis. Because epithelium has no blood supply, these epithelial cells are nourished by capillaries in the upper dermis. As the epidermal cells are pushed away from these capillaries, the cells weaken and die. This gradual dying process changes the appearance of the cells, resulting in visible layers in the epidermis.

The top layer of the epidermis, the **stratum corneum**, is composed of dead cells joined by strong cell-to-cell junctions. These cells are filled with **keratin**, a waterproof substance that accumulates in the epidermal cells as they progress toward the skin surface. This dead layer provides the skin's nonspecific defense against invasive pathogens. Few pathogens are attracted to dead cells, and keratin repels waterborne pathogens along with water.

Skin color results from the brown pigment **melanin**, which is produced by **melanocytes** in the deepest epidermis (FIGURE 9.6). A UV light stimulates production of a hormone that in turn stimulates the melanocytes

Melanocytes
Cells that produce melanin, a brown, light-absorbing pigment.

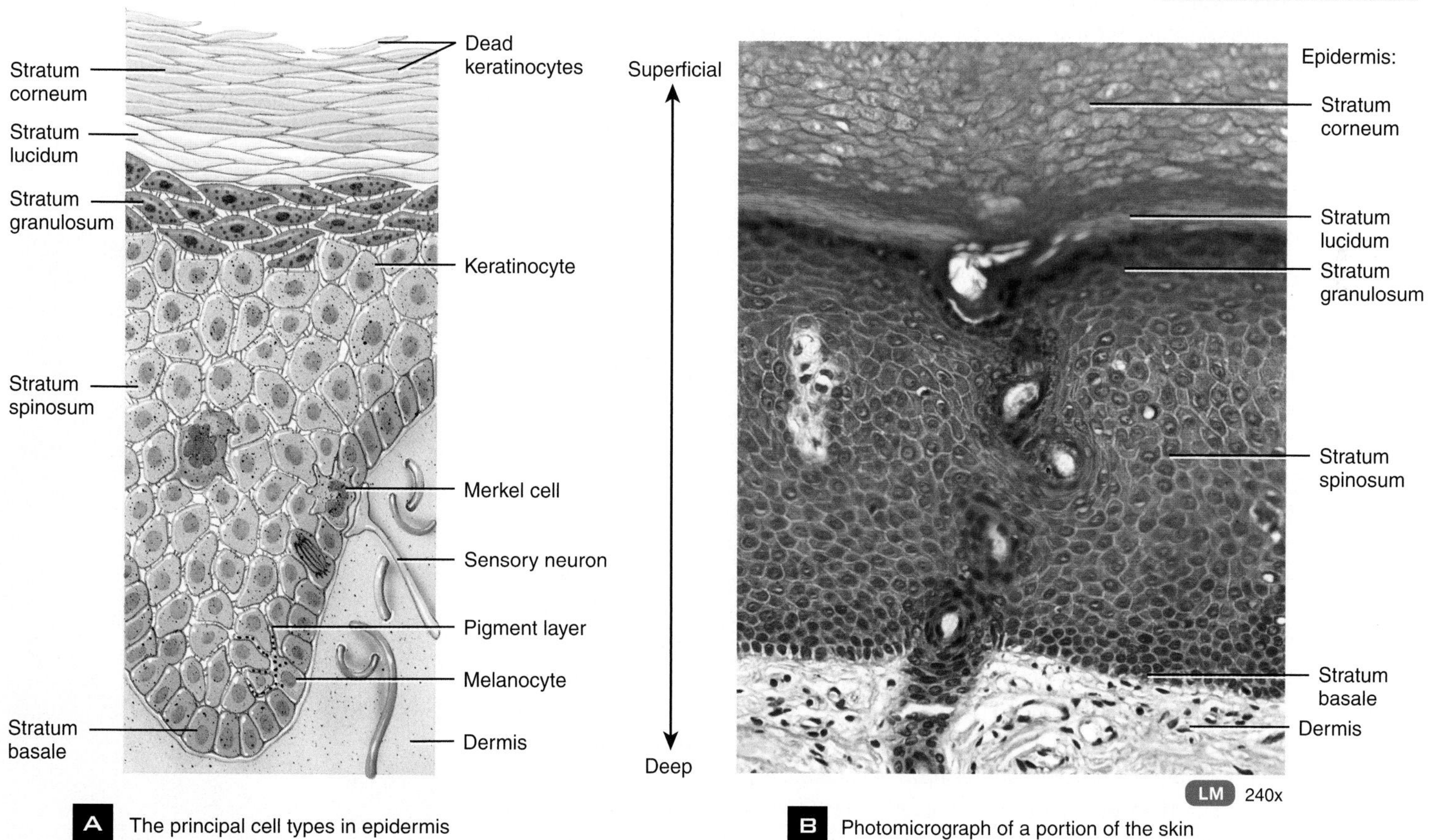

A The principal cell types in epidermis

B Photomicrograph of a portion of the skin

Pigmented epidermis FIGURE 9.6

Ethics and Issues

Skin and society: Beauty is only skin deep

"Your epidermis is showing!" This bit of elementary-school silliness usually embarrasses kids who cannot define "epidermis." But it is true! Our epidermis is always showing, along with the skin's accessory organs—hair and nails. Because it's always showing, many people diligently adorn, sculpt, and color their skin and hair. We push gold and silver through the skin and suspend adornments from our ears, noses, navels, lips, and eyebrows. We mark, scar, and draw on skin to indicate social status or personal expression. We judge others by the appearance of their skin and hair, and we constantly search our own skin for blemishes and discolorations.

All this time and effort has created a huge market for products to improve skin and hair. Drug stores carry creams devoted to moisturizing, firming, bleaching, fixing wrinkles, and reversing the aging process—as if that were possible! The cosmetic aisle is chock-a-block with powders and creams that claim to perfect skin color, powders to mimic rosy cheeks, and potions to thicken and darken eyelashes and enhance the eyebrows. You can paint almost any color of the rainbow on your eyelids or fingernails. Hair products include dyes, curling creams, straightening oils, and various gels, mousses, and sprays. In the United States, the cosmetics industry accounts for billions in revenues, and we have not even mentioned the popularity of tattooing and piercing, especially among the young.

This fascination with skin and hair is not a modern phenomenon. Throughout history, cultures have defined beauty and desirability, then altered the appearance of their skin and hair to achieve it. Egyptians dyed their hair and lips, and applied coal and clay to their eyes. In medieval Europe, extremely white skin was a sign of beauty, so whitening powders and creams were popular. African tribes mark their chiefs with tattoos, scarification, and clay designs to symbolize strength and power. Traditional Polynesians indicated social rank, sexual maturity, and genealogy with tattoos. These activities have continued through the ages. Modern cultures are at least as active as ancient cultures.

Tattoos are permanent ink markings that are injected into the skin with hollow needles. Although the tattoo appears in the epidermis, the ink is injected into the dermis (otherwise the tattoo would disappear, as the epidermal cells are continually sloughed off). Tattoos fade when the ink particles gradually diffuse into the surrounding dermis. Black tattoos are distinctly visible for a longer period because black ink contains suspended particles that diffuse more slowly than the dissolved chemicals used in other inks. One way to remove a tattoo is to speed up the dispersal of the ink with a precise beam of laser energy. The energy helps break up the ink, which is then removed through the cardiovascular and urinary systems.

The obsession with ideals of beauty raises medical issues, as women and increasing numbers of men suffer eating disorders such as anorexia nervosa and bulimia, or undergo elective surgery to meet—or exceed—the current standards of beauty. Cosmetic surgery is commonly used in an effort to meet standards that actually transcend "normal" proportions. How many Western women have been tempted to match the ultra-slim (and probably digitally enhanced) models in magazines and catalogs? Is this going too far? Would it be better to define beauty as natural appearance? Can you abandon the beauty fantasies of fashion magazines and movie screens and adorn yourself only to the point that makes you comfortable?

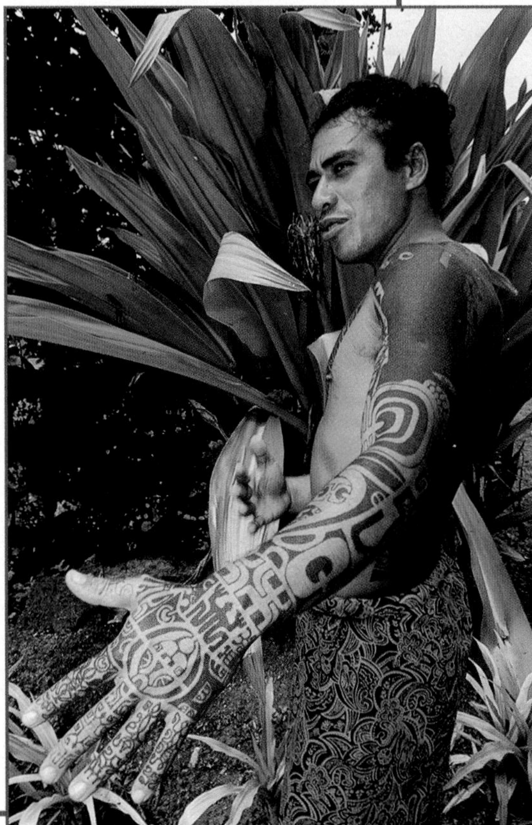

to produce more melanin, resulting in a tan. Interestingly, humans, regardless of race, have the same number of melanocytes; different levels of melanin production account for the different skin colors. Melanocytes are less active in those with pale skin. In those with dark skin, highly active melanocytes produce lots of melanin, even with low sunlight exposure. In evolutionary terms, dark skin is an adaptation that protects tropical people from the intense sun. White skin is adaptive closer to the poles because it allows the entry of enough ultraviolet light to produce vitamin D. Skin cancer is a concern for anyone who has exposed their skin to sunlight. See the Health, Wellness, and Disease feature "Skin cancer" on page 266 for an in-depth discussion of this disease.

THE DERMIS IS THE SOURCE OF NUTRITION FOR THE EPIDERMAL CELLS

The bottom layer of skin, the dermis, is composed of loose, irregular connective tissue. The dermis has a large blood supply and extensive innervation. The **accessory organs** of the skin (hair, glands, and nails) lie in the dermis. The top portion of the dermis is arranged in ridges and whorls (FIGURE 9.7). These ridges on the fingertips, palms, and toes make fingerprints (see the I Wonder . . . box on p. 270).

All the sensory organs of the skin reside in the dermis. Free (exposed) nerve endings register the sensation of pain (**nociceptors**), whereas

> **Nociceptors**
> Non-adapting pain receptors in the skin (noci = pain).

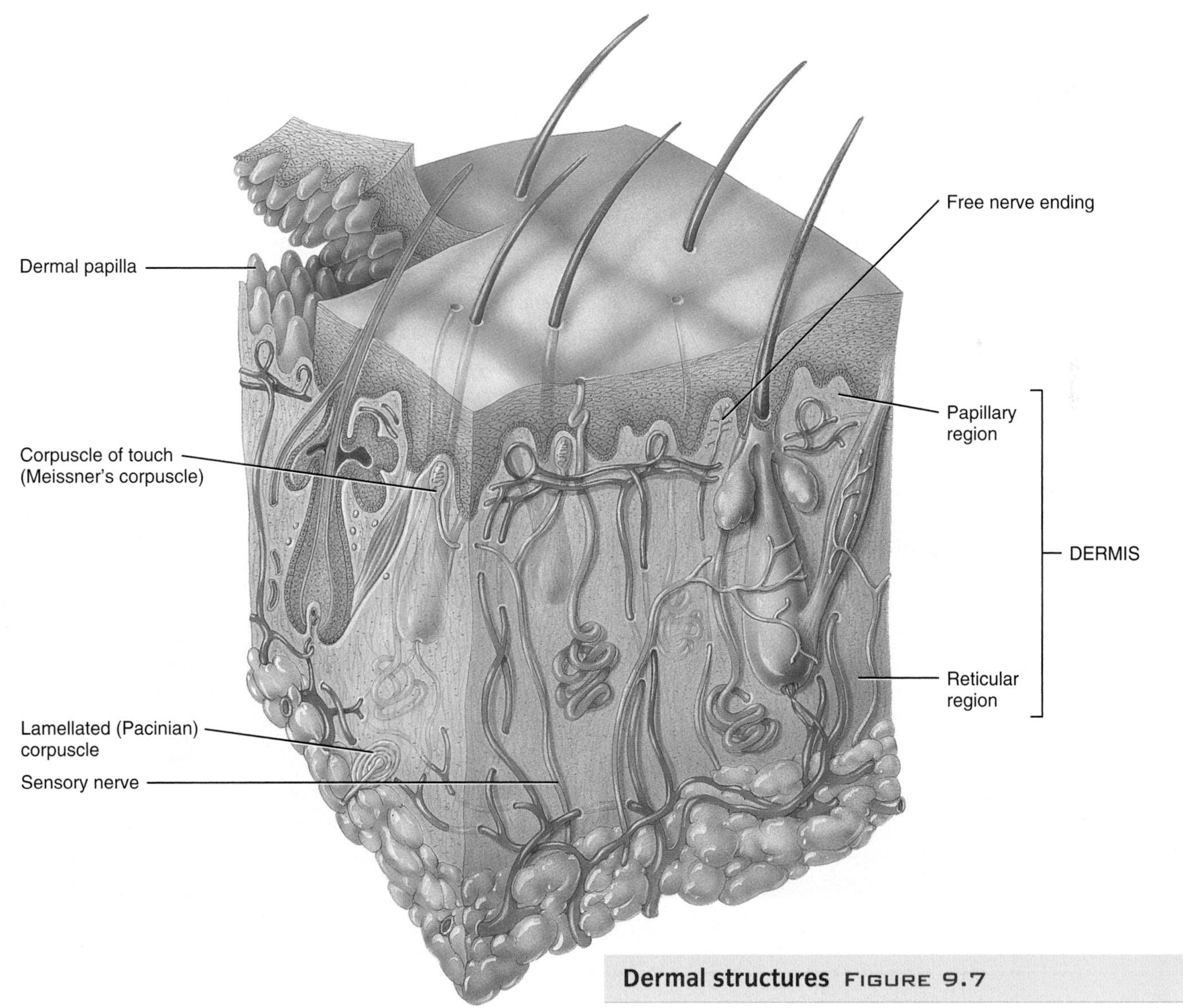

Dermal structures FIGURE 9.7

Health, Wellness, and Disease

Skin cancer

Skin cancer is common in the United States. In 2004, 1 in 65 Americans was diagnosed with some form of skin cancer. The good news is that skin cancer occurs in the epidermal cells and is easily detected at an early stage. As with all cancers, these tumor cells eventually begin to multiply rapidly and uncontrollably. Skin cancer is related to sun exposure because the ultraviolet radiation in sunlight damages the DNA in skin cells.

To avoid skin cancer, reduce sun exposure by using a lotion with an SPF of at least 15. SPF, which stands for Sun Protection Factor, indicates the degree of protection; higher numbers offer more protection. The two basic types of SPF are zinc ointments that reflect the sun's rays and chemicals that absorb the UV rays. Dermatologists prefer the zinc ointments because reflection is safer than absorption close to the skin's surface.

Basal cell carcinoma (BCC) is the most common cancer in humans, accounting for over 1 million cases per year in the United States alone. This cancer develops in the basal or deepest cells of the epidermis, usually in places that are routinely exposed to the sun. The appearance can vary, but the tumor is usually a slow-growing, shiny or scaly bump. A wound that repeatedly heals and opens may be a form of BCC. These cancers rarely metastasize, or spread to other tissues, but dermatologists still recommend that they be removed.

Squamous cell carcinoma (SCC) is a tumor of the upper layers of the skin. These cancers usually develop a crusty or scaly covering and grow rapidly. The threat of metastasis is much higher with SCC than with basal cell carcinoma, so SCC tumors should be removed as soon as possible. Approximately 16 percent of skin cancer cases are SCC.

Melanomas are the most aggressive skin cancer, rapidly spreading to the lymph nodes and other tissues, but they comprise only 4 percent of all diagnosed skin cancers. The cancerous cells are melanocytes—ironically, the same cells that protect us from harmful UV radiation. Cancerous melanocytes divide rapidly and spread to the dermis. A melanoma is a dark spot on the skin that can be identified with the "ABCD" guidelines:

A = asymmetry. A noncancerous mole is usually round, and both sides match each other. Melanoma grows in all directions, but at different speeds, creating an irregular (asymmetrical) appearance.

B = border. A noncancerous mole has a distinct border that you could easily trace with a pen. Melanomas often have scalloped borders, or areas where they fade into the surrounding skin.

C = color. A noncancerous blemish has a uniform color, but melanomas tend to have several colors in one blemish. One tumor may have areas that are dark blue-black, brown, red, or even white.

D = diameter. Melanomas are often larger than noncancerous blemishes. A spot that is larger than a typical pencil eraser may be a melanoma.

You should know your own skin well enough to recognize suspicious blemishes. If any of these descriptions apply, consult a medical professional for evaluation, biopsy (cellular examination), and treatment if necessary.

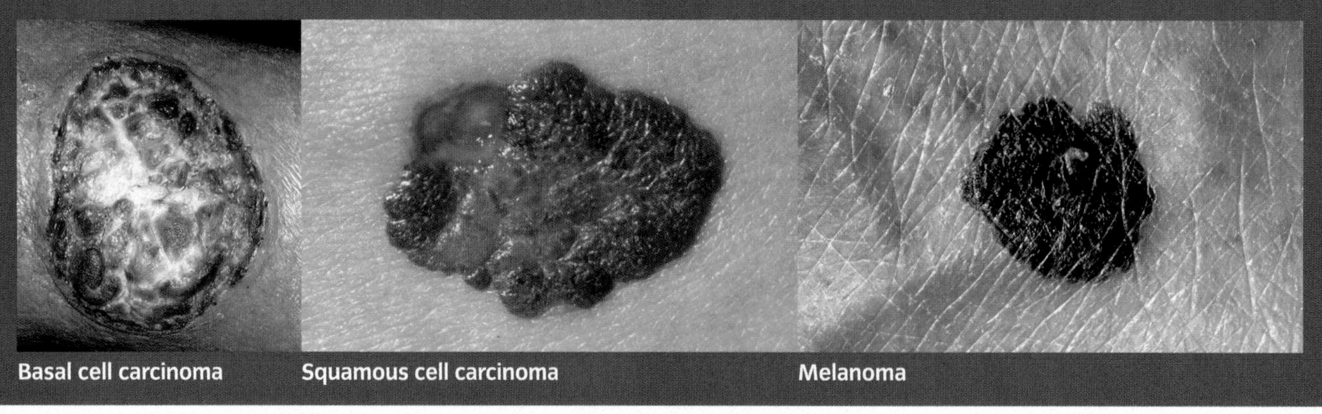

Basal cell carcinoma | Squamous cell carcinoma | Melanoma

specialized structures attached to cutaneous nerves respond to light touch and pressure. **Meissner's corpuscles** and Merkel discs in the upper dermis register light touch and fine sensations, respectively. Deeper in the dermis, near the hypodermis, the **Pacinian corpuscles** register pressure. Meissner's corpuscles and Pacinian corpuscles **adapt**, meaning the sensory structures will begin to ignore light touch or pressure that does not change. When you put on a shoe in the morning, the Pacinian corpuscles in your foot register the shoe's pressure. During the day, the pressure doesn't change, and you are no longer aware that you are wearing shoes. Should the pressure become painful, however, pain receptors remind you of your shoes. Unfortunately, pain receptors do not adapt, so your discomfort will remain until you somehow remove the excess pressure.

ACCESSORY STRUCTURES OF THE SKIN LUBRICATE AND PROTECT

The accessory structures of the skin are the glands, hair, and nails. The glands produce sweat for thermal homeostasis, or oils to keep the skin flexible. The hair and nails are protective structures.

Oil (**sebaceous**) glands are found within hair follicles. Oil is secreted onto the hair shaft, helping to keep the hair and surrounding skin supple. The hormones of puberty increase the output of these glands, often leading to acne, defined as a physical change in the skin because of a bacterial infection in the sebaceous glands. Acne causes the development of lesions, cysts, blackheads, or whiteheads, common terms for various combinations of dirt, infection, and skin oils. Fortunately, doctors can now treat virtually all types of acne and usually prevent scarring that can follow uncontrolled infections.

Sebaceous glands are located wherever there is hair (FIGURE 9.8). This means we have oil glands everywhere on our bodies except in hairless skin, such as on the lips. This absence of oil glands explains the need for lip balms to alleviate drying and chapping in this oil-less skin.

THE SKIN PLAYS A CRUCIAL ROLE IN TEMPERATURE CONTROL

Sweat glands are active in maintaining thermal homeostasis. They are found all over the body, with the exception of the lips and the tip of the penis. Sweat glands

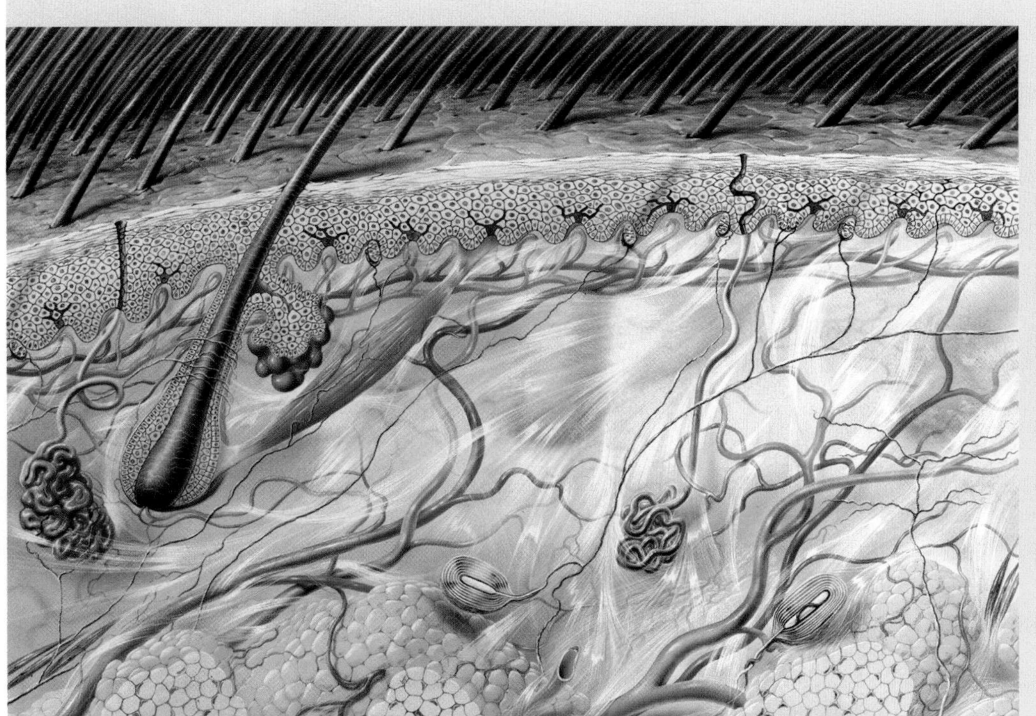

Sebaceous gland

FIGURE 9.8

Note that these glands are always associated with hairs, lying next to the hair with their ducts opening directly onto the hair (colored purple in this figure to match the epidermal cells from which they originate). When the hair is moved, oils are secreted from these ducts.

Apocrine
A cellular secretion that pinches off the upper portion of the cell with the secretion.

Eccrine
A cellular secretion that does not include any portion of the secreting cell.

are basically a tube from the surface of the skin into the dermis. At the base of the dermis, the tube coils into a knot (FIGURE 9.9). Most sweat glands open to the surface at a pore, with no hair associated. The larger sweat glands of the axillary region, the **groin** region, and the areolae of the breasts become active during puberty. These glands secrete fluids using the **apocrine** pathway rather than the **eccrine** pathway used for the many sweat glands found on the surface of the body. Consequently, apocrine glands secrete a protein mix along with the water and ions. At the surface of the skin, bacteria break down these proteins, creating the body odor associated with sexual maturity.

Sweat is produced in response to rising internal temperature. Blood vessels in the dermis dilate, allowing a larger volume of blood to flow from the core of the body to the skin. This blood transports excess heat to the skin where it activates thermoreceptors that send impulses to the brain to activate the sweat glands. The blood, having transferred its heat to the skin, returns to the heart somewhat cooler. This additional heat does not remain at the skin because when sweat evaporates, it reduces core body temperature by removing the energy required to vaporize water (water's heat of vaporization = 540 calories/gram). During average activity, your sweat glands produce approximately a coffee cup (150 ml) of fluid per day. Athletic activity increases this volume tremendously; up to 2.5 liters of fluid per hour can be lost during strenuous activity in hot weather. In the 2003 Tour de France, Lance Armstrong lost a full 6 percent of his body weight during a hot, intense, one-hour race. This extreme fluid loss took a toll on his performance and overall health, and Armstrong needed two days to recover. For optimal performance and general health, endurance athletes must hydrate before and during competition.

HAIR—AN EVOLUTIONARY RELIC?

What is hair, and why does it grow where it does? Although we think of hair mainly as the coarse structures projecting from and protecting our head, hair actually covers most of our bodies, including our face, shoul-

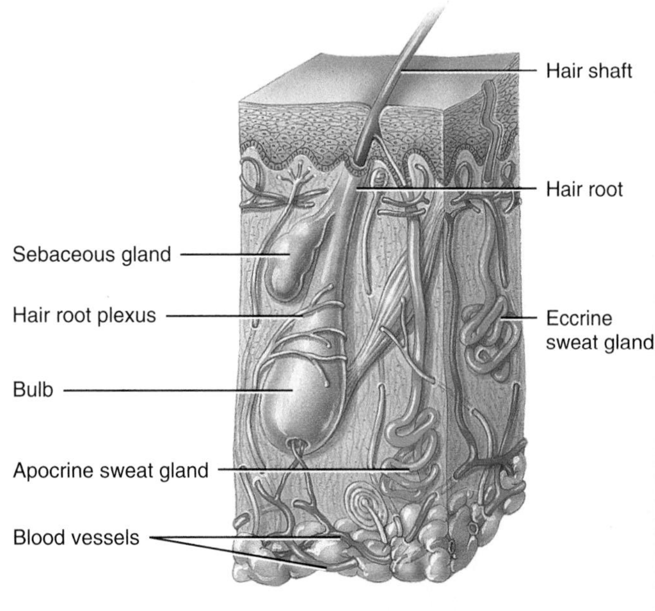

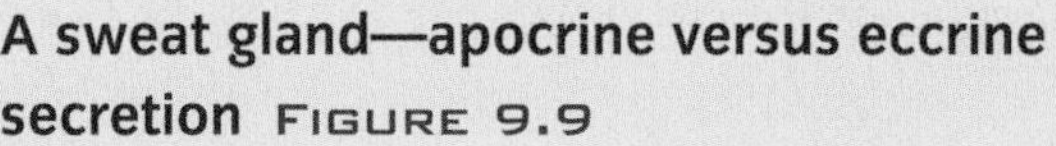
A sweat gland—apocrine versus eccrine secretion FIGURE 9.9

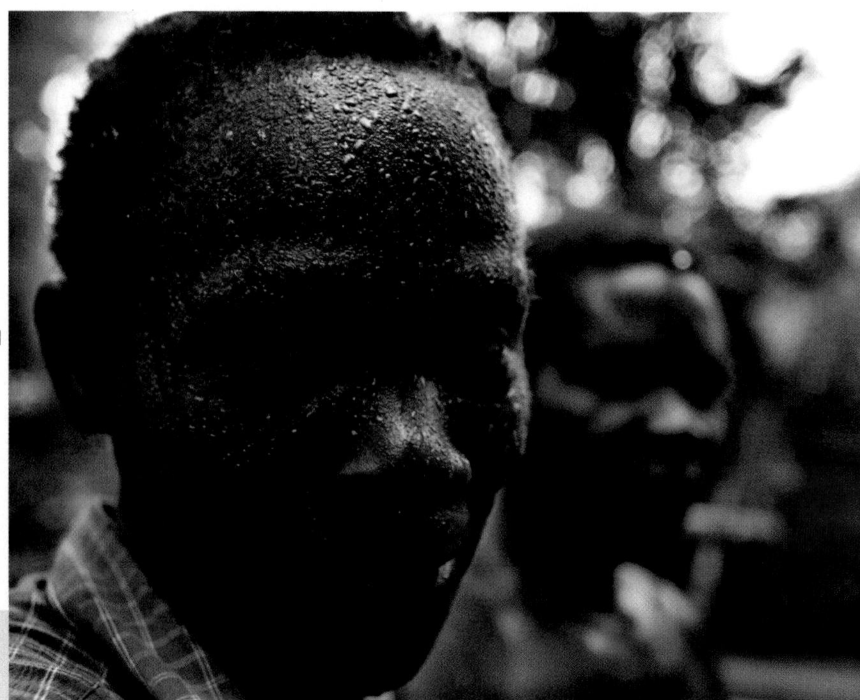

ders, back, and belly. Humans are not really "hairless apes," although most of our hair is fine and sparse compared to that of the other apes. Hair serves as an insulator as well as protection for the eyes, nostrils, and ear openings. On our heads, hair prevents loss of heat from blood flowing beneath the scalp. On a man's face, hair indicates sexual maturity.

Follicle
A small cavity or cul-de-sac; hair originates in a hair follicle.

Hair is formed from the division of specialized epidermal cells in the hair **follicle**, located in the dermis. Just as new epidermal cells push older cells outward, the growing hair shaft pushes older cells away from the blood supply. Beyond the epidermis, the hair shaft is composed of dead cells.

Human hairs are attached to a small slip of skeletal muscle called the **arrector pili** muscle (FIGURE 9.10), which can lift the hair erect, away from the skin. In fur-bearing animals, the arrector pili muscles raise the hairs to trap an insulating layer of air against the skin. Our body hair is too meager to maintain a layer of insulation. Raised fur along the spine of most mammals signals aggression, and we are taught to move slowly away from a dog whose "hackles are raised." But in humans, arrector pili muscles produce "chicken skin" or "goose bumps," which are no better at preserving heat than they are at signaling aggression! When our "hackles are raised," we do not look large or menacing, but we may feel our skin "crawl." This phenomenon is associated with fear because the arrector pili muscles are innervated only by the sympathetic nervous system.

Hair FIGURE 9.10

Hair is formed from pockets of epithelium that dive deep into the dermis. The hair follicle produces a hair shaft, composed of epithelial cells arranged in many layers. The innermost layer of the hair shaft contains the pigments that color our hair. The papilla of the hair follicle is what keeps the cells that form the hair shaft alive. Decreasing the blood flow through the papilla results in losing the hair shaft.

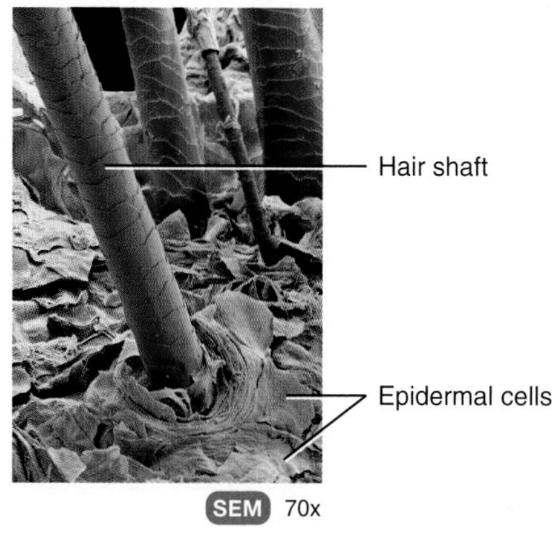

A Several hair shafts showing the shinglelike cuticle cells

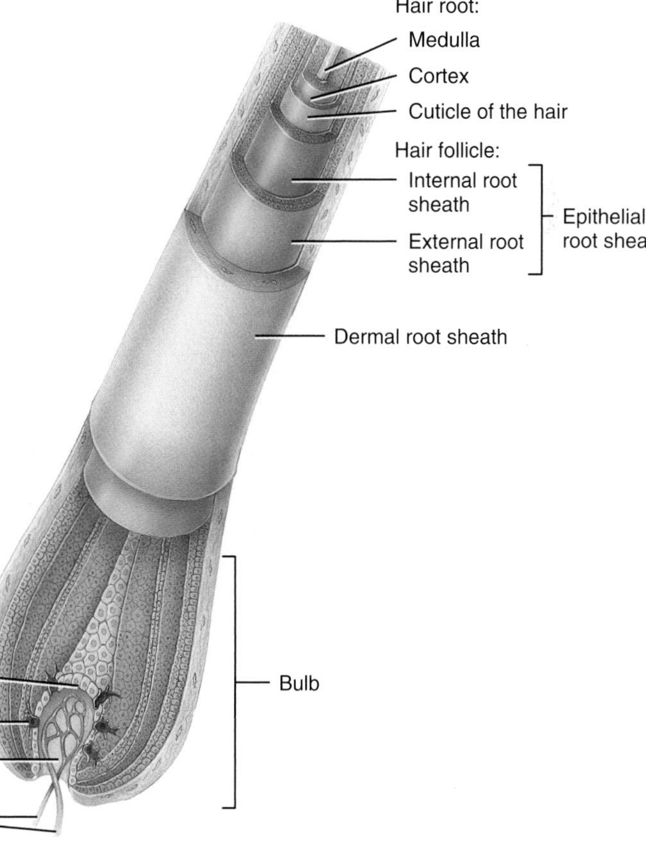

B Frontal section of hair root

I WONDER . . .

Could I be falsely accused based on fingerprint identification?

From an early age, we leam that we are individuals, unique in our own right. For example, no one else has the same pattern of whorls and ridges on their fingertips.

Fingerprints interested our ancestors. At least one cave painting shows the whorls and patterns on a hand. Babylonian merchants used thumbprints for identification in business transactions, and ancient Chinese merchants signed pots and plates with their thumbs. The first scientific mention of fingerprints was in 1686, when anatomist Marcelo Malphighi noted the intricate patterns of whorls and ridges. In 1823, another anatomist described nine patterns of fingerprints, but he did not claim that each person's prints are unique.

Fingerprints were first used to identify individuals in 1856 in India. A European magistrate, Sir William Hershel, began asking native Indians to press their inked hand to the back of legal contracts, such as bills of sale and court documents. Ironically, the Indians believed that touching the paper instilled some personal connection to the contract, which was, to them, more convincing than their signatures. Sir Hershel eventually dropped the palm print in favor of the middle and index finger, which seemed just as persuasive. In reviewing these "signatures," he found he could match people to their prints and also identify forgeries. Although he privately believed that fingerprints were unique, he did not publish his theory.

Not until 1880 did the scientific community accept the idea that individual fingerprints are unique. CSI (Crime Scene Investigation) was not far behind. Fingerprints were first used in criminal investigations in 1891 in Argentina, 1901 in England, and 1902 in the United States. By 1903, the United States was fingerprinting criminals for identification. The Federal Bureau of Investigation (FBI) operates a national fingerprint registry. Most civilian fingerprints are on paper, whereas military and criminal prints are now digitally stored.

To make a fingerprint identification, an examiner will compare two or more fingerprints and try to match a certain number of points on them. However the standards for identification are vague. In 1918, experts decided that finding 12 matching points on a fingerprint was conclusive, but this was not written into law, and no state requires a minimum number of matches. In some criminal cases, mismatches have caused wrongful accusations and convictions. After a 2004 terrorist bombing in Madrid, Spain, for example, an Oregon attorney was erroneously jailed due to a partial fingerprint identification.

With a large number of matching points, a fingerprint identification is likely to be correct, but because there is no absolute, scientific standard for fingerprint identification, courts should still require corroborating evidence to convict. Despite the claims that each person's fingerprints are unique, fingerprint identification is not nearly as scientific or as foolproof as DNA fingerprinting, a process that grew directly from the science of genetics.

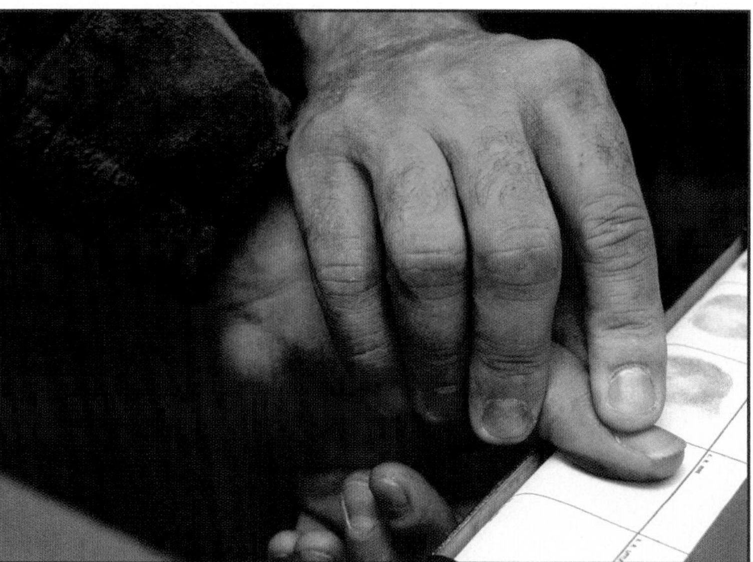

Axillary and groin hair that appears after puberty is thought to indicate reproductive status. Although we associate body odor with unhygienic practices, among many animals, scent can signal a readiness to reproduce. Our cultural practices of washing and perfuming our bodies remove these natural smells, making these patches of postpubescent hair more important as a visual reproductive cue than as an odor catcher.

NAILS REINFORCE THE FINGERS AND TOES

Keratinized Filled with keratin and therefore waxy.

Nails are flattened sheets of **keratinized** cells that protect the ends of the digits (FIGURE 9.11). Nails arise from a thick layer of specialized epithelial cells at the nail root called the **lunula** at the base of the nail bed. The **cuticle** is a layer of epidermis that covers the base of the nail. Nails protect the ends of the digits from physical damage as we move them through the environment.

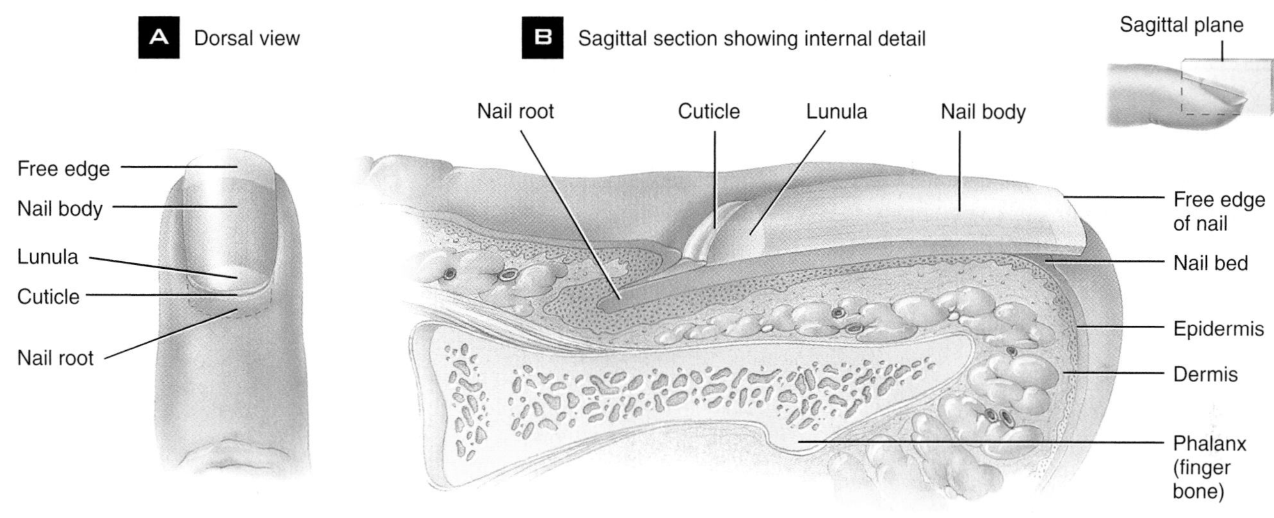

Nails FIGURE 9.11

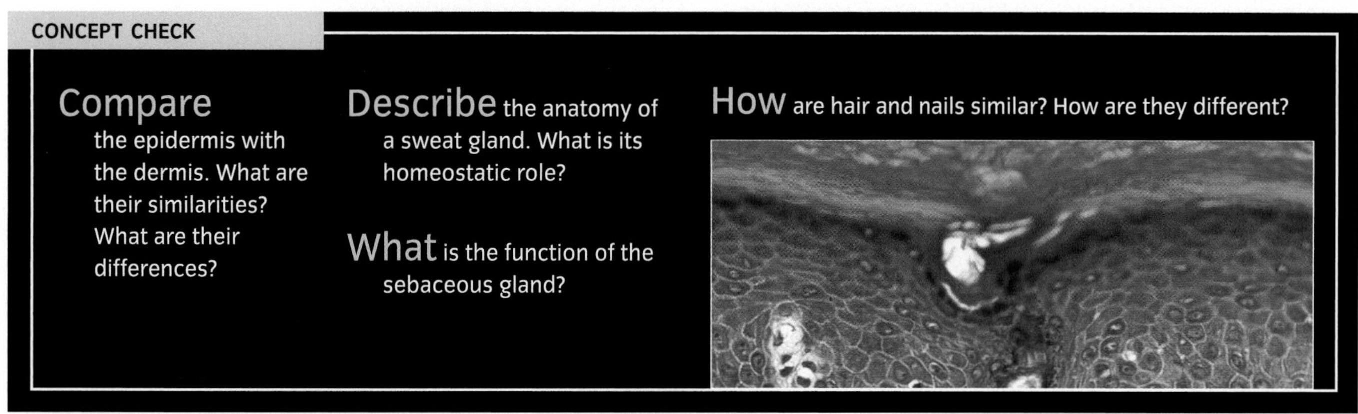

CONCEPT CHECK
Compare the epidermis with the dermis. What are their similarities? What are their differences?
Describe the anatomy of a sweat gland. What is its homeostatic role?
What is the function of the sebaceous gland?
How are hair and nails similar? How are they different?

We Have Other Innate Physical Barriers

LEARNING OBJECTIVES

Describe the four types of membranes in the body.

Explain how each membrane functions as a physical barrier.

The skin is our first line of defense against pathogenic invasions, but other **membranes** also serve as physical barriers against invasion. A membrane is a simple organ composed of a layer of simple or stratified epithelium supported by connective tissue (FIGURE 9.12).

Like the cutaneous membrane, mucous membranes provide nonspecific immunity. This is essential because mucous membranes line any cavity open to the exterior, including the mouth and digestive tract, the respiratory tract, the urinary tract, and the reproductive tract. Instead of being covered in keratinized dead cells, these tracts are covered in mucus that retards pathogens. The mucus, secreted by the epithelial cells of the membrane, constantly washes the membrane. Often, larger volumes of fluid wash these membranes as well. Urine flows across the urinary tract membrane; vaginal secretions flow out of the body across the mucous membranes of the female reproductive tract; and saliva continuously washes the oral cavity.

Serous membranes are found within the ventral body cavity and include the **peritoneum** lining the abdominal cavity, the **pericardium** lining the heart, and

Four types of membranes FIGURE 9.12

One of the four types of membranes lines every surface of the body, internally and externally. **Mucous** membranes line openings that communicate with the exterior. **Serous** membranes line sealed cavities not connected to the external environment. The **cutaneous** membrane is the skin. **Synovial** membranes line joints.

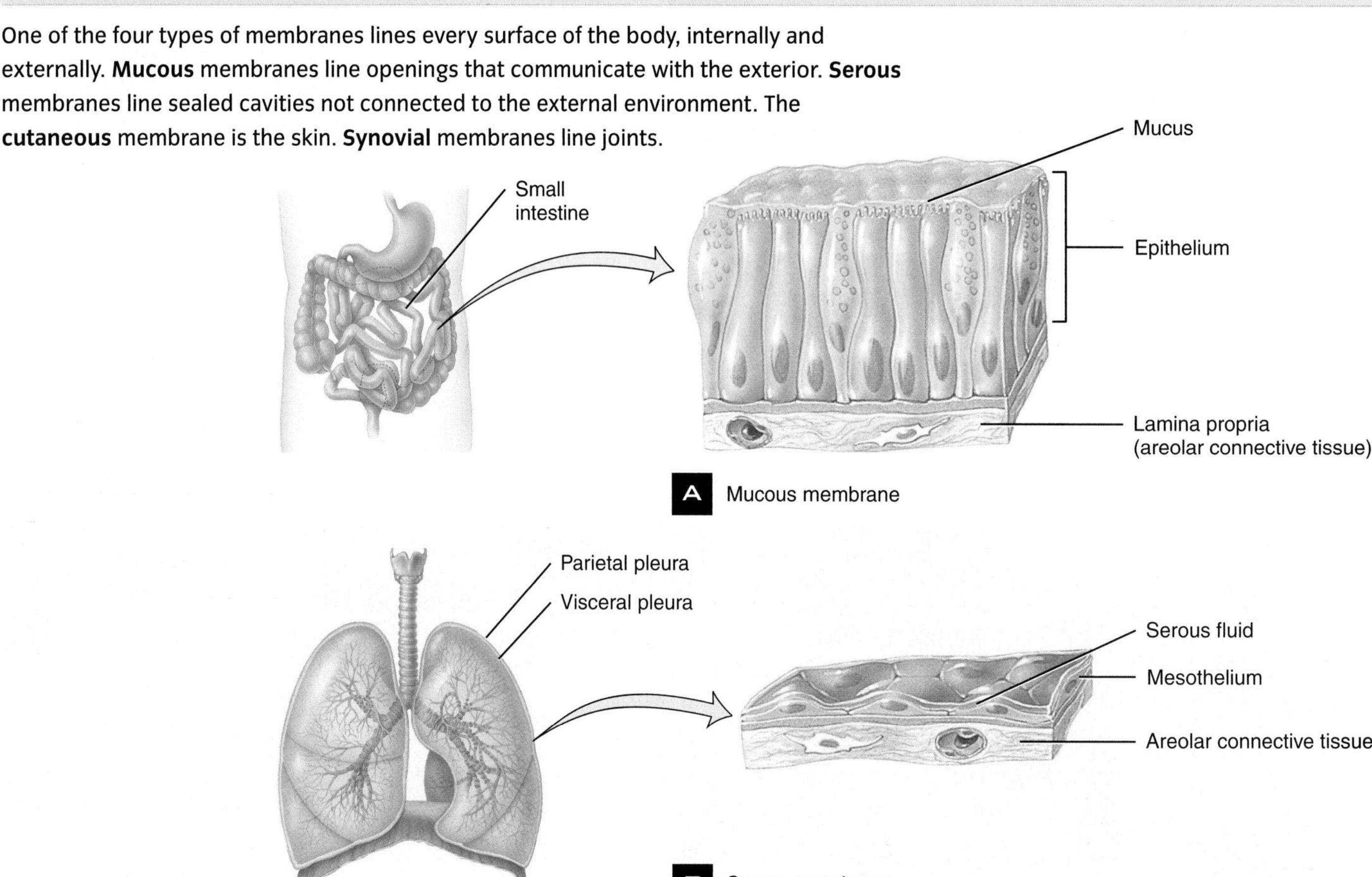

the **pleural** membrane lining the lungs. Serous membranes are double membranes, with one layer attached to the organ and the other to the wall of the cavity. Imagine punching a half-filled punching ball. The walls of the bag that surround your hand represent the inner membrane, and the rest of the bag becomes the outer membrane. In the body, serous fluid would lie between these layers, permitting smooth movement of the covered organ.

All movable joints, such as the knee or elbow, are lined with a synovial membrane. Like the serous membrane, synovial membranes secrete slippery, lubricating fluid. However, synovial membranes secrete fluid into the space the membrane surrounds rather than between the two layers of the membrane, as in serous membranes. The main function of both of these slippery membranes is to permit movement of underlying organs (the beating heart, the expanding lungs, the moving bones at a joint) without damaging nearby tissues. Synovial and serous membranes, unlike cutaneous and mucous membranes, do not function as physical barriers. (Synovial membranes were discussed in greater detail in Chapter 4.)

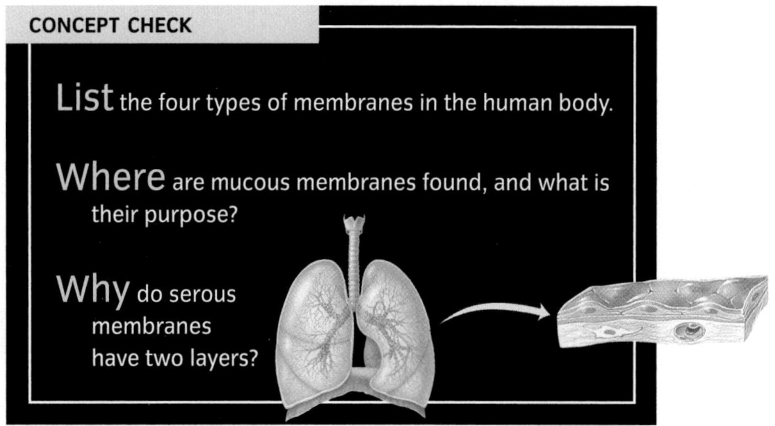

CONCEPT CHECK
List the four types of membranes in the human body.
Where are mucous membranes found, and what is their purpose?
Why do serous membranes have two layers?

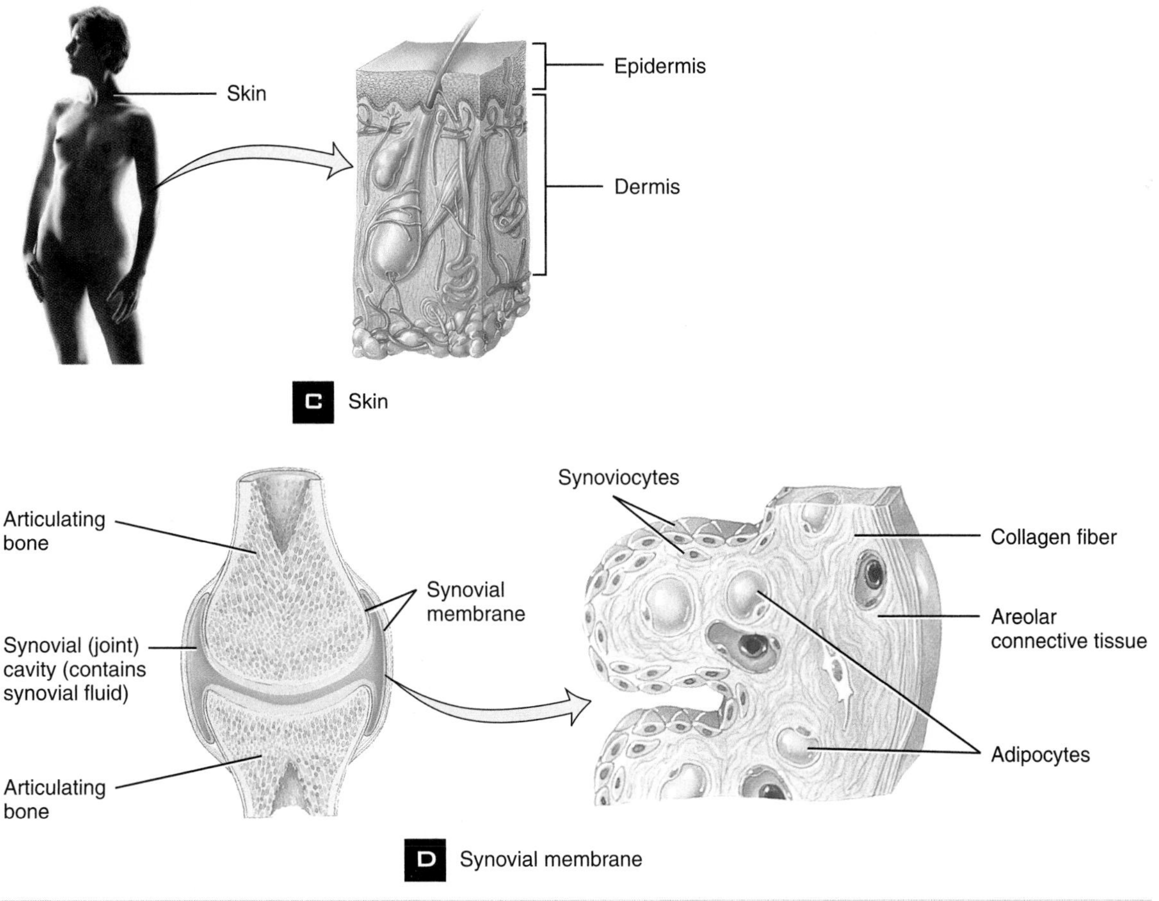

C Skin

D Synovial membrane

Chemical Barriers Can Defeat Bacteria

LEARNING OBJECTIVES

Describe the activities of the complement system and interferon.

Explain the function of local hormones as they relate to innate defenses.

espite the "fortress wall" of skin and mucous membranes, bacteria and other pathogens can often enter the body and cause homeostatic imbalances. When this happens, internal defenses immediately try to combat the pathogen. Innate defenses destroy pathogens without distinguishing between—or even recognizing—them. In contrast, specific immunity protects against

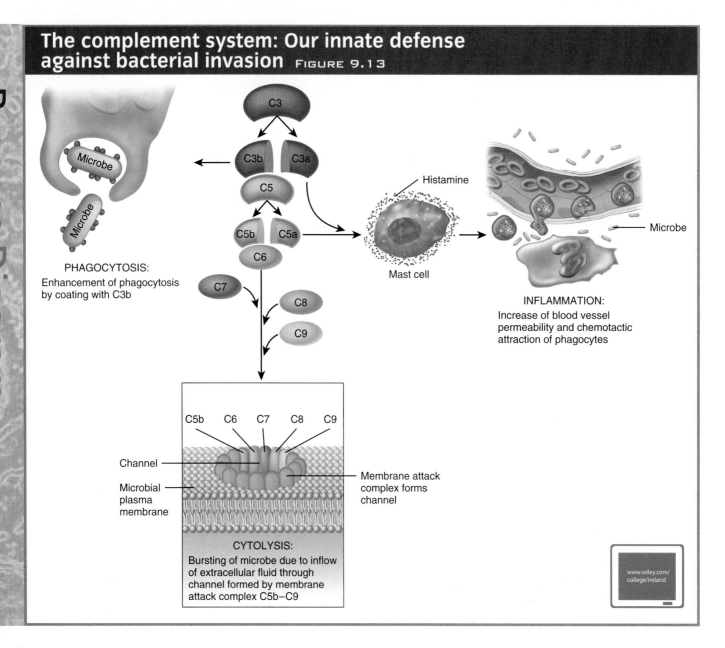

Process Diagram

The complement system: Our innate defense against bacterial invasion FIGURE 9.13

C3

C3b

C3a

C5

C5b

C5a

C6

C7

C8

C9

Microbe

Microbe

PHAGOCYTOSIS:
Enhancement of phagocytosis by coating with C3b

Histamine

Mast cell

Microbe

INFLAMMATION:
Increase of blood vessel permeability and chemotactic attraction of phagocytes

C5b C6 C7 C8 C9

Channel

Microbial plasma membrane

Membrane attack complex forms channel

CYTOLYSIS:
Bursting of microbe due to inflow of extracellular fluid through channel formed by membrane attack complex C5b–C9

www.wiley.com/college/ireland

Interferon fights viral invasion FIGURE 9.14

Cells produce interferon to help ward off a viral infection.

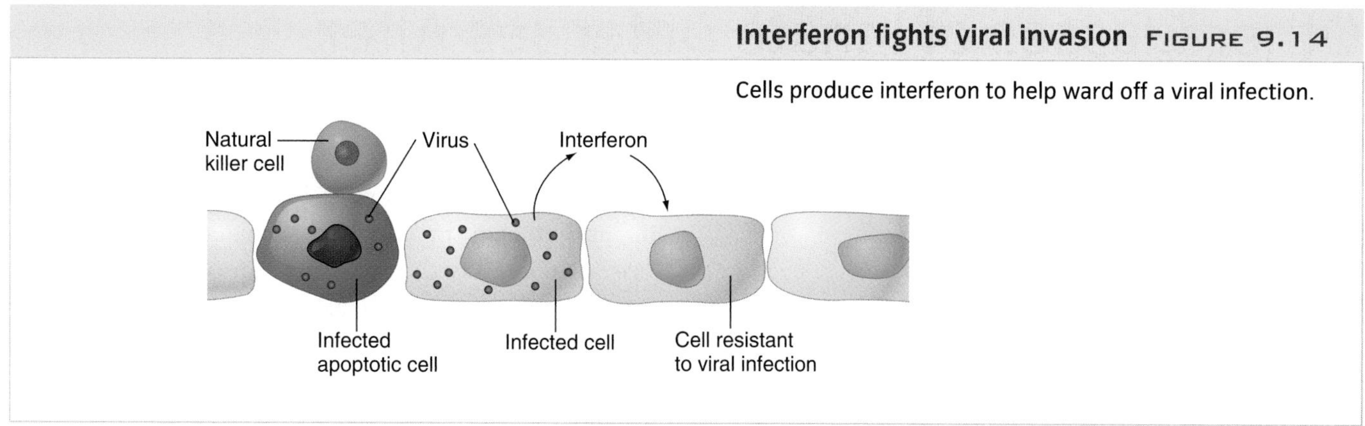

particular threats. Specific immunity must be acquired through contact with the pathogen or purposeful introduction of the pathogen to the immune system with *vaccination*.

Our nonspecific chemical defense against bacteria is called the **complement system** (FIGURE 9.13). This series of chemical reactions brings together a group of proteins that are usually floating freely in the plasma. These proteins are stacked in a specific order to create a "complement" of proteins that functions like an antibacterial missile. When a bacterial invasion is encountered, the complement complex assembles, attaches to the bacterial walls, and impales the cell with the protein complex. With the bacterial wall breached, osmotic pressure forces water into the bacterium, destroying its chemistry and killing it.

Complement is effective against bacteria but not viruses. When cells are infected with a virus, one defensive response is to produce **interferon** (FIGURE 9.14). Interferon is a "local" or paracrine hormone that is secreted to affect nearby cells. It is a chemical warning, similar to the tornado warning sirens of the Midwest or the tsunami warnings in coastal communities. When cells detect interferon in the extracellular fluid, they prepare for viral invasion. Ideally, the viral infection can then be limited to a small area, allowing it to run its course with little effect on overall body functioning.

CONCEPT CHECK

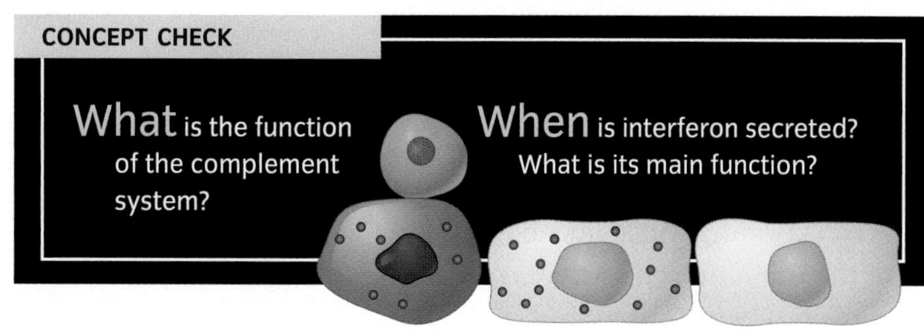

What is the function of the complement system?

When is interferon secreted? What is its main function?

Other Classes of Innate Defenses Alter the Environment Around the Pathogen

LEARNING OBJECTIVES

List the innate defenses not categorized as physical or chemical barriers.

Briefly explain the physiology of fever.

Three other classes of nonspecific defenses also function to destroy pathogens without distinguishing among them: fever, inflammation, and phagocytes.

Fever is defined as a change in the body's temperature set point, resulting in an elevation in basal body temperature above 37.0°C (98.6°F). Proteins called **pyrogens** reset the body's thermostat to a higher temperature. Fever may harm the pathogen directly, but more likely it aids defensive mechanisms by raising the metabolic rate. For every 1°C rise in body temperature, your metabolic rate increases by 10 percent. At elevated temperatures, enzymes and repair processes work faster, cells move more quickly, and specific immune cells are mobilized more rapidly. In addition, your spleen **sequesters** (holds) more iron at higher temperatures, which many bacteria require to reproduce.

The adage "feed a fever" is correct. Fever elevates your basal metabolic rate, increasing your use of energy. Unless you replenish your energy supplies, you will tire quickly, which will increase the homeostatic imbalance created by the pathogen. Feeding your body will aid the recovery process by providing nutrients necessary for the functioning of the immune cells, whereas not eating may deplete your reserves and give the pathogen the upper hand.

Inflammation is similar to fever in its goal, but it is a localized, not whole-body, method for increasing enzyme function. **In situ** (in place) swelling, redness, heat, and pain are associated with inflammation. Damaged or irritated cells release **prostaglandins**, proteins, and potassium, which trigger inflammation when released into the interstitial fluid. The benefits of inflammation include temporary tissue repair, blockage of continued pathogen entry, slowing of pathogen spreading, and quicker repair of the damaged tissue. The redness associated with inflammation of the skin shows how capillaries become "leaky," allowing blood to bring immune-system cells and various compounds to injured or diseased tissues.

Inflammation can be triggered by many factors, including pathogen entry, tissue abrasion, chemical irritation, or even extreme temperature (FIGURE 9.15). For example, mosquito bites stimulate inflammation in almost everyone. The red, hot, itchy welt ac-

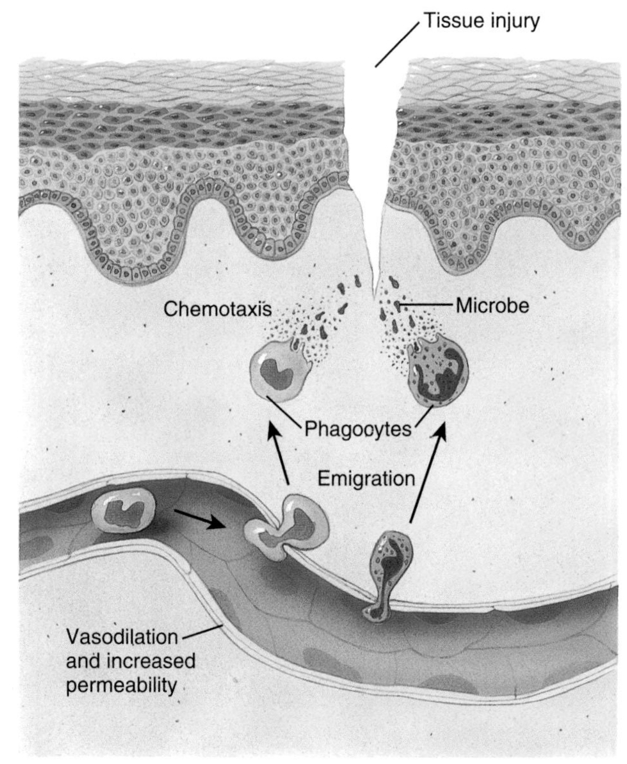

Phagocytes migrate from blood to site of tissue injury

Inflammation FIGURE 9.15

Inflammation results from the combined effects of several chemicals released at the injury site. Chemical messengers bring the phagocytes to the injury, where they either release compounds that destroy the pathogen or **phagocytose** (consume) the pathogen.

tually represents a local inflammation resulting from the lady mosquito's poor table manners. As she completes her meal and withdraws her proboscis, she spits into the skin, releasing cellular debris and salivary chemicals that initiate an inflammatory response.

Phagocytes are a final nonspecific defense for dealing with stressors. The root *phago* means "to eat"; and you already know that *cyte* translates to "cell." Phagocytes, therefore, are eating cells, or cells that wander through the tissues, engulfing and removing anything that does not belong there. Phagocytes, the first cellular line of defense against pathogens, remove all dead or dying cells, cellular debris, and foreign material, which makes them a nonspecific defense.

Phagocytes come in different sizes. **Microphages** are quite small and are mainly found in the nervous system. **Macrophages** are large, actively patrolling cells. They arise from blood cells and travel through every tissue looking for foreign material (FIGURE 9.16). Macrophages escape the bloodstream by squeezing between the cells of the vessel wall, a process called **diapedesis**. Some tissues have resident, or "fixed," macrophages, whereas other tissues get patrols of wandering macrophages passing through, like security guards making the rounds at a mall.

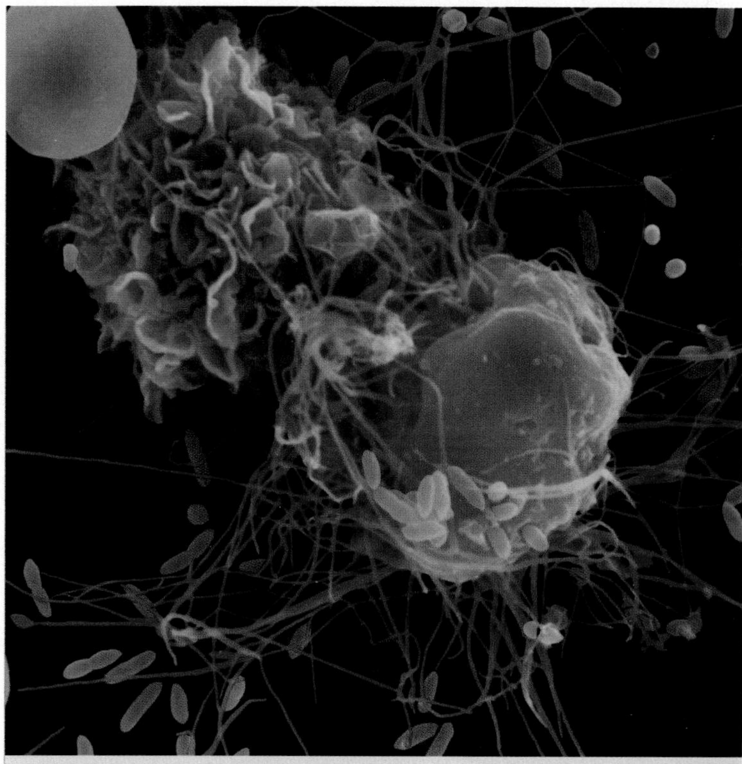

Macrophage eating technique FIGURE 9.16

Both microphages and macrophages are attracted to pathogens and damaged cells via chemical messengers. Once they locate a pathogen or damaged cell, they surround, engulf, and destroy it. Some phagocytes are capable of continuous removal of pathogens and cellular debris, whereas others have a limit on how much they can ingest. Once they reach that limit, the phagocyte itself dies and must be gotten rid of. Pus is actually dead phagocytes, filled with cellular debris from the wound they were helping to clean.

In the opening of this chapter, we took a look at the stresses associated with war. Fortunately, not all of us will experience that level of stress, but we all face stress in our lives, and most of us cope with it successfully. The routine activities of the body dealing with stress and pathogens have amazing complexity and efficiency. Usually we are conscious only of our nervousness or tiredness as we deal with stresses, not the myriad defensive activities taking place in our bodies.

Although these defenses seem able to cover any pathogen that comes along, humans still succumb to illness. Stresses can overwhelm our defense mechanisms, preventing our nonspecific defenses from protecting us. Pathogens can slip past our protective membranes and overcome the effects of complement or interferon, inflammation, or fever, escaping detection until the problem is too large for nonspecific responses. At that point, the body requires specialists from the immune system, which is the topic of the next chapter.

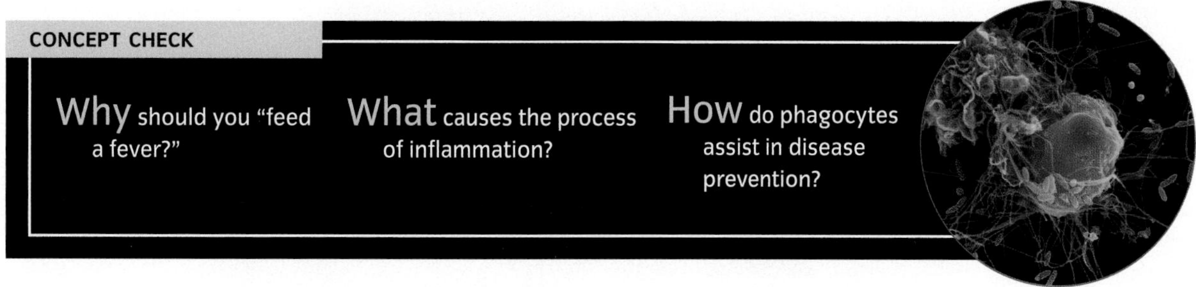

CONCEPT CHECK
Why should you "feed a fever?"
What causes the process of inflammation?
How do phagocytes assist in disease prevention?

CHAPTER SUMMARY

1 What Is Stress?

Humans face many types of stress from physical, emotional, social, or microbial sources, and we have many systems to deal with them, including the skin, whole-body and localized reactions, and a variety of chemical and physical mechanisms to reduce, eliminate, or survive stress.

2 The General Adaptation Syndrome Helps Overcome Stress

The body responds to stress with the three stages of the General Adaptation Syndrome: alarm, resistance, and exhaustion. During alarm, the fight-or-flight mechanism predominates. This stage tries to remove the body from the stressor. If unsuccessful, the resistance phase begins. During this phase, blood ion concentrations are pushed far from homeostasis in an attempt to maintain elevated blood glucose. Should resistance continue for a prolonged period, the body will reach exhaustion. During exhaustion, the body retreats from the fight and tries to recover from the altered ion balances created in the previous stage. At this stage, organ systems fail and the organism can die.

3 The Skin Is the Primary Physical Barrier

The skin is composed of the stratified squamous cells of the epidermis and underlying connective tissues of the dermis. Hair, nails, and glands are accessory organs. Sensory structures in the dermis detect pressure, temperature, and pain. Glands secrete either oils or sweat onto the surface of the skin and hairs. The sweat glands help maintain thermal homeostasis. Nails and hair serve protective functions.

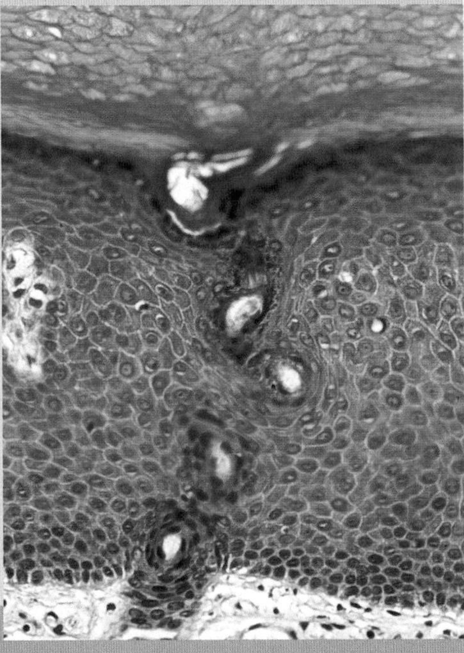

4 We Have Other Innate Physical Barriers

Four types of membranes provide an important part of innate immunity. The cutaneous membrane is our skin. Serous membranes in the ventral body cavity permit movement of underlying organs, and include the peritoneum lining the abdominal cavity, the pericardium lining the heart, and the pleural membrane lining the lungs. Mucous membranes provide nonspecific immunity in cavities open to the exterior, including the mouth, digestive tract, respiratory tract, urinary tract, and reproductive tract. Mucus, secreted by the epithelial cells of the membrane, retards pathogens on mucous membranes. Synovial membranes secrete a slippery, lubricating fluid to permit movement of underlying bones at joints.

5 Chemical Barriers Can Defeat Bacteria

The complement system fights bacteria by destroying their cell walls. Interferon, secreted by cells that are infected by a virus, is a chemical warning that helps nearby cells prepare for viral invasion.

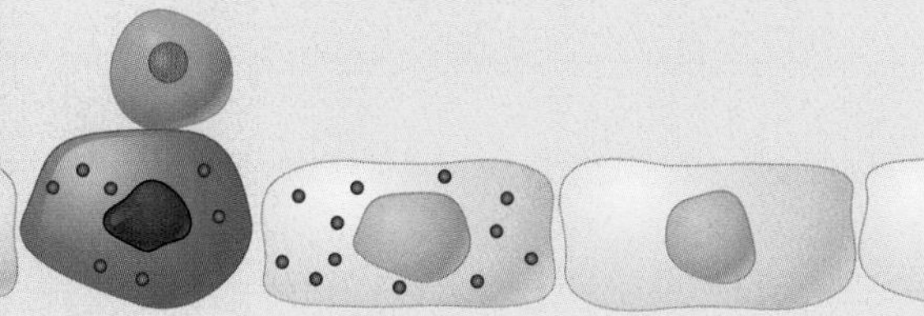

6 Other Classes of Innate Defenses Alter the Environment Around the Pathogen

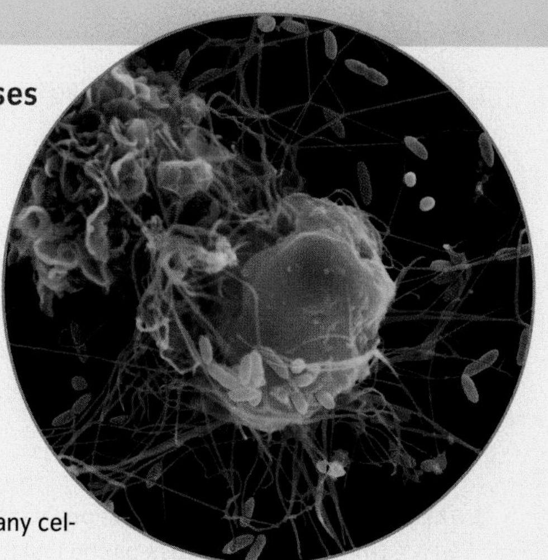

Fever raises the body temperature so that chemical reactions will act more quickly, and it is therefore effective against a wide range of threats. Inflammation is a series of reactions that allow more blood to reach the site of infection to help with tissue repair, block the entry of more pathogens, and slow the spread of pathogens. Phagocytes are cells that remove circulating pathogens, as well as any cellular debris created during infections.

KEY TERMS

- **apocrine** p. 268
- **complement system** p. 257
- **dermis** p. 262
- **eccrine** p. 268
- **epidermis** p. 262
- **epinephrine** p. 258
- **follicle** p. 269
- **interferon** p. 257
- **keratinized** p. 271
- **melanocytes** p. 263
- **nociceptors** p. 265
- **pathogen** p. 257
- **phagocytes** p. 257

CRITICAL THINKING QUESTIONS

1. Marie sat quietly in the back of the class feeling relaxed, even though this was her first college class. "Here goes; this is the beginning of my future," she excitedly thought. As the teacher walked to the front of the room, Marie suddenly felt dizzy and broke into a cold sweat. What was happening to her? What is the natural course of these events?

2. Swimming in the ocean may expose a bather with an open wound to staphylococcus infection. What characteristics of the skin normally prevent these infections? How does an open wound compromise these defenses?

3. Everyone gets a common cold once in a while. Usually, this is caused by a rhinovirus (nasal virus) that may infect the nose, sinuses, ears, and/or bronchial tubes (lungs). What defenses must this virus overcome to cause an infection? If you wanted to manufacture a compound to defend against the common cold, which natural biochemical defense(s) would you try to mimic?

4. One of the first symptoms of menopause in women reaching the end of their reproductive years is "hot flashes." Without warning, hot flashes raise the body temperature and cause sweating and thermal discomfort. From what you understand of fever, what is happening biochemically during a hot flash? Could a hot flash help defend against pathogens? What is the difference between the raised temperature of a hot flash and a true fever?

5. Suppose you lacked all innate or nonspecific defenses. First, list exactly what you would be missing. Second, for each item, describe how life would be different without that mechanism. For as many of the listed items as possible, invent some behavioral changes that would promote your survival.

SELF TEST

1. Which of the following can be classified as stressors?
   a. Eating a heavy meal
   b. Coming down with strep throat
   c. Beginning a new college semester
   d. All of the above are stressors.

2. Innate immunity includes all of the following EXCEPT
   a. skin and mucous membranes.
   b. phagocytes.
   c. antibodies and immune cells.
   d. complement system.

3. The phase of the General Adaptation Syndrome that begins with a large dumping of epinephrine into the system is
   a. the alarm phase.
   b. the resistance phase.
   c. the exhaustion phase.
   d. All of the phases include dumping epinephrine.

The following six questions all relate to this figure.

4. Identify the structure labeled B on this diagram.
   a. Epidermis
   b. Hypodermis
   c. Dermis
   d. Adipose tissue

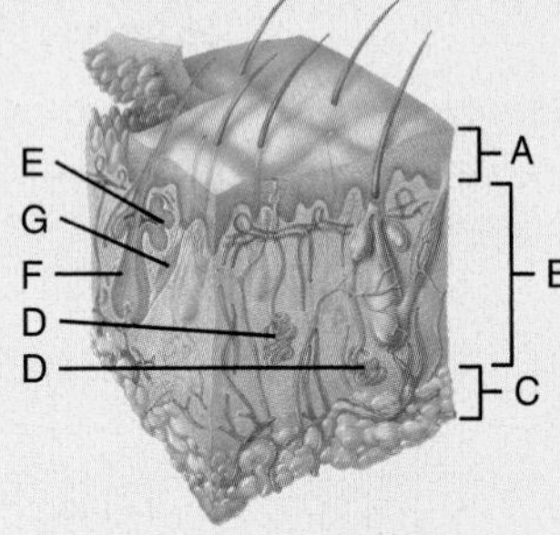

5. The function of the structure indicated by E is to
   a. produce sweat.
   b. raise the hair follicle.
   c. produce oil.
   d. protect the dermis.

6. Which structure is directly responsible for thermal homeostasis?
   a. A
   b. C
   c. D
   d. G

7. The nonvascular layer of skin is indicated by the letter
   a. A
   b. B
   c. C
   d. E

8. The sebaceous gland, which produces oil that lubricates and softens, is indicated by the letter
   a. B
   b. C
   c. D
   d. E

9. Dead cells filled with keratin would be found in which area of the skin?
   a. A
   b. B
   c. C
   d. G

10. The function of melanocytes is to
    a. produce keratin.
    b. maintain internal temperature.
    c. produce dark pigments to absorb light.
    d. store energy for later use.

11. The sensory organs in the dermis that detect light touch are the
    a. Meissner's corpuscles.
    b. Merkel discs.
    c. Pacinian corpuscles.
    d. nociceptors.

12. True or False? Oil glands are located everywhere on the skin, including the face and lips.

13. This image shows
    a. an oil gland.
    b. an adipose cell.
    c. a sebaceous gland.
    d. an eccrine sweat gland.

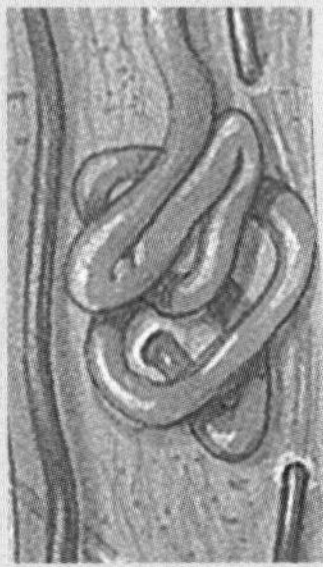

14. The two structures that protect and reinforce the skin are the
    a. hair follicles and nails.
    b. sweat glands and hair follicles.
    c. nails and oil glands.
    d. sweat and oil glands.

15. Which type of membrane is shown in figure A below?

a. Mucous membrane
b. Serous membrane
c. Cutaneous membrane
d. Synovial membrane

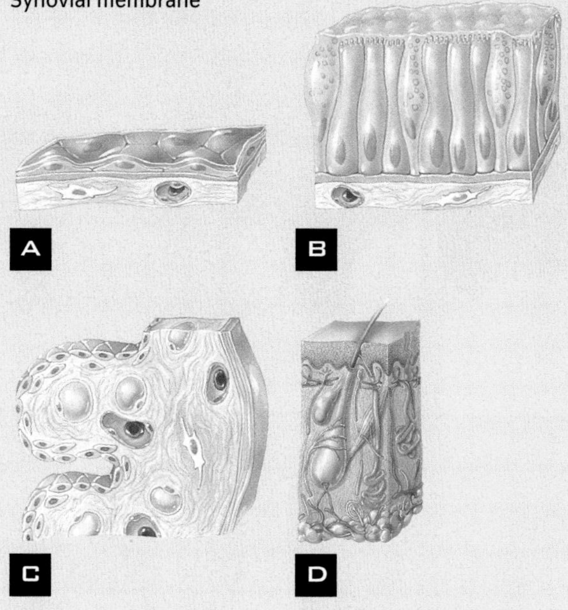

16. The chemical defense that destroys bacteria is called

a. immunity.
b. the complement system
c. interferon.
d. phagocytosis.

17. The idea behind ___________ is that the temperature increase they cause will raise metabolic rate and speed activity of the immune system.

a. prostaglandins
b. interferons
c. pyrogens
d. complement systems

18. The type of innate defense against pathogens seen in this figure is

a. inflammation.
b. fever.
c. interferon.
d. phagocytosis.

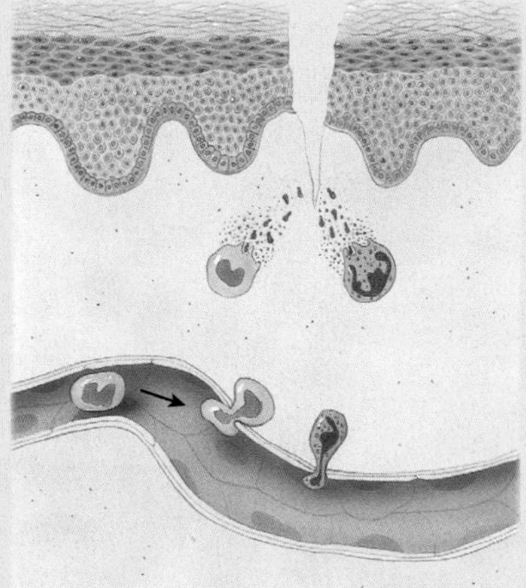

19. The job of a phagocyte is to

a. patrol tissues and remove pathogens.
b. patrol and monitor the health of the nervous system.
c. remove dead or dying cells.
d. All of the above.

20. The innate defense classified as a chemical barrier is

a. the cutaneous membrane.
b. phagocytes.
c. fever.
d. interferon.

The Lymphatic System and Specific Immunity

10

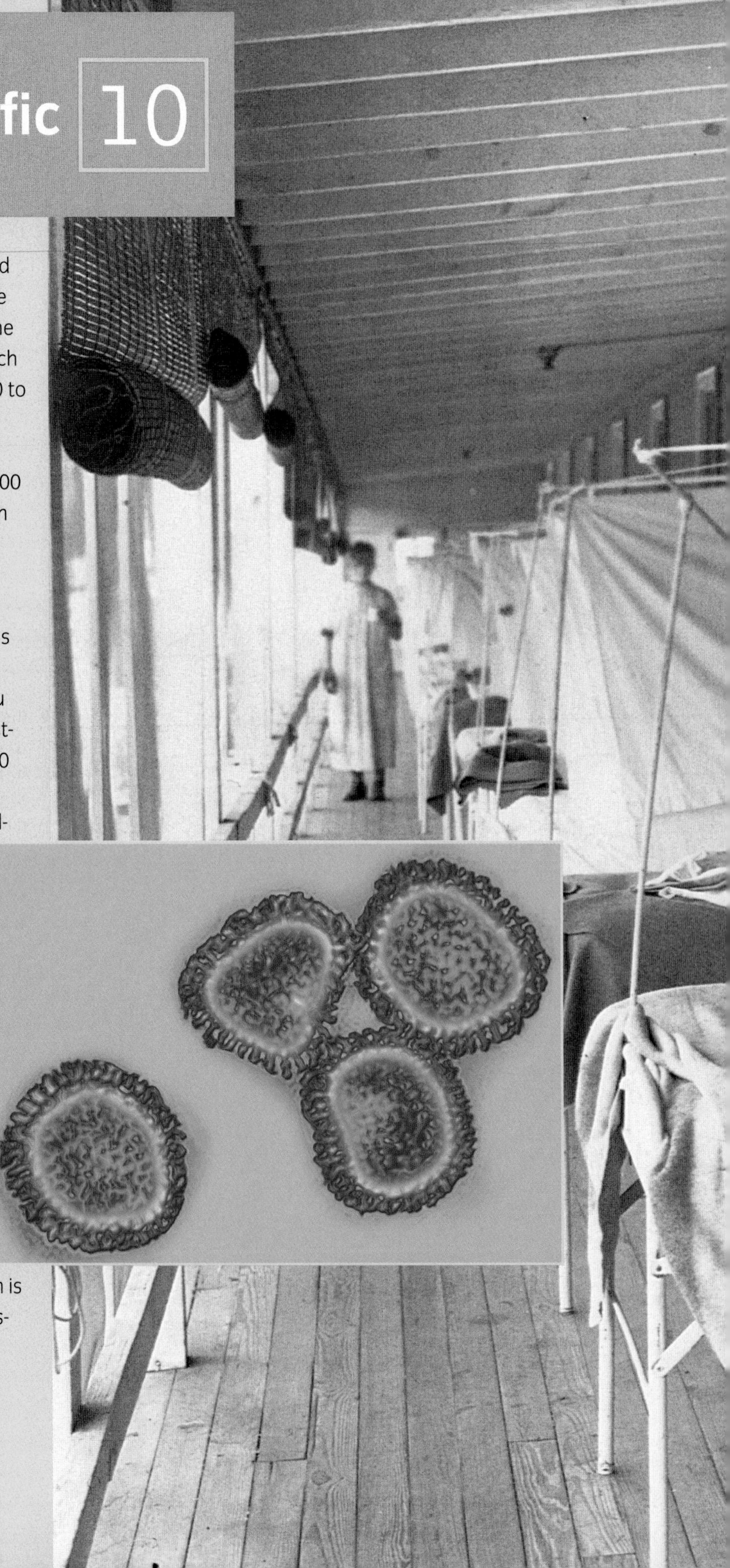

World War I was such a horrific event in world history that it continues to overshadow the worst disease outbreak of the twentieth century. The 1918–1919 worldwide epidemic of Spanish flu, which broke out toward the end of the Great War, killed 20 to 50 million people, compared to an estimated 15 million civilians and soldiers killed in WW I. In the United States, Spanish flu killed an estimated 675,000 people. Scientists think the flu virus was harbored in birds and then, for reasons unknown, "jumped" to humans and began to spread.

The 1918 Spanish flu disaster is a history that could be repeated: A new and deadly influenza virus is now circulating among wild and domestic birds. Traveling with migratory birds, this so-called bird flu has already reached Turkey, Egypt, Nigeria, and Eastern Europe. The avian virus has killed more than 100 people, mainly in Southern Asia. To prevent its spread, public health authorities have had to kill millions of chickens, ducks, and other fowl.

As we write, the avian influenza virus does not have the ability to spread from one person to another, so all patients have been infected through close contact with an infected bird. But if avian flu becomes able to spread directly from one person to another, millions could die before a vaccine is developed.

Luckily, humans are not completely defenseless against disease. We have already seen that all of us have general defenses against pathogens. A second, more specific, arm of this defense is found in the lymphatic system. The intricate interplay of antibodies, killer cells, and memory cells in the immune system is the focus of this chapter. Without the lymphatic system, humans would have died out long before the 1918 pandemic.

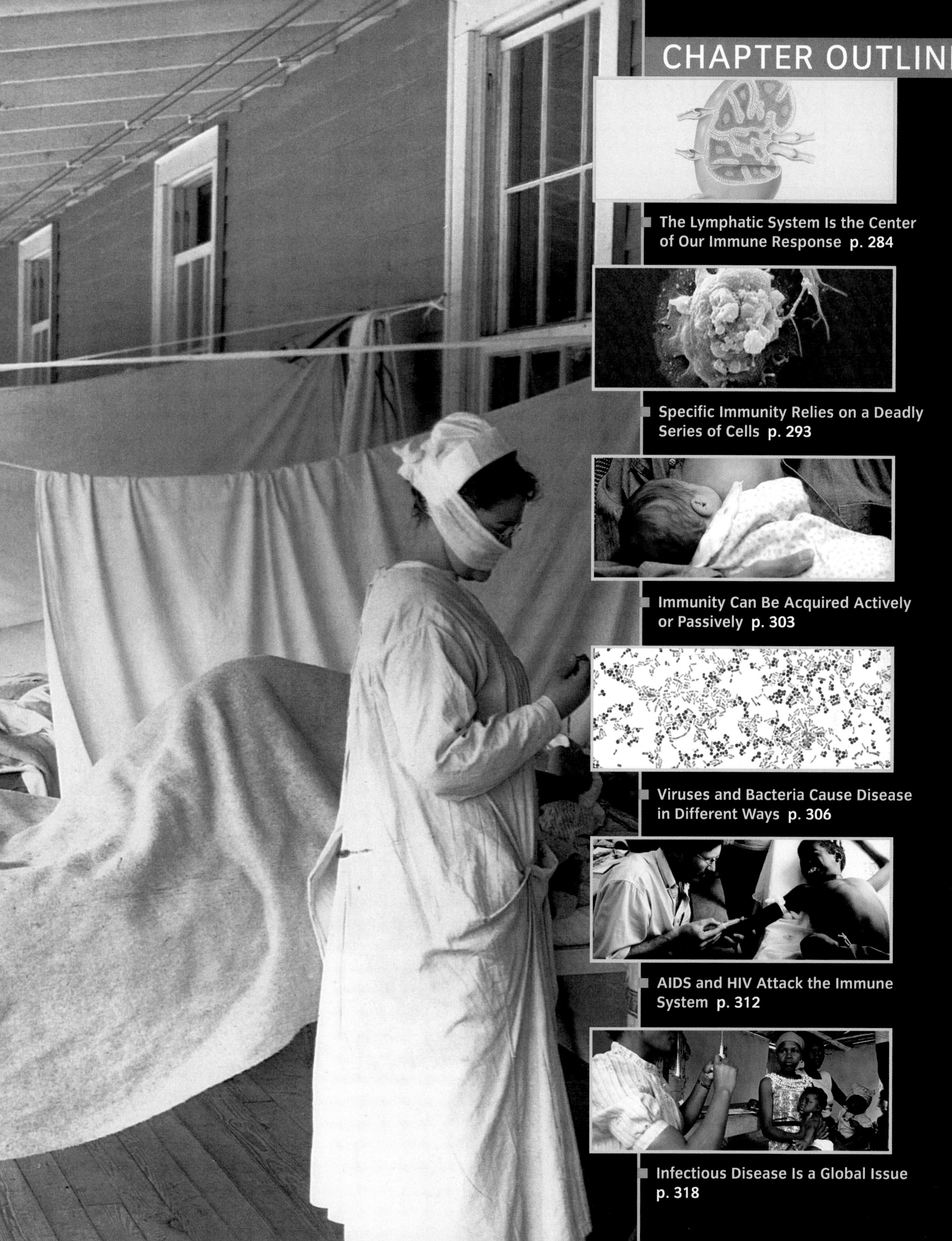

CHAPTER OUTLINE

The Lymphatic System Is the Center of Our Immune Response

Learning Objectives

Identify the structures of the lymphatic system.

Describe the flow of lymph.

To a pathogenic bacterium, humans are a walking meal of proteins, sugars, fats, and other good things to eat. To a virus, we are an uncountable number of cells that can be converted into "factories" for making thousands of new viruses. But despite the huge array of pathogens waiting to infect us, most of the time, most of us are healthy.

Immune response The disease-fighting activity of an organism's immune system.

Lymphatic system The tissues, vessels, and organs that produce, transport, and store cells that fight infection.

When nonspecific defenses such as those discussed in Chapter 9 prove inadequate, our body can employ more selective defenses against disease. This defense, called our **immune response**, is governed by the **lymphatic system**. The immune response is acquired, not innate, meaning that it is a conditioned or "learned" reaction of the lymphatic system. Whereas the innate defenses function the same way regardless of the pathogen, the immune response is specific. Each pathogen triggers a slightly different reaction, and the immune system must "learn" to identify each pathogen through experience.

The lymphatic system helps explain why we rarely need medical help to combat infectious disease, and how we benefit from vaccinations. The lymphatic system is complicated but lovable. Without its good offices, you likely would not be studying human biology today. Instead, you would be long gone.

THE LYMPHATIC SYSTEM REACHES MOST OF THE BODY

The organs of the lymphatic system include the tonsils, spleen, thymus, lymph nodes, and the Peyer's patch glands of the digestive system. Connecting these organs is a network of lymphatic vessels that collect lymph from the tissues and deposit it in the bloodstream.

The lymphatic system is composed of lymphatic vessels and lymphoid organs (FIGURE 10.1). Like the circulatory system, the lymphatic system touches most of the body and carries out both transportation and homeostatic services. You are probably familiar with the lymph nodes, those small, bean-shaped structures that you may feel alongside your Adam's apple when you have a sore throat. You may be surprised to learn that you have lymph nodes elsewhere, including your intestinal tract and chest. These lymph nodes function in concert with lymphatic tissue, organs, and vessels to (1) return excess fluid from the tissues to the bloodstream, (2) absorb fats from the intestine and transport them to the bloodstream, and (3) defend the body against specific invaders.

Your tissues are bathed in **lymph**, a clear fluid that is called interstitial fluid when it is found in the **interstices** between cells. Chemically, lymph is quite similar to blood plasma, which makes sense because lymph originates in fluid that diffuses from the capillaries into the tissue. If you scrape your epidermis, say when you "skin your knee," clear interstitial fluid will bead up on the exposed dermis. Normally, lymphatic vessels collect this fluid

Interstices The small fluid-filled spaces between tissue cells.

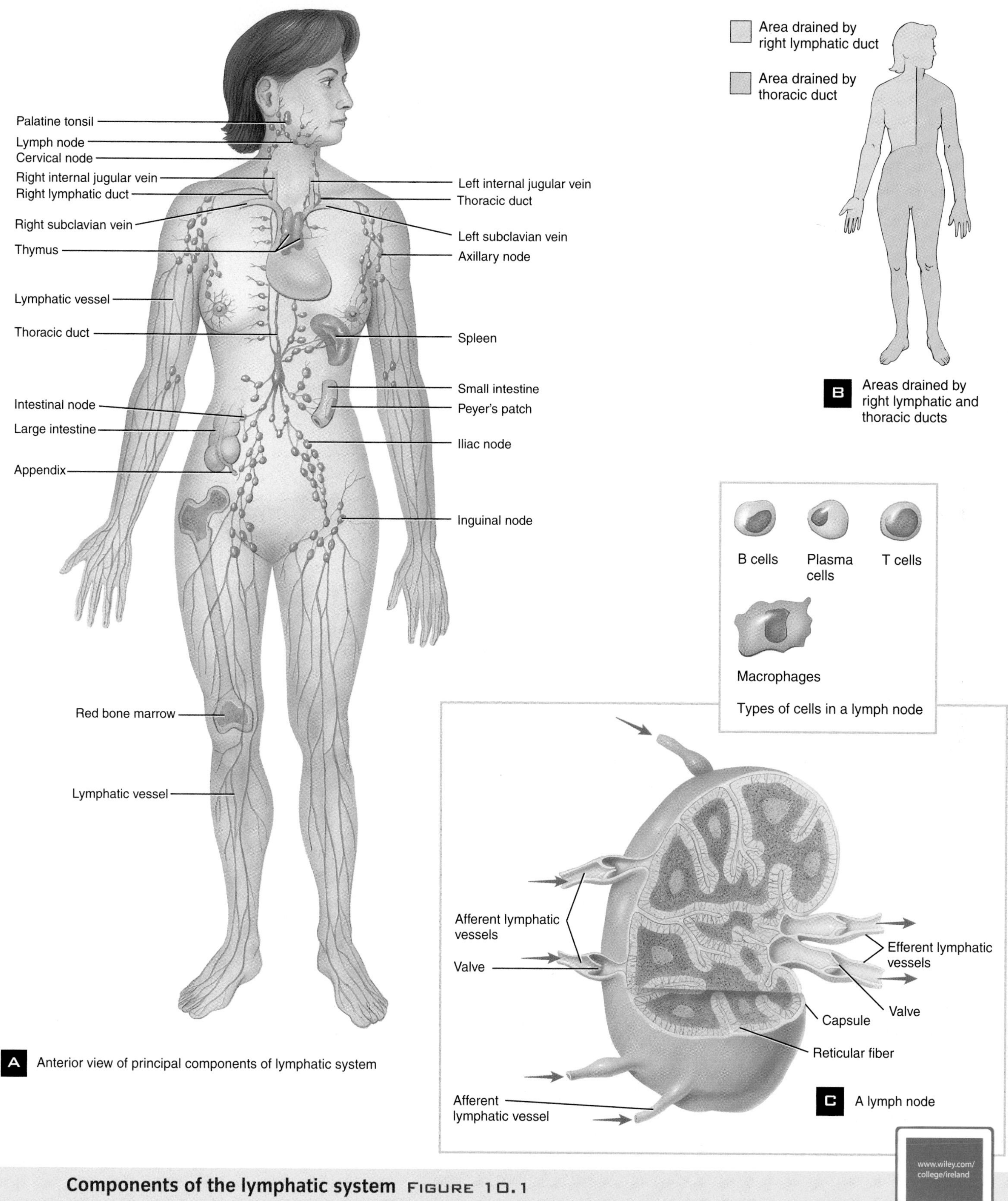
Palatine tonsil
Lymph node
Cervical node
Right internal jugular vein
Right lymphatic duct
Right subclavian vein
Thymus
Lymphatic vessel
Thoracic duct
Intestinal node
Large intestine
Appendix
Red bone marrow
Lymphatic vessel
Left internal jugular vein
Thoracic duct
Left subclavian vein
Axillary node
Spleen
Small intestine
Peyer's patch
Iliac node
Inguinal node
A Anterior view of principal components of lymphatic system
Area drained by right lymphatic duct
Area drained by thoracic duct
B Areas drained by right lymphatic and thoracic ducts
B cells
Plasma cells
T cells
Macrophages
Types of cells in a lymph node
Afferent lymphatic vessels
Valve
Efferent lymphatic vessels
Valve
Capsule
Reticular fiber
Afferent lymphatic vessel
C A lymph node
www.wiley.com/college/ireland

Components of the lymphatic system FIGURE 10.1

Process Diagram

Formation of lymph FIGURE 10.2

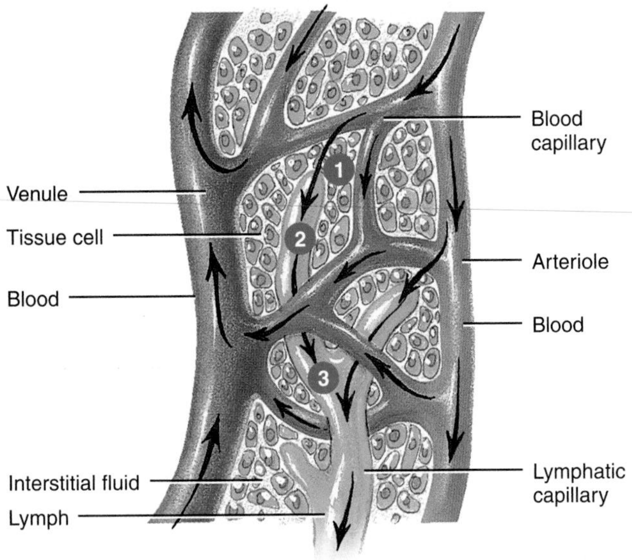

Relationship of lymphatic capillaries to tissue cells and blood capillaries

1. Blood pressure forces the fluid portion of the blood out at the capillaries, bathing the tissues.
2. The excess fluid is then forced into the lymphatic capillaries from the tissues by fluid pressure and osmotic pressure.
3. The fluid already in the lymphatic vessel opposes the mass movement of tissue fluid into the lymphatic system, helping to keep the tissues moist.

www.wiley.com/college/ireland

for return to the bloodstream. When interstitial fluid is inside lymph vessels, we call it lymph (FIGURE 10.2).

LYMPHATIC CAPILLARIES AND VESSELS RESEMBLE A PARALLEL CIRCULATORY SYSTEM

The lymphatic system has many similarities to the circulatory system because both systems reach almost every cell in the body. Because interstitial fluid is so widespread, lymphatic **capillaries** (very small vessels) are also found throughout the body. Often the lymph in these capillaries is filled with ingested fats, turning the vessel milk white. When lymphatic veins and capillaries were first discovered, they were called "wee milked veins" for this reason.

In the circulatory system, capillaries are part of a closed system that takes blood from the heart to the body and back to the heart. In contrast, lymphatic capillaries are one-way tubes. They are part of an **open system** where vessels lead from the tissues to the bloodstream but not in the opposite direction.

Open system
A system with a starting point and an ending point rather than a continuous circular flow.

Unlike the circulatory system, the lymphatic system has no central pump. Lymph flows through tissues and into lymphatic capillaries mainly because of the squeezing action of skeletal muscles. As muscles contract, they shorten and thicken, forcing excess fluid from the muscular tissue and surrounding organs into the lymphatic capillaries. Lymphatic capillaries allow fluid to enter but not to exit because their walls are composed of cells positioned with slight overlaps (FIGURE 10.3). Pressure from outside the vessel parts the cells so that fluid can enter the lumen (center) of the capillary. Fluid pressure inside the capillary presses the cells shut so that the fluid cannot escape. This action is rather like your front door. If you push on one side, the door will open, but if you push from the other side, it will only close tighter.

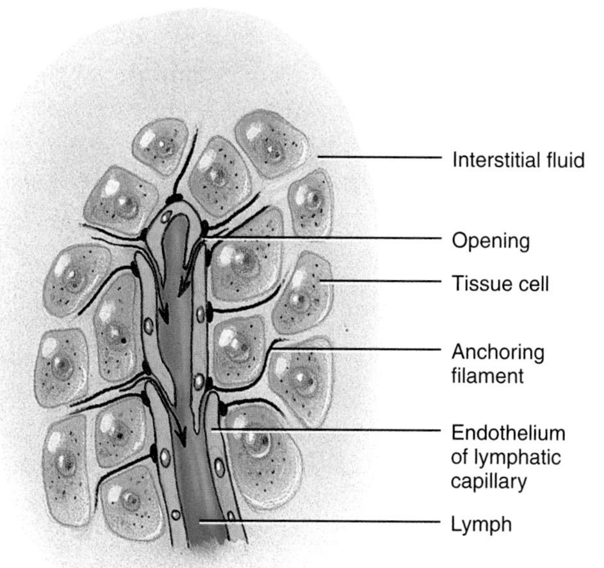

Lymphatic capillary walls
FIGURE 10.3

Lymphatic vessels are similar to the veins, which are thin-walled, flexible, and not built to withstand much pressure. Because lymph flows through the lymphatic system without being pumped, larger lymphatic vessels require **valves** to prevent backflow (FIGURE 10.4). The long connective tissue flaps of lymphatic valves prevent lymph from flowing backwards into the lymphatic capillaries.

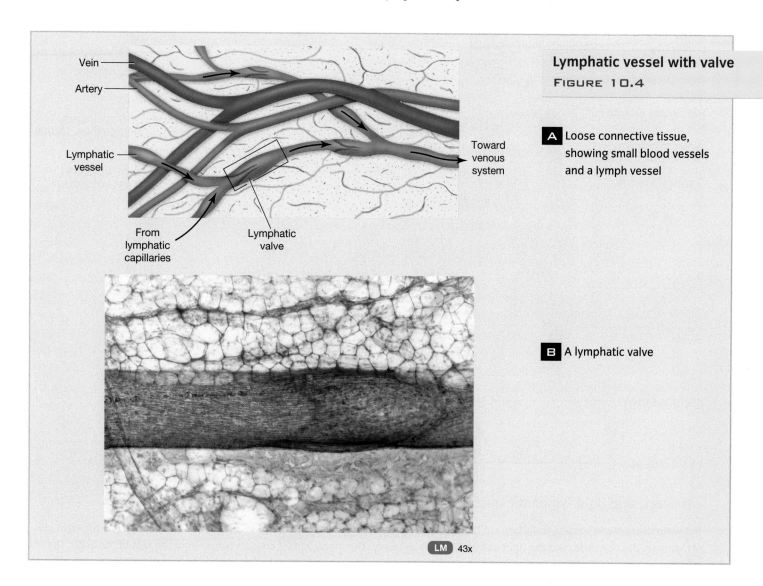

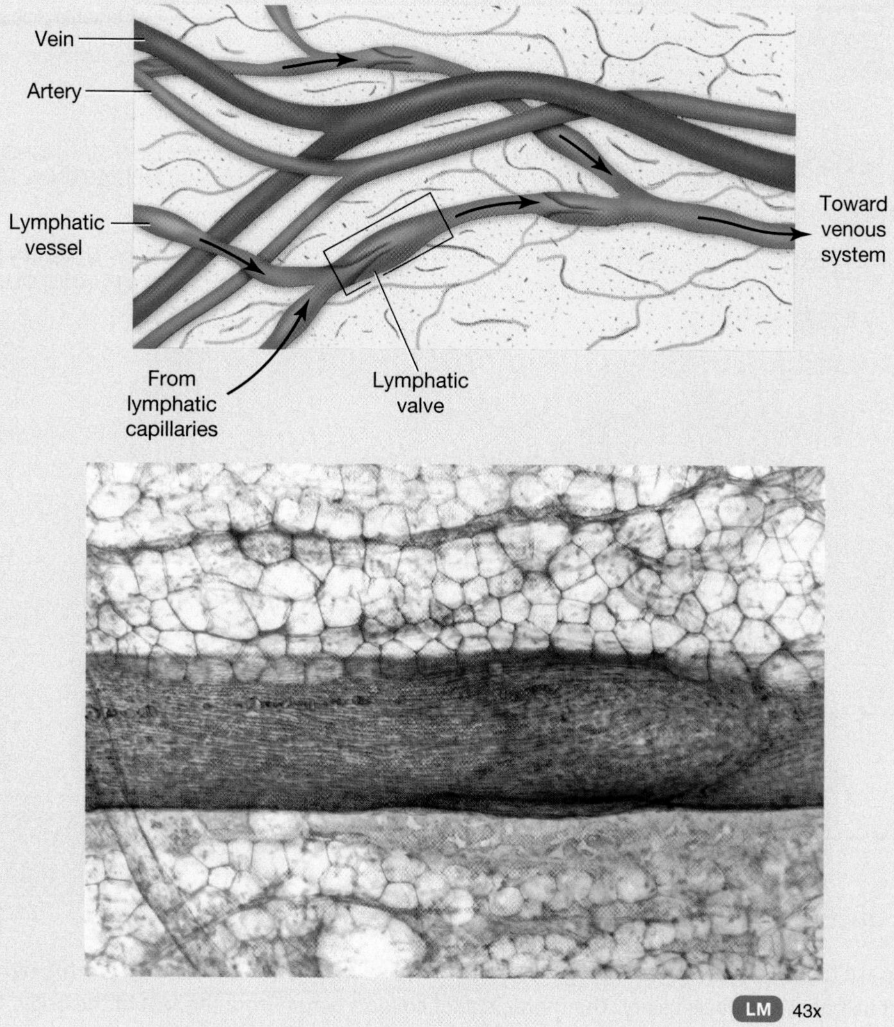

Lymphatic vessel with valve
FIGURE 10.4

A Loose connective tissue, showing small blood vessels and a lymph vessel

B A lymphatic valve

LM 43x

Mediastinum
The central portion of the thoracic cavity between the lungs, housing the heart, major blood vessels, and lymphatics.

Lymphatic vessels transport their lymph to either the thoracic duct in the **mediastinum**, or to the right lymphatic duct, just posterior to the right clavicle (FIGURE 10.5). Both ducts drain into the subclavian veins, allowing lymph to return to the bloodstream.

The right lymphatic duct drains the right side of the head, the right shoulder, and the upper portion of the right chest. Lymph collected from the rest of the body is drained into the thoracic duct. This arrangement causes concern for breast cancer patients, whose cancer may metastasize into the lymph. If this happens, it is easy to see how quickly those cells can be spread throughout the body via the lymphatic system.

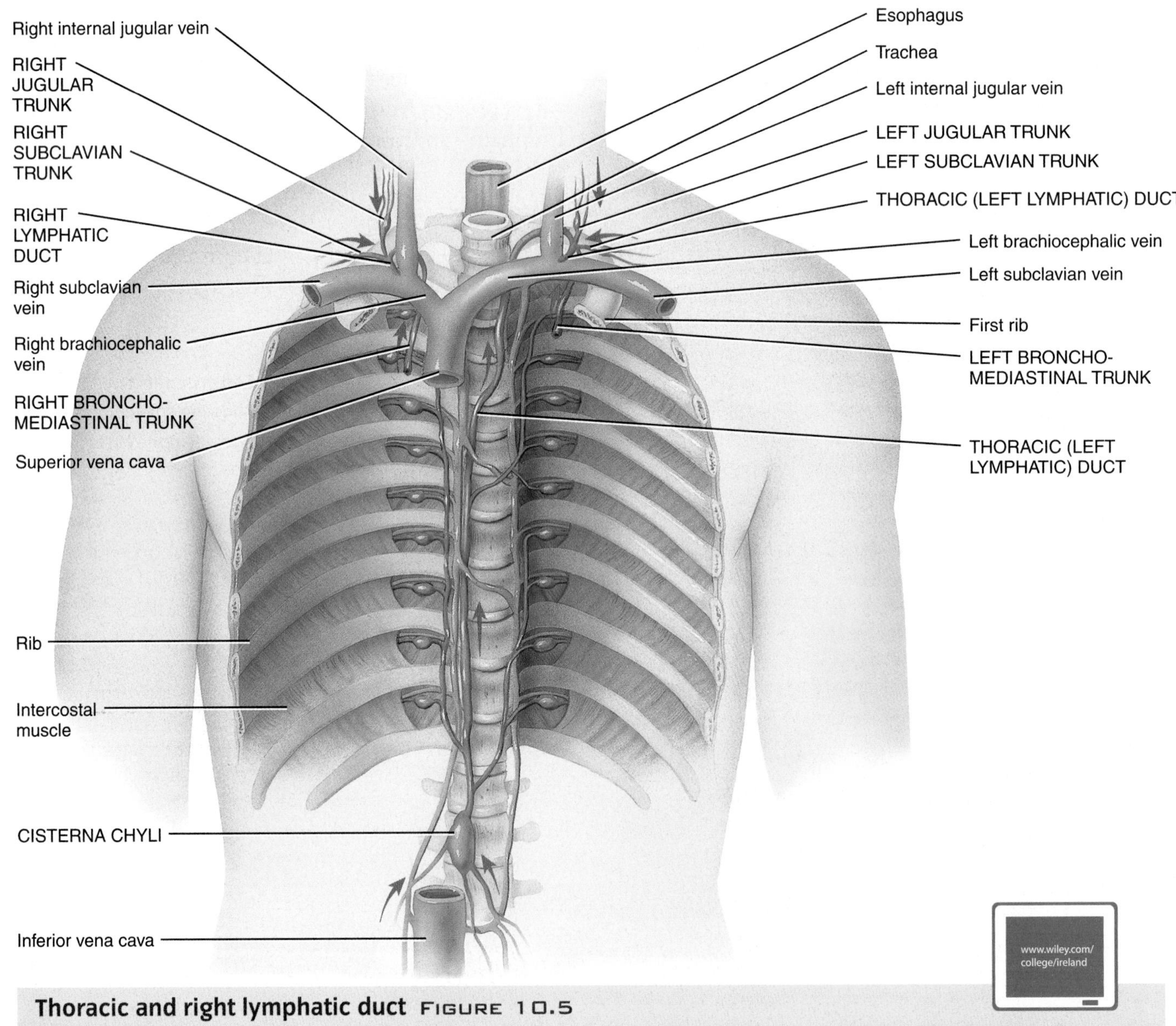

Thoracic and right lymphatic duct FIGURE 10.5

The main ducts of the lymphatic system are the thoracic and the right lymphatic ducts. The right lymphatic duct drains the right side of the head, the right arm, and the right half of the thoracic cavity. The thoracic duct collects lymph from the rest of the body. The veins listed here are described in more detail in Chapter 11.

LYMPHOID ORGANS FILTER AND PROTECT

Before lymph returns to the bloodstream, it must be filtered and cleaned. Otherwise, the lymph would dump the cellular debris and waste products it has picked up while traveling through the tissues into the bloodstream. This cleaning occurs in the lymphoid organs—the lymph nodes, tonsils, spleen, thymus gland, and bone marrow.

Lymph nodes are cleansing units

Lymph nodes are small, encapsulated glands that are strategically located to filter large volumes of lymph. The **inguinal** nodes are in the groin, the **axillary** nodes are in the armpit, the **cervical** nodes are in the neck, and the **mesenteric** lymph nodes form a chain at the center of the abdominal cavity.

Mesenteric
Pertaining to the membranous fold in the abdominal cavity attaching many of the abdominal organs to the body.

Nodes are filtering stations for lymph (FIGURE 10.6). Lymph enters a node via many passages but can leave by only one or two exits, forcing lymph to flow through the nodes in one direction.

Lymph nodes filter lymph that has been collected from nearby tissues, and they can tell us a good deal about the health of that region of the body. "Swollen glands" are lymph nodes that are enlarged due to localized or systemic infection, abscess formation, malignancy, or other, rarer causes. A bacterial infection can often be detected in the lymph because immune cells in lymph nodes will increase in number and produce antibodies. The population of lymphocytes will rise in the lymph node as these cells attack the pathogens. Swelling might be present even when the infection is not producing other symptoms. The particular lymph nodes that are swollen depend on the type of problem and the body parts involved; identifying the location can help locate the infection.

Many infections can cause swollen lymph nodes, including mononucleosis, German measles, tuberculosis, mumps, ear infections, tonsillitis, an abscessed tooth, gingivitis (infection of the gums), large, untreated dental cavities, and various sexually transmitted diseases. Immune disorders that can cause swollen lymph nodes include rheumatoid arthritis and HIV.

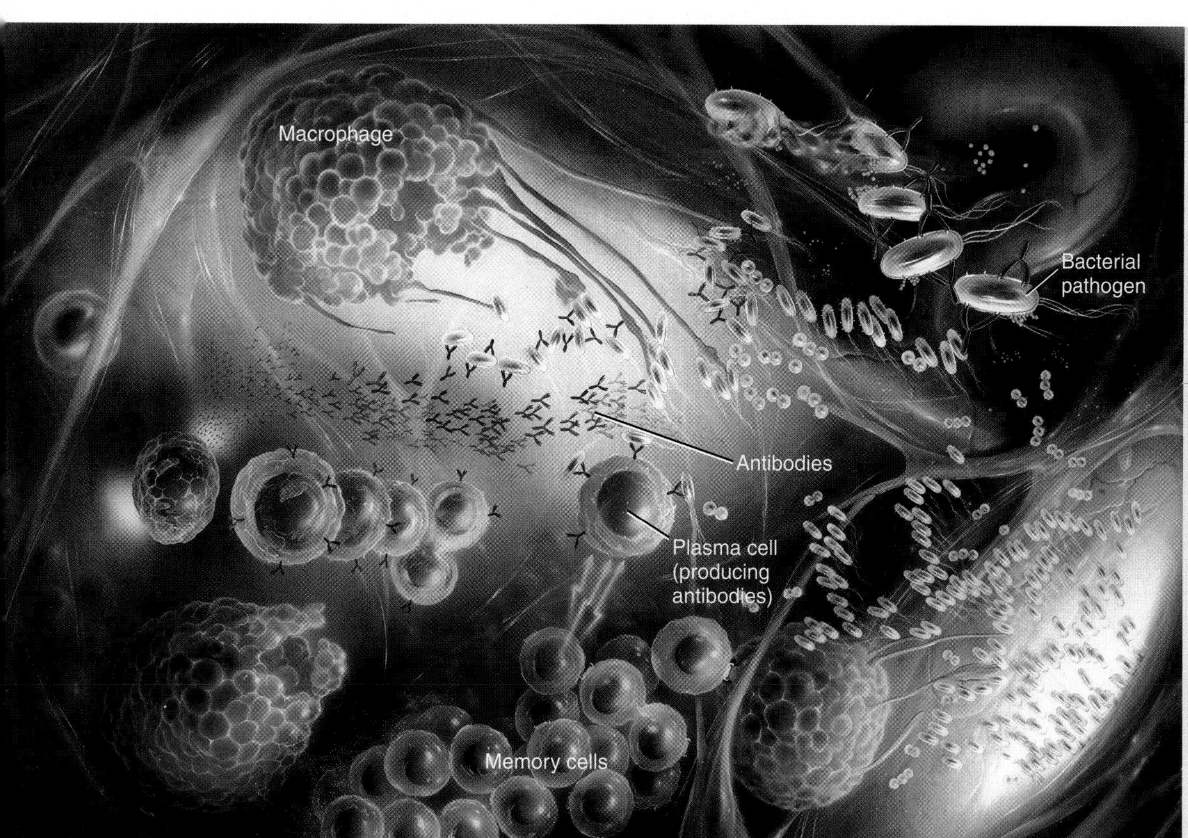

Lymph node
FIGURE 10.6

Lymph slowly flows through a maze inside the node, giving phagocytic cells in the lymph node time to interact with the fluid and remove and destroy infectious agents and debris.

I WONDER . . .

When I was diagnosed with mononucleosis, why was I told not to jump around?

Mononucleosis ("mono") is an infection caused by the Epstein-Barr virus. Many children get a mild case, which can be mistaken for an ordinary cold. In a young adult, mono usually produces symptoms four to seven weeks after exposure. These symptoms can include sore throat, fever, headache, fatigue, and loss of appetite. Many symptoms concern the lymphatic system: swollen tonsils, white patches on the back of the throat, and swollen glands (lymph nodes) in the neck.

Epstein-Barr virus infects cells in the salivary gland, causing the saliva to carry a large quantity of virus. This explains why Epstein-Barr is often transmitted through kissing and is called the "kissing disease." Other, less romantic transfers can occur through coughing, sneezing, or sharing food utensils.

Mononucleosis is usually treated with rest and fluids. Because it is viral, antibiotics are not helpful, although they may be used to defeat any bacterial infections that accompany mono.

The complications of mono include hepatitis (liver inflammation), jaundice, and a reduction of platelets, a formed element of the blood that initiates clotting. But the most serious complication is a ruptured spleen. The spleen is the lymphatic organ that filters and cleans whole blood, and it expands during an infection, just as the lymph nodes do while normally fighting infection.

B-cells (B-lymphocytes) are a second infection site of Epstein-Barr, and one job of the spleen is to remove abnormal blood cells. During mononucleosis, the spleen must deal with a large number of abnormal blood cells, which causes the organ to swell and become tender. As with most swollen organs, stretching stresses the membranes and capsule of the spleen, making it susceptible to bursting upon impact or injury.

A burst spleen causes internal bleeding. Symptoms of rupture include pain in the left upper part of the abdomen, a lightheaded feeling, a racing heart, abnormal bleeding elsewhere, and trouble breathing. A ruptured spleen is a medical emergency, entailing either quick repair surgery or a spleen removal.

Although few people with mononucleosis suffer a ruptured spleen, this life-threatening complication is worth avoiding. As the fatigue starts to abate, light exercise will help recovery, but the patient should avoid sports or other activities until a doctor grants permission. It's usually safe to resume activities about four to eight weeks after the first symptoms appear.

If you avoid a ruptured spleen, Epstein-Barr virus is unlikely to cause any long-term harm, and your new antibodies are quite likely to ensure that you never get the disease again.

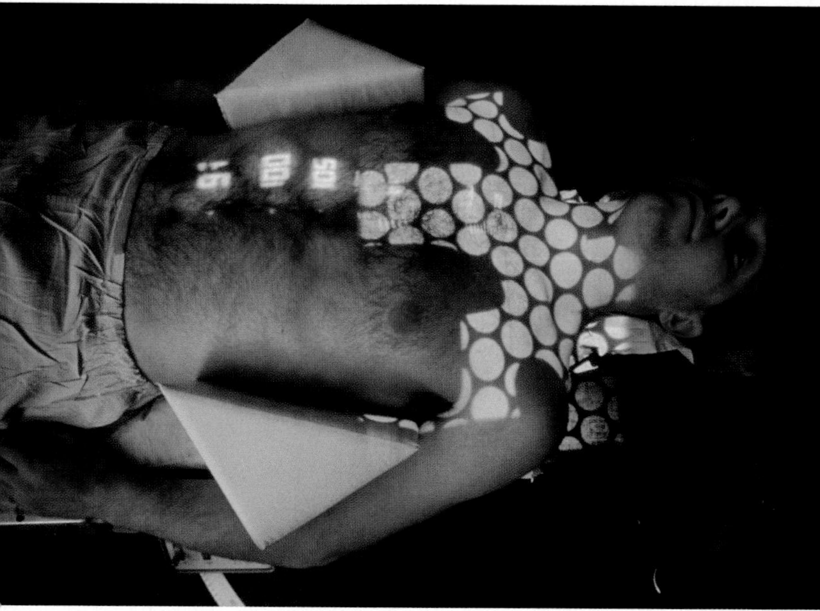

Hodgkin's disease FIGURE 10.7

Hodgkin's disease is a cancer of the lymph nodes, usually contracted in people who have had mononucleosis or measles. Not much is yet understood about the causes of this disease, but once diagnosed, survival rates are high. In this photo, a patient is undergoing radiotherapy on a linear accelerator.

Cancers that can cause swollen glands include leukemia, Hodgkin's disease (FIGURE 10.7), and non-Hodgkin's lymphoma. Swollen lymph nodes may also be caused by certain medications or vaccinations.

Cells of certain cancers, especially breast cancer, can be found in lymph nodes near the site of the primary tumor. As these cells metastasize, or migrate, to form new tumors, the number of lymph nodes containing cancer cells increases. This then is a good indicator of how advanced the cancer is.

Tonsils and MALT are patches of unencapsulated lymphatic tissue

The **tonsils** are similar to lymph nodes in their organization and function. You were born with two sets of tonsils: the pharyngeal tonsils in the nasopharynx, and the palatine tonsils, which are visible on either side of the pharyngeal opening. The main difference between tonsils and lymph nodes is that the tonsils are not entirely encapsulated. Instead, they are open to the fluids that pass through the pharynx. Infectious agents can be trapped in these organs, swelling the tonsil enough to almost shut off the pharynx.

Similar patches of lymphoid tissue are found in the lining of the small intestine. These egg-shaped masses, called mucosa-associated lymphoid tissue, or **MALT**, help filter fluid absorbed from the intestinal lumen.

The largest lymphatic organ is the spleen

The largest collection of lymphoid tissue in the body is the fist-sized spleen (FIGURE 10.8), which is much larger than any lymph node. The spleen has a strong outer capsule surrounding red and white pulp. Red pulp, containing red blood cells and macrophages, purifies blood by removing bacteria and damaged or exhausted red blood cells. The white pulp contains **lymphocytes** and is involved in specific immunity. For this reason, the spleen is considered a lymphatic organ, even though it filters whole blood rather than lymph. Although when healthy we are rarely aware that we even have a spleen, during some diseases careful treatment of it becomes essential and critical. See the I Wonder . . . feature for more information on this organ's reaction to disease.

Lymphocytes White blood cells that patrol the body, fight infection, and prevent disease.

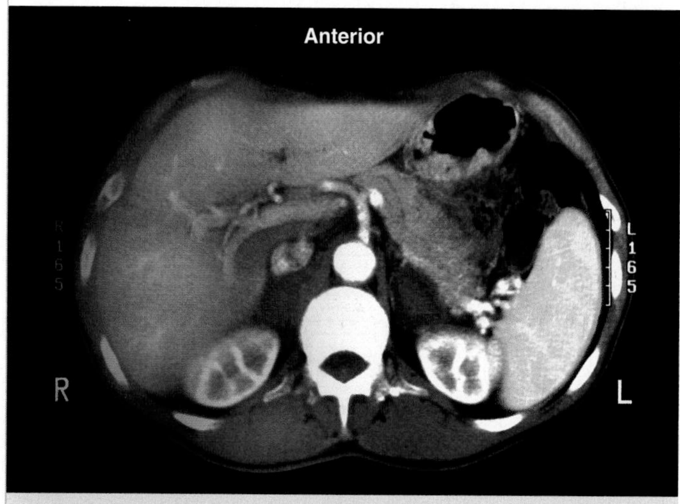

Spleen FIGURE 10.8

The spleen is highlighted in yellow in this CT scan.

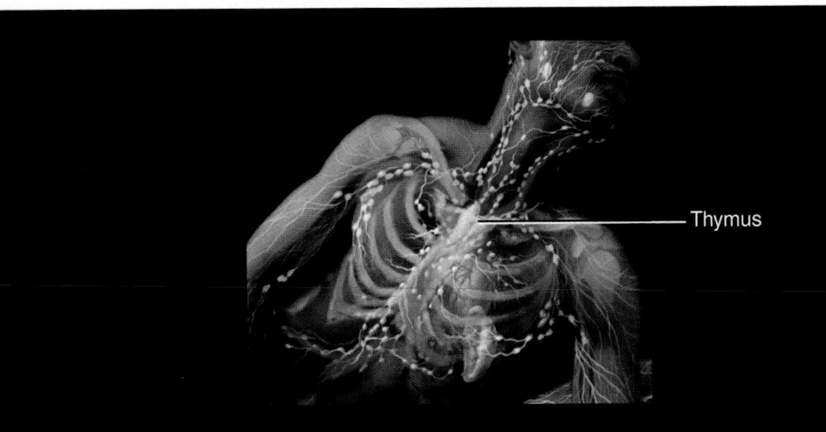

Thymus FIGURE 10.9

The thymus is largest at puberty and shrinks with age, losing function as it shrinks. One reason your parents or grandparents probably suffer more than you from a common cold or a passing virus is thymic atrophy.

The thymus produces mature immune cells

The thymus gland is located in the mediastinum of the thoracic cavity, behind the sternum and draping over the upper portion of the heart. It is composed of two lobes held together by connective tissue (FIGURE 10.9).

The primary function of the thymus is to produce mature, functional T cells, a distinct group of immune cells. The cortex of the thymus gland is involved in "training" T cells to distinguish self from pathogens. It also produces thymic hormones that promote maturation of T cells.

The changing activity of the immune system with age helps explains the recommendations the U.S. Centers for Disease Control and Prevention (CDC) has made for flu shots (FIGURE 10.10). The CDC currently recommends that flu shots be given to people age 50 and over, nursing home residents, children 6 months to 5 years, pregnant women, people with chronic health problems, and certain healthcare and daycare workers. But when vaccine becomes scarce, healthy people under 65 are urged to forgo the shot.

Bone marrow also produces mature immune cells

The final type of lymphatic tissue is red bone marrow. In children, red bone marrow is found in the center of virtually all the bones. When we reach adulthood, only the skull bones, sternum, ribs, clavicle, pelvic bones, and the vertebral column retain red marrow. The remaining bones contain yellow marrow in their marrow cavities. Red bone marrow includes blood **stem cells** that can produce the red and white blood cells. The cells involved in specific immunity are a subset of the white blood cells, produced in the red marrow.

Stem cells
Undifferentiated cells that remain able to divide and specialize into functional cells.

As we now understand, the lymphatic system cleans and returns excess fluid to the circulatory system. It is also of paramount importance in maintaining homeostasis through its role in specific immunity.

Flu shot FIGURE 10.10

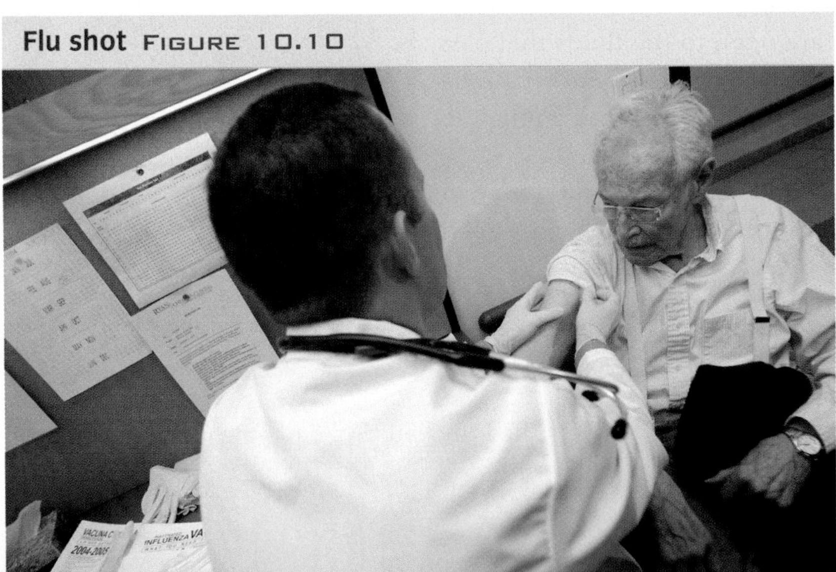

CONCEPT CHECK

List the functions of the lymphatic system.

Compare the open lymphatic system to the closed circulatory system. What structures are similar? How are these systems different?

Explain the function of the spleen.

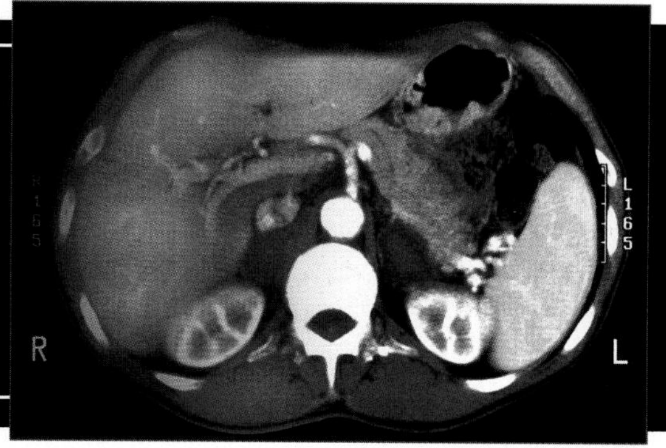

Specific Immunity Relies on a Deadly Series of Cells

LEARNING OBJECTIVES

Identify the cells involved in specific immunity.

Define humoral immunity and compare it to cellular immunity.

List the five classes of antibodies.

The immune system has two methods for combating pathogens, both of which are carried out by specific cells in the bloodstream, called lymphocytes. In one method, the cells function directly in any pathogen attack, while the other uses compounds produced by immune cells. The first type is referred to as cellular or cell-mediated immunity, while the second is humoral, or fluid, immunity. Cellular immunity uses a variety of deadly cells to do its work of destroying invading pathogens. These cells include macrophages and natural killer cells. Humoral immunity, on the other hand, employs cells that produce disease-fighting compounds, including T cells and B cells.

SPECIFIC IMMUNITY IS CONTROLLED BY CELLS THAT RECOGNIZE AND REMEMBER PATHOGENS

When a pathogen slips past our nonspecific defenses, the battle is not over. Rather than immediately succumb to the disease, we rely on our specific defense—the immune system. This system is composed of a set of blood cells collectively called lymphocytes. The various subtypes of lymphocytes look alike but have subtly different functions. All immune cells share common characteristics, including:

- The ability to distinguish self from nonself (otherwise, immune cells would destroy the very fabric on which they depend).
- Specificity, meaning they react only to a particular antigen (a component of a disease-causing agent).
- The ability to remember certain pathogens and react more quickly the second or subsequent times the pathogen is encountered. The immune system responds far faster the second time it encounters a particular antigen, which is the basis for **immunization**.

Immunization Stimulating resistance to a specific disease through exposure to a nonpathogenic form of the disease-causing organism.

Regardless of these similarities, each immune cell has critical and unique traits and tasks to perform, as we'll see shortly.

THE IMMUNE SYSTEM IS A WORLD OF ASSASSINS

> **Antibodies**
> Proteins produced by B lymphocytes and directed against specific pathogens or foreign tissue.

Two main classes of lymphocytes are involved in immunity, B cells and T cells. B cells (B lymphocytes) mature in the bone marrow and spend most of their time inside lymph nodes. B cells produce **antibodies** that are specific to a particular pathogen. T cells (T lymphocytes) mature in the thymus gland in response to thymic hormones. T cells make up about half of the circulating lymphocytes in the blood. T cells are responsible for stimulating B cells, as well as the direct destruction of antigens.

Lymphocytes have receptors on their cell membranes waiting to detect the exact antigen, which fits the receptor like a lock and key (FIGURE 10.11). Each lymphocyte is specific to one antigen; it will ignore all others. During our lives, we are exposed to thousands of antigens. Amazingly, our lymphocytes develop a specific response to every one of them by mixing and matching receptor proteins that are created by genes of the immune system. Small changes in receptor shape on the surface of a T cell or B cell will cause that cell to react to a different antigen.

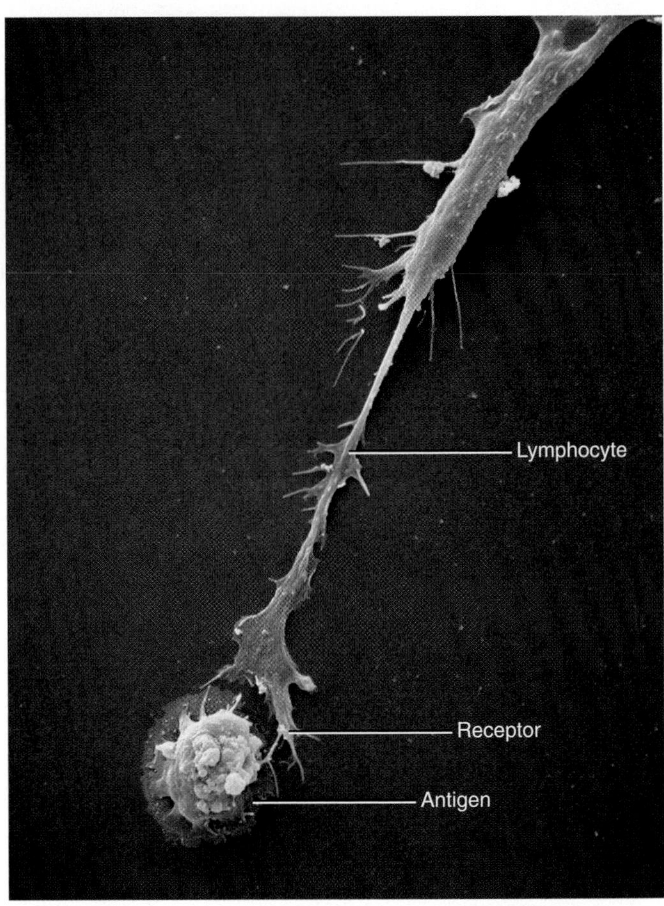

Lymphocyte with antigen attached to receptor
FIGURE 10.11

When a lymphocyte encounters the matching antigen, it bonds to that antigen and the lymphocyte is stimulated. Depending on the type of lymphocyte, stimulation results in either humoral immunity or cell-mediated (cellular) immunity. T cells are responsible for cellular immunity, whereas both B cells and T cells are involved in humoral immunity.

HUMORAL IMMUNITY IS MEDIATED BY ANTIBODIES

> **Agglutinate**
> To clump with other cells due to the adhesion of surface proteins.

Humoral immunity involves B cells and antibodies; the name "humoral" reflects the fact that this immunity takes place in the fluids of the body. Antibodies are proteins that remove antigens from the bloodstream, usually by causing them to **agglutinate**. Each B cell produces a different antibody that is directed toward a specific antigen. Because the B cell "wears" this antibody on its surface, the antibody is called a marker. When the surface antibody reacts with its specific antigen, the B cell is activated and begins to divide, making clones of itself. Because the antigen in effect "chooses" or selects which B cell will be cloned, this process is called clonal selection.

The cloned B cells produced during clonal selection are identical to the original, so they will link to the same antigen that started the cloning in the first place. As the cloned B cells are produced, two populations are created: plasma cells and memory cells. Mature antibody-producing B cells, called plasma cells, pump out an arsenal of antibodies, ensuring that the antigen is removed from the body (FIGURE 10.12). When the antigen is gone, the plasma cells undergo **apoptosis** and die.

Apoptosis
Programmed cell death.

The second variety of cloned B cells, called memory cells, contributes to a library of long-term immunity that we call the secondary immune response. For as long as 10 years, memory cells stand ready to go into action. If the pathogen reappears within that period, the memory cells quickly produce antibodies, ready to combat the pathogen before it can cause harm. Vaccinations and booster shots trigger the formation of memory cells, thus allowing us to fight pathogens that have never actually caused us to get sick. We have memory cells for a disease whose symptoms we have never actually experienced.

T cells are traveling fighters Because B cells reside in lymph nodes where they may not contact antigens, they have an alternative route to stimulation. Recall that T cells travel throughout the body in the bloodstream. Helper T cells patrol the blood in concentrations from 500 to 1,500 cells per cubic millimeter. The surface of helper T cells carries the same type of antigen-binding antibodies as B cells. When the proper antigen binds to a helper T cell, it is stimulated to clone. This cloning of helper T cells amplifies the message that

B cell activation FIGURE 10.12

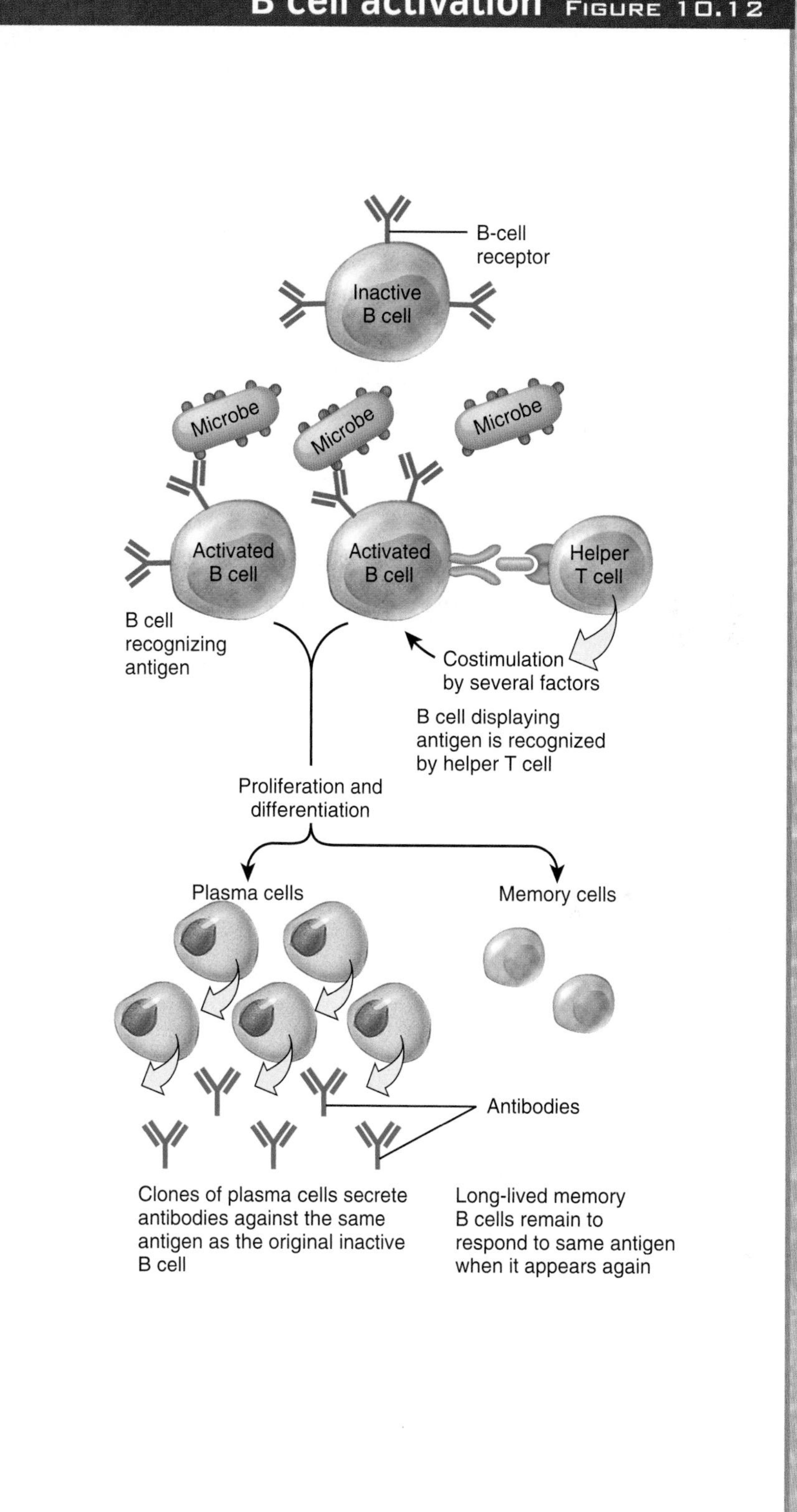

T cell activation

FIGURE 10.13

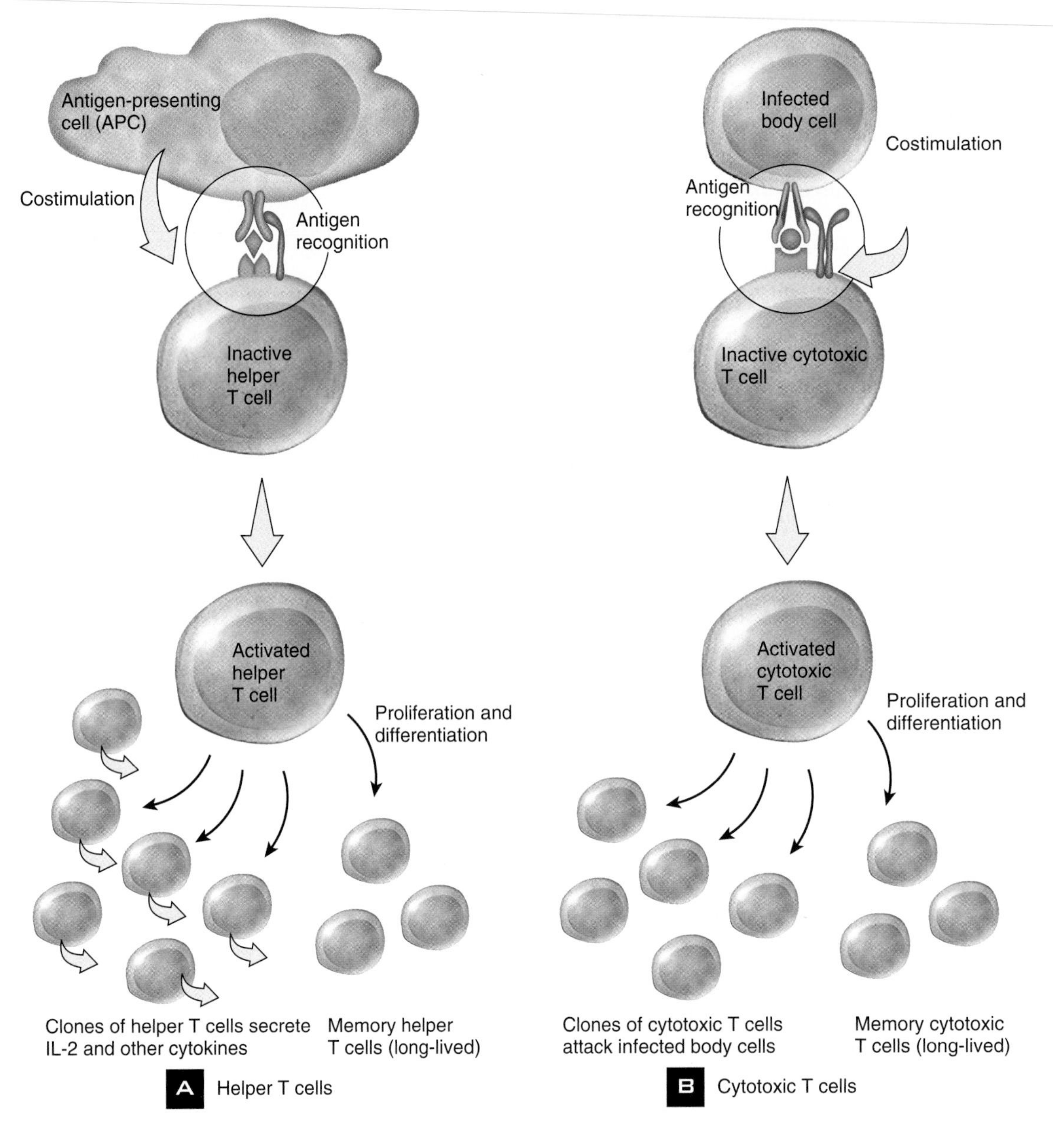
Antigen-presenting cell (APC)
Costimulation
Antigen recognition
Inactive helper T cell
Activated helper T cell
Proliferation and differentiation
Clones of helper T cells secrete IL-2 and other cytokines
Memory helper T cells (long-lived)
A Helper T cells
Infected body cell
Costimulation
Antigen recognition
Inactive cytotoxic T cell
Activated cytotoxic T cell
Proliferation and differentiation
Clones of cytotoxic T cells attack infected body cells
Memory cytotoxic T cells (long-lived)
B Cytotoxic T cells

there is a pathogen to be taken care of, greatly increasing the chances of stimulating the proper B cell. Much like B cells, cloned helper T cells may be activated to fight the invading antigen, or they may produce memory cells.

Activated helper T cells secrete compounds that stimulate other lymphocytes. Interleukin-2 (IL-2), the best known of these compounds, promotes T cell growth and stimulates macrophage functioning (FIGURE 10.13). An activated helper T cell will activate the specific B cells that are directed against that same antigen. These B cells then begin to produce antibodies, just as if they had encountered the antigen directly.

CD4 Recognition element in major histocompatibility complex (MHC) class II immune responses; identifies certain T cells.

Class II MHC (major histocompatibility complex) Recognition proteins present on the membranes of antigen-presenting cells and lymphocytes.

On the surface of each helper T cell is a protein complex called **CD4**, which serves as a docking station for a B cell while it is being stimulated to clone and produce antibodies. CD4 is attracted to **Class II MHC (major histocompatibility complex)** molecules, which occur on the surface of B cells and other immune cells. The CD4/MHC interaction holds the T cell to the B cell, enhancing stimulation (FIGURE 10.14).

Memory helper T cells, produced when helper T cells clone and differentiate, lie in wait in the blood, ready to jump quickly into action should the same antigen again threaten the body.

Some T lymphocytes differentiate into natural killer (NK) cells, which are actually part of our innate defense system. They are introduced here because they are produced exactly like the helper T cells of our specific immune defenses. NK cells function as a natural cancer screen, patrolling the body and identifying virally infected cells and tumor cells. After detection, NK cells kill the diseased cell via cell-to-cell contact. NK cells are not specific because they remove all foreign or infected cells in exactly the same way. They do not respond to immunization, and they do not seem to produce clones of memory cells.

CD4/MHC complex interaction

FIGURE 10.14

The binding of the CD4 protein to MHC-II helps to secure the T cell receptor (TCR) and antigen to the inactive helper or cytotoxic T cell. Once this complex is secure, the T cell will become activated and produce clones and memory cells. Helper T cells then produce chemicals that stimulate the immune response, while cytotoxic T cells actively attack infected body cells.

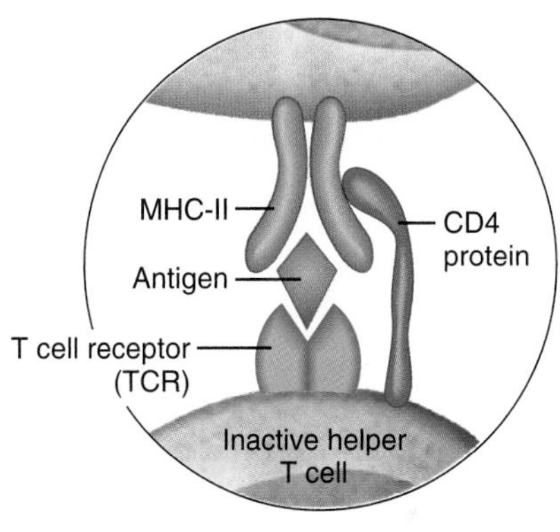

Process Diagram

ANTIBODIES ARE MORE SPECIFIC THAN YOUR SOCIAL SECURITY NUMBER

Antibodies are proteins secreted by plasma cells in response to antigen binding. Antibodies all have the same general shape: a doubled, Y-shaped protein with one heavy chain and one light chain polypeptide. Because the vast majority of heavy chains are identical among antibodies of one class, the heavy chain is called the constant region.

The upper tips of the heavy chain and the corresponding tips on the light chain identify each antigen, and because they change so much, they are called the variable region. It is the variable region that interacts with the antigen and causes agglutination. A large conglomeration of antigen and antibody marks the antigens for destruction by the macrophages. The antigen/antibody complex often activates complement as well, enhancing antigen destruction.

The five classes of antibodies (also called immunoglobulins) are IgG, IgM, IgA, IgD, and IgE (FIGURE 10.15).

- IgG, by far the most common antibody, occurs in the circulating blood, lymph, and extracellular fluid. IgG immunoglobulins bind directly to an antigen, inactivating it almost immediately (FIGURE 10.16).
- IgM is the first immunoglobulin released in any immune response and is also the predominant immunoglobulin produced in infants. IgM is a large polymer of five Y-shaped molecules that causes infected or foreign cells to clump together when IgM binds to them. Like IgG, IgM also aids in the release of complement.
- IgA can be a monomer, dimer (two subunits), or larger molecule composed again of Y-shaped units. One form of IgA, found in secretions such as saliva, can bind to pathogens before they enter the bloodstream.
- IgD, found on mature B cells, binds antigens that stimulate B-cell activation.
- IgE, the immunoglobulin responsible for immediate allergic reactions, appears on the surface of basophils and mast cells, both of which release histamines and other chemicals implicated in allergic symptoms.

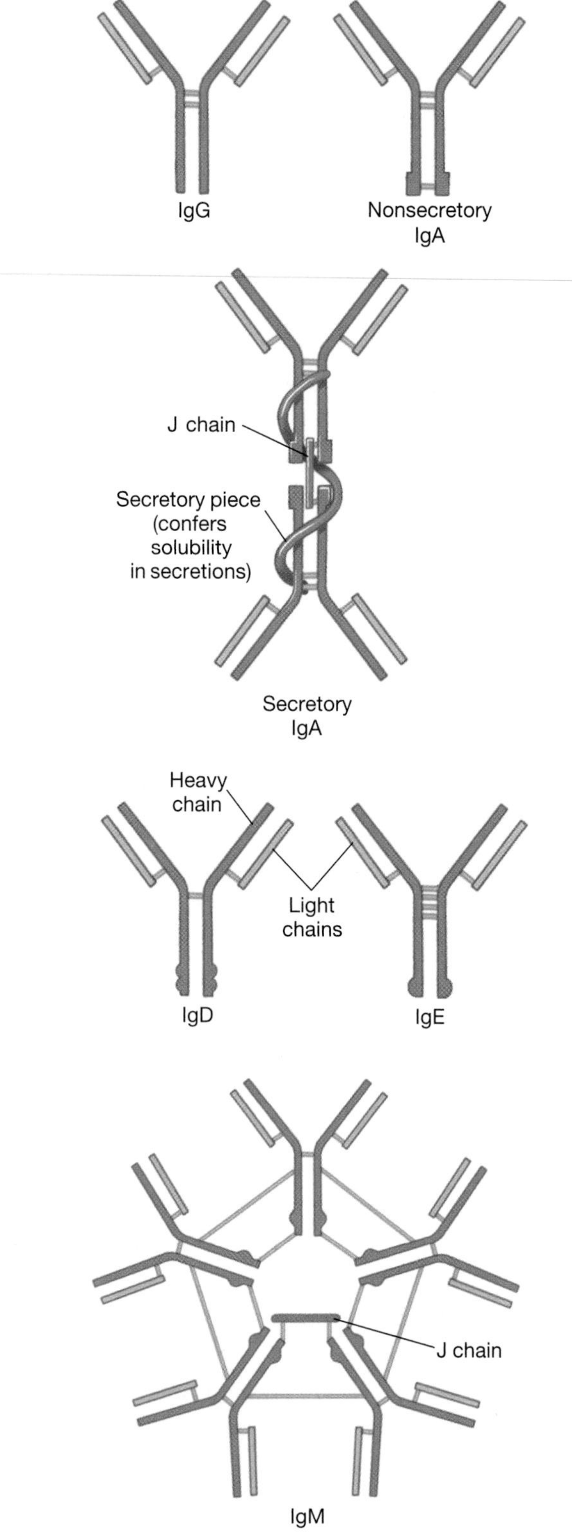

Five classes of antibodies FIGURE 10.15

In the body, natural humoral immunity results when many different plasma cells are simultaneously stimulated to form antibodies. Each clone of plasma

Elimination of bacteria by IgG and macrophage FIGURE 10.16

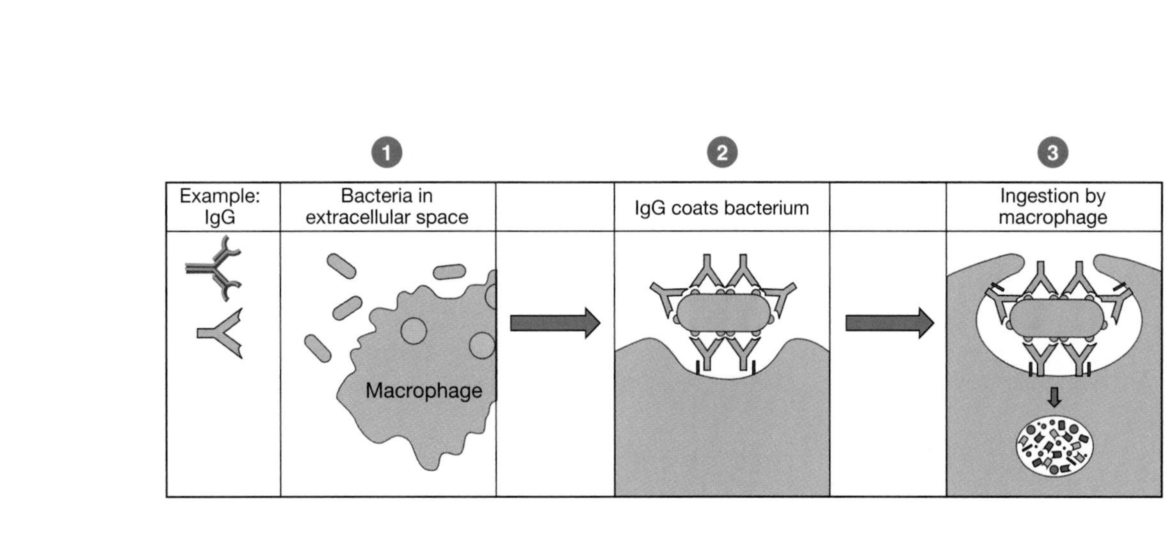

Process Diagram

cells originates from a different B cell. Each of these plasma cells produces an antibody that responds to a slightly different portion of the invading pathogen. The resulting soup of antibodies is polyclonal, meaning that the antibodies are produced by many different plasma cells. Polyclonal antibodies are directed against one specific antigen, but they link to many different antigenic sites on that antigen. Directing so many slightly different antibodies against differing portions of the same antigen ensures that no antigen will be left in the bloodstream.

Because antibodies are specific, they are an interesting source of precisely targeted drugs. Most of these cutting-edge medical treatments propose to use "monoclonal antibodies." As the words imply, monoclonal antibodies are antibodies that are formed from clones of a single activated cell. To make a monoclonal antibody, a plasma cell is removed from the body, then cloned in the laboratory, creating a large number of identical cells that all produce the same antibody (FIGURE 10.17, p. 300). Monoclonal antibodies are specific, reacting to only one particular antigenic portion of the pathogen. It may be possible to use them to target cancer cells by combining an antibody specific to the tumor cell with either a radioactive particle or a cell-killing medicine. In either case, the idea would be to deliver the death knell directly to the tumor cells, without harming healthy cells.

The specificity of monoclonal antibodies is often used in medical tests. The pregnancy tests sold in drugstores use a monoclonal antibody directed against a protein found only in the urine of pregnant women. Because monoclonal antibodies are so specific, any reaction in the test proves that the woman is pregnant. (If there is no reaction, the test should be repeated within a few days because the protein level could be too low to detect on the first test.)

CELLULAR IMMUNITY IS MEDIATED BY CELLS

Cellular immunity is governed by the T cells that are carried through the tissues in the blood. There are two large populations of T cells: helper T cells and **cytotoxic T cells**. Unlike B cells, which can directly detect the presence of an antigen using the antibodies on their

Cytotoxic T cells Subset of T lymphocytes responsible for killing virally infected cells.

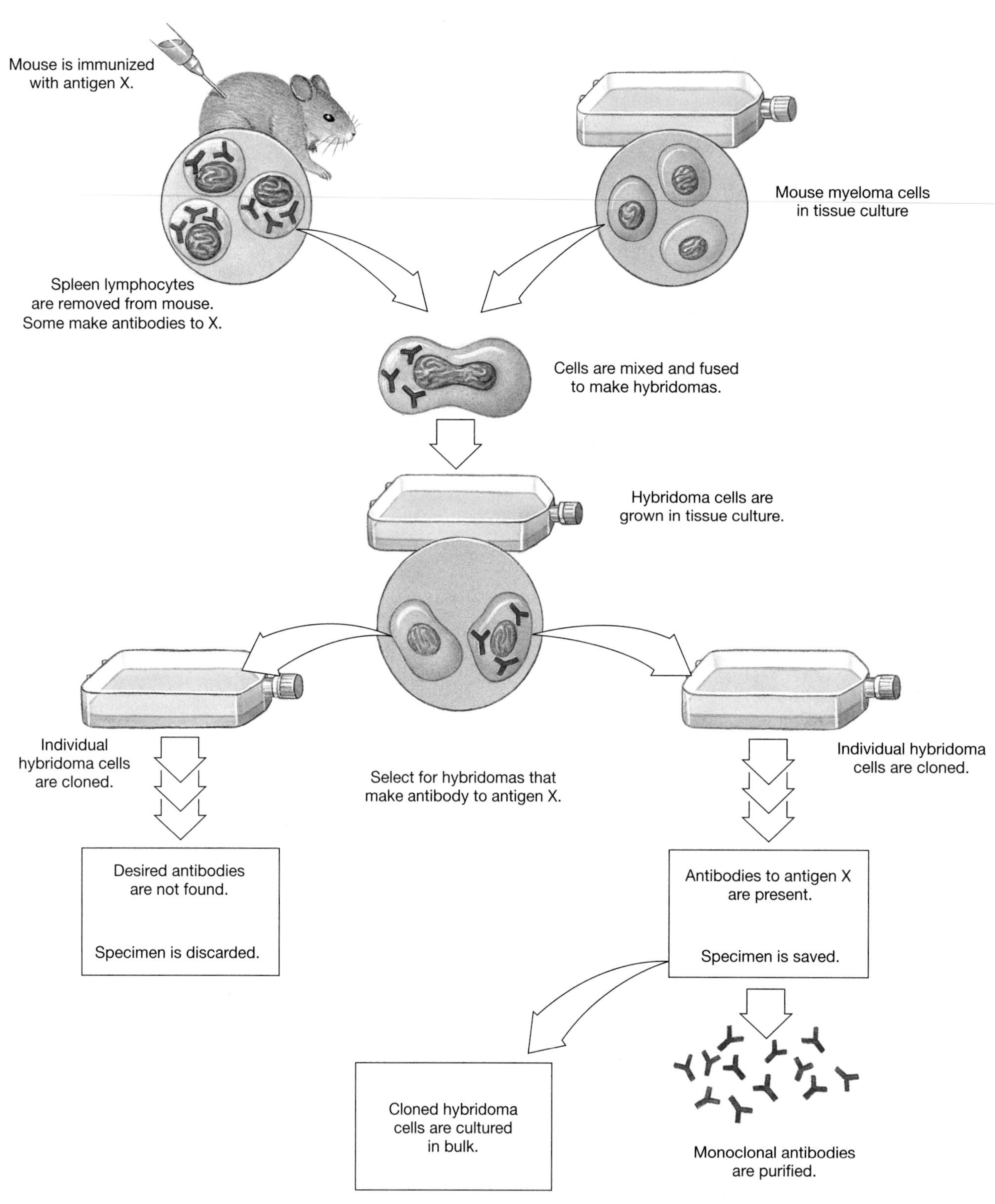
Mouse is immunized
with antigen X.
Mouse myeloma cells
in tissue culture
Spleen lymphocytes
are removed from mouse.
Some make antibodies to X.
Cells are mixed and fused
to make hybridomas.
Hybridoma cells are
grown in tissue culture.
Individual
hybridoma cells
are cloned.
Select for hybridomas that
make antibody to antigen X.
Individual hybridoma
cells are cloned.
Desired antibodies
are not found.
Specimen is discarded.
Antibodies to antigen X
are present.
Specimen is saved.
Cloned hybridoma
cells are cultured
in bulk.
Monoclonal antibodies
are purified.

Monoclonal antibody production
FIGURE 10.17

Monoclonal antibodies are produced by fusing normal mouse spleen cells with myeloma cells. The resulting hybridoma cells are then grown and those producing the antibody are selected. Antibody can be harvested in large quantities using this method.

surface, T cells must have the antigen presented to them. This is done by antigen-presenting cells (APC), usually macrophages.

Helper T cells stimulate the production not only of activated B cells, but also of APC macrophages. APC macrophages engulf large antigens and "present" them, removing the antigen from circulation and literally wearing a portion of the antigen on their surface (FIGURE 10.18). This is the ultimate example of "You are what you eat!"

Cytotoxic T cells are mainly responsible for cellular immunity. Just like B cells, cytotoxic T cells are activated when antigens bind to receptors on their surface. They are also stimulated to divide by **cytokines** released from helper T cells. Cytotoxic T cells respond specifically to altered HLA (human leukocyte antigen) proteins. The

> **Cytokines**
> Chemical signals released by immune cells during the immune response.

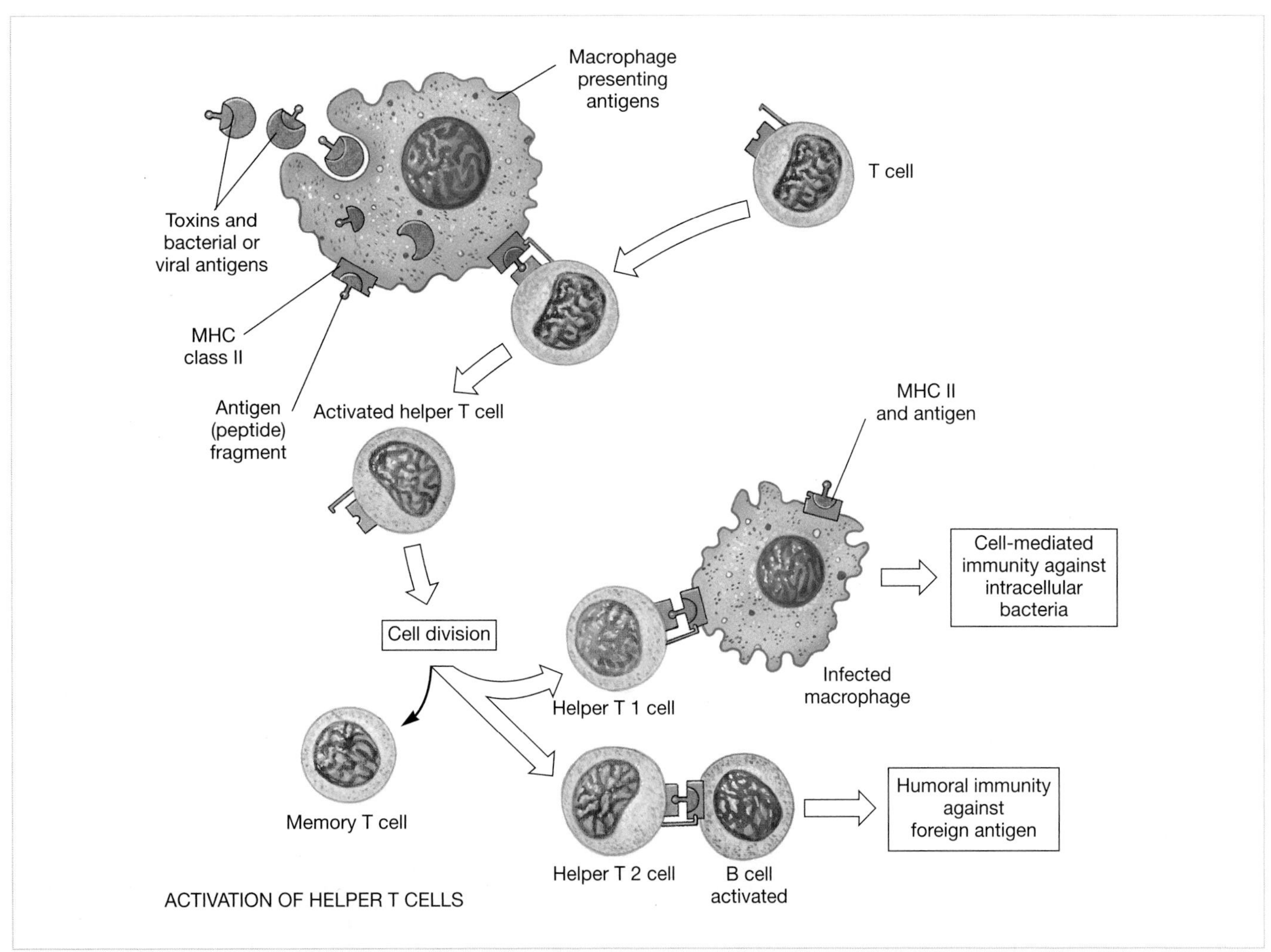

Macrophage presenting antigen FIGURE 10.18

When the macrophage phagocytoses an antigen, it breaks the antigen into subunits, which are attached to the MHC of the macrophage and presented on the surface of the macrophage membrane. Traveling T cells can easily locate these antigenic fragments and respond to them. In this way, the MHC increases the probability that T cells will encounter and respond to the antigen.

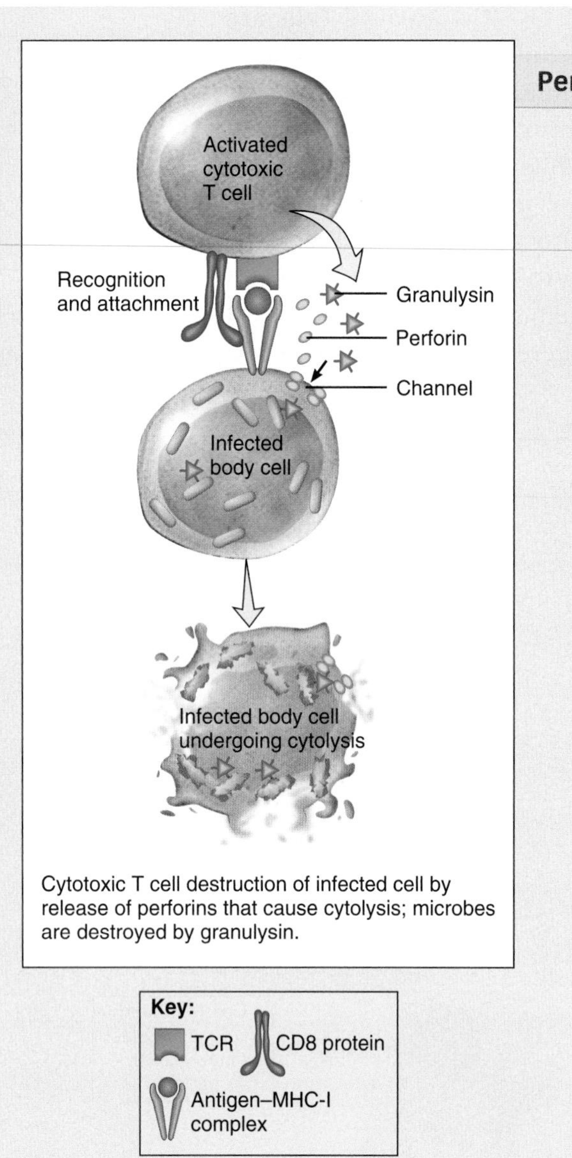

Cytotoxic T cell destruction of infected cell by release of perforins that cause cytolysis; microbes are destroyed by granulysin.

Perforin FIGURE 10.19

HLA complex is a marker that identifies the cell as belonging to the body and is what we identify when we "tissue type" a person and an organ before an organ transplant. HLA mismatches can trigger a rejection reaction that can destroy poorly matched transplanted organs.

Most cells with foreign HLA complexes are cancerous or virally infected, but cytotoxic T cells will remove any cell without the proper HLA antigens, even cells that are beneficial to the body. Cytotoxic T cells, or killer cells, physically attach to the foreign HLA-carrying cell and release **perforin** molecules from their vacuoles. Perforin molecules are like little molecular darts that poke through the plasma membrane of the infected cell (FIGURE 10.19). A pore forms in the cell membrane, allowing salts and water to enter the cell, causing it to swell and burst.

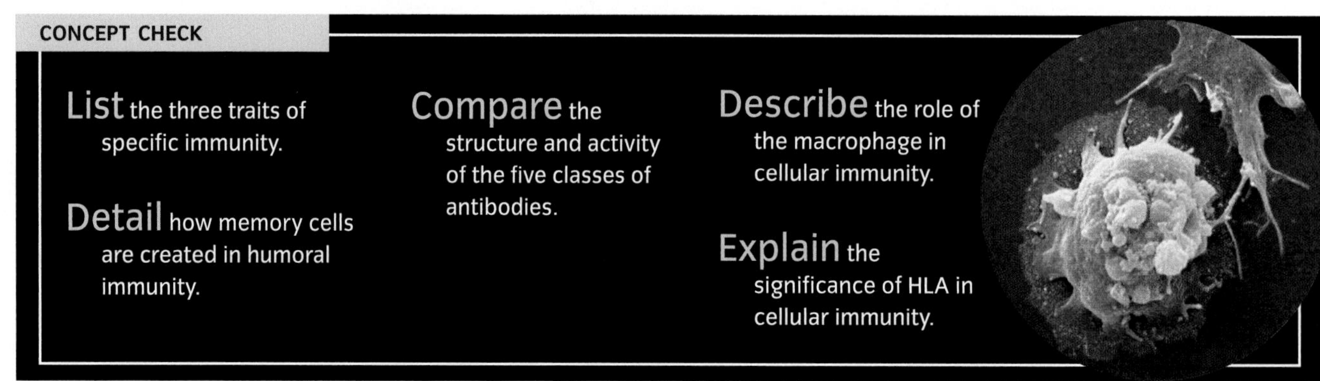

CONCEPT CHECK

List the three traits of specific immunity.

Detail how memory cells are created in humoral immunity.

Compare the structure and activity of the five classes of antibodies.

Describe the role of the macrophage in cellular immunity.

Explain the significance of HLA in cellular immunity.

Immunity Can Be Acquired Actively or Passively

Learning Objectives

Define the primary and secondary immune responses.

Compare active and passive immunity.

ACTIVE IMMUNITY IS THE "TRAINABLE" IMMUNE SYSTEM

Most of us acquire immunity from experience. We are exposed to a pathogen, it invades our tissues, and our immune system counterattacks by making antibodies (as just described). This is natural **active immunity**: Your immune system is exposed to the antigen in the natural course of your life; it adapts and actively combats the pathogen. **Passive immunity**, in contrast, occurs when antibodies are transferred without stimulating the immune system.

The primary advantage of active immunity comes from the creation of memory cells, which arise many hours after the initial reaction to the pathogen. The body needs days to respond to the pathogen, stimulate the proper cells, and follow the chain through helper T cells to B cells to plasma cells to antibody production. Then the body needs a few more days of antibody production to elevate the antibody **titer** to an effective level.

Titer
Level of a compound or antibody in the blood.

Memory cells produced during the primary response remain in the body for years, lying dormant until the same antigen reappears, when they will start the secondary response. This secondary response to a particular antigen happens far faster than the first response because the immune system needs to stimulate and clone only the memory cells (FIGURE 10.20). Secondary responses also require less energy from the body.

Although active immunity can prevent illness from a second exposure to a pathogen, the process we have described requires that you have previously been exposed to the pathogen, gotten sick, and recovered. It's preferable to prevent illness from the outset, so we never get the disease; some pathogens, after all, are extremely fatal! Fortunately, immunity can be obtained through artificial means as well. In this case, we intentionally introduce a pathogen to the body rather than allow you to contract the pathogen naturally. These pathogens are **attenuated** so that they can stimulate a primary immune response without causing disease.

Attenuated
Reduced capability of a pathogen to cause disease.

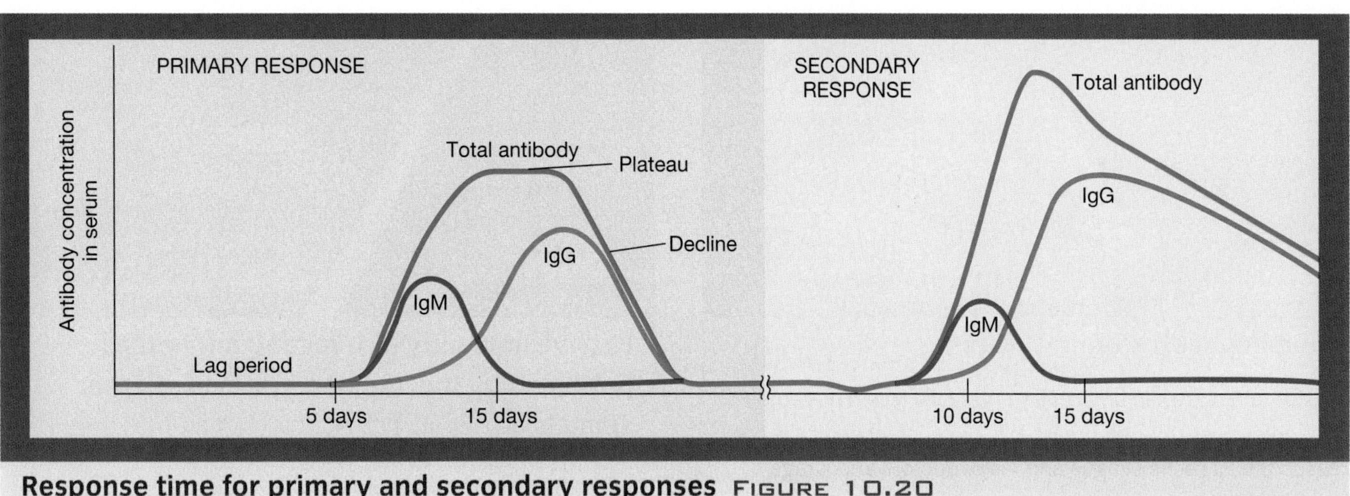

Response time for primary and secondary responses FIGURE 10.20

PASSIVE IMMUNITY GETS HELP FROM THE OUTSIDE

As noted in the previous section, passive immunity is the transfer of antibodies without stimulating the immune system. Although active immunity is helpful because the memory cells can launch a quick secondary response, passive immunity is also beneficial because you do not expend energy creating antibodies or producing clones. However, passive immunity is like giving an infantryman a gun with only one magazine. Introduced antibodies provide the recipient with immediate resistance to specific antigens. Once the antibodies are used or broken down, however, the body cannot create more, and the immune protection is lost. There are no memory cells because the antibodies were not created by active stimulation of the immune system.

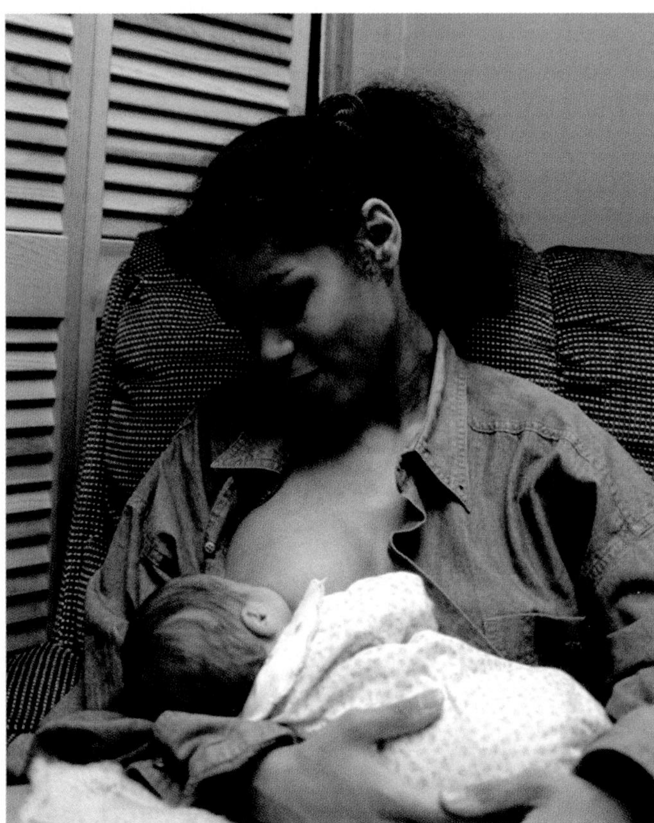

Nursing baby FIGURE 10.21

Passive immunity can be acquired naturally, when maternal antibodies pass through breast milk to an infant, which is one reason the La Leche League and many doctors encourage breast feeding.

La Leche League is a nonprofit organization that promotes healthy prenatal and postnatal care for both the infant and the mother. Their most well-known campaign is designed to educate women on the advantages of breast feeding until at least age 6 months. The antibodies received from the maternal blood in utero sustain the infant for approximately two to three months. Soon after, these antibodies begin to break down, and the infant must either produce antibodies via active immunity or receive maternal antibodies via breast milk (FIGURE 10.21). Breast-fed infants continue to gain passive immunity from their mothers and are therefore more able to resist disease. Infant formula may have a nutrient content similar to human milk, but it does not contain any antibodies.

Passive immunity can be used to fight diseases that cannot be fought in any other way. Horses, goats, rats, mice, and rabbits have all been used to generate antibodies against specific diseases (FIGURE 10.22). These antibodies are harvested, purified, and administered to humans for treating diseases such as diphtheria, botulism, and tetanus.

Passive immunity: Harvesting antibodies produced by the immune systems of other animals FIGURE 10.22

Passive immunity can also be administered artificially in gamma globulin shots, which are mixtures of many antibodies designed to match the pathogens the patient may contact. These are often given before travel to foreign countries, where new diseases may be encountered. Passive immunity generally lasts three to six months, long enough for most foreign vacations.

IN AUTOIMMUNE DISEASES, DEFENSE BECOMES OFFENSE

The immune system is a complicated network of cells and cell components that normally defend the body and eliminate bacterial, viral, and other pathogenic infections. This sophisticated mechanism goes awry in autoimmune disease, when the immune system mistakenly attacks the body's own cells, tissues, and organs. *Auto* is Greek for "self," so an autoimmune response is an immune response in which the body attacks itself.

The many autoimmune diseases have different effects depending on what tissue is under attack. In multiple sclerosis, the autoimmune attack is directed against nervous tissue. Immune cells break down the myelin sheath on neurons of the CNS, resulting in the buildup of scar tissue that impedes normal impulse transmission. Crohn's disease is an autoimmune disease directed against the gut. Type I diabetes mellitus is an autoimmune disease that attacks the pancreas. If the pancreas is not functioning properly, cells of the body cannot absorb glucose as they should, resulting in the myriad symptoms of diabetes. In diseases like systemic lupus erythematosus (lupus), the site of the attack may vary. In one person, lupus may affect the skin and joints, whereas in another it may affect the skin, kidney, and lungs. Rheumatoid arthritis is an extremely common autoimmune disease, attacking the joint capsules of the body, causing painfully deformed joints (**FIGURE 10.23**). Although this type of arthritis is usually considered a disease of older people, 1 in 1,000 children under the age of 16 show signs of juvenile rheumatoid arthritis.

The damage of autoimmune disease may be permanent. Once the insulin-producing cells of the pancreas are destroyed in Type I diabetes, they do not regenerate. Autoimmune diseases afflict millions of Americans, and for reasons not understood, they strike more women than men. Some autoimmune diseases are also more frequent in certain minority populations. For example, lupus is more common in African American and Hispanic women than in Caucasian women of European ancestry. Rheumatoid arthritis and scleroderma, another autoimmune disease, affect a higher percentage of some Native American communities than the general U.S. population.

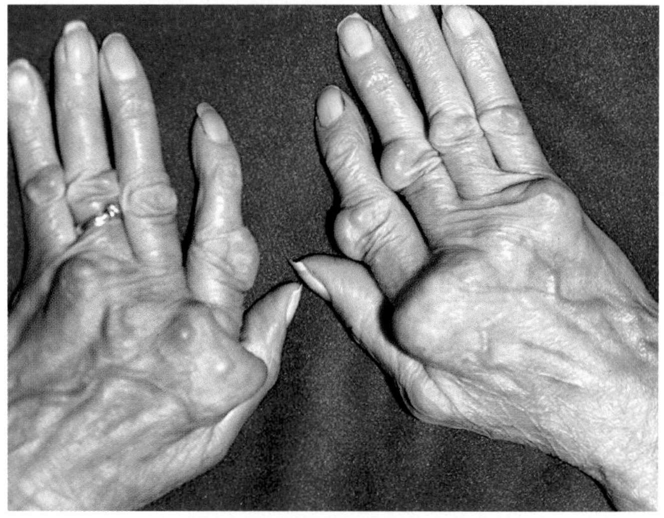

Hands of a patient suffering from rheumatoid arthritis FIGURE 10.23

Note the twisted appearance of the fingers above the affected joints in this patient suffering from rheumatoid arthritis. In this autoimmune disease, immune cells attack the synovial lining of the joints, causing them to become inflamed and swollen. As the disease progresses, a thickening develops in the joint, further hindering movement. In the final stages of rheumatoid arthritis, the swollen joint lining produces enzymes that digest the bone of the joint, resulting in the deformities shown here.

CONCEPT CHECK

Why is the secondary immune response so much more effective than the primary immune response?

Define passive immunity and give one example.

Define active immunity and give one example.

Viruses and Bacteria Cause Disease in Different Ways

Learning Objectives

Define prokaryotic and eukaryotic.

Outline the structure of a typical bacterium.

Relate the structure of a virus to its function.

Define the lysogenic and lytic life cycles of viruses.

Throughout this chapter, we have been discussing viral and bacterial infection. It is important to recognize the differences between these two categories of pathogens. These differences emerge from the fact that one is a true cell, whereas the other is a bit of protein surrounding a few genes.

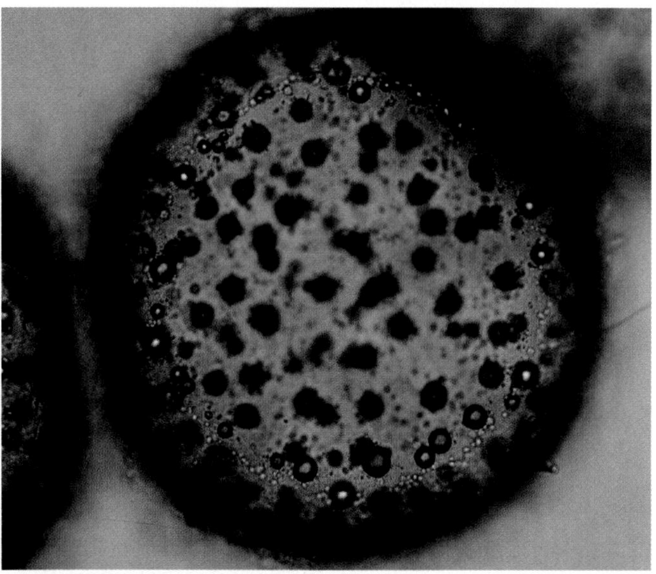

***Thiomargarita namibiensis* cell**

FIGURE 10.24

Scientists used to think that the reliance of a bacterial cell on diffusion limited its size, but the discovery of this giant cell disproved that idea. *T. namibiensis* can grow so large because it fills its center with a nitrogen-containing vacuole. Nutrients and waste are diffused between the cell membrane and the exterior, and between the central vacuole and the bacterial cytoplasm.

BACTERIA ARE SINGLE-CELLED WONDERS

Bacteria are **prokaryotic** cells that can be found in the ground, in the water, even in the air, not to mention inside humans and our fellow animals. Bacteria are generally smaller than eukaryotic cells, ranging in size from the 100-nanometer mycoplasma to the average-sized 7-micron cyanobacterium. A bacterial giant was recently discovered in the seafloor off Namibia. *Thiomargarita namibiensis* (FIGURE 10.24), which means "Sulfur pearl of Namibia," was discovered in 2000 by Dr. Andreas Teske of Woods Hole Oceanographic Institute. This spherical bacterium is roughly the size of a period in a 12-point font. (Most bacteria are barely visible with a light microscope.)

Prokaryotic Referring to single-celled organisms with no membrane-bound organelles, a single circular piece of DNA, and few specialized structures.

Photosynthesis Process of producing carbohydrates with sunlight, chlorophyll, carbon dioxide, and water.

Like all prokaryotes, bacteria have no internal membranes, no division of labor, and no specialized area where DNA is stored. Any special function, such as **photosynthesis** or isolating the single, circular strand of DNA, is carried out by the cell membrane. Bacteria do have one organelle in common with eukaryotic cells, however. Bacterial cells transcribe and translate DNA just like eukaryotes, so they have ribosomes in their cytoplasm. These ribosomes are so similar to those in eukaryotic cells that some scientists think all cells may have a common origin.

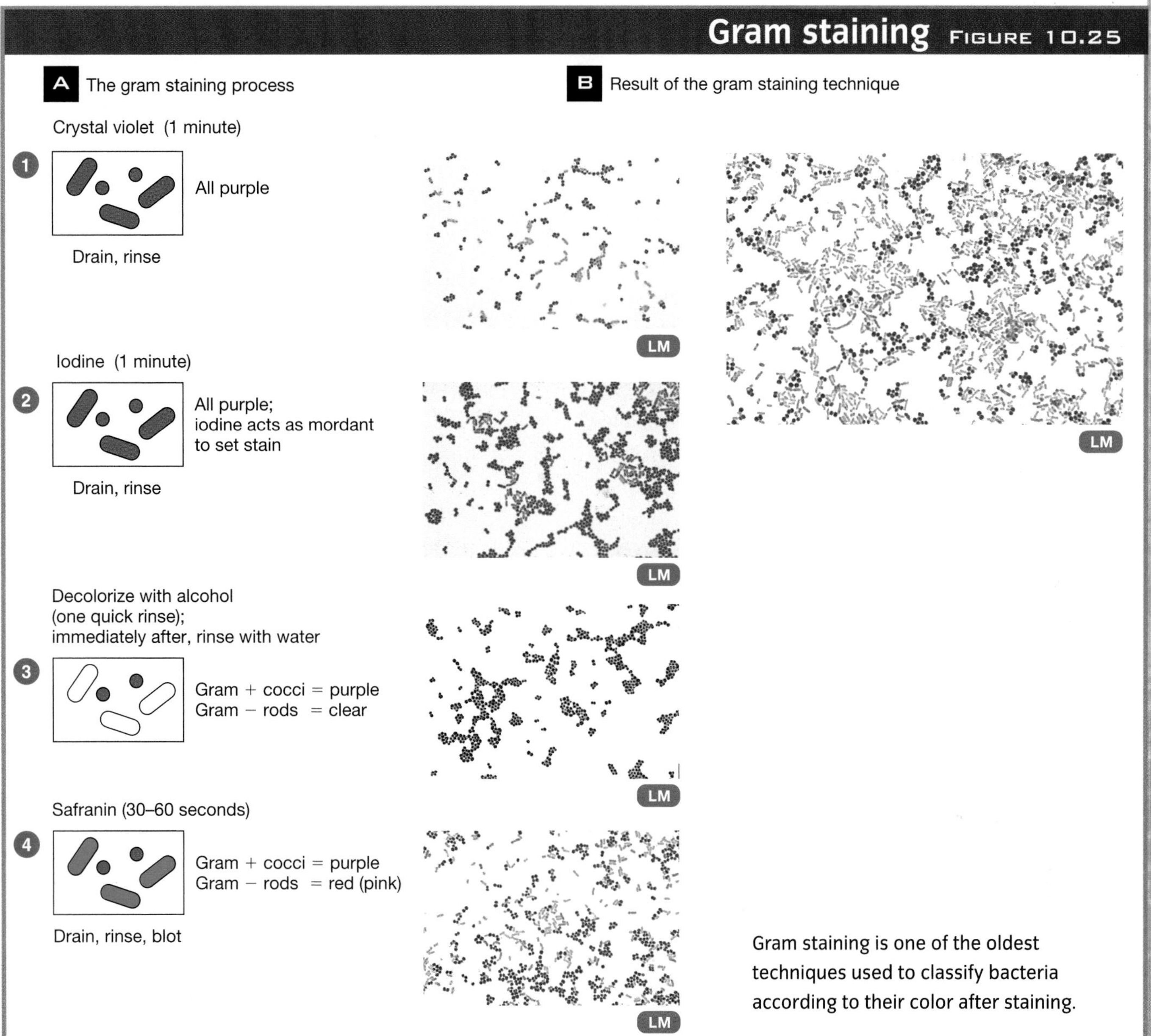

Gram staining is one of the oldest techniques used to classify bacteria according to their color after staining.

Bacteria are classified by shape, staining, or genetics Being relatively simple organisms, bacteria were traditionally classified by shape and by the staining patterns of their cell wall. The shape of bacteria falls into two broad categories: spherical and rod-shaped. Terms for spherical bacteria include cocci for single spherical bacteria, streptococci for those that live in chains, and staphlococci for those that grow in large masses. Bacilli (singular, *bacillus*) are rod-shaped cells that can be oval, tapered, or curved. Spirochetes are long rod-shaped bacterial cells that twist about their long axis. The bacterium that causes Lyme disease is an example of a spirochete.

Gram stain, the most common bacterial staining technique, was developed by Hans Christian Gram to distinguish two types of bacterial infections in the lungs. Bacteria are either gram positive or gram negative. Gram-positive bacteria retain a purple color from the gram stain, whereas gram-negative bacteria pick up a red dye, safranin (FIGURE 10.25). *Staphylococcus aureus* (staph infections) and *Streptococcus pneumoniae* (strep infections), are both gram positive, whereas *Escherichia coli* (*E. coli*) is gram negative.

A third, more scientific way to classify bacteria reflects their genetics, not their appearance. Two strains can be compared with DNA-DNA hybridization, which measures how closely the DNA of one species resembles that of the other. Alternatively, the study could focus on a particular common gene that changes slowly through time. A third technique can look at ribosomal RNA; in fact, a particular type of RNA, called 16S RNA, is used to track the evolutionary relationships of the entire tree of life, not just bacteria.

Before leaving the subject of bacteria, here's something to ponder: Humans live in concert with 200–1,000 types of bacteria. If you could count the bacteria in your GI tract, their number would exceed the number of cells in your body. Your mouth probably houses more than 400 species of bacteria all by itself. Clearly, most of these bacteria are harmless or even helpful. Bacteria in your gut, to take just one example, produce vitamin K, which is essential in blood clotting. Without bacteria in your body, you would die. So before you spend money on antibacterial soap or cleanser, consider that most of the microbes you encounter are harmless, helpful, or easily controlled by your nonspecific and adaptive defenses.

Antibiotics kill bacteria

When we need to kill bacteria, we turn to **antibiotics**, drugs that interfere with cellular processes that bacterial cells undergo every day. Various antibiotics prevent protein synthesis by binding to bacterial ribosomal RNA; others destroy essential metabolic pathways; and still others block DNA and RNA synthesis. Antibiotics also affect cell walls, which are found in bacteria but not in mammals, either breaking them down or preventing new cell walls from forming.

Bacteria have fiendishly clever defenses against antibiotics. Although bacteria sometimes mutate to acquire antibiotic resistance, more commonly, they acquire a resistance gene from other bacteria. This antibiotic resistance is developing into a huge problem, as bacteria are rapidly becoming immune to many modern antibiotics. One gene, or a ring of genetic material including a few genes, may carry resistance to several antibiotics, and it may be transferred from one species of bacterium to another, not just among bacterial cells of a single species. For a discussion of one source of antibiotic resistance, see the Ethics and Issues box.

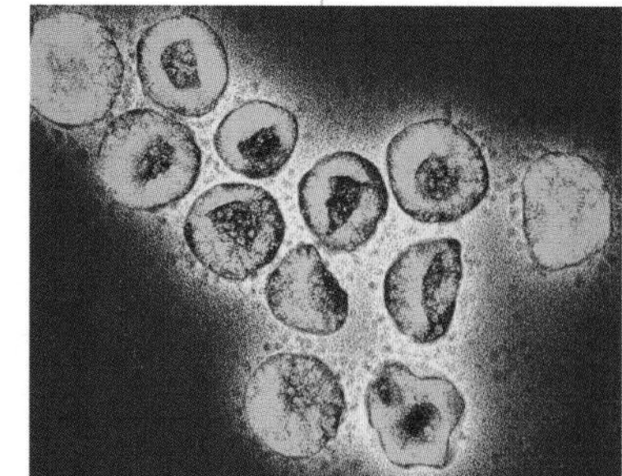

Bacteria can become resistant to specific antibiotics through several mechanisms:

- The bacterial membrane permeability changes so the antibiotic cannot enter.
- The antibiotic receptor protein on the bacterial surface changes so the antibiotic cannot attach.
- The bacterial metabolism alters and starts pumping the antibiotic out of the cell.
- The bacteria produce enzymes that destroy the antibiotic.

VIRUSES CAN REPRODUCE, BUT THEY ARE NOT ALIVE

Viruses are different from bacteria. Not only are they far smaller, but they lack most characteristics of life. Viruses cannot reproduce without a **host cell**, they do not metabolize, and they are not composed of cells. A virus is merely a snippet of nucleic acid (either DNA or RNA) contained inside a protein coat, called a capsid. Enzymes may be carried within the protein coat as well. Ebola, AIDS, smallpox, chickenpox, influenza, shingles, herpes, polio, rabies, and hantavirus are all viral diseases. Some viruses, called bacteriophages, attack bacteria (FIGURE 10.26, p. 310). Because of their small size and ease of purification, bacteriophages are used in research and medicine to introduce genes into cells.

Host cell
A cell that harbors a virus.

Are antibiotics being overused on the farm?

Americans love meat; in 2001, the average American ate about 220 pounds of beef, poultry, and pork. Meat consumption is rising in other countries as well; beef consumption in Brazil, for example, is increasing by nearly 5 percent per year.

Raising meat animals is a multibillion dollar industry, where profits rest on tiny margins. The cost of feed and the time to reach market weight must be minimized, and producers are eager to decrease costs by growing animals faster. Decades ago, scientists observed that feeding small, sub-therapeutic doses of antibiotics to farm animals would speed weight gain. Today, low doses of antibiotics are fed to most meat animals in developed countries. According to one estimate, such use accounts for 70 percent of all antibiotics used in the United States. Beyond speeding growth, these antibiotics also prevent disease in crowded animal facilities.

But how does routine feeding of antibiotics affect the bacteria that are the target of antibiotics? A sub-therapeutic dose of antibiotic does not kill all bacteria in an animal, and the ones that survive are more likely to resist the drug. When these bacteria multiply, they may form antibiotic-resistant strains of common, and sometimes deadly, bacteria.

In human medicine, low doses are rare: We take enough antibiotics to kill all the microbes, and we are told to finish all our pills in order to prevent antibiotic-resistant microbes from emerging.

Antibiotic resistance is a real danger. More than 50 years after antibiotics were labeled "miracle drugs" for killing common bacteria like tuberculosis, streptococcus, and staphylococcus, resistant bacteria are overcoming antibiotics. Could we return to the dreadful days of untreatable bacterial epidemics?

Research indicates that feeding antibiotics does undermine the drugs. After poultry started receiving fluoroquinolone antibiotics in 1995, doctors began seeing patients with resistant *campylobacter* and *salmonella* infections. The Food and Drug Administration found that 20 percent of ground beef in a supermarket was contaminated with *salmonella*, and 84 percent of those bacteria were resistant to at least one antibiotic. Antibiotic-resistant bacteria have been found in groundwater, surface water, and air near large animal operations.

The medical profession has repeatedly warned about overusing antibiotics. According to a 2002 analysis in the journal *Clinical Infectious Diseases*, "Many lines of evidence link antimicrobial-resistant human infections to foodborne pathogens of animal origin." Ceasing to feed sub-therapeutic doses to food animals "will lower the burden of antimicrobial resistance in the environment, with consequent benefits to human and animal health."

Farming trade organizations, however, argue that many factors are involved in creating antibiotic-resistant bacteria, including medical practices and evolution through natural selection. Furthermore, they say that after antibiotic feeding was banned in Denmark, animals got sicker, and farmers had to use more antibiotics to treat disease.

At root is an ethical question: Feeding antibiotics to animals may be profitable, but is this worth the hazard of resistance? Are we overusing antibiotics on the farm, or are public health officials overreacting to a reasonable use of cost-saving technology?

www.wiley.com/college/ireland

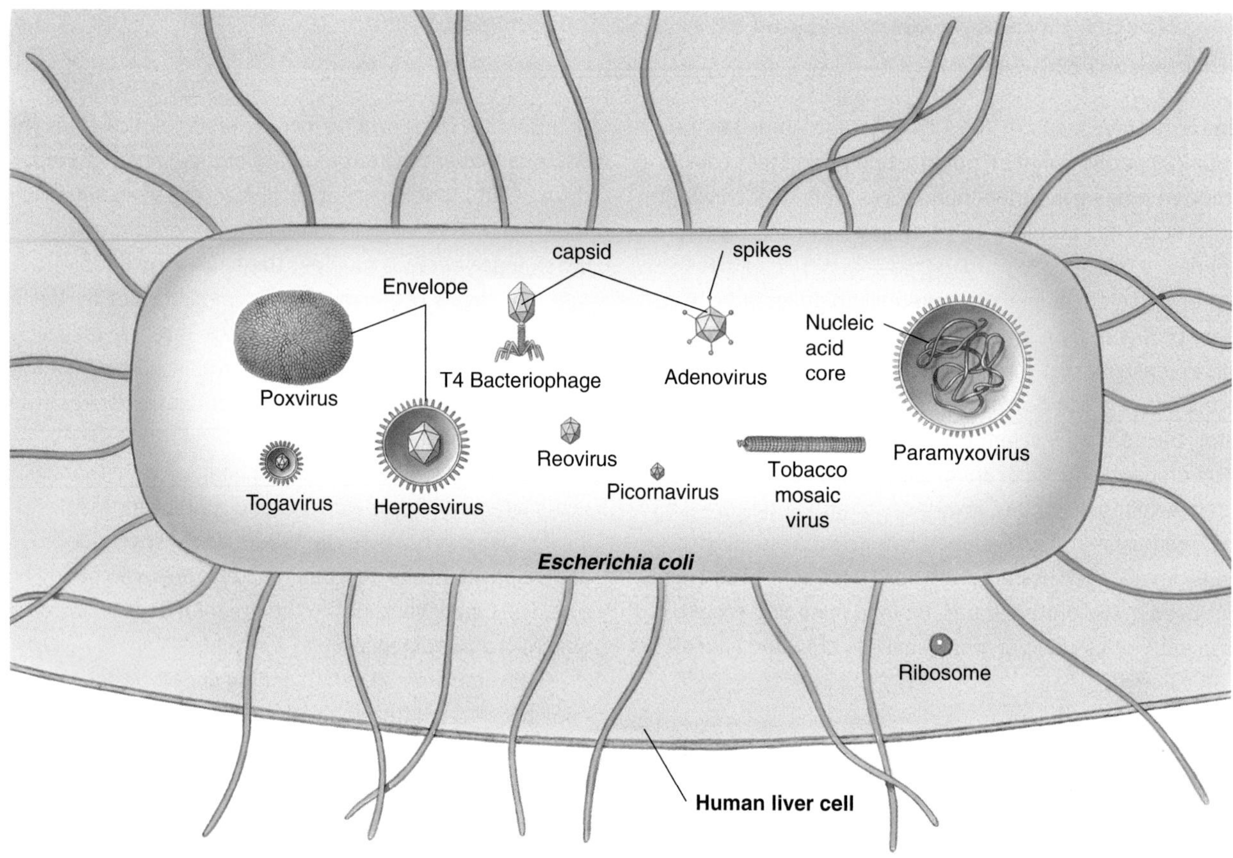

Adenovirus and bacteriophage FIGURE 10.26

Note the comparative sizes of the bacterium and viruses pictured here inside a typical human liver cell. Both the virus and the bacterium are simple structures, with no organelles or specialized compartments. The liver cell is far more complex.

Viruses are cellular parasites. When they contact their preferred host cell, they inject their nucleic acid into the host and take over its functioning. The host cell becomes a viral factory, producing new viruses at an alarming rate.

Viral DNA may remain dormant in the host cell, as happens in viruses that have a "lysogenic cycle" of replication, or it may imediately affect the cell, as occurs in the lytic cycle (FIGURE 10.27). When viral DNA takes over the host cell, the viral DNA governs the functioning of the host cell. With the proper environmental cue, the dormant virus is stimulated and begins to form new virus particles within the host cell. Eventually the host cell will fill with virus particles and burst, releasing new viruses into the body. Other viruses, like the adenovirus that is one cause of the common cold, have a lytic life cycle. After infection, there is no dormant phase. Lytic viruses cause the host cell to immediately become a viral factory, pumping out more viruses almost instantaneously.

Because viruses are not living, they are not susceptible to antibiotics. There is no cell wall to break down, no metabolic pathways to destroy, and no protein synthesis to disrupt. This is why you are not given antibiotics when you are suffering from the flu. However, a few drugs can counteract specific viruses. Acyclovir, for example, breaks down into a compound that inhibits the replication of herpes simplex virus. A wide range of compounds are being used to prevent the replication of HIV, the virus that causes AIDS. However, in most cases, when you contract a virus, all that modern medicine can do is treat the symptoms and wait for

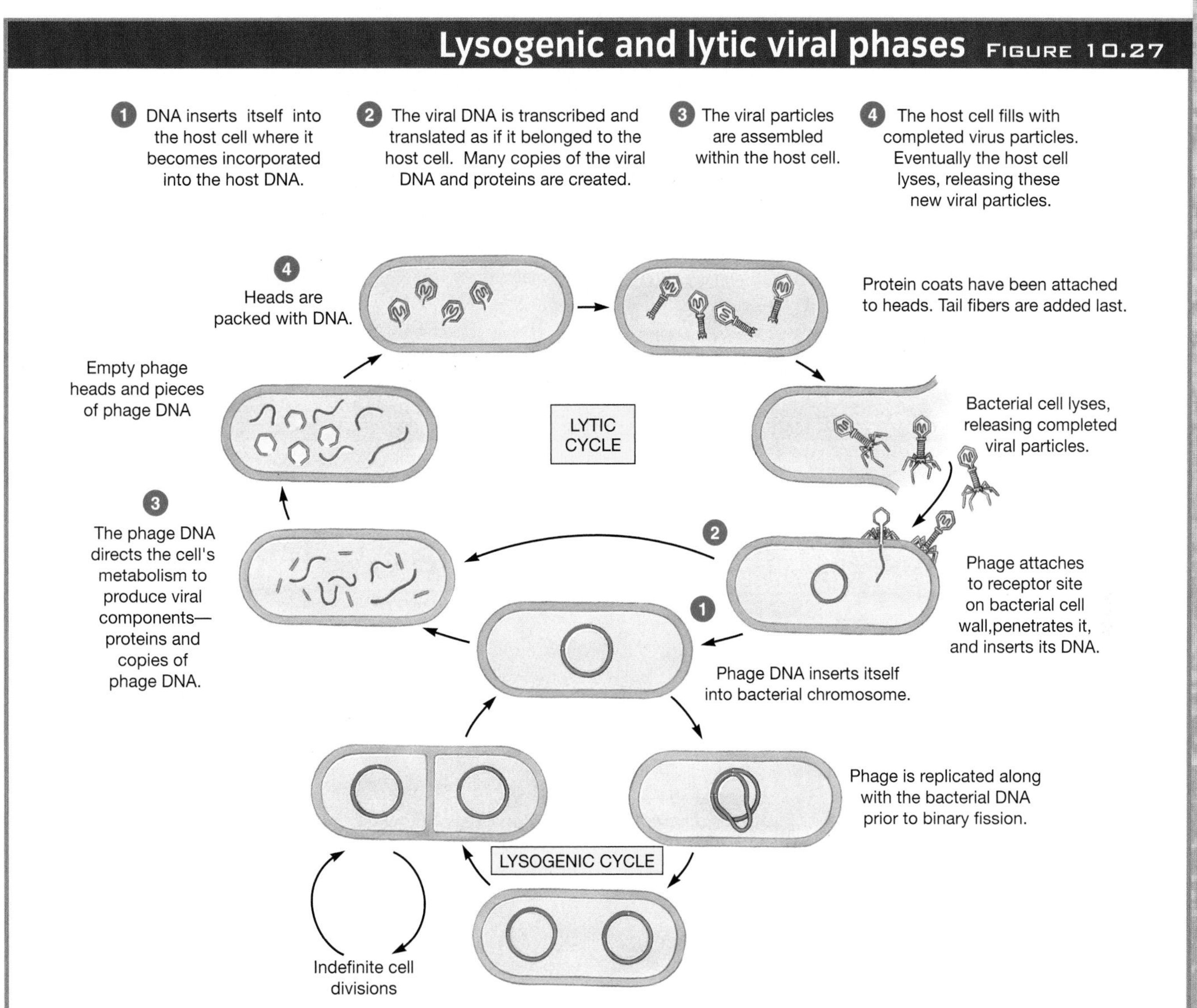
1 DNA inserts itself into the host cell where it becomes incorporated into the host DNA.
2 The viral DNA is transcribed and translated as if it belonged to the host cell. Many copies of the viral DNA and proteins are created.
3 The viral particles are assembled within the host cell.
4 The host cell fills with completed virus particles. Eventually the host cell lyses, releasing these new viral particles.
4 Heads are packed with DNA.
Protein coats have been attached to heads. Tail fibers are added last.
Empty phage heads and pieces of phage DNA
LYTIC CYCLE
Bacterial cell lyses, releasing completed viral particles.
3 The phage DNA directs the cell's metabolism to produce viral components— proteins and copies of phage DNA.
2
Phage attaches to receptor site on bacterial cell wall,penetrates it, and inserts its DNA.
1
Phage DNA inserts itself into bacterial chromosome.
Phage is replicated along with the bacterial DNA prior to binary fission.
LYSOGENIC CYCLE
Indefinite cell divisions

Lysogenic and lytic viral phases FIGURE 10.27

your immune system to contain and destroy the virally infected cells.

OTHER PATHOGENS CARRY OTHER DANGERS

Two other categories of pathogens can attack human beings in the proper conditions: fungi and prions. Fungi are eukaryotic organisms that play a major role in decay processes in the natural world. Those that you are most familiar with include mushrooms and molds. In general, fungal diseases are more common in warm, moist conditions. They can range from athlete's foot, a skin infection, to yeast infections of the female reproductive tract. Aspergillosis is a fungal infection of the respiratory tract that can cause asthmatic symptoms. Zygomycosis is a fungal infection of the blood vessels that is predominantly found in patients with a compromised immune system due to an underlying disease.

Prions are misshapen proteins that cause mad cow disease (spongiform encephalopathy) in cattle and Creutzfeldt-Jacob disease in humans. Prions have even fewer of the characteristics of life than viruses, but they do cause similar proteins to become deformed, resulting in a chain reaction of destruction. Prions can attack the brain in a wide range of mammals, ranging from deer to cats to humans. These diseases are fatal and untreatable but extremely rare.

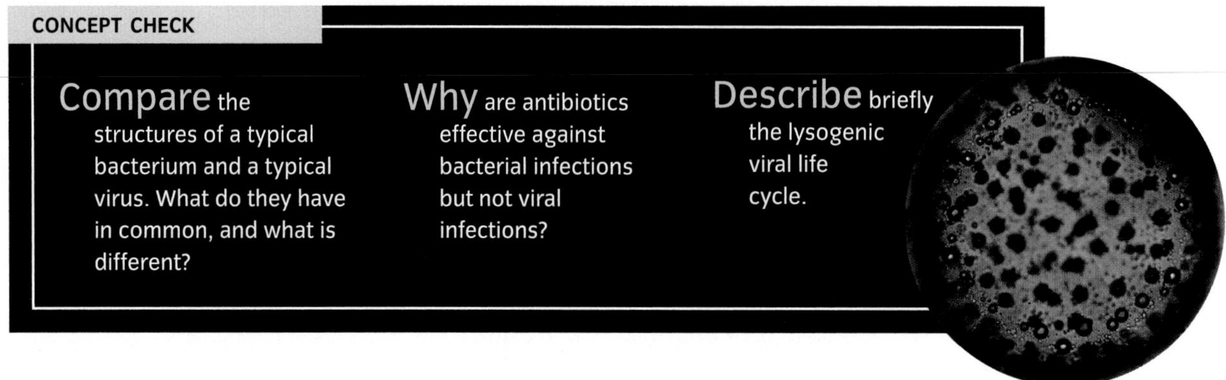

CONCEPT CHECK
Compare the structures of a typical bacterium and a typical virus. What do they have in common, and what is different?
Why are antibiotics effective against bacterial infections but not viral infections?
Describe briefly the lysogenic viral life cycle.

AIDS and HIV Attack the Immune System

LEARNING OBJECTIVES

Describe the structure of the AIDS virus.

Understand the transmission modes of HIV.

Explain the difference between HIV and AIDS.

Describe the problems that AIDS vaccines have encountered.

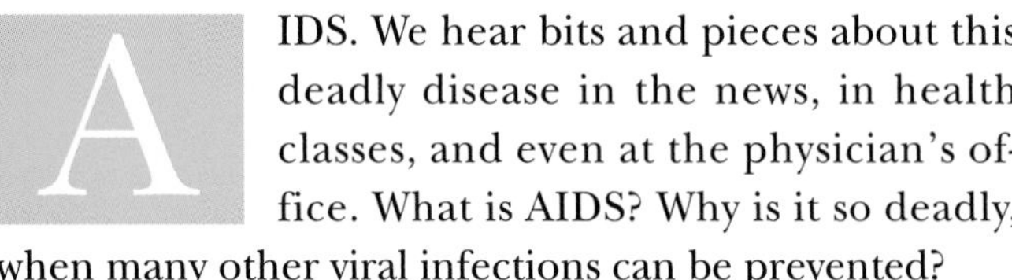

AIDS. We hear bits and pieces about this deadly disease in the news, in health classes, and even at the physician's office. What is AIDS? Why is it so deadly, when many other viral infections can be prevented?

AIDS, acquired immune deficiency syndrome, is not actually a viral infection so much as the name for a series of diverse symptoms associated with long-term infection by the human immunodeficiency virus (HIV). These symptoms include extreme loss of weight, cancerous blotches on the skin, opportunistic infection with anything that is going around, persistent fevers with accompanying night sweats, chronically swollen lymph nodes, and extreme fatigue not associated with exercise or drug use.

To understand is to protect To avoid contracting AIDS, we must understand the biology of HIV. Unlike many viruses, HIV is unstable outside of body fluids and can survive for only approximately 20 minutes when in contact with drying air and oxygen. This means most HIV transmission occurs through body fluids. Live viruses can exist in semen, blood, vaginal secretions, saliva, and tears. Thus far, transmission of HIV has been documented only through blood, semen, and vaginal secretions. Even then, transmission often requires an open wound or other tear in the epithelial lining, which gives the viral particles access to the bloodstream.

Unprotected sex is the primary mode of HIV transmission. Small tears in the vaginal and anal lining that occur during intercourse give HIV particles present in the semen easy access to the second individual's bloodstream. Sexually transmitted diseases can cause open wounds in these membranes that facilitate the spread of HIV as well. The virus is also prevalent among intravenous drug users who share needles and directly transfer small quantities of blood between bloodstreams. Back when we understood little about HIV, our blood supply was tainted with the virus, and recipients

HIV structure FIGURE 10.28

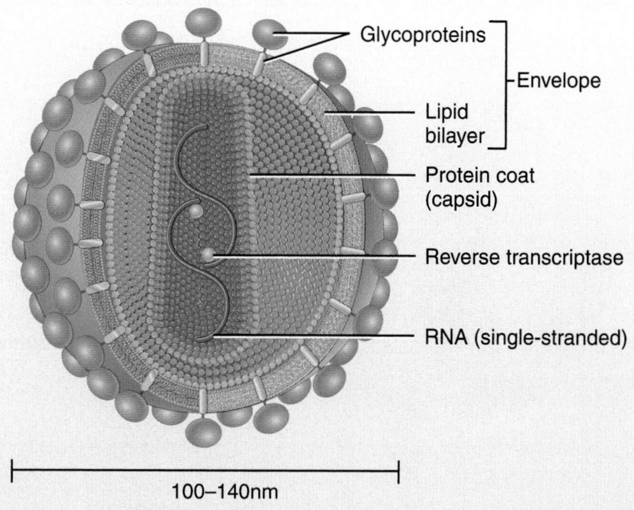

of blood transfusions occasionally got AIDS. Since the mid-1980s however, antibody tests have been used to screen out blood contaminated with HIV, essentially eliminating infection through transfusion.

In the early years of the AIDS **epidemic** in the United States, the first cases of the disease were contracted through homosexual sex or blood transfusions. Today, a growing number of heterosexual women carry HIV, but the rate of infection in children under 13 is falling. The virus can pass across the placenta and through breast milk. The possibility of an HIV-positive mother giving HIV to her child is greatly reduced by the use of anti-HIV medications.

Epidemic
Disease outbreak.

The best way to avoid HIV is to refrain from risky behaviors. Know your partner before engaging in sexual relations. Use a condom for protection. Avoid intravenous drug use, and be aware of any accidental blood contact. If you come into contact with another's blood, wash immediately and inspect the skin for cuts or scrapes. Mucous membranes are susceptible because they are penetrable by HIV. Take extra care not to introduce blood or body fluids to mucous membranes.

HIV targets the helper T cell

The scientific community needs to know more than just the mode of transmission in order to combat the AIDS epidemic. We must also understand what the virus does once it enters the body. We know that HIV targets the helper T cell, also called the CD4 T cell, eventually turning it into a virus factory. We also know the general anatomy of HIV (FIGURE 10.28).

Because HIV is a lysogenic virus, years can pass between infection and the onset of symptoms. Once HIV enters the body, it travels in the blood, where it eventually contacts a CD4 T cell. The virus attaches to the T cell at the CD4 receptor and fuses with the cell membrane, releasing its components into the host cell.

HIV uses RNA to encode its genetic instructions, so it is classified as a **retrovirus**. In order to infect a human cell, this RNA must be converted to DNA and inserted into the host cell's genetic material. Once inside the host cell, a viral enzyme called reverse transcriptase makes a DNA copy, called cDNA, of the viral RNA. A second viral enzyme then duplicates and inserts this cDNA into the host cell's DNA, so the HIV genetic material becomes part of the host DNA. The genes that code for HIV are called a provirus at this point and are indistinguishable from the host cell DNA.

At some point, perhaps 10 to 15 years later, an environmental change occurs in this infected CD4 T cell, and the provirus activates. The provirus then directs the transcription and translation of the HIV genes, shutting off the CD4 T cell's normal functions and turning it into a virus factory.

Process Diagram

HIV reproduction FIGURE 10.29

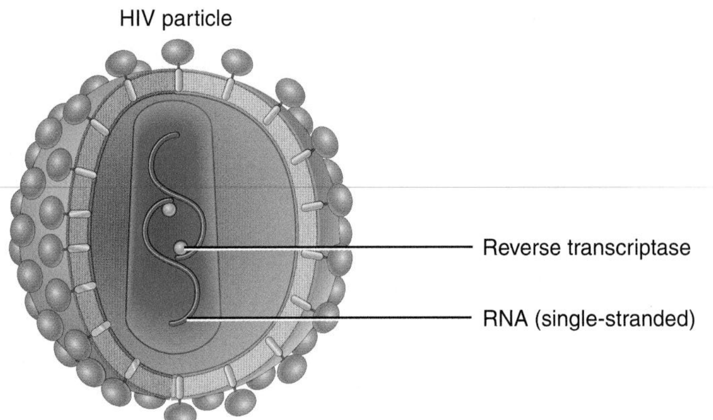

1. The virus attaches to the CD4 receptor of the cell.
2. The attached viral particle injects both viral RNA and reverse transcriptase into the host cell.
3. HIV reverse transcriptase makes a cDNA copy of its genetic information from the viral RNA.
4. The viral cDNA is incorporated into the host cell DNA, integrating it into the genome of the helper T cell, where it lies dormant.
5. When triggered, the host cell begins to manufacture more viral particles, copying the cDNA and producing viral RNA and viral proteins.
6. New HIV particles are assembled from the viral RNA and viral proteins, and leave the host cell to infect other cells of the body.

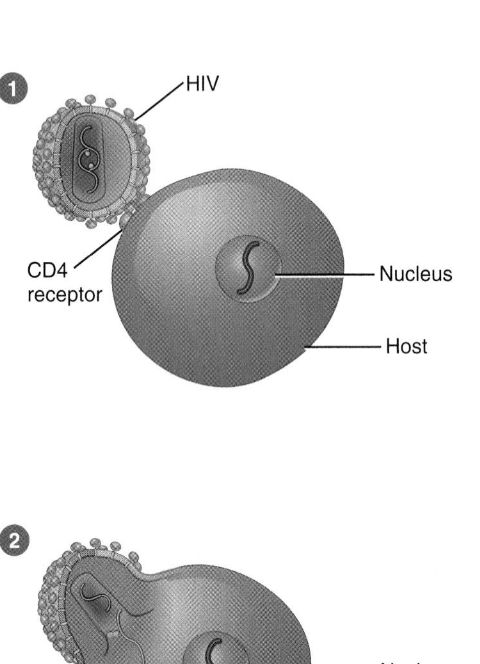

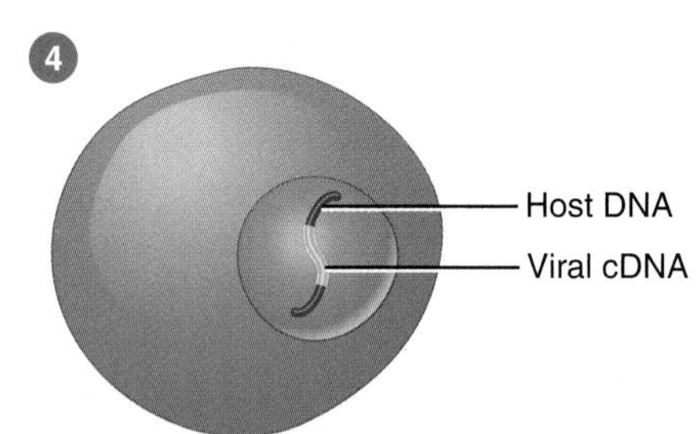

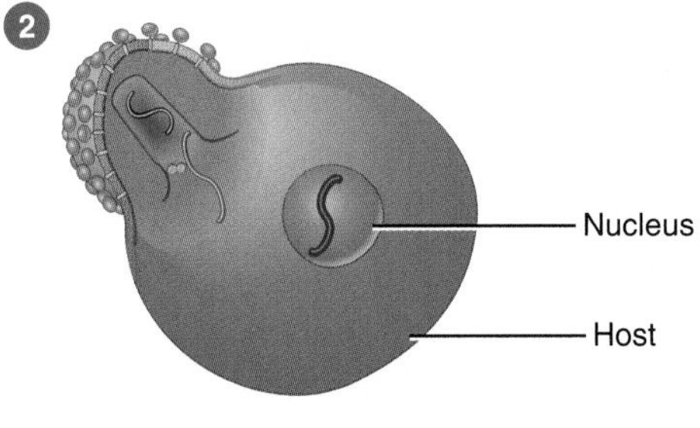

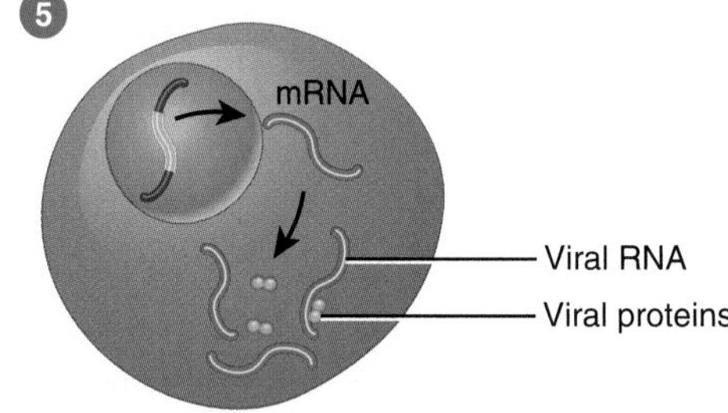

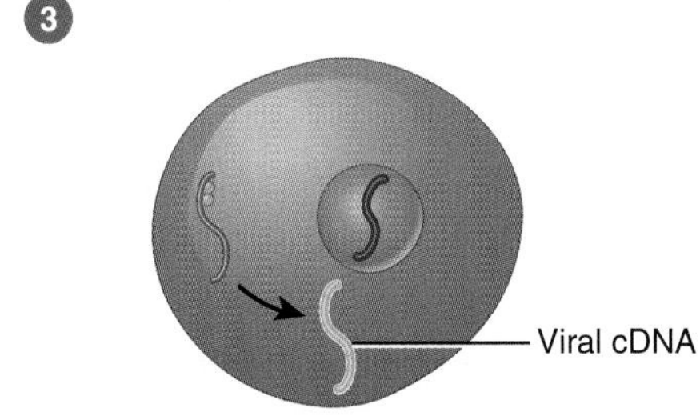

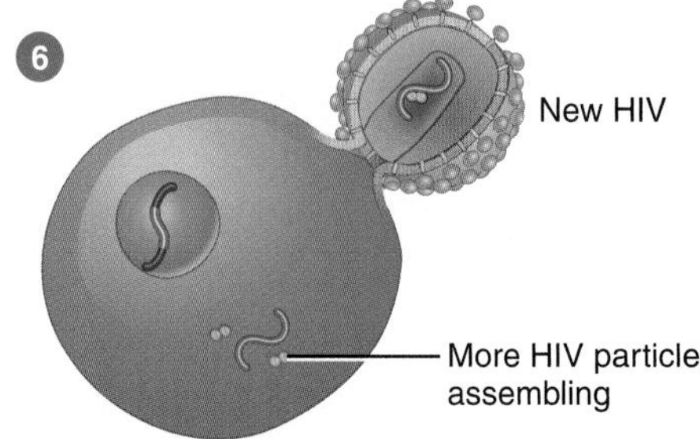

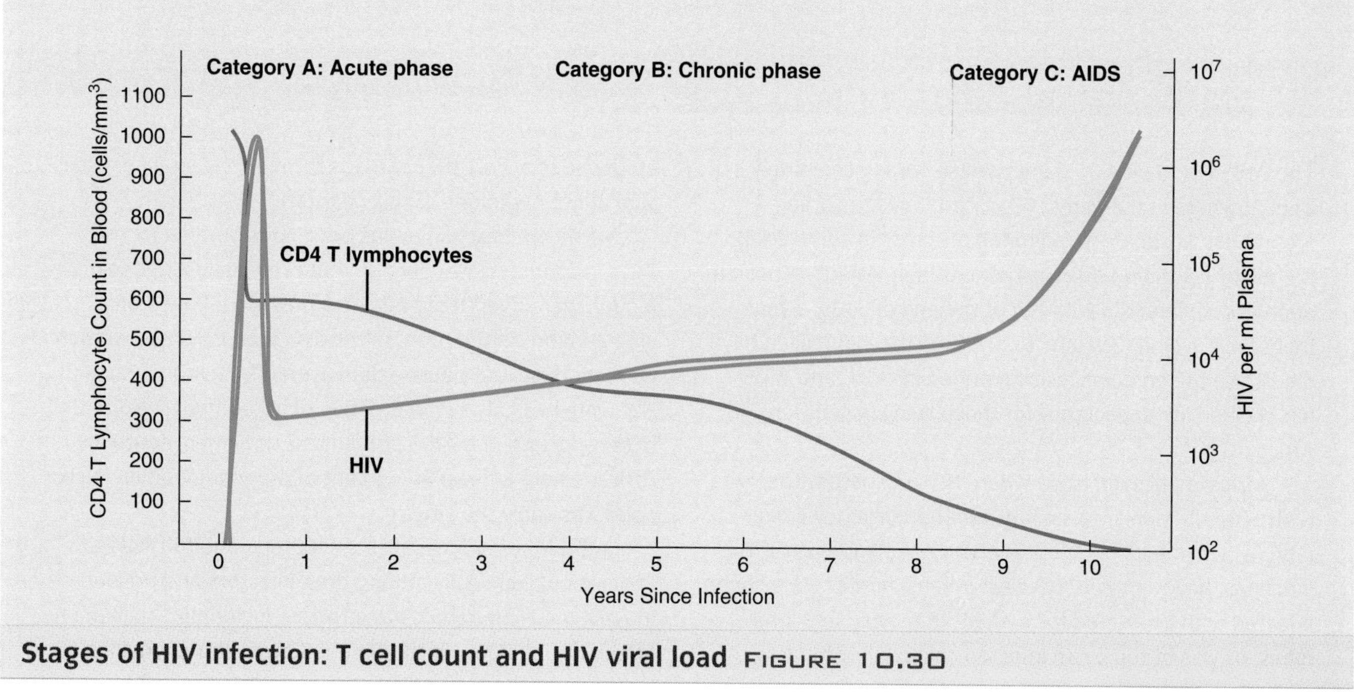

Stages of HIV infection: T cell count and HIV viral load FIGURE 10.30

The number of viral particles in the blood is called the viral load. The viral load soars after the infection, then it drops as the CD4 T cells are infected. The viral load increases again when the infected T cells start producing more virus. See FIGURE 10.29 for a summary of the process of HIV reproduction.

The infection pattern of HIV causes recognizable stages for patients (FIGURE 10.30). During the acute phase of HIV infection, the patient has a high viral load. The CD4 T cell count is normal (500+ per mm^3), and the immune system is functioning normally. A small proportion of people complain of flulike symptoms during this stage, but the majority of patients have no symptoms as the HIV virus is attacking their T cells.

The number of T cells remains high during this first attack of HIV. Eventually, however, the virus will gain the upper hand. Viral load will rise, the CD4 T cell count will fall, and the patient will suffer chronic infections. The CD4 T cell count drops below 500 per mm^3, to as low as 200 per mm^3, as infection after infection attacks the body. The lymph nodes swell with each infection and remain swollen for prolonged periods, damaging the node tissue. With fewer CD4 cells to initiate the immune response, the patient is susceptible to many diseases that a healthy immune system defeats daily. One indicator disease for this stage of HIV infection is thrush, a yeast infection in the throat and mouth. Uninfected patients easily combat this fungus but not those with lowered T cell counts.

It can take anywhere from 1 to 15 years for HIV to develop into AIDS. Once chronic infection sets in, full-blown AIDS, defined as a CD4 count below 200 per mm^3—is not far behind. The patient suffers a dramatic weight loss, the lymph nodes are damaged beyond their ability to function, and opportunistic infections like *Pneumocystis carinii* (an otherwise rare form of pneumonia), tuberculosis, or Kaposi's sarcoma attack the body (FIGURE 10.31). The patient usually succumbs to one of these infections, so death is an indirect result of the HIV infection.

AIDS patient
FIGURE 10.31

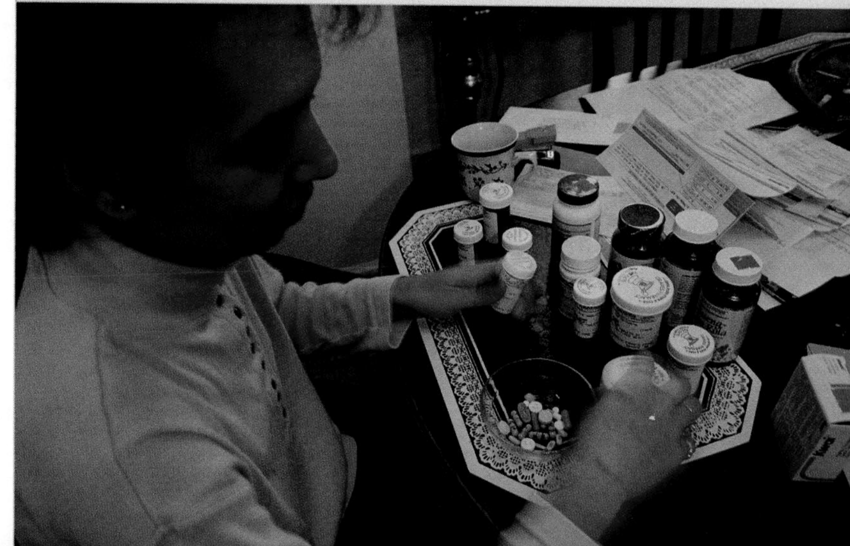

Lessons of the AIDS epidemic

The AIDS worldwide epidemic, which sprang seemingly from nowhere in the early 1980s, continues to spread around the world. Despite limited progress in some areas, the toll of death and disease remains high. About 40 million people were infected with HIV at the end of 2005, a year that saw 5 million new infections. More than 25 million have died since the epidemic was identified in 1981, and AIDS has slashed life expectancy for those living in southern Africa.

Africa is believed to be where HIV originated. The virus is structurally similar to simian immunodeficiency virus (SIV), which infects our closest relatives, chimpanzees. The virus may have "jumped species" when a hunter ate a chimp infected with a mutated form of SIV that could infect humans, or was bitten by an infected chimp while hunting. A similar danger exists today. Scientists say a deadly avian flu virus may spread from poultry to people, starting a global epidemic of a virus that spreads through the air (see discussion at the beginning of this chapter).

Although East Africa has been considered a focus of AIDS, scientists now concede that previous estimates of infection were overstated because they were based on women who visited prenatal clinics. Because these women were more sexually active than average, their high rates of HIV were not fully representative of all women. Despite the over-reporting, the AIDS epidemic is thriving in southern Africa, where at least 20 percent of pregnant women in six countries carry the virus.

AIDS is spreading in Eastern Europe and the Russian Federation, especially among drug injectors and prisoners. In Asia, 8 million people are infected, including 1 million infected during 2005. In the United States, about 1 million people are living with HIV, and about 42,000 people have AIDS.

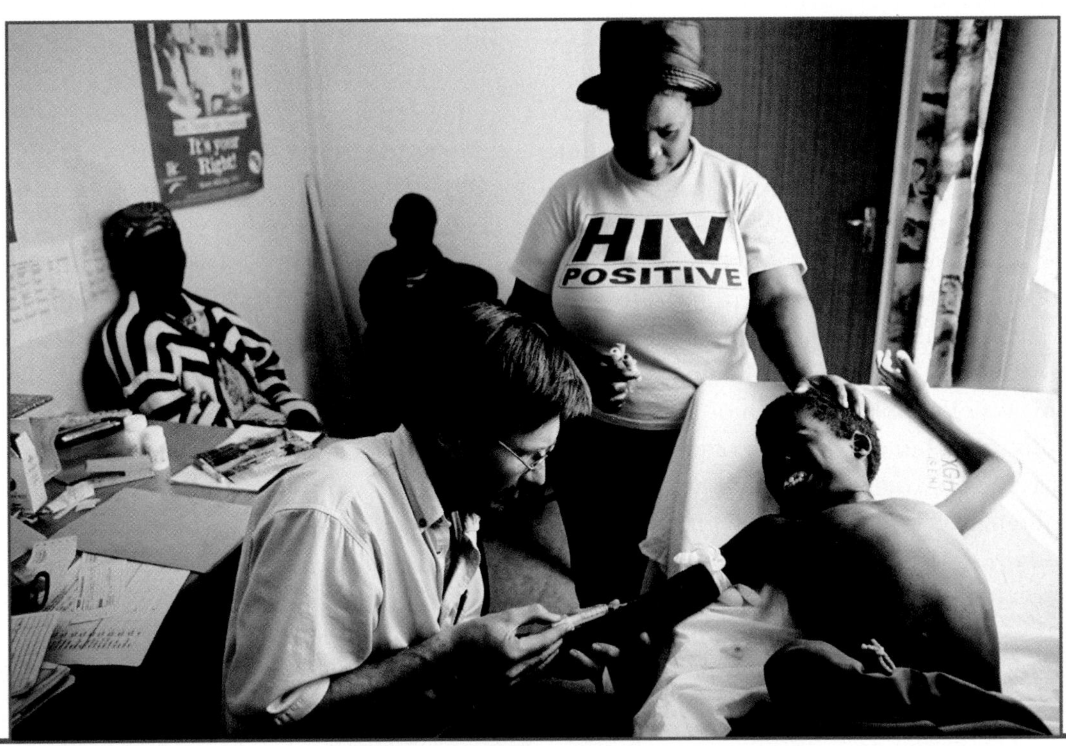

HIV treatment remains an uphill battle Although AIDS cannot be cured, we are getting better at controlling the virus and its symptoms. The latest state-of-the-art treatment is called highly active antiretroviral therapy (HAART), which includes nucleotide analogs and protease inhibitors. Protease inhibitors block the enzyme protease needed to produce new viral particles. Nucleotide analogs like AZT are structurally similar to one of the four DNA nucleotides, and they prevent translation of the HIV proviruses in infected cells. The analogs are picked up during transcription and added to the growing mRNA molecule. The analogs either stop the formation of the new chain, inhibit reverse transcriptase from completing the chain, or prevent translation of the cDNA. When nucleotide analogs are present, the message is not usable, so production of viral proteins stops.

These treatments are effective but demanding. The patient must take a complicated regime of pills throughout the day, and the side effects of these medications commonly include diarrhea, hepatitis, and diabetes.

HIV vaccines are hard to make Many viral pathogens, including smallpox, polio, and chickenpox, are controlled by vaccines, so it is logical to think a vaccine would control the AIDS epidemic as well. Medical experts are working on a preventative vaccine for those not yet infected with the virus and on a therapeutic vaccine for those already infected, but HIV vaccines do not yet work. Traditional vaccines use an attenuated viral particle, with an intact protein coat but no capability of causing infection. Injecting attenuated virus into a healthy person triggers the production of antibodies toward the viral coat. Unfortunately, HIV mutates too quickly for this tactic to work. Even if a vaccine did work against one strain of the virus, the virus changes so quickly that the vaccine would be useless in a very short time. Those vaccinated against the original strain could succumb to the newly mutated one. Scientists are looking into vaccines that stimulate the immune system using an integral part of the viral coat, such as the portion that initiates contact with the T cell. Thus far, several dozen vaccines have been tested in the United States or overseas. In July 2005, two vaccines reached phase-three trials, the last hurdle before licensing, but neither worked well enough to proceed. See the Health, Wellness, and Disease box for information on the understandings gained from HIV research. At present, the only good advice regarding HIV is this: The disease is fairly easy to prevent and impossible to cure. Prevention matters, and it works.

We can learn from disease The AIDS **pandemic** carries many lessons. When a new virus breaks out, neither vaccine nor cure is likely to be available. International scientific and public health cooperation is needed to combat diseases that often originate and survive in regions where the necessary scientific, social, and financial resources are in short supply.

> **Pandemic**
> An epidemic in a wide geographic region.

Epidemics can cause fear, resentment, and rumor. Some conspiracy theorists have blamed AIDS on plots by spy agencies or on failed vaccination campaigns. The government of South Africa, with perhaps the worst infection rate in the world, has refused to admit that HIV causes AIDS. This anti-scientific attitude makes prevention campaigns nearly impossible.

The first step in confronting an epidemic is to understand the science of the pathogen. But scientific knowledge becomes useful only when we use it to identify the economic and social practices that spread the disease, and then act to change those practices. The AIDS pandemic has shone a light on social "customs" that spread deadly pathogens. Prostitution, polygamy, unsafe sex, intravenous drug use, unfaithful spouses, and rampant sexuality in gay sex clubs have all been responsible for transmitting HIV in various places.

In many countries, more women than men are infected. Even if these women know how to protect themselves against infection, many lack the social power to enforce monogamy or condom use. Thus, educating and empowering women becomes a key strategy in slowing a pandemic that is undoing decades of hard-won economic progress in poor countries.

Pandemics may force a change in familiar economic arrangements. At some point, does the reality that poor countries need access to life-saving medicines overcome the patent rights of drug companies? Much of the recent progress against AIDS has come from broader use of antiviral medicines. India, for example, chose to bend patent laws to slow the AIDS epidemic by manufacturing generic versions of patented medicines. For too many years after expensive antivirals had begun saving lives in rich countries, AIDS remained a death sentence in poor countries. But the United Nations says that is changing: "More than one million people in low- and middle-income countries are now living longer and better lives because they are on antiretroviral treatment." These drugs saved an estimated 250,000 to 350,000 lives in 2005.

In retrospect, many governments bungled the initial response to AIDS by denial or by staging lame, uncoordinated campaigns against infection. To date, no HIV vaccines work. Even though we have relied on vaccines to control viruses for a century, for the foreseeable future the battle against AIDS will focus on changing behavior and maximizing the use of imperfect medicines.

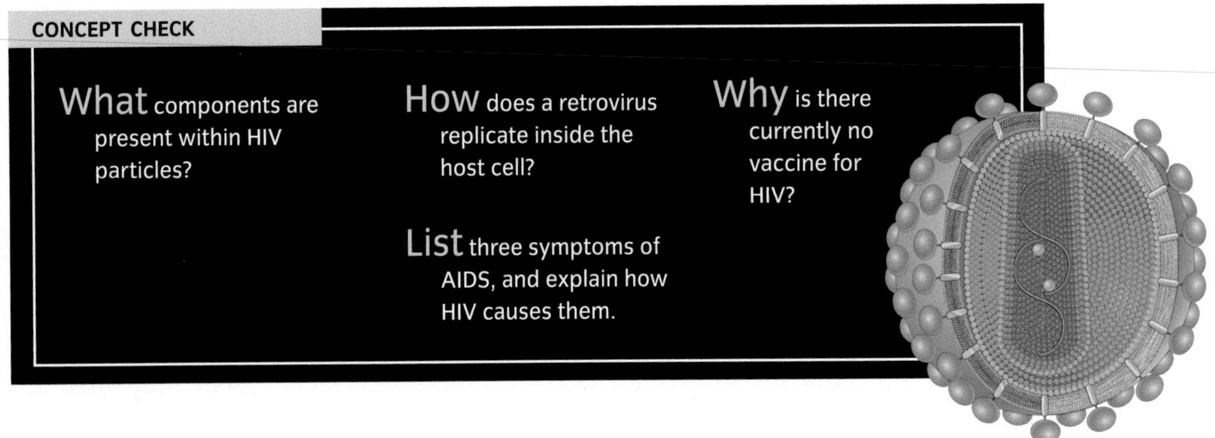

CONCEPT CHECK
What components are present within HIV particles?
How does a retrovirus replicate inside the host cell?
List three symptoms of AIDS, and explain how HIV causes them.
Why is there currently no vaccine for HIV?

Infectious Disease Is a Global Issue

LEARNING OBJECTIVES

Explain the importance of disease surveillance.

Describe one major program of WHO during 2005.

Although our bodies have an excellent series of defenses against disease, epidemics still occur. Because epidemics can cross borders, combating them requires international leadership. Since 1948, the World Health Organization (WHO) has been the branch of the United Nations dedicated to ensuring that every human attains the highest possible level of health (FIGURE 10.32). The policies of this organization are designed to enhance quality of life through improvements in physical, mental, and social well-being.

In collaboration with national health organizations such as the U.S. Centers for Disease Control and Prevention, WHO tries to keep tabs on epidemics. Researchers from WHO track the spread of epidemics in an attempt to predict viral outbreaks. WHO helps transfer samples of new diseases to safe labs where they can be quickly identified quickly.

The WHO also helps predict which strains of influenza (the "flu") are most likely to appear among

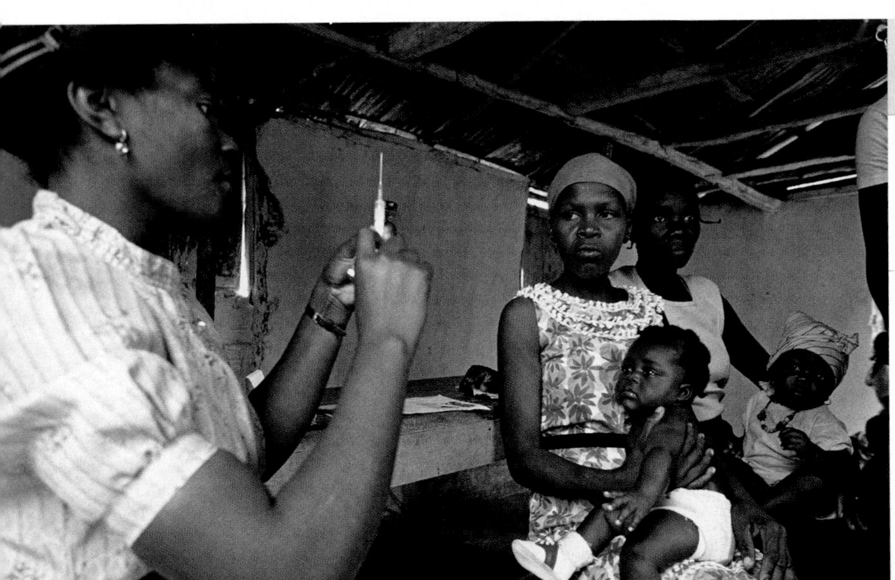

A rural clinic in a developing country

FIGURE 10.32

The World Health Organization provides vaccines for children in developing countries as part of their effort to eradicate crippling diseases. Often these vaccines are given in free health clinics such as this one in Haiti.

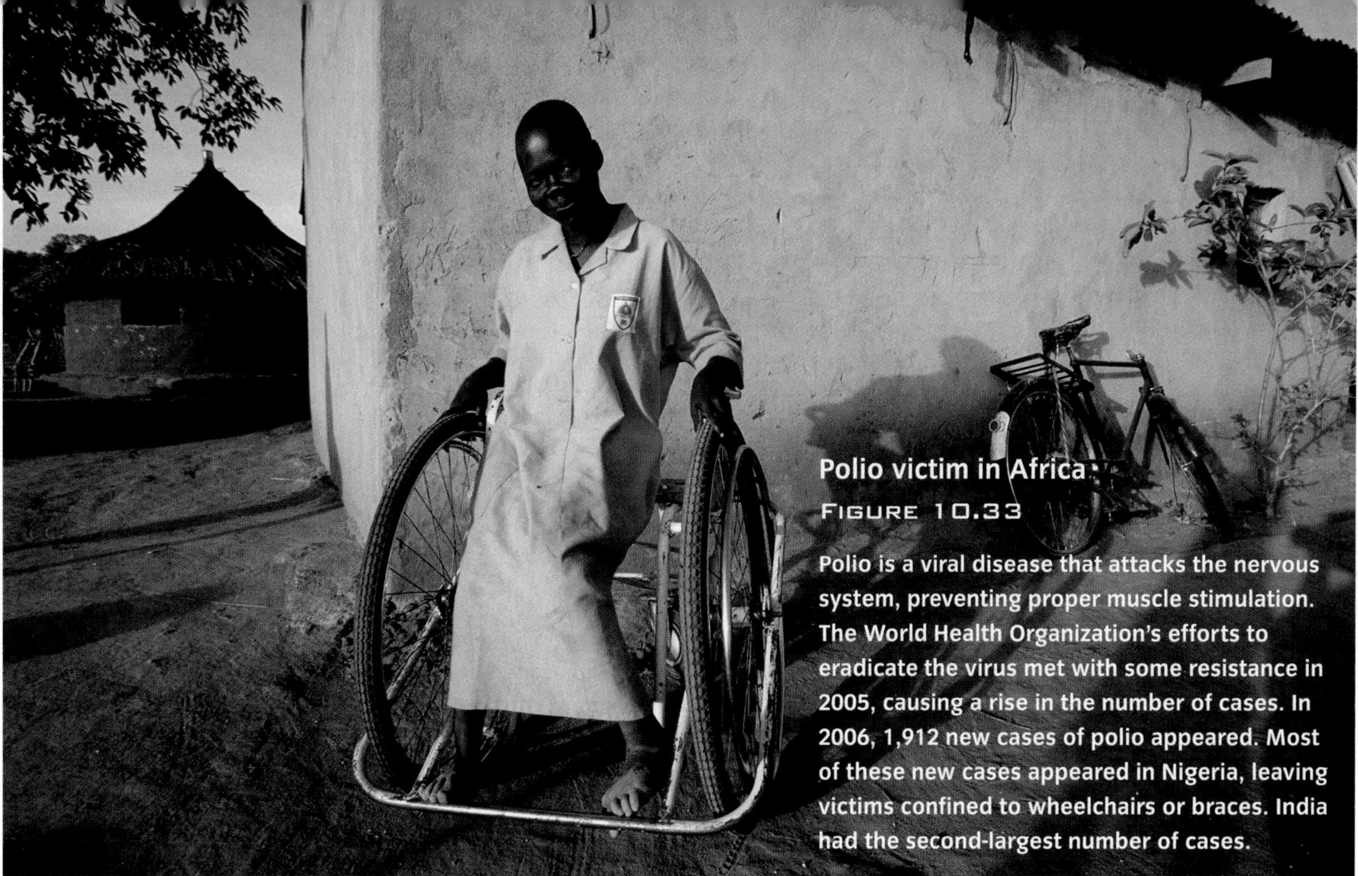

Polio victim in Africa FIGURE 10.33

Polio is a viral disease that attacks the nervous system, preventing proper muscle stimulation. The World Health Organization's efforts to eradicate the virus met with some resistance in 2005, causing a rise in the number of cases. In 2006, 1,912 new cases of polio appeared. Most of these new cases appeared in Nigeria, leaving victims confined to wheelchairs or braces. India had the second-largest number of cases.

humans each winter. Their predictions are based on past influenza strains and on hypotheses of how changing environmental conditions may affect the competitive advantages of particular strains. On this basis, the organization then selects which antigens to include in the "flu shot," and corporations and national medical systems make and administer the shots to at-risk individuals.

Common influenza remains a deadly nuisance, but smallpox was one of the greatest killers in history. At the end of the twentieth century, the WHO directed a worldwide campaign to eradicate smallpox, the only viral disease ever eradicated from the human population. WHO is now in the midst of a campaign to eradicate polio, which attacks the motor neurons of the brain stem and spinal cord and causes paralysis in 1 of 200 cases (FIGURE 10.33).

In 2002, the eradication program was working well, with only three Asian and three African countries reporting cases. Since then, the program has been undermined by fear, rumors, and political manipulation. For 16 months, northern Nigeria refused to administer polio vaccines because Islamic leaders charged that the vaccine was a disguised sterilization campaign. The lapse in vaccinations caused a resurgence of polio. Between April 2005 and April 2006, 1,876 cases were seen worldwide, including 769 in Nigeria. The other cases in Africa were traced via genetic fingerprinting to the viral strain of polio in Nigeria. Nigeria rescinded its ban on vaccination at the end of 2004, but 21 countries have been reinfected since 2003.

CONCEPT CHECK

What is one key role of the World Health Organization?

How does WHO determine which strains of influenza are most threatening each year?

CHAPTER SUMMARY

1 The Lymphatic System Is the Center of Our Immune Response

The lymphatic system returns interstitial fluid to the cardiovascular system, absorbs and transports fats, and provides specific immunity. The system is composed of lymphatic organs, lymphatic tissue, and lymphatic vessels. Lymph forms when portions of blood are forced through the capillary wall. This lymph fluid bathes and cleans the tissues.

2 Specific Immunity Relies on a Deadly Series of Cells

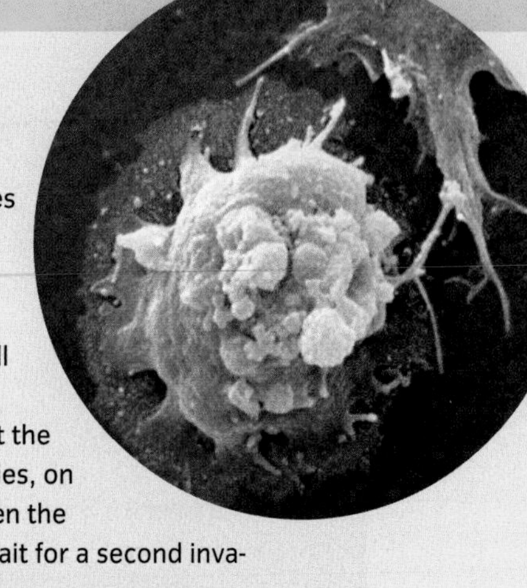

Cellular immunity is embodied in lymphocytes in the bloodstream and lymph nodes. Humoral immunity is carried out by B cells residing in the lymph nodes. Helper T cells detect a specific antigen and stimulate the proper B cell. That B cell then clones, producing plasma cells and memory cells. The plasma cells produce antibodies against the specific antigen. There are five classes of antibodies, on the basis of shape and timing of appearance. When the pathogen has been cleared, memory cells lie in wait for a second invasion by the same pathogen.

3 Immunity Can Be Acquired Actively or Passively

Active immunity refers to immunity obtained through activating your immune system and creating memory cells. Both immunizations and the natural course of recovering from disease cause a population of memory cells to form in the body. When the same antigen reappears, the memory cells immediately clone and eliminate the antigen. This secondary response is far faster than the primary immune response. Passive immunity occurs when antibodies are given to an individual rather than formed by that individual. Natural passive immunity occurs when a fetus or infant receives antibodies from the mother, through diffusion across the placenta and then via breast milk.

4 Viruses and Bacteria Cause Disease in Different Ways

Bacteria and viruses cause most infectious disease in humans. Bacteria are prokaryotic cells. They have a cell wall, a cell membrane, ribosomes, a circular piece of DNA anchored to the cell wall, and some intracellular fluid. Bacteria are classified by shape, gram staining, and genetics. Viruses are small bits of nucleic acid covered in a protein coat, but they are not considered alive. Antibiotics kill bacteria by disrupting their cell membranes or other metabolic processes, but they have no effect on viruses. Prions and fungi also cause disease.

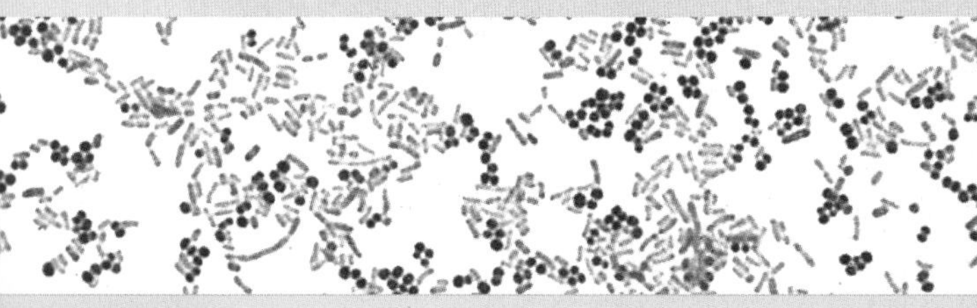

5 AIDS and HIV Attack the Immune System

HIV is a blood-borne virus that causes death via AIDS. It is a retrovirus, infecting individuals through blood-to-blood contact that usually occurs during unprotected sex or use of contaminated needles. The cycle of HIV begins with introduction to the bloodstream. It then attaches to and invades a host CD4 T cell, where it copies its own RNA into cDNA. Next, the viral genes are inserted into the host cell DNA. Symptoms are negligible at this point. Years later, the infected CD4 T cells begin to produce virus, increasing the viral load of the patient and decreasing the CD4 T cell count. AIDS is diagnosed when the CD4 count drops below 200 per mm^3 and the patient is suffering from opportunistic infections that healthy individuals' immune systems easily fight. Vaccine treatment for HIV remains out of reach, but researchers are getting closer to success.

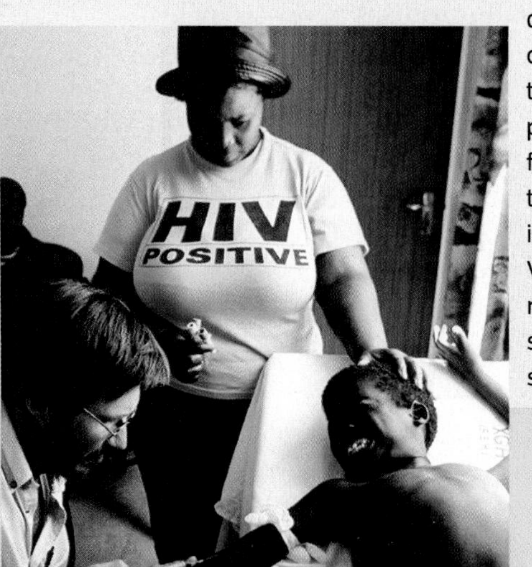

6 Infectious Disease Is a Global Issue

The World Health Organization (WHO) is responsible for monitoring and predicting pandemics and for helping national health organizations coordinate healthcare worldwide. Epidemics are often easier to prevent than to treat, so global monitoring is needed to track them to their source and start containment.

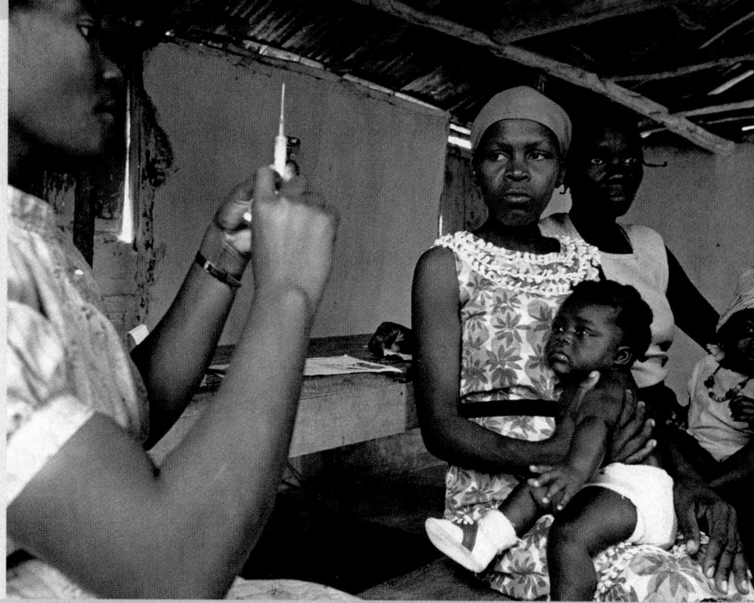

KEY TERMS

- **agglutinate** p. 294
- **antibodies** p. 294
- **apoptosis** p. 295
- **attenuated** p. 303
- **CD4** p. 297
- **class II MHC (major histocompatibility complex)** p. 297
- **cytokines** p. 301
- **cytotoxic T cells** p. 299
- **epidemic** p. 313
- **host cell** p. 308
- **immune response** p. 284
- **immunization** p. 293
- **interstices** p. 284
- **lymphatic tissue** p. 284
- **lymphocytes** p. 291
- **mediastinum** p. 288
- **mesenteric** p. 289
- **open system** p. 286
- **pandemic** p. 317
- **photosynthesis** p. 306
- **prokaryotic** p. 306
- **stem cells** p. 292
- **titer** p. 303

CRITICAL THINKING QUESTIONS

1. Rheumatoid arthritis is an autoimmune disease. In autoimmune diseases, your immune system loses its ability to differentiate self from non-self and begins to attack your body. In rheumatoid arthritis, the attack affects cartilage in the joints. Using what you have learned about the immune response, what symptoms would you predict? How would the normal functioning of the immune system lead to these symptoms? What might a physician prescribe for rheumatoid arthritis?

2. Dengue fever is a tropical disease that, by 2005, had reached epidemic proportions in Malaysia and Vietnam. The disease spreads quickly by the *Aedes aegypti* mosquito. Explain how a vaccine might slow this epidemic. What characteristics would the vaccine need? What are the differences between the primary and secondary immune responses in terms of a dengue vaccine?

3. Compare a bacterium to a virus. Which is larger? How do their internal structures compare? How do their outer casings, or membranes, compare? What is the infectious pathway of most bacteria? How do viruses infect cells?

4. *Herpes simplex (HS)* is the name for a group of viruses that attack human cells. This virus is lysogenic, causing cold sores (HS I) or genital warts (HS II). Both of these varieties display as open canker sores that periodically reappear. Reviewing the lysogenic cycle of viral infection, predict what is happening within an infected cell during the appearance of a cold sore.

5. The flu is a serious problem for the WHO. Why is this so? Flu seems like a minor inconvenience, leaving most of us ill for a mere few days. Why is influenza still a number one priority of the WHO? What can you say about the origin of a serious influenza epidemic?

SELF TEST

1. True or False? The lymphatic system is anatomically similar to the circulatory system, with a series of vessels that transport lymph to and from the heart.

2. Functions of the lymphatic system include all of the following EXCEPT
   a. maintaining tissue fluid homeostasis.
   b. absorbing fats from the intestinal tract.
   c. defending against bacterial invasion via fever.
   d. defending against specific invaders.

3. Identify the structure indicated as A on the diagram.
   a. Lymph node
   b. Tonsil
   c. Peyer's patch
   d. Spleen

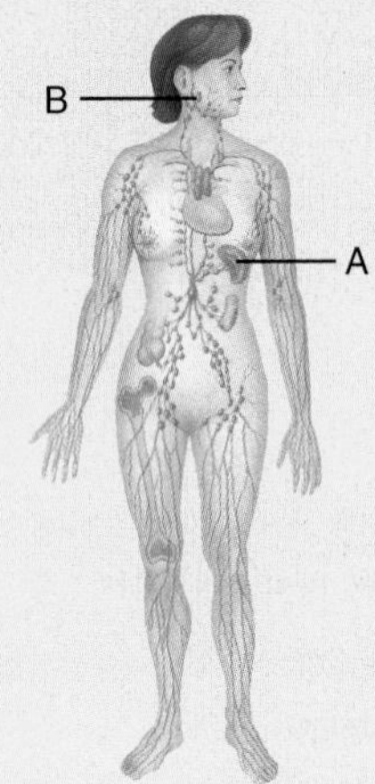

4. The structure indicated as B on that same diagram
   a. produces lymphocytes.
   b. filters blood.
   c. cleans body fluids passing the organ.
   d. has no known function, and can easily be removed without damage to the individual's health.

5. The lymphatic organ labeled A in the photograph below is the
   a. spleen.
   b. thymus.
   c. tonsil.
   d. inguinal lymph node.

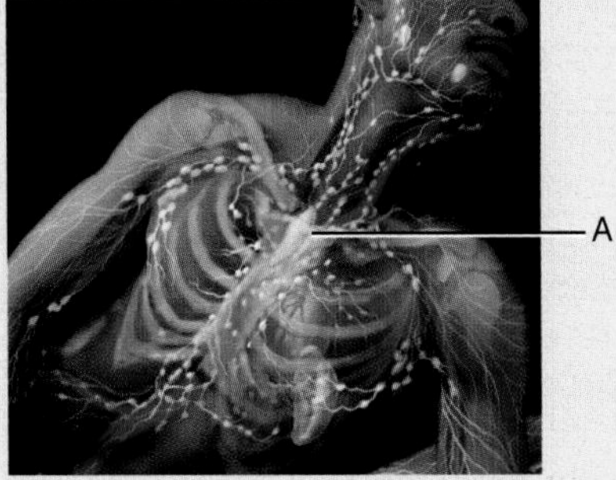

6. The stem cells in red bone marrow produce
   a. both red and white blood cells.
   b. B cells only.
   c. T cells only.
   d. macrophage only.

7. Humoral immunity employs
   a. T cells.
   b. B cells.
   c. antibodies.
   d. all of the above.

8. True or False? Specific immunity requires cells to demonstrate specificity, memory, and self-recognition.

9. The portion of this image labeled A is the
   a. antigen.
   b. random lymphatic cell.
   c. specific receptor on lymphatic cell.
   d. specific pathogen.

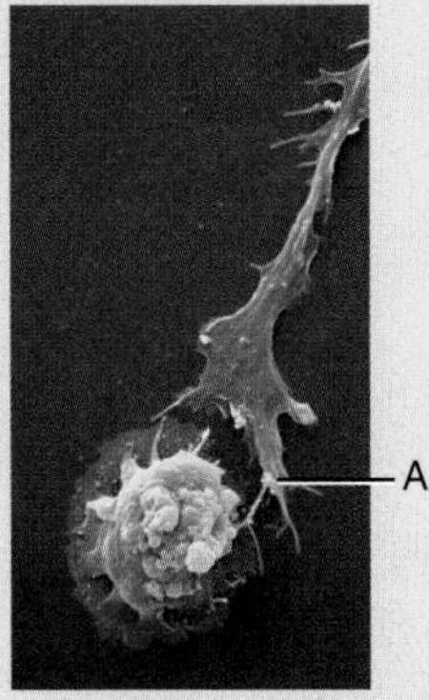

10. Vaccinations and booster shots are designed to assist in the formation and maintenance of
    a. antibodies in the bloodstream.
    b. cloned B memory cells.
    c. cloned T memory cells.
    d. plasma cells.

11. The type of T cell that binds an antigen, clones to amplify the signal, and then stimulates the B cell that will produce the matching antibody is the
    a. natural killer T cell.
    b. thymic cell.
    c. APC cell.
    d. helper T cell.

12. The activation of a T cell requires direct interaction between the pathogen and
    a. the CD4 protein complex of the T cell.
    b. the class II MHC molecules on the surface of the T cell.
    c. the CD4 protein complex of the B cell.
    d. the antigen-presenting cell.

13. The function of the antibody shown here is to
    a. bind directly to the antigen.
    b. trigger allergic responses.
    c. agglutinate the pathogen.
    d. stimulate B cell activation.

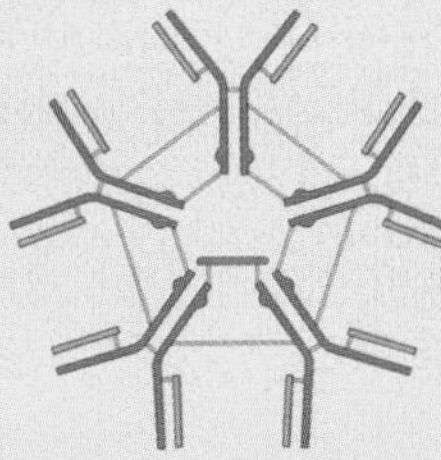

14. ___________ are directed against one specific antigen but link to many different antigenic sites on that antigen.

a. Monoclonal antibodies
b. Polyclonal antibodies
c. Genetically engineered antibodies

15. The type of immune cell causing the reaction seen here, where the pathogenic cell is attacked by released perforin, is the

a. helper T cell.
b. cytotoxic T cell.
c. HLA cell.
d. antigen-presenting cell.

16. True or False? The secondary immune response, occurring after initial exposure to the pathogen, is much faster than the primary response.

17. The type of immunity achieved by the infant via breast milk is

a. active artificial immunity.
b. passive artificial immunity.
c. passive natural immunity.
d. active natural immunity.

18. Which of the following is NOT characteristic of the activity of immune cells in an autoimmune disease?

a. Attack the body's own cells
b. Cause erratic degeneration of joint tissue
c. Destroy pancreatic cells
d. Increase in number and specificity in the blood

19. The type of bacteria found in long chains of spherical organisms is

a. staphylococcus.
b. coccus.
c. bacillus.
d. streptococcus.

20. The phase of the viral life cycle depicted below is the

a. lytic phase.
b. lysogenic phase.
c. replication phase.
d. dormant phase.

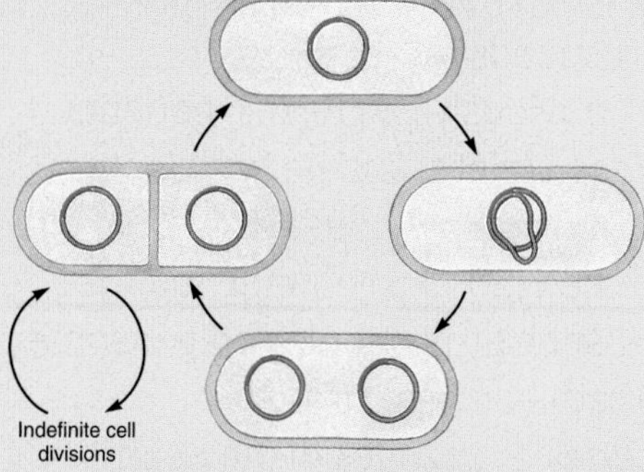

21. HIV attaches to the CD4 protein complex of the ___________, obtaining entry to the cell where it may lie dormant for many years.

a. cytotoxic T cell
b. helper T cell
c. B cell
d. macrophage

The Cardiovascular System

11

Lance Armstrong, seven-time winner of the Tour de France, is not like the rest of us, and one of his most unusual features is his heart. At rest, for example, Armstrong's ticker is said to beat 32 times a minute, less than half of the average for people his age. That slow rhythm provides a big clue to the efficiency of his heart (and musculature) because it suggests how much power he can produce when his heart speeds up.

Armstrong obviously has terrific genetics for an endurance athlete, and he has also perfected the arduous training needed to beat all contenders in the 3,200+ kilometer Tour de France. But most healthy people can get some benefits from "athletic heart syndrome," which is the series of changes in the heart's anatomy and physiology that result from regular, strenuous exercise. This training allows more dilation in the heart's four chambers so they accept and discharge more blood on each stroke, thereby delivering more oxygen and nutrients to the body. Athletic heart syndrome, combined with the many benefits that emerge from a combination of genetics and training, explain why Armstrong had the fastest times on the Tour, seven years running.

In this chapter, we will introduce the structure and function of the heart as well as the rest of the cardiovascular system, and then look at some things that can go awry with this essential delivery and trash-removal system. The cardiovascular (CV) system is a triumph of sophisticated design, but cardiovascular problems are also the predominant cause of death in many developed countries.

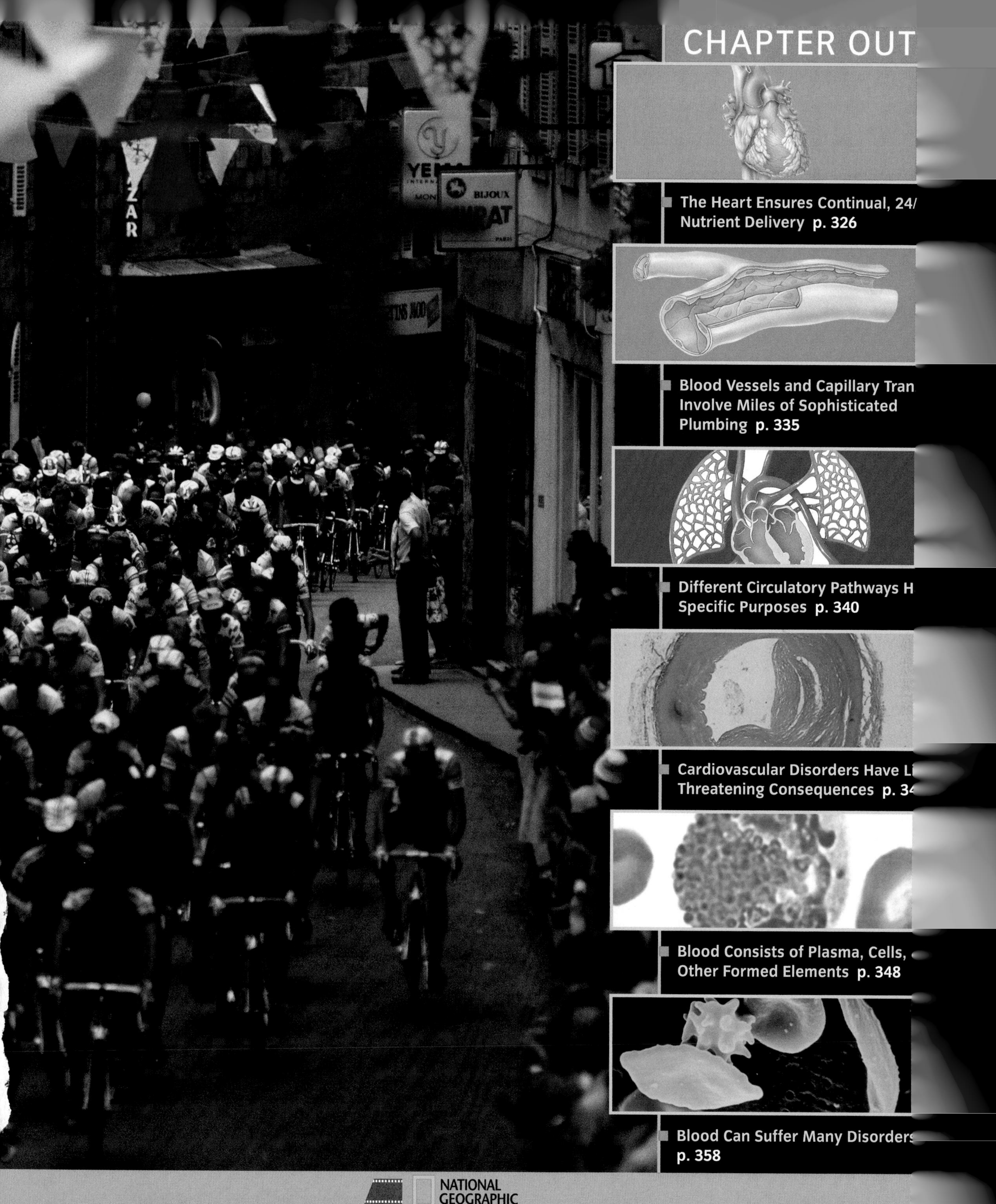

CHAPTER OUT

NATIONAL GEOGRAPHIC

The Heart Ensures Continual, 24/7 Nutrient Delivery

LEARNING OBJECTIVES

Describe the structure of the heart.

Trace the flow of blood through the heart.

Explore the electrical signaling that produces contraction.

Describe how the tracings on an ECG reflect the heart's beat.

On the day Dr. Seuss's Grinch discovered the true meaning of Christmas, his heart grew three sizes. The tin man in the *Wizard of Oz* wanted a heart so he could have emotions. We've all heard the heart described as our emotional center, but the heart is literally the center of the cardiovascular system. The heart is a pump that pushes blood through miles of blood vessels. The blood pressure generated by each heartbeat ensures that nutrients and oxygen reach every cell, directly or indirectly.

To understand the importance of the CV system, look at any large city. Vehicles transport food, goods, and raw materials into the city and deliver them to residents and institutions. After the goods are consumed, waste that is left over must be recycled, burned, reused, or shipped away. Any obstruction to this flow is likely to damage the city. Within days after trash collectors went on strike in the 1980s, garbage was piling up along the streets of New York City, blocking traffic, impeding business, and offending millions of noses. The city almost ground to a halt until a new contract was signed and trash removal resumed. Similarly, if the human body cannot move water, nutrients, and oxygen into the tissues, and remove wastes from them, tissues will die, and the organism as well.

In delivering oxygen and removing carbon dioxide, the cardiovascular and respiratory systems work together. The **respiratory** system (Chapter 12) brings oxygen to the blood and removes carbon dioxide from it. The **cardiovascular** system transports that blood, carry-

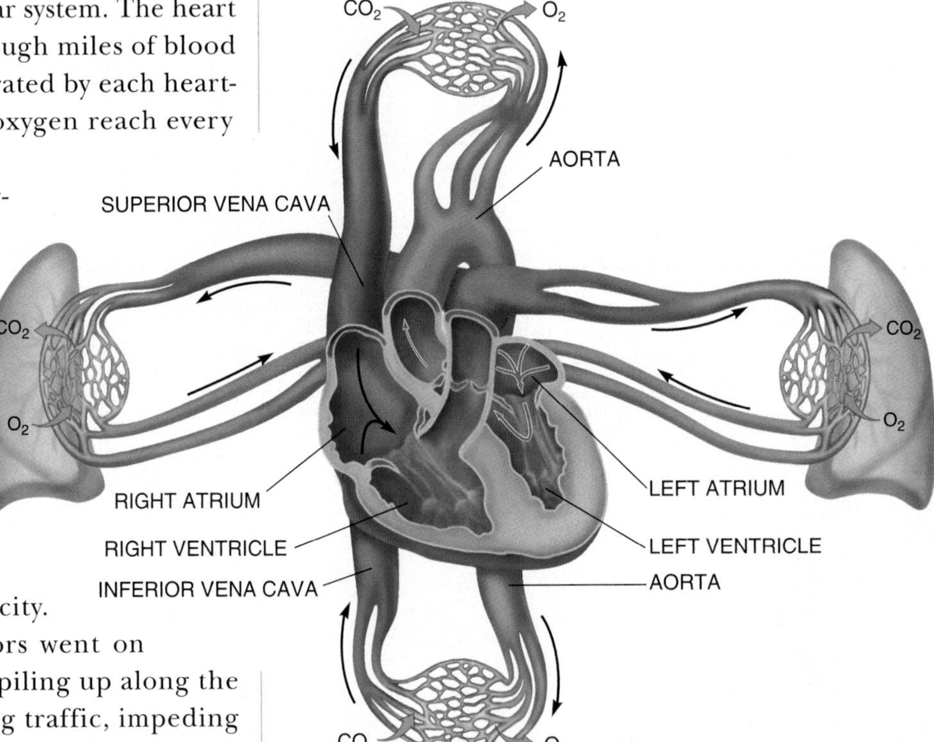

Basic schematic of the CV system

FIGURE 11.1

The heart and blood vessels form a closed circuit that transports blood from the heart to various parts of the body and back to the heart.

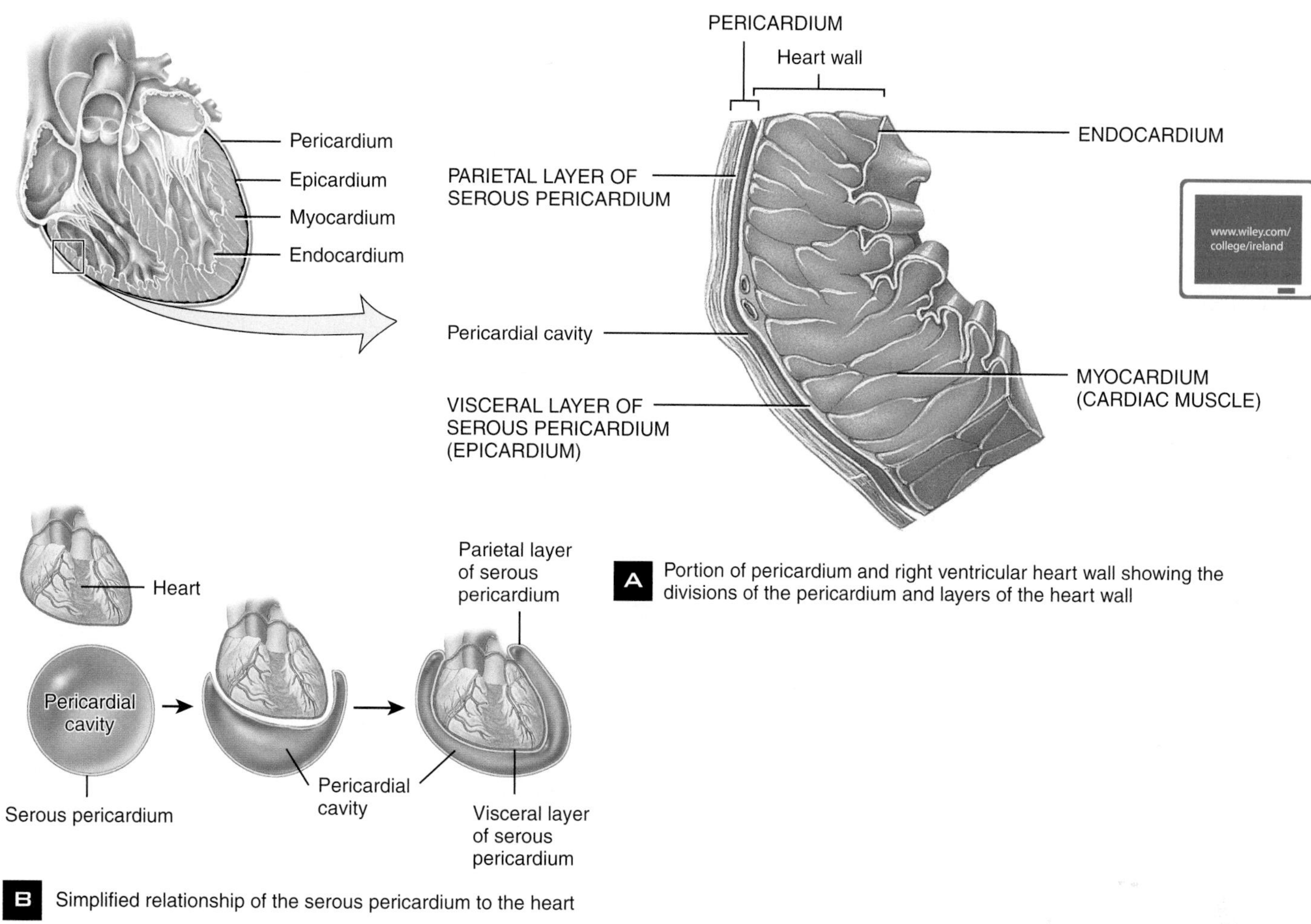

Pericardium FIGURE 11.2

The pericardium includes two layers, one lining the walls of the thoracic cavity and the other attached to the cardiac muscle of the heart. Between these two membranes is a thin, slippery layer of serous fluid, allowing the heart to move within the cavity without damaging itself or the thoracic area.

ing nutrients, wastes, and dissolved gases to and from the tissues. The cardiovascular system includes the **heart, blood vessels**, and **blood**. We will look first at the heart and the blood vessels, and then at the blood that flows through that closed circuit (FIGURE 11.1).

The heart resides in the center of the thoracic cavity, hanging by the great blood vessels that deliver and remove blood. The **pericardium** (FIGURE 11.2) is a serous membrane that surrounds the heart and allows it to beat without causing damage to itself—beating causes the heart to jump around in the **mediastinum**.

Mediastinum The area between the lungs, containing the heart, lymphatics, and vessels of the thoracic region.

The heart is composed of four chambers—two **ventricles** and two **atria**. The atria are smaller, thin-walled chambers sitting atop the thick-walled, muscular ventricles. The atria receive blood from large veins and direct it into the ventricles, which expel the blood under great pressure toward the lungs or body.

During development, the heart forms from two adjacent vessels. By the third week of development, these two vessels fuse to form one large chamber (a ventricle) with two smaller vessels delivering blood to it. The heart is upside down at this point, with the ventricle above the two incoming vessels. Around week 5, the heart curves back on itself in an "S" turn, creating the familiar anatomy, with the atria on top (FIGURE 11.3).

The adult heart is shown in FIGURE 11.4. Note the thick ventricular walls, especially in the left ventricle. It is the left ventricle that must generate enough force to push blood throughout the body. The less muscular right ventricle pushes blood only to the nearby lungs. The walls of the atria are even less muscular because these chambers are essentially holding tanks for blood after it returns from the body or lungs, rather than pumping chambers.

> **Tricuspid**
> The valve between the right atrium and right ventricle, composed of three points (cusps) of connective tissue.

Each heart chamber contains one valve that opens to allow blood to pass and then closes when the chamber contracts to pump. Because these valves are found between the atria and the ventricles, they are called atrioventricular valves. The atrioventricular valves are the **tricuspid** valve in the

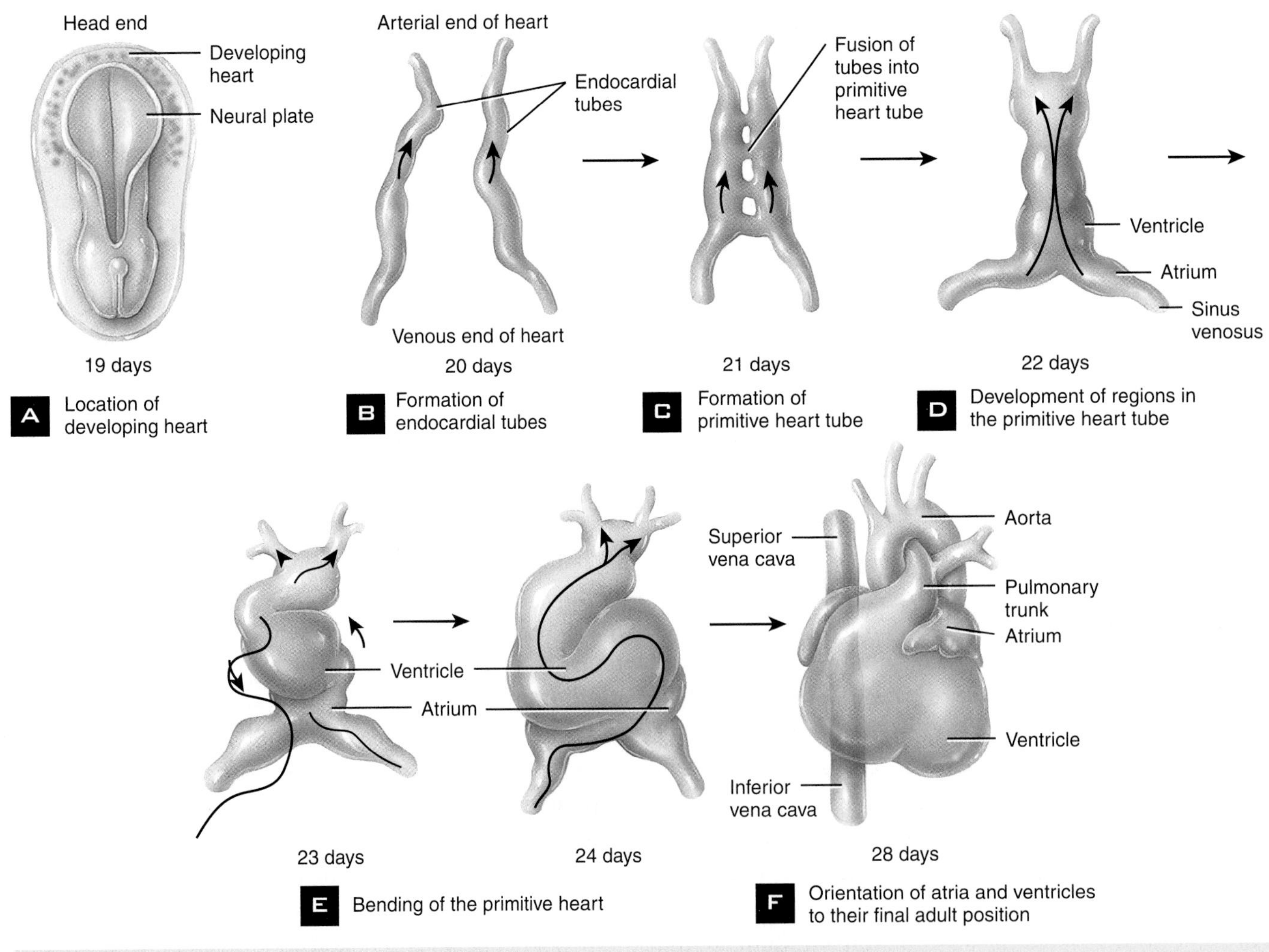

Development of the heart FIGURE 11.3

Bicuspid
The valve between the left atrium and left ventricle, composed of two opposing cusps or flaps of connective tissue.

Prolapse
Movement of an organ from its original position in the body, usually because of gravity or pressure.

right ventricle and the **bicuspid**, or **mitral** valve, in the left ventricle. Valves are composed of dense, irregular connective tissue and are held in place by the **chordae tendineae** (literally chords of tendons). These "heart strings" anchor the cusps of the valves to the **papillary muscles**. When we listen to a heart beat, even without a stethoscope, part of what we hear is the thrumming of the heartstrings as they are pulled tight and pressurized blood flows past them.

If the opposing surfaces of a valve fit poorly, blood can slip past, causing the valve to flutter. This fluttering creates a murmur (an audible change in heart sound) and can possibly lead to valve **prolapse**. The most leak-prone valve is the mitral valve in the left ventricle. When the mitral valve fits poorly, the condition is called mitral valve prolapse (MVP). MVP runs in families and affects more women than men. Although MVP saps heart efficiency, patients rarely report symptoms and require no medical treatment.

At the base of the great arteries leaving the heart are the **pulmonary** and **aortic valves**. These valves are shaped like three flexible bowls, anchored to the walls of the great vessel (FIGURE 11.5, p. 330). When the heart pushes blood into the pulmonary or aortic artery, the bowls flatten against the artery walls so the blood can flow freely. When pressure drops inside the heart, blood in the arteries pushes back, ballooning the three bowls so they open and contact one another, closing the arterial opening leading back to the heart. Because these valves lack the chordae tendineae, these valves make no humming sound when they close. Instead, they produce a tapping noise as they fill and knock against one another. This sharper noise can be heard when listening to the heart beat.

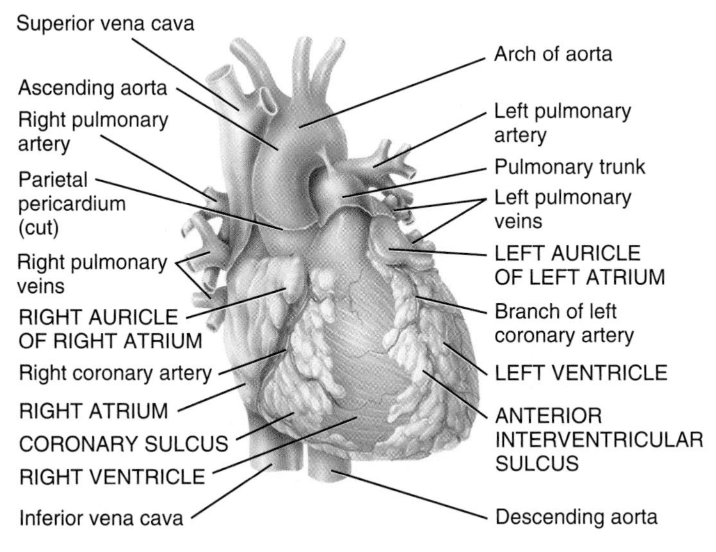

A Anterior external view showing surface features

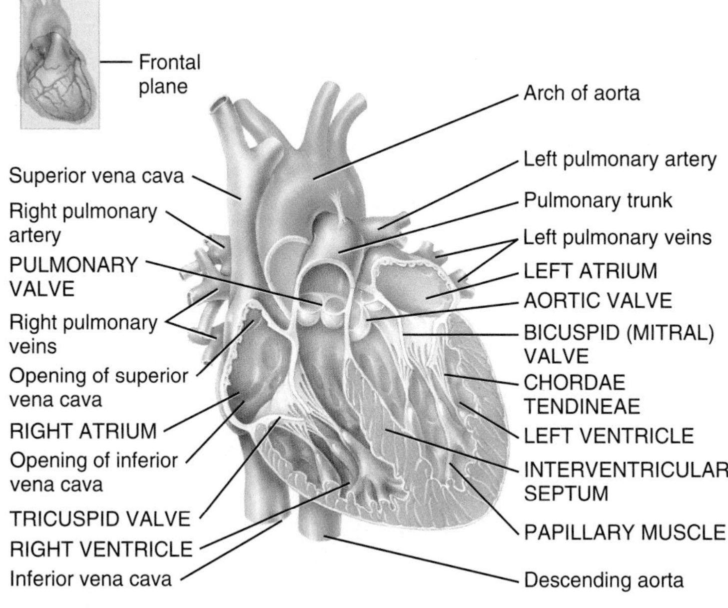

B Anterior view of frontal section showing internal anatomy

The adult heart FIGURE 11.4

Thinking about the heartbeat, we immediately imagine the characteristic "lubb-dupp" sound (FIGURE 11.6). This sound is generated by the valves, and it can have clinical significance. Normal heart sounds are called S1 ("lubb") and S2 ("dupp"). S1 is a loud, resonating sound caused by blood pressure against the atrioventricular valves. This pressure closes the bicuspid and tricuspid valves, pulling the chordae tendinae and the entire supporting framework of cardiac muscle. The second sound forms when the ventricles relax and blood in the pulmonary artery and the aorta flows back toward the ventricles. The arterial valves catch the backflow and snap against one another—"dupp." If the two ventricles are slightly out of sequence, so that one closes first, S2 may "stutter" or "split." An occasional split S2 is normal, but a constant split may indicate **hypertrophy** of one ventricle, a serious cardiac disorder. Listening to these heart sounds, or any internal body sounds for that matter, is termed **auscultation**.

> **Hypertrophy**
> Enlargement of an organ owing to enlarged cells rather than an increasing number of cells.

A heart "murmur" can indicate valve malfunction. This whooshing, blowing, or rasping noise occurs when blood passes the valves in a turbulent flow. Murmurs may signal serious valve trouble, but not all murmurs are cause for alarm. Children often develop a murmur because the cardiac muscle grows much faster than the valves, which are made of connective tissue. For a while, the valves are simply too

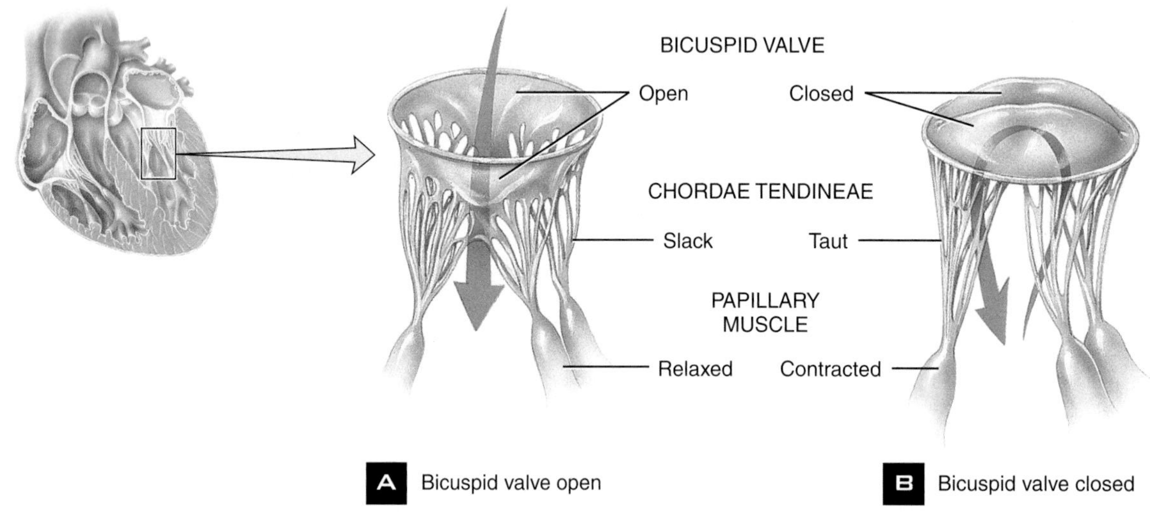

A Bicuspid valve open

B Bicuspid valve closed

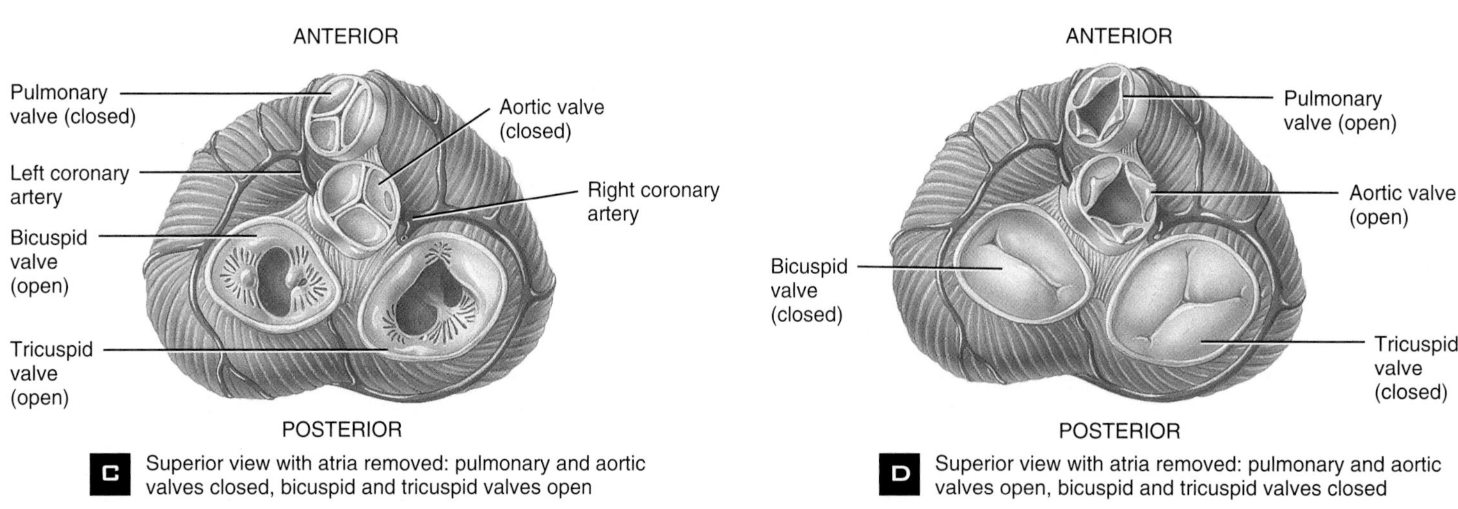

C Superior view with atria removed: pulmonary and aortic valves closed, bicuspid and tricuspid valves open

D Superior view with atria removed: pulmonary and aortic valves open, bicuspid and tricuspid valves closed

Heart valves FIGURE 11.5

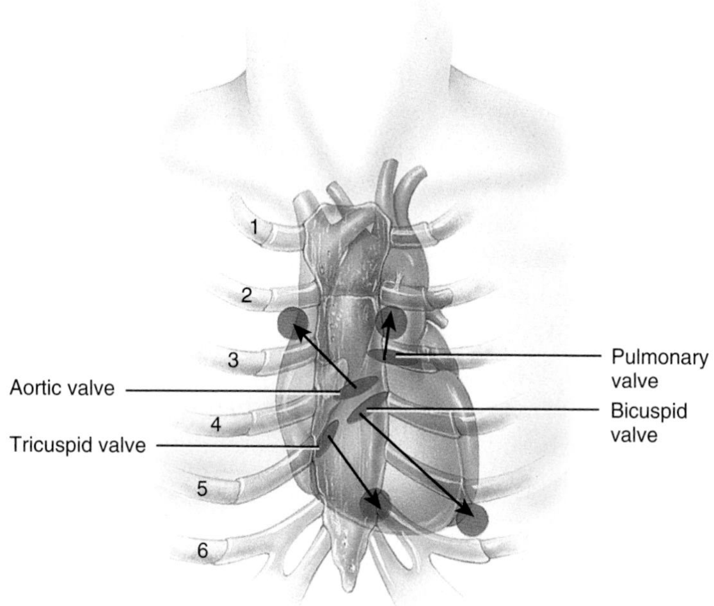

Anterior view of heart valve locations and auscultation sites

Heart sound auscultation spots FIGURE 11.6

When listening to the valves functioning in the heart, it is best to position the stethoscope bell immediately over those valves.

small for the heart! Many women develop a murmur during pregnancy as a result of the dramatic increase in blood volume. Pumping the extra volume exaggerates any small murmurs that are present.

Blood flows twice through the heart

Blood enters the heart twice during one complete circuit of the cardiovascular system—once through the right side and then again through the left. Blood enters the right atrium from the **superior** and **inferior vena cavae** and the **cardiac sinus**, and then drops through the tricuspid atrioventricular valve into the right ventricle, which pumps blood to the lungs.

> **Cardiac sinus**
> Large vein on the dorsal surface of the right atrium that collects blood from the cardiac veins and returns it to the chambers of the heart.

Blood that returns from the lungs enters the left atrium and drops through the mitral valve into the left ventricle, which pumps the blood throughout the body (with the exception of the respiratory membranes of the lungs). This cyclic movement of blood through the heart and body is propelled by the cardiac cycle of the heart.

> **Diastole**
> Relaxation of the heart.

> **Systole**
> Contraction of the heart.

At the beginning of the cardiac cycle (FIGURE 11.7, p. 332), the heart is in **diastole**. The ventricles have relaxed after their recent contraction, and their volume has increased. This increase in volume quickly decreases the pressure in the ventricles below that of the atria, drawing in blood through the atrioventricular (AV) valves. The majority of ventricular filling occurs as these AV valves open. Then the two atria undergo **systole** and force the remaining atrial blood into the ventricles. This step takes approximately 0.15 second.

After atrial systole, the ventricles contract, taking another 0.30 second. The rapid pressure increase inside the ventricles forces the atrioventricular valves closed, and the semilunar valves open. The blood trapped in the ventricles cannot escape back to the atria through the atrioventricular valves, so it is forced through the semilunar valves, into the great arteries. Blood leaves the right side of the heart via the pulmonary valve and enters the **pulmonary trunk**, which takes it to the lungs. The blood exiting the left ventricle passes through the aortic valve and reaches the rest of the body.

As the ventricles contract, the atria relax. After a brief ventricular contraction, the entire heart relaxes. Most of the cardiac cycle (an average of 0.40 second) is spent in diastole.

The meaning of blood pressure

The heartbeat propels blood through the closed cardiovascular system. As the ventricles undergo systole, they exert pressure on the blood in the entire cardiovascular system. The volume of the blood in the system does not change, but its pressure does. The force created by the left ventricle generates the pulse we can feel and the blood pressure that is measured at the doctor's office. You may be able to recite your blood pressure, which is presented in standard form as systolic pressure over diastolic pressure, such as 110/60 or 193/85.

These numbers have physiological meaning. Systolic pressure measures the force of left ventricle

Process Diagram

The cardiac cycle Figure 11.7

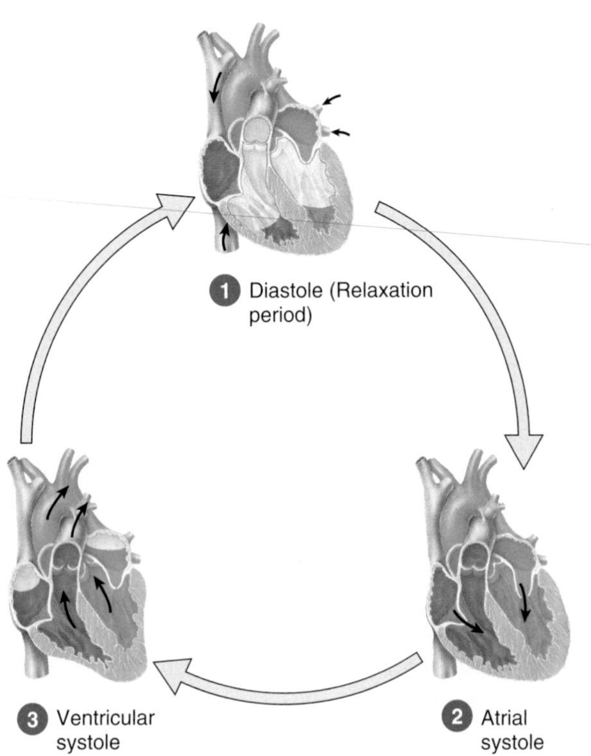

The majority of the cardiac cycle is spent in diastole.

1. At the beginning of the cycle, the heart is completely relaxed, with blood entering both the left and right atria.
2. As the heartbeat begins, the atria contract. This forces blood from the atria into the ventricles.
3. Soon after atrial systole, ventricular systole occurs. The atria relax during ventricular systole. The ventricles remain contracted for a measurable time, and then the entire heart returns to diastole.

contraction, which pushes blood through the circulatory system. This number is low in children and creeps up with age, as the blood vessels become less elastic. Diastolic pressure is the force your blood exerts on the walls of your closed circulatory system while the heart is in complete diastole (relaxation). Contrary to popular belief, the diastolic number cannot be zero unless all the blood has been drained from the organism. High blood pressure is loosely defined as a blood pressure reading of 140/90 or above (see the section, "Cardiovascular Disorders Have Life-Threatening Consequences," later in this chapter).

The heartbeat is under intrinsic and extrinsic control Your heart began beating during your third week of development, and it must continue beating to supply your body's oxygen and nutrient demands until the last minutes of your life. The rate of heartbeat is under two types of control: **Intrinsic controls** establish the usual, day-in, day-out pace of heart beats; **extrinsic controls** modulate the baseline rate to meet the body's immediate demands.

Unlike other muscle cells, cardiac muscle cells (Figure 11.8) undergo rhythmic contractions without receiving nerve impulses. The particular rhythm of each cell is based on the rate of calcium ion leakage from the sarcoplasmic reticulum. Recall that the trigger for skeletal muscle contraction is the calcium ion. Cardiac muscle cells are constantly leaking this important ion, and when the intercellular calcium concentration reaches threshold, the cell spontaneously contracts. When two or more cardiac muscle cells touch one another, they begin to beat in unison, following the pace of the faster cell.

A group of cells in the upper wall of the right atrium has the fastest intrinsic beat, and it serves as the heart's pacemaker. Because these pacemaker cells are near the entrance of the **coronary sinus**, they are called the SA (sinoatrial) node. When the SA node initiates the heartbeat, the signal to contract passes in wavelike fashion from cell to cell through the right and then the left atrium, causing both to contract.

At the base of the left atrium, near the ventricle, lies a group of cells called the **AV node**. AV stands for atrioventricular, reflecting the placement of this node. These cells are a relay station that delays the contraction impulse before sending it on. (Like the SA node, these cells cannot be distinguished visually.) The delay allows the atria to complete their contraction before the ventricles are stimulated to begin contracting. After the delay, the impulse passes through a series of conductive tissues before reaching the cells of the ventricles.

From their relative sizes, it's obvious that ventricles have far more cells than atria. Although the impulse to contract could spread from cell to cell in the ventricles just as in the atria, the contraction would be ineffective because closer cells would have finished contracting before the contraction impulse reached more distant cells. Instead of producing the forceful contrac-

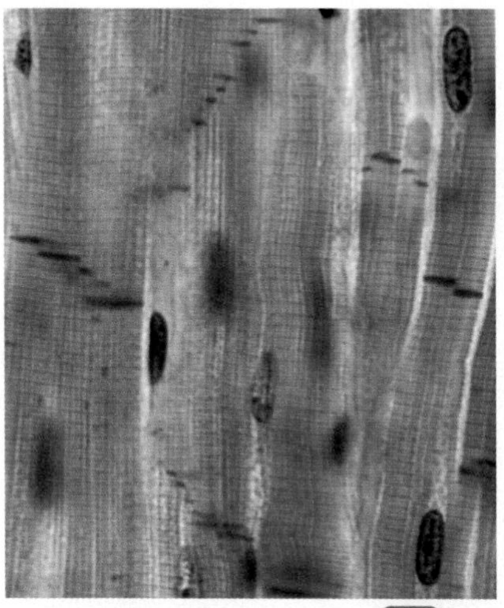

An individual cardiac cell in a culture
FIGURE 11.8

tion needed to build up pressure and open the semilunar valves, blood would just slosh around in the ventricle. If you try to pop a water balloon by grabbing the top, the middle, and bottom in order, the water will simply move away from your hands without breaking the balloon. But if you can grab the balloon everywhere at once, the dramatic rise in internal pressure will pop it.

Purkinje fibers Conduction myofibers that reach individual cells of the ventricles.

To obtain simultaneous contraction, the ventricles require a conduction system for the contraction impulse (FIGURE 11.9). This system starts at the AV node and goes to the **AV bundle** at the center of the heart, near the interventricular septum. Here the system splits into the **left** and **right bundle branches,** which carry the impulse to the apex of the heart, and then up the outer walls. From the bundle branches, the impulse travels on smaller **Purkinje fibers**, which end at clusters of ventricular cells. Using this system, all ventricular cardiac muscle cells contacted by Purkinje fibers get the impulse simultaneously, resulting in synchronous contraction of the entire ventricle.

Conduction system of the heart FIGURE 11.9

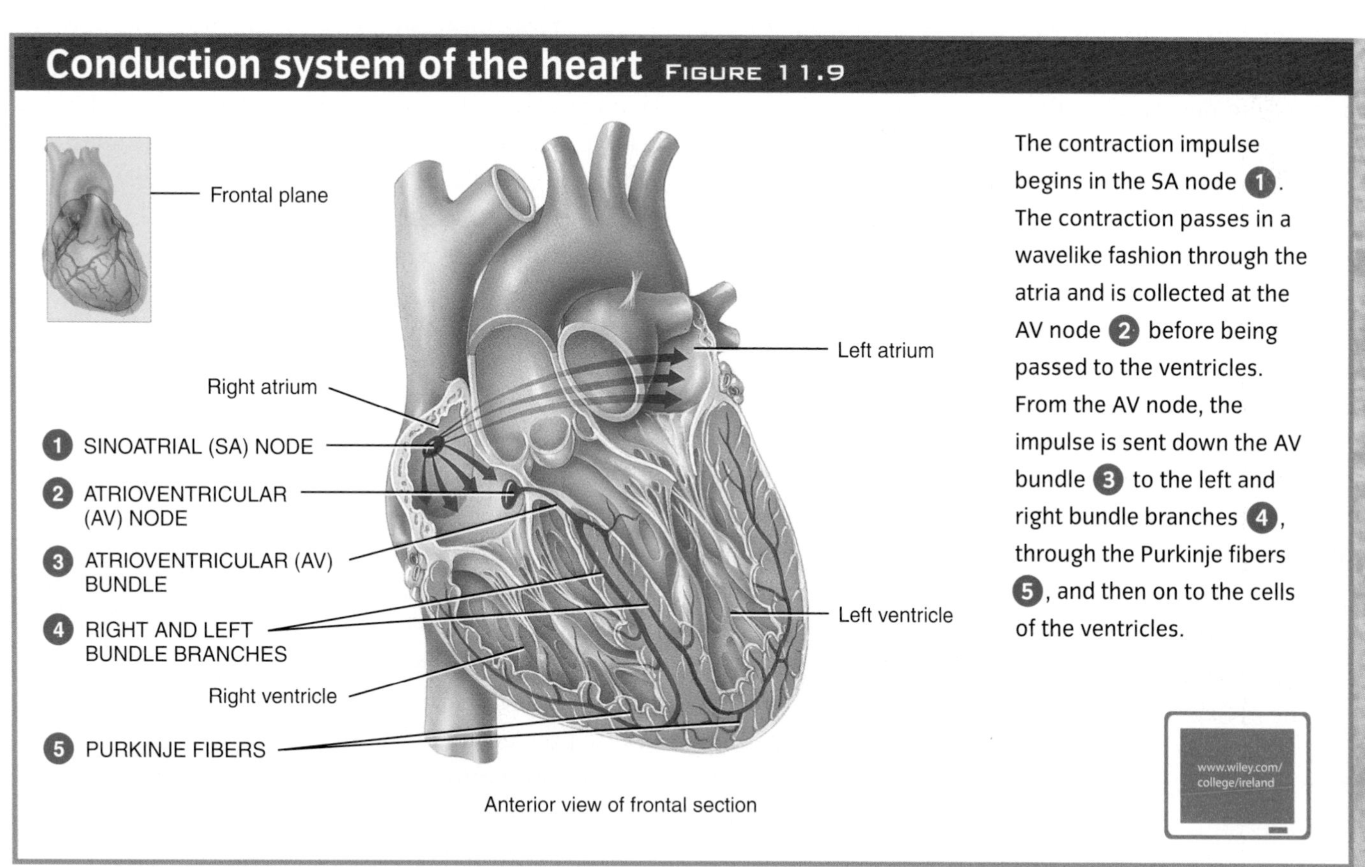

The contraction impulse begins in the SA node 1. The contraction passes in a wavelike fashion through the atria and is collected at the AV node 2 before being passed to the ventricles. From the AV node, the impulse is sent down the AV bundle 3 to the left and right bundle branches 4, through the Purkinje fibers 5, and then on to the cells of the ventricles.

The SA node and the conduction system govern the baseline, or resting, heart rate. But if the body needs more blood than this heart rate can deliver, several **extrinsic heart rate controls** may enter the picture. One extrinsic control resides in the cardiac control center in the medulla oblongata. This center can override the intrinsic heartbeat, increasing or decreasing the rate as necessary. When the sympathetic division of the autonomic nervous system is active, heart rate increases significantly. Similarly, heart rate immediately rises in cardiac cells that are exposed to norepinephrine, the sympathetic division neurotransmitter.

The heart itself can also affect contraction rate and strength. **Starling's law** states that when the ventricles are stretched by increased blood volume, they recoil with matching force. Thus increased blood flow to the heart, which occurs when we start hard physical work or exercise, causes the heart to respond with more forceful pumping—just what we need to move oxygenated blood to the active muscles.

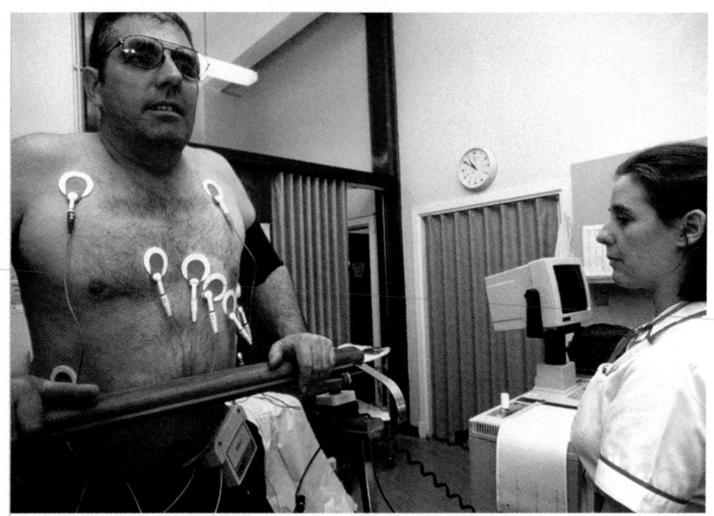

Man hooked up to an ECG machine FIGURE 11.10

The electrical changes associated with cardiac systole and diastole can be picked up on the surface of the body with electrodes attached to the skin. A conductive gel between the electrode and the skin enhances sensitivity. The resulting tracing is used to diagnose cardiac function.

The electrocardiogram records electrical activity

Regardless of what is controlling the heart rate, the cardiac muscle cells generate a pattern of electrical signals as they go through the cardiac cycle. The cells of the myocardium depolarize immediately before they contract and repolarize as they relax. Because so many cells are involved in this cycle, the electrical signals are strong enough to be detected on the skin, where they can be recorded on an **electrocardiogram**, or ECG (FIGURE 11.10).

The ECG tracing (FIGURE 11.11) has a defined series of peaks and valleys. As the SA node fires, the atrial cells depolarize, causing a hill-shaped upward deflection called the **P wave**. Within 100 milliseconds, atrial systole follows. The ECG tracing briefly flattens, then starts a large upward deflection. This **QRS complex** is created by the simultaneous depolarization of the many ventricular cells. As the ventricles briefly remain in systole, the ECG is momentarily flat. As the ventricles relax, the cells repolarize, creating the deflection called the **T wave**, which marks the return of cardiac diastole.

Electrocardiogram A graphic representation of the electrical conditions during a heartbeat.

These deflections can help clinicians evaluate cardiac function. During the **P-R interval** (from atrial depolarization to ventricular depolarization), the contraction impulse is transmitted from the SA node, through the atria, to the AV node, and finally through the conduction system. An interval longer than 0.2 second may indicate damage to the conduction system or the AV node. A long **Q-T interval** (the total time of ventricular contraction and relaxation) may indicate **congenital** heart defects, conduction problems, coronary **ischemia**, or even cardiac tissue damage from a previous heart attack. If the problems seen in the ECG are severe, the heart muscle may stop functioning properly or it may be too weak to be effective. We have the technology and medical skill to replace the heart, but the operation raises complex issues. The Ethics and Issues box, "Ethics of heart replacement" on page 338, further explores the controversy surrounding this complex procedure.

Congenital Present at birth.

Ischemia Lack of oxygen to a tissue because of constriction or blockage of the blood vessels.

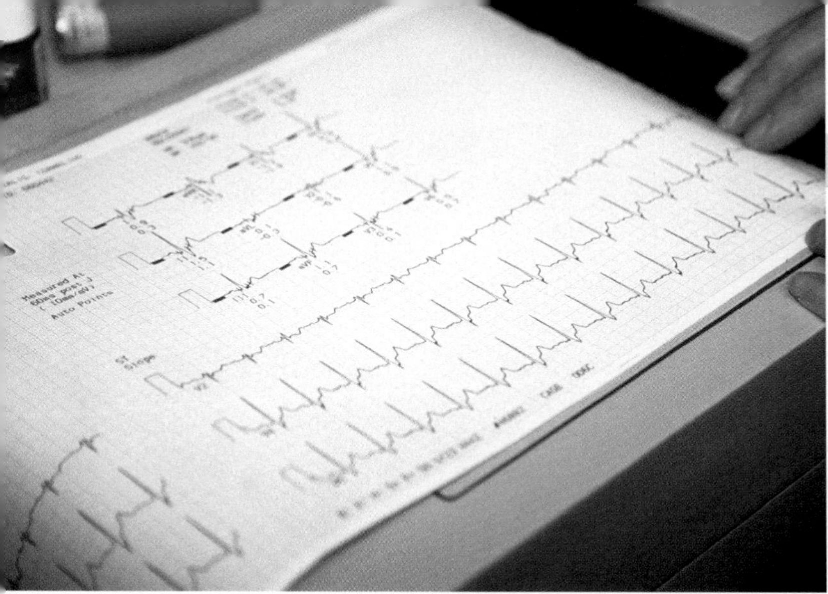

Typical ECG tracing FIGURE 11.11

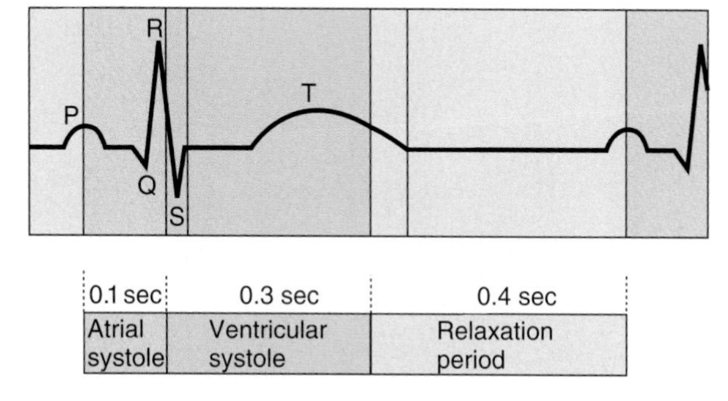

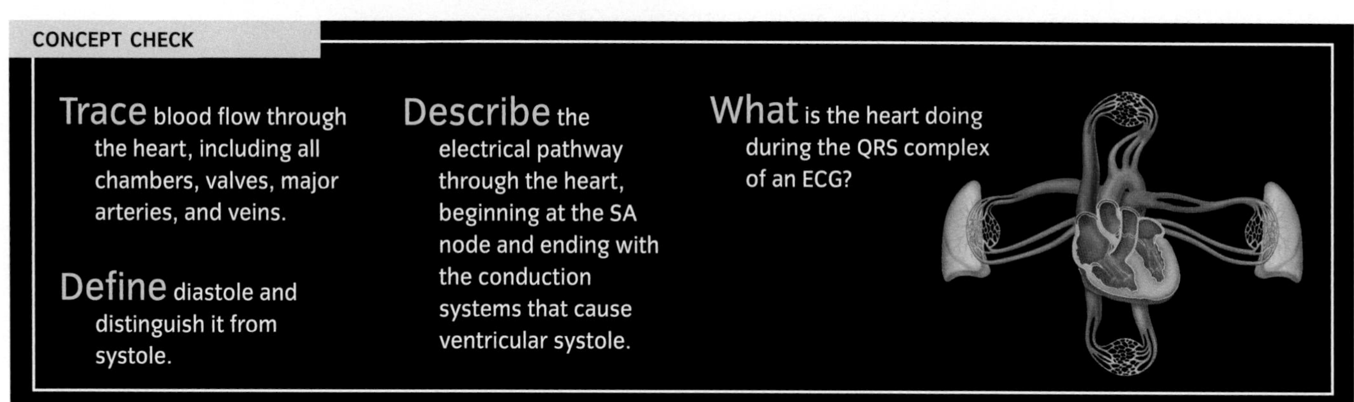

CONCEPT CHECK
Trace blood flow through the heart, including all chambers, valves, major arteries, and veins.
Define diastole and distinguish it from systole.
Describe the electrical pathway through the heart, beginning at the SA node and ending with the conduction systems that cause ventricular systole.
What is the heart doing during the QRS complex of an ECG?

Blood Vessels and Capillary Transport Involve Miles of Sophisticated Plumbing

LEARNING OBJECTIVES

Compare the structure and function of the three types of blood vessels.

Discuss the function of capillary beds and venous valves.

The cardiovascular system has three categories of vessels that are strung together in a large web that begins at the heart, reaches the tissues, and returns to the heart. The vessels in this continuous circuit are the **arteries, capillaries**, and **veins** (FIGURE 11.12, p. 336). Each type of vessel has a different function.

Arteries are blood vessels on the output (ventricular) side of the heart. Arteries closest to the heart have large diameters and thick walls because the heart's pumping causes them to stretch and recoil with each beat. Further from the heart, diameter and wall thickness both decrease because this distance reduces the fluid pressure from the heart. **Arterioles** are small vessels that branch from larger arteries and are structurally similar. In the arterioles, the total cross-sectional area of the blood vessels increases, even though each vessel is smaller in diameter. This larger cross section allows

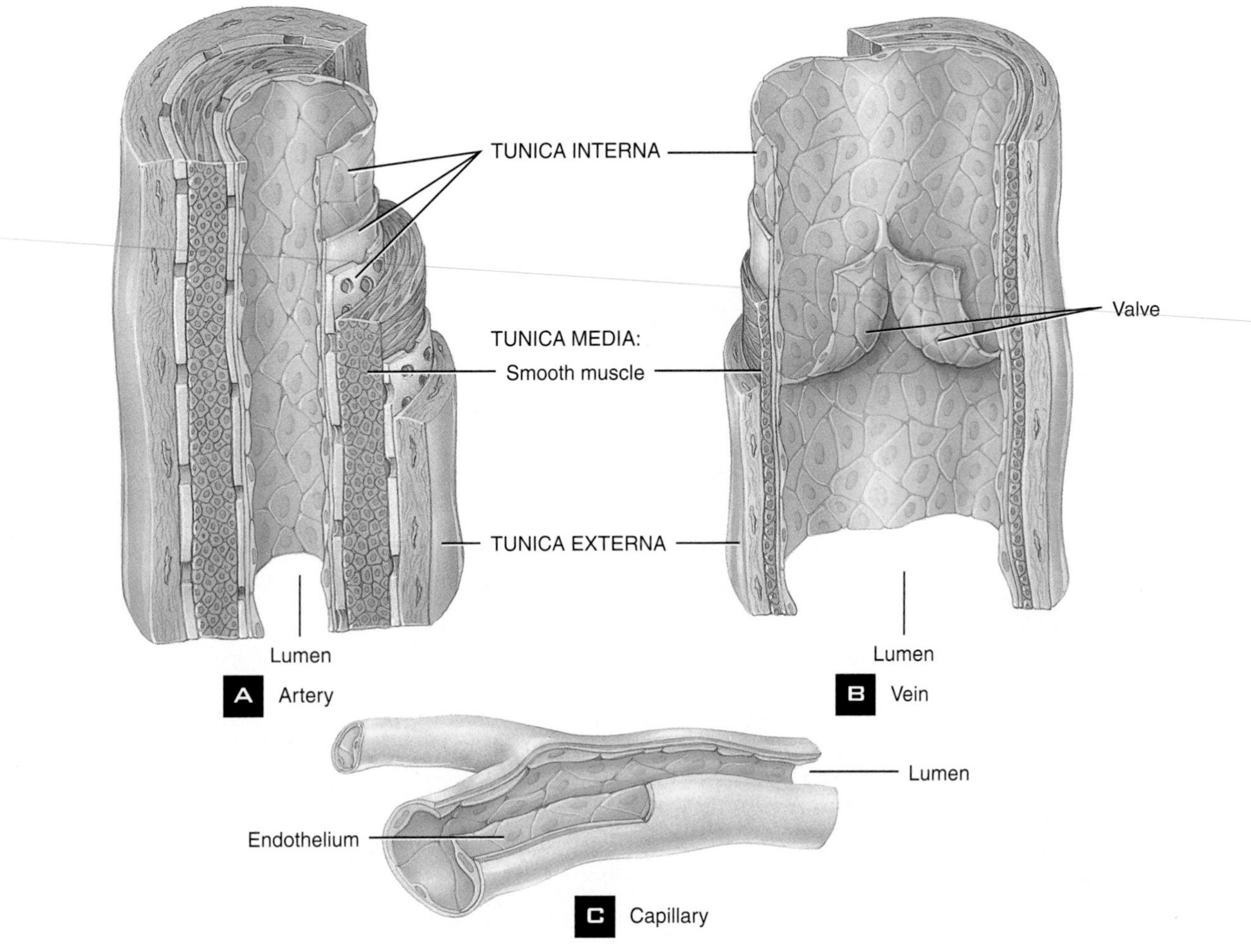

Artery, capillary, and vein structure FIGURE 11.12

Note that the artery is the thickest of the vessels. Arteries take blood from the heart to the tissues of the body and are subjected to the largest pressures. They have a layer of resilient muscle in their walls that allows for the bouncing pulse we can feel through the skin. Capillaries are extremely thin-walled, usually only one cell thick. They are the diffusion vessels of this system. Veins are thinner than arteries but have more substance than the capillaries. Valves prevent backflow of blood in these weak-walled vessels.

Lumen
The inner, hollow portion of a tubular structure; the center of the blood vessel.

the blood to slow. The smaller **lumen** exposes the blood to more surface area, which creates friction, further slowing the flow (FIGURE 11.13).

Arterioles lead to **capillaries**, the smallest blood vessels. The wall of a capillary is one cell layer thick, and the lumen is barely big enough for one blood cell. Capillaries are exchange vessels that reach to almost every cell, and they are the only vessels that permit the vital exchange of gases, nutrients, and waste across the blood vessel wall. The slow blood flow and high cross-sectional area provide enough time and surface area for exchange to occur. Capillaries form large **capillary beds** within the tissues, where blood flow is regulated by **precapillary sphincters**. These small, ringlike muscles can close or open parts of a capillary bed, depending on the oxygen and nutrient demands of the tissue (FIGURE 11.14).

Capillary bed
Interwoven mat of capillaries threading through a tissue.

Blood pressure and blood flow in arteries, capillaries, and veins FIGURE 11.13

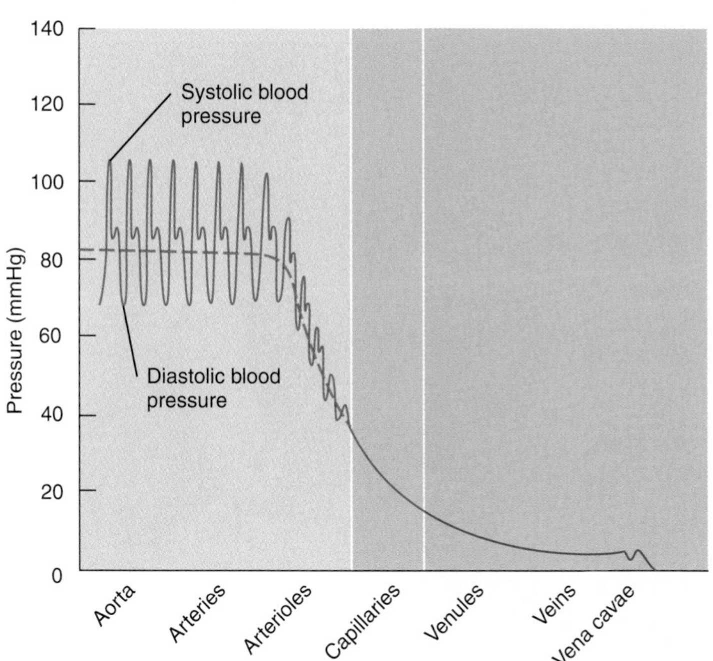

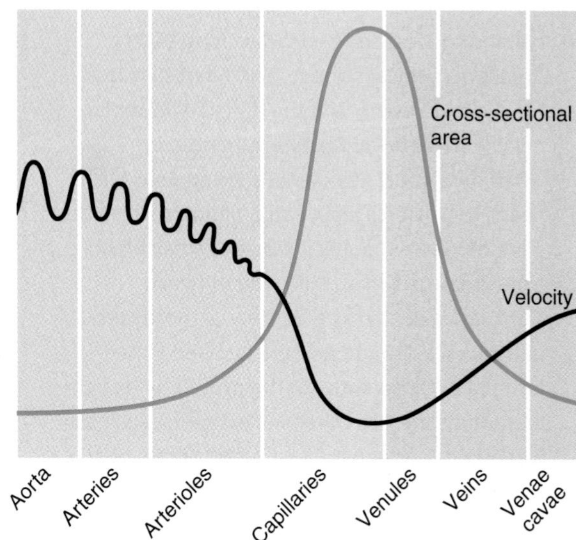

Capillary bed and exchange flow FIGURE 11.14

NFP is the net filtration pressure of the blood. It is based on the force of the blood in the capillary, the pressure of the interstitial fluid, and the osmotic pressure differences between these two. NFP moves fluid out of the capillaries at the arteriole end, and allows fluid absorption into the capillaries at the venule end.

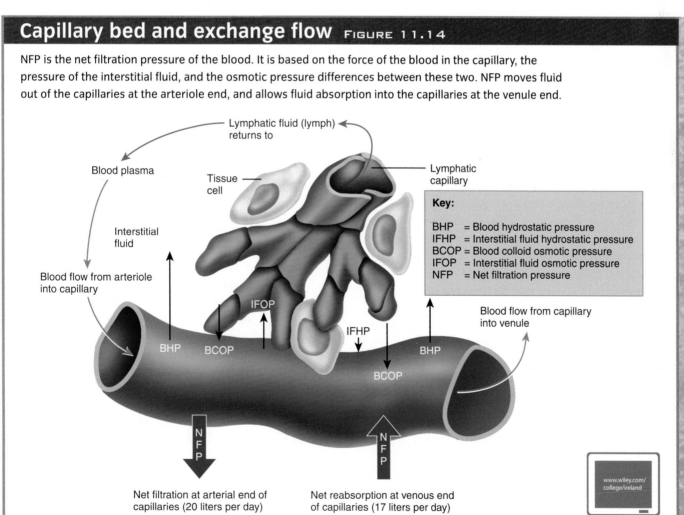

Lymphatic fluid (lymph) returns to

Blood plasma

Tissue cell

Lymphatic capillary

Interstitial fluid

Blood flow from arteriole into capillary

Key:

BHP = Blood hydrostatic pressure
IFHP = Interstitial fluid hydrostatic pressure
BCOP = Blood colloid osmotic pressure
IFOP = Interstitial fluid osmotic pressure
NFP = Net filtration pressure

IFOP

IFHP

BHP

BCOP

BHP

BCOP

Blood flow from capillary into venule

NFP

NFP

Net filtration at arterial end of capillaries (20 liters per day)

Net reabsorption at venous end of capillaries (17 liters per day)

www.wiley.com/college/ireland

Ethics and Issues

Ethics of heart replacement

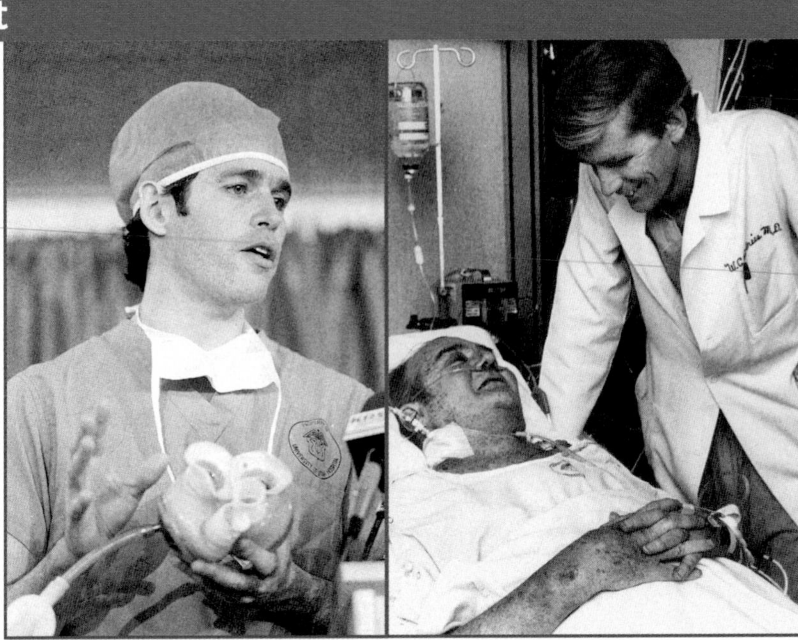

Replacing diseased hearts has long been a dream of medical science. The heart may look like a simple pump, but the difficulty of replacing it with metal and plastic emerged in 1983, when dentist Barney Clark received an artificial heart, then died a slow and painful death (in the glare of massive publicity) due to small blood clots created by the replacement heart.

Clark's death raised a slew of thorny ethical questions. Were the doctors justified in performing a transplant with the prototype heart? Were they and Clark medical pioneers, selflessly working to perfect and test a technology that would later benefit thousands of others with failing hearts? Or were the heart developers self-promoters who took advantage of a dying man in their search for fame and fortune?

The Clark experience put a damper on the quest for an artificial heart—although some machines are now used as a "bridge" to sustain patients until a suitable heart becomes available for transplant. Another artificial heart was tested—with much less publicity—in 2001; it also had problems.

It's now clear that the best replacement for a human heart is another human heart. Many early heart transplants were rejected by the body's immune system, but the use of immune-suppressing drugs after 1980 vastly improved success rates. However, transplant organs must still be tissue matched to the recipient to reduce the chance of rejection. According to the American Heart Association, about 2,000 heart transplants are performed each year in the United States, and many heart transplant recipients enjoy 10 or more fairly normal years after the surgery.

Sadly, despite our ability to successfully transplant organs, patients are still dying of diseases a transplant would "cure." With too few donated organs available, patients with kidney, liver, or heart disease are dying on waiting lists. Transplants raise their own set of ethical problems:

- Should people wait their turn on the list and receive transplants when they are near death? Or should they get transplants sooner, when they are more likely to benefit from the procedure?
- Should states that actively solicit organ donors (through check-offs on drivers' licenses, for example) be permitted to keep organs, or should they be shared regionally or nationally?
- Is the immense cost of a transplant always worth it, or only when a certain life span can be expected?

Here's a final question. Some argue that modern medicine, with its focus on the individual patient, has become too conservative. In 1900, Dr. Walter Reed and colleagues proved that mosquitoes carry the dreaded yellow fever virus by exposing volunteers to mosquito bites. One of those volunteers died from the experiment, but the results were used to support mosquito eradication campaigns that brought yellow fever to its knees. Would a similar experiment be permitted today, given the potential for massive social benefits?

The possible use of replacement tissues derived from stem cells raises a related set of ethical questions that were discussed in Chapter 3's Ethics and Issues box on stem cell research. Finally, all cases of organ replacement or transplant raise certain key ethical issues: Who will pay for these elaborate procedures? Can desperately sick patients make informed decisions about their own treatment? And can doctors and scientists who are enthusiastically working on potential "medical miracles" make wise decisions about the risks and benefits of untried procedures?

Skeletal muscle squeezing vein, assisting in return FIGURE 11.15

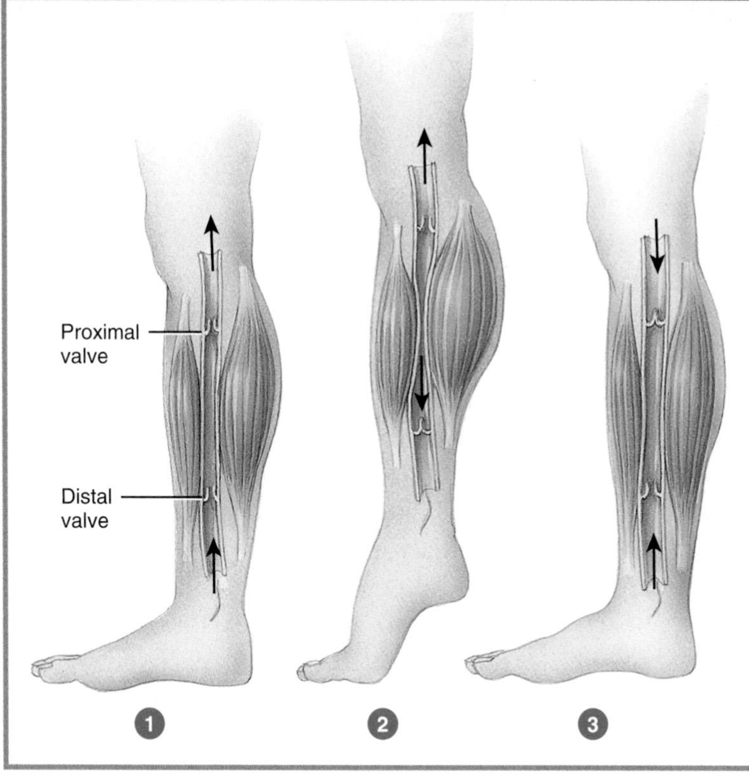

1 At rest, skeletal muscles do not impede the flow of blood back to the heart through veins. 2 When the muscles of the leg are contracted, however, they push against and flatten portions of the veins. This rhythmic flattening and 3 releasing "milks" the veins, moving the blood more efficiently toward the heart.

Process Diagram

Blood leaving capillaries collects in larger vessels called **venules** and veins heading back toward the heart. At this point, circulation resembles the flow of water from rivulets into creeks into rivers and eventually to the sea. As the veins get bigger, the walls thicken slightly. Because the veins are beyond the capillaries, the heart's pumping cannot put much pressure on venous blood. Therefore, the veins are not as thick as arterial walls. The blood in the veins is moving with barely any pressure, so the veins do not need to be terribly strong.

Venules Small veins that drain blood from capillaries to larger veins.

Despite the low pressure, the blood continues to flow toward the heart. Part of the reason is fluid dynamics: Fluids flow easily from a smaller vessel to a larger one, where there is less friction from the vessel walls. Returning the blood from the legs to the heart poses a special challenge because the flow must counteract gravity, with almost no help from the heart. Blood does not pool or flow backwards in the legs because a series of valves in the large veins prevent reverse flow. Also, the contraction of skeletal muscle squeezes the veins and creates a pumping action, pushing blood up toward the heart (FIGURE 11.15). Exercise is often prescribed to move blood and prevent **edema** of the lower extremities.

Edema Abnormal swelling in tissues.

CONCEPT CHECK

What are the three types of cardiovascular vessels?

Why do veins have valves? Are valves necessary in arteries?

How does the large surface area of the arteriole affect blood pressure in the arterioles?

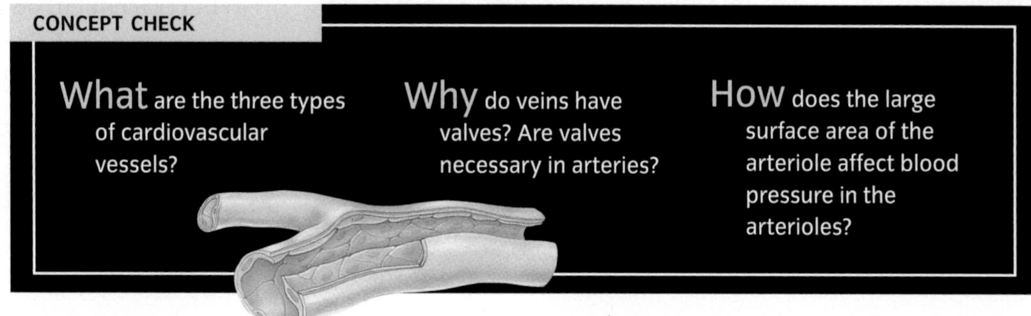

Different Circulatory Pathways Have Specific Purposes

LEARNING OBJECTIVES

Define a closed circuit.

Compare the function and flow of the systemic, pulmonary, and hepatic portal pathways in humans.

Blood can take one of two pathways from the heart: the **pulmonary circuit** toward the lungs or the **systemic circuit** toward the tissues. The purpose of the pulmonary circuit is to exchange carbon dioxide in the blood for oxygen from the environment. The systemic circuit brings this oxygen (and nutrients) to the tissues, then removes carbon dioxide from them.

The **pulmonary circuit** extends from the right side of the heart to the capillary beds of the lungs and on to the left atrium (FIGURE 11.16). Blood entering the right atrium is low in oxygen, having just returned from the body. This deoxygenated, carbon-dioxide-rich blood drops to the right ventricle, and is propelled to the lungs, where it picks up oxygen, releases carbon dioxide, and returns to the left atrium.

The **systemic circuit** begins when oxygen-rich blood enters the left atrium. This oxygen-rich blood then enters the left ventricle, and, during ventricular systole, is pumped through the aortic arch to the body. After passing through the capillaries, venous blood returns to the superior and inferior vena cavae. These large veins drain into the right atrium, where blood reenters the pulmonary circuit. The systemic circuit includes most blood vessels in the body.

The first branches from the aortic arch are the **coronary arteries**, which deliver oxygen-rich blood to the cardiac muscle. Although the left side of the heart is full of oxygen-rich blood, that blood and its oxygen are not available to the heart tissue because the inner lining of the heart, the **endocardium**, is not a diffusion membrane. Therefore, cardiac tissue must obtain oxygen through a capillary bed, just like every other tissue. These coronary arteries are narrow and prone to clogging. If they are blocked, less oxygen-rich blood is delivered to the heart, causing a heart attack. Heart attacks are discussed later in this chapter.

Although blood usually flows from arteries to capillaries to veins, this pattern is modified in a few places. In **portal systems,** blood flows from arteries to capillaries to veins, as usual (FIGURE 11.17, p. 342). The veins, however, break up into another set of capillary beds before the blood returns to the heart. This allows the blood to slow in the organ before being pushed back to the heart.

Additionally, there is an altered blood flow in the fetus, which gets its oxygenated blood through the placenta, or umbilicus, not from its lungs. The lungs are not yet functioning and will not be needed to diffuse oxygen until after birth. The umbilicus carries oxygenated, nutrient-rich blood from the placenta to the fetal liver, where the blood then continues through the fetal systemic circuit. The umbilical arteries carry fetal blood from the fetus to the placenta to be cleansed. The complete set of fetal circulatory modifications is discussed in more detail in Chapter 17.

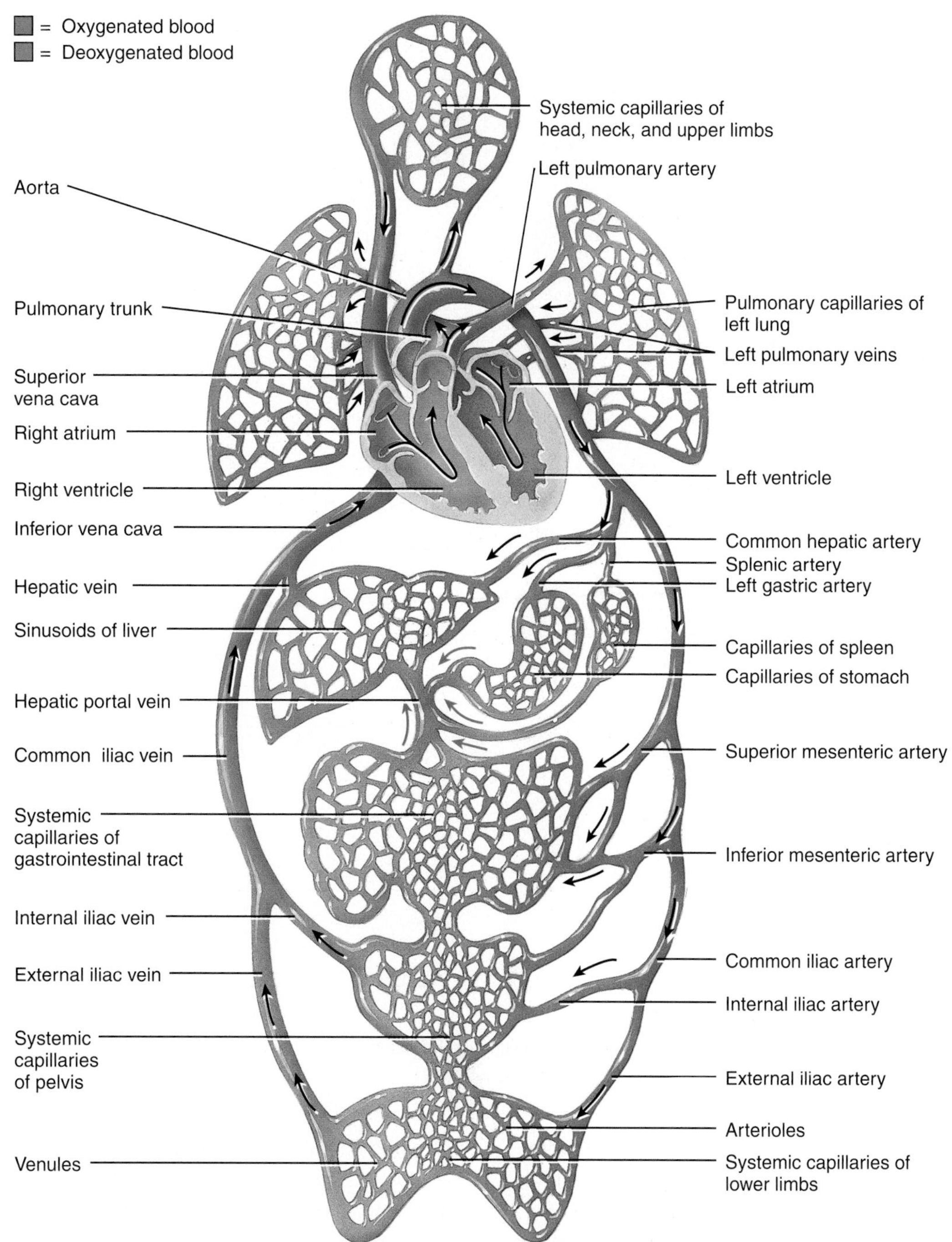

Pulmonary and systemic circulatory routes FIGURE 11.16

The two main circulatory routes in the body are seen here. The pulmonary circulatory route takes blood from the heart to the respiratory surface of the lungs and back to the heart (short black arrows). The much more complicated systemic circuit delivers blood to the other organs, and then back to the heart (longer black arrows). The red arrows represent the hepatic portal circulation (see FIGURE 11.17, p. 342).

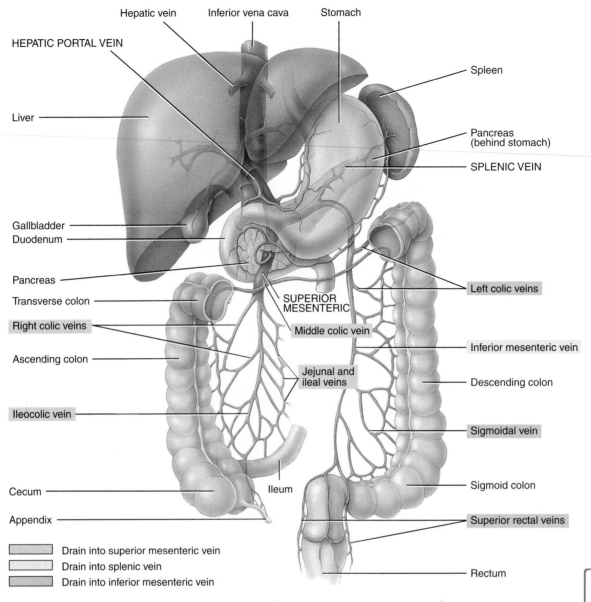

Hepatic vein
Inferior vena cava
Stomach
HEPATIC PORTAL VEIN
Spleen
Liver
Pancreas (behind stomach)
SPLENIC VEIN
Gallbladder
Duodenum
Pancreas
Left colic veins
Transverse colon
SUPERIOR MESENTERIC
Right colic veins
Middle colic vein
Inferior mesenteric vein
Ascending colon
Jejunal and ileal veins
Descending colon
Ileocolic vein
Sigmoidal vein
Cecum
Ileum
Sigmoid colon
Appendix
Superior rectal veins
Drain into superior mesenteric vein
Drain into splenic vein
Drain into inferior mesenteric vein
Rectum

Anterior view of veins draining into the hepatic portal vein

www.wiley.com/college/ireland

Hepatic portal system FIGURE 11.17

In the hepatic portal system, blood flow begins and ends with capillaries. Before blood enters the liver, it absorbs nutrients in capillary beds in the small intestine. This blood collects in the hepatic portal vein, which drains from the small intestine to the liver, and then passes through another capillary bed. The blood flow slows in the capillary bed of the liver so *hepatocytes* (liver cells) can remove detrimental ions and compounds that were picked up by the digestive tract. The cleansed blood collects in the hepatic vein and drains to the inferior vena cava.

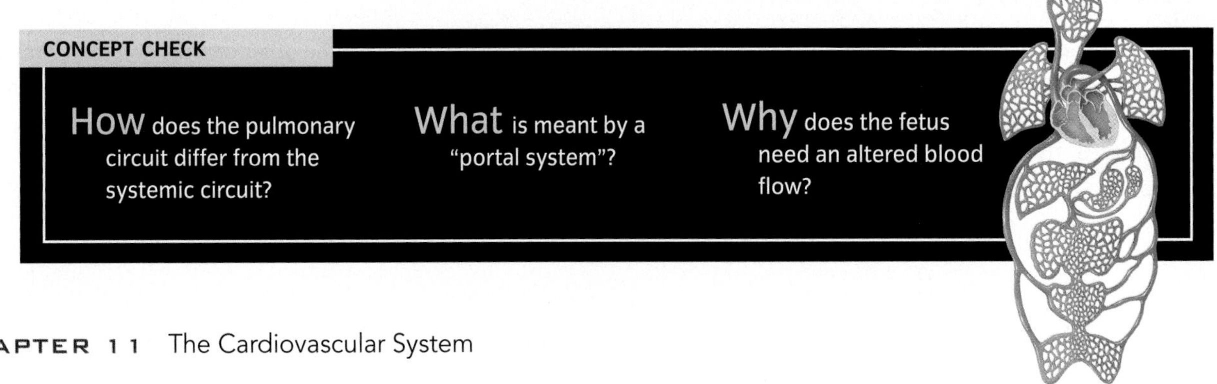

Cardiovascular Disorders Have Life-Threatening Consequences

LEARNING OBJECTIVES

List the most common disorders of the heart and vessels.

Probe the risk factors for heart attack.

Explore the role of lifestyle in cardiovascular disease.

Many cultures equate great emotional pain with a "broken heart," but in reality, love gone awry does not interfere with cardiac function. However, **cardiovascular disease** (CVD) does, and in fact is the leading cause of death in Western countries. CVD takes many forms, each with its own symptoms and treatments.

The most common cardiovascular diseases include hypertension, atherosclerosis, heart attack, heart failure, embolism, stroke, and varicose veins. You probably know someone who has suffered from one of these conditions. The risk factors for cardiovascular disease can be genetic or environmental. Genetic risk factors include family history, gender, and ethnic background. A family history of heart attack prior to age 55 indicates a genetic predisposition to heart disease. Males suffer from CVD more frequently than females, although this gap is closing. The reasons for this change could include the advancement of women into the higher-stress jobs once dominated by men, women's longer life spans, and the postmenopausal reduction in estrogen levels. The incidence of CVD is higher in African Americans than in Americans of European descent, indicating a genetic predisposition.

Even if you have genetic risk factors, all is not lost. Many risk factors, such as smoking, overeating, and spending too much time on the sofa rather than exercising, are fairly easy to control. Monitoring your diet and getting a modicum of exercise will decrease your chance of CVD. Some studies indicate that 40 minutes a day of low-impact exercise reduces the chance of heart disease by one-third. And the exercise need not be terribly strenuous—gardening, yoga, or ballroom dancing can all improve cardiovascular performance. In many cases, preventing or controlling CVD is not difficult; it just requires some dedication and understanding.

HIGH BLOOD PRESSURE STRESSES THE ENTIRE BODY

One of the most prevalent CVDs is **hypertension**, or high blood pressure. Hypertension is often called the "silent killer" because it may produce no symptoms before disaster strikes. As mentioned, hypertension is diagnosed when systolic blood pressure is above 140 mmHG, or diastolic pressure is above 90 mmHG. A high diastolic number indicates a decline in blood-vessel elasticity that increases the chance that the force of systolic contraction will exceed the capacity of the circulatory system. In chronic hypertension, capillary beds leak blood into the surrounding tissues, or break entirely, causing internal bruising. Although hypertension is harmful to many organs, the key risk is stroke. Dietary restrictions, moderate exercise, reducing smoking and drinking, and medications can all control hypertension.

Recent research is discovering a genetic link to some forms of high blood pressure. In particular, two genes are involved in the conversion and activation of the protein angiotensinogen. This enzyme converts angiotensin I in the plasma into angiotensin II, which constricts the blood vessels. If this pathway is hyperactive, constricted vessels will reduce the blood-flow capacity, and the noncompressible blood will push harder against the vessel walls, increasing blood pressure.

ARTERY DAMAGE IS A MAJOR CAUSE OF MORTALITY AND DISABILITY

Atherosclerosis (literally "hardened vessels") is another disease of the blood vessels. When **plaques,** fatty deposits of cholesterol, accumulate inside the vessel walls, they **occlude** (slow or block) the lumen, reducing blood flow (FIGURE 11.18). More serious complications can arise if the plaque causes a clot to form within the vessel. A clot that is attached to the vessel wall is called a thrombus; if it loosens and floats in the bloodstream, it is called an **embolism**. This floating clot can lodge in a smaller vessel, completely blocking blood flow and causing tissue death.

An **aneurysm** occurs when a vessel wall balloons under pressure, forming a weak spot that can be burst by the increased blood pressure generated by each heartbeat (FIGURE 11.19). Burst aneurysms are usually fatal, because they tend to develop in large, high-volume arteries in the abdomen or brain. Because arteries are not exchange vessels, aneurysms can sometimes be repaired before they burst by replacing the ballooned area with plastic tubing, but only if they are detected early enough.

An embolism or aneurysm in the brain causes stroke. Whether the problem is a blockage or excess bleeding, stroke starves the tissues fed by the blocked or broken artery of oxygen and nutrients. Although quick removal of an embolism can control the damage, brain tissue usually dies. Initial symptoms of stroke include sudden difficulty speaking, blindness in one eye, and numbness and/or weakness, usually on one side of the body. Stroke can also cause **aphasia** (loss of speech), loss of fine motor control, paralysis, or even death. New emphasis on quick treatment of strokes has reduced the disability, but many of the 700,000 Americans who suffer a stroke each year do suffer widespread brain damage.

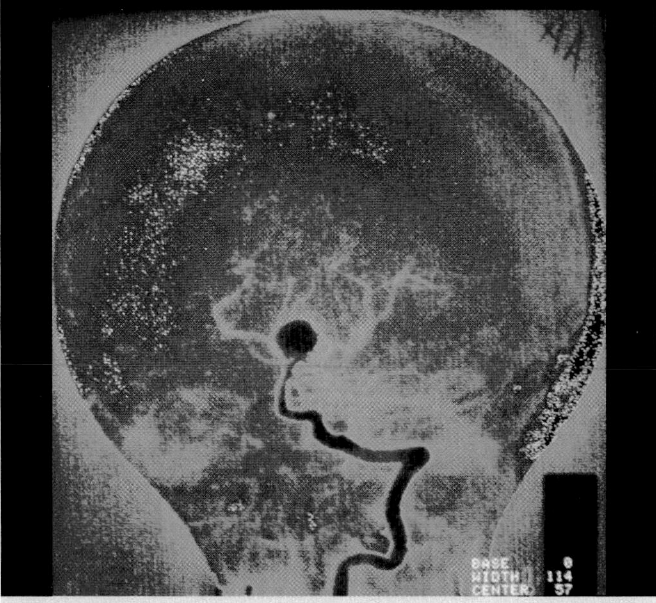

Aneurysm FIGURE 11.19

In this MRI image, a ballooning of the cerebral artery is clearly visible. Aneurysms may occur in high-pressure arteries such as the aorta or cerebral arteries. A burst aneurysm can be deadly.

THE CAUSES AND CONSEQUENCES OF HEART ATTACK

Perhaps the most fearsome cardiovascular disorder is **heart attack,** the death of some heart muscle due to a lack of oxygen. Heart attack, or **myocardial infarction**

Plaque formation FIGURE 11.18

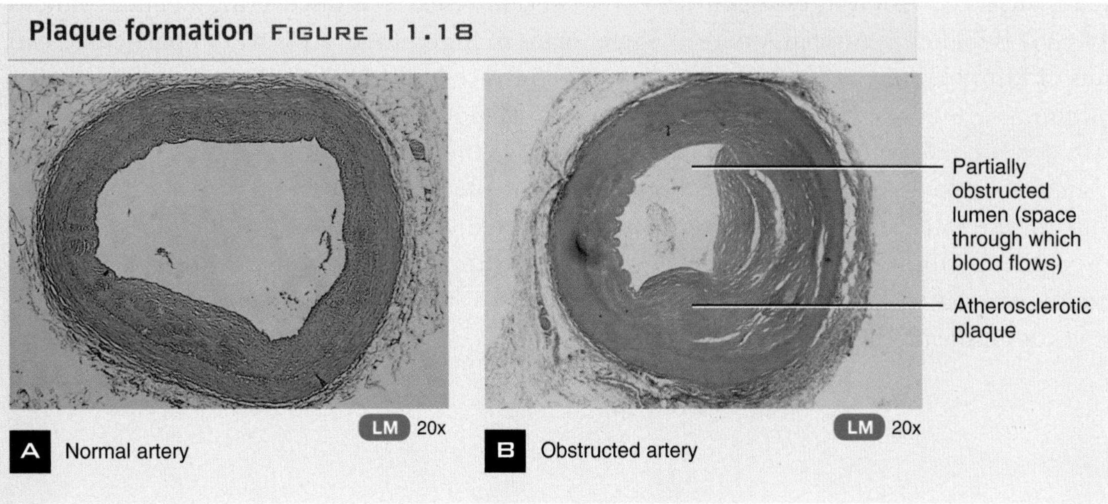

(MI), causes one in five deaths in the United States. Each year, the population of the United States suffers more than 1.2 million nonfatal heart attacks; of those patients, 40 percent will die within one year. (See the I Wonder . . . box on p. 347.)

Dead cardiac tissue ceases to conduct electricity, so the contraction impulse cannot pass. A ventricle that cannot contract completely cannot move blood efficiently, and the result is reduced cardiac output.

MI usually occurs when plaque in a coronary artery occludes the blood flow. While the plaque is forming and blood flow is diminishing, the heart tissue may act like a cramped muscle. The pain from this temporary loss of oxygen is usually described as a crushing feeling in the chest, pain that radiates through the chest and left arm, or a numbness in the left arm. This condition is called **angina pectoris**, or simply angina.

Angina can arise when the heart is working hard, such as during strenuous exercise, or when it is stressed by, for example, smoking cigarettes. This "**stable angina**" can be treated by reducing activity and/or quitting smoking. **Unstable angina**, in contrast, appears with no apparent stimulus and is often an early warning of impending heart attack. Fortunately, people with unstable angina often think they are having a heart attack and seek immediate medical attention.

Angina may be controlled with nitrate drugs, such as nitroglycerine and isosorbide, which relax cardiovascular smooth muscle. As the smooth mucle in the walls of the coronary arteries relax, blood pressure decreases and blood can flow more smoothly past the obstructive plaque.

If medication does not restore a normal lifestyle, surgical procedures may be recommended, including **balloon angioplasty**, placement of a **stent**, or **bypass surgery** (FIGURE 11.20). Balloon angioplasty pushes soft, fatty plaque against the vessel wall, reopening the lumen. The physician inserts a catheter with a deflated balloon at the end into the femoral artery and

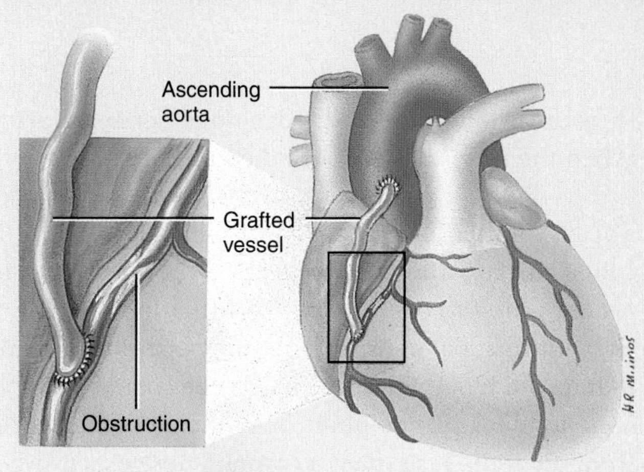

A Bypass surgery

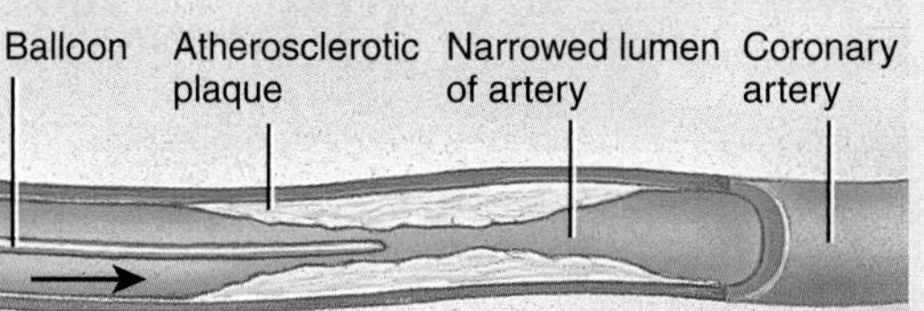

Balloon catheter with uninflated balloon is threaded to obstructed area in artery

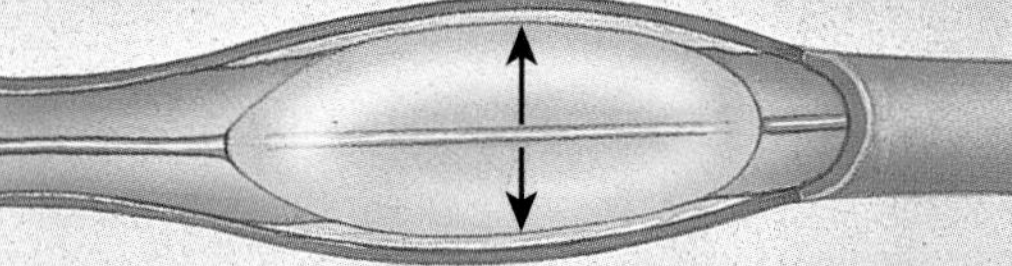

When balloon is inflated, it stretches arterial wall and squashes atherosclerotic plaque

After lumen is widened, balloon is deflated and catheter is withdrawn

B Balloon angioplasty

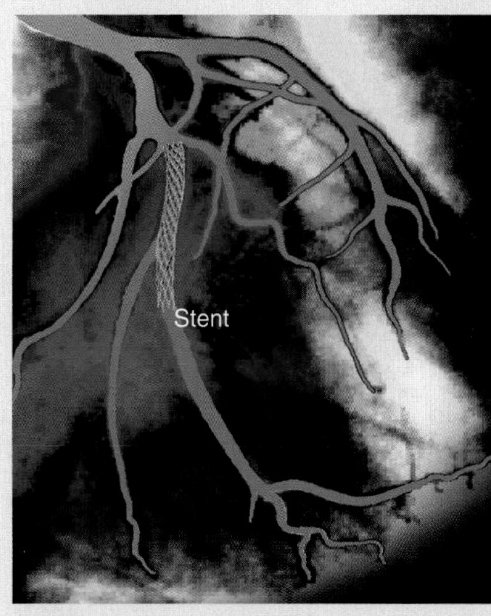

C Angiogram showing a stent in a coronary artery

Surgical procedures for reestablishing blood flow in occluded coronary arteries FIGURE 11.20

threads the catheter to the occluded coronary artery. When the balloon is inflated, the lumen expands.

Balloon angioplasty can fail if the plaque does not stick to the vessel wall. Stents, which look like a tiny roll of chicken wire (see FIGURE 11.20c, on p. 345), are designed to overcome this difficulty. A stent supports the arterial walls, permanently opening the vessel to improve blood flow. A stent may be coated with medicine to block plaque buildup; as the medicine leaches from the stent, it supplies a constant dose exactly where it is needed.

If the coronary artery is physically damaged, or the plaque buildup is severe, bypass surgery may be required. This is open-heart surgery and is obviously much more invasive than angioplasty. Surgeons break the breastbone to reach the heart, and periodically stop the heart to perform delicate suturing. A section of blood vessel, usually from the femoral vein, is removed to serve as the bypass vessel. (Blood return from the legs is not hindered because the venous system includes many **anastomoses** that provide alternate pathways for blood return.) The surgeons suture a small length of femoral vein around the blockage in the coronary artery, creating something that works like a highway detour: Blood bypasses the congestion and returns to normal circulation beyond the blockage. Each detour counts as one bypass, so a triple bypass surgery involves three detours. Bypass surgery is usually highly effective, but recovery is much slower than after balloon angioplasty or stent placement, owing to the major thoracic surgery and the healing required in the leg.

> **Anastomoses**
> Networks or connections between two or more vessels.

Congestive heart failure is due to a weak heart

With age, many hearts simply weaken and fail to push enough blood through the circulatory system. Such a condition, called **congestive heart failure**, is increasingly common now that more people are living deep into old age. A weakened left ventricle fails to move the blood, allowing fluid to back up and leak into the lungs, causing **pulmonary edema**. As the blood flow slows, fluid also builds up elsewhere. Congestive heart failure is named for the resulting congestion in the thoracic cavity. When this fluid presses against the heart, beating becomes even more difficult.

Heart failure, unlike angina or MI, is a gradual disease. As the heart weakens, the body attempts to compensate. The heart itself expands, enlarging the volume of each ventricle. Enlarged volume requires more muscle mass to push the blood, causing a further expansion in the heart. As these two remedies fail, **tachycardia** pushes more blood through the body, until the heart cannot maintain the rapid pulse. Symptoms of congestive heart failure include fatigue, difficulty breathing, tachycardia, and possibly even death as fluid builds up in the lungs and drowns the respiratory membranes.

> **Tachycardia**
> Resting heart rate above 100 beats per minute.

When veins become visible they function less effectively

Not all cardiovascular diseases are fatal. **Varicose veins** and **spider veins** are unsightly and can be painful (FIGURE 11.21). Varicose veins are distensions of the venous walls near valves. As the blood

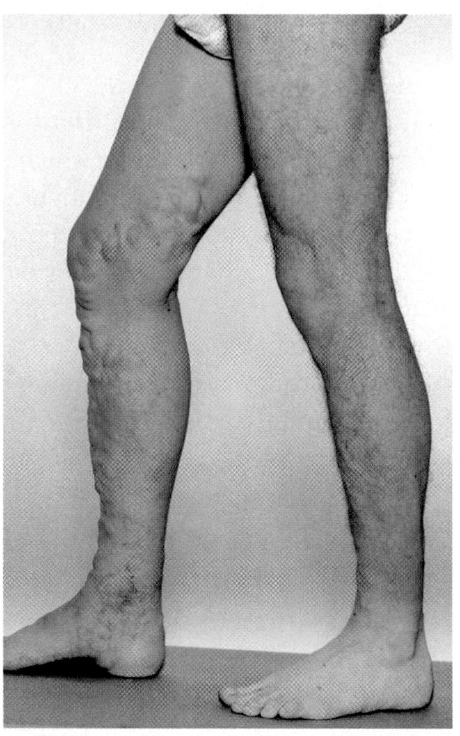

Varicose veins FIGURE 11.21

Why do trained athletes have heart attacks?

It's a paradox. People exercise to improve their overall health and their CV systems. Yet once in a while, trained athletes keel over from heart attacks. Why? Is exercise dangerous?

A series of studies over the past 20 years found that extreme athletic effort does seem to raise the risk of heart attack, at least for about 24 hours after the event. But they also show that most of the stricken athletes already had heart disease to begin with.

Among youths, sudden cardiac death during exercise is quite rare. The key danger seems to be hypertrophic cardiomyopathy—a syndrome of heart "overgrowth" that usually can be detected with ultrasound examination.

A wide variety of preexisting conditions can cause athletic heart failure among people over 40. Most common is atherosclerosis—hardened arteries that suffer an embolism or aneurysm during the event. But spasms of the coronary arteries, deformed valves, and other common causes of MI are also to blame. Arrythmias—erratic heart rhythms—can also cause athletic heart failure because extreme exertion can sometimes trigger a fatal arrhythmia.

The number of sudden cardiac deaths is low among athletes. In 1998, a South African scientist reported that "6 in 100,000 middle-aged men die during exertion per year. . . . the evidence suggests that exercise seems to trigger or cause sudden cardiac death in athletes with underlying heart disease." (*Encyclopedia of Sports Medicine and Science*, 1998, Dr. Timothy D. Noakes)

B.J. Maron, L.C. Poliac, and W.O. Roberts of the Minneapolis Heart Institute found that of 215,413 competitive marathon runners studied over a 30-year period, only four suffered sudden cardiac deaths. Three deaths resulted from coronary artery disease, and the other from unusual coronary artery anatomy. This study indicated that adults who were fit enough to run 26-mile marathon races faced a 1 in 50,000 risk of sudden cardiac death during the event, or immediately afterward.

The stricken athletes may simply be people who were unwittingly close to a heart attack. After a severe snowfall in the United States, one study found a temporary spike in heart-attack deaths during the storm but no change in the monthly toll. The researchers concluded that the people who had heart attacks while shoveling snow probably would have died within a few days anyway.

Most experts say exercise helps the heart and overall CV system by strengthening cardiac muscle and the vasculature, and by improving the efficiency of circulation and respiration. Although athletes can get MI (myocardial infarction) from many underlying causes, "none of these conditions is caused by exercise, however vigorous," wrote Timothy Noakes, of the University of Cape Town's physiology department. "Rather, the evidence is clear that regular exercise acts against the development especially of coronary atherosclerosis."

But one line of evidence suggests that exercise can damage the heart: A few recent studies have found the enzyme troponin I in athletes' blood after endurance events. Troponin I in the blood is considered an indication of death of heart muscle cells. Nevertheless, a 2005 study of data from a long research project in Massachusetts associated working out almost every day with an extended life span. Among about 5,000 Americans, middle-aged or older, a moderate or high level of activity added 1.3 to 3.7 years to the life span, compared to those with little exercise. The major benefit? A reduction in heart disease.

What is the best advice for an adult embarking on an exercise program? See the doctor first, and be sure to report any cardiac symptoms. Although there is some dispute on the matter, athletes live longer and healthier lives than nonathletes, and for most people, exercise helps control weight, improves mood and the immune system, and keeps the cardiovascular system in tip-top condition.

The Great North Run in London attracts many athletes, both well trained and amateur.

moves into the veins, it pools against the valves. If the walls of the vein are weak, owing to disease or genetics, the vessel will expand. More blood will move into the distended area with each heartbeat without moving up toward the heart, pushing the walls out even further. The vein eventually pops out of the musculature and becomes visible as a bluish wrinkled cord directly beneath the skin. The varicose portion of the vein can be removed surgically if it becomes too painful or unsightly.

Spider veins are less visible because they involve venules, not veins. These surface venules fill with blood but do not empty. They are visible through the skin as pale purple or blue tracings and usually occur in small to large patches on the face or thighs. Treatment for spider veins is purely cosmetic, and involves injecting the blocked venule with sterile saline solution to displace the pooled blood.

The exact cause of these two venous disorders is not known, but they do run in families. More women suffer varicose veins than men, and hormonal changes are often implicated. Many women develop varicose veins or spider veins during pregnancy.

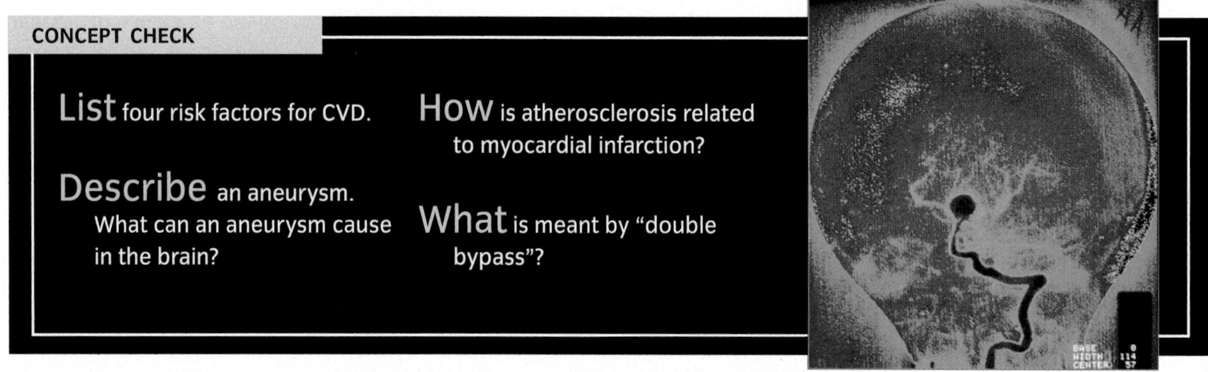
CONCEPT CHECK

List four risk factors for CVD.

Describe an aneurysm. What can an aneurysm cause in the brain?

How is atherosclerosis related to myocardial infarction?

What is meant by "double bypass"?

Blood Consists of Plasma, Cells, and Other Formed Elements

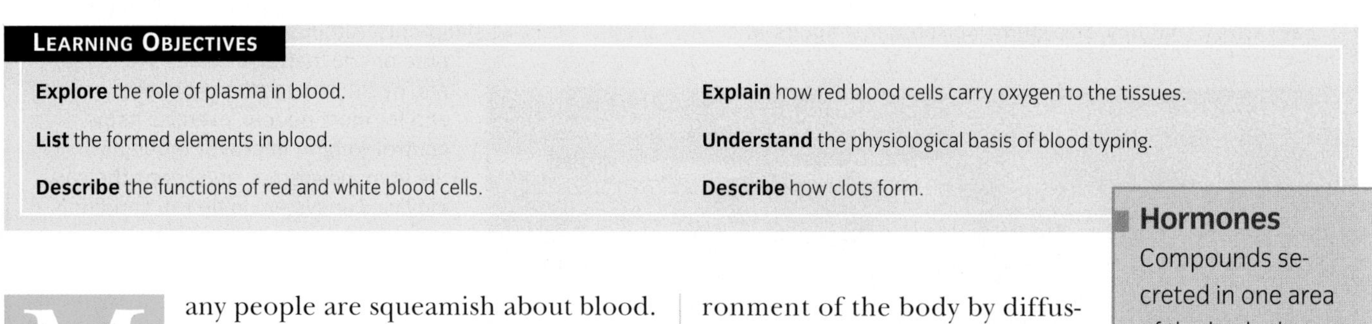
LEARNING OBJECTIVES

Explore the role of plasma in blood.

List the formed elements in blood.

Describe the functions of red and white blood cells.

Explain how red blood cells carry oxygen to the tissues.

Understand the physiological basis of blood typing.

Describe how clots form.

Many people are squeamish about blood. They do not like to see it outside the circulatory system, and the mere thought of it can weaken their knees. This is unfortunate because blood is a unique and essential connective tissue. It is composed of a liquid portion, the **plasma**, and a solid portion, the **formed elements**, which are mainly cells (FIGURE 11.22).

The functions of blood are all critical to maintaining homeostasis. Blood regulates the internal environment of the body by diffusing ions and other materials into the interstitial fluid. It forms clots to prevent blood loss at injuries. Blood also transports heat between the body core and the skin. Dissolved in the plasma are **hormones**, nutrients and gases that are needed in other areas, so blood serves as a mode of transport for these compounds. In addition, the formed elements in the

Hormones Compounds secreted in one area of the body that are active in another, usually carried by the blood.

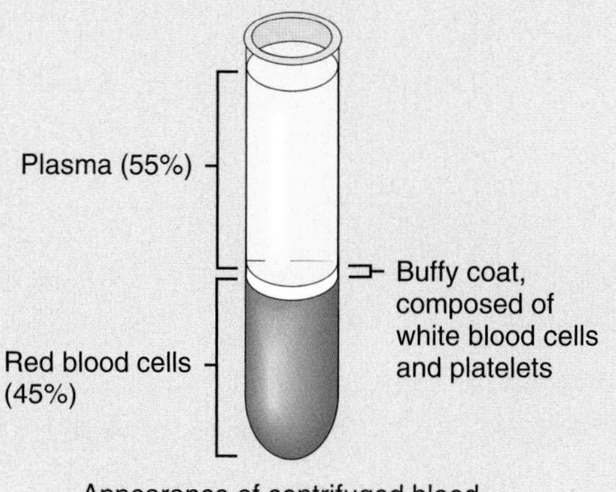

Centrifuge tube of spun blood FIGURE 11.22

When blood is spun in a centrifuge, the formed elements settle to the bottom of the tube, leaving the plasma on the top. The red blood cells (RBC) are heavier than the white blood cells, so the white blood cells are found in a small "buffy coat" above the packed red blood cells.

blood deliver oxygen and patrol the body to destroy pathogens. Both specific and nonspecific immunity occur within the blood. None of these functions is reason to fear blood; instead they indicate just how remarkable this tissue is.

The formed elements in the blood are the cells and cellular fragments—99.9 percent of the formed elements are red blood cells, which give blood its red color; the other 0.1 percent are white blood cells and platelets.

Plasma comprises approximately 46 to 63 percent of total blood volume. Plasma is 92 percent water, 7 percent dissolved proteins, and 1 percent **electrolytes**, nutrients, and wastes. The proteins help maintain blood's osmotic pressure, so water will remain inside the vessels instead of diffusing into tissues. The protein albumin is particularly important in maintaining osmotic pressure. If the albumin level drops, osmotic pressure of the blood shifts, forcing water from the blood into the tissues, causing edema. Edema can also be caused by many other factors that alter the osmotic balance between blood and tissue.

Electrolytes Compounds that form a solution that can conduct electricity.

Colony stimulating factors Blood-borne compounds that cause cells in the bone marrow to produce new blood cells.

The formed elements of the blood are cells or bits of cells that originate in the red bone marrow (FIGURE 11.23, p. 350). In adults, red marrow is located within the epiphyses of the long bones, in the hip and sternum. Under the direction of hormones and **colony stimulating factors**, blood stem cells differentiate into **erythrocytes**, platelets, or **leukocytes**. Leukocytes further differentiate into five types of white blood cells.

Erythrocytes Red blood cells.

Leukocytes White blood cells.

LEUKOCYTES ARE DEFENSIVE CELLS

Leukocytes are specialized for defense, and though not abundant, they are critical to the immune system. There are approximately 5,000 to 11,000 white blood cells per mm^3 of blood, compared to the 4 to 6 million red blood cells per mm^3.

The five types of white blood cell (WBC) include three granular cells: **neutrophils, eosinophils**, and **basophils**, and two agranular cells: **lymphocytes** and **monocytes**. "Granular" means that when the cells are stained, dark granules appear in the cytoplasm under a microscope.

The odd names of the granulocytes (neutrophil, eosinophil, and basophil) reflect what happens when they are placed in Wright's stain, a mixture of stains used to identify white blood cells. Neutrophil granules become stained with the neutral stain (their granules "like" neutral stain, which is the literal translation of "neutrophil"). Eosinophil granules stain a bright orange-pink, the color of the eosin portion of the stain. Basophil granules take on the basic (pH 11) stain color, nearly black.

Agranulocytes contain no granules in their cytoplasm. Lymphocytes are small, round cells with little visible cytoplasm, whereas the monocytes are the largest of the white blood cells, with quite a bit

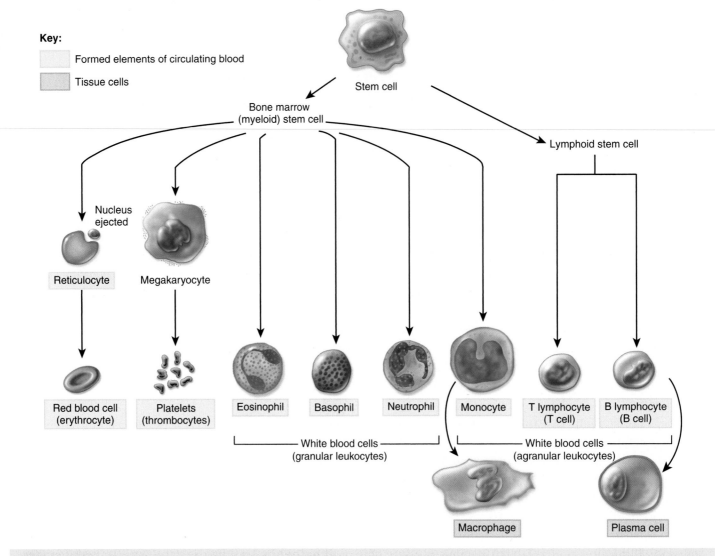

Key:
Formed elements of circulating blood
Tissue cells

Stem cell
Bone marrow (myeloid) stem cell
Lymphoid stem cell
Nucleus ejected
Reticulocyte
Megakaryocyte
Red blood cell (erythrocyte)
Platelets (thrombocytes)
Eosinophil
Basophil
Neutrophil
Monocyte
T lymphocyte (T cell)
B lymphocyte (B cell)
White blood cells (granular leukocytes)
White blood cells (agranular leukocytes)
Macrophage
Plasma cell

Blood cell formation FIGURE 11.23

Blood cell production, called **hematopoiesis**, occurs in bone marrow after birth. All the different types of blood cell arise from one type of pluripotent stem cell.

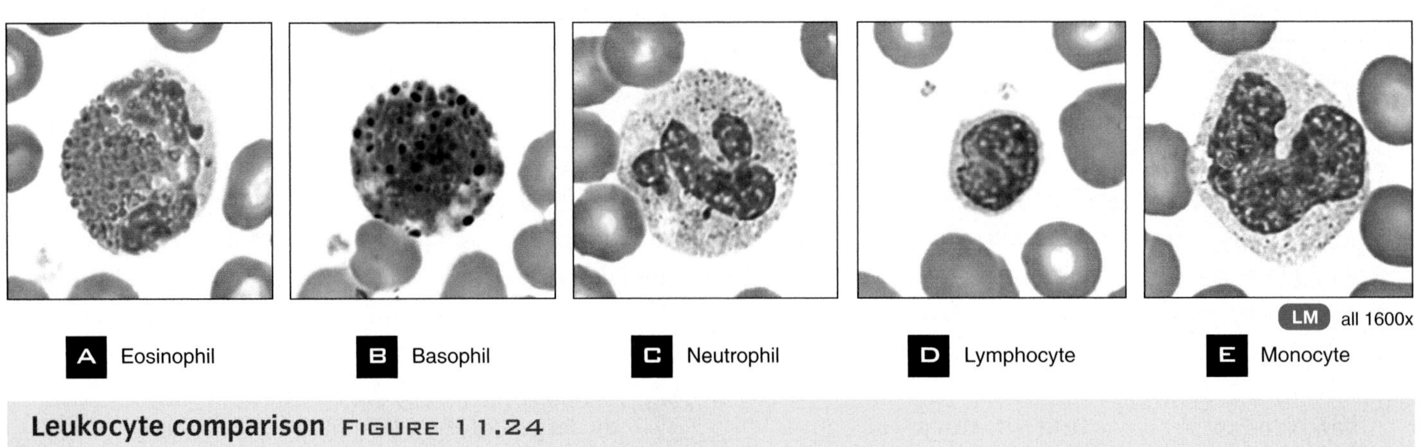

Leukocyte comparison FIGURE 11.24

of cytoplasm surrounding a large, kidney-shaped nucleus (FIGURE 11.24).

The proportions of leukocytes remain fairly constant in a healthy individual. Neutrophils make up the majority of circulating WBCs, with lymphocytes a close second. Monocytes are the third most common WBC, followed by eosinophils and lastly basophils. You can remember this order with this catchphrase: "Never let monkeys eat bananas." (n = neutrophils, l = lymphocytes, m = monocytes, e = eosinophils, and b = basophils)

Each cell has a specific function in warding off pathogens. When necessary, specific populations of WBCs increase, altering the overall proportions. The proportion of white blood cells gives an indication of what type of pathogen is present. See TABLE 11.1 for a physical description and the main function of each type of WBC.

Summary of formed elements in blood TABLE 11.1

Name and Appearance	*Number*	*Characteristics**	*Functions*
Red Blood Cells (RBCs) or Erythrocytes	4.8 million/μL in females; 5.4 million/μL in males	7–8 μm diameter, biconcave discs, without nuclei; live for about 120 days.	Hemoglobin within RBCs transports most of the oxygen and some of the carbon dioxide in the blood.
White Blood Cells (WBCs) or Leukocytes	5,000–10,000/μL	Most live for a few hours to a few days.[†]	Combat pathogens and other foreign substances that enter the body.
Granular leukocytes			
Neutrophils	60–70% of all WBCs	10–12 μm diameter; nucleus has 2–5 lobes connected by thin strands of chromatin; cytoplasm has very fine, pale lilac granules.	Phagocytosis. Destruction of bacteria with lysozyme, defensins, and strong oxidants, such as superoxide anion, hydrogen peroxide, and hypochlorite anion.
Eosinophils	2–4% of all WBCs	10–12 μm diameter; nucleus usually has 2 lobes connected by a thick strand of chromatin; large, red-orange granules fill the cytoplasm.	Combat the effects of histamine in allergic reactions, phagocytize antigen–antibody complexes, and destroy certain parasitic worms.
Basophils	0.5–1% of all WBCs	8–10 μm diameter; nucleus has 2 lobes; large cytoplasmic granules appear deep blue-purple.	Liberate heparin, histamine, and serotonin in allergic reactions that intensify the overall inflammatory response.
Agranular leukocytes			
Lymphocytes (T cells, B cells, and natural killer cells)	20–25% of all WBCs	Small lymphocytes are 6–9 μm in diameter; large lymphocytes are 10–14 μm in diameter; nucleus is round or slightly indented; cytoplasm forms a rim around the nucleus that looks sky blue; the larger the cell, the more cytoplasm is visible.	Mediate immune responses, including antigen–antibody reactions. B cells develop into plasma cells, which secrete antibodies. T cells attack invading viruses, cancer cells, and transplanted tissue cells. Natural killer cells attack a wide variety of infectious microbes and certain spontaneously arising tumor cells.
Monocytes	3–8% of all WBCs	12–20 μm diameter; nucleus is kidney shaped or horseshoe shaped; cytoplasm is blue-gray and has foamy appearance.	Phagocytosis (after transforming into fixed or wandering macrophages).
Platelets (Thrombocytes)	50,000–400,000/μL	2–4 μm diameter cell fragments that live for 5–9 days; contain many vesicles but no nucleus.	Form platelet plug in hemostasis; release chemicals that promote vascular spasm and blood clotting.

*Colors are those seen when using Wright's stain.

[†]Some lymphocytes, called T and B memory cells, can live for many years once they are established.

ERYTHROCYTES CARRY OXYGEN

Erythrocytes, or red blood cells, transport oxygen to the tissues and are by far the most common blood cells. Red blood cells (RBCs) are little more than a membrane-bound sac of **hemoglobin**, a protein that contains the pigment heme (FIGURE 11.25). Each RBC carries approximately 200 million hemoglobin molecules. Each of these molecules has at its center an atom of iron. This iron picks up oxygen (it rusts, in essence) in an environment where the oxygen content is high, and releases oxygen where oxygen is scarce. The concentration of oxygen in these areas is measured in bars and is written as the partial pressure of the gas. Often the total gas pressure is due to more than one gaseous element. Each individual gas exerts a portion of the total, or a partial pressure. The partial pressure of oxygen is annotated as P_{O_2}, and the partial pressure of carbon dioxide is indicated P_{CO_2}. Hemoglobin responds to the P_{O_2} in tissues and blood. This is a perfect setup because the body needs to transport oxygen from the lungs (where oxygen concentration is high) to the tissues (where the concentration is low). Hemoglobin is so perfect, in fact, that it is the only respiratory protein found in vertebrates; the same protein also carries oxygen for fish, whales, and frogs. Hemoglobin also appears throughout the **invertebrates**, where it floats in the blood, or **hemolymph**, of some insects, clams, and worms.

Invertebrate
Organism without a vertebral column, such as an earthworm, crab, or starfish.

Hemolymph
An oxygen-carrying fluid that circulates through the tissues of many invertebrates with open circulatory systems.

Hemoglobin also responds to changes in pH and temperature. In low pH or high temperature, both of which occur in active muscle, hemoglobin drops its oxygen more readily, so that the RBC delivers the oxygen exactly where it is needed (FIGURE 11.26). No wonder this respiratory protein is ubiquitous.

Erythrocytes are unique in several ways. As the immature red blood cell develops, it kicks out the nucleus to make room for more hemoglobin. Without a nucleus, the cell can neither repair itself nor direct cellular activities, including such basics as cellular respiration. Red blood cells do not survive long in the circulatory system. All the pressure from the left ventricle of the heart races the RBCs through the vessels and squishes them, one cell at a time, through the capillary beds (FIGURE 11.27). While passing through these beds, RBCs not only drop their oxygen, but they also suffer physical damage, which cannot be repaired in the absence of a nucleus.

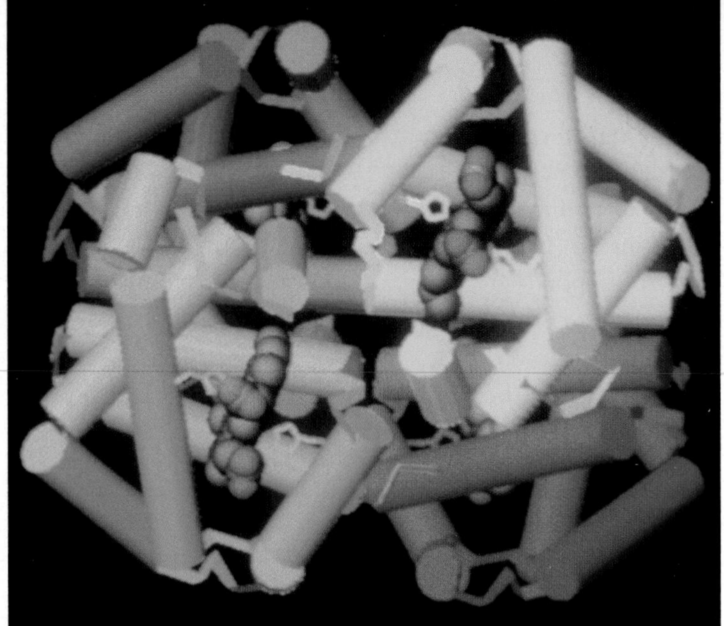

Hemoglobin molecule FIGURE 11.25

RBCs circulate for approximately 120 days before they are damaged enough to need removal from the circulatory system. The spleen and liver are responsible for removing these cells, breaking them down, and recycling their constituent minerals and proteins. An estimated 2 million RBCs are broken down per second. Because we do not run out of RBCs, we must produce them at the same rate: an incredible 2 million cells per second.

The rate of **erythropoiesis** is affected by hormones and environmental need. When blood oxygen drops, the kidneys are stimulated to produce erythropoietin, a hormone that stimulates RBC production. Because the presence of more red blood cells translates into more oxygen-carrying capacity, athletes can use this physiological fact to

Erythropoiesis
The formation of red blood cells (*poiesis* = to form, *erythro* = red).

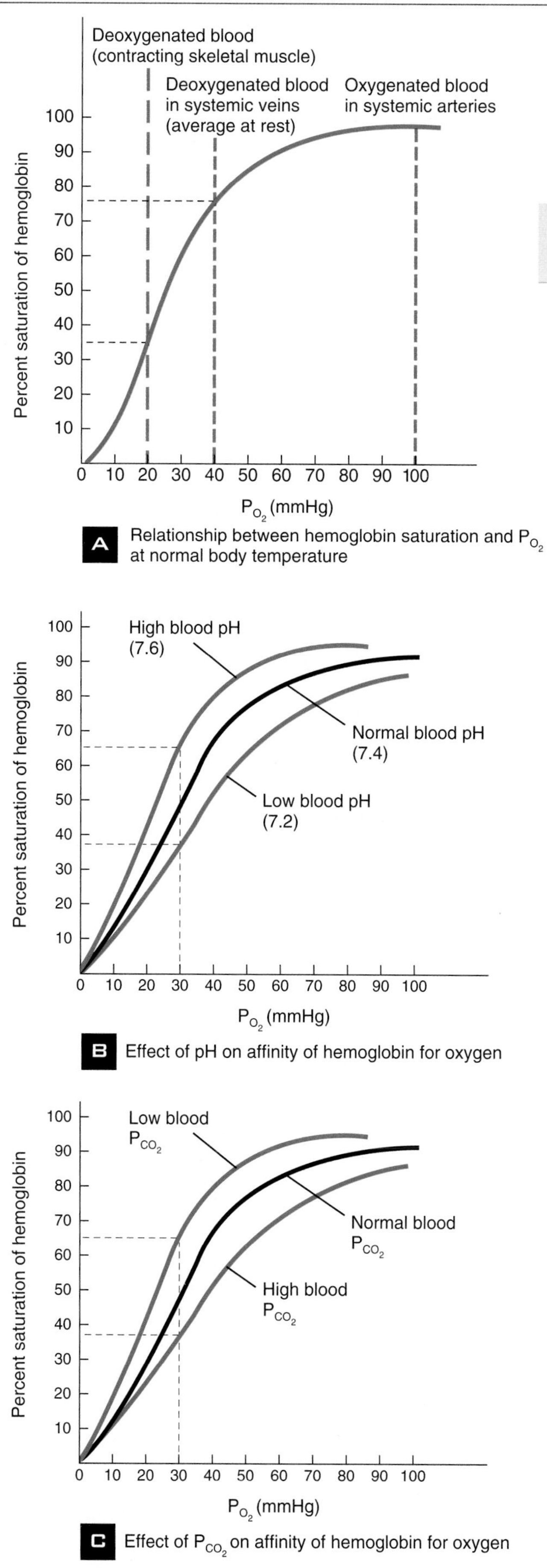

Graphs of hemoglobin uptake and release in varying conditions FIGURE 11.26

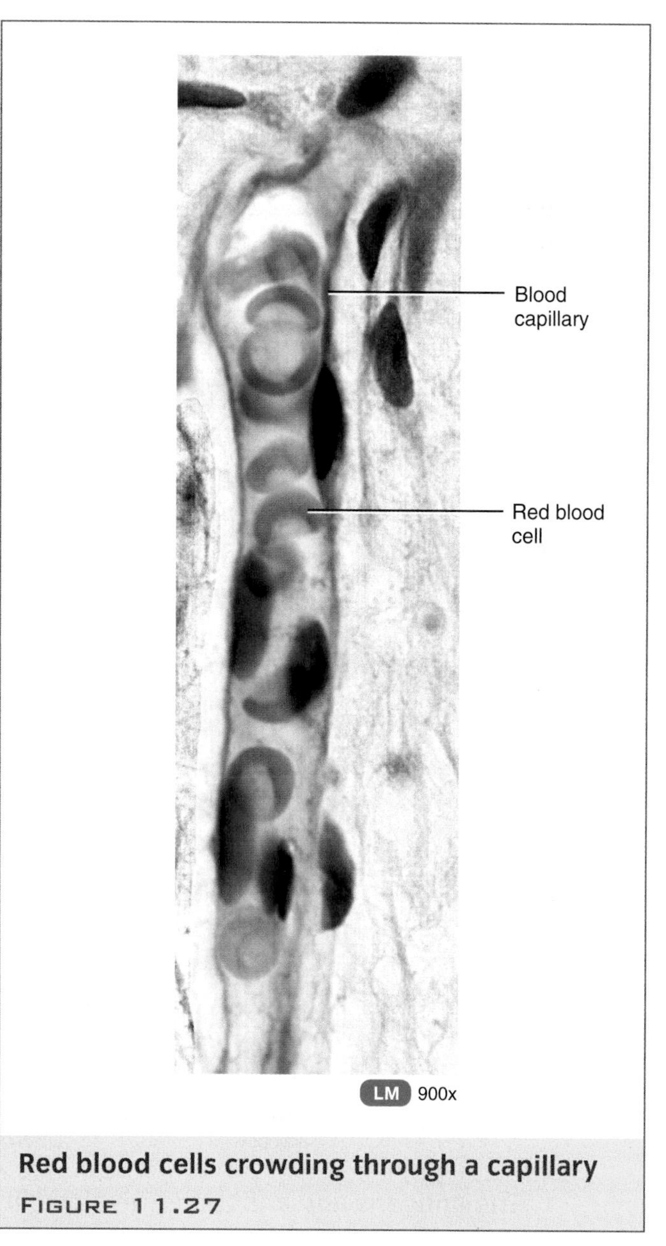

Red blood cells crowding through a capillary
FIGURE 11.27

improve their training. Because oxygen is scarce at higher altitudes, many athletes train at higher elevations just to stimulate RBC production. Some athletes have also used commercial erythropoietin, or EPO, to do the same thing. Although this hormone does increase RBC production, the performance advantage is unproven, and EPO is banned in many sports.

RED BLOOD CELL SURFACE PROTEINS DETERMINE BLOOD TYPE

Red blood cells, like other somatic cells, have many marker proteins on their surfaces, but the most important set is the markers that determine blood type. Blood type is described as A, B, AB, or O. Although you probably know your blood type, you may have no idea what those letters mean.

Codominant
Neither form of a gene will overshadow the other; when both forms are present, the individual will express both equally.

A, B, and O were arbitrarily chosen to identify the protein markers on the surface of your red blood cells. People with the "A" marker have type A blood; people with the "B" marker have type B. Because these traits are **codominant**, some people have both A and B markers, which we call type AB blood. Those with neither A nor B markers have type O blood, which represents the condition described as "no markers."

Agglutinin
Agent that causes cells to clump together or agglutinate.

Recent findings, however, indicate that type O blood has the precursor to the A and B markers on its surface. This precursor, called H substance, is modified to form the A and B antigens on the surface of types A, B, and AB blood (FIGURE 11.28). Apparently, people with type O do not modify the H substance, leaving it in its original form, able to trigger antibodies to both A and B antigens.

Despite certain dieting fads, the A, B, and O blood markers are important only when we must receive blood. The plasma of people with type A blood contains an anti-B **agglutinin** that will clump B blood. Similarly, those with type B blood have plasma that contains an anti-A agglutinin, which clumps type A blood. Type O blood carries both anti-A and anti-B agglutinins. This does not harm the individual because their RBCs have neither marker. It stands to reason that those with type AB blood have neither anti-A nor anti-B agglutinins because either would agglutinate their blood, with fatal results. If the blood type is not

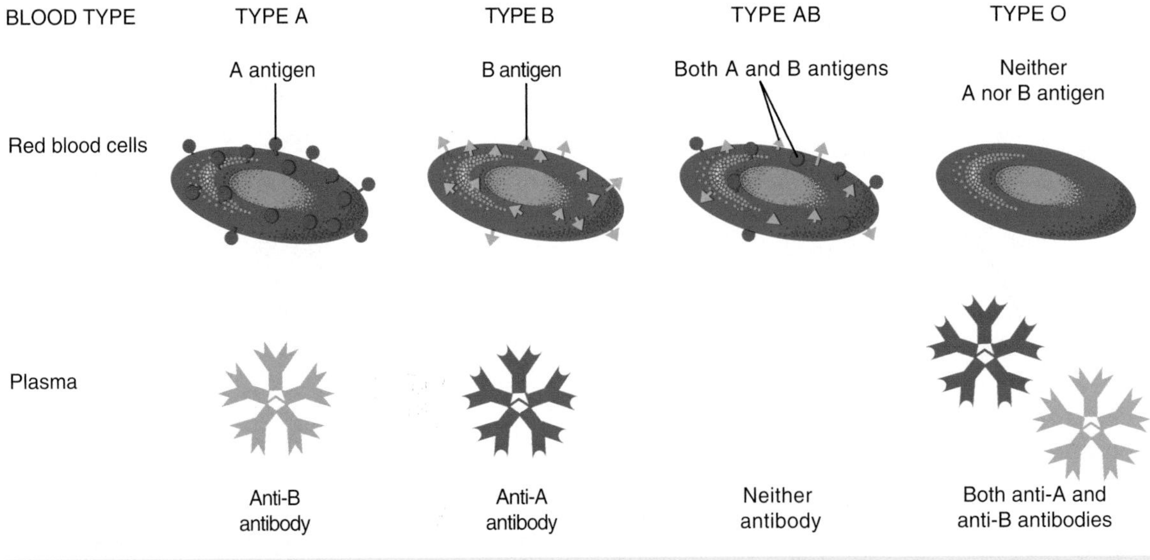

The antigens and antibodies involved in blood typing FIGURE 11.28

matched before a blood transfusion, a recipient's agglutinins will clump the introduced blood, negating any benefits of the additional blood volume and possibly causing life-threatening problems.

To determine blood type, technicians mix small samples of blood with each type of agglutinin and observe any reaction. If your blood clumps up when mixed with anti-A agglutinin, your RBCs have the surface marker A, and your blood is type A. If no clumping occurs when your blood is mixed with anti-A, you have type B or type O. Samples that clump when mixed with anti-B agglutinins have the B marker, and are type B. A sample that reacts with both anti-A and anti-B must have both A and B markers, and is thus type AB. Conversely, if no reaction appears with either agglutinin, the sample is type O.

In the United States, type O blood is more common than type A; however, the proportions of each type differ among ethnic groups. Slightly more than 46 percent of the Caucasian U.S. population is type O, whereas 38.8 percent is type A. Types B and AB are much rarer, comprising 11.1 and 3.9 percent of the population, respectively. Among African Americans, A and B are more evenly distributed: 49 percent are type O, 27 percent are type A, 20 percent are type B, and 4 percent are type AB. As a comparison, the Native American population is largely type O (79 percent). Very few Native Americans have type A blood (16 percent), and even fewer have type B (4 percent) or type AB (1 percent).

Because blood types are genetically based, they can be used to identify fathers in paternity suits, to eliminate or incriminate suspects in criminal investigations, and even to study ancient population migrations. A total of 26 blood groups other than ABO have been identified for these pursuits, including the MNS's, P, Lutheran (LU), and Kell (KEL).

Another blood-cell antigen, called **Rh factor**, is either present (Rh^+) or absent (Rh^-) on RBCs. On a blood-type card, Rh factor is the plus or minus sign after the A-B-O designation: "A^+" or "O^-." Rh factor works differently than ABO because people who are Rh^- do not ordinarily have anti-Rh agglutinin. But once they are exposed to Rh^+ blood, their immune system starts to produce anti-Rh antibodies. To prevent complications from anti-Rh^- antibodies, the Rh factor must be matched in an Rh^- person who has previously received an unmatched transfusion.

Eighty-five percent of the U.S. population is Rh^+. Rh^- mothers are at risk of **hemolytic disease of the newborn**, or HDN (FIGURE 11.29), if the father is Rh^+. As an Rh^+ child develops in the uterus, fetal blood and maternal blood do not mix, so the mother will not make anti-Rh agglutinin. But during birth, these two blood supplies may contact each other, in which case the mother will begin to produce anti-Rh

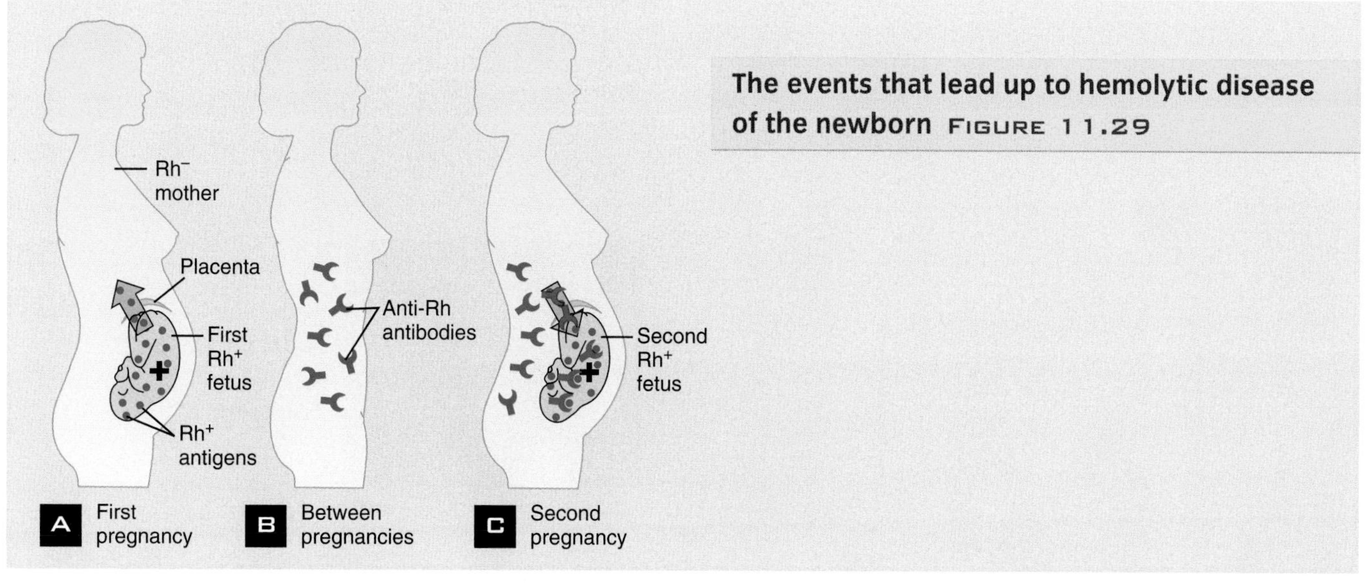

The events that lead up to hemolytic disease of the newborn FIGURE 11.29

Process Diagram

Clot formation FIGURE 11.30

1. Prothrombin is formed by the extrinsic pathway or the intrinsic pathway.
2. The enzyme prothrombinase converts prothrombin into the enzyme thrombin.
3. Thrombin converts soluble fibrinogen into insoluble fibrin. Fibrin forms the threads of the clot.

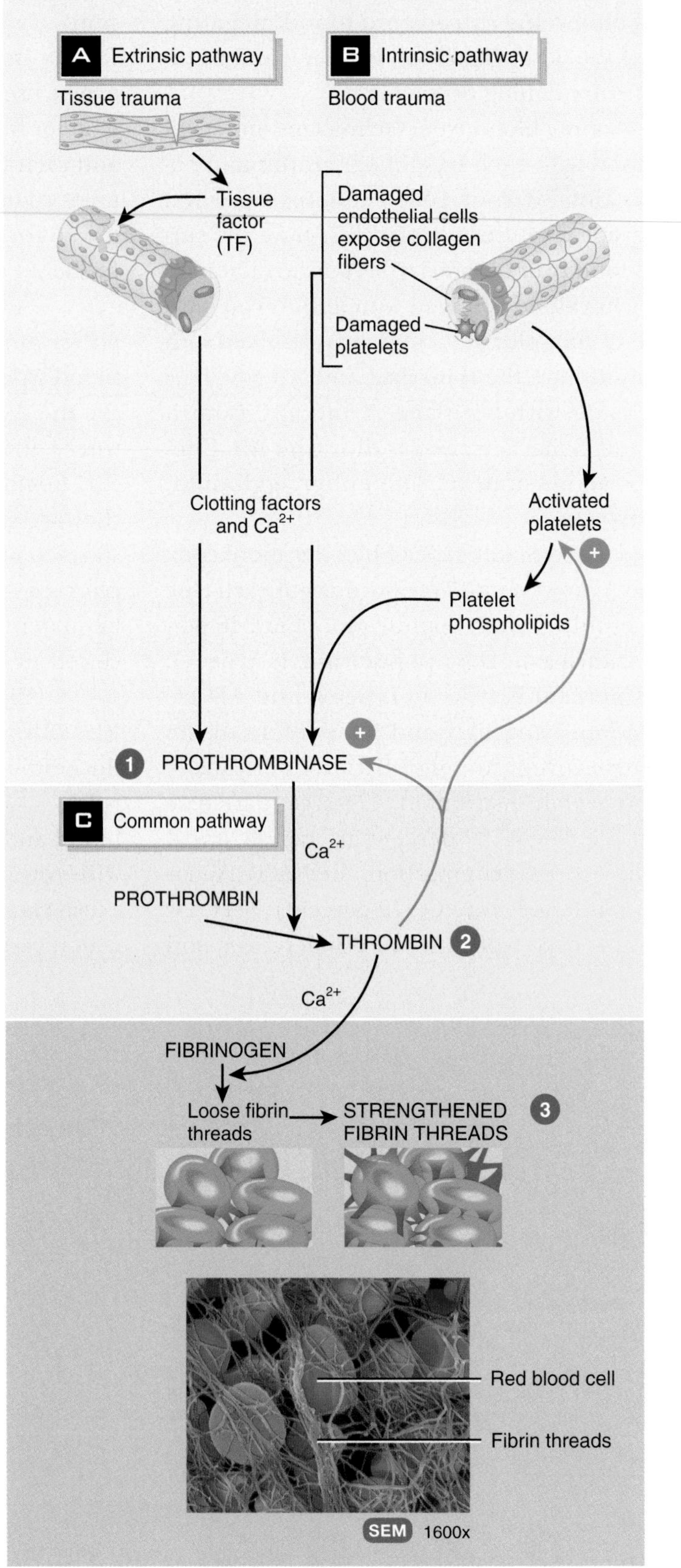

antibodies. These antibodies will not affect her, or her first child, but if she becomes pregnant with a second Rh^+ child, her Rh^+ antibodies will cross the placenta and cause agglutination and destruction of this second baby's blood. Rh^- mothers can be prevented from producing these antibodies **during the second birth** by inoculation with a dose of anti-Rh^+ antibodies **immediately after the first birth**. These antibodies clump the Rh^+ blood and remove it from the mother's blood supply before her immune response is launched.

PLATELETS GOVERN BLOOD CLOTTING

Platelets, the final type of formed element, are not even complete cells, but rather fragments of large cells called **megakaryocytes** that remain in the bone marrow. These huge cells bud pieces from their cytoplasm and release them into the bloodstream, forming more than 200 billion platelets per day. The fragments lack organelles and energy stores, but they do contain packets of physiologically active compounds. Once these compounds are released from the platelet into the surrounding plasma, they begin a series of events leading to the formation of a blood clot, in a process called hemostasis.

Clotting is necessary for maintaining fluid homeostasis; as we know, severe bleeding is a life-threatening emergency. Clotting is a complicated process in which a series of plasma proteins interact with clotting factors released by the platelets (FIGURE 11.30). Clotting begins when a blood vessel is damaged, turning its normally smooth interior rough. These rough edges catch platelets flowing past, forming a **platelet plug** that may seal the wound without need for a true clot. A platelet plug is what prevents bleeding from a paper cut.

If the rip is too large for a platelet plug, a clot will form as the stuck platelets rupture with the pressure of the passing blood and release compounds that react with plasma components. These interactions begin a series of events that will continue until blood flow ceases. The damaged tissue and trapped platelets release **prothrombinase activator,** which converts the plasma protein **prothrombin** into its active form, **thrombin**. Thrombin, in turn, activates the plasma protein **fibrinogen**, forming long thin fibers of **fibrin**. The fibrin threads get caught in the rough edges of the torn vessel, creating a net. As blood flows through the fibrin net, red blood cells get trapped. More fibers are delivered by fresh plasma that reaches the wound. The new fibrinogen interacts with fresh thrombin, and the clotting cascade continues until the plasma stops flowing and ceases bringing more protein. When plasma stops flowing, clotting has succeeded at stopping the bleeding. Clotting is a rare example of positive feedback in the body (see the green arrows in FIGURE 11.30).

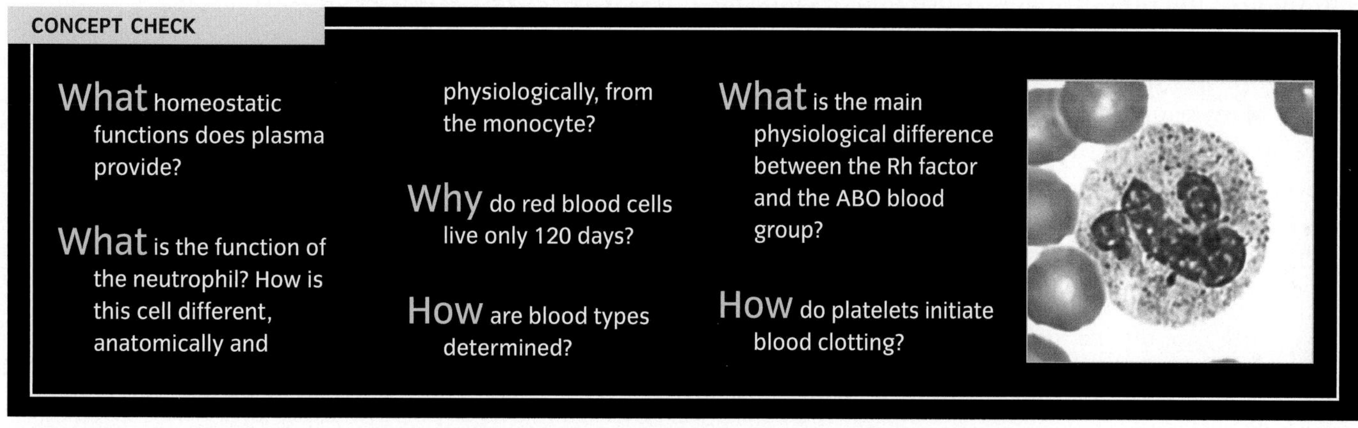

CONCEPT CHECK
What homeostatic functions does plasma provide?
What is the function of the neutrophil? How is this cell different, anatomically and physiologically, from the monocyte?
Why do red blood cells live only 120 days?
How are blood types determined?
What is the main physiological difference between the Rh factor and the ABO blood group?
How do platelets initiate blood clotting?

Blood Can Suffer Many Disorders

LEARNING OBJECTIVES

Describe the general pathology of leukemia.

Define anemia and list its symptoms.

List common blood-borne pathogens.

Because blood is so vital, when something goes wrong with it, our quality of life is severely diminished. Disorders of the blood can be life-threatening, as in leukemia, or they can cause acute disabilities, as in anemia. **Infectious mononucleosis**, commonly called mono, is a common blood disease on college campuses. Mono is caused by the **Epstein-Barr virus** and is transmitted through saliva—hence the popular name of "kissing disease." (We discussed mono in detail in Chapter 10.)

BLOOD CANCER: LEUKEMIA

Perhaps the most frightening blood disorder is **leukemia,** literally: "white blood." Leukemia is a general term for several cancers of the bone marrow. In most leukemias, the white blood cells are shaped abnormally and do not function properly. More than 2,000 children and 27,000 adults in the United States are diagnosed with leukemia every year.

Many symptoms of leukemia are flu-like, and all are related to those nonfunctional white blood cells. Infections take hold more readily and are more persistent. Lymph nodes and the spleen swell in an effort to rid the body of these defective leukocytes. To add to the difficulty, when the bone marrow is pushing out too many white blood cells, it often reduces its output of red blood cells, which reduces the blood's oxygen-carrying capacity, causing fatigue and weakness.

The causes of leukemia are unknown, and although some risk factors have been identified, having these factors does not mean you will necessarily develop leukemia any more than having the risk factors associated with cardiovascular disease means you will have a heart attack. Beyond a family history of the disease, the risk factors include exposure to ionizing radiation from nuclear weapons and nuclear waste and exposure to carcinogens such as benzene.

Leukemia can be classified by its pattern of onset or by the specific cells affected. **Acute** leukemia appears quickly, filling the blood with extremely immature white blood cells called **blasts**. **Chronic** leukemia appears far more slowly, with blood cells that are more developed, but still immature. Both acute and chronic leukemia can affect either **myeloid** or **lymphoid** cells. Both cells mature in the bone marrow, but myeloid cells become the granulocyte form of white blood cells, whereas lymphoid cells mature in the lymph glands and become lymphocytes (see FIGURE 11.23).

Treatments for leukemia vary depending on the stage of disease and the type of affected leukocytes, but the goal is to move the patient into **remission**—where the disease may remain in the patient's bone marrow, but the leukocytes are functionally normal. Treatment—chemotherapy, radiation therapy, bone marrow transplant, or biological therapy—is often successful for a period. Because the disease often reappears, leukemia patients need continual medical monitoring.

ANEMIA MEANS A SHORTAGE OF ERYTHROCYTES

Anemia is a reduction in the red blood cell population and thus in the blood's oxygen-carrying capacity. The symptoms of anemia include fatigue, weakness, shortness of breath, and sometimes chest pains like angina. Anemia is easily diagnosed via **hematocrit** (FIGURE 11.31). A packed cell level below 42 percent in adult males, or 38 percent in adult females, often indicates some form of anemia.

Anemia is classified by the cause of the red blood cell deficiency. A shortage of iron, vitamin B12, intrinsic factor (a hormone that allows for iron absorption), or other essential proteins all inhibit RBC production. Excessive bleeding will reduce RBC counts.

Hematocrit tubes

FIGURE 11.31

The level of packed cells in the hematocrit tube after spinning in a centrifuge indicates the amount of oxygen-carrying capacity of that sample of blood. Adults should have a packed cell volume above 38 percent. If the packed cell volume is lower, anemia is suspected.

Improper formation of red blood cells and nutritional deficits can also cause anemia. Sickle cell anemia (see the Health, Wellness, and Disease box, on p. 360) is a special type of anemia in which the hemoglobin molecule is not properly formed.

CARBON MONOXIDE PREVENTS THE BLOOD FROM CARRYING OXYGEN

Carbon monoxide is an odorless environmental poison that prevents the blood from carrying oxygen and can cause death or disability. Carbon monoxide (CO) molecules establish an irreversible bond to hemoglobin, thereby preventing the hemoglobin molecule from carrying oxygen. Red blood cells contaminated with CO float uselessly through the blood until they wear out and are destroyed. Normally, air contains almost no CO, so this irreversible binding is irrelevant. But the CO concentration increases dramatically in some environments, primarily when fossil fuels are burned and the exhaust fumes are returned to the combustion zone. This can happen if a car runs in a closed garage or a malfunctioning furnace recycles fumes into a residence. Because severe CO poisoning can starve the tissues of oxygen, causing brain damage, myocardial infarction, or death, carbon monoxide detectors (much like smoke detectors), are an affordable and sensible precaution. The first symptoms of CO poisoning are drowsiness and headache. If you suspect carbon monoxide poisoning, move to fresh air and seek medical help. Blood transfusions may be needed to replace carbon monoxide-polluted red blood cells with functional erythrocytes.

PATHOGENS CAN LIVE IN THE BLOOD

Though not necessarily a disorder of the blood itself, many pathogens travel in the blood, including **hepatitis, HIV**, and other sexually transmitted diseases. The best defense against blood-borne pathogens is to prevent your blood from contacting another person's blood. A key source of infection is unprotected sex, which can tear mucous membranes, causing unintentional contact between the two bloodstreams. Healthcare workers are constantly reminded to take precaution around all "sharps," because an inadvertent "stick" with a used needle can spread blood-borne pathogens. To prevent infection through transfusion of tainted blood, blood banks routinely test their stocks of blood for viral contamination. Blood-borne pathogens include a wide variety of diseases; in each case treatment aims to eliminate the pathogen from the blood.

Health, Wellness, and Disease

Sickle cell anemia

About 72,000 Americans have sickle cell anemia, a genetic defect that deforms red blood cells and impairs their ability to circulate and transport oxygen. The sickle cell mutation changes hemoglobin, the oxygen-transport protein. Normally, red blood cells (RBCs) are shaped like donuts. In sickle cell anemia, they are hard and crescent or "sickle" shaped in environments with a low P_{O_2}. The deformed cells get caught in capillaries, forming a plug that prevents other RBCs from delivering oxygen to these tissues. And because these abnormal RBCs live only 10 to 20 days, in contrast to the normal 120-day life span, patients with sickle-cell disease also have general anemia, caused by a shortage of RBCs.

Sickle cell disease is an inherited defect in the gene that forms hemoglobin. A person who inherits one "sickle" gene and one normal gene becomes a carrier for sickle cell disease; they have "sickle cell trait." These carriers do not have symptoms of sickle cell disease, but their children are at risk. If both parents have sickle cell trait, each child has a 1 in 4 chance of having the disease and a 2 in 4 chance of being carriers (having the sickle cell trait), leaving only a 1 in 4 chance of being free of sickle cell trait.

Why does the sickle cell trait persist among humans? Evolution, after all, tends to remove defective genes from the population, and they should eventually disappear. The answer is that the sickle cell trait protects people from the deadly consequences of malaria, a parasite which attacks red blood cells. This resistance to malaria explains why the sickle cell gene is common in Africa, South and Central America, some Mediterranean countries, and India. In these places, protection against malaria is an adaptive trait, even if it does reduce the fitness of the individual by placing their offspring at risk of inheriting sickle cell anemia.

Although sickle cell disease is inherited and present at birth, symptoms usually don't occur until after 4 months of age. Beyond the problem with oxygen transport, the deformed cells also cause small blood clots and recurrent painful episodes called "sickle cell crises." Other problems include potentially life-threatening *hemolytic crises* when damaged red blood cells break down, *splenic sequestration crises* as the spleen enlarges and traps the damaged blood cells, and *aplastic crises* if a certain type of infection causes the bone marrow to stop producing red blood cells.

Repeated crises can damage the kidneys, lungs, bones, eyes, and central nervous system. Blocked blood vessels and damaged organs cause acute painful episodes, which most patients suffer at some point, that can last hours to days. These acute painful episodes affect the bones of the back, the long bones, and the chest. The crises may require hospitalization for pain control, oxygen, and intravenous fluids.

Sickle cell disease can cause death by organ failure and infection. Some patients experience minor, brief, and infrequent episodes, while others endure severe, prolonged, and frequent episodes with many complications. People with sickle cell disease need treatment to prevent and reduce symptoms. Blood transfusions can treat the anemia portion of the disease. In the past, death from organ failure usually occurred between ages 20 and 40. More recently, because of better understanding and management, affected people live into their 40s and 50s.

www.wiley.com/college/ireland

Beginning to sickle
Crenated
Normal
Sickled

SEM 3310x

Red blood cells

Again and again in this discussion, we have returned to one of the primary roles of the blood: to distribute oxygen and remove carbon dioxide. To do its work, the cardiovascular system must interact closely with the respiratory system, which is the point of entry for oxygen and the point of departure for carbon dioxide. If you need more evidence of the tight interaction between the CV and respiratory systems, pay attention to your own body. Take your pulse while resting, and simultaneously count your breaths. Then run up some stairs and repeat. Notice that both your pulse and your breathing have accelerated. To understand what is happening during this interaction of heartbeats and breaths, we must move on to the respiratory system.

CONCEPT CHECK

Differentiate between chronic and acute leukemia.

Why is sickle cell anemia still present in the human gene pool?

How does carbon monoxide reduce the oxygen-carrying capacity of blood?

List two blood-borne pathogens.

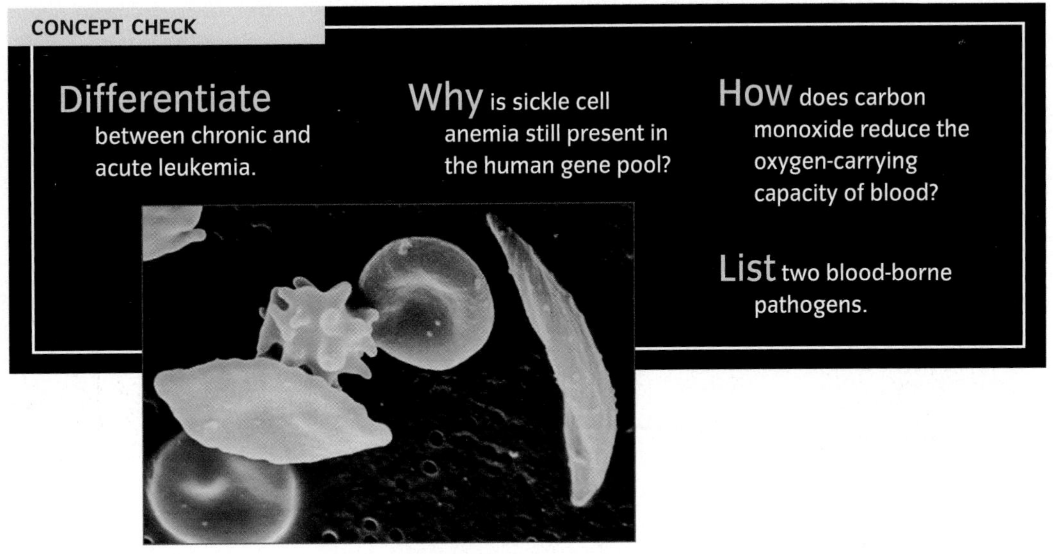

CHAPTER SUMMARY

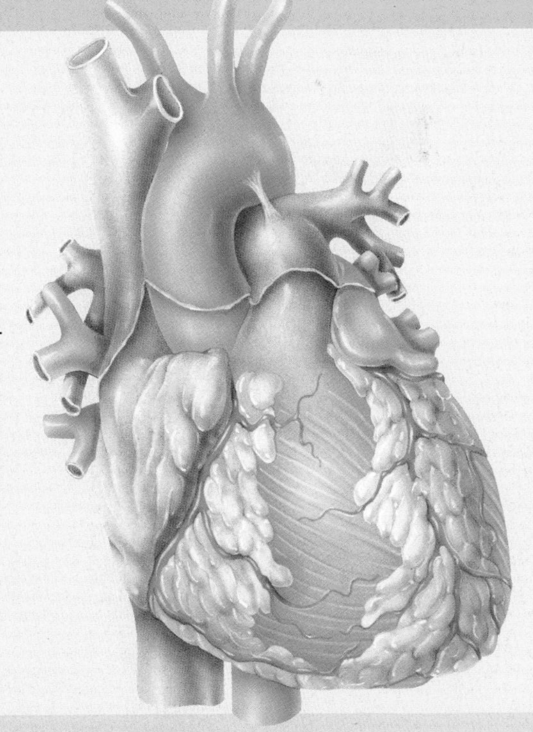

1 The Heart Ensures Continual, 24/7 Nutrient Delivery

The cardiovascular system is responsible for the transport of nutrients, gases, and waste products in the body. It is a closed system, consisting of the heart, arteries, veins, and capillaries. The heart serves as the pump for the cardiovascular system, pushing blood through the body. The heart has four chambers: two atria and two ventricles. The ventricles generate the force needed to move the blood. The bicuspid, tricuspid, aortic, and pulmonary valves in the heart prevent backflow. Cardiac output is the amount of blood pumped by the heart in one minute.

The conduction system of the heart consists of the SA node, AV node, AV bundle, bundle branches, and the Purkinje fibers. The contraction impulse follows this pathway, ensuring the heart contracts effectively. The ECG records the changes in electrical charge as the cardiac cells contract. The P wave is the depolarization of the atria, the QRS complex is the depolarization of the ventricles, and the T wave is the repolarization of the ventricles. The time between events measures the speed of transfer of the contraction impulse.

CHAPTER SUMMARY

2 Blood Vessels and Capillary Transport Involve Miles of Sophisticated Plumbing

Arteries carry blood from the heart, capillaries are the exchange vessels, and veins return the blood to the heart. The walls of these vessels differ according to the differing pressures they carry. Veins, with extremely low-pressure flow, require valves in order to prevent backflow.

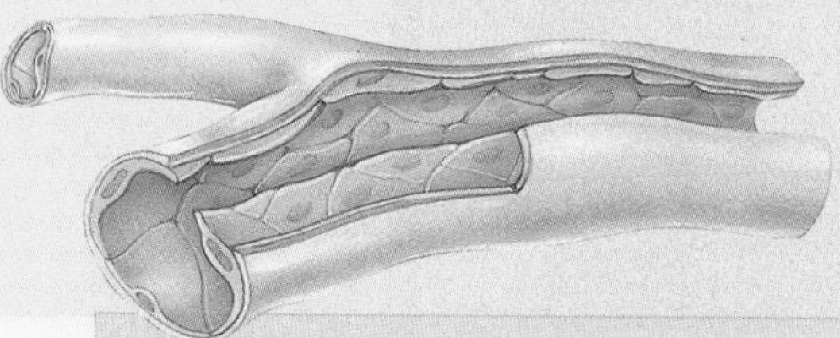

3 Different Circulatory Pathways Have Specific Purposes

Vessels that lead from the heart to the lungs and back to the heart comprise the pulmonary circuit. The systemic circuit includes vessels that leave the heart, travel through the tissues, and return to the heart. Portal systems, like the hepatic portal system, contain two capillary beds.

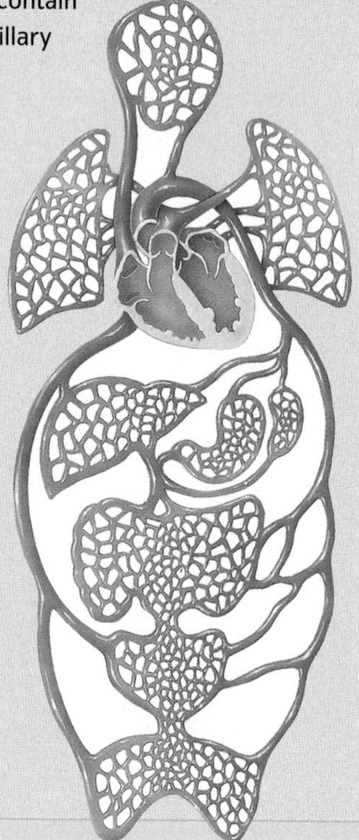

4 Cardiovascular Disorders Have Life-Threatening Consequences

Cardiovascular disease is the leading cause of death in the United States. It includes many different problems, such as hypertension, atherosclerosis, heart attack, heart failure, embolism, and stroke. Genetic factors play a role in hypertension, atherosclerosis, heart attack, and heart failure. High blood pressure affects body tissues by damaging or destroying capillary beds. Vessels become clogged with fatty deposits in atherosclerosis. Heart attack, or myocardial infarction, is due to a lack of blood flow to a region of the heart. Angioplasty, stent placement, or bypass surgery may correct cardiac atherosclerosis and prevent heart attack. Heart failure is an inability of the heart to pump blood from the left ventricle through the body.

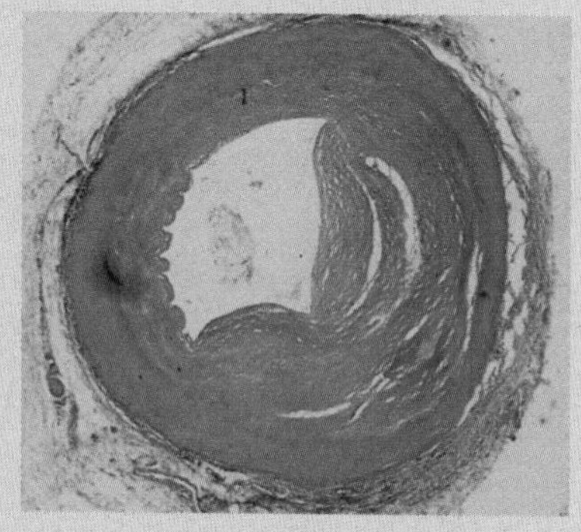

5 Blood Consists of Plasma, Cells, and Other Formed Elements

Blood is a liquid connective tissue composed of plasma, red blood cells, white blood cells, and platelets. The plasma serves to hydrate the body and dissolve nutrients. The red blood cells transport oxygen, using hemoglobin, which drops oxygen in areas of low oxygen concentration and picks it up in areas of high concentration. RBCs carry marker substances on their surface, designating blood as type A, B, AB, or O. In addition, there is an Rh factor on most people's RBCs. The ABO blood groups are genetically determined and can be used to trace lineage. Type A blood has anti-B agglutinins; Type B blood has anti-A agglutinins, and Type O blood has both agglutinins. Type AB blood has neither agglutinin because that would harm the individual. Many other blood groups are based on proteins and glycoproteins on the surface of the RBCs.

White blood cells provide immunity and nonspecific defense. There are five types of white blood cells: neutrophils, lymphocytes, monocytes, eosinophils, and basophils. Each has a specific job and occurs in a specific percentage in a healthy individual. Platelets maintain fluid homeostasis. They either form a platelet plug to block the loss of blood in small tears, or release factors that initiate clotting. Clot formation is a positive feedback loop, continuing until blood no longer flows past the injured area.

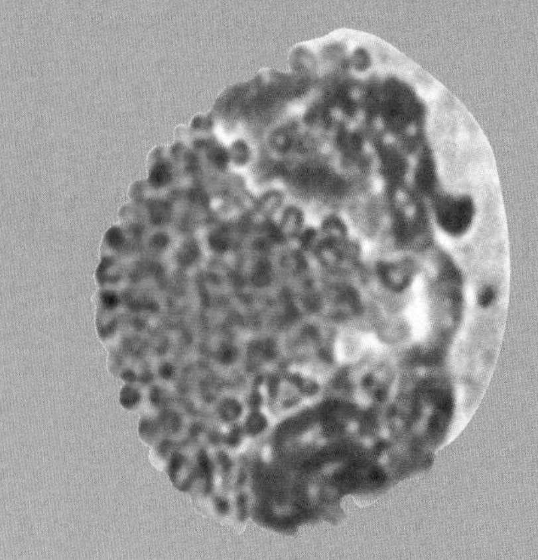

6 Blood Can Suffer Many Disorders

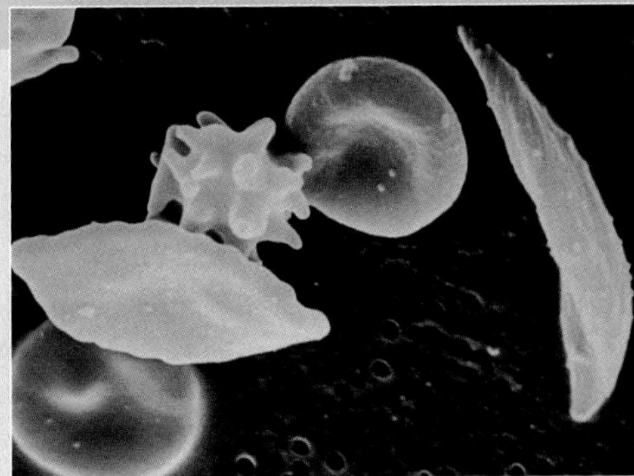

In anemia, the most common blood disorder. RBC numbers decline and oxygen-carrying capacity of the blood drops. Causes range from lack of iron in the diet to inadequate protein formation, to bleeding and loss of blood volume. Sickle cell anemia is a special type of anemia in which the hemoglobin is incorrectly formed, causing a drop in RBC levels. Leukemia is another blood disorder that affects the white blood cells. Causes of leukemia may include exposure to carcinogens or nuclear radiation. Leukemia is treated with chemotherapy, or bone marrow transplant surgery. Blood can carry a wide range of pathogens. Many are spread by contact with contaminated blood, which is one reason for using caution in sexual activity. Blood banks must test blood for viruses before distributing blood for transfusions. HIV, herpes, and other STDs are examples of blood-borne pathogens.

KEY TERMS

- **agglutinin** p. 354
- **anastomoses** p. 346
- **bicuspid** p. 329
- **capillary bed** p. 336
- **cardiac sinus** p. 331
- **codominant** p. 354
- **colony stimulating factors** p. 349
- **congenital** p. 334
- **diastole** p. 331
- **edema** p. 339
- **electrocardiogram** p. 334
- **electrolyte** p. 349
- **erythrocytes** p. 349
- **erythropoiesis** p. 352
- **hemolymph** p. 352
- **hormones** p. 348
- **hypertrophy** p. 330
- **invertebrate** p. 352
- **ischemia** p. 334
- **leukocytes** p. 349
- **lumen** p. 336
- **mediastinum** p. 327
- **prolapse** p. 329
- **Purkinje fibers** p. 333
- **systole** p. 331
- **tachycardia** p. 346
- **tricuspid** p. 328
- **venules** p. 339

CRITICAL THINKING QUESTIONS

1. Reptiles and amphibians have a three-chambered heart, with only one ventricle. Blood flows from the lungs and body into this single pumping chamber, which pushes it to the body or the lungs. How does this compare with the functioning of the four-chambered heart of mammals? Explain the physiological advantage of separate left and right ventricles.

2. Artificial pacemakers can override the natural heartbeat set by the SA node. These electronic devices set a constant heartbeat that is not sensitive to the body's demands. List some activities that would be challenging for a patient with an artificial pacemaker. What innovations could improve pacemaker technology?

3. Most capillaries are diffusion vessels, meaning that nutrients, oxygen, waste material, and hormones can pass through their walls and into surrounding cells (or vice versa). What features of the structure of a capillary wall raise diffusion capacity—how does structure relate to function in this case? What special modifications would you expect to see in areas where diffusion is prevented, as in capillaries of the brain?

4. Marie was born and raised in Denver, Colorado, the "mile high" city. She has been a cross-country runner since grade school. When Marie went to college in Florida, her running times improved. What might explain this sudden improvement?

5. Hemophilia is an inherited clotting disorder attributable to the absence of one necessary blood-clotting factor. How might the inability to form blood clots affect daily life? Why is homeostasis vital to the blood?

SELF TEST

1. The correct pattern of blood flow through the cardiovascular system is as follows:
   a. heart → veins → arteries → capillaries → heart.
   b. heart → arteries → capillaries → veins → heart.
   c. heart → veins → capillaries → arteries → heart.
   d. heart → capillaries → veins → arteries → heart.

For questions 2,3 and 4, refer to the following figure.

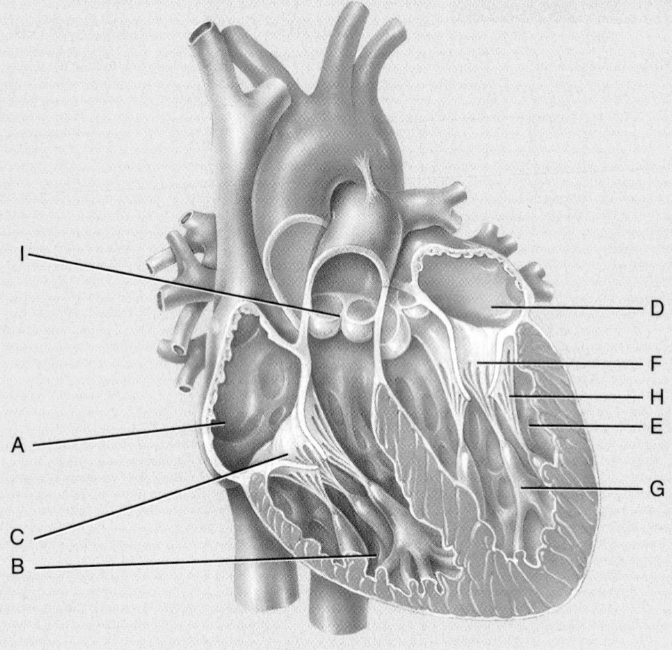

2. The chamber of the heart that receives blood from the lungs is
   a. A.
   b. B.
   c. D.
   d. E.

3. The valve that prevents backflow of blood returning from the body is
   a. A.
   b. C.
   c. F.
   d. I.

4. The structure(s) responsible for supporting and stabilizing the interventricular valves is (are)
   a. C.
   b. F.
   c. G.
   d. H.
   e. both G and H.

5. True or False? When the heart is relaxed, it is said to be in diastole.

6. During the cardiac cycle, the stage that immediately follows atrial systole is
   a. atrial diastole.
   b. ventricular systole.
   c. ventricular diastole.
   d. whole heart diastole.

The following figure should be used to answer questions 7, 8, and 9.

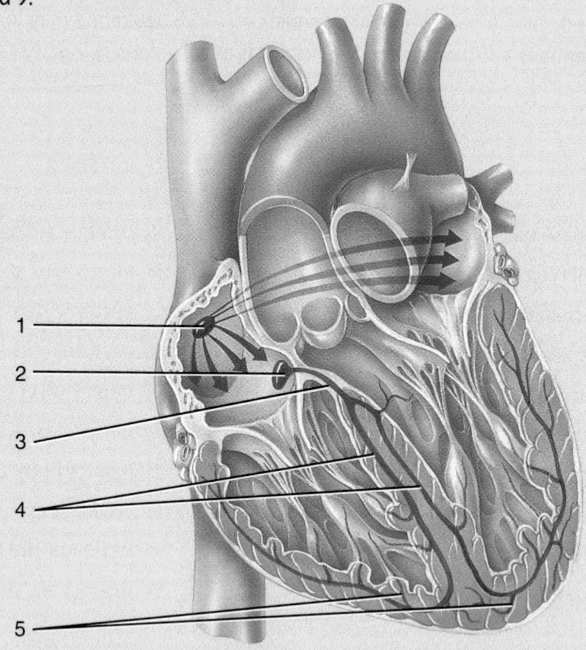

7. The structure that initiates the heartbeat, indicated by the number 1, is the
   a. Purkinje fibers.
   b. AV node.
   c. bundle branches.
   d. SA node.

8. Once the heartbeat begins, the function of the structure labeled 2 is to
   a. spread the impulse to contract to the cells of the atria.
   b. slow the impulse to contract and pass it to the AV bundle and on to the ventricles.
   c. allow the impulse to reach all the cells of the ventricles simultaneously.
   d. send the impulse to contract on to the bundle branches.

9. The structure responsible for the P wave on an ECG is number

a. 1.
b. 2.
c. 3.
d. 4.

10. The blood vessel that is thin-walled, includes valves, and carries blood under little pressure is the

a. artery.
b. capillary.
c. vein.
d. All of the above fit this description.

11. The main difference between the pulmonary circuit and the systemic circuit is that in the pulmonary circuit,

a. oxygen-rich blood leaves the heart for the lungs.
b. pulmonary veins carry oxygen-poor blood.
c. pulmonary arteries carry oxygen-poor blood.
d. blood in the pulmonary circuit goes to the brain only.

12. When a blood vessel of the leg becomes occluded (blocked) by a fatty deposit, the resulting condition is

a. stroke.
b. aneurysm.
c. myocardial infarction.
d. atherosclerosis.

13. Congestive heart failure

a. causes a build-up of fluid in the lungs and pericardium.
b. is more common in the elderly than the young.
c. is due to a weakened left ventricle.
d. All of the above are true.

14. The most common cell in the blood is the

a. neutrophil.
b. leukocyte.
c. erythrocyte.
d. platelet.

15. True or False? The liquid portion of the blood, the plasma, contains water, proteins, and cells.

16. Which of the cells shown in the figure below is least common in the blood?

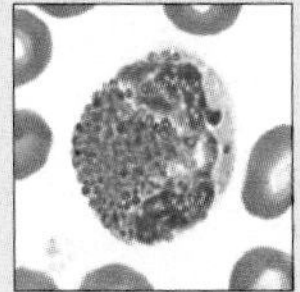
(a) Eosinophil

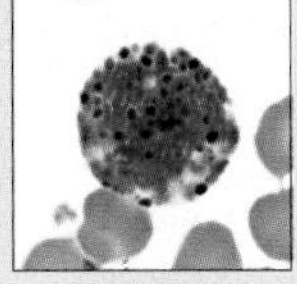
(b) Basophil

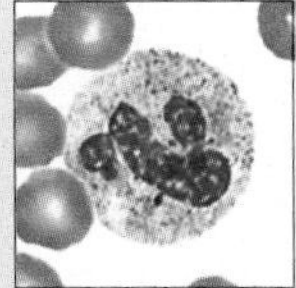
(c) Neutrophil

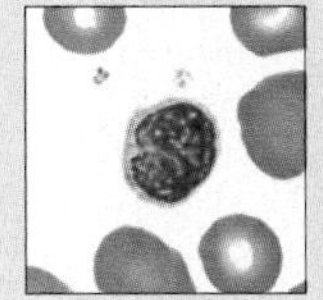
(d) Lymphocyte

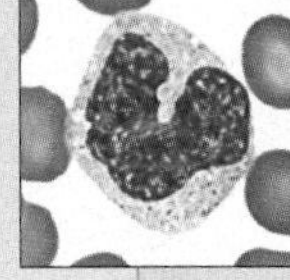
(e) Monocyte

a. Neutrophil
b. Eosinophil
c. Basophil
d. Monocyte
e. Lymphocyte

17. Hemoglobin is specialized to ___________ oxygen where pH is low, oxygen concentration is low, or temperatures are high.

a. release
b. pick up

18. What is the blood type of the cell shown below?

a. A
b. B
c. O
d. AB

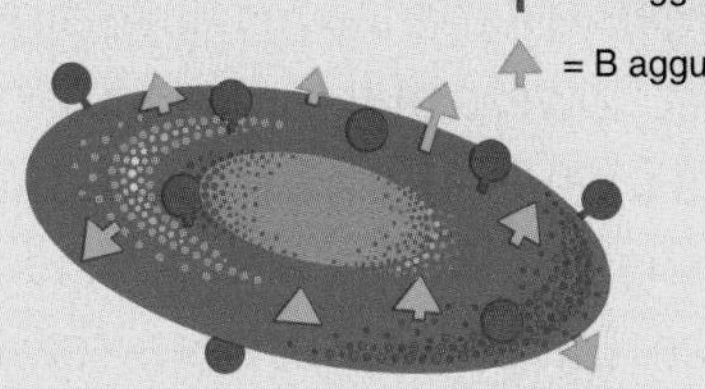

19. The plasma protein that is activated and forms a network of fibers across a wound to trap RBCs is

a. prothrombin.
b. thrombin.
c. fibrinogen.
d. fibrin.

20. The test shown below is used to diagnose

a. anemia.
b. acute leukemia.
c. infectious mononucleosis.
d. angina.

The Respiratory System: Movement of Air

12

It is a clear, crisp fall morning. The leaves are turning, and the air carries a hint of the chill to come. Many of the leaves have already fallen, and the last flowers of the season are in bloom. For Marie's friends, the heat of the summer is past, and this is a time of energy and exuberance. But for Marie, it is a season fraught with threat—her very survival is in danger. The air is filled with mold spores, leaf dust, and other fragments of dead plants. For her, those piles of leaves are simply huge colonies of mold and heaps of dust that make breathing, the essential act of life, difficult. Most people don't give a second thought to breathing; but Marie does in the moldy season. She has so little energy that she must stop to rest on a flight of stairs. Oxygen is so scarce in her blood that her lips and even her fingernails sometimes turn blue.

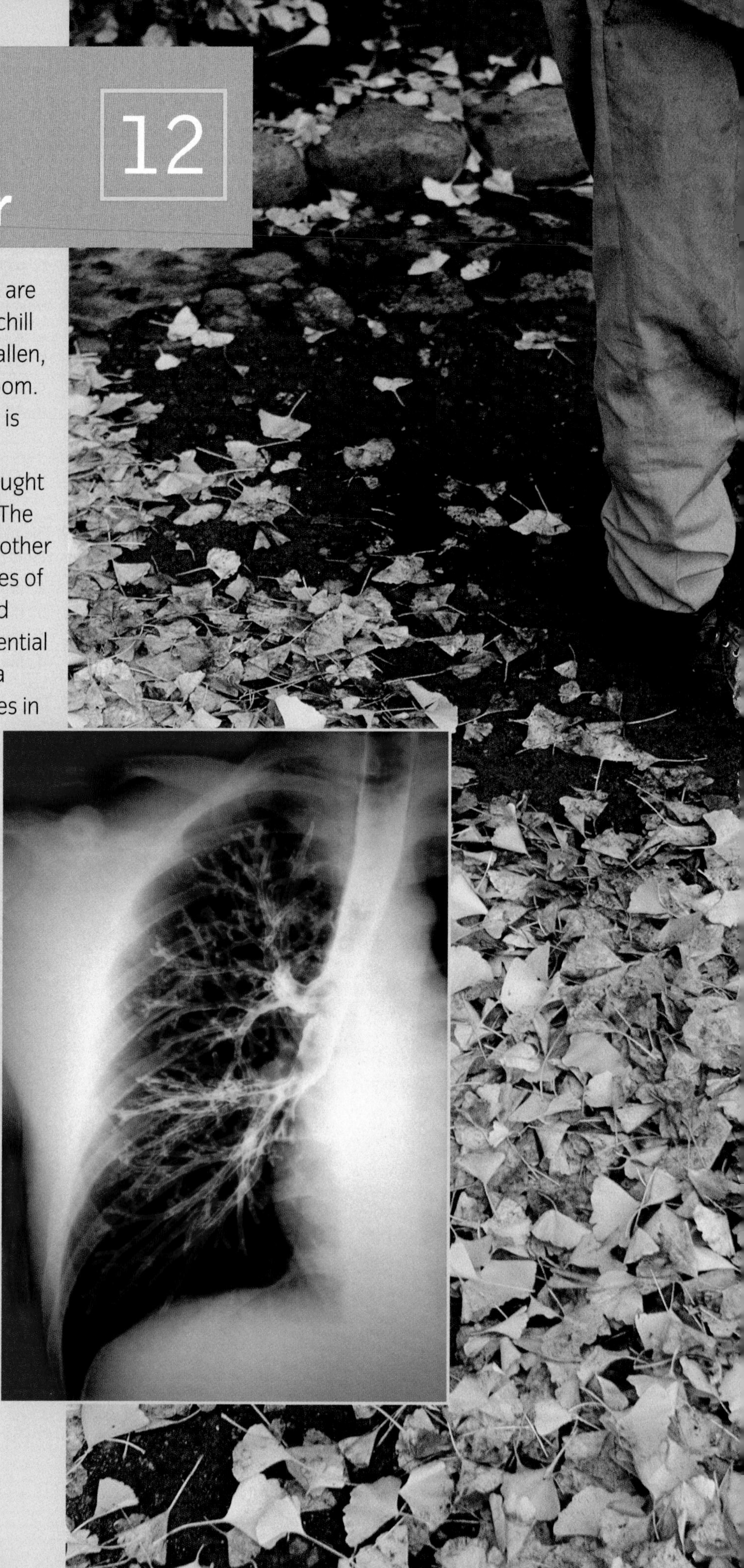

Marie is an asthmatic. Her lungs respond to mold and airborne dust with an inflammatory reaction that threatens her very survival during allergy season. Every minute, humans must transport oxygen to their cells and remove the carbon dioxide produced by cellular respiration. Human life depends on the integrated functioning of the cardiovascular and respiratory systems because neither system, by itself, can supply what we need to survive. And for Marie and millions of other asthmatics, this gas exchange process cannot be taken for granted. But to understand how asthma and other diseases affect the respiratory system, we must first get a grasp of the intricate anatomy and physiology of the lungs and airways.

CHAPTER OUT
The Respiratory System Provides
Us with Essential Gas Exchange
well as Vocalization p. 368
In Order to Respire, Air Must Be
Moved in and out of the
Respiratory System p. 382
External Respiration Brings Supp
for Internal Respiration p. 387
Transport of Oxygen and Carbo
Dioxide Requires Hemoglobin a
Plasma p. 390
Respiratory Health Is Critical to
Survival p. 393

The Respiratory System Provides Us with Essential Gas Exchange as well as Vocalization

LEARNING OBJECTIVES

Explore the overall function of the respiratory system.

Identify the structures of the upper and lower respiratory tracts.

Differentiate the conducting zone from the respiratory zone.

Discuss the anatomy and physiology of the alveolar sac.

Thus far in our treatment of survival, we have talked about protecting ourselves from the environment, moving through the environment, and sensing and reacting to external and internal changes. We explored how the cardiovascular system (Chapter 11) moves nutrients, gases, and waste through the body. Now it is time to discuss how the cardiovascular system cooperates with the respiratory system, which delivers oxygen and expels carbon dioxide. The respiratory system also filters incoming air, maintains blood pH, helps control fluid and thermal homeostasis, and produces sound. Otherwise, speech (and biology lectures!) would be impossible.

Upper respiratory tract Respiratory organs in the face and neck.

Lower respiratory tract Respiratory organs within the thoracic cavity, including the bronchial tree and the lungs.

The respiratory system has two anatomical divisions, the **upper respiratory tract** and the **lower respiratory tract**, with separate but related functions (FIGURE 12.1). The upper tract conditions air as it enters the body, and the lower respiratory tract allows oxygen to enter the blood, and waste gases to leave it.

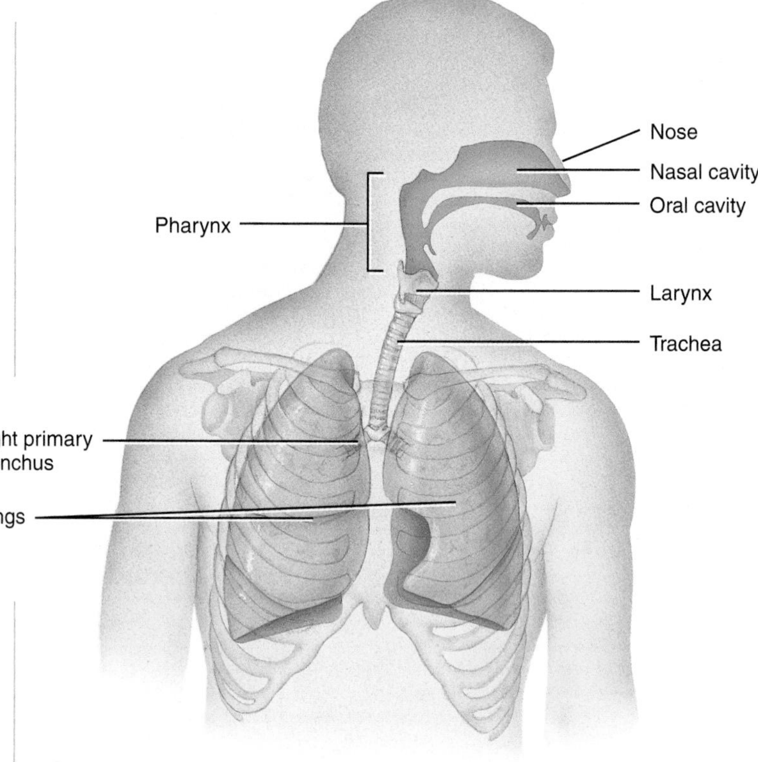

Respiratory tract anatomy FIGURE 12.1

THE UPPER RESPIRATORY TRACT HAS AN INSPIRING ROLE

The structures of the upper respiratory tract—the **nose**, **pharynx**, and **larynx**—warm, moisten, and filter the incoming air (FIGURE 12.2). The nose is one of the first body parts that small children can identify. We are familiar with the external portion of the nose, consisting of the nasal bone and hyaline cartilage, covered by skin and muscle. The division between the two nostrils, or **external nares**, is a plate of hyaline cartilage called the **septum**. The septum is attached to the vomer bone at its base. Both the septum and the cartilages that make up the sides of the nose serve to support the nasal openings. If a blow to the nose moves

Pharynx Throat.

Larynx Voice box (Adam's apple).

Bony framework:
Frontal bone
Nasal bones
Maxilla
Cartilaginous framework:
Lateral nasal cartilages
Septal cartilage

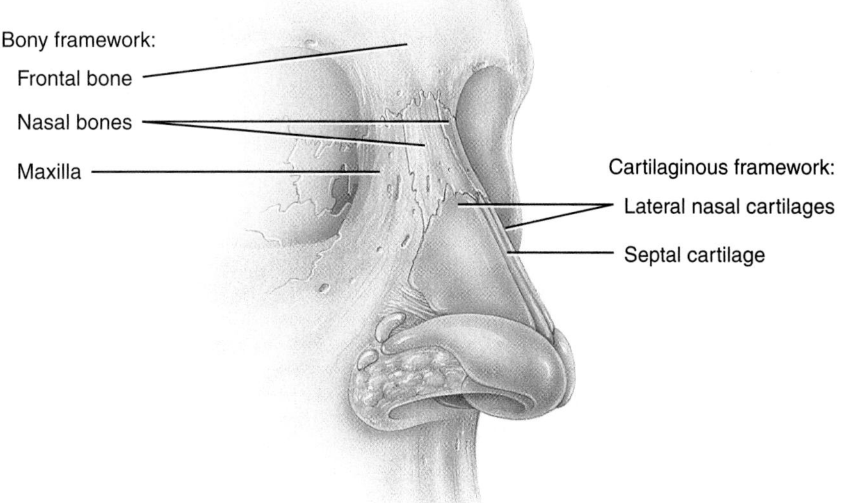

A Anterolateral view of external portion of nose

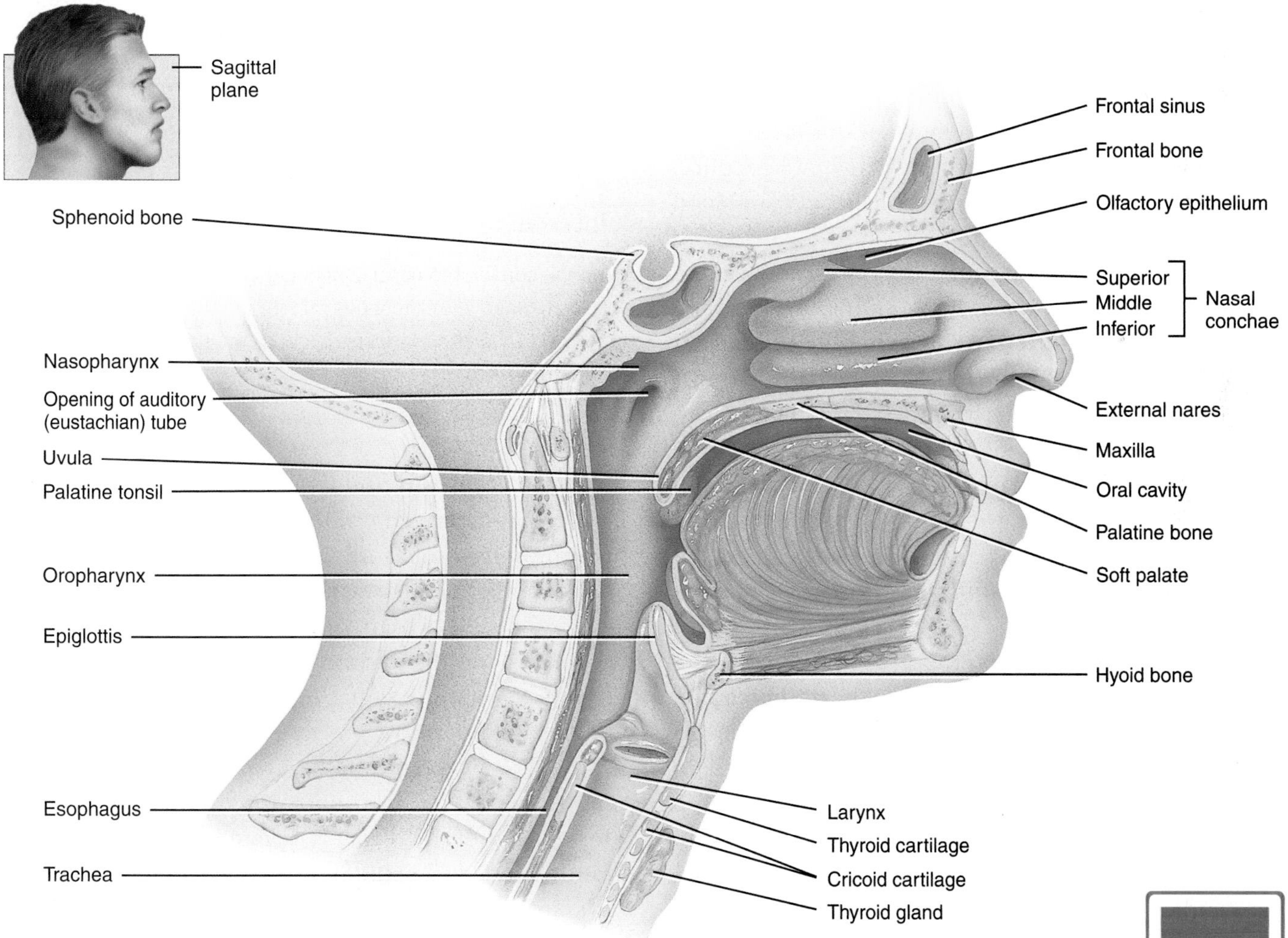

Sagittal plane
Sphenoid bone
Nasopharynx
Opening of auditory (eustachian) tube
Uvula
Palatine tonsil
Oropharynx
Epiglottis
Esophagus
Trachea
Frontal sinus
Frontal bone
Olfactory epithelium
Superior
Middle
Inferior
Nasal conchae
External nares
Maxilla
Oral cavity
Palatine bone
Soft palate
Hyoid bone
Larynx
Thyroid cartilage
Cricoid cartilage
Thyroid gland

B Sagittal section of the left side of the head and neck showing the location of respiratory structures

www.wiley.com/college/ireland

External nose anatomy including septum FIGURE 12.2

these cartilages to the side of the vomer, airflow is blocked. To treat this "deviated septum," surgeons restore the septum into position and open both nasal passageways. Surgery on the nose (called **rhinoplasty**) can also be done for cosmetic reasons, usually by breaking the nasal bone and reshaping the nasal cartilages.

As mentioned, the nasal cavity warms, filters, and moistens incoming air, and does so far better than the mucous membranes of the mouth. Swirls and ridges in the nasal cavity slow the air. As inhaled air moves through this convoluted space, it contacts the nasal epithelium. The epithelium in the upper respiratory tract is **pseudostratified ciliated columnar** epithelium. In the nasal region, this tissue is covered in mucus and constantly washed by tears draining from the eyes.

A large blood supply warms the nasal epithelium, and both the warmth and moisture are transferred to the inhaled air (FIGURE 12.3). If you have ever bumped your nose, you know of this large blood supply. Most of us have had a bloody nose at least once, and were surprised by the remarkable quantity of blood that leaked out.

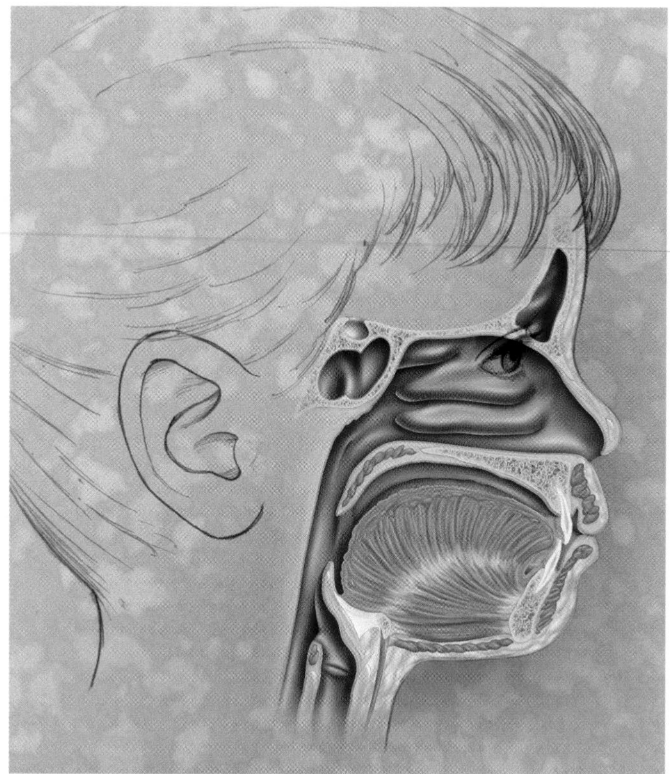

Internal nasal passages FIGURE 12.3

Note the convoluted nasal conchae that swirl the incoming air. These are lined with a mucous membrane that adds moisture and heat to the air before sending it to the lower respiratory tract.

Filtering is a vital function of the nose because inhaled particles would seriously inhibit airflow in the lower respiratory tract. Coarse hairs in the nostrils filter out larger particles, and the mucus of the nasal passages further filters incoming air by trapping small particles.

A final function of the nasal epithelium is the sense of smell (as described in Chapter 8). To smell something more clearly, we often inhale deeply to ensure that airborne compounds reach the patch of nasal epithelium that is studded with chemosensory neurons.

The **internal nares,** the twin openings at the back of the nasal passageway, lead to the **nasopharynx**, or upper throat (FIGURE 12.4). The passageway between the nose and throat is normally open for breathing, but it must close when we swallow. The **uvula,** a fleshy tab of tissue that hangs down in the back of the throat, contracts when touched by solids, moving upward and closing the internal nares. When your doctor asks you to say "Ahh" during a throat examination, you contract the uvula and move it up so the doctor can see the nasopharynx and the tonsils on the posterior wall of the pharynx. If you laugh or cough while drinking, the uvula may spasm, and liquids may leak past it. These liquids may be forced out the external nares, causing a burning sensation as they travel the nasal passages—and some slight embarrassment.

The eustachian, or auditory, tubes link the nasopharynx and the middle ear. When your ears "pop," these tubes open to equalize air pressure between the middle and outer ear.

The **oropharynx,** the area directly behind the tongue, is covered by the uvula when it hangs down. This portion of the throat is devoted to activities of the mouth. Food and drink pass through the oropharynx with each swallow, so the mucous membrane and epithelium lining this region are usually thicker and more durable than elsewhere in the pharynx. The palatine and lingual tonsils are found in the oropharynx as well.

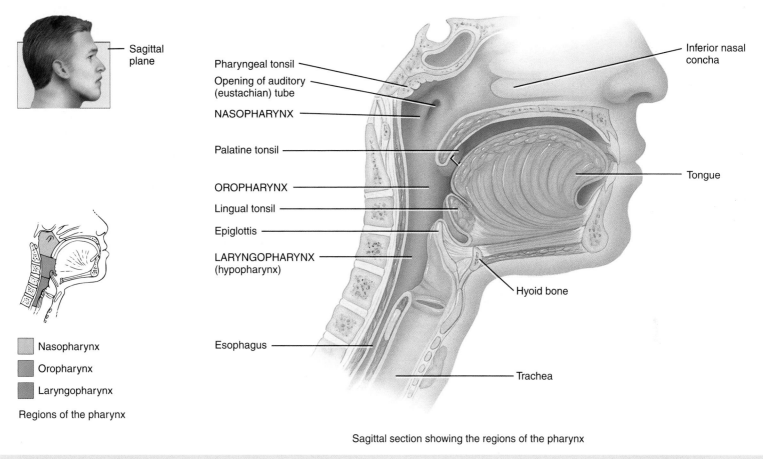

Sagittal plane

Pharyngeal tonsil
Opening of auditory (eustachian) tube
NASOPHARYNX
Palatine tonsil
OROPHARYNX
Lingual tonsil
Epiglottis
LARYNGOPHARYNX (hypopharynx)
Esophagus
Inferior nasal concha
Tongue
Hyoid bone
Trachea

Nasopharynx
Oropharynx
Laryngopharynx

Regions of the pharynx

Sagittal section showing the regions of the pharynx

Pharynx FIGURE 12.4

The lowest level of the pharynx, called the **laryngopharynx,** is the last part of the respiratory tract shared by the digestive and respiratory systems. The end of the laryngopharynx has two openings. The anterior opening leads to the larynx and the rest of the respiratory system. The posterior opening leads to the esophagus and the digestive system.

The larynx divides the upper and lower respiratory tracts. This structure, composed entirely of cartilage, holds the respiratory tract open, guards the lower tract against particulate matter, and produces the sounds of speech. The larynx is composed of nine pieces of hyaline cartilage: three single structures and three paired structures. The single pieces are the **thyroid cartilage,** the **epiglottis,** and the **cricoid cartilage** (FIGURE 12.5, p. 372).

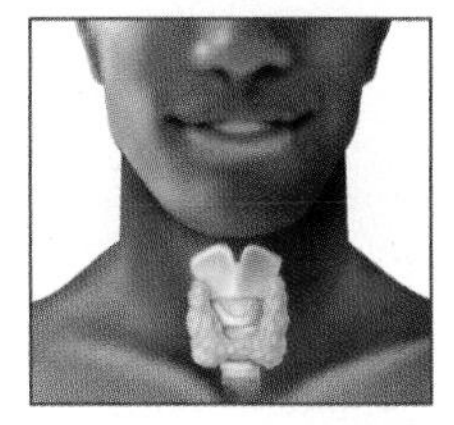

The **thyroid cartilage** lies in the front of the larynx. It is shield-shaped and often protrudes from the throat. Because males produce more testosterone than females, and testosterone stimulates cartilage growth, the thyroid cartilage in men is usually larger than in females. One common name for the larynx, "Adam's apple," refers to the larger larynx in men.

The **epiglottis** covers the opening to the lower respiratory tract to prevent food from entering (*epi* means "on top of" and *glottis* means "hole"). The epiglottis is a leaflike flap of cartilage on the superior (upper) aspect of the larynx, covering the hole leading to the lungs. A pair of small **corniculate cartilages** hold the epiglottis in position above the glottis. The larynx is attached to the tongue muscles. When the tongue pushes against the roof of the mouth in preparation for swallowing, the larynx moves up toward the epiglottis. Food particles hitting the top of the epiglottis complete

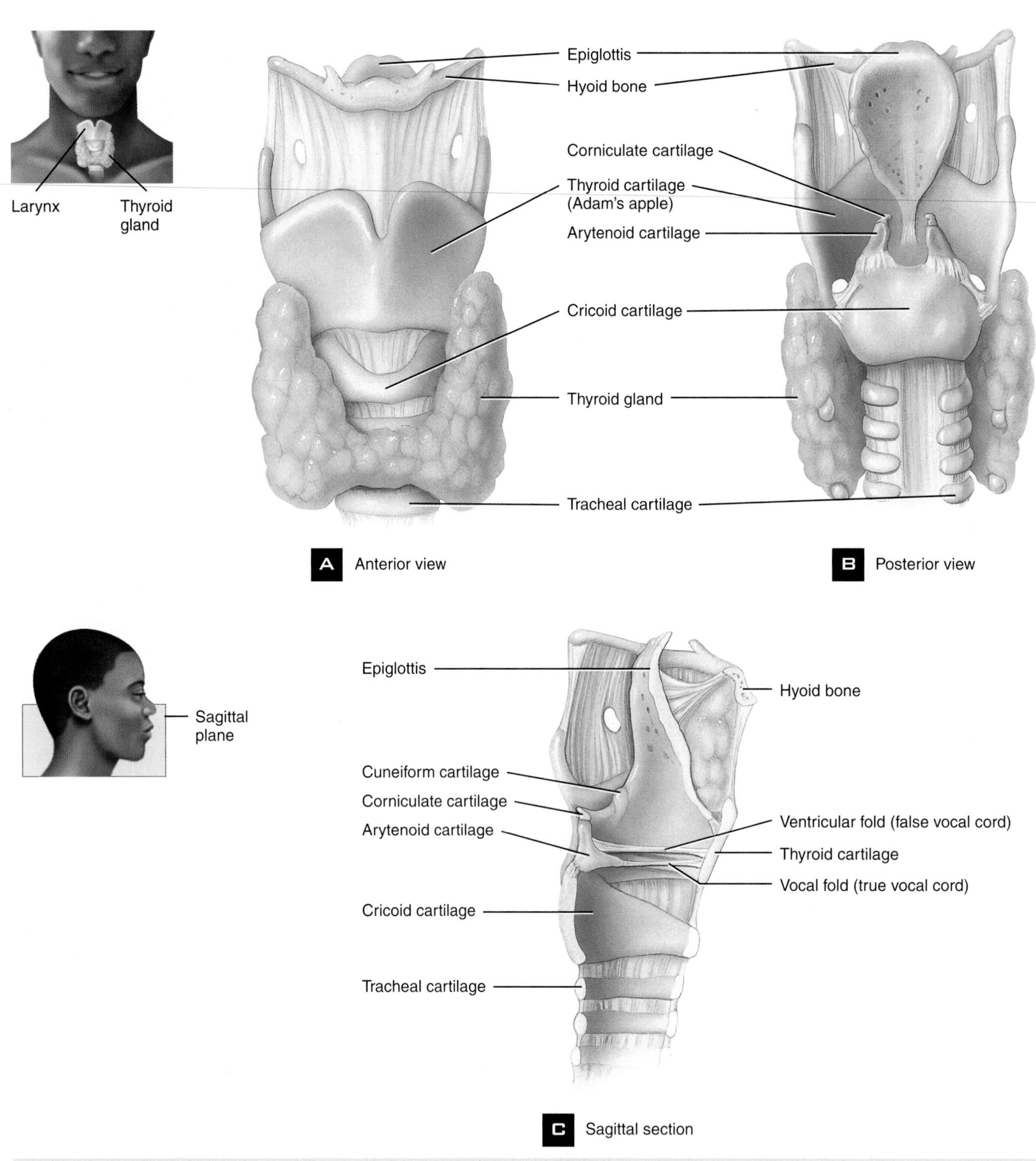
Larynx
Thyroid gland
Epiglottis
Hyoid bone
Corniculate cartilage
Thyroid cartilage (Adam's apple)
Arytenoid cartilage
Cricoid cartilage
Thyroid gland
Tracheal cartilage
A Anterior view
B Posterior view
Sagittal plane
Epiglottis
Hyoid bone
Cuneiform cartilage
Corniculate cartilage
Arytenoid cartilage
Ventricular fold (false vocal cord)
Thyroid cartilage
Vocal fold (true vocal cord)
Cricoid cartilage
Tracheal cartilage
C Sagittal section

Larynx FIGURE 12.5

the closure by causing the epiglottis to rest against the top of the larynx. You can feel this movement by touching your "Adam's apple" and swallowing. You will feel the entire larynx move up with your tongue.

The **cricoid cartilage** is the only complete ring of cartilage in the respiratory system. It is narrow in front but thick in the back of the larynx. The cricoid cartilage holds the respiratory system open. If it is crushed, airflow is impeded and breathing becomes nearly impossible. In an emergency, it may be necessary to surgically open the airway below a crushed cricoid cartilage.

The larynx is called the "voice box" because it is the location of the vocal cords. As seen in FIGURE 12.5C, you have two pairs of vocal cords: false vocal cords, or ventricular folds, and true vocal cords, or **vocal folds** (FIGURE 12.6). The vocal fold are covered by mucous membrane and held in place by elastic ligaments stretched across the glottis. These folds vibrate as air moves past them, producing sound. High-pitched sounds occur when tension on the vocal folds increases, and low-pitched sounds occur when the tension is reduced. We unconsciously adjust tension on the vocal folds by moving the paired laryngeal cartilages. The **arytenoid** and **cuneiform cartilages** both pull on the vocal folds to alter pitch. The amplitude, or amount the cords are vibrating, determines sound volume.

Vocal folds
A pair of cartilaginous cords stretched across the laryngeal opening that produce the tone and pitch of the voice.

As boys reach puberty and their testes produce more testosterone, their voices change. Testosterone stimulates the growth of cartilage in the larynx, thickening the vocal folds. Boys train their voices through daily use to adjust their vocal fold tension based on the size of the larynx. As the larynx grows, the tension needed to produce the same sounds changes. In effect, the male must retrain his voice to maintain vocal tone. When the larynx is growing quickly, the male voice will often "crack" or "squeak" due to his inability to adjust the tension on his changing vocal folds.

The lower respiratory tract routes air to the lungs The main function of the lower tract is to move inhaled air to the **respiratory membrane**. Physiologically, the upper tract and the first portion of the lower tract make up the **conducting zone** of the respiratory system, which conducts air from the atmosphere to the **respiratory zone** deeper in the body, where the actual exchange of gases takes place. The

Respiratory membrane
The thin, membranous "end" of the respiratory system, where gases are exchanged.

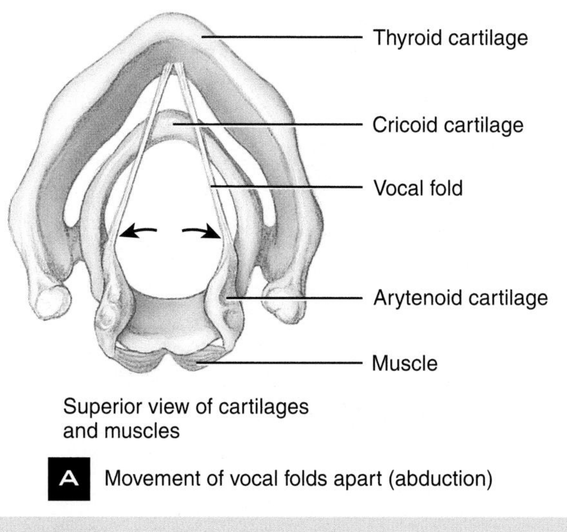

A Movement of vocal folds apart (abduction)

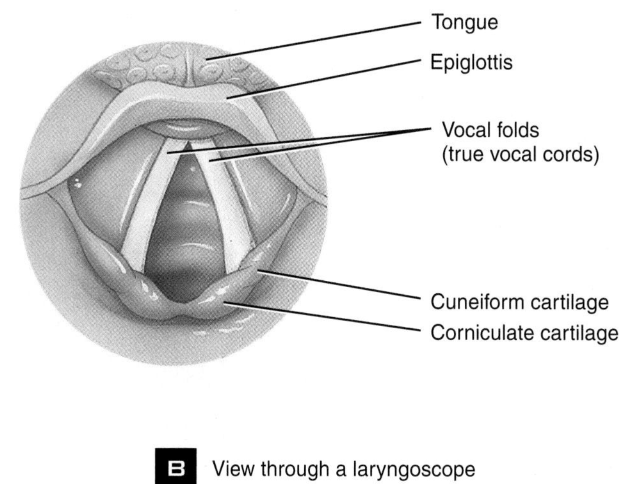

B View through a laryngoscope

Vocal folds FIGURE 12.6

conducting zone includes all the structures of the upper respiratory tract, as well as the **trachea, bronchi, bronchioles**, and **terminal bronchioles**. The respiratory zone lies deep within the lungs and includes only the **respiratory bronchioles** and the **alveoli**. The lower portion of the conducting zone and the respiratory zone are collectively referred to as the bronchial tree (FIGURE 12.7).

The trachea connects the larynx to the bronchi Beyond the larynx, air enters the trachea, a 12-centimeter tube extending from the base of the larynx to the fifth thoracic vertebra (FIGURE 12.8). The trachea is approximately 2.5 centimeters in diameter, and is composed of muscular walls embedded with 16 to 20 "C"-shaped pieces of hyaline cartilage. (Remember that the cricoid cartilage of the larynx is the only complete ring of cartilage in the respiratory system.) The opening of each "C" is oriented toward the back. You can easily feel the tracheal rings through the skin of your throat, immediately below your larynx.

These cartilage "C" rings support the trachea so it does not collapse during breathing, while also allowing the esophagus to expand during swallowing. When you swallow a large mouthful of food, the esophagus pushes into the lumen of the trachea as the mouthful passes. If the tracheal cartilages were

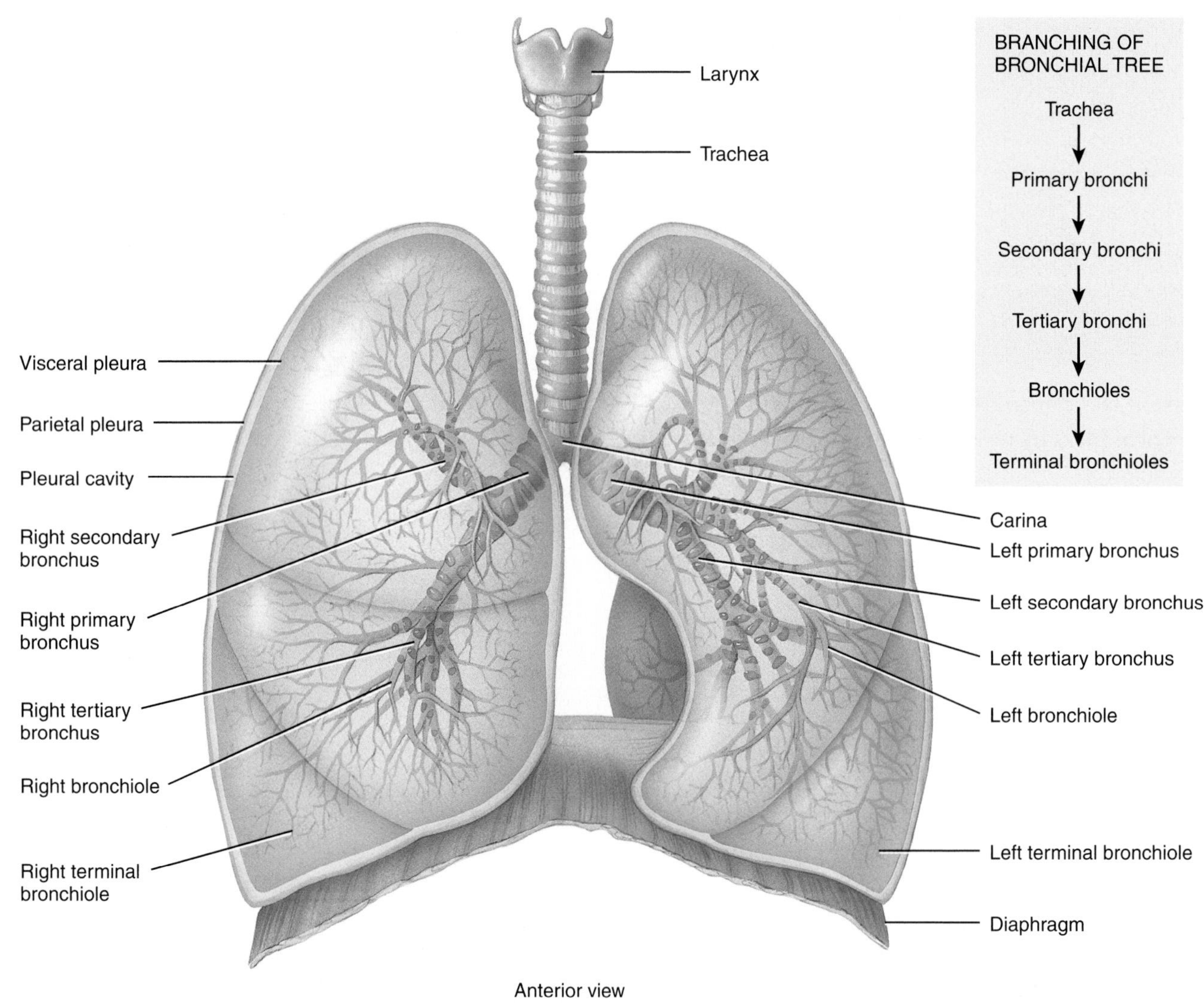

Bronchial tree FIGURE 12.7

circular, the food would push the entire trachea forward. But the trachea is attached to the bronchi of the lungs, so this would also move the lungs upward in the thoracic cavity—a structural nightmare! The C-shaped cartilage allows the back of the trachea to compress, so the lungs can remain in the thoracic cavity and the trachea can get the support it needs to remain open.

In advanced first-aid classes, you learn to locate these rings and identify a position between two rings. You can save someone with a crushed larynx from suffocating by opening the trachea between "C" rings and inserting a temporary breathing tube so air can flow to the lungs, bypassing the crushed larynx. This is called a **tracheotomy**. Another way to restore breathing is to **intubate**—to insert a tube through the mouth or nose, through the larynx and into the trachea. The tube pushes obstructions aside and/or helps suction them out.

At the lower base of the trachea is an extremely sensitive area called the **carina**. The mucous membrane of the carina is more sensitive to touch than any other area of the larynx or trachea, so this spot triggers a dramatic cough reflex when any solid object touches it.

At the level of the fifth thoracic vertebra, the trachea splits into two tubes called the **primary bronchi**, which lead to each lung. Despite their common function, the two bronchi are slightly different. The right primary bronchus is shorter, wider, and more vertical than the left. For this reason, inhaled objects often get lodged in the right primary bronchus. These two bronchi are constructed very much like the trachea and are held open with incomplete rings of cartilage in their walls.

Inside the lungs, the primary bronchi divide into the **secondary bronchi** (see FIGURE 12.7). The right bronchus divides into three secondary bronchi, whereas the left splits into two. This branching pattern continues getting smaller and smaller as the tubes extend farther from the primary bronchus. The sequentially smaller tubes are called **tertiary bronchi**, **bronchioles**, **terminal bronchioles**, and **respiratory bronchioles**. The respiratory system looks like an upside-down tree, with the base at the nasal passages and the tiniest branches leading to the "leaves" deep within the lungs.

The bronchial tree undergoes two major changes as it reaches deeper into the body:

Trachea FIGURE 12.8

1. The cells of the mucous membrane get smaller. The epithelium of the upper and beginning portion of the lower respiratory tract is pseudostratified ciliated columnar epithelium; these fairly large cells secrete mucus, which the cilia sweeps upward and outward with any inhaled particles. The epithelium changes to the slightly thinner, ciliated columnar epithelium in the larger bronchioles. The smaller bronchioles are lined with smaller ciliated cuboidal epithelium. Terminal bronchioles have no cilia and are lined with simple columnar epithelium. If dust reaches all the way to the terminal bronchioles, it can be removed only by **macrophages**.

2. The composition of the walls of the bronchi and bronchioles changes. Smaller tubes need less cartilage to hold them open, so the incomplete rings of cartilage supporting the bronchi are gradually replaced by plates of cartilage in the bronchioles. These plates diminish in the smaller bronchioles, until the walls of the terminal bronchioles have virtually no cartilage. As cartilage decreases, the percentage of smooth muscle increases. Without cartilage, these small tubes can be

Macrophage
Large, phagocytic immune cell that patrols tissue, ingesting foreign material and stimulating immune cells.

completely shut by contraction of this smooth muscle. In asthma and other constrictive respiratory disorders, this smooth muscle becomes irritated and tightens, reducing the tube diameter, sometimes even effectively closing it.

Epinephrine, a hormone that is released into the bloodstream when we exercise or feel fright, relaxes smooth muscle. In the lungs, epinephrine relaxes the smooth muscle of the terminal bronchioles, enlarging the lumen and allowing greater airflow. This in turn increases the oxygen content of the blood and allows the muscles to work more efficiently. Someone you know who has asthma probably carries an "inhaler" filled with "rescue medication." If you get a look at the label, the active ingredient is probably epinephrine, norepinephrine, or a derivative. Spraying these drugs on the walls of the bronchioles immediately relaxes the smooth muscle, dramatically increasing tubule diameter.

The thoracic cavity houses the two organs of respiration, the **lungs** (FIGURE 12.9). These lightweight organs extend from just above the clavicle to the twelfth thoracic vertebra and fill the rib cage. The base of the lungs is the broad portion sitting on the diaphragm. The apex is the small point extending above the clavicles.

Although the lungs are paired, they are not identical. The right lung is shorter and fatter, and it has three lobes, whereas the left lung has only two lobes. The left lung is thinner and has a depression for the heart, called the cardiac notch, on the medial side. The central portion of the thoracic cavity is called the **mediastinum;** therefore, the medial portion of the lungs is

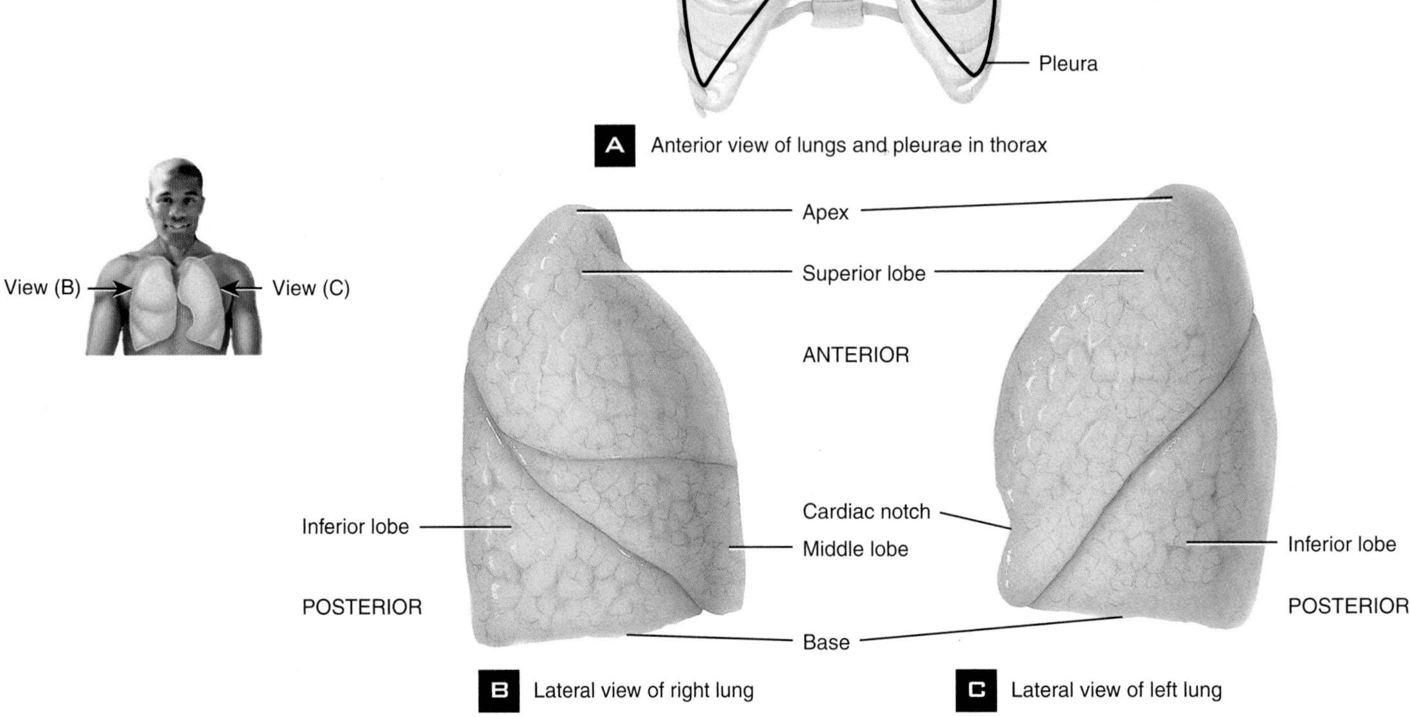

External anatomy of the lungs FIGURE 12.9

Hilum
Site of entry and exit for the nerves, blood vessels, and lymphatic vessels on most organs.

the mediastinal surface. On this surface lies the **hilum** of the lung. Entering and exiting the lung at the hilum are the bronchi, along with the major blood vessels, lymphatics, and nerve supply for the organ.

The pleura wraps the lungs The lungs are covered in a serous membrane called the **pleura** that allows the lungs to expand and contract without tearing the delicate respiratory tissues (FIGURE 12.10). The pleura is anatomically similar to the pericardium around the heart in that they are both composed of two membranous layers separated by serous fluid. The **visceral pleura** is snug against the lung tissue, and the **parietal pleura** lines the thoracic cavity. The **pleural cavity** between the two pleural membranes contains serous fluid. The surface tension of the fluid between these two membranes creates a slight outward pull on the lung tissue. Have you noticed that a thin layer of water on a glass table holds other glass objects to it? In the lungs, this same phenomenon causes adhesion between the visceral and parietal pleura. There is also a slight vacuum in the pleural space, created during development of the lungs and thoracic cavity. This vacuum is essential to proper lung functioning.

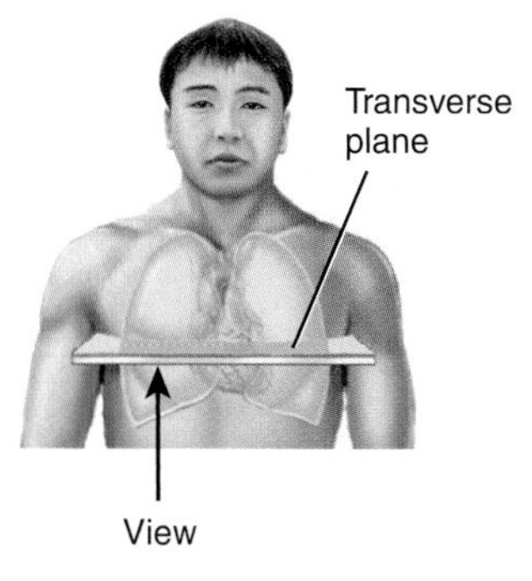

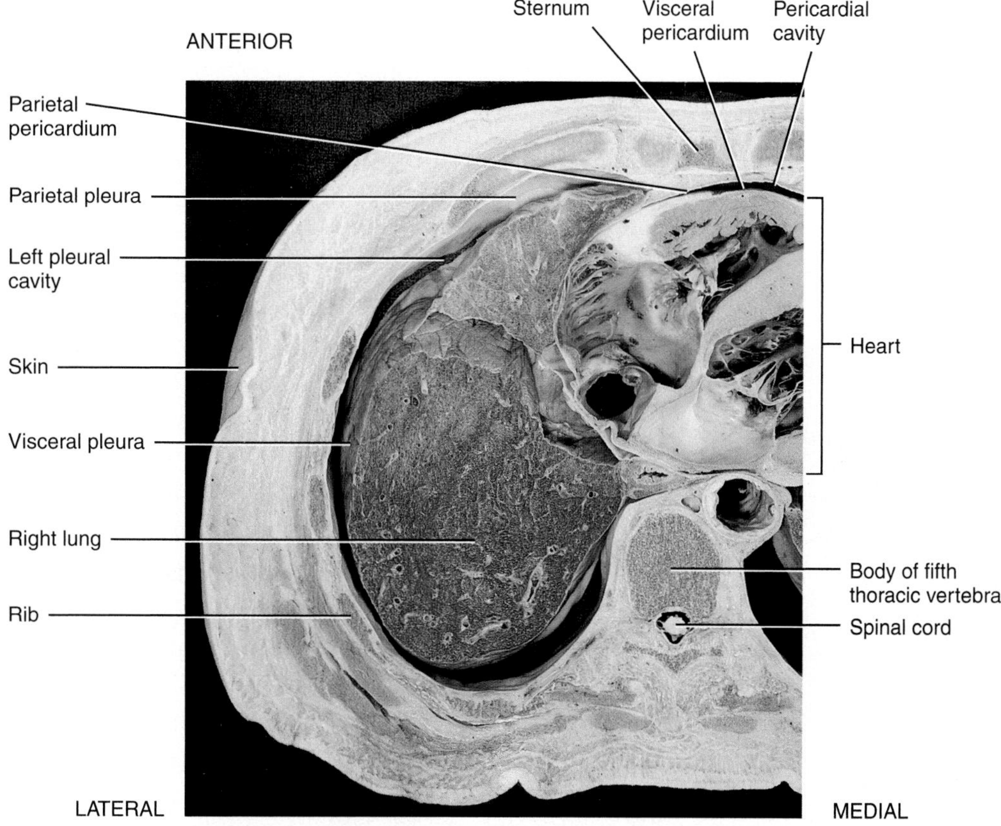

Inferior view of a transverse section through the thoracic cavity showing the pleural cavity and pleural membranes

Pleura FIGURE 12.10

If the partial vacuum within the pleural space is lost, inhalation becomes difficult. This can happen if the thoracic cavity is punctured through injury or accident, causing either a **pneumothorax** (air in the pleural space) or a **hemothorax** (blood in the pleural space). If enough air or blood enters the pleural space, lung tissue in that area can collapse (FIGURE 12.11). The air or blood must be evacuated, and pleural integrity restored, to reinflate the lung and reestablish normal breathing. **Pleurisy** is less devastating and more common than a collapsed lung. In pleurisy, the pleural membranes swell after being inflamed or irritated, and they rub against each another. Every breath is painful, and deep breathing, coughing, or laughing may be excruciating. Anti-inflammatory drugs can reduce these symptoms.

Pleurisy Inflammation of the covering surrounding the lungs, causing painful breathing.

The **lobes** of each lung are separate sections of the organ that can be lifted away from the other lobes, just as a butcher might separate lobes of beef liver. Air enters each lobe through one secondary bronchus. Despite having different numbers of secondary bronchi, each lung has ten terminal bronchioles, each supplying one **bronchopulmonary segment**.

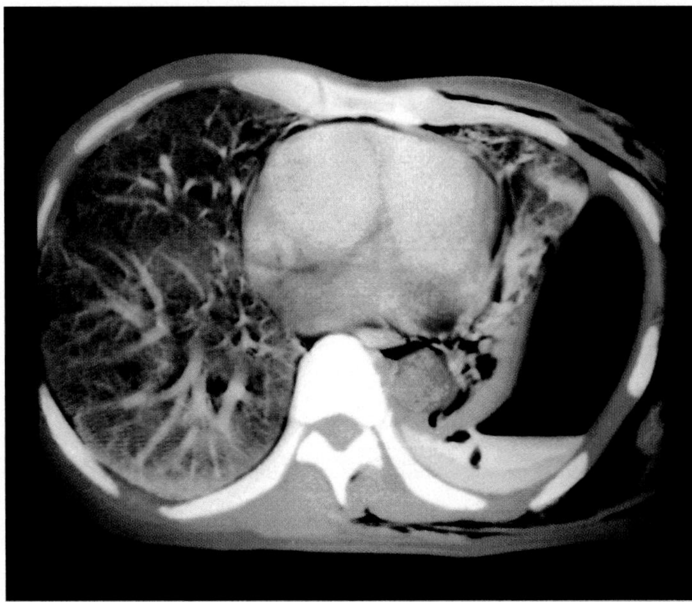

Collapsed lung FIGURE 12.11

Gases are exchanged in the respiratory zone

A bronchopulmonary segment looks somewhat like a bunch of grapes on a vine (FIGURE 12.12). One terminal bronchiole feeds all the respiratory membranes of each bronchopulmonary segment. One pulmonary arteriole runs to each segment, and one pulmonary venule returns from it. Small groups of respiratory membranes, called lobules, extend off the terminal bronchiole. These lobules are wrapped in elastic tissue and covered in pulmonary capillaries. Lobules are attached to the terminal bronchiole by a respiratory bronchiole.

The respiratory bronchiole leads to alveolar ducts, which finally conduct air to the alveoli, the respiratory membranes for the entire system. Only here, in the alveoli, after traveling through the entire set of tubes in the conducting zone, can gases diffuse. It is here, and here alone, that oxygen enters the bloodstream and carbon dioxide exits.

The **alveolus** is a cup-shaped membrane at the end of the terminal bronchiole. Alveoli are clustered into an **alveolar sac** at the end of terminal bronchiole. The key to respiration is diffusion of gases, and diffusion requires extremely thin membranes. The walls of the alveolar sac are a mere two **squamous epithelial cells** thick—one cell from the alveolar wall and one from the capillary wall (FIGURE 12.13). As mentioned in the beginning of the chapter, asthma impedes air flow to these respiratory membranes. (See also the I Wonder . . . feature, "Why are asthma rates going up?" on p. 380.)

Diffusion of gases across the cell membrane requires a moist membrane, but moist membranes have a tendency to stick together much like plastic food wrap. **Septal cells**, scattered through the lung, produce **surfactant**, a detergent-like fluid that moistens the alveoli but prevents the walls from sticking together during exhalation. (Imagine how a thin layer of watery detergent would release the bonding of a ball of plastic wrap.) The surfactant also serves as a biological detergent, solubilizing oxygen gas to promote uptake.

Because septal cells begin secreting surfactant only during the last few weeks of pregnancy, **premature babies** often

Premature babies Infants born prior to the normal gestational period of 40 weeks.

Terminal bronchiole
Pulmonary venule
Pulmonary arteriole
Lymphatic vessel
Elastic connective tissue
Respiratory bronchiole
Alveolar ducts
Pulmonary capillary
Alveolar sac
Visceral pleura
Alveoli

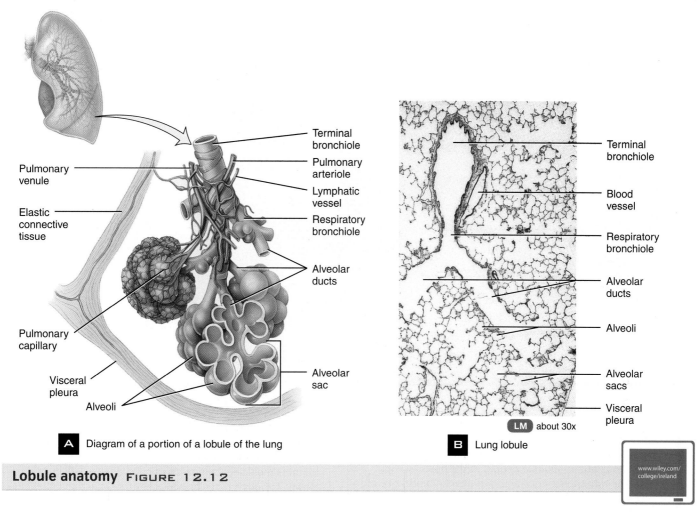

A Diagram of a portion of a lobule of the lung

Terminal bronchiole
Blood vessel
Respiratory bronchiole
Alveolar ducts
Alveoli
Alveolar sacs
Visceral pleura
LM about 30x

B Lung lobule

Lobule anatomy FIGURE 12.12

www.wiley.com/college/ireland

have difficulty breathing. Every inhalation requires a gasp to reinflate the collapsed alveoli because their walls stick together. In the late 1980s, artificial surfactant was first administered to premature infants as an inhalant during the first week of life. Now this drug is routinely given to premature infants to enhance respiration before septal cells begin producing surfactant.

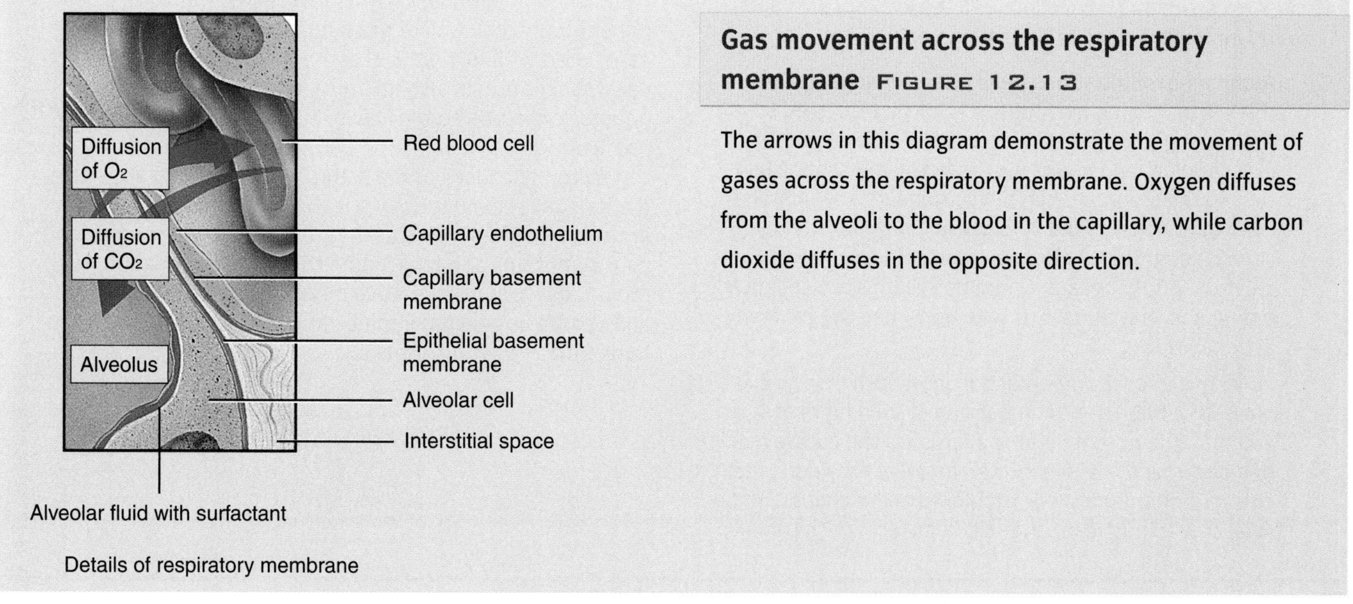

Gas movement across the respiratory membrane FIGURE 12.13

The arrows in this diagram demonstrate the movement of gases across the respiratory membrane. Oxygen diffuses from the alveoli to the blood in the capillary, while carbon dioxide diffuses in the opposite direction.

I WONDER . . .

Why are asthma rates going up?

Asthma, a constriction of the bronchi that causes wheezing and shortness of breath, has become an epidemic. An estimated 14 to 30 million Americans have asthma, including at least 6 million children, many in inner cities. The disease sends about 500,000 people to the hospital each year. The rate of asthma diagnoses has doubled since the 1980s.

Part of the increase may be due to better diagnosis, but could something else be increasing this disease? The answer would lie in the environmental causes of asthma and/or in the human response to those causes. Asthma can result from exposure to irritants and allergens, including pollen, cockroaches, mold, cigarette smoke, air pollutants, respiratory pathogens, exercise, cold air, and some medicines. Researchers have examined these exposures and found important clues to the asthma epidemic:

- Asthma hospitalizations peak just after school starts in the fall. In a Canadian study, schoolchildren aged 5 to 7 were going to emergency rooms a few days before preschoolers and adults. Of the wheezing schoolchildren, 80 to 85 percent had active rhinovirus (common cold) infections, as did 50 percent of adults. The research suggests that the common cold is spread by children (partly because immune systems are still developing) and that rhinovirus infections may trigger many asthma attacks.
- Poverty and environmental pollution both help to explain why inner-city Americans have such high rates of asthma. One potent asthma allergen is the cuticle (shell) of a cockroach, an insect often found in crowded inner cities. Compounding this problem are the chaotic home lives characteristic of inner-city families, which can also interfere with timely administration of medicines to control asthma symptoms.
- Children who lived on a farm before age 5 have significantly lower rates of asthma, wheezing, and use of asthma medicine, compared to children who live in town. Although allergies play a key role in asthma, the farm children did not have lower rates of hay fever, an allergic reaction to pollen.

None of these studies exactly explains the surge in asthma diagnoses, but the last one does offer a clue. Some scientists suspect that early exposure to dirt and/or infectious disease somehow "tunes" the immune system to reduce the hyperactive reaction that contributes to the inflammation of asthma. Early exposure to endotoxin, a component of the cell wall of gram-negative bacteria, has been associated with low rates of asthma. But the picture is complicated: Endotoxin also inflames lung tissue in healthy people, and some studies have linked it to more wheezing, not less.

With the causes of the asthma epidemic still uncertain, the best take-home message is this: Most cases of asthma are controllable. If you suffer from it, know what triggers your symptoms and take action to reduce your exposure to them. Take your preventative medications as prescribed and get the suggested immunizations to prevent viral infections from triggering attacks.

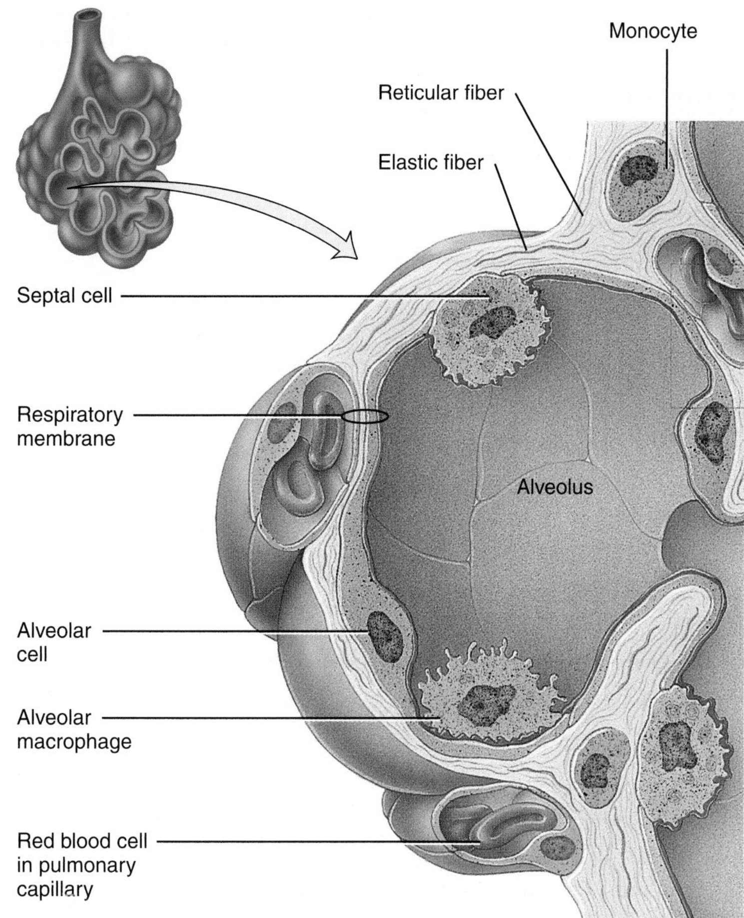

Section through an alveolus showing its cellular components

Anatomy of an alveolar sac FIGURE 12.14

The respiratory membrane consists of a layer of alveolar cells, an epithelial basement membrane, the capillary basement membrane, and the endothelium of the capillary. These membranes are found at the end of the respiratory tree.

Alveolar macrophages, or **dust cells**, patrol the alveoli (FIGURE 12.14). These immune cells remove any inhaled particles that escaped the mucus and cilia of the conducting zone.

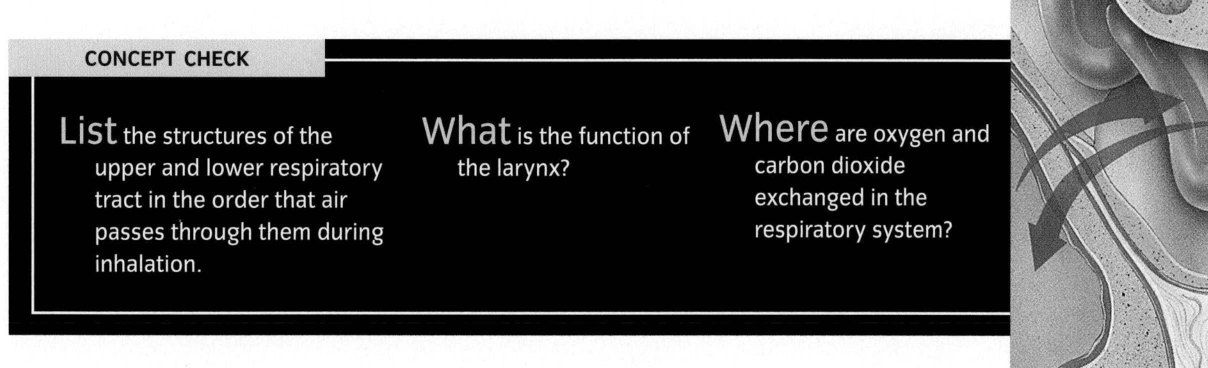

CONCEPT CHECK

List the structures of the upper and lower respiratory tract in the order that air passes through them during inhalation.

What is the function of the larynx?

Where are oxygen and carbon dioxide exchanged in the respiratory system?

In Order to Respire, Air Must Be Moved in and out of the Respiratory System

LEARNING OBJECTIVES

Explain inhalation and exhalation in terms of muscle activity.

Describe the various lung volumes, and explain their relationship.

The anatomy of the respiratory system eases the exchange of gases between the air and the body. But how is external air brought into the depths of the respiratory system during inhalation (or inspiration)? Inhalation (and the opposite movement, called exhalation or expiration) are governed by muscular movements of the thoracic cavity. Inhalation is an active process, requiring muscle contractions, but exhalation requires only that those muscles relax. The combined inflow and outflow of air between atmosphere and alveoli is called pulmonary ventilation. Pulmonary ventilation is governed by Boyle's law, which states that the volume of a gas varies inversely with its pressure (FIGURE 12.15). In other words, if you increase the size of a container of gas without adding gas molecules, the pressure must decrease.

When you inhale, your muscles expand your thoracic cavity (FIGURE 12.16). Your diaphragm contracts, dropping the bottom from the thoracic cavity. This dropping of the diaphragm causes most of the size increase in the thoracic cavity during an inhalation. The intercostal muscles also contract, raising the ribs slightly. (You can feel this by holding your sides as you breathe and feeling your ribs expand and contract.) The lungs connect to the walls of the thoracic cavity through the pleura, so the lungs must follow the moving walls of the thoracic cavity. The increasing volume of the lungs during inhalation causes the pressure to drop, causing gas molecules to rush in from the environment outside your nostrils. Because air moves from high-pressure zones to low-pressure zones, air moves into your lungs to equilibrate this pressure gradient. This is how inhalation occurs.

Drowning occurs when water, which is too heavy to be removed from the lungs, is pulled into them. Our respiratory muscles cannot expel the water, and water carries too little oxygen to diffuse into our blood. In fact, oxygen will diffuse in the opposite direction, from blood to the water!

When the muscles that expanded the thoracic cavity relax, the thoracic cavity returns to its original

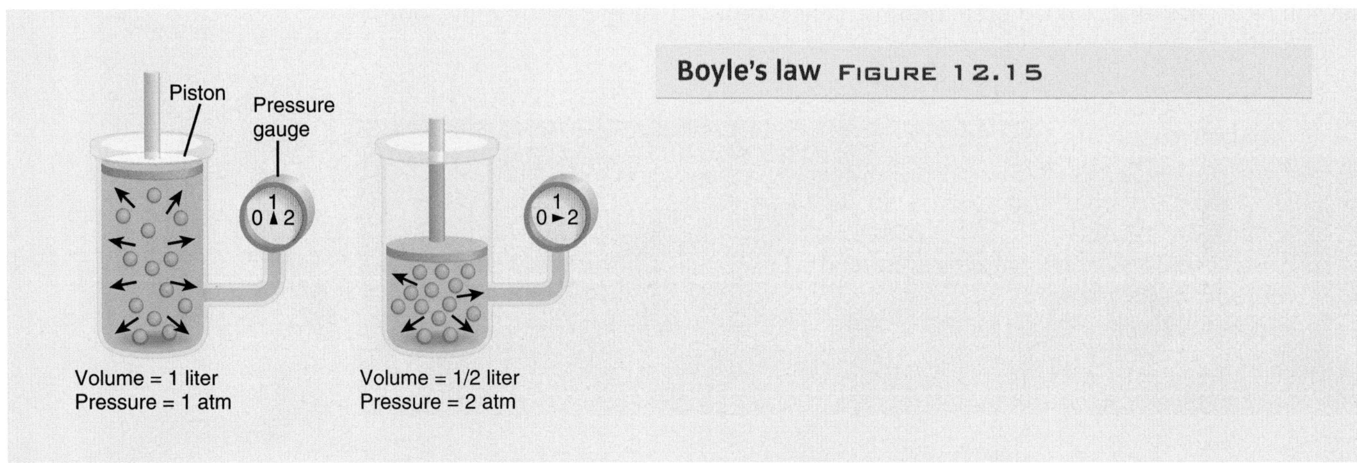

Boyle's law FIGURE 12.15

Inhalation: diaphragm drops, volume increases FIGURE 12.16

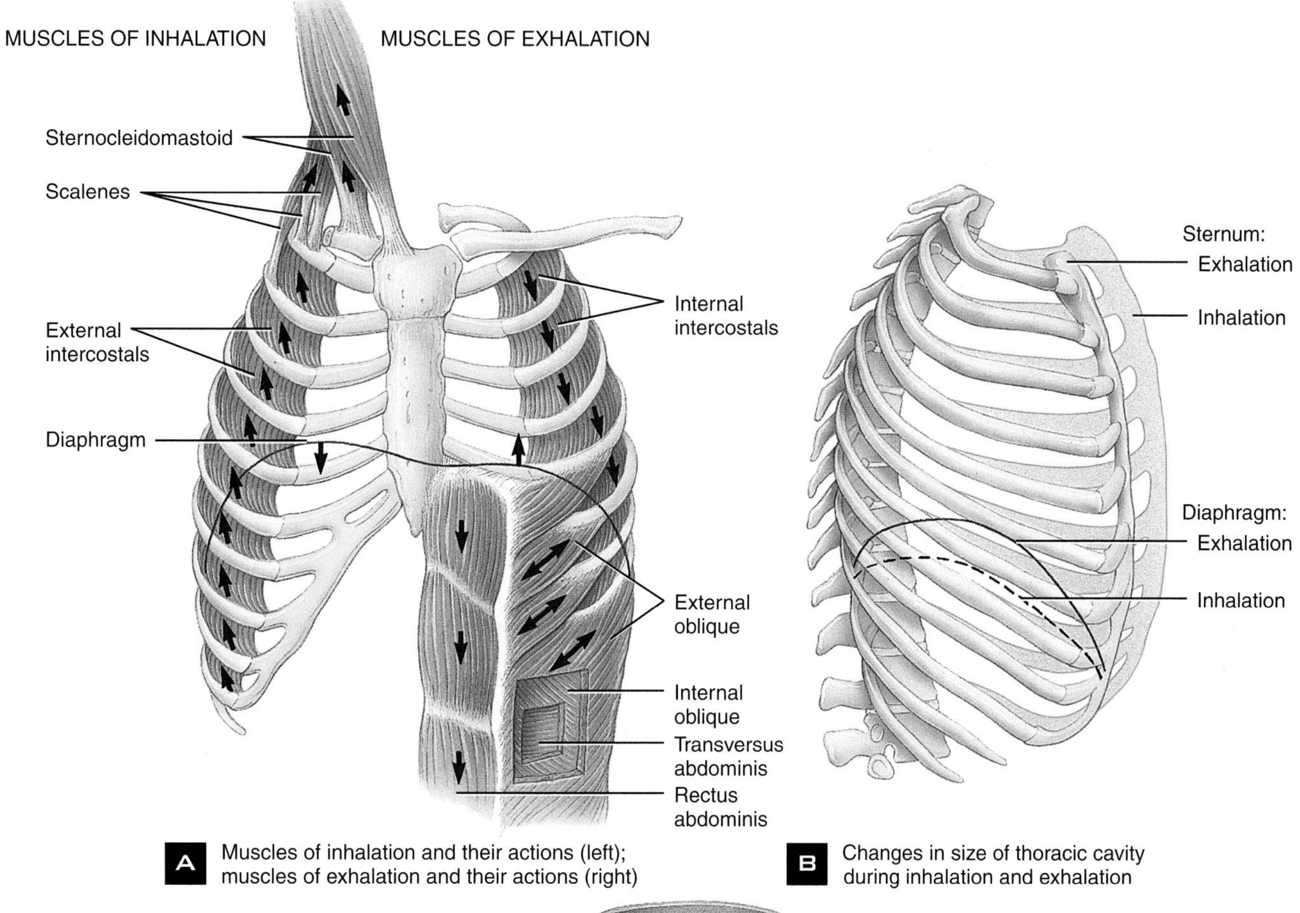

A Muscles of inhalation and their actions (left); muscles of exhalation and their actions (right)

B Changes in size of thoracic cavity during inhalation and exhalation

C During inhalation, the ribs move upward and outward like the handle on a bucket

1. The diaphragm performs 75 percent of the work in normal respiration, with help from the intercostal, sterno-cleidomastoid, serratus anterior, pectoralis minor, and scalene muscles. You can identify these other muscles by watching in a mirror while inhaling deeply. Neck and shoulder muscles will appear as they contract.
2. The lungs increase in volume as they follow the walls of the thoracic cavity, decreasing the pressure within them.
3. This pressure decrease sets up a pressure gradient, with the atmosphere outside the nose higher than pressure deep within the lungs. Air moves into the lungs to equalize the pressure.
4. The diaphragm relaxes, the intercostals relax, and the thoracic cavity returns to its former size. The volume of the cavity decreases, increasing the pressure on the gases within the cavity.
5. The gases within the cavity rush outward through the nostrils to again equalize pressure between the lungs and the environment. One complete cycle of pulmonary ventilation includes an inhalation and an exhalation.

size, which raises pressure in the thoracic cavity above that outside the nostrils (FIGURE 12.17). Again, because air moves toward areas of low pressure, the respired air exits the respiratory tract. During exhalation, the lungs act like a bicycle pump: The container holding the air shrinks, gas pressure rises, exceeding pressure outside the pump, so air must leave the container.

Exhalation is a passive process, mainly involving muscular relaxation. If we forcibly exhale, as in sighing or yelling, we contract muscles that directly and indirectly shrink the thoracic cavity. In forcible exhalation, the abdominal muscles contract, pushing the abdominal organs and the diaphragm. You can prove this by placing a hand on your abdomen and forcefully exhaling. You will feel these muscles contract as you force out the air.

Recall that the alveoli are thin and moist. Surfactant helps prevent these membranes from gumming up and sticking together during exhalation.

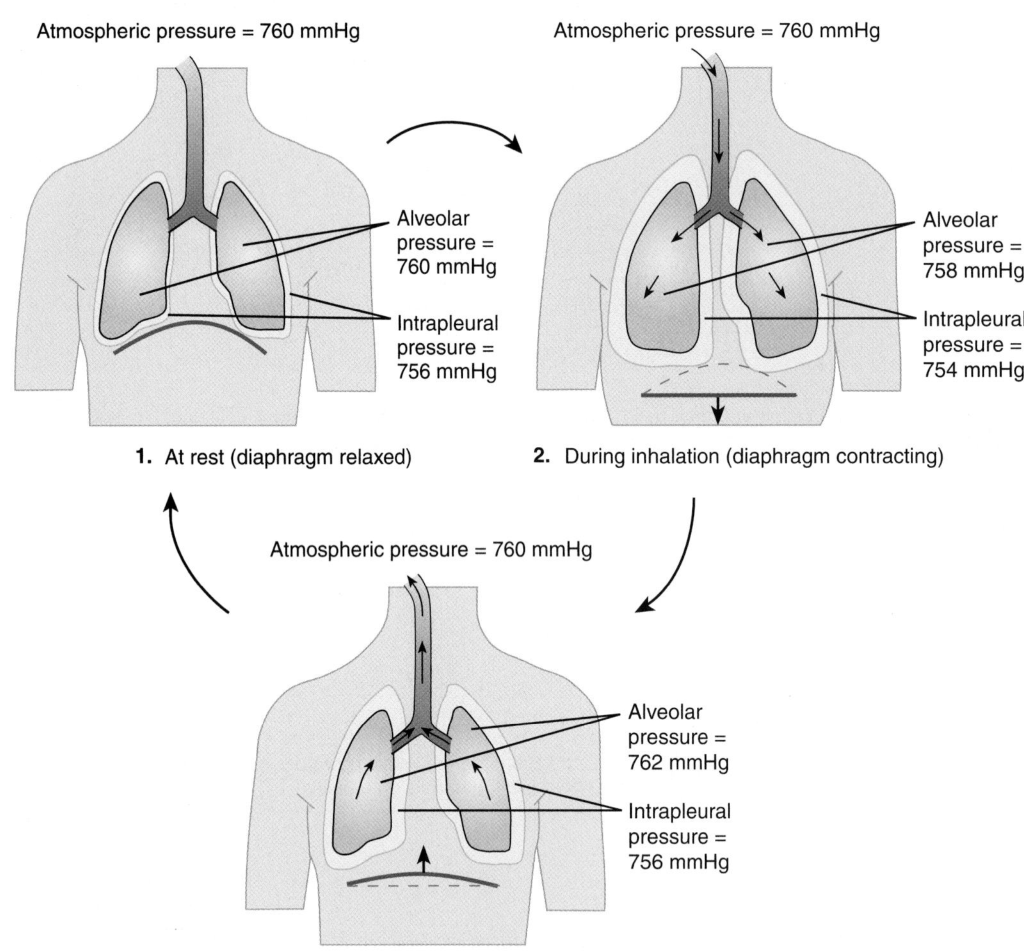

Pressure changes in pulmonary ventilation FIGURE 12.17

Pressure changes within the thoracic cavity occur as the volume of the cavity increases and decreases. During inhalation, the diaphragm contracts, the chest expands, and the lungs are pulled outward. All of these decrease pressure within the lungs, allowing external air to rush in. Relaxing the diaphragm and the intercostals drops the volume of the lungs, increasing their pressure and forcing the air back out.

A second factor is the vacuum between the two layers of pleura. This vacuum forms during fetal development, when the walls of the thoracic cavity enlarge faster than the lungs. The parietal pleura is pulled outward with the expanding walls while the visceral pleura remains attached to the lungs. The resulting vacuum is essential to respiration because it prevents collapse of the thin alveoli during exhalation. The walls of the alveoli spring inward as the air leaves the respiratory tract, but the alveolar walls do not collapse and stick together, partly because of the outward pull of the vacuum between the pleura. In addition, the slight vacuum helps the lungs enlarge and fill with air on the next inhalation.

YOUR BRAIN STEM SETS YOUR RESPIRATORY RATE

As you read this text, you are breathing at a steady rate. These constant, day-in, day-out breaths are called your **resting rate**. Respiratory rate is governed by the medulla oblongata and the pons in the brain stem. The respiratory center in the medulla oblongata causes rhythmic contractions of the diaphragm, stimulating contraction for two seconds and allowing three seconds of rest. This cycle repeats continuously unless overridden by higher brain function (FIGURE 12.18). You can override the medullary signal by holding your breath or by forcibly exhaling, but you cannot hold your breath until you die. Many small children use this threat to blackmail adults, but let them try! The pons will not let anybody "forget" to breathe. Once the carbon dioxide level builds to a critical point, the child will pass out, and the pons will regain control of breathing. You can bet that child will resume breathing.

The body can sense the levels of carbon dioxide and oxygen in the blood through **chemoreceptors** in the carotid artery and aorta. High carbon dioxide levels immediately trigger an increase in the depth and rate of respiration. These chemoreceptors respond to a 10 percent increase in carbon dioxide levels by doubling the respiratory rate. In contrast, a much larger decrease in oxygen level is needed before these receptors will cause the respiratory rate to rise.

Chemoreceptors
Sensory receptors that detect small changes in levels of specific chemicals, such as carbon dioxide.

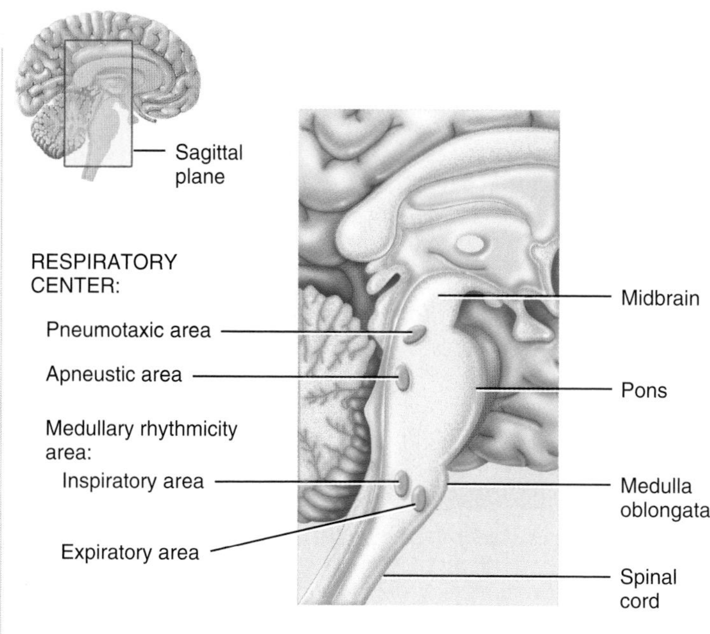

Respiratory centers in the brain
FIGURE 12.18

Different respiratory volumes describe different types of breath

During normal breathing, the volume of air inhaled per minute reflects the respiratory rate and the volume of each normal breath, called the **tidal volume (TV)**. Tidal volume, approximately 500 ml, is somewhat more than the amount of air that is actually exchanged because the trachea, larynx, bronchi, and bronchioles are "anatomic dead spaces" that do not participate in gas exchange. These dead spaces have a volume of about 150 ml. So each tidal breath delivers about 350 net ml of air to the respiratory membranes.

Just as you can consciously control your breathing rate, you can increase the volume of breath by contracting more muscles during inhalation. During a "forced inhalation," the average adult male can inhale approximately 3,300 ml of additional air, and the average adult female can force in approximately 1,900 ml. This volume is called **inspiratory reserve volume (IRV)**.

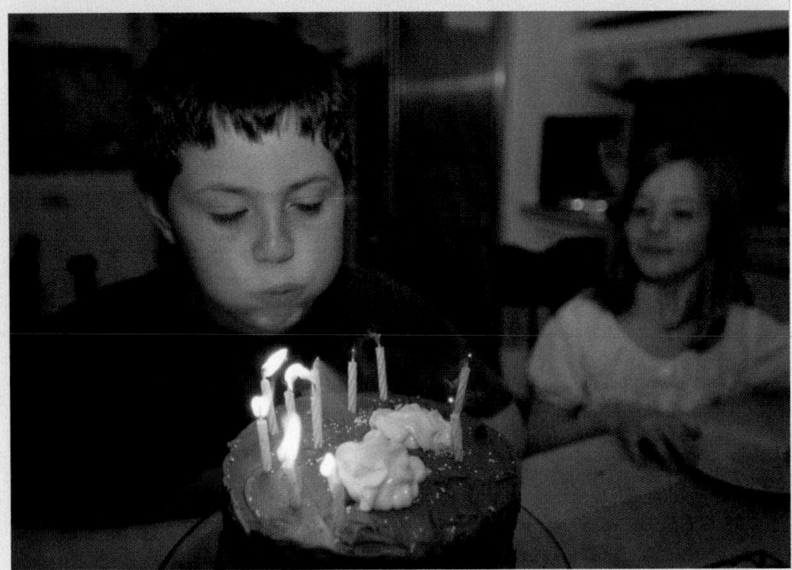

Using expiratory reserve volume FIGURE 12.19

Normal tidal volume moves air in and out of the lungs without taxing the respiratory muscles. When a larger volume of air must be exchanged, the intercostals, the scalenes, and the abdominal muscles are used as well. In adults, the volume of air exhaled can increase from approximately 500 ml to over 3,000 ml. Even children have plenty of air for birthday rituals!

Similarly, we can exhale much more than the 500 ml tidal volume after a normal tidal inhalation, up to about 1,000 ml for males and 700 ml for females, in the **expiratory reserve volume (ERV)**. This volume is lower than IRV because exhalation is largely passive; we have no muscles that directly compress the thoracic cavity beyond that used in a tidal breath. The best we can do is indirectly pressurize the thoracic cavity by contracting the abdominal muscles, forcing the contents of the abdominal cavity up against the diaphragm (FIGURE 12.19).

Vital capacity (VC) measures the total volume of air your lungs can inhale and exhale in one huge breath, which is essentially the maximum amount of air your lungs can move in one respiratory cycle (FIGURE 12.20). VC is the sum of inspiratory reserve volume, tidal volume, and expiratory reserve volume. For most people, the VC is between 3,100 and 4,800 ml; males generally have the larger volume.

The amount of air that remains in the lungs after forced expiration is called **residual volume (RV)**. The residual volume holds the alveoli open and fills the "anatomical dead spaces." The RV is usually between 1,100 and 1,200 ml. You can add your RV to your VC to find your total lung capacity.

Have you ever fallen from a tree or a swing and landed on your back? Perhaps you could not breathe for a minute because you had "gotten the wind knocked out of you." In our terms, your problem was a loss of

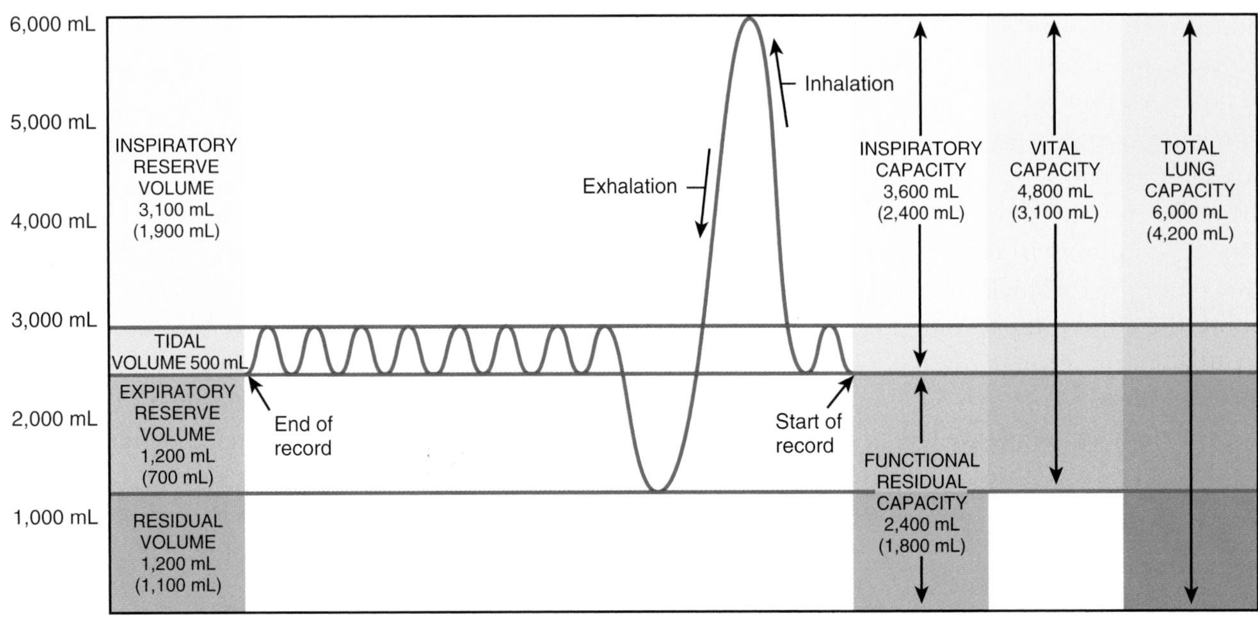

Respiratory volumes FIGURE 12.20

Average respiratory volumes for both males and females. Female volumes are slightly smaller, and given in parentheses.

residual volume. The force of impact momentarily shrank the thoracic cavity beyond what muscular contractions could achieve and forced out some of the residual volume. Your first breath was painful and may even have produced awkward noises as you reinflated the empty alveoli to refill your RV. This is just what infants do with their first few gasps after birth (which are commonly mistaken for crying).

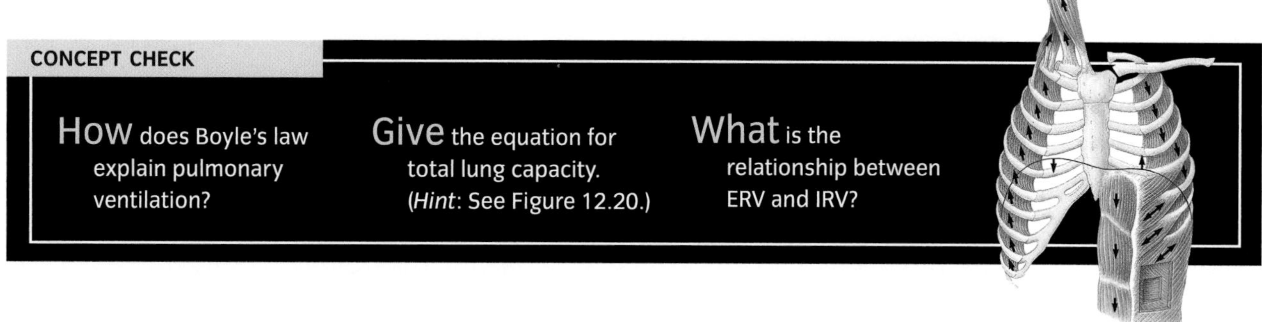

CONCEPT CHECK
How does Boyle's law explain pulmonary ventilation?
Give the equation for total lung capacity. (Hint: See Figure 12.20.)
What is the relationship between ERV and IRV?

External Respiration Brings Supplies for Internal Respiration

LEARNING OBJECTIVES

Define internal and external respiration.

Discuss the movement of gas from air to blood and from blood to tissues.

Thus far we have discussed only pulmonary ventilation—the moving of air into the respiratory system. Once gases are in the alveoli, **external respiration** occurs. **External respiration** is the exchange of gases between the air in the alveoli and the blood in the respiratory capillaries. A second respiratory process—**internal respiration**—is the exchange of gases between body cells and blood in the systemic capillaries.

EXTERNAL RESPIRATION SECURES OXYGEN, DISPOSES OF CARBON DIOXIDE

The exchanges during external and internal respiration are driven by the **partial pressures** of oxygen and carbon dioxide. In external respiration, the driving force is the difference in the partial pressures in the alveolar air and the capillary blood. In internal respiration, the driving force is the partial pressure difference in the capillary blood and the tissue fluid.

Partial pressure The percentage of total gas pressure exerted by a single gas in the mixture.

The air we breathe is composed of many gases. Nitrogen is the most common, making up 78.1 percent of the atmosphere by volume. Oxygen is the second most common gas, occupying 20.9 percent of total volume. Water vapor varies by location and weather, ranging from 0 to 4 percent of volume. And finally, carbon dioxide makes up a measly 0.4 percent of air by volume. The air pressure in any mass of air is a sum of the partial pressures of each constituent gas, so the pressure exerted by each gas is directly related to its proportion in the atmosphere. Thus, in air, 78.1 percent of the pressure is generated by nitrogen molecules, 20.9 percent by oxygen, and 0.4 percent by carbon dioxide. Knowing that atmospheric pressure is usually close to 760 mmHg, we can calculate the partial pressures of each gas.

Why discuss partial pressure? Because it explains the movement of oxygen and carbon dioxide in respiration. **Dalton's law** states that gases move

Atmospheric pressure is the sum of the pressures of all these gases:

$$\text{Atmospheric pressure (760 mmHg)} = P_{N_2} + P_{O_2} + P_{H_2O} + P_{CO_2} + P_{\text{other gases}}$$

We can determine the partial pressure exerted by each gas in the mixture by multiplying the percentage of the gas in the mixture by the total pressure of the mixture. Atmospheric air is 78.6% nitrogen, 20.9% oxygen, 0.04% carbon dioxide, and 0.06% other gases; a variable amount of water vapor is also present, about 0.4% on a cool, dry day. Thus, the partial pressures of the gases in inhaled air are as follows:

$$\begin{aligned}
P_{N_2} &= 0.786 \times 760\text{ mmHg} = 597.4\text{ mmHg}\\
P_{O_2} &= 0.209 \times 760\text{ mmHg} = 158.8\text{ mmHg}\\
P_{H_2O} &= 0.004 \times 760\text{ mmHg} = 3.0\text{ mmHg}\\
P_{CO_2} &= 0.0004 \times 760\text{ mmHg} = 0.3\text{ mmHg}\\
P_{\text{other gases}} &= 0.0006 \times 760\text{ mmHg} = 0.5\text{ mmHg}\\
\text{Total} &= 760.0\text{ mmHg}
\end{aligned}$$

Dalton's law FIGURE 12.21

Each gas in the atmosphere exerts its own partial pressure, which add up to total atmospheric pressure. Each gas can independently diffuse from areas of high concentration to areas of low concentration.

independently down their pressure gradients, from higher to lower pressure (FIGURE 12.21). So oxygen will diffuse from the air in the alveoli into the blood, whereas carbon dioxide will diffuse from blood to the alveoli. Each gas independently moves toward an area of lower pressure without affecting any other gas.

The partial pressure of oxygen in the air of the alveoli is approximately 100 mmHg, whereas the partial pressure of oxygen in the tissues hovers near 40 mmHg. Through simple diffusion, oxygen moves from the air in the alveoli through the thin respiratory membrane and into the blood. By the time blood in the respiratory capillaries completes its journey through the lungs, the partial pressure of oxygen in the blood has equilibrated with that of the air. Blood returning to the heart's left atrium carries oxygen with a partial pressure of 100 mmHg, ready to be pumped to the tissues (FIGURE 12.22).

While oxygen is diffusing into the blood, carbon dioxide is leaving it. The partial pressure of carbon dioxide in the blood returning to the left side of the heart is about 40 mmHg. Blood picks up carbon dioxide as it courses through the tissues, and by the time it reaches the alveoli, the partial pressure of carbon dioxide is 45 mmHg, higher than the 40 mmHg in the alveolar air. This CO_2 pressure gradient causes carbon dioxide to diffuse from the blood to the alveolar air. When the blood leaves the lungs and enters the left atrium, its carbon dioxide partial pressure has dropped to 40 mmHg. The difference between 40 and 45 mmHg tells us how much of this waste gas was removed from the body.

INTERNAL RESPIRATION SUPPLIES OXYGEN TO THE CELLS AND REMOVES THEIR GASEOUS WASTE

Internal respiration is the exchange of gases between the blood and the cells (see FIGURE 12.22). For survival, oxygen in the arteries must reach the tissues, and carbon dioxide generated in the cells must leave the body. In the capillaries of the systemic circulation, the two gases again diffuse in opposite directions. Oxygen enters the tissues, and carbon dioxide diffuses out, again based on partial pressure. The partial pressure of oxygen in the capillary beds of the systemic circuit is approximately 95 mmHg, whereas the partial pressure of oxygen in most tissues is about 40 mmHg. This gradient allows oxygen to leave the

Summary of external and internal respiration

FIGURE 12.22

Most oxygen is transported by the hemoglobin (Hb) of the red blood cells (RBCs) as HB–O_2. Carbon dioxide is carried in the blood plasma by RBC hemoglobin as HB–CO_2 and as bicarbonate ions (HCO_3^-).

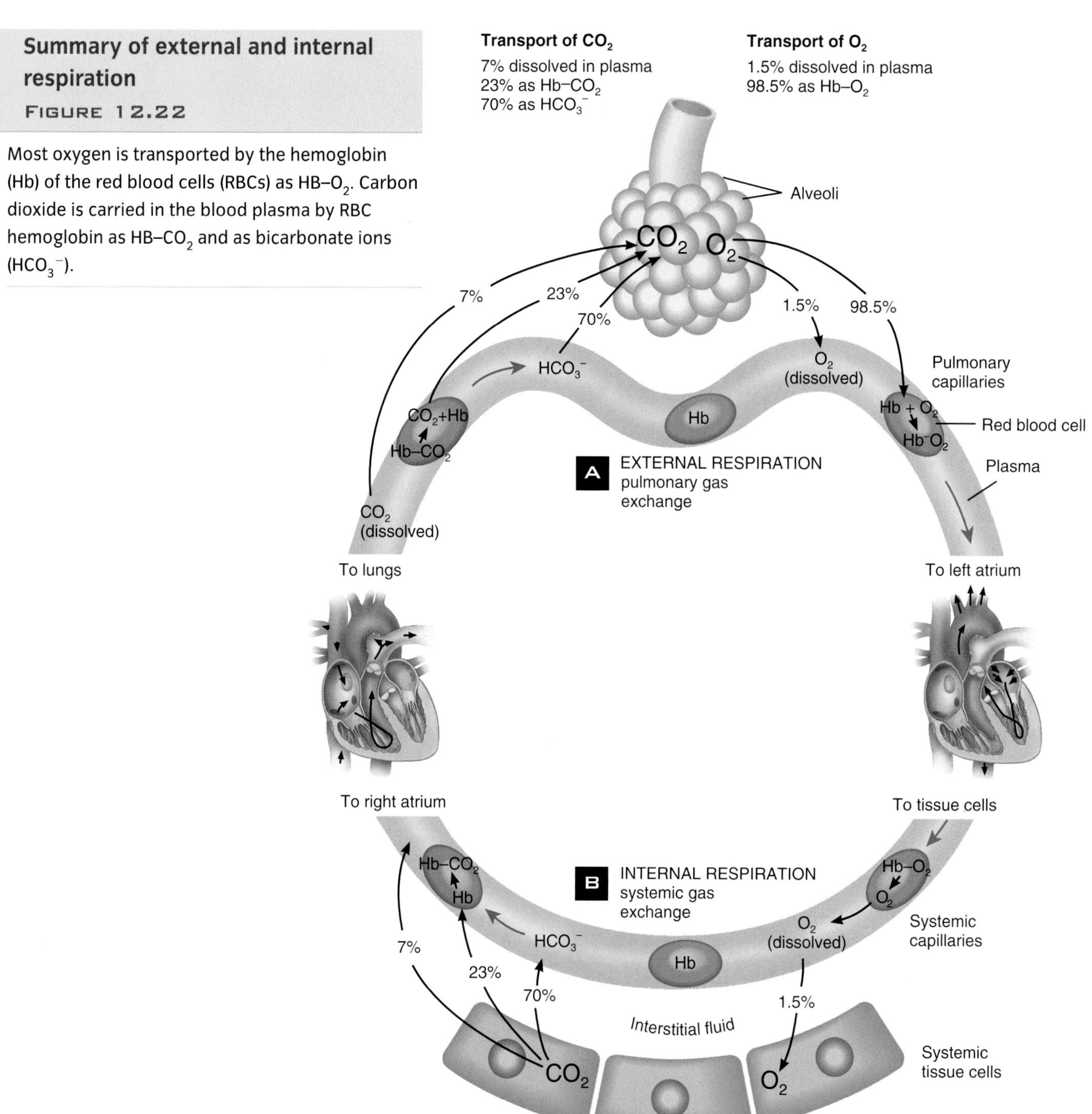

blood and enter the respiring cells without requiring energy from the body.

Cellular respiration produces carbon dioxide, and the partial pressure of carbon dioxide in the tissues is about 45 mmHg. Blood in the capillary beds has a carbon dioxide partial pressure of 40 mmHg. This small gradient is still enough to cause carbon dioxide to diffuse from the cells to the blood, which carries it off to the lungs for release into the alveolar air.

CONCEPT CHECK

What is the difference between external and internal respiration?

Explain how Dalton's law governs the movement of oxygen and carbon dioxide in respiration.

How does the blood transport oxygen and carbon dioxide?

Transport of Oxygen and Carbon Dioxide Requires Hemoglobin and Plasma

LEARNING OBJECTIVES

Understand the role of hemoglobin in respiration.

Recognize the role of carbon dioxide in maintaining blood pH.

Respiration involves not only the structures of the respiratory system, but also the functioning of the cardiovascular system. The respiratory system moves the gases in and out of the body, while the cardiovascular system transports them within the body. The pulmonary capillaries exchange gases in the lungs, while the systemic capillaries exchange gases in the body. The final piece to this puzzle is to determine how these gases are carried through the cardiovascular system between these two capillary beds.

HEMOGLOBIN TRANSPORTS OXYGEN

As we know, the **hemoglobin** molecule carries oxygen in the blood (FIGURE 12.23). Hemoglobin picks up oxygen through a bond between the oxygen molecule and the iron atom of the heme molecule. Hemoglobin has a high **affinity** for oxygen

Affinity
An attraction between particles that increases chances of combining.

Hemoglobin with oxygen binding site indicated FIGURE 12.23

Illustration, Irving Geis. Image from the Irving Geis Collection/Howard Hughes Medical Institute. Rights owned by HHMI. Reproduction by permission only.

under some conditions but will release it under other conditions. The oxygen-hemoglobin dissociation curves discussed in Chapter 11 and reviewed below show hemoglobin's unique characteristics.

The bond between oxygen and hemoglobin is reversible. Oxygen binds to the iron atom in the hemoglobin molecule when the partial pressure of oxygen is high, the pH is high, and the temperature is low. In areas where these conditions do not exist, hemoglobin releases oxygen. Even minute changes in temperature or pH will cause oxygen release (FIGURE 12.24). Such differences exist in active tissue—muscles generate heat while contracting, which warms the muscle. Contraction requires oxygen to fuel ATP production, which produces lactic acid, which lowers the pH. Both of these factors increase oxygen delivery to the muscle cells.

Several mechanisms transport carbon dioxide Hemoglobin is best known for carrying oxygen, but it also conveys about 23 percent of total carbon dioxide through the bloodstream. This carbon dioxide binds to the protein portion of hemoglobin, forming **carbaminohemoglobin (Hb–CO_2)** (see FIGURE 12.23).

Another 7 percent of the blood-borne carbon dioxide is carried as dissolved CO_2 gas. The major share of blood-borne carbon dioxide (about 70 percent of total carbon dioxide) moves as a **bicarbonate ion** in plasma. A bicarbonate ion is produced in steps. First, carbon dioxide and water combine to form carbonic acid inside red blood cells. The enzyme **carbonic anhydrase** speeds this reaction, allowing red blood cells to remove most of the carbon dioxide from the blood. This carbonic acid then dissociates into a hydrogen ion and a bicarbonate ion. The hydrogen ion is picked up by hemoglobin, forming **reduced hemoglobin**. The bicarbonate ion is transferred out of the RBC in exchange for a chloride ion entering the RBC. The large transport of chloride ions into the RBCs, called the **chloride shift,** is an exchange reaction that requires no ATP because it merely switches the positions of the anions. The bicarbonate ion in the plasma then serves as a **buffer**, helping to

Bicarbonate ion
HCO_3^-, a buffering ion.

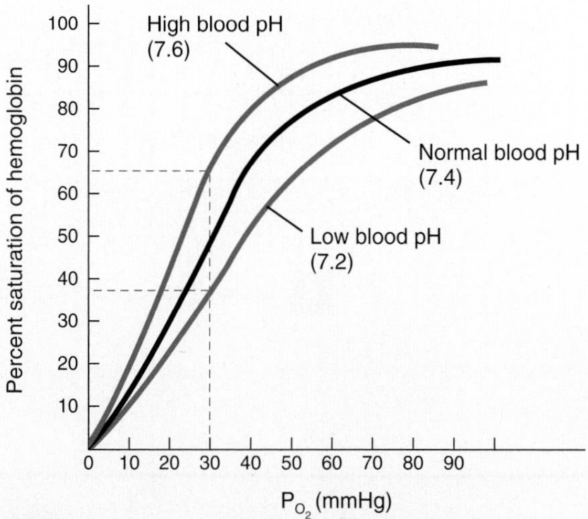

A Effect of pH on hemoglobin's affinity for oxygen

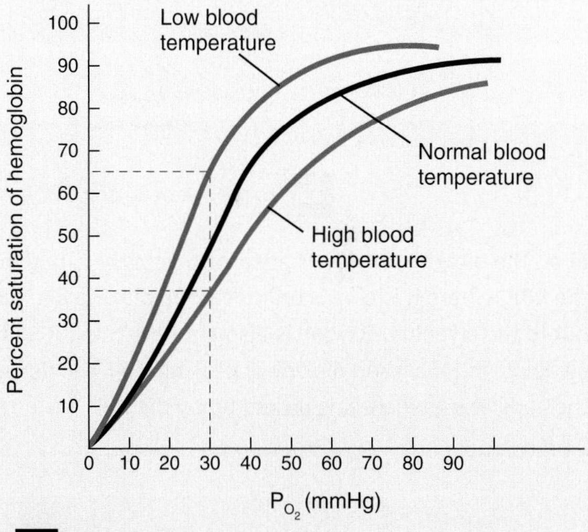

B Effect of P_{CO_2} on hemoglobin's affinity for oxygen

Effects of pH and temperature on hemoglobin binding FIGURE 12.24

Process Diagram

Carbon dioxide transport in blood FIGURE 12.25

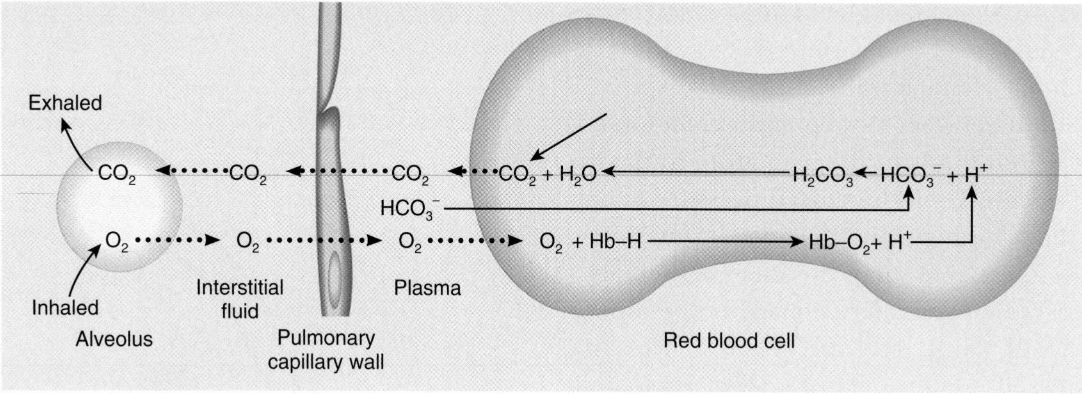

A Exchange of O_2 and CO_2 in pulmonary capillaries (external respiration)

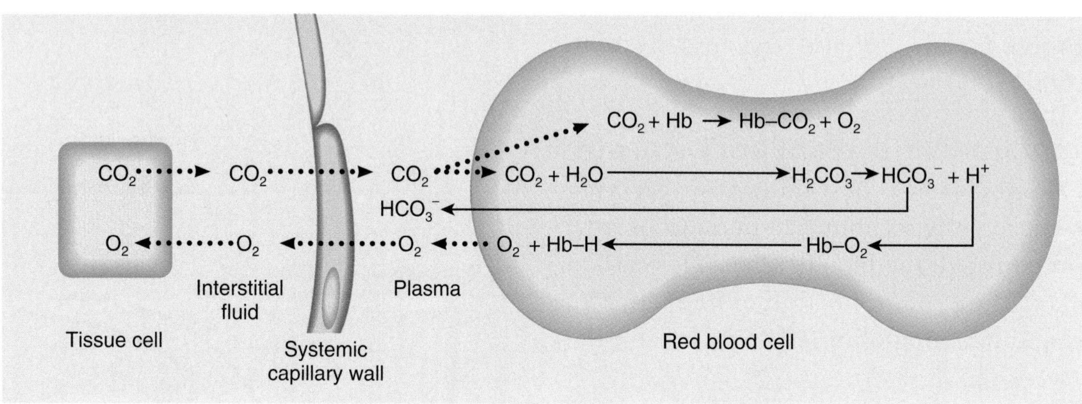

B Exchange of O_2 and CO_2 in systemic capillaries (internal respiration)

In **A**, the bicarbonate ion is absorbed from the blood into the RBC, where it is converted to carbon dioxide and passed out to the alveolus. Oxygen is also entering the RBC at the alveolus. In **B**, carbon dioxide is passing from the tissues to the capillaries, where it is picked up by the RBC. Inside the RBC, the carbon dioxide is converted to bicarbonate ions that are then pumped back out to the blood where they serve as a buffer. Oxygen is seen leaving the RBC and diffusing into the tissues, where it is used to drive cellular activities.

Buffer
A compound that absorbs hydrogen ions or hydroxide ions, stabilizing pH.

maintain blood pH (FIGURE 12.25). Without this buffering, we could not control our internal pH, and we would perish.

Reduced hemoglobin has a deep crimson, almost purple color, which is why vienous blood looks so blue when viewed through our skin. The red color of arterial blood is due to a high concentration of **oxyhemoglobin** (Hb–O_2). But blood inside your body is never as crimson as what is spilled when you cut yourself. The partial pressure of oxygen in the atmosphere is far higher than anywhere in your body, so hemoglobin quickly picks up more oxygen when you bleed.

Oxyhemoglobin
Hemoglobin molecule with at least one oxygen molecule bound to the iron center.

CONCEPT CHECK

How is oxygen carried in the blood?

What is the role of hemoglobin in gas transport?

What is one positive role of carbon dioxide in the blood?

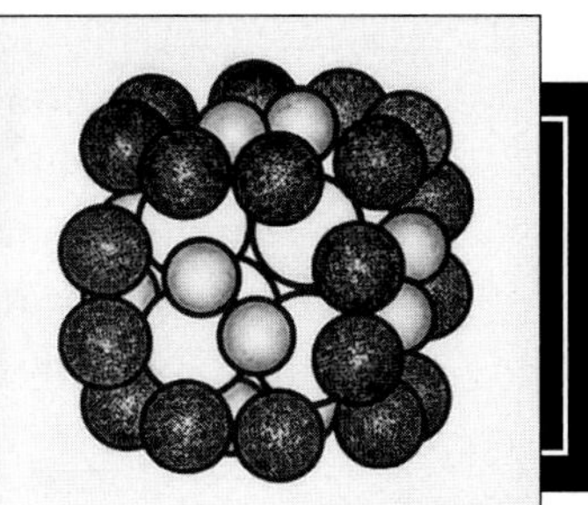

Respiratory Health Is Critical to Survival

LEARNING OBJECTIVES

Discuss two common disorders of the upper respiratory tract.

Identify the symptoms of obstructive respiratory disorders.

Understand the main disorders of the lower respiratory tract.

The previous chapter introduced cardiovascular disorders and outlined their obvious impact on respiration. If the blood does not circulate properly, or if it does not carry enough oxygen, external and internal respiration are impaired.

The upper respiratory tract is susceptible to infection and inflammation of the nasal passages, sinuses, and larynx. One of the most common upper respiratory diseases is **sinusitis,** an inflammation or swelling of the sinuses ("-itis" means inflammation). **Sinuses** are cavities in the skull, lined with the same type of mucous membrane as the nasal passages (FIGURE 12.26). Sinuses exist in the frontal bone, ethmoid, sphenoid, and maxillary bones, but the largest are in the frontal

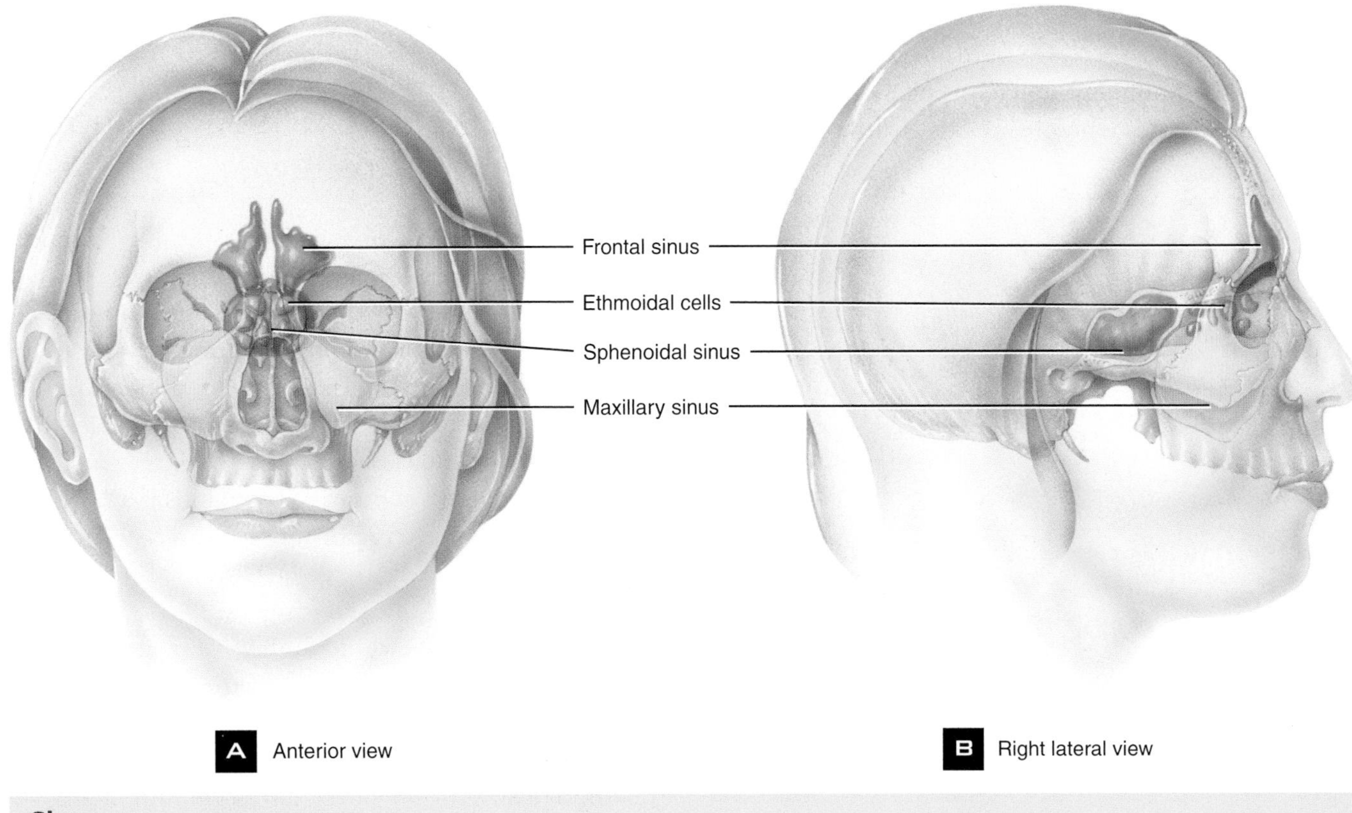

Sinuses FIGURE 12.26

Histamine A compound involved in allergic reactions that causes capillary leakage and increased fluid movement to affected tissues.

Acute sinusitis Inflammation of the sinuses with sudden onset and usually of short duration.

Chronic sinusitis Inflammation of the sinuses that persists for long periods of time.

bone. When you succumb to the common cold or flu, viruses swell the nasal membranes. **Histamines** are released, and mucus production increases as the membranes try to rid the body of the virus. If the membrane lining a sinus swells, the opening can shut, preventing mucus produced in the sinus from draining and causing it to build up pressure in the closed sinus. Resident populations of streptococcus or staphylococcus bacteria can also grow unchecked in the closed sinus. **Acute sinusitis** is usually caused by a common cold and goes away on its own within two to three weeks. **Chronic sinusitis**, in contrast, is more severe and its causes are less clear. Most people who suffer from chronic sinusitis also have allergies, asthma, or a compromised immune system owing to a disease like AIDS. Treating this type of sinusitis is also more difficult; antibiotics, inhalant steroids, or even oral steroids may be used, depending on the case.

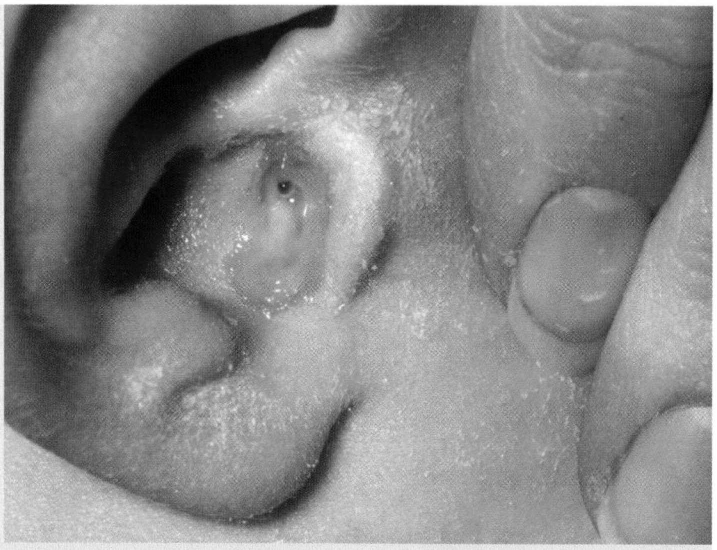

Distended eardrum caused by otitis media FIGURE 12.27

If you have a young child, you probably know about **otitis media** (FIGURE 12.27). This inflammation of the middle ear fills the middle ear with fluid, distending the eardrum. A stretched eardrum can cause severe pain, and the eardrum can rupture as bacteria within the trapped fluid multiply. Otitis media is usually caused by a bacterial infection that can be treated with antibiotics. The pathogens most often arrive through the eustachian tube, with its open connection between the middle ear and the nasopharynx. In small children, the tube is almost horizontal, so fluids in the mouth can easily travel to the middle ear, especially since the bottom of the tube opens with each swallow. As we age, our facial bones expand, tilting the eustachian tubes toward the vertical, so fluids do not flow so readily to the middle ear. For this reason, ear infection rates drop with age.

Diseases of the lower respiratory tract are usually either **obstructive**, meaning that something is obstructing the normal flow of gases through the lungs, or **constrictive**, indicating that the airways have been narrowed in some way.

CONSTRICTIVE DISEASES ARE SERIOUS BUT OFTEN SPORADIC

As the name implies, constrictive respiratory diseases constrict the airways. One common constrictive disease of the lower respiratory tract is **bronchitis,** an inflammation of the mucous membrane lining the bronchi. When this membrane swells, the lumen of the bronchiole constricts. Often these infected bronchioles also produce more mucus, which can block air passages. The most common symptom of bronchitis is a deep, often painful, cough. Acute bronchitis can be caused by viruses and occasionally bacteria. Chronic bronchitis is most often caused by smoking and can last from months to years, depending on the severity of the reaction to smoke and the duration of the smoking habit. The main symptom of acute and chronic bronchitis is a productive cough. In acute bronchitis, shortness of breath, tightness of the chest, and a general feeling of illness often accompany the cough. Treatment for bron-

Chronic obstructive pulmonary disease: Why are chronic bronchitis and emphysema so deadly?

Chronic obstructive pulmonary disease (COPD) is actually two diseases—emphysema and chronic bronchitis—that both obstruct airflow. (Doctors use *COPD* because individual patients often have both diseases.) In the United States, the death rate from COPD has doubled in the past 30 years, to an estimated 120,000 annually. Globally, scientists predict that COPD will be the third-largest cause of death by 2020. The major cause is cigarette smoking, but other airborne toxins and pollutants are also to blame.

About 8.6 million Americans have been diagnosed with chronic bronchitis, which starts when the bronchi get inflamed. Scarring of the bronchi is accompanied by heavy mucus flow and chronic cough. Mucus and thick bronchial walls obstruct airflow, and bacterial infections can fester in the gathered mucus.

Emphysema begins when a pollutant or cigarette smoke damages the alveoli, forming holes that cannot be repaired. Delicate lung structures become fibrotic (filled with fibers) and stiff, reducing their elasticity, making exhaling difficult. The disease starts gradually, with a shortness of breath, and gets worse with age. More than 80 percent of cases are caused by smoking. About 5 percent of Americans suffer from genetic emphysema caused by the lack of a protein necessary for lung function.

Both types of COPD reduce gas transfer in the lungs, causing shortness of breath. Exercise, and even daily activity, become difficult or impossible. COPD may be treated with antibiotics, anti-inflammatories, and bronchodilators, which open the airways to ease breathing. Advanced emphysema patients need supplemental oxygen. An increasing number of COPD patients are receiving lung transplants. Although transplants can prolong survival, the lungs often fail much sooner than other transplanted organs.

Given the increasing death rate, and the fact that emphysema is invariably fatal, new perspectives are needed on COPD. An intensified battle against smoking is an obvious first step that could bring many other benefits (see the Ethics and Issues box, "Tobacco, the universal poison," on p. 397). Researchers have found other clues that could help explain and treat COPD. For example, a 20-year study found that asthmatics were 12 times as likely to develop emphysema as other people. Asthma and emphysema are considered separate diseases, but this evidence suggests that the emphysema epidemic may be part of the asthma epidemic. (For more on asthma, see the I Wonder box, "Why are asthma rates going up?" on p. 380.)

Other research examines what happens after a diagnosis of COPD. Stopping smoking can greatly extend one's life span, and exercise makes a difference. A study at Ohio State University found that a 10-week exercise program increased cognitive, psychological, and physical function in COPD patients. After a COPD patient spends time in the hospital to treat a medical crisis, physical rehabilitation can reduce symptoms and restore some quality of life.

COPD is not a pretty picture. In most cases, the disease is much easier to prevent than to cure. For smokers, prevention should start now.

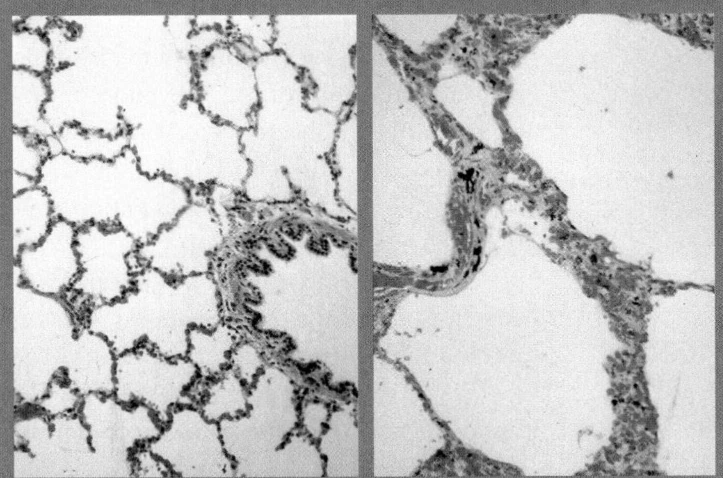

Left: a normal lung, with intact alveolar walls. Right: In an emphysemic lung, the alveolar membrane is destroyed.

chitis includes rest, plenty of fluids, and perhaps over-the-counter cough medicine. If the cough persists, an inhalant **bronchodilator** may be prescribed to relax the smooth muscle of the bronchi, open the constricted tubes, and help clear the mucus (FIGURE 12.28).

Asthma is a constrictive pulmonary disease that can be life-threatening. During an asthma attack, the smooth muscle of the bronchi contracts, mucus production increases in these tubes, and the bronchi swell, interfering with the passage of air. Breathing grows laborious, and wheezing is common during exhalation. Asthma attacks are usually triggered by an external source, such as exercise, viral infection, or inhalation of cold air or an allergen, or by high levels of ozone in the air (see the I Wonder . . . box on the increasing prevalence of asthma on p. 380).

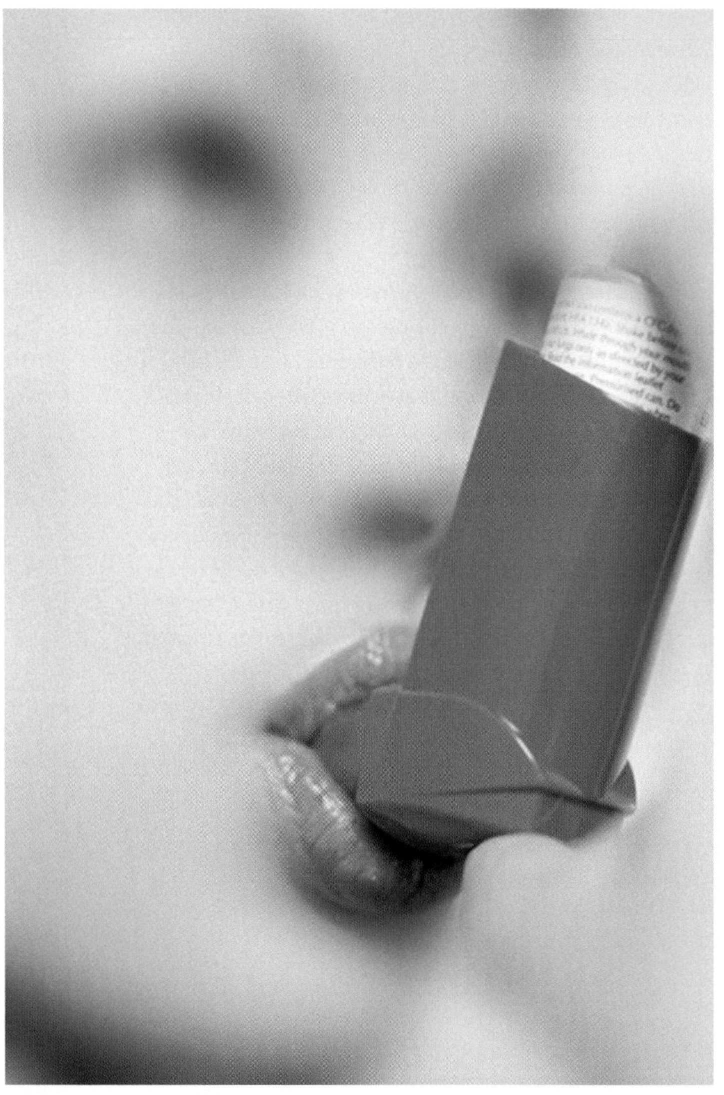

Inhalers contain bronchodilator drugs
FIGURE 12.28

Asthma may result from an overactive immune system, and to many people, inhaling an allergen can cause an immediate, dangerous airway constriction. Many asthma patients carry inhalers containing bronchodilator drugs to quickly open the airways during an attack. As a preventive measure between attacks, many chronic asthmatics inhale corticosteroids to reduce the number and severity of asthma attacks. Despite these medicines, however, asthma still kills up to 5,000 people every year in the United States.

OBSTRUCTIVE DISEASES CAUSE PERMANENT LUNG DAMAGE

Although asthma is a serious disease, it does not permanently damage lung tissue. In contrast, the chronic obstructive pulmonary diseases, including **emphysema** and fibrosis, do damage or destroy the terminal and respiratory bronchioles. The most common obstructive pulmonary diseases are **pneumonia, tuberculosis, emphysema**, and **lung cancer** (see the Health, Wellness, and Disease box on p. 395). After exhalation in all of these diseases, the tubes of the airway do not spring back open because the elastic tissue is destroyed. Pressure builds in the lungs as the patient tries to force air through the collapsed tubes, damaging the delicate alveoli and reducing the respiratory surface area. The most common cause of emphysema is smoking, but environmental pollutants and even genetic factors can also be to blame. Pulmonary fibrosis, a destructive increase in collagen that also makes the lungs less elastic, often results from occupational exposure to silicon or other irritants.

Lung tissue must remain warm and moist because gases cannot diffuse across a dry membrane. Unfortunately, these same conditions are perfect for bacterial growth. Bacteria living in the warm, moist, lung tissue cause two of the more common obstructive respiratory diseases: **pneumonia** and **tuberculosis**.

Tobacco, the universal poison

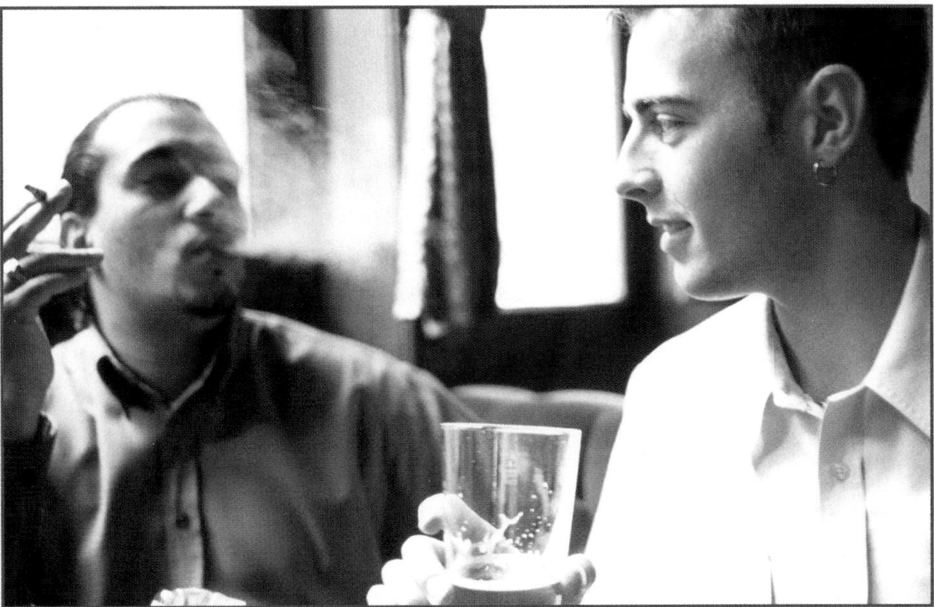

In 1964, the U.S. Surgeon General issued an influential Report on Smoking and Health. The report looks tame today, given how much we now know about the toxicity of tobacco smoke, but it was an early acknowledgment that smoking causes lung cancer. Today, smoking-related lung cancer kills an estimated 174,000 people in the United States per year, and the number is rising.

Some of the approximately 4,000 compounds in tobacco smoke attack the delicate epithelial cells lining the respiratory tract and allow them to grow without control—the hallmark of cancer. Because early tumors are invisible, lung cancer is not usually detected until it has spread; therefore, the five-year survival rate is only 15 percent. Smoking and tobacco smoke also:

- are the major cause of emphysema.
- increase the risk of acute myeloid leukemia and cancer of the throat, mouth, bladder, kidney, stomach, cervix, and pancreas, according to the American Cancer Society.
- impair several functions of the uterine tube, which conducts both gametes and the embryo, and alters female hormone effectiveness. Both effects could explain why smoking women have higher rates of reproductive problems, including undersized and/or premature infants.
- kill nerve cells, interfering with smell and taste.
- elevate pulse and body temperature.
- increase the risk of heart disease by a factor of 2 to 4.
- raise the level of carbon monoxide and reduce the level of oxygen in the blood, which in turn reduce the ability to exercise or even move about comfortably.
- destroy cilia in the airways, reducing the ability to expel mucus.
- promote heartburn and peptic ulcers by increasing stomach acid production.

Nicotine causes its own set of problems. Nicotine is a vasoconstrictor, which forces the heart to work harder. Nicotine's neurological effects include increased concentration, a reduction in hunger, and a subtle boost in mood. Nicotine triggers the release of dopamine, a "feel-good" neurotransmitter, making "coffin nails," or cigarettes, highly addictive.

The tobacco industry is expert at promoting the delusion that smoking is "cool." Many of their campaigns are targeted, subtly or not, at young people. It's only logical. With so many customers dying each year, they need to replace them with young, healthy smokers.

But many people are quitting, and fewer people are starting. The number of cigarettes smoked per capita has declined 59 percent from 1963 to 2004. Still, an estimated 45 million Americans smoke. Smoking causes about 30 percent of all cancers and an estimated 438,000 premature deaths.

Lethargy
Tiredness.

Pneumonia is a general term for a buildup of fluid in the lung, often as a response to bacterial or viral infection. When the delicate membranes in the alveoli become inflamed, they secrete fluid in an attempt to eradicate the pathogen, but this fluid inhibits gas exchange across the membrane. Symptoms of pneumonia include a productive cough, **lethargy**, fever, chills, and shortness of breath. Treatment depends on the underlying cause of the fluid buildup. Although pneumonia usually can be treated, it can be fatal, especially in patients with weak immunity owing to other serious illnesses.

Tuberculosis (TB) is a disease caused by *Mycobacterium tuberculosis* infection. This tiny bacterium can pass from person to person in airborne droplets generated by a sneeze and cough. The inhaled bacteria multiply from one small region of the infected organ, called the "focus." Because it is airborne, the focus in humans is usually in the lung tissue. If the immune system can combat the disease, scar tissue may form at the focus. In those rare instances where the body does not eliminate the infection, the bacteria can enter the lymphatic system and infect just about any organ. The bacterium can also remain dormant for years and then reappear in the lungs without warning. Symptoms of TB resemble those of pneumonia, including a productive (and often bloody) cough, fever, chills, and shortness of breath. TB also causes weight loss and night sweats. TB is usually diagnosed if a focus appears on a chest X-ray. Previous exposure can be detected with a simple skin test, which is mandatory for children entering U.S. public schools (FIGURE 12.29).

A century ago, TB was a major deadly health threat, but antibiotics have reduced the incidence in industrialized nations. Unfortunately, TB is on the rise again because antibiotic-resistant strains have now appeared, and many patients must take multiple antibiotics for many months to clear the infection. TB is one of several cases where bacteria are starting to evade antibiotics that once controlled them. This shows how misuse of antibiotics, combined with their widespread use in animal agriculture, may help breed antibiotic-resistant strains of bacteria.

Cancer can attack just about any organ system, but lung cancer causes one-third of all cancer deaths in the United States. Lung cancer can affect the bronchi or the alveoli. In either case, the cells proliferate, obstruct airflow, and prevent gas exchange. Lung cancer is primarily due to tobacco smoking; nearly 90 percent of all patients in the United States are current or former smokers. The Ethics and Issues box on page 397 discusses this in greater detail. Lung cancer takes years to develop, and the risk increases with each year of smoking. The good news is that quitting smoking re-

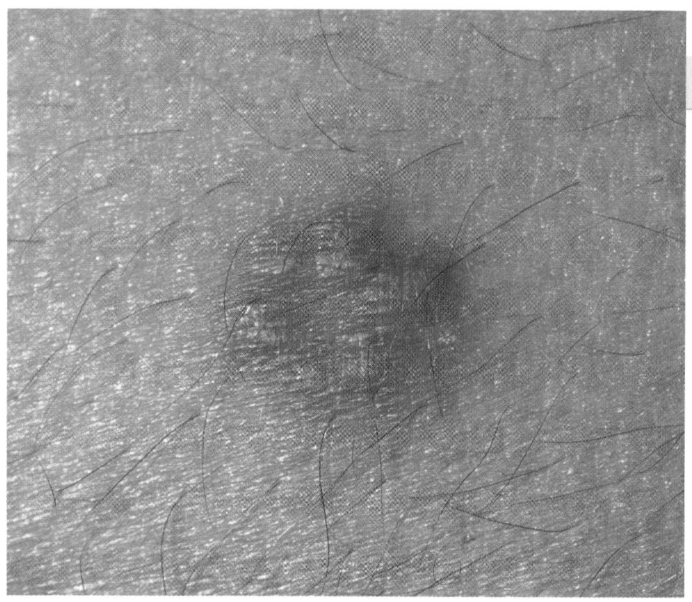

Positive TB test result FIGURE 12.29

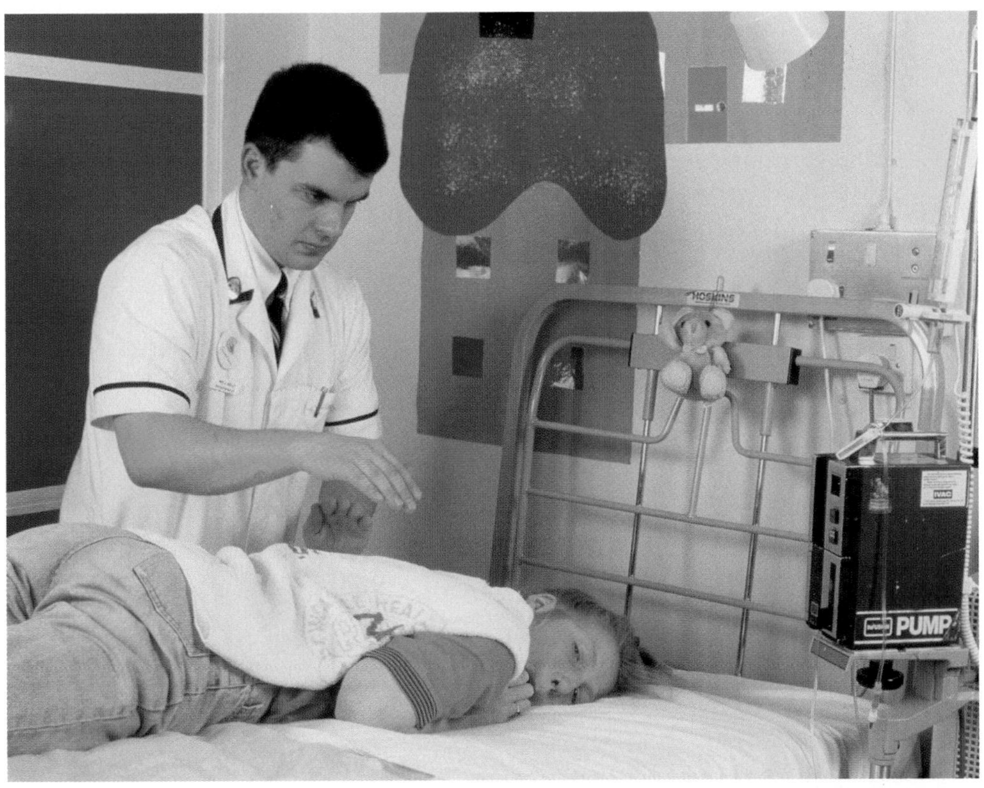

CF patient receiving physical therapy "clapping" to mobilize mucus in the lungs FIGURE 12.30

duces the risk, even for long-term smokers. As with other respiratory illnesses, the symptoms include a chronic cough, possibly with bleeding, wheezing, and chest pain. Treatment may include surgical removal of the tumor or destruction of the cancer with radiation or chemotherapy. Unlike other cancers, lung cancer is relatively easy to prevent. Avoid smoking and exposure to environmental carcinogens such as asbestos, silicon, coal dust, and radon gas.

Cystic fibrosis (CF) results from a defective gene that controls the consistency of mucus in the lungs. The CF version of this gene causes thick, sticky mucus to be produced, rather than thin, fluid mucus that is conducive to diffusion. This thick mucus traps bacteria and slows airflow through the bronchial tree, and it may also block the pancreas and bile duct. Treatment for the lung obstruction includes physical therapy to dislodge the mucus (FIGURE 12.30), and new drugs that may make the mucus more fluid. Approximately 30,000 people in the United States are currently living with cystic fibrosis. Another 1,000 are diagnosed yearly, usually before age 3. One promising line of research would use gene therapy to correct the defect that causes CF.

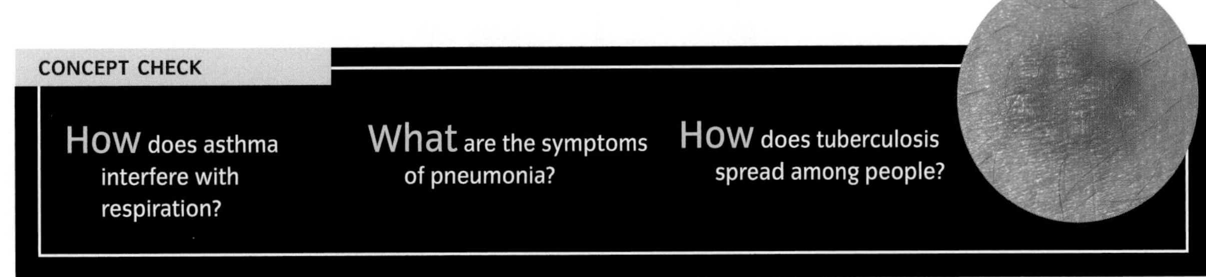

CONCEPT CHECK

How does asthma interfere with respiration?

What are the symptoms of pneumonia?

How does tuberculosis spread among people?

CHAPTER SUMMARY

1 The Respiratory System Provides Us with Essential Gas Exchange as well as Vocalization

The respiratory system delivers oxygen and removes carbon dioxide, helps balance blood pH, sustains fluid and thermal homeostasis, and produces speech in the larynx. The upper respiratory tract warms, moistens, and filters incoming air. The lower tract exchanges gas with the environment. The bronchial tree reaches into the lobes of the lungs. At the end of the respiratory bronchioles are the alveoli, the thin membranous sacs where gas exchange occurs. Septal cells produce surfactant to prevent the alveolar membranes from sticking together. Dust cells patrol the respiratory membrane to remove foreign particles.

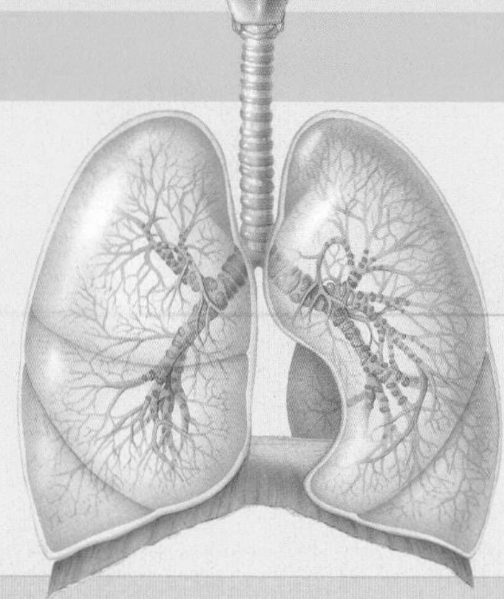

2 In Order to Respire, Air Must Be Moved in and out of the Respiratory System

Pulmonary ventilation is the movement of air in and out of the lungs, based on Boyle's law of gases. Tidal volume is the amount of air you inspire during a normal, 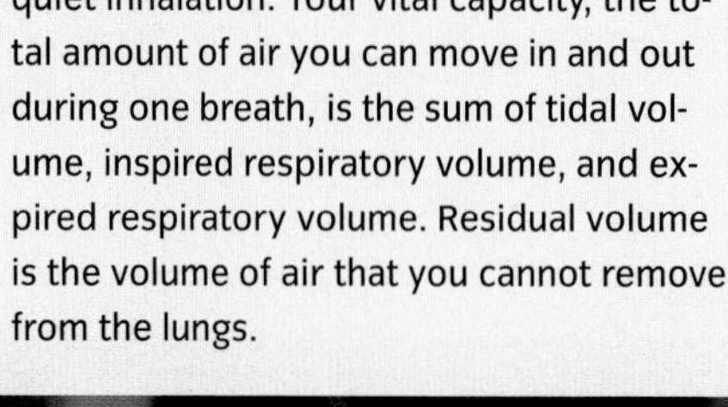
quiet inhalation. Your vital capacity, the total amount of air you can move in and out during one breath, is the sum of tidal volume, inspired respiratory volume, and expired respiratory volume. Residual volume is the volume of air that you cannot remove from the lungs.

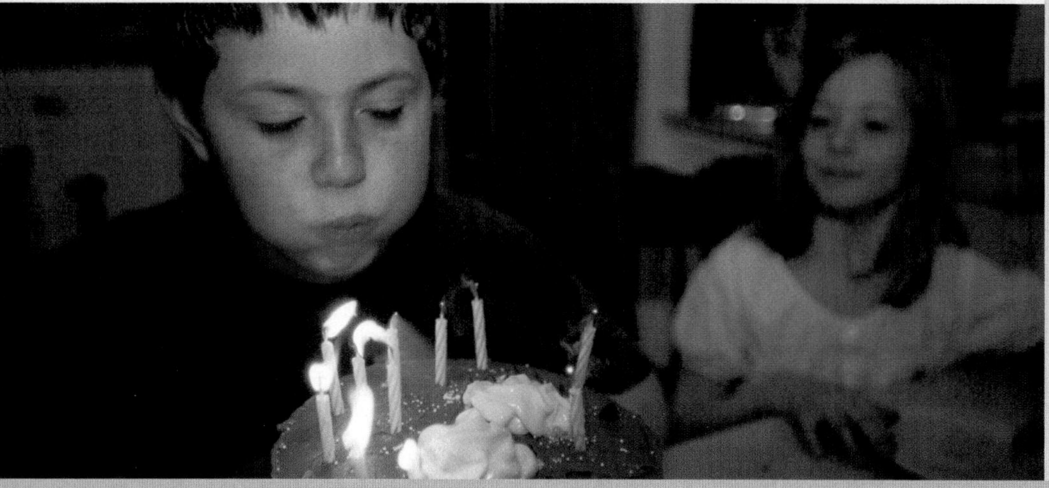

3 External Respiration Brings Supplies for Internal Respiration

External respiration is the exchange of gases between air in the alveoli and blood in the pulmonary capillaries. Oxygen enters the red blood cells, while carbon dioxide exits. Internal respiration is the transfer of gases between systemic capillaries and body cells. Oxygen diffuses into the cells, while carbon dioxide diffuses into the blood. The diffusion in both types of respiration is based on Dalton's law of partial pressures.

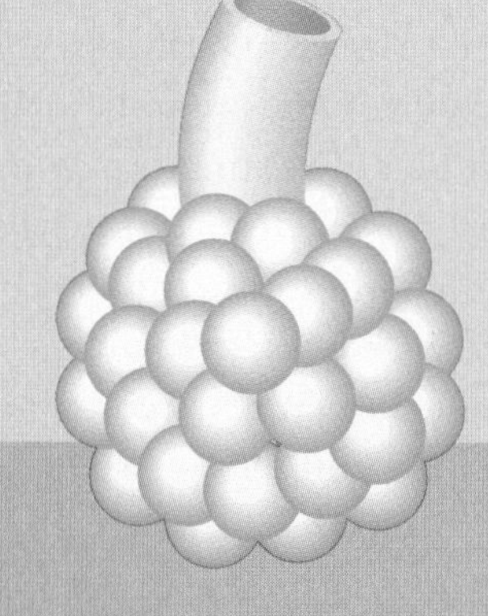

4 Transport of Oxygen and Carbon Dioxide Requires Hemoglobin and Plasma

Oxygen is carried bound to iron in hemoglobin molecules in red blood cells. Most carbon dioxide is moved as bicarbonate ions in plasma. Bicarbonate also serves as a buffer that stabilizes pH in the blood. Some carbon dioxide is carried by the protein portion of hemoglobin, turning venous blood blue.

5 Respiratory Health Is Critical to Survival

In constrictive respiratory diseases like asthma and bronchitis, airway diameter is reduced. Obstructive diseases, including emphysema, fibrosis, tuberculosis, pneumonia, and lung cancer, feature physical obstructions to airflow. The death toll due to lung cancer in the United States is high, but the disease is preventable because most cases are caused by smoking.

KEY TERMS

- **acute sinusitis** p. 394
- **affinity** p. 390
- **bicarbonate ion** p. 391
- **buffer** p. 392
- **chemoreceptors** p. 385
- **chronic sinusitis** p. 394
- **hilum** p. 377
- **histamine** p. 394
- **larynx** p. 368
- **lethargy** p. 398
- **lower respiratory tract** p. 368
- **macrophage** p. 375
- **oxyhemoglobin** p. 392
- **partial pressure** p. 387
- **pharynx** p. 368
- **pleurisy** p. 378
- **premature babies** p. 378
- **respiratory membrane** p. 373
- **upper respiratory tract** p. 368
- **vocal folds** p. 373

CRITICAL THINKING QUESTIONS

1. We know humans cannot breathe under water, and yet fishes can. One difference between fish gills and human lungs is that the blood in the gill flows in a countercurrent pattern. This means the water and blood flow across the respiratory surface in opposite directions. How might this speed oxygen removal from the water? Draw a schematic of this arrangement with arrows to show how countercurrent flow works. What else do humans lack for breathing under water? How might our physiology be "improved" to allow us to extract oxygen from water?

2. Although lung cancer is the most common cancer associated with smoking, the larynx is also susceptible to tobacco smoke. When cancer is detected in the larynx, the affected area is removed. What problems would you expect if the entire larynx was removed? Often the tumors appear on the vocal folds. How might removal of these growths affect vocalization? What alternative methods of sound production might be available to victims of laryngeal cancer?

3. In Chapter 11, we discussed carbon monoxide poisoning. How would the respiratory system change if red blood cells were saturated with CO? What might happen to the respiratory rate? To airway diameter? Death occurs after the patient slips into unconsciousness. Physiologically, what is causing that unconsciousness?

4. Chapter 7 explained the sympathetic nervous system. How does activation of the "fight-or-flight" nervous system affect the respiratory system? What neurotransmitter is released, and how does it affect the functioning of the upper and lower respiratory tracts? What happens to pulmonary ventilation when the sympathetic nervous system is in control? Are there any changes in external or internal respiration?

SELF TEST

1. The function of the upper respiratory system is to

a. warm incoming air.
b. vocalize.
c. exchange gases with blood.
d. prevent lung infections.

2. The portion of the pharynx that is lined with thick mucus and includes the tonsils is the

a. laryngopharynx.
b. oropharynx.
c. nasopharynx.
d. nares.

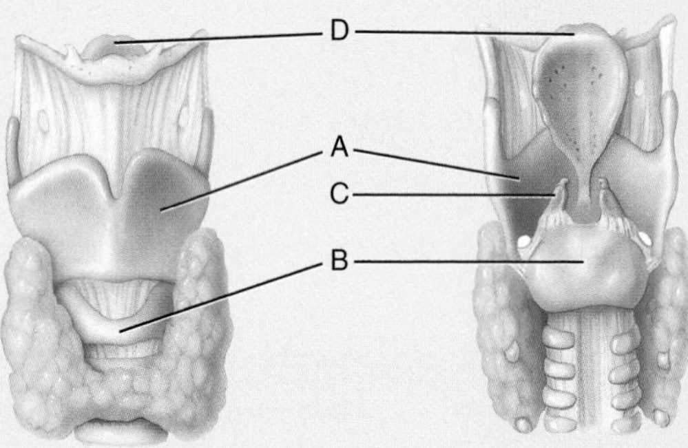

3. The cartilage that varies the pitch of the voice is indicated by which letter in the above diagram?

a. A
b. B
c. C
d. D

4. On the diagram above, the cartilage that prevents food and liquids from entering the lower respiratory system is labeled

a. A.
b. B.
c. C.
d. D.

5. The proper sequence of structures in the lower respiratory tract is

a. trachea → bronchioles → bronchi → respiratory bronchioles.
b. trachea → respiratory bronchioles → bronchioles → bronchi.
c. trachea → bronchi → bronchioles → respiratory bronchioles.
d. trachea → bronchi → respiratory bronchioles → bronchioles.

6. The only complete circle of cartilage in the respiratory system lies in the

a. trachea.
b. larynx.
c. bronchi.
d. bronchioles.

7. The most touch-sensitive area in the respiratory system is the

a. carina.
b. respiratory membrane.
c. uvula.
d. bronchial tree.

8. A side effect of the respiratory tubes getting smaller and smaller is that

a. cartilage support lessens.
b. the proportion of smooth muscle increases.
c. the surface area of the respiratory system increases.
d. All of the above are true.

9. The structure indicated by the arrow on the figure below is the

a. upper lobe of the left lung.
b. carina.
c. hilus of the left lung.
d. cardiac notch of the left lung.

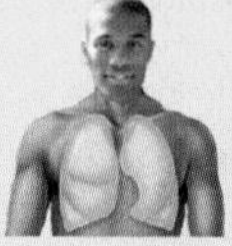
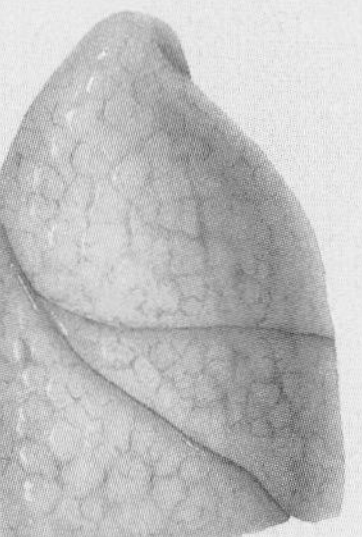
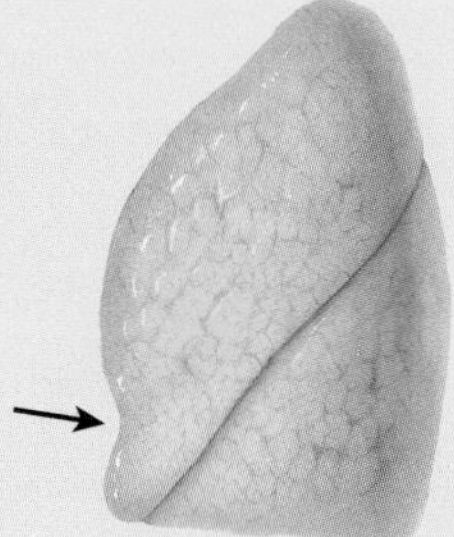

10. The function of the structure indicated by the arrow in the figure below is to

a. serve as a diffusion membrane for gases.
b. produce surfactant.
c. patrol the alveoli, removing debris and bacteria.
d. support the delicate walls of the alveolus.

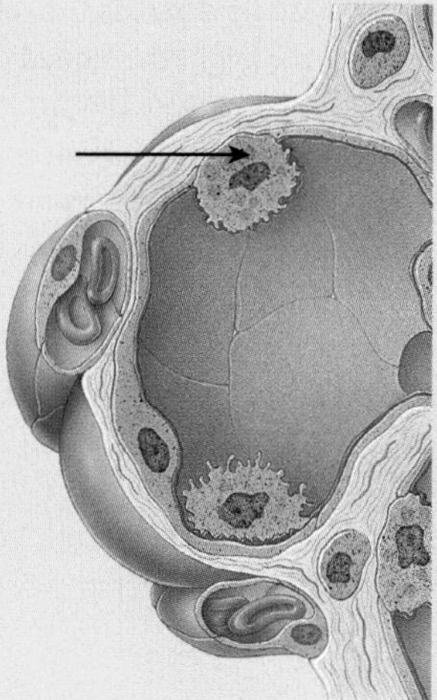

11. The function of the entire area depicted in the figure at the bottom of the previous page is
   a. diffusion of gases into and out of the blood.
   b. infection fighting within the lungs.
   c. movement of air into the deeper tissues of the respiratory system.
   d. thermal homeostasis.

12. During inspiration, the diaphragm __________, __________ the volume of the thoracic cavity.
   a. contracts, increasing
   b. contracts, decreasing
   c. relaxes, increasing
   d. relaxes, decreasing

13. True or False? The gas law that dictates the differential movement of carbon dioxide and oxygen into and out of the tissues of the body is Boyle's law.

14. Identify the volume indicated as A on this diagram:

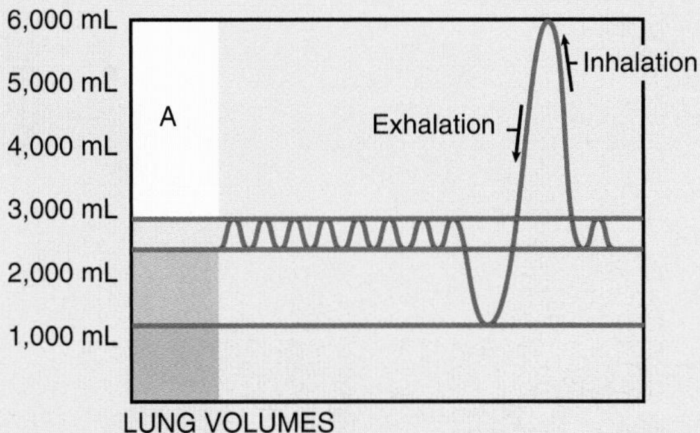

   a. Vital capacity
   b. Tidal volume
   c. Expiratory reserve volume
   d. Inspiratory reserve volume

15. The movement of oxygen from the blood into the tissues is referred to as
   a. internal respiration.
   b. external respiration.
   c. Dalton's law.

16. Carbon dioxide moves from the tissues of the body into the blood because
   a. the partial pressure of oxygen is lower in the tissues.
   b. the partial pressure of carbon dioxide is lower in the blood.
   c. the volume of carbon dioxide decreases in the blood.
   d. carbon dioxide floats in the blood, and will always travel upwards.

17. Oxygen is carried on the
   a. plasma proteins of the blood.
   b. protein portion of the hemoglobin molecule.
   c. iron portion of the hemoglobin molecule.
   d. white blood cells.

18. Hemoglobin binds oxygen more tightly when oxygen concentrations are __________ and pH is __________.
   a. low, low
   b. high, low
   c. high, high
   d. low, high

19. The structures indicated by the letter A are often susceptible to
   a. flooding with mucus.
   b. bronchitis.
   c. otitis media.
   d. sinusitis.

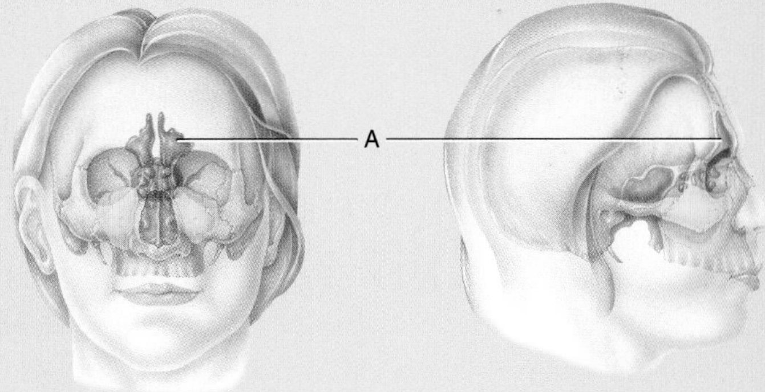

20. True or False? Bronchitis is an example of a constrictive disease.

13 The Digestive System

Have you seen *Super Size Me*—the movie by the man who ate nothing but McDonald's for one excruciating month? Part of the delicious delight of watching Morgan Spurlock work his way through endless Big Macs stems from pure contrariness. Your mother, after all, told you not to eat junk food, and here is Spurlock, gobbling like mad. The other delight comes from mother's vindication. Sure enough, Spurlock suffers mightily for his excess.

Long ago, when the Beatles sang, "You know that what you eat, you are," the idea that food might affect health was revolutionary. But not anymore. Nowadays, the idea that the food that you consume can affect your health is commonplace, and indeed many are surprised by a study that finds, for example, that eating less fat may *not* reduce the incidence of breast cancer, or that calcium supplements may not ward off osteoporosis.

At the center of all this concern is the digestive system, an essential series of organs that are designed to extract every last gram of nutrition from whatever goes down the gullet. In an era of rising obesity, such efficiency is not necessarily a good thing: Some designer fats are being deliberately concocted to avoid digestion. But that's the exception. In general, the goal of the digestive system is to convert food into simple compounds that the body can use for making, cellular energy, adenosine triphosphate (ATP), adipose tissue, and the building blocks of cells and tissues.

CHAPTER OUT

Nutrients Are Life-Sustaining

LEARNING OBJECTIVES

Differentiate macronutrients from micronutrients.

Describe how nutrients enter our cells.

All **aerobic** cells, and therefore all humans, need oxygen to survive. Oxygen drives cellular respiration by serving as the ultimate electron "pull," creating the hydrogen ion concentration gradient required to form ATP. However, one cannot live by oxygen alone!

The cells of our body require **nutrients** in usable form to maintain homeostasis and create ATP. Because we are **heterotrophs,** we cannot manufacture our own organic compounds, so we must obtain them from the environment. Consequently, we spend an awful lot of our time locating, preparing, and ingesting food.

Aerobic Requiring oxygen to metabolize.

Nutrients Ingredients in food that are required by the body.

Eating is so important that virtually every culture has elaborate rituals surrounding food. Think of your last Thanksgiving celebration, or even your birthday. Both of these events traditionally include a specific celebratory food: turkey with all the trimmings, or a cake with candles. And in both cases, there were rituals surrounding the food. We take a moment to reflect on all the good things in our life before eating Thanksgiving dinner, and we sing "Happy Birthday" and blow out candles before cutting the cake.

Although we may not understand why, we intuitively know we need nutrients in order to survive. But exactly what are nutrients? A nutrient is defined as any compound required by the body. The two main types of nutrients are **macronutrients** (carbohydrates, lipids, and proteins) and **micronutrients** (vitamins and minerals). These are organic and inorganic compounds that we obtain from food rather than synthesizing. We ingest carbohydrates, lipids, and proteins to provide the necessary energy and starting materials for us to create our own carbohydrates, lipids, and proteins. From these macronutrients, we synthesize cellular components such as the cell membrane, enzymes, organelles, and even entirely new cells during mitosis and meiosis. Micronutrients are required for the proper functioning of essential compounds, such as the enzymes of cellular respiration. Review Chapter 2 to refresh your understanding of carbohydrates, lipids, and proteins.

THERE ARE THREE CLASSES OF MACRONUTRIENTS

The average supermarket contains more than 20,000 food products, but these all come down to three macronutrient groups: carbohydrates, fats, and proteins. These groupings are distinct from the six major food groups, which are classified by food type rather than biochemical makeup. For example, fruits, a food group, provide us with carbohydrates in the form of fructose, and meats, another food group, are rich in protein.

One macronutrient we often hear about in diet discussions is the **carbohydrate**, and for good reason. Carbohydrates are our most efficient source of energy. Carbohydrates are composed of carbon, hydrogen, and oxygen in a 1:2:1 ratio. The most common carbohydrate, glucose, has a chemical formula of $C_6H_{12}O_6$. Our cells are excellent at breaking down glucose to produce ATP or to synthesize amino acids, glycogen, or triglycerides. Carbohydrate digestion is so efficient that we can ingest glucose and break it down completely into energy, carbon dioxide, and water. Although we are efficient carbohydrate burning machines, sometimes fad diets encourage us to avoid this energy source. The Health, Wellness, and Disease box on the Atkins diet on page 410 takes a closer look at this way of thinking.

Glycolysis, the Krebs cycle, and electron transport FIGURE 13.1

Glycolysis occurs in the cytoplasm, requiring two molecules of ATP to begin, but generating a total of four ATP molecules in the conversion of glucose to pyruvate. With oxygen present, the two pyruvate molecules are shuttled to the mitochondrion, where they are passed through a series of chemical reactions, each step of which releases energy that is harvested in ATP, NADH, and $FADH_2$. These reactions are referred to as the Krebs, or TCA, cycle. The NADH and $FADH_2$ created in the Krebs cycle then drive the reactions of the electron transport chain, where hydrogen ions are moved to the center of the mitochondrion, creating a hydrogen ion gradient. This gradient drives chemiosmosis, the final step in this process. At this point, the energy harvested from the original glucose molecule is finally converted to about 32 ATP molecules.

1. Glycolysis. Oxidation of one glucose molecule to two pyruvic acid molecules yields 2 ATPs.
2. Formation of two molecules of acetyl coenzyme A yields another 6 ATPs in the electron transport chain.
3. Krebs cycle. Oxidation of succinyl CoA to succinic acid yields 2 ATPs, 2 molecules of $FADH_2$ and 6 molecules of NADH.
4. The 6 NADH + $6H^+$ produced in the Krebs cycle yields 18 ATPs in the electron transport chain. The similarly produced $FADH_2$ yields 4 ATPs in the electron transport chain.

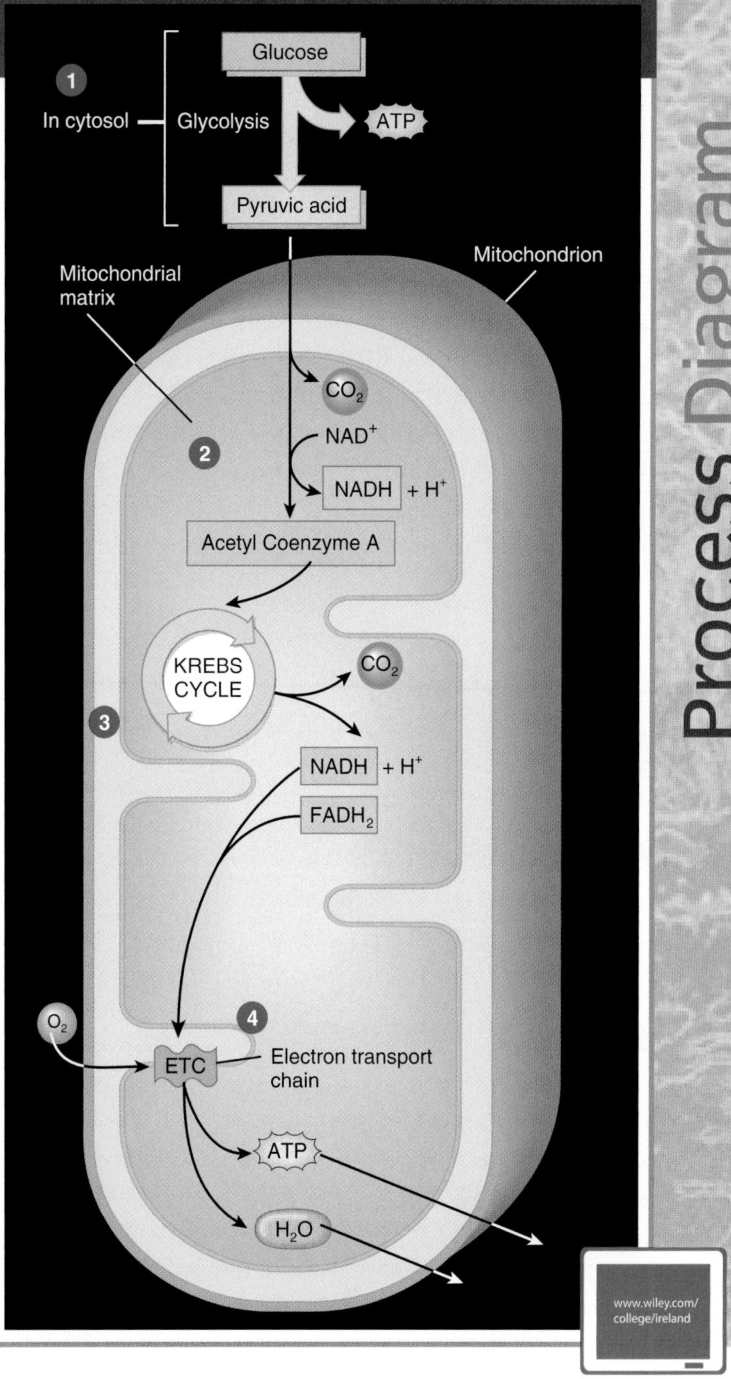

Process Diagram

Glycolysis
The enzymatic breakdown of glucose to pyruvate, occurring within the cytoplasm.

Chemiosmosis
The diffusion of hydrogen ions across a membrane, generating ATP as they move from high concentration to low.

Carbohydrate digestion, or cellular respiration, is actually a controlled burning of the glucose molecule through a series of enzymatic reactions. Burning releases energy all at once, whereas carbohydrate metabolism releases that same energy gradually. The first reaction is **glycolysis**, which converts one glucose molecule into two **pyruvate** molecules, releasing a bit of energy. Assuming oxygen is present, the pyruvates are then passed to a mitochondrion where oxidation continues. The mitochondrion completes the enzymatic burning of glucose by passing the compounds through the **Krebs cycle**, where energy-rich compounds are created, and then passing these energy-rich compounds through the **electron transport chain**. During these steps, the carbon dioxide we exhale is produced. **Chemiosmosis** within the inner membrane of the mitochondrion produces most of the ATP for the cells (FIGURE 13.1).

Lipids—fats—are a second class of macronutrient. Fats are long chains of carbon molecules, with many more carbon atoms and far fewer oxygen atoms than carbohydrates. We need a little fat in our diet; however, fats are added to many dishes in one form or another. They carry flavor and add texture to food. According to marketing tests, they coat our mouths and provide a much-craved oral gratification. Fats can be either **saturated**, meaning the carbon chain has every space occupied with hydrogens, or **unsaturated**, meaning there are some double bonds in the carbon chain (FIGURE 13.2). Because double bonds kink the long carbon chains, unsaturated fats cannot pack tightly together. Unsaturated fats, including vegetable oils, are liquid at room temperature. Saturated fats are solid at room temperature and are usually derived from animals, but coconut oil is also a saturated fat.

The American Cancer Society reports that diets high in fat can increase the incidence of cancer and gives recommendations for minimizing your risk (TABLE 13.1). They reason that high-fat diets are high in **calories**, leading to obesity, which is associated with increased cancer risks. Saturated fats may increase cancer risk, whereas other fats, such as **omega-3** fats from fish oils, may reduce the risk of cancer.

Calories
The amount of heat stored in food. One calorie is the amount of heat needed to raise the temperature of 1 kilogram of water 1 degree Celsius.

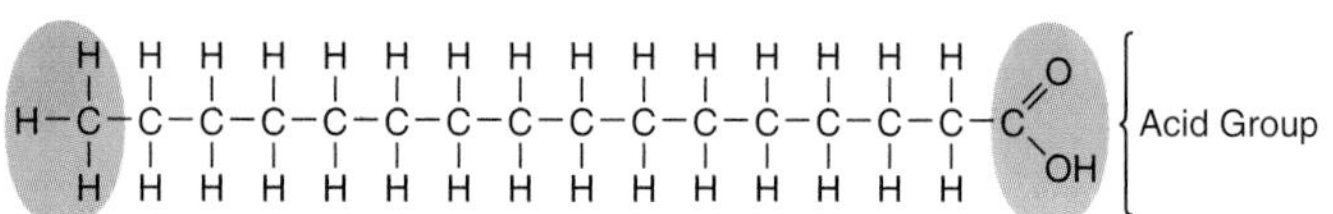

Saturated fatty acid: palmitic acid

Carbon-carbon double bonds

Monounsaturated fatty acid: oleic acid (omega-9)

Polyunsaturated fatty acid: linoleic acid (omega-6)

Polyunsaturated fatty acid: alpha-linoleic acid (omega-3)

Saturated and unsaturated fats
FIGURE 13.2

Almost all animal fats are saturated fats, especially those found in beef and dairy products. Most plants produce unsaturated fats, the notable exceptions being coconuts, cocoa butter, and palm kernel oils. For this reason, vegetable oil is liquid at room temperature, whereas butter or cocoa butter is solid.

Good and bad fats TABLE 13.1

To limit your intake of cholesterol, *trans* fat, and saturated fat:

- Trim the fat from your steak and roast beef.
- Serve chicken and fish, but don't eat the skin.
- Try a vegetarian meal once a week.
- Limit your eggs to once or twice a week.
- Choose low-fat milk and yogurt.
- Use half your usual amount of butter or margarine.
- Have only a small order of fries or share them with a friend.

To increase your intake of polyunsaturated and monounsaturated fats:

- Use olive, peanut, or canola oil for cooking and salad dressing.
- Use corn, sunflower, or safflower oil for baking.
- Snack on nuts and seeds.
- Add olives and avocados to your salad.

To increase your omega-3 intake:

- Sprinkle flax seed on your cereal or yogurt.
- Add another serving of fish to your weekly menu.
- Have a leafy green vegetable with dinner.
- Add walnuts to your cereal.

Vegan
A vegetarian who consumes only plant products, eating no animal products whatsoever.

The last class of macronutrients is **protein**. Proteins are an essential part of our daily diet because amino acids are not stored in the body. Instead of completely breaking down the amino acids of ingested proteins for energy, we usually recycle them into proteins of our own. Of the 20 amino acids that make up living organisms, we can manufacture only 11. The remaining 9 **essential amino acids** must come from our diet (TABLE 13.2A). This presents a problem only for those individuals who choose not to consume red meat.

Complete proteins, such as red meat and fish, contain all 20 amino acids. Unlike meat, no single vegetable or fruit contains all eight essential amino acids. But for those who choose to restrict meat intake, eating legumes and grains, or combining cereal with milk, will provide a full complement of amino acids. **Vegans** and vegetarians can be quite healthy, assuming they monitor their protein intake. See TABLE 13.2B for a list of food combinations that contain complementary amino acids.

MYPYRAMID IS A DIETARY GUIDELINE

Food groups are not nutrient classes. Rather, food groups are the major categories of foods: meats, dairy, breads and pastas, vegetables, and oils or fats. Each group is important to overall health, and each group has a different daily caloric intake recommendation. For example, the recommended daily allowance (RDA) for meats is quite low, at two servings per day, or 50 grams for women and 63 for men. Most Americans get far more than that in their diet.

You may be familiar with the traditional **food guide pyramid**, which suggests healthy proportions of the food groups, based on the eating habits of healthy people in the United States and around the world. The pyramid offers guidelines on the number of servings of each type of food that should be eaten each day. The bottom of the pyramid is breads, cereals, and pastas, with a recommended 6 to 11 servings per day. Fruits and vegetables are next, with a recommended 3 to 5 servings of each daily. Milk and cheeses, proteins and beans both fill the next level at 2 to 3 servings of each a day. The top of the pyramid is fats, with a recommendation that they be used "sparingly."

Essential and nonessential amino acids
TABLE 13.2A

Essential amino acids	*Nonessential amino acids*
Histidine	Alanine
Isoleucine	Arginine*
Leucine	Asparagine
Lysine	Aspartic acid (aspartate)
Methionine	Cysteine (cystine)*
Phenylalanine	Glutamic acid (glutamate)
Threonine	Glutamine*
Tryptophan	Glycine*
Valine	Proline*
	Serine
	Tyrosine*

* These amino acids are considered conditionally essential by the Institute of Medicine, Food and Nutrition Board *(Dietary Reference Intakes for Energy, Carbohydrates, Fiber, Fat, Protein and Amino Acids.* Washington, DC: National Academy Press, 2002).

Complementary proteins
TABLE 13.2B

Rice and beans
Rice and lentils
Bread with peanut butter
Tofu and cashew stir-fry
Bean burrito in corn tortilla
Hummus/chickpeas and sesame seeds
Black-eyed peas and corn bread
Tahini (sesame seeds) and peanut sauce
Trail mix (soybeans and nuts)
Rice and tofu

Atkins diet: Will eliminating carbohydrates help me lose weight?

In 1972, cardiologist Dr. Robert Atkins rocked the diet world with his book on a "diet revolution" that placed extreme emphasis on protein and fat, and discouraged eating vegetables or carbohydrates. When a revised version of the diet was published in 1992, the book became an instant best-seller. Dieters waxed rhapsodic about the quick and persistent weight loss they obtained by cutting carbs and preferring protein.

The physiology is pretty simple. Lacking carbohydrates, the normal source for glucose needed to produce ATP, the body mobilizes fat stores and converts fat into small molecules called ketones. As ketones are oxidized to produce ATP, the body enters a metabolic state called ketosis. The quick weight loss of the first week is caused by water loss, and that loss cannot be sustained. Starting the second week, weight loss slows drastically, because the only way to lose weight is to expend more energy than we take in, and Atkins is a calorie-rich diet.

As the Atkins diet sold millions of copies, it attracted a storm of criticism from researchers and organizations concerned with nutrition and obesity. For starters, they wanted to see evidence that the diet worked. Although the Atkins organization offered anecdotal evidence, independent researchers could not find proof. For example, the U.S. National Institutes of Health keeps track of people who have successfully kept off at least 13.6 kg for five years on its National Weight Control Registry (NWCR). The Registry's first study showed that its "successful losers" were eating a low-calorie, low-fat diet—the opposite of Atkins.

Other concerns focused on safety. With heart disease still the number one killer, did it make sense to promote eating fat, which gathers in the arteries and contributes to atherosclerosis? With the antioxidants in vegetables playing an ever-clearer role in health, should dieters abandon the antioxidant-laden broccoli for high-fat meat? Doctors also pointed to the known side effects of a high-protein, high-fat diet, including kidney failure, high blood cholesterol, osteoporosis, kidney stones, and cancer. The word from established medical organizations was unequivocal: "The American Heart Association does not recommend high-protein diets for weight loss."

It's hard to know whether the Atkins diet failed under a shower of expert criticism, or through the simple fact that people could not stay with it. At any rate, Atkins blazed bright and fizzled like a comet zooming across the night sky. After selling millions of books, Atkins Nutritionals, Inc., filed for bankruptcy in 2005.

But the death of the Atkins diet did not mark the death of the frenzy over being fat. The national obesity epidemic continues, and it's safe to predict that another quack diet cannot be far off. We can only hope that your knowledge of human biology will protect you from getting suckered by an unhealthy diet. In health, as in jobs, lovers, and promises in general, the same rule applies: If it sounds too good to be true, it probably is.

The U.S. Department of Agriculture recently updated its food pyramid with MyPyramid, found online at http://www.mypyramid.gov (FIGURE 13.3). Although this pyramid is more in tune with current research, it is based on the same principles as the traditional pyramid. It still recommends that we get most of our caloric value from carbohydrates and that we limit our fat intake. Rather than arrange the food groups horizontally, however, they are arranged vertically. This gives a more accurate visual picture because we require all the food groups in order to be healthy. We should not base our caloric intake on carbohydrates, but we do get a majority of our calories from them. This site is also more personal, giving recommendations for serving size and number based on age, gender, and activity level. When you submit your personal statistics to the MyPyramid Web site, you receive food intake guidelines specific to your lifestyle. Underneath your MyPyramid are a few suggestions for improving your choices within each group. The suggested amount of whole grains is listed as a portion of the carbohydrates, and the vegetable group is divided into dark greens, orange vegetables, dry beans and peas, starchy vegetables, and others. Although this is by no means an exhaustive view of good eating, it does provide enough of a base for you to begin making healthier choices.

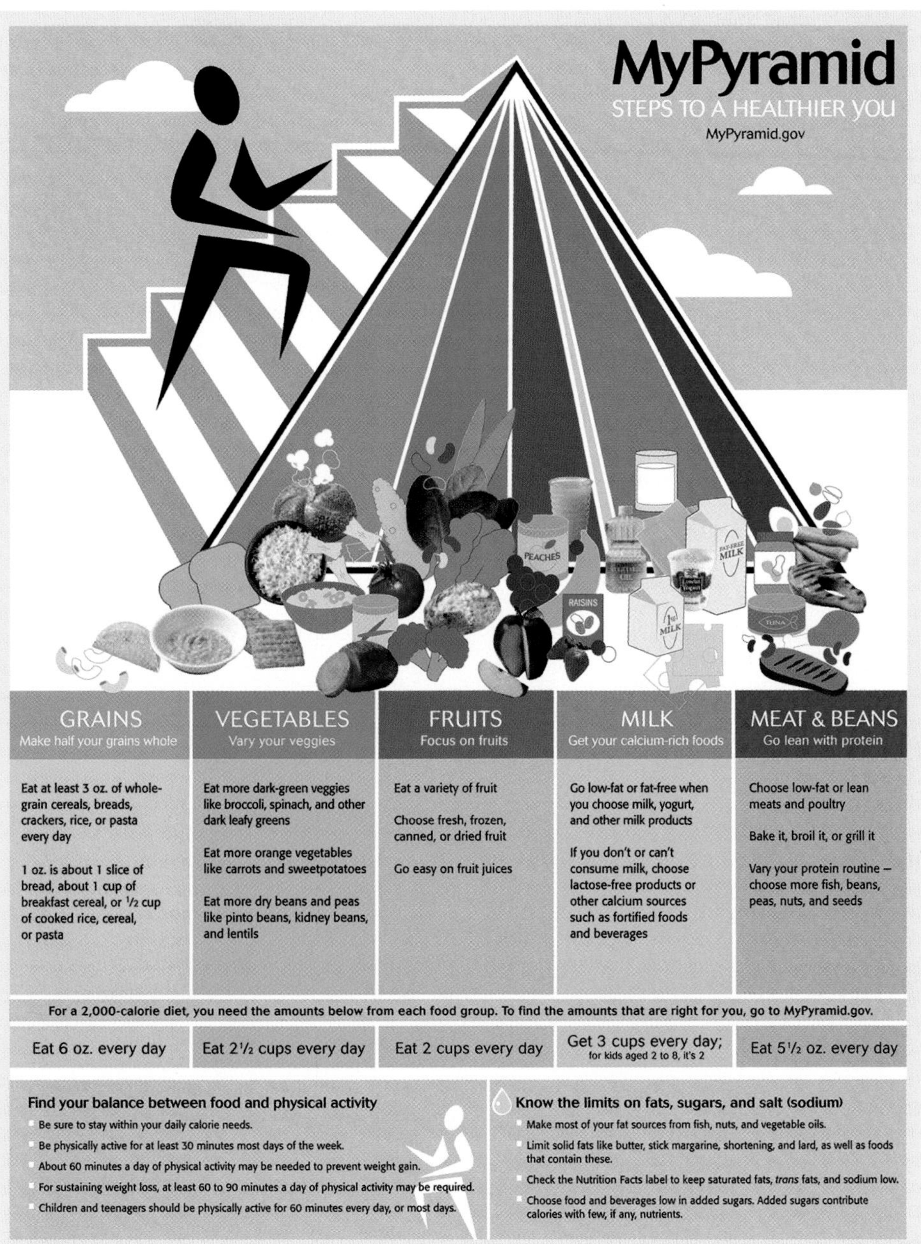

MyPyramid FIGURE 13.3

It is important to note that carbohydrates remain our best source of energy.

VITAMINS AND MINERALS ARE MICRONUTRIENTS

A healthy diet must include vitamins and minerals. Unlike macronutrients, these micronutrients are not broken down, but instead are required for enzyme function or specific protein synthesis. Vitamins are organic substances, such as thiamine, riboflavin, and vitamin A (TABLE 13.3, pp. 412–413). Minerals are inorganic substances such as calcium, zinc, and iodine (TABLE 13.4, pp. 414–415).

A healthy diet with plenty of fruit and vegetables will give you most of the necessary vitamins and minerals.

Vitamins TABLE 13.3

Vitamin	Comment and source	Functions	Deficiency symptoms and disorders
Fat-soluble	All require bile salts and some dietary lipids for adequate absorption.		
A	Formed from provitamin beta-carotene (and other provitamins) in GI tract. Stored in liver. Sources of carotene and other provitamins include orange, yellow, and green vegetables; sources of vitamin A include liver and milk.	Maintains general health and vigor of epithelial cells. Beta-carotene acts as an antioxidant to inactivate free radicals. Essential for formation of light-sensitive pigments in photoreceptors of retina. Aids in growth of bones and teeth by helping to regulate activity of osteoblasts and osteoclasts.	Deficiency results in atrophy and keratinization of epithelium, leading to dry skin and hair; increased incidence of ear, sinus, respiratory, urinary, and digestive system infections; inability to gain weight; drying of cornea; and skin sores. **Night blindness** or decreased ability for dark adaptation. Slow and faulty development of bones and teeth.
D	Sunlight converts 7-dehydrocholesterol in the skin to cholecalciferol (vitamin D_3). A liver enzyme then converts cholecalciferol to 25-hydroxycholecalciferol. A second enzyme in the kidneys converts 25-hydroxycholecalciferol to calcitriol (1,25-dihydroxycalciferol), which is the active form of vitamin D. Most is excreted in bile. Dietary sources include fish-liver oils, egg yolk, and fortified milk.	Essential for absorption of calcium and phosphorus from GI tract. Works with parathyroid hormone (PTH) to maintain Ca^{2+} homeostasis.	Defective utilization of calcium by bones leads to **rickets** in children and **osteomalacia** in adults. Possible loss of muscle tone.
E (tocopherols)	Stored in liver, adipose tissue, and muscles. Sources include fresh nuts and wheat germ, seed oils, and green leafy vegetables.	Inhibits catabolism of certain fatty acids that help form cell structures, especially membranes. Involved in formation of DNA, RNA, and red blood cells. May promote wound healing, contribute to the normal structure and functioning of the nervous system, and prevent scarring. May help protect liver from toxic chemicals such as carbon tetrachloride. Acts as an antioxidant to inactivate free radicals.	May cause oxidation of monounsaturated fats, resulting in abnormal structure and function of mitochondria, lysosomes, and plasma membranes. A possible consequence is hemolytic anemia.
K	Produced by intestinal bacteria. Stored in liver and spleen. Dietary sources include spinach, cauliflower, cabbage, and liver.	Coenzyme essential for synthesis of several clotting factors by liver, including prothrombin.	Delayed clotting time results in excessive bleeding.
Water-soluble	Dissolved in body fluids. Most are not stored in body. Excess intake is eliminated in urine.		
B_1 (thiamine)	Rapidly destroyed by heat. Sources include whole-grain products, eggs, pork, nuts, liver, and yeast.	Acts as coenzyme for many different enzymes that break carbon-to-carbon bonds and are involved in carbohydrate metabolism of pyruvic acid to CO_2 and H_2O. Essential for synthesis of the neurotransmitter acetylcholine.	Improper carbohydrate metabolism leads to buildup of pyruvic and lactic acids and insufficient production of ATP for muscle and nerve cells. Deficiency leads to: (1) **beriberi**, partial paralysis of smooth muscle of GI tract, causing digestive disturbances; skeletal muscle paralysis; and atrophy of limbs; (2) **polyneuritis**, due to degeneration of myelin sheaths; impaired reflexes, impaired sense of touch, stunted growth in children, and poor appetite.

Vitamin	*Comment and source*	*Functions*	*Deficiency symptoms and disorders*
B_2 (riboflavin)	Small amounts supplied by bacteria of GI tract. Dietary sources include yeast, liver, beef, veal, lamb, eggs, whole-grain products, asparagus, peas, beets, and peanuts.	Component of certain coenzymes (for example, FAD) in carbohydrate and protein metabolism, especially in cells of eye, integument, mucosa of intestine, and blood.	Deficiency may lead to improper utilization of oxygen resulting in blurred vision, cataracts, and corneal ulcerations. Also dermatitis and cracking of skin, lesions of intestinal mucosa, and one type of anemia.
Niacin (nicotinamide)	Derived from amino acid tryptophan. Sources include yeast, meats, liver, fish, whole-grain products, peas, beans, and nuts.	Essential component of NAD and NADP, coenzymes in oxidation-reduction reactions. In lipid metabolism, inhibits production of cholesterol and assists in triglyceride breakdown.	Principal deficiency is pellagra, characterized by dermatitis, diarrhea, and psychological disturbances.
B_6 (pyridoxine)	Synthesized by bacteria of GI tract. Stored in liver, muscle, and brain. Other sources include salmon, yeast, tomatoes, yellow corn, spinach, whole-grain products, liver, and yogurt.	Essential coenzyme for normal amino acid metabolism. Assists production of circulating antibodies. May function as coenzyme in triglyceride metabolism.	Most common deficiency symptom is dermatitis of eyes, nose, and mouth. Other symptoms are retarded growth and nausea.
B_{12} (cyanocobalamin)	Only B vitamin not found in vegetables; only vitamin containing cobalt. Absorption from GI tract depends on intrinsic factor secreted by gastric mucosa. Sources include liver, kidney, milk, eggs, cheese, and meat.	Coenzyme necessary for red blood cell formation, formation of the amino acid methionine, entrance of some amino acids into Krebs cycle, and manufacture of choline (used to synthesize acetylcholine).	Pernicious anemia, neuropsychiatric abnormalities (ataxia, memory loss, weakness, personality and mood changes, and abnormal sensations), and impaired activity of osteoblasts.
Pantothenic acid	Some produced by bacteria of GI tract. Stored primarily in liver and kidneys. Other sources include kidney, liver, yeast, green vegetables, and cereal.	Constituent of coenzyme A, which is essential for transfer of acetyl group from pyruvic acid into the Krebs cycle, conversion of lipids and amino acids into glucose, and synthesis of cholesterol and steroid hormones.	Fatigue, muscle spasms, insufficient production of adrenal steroid hormones, vomiting, and insomnia.
Folic acid (folate, folacin)	Synthesized by bacteria of GI tract. Dietary sources include green leafy vegetables, broccoli, asparagus, breads, dried beans, and citrus fruits.	Component of enzyme systems synthesizing nitrogenous bases of DNA and RNA. Essential for normal production of red and white blood cells.	Production of abnormally large red blood cells (macrocytic anemia). Higher risk of neural tube defects in babies born to folate-deficient mothers.
Biotin	Synthesized by bacteria of GI tract. Dietary sources include yeast, liver, egg yolk, and kidneys.	Essential coenzyme for conversion of pyruvic acid to oxaloacetic acid and synthesis of fatty acids and purines.	Mental depression, muscular pain, dermatitis, fatigue, and nausea.
C (ascorbic acid)	Rapidly destroyed by heat. Some stored in glandular tissue and plasma. Sources include citrus fruits, tomatoes, and green vegetables.	Promotes protein synthesis including laying down of collagen in formation of connective tissue. As coenzyme, may combine with poisons, rendering them harmless until excreted. Works with antibodies, promotes wound healing, and functions as an antioxidant.	Scurvy; anemia; many symptoms related to poor collagen formation, including tender swollen gums, loosening of teeth (alveolar processes also deteriorate), poor wound healing, bleeding (vessel walls are fragile because of connective tissue degeneration), and retardation of growth.

Minerals TABLE 13.4

Mineral	Comments	Importance
Calcium	Most abundant mineral in body. Appears in combination with phosphates. About 99% is stored in bone and teeth. Blood Ca^{2+} level is controlled by parathyroid hormone (PTH). Calcitriol promotes absorption of dietary calcium. Excess is excreted in feces and urine. Sources are milk, egg yolk, shellfish, and leafy green vegetables.	Formation of bones and teeth, blood clotting, normal muscle and nerve activity, endocytosis and exocytosis, cellular motility, chromosome movement during cell division, glycogen metabolism, and release of neurotransmitters and hormones.
Phosphorus	About 80% is found in bones and teeth as phosphate salts. Blood phosphate level is controlled by parathyroid hormone (PTH). Excess is excreted in urine; small amount is eliminated in feces. Sources are dairy products, meat, fish, poultry, and nuts.	Formation of bones and teeth. Phosphates ($H_2PO_4^-$, HPO_4^-, and PO_4^{3-}) constitute a major buffer system of blood. Plays important role in muscle contraction and nerve activity. Component of many enzymes. Involved in energy transfer (ATP). Component of DNA and RNA.
Potassium	Major cation (K^+) in intracellular fluid. Excess excreted in urine. Present in most foods (meats, fish, poultry, fruits, and nuts).	Needed for generation and conduction of action potentials in neurons and muscle fibers.
Sulfur	Component of many proteins (such as insulin and chrondroitin sulfate), electron carriers in electron transport chain, and some vitamins (thiamine and biotin). Excreted in urine. Sources include beef, liver, lamb, fish, poultry, eggs, cheese, and beans.	As component of hormones and vitamins, regulates various body activities. Needed for ATP production by electron transport chain.
Sodium	Most abundant cation (Na^+) in extracellular fluids; some found in bones. Excreted in urine and perspiration. Normal intake of NaCl (table salt) supplies more than the required amounts.	Strongly affects distribution of water through osmosis. Part of bicarbonate buffer system. Functions in nerve and muscle action potential conduction.
Chloride	Major anion (Cl^-) in extracellular fluid. Excess excreted in urine. Sources include table salt (NaCl), soy sauce, and processed foods.	Plays role in acid-base balance of blood, water balance, and formation of HCl in stomach.
Magnesium	Important cation (Mg^{2+}) in intracellular fluid. Excreted in urine and feces. Widespread in various foods, such as green leafy vegetables, seafood, and whole-grain cereals.	Required for normal functioning of muscle and nervous tissue. Participates in bone formation. Constituent of many coenzymes.
Iron	About 66% found in hemoglobin of blood. Normal losses of iron occur by shedding of hair, epithelial cells, and mucosal cells, and in sweat, urine, feces, bile, and blood lost during menstruation. Sources are meat, liver, shellfish, egg yolk, beans, legumes, dried fruits, nuts, and cereals.	As component of hemoglobin, reversibly binds O_2. Component of cytochromes involved in electron transport chain.

However, many Americans now supplement their diets with moderate levels of vitamins and minerals, just to ensure they receive what they need on a daily basis. The usual supplement taken is an over-the-counter (OTC) multivitamin supplement. These often include vitamins E, C, and A, which help remove free radicals, thereby boosting the immune system and perhaps prolonging cell life. As with anything, excess is not healthy. Taking too large a quantity of fat-soluble vitamins can cause them to build up in the liver, hampering its function.

Selected minerals are usually also found in OTC multivitamins, such as calcium, phosphorus, iodine, magnesium, and zinc, among many other micronutrients. Some minerals are found in high concentration in foods, especially prepared foods. Sodium, for example, is extremely high in most frozen and prepared foods. Because a large quantity of these convenience foods is consumed by the general population, sodium supplements are seldom advisable, because too much sodium in the diet may lead to hypertension.

By eating mostly whole grains, we obtain vitamins and minerals as well as glucose. Whole grain also provides **fiber**, which helps move feces along the large intestine and decreases the risk of colon cancer.

Mineral	Comments	Importance
Iodine	Essential component of thyroid hormones. Excreted in urine. Sources are seafood, iodized salt, and vegetables grown in iodine-rich soils.	Required by thyroid gland to synthesize thyroid hormones, which regulate metabolic rate.
Manganese	Some stored in liver and spleen. Most excreted in feces.	Activates several enzymes. Needed for hemoglobin synthesis, urea formation, growth, reproduction, lactation, bone formation, and possibly production and release of insulin, and inhibition of cell damage.
Copper	Some stored in liver and spleen. Most excreted in feces. Sources include eggs, whole-wheat flour, beans, beets, liver, fish, spinach, and asparagus.	Required with iron for synthesis of hemoglobin. Component of coenzymes in electron transport chain and enzyme necessary for melanin formation.
Cobalt	Constituent of vitamin B_{12}.	As part of vitamin B_{12}, required for erythropoiesis.
Zinc	Important component of certain enzymes. Widespread in many foods, especially meats.	As a component of carbonic anhydrase, important in carbon dioxide metabolism. Necessary for normal growth and wound healing, normal taste sensations and appetite, and normal sperm counts in males. As a component of peptidases, it is involved in protein digestion.
Fluoride	Components of bones, teeth, other tissues.	Appears to improve tooth structure and inhibit tooth decay.
Selenium	Important component of certain enzymes. Found in seafood, meat, chicken, tomatoes, egg yolk, milk, mushrooms, and garlic, and cereal grains grown in selenium-rich soil.	Needed for synthesis of thyroid hormones, sperm motility, and proper functioning of the immune system. Also functions as an antioxidant. Prevents chromosome breakage and may play a role in preventing certain birth defects, miscarriage, prostate cancer, and coronary artery disease.
Chromium	Found in high concentrations in brewer's yeast. Also found in wine and some brands of beer.	Needed for normal activity of insulin in carbohydrate and lipid metabolism.

Milled
Grain ground into flour.

Milled grains lose their fibrous, mineral-rich outer husk, diminishing their nutritional value. Simple carbohydrates, such as sucrose, usually provide energy and nothing else. These are sometimes called "empty calories" on the theory that they contribute more to weight gain than to homeostasis.

CONCEPT CHECK

What are the major macronutrients?

What is a micronutrient?

Describe the differences between the traditional food pyramid and MyPyramid.

Differentiate between vitamins and minerals.

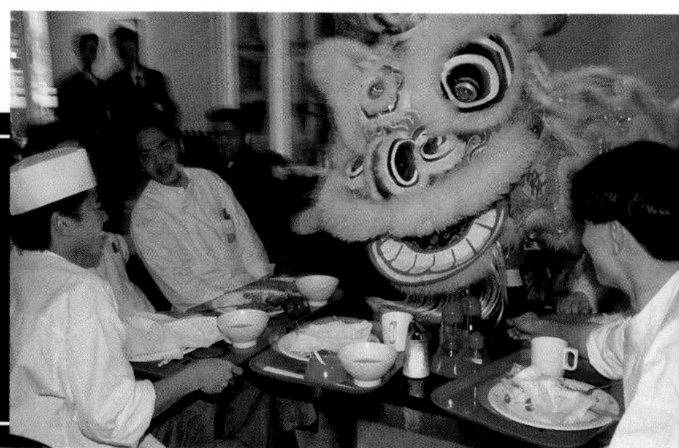

The Digestive System Processes Food from Start to Finish

Learning Objectives

Describe the general anatomy of the digestive tract.

List the digestive organs in order from mouth to anus.

Briefly explain the function of each organ in the digestive system.

THE GI TRACT REMAINS THE SAME THROUGHOUT ITS LENGTH

The digestive system is sometimes called a "tube within a tube," because it is a hollow structure with two openings that runs the length of your body. The digestive system, also called the "gastrointestinal system" or **GI tract**, begins at the oral cavity, winds through the abdominal cavity, and ends at the anus (FIGURE 13.4).

The structure of the GI tract is essentially the same along its entire length. The innermost layer is composed of a mucous membrane, or **mucosa**. This slippery, smooth layer allows ingested food to move along the tract without tearing it. Under the mucosa, the **submucosa** includes the glands, nerves, and blood supply for the tract itself. The **muscularis** gives the tract the ability to move substances lengthwise. For most of the tract, the muscularis is composed of one layer of longitudinal muscle above another layer of circular muscle (FIGURE 13.5). These layers work in unison to create the **peristaltic wave** (FIGURE 13.6) that propels food through the tube.

> **Peristaltic wave** Rhythmic muscular contractions of a tube that force contents toward the open end.

The outer layer of the GI tract, the **serosa,** is a slippery membrane that permits the tract to move inside the abdominal cavity without catching or causing discomfort. Your digestive system is always active, as muscular contractions shift, lengthen, and shorten the tube. Despite this constant movement, you normally neither see nor feel the movement.

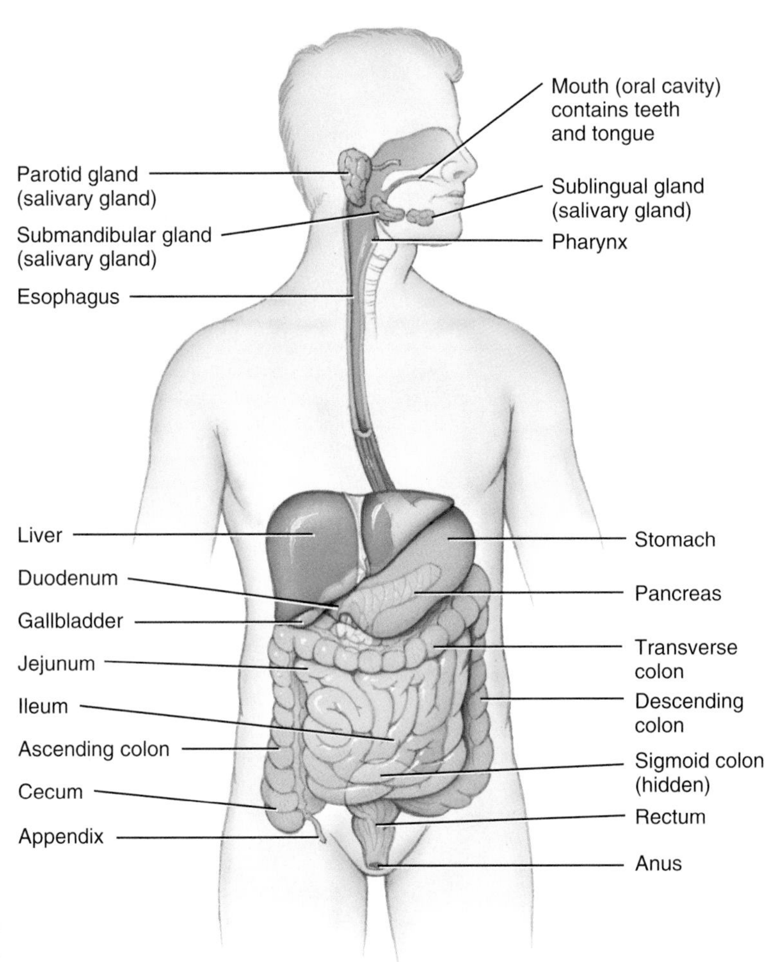

Digestive system overview FIGURE 13.4

The tubular structure of the GI tract is obvious when looking at it in its entirety. The tube begins at the esophagus, and with slight modifications, travels the length of the tract, ending at the anus. These modifications alter the function of the tract at various points, which we describe as different organs.

www.wiley.com/college/ireland

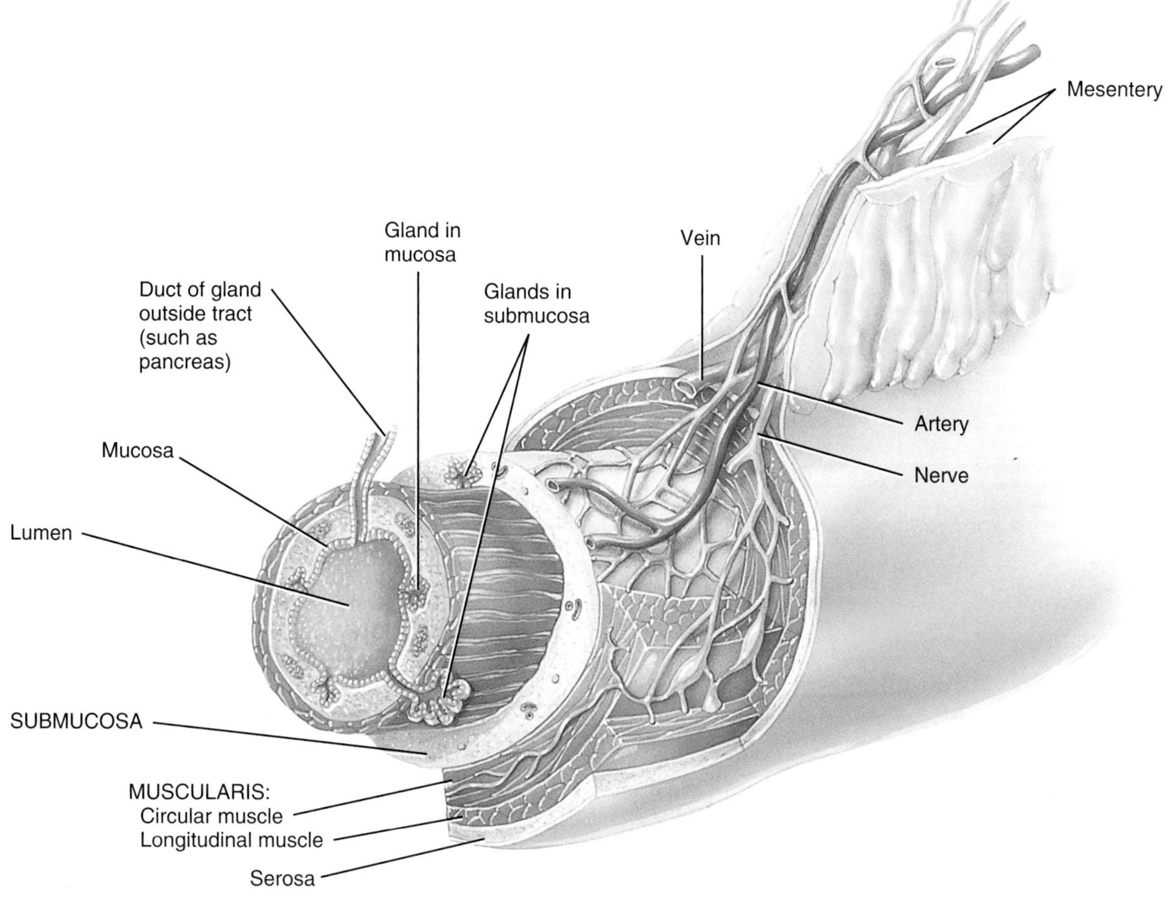

Layers of the GI tract FIGURE 13.5

The serosa allows the GI tract to move as food passes within it. The muscularis is responsible for generating the movement of the tube, whereas the mucosa and submucosa come into contact with the food and provide the blood supply and innervation for the inner lining of the tract.

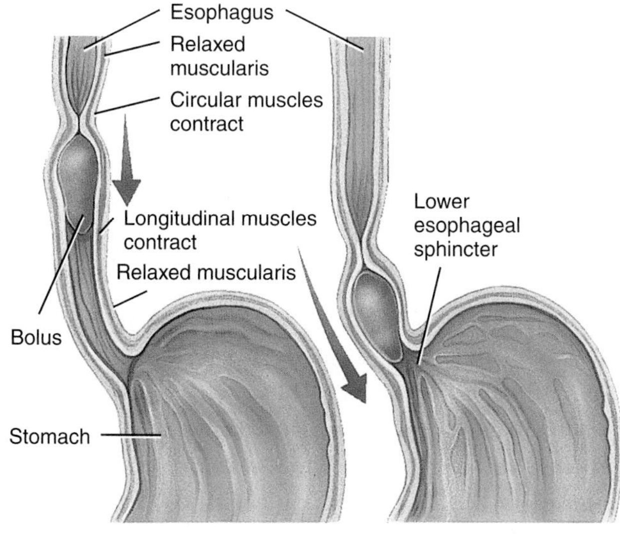

Anterior view of frontal sections of peristalsis in esophagus

Peristaltic wave generation FIGURE 13.6

The peristaltic wave is generated as you consciously swallow food. Movement of the tongue initiates the muscularis to begin a ring of contraction that is passed throughout the entire tract. Once you swallow food, the peristaltic wave travels the length of the tube; you no longer have conscious control over those smooth muscle contractions.

DIGESTION BEGINS IN THE ORAL CAVITY

The best way to understand the actions of the digestive system is to follow some food through the GI tract, starting at the oral cavity, or mouth. Think about a hot slice of pizza. How does it provide energy and nutrients? Let's follow that slice along the digestive tract, and see how the body pulls nutrients from it and how its energy is used to create adipose tissue for energy storage or ATP for immediate use.

The pizza enters the digestive tract through the oral cavity. We tear off a bite of pizza with **incisors**, and then crush it with the **molars** and **premolars**. Teeth function as cutting tools (incisors), piercing and ripping utensils (canines), or grinding instruments (molars and premolars). Although we are not born with teeth extending through the gums, they erupt soon after birth in a predictable pattern. Incisors appear first, allowing food to be bitten off, often by 8 months of age. The premolars and molars appear last, with "wisdom teeth," our final set of grinding molars, appearing sometimes as late as our mid-twenties or early thirties.

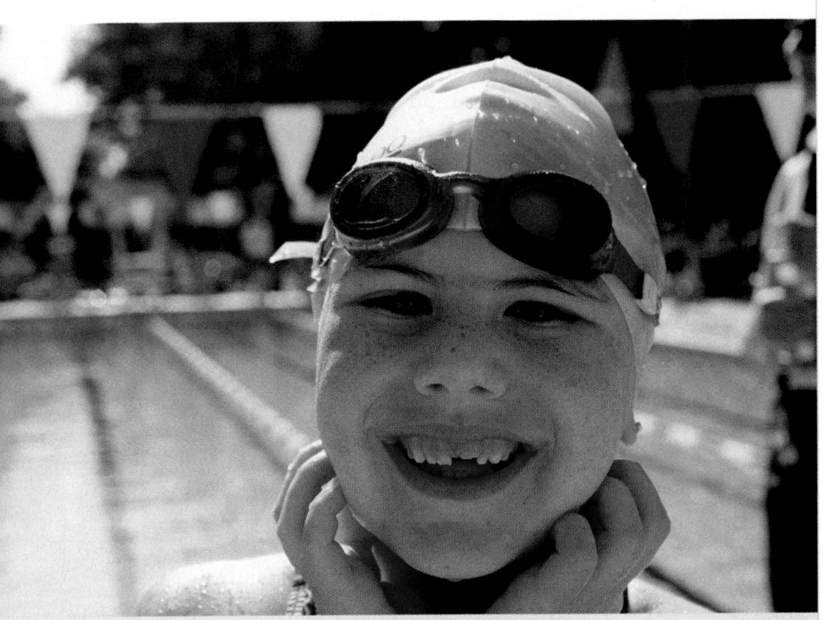

The transition from baby teeth to permanent teeth FIGURE 13.7

Teeth erupt from the gums in a specific order as we mature. They may appear more slowly in some individuals, but the pattern of eruption is predictable.

We first obtain 20 primary, deciduous, or baby teeth (FIGURE 13.7). These are replaced by our 32 permanent teeth, usually by age 21 (FIGURE 13.8).

The small bits of pizza are **macerated** with saliva. **Mechanical digestion** increases the efficiency of enzymes in the stomach and small intestine by creating small bits with a great deal of surface area where enzymes can carry out the process of **chemical digestion**.

Macerated Soaked until soft and separated into constituent parts.

Mechanical digestion Physically crushing, chopping, and cutting food.

Chemical digestion Breaking down food using enzymes that alter the chemical structure of the food.

Most people try to take good care of their teeth, with regular brushing, flossing, and visits to the dentist. Why do we bother with such dental cleanliness? Your mouth contains hundreds of species of bacteria, which live on the oral surfaces and multiply rapidly when sugar is available. These bacteria excrete wastes as they grow and metabolize. The wastes are usually acidic, and if the acid remains on tooth surfaces, it can eat through the enamel to the softer **dentin** at the center of the tooth. **Plaque** is a combination of the bacterial colonies, their bacterial wastes, leftover sugars from chewed up food, epithelial cells from the host, and saliva. Plaque begins as a sticky substance on the surfaces of the teeth but can calcify with time into the tough layer of tartar your hygienist must scrape off.

The largest increase in bacterial growth occurs 20 minutes after eating. The bacterial colonies are metabolizing the food from your last meal, growing and dividing at their highest rate. As the bacteria are multiplying rapidly, they are digesting the sugar in your mouth and creating large quantities of acidic waste. Once the food is removed, the bacterial division slows. If you do not thoroughly and routinely remove this buildup of bacteria and acid, the acid may decay the enamel on the teeth, causing cavities. A cavity does not cause pain at first, but as the acids reach farther into the tooth, they eventually hit softer tissue near the

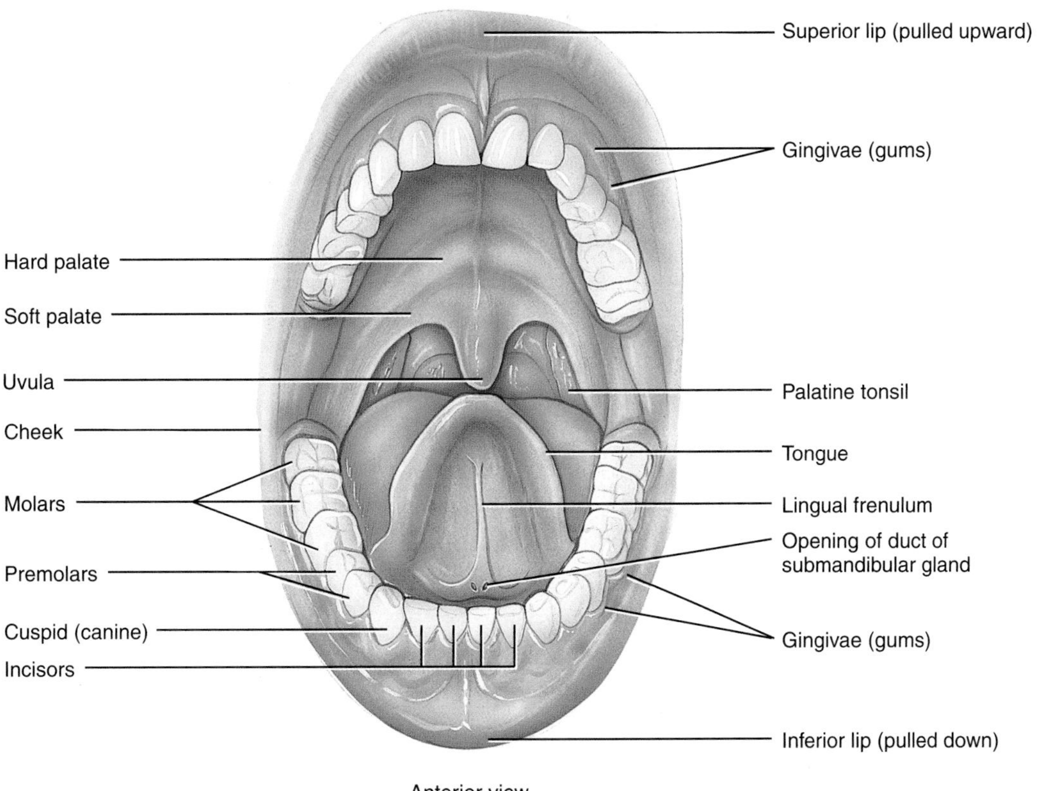

Oral cavity FIGURE 13.8

The teeth and tongue in the oral cavity are ideal for mechanical digestion. The food is rolled around with the tongue and broken into smaller pieces with the teeth.

tooth's nerve, called the **pulp**. By this time, the cavity is quite large and will require dental repair.

The recommended biannual dental cleaning is a great way to monitor plaque buildup and cavity formation. While removing plaque, the hygienist may spot any small cavities, which the dentist can repair before they destroy the pulp of the tooth. The repair process involves drilling out all rotten material and replacing it with an air-tight seal made of gold, silver alloy, or composite resin. Mercury amalgam is no longer used to fill cavities due to the health risks of mercury, which is a potent neurotoxin. Some dentists recommend replacing old amalgam fillings with composite resin, to avoid later complications.

The tongue balls things up

The tongue manipulates the now-crushed pizza into a **bolus** and positions that bolus at the back of the oral cavity so it can be swallowed. The tongue is a muscle that can move in almost any direction in the oral cavity. On its surface, keratinized epithelium covers each **papilla**, creating a rough texture to help move the slippery food into position where the teeth can masticate it. Taste buds reside along the sides of these papillae. The tongue also secretes watery mucus containing a digestive enzyme, **lingual lipase,** from **sublingual salivary glands** on its undersurface. This enzyme begins the chemical digestion of lipids by breaking down triglycerides, such as those in the pizza's cheese.

Bolus
A round, soft mass of chewed food within the digestive tract.

Papilla
Any small rounded projection extending above a surface.

Lingual
Relating to speech or the tongue.

The tonsils are the first line of defense against microbes

The **uvula** hangs from the top of the oral cavity at the back of the mouth. This structure functions as a trap door, swinging upward and closing the entrance to the nasal cavity when solid or liquid is forced to the back of the throat. The **tonsils**, at the back of the oral cavity, are your first line of defense against any microbes that may enter your mouth along with the pizza. When bacteria invade the oral cavity, the tonsils swell as they attempt to destroy the pathogen through the action of specific immune tissues.

MALT is a disease-prevention tissue

Food is rarely sterile, and yet we almost never suffer disease from ingesting it. Starting with the tonsils, the mucosa of the GI tract contain a disease-prevention tissue called **MALT** (mucosa-associated lymphatic tissue). MALT is also prevalent in the small intestine, large intestine, and appendix. These nodules of lymphatic tissue prevent pathogens from taking over the **lumen** of the digestive tract and are important for preserving homeostasis. MALT tissues represent a large percentage of the entire immune system, including about half of the body's total lymphocytes and macrophages. Without MALT, pathogens could grow within the digestive tract, penetrate the epithelial lining, and cause serious internal infections.

Although MALT is effective, it can be overrun. Bacteria ingested with food suddenly enter a warm, moist, nutrient-rich environment, and they can bloom and overwhelm the body's ability to combat them. Often the acid environment of the stomach will kill these blooming bacteria, but sometimes even that is not enough. If the bacterial colony survives the stomach, the body may flush the entire tract with diarrhea or vomiting to help the immune system rid the body of the invading bacterium.

The salivary glands aid in digestion

The **salivary glands,** located within the oral cavity, secrete watery saliva, normally in small quantities to moisten the oral mucosa. As soon as we smell the pizza, however, salivary production increases. Even the thought of food can increase saliva production. When food is in the mouth, excess saliva is needed to mix with the food and form the slippery bolus required for swallowing.

The major salivary glands are the **parotid** glands, located below and in front of the ears, and the **submandibular** glands under the tongue. The parotid glands produce watery saliva that includes some ions (sodium, potassium, chloride, bicarbonate, and phosphate) and organic substances. The submandibular glands produce thicker, ropey saliva with similar ion content but a larger concentration of mucus. When the sympathetic nervous system is active, watery secretion from the parotid glands is inhibited, whereas the sticky submandibular secretion is not. This leaves us with the familiar "cotton mouth" feeling that we associate with nervousness.

In addition to water and ions, saliva contains **lysozyme**, a **bacteriolytic** enzyme that helps destroy bacteria in the oral cavity. Another important component of saliva is **salivary amylase**, a digestive enzyme that breaks carbohydrate polysaccharides into monosaccharides. Amylase occurs in low levels in saliva and in larger quantities in pancreatic secretions. As we chew the pizza crust, salivary amylase begins breaking the large carbohydrates down into the small monosaccharides that cells can absorb further down the GI tract.

Bacteriolytic
Agent that lyses or destroys bacteria.

Mumps, a common disease of the salivary glands, causes swelling of the glands, sore throat, tiredness, and fever (FIGURE 13.9). Mumps spreads from

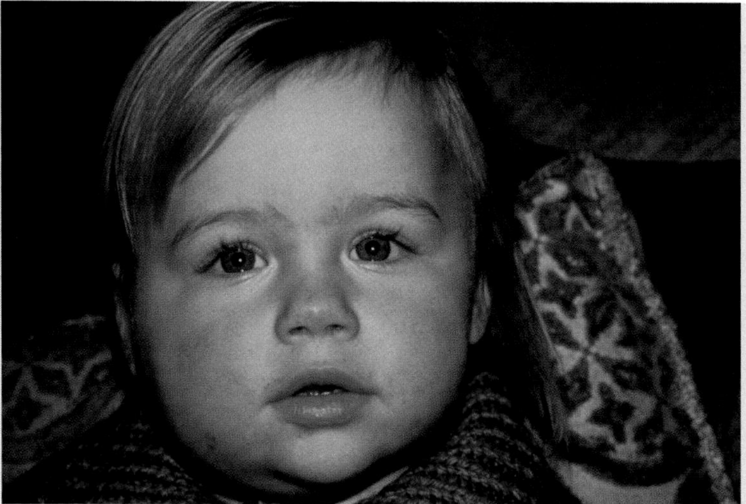

Mumps in a young child FIGURE 13.9

No longer the threat it was in the 1950s, mumps causes painful swelling of the salivary glands, most often the parotid glands. In older children and adults, mumps is far more serious, and can also cause swelling of the brain, pancreas, testes, or ovaries.

person to person in saliva, either by inhaling small bits of sneezed saliva or by sharing utensils or food contaminated with droplets of saliva. Cases of mumps have dropped steadily since 1967, when the mumps, measles, and rubella (MMR) vaccine was introduced. MMR is now part of routine infant vaccinations.

Although the mumps virus is uncomfortable in young children, it can be severe in postpubescent individuals. The virus usually settles in the parotid salivary glands, causing them to swell and feel jellylike. In adolescent males, the testes are often affected, leading to painful swelling but rarely sterility. Mumps may also cause swelling or inflammation of the pancreas, brain, meninges, or ovaries. Encephalitis (swelling of the brain tissue) can be life-threatening and may result in permanent damage. Fortunately, this is a rare complication of mumps. Hearing loss may also occur in mumps, but it is often temporary. As we vaccinate more infants, mumps could become a disease of the past, following the same pattern as German measles and polio.

Deglutition occurs in stages Swallowing, or **deglutition**, occurs as the bolus of macerated, saliva-mixed pizza is moved to the back of the throat. The tongue positions the bolus at the opening to the esophagus, where you consciously decide to swallow the pizza. This is the last muscular movement you control until the pizza has worked its way to the other end of your GI tract. The tongue is composed of voluntary, consciously controlled skeletal muscle. The muscularis of the GI tract is smooth muscle, controlled by the autonomic nervous system. At the very end of the tract, the anal sphincter is again skeletal muscle.

Swallowing has three stages. During the **voluntary stage**, you consciously swallow the pizza. During the **pharyngeal stage** (FIGURE 13.10), the bolus involuntarily passes through the pharynx. The trachea is closed to allow the bolus to pass the larynx and enter the esophagus. It is here that the uvula covers the nasal opening and the larynx moves upward against the epiglottis. The epiglottis covers the opening to the

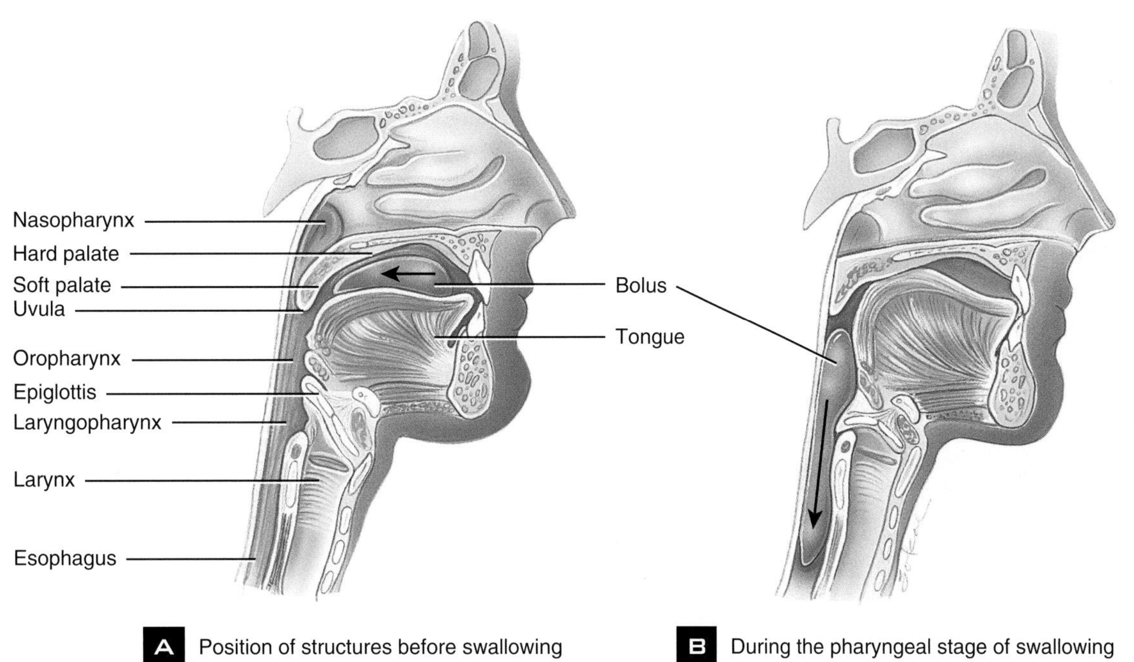

Swallowing and the pharynx FIGURE 13.10

As the bolus of food is swallowed, the larynx moves up, in turn shifting the position of the epiglottis. The bolus of food then slides past the larynx and on to the esophagus. The wave of contraction begun here continues through the entire system, pushing this mouthful into the stomach and eventually on to the remaining organs of the GI tract.

respiratory system, and the bolus slides back toward the esophagus instead of dropping into the respiratory system. Talking while eating can cause the epiglottis to spasm because it must be opened to allow air to escape in order to vocalize, but must be closed to prevent the bolus from sliding into the respiratory tract. Because the epiglottis cannot be opened and closed at the same time, it spasms. Choking can result, and we may require assistance to remove the misplaced bolus.

THE ESOPHAGUS CONNECTS THE ORAL CAVITY WITH THE STOMACH

The esophagus is a collapsible 20- to 25-centimeter long conduit that connects the oral cavity with the stomach (FIGURE 13.11). Once the bolus of pizza arrives at the top of the esophagus, a peristaltic wave begins. In this third stage of swallowing, the **esophageal stage**, food moves through the esophagus into the stomach via peristalsis. This wave will push the bolus along the esophagus in a controlled manner (neither food nor drink free-fall into the stomach). The esophagus terminates at its lower end with a sphincter muscle. A sphincter muscle is a circular muscle that closes off a tube, functioning like a rubber band pulled tightly around a flexible straw. They appear many times along the GI tract, dividing one organ from the next. The **lower esophageal sphincter** (LES) at the base of the esophagus opens as the pizza bolus touches it, dropping the bolus into the upper portion of the stomach. You can listen to water traveling through the esophagus and hitting the LES if you have a stethoscope. Place the bell of the stethoscope near your xyphoid process and swallow a mouthful of water. You should be able to count to 10, then hear the water splash against the lower esophageal sphincter. If you are lucky, you might hear the water splash again as it enters the stomach when the LES opens.

The esophagus runs right through the diaphragm at the **esophageal hiatus**. Occasionally a portion of the upper stomach can protrude through this opening, resulting in a hiatal hernia. This condition can be painful and often requires medical intervention.

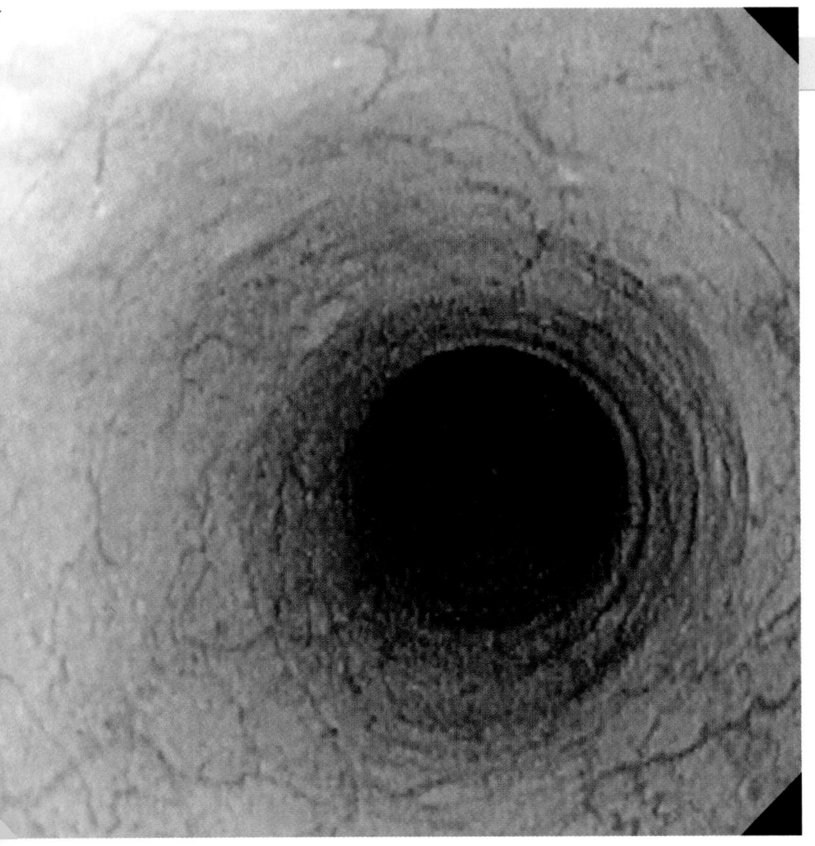

The esophagus FIGURE 13.11

The esophagus is a straight tube, most representative of the four layers of the GI tract. There are no modifications of the tract in this organ, which ends with the lower esophageal sphincter.

THE STOMACH PUTS FOOD TO THE ACID TEST

The next organ the pizza encounters in the digestive system is the **stomach**, a J-shaped organ that lies beneath the esophagus. The stomach is separated from the esophagus and the small intestine by two sphincter muscles. The lower esophageal sphincter is the upper boundary of the stomach, and the **pyloric sphincter** marks the lower end of the stomach. The pyloric sphincter, the strongest sphincter muscle of the digestive tract, opens to allow **chyme** to enter the small intestine only when chemically ready. This sphincter is so powerful that it can cause projectile vomiting in infants. The stomach contracts forcefully to push the food into the small intestine, but the pyloric sphincter remains closed until the chyme is fluid enough to be passed on. If the pyloric sphincter refuses to open, the contents of the stomach are instead ejected through the weaker lower esophageal sphincter, leaving the body at impressive speed.

> **Chyme**
> The thick, partially digested fluid in the stomach and small intestine.

Histologically speaking, the stomach is "the pits" The typical structure of the gastrointestinal tract undergoes modification at the stomach (FIGURE 13.12). The muscularis is usually composed of two layers of muscle, one longitudinal and one circular. The stomach has a third layer of muscle, called the **oblique** layer. The function of the stomach is to churn and mix the bolus of pizza with the acid environment of the stomach and begin protein digestion. The oblique layer helps this churning and mixing. Because the stomach is a holding area for ingested food, it must be able to expand. The walls of the stomach contain folds, or **rugae**, that permit expansion somewhat like a deflated punching ball.

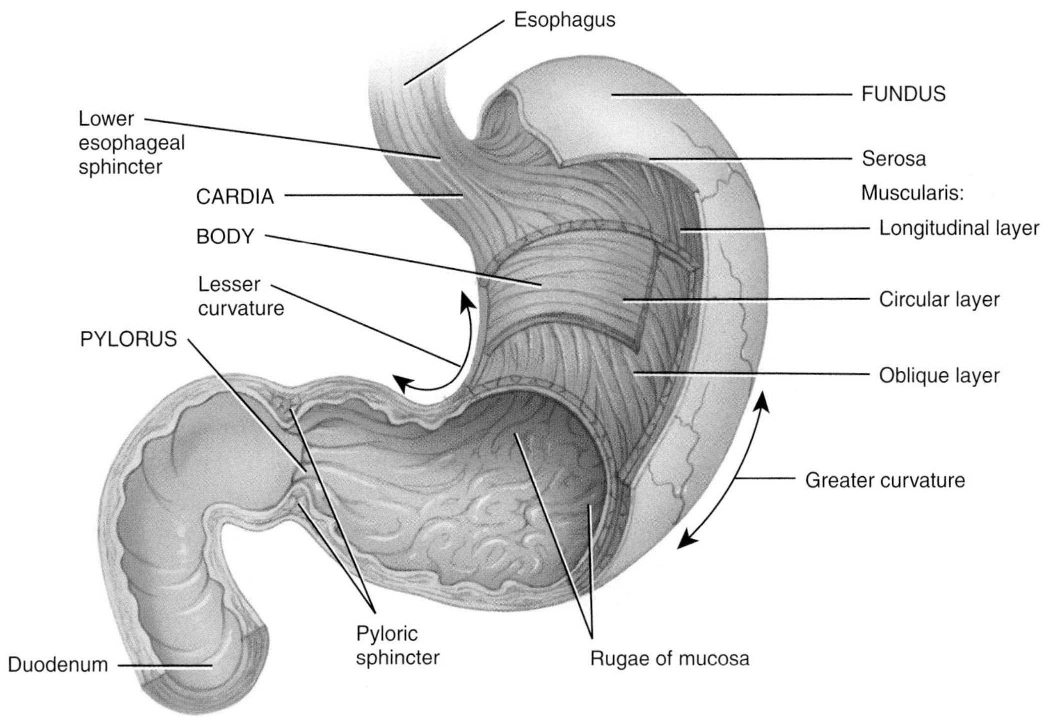

The stomach FIGURE 13.12

A final modification of the stomach is due to the chemical environment in the organ, where the pH is only 2. Such high acidity breaks down large macromolecules and destroys many microbes, but it can also harm the stomach lining. Furthermore, the stomach also secretes enzymes that digest protein, which is what the stomach walls are composed of. Therefore, the stomach must be protected from its own contents. The stomach does this by producing a protective layer of thick, viscous, alkaline mucus. Nowhere else does the digestive tract need, or produce, such a mucus coating.

Gastric
Indicates a relationship to the stomach.

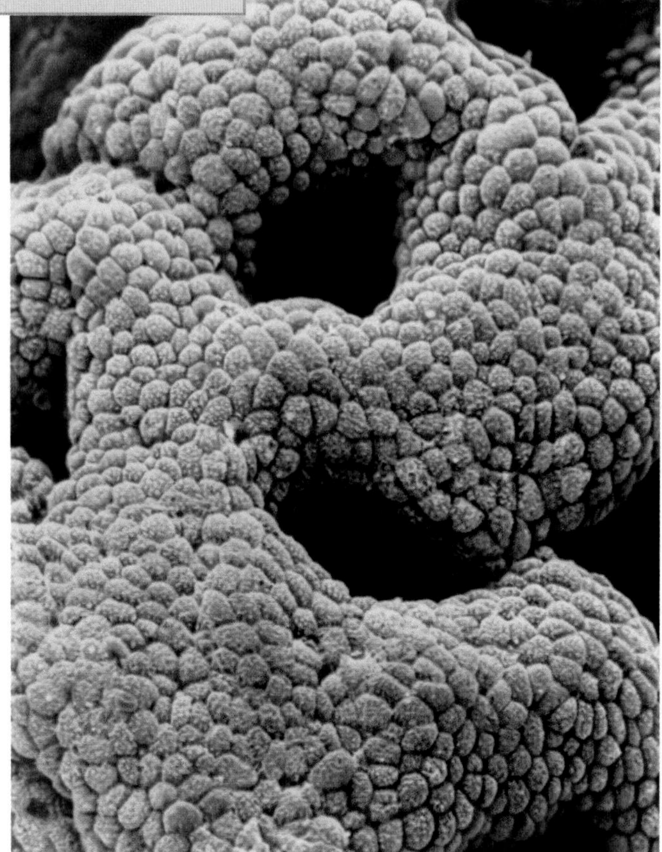

Gastric pits FIGURE 13.13

Gastric pits are composed of chief cells and parietal cells. These cells are responsible for creating the specialized environment of the stomach.

The walls of the stomach contain **gastric** pits, which secrete 2 to 3 quarts of **gastric juice** each day (FIGURE 13.13). These pits are composed of **chief** cells and **parietal** cells. The chief cells secrete **pepsinogen** and **gastric lipase**. Pepsinogen is an inactive precursor of the enzyme **pepsin**, which digests proteins, and therefore must be secreted in inactive form. (If pepsin itself were produced in stomach cells, it would digest the proteins of those cells.) Pepsinogen forms pepsin only under pH 2. The parietal cells produce **hydrochloric acid** and **intrinsic factor**. The hydrochloric acid is responsible for the acidic pH of the stomach, which both activates pepsin and kills microbes. Intrinsic factor is necessary for the absorption of vitamin B_{12}, a micronutrient that helps produce blood cells. Although intrinsic factor is produced in the stomach, it is active in the small intestine.

As the pizza is churned in the stomach, **gastric lipase** will continue the chemical breakdown of fats that began in the mouth. This enzyme specializes in digesting short-chian fatty acids such as those found in milk, but works at an optimum pH of 5 or 6. In adults, both gastric lipase and lingual lipase have limited roles.

In the stomach, the pizza bolus is converted to a pasty, liquid chyme. Pepsinogen is converted to pepsin and digests the proteins of the tomato sauce and the cheese. The low pH assists in denaturing proteins and breaking down the remaining macromolecules, providing an easy substrate for digestion in the small intestine.

The stomach is an active organ. As the bolus of food reaches the stomach, small **mixing waves** are initiated. These waves occur every 15 seconds or so and help to break up the pizza. Even with these mixing waves, the pizza may stay in the **fundus** of the stomach for as long as an hour before being moved into the body of the stomach. There the pizza mixes with the gastric secretions and becomes soupy and thin. The mixing waves of the stomach become stronger, intensifying as they reach the pyloric sphincter. With each wave, a small portion of the chyme is forced through the pyloric sphincter and into the small intestine. The rest of the chyme washes back toward the body of the stomach to be churned further with the next mixing wave.

Fundus
The bottom portion of any hollow organ.

Phases of gastric digestion FIGURE 13.14

The activation of the stomach includes three phases.

1. **Cephalic phase.** In the first phase, thoughts of food and the feel of food in the oral cavity stimulates increased secretion from the gastric pits. The stomach also begins to churn more actively in preparation for the incoming food.
2. **Gastric phase.** When the bolus reaches the stomach, the second phase of gastric digestion begins. Here the stomach produces gastrin as well as continuing the production of pepsin and HCl. Gastrin aids in stimulation of the gastric pits, providing a feedback system that speeds digestion. Impulses from the stomach also go back to the brain, maintaining contact with the nervous system.
3. **Intestinal phase.** In the final phase of gastric digestion, the chyme begins to leave through the pyloric sphincter. As the chyme leaves the stomach, gastrin production decreases, the impulses to the brain indicate a lessening of chyme, and the brain begins to slow the stimulation of the gastric pits. At the same time, hormones from the beginning portion of the small intestine initiate activation of the small intestine.

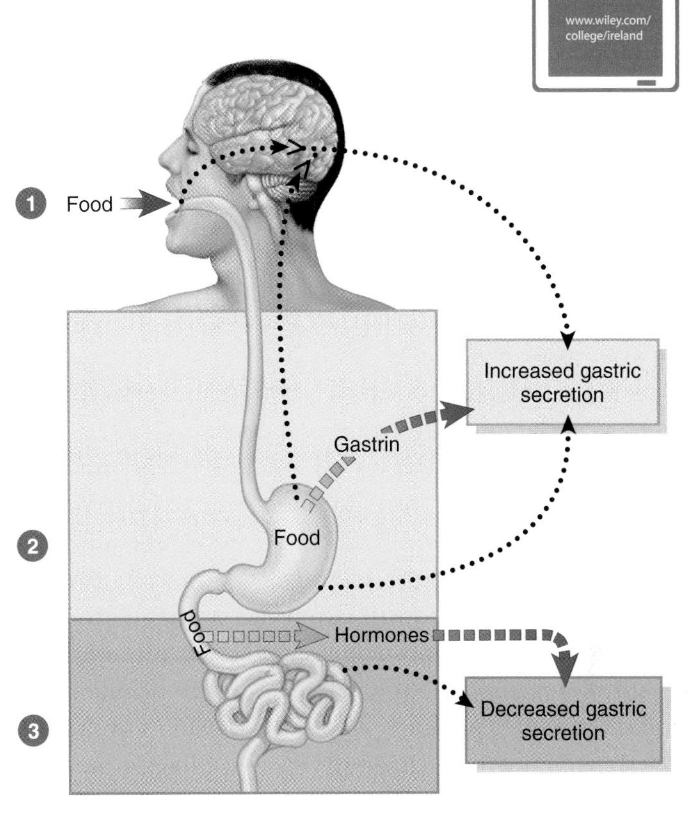

Process Diagram

Gastric digestion includes three phases

Digestion occurs in three phases in the stomach (FIGURE 13.14). During the **cephalic phase**, digestion consists of reflexes initiated by the senses, as the name implies. This phase started when you ordered the pizza, intensified as you got out the utensils to eat it, and peaked as you smelled the pizza after delivery. The scents and sounds associated with eating stimulate specific portions of the medulla oblongata, which in turn trigger secretion of the gastric pits. The parasympathetic nervous system is activated, increasing stomach movement. Interestingly, these reflexes can be dampened by stimulation of the sympathetic nervous system. Anger, fear, or anxiety opposes the parasympathetic nervous system, shutting down the cephalic phase and reducing your feelings of hunger.

Once food enters the stomach, stretch receptors and chemoreceptors are activated, initiating the **gastric phase**. Hormonal and neural pathways are set in motion, causing an increase in both gastric wave force and secretion from the gastric pits. As chyme is pushed past the pyloric sphincter, stomach volume decreases and stretch receptors begin to relax. This in turn diminishes the intensity of the gastric phase.

The final phase of gastric digestion is the **intestinal phase**. As chyme passes through the pyloric sphincter, intestinal receptors are stimulated. These receptors inhibit the actions of the stomach, causing it to return to rest. At the same time, these receptors stimulate digestion in the small intestine.

Once in the small intestine, the chyme itself stimulates the release of hormones. Chyme containing glucose and fatty acids, such as the chyme from the pizza, causes the release of **cholecystokinin** (CCK) and **secretin**. CCK inhibits stomach emptying, whereas secretin decreases gastric secretions. Both of these also affect the **liver, pancreas**, and **gallbladder**, the accessory organs of the gastrointestinal tract. The combined action of these hormones holds the pizza in the stomach for a prolonged period, ensuring the pizza is sufficiently broken down, despite its high level of hydrophobic fats.

After 2 to 4 hours, the stomach has emptied, and all the chyme has entered the small intestine. Because the pizza has a high fat concentration, it will move rather slowly through the stomach, taking closer to 4 hours. Had you eaten stir-fried vegetables with their much lower fat content, your stomach would have emptied much more quickly, leaving you feeling hungry again after just a few hours.

Sometimes food in the stomach does not "agree" with the stomach because it contains bacteria or toxins that irritate the stomach lining. This situation may cause vomiting. Although not an easy task from a physiological standpoint, reversing the peristaltic wave and churning the stomach violently while holding the pyloric sphincter closed will expel the stomach contents. The esophageal sphincter is weaker than the pyloric sphincter and will open first when the stomach contents are under pressure. The entire contents of the stomach then return through the esophagus and the mouth. The acidity of the stomach is not buffered, causing some burning as the fluid passes the mucous membranes of the mouth and throat. Repeated vomiting can be detrimental to the lining of the mouth as well as the tooth enamel. In addition, replacing the hydrogen ion concentration in the stomach can deplete the hydrogen content of the blood, leading to electrolyte imbalances.

THE SMALL INTESTINE COMPLETES THE NUTRIENT-EXTRACTION PHASE

Once in the small intestine, the pizza's nutrients are finally ready for absorption. This organ is the only portion of the GI tube where nutrients are taken into the cells. Prior to reaching the small intestine, the food was cut up, broken down, and denatured. Some enzyme activity was initiated to break down large macromolecules. Here in the small intestine, the nutrients from the pizza are finally absorbed into the body.

The small intestine has three regions: the **duodenum**, the **jejunum**, and the **ileum**. The duodenum is the shortest of the regions, extending approximately 25 centimeters from the pyloric sphincter. The name duodenum means 12, reflecting the fact that the region is approximately 12 fingers long. The jejunum encompasses the next meter or so. Jejunum means empty, and this region is characteristically empty during autopsy. The longest portion, the ileum, is about 2 meters long. The entire length of the small intestine is 3 meters, making it the longest digestive organ. This structure is packed into the abdominal cavity by twisting and winding around the central **mesenteries**.

Small intestine FIGURE 13.15

The small intestine is characterized by its velvet-like mucosa. The sole purpose of this organ is to absorb nutrients, requiring a large surface area. The mucosa is folded, and cells are lined with microvilli and even covered in individual eyelash-like extensions to provide as much surface area as possible.

Mesenteries
Folds in the lining of the abdominal cavity that help to secure the digestive organs.

How large is the surface of the small intestine?

Within the small intestine, the mucosa is shaped into permanent circular folds, which add important surface area to the organ (FIGURE 13.15). Not only do these folds increase absorption, they also force the chyme to move in spiral fashion, which creates a longer pathway through the intestine, allowing more time to absorb nutrients.

The small intestine has an interesting histology. Because the whole point of the organ is to provide a surface area for absorption, the small intestine has many microscopic projections. The mucosa has finger-like extensions, or **villi**, each one approximately 0.5 to 1 mm long (FIGURE 13.16). These villi give the inner surface of the small intestine the look and feel of velvet. Areolar connective tissue is located at the center of each villus. This connective tissue supports an arteriole, a venule, a blood capillary network connecting the two, and a **lacteal**.

Beyond the villi, the small intestine also has **microvilli** on each **apical membrane** of the small intestinal mucosa. These hairlike projections of the cell membrane increase the cell's surface area. The microvilli are small and difficult to resolve under a light microscope, where they look like a fuzzy line, not individual structures. The entire surface of the cell is called a **brush border**. Through an electron microscope, scientists have discovered even smaller projections on the surface of these brush borders, which again increase surface area.

Apical membrane
Membrane at the free end, or top, of the intestinal cells.

The walls of the small intestine are also dotted with intestinal glands, which secrete intestinal juice to help digestion. The small intestine has an abundance of MALT, in the form of **Peyers patches**. These nodules of lymphatic tissue are akin to tonsils and embedded in the intestinal walls (FIGURE 13.17).

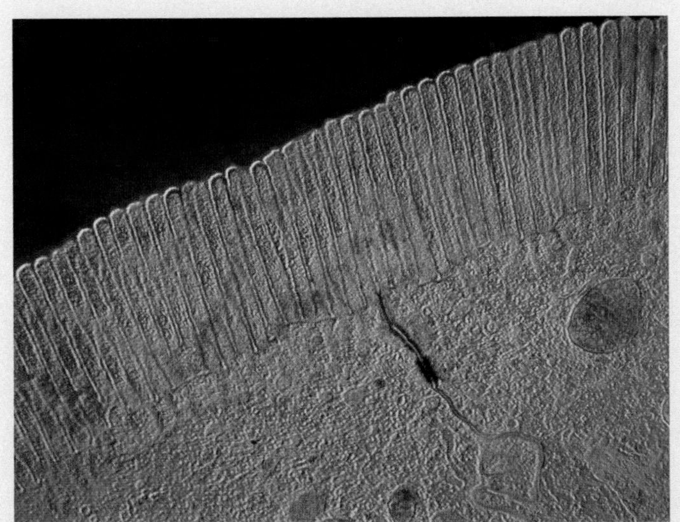

Microvilli in the small intestine

FIGURE 13.16

The cells of the small intestine are the only nutrient-absorbing structures in the digestive system. The larger their surface area, the greater the chances that nutrients taken in with the original food will be absorbed before they pass through the small intestine. The incredibly extensive surface area of these cells allows fats and nutrients to diffuse or be actively absorbed at a high rate.

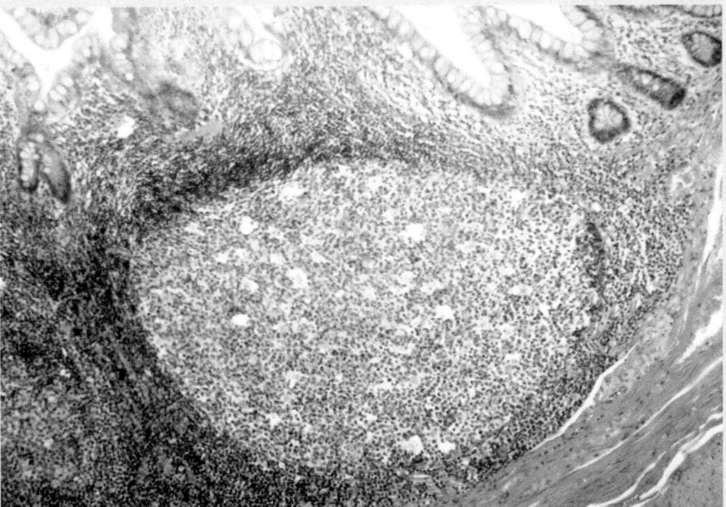

Intestinal wall with Peyers patches

FIGURE 13.17

Peyers patches are an important part of the immune system, protecting the lumen of the digestive tract from bacterial invasion. If even one bacterium escaped the stomach, it could potentially cause serious problems here in the nutrient-rich, warm, moist environment of the small instestine. It is the job of these Peyers patches to prevent these problems.

Digestion occurs in the small intestine Both mechanical and chemical digestion occur in the small intestine. Mechanically, the peristaltic wave is modified into **segmentations** and **migrating motility complexes**. Segmentations are localized mixing contractions that swirl the chyme in one section of the intestine. They allow the chyme to interact with the walls of the small intestine but do not move it along the tract. Migrating motility complexes move the chyme along the length of the small intestine. These movements strengthen as the nutrient level in the chyme decreases.

When soupy chyme enters the duodenum, digestion of proteins, lipids, and carbohydrates has just begun. **Pancreatic juice** is added to the chyme as it enters the small intestine, adding a suite of digestive enzymes that are specific for different macromolecules. For example, **sucrase, lactase, maltase**, and **pancreatic amylase** all digest carbohydrates.

Pancreatic juice The fluid produced by the pancreas and released into the small intestine.

The pH buffers of the pancreatic juice immediately bring the pH of the chyme from 2 back to 7 in the small intestine, protecting the lining of the duodenum. Raising the pH up to 7 protects the walls of the small intestine; however, it renders pepsin inactive. Protein digestion continues using **trypsin, chymotrypsin, carboxypeptidase**, and **elastase**, all secreted from the pancreas. Protein digestion is completed on the exposed edges of the intestinal cells themselves, using the enzymes **aminopeptidase** and **dipeptidase**.

In adults, most lipid digestion occurs in the small intestine because lingual lipase and gastric lipase are barely effective in adults. **Pancreatic lipase** is the main enzyme causing the breakdown of fats in adults, removing two of the three fatty acids from ingested triglycerides.

In the cells of the small intestine, carbohydrates, short-chain fatty acids, and amino acids are absorbed from the chyme and transported to the capillaries of the lacteal (FIGURE 13.18). Absorbed triglycerides are too large to pass directly into the bloodstream. They are converted to **chylomicrons** and transported in the lymphatic capillary of the lacteal. From here, the fats flow with lymph to the subclavian vein. Once in the bloodstream, **lipoprotein lipase** breaks chylomicrons down to short-chain fatty acids and glycerol.

Chylomicrons Small lipoproteins carrying ingested fat from the intestinal mucosa to the liver.

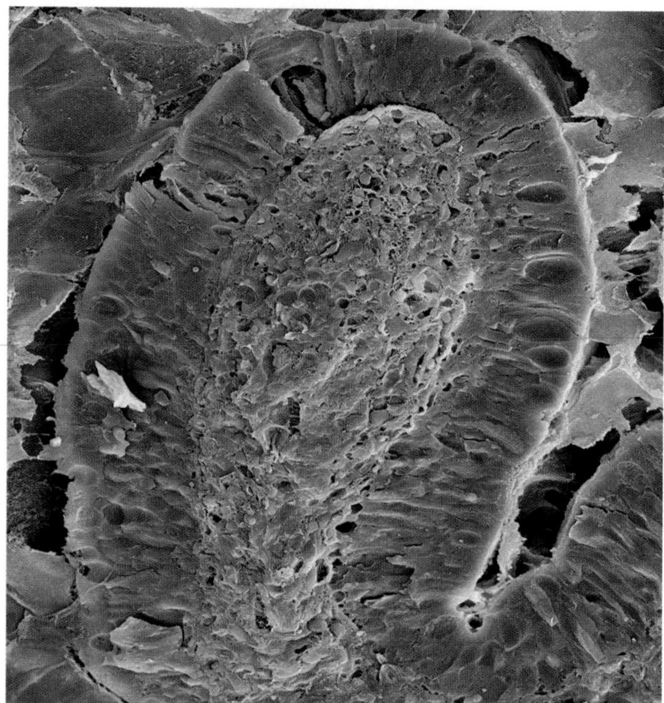

A villus FIGURE 13.18

Nutrients absorbed by the cells of the intestinal wall are passed through the cell and into the capillary network or the lymphatic vessel of the lacteal. Lacteals include blind-ended lymphatic capillaries that permit absorption of ingested fats. Nutrients are usually absorbed directly into the lacteal capillary system, part of the systemic circulatory system.

You have probably heard of LDL (low-density lipoprotein) and HDL (high-density lipoprotein) cholesterol, but you may not be aware of their functions. Transporting insoluble fats through the aqueous bloodstream requires a protein carrier. Initially this carrier is LDL, which is essentially a small protein carrying a large fat droplet, hence the term *low density*. LDL is often called "bad" cholesterol because higher levels of LDL in your blood indicate a greater proportion of large fat droplets being carted from the lacteals to the liver for degradation. LDL can often "drop its load" along the way, leading to plaque formation and atherosclerosis.

HDL, on the other hand, is sometimes called "good" cholesterol. High-density lipoprotein uses a large protein to carry a small fat. HDL carries small fats from storage to the muscles and liver, where they are metabolized. Because of the opposing roles of these two carrier molecules, the LDL-to-HDL ratio helps assess cholesterol levels and heart disease risk.

ACCESSORY ORGANS HELP FINISH THE JOB

Although the gastrointestinal tract provides both a location for nutrient digestion and the surface required to absorb those nutrients, it cannot complete the job alone. Along the length of the tract several accessory organs assist in digestion, including the pancreas, liver, and gallbladder.

The pancreas is an enzyme factory

The pancreas functions as an exocrine gland in the digestive system, producing enzymes that are released via the **pancreatic duct**. Almost all of the enzymes that act in the small intestine are made in the pancreas. Pancreatic juice also buffers the acidity of the chyme as it leaves the stomach. The small intestine does not have the protective layer of mucus found in the stomach, so it has no protection from the corrosive pH 2 solution being released from the pyloric sphincter. The pancreas secretes pancreatic juice into chyme immediately as it enters the duodenum, largely neutralizing the chyme to safeguard the duodenum from acid burns.

In addition to secreting digestive enzymes into the digestive tract, the pancreas is also responsible for secreting hormones into the bloodstream. The pancreas makes insulin and glucagon, which are responsible for regulating glucose uptake by the cells. Insulin stimulates glucose uptake, whereas glucagon causes glucose to be released into the bloodstream by those muscle and liver cells sequestering it. These hormones will be covered in Chapter 15.

Ulcers are holes in the GI tract

Ulcers are open wounds that remain aggravated and painful instead of healing. A gastric or duodenal ulcer is such a wound in the lining of the GI tract (FIGURE 13.19). Gastric ulcers occur in the stomach, whereas duodenal ulcers are in the duodenum of the small intestine.

The mucous lining that normally protects the stomach from digestion must be compromised for an ulcer to develop. This can happen when alcohol or aspirin enters the stomach because these compounds can degrade the mucous lining. Aspirin labels direct you to take them with a full glass of water so that the pill is washed through the stomach, or dissolved rather than left sitting on the mucous layer. If the mucous layer is worn away, acidity in the lumen begins to burn the stomach lining, and pepsin will digest proteins of the stomach cells, creating an ulcer. Although in the past ulcers were commonly blamed on stress that caused the release of excess stomach acid, many gastric ulcers are actually caused by infection with *Helicobacter pylori,* a spiral bacterium that thrives in the highly acidic stomach. People who are susceptible to this bacterium often develop gastric ulcers due to bacterial colonies that live on the mucus. Rather than counsel these patients to reduce their stress level, the old-time ulcer treatment, they are given antibiotics to cure their ulcers.

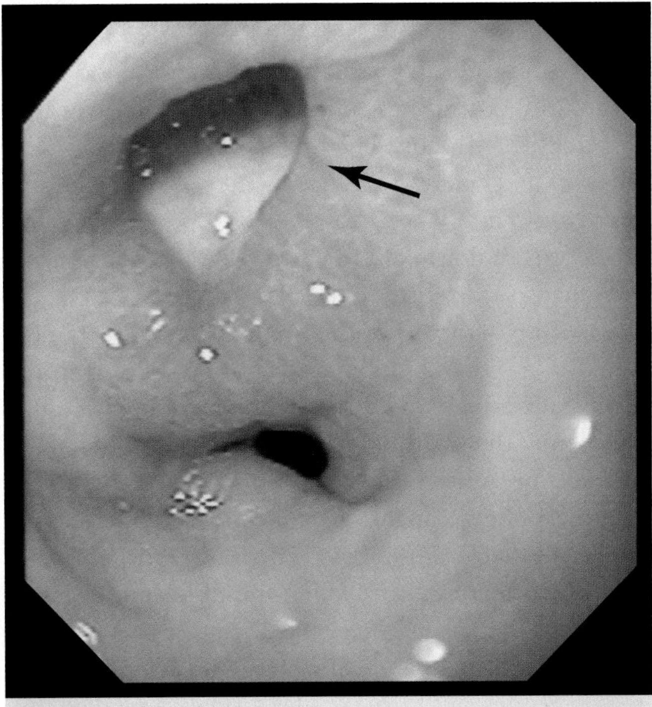

Gastric ulcer FIGURE 13.19

The liver detoxifies what we add to the bloodstream

The liver is the largest organ aside from the skin and usually weighs about 1,450 grams. The liver has two lobes, and sits mostly on the right side of the body. Within the lobes of the liver, the **hepatocytes** are arranged in **lobules** (FIGURE 13.20), designed to allow maximum contact between hepatocytes and venous blood. The lobules monitor blood collected from the small intestine, adding and subtracting materials to maintain fluid homeostasis.

Hepatocytes
Liver cells (*hepato* = liver; *cyte* = cell).

The liver is served by a **portal system**. The veins of the small intestine drain into the liver, where they break into capillaries again before being collected into a larger vein and returned to the heart. Blood flows through the digestive organs, travels from arteries to capillaries to veins, and proceeds on to the liver, where it moves back to capillaries, then to the veins that return to the heart. This portal system gives the hepatocytes access to the blood coming from the small intestine. This blood includes all absorbed compounds, and nutrients, as well as toxins, from the small intestine. The hepatocytes must cleanse the blood before it reaches the heart, removing toxins and storing excess nutrients, such as iron, and fat-soluble vitamins such as A, D, and E.

Cholesterol, plasma proteins, and blood lipids are manufactured in the hepatocytes. The liver also monitors the glucose level in the blood; when it exceeds 0.1%, hepatocytes remove and store the excess as glycogen. When the glucose level drops, stored glycogen is broken down and released from the hepatocytes, and glucose again rises in the blood.

Bile is formed by the liver as a by-product of the breakdown of hemoglobin and cholesterol. It is stored in the gallbladder, under the right lobe of the liver. Bile salts from the gallbladder are released when fatty chyme is present in the duodenum, such as that from the greasy cheese pizza. The concentrated bile salts act as an **emulsifier** or biological detergent, breaking larger fat globules into smaller ones. Bile aids in fat digestion by increasing the surface area on which the digestive activities of pancreatic lipase can act.

Stones can form in bile. A small crystal of cholesterol that forms in the gallbladder may attract calcium ions from the concentrated bile, resulting in the formation of a stone. Stones can grow big enough to get stuck in the bile duct when the gallbladder releases its contents. This causes pain and blocks the flow of bile. The gallbladder is often removed if stones are a chronic problem. After removal, bile is still produced but not stored. The patient should not eat fatty meals, because there is no store of bile to aid in lipid digestion.

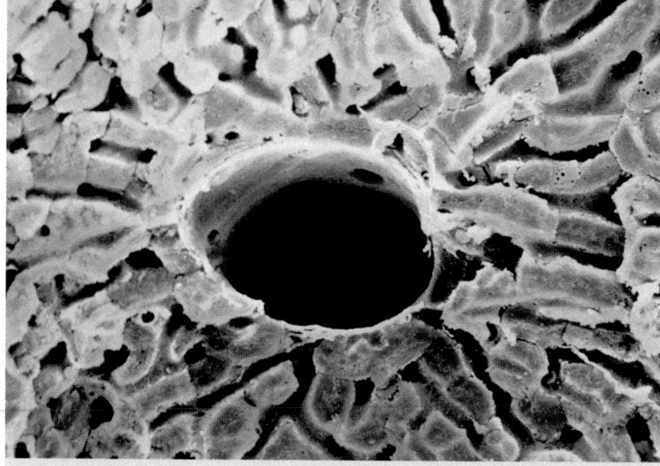

Liver lobule FIGURE 13.20

Each lobule is composed of a *triad* consisting of a hepatic portal vein, a hepatic artery, and a bile duct. These structures are found in the center of the lobule, with small channels that radiate to the individual cells of the lobule, like the spokes on a bicycle wheel. Fluid within the lobules is cleansed by the hepatocytes and sent on to the vena cavae.

Liver diseases can be deadly

Because the liver serves as a detoxification center for blood coming from the intestinal tract, it is exposed to many toxic substances, and the liver can be damaged by the very substances it is detoxifying. This occurs in the disease **cirrhosis** of the liver. Cirrhosis is a general term for a series of events that cause scar tissue to build up in the liver. The scar tissue impedes the blood flow through the lobules, preventing hepatocytes from doing their job. Detoxification of intestinal blood is impaired. Cirrhosis can be caused by alcohol consumption, chronic hepatitis infection, autoimmune diseases that attack the liver, or even congenital defects.

Other common liver diseases are viral, including hepatitis A, B, and C.

Hepatitis A is transmitted through drinking water or eating foods contaminated with infected fecal matter. Unlike other hepatitis viruses, hepatitis A virus remains intact as it passes through the stomach and the intestinal tract. It is still virulent after leaving the body in the feces. Hepatitis A also can be transmitted directly through kissing or sexual contact. The virus can live for 3 to 4 hours on hard surfaces, such as eating utensils. Symptoms of infection include fatigue, nausea, vomiting, liver pain, dark urine, and light colored stools. Recovery usually takes approximately 6 months, during which time it is difficult to work or carry out daily duties.

Hepatitis B is passed from person to person in body fluids. This form of hepatitis is serious because it has few symptoms, so individuals can unknowingly transmit the virus. After some years, hepatitis B causes permanent liver damage, liver failure, or liver cancer.

Hepatitis C is the most common chronic bloodborne virus in the United States. It is transmitted through blood-to-blood contact, like HIV. After suffering through a flu-like illness just after infection, people infected with hepatitis C are almost completely symptom-free. After many years, liver damage begins to show up, and the damage progresses. It is most accurately diagnosed via tests for antibodies. Although there is no cure, treatment includes maintaining a healthy diet and exercise program. Hepatitis C is a major cause of liver transplants.

Chyme passes into the large intestine

Once the pizza that we ate hours ago reaches the end of the small intestine, the body cannot pull any more nutrients from it. The chyme now passes from the small intestine into the next portion of the GI tract, the large intestine (FIGURE 13.21). The overall function of the large intestine is to reabsorb the water that was added to the chyme to begin digestion. Along with the water, the large intestine absorbs many dissolved minerals and some vitamins. The valve that makes the transition from the ileum of the small intestine to the **cecum** of the large intestine is called the **ileocecal valve**. The ileum joins the large intestine a few centimeters from the bottom. The cecum hangs below the junction, forming a blind pouch that ends in the **vermiform appendix**.

Although the function of the appendix is unclear, it may play a role in the immune system. When the appendix acts up, we get **appendicitis**, which presents as pain near the belly button that migrates to the lower right side. Other symptoms include nausea, vomiting, low fever, constipation or diarrhea, inability to pass gas, and abdominal bloating. The abdomen becomes increasingly tender, and simple movements cause pain. These are all symptoms of a blockage in the appendix that prevents normal flow through the large intestine. Feces may be blocking the entrance, or lymph nodes in the surrounding walls may be swollen due to infection. In either instance, the contents of the appendix cannot move, leading to a buildup of pressure, decreased blood flow, and inflammation. If the pressure is not relieved quickly, the entire organ can rupture or suffer **gangrene**. For unknown reasons, most cases of appendicitis occur in people aged 10 to 30. As soon as inflammation is diagnosed, the appendix is surgically removed to prevent it from rupturing and releasing pathogens into the intestine or the abdominal cavity.

Gangrene
Tissue death due to lack of blood flow.

The remainder of the large intestine is commonly called the **colon**. The four divisions of the colon describe the direction of flow within them. The

Large intestine FIGURE 13.21

The four parts of the colon can be easily seen in this image. A substance was given to the patient that reflects X rays. Exposure to X rays then provides a clear view of the colon. The ascending colon is on the left, the large downward loop is the transverse colon, and the very densely stained descending colon runs along the right side. The sigmoid colon makes its characteristic "S" turn at bottom center.

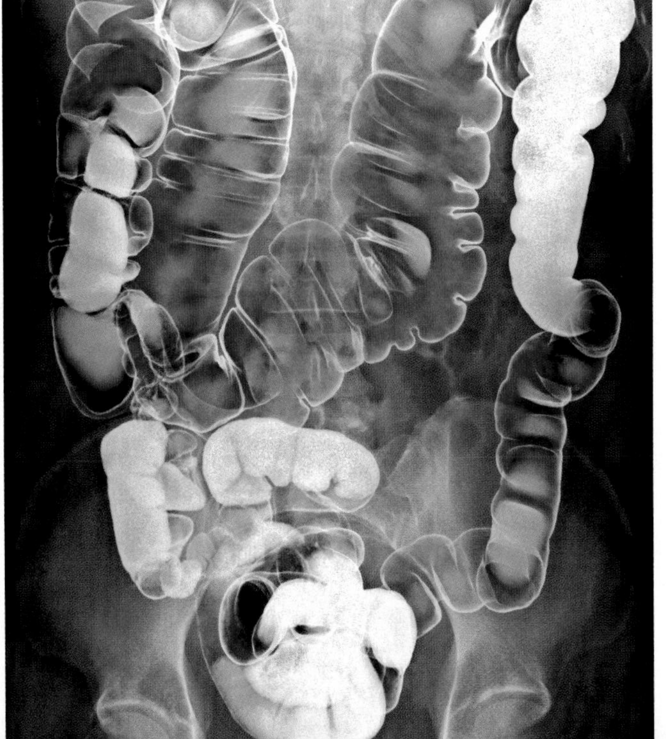

Polyps
Growths protruding from a mucous membrane.

ascending colon runs up the right side of the abdominal cavity. The **transverse colon** cuts across the top of the abdominal cavity, posterior to the stomach. At the left side of the abdominal cavity the colon turns back down, in the **descending colon**. At the lower left of the abdominal cavity, the colon makes an S turn to wind up in the center of the body. This turn is called the **sigmoid colon** and is the portion of the colon where feces often sit for long periods of time before moving out the rectum. Often, **polyps** can develop in the colon as feces rest against the mucosa.

The walls of the large intestine have **haustra**, pouches created by strands of muscle in the walls. These pouches fill with undigested material, which moves from pouch to pouch via **mass movements**.

Diarrhea results from an irritation of the colon. The chyme moves through the colon far too quickly for water or minerals to be absorbed. Medicines that prevent mass movements are often helpful in slowing the movement of chyme through the large intestine, giving the walls of the organ ample time to return the excess water to the bloodstream. To combat severe diarrhea, remedies that contain minerals and fluid are ingested to replace what is lost in the diarrhea.

The last 20 centimeters of the colon are the **rectum** and **anus**. Chyme remains in the colon for 3–10 hours, during which time it becomes progressively drier. Compacted chyme is called feces. When feces enter the upper portion of the rectum, they trigger the opening of the internal anal sphincter, a smooth muscle. The feces move into the rectum and press against the external anal sphincter. This triggers **defecation**, a skeletal muscle action. As with all skeletal muscles, control over defecation is voluntary. On average, by age $2\frac{1}{2}$ children are mature enough to control defecation.

Summary of the functions of the digestive organs TABLE 13.5

Organ	*Functions*
Mouth	See other listings in this table for the functions of the tongue, salivary glands, and teeth, all of which are in the mouth. Additionally, the lips and cheeks keep food between the teeth during mastication, and buccal glands lining the mouth produce saliva.
Tongue	Maneuvers food for mastication, shapes food into a bolus, maneuvers food for deglutition, detects taste and touch sensations, and initiates digestion of triglycerides.
Salivary glands	Produce saliva, which softens, moistens, and dissolves foods; cleanses mouth and teeth; and initiates the digestion of starch and lipids.
Teeth	Cut, tear, and pulverize food to reduce solids to smaller particles for swallowing.
Pharynx	Receives a bolus from the oral cavity and passes it into the esophagus.
Esophagus	Receives a bolus from the pharynx and moves it into the stomach. This requires relaxation of the upper esophageal sphincter and secretion of mucus.
Stomach	Mixing waves macerate food, mix it with secretions of gastric glands (gastric juice), and reduce food to chyme. Gastric juice activates pepsin and kills many microbes in food. Intrinsic factor aids absorption of vitamin B_{12}. The stomach serves as a reservoir for food before releasing it into the small intestine.
Pancreas	Pancreatic juice buffers acidic gastric juice in chyme (creating the proper pH for digestion in the small intestine), stops the action of pepsin from the stomach, and contains enzymes that digest carbohydrates, proteins, triglycerides, and nucleic acids.
Liver	Produces bile, needed for emulsification and absorption of lipids in the small intestine; detoxifiles blood containig absorbed nutrients and other substances.
Gallbladder	Stores and concentrates bile and releases it into the small intestine.
Small intestine	Segmentations mix chyme with digestive juices; migrating motility complexes propel chyme toward the ileocecal sphincter; digestive secretions from the small intestine, pancreas, and liver complete the digestion of carbohydrates, proteins, lipids, and nucleic acids; circular folds, villi, and microvilli increase surface area for absorption; site where nutrients and water are absorbed.
Large intestine	Haustral churning, peristalsis, and mass peristalsis drive the contents of the colon into the rectum; bacteria produce some B vitamins and vitamin K; absorption of some water, ions, and vitamins; defecation.

Material moves through the large intestine in mass movements, created using a peristaltic wave. In the colon, water is reabsorbed from the soupy chyme, concentrating the waste material and conserving fluid. As the water is pulled back into the bloodstream across the lining of the colon, so too are minerals and vitamins. The removal of water leaves undigested remains of food and fiber in the colon, as well bacteria, such as *E. coli* and other **obligate anaerobes** that naturally live in the large intestine. These colonies are necessary in the colon because they break down indigestible material and often produce essential vitamins. Sometimes these colonies can be embarrassing because they generate gas when fermenting solids.

Obligate anaerobes Bacteria that require an oxygen-free environment.

See TABLE 13.5 for a summary of the organs involved in digestion.

CONCEPT CHECK

List the three phases of deglutition.

Detail the function of the gastric pits. List the cells of these pits and their secretions.

What is the function of the pancreas?

List the four parts of the colon.

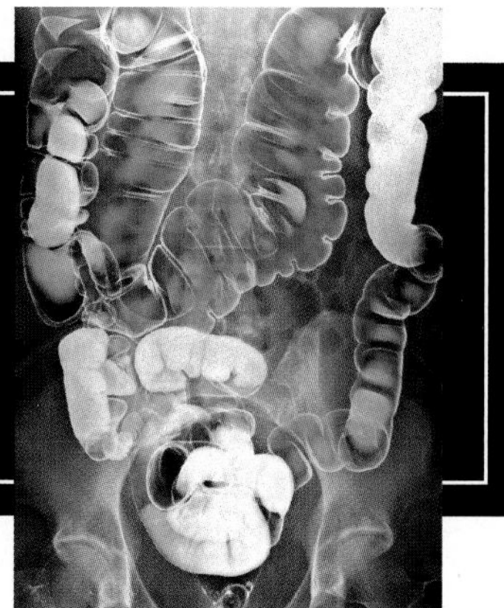

Digestion Is Both Mechanical and Chemical

LEARNING OBJECTIVES

Define mechanical and chemical digestion.

List the major enzymes of chemical digestion, and note their substrates.

Throughout this look at the digestive system, we have discussed various organs and their contribution to the process of digestion. Now it's time to summarize, so that we can view digestion as one continuous process.

Digestion is the breaking down of food into substances that can be absorbed and used by the body. This is accomplished through two processes: mechanical digestion and chemical (or enzymatic) digestion. **Mechanical digestion** refers to the chopping, cutting, and tearing of large pieces of food into smaller ones. Bites of apple, for example, are crushed and torn into pieces in your mouth, but these pieces are still recognizable as apple pieces, and no chemical alteration has occurred. They have all the properties and chemical bonds of the original apple, but with a larger surface area needed for chemical digestion.

Mechanical digestion occurs mainly in the mouth. Once the bolus of food is passed to the esophagus, a small amount of mechanical digestion occurs in the stomach, as it rolls and churns the food into chyme. The chyme then moves through the pyloric sphincter into the duodenum, where large droplets of fat are emulsified via bile. The action of bile is a form of mechanical digestion, breaking larger fat droplets into smaller ones without altering the chemical structure of the fats. At this point, the chyme is ready for enzymatic degradation, and mechanical digestion is finished.

Unlike mechanical digestion, enzymatic digestion alters chemical bonds. Most of the food we ingest is composed of **polymers**, long chains of repeating subunits, which our digestive enzymes must break into short chains or monomers. It is these shorter units that are absorbed in the small intestine and used to produce the proteins and energy needed for survival.

Process Diagram

Hydrolase activity FIGURE 13.22

Most of our digestive enzymes are *hydrolases*, meaning that they catalyze the breakdown of large polymers by inserting water molecules between monomers. We unconsciously know that digesting requires water because we find it uncomfortable to eat without drinking.

1. Large molecule of food enters the digestive system.
2. Enzyme binds to food (substrate) molecule.
3. Enzyme uses H_2O to split the substrate molecule in half, leaving an OH^- on one product molecule and an H^+ on the other.

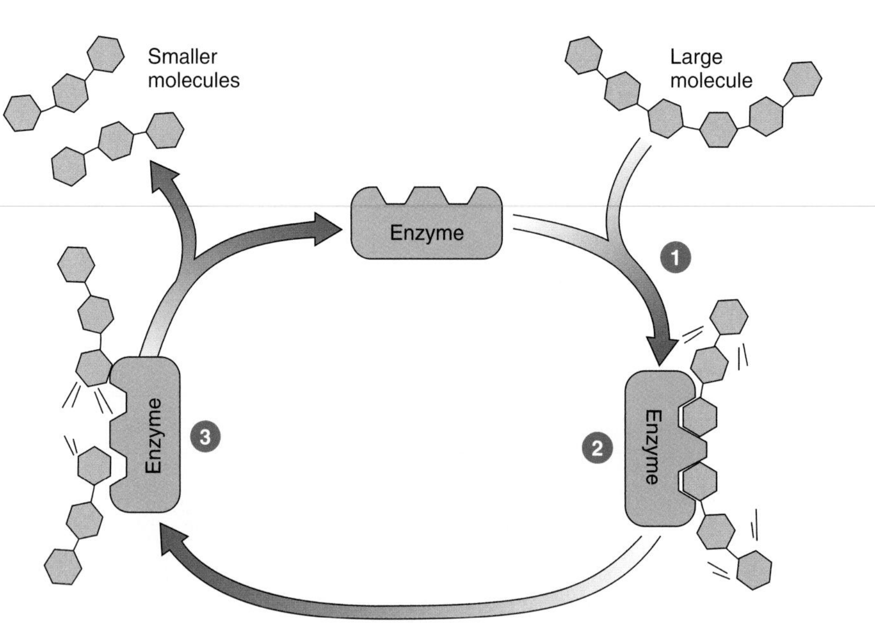

In order to digest our myriad foodstuffs, we need several digestive enzymes (FIGURE 13.22). As you know, enzymes are functional proteins that work best under a set of optimal conditions of pH, temperature, substrate, and product levels. (The substrate is the compound the enzyme acts upon, and the product is the result of that enzymatic action.) All enzymes are specific for a particular substrate and catalyze only one reaction.

Enzyme names are usually built from the name of the substrate, followed by the suffix "-ase." For example, lipase digests lipids, and nucleases digest nucleic acids. The major digestive enzymes, along with their substrates, products, and sources, are listed in TABLE 13.6.

All digestive enzymes except two act in the small intestine. Salivary amylase begins to digest carbohydrates in the mouth and continues in the bolus of food entering the stomach. Pepsin, in the stomach, works best at pH 2. The rest of the digestive enzymes operate best at pH 7, and are found inside the small intestine.

For some organisms, locating and ingesting nutrients is relatively simple. The single-celled **amoeba** oozes through the environment, constantly searching for nutrients. When it runs across a bit of organic material, the amoeba engulfs the particle and brings it into its body via **phagocytosis**. Once inside the amoeba, the particle is broken into its building blocks by digestive enzymes in the lysosome (FIGURE 13.23). Monosaccharides are

Amoeba A single-celled organism that moves using pseudopods (false feet formed by oozing a portion of the body forward).

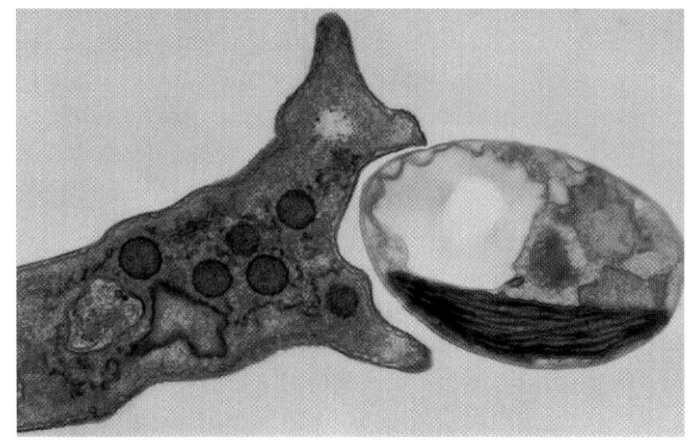

Amoeba eating FIGURE 13.23

Digestive enzymes TABLE 13.6

Enzyme	Source	Substrates	Products
SALIVA			
Salivary amylase	Salivary glands.	Starches (polysaccharides).	Maltose (disaccharide), maltotriose (trisaccharide), and α–dextrins.
Lingual lipase	Lingual glands in the tongue.	Triglycerides (fats and oils) and other lipids.	Fatty acids and diglycerides.
GASTRIC JUICE			
Pepsin (activated from pepsinogen by pepsin and hydrochloric acid)	Stomach chief cells.	Proteins.	Peptides.
Gastric lipase	Stomach chief cells.	Triglycerides (fats and oils).	Fatty acids and monoglycerides.
PANCREATIC JUICE			
Pancreatic amylase	Pancreatic acinar cells.	Starches (polysaccharides).	Maltose (disaccharide), maltotriose (trisaccharide), and α–dextrins.
Trypsin (activated from trypsinogen by enterokinase)	Pancreatic acinar cells.	Proteins.	Peptides.
Chymotrypsin (activated from chymotrypsinogen by trypsin)	Pancreatic acinar cells.	Proteins.	Peptides.
Elastase (activated from proelastase by trypsin)	Pancreatic acinar cells.	Proteins.	Peptides.
Carboxypeptidase (activated from procarboxypeptidase by trypsin)	Pancreatic acinar cells.	Amino acid at carboxyl end of peptides.	Amino acids and peptides.
Pancreatic lipase	Pancreatic acinar cells.	Triglycerides (fats and oils) that have been emulsified by bile salts.	Fatty acids and monoglycerides.
Nucleases			
Ribonuclease	Pancreatic acinar cells.	Ribonucleic acid.	Nucleotides.
Deoxyribonuclease	Pancreatic acinar cells.	Deoxyribonucleic acid.	Nucleotides.
BRUSH BORDER			
α–Dextrinase	Small intestine.	α–Dextrins.	Glucose.
Maltase	Small intestine.	Maltose.	Glucose.
Sucrase	Small intestine.	Sucrose.	Glucose and fructose.
Lactase	Small intestine.	Lactose.	Glucose and galactose.
Enterokinase	Small intestine.	Trypsinogen.	Trypsin.
Peptidases			
Aminopeptidase	Small intestine.	Amino acid at amino end of peptides.	Amino acids and peptides.
Dipeptidase	Small intestine.	Dipeptides.	Amino acids.
Nucleosidases and phosphatases	Small intestine.	Nucleotides.	Nitrogenous bases, pentoses, and phosphates.

released from carbohydrates, amino acids are released from proteins, and small carbon compounds are released from fatty acids. These small, organic compounds are then used by the amoeba to generate essential enzymes, cellular structures, and energy. Micronutrients are obtained by the amoeba in a similar fashion, via **pinocytosis**. Often micronutrients are released from larger compounds during lysosomal digestion.

The human body is far more complex than the amoeba, but each cell still needs nutrients in order to survive. Interestingly, human cells absorb nutrients in exactly the same manner as the amoeba: through diffusion, osmosis, facilitated diffusion, and active transport (including both phagocytosis and receptor-mediated endocytosis). However, the cells cannot leave their positions in the tissues to ooze through the environment in search of nutrients. Although that would make a wonderful B-movie plot, our cells must remain organized and in position! Therefore, the digestive system's job is to prepare nutrients for circulation through the blood, which reaches every cell.

Regulation of our digestive activities is based on blood sugar levels. Normally, blood sugar is kept at approximately 70 to 110 mg glucose per 100 ml blood. This level is essential to keep neurons functioning. When blood glucose drops, we feel hungry. If we eat, blood sugar levels rise from the absorption of ingested glucose. If we do not eat, we begin to break down glycogen stores, where excess glucose has been stored in liver and skeletal muscles. Glycogen can break down to glucose relatively quickly. Fats and proteins can also be converted to glucose, but at a higher energy expense. During starvation, the protein of skeletal muscle, and even heart muscle, is broken down to provide glucose for the brain, as described in the coverage of the general adaptation syndrome in Chapter 9.

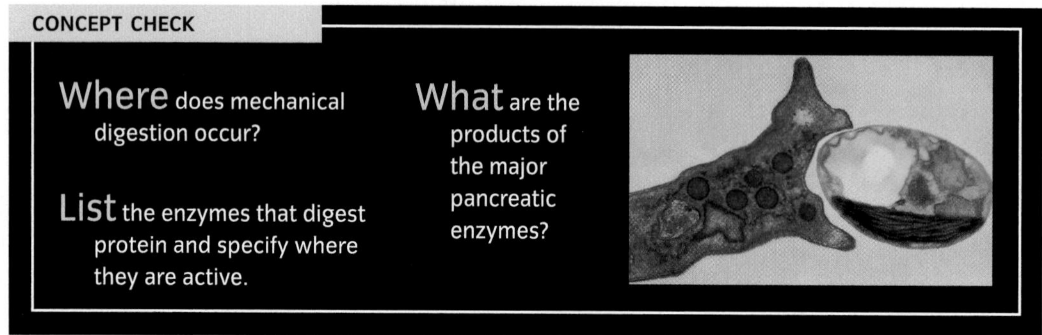

CONCEPT CHECK
Where does mechanical digestion occur?
List the enzymes that digest protein and specify where they are active.
What are the products of the major pancreatic enzymes?

Nutritional Health and Eating Disorders: You Truly Are What You Eat

Learning Objectives

Understand how to make healthy eating choices.

List the common eating disorders and their symptoms.

Diet and nutrition are important aspects of overall health because most of the compounds that enter the body enter through the digestive system. If we put nothing useful into the digestive system, our bodies will not have a good source of raw material for the proteins, enzymes, and energy required for life. Conversely, if we fill our digestive system with foods high in necessary nutrients, our bodies will function at peak levels. Of course, we can get too much of a good thing. If we ingest more calories than we "spend," regardless of their quality, we will store the excess in adipose tissue as fat (triglycerides).

Much attention is given to our diets, and its effect on our body, both in the media and in society. Our society is obsessed with being thin. For some, this obsession takes an unhealthy turn, in the form of two common eating disorders, **anorexia nervosa** and **bulimia nervosa**. Both of them stem from the desire to be thin and therefore, "beautiful" and are described in the Ethics and Issues box. Another problem related to eating is even more common: **obesity**. All of these eating

What is an ideal weight? How far should you go to look skinny?

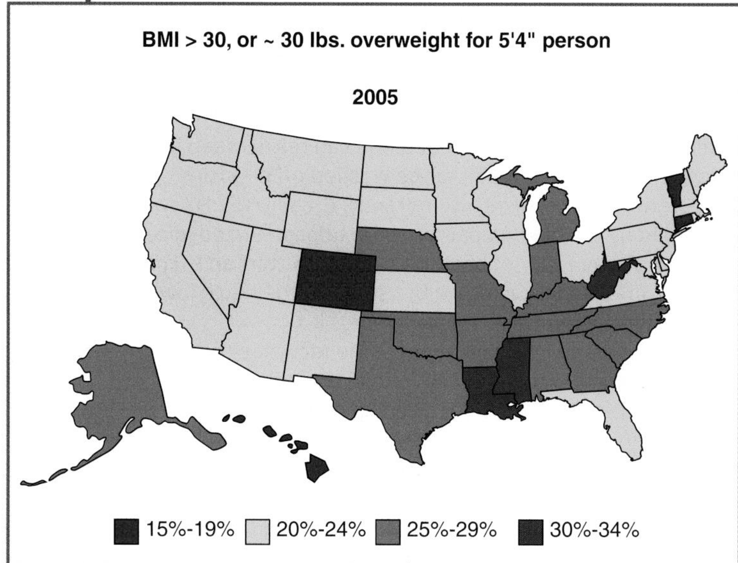

Need to lose weight? In the United States, the rate of obesity is soaring; 30 percent of U.S. residents at least 20 years of age are considered obese. Since 1980, the rate of obesity has tripled among young people. But even people with normal weight seem to feel they would be smarter, sexier, and more lovable if they could dump a few pounds.

Certainly, being overweight and especially obese is associated with high rates of hypertension, some cancers, and type 2 diabetes. But an obsession with being overweight can take its own toll: Often those with this obsession have a general feeling that they are not okay in their appearance or performance. Some of these people develop eating disorders.

Eating disorders primarily affect young women; between 1 and 4 percent of women aged 14–25 suffer from one eating disorder or another, according to federal figures. The most common disorders are:

- Anorexia nervosa, a form of self-starvation. Thin people think they are fat and use severe diets, intense exercise, or purging in an attempt to lose weight. The physical symptoms resemble starvation: osteoporosis, brittle hair, intolerance of cold, and muscle wasting, among many other possible problems.
- Bulimia nervosa, secretive eating binges followed by vomiting or enemas to clear the food before it can be digested. Because bulimics may not be severely thin, the disorder can remain undiagnosed for a long while. Complications include electrolyte imbalance and acid damage to the upper GI tract.
- Binge eating disorder also features bouts of uncontrolled eating, without the purging phase. For this reason, binge eating disorder can cause obesity.

Eating disorders are not merely "lifestyle diseases." According to federal statistics from 1994, an estimated 10 percent of anorexia patients will die of complications, an extremely high rate for a psychiatric illness. Although the exact causes of eating disorders are uncertain, it's common to see a history of depression, anxiety, or substance abuse. Eating disorders are also more common among people like models or actors, who need to be thin for occupational reasons. The ideal therapy is likewise uncertain and can range from patient education to in-patient treatment in psychiatric hospitals. Treatment seems to work better if begun early.

Eating disorders are largely a hidden epidemic, but two social factors might make them less common: a reduction in the overall level of overweight and obese individuals, and an emphasis on finding a healthy weight, instead of following the thin-is-beautiful approach of fashion magazines and many movie actresses.

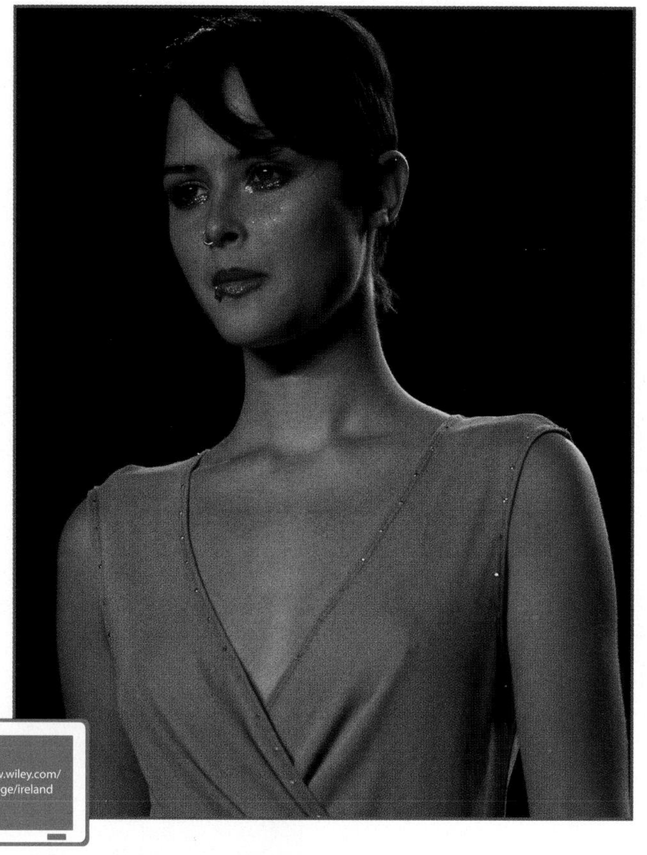

www.wiley.com/college/ireland

I WONDER . . .

Are *E. coli* bacteria hazardous to my health?

The healthy human colon is a sea of bacteria, including vast numbers of *Escherichia coli* (*E. coli*). A large percentage of the mass of feces consists of billions of *E. coli* cells. Inside the GI tract, almost all *E. coli* are helpful or at worst harmless. However, *E. coli* can infect the urinary tract (causing UTI, a urinary tract infection). If it escapes the colon and enters the abdomen, it can cause peritonitis, a serious abdominal infection. We often hear about outbreaks of *E. coli infections* that cause serious illness for a few unfortunate victims, or outbreaks that sweep entire small towns. Recently 146 U.S. citizens in 23 states suffered from *E. coli* poisoning, and one person died. All of these cases have been traced to tainted spinach crops, causing a crisis in the spinach and lettuce industry. What exactly causes these problems? How do the bacteria get into the food supply in the first place?

Bacteria are divided into "strains" according to some genetic trait, and one of the many strains of *E. coli* causes a severe form of food poisoning. Called *E. coli* O157:H7, this genetic variant releases toxins that cause severe, bloody diarrhea. The Centers for Disease Control and Prevention estimate that *E. coli* O157:H7 causes about 73,000 illnesses per year. Most infections clear up after 5 to 10 days. Antibiotics are not needed and may contribute to kidney damage.

In rare cases, *E. coli* O157:H7 can cause a far more serious disease, called hemolytic uremic syndrome. This syndrome kills red blood cells and causes an average of 61 fatalities each year in the United States, mainly through kidney failure. The syndrome is most severe among children, the elderly, and people with immune deficiencies. Some survivors require dialysis, others can suffer blindness or paralysis.

E. coli O157:H7 normally live in the intestines of healthy cattle, and some other ruminants. Many human infections come from meat contaminated by the contents of cattle intestines at the slaughterhouse. Ground beef is a common carrier because the bacteria can reside deep inside the meat, where it cannot be washed off or easily heat-sterilized by cooking. More rarely, *E. coli* O157:H7 can be spread by an infected person, or in unpasteurized juice or water. The number of bacteria needed to start an *E. coli* O157:H7 infection is unknown but seems to be much lower than for typical food-borne pathogens.

The routes of infection suggest the tactics for self-defense against *E. coli* O157:H7. While preparing meat, segregate it from other food, clean up carefully, and wash hands often. Cook hamburger to at least 72°C (160°F). If you cannot check the temperature with a digital thermometer, cook until the pink inside turns brown. If children develop this infection, they must be instructed to observe sanitary procedures, especially frequent hand-washing, so they do not pass it along.

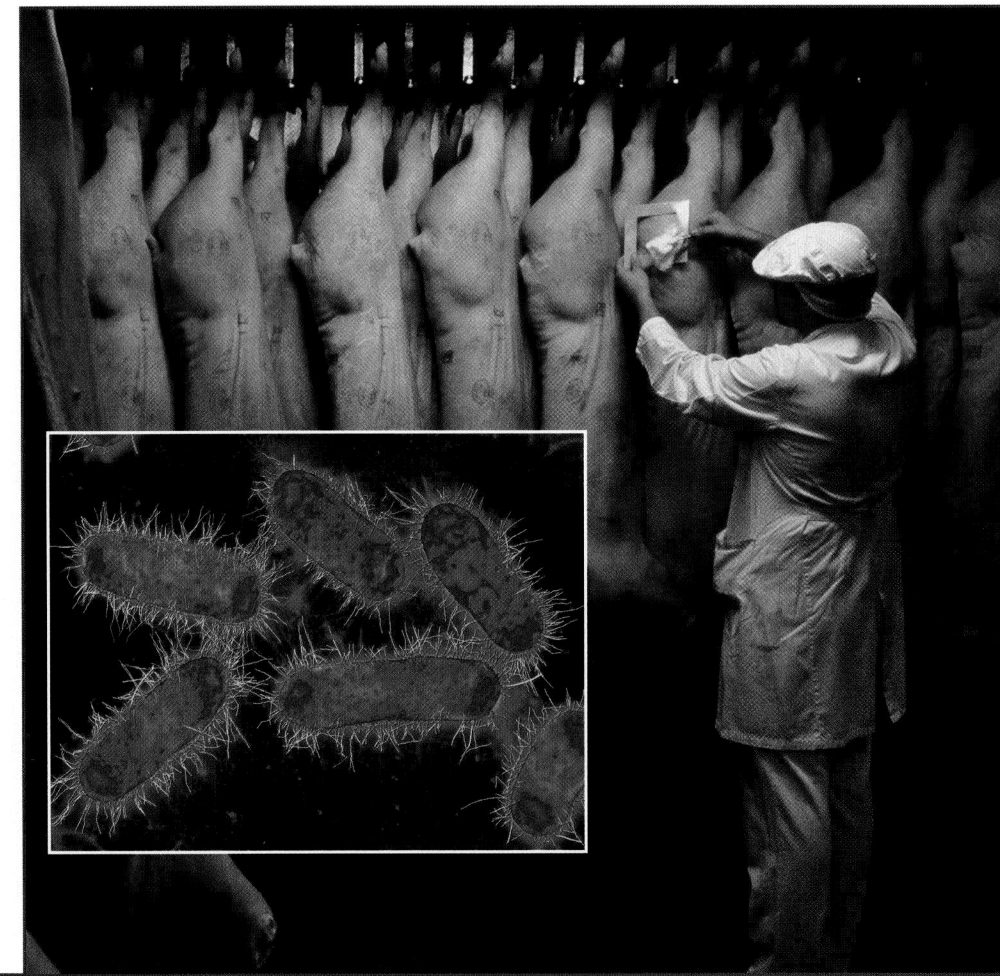

disorders grow out of a culture that is obsessed with beauty and will be hard to resolve without changing the social view of beauty.

FOOD IS LIFE-SUSTAINING, BUT SOMETIMES IT CAN BE LIFE-THREATENING AS WELL

Eating disorders are not the only pathologies of the digestive system. There are almost as many food-borne diseases as there are foods to carry them. More than *250 food-borne diseases* are known, ranging from bacteria and viruses to parasites and toxins from the foods themselves. The many types of food poisoning share a common thread. They are usually found growing in or on the foods we eat. All of these diseases enter the body through the digestive tract. Symptoms can vary, but the immediate symptoms usually include nausea, vomiting, abdominal cramps, and/or diarrhea. These symptoms represent the body's attempt to rid itself of the pathogen or toxin. If these flushing techniques fail, we will experience the specific symptoms of the invading organism.

Three common bacterial food poisonings are *Campylobacter, Salmonella,* and *Escherichia coli (E. coli)*. *Campylobacter* is a normal resident of the intestinal tract of chickens and other fowl. Commonly ingested in undercooked poultry, *Campylobacter* is the number one cause of bacterial diarrhea in the world. *Salmonella,* found in the intestines of birds, reptiles, and mammals, causes the usual food poisoning symptoms, but can become much more serious if untreated. *Salmonella* can escape the intestinal tract and enter the bloodstream, leading to **septicemia**, a life-threatening condition in which the blood carries a poison throughout the body. *E. coli* is normally present in the colon of cattle, pigs, and humans. A toxic form of *E. coli* is described in the I Wonder . . . box.

The most common viral food contaminant is Calicivirus or Norwalk-like virus, which causes vomiting that lasts for approximately two days, with little diarrhea or fever. Norwalk-like virus has even spread from infected fishermen through their oyster catch. Stomach flu has similar symptoms; it is actually not influenza but rather a viral infection that attacks and irritates the stomach and small intestine. Stomach flu is transmitted through kissing, touching, or sharing food, drinks, or utensils. Food preparation workers who carry the virus can spread it through the food they handle.

Whether the homeostatic balance of the body is disrupted by food-borne illness or merely by eating, we must have a system in place to restore it. These changes must be rectified to keep the blood and other body fluids within their narrow ranges. Maintaining fluid homeostasis is a matter of survival. Monitoring and maintaining the composition of the blood and the entire internal environment is the job of yet another system, the urinary system, which is covered in Chapter 14.

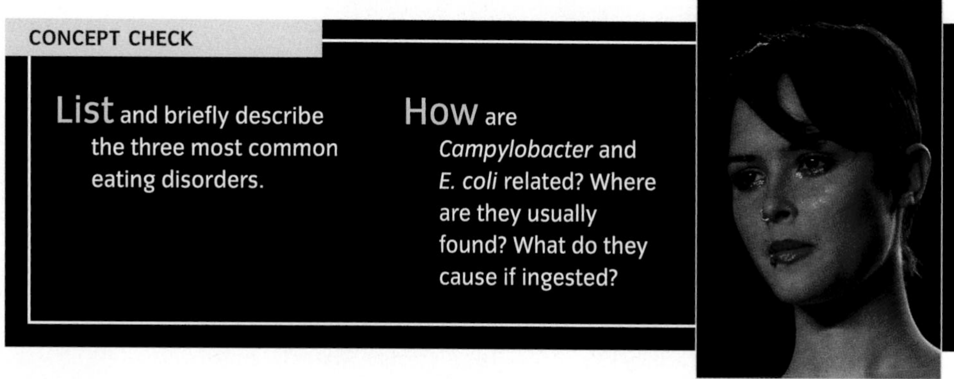

CONCEPT CHECK

List and briefly describe the three most common eating disorders.

How are *Campylobacter* and *E. coli* related? Where are they usually found? What do they cause if ingested?

CHAPTER SUMMARY

1 Nutrients Are Life-Sustaining

Food contains macronutrients—carbohydrates, fats, and proteins—and micronutrients—vitamins and minerals. Vitamins are organic substances and minerals are inorganic. Both are necessary for maintaining homeostasis, and both can be obtained safely from over-the-counter supplements, but the daily diet should be rich in fruits and vegetables. How much and what type of food we ingest plays a large role in our health. The U.S. Food and Drug Administration has recently upgraded the basic food pyramid to factor in age, activity levels, and gender.

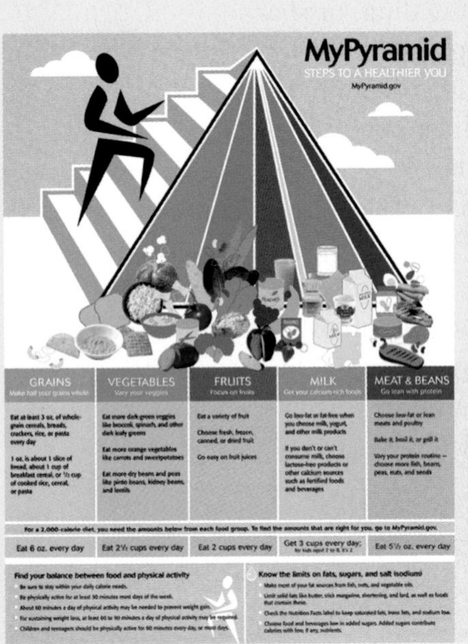

2 The Digestive System Processes Food from Start to Finish

The digestive system ingests food, mixes and propels that food through the digestive organs, mechanically and chemically breaks down the food, absorbs nutrients from the food, and releases the undigested wastes. The digestive system, or GI tract, is one continuous tube, divided by sphincter muscles. Each organ has anatomical alterations that allow it to perform a specific function. The organs, in order, are the oral cavity, esophagus, stomach, small intestine, large intestine, and rectum. Accessory organs, including the salivary glands, liver, gallbladder, and pancreas, assist in digestion. The salivary glands release salivary amylase and lubricate the bolus of food. The liver cleanses the blood as it drains from the small intestine. The gallbladder stores and releases bile. The pancreas produces digestive enzymes and buffers that control the pH of the digesting chyme in the small intestine.

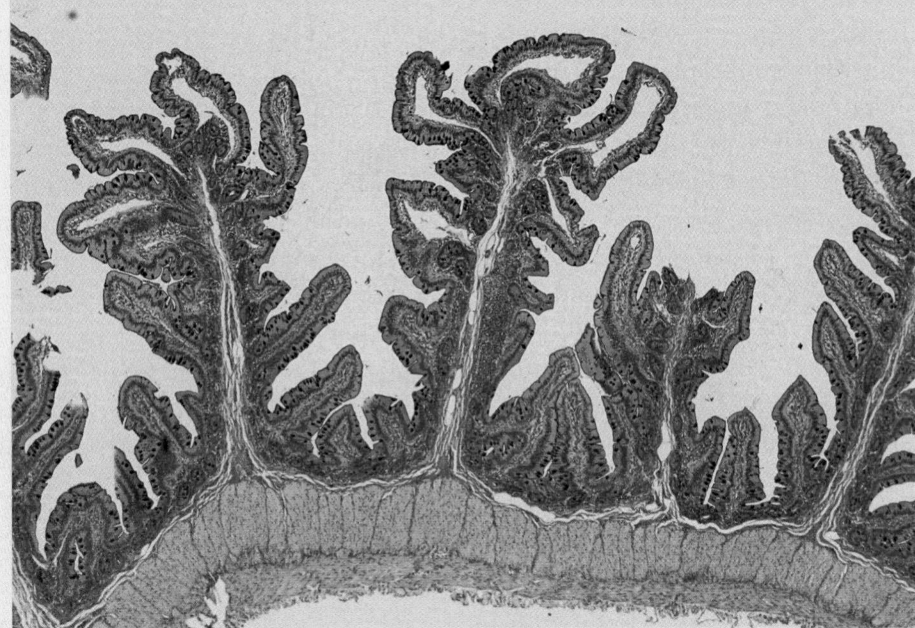

3 Digestion Is Both Mechanical and Chemical

Mechanical digestion starts in the mouth, where the teeth grind and crush the food. Saliva moistens the food, forming a bolus that can be swallowed. Muscular contractions push the bolus through the esophagus into the stomach, where high acidity starts to break it down. This acidity kills most pathogens but can attack the stomach wall if the mucous lining is damaged. In the small intestine, enzymes continue to break down the material called chyme. Macromolecules are absorbed through the highly convoluted lining of the small intestine and into the blood supply. As the now-nutrient depleted chyme moves through the large intestine, water is removed. The waste material, including a large proportion of harmless bacteria, is moved into the rectum and excreted.

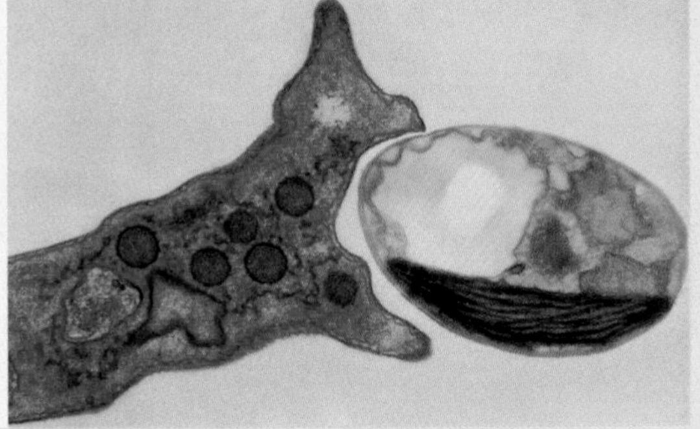

4 Nutritional Health and Eating Disorders: You Truly Are What You Eat

The primary nutritional disease in the United States is obesity. The major eating disorders are anorexia nervosa, bulimia nervosa, and binge eating disorder. All can be treated with a combination of proper diet and professional mental-health care. A number of food-borne pathogens, both bacterial and viral, can cause disease, but good sanitation can prevent many of them from being spread.

KEY TERMS

- **aerobic** p. 406
- **amoeba** p. 434
- **apical membrane** p. 427
- **bacteriolytic** p. 420
- **bolus** p. 419
- **calories** p. 408
- **chemical digestion** p. 418
- **chemiosmosis** p. 407
- **chylomicrons** p. 428
- **chyme** p. 423
- **fundus** p. 424
- **gangrene** p. 431
- **gastric** p. 424
- **glycolysis** p. 407
- **hepatocytes** p. 430
- **lingual** p. 419
- **macerated** p. 418
- **mechanical digestion** p. 418
- **mesenteries** p. 426
- **milled** p. 415
- **nutrients** p. 406
- **obligate anaerobes** p. 433
- **pancreatic juice** p. 428
- **papilla** p. 419
- **peristaltic wave** p. 416
- **polyps** p. 432
- **vegan** p. 409

CRITICAL THINKING QUESTIONS

1. Go to the MyPyramid Web site (http://www.mypyramid.gov/) and obtain your personal food guide. Alter your personal characteristics, and compare the results. Describe what happens to the recommended guidelines as you age. What happens as if your exercise level increases? Are these changes the same for males and females, or does gender alter the caloric recommendation?

2. Starting at the esophagus, trace the pathway of food through the system. At each organ, indicate anatomical adaptations to the general GI tract tube structure that enhance the specific functions of that organ.

3. One of the more drastic solutions for overeating is to "staple" the stomach, a procedure called gastric bypass surgery. This surgery reduces stomach size, preventing it from holding so much. How would this affect the functioning of the stomach? What essential hormone will decrease in the blood as the surface area of the stomach decreases?

4. Give a brief review of the structure of a liver lobule. Explain why cirrhosis of the liver can lead to jaundice and eventual liver failure. What exactly prevents the liver lobule from functioning?

5. Some emotions curb the appetite. How does this happen?

SELF TEST

1. Macronutrients include all of the following EXCEPT
   a. carbohydrates.
   b. lipids.
   c. vitamins.
   d. proteins.

2. The reactions in this diagram are collectively referred to as
   a. chemiosmosis.
   b. the Krebs cycle.
   c. mitochondrial reactions.
   d. cellular respiration.

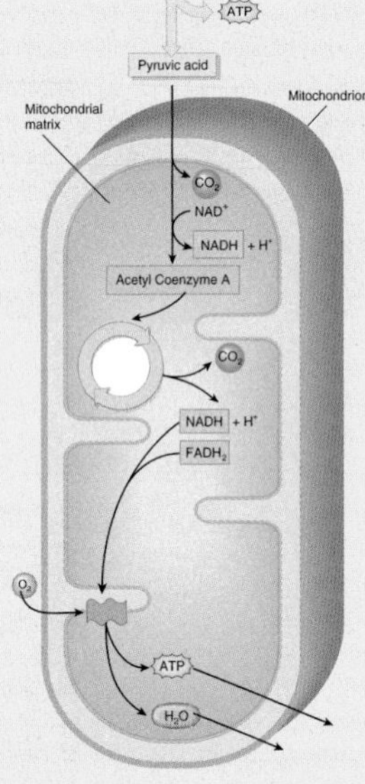

3. The first step in the reaction shown above
   a. is called glycolysis.
   b. converts one glucose molecule to two pyruvate molecules.
   c. releases a net of 2 ATP molecules.
   d. All of the above describe the first reaction shown.

4. In the figure below, the molecule of unsaturated fat is indicated as
   a. A.
   b. B.
   c. Neither of these molecules are unsaturated fats.
   d. Both of these molecules are unsaturated fats.

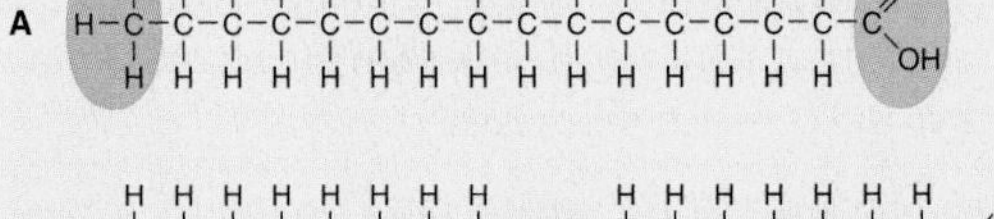

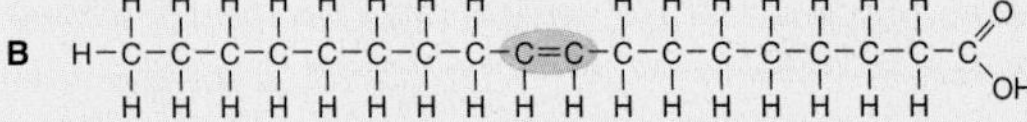

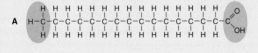

5. The MyPyramid Web site is designed to give you
   a. tips on healthy eating in general.
   b. easy access to the caloric content of most common foods.
   c. tips on healthy eating based on your gender, age, and activity level.
   d. assistance in reducing obesity.

6. True or False? Calcium, zinc, and iodine are all examples of vitamins.

7. The correct order of layers in the GI tract from external surface to lumen is
   a. serosa → muscularis → submucosa → mucosa.
   b. mucosa → submucosa → serosa → muscularis.
   c. muscularis → submucosa → mucosa → serosa.
   d. submucosa → mucosa → muscularis → serosa.

8. The muscularis of the GI tract is responsible for
   a. protecting the lumen.
   b. creating the peristaltic wave.
   c. absorbing water and nutrients.
   d. allowing the tract to slide around inside the abdominal cavity.

9. The teeth responsible for grinding and crushing are the
   a. incisors.
   b. canines.
   c. premolars.
   d. All types of teeth grind food.

10. Immune defenses in the digestive system include all of the following EXCEPT
    a. MALT.
    b. Peyers patches.
    c. liver.
    d. tonsils.

11. The stage of swallowing that involves the rising of the larynx is shown in this figure as
    a. A.
    b. B.

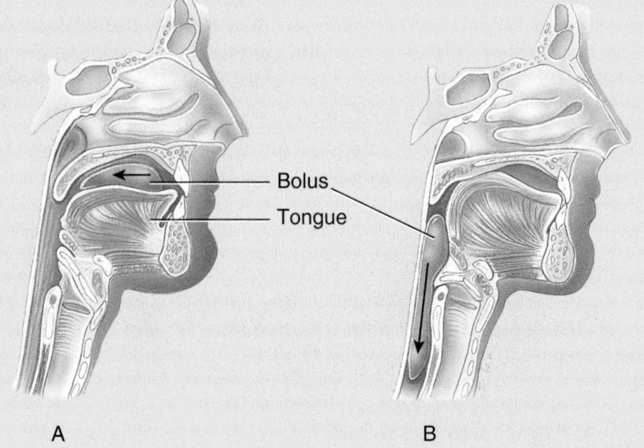

12. One function of the organ containing these structures is

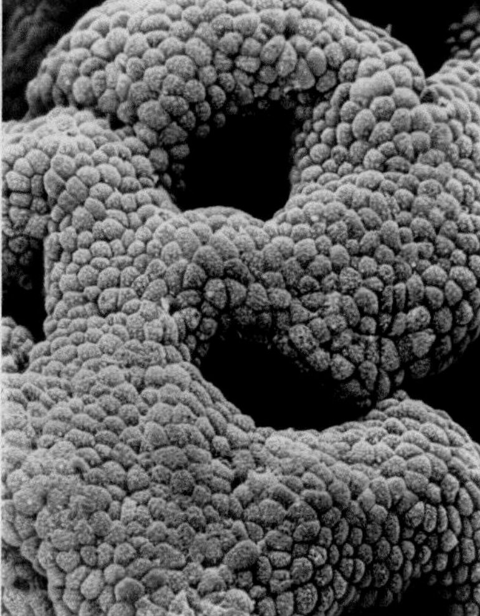

a. chemical digestion of carbohydrates.
b. mechanical digestive action of bile.
c. chemical digestion of proteins.
d. nutrient absorption.

13. The organ that is responsible for producing digestive enzymes is the

a. liver.
b. gallbladder.
c. pancreas.
d. sublingual salivary gland.

14. Most stomach ulcers are caused by

a. stress.
b. aspirin eroding the mucous lining of the stomach.
c. a spiral bacterium.
d. alcoholism.

15. The phase of gastric digestion that is initiated simply by the smell of food is the

a. cephalic phase.
b. gastric phase.
c. intestinal phase.
d. All three phases are triggered by the smell of food.

16. The function of the organ containing the structures shown below is to

a. chemically digest food.
b. mechanically digest food.
c. absorb nutrients.
d. All of the above are true of this organ.

17. The structure shown below is found in the ___________ and serves to ___________.

a. large intestine, decrease surface area
b. small intestine, increase surface area
c. stomach, produce HCl
d. liver, produce and store bile

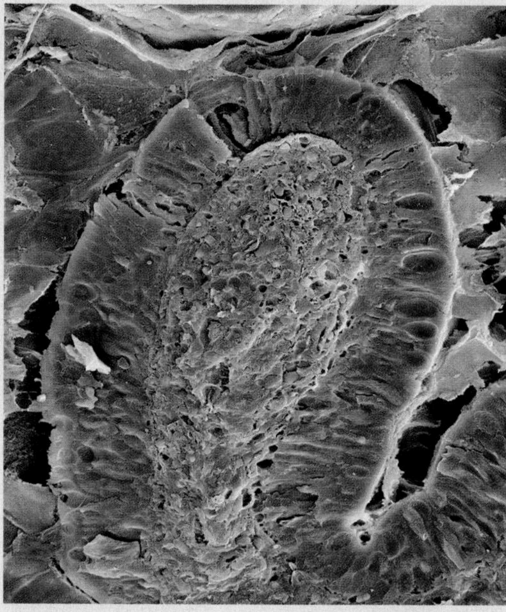

18. The most common viral liver disease in the United States is

a. hepatitis A.
b. hepatitis B.
c. hepatitis C.
d. All three are equally uncommon.

19. The common eating disorder, anorexia nervosa, can be described as

a. the binge-purge disease.
b. food poisoning.
c. overeating.
d. severe under-eating.

20. The bacterium *E. coli* is normally found

a. in the colon.
b. in the small intestine.
c. in the stomach.
d. throughout the digestive system.

The Urinary System

14

Peggy and Rob recently moved from their college town in coastal Washington State to northern New Mexico. Not long afterward, Peggy began to feel dizzy and noticed that she was urinating less and didn't feel quite as sharp mentally. To herself, she blamed these symptoms on the stress of moving and setting up their home in a strange community far from their families. But Peggy did not feel any better as she got established in her new life. In fact, within weeks, she had some near-blackouts when rising from a chair.

Eventually, Rob asked Peggy about her fluid intake. Their new environment was much warmer and drier than their old one, and although he did not notice much extra perspiration, he felt constantly thirsty and was toting a water bottle along on errands. In the middle of the night, he'd even wake up and take a long guzzle from the water glass. When he thought about it, he noticed that his urine was more colorful than what he remembered from Washington. And he realized that when he wore a backpack, his shirt got soaked in the areas where air could not circulate.

Peggy, however, was not drinking more than usual and had unwittingly pushed herself toward dehydration. By changing her body's fluid balance, she had opened herself up to systemic symptoms that could even be life-threatening. What organ system is responsible for maintaining fluid balance? How does the body know when to conserve water and when to excrete it? These are the focus of our chapter on the urinary system.

CHAPTER OUTLINE

The Urinary System Filters, Transports, and Stores Waste Products

LEARNING OBJECTIVES

Define the functions of the urinary system.

Identify the organs of the urinary system.

Diagram the anatomy of the kidney and the nephron.

The urinary system excretes aqueous waste as it maintains fluid balance and blood volume (FIGURE 14.1). It also regulates blood composition, helps to maintain blood pressure, monitors and maintains red blood cell levels, and assists in vitamin D synthesis. In addition, the urinary system is responsible for monitoring and adjusting the ionic composition of the blood, regulating the pH of the blood, regulating blood volume, maintaining blood glucose levels, and producing hormones that regulate calcium levels. And it does all this with four organs: pairs of kidneys and ureters, and the urinary bladder and urethra.

Listing all of these functions at once, it becomes obvious that the four organs of the urinary system are responsible for regulating the fluid environment of the body. As a whole, these are such vital functions that if the urinary system fails, the body will shut down within

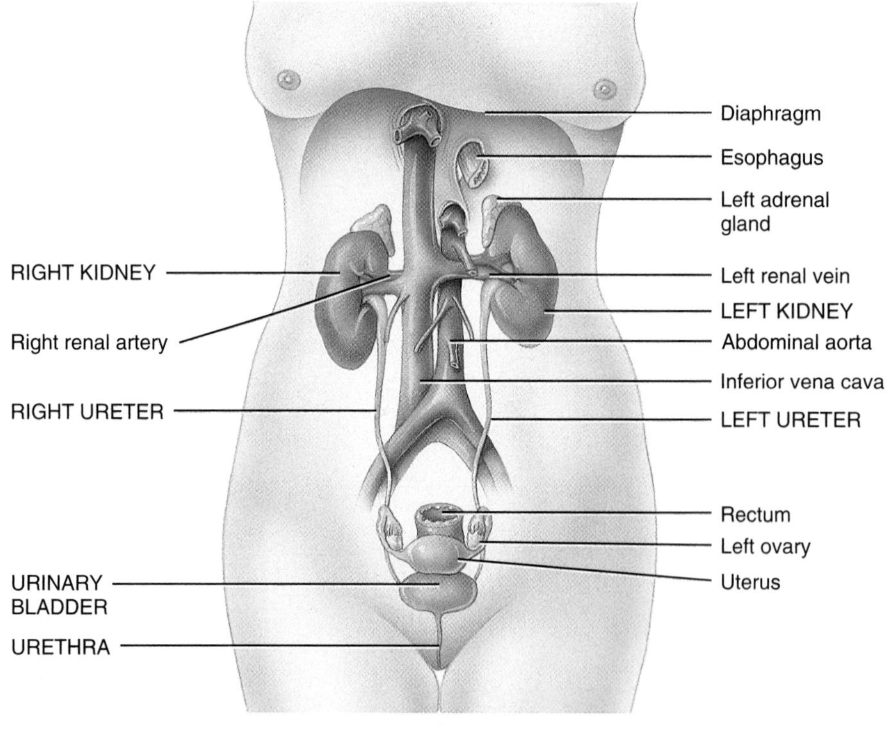

Urinary system FIGURE 14.1

The urinary system lies behind the peritoneum, protected by strong back muscles and fat. The kidneys are the organs responsible for filtering the blood, whereas the other three organs transport and excrete the resulting urine.

a few days. Urine is formed as a by-product of the system's functions. All waste materials removed from the blood by the urinary system leave the body in urine.

THE KIDNEYS ARE FILTERING ORGANS

The kidneys filter blood and produce hormones. These two fist-sized, bean-shaped organs lie immediately beneath the back musculature, embedded in a protective layer of fat (FIGURE 14.2). The kidneys are retroperitoneal, meaning they lie posterior to the peritoneal membrane. Because of this relatively unprotected placement, the kidneys are susceptible to injury from an external blow. Consequently, football pads are designed to cover the kidney area, and boxers are not permitted to punch opponents in the back. Due to the liver, the right kidney is slightly lower than the left.

The kidneys themselves are covered with a tough outer membrane, the renal capsule (FIGURE 14.3, p. 448). A large renal artery enters the kidney at the **hilus**. One quarter of the blood from every heartbeat gets shunted through the renal arteries to the kidneys. The hilus provides exit for the equally large renal vein and the kidney's nerves and lymphatic vessels. The ureters also pass through the hilus.

A sagittal section through a kidney reveals a uniform outer cortex and an irregular inner medulla. The cortex appears grainy and solid, and portions of it dip between the **renal pyramids** of the medulla. The renal pyramids are cone-shaped structures formed from an accumulation of collecting ducts filled with urine. The area adjacent to the hilus is the **renal pelvis**, where the formed urine is collected and passed to the ureters. The renal pelvis is coated in a protective mucous membrane because the urine it contains is toxic to the cells. Among other substances, this urine contains **nitrogenous wastes** filtered from the blood.

> **Nitrogenous wastes**
> Compounds containing nitrogen, such as urea, that are produced during protein metabolism.

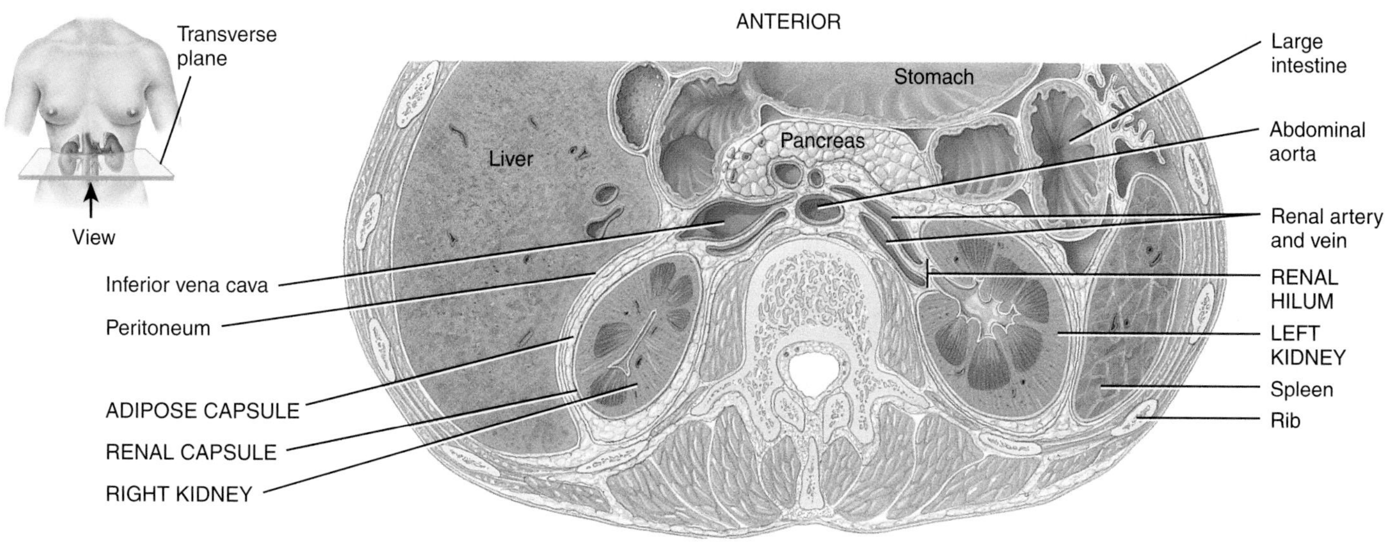

Kidney placement in the body FIGURE 14.2

The kidneys lie in the posterior portion of the abdominal cavity, behind the digestive organs. They are not symmetrically positioned; the right kidney is displaced downward by the liver.

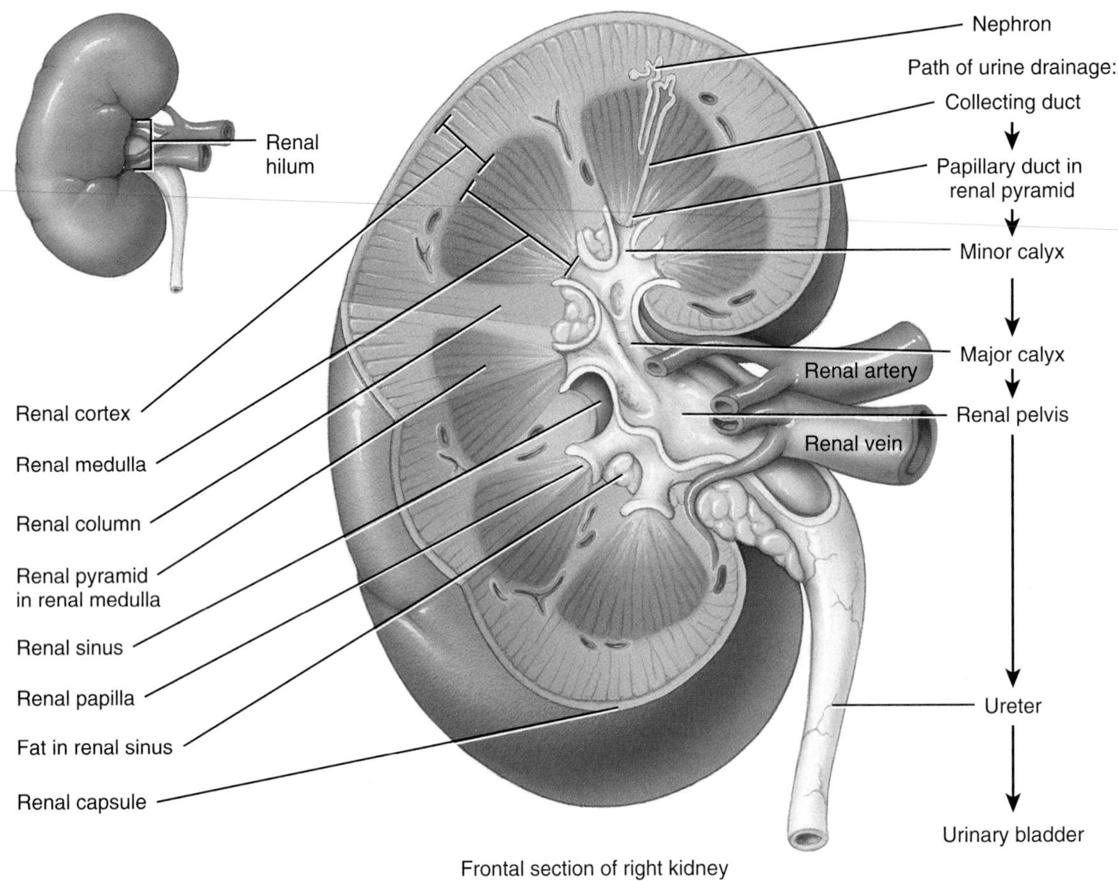

Internal anatomy of the kidney FIGURE 14.3

The renal cortex contains a large blood supply, and it is here that filtration occurs. The renal medulla is involved in the fine tuning of this filtrate, and the renal pelvis transports the final waste product, the urine, from the kidneys.

Nephron The filtering unit of the kidney.

Peritubular capillaries (*peri* = around; *tubular* = nephron tubules) Capillaries that surround the nephron.

The kidneys are composed of millions of **nephrons**, packed together under the renal capsule (FIGURE 14.4). The large blood supply that enters the kidneys is diverted through ever-smaller arteries and arterioles until it winds its way to a knotted vessel at the beginning of each nephron. The blood vessel leaving each nephron then breaks into **peritubular capillaries**, which wind around the entire nephron before collecting into venules. The venous system of the kidneys follows the same route as the arterial system, eventually leaving the kidney in the large renal veins. At the nephron, the blood is filtered, and the necessary ions and nutrients are returned to the circulatory system. The waste material remains in the fluid within the tubules of the nephron.

Beyond cleaning blood, the kidneys also produce the hormones calcitriol and erythropoietin, which regulate the concentration of calcium and formed elements in blood. Calcitriol, the active form of vitamin D, helps maintain blood calcium levels. Erythropoietin stimulates production of new red blood cells.

Kidney anatomy and kidney blood flow FIGURE 14.4

Process Diagram

Blood flow through the kidneys must be highly regulated. A full quarter of blood flow is sent to the kidneys rather than the body tissues. 1 Blood enters the kidneys via the renal artery. 2 The renal artery branches into the segmental arteries that supply each renal pyramid of the kidneys. 3 Segmental arteries give rise to the interlobar arteries that dive between renal pyramids. 4 These arteries then loop over the renal pyramids in arcuate arteries. 5 The interlobular arteries then take the blood to the renal cortex, where 6 it is further divided into afferent arterioles leading to the glomerular capillaries. 7 Filtered blood leaves the glomerulus through the efferent arterioles, where it moves to the peritubular capillaries and is then collected by the 8 interlobular veins. From here, the pathway reverses, moving consecutively through 9 arcuate veins, 10 interlobar veins, and finally 11 leaving the kidneys via the renal vein.

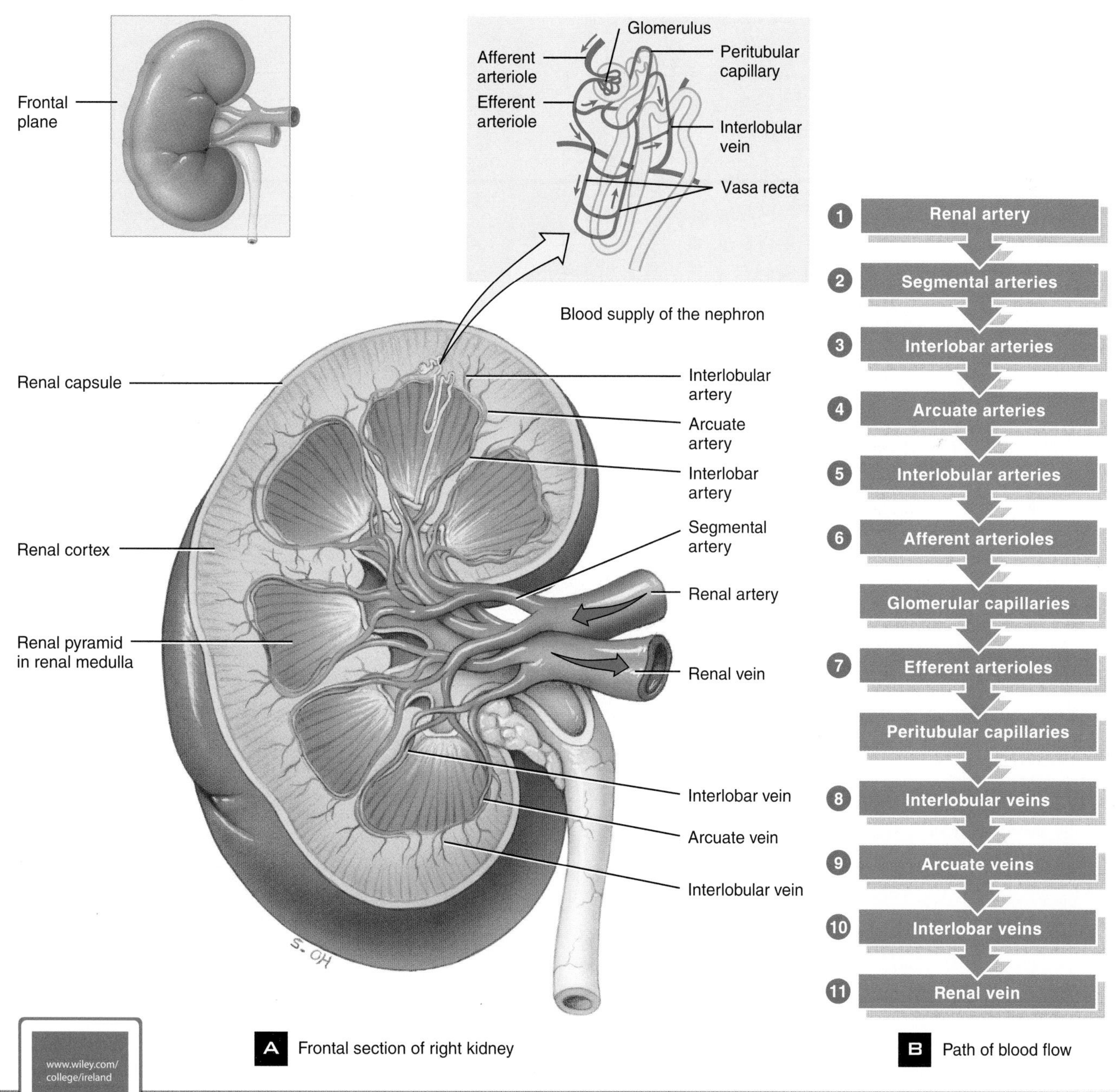

www.wiley.com/college/ireland

A Frontal section of right kidney

B Path of blood flow

MILLIONS OF NEPHRONS DO THE FILTERING WORK

When observing a kidney under a light microscope, it becomes obvious that the organ is in fact a large collection of small nephrons, each responsible for filtering a portion of the blood that passes through the kidney (FIGURE 14.5). Each nephron is encircled by a separate capillary bed, providing the link between the urinary and cardiovascular systems. This link begins at a glomerulus, or vessel knot. The glomerulus is formed from an incoming arteriole. Leaving the glomerulus,

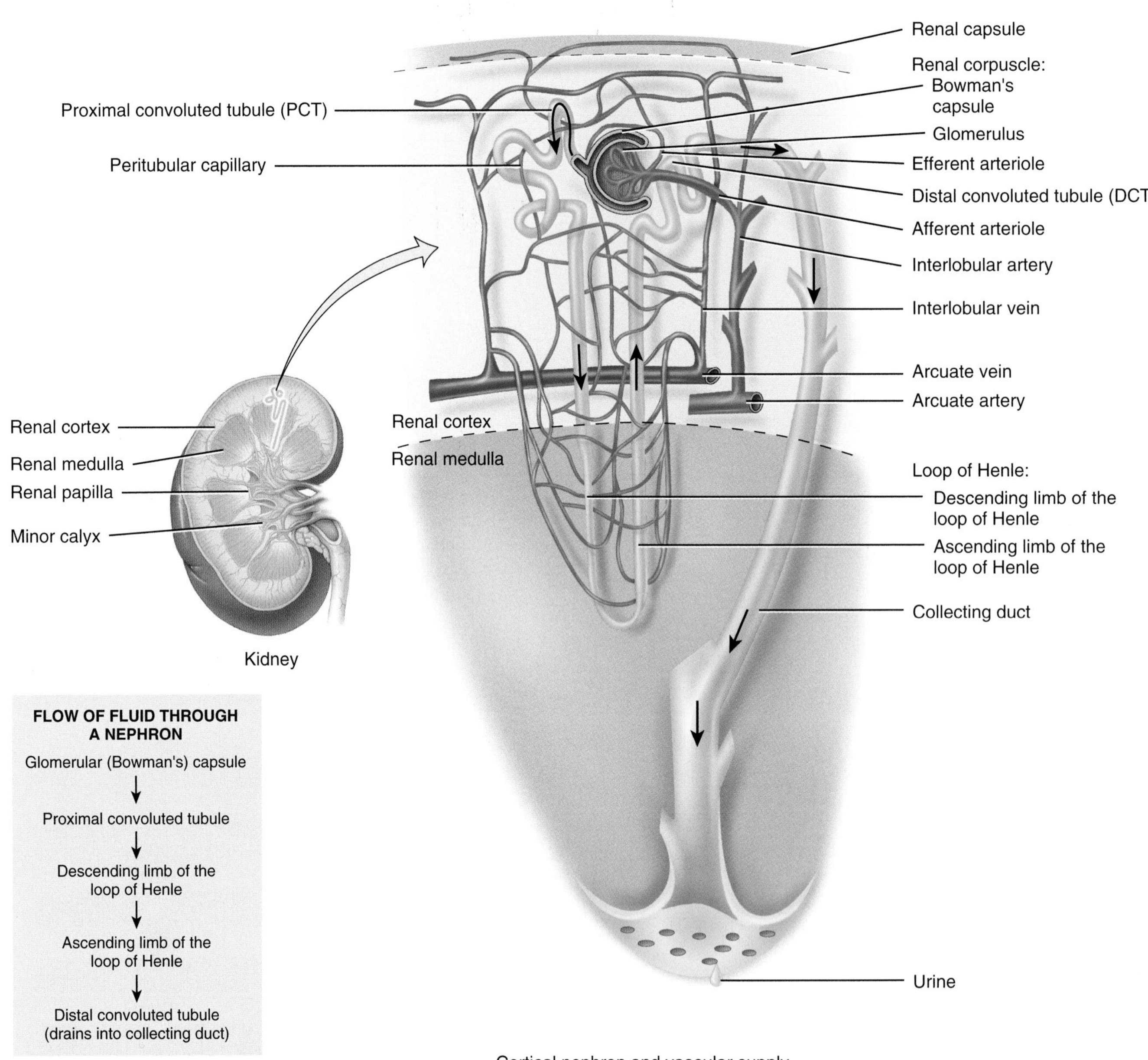

Nephron with capillary bed FIGURE 14.5

The nephron is the filtering unit of the urinary system. It is here that the homeostatic fluid balance of the entire system is carried out.

the efferent arteriole leads to a capillary bed surrounding the nephron. The blood from the capillary bed is collected in a renal venule and transferred to progressively larger veins until it leaves the kidney at the renal vein.

The nephron itself is composed of a glomerular capsule surrounding the glomerulus, a proximal convoluted tubule, a loop, and a distal convoluted tubule connected to a collecting duct. Each portion of the tubule has a distinct role in filtering blood, balancing ions and pH, and removing wastes.

The initial portion of the nephron, called Bowman's capsule, or the glomerular capsule, surrounds the glomerulus. The tubule that extends from the glomerular capsule is the proximal convoluted tubule (PCT). Proximal means "close to," and convoluted means "having twists or coils." This tubule is the one closest to the glomerulus, and it does have plenty of twists and turns.

Most of the nephrons begin in the cortex of the kidneys. The collection of many, many capsules and associated PCTs make up the outer cortex of the kidneys.

From the PCT, the newly filtered fluid is transported into the loop of the nephron. This portion of the nephron extends from the cortex into the medulla of the kidneys, making up a portion of the renal pyramid. The loop dives down into the medulla and back up to the cortex, where it joins with the distal convoluted tubule (DCT). Distal means "further from," indicating that this tubule lies some distance from the glomerulus. The DCT leads directly to the collecting duct. One collecting duct gathers newly formed urine from a series of nephrons and drains it to the renal pelvis. These collecting ducts comprise the majority of the renal pyramids.

The urine that reaches the renal pelvis is almost ready for excretion from the body. As it travels through the rest of the urinary system, it is subjected to small adjustments in composition before it is voided, or released. Owing to the action of the nephron, blood leaving the peritubular capillaries is cleansed, balanced, and ready to be transported to the rest of the body.

Amazingly, the body maintains more nephrons than it needs. This is not characteristic of the human body—usually when there is an excess of proteins, compounds, or structures, the body will break down the excess and retain only the bare minimum needed for survival. Recall that unused muscular tissue atrophies, leaving no sign of its existence. Literally millions of extra nephrons are maintained. We have enough filtering capacity in one kidney to provide all the cleansing and monitoring of fluid balance necessary for life. Having two kidneys allows us to donate a kidney for transplant and not suffer adverse effects on either fluid balance or general well-being.

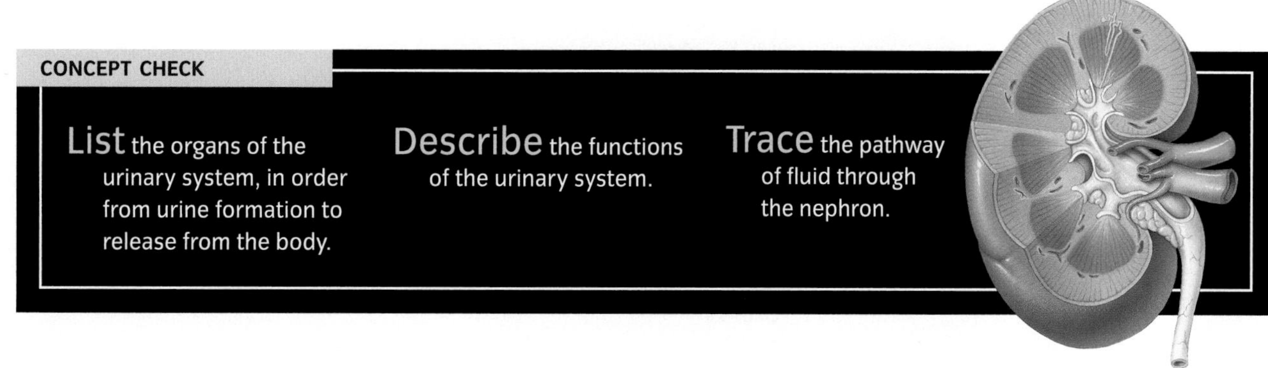

CONCEPT CHECK
List the organs of the urinary system, in order from urine formation to release from the body.
Describe the functions of the urinary system.
Trace the pathway of fluid through the nephron.

Urine Is Formed through Filtration and Osmosis

LEARNING OBJECTIVES

Define glomerular filtration.

Explain the functions of the PCT, loop, and DCT.

Urine formation begins in the glomerulus and is finalized in the renal pelvis, through the processes of filtration, active transport, and osmosis. As blood passes through the glomerulus of the nephron, most of the liquid is forced out of the arteriole and into the lumen of the nephron. This is glomerular filtration. The nephron capsule completely surrounds the glomerulus, looking like a sock balled up around the blood vessel. Water, nitrogenous wastes, nutrients, and salts are all forced from the blood at this point.

LIQUID IS FORCED INTO THE LUMEN OF THE GLOMERULUS

To understand how this filtration occurs, it may help to review the material on osmosis and pressure from Chapter 3. Glomerular blood pressure is higher than

Process Diagram

Glomerular pressure FIGURE 14.6

1 The pressure of blood in the glomerular vessel causes fluid to leak into the nephron.
2 The fluid already in the nepron exerts an opposite force. This back pressure prevents a huge influx of filtrate.
3 The blood, now thicker, pulls water back into it from the more watery filtrate in the nephron. The net filtration pressure is the sum of three forces: #1 pushing fluid into the nephron, while #2 and #3 pull water back from the nephron.
Afferent arteriole
Proximal convoluted tubule
NET FILTRATION PRESSURE (NFP)
Efferent arteriole
Bowman's capsule
Capsular space

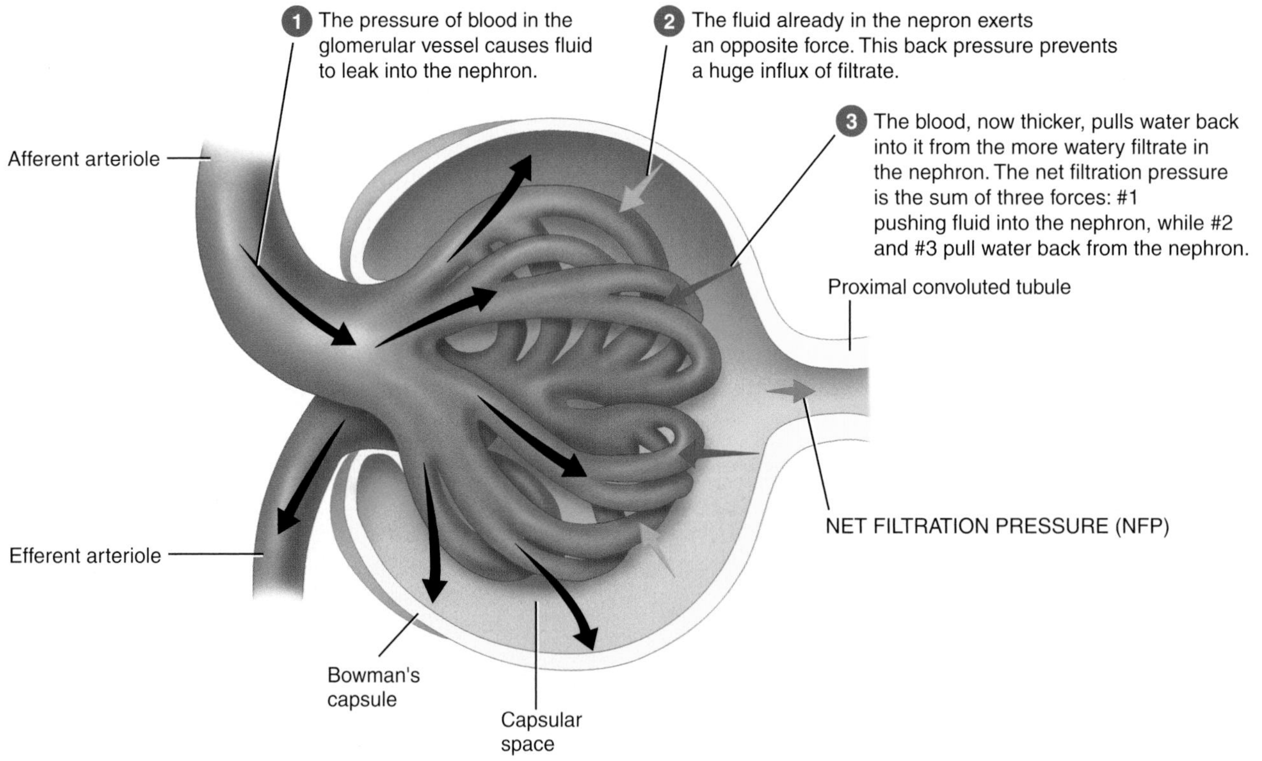

Three criteria must be met in order to filter the constituents of the blood plasma through the glomerulus. 1 Blood pressure must be high enough to force plasma out of the glomerular vessel walls; 2 the fluid already in the glomerulus must have a low enough pressure so more fluid can be forced into the nephron tubules; and 3 the osmotic pressure of blood in the peritubular capillaries must be high enough to draw water back into the capillaries from the nephron tubule. If these conditions are not met, the nephron cannot filter the blood, and the urinary system will fail.

www.wiley.com/
college/ireland

systolic blood pressure. This increase is partially caused by the kinking and twisting of the glomerular vessels. You have experienced this in a garden hose if you have ever bent it. The pressure increases because the water must travel past the obstructions. A similar phenomenon occurs in the glomerular vessels. In addition, the incoming (afferent) arterioles have a larger diameter than the outgoing (efferent) glomerular arterioles. This increases pressure in the glomerulus by creating back pressure. The total pressure on the blood forces most of the fluid into the capsule. To filter the blood, the blood pressure must overcome the pressure of the fluid already in the capsule (capsular pressure) as well as the osmotic pressure of the blood itself.

Because the glomerular system relies on pressure, there is a lower limit to its functioning (FIGURE 14.6). If your systolic pressure drops below 60 millimeters Hg, blood in the glomerulus will not be forced through the glomerular wall because glomerular pressure will not rise high enough to force plasma from the blood vessels. This leads to serious complications because the aqueous portion of the blood cannot filter into the nephron and therefore cannot be cleansed.

During filtration, the formed elements and plasma proteins remain in the glomerular vessel because they are too large to pass through the **fenestrations** of the cells that line the glomerulus (FIGURE 14.7). The proteins left in the capillary blood are essential because they set up the osmotic gradient that later pulls most of the water from the filtrate back into the blood. Every day, approximately 180 liters of fluid are filtered from the blood, but only a small fraction of that is excreted. Imagine how different life would be if we lost 180 liters of fluid every day! That is equal to 60 times the total plasma volume of the body. Not only would we have to drink constantly, but we would most likely also have a different social custom surrounding the need to urinate, because it would occur almost constantly. In the body as in the biosphere, recycling makes a real difference.

Fenestrations
Windows or openings between cells in the lining of the glomerulus.

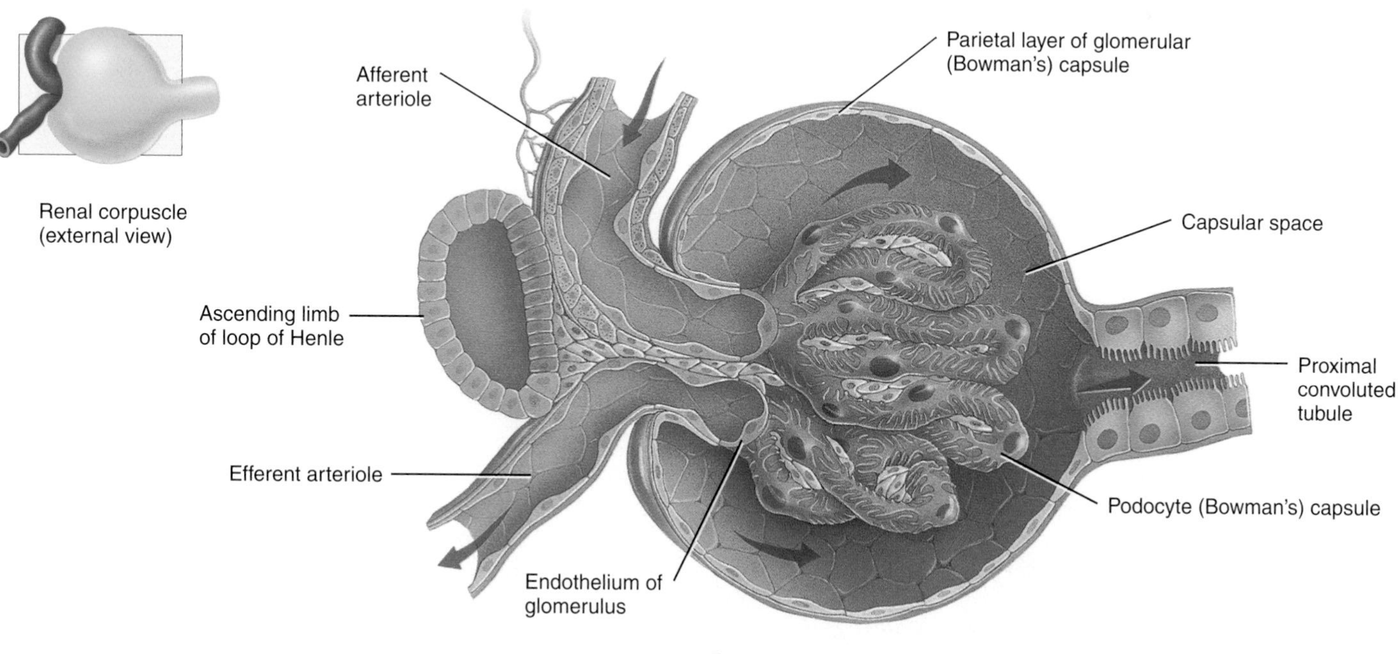

Renal corpuscle histology FIGURE 14.7

The blood is forced through the walls of the glomerulus by increased blood pressure. Water, ions, nutrients, and waste materials pass through the fenestrations of the podocytes (specialized cells comprising the glomerular walls).

How much water should I drink?

Water. It's just about the most popular nutrient around, and these days, hydration is almost a personal virtue. Adequate hydration is necessary for the functioning of the kidneys and most other body systems. Water maintains the blood's volume, helping it transport nutrients and remove waste. When water is scarce, homeostasis is threatened, especially during illness or physical exertion.

Hydration, combined with adequate sodium levels in the interstitial fluids, helps keep the osmolarity the same on both sides of the cell membrane. Underhydration, if not corrected by the urinary system through mechanisms discussed in the main text, can cause water to osmose from the cells into interstitial fluid. Overhydration, sometimes called "water intoxication," can cause water to osmose into cells, creating swelling, or edema.

So how much water should we drink? Historically, the standard advice was to drink eight 8-ounce glasses of water each day. Oddly, this widely trumpeted health advice did not have a scientific basis. In 2004, the National Academy of Sciences (NAS) came up with some scientifically driven drinking advice. Instead of performing experiments, the academy's experts concluded that healthy people who have access to food and water are almost never chronically dehydrated. So they set the water requirement by measuring the water intake of healthy, well-hydrated people.

The NAS researchers found that each day, healthy women in the United States consume about 2.7 liters of water from food and beverages, whereas men consume about 3.7 liters. About 80 percent of that amount comes from water and other beverages, and the rest from moisture in food. In general, the NAS experts said, hydration is not something to worry about because "thirst provides the body feedback that we are getting dehydrated, so that we can consume more fluids."

If this recommendation is accurate, the "8 by 8" advice (1.9 liters) seriously understated the need for water. According to the NAS recommendations, a man needs to ingest 3.0 liters of fluids (not counting moisture from food) to stay hydrated!

In hot, dry conditions, or during strenuous exercise, fluid consumption must increase. And that raises one final question: When is water a curse rather than a cure? In rare situations, athletes have died from hyponatremia, a low blood sodium concentration, after drinking too much water during long endurance competitions. The condition resembles water intoxication: Cells swell and can burst, as water responds to the higher sodium concentration by moving via osmosis from interstitial fluid into the cells. So instead of guzzling gallons of water during athletic events, it may be better to swill sports drinks, which contain sodium and other electrolytes, or even eat some potato chips as you drink.

TUBULAR REABSORPTION RECYCLES WATER TO THE BLOOD

As filtrate passes through the nephron, ions and water are returned to the peritubular capillaries in a process called tubular reabsorption. Approximately 80% of the filtered water is returned to the blood immediately at the PCT. Glucose, amino acids, and salts are also returned to the bloodstream. The walls of the proximal convoluted tubule have a large surface area to accommodate all this reabsorption. The cells that line the PCT are covered with **microvilli**. These cells are adjacent to the endothelial cells of the peritubular capillaries, creating a thin layer that allows diffusion from the tubule to the blood.

Essential ions and water are sent back to the blood via osmosis and diffusion (FIGURE 14.8). Glucose returns using **facilitated diffusion**. The walls of the PCT have a finite number of glucose receptors to pick up glucose from the filtrate. Normally, there are enough receptors to remove all the glucose from the filtrate and return it to the blood. But excess glucose in the blood will overrun these receptors and drop from the PCT into the loop of the nephron. Once beyond the PCT, glucose cannot be returned to the bloodstream. It is said to "spill" into the urine because it literally spills into the loop. One symptom of diabetes mellitus is glucose spilling as a result of very high levels of glucose in the original filtrate.

Waste products and other unwanted substances too large to filter from the blood at the glomerulus, such as steroids and drug breakdown products, are actively **secreted** into the filtrate at the distal convoluted tubule (FIGURE 14.9).

Microvilli Small hair-like projections extending from the free surface of epithelial cells.

Facilitated diffusion Moving substances from high concentration to low with the assistance of a carrier molecule.

Secreted Moved from the blood to the filtrate, using energy.

PCT function FIGURE 14.8

Reabsorption of water, bicarbonate ions, organic solutes including glucose and amino acids, and important ions such as sodium, calcium, and potassium, occur in the PCT.

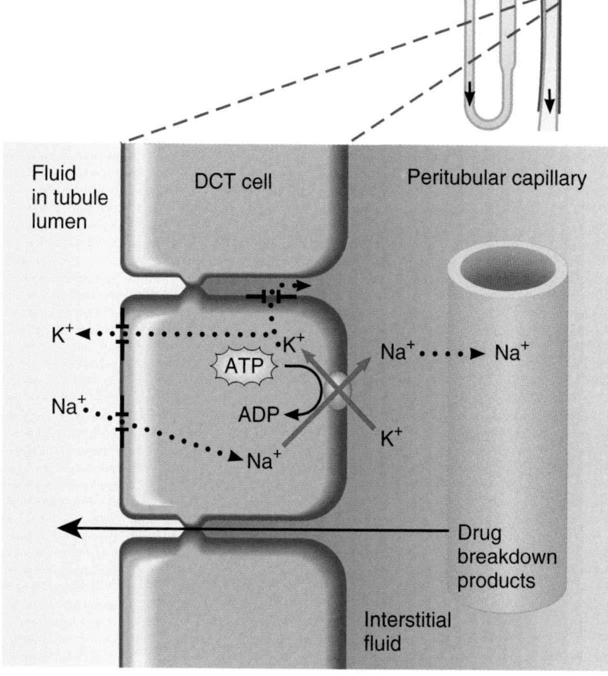

Key:
Diffusion
Leakage channels
Sodium-potassium pump

DCT functioning FIGURE 14.9

The cells surrounding the DCT are especially susceptible to hormonal controls, allowing a final adjustment of blood composition.

Tubular secretion requires ATP, and provides delicate control over fluid homeostasis. Tubular secretion provides a final fine-tuning of the dissolved compounds in the blood. This process also provides clues as to the amount and type of drugs that are traveling through the body. Most of the breakdown products of drugs, both pharmaceutical and recreational, are large and must be secreted into the nephron. The Ethics and Issues box on page 469 discusses this in detail.

The loop of the nephron and the collecting duct remove even more water from the filtrate, serving to precisely regulate fluid loss (FIGURES 14.10 and 14.11). Interestingly, the descending arm of the loop of the nephron is permeable to water, but the ascending limb is not. Therefore, water leaves the filtrate as it moves down the loop of the nephron, and salts leave the filtrate as it flows up the ascending arm, creating a salt gradient in the medulla of the kidney. This is referred to as **countercurrent multiplication**, or **CCM**, because the ascending and descending loop flow opposite one another. Each current affects the water and salt concentration in the other. The collecting ducts pass right through the salt gradient set up by the CCM, providing one last opportunity to remove water from the urine before sending it on to the ureters. Water is a vital fluid and as such must be carefully monitored. The Health, Wellness, and Disease box, "How much water should I drink?" on page 454 takes a look at the role of water in personal health.

Countercurrent multiplication (CCM)
Increasing the diffusion rate by flowing solutions in opposite directions on either side of the diffusion membrane.

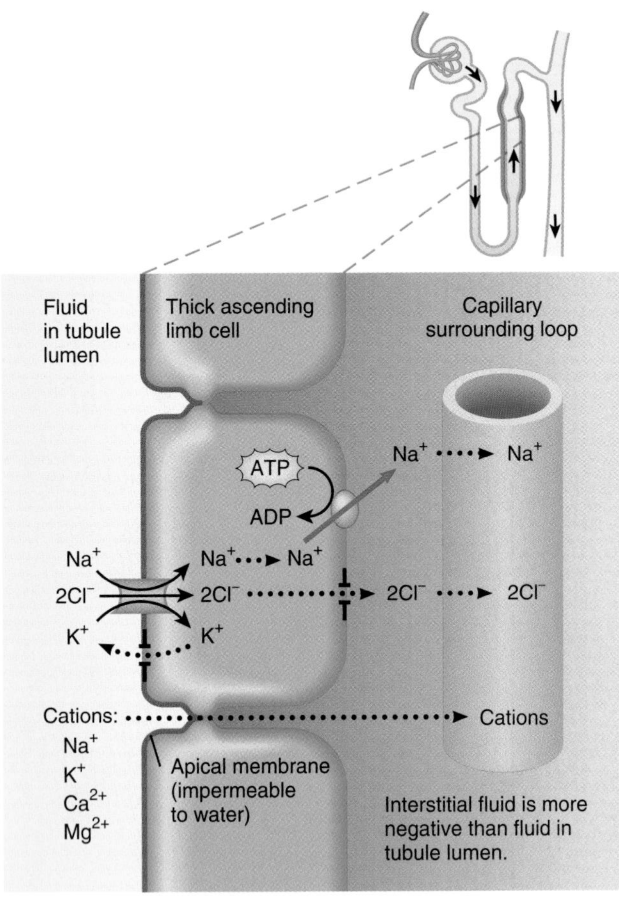

Key:

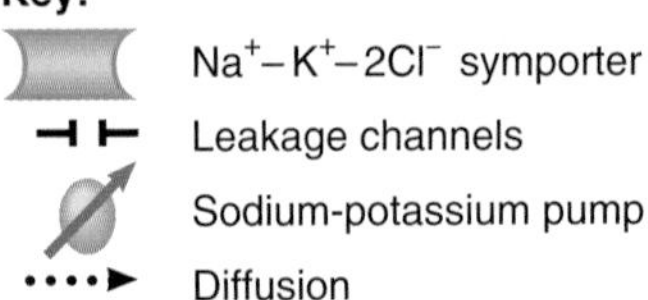

Loop functioning FIGURE 14.10

The countercurrent multiplication activity of the nephron loop can be easily seen in this diagram. Because of this action, the medulla of the kidney is a very salty area.

Urine formation FIGURE 14.11

Urine formation begins in the glomerulus, where blood plasma is filtered and collected by the renal corpuscle. At the PCT, most of the water and many ions and nutrients are reabsorbed by the blood. The loop removes more water and ions, setting up a salt gradient in the medulla of the kidney. The distal convoluted tubule allows for final adjustments in the composition of urine and blood by actively secreting larger substances from the blood to the urine.

RENAL CORPUSCLE

Glomerular filtration rate: 105–125 mL/min

Filtered substances: water and all solutes present in blood (except proteins) including ions, glucose, amino acids, creatinine

DISTAL CONVOLUTED TUBULE

Reabsorption (into blood) of:

Water	10–15% (osmosis)
Na^+	5%
Cl^-	5%
Ca^{2+}	variable

PROXIMAL CONVOLUTED TUBULE

Reabsorption (into blood) of filtered:

Water	65% (osmosis)
Glucose	100%
Amino acids	100%
Na^+	65%
K^+	65%
Cl^-	50%
HCO_3^-	80–90%
Ca^{2+}, Mg^{2+}	variable
Urea	50%

Secretion (into urine) of:

H^+	variable
Ammonia	variable
Urea	variable
Creatinine	small amount

LAST PART OF DISTAL TUBULE AND COLLECTING DUCT

Reabsorption (into blood) of:

Water	5–9% (insertion of water channels stimulated by ADH)
Na^+	1–4%
HCO_3^-	variable amount
Urea	variable

Secretion (into urine) of:

K^+	variable amount to adjust for dietary intake (leakage channels)
H^+	variable amounts to maintain acid–base homeostasis (H^+ pumps)

Tubular fluid leaving the collecting duct is dilute when ADH level is low and concentrated when ADH level is high.

Urine

LOOP OF HENLE

Reabsorption (into blood) of:

Water	15% (osmosis in descending limb)
Na^+	20–30% (ascending limb)
K^+	20–30% (ascending limb)
Cl^-	35% (ascending limb)
HCO_3^-	10–20%
Ca^{2+}, Mg^{2+}	variable

Secretion (into urine) of:

Urea	variable

CONCEPT CHECK

List two ways that blood pressure is increased in the glomerulus.

What is the main function of the PCT?

How does the DCT help monitor and adjust blood composition?

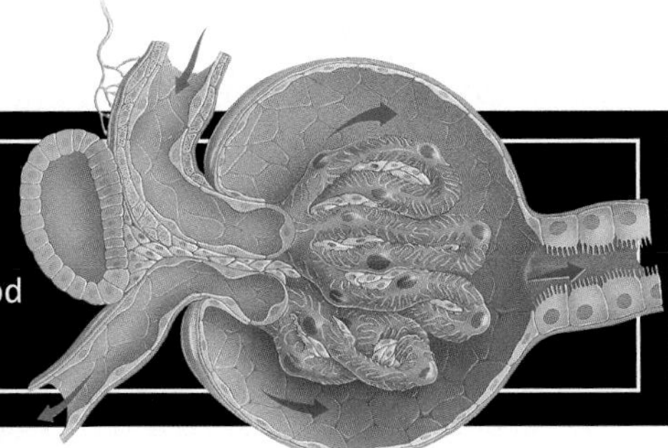

Urine Is Transported to the Bladder for Storage

LEARNING OBJECTIVES

Outline the functions of the bladder and the urethra.

Understand why females are more susceptible to urinary tract infections.

Once the filtrate has passed through the nephron and collecting ducts and reaches the renal pelvis, it is referred to as urine. Most of the fine-tuning of ion concentration and water content is completed by this point (TABLE 14.1). Water can still be removed as the urine sits in the remaining organs of the urinary system, but the salt content is relatively stable.

While in the renal pelvis, water can continue to leave the urine, concentrating the salts in the urine, which can lead to the formation of kidney stones. These rock like masses, usually composed of **calcium oxalate**, can grow large enough to block renal flow (FIGURE 14.12). Kidney stones are extremely painful as they move through the urinary pelvis and can become lodged in the kidney or the ureters. Some kidney stones are jagged or pointy, making them even more likely to jam. Removal of kidney stones rarely requires medical assistance. Drinking lots of water and resting as the stone moves through the renal pelvis and ureter often do the trick, but some stones are too large to pass. These may be broken apart by ultrasound waves so the fragments can be excreted. Because kidney stones often reappear, patients are advised to avoid foods high in calcium, eat less protein (to decrease urine acidity), and drink more fluids, especially water.

Calcium oxalate A chemical compound composed of calcium ions bound to the oxalate ion ($C_2O_4^{2-}$).

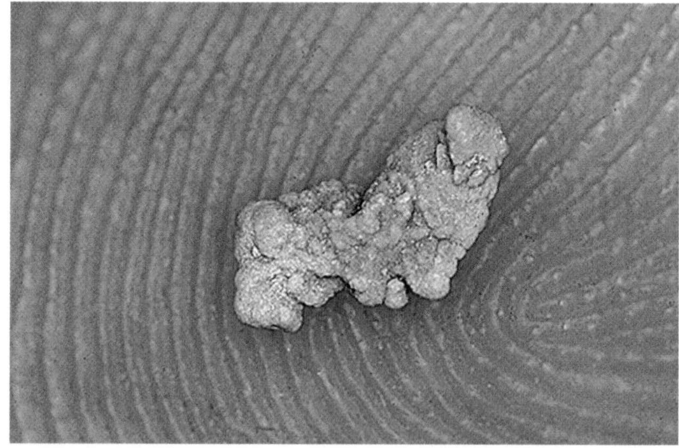

Kidney stone FIGURE 14.12

A typical small kidney stone is seen here on the tip of a finger. Stones can be as large as a pearl or even, rarely, a golf ball.

Substances filtered, reabsorbed, and excreted in urine per day TABLE 14.1

Substance	*Filtered* (enters renal tubule)*	*Reabsorbed (returned to blood)*	*Excreted in Urine*
Water	180 liters	178–179 liters	1–2 liters
Chloride ions (Cl^-)	640 g	633.7 g	6.3 g
Sodium ions (Na^+)	579 g	575 g	4 g
Bicarbonate ions (HCO_3^-)	275 g	274.97 g	0–0.03 g
Glucose	162 g	162 g	0
Urea	54 g	24 g	30 g†
Potassium ions (K^+)	29.6 g	29.6 g	2.0 g‡

* Assuming glomerular filtration is 180 liters per day.

† In addition to being filtered and reabsorbed, urea is secreted.

‡ After virtually all filtered K^+ is reabsorbed in the convoluted tubules and loop of Henle, a variable amount of K^+ is secreted in the collecting duct.

From the renal pelvis, urine travels down the ureters (FIGURE 14.13) to the urinary bladder. The ureters are long, thin muscular tubes lined with mucosa. The ureters loop behind the urinary bladder and enter it at the base. This allows the bladder to expand upward without dislodging the ureters.

With every heartbeat, blood is pushed into the glomerulus and filtered. The nephrons constantly form urine, so the tubes and ducts of the urinary system are always full of fluid. As more urine is produced, it pushes what is already formed down the ureters and into the bladder, where small contractions move the urine toward the bladder.

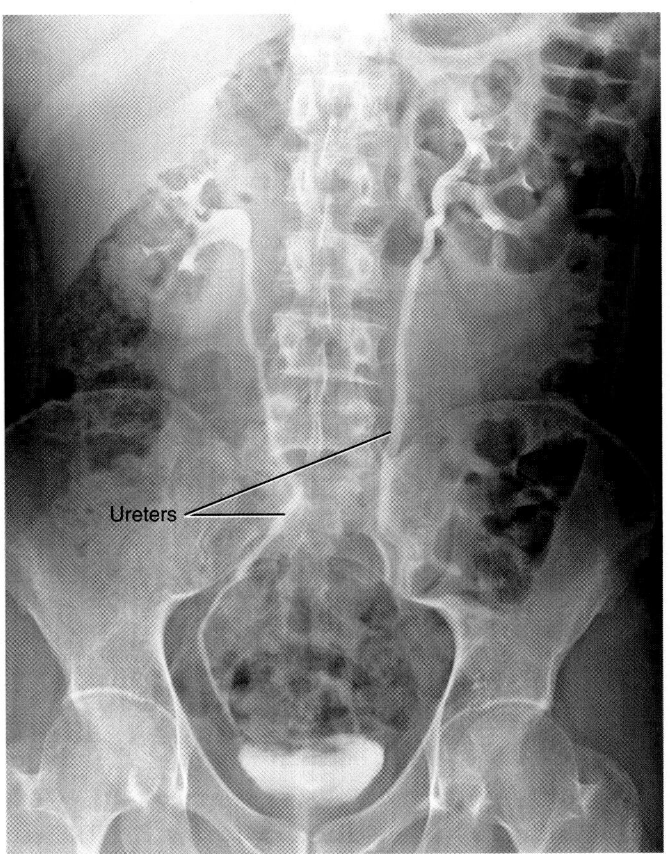

Ureters FIGURE 14.13

Ureters carry urine from the renal pelvis to the urinary bladder. They are approximately 20 centimeters long and curl behind the urinary bladder to enter from the trigone, or base of the bladder. These tubes are ringed with smooth muscle, which helps propel the urine to the bladder.

THE URINARY BLADDER STORES URINE BEFORE RELEASE

The urinary bladder is a hollow, variable-sized organ (FIGURE 14.14). It lies in the pelvic cavity, posterior to the pubic bones and the pubic symphysis. The base of the bladder has a triangular area where the two ureters enter and the urethra exits. This area is called the trigone. The bladder is lined with transitional epithelium to allow for expansion without tearing or

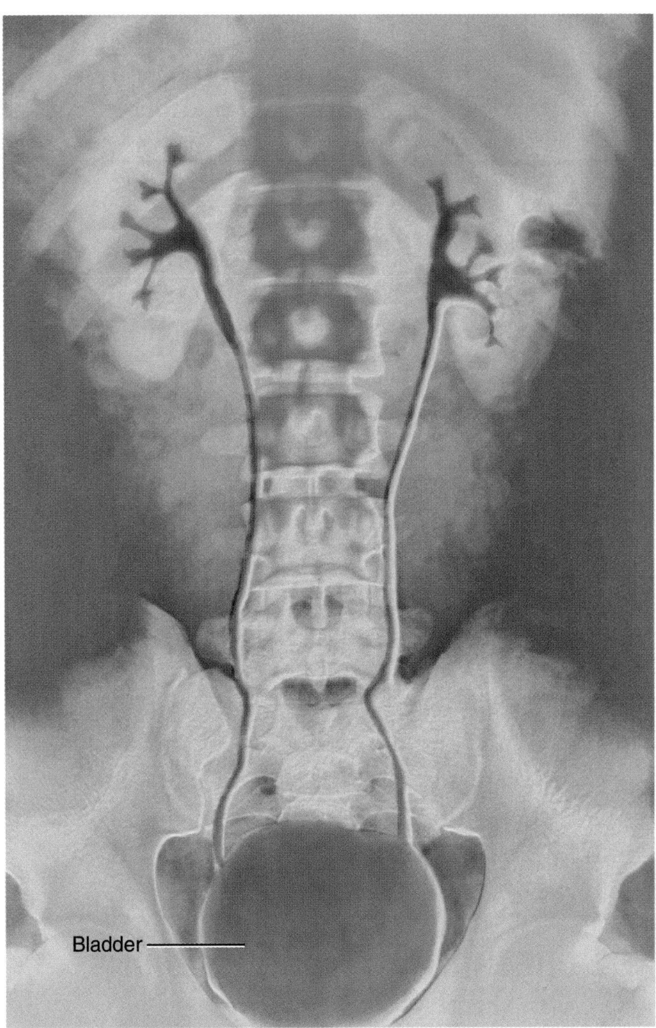

Bladder FIGURE 14.14

Transitional epithelium lines the walls of this distensible organ. These cells also secrete mucus to protect the bladder from toxic compounds in the urine.

destroying the integrity of the inner lining. The empty bladder is the size of a walnut, but can stretch to hold up to 800 ml of fluid in males, and slightly less in females.

Discharging urine from the bladder is called urinating, voiding, or micturition. This reflex involves both smooth and skeletal muscles. Urine is constantly forming and draining into the bladder. When the bladder contains approximately 300 ml, pressure in the bladder stimulates stretch receptors that send nerve impulses to the micturition center. The micturition reflex causes contraction of the walls of the bladder and relaxation of the **internal urethral sphincter** muscle. Urine moves down into the urethra, pressing on the **external sphincter muscle**. At this point, you can consciously control the opening of the external urinary sphincter. Should you choose not to empty the bladder, the urge to urinate will subside until the next 300 ml collects in the bladder.

Internal urethral sphincter Ring of involuntary smooth muscle that keeps the urethra closed.

External sphincter muscle Ring of voluntary skeletal muscle that closes the urethra.

As we mature, we learn to anticipate and control this reflex, but we cannot delay micturition indefinitely. The bladder continues to expand, and a second reflex will begin shortly. Just as we are not able to hold our breath until we die, we cannot retain urine until the bladder bursts. When the bladder reaches 700 to 800 ml, micturition occurs despite our best efforts to control the external urethral sphincter.

THE URETHRA TRANSPORTS URINE OUT OF THE BODY

When micturition occurs, the urine leaves the body via the urethra, a single tube extending from the trigone of the bladder to the exterior. In females, the urethra is a short 5 centimeters, emptying in front of the vaginal opening. The male urethra is almost four times longer because it runs the length of the penis (FIGURE 14.15). The urinary and reproductive systems join in the male, sharing the male urethra. In the female, the two systems are separate. The female urethra carries only urine, and the female reproductive tract opens at the vagina.

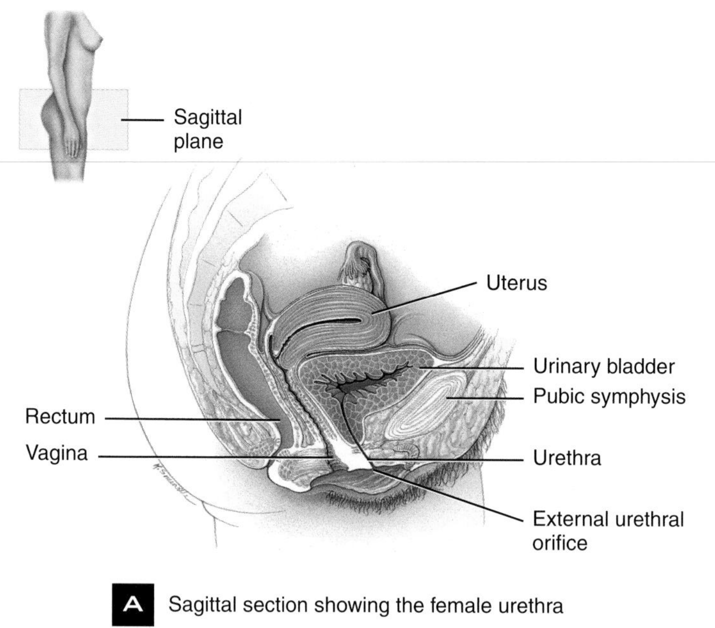

A Sagittal section showing the female urethra

Comparison of female and male urethras
FIGURE 14.15

Because the distance from the exterior to the bladder is shorter in females, they suffer far more urinary tract infections (UTIs). Bacteria outside the body can travel the short distance up the urethra and colonize the bladder, resulting in painful urination, often accompanied by bleeding from the irritated bladder walls. (If the urine contains glucose, the bacteria multiply even faster.) UTIs are serious infections that must be cleared up. If the bacteria remain in the bladder, they will eventually travel up the ureters and colonize the pelvis and tubules of the kidney. Kidney infections are painful and serious because they block normal kidney function and can lead to kidney failure.

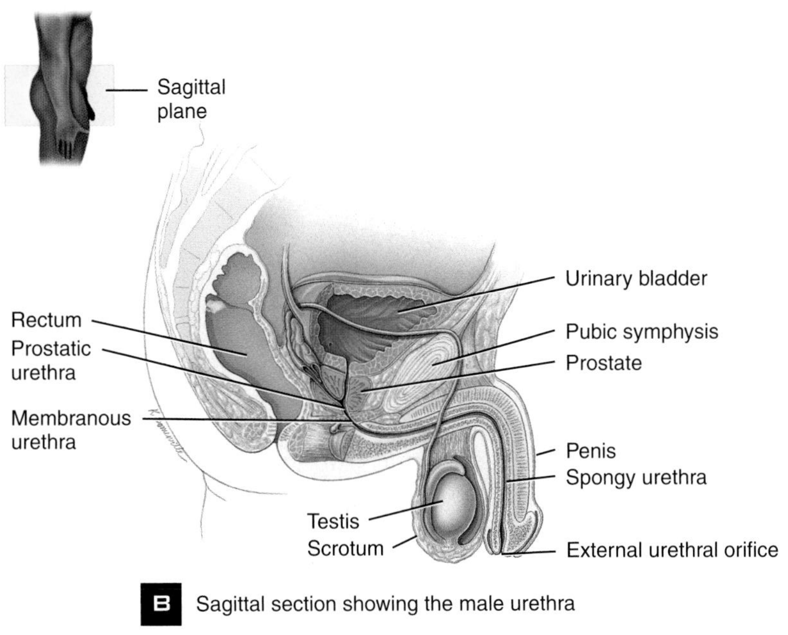

B Sagittal section showing the male urethra

INCONTINENCE IS THE LOSS OF CONTROL OVER VOIDING

As we age, many things change, including our ability to control micturition when the urge arises. **Incontinence** can and does occur in all age brackets, genders, and social levels, but it is far more common in elderly women. Perhaps the stress of bearing children weakens the muscles of the pelvic floor, leading to greater difficulty controlling these muscles in later years.

An estimated 12 million Americans suffer incontinence, and most do not require surgery. Incontinence can be a symptom of many different pathologies but is not a pathology in its own right. Causes include chronic urinary tract infections, side effects of medication, muscular weakness, an enlarged prostate gland in males, **constipation**, or neuromuscular disease. There are three types of incontinence determined by the underlying cause of the problem, each with the same result. **Stress incontinence** is the leaking of urine during physical exertion. **Urge incontinence** is the inability to quell the urge to urinate. **Overflow incontinence** is the overflowing of the urinary bladder caused by waiting too long before urinating, as happens in young children who are learning to control their sphincter muscles. Treatment for incontinence is tailored to the cause. Muscular strengthening exercises or behavioral modification may be recommended.

Incontinence The inability to prevent urine leakage.

Constipation Difficult or infrequent defecation, leading to dry, potentially painful fecal evacuation.

CONCEPT CHECK

What is the micturition reflex?

How does the urethra differ in males and females?

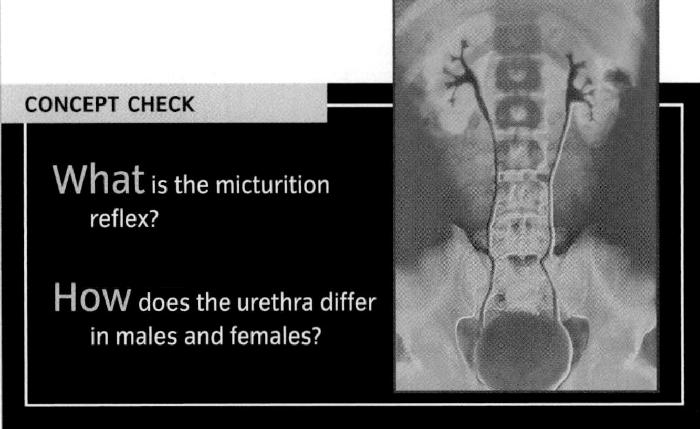

The Urinary System Maintains the Body's Water–Salt Balance

LEARNING OBJECTIVES

Describe the hormones used to regulate fluid balance.

Explain the function of ADH, aldosterone, ANP, and BNP.

A key function of the kidneys is to maintain the body's water and salt balance. Excreted urine usually has a much different **osmolarity** than the blood. When originally filtered from the blood, the fluid in the nephrons has the same water-to-solute ratio as the blood. As it moves through the nephron, this ratio changes to produce concentrated or dilute urine, depending on the body's demands. Dilute urine is produced by removing **solutes** from the forming urine leaving the nephron. Water cannot pass back across the walls of the DCT or the collecting tubule. As the ion concentration drops in the urine, the water proportion increases, so fluid reaching the collecting duct is far less concentrated than blood plasma, resulting in dilute urine (FIGURE 14.16).

Osmolarity Osmotic pressure of a solution.

Solutes Substances dissolved in the solvent.

Concentrated urine is produced by the reabsorption of water at the loop of the nephron and the collecting duct (FIGURE 14.17). The cells of the DCT can be controlled to reabsorb water by the presence of certain hormones. Water can also be reabsorbed across the walls of the urinary bladder. This reabsorption explains why the first morning urination is more concentrated than urine produced and passed later in the day.

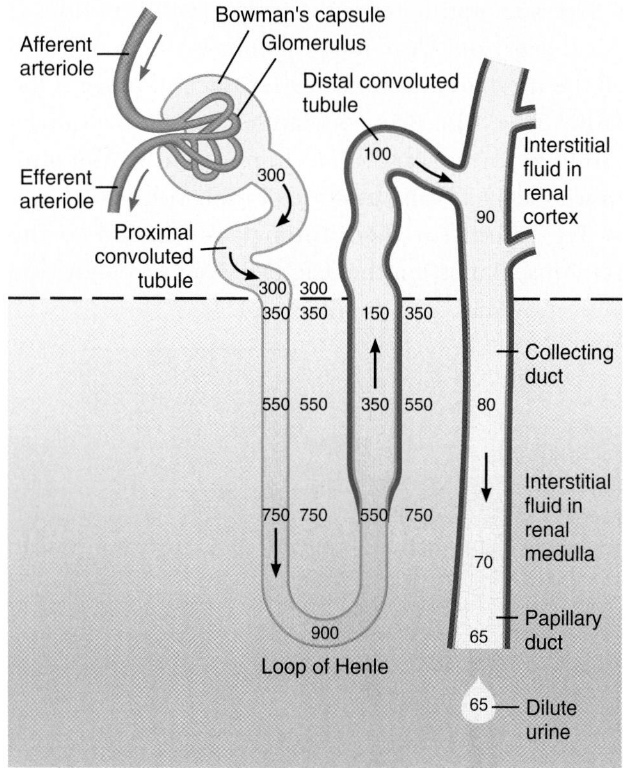

Dilute urine formation FIGURE 14.16

The numbers shown here represent the osmolarity of the filtered fluid in milliosmoles per liter. The brown lines surrounding the ascending loop of the kidney and the DCT show where water cannot leave the tubule. The blue lines indicate areas of the collecting duct that are impermeable to water only when ADH is not present. If ADH is absent, urine osmolarity can be as low as 65 mOsm/liter, which is very dilute.

Formation of concentrated urine FIGURE 14.17

The area of the nephron outlined in green indicates where ions are reabsorbed into the interstitial fluid. The nephron here is also relatively impermeable to water, creating a salty area in the medulla of the kidney through which the collecting ducts travel. As fluid moves through this salty area, water is pulled from the collecting ducts, resulting in concentrated urine. All numbers are again in milliosmoles per liter.

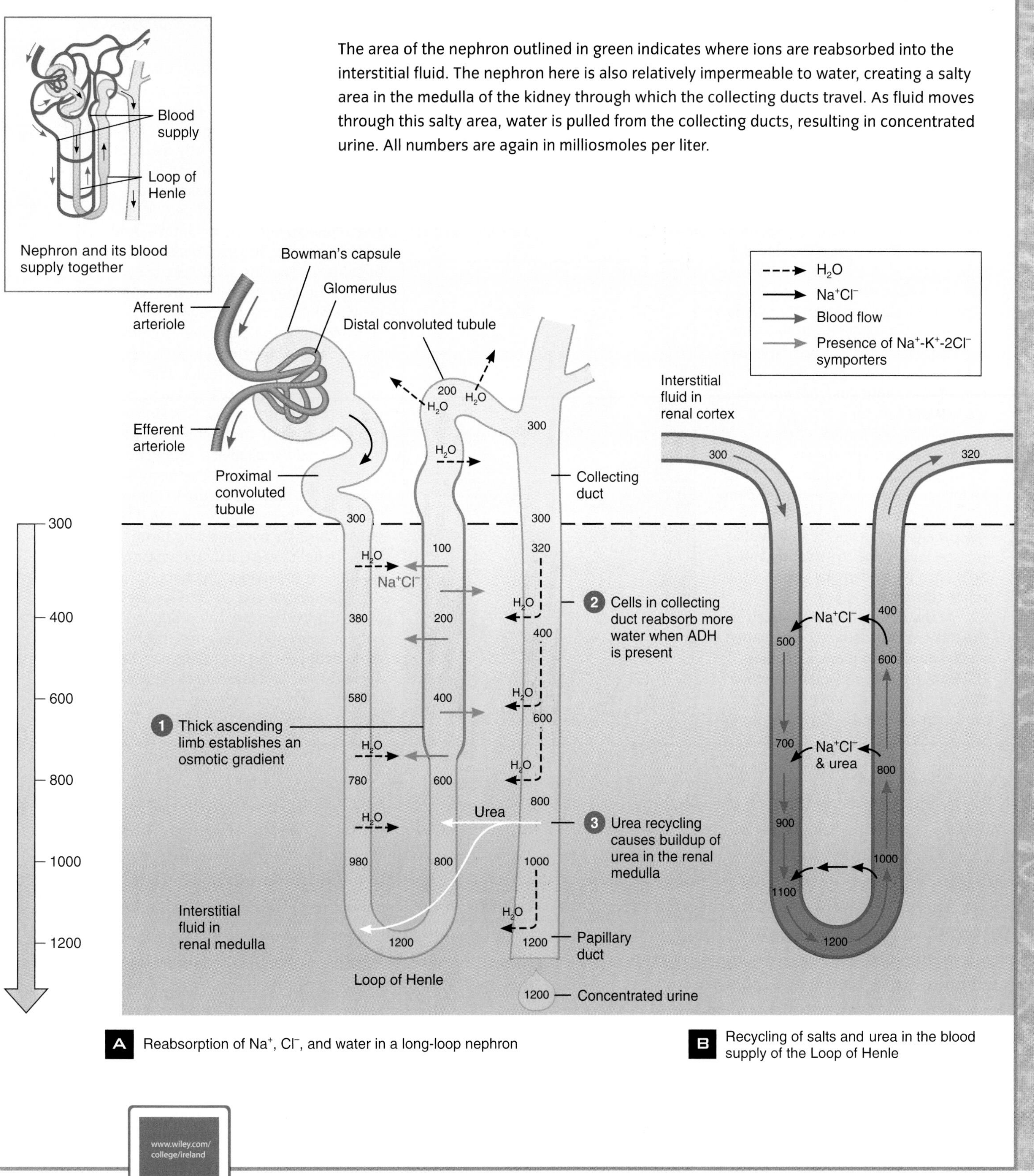

A Reabsorption of Na^+, Cl^-, and water in a long-loop nephron

B Recycling of salts and urea in the blood supply of the Loop of Henle

www.wiley.com/college/ireland

I WONDER . . .

Why do they keep telling me to cut down on salt?

Table salt is sodium chloride, and sodium is about as essential as electrolytes get because it helps control osmosis throughout the body. But eating a lot of salt can raise the blood pressure by causing a subtle swelling of the tissues. Over long periods, hypertension can cause deadly or disabling strokes, heart attacks, heart failure, or kidney failure.

For years, we've been told to cut down on the salt. But that is easier said than done. As mentioned in the text, the kidney is built to recycle salt, enabling some people to survive on less than 1 gram per day. Salt improves the taste of food, and evolution has forced us to crave salt. Sodium chloride is so vital that in the hot, dry Sahel region bordering the Sahara Desert, salt was once traded—gram for gram—for gold. You can certainly live without gold, but you will die without salt.

In 2004, the National Academy of Sciences waded into the salt debate by suggesting that people aged 19 to 50 ingest at least 1.5 grams of sodium (3.75 grams of salt) per day. This intake, the group's experts said, would replace losses due to sweating and help ensure an adequate supply of other nutrients.

At the other end of the spectrum, the National Academy recommended a maximum of 2.3 grams of sodium (5.75 grams salt), and mentioned the danger of hypertension associated with increased levels. For many Americans and Canadians who eat processed food, that maximum poses a problem. Salt is liberally added to most processed food because it improves shelf life and people like the taste.

Interestingly, the "cut the salt" advice is hotly debated. Some large clinical trials have associated lower salt levels with lower blood pressure, which is usually desirable. But some experts note that only about one-third of people get a strong hypertensive response to high sodium intake. And in some studies, the cardiovascular benefits of reducing salt intake show up only among people who were overweight when the study began.

Adding to the confusion, a study reported in 2006 found a higher, not lower, rate of cardiovascular deaths among people who ate less than 2.3 grams of sodium per day. This kind of result leads some researchers to question the traditional advice about cutting down on salt. Even if reducing salt does reduce blood pressure in the short term, there is no proof that it reduces mortality over the long term. If these benefits exist, a definitive clinical trial has yet to identify them.

Obviously if you want to reduce chronic hypertension, salt reduction is not your only option. Exercise helps, as do antihypertensive medicines. Reducing stress and learning to relax will also reduce blood pressure, without making your food taste bland.

Water can be reabsorbed into the bloodstream at the DCT and collecting duct with help from the hormone ADH (antidiuretic hormone) (FIGURE 14.18). A diuretic increases the volume of urine produced, whereas an antidiuretic has the opposite effect. ADH will therefore decrease urine volume. ADH is secreted by the posterior lobe of the pituitary gland, located on the undersurface of the brain, in response to blood volume. ADH in the blood causes the cells surrounding the collecting duct and the DCT to remove more water from the urine, returning it to the depleted bloodstream.

Surprisingly, sodium is conserved almost as stingily as water. It is important to remember that where sodium goes, water follows. If you live in a humid area, you already know this. In the Deep South, the Midwest in summer, the southern shorelines of the East and West Coasts, and on Pacific islands, humidity can clog salt shakers because sodium chloride draws water molecules from the air, causing clumps in the salt shaker. If you add a few grains of uncooked rice to the salt shaker, they will absorb the water from the salt and prevent the salt crystals from sticking together.

In the nephron, this attraction between water and sodium is used to good advantage. More than 99 percent of the sodium filtered from the blood at the glomerulus is returned before the urine leaves the nephron. Two-thirds of this reabsorption occurs at the PCT. Another 25 percent of the filtered sodium is removed from the forming urine at the ascending limb of

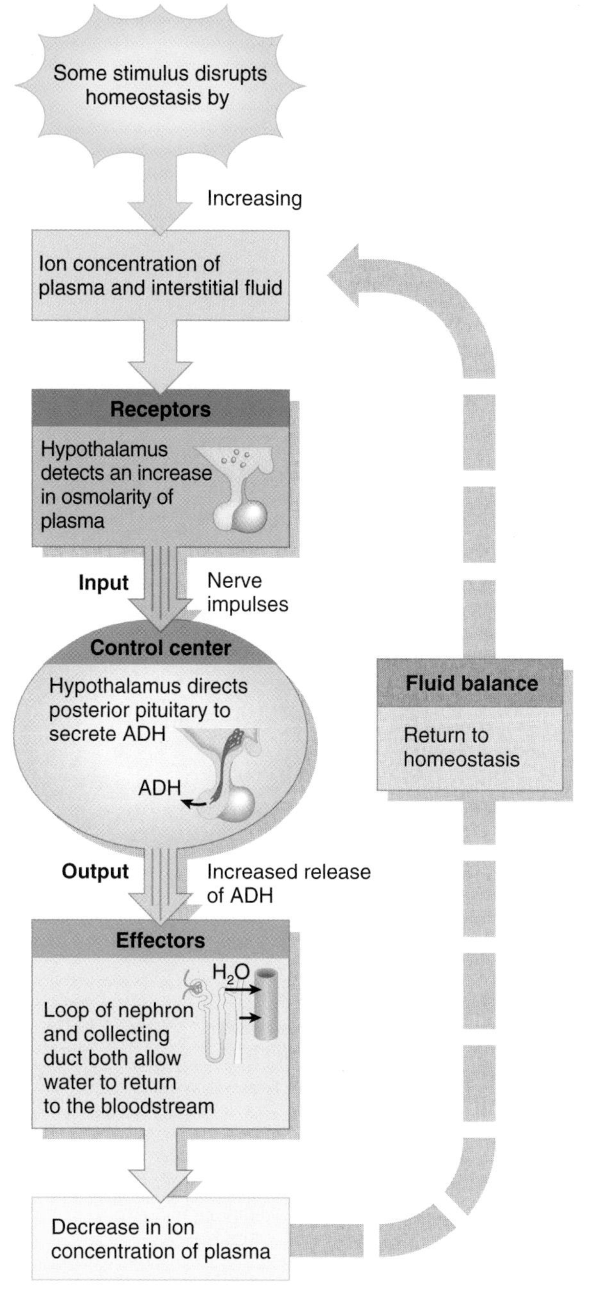

ADH feedback system FIGURE 14.18

If you drink less water than you need, ADH will be secreted to preserve the volume of water in your body. A small volume of more concentrated urine will be produced. Conversely, if you drink a lot of water, ADH will not be secreted and more fluid will be lost through the urinary system.

the loop of the nephron. This loop sets up a sodium gradient in the medulla of the kidney by removing sodium from the filtrate. Sodium is also reabsorbed from the DCT and the collecting duct, so sodium levels are strictly maintained. (For more on salt consumption, see the I Wonder . . . box.)

Several hormones are involved in salt regulation. Aldosterone, atrial natriuretic peptide (ANP), and brain natriuretic peptide (BNP) all regulate sodium reabsorption at the distal convoluted tubule. Aldosterone causes the excretion of potassium ions and the reabsorption of sodium ions, so water will leave the filtrate with the sodium ions rather than the potassium ions. ANP and BNP both oppose ADH. When either is present, the kidneys produce lots of dilute urine.

Ingested chemicals can also affect nephron function. Caffeine and alcohol both increase urine production, apparently through decreased ADH production. When caffeine is ingested in quantities below 350 mg, we experience central nervous system stimulation, decreased sleepiness, and possible increases in athletic performance. But the side effects include headache and drowsiness as the caffeine wears off, and insomnia. The diuretic effects of caffeine are not particularly helpful. Dehydrated muscles cramp more easily and are less likely to be repaired after injury. Your body cannot achieve peak function without good hydration.

CONCEPT CHECK

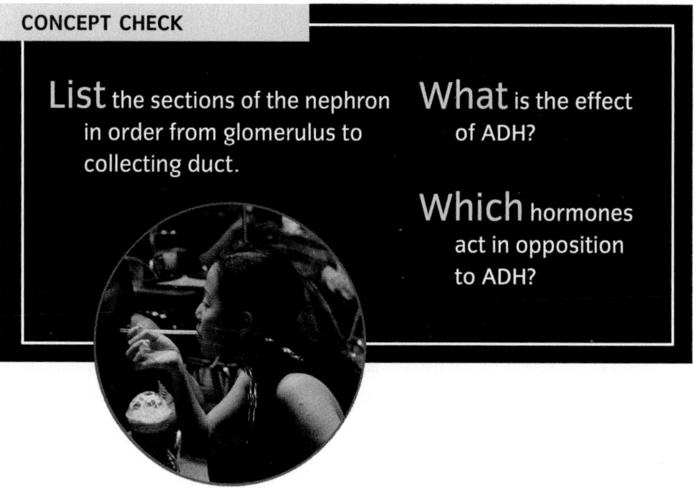

List the sections of the nephron in order from glomerulus to collecting duct.

What is the effect of ADH?

Which hormones act in opposition to ADH?

The Kidneys Help Maintain the Blood's Acid–Base Balance

LEARNING OBJECTIVES

Define the carbonate buffering system of the blood.

Explain the role of the kidneys in maintaining blood pH.

Body pH must be held within a narrow range (7.35 to 7.45). This is done primarily through the bicarbonate buffer system of the respiratory system, with help from the urinary system. This pH stability is achieved through chemical equilibrium. In the body, all three of the product sets shown below are in equilibrium, balanced as if on a teeter-totter. Adding water to the body increases the reactants at the right (H_2O + CO_2), causing the amount of H_2CO_3 and $H^+ + HCO_3^-$ to increase proportionately as the reactions return to equilibrium. This is called "pushing the reaction to the left."

$$H^+ + HCO_3^- \Rightarrow H_2CO_3 \Rightarrow H_2O + CO_2$$

When carbon dioxide is exhaled, the above reactions are "pushed to the right." In order to maintain an equal concentration of reactants on either side of the arrows, more hydrogen ions are picked up by the bicarbonate ion (HCO_3^-) and removed from the blood, returning the reaction to a point where all three compartments are equal (FIGURE 14.19). This homeostatic function is so vital that the rate and depth of breathing respond to the level of carbon dioxide in

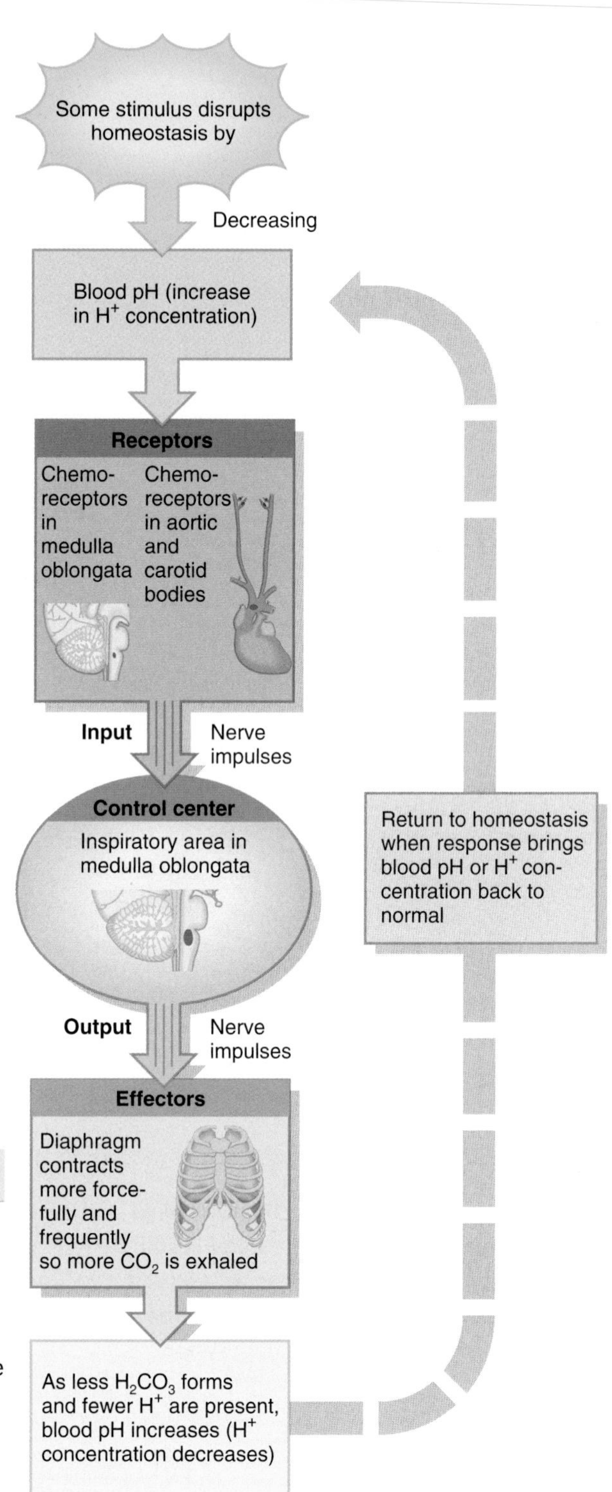

Respiratory system regulation of blood pH FIGURE 14.19

The breathing rate is controlled by the level of bicarbonate ion (carbon dioxide) in the blood. This in turn maintains the buffering capacity of the blood, so the fluid balance of the body is in homeostasis. As the breathing rate increases, carbon dioxide is washed out of the blood, decreasing the bicarbonate ion level. This in turn can cause dizziness as the body responds to the loss of buffering capacity. The kidneys must work harder to maintain fluid homeostasis and ion concentrations when this happens. There is a time lag in rectifying the situation because the alteration of blood composition by the kidneys is a bit slower than the removal of carbon dioxide via breathing.

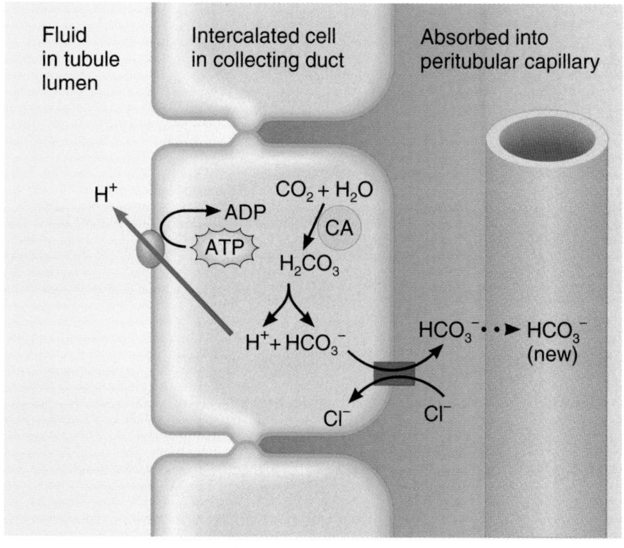

A Secretion of H^+

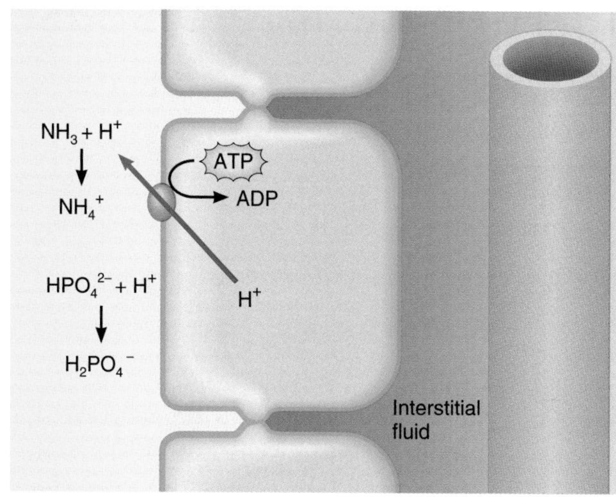

B Buffering of H^+ in urine

Key:

 Proton pump (H^+ ATPase) in apical membrane

HCO_3^-/Cl^- antiporter in basolateral membrane

Diffusion

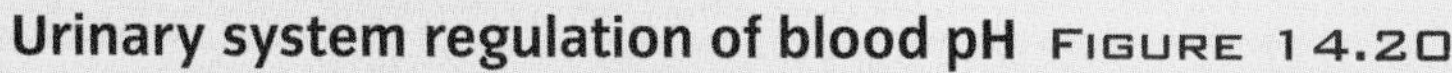

Urinary system regulation of blood pH FIGURE 14.20

Urine is usually acidic, meaning the excess of hydrogen ions typically found in the blood is filtered into the urine. There is usually not an excess of bicarbonate ions in the blood to filter into the urine; however, urine pH fluctuates according to the physical state of the body and is considered normal between pH 4.6 and 8.0. High-protein diets can increase acidity, whereas vegetarian diets increase alkalinity.

the blood, not the level of oxygen. When blood pH decreases (due to increased of H^+ concentration), the breathing rate increases. When blood pH rises, breathing is depressed, providing an instantaneous "fix" for blood pH.

Although the respiratory system is the main regulator of blood pH, the kidneys also play a role (FIGURE 14.20). Whereas the bicarbonate system uses an equilibrium reaction, the urinary system removes acidic and basic substances from the fluid and literally flushes them out. If the blood is too acidic, the kidneys can excrete hydrogen ions and send bicarbonate ions back to the blood. Conversely, if the blood is too basic, the kidneys will return hydrogen ions to the blood and excrete bicarbonate ions. This may adjust pH more slowly than the respiratory system, but the results are permanent.

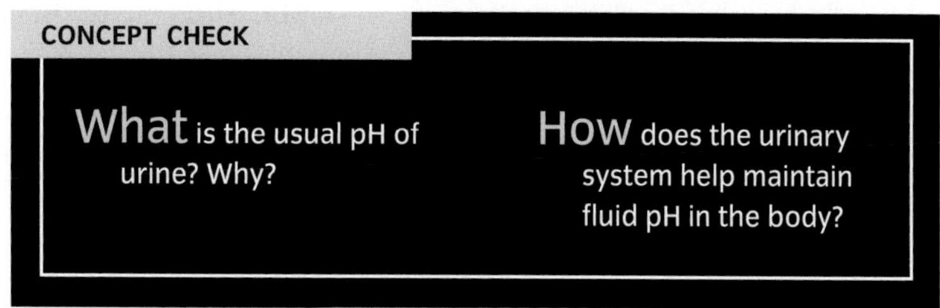

CONCEPT CHECK

What is the usual pH of urine? Why?

How does the urinary system help maintain fluid pH in the body?

Life-Threatening Diseases Affect the Urinary System

LEARNING OBJECTIVES

List the health facts that can be learned from routine urinalysis.

Explain the processes of conventional dialysis and hemodialysis.

Chemical analysis of urine can reveal a number of serious diseases as well as the use of illegal drugs (see the Ethics and Issues box). Urinalysis (UA) is a simple, common test that is routinely done in the doctor's office to produce a view of your internal health. It is noninvasive, meaning that instruments or sensing equipment are not placed in or on your body. Because urine is the by-product of filtered blood, any unusual compounds or incorrect levels of normal blood constituents will appear in the urine.

Casts Small structures formed by mineral or fat deposits on the walls of the renal tubules.

WARNING SIGNALS FROM URINALYSIS

Abnormal components in urine can include albumin, hemoglobin, red blood cells, white blood cells, glucose, and **casts**. Each can indicate a specific problem. See TABLE 14.2 on this page and on page 470 for listings of the normal and abnormal constituents of urine.

- **Albumin** is a small protein that, if present in the urine, must be entering the nephrons at the glomerulus. This could reflect high blood pressure in the glomerulus that forces proteins through the podocyte walls, or tears in glomerular arterioles. Normally, albumin remains in the blood to provide an osmotic force to draw excess water back from the filtrate to the blood. Albumin or other less common proteins in the urine are diagnosed as protein urea, but this may not indicate pathology. Serious weight training puts tremendous pressure on the capillaries and can force protein into the urine.

- **Hemoglobin** indicates bleeding in the upper urinary tract because the **red blood cells** have been present in the urine long enough to break open and release hemoglobin. Intact red blood cells would indicate bleeding closer to the lower end of the urinary tract, perhaps in the urethra. **White blood cells** in the urine indicate that an immune response is occurring, usually in response to an infection of the urinary tract, or occasionally the kidney.

Normal and abnormal constituents of urine

TABLE 14.2A

Normal Constituent	*Description*
Volume	One to two liters in 24 hours but varies considerably.
Color	Yellow or amber but varies with urine concentration and diet. Color is due to urochrome (pigment produced from breakdown of bile) and urobilin (from breakdown of hemoglobin). Concentrated urine is darker in color. Diet (reddish colored urine from beets), medications, and certain diseases affect color. Kidney stones may produce blood in urine.
Turbidity	Transparent when freshly voided but becomes turbid (cloudy) upon standing.
Odor	Mildly aromatic but becomes ammonia–like upon standing. Some people inherit the ability to form methylmercaptan from digested asparagus that gives urine a characteristic odor. Urine of diabetics has a fruity odor due to presence of ketone bodies.
pH	Ranges between 4.6 and 8.0; average 6.0; varies considerably with diet. High-protein diets increase acidity; vegetarian diets increase alkalinity.
Specific gravity	Specific gravity (density) is the ratio of the weight of a volume of a substance to the weight of an equal volume of distilled water. In urine, it ranges from 1.001 to 1.035. The higher the concentration of solutes, the higher the specific gravity.

How does a urine test prove drug abuse?

Urinalysis (UA) is a noninvasive way to get a holistic picture of the immediate physiological events occurring in the body. Urine tells us about the chemical processes in the body, just as examining the chemicals in a river tells us about events in the river's watershed. If a person has been taking illicit drugs, prescribed medications, or even diet supplements, those compounds, or their breakdown products, will show up in the urine.

Urinalysis is most often used as a screening and diagnostic tool. UAs are performed when people complain of abdominal pain, back pain, painful or frequent urination, blood in the urine, or other symptoms of a urinary tract infection, which may show up as elevated levels of white blood cells. It is also a routine part of regular physical examinations.

The test also detects substances associated with many metabolic and kidney disorders. An abnormal UA can be an early warning of trouble, because substances like protein or glucose begin to appear in the urine before a person is aware of a problem. The health care provider must correlate the urinalysis results with physical complaints and clinical findings to make a diagnosis.

Have you ever wondered what happens to your urine sample? The medical professional first examines its physical characteristics, such as clarity, color, odor, and **specific gravity**.

Specific gravity A ratio of the density of a substance to the density of pure water.

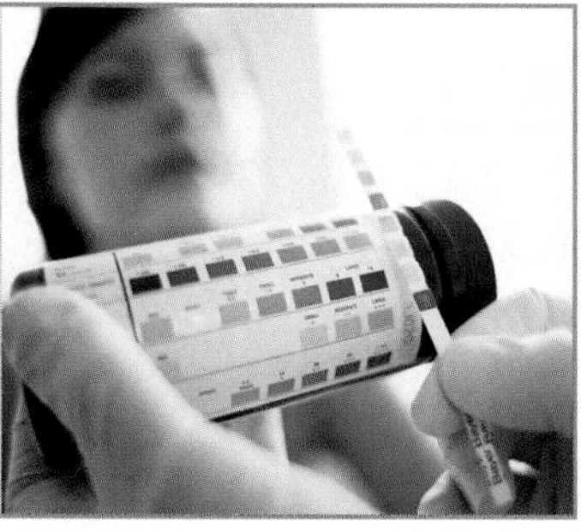

The next step is a chemical analysis, usually with a "dipstick" test that includes many pads soaked with indicator substances. The strip is dipped into the sample, and the urine interacts with the chemicals in each pad. After a specific time (usually 30 seconds), the color of each pad is compared to a reference chart. This comparison is now automated for greater accuracy.

Urine usually contains urochrome, which gives it that yellow color; nitrogenous wastes like ammonia and urea from metabolic processes; water; ions; and cast-off cells from the epithelial lining of the bladder and urethra. In addition, many large molecules enter the urine when blood in the peritubular capillaries passes the distal convoluted tubule. These molecules can include breakdown products of legal and illegal drugs, vitamin and mineral supplements, or even various environmental contaminants.

The pH of urine should be between 4.6 and 8.0, and the specific gravity between 1.002 and 1.028. For reference, distilled water has a specific gravity of 1.000, and the Pacific Ocean has an average specific gravity of 1.023.

When urinalysis is used to test for the presence of drugs, the sample is first put through a fast, inexpensive, and inexact screening test. Samples that test positive are then put into an analytical machine called the gas chromatograph-mass spectrometer (GC/MS). The GC/MS first separates compounds based on their mass, and then uses detector chemicals to identify certain compounds. The machine is expensive but is so sensitive that it can easily detect traces of compounds at concentrations of 1 part per billion, or even less.

The GC/MS produces a graph showing all compounds detected in the sample. If peaks on the graph indicate the presence of illicit drugs or their metabolites, the test is said to be positive and the person is considered a user of illicit drugs.

Drug testing is often sold as a cure-all for detecting drug use, especially among students, athletes, or potential employees. But urinalysis is not perfect. A test result can mistake metabolites of over-the-counter or prescription drugs for those of illicit drugs, forcing test administrators to interpret results carefully. Poppy seeds, found on bagels and pastries, can break down into compounds that resemble metabolites from opiate drugs, because opiates and poppy seeds both come from the opium poppy. Urinalysis is more effective at detecting some drugs than others. Marijuana and other drugs are detectable in the urine for long periods, whereas alcohol and cocaine are cleared quickly from the body. Finally, drug testing is expensive, and some studies suggest that the knowledge that urinalysis will be performed on a regular basis has little effect on employee performance.

The most common "street" advice for fooling a urinalysis, or passing as "clean" despite having recently taken illicit drugs, is to dilute the urine by drinking massive quantities of water. But the GC/MS is so sensitive that it can usually pick up traces of drug metabolites in very dilute urine. Deliberately ingesting compounds that will interfere with drug tests also raises moral questions. And these "interferences" are identified in the test, raising a red flag. The best, most surefire way to test drug-free is to live drug-free.

Normal and abnormal constituents of urine TABLE 14.2B

Abnormal Constituent	*Description*
Albumin	A normal constituent of plasma, it usually appears in only very small amounts in urine because it is too large to pass through capillary fenestrations. The presence of excessive albumin in the urine—**albuminuria**—indicates an increase in the permeability of filtration membranes due to injury or disease, increased blood pressure, or irritation of kidney cells by substances such as bacterial toxins, ether, or heavy metals.
Glucose	The presence of glucose in the urine is called **glucosuria** and usually indicates diabetes mellitus. Occasionally it may be caused by stress, which can cause excessive amounts of epinephrine to be secreted. Epinephrine stimulates the breakdown of glycogen and liberation of glucose from the liver.
Red blood cells (erythrocytes)	The presence of red blood cells in the urine is called **hematuria** and generally indicates a pathological condition. One cause is acute inflammation of the urinary organs as a result of disease or irritation from kidney stones. Other causes include tumors, trauma, and kidney disease, or possible contamination of the sample by menstrual blood.
Ketone bodies	High levels of ketone bodies in the urine, called **ketonuria**, may indicate diabetes mellitus, anorexia, starvation, or simply too little carbohydrate in the diet.
Bilirubin	When red blood cells are destroyed by macrophages, the globin portion of hemoglobin is split off and the heme is converted to biliverdin. Most of the biliverdin is converted to bilirubin, which gives bile its major pigmentation. An above-normal level of bilirubin in urine is called **bilirubinuria**.
Urobilinogen	The presence of urobilinogen (breakdown product of hemoglobin) in urine is called **urobilinogenuria**. Trace amounts are normal, but elevated urobilinogen may be due to hemolytic or pernicious anemia, infectious hepatitis, biliary obstruction, jaundice, cirrhosis, congestive heart failure, or infectious mononucleosis.
Casts	**Casts** are tiny masses of material that have hardened and assumed the shape of the lumen of the tubule in which they formed. They are then flushed out of the tubule when filtrate builds up behind them. Casts are identified by either the cells or substances that compose them or their appearance.
Microbes	The number and type of bacteria vary with specific infections in the urinary tract. One of the most common is *E. coli*. The most common fungus to appear in urine is the yeast *Candida albicans*, a cause of vaginitis. The most frequent protozoan seen is *Trichomonas vaginalis*, a cause of vaginitis in females and urethritis in males.

- **Glucose** in the urine signifies diabetes mellitus. As described previously, glucose spills into the urine due to a high concentration in the blood.
- **Casts** are plugs of material, shaped like the nephron tubules, that build up in the tubules and then get forced out by pressure. Casts can be formed from minerals that enter the filtrate and clog the PCT and nephron loop, or they can be composed of proteins and cells that find their way into the system. Casts always indicate serious kidney trouble, such as nephritis or glomerulonephritis.

KIDNEY DISEASE IS LIFE-THREATENING

Without functioning kidneys, blood composition cannot be maintained and homeostasis will be lost. Three of the most common kidney diseases are nephritis, glomerulonephritis, and polycystic kidney disease. Of these, only polycystic kidney disease is inherited. This disease causes cysts to form in the kidneys, destroying normal kidney tissue. In severe cases, the patient may require **dialysis** or even a kidney transplant.

Dialysis
Substance exchange via diffusion across a membrane, artificially mimicking the kidney.

Nephritis and glomerulonephritis are both inflammations of the filtering unit of the kidney. Because the kidney is covered by the renal capsule, any inflammation within the kidney increases pressure, compresses kidney tissues, and slows or halts filtration at the glomerulus. The kidney is shut down until the swelling is resolved.

Glomerulonephritis is a general term for blockage of kidney circulation, with subsequent swelling and shutdown of the nephrons. When the kidneys cannot filter blood, toxins build up and the blood becomes filled with metabolic wastes. Blood volume and composition are disturbed, and the patient becomes ill. Symptoms include nausea, dizziness, fatigue, and memory loss. If the ion and fluid balance is not restored, death can result.

As noted, dialysis is the exchange of aqueous substances between two solutions through a membrane. In effect, the nephron performs dialysis with the peritubular capillaries on a continuous basis. When the kidneys shut down, dialysis must continue somehow, or the blood will become toxic to the body cells. Dialysis machines permit dialysis to occur outside the body.

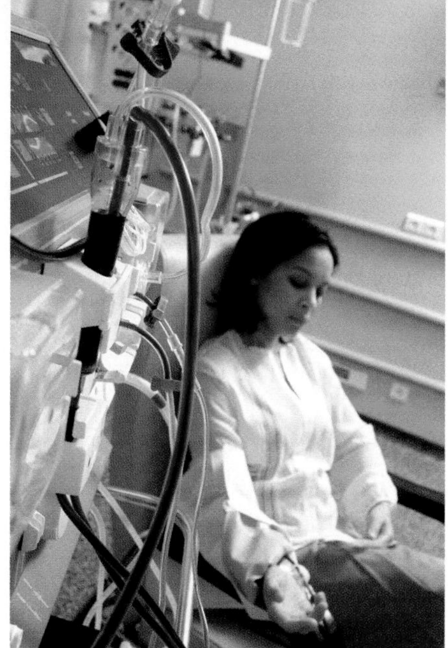

A patient undergoing dialysis
FIGURE 14.21

Hemodialysis is dialysis between blood and another fluid (FIGURE 14.21). This is a relatively common procedure used to compensate for impaired kidney function. It can be done for extended periods, such as when the kidneys have failed and no matching donor kidney is available.

In hemodialysis, blood is withdrawn from an artery and passed across a dialysis membrane. Toxins in the blood diffuse into the prepared solution, while necessary blood plasma components are either (1) prevented from diffusing by putting the same concentration of these components in the dialysis fluid as in the blood or (2) added to the blood by increasing their levels in the dialysis fluid. The dialyzed blood is then sent back to the body. The procedure takes three to four hours and must be done three times a week.

Hemodialysis is tough on the blood cells because they are passed through tubes and across membranes under pressure. If the patient requires dialysis for a long period, peritoneal dialysis may be recommended. In this procedure, two liters of dialysis fluid are put directly into the abdominal cavity, left to diffuse for a period, and then removed. The **peritoneum** serves as the dialysis membrane. As with hemodialysis, this procedure must be performed regularly; peritoneal dialysis is completed several times a day to sustain life.

Peritoneum
Membrane lining the abdominal cavity.

Kidney transplants are relatively common, second only to corneal transplants in numbers performed per year. Because of the kidneys' retroperitoneal placement, they are easy to reach surgically. The kidneys have essentially one artery and one vein. These large vessels are easily cut and sutured to the donor kidney. The suturing can be completed using laparoscopes, and the kidney removed and/or replaced through a three-inch incision on the side of the abdomen. Kidney transplants are highly successful transplant operations, with almost 80 percent patient and organ survival rate after one year. Living transplants, transplanting organs obtained by removing one kidney from a living, healthy donor rather than an accident victim, have success rates above 90 percent.

Obtaining nutrients, absorbing the necessary compounds, getting rid of waste products, and maintaining fluid homeostasis are all imperative to survival. With the digestive and urinary systems handling these vital functions, we humans can turn our attention to other occupations. One of the more interesting of these is reproduction, the topic of the next chapter. Perpetuating the species is the underlying biological drive that keeps us going on this planet.

CONCEPT CHECK

List four abnormalities that may appear in a urinalysis, and describe what each indicates.

Differentiate between polycystic kidney disease and nephritis.

How does hemodialysis work?

CHAPTER SUMMARY

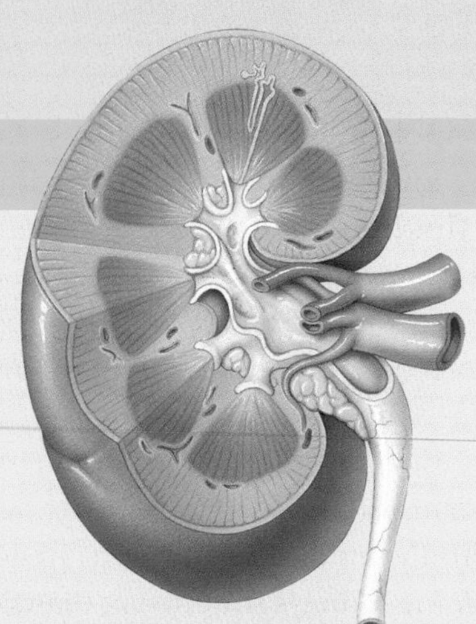

1 The Urinary System Filters, Transports, and Stores Waste Products

The urinary system is responsible for maintaining fluid homeostasis, ion balance, and blood calcium concentration and for removing fluid waste from the body. The system includes the paired kidneys, the paired ureters, the urinary bladder, and the urethra. A nephron is composed of a glomerular capsule surrounding the glomerulus, a proximal convoluted tubule, a loop, and a distal convoluted tubule connected to a collecting duct. Each portion of the tubule has a distinct role in filtering blood, balancing ions and pH, and removing wastes.

2 Urine Is Formed through Filtration and Osmosis

The nephrons are the functional units of the urinary system. It is here that 180 liters of fluid are filtered and maintained per day. Glomerular filtration depends on blood pressure, capsular hydrostatic pressure, and osmotic pressure of the blood. The filtrate is captured in Bowman's capsule and passed to the PCT, where most of the necessary nutrients and water are reabsorbed. The loop of the nephron extends into the middle of the kidney and assists in removal of salts and water. The DCT is involved in secretion from the blood to the forming urine. The collecting ducts remove urine from the kidney.

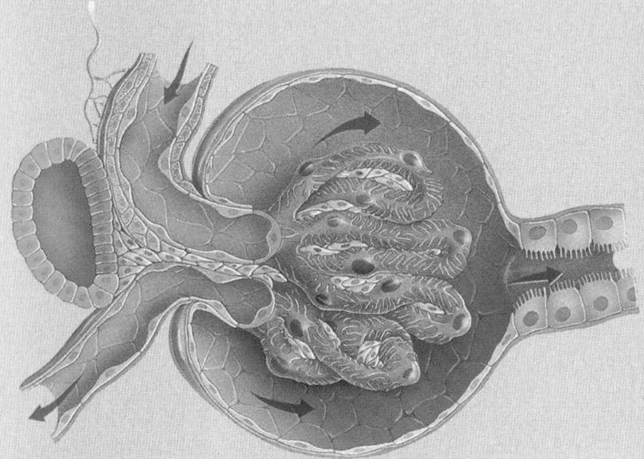

3 Urine Is Transported to the Bladder for Storage

Urine is stored in the bladder until voided. After 300 ml of urine fills the bladder, the micturition reflex is stimulated. The voluntary, external urinary sphincter determines when voiding takes place. Incontinence is the loss of this control. The female urethra is shorter than the male urethra, leading to a higher incidence of urinary tract infection in females.

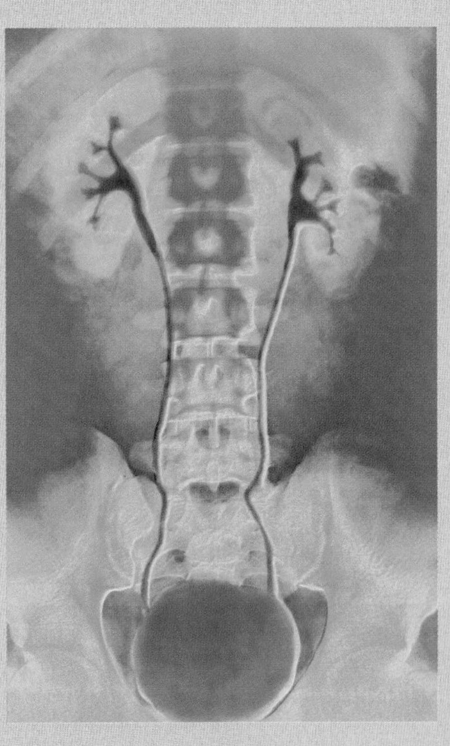

4 The Urinary System Maintains the Body's Water–Salt Balance

Hormones regulate the amount of water and ions excreted with the urine. ADH prevents the loss of water, causing the production of concentrated urine. Aldosterone regulates sodium reabsorption, effectively removing water and sodium from the body. ANP and BNP work in opposition to ADH, causing the formation of dilute urine.

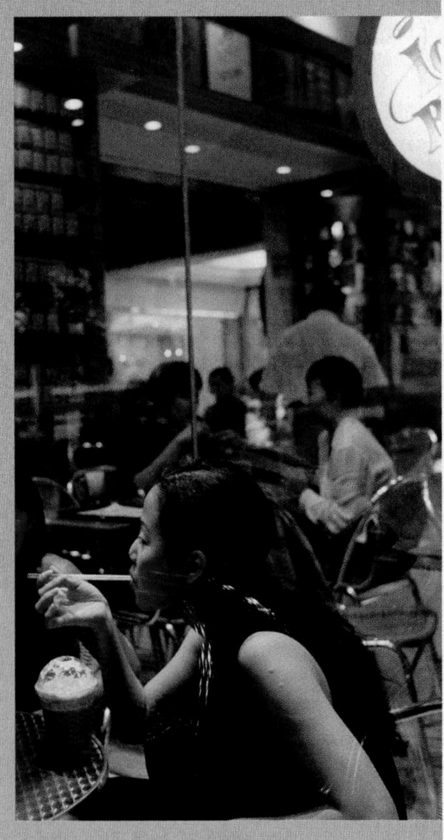

5 The Kidneys Help Maintain the Blood's Acid–Base Balance

The kidneys and respiratory system combine to control blood pH. The urinary system can remove acidic and basic substances from the fluid and flush them from the body, with permanent results. If the blood is too acidic, the kidneys can excrete hydrogen ions and send bicarbonate ions back to the blood. If the blood is too basic, the kidneys will return hydrogen ions to the blood, and excrete bicarbonate ions.

6 Life-Threatening Diseases Affect the Urinary System

Dialysis is the exchange of aqueous substances between two solutions through a membrane. In effect, the entire nephron performs dialysis with the peritubular capillaries on a continuous basis. When the kidneys shut down, dialysis must continue somehow, or the blood will become toxic to the cells of the body. Dialysis machines permit dialysis to occur outside the body.

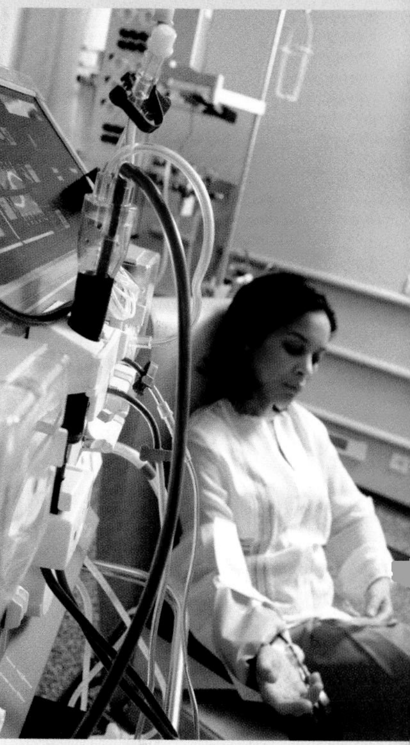

KEY TERMS

- **calcium oxalate** p. 458
- **casts** p. 468
- **constipation** p. 461
- **countercurrent multiplication** p. 456
- **dialysis** p. 470
- **external sphincter muscle** p. 460
- **facilitated diffusion** p. 455
- **fenestrations** p. 453
- **incontinence** p. 461
- **internal urethral sphincter** p. 460
- **microvilli** p. 455
- **nephron** p. 448
- **nitrogenous wastes** p. 447
- **osmolarity** p. 462
- **peritoneum** p. 471
- **peritubular capillaries** p. 448
- **secreted** p. 455
- **solutes** p. 462
- **specific gravity** p. 469

CRITICAL THINKING QUESTIONS

1. Imagine that you contracted a urinary tract infection and did not treat it. Trace the pathway of the bacteria as it moves up the urinary system. What structures in the kidneys would you expect to be damaged by the bacteria?

2. Many home pregnancy tests look for a specific protein in the urine. This compound is present only in pregnant women. Why do the tests recommend using first morning urine? How is that different from urine produced and excreted at midday?

3. Caffeine and alcohol both block the secretion of ADH from the posterior pituitary gland. Explain what this does to fluid balance. Does it make sense to drink caffeine before an athletic event? Explain why a cold beer might not be such a great idea on a hot afternoon.

4. What are the differences between hemodialysis and peritoneal dialysis? What are the benefits of each? What are the drawbacks?

5. Assume you were given the following results from a series of urinalysis tests. What would each test indicate?

   Cloudy urine, above-normal specific gravity, high white blood cell count, many transitional epithelial cells.

   Presence of protein, casts, and hemoglobin.

   Presence of glucose and ketones (ketones are a by-product of the digestion of body proteins).

   Pale yellow color, pH 6.3, specific gravity 1.015, no RBCs, no proteins.

SELF TEST

1. Which of the following is NOT a specific function of the urinary system?

a. Production of urine
b. Maintenance of blood pH
c. Maintenance of blood volume
d. Maintenance of red blood cell levels

2. The function of the structure indicated as C is

a. filtration of blood.
b. transport of urine within the body.
c. transport of urine from the body.
d. storage of produced urine.

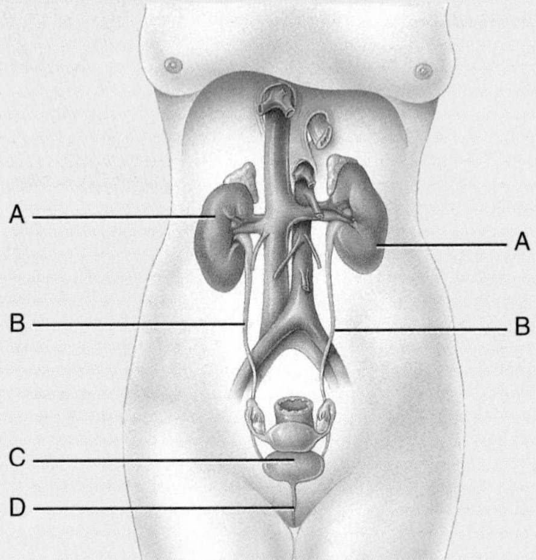

3. In the above diagram, the urethra is indicated by the label

a. A.
b. B.
c. C.
d. D.

4. The renal pyramids are composed of

a. glomeruli.
b. peritubular capillaries.
c. collecting ducts.
d. renal capsules.

5. The correct sequence of blood vessels through the kidney is:

a. renal artery → arcuate artery → afferent arteriole → efferent arteriole → peritubular capillaries → renal vein.
b. renal vein → renal artery → peritubular capillaries → arcuate artery → interlobar veins.
c. renal artery → peritubular capillaries → efferent arteriole → afferent arteriole → renal vein.
d. renal artery → efferent arteriole → peritubular capillaries → interlobar artery → renal vein.

6. The function of the structure labeled B is to

a. filter blood.
b. collect filtrate.
c. reabsorb necessary nutrients.
d. secrete unwanted large waste products.

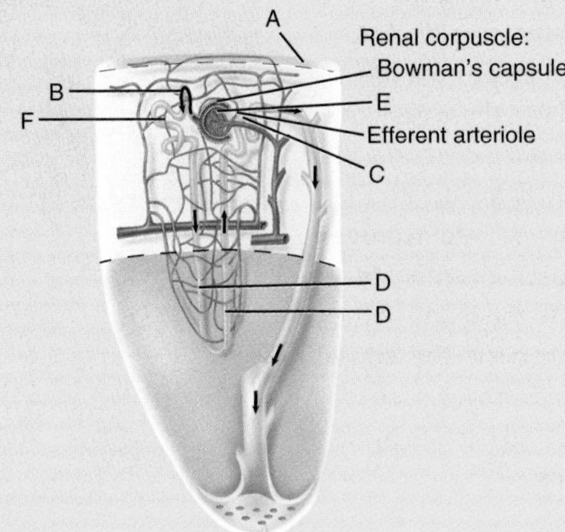

7. In the above figure, label E indicates the

a. PCT.
b. loop of Henle.
c. DCT.
d. glomerulus.

8. The portion(s) of the nephron that are usually found in the cortex of the kidney are the

a. PCT, capsule, and loop of Henle.
b. PCT and DCT.
c. loop of Henle and collecting duct.
d. PCT, capsule, and DCT.
e. capsule only.

9. When blood is filtered through the glomerulus, the two forces opposing movement into the nephron are

a. A and B.
b. B and C.
c. A and C.
d. All of these forces oppose movement into the capsule.

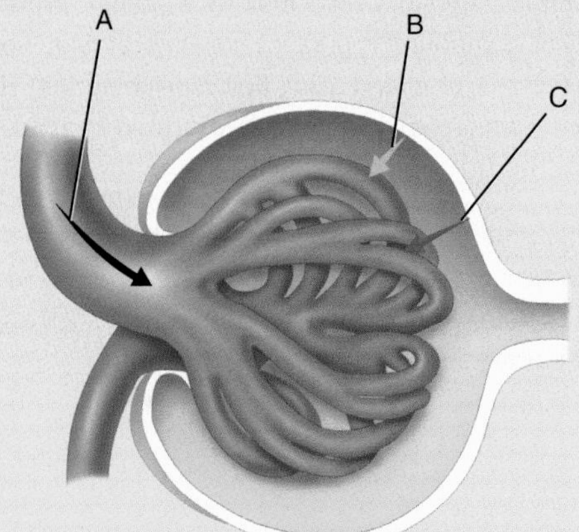

10. Most of the filtered water, and hopefully all of the filtered glucose is returned to the bloodstream at the
   a. PCT.
   b. DCT.
   c. loop of Henle.
   d. collecting duct.

11. The portion of the nephron that sets up the salt gradient in the medulla of the kidney is the
   a. PCT.
   b. DCT.
   c. loop of Henle.
   d. collecting duct.

12. The structure shown below is formed when
   a. overly dilute urine is produced.
   b. overly concentrated urine is produced.
   c. kidney failure is experienced.
   d. too many calcium-rich foods are consumed.

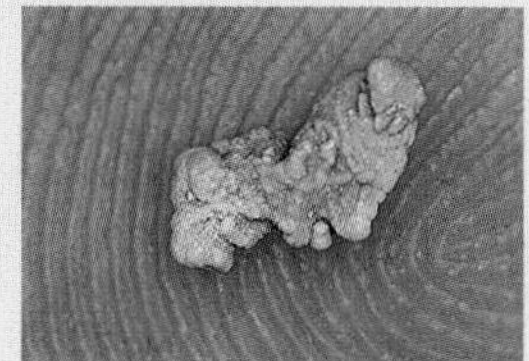

13. True or False? The urethra in the image below carries both urine and reproductive fluids.

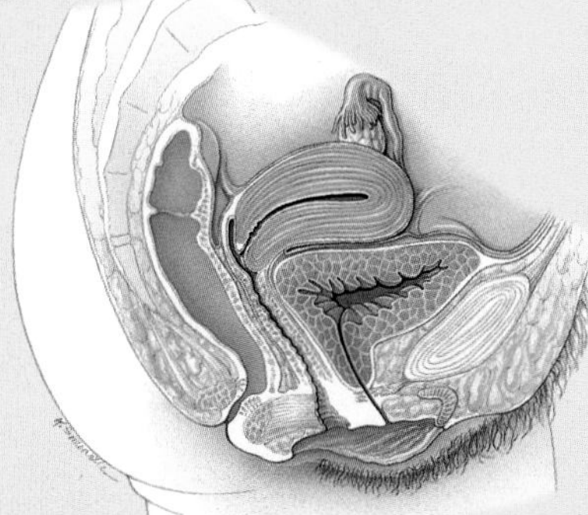

14. Dilute urine is produced when
   a. ADH is present.
   b. ADH is absent.
   c. solutes are added to the forming urine.
   d. water is removed from the collecting ducts.

15. If you drink more water than you need, ___________ will be secreted and you will lose water through the urinary system.
   a. ADH
   b. ANP
   c. aldosterone
   d. Both b and c will be secreted.

16. True or False? Both caffeine and alcohol serve as diuretics, causing the production of copius dilute urine.

17. The compound most important in driving respiration rates and depth of breathing shown in the feedback loop in this figure is
   a. oxygen.
   b. bicarbonate ions.
   c. carbon dioxide.
   d. hydrogen ions.

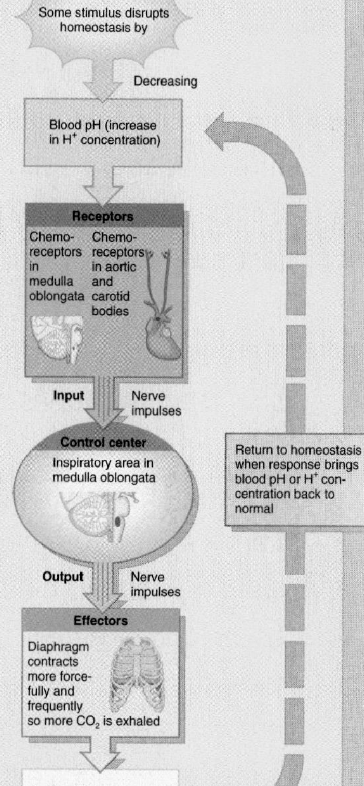

18. Urinalysis is able to detect all of the following EXCEPT
   a. vitamin supplementation.
   b. illegal drug use.
   c. viral infection.
   d. metabolic kidney disorders.

19. White blood cells in the urine indicate
   a. normal urinary tract functioning.
   b. a possible UTI.
   c. bleeding in the kidneys or ureters.
   d. diabetes mellitus.

20. During the procedure shown at right, if a physician wants to remove excess potassium from the blood, she or he must include ___________ of potassium in the dialysis fluid than those that occur in the blood.
   a. higher levels
   b. lower levels

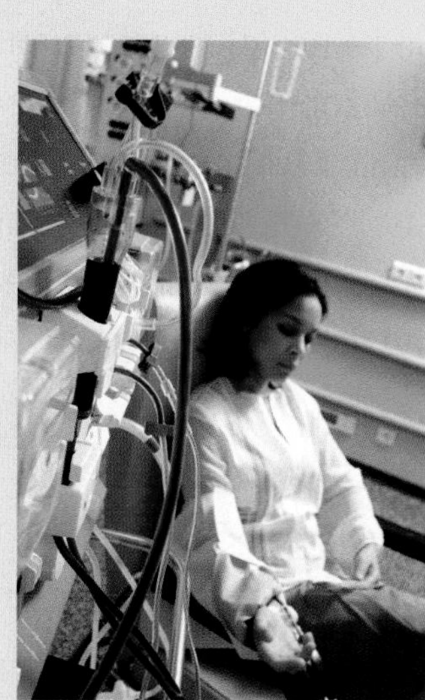

The Endocrine System and Development

15

"Mom, I need some new jeans," Marc called from his room. The fifteen-year-old opened his bedroom door and handed out a pair of jeans that were new just two months earlier. "These are too small," he claimed.

"Is that possible? And why does his voice sound so full—does he have a sore throat?" his mother fretted, folding the pants. She sat back and thought about her little boy. As recently as last summer, he was a small guy, running after his older brother, scraping his knees and relentlessly trying improbable jumps on his skateboard. Now just a few months later, he seems tremendously tall, with thicker limbs and a deeper voice.

In fact, the rapid growth of bone and muscle at this age is perfectly normal, as is the alteration in the boy's voice. Marc is maturing quickly, changing in appearance and physical ability. What is causing this predictable sequence of transformations in growth and development? "Maybe I am feeding him too much," his mother sighed as she got up, put away the undersized jeans, and began to prepare yet another oversized meal for her growing teenager.

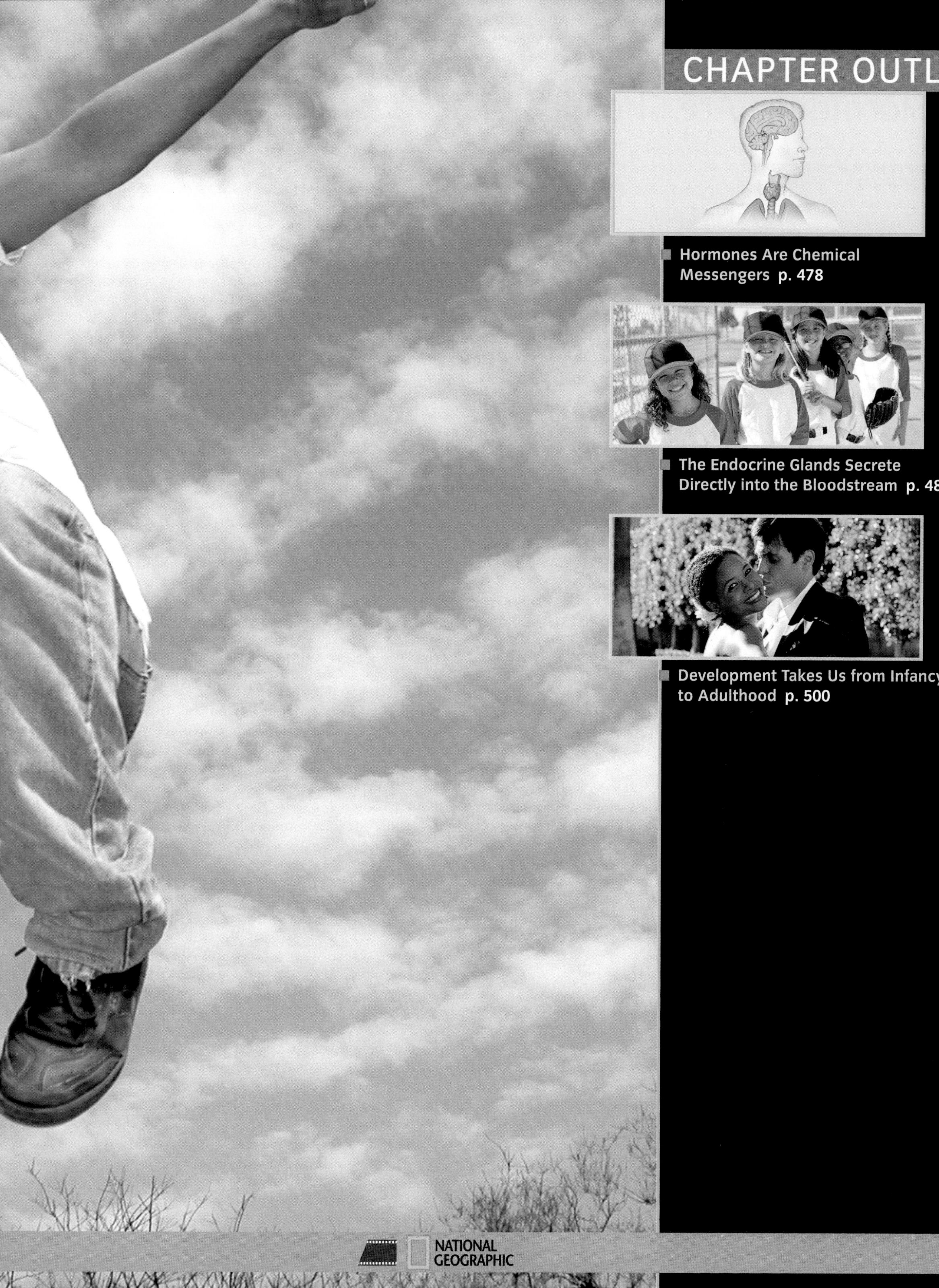

CHAPTER OUTLINE

Hormones Are Chemical Messengers

Learning Objectives

Define a hormone.

Differentiate between steroid and nonsteroid hormones.

Briefly explain how hormones are controlled.

Life is a series of precisely timed processes. We are born, we grow, we become sexually mature, we reproduce, and we age. All these changes require precise, long-range timing of events over years, and permanent, predictable interactions among body systems. Controlling these stages is the job of the endocrine system. This odd collection of organs communicates with the cells using chemical messengers called **hormones**.

At first blush, the endocrine system seems simple enough. When we are ready for a protein to be formed, or a series of reactions to begin, we merely activate a hormone. But when you begin to focus on hormonal control, plenty of questions arise. How do hormones work at the cellular level? Why do some hormones affect the output of other hormones or affect only certain tissues? How can one hormone have many different effects?

The endocrine system is built quite differently from the other systems we have viewed, since it is a group of separate structures, called endocrine glands, that are connected by the cardiovascular system. An endocrine gland is a gland that secretes its products directly into the bloodstream rather than through ducts to the surface of the gland. The main glands of the system are shown in **Figure 15.1**.

The products of endocrine glands are hormones. You already know the names of several hormones, such as testosterone, estrogen, and adrenaline. Hormones are chemically active compounds that are produced in one area of the body but have their effect elsewhere. The cells that hormones act upon are called their **target cells**. Hormones are responsible for the many sequential changes of growth and maturation. They are also agents of response when homeostasis is disrupted. Hormones maintain fluid balance, control calcium and glucose levels in the blood, assist in tissue repair, maintain basal metabolic rate, and assist in digesting food. They can do

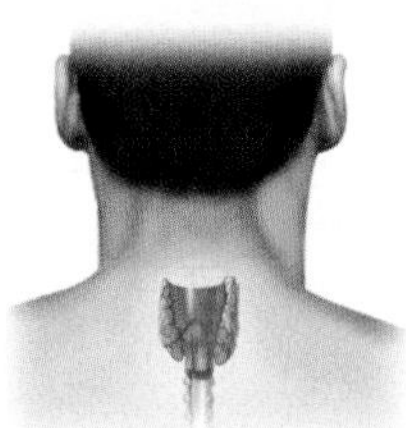

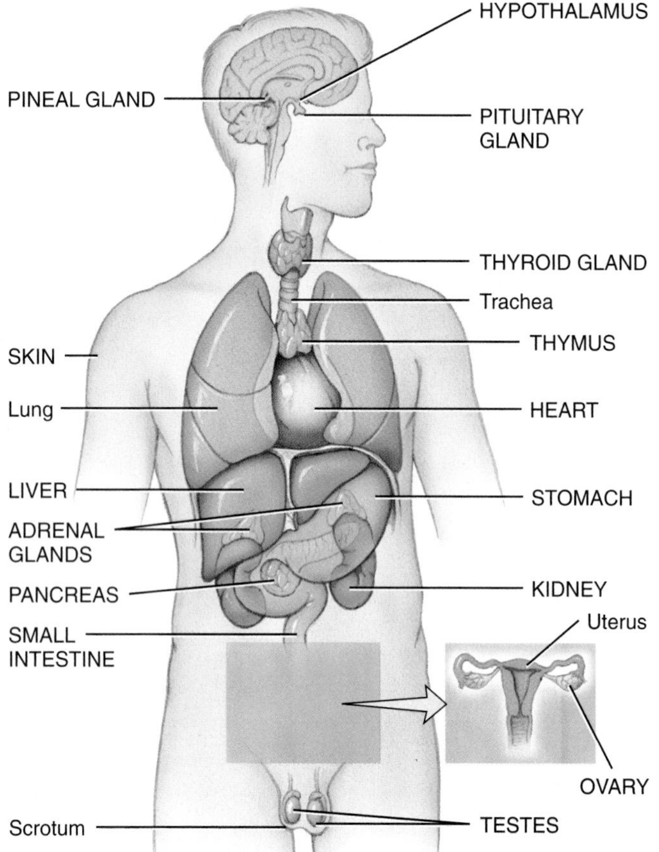

Overview of the endocrine system
Figure 15.1

all of this because hormones are carried to virtually all cells of the body via the bloodstream. They cause an effect only in their specific target cells. Interestingly, one hormone can have many different target cells; however, the overall effect of the hormone will be the same. For example, growth hormone causes muscle tissue to enlarge, bone tissue to increase matrix production, and glycogen stores to be released to fuel the increased protein production. All of these effects work together to increase body size.

The two main classes of hormone are steroid hormones, which are structurally related to cholesterol, and nonsteroid hormones, which are composed mainly of amino acids. The main difference between these classes is solubility. Steroid hormones, including testosterone and estrogen, are lipid soluble, so they can pass directly through the phospholipid bilayer of cell membranes. Nonsteroid hormones are not lipid-soluble, so they cannot penetrate the cell membrane. This single difference translates into completely different modes of action.

STEROID HORMONES ENTER THE CELL WITHOUT HELP

Because steroid hormones are lipid soluble, they can pass directly through cell and nuclear membranes of their target cells, reaching specific receptors in the cytoplasm or nucleoplasm. Once the hormone binds to its receptor, it forms a hormone–receptor complex. This complex affects the transcription of genes, and either upregulates (increases) or downregulates (decreases) the production of specific proteins. This change in production rate shifts the complement of proteins inside the cell (FIGURE 15.2).

Give a toddler a set of building blocks and you will usually be rewarded with a proudly displayed, three-block tower. If you then give the child blocks of different shapes, you will be presented with an entirely new building. The shape of the blocks determines the shape of the building. In much the same way, the protein complement of the cell determines its function. Directing the construction of proteins, after all, is what genes do, so in this sense, hormones rule protein production.

This process, from gene activation to protein production to final effect, can take anywhere from a few minutes to many hours. The process requires transcription and translation (as described in Chapter 3), and then the newly created protein must be used to alter the activities of the target cell. For this reason, steroid hormones act relatively slowly when compared with nonsteroid hormones or neural impulses.

Process Diagram

Steroid hormone activity

FIGURE 15.2

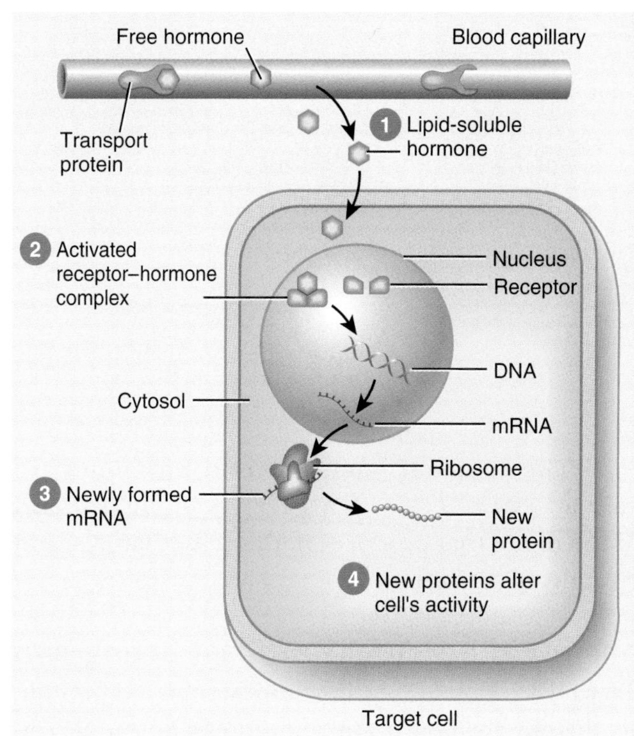

Lipid-soluble hormones act on receptors inside the cell. 1 The lipid-soluble hormone diffuses into the cell and binds to a receptor in the cytoplasm or the nucleoplasm. 2 This activated receptor–hormone complex alters gene expression. 3 Messenger RNA initiated by the hormone leaves the nucleus and directs the formation of a new protein. 4 The cell uses these new proteins, which alter the activity of the cell. All of this takes a measurable time—more time than is needed to simply alter the function of a protein already in the cell via nonsteroid hormones or neural impulses.

Steroid hormones TABLE 15.1

Hormone	*Organ that produces it*	*Target cells*	*Effects*
Androgens and estrogens	Ovaries, testes, adrenal cortex	Most cells of the body	Development of male or female secondary sexual characteristics
Mineralocorticoids	Adrenal cortex	Kidneys	Increase absorption of sodium and water by the kidneys, accelerate potassium loss
Glucocorticoids	Adrenal cortex	Most cells of the body	Promote liver formation of glucose and glycogen, release amino acids from muscle, anti-inflammatory effects
Calcitriol	Kidneys	Intestinal lining	Stimulates calcium and phosphate absorption, inhibits PTH release

The common human steroid hormones are listed in TABLE 15.1. For each hormone, you will find the organ that produces it, the cells it targets, and the results of its actions.

Nonsteroid hormones are fast acting and powerful

Nonsteroid hormones, such as epinephrine, thyroid hormones, and antidiuretic hormone, can affect target cells much more quickly than steroid hormones because they affect the activity of proteins that are already present in that cell. Nonsteroid hormones are water-soluble, so they are easily transported to the cell in blood or interstitial fluid. However, water-soluble hormones cannot diffuse across the phospholipid bilayer of the target cell.

To overcome this obstacle, nonsteroid hormones act on specific receptors that stud the surface of target cells. These receptors are integral membrane proteins, often with an associated but inactive molecule attached to the cytoplasmic side of the protein. Hormone binding to the exterior of the integral protein changes the receptor and activates the associated inactivated molecule, releasing it into the cytoplasm. The released molecule, usually **cyclic AMP (cAMP)**, becomes a second messenger that carries information from the hormone (the first messenger) to the machinery of the cell. cAMP in turn activates an enzyme, often a **kinase**, that can alter various biochemical and cellular pathways.

Cyclic AMP (cAMP)
A form of adenosine monophosphate in which the phosphate appears in ring formation, carrying little energy (not enough to harness for metabolic processes).

Kinase
A group of enzymes, all of which carry a phosphate from one compound to another.

Frequently, a series of enzyme activations occurs after an aqueous hormone binds. This activation takes only seconds or at most less than a minute, compared to the minutes to hours needed to produce a new protein via a steroid hormone.

Another benefit of nonsteroid hormone activity is that at each step, the original signal is amplified (FIGURE 15.3). One hormone can eventually cause the activation of many enzymes. Because the effects of a small amount of hormone can be greatly exaggerated, nonsteroid hormones are quite potent.

Nonsteroid hormones include very small compounds derived from single amino acids, as well as compounds composed of short chains of amino acids. These larger hormones are called peptide hormones. Despite the size difference, both varieties are water soluble and therefore use the same general pathway.

The common nonsteroid hormones in the body are listed in TABLE 15.2 on page 482. We will return to these hormones when discussing the endocrine glands.

Nonsteroid hormone activity FIGURE 15.3

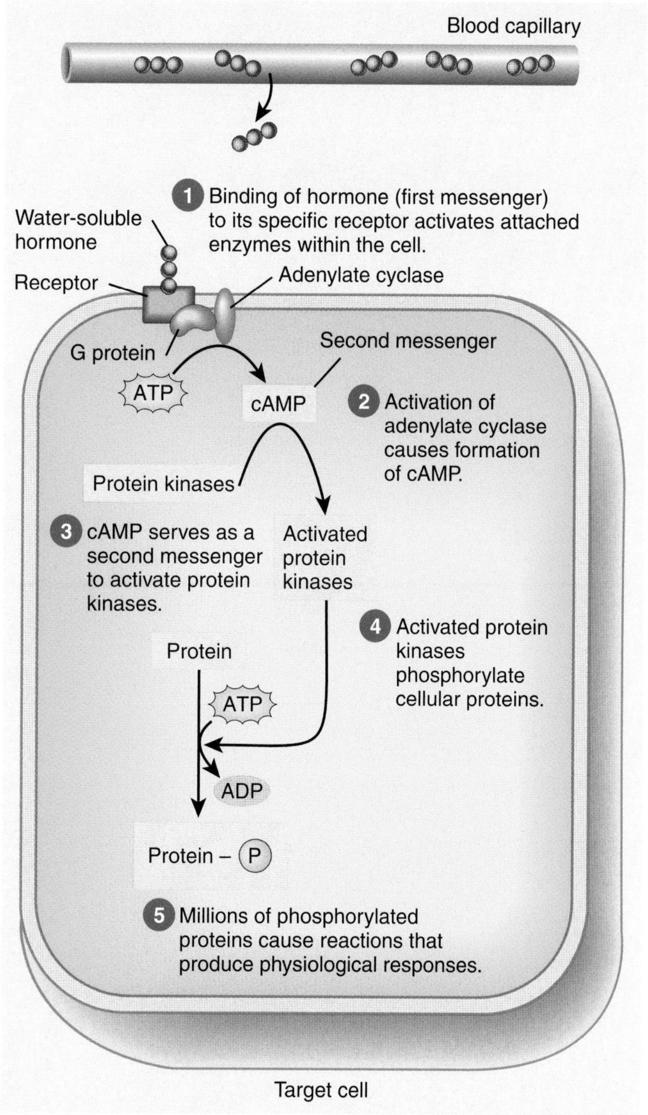

Blood capillary

Water-soluble hormone

Receptor

1 Binding of hormone (first messenger) to its specific receptor activates attached enzymes within the cell.

Adenylate cyclase

G protein

Second messenger

ATP

cAMP

2 Activation of adenylate cyclase causes formation of cAMP.

Protein kinases

3 cAMP serves as a second messenger to activate protein kinases.

Activated protein kinases

4 Activated protein kinases phosphorylate cellular proteins.

Protein

ATP

ADP

Protein – P

5 Millions of phosphorylated proteins cause reactions that produce physiological responses.

Target cell

Somatostatin A water-soluble hormone that prevents the secretion of growth hormone; literally to "keep the body the same" (*soma* = body).

Hormones need tight controls The action of a hormone is entirely dependent on the target cell, so a particular hormone can have widely varying effects on different tissues. For example, **somatostatin** prevents the secretion of growth hormone in the hypothalamus, helps regulate digestive function in the abdominal cavity, and prevents the pancreas from releasing hormones that regulate blood sugar. Notice that when somatostatin is released, this entire suite of effects is likely, unless something changes in the target tissues.

Regardless of the hormone class, an endocrine gland's activity must be controlled. Recall from Chapter 1 that homeostasis is usually maintained via negative feedback, and this regulation applies to most hormones. The hormone's effect on the body may diminish the trigger that stimulated its production, or a second hormone may oppose the action of the first. In the

Nonsteroid hormones TABLE 15.2

Hormone	*Organ that produces it*	*Target cells*	*Effects*
Epinephrine and norepinephrine	Adrenal medulla	Most cells of the body	Increases cardiac activity, blood pressure, and blood glucose levels, releases stored lipids
Thyroid-stimulating hormone (TSH)	Anterior pituitary	Thyroid gland	Secretion of thyroid hormones
Luteinizing hormone (LH)	Anterior pituitary	Immature egg cells of ovary; interstitial cells of testes	Ovulation in ovary; secretion of testosterone in testes
Follicle-stimulating hormone (FSH)	Anterior pituitary	Immature egg cells of the ovary; immature sperm cells of the testes	Development of eggs and production of estrogen; development of sperm
Thyroxine (T4)	Thyroid gland	Most cells of the body	Increases energy utilization, oxygen consumption, growth, and development
Melatonin	Pineal gland	Hypothalamus	Inhibits secretion of GnRH (gonadotropin-releasing hormone, which governs the release of FSH and LH)
Oxytocin	Hypothalamus	Uterus and mammary glands in females, vas deferens and prostate gland in males	Smooth muscle contractions during labor, milk release, and male ejaculatory event
Antidiuretic hormone	Hypothalamus	Kidneys	Reabsorption of water, elevation of blood pressure
Adrenocorticotropic hormone (ACTH)	Anterior pituitary	Adrenal cortex	Secretion of glucocorticosteroids
Growth hormone (hGH)	Anterior pituitary	All cells of the body	Growth, protein synthesis, lipid movement
Melanocyte-stimulating hormone (MSH)	Anterior pituitary	Melanocytes	Increased melanin synthesis in epidermis
Prolactin (PRL)	Anterior pituitary	Mammary glands	Production of milk
Insulin	Pancreas	Most cells of the body	Promotes uptake of glucose, stimulates storage of lipids
Glucagon	Pancreas	Liver, adipose tissues	Activates lipid reserves, elevates blood glucose levels
Parathyroid hormone	Parathyroid glands	Bone, kidneys	Increases calcium concentration in body fluids
Calcitonin	Thyroid gland	Bone, kidneys	Decreases calcium concentration in body fluids
Erythropoietin	Kidneys	Red bone marrow	Stimulates the production of RBCs
Inhibin	Testes, ovaries	Anterior lobe of the pituitary gland	Inhibits secretion of FSH

simplest example of negative feedback, the endocrine system acts as the control center, responding to changes in blood or interstitial fluid chemistry. The hormone connects the control center and the effector (its target cells). Activation of the target cells shuts down the original stimulus, and homeostasis is restored.

The situation becomes more complicated when several hormones interact. These hormones can either directly or indirectly stimulate or inhibit one another, resulting in webs of interaction. The net result is the same, however: Homeostasis is maintained via negative feedback.

The Endocrine Glands Secrete Directly into the Bloodstream

LEARNING OBJECTIVES

List the major endocrine glands and their hormones.

Define the relationships among endocrine glands.

MASTERS OF THE GLANDULAR UNIVERSE: THE HYPOTHALAMUS AND THE PITUITARY

The endocrine system is directly tied to the nervous system through the hypothalamus. This bit of the forebrain monitors water and ion balance, body temperature, and carbohydrate metabolism. The hypothalamus physically connects with the pituitary gland and shares a portal circulation route. The hypothalamus secretes releasing and inhibiting factors into a portal system of capillaries that are connected directly to the capillaries of the anterior pituitary gland. This portal system (circulation that flows from a capillary bed to veins to another capillary bed) allows quick delivery of hypothalamic regulatory factors to the pituitary gland and also permits rapid response of the pituitary cells through the release of pituitary hormones.

The pituitary gland hangs from the hypothalamus into a depression in the sphenoid bone (FIGURE 15.4 on p. 484). It is composed of two sections, the anterior pituitary gland and the posterior pituitary gland. The two parts of the pituitary gland are suspended from the hypothalamus by a thin stalk. The pituitary gland secretes **endorphins** and **enkephalins** as well as nine hormones, which in turn stimulate other endocrine glands. The posterior pituitary gland contains the axons of neurons that originate in the hypothalamus. These neurons secrete two hormones at the posterior pituitary gland. The anterior pituitary gland is composed of epithelial tissue and produces seven hormones, all under control of the hypothalamus.

Endorphins and enkephalins Naturally-occurring compounds that reduce the sensation of pain and produce a feeling of well-being.

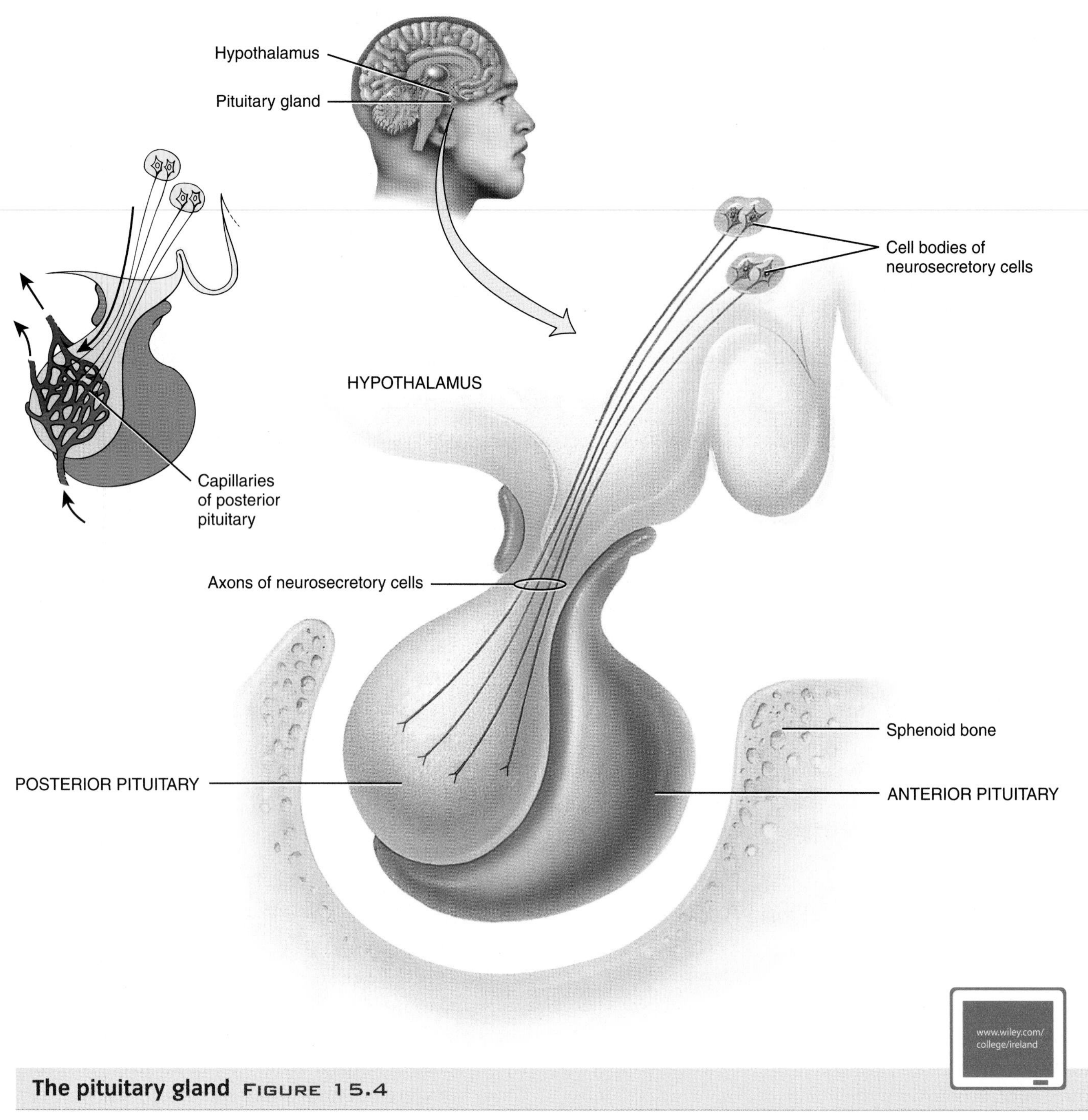

The pituitary gland FIGURE 15.4

Because the pituitary hormones affect other endocrine glands, the pituitary gland used to be called the master gland of the endocrine system. Now that we understand how the hypothalamus governs the pituitary gland, that terminology is obsolete.

Posterior pituitary gland acts in childbirth and water regulation The posterior pituitary gland is composed of **neuroendocrine** neurons. These cells can carry nerve impulses and produce two hormones for release into the bloodstream. The posterior pituitary gland produces **oxytocin** and **antidiuretic hormone**.

The hormone oxytocin has important roles in childbirth and lactation. This hormone initiates labor and causes cells in the mammary gland ducts to contract during the "milk let-down" response. This is one of the few examples of positive feedback in the human body. As labor nears, the uterus becomes more sensitive to oxytocin, reacting to small amounts of the hormone with larger contractions. During nursing, the "milk let-down" response is triggered by a neuroendocrine reflex. As the newborn suckles, sensory receptors in the nipple send impulses to the hypothalamus, which respond by increasing oxytocin production.

Antidiuretic hormone, or ADH, affects nephrons of the kidney. As discussed in the previous chapter, ADH prevents water loss by altering the permeability of the distal convoluted tubule cells to water. The hypothalamus initiates production of ADH when it detects low water levels in the blood. As ADH triggers target cells in the kidney, water level in the blood increases. When the hypothalamus ceases detecting low water levels, it quits producing ADH. You can get an indication of the strength of ADH when you drink caffeine or alcohol, both of which inhibit the ADH release. You may have noticed a need to urinate soon after drinking a cup of coffee. Because ADH is absent, all the water collected in the distal portion of the nephron leaves the kidney for the bladder.

In diabetes insipidus, either the posterior pituitary gland does not produce enough ADH or the ADH receptors in the kidney fail. People with diabetes insipidus produce large quantities of very dilute urine. In severe cases, fluid loss can exceed 10 liters a day. If these people do not drink enough water, they may die of dehydration.

The anterior pituitary gland produces seven hormones The anterior pituitary gland produces hormones that stimulate growth, metabolic rate, milk production, and **glucocorticoid** production. Glucocorticoids are steroid hormones that maintain mineral balance and control inflammation and stress. Once puberty begins, the anterior pituitary gland also secretes the hormones that maintain reproductive ability. These hormones are listed in TABLE 15.3.

Four anterior pituitary hormones—ACTH, FSH, LH, and TSH—are messenger hormones that cause target cells to secrete other hormones. These messengers travel through the bloodstream to a second endocrine gland. Once they interact with target cells on these second endocrine glands, those glands secrete hormones that will alter homeostatic balance.

Hormones of the pituitary gland TABLE 15.3

Hormone	*Target cells*	*Primary Action*
POSTERIOR PITUITARY GLAND		
Antidiuretic hormone (ADH)	Kidneys	Promotes water retention
Oxytocin	Uterus, mammary gland ducts	Labor and milk let-down contractions
ANTERIOR PITUITARY GLAND		
Thyroid-stimulating hormone (TSH)	Thyroid	Secretion of T3 and T4
Adrenocorticotropic hormone (ACTH)	Adrenal cortex	Secretion of glucocorticoids
Follicle-stimulating hormone (FSH)	Ovaries and testes	Promotes gamete development
Luteinizing hormone (LH)	Ovaries and testes	Ovulation, and testosterone production
Prolactin (PRL)	Mammary glands	Milk production
Growth hormone (hGH)	Most cells of the body	Promotes growth
Melanocyte-stimulating hormone (MSH)	Melanocytes	Increases melanin synthesis

Mineralo-corticoids Steroid hormones involved in maintaining water and ion balance.

Gonadotropins Hormones that stimulate activity in the gonads (ovary and testes).

Adrenocorticotropic hormone, ACTH, has a long name that explains the hormone's action quite nicely. "Adrenocortico" indicates the cortex of the adrenal glands (FIGURE 15.5), those small bits of endocrine tissue atop each kidney, and the suffix *"-tropic"* means "acting upon." ACTH stimulates the adrenal cortex to produce glucocorticoids and **mineralocorticoids**. These two classes of hormone maintain homeostasis during stress and control glucose metabolism.

Follicle-stimulating hormone (FSH) and **lutenizing hormone** (LH) are both **gonadotropins**. They stimulate the growth and functioning of the ovaries and testes, which in turn produce estrogen and testosterone. These hormones are usually not produced until age 10 to 13. A surge in production of FSH and LH initiates puberty and the graduation from childhood to adolescence.

Thyroid-stimulating hormone (TSH) activates the thyroid to produce T3 (triiodothyronine) and T4 (thyroxin), which will be covered later in this chapter. Both hormones are involved in maintaining your basal metabolic rate. They determine how quickly and efficiently your body uses the energy you consume.

Two hormones from the anterior pituitary gland, prolactin (PRL) and human growth hormone (hGH), act directly on target tissue instead of serving as messenger hormones.

Prolactin (PRL) stimulates milk production in females. Males also produce prolactin, but the exact function is uncertain. In sexually mature male birds, prolactin is important in attaining brightly colored plumage. Prolactin is also thought to play a role in sexual dimorphism (difference between the sexes) in amphibians.

Human growth hormone (hGH) affects almost every body system. hGH stimulates the growth of muscle, cartilage, and bone, and causes many cells to speed up protein synthesis, cell division, and the burning of fats for energy. This hormone is active at birth, then goes into overdrive during childhood and adolescence, causing bone elongation, muscle growth, and an overall increase in body mass. Although less effective after puberty, hGH is still essential to adult health. During this time, the bones have finished growing and the epiphyseal plates are sealed, so height cannot increase. However, muscles and cartilage can and do continue to enlarge. Bones can also increase in girth, strengthening the skeleton. hGH also assists in the burning of fats and amino acids when glucose stores are low.

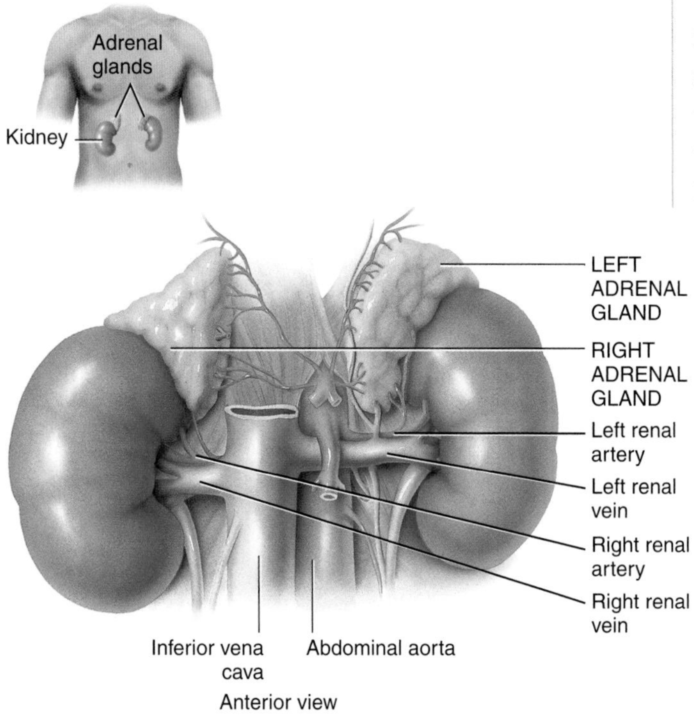

Adrenal glands FIGURE 15.5

Gigantism FIGURE 15.6

Growth hormone noticeably stimulates growth of skeletal muscle and bone. While discussing the muscular system, we mentioned that this hormone is one substance that unscrupulous athletes abuse to chemically enhance their training. Abnormal levels of growth hormone are also produced naturally in some diseases.

If growth hormone is produced in large amount prior to puberty, the bones and muscles will continue to grow, causing gigantism (FIGURE 15.6). The tallest person ever measured was Robert Wadlow, the Alton Giant, who was 8′11″ (2.71 M) tall when he died. Andre the Giant, another victim of growth hormone hypersecretion, was 6′3″ (1.9 M) at age twelve and reached 7′4″ (2.23 M) by adulthood.

While gigantism is rare, a different type of growth hormone hypersecretion is more common. Acromegaly is the secretion of excess growth hormone after the closure of the epiphyseal plates, when further increase in height is impossible. Acromegaly typically enlarges cartilage, causing an enlarged chin, accelerated growth of the nose, ears, and voice box, as well as a coarsening of the skin, and an enlargement of the hands and feet (FIGURE 15.7).

Growth hormone can also be hyposecreted during development, causing pituitary dwarfism. The growth plates close too early, organs stop growing, and childlike proportions remain throughout adulthood. Pituitary dwarfism can be treated with artificial injections of hGH (see the Ethics and Issues box on p. 488 for a more detailed discussion of the use of hGH injections).

Acromegaly FIGURE 15.7

Ethics and Issues

Is being short a disease? What are the ethics of human growth hormone?

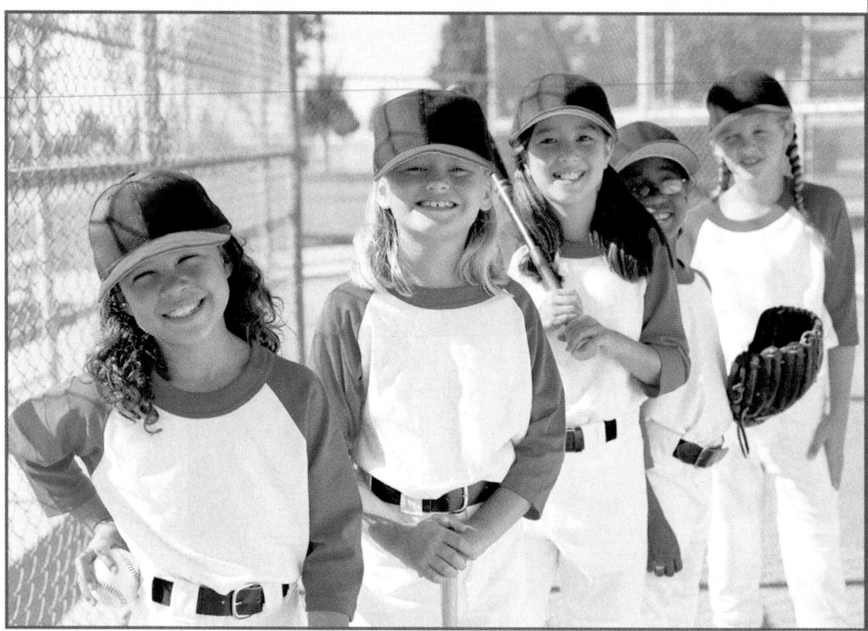

Human growth hormone (hGH) is manufactured by the anterior pituitary gland—and by large pharmaceutical companies. In the body, hGH stimulates growth of the muscles and long bones, making people taller and more muscular. And so in 1985, when a drug company began using genetically engineered bacteria to synthesize hGH, the compound was used to treat people who made insufficient hGH. A 2002 study found that additional hGH can increase final height by an average of 7 centimeters in those whose pituitary gland does not produce enough hGH.

Gradually, the hormone saw wider use and is now being used to treat something that's not even a disease: idiopathic short stature, or ISS. "Idiopathic" means "we don't know why" and being short, well, everybody understands that part.

The Food and Drug Administration has now approved the use of hGH in children who have a normal hGH profile, but who are in the bottom 1–2 percentile for height. These are the children with ISS.

By opening the market to ISS, the FDA helped fuel the ongoing debate over hGH treatment. Nobody questions using the hormone to treat children who do not produce enough of the hormone naturally. But should other children be treated?

That depends on your answer to this question: Is being short a medical condition? Nobody questions that short children get teased in school or that taller men have higher lifetime earnings. But scientists have tried and failed to identify psychological damage from shortness. Another 2002 study, for example, found that "the psychological adaptation of individuals who are shorter than average is largely indistinguishable from others, whether in childhood, adolescence or adulthood."

Curiously, far more boys than girls are asking doctors to treat them for shortness, even though shortness affects both sexes equally. This suggests a cultural stereotype: height is more important for boys.

Artificial hGH is produced through genetically engineering a bacterial cell, and as biotechnology becomes ever more powerful, more ethical questions are arising. Many of these questions concern the use of genetic technologies to perfect the human. What is the price of perfection? Do parents have a duty to accept healthy children as they are, or do they have the freedom—even the obligation—to "improve" their children's anatomy and physiology?

And what about cost? Until 1985, hGH was extracted from cadavers at great cost and used to treat the microscopic percentage of children who lacked the hormone. Now, short children are trooping to the endocrinologists, hoping for a diagnosis of ISS and a miracle from a bottle. The growth hormone in these bottles is mass-produced by transformed bacterial cells or created through laboratory synthesis. These technologies are expensive. hGH costs about $20,000 per year, and several years of treatment may be needed. As health costs skyrocket, many argue that spending so much to treat something that is not even a disease will siphon funds from more pressing, and more beneficial, medical treatments.

Curiously, short stature, one scientist observed, was not even a medical diagnosis until artificial growth hormone became available. In other words, the technology and the economics created the medical syndrome. Is that putting things backwards?

The pancreas is both an endocrine and exocrine gland The pancreas plays a dual role: as an exocrine gland that secretes digestive enzymes through ducts and as an endocrine gland that secretes a number of endocrine hormones involved in maintaining blood glucose levels. Embedded within the exocrine structures of the pancreas are specialized clusters of cells called the islets of Langerhans, which secrete hormones directly into the blood (FIGURE 15.8).

The pancreas FIGURE 15.8

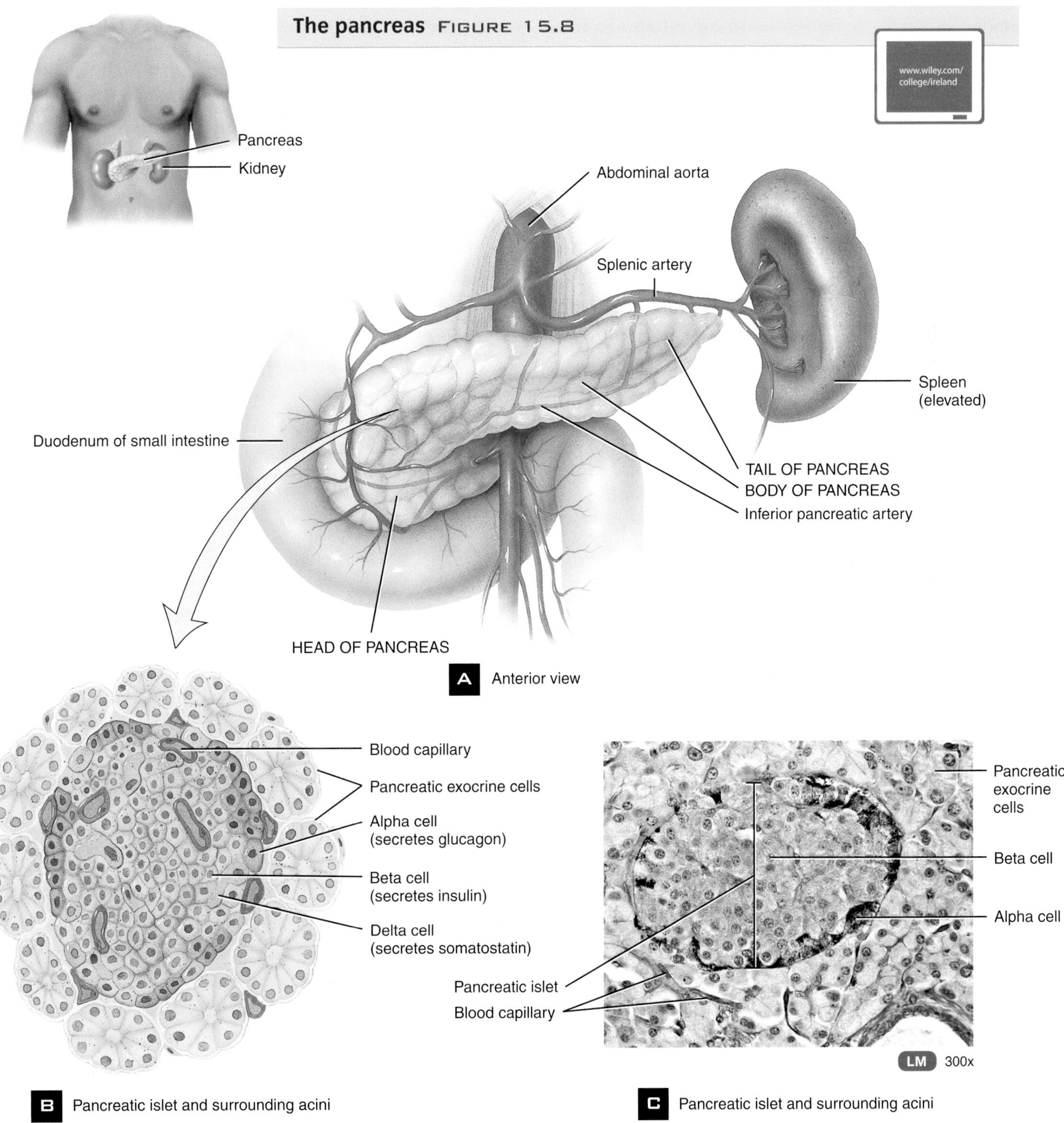

A Anterior view

B Pancreatic islet and surrounding acini

C Pancreatic islet and surrounding acini

The islets include alpha, beta, and delta cells. The alpha cells secrete **glucagon** when glucose levels are low (the name, which sounds like "glucose gone," suggests the function). Glucagon stimulates liver cells to break down stores of glycogen, releasing glucose into the blood. It also causes the breakdown of glycogen in muscle and the production of glucose from amino acids. Glucagon increases blood sugar between meals, supplying energy to the brain and active muscles.

Beta cells of the pancreatic islets secrete **insulin**, a hormone that opposes glucagon. Insulin lowers blood sugar by stimulating liver, muscle, and fat cells to take up glucose. It is the hormone responsible for clearing from the blood all the glucose you get from a meal. If insulin fails, blood glucose levels will rise, causing osmotic balance problems in all tissues. Lack of proper insulin functioning is easily detected by finding sugar in the urine, which immediately suggests the presence of diabetes (see the Health, Wellness, and Disease box on pp. 494–495 for more on this disorder).

The delta cells of the islets of Langerhans secrete the hormone, somatostatin. This hormone seems to inhibit the production of insulin, glucagon, hGH, and a host of other hormones from other glands. The receptors for this hormone are coupled to G-proteins inside the target cells. When somatostatin binds to these receptors, it prevents further processing of hormones through G-protein/cAMP second messengers.

THE ADRENAL GLANDS PLAY MULTIPLE HORMONAL ROLES

The adrenal glands, atop the kidneys, secrete a number of hormones (FIGURE 15.9). These glands have an essential **cortex** and a nonessential **medulla**. The cortex secretes glucocorticoids, mineralocorticoids, and small amounts of estrogen and testosterone. The adrenal medulla secretes epinephrine and norepinephrine, which cause the fight-or-flight reaction discussed in Chapter 7, in response to real or perceived stress.

Glucocorticoids are a group of hormones involved in glucose metabolism. The glucocorticoid secretion of the adrenal cortex is **cortisol**. Cortisol is similar to glucagon in that it promotes the use of fats and proteins as energy sources. Specifically, it causes muscle tissue to break down proteins to amino acids, which the liver can convert to glucose. Cortisol is also an anti-inflammatory. You may have topical cortisol in your medicine cabinet, labeled hydrocortisone, to control itches and rashes. You should not use these products for long, however, as a high level of cortisol can suppress your immune system.

Cortex Outer portion of the organ.

Medulla Inner portion of the organ.

Cortisol-releasing hormone A compound secreted by the hypothalamus into the portal system, leading to the pituitary gland; causes release of ACTH, a pituitary hormone.

The feedback control on cortisol production is typical of the endocrine system. ACTH is released from the anterior pituitary gland when blood cortisol is low. CRH (**cortisol-releasing hormone**) is released from the hypothalamus when the cortisol level drops, causing ACTH to be produced from the anterior pituitary gland. As ACTH level increases, cortisol is produced. Rising blood levels of cortisol inhibit ACTH and CRH. These hormones fluctuate constantly around their ideal, keeping cortisol levels within a narrow range. Cortisol secretion is also affected by physical injury or emotional stress, both of which cause a marked increase in cortisol. The resulting rise in blood glucose is useful during injury repair, and the anti-inflammatory activities decrease fluid loss by capillaries, reducing tissue water retention during stress.

Mineralocorticoids are also secreted by the adrenal cortex. These hormones monitor and maintain ion balance. Sodium and potassium are closely regulated by the hormone **aldosterone**, which also affects water balance. Recall that where sodium goes, water follows. By maintaining proper sodium concentrations, aldosterone assists in maintaining correct fluid levels inside and outside cells. Aldosterone is produced when sodium and water levels are too low, or potassium levels

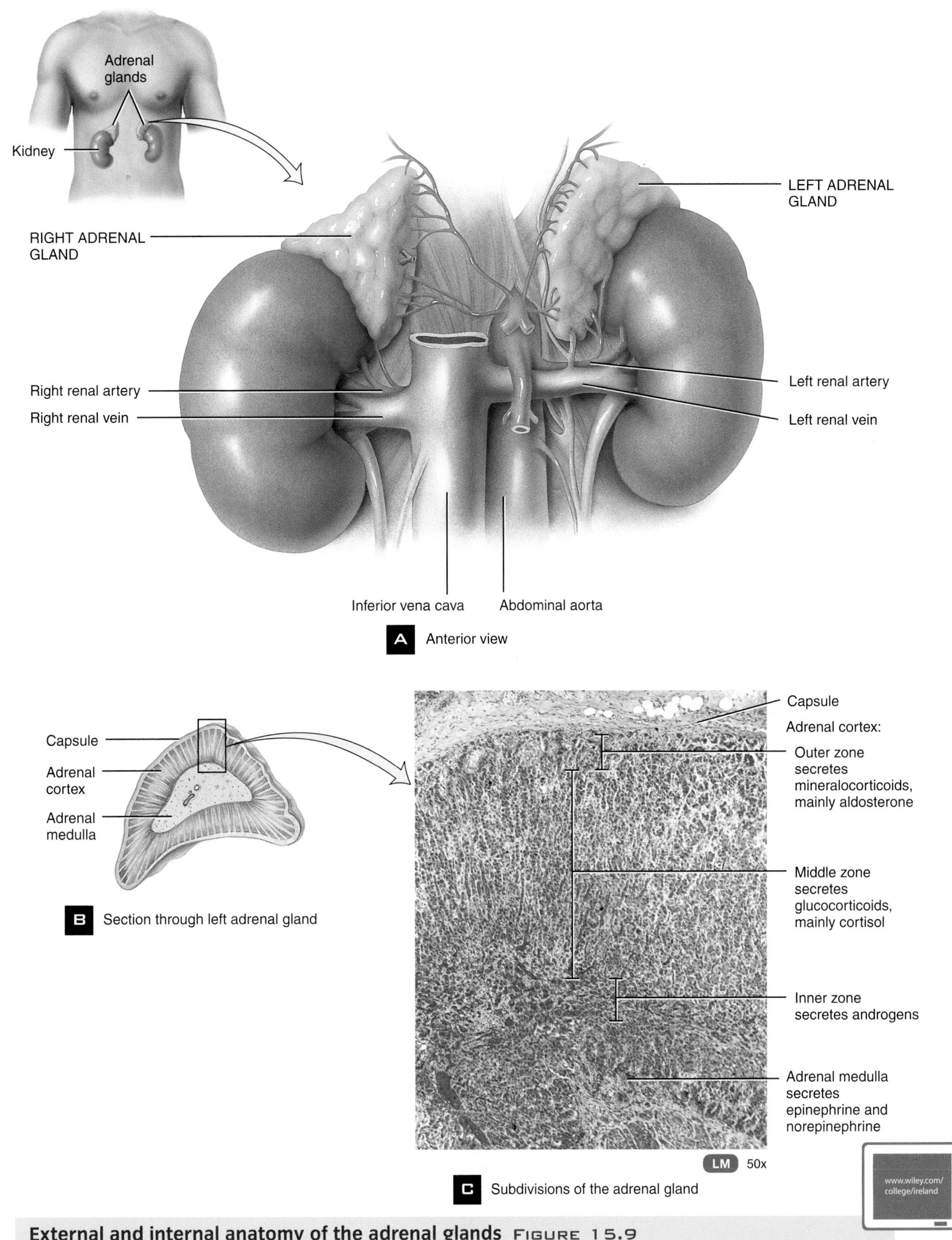

External and internal anatomy of the adrenal glands FIGURE 15.9

are too high (FIGURE 15.10). Aldosterone causes retention of sodium, and therefore water, in the kidneys by exchanging sodium ions, destined for excretion, with potassium ions. Sodium levels increase, while potassium levels decrease. The action of this hormone is one reason athletes are told to ingest more potassium during the summer. In an attempt to retain the water lost during practice, the body produces aldosterone, which drastically lowers the potassium level, raising the risk of muscle cramps.

Adrenal diseases include Cushing's syndrome and Addison's disease. Cushing's syndrome is caused by hypersecretion of the adrenal cortex, which puts excess cortisol in the blood. The cortisol breaks down muscle proteins and redistributes body fat (causing the typical round, flushed "moon face"), a deposit of fat at the back of the neck, and small, thin arms and legs (FIGURE 15.11). Patients also suffer blood chemistry imbalances, primarily excess glucose. Their bones become weak, and they suffer from hypertension and mood swings.

Addison's disease, the hyposecretion of glucocorticoids and aldosterone, is usually due to autoimmune destruction of the adrenal cortex. The resultant lack of glucocorticoids causes mental slowness, anorexia, weight loss, and a bronzing of the skin. President J. F. Kennedy suffered from Addison's disease while in office, but few related his tanned appearance (FIGURE 15.12) to a disease, and his quick-witted performance at press conferences showed no sign of mental slowness. The lack of aldosterone raises blood potassium and reduces blood sodium, which can cause low blood pressure, dehydration, and irregular heartbeat. Treatment for Addison's disease involves replacing the lost hormones and increasing dietary sodium.

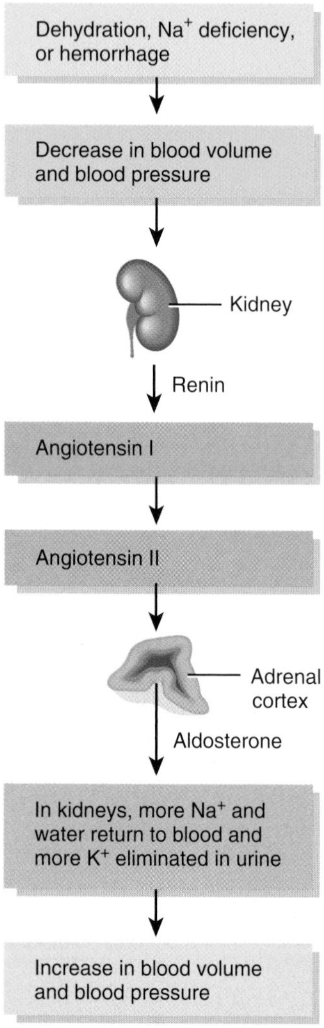

Dehydration, Na^+ deficiency, or hemorrhage

↓

Decrease in blood volume and blood pressure

↓

Kidney

↓ Renin

Angiotensin I

↓

Angiotensin II

↓

Adrenal cortex

↓ Aldosterone

In kidneys, more Na^+ and water return to blood and more K^+ eliminated in urine

↓

Increase in blood volume and blood pressure

Aldosterone functioning
FIGURE 15.10

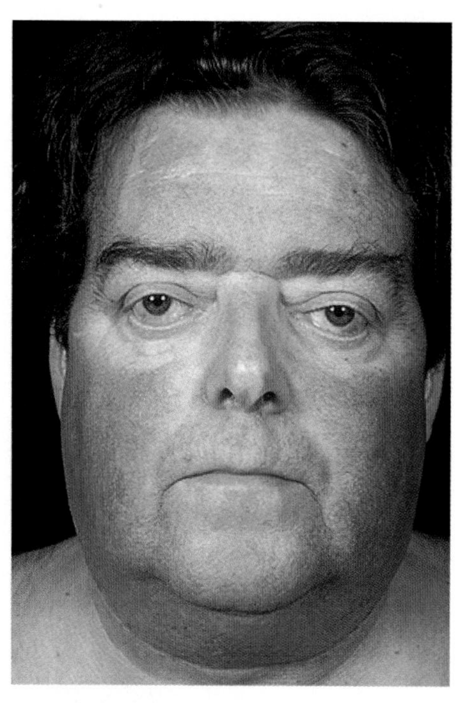

Cushing's syndrome
FIGURE 15.11

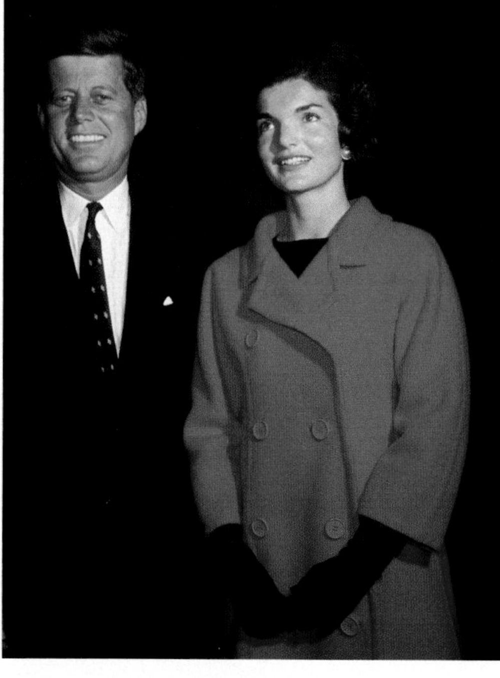

JKF and wife Jacqueline
FIGURE 15.12

THE THYROID AFFECTS ENERGY AND CALCIUM METABOLISM

Basal metabolic rate
Rate of energy usage when body is quiet, resting, and fasting.

Under the influence of thyroid-stimulating hormone from the anterior pituitary gland (discussed earlier), the thyroid gland secretes two structurally similar hormones, T3 and T4 (T3 is often converted to T4 in the body). These hormones are involved in **basal metabolic rate** and help determine how quickly and efficiently you use energy. You may have noticed that some friends can devour enormous helpings of food without gaining weight, while others seem to gain weight from just eye-balling a slice of cake. These differences in energy use and storage are partly regulated by your friends' thyroid hormones and their resulting basal metabolic rate (see the I Wonder . . . box on p. 498 for more about basal metabolic rate). Thyroxin (T4) is responsible for the cellular conversion of glucose to ATP. Higher T4 production increases basal metabolic rate, meaning more work is done, more heat is produced, and more energy is expended. Too much or too little of these hormones can cause abnormal growth and development.

Hypothyroidism occurs when the thyroid secretes too little T3 and T4. Congenital hypothyroidism occurs from birth and can lead to mental retardation and stunted bone growth unless treated immediately. Myxedema results when the thyroid works normally at birth but fails to secrete enough hormone in adult life, causing slow heart rate, low body temperature, dry hair and skin, muscular weakness, general tiredness, and a tendency to gain weight. This condition is more prevalent in females than males. Oral hormone replacement can treat either form of hypothyroidism.

Hyperthyroidism is the opposite condition: the oversecretion of thyroid hormones. The metabolic rate can be 60 to 100 percent above normal. Exophthalmos, fluid buildup behind the eyes, may cause the eyes to "pop" from their sockets and make the whites of the eyes visible all around the iris (FIGURE 15.13). Graves disease, another common hyperthyroidism disease, again occurs more often in females. Graves disease may be treated with surgical removal of part of the thyroid or the application of radioactive iodine to the thyroid. As the gland absorbs the iodine, some of its tissue dies, which reduces its output.

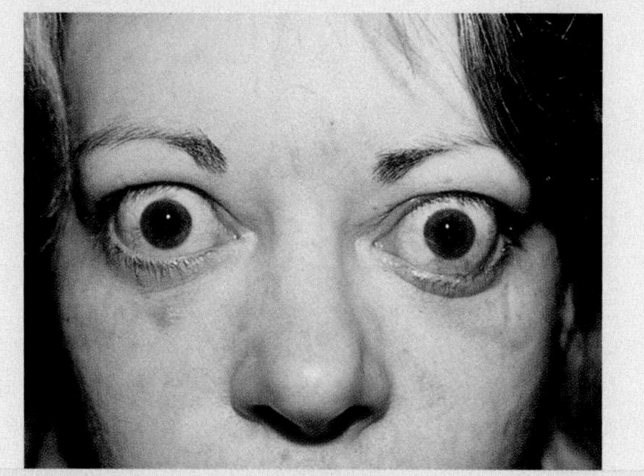

Exophthalmos FIGURE 15.13

T3 and T4 require three and four atoms of iodine, respectively, to complete their production. When iodine is scanty in the diet, the precursors to T3 and T4 cannot be converted to completed hormones, so they are held in the thyroid. TSH is continually produced by the anterior pituitary gland because thyroid hormone levels cannot rise in the blood. This causes the thyroid to enlarge enough to appear on the surface of the larynx, a condition called *goiter* (FIGURE 15.14). Goiter can be prevented by simply adding iodine to the diet, so T3 and T4 formation can be completed. Seawater contains iodine, so when waves crash against the shore iodine is aerated with the spray. This is sufficient to maintain healthy thyroid functioning for people living near a coast. In the United States, goiter is prevented by adding iodine to table salt. In a supermarket, you can buy plain salt (NaCl) or iodized salt for the same price. The government subsidizes the addition of iodine so people living inland will not suffer goiters.

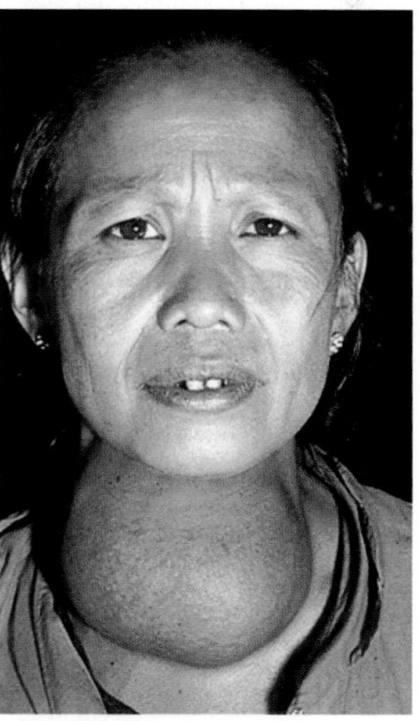

Goiter FIGURE 15.14

Health, Wellness, and Disease

Diabetes: What is it, and why is it so common?

Diabetes mellitus, or simply diabetes, is a common and serious impairment of glucose homeostasis. In essence, the body loses control of the level of blood glucose. Diabetes may be the most serious chronic disease in the United States, with a medical cost of $92 billion a year. About 7 percent of the U.S. population has diabetes, including 6.2 million people who don't know that they have the disease. Diabetes is the sixth-leading cause of death and contributes to about 225,000 U.S. deaths per year. A person with diabetes is twice as likely to die within a year as a demographically similar person without diabetes.

Diabetes mellitus centers on insulin, the hormone that allows glucose to leave blood and enter the cells. Diagnosis of diabetes requires an observation of hyperglycemia (high blood glucose) on at least two occasions. High blood glucose is itself a problem because body cells cannot absorb and utilize the energy they need. But chronic hyperglycemia also damages the kidneys, eyes, nerves, heart, and blood vessels.

Diabetes has two varieties:

- Type I: Usually appears before age 25, when the pancreas suddenly stops making functional insulin.
- Type 2: Usually appears during adulthood. The patient has some insulin but maintains an excessive level of blood glucose.

Type 1 diabetes

Type 1 diabetes used to be called "juvenile diabetes," because of its early onset. An **autoimmune** attack destroys islet cells, rendering the patient unable to make insulin. Type 1 is more prevalent among Caucasians and seems to have a genetic link, as the risk increases if you have a relative with diabetes or another autoimmune disease. When type 1 diabetes strikes, the patient must take over the normal responsibilities of the pancreatic islets, monitoring blood glucose and injecting insulin when the level climbs.

Autoimmune
An immune response launched against healthy tissues of the individual's body, destroying normal organs.

Type 2 diabetes

Type 2 diabetes, once called "adult-onset diabetes," accounts for at least 90 percent of cases. For some reason, cells cease responding to insulin, a phenomenon called insulin resistance. In some cases, the beta cells also fail to produce enough insulin. Even if the blood has an abnormally high level of insulin, the cells still cannot absorb glucose properly. Type 2 seems to combine genetic and behavioral components, since it is strongly associated with a family history of diabetes, older age, obesity, and lack of exercise. Type 2 diabetes is more common in women, especially those with a history of gestational diabetes (diabetes during pregnancy), and among Hispanic, Native American, and African American populations.

The many complications of diabetes

Not getting enough glucose inside the cells is only one of the problems caused by diabetes. It is not completely clear why one defect, high blood glucose levels, causes such wide-ranging and serious complications, but a large part of the reason is that high blood glucose damages small blood vessels. Diabetes can harm:

- The cardiovascular system. Sixty-five percent of diabetics die of heart attack or stroke, a much higher percentage than among the general population, and patients must pay close attention to blood cholesterol and blood pressure. Impaired blood circulation, especially in the legs, is common and often severe enough to force amputations. Aside from those caused by accidental trauma, diabetes is the number-one cause of leg and foot amputations. Poor circulation also impairs wound healing.
- The eyes. Fine blood vessels in the retina leak, causing a type of blindness called diabetic retinopathy. Among people aged 20 to 74, diabetic retinopathy is the major cause of new cases of blindness.
- The nerves. The majority of diabetes patients have some nerve impairment, which can harm digestion and cause numbness in the extremities.
- The kidneys. Diabetes caused 44 percent of the reported kidney failures in 2002, making it the largest single cause of kidney failure.

Prevention and treatment are the keys to diabetes

At present, no one knows how to prevent type 1 diabetes, although researchers are actively investigating how to block the immune attack on the islet cells. Type 2 is a different story. If you have a family history of diabetes, are overweight, and/or sedentary, it may pay to find out if you have "prediabetes," an elevated level of blood glucose that is not yet high enough to signify true diabetes. Those with a high risk of diabetes can control their blood glucose levels through diet and exercise, reducing their body's exposure to the toxicity of high blood glucose, and slowing the onset of diabetes.

The key first step in treating diabetes is to control blood glucose through a combination of insulin, other medications, diet, exercise, and close monitoring of blood glucose. Insulin may be injected or supplied through a pump implanted beneath the skin. Inhalable insulin is about to reach the market, providing an even easier method of administration.

Type 2 diabetes, once considered an adult-onset disease, is now appearing in younger people, apparently as a result of unhealthy diet and lack of activity. Type 2 diabetes is a major concern of the U.S. obesity epidemic. Unlike type 1 diabetes, type 2 can often be controlled by changing diet and lifestyle. Eating smaller portions and increasing exercise can reduce symptoms. Patients with either form of diabetes need to adopt a healthy low-sugar diet and may need advice and medication to control blood lipids and blood pressure. Following the best medical practices can reduce the extent and cost of this deadly and disabling epidemic.

www.wiley.com/college/ireland

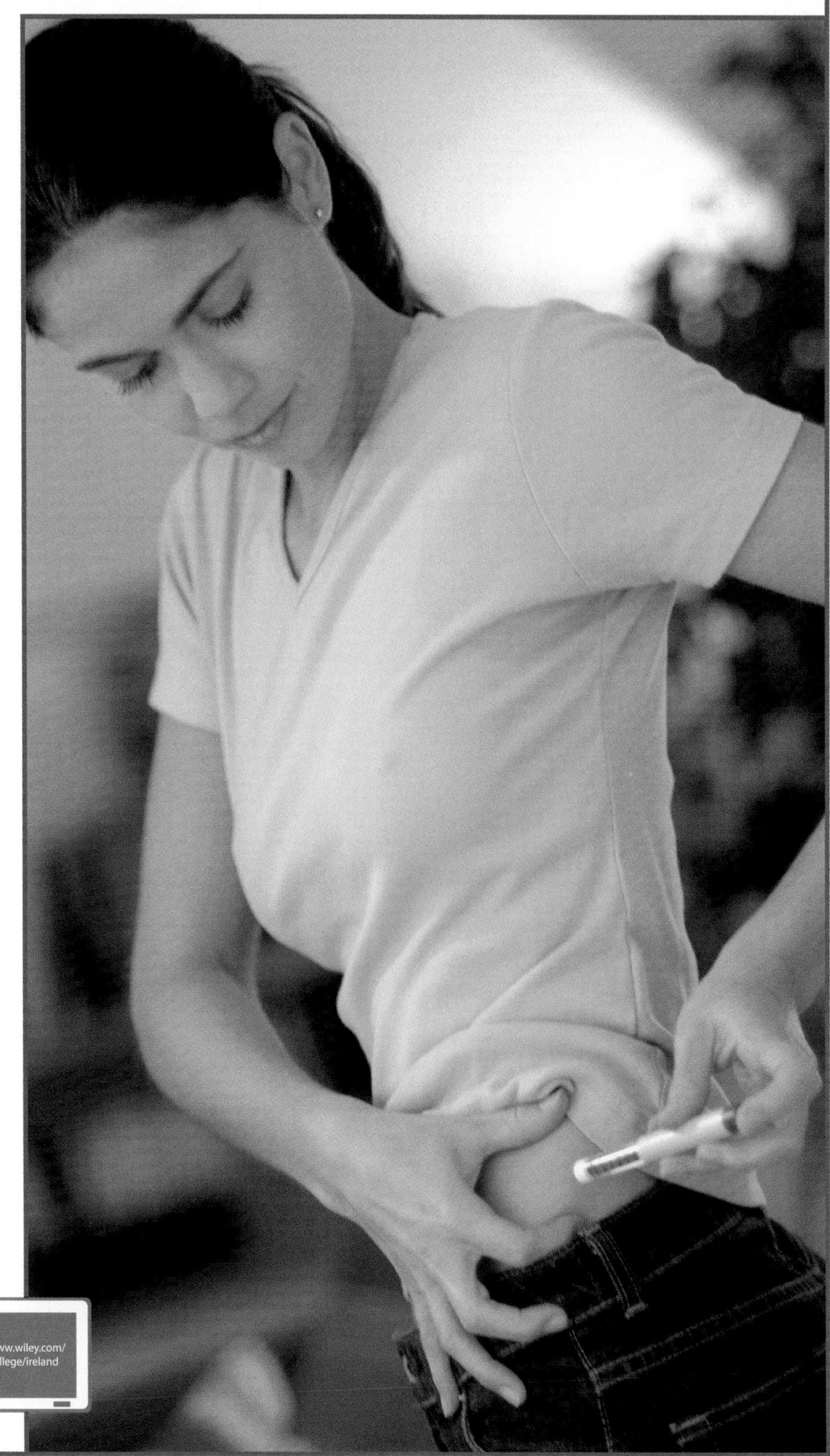

The thyroid also plays a role in calcium regulation. Thyroid cells not involved in producing T3 or T4 produce calcitonin. This hormone stimulates calcium uptake by osteoblasts, putting "calcium in" the bones, as the name suggests. Calcitonin also inhibits osteoclasts, preventing bone from being destroyed. In short, calcitonin causes increased bone mass. The feedback control on calcitonin is simple. When blood calcium is high, calcitonin is produced. When the blood calcium level drops dangerously low, calcitonin is inhibited.

THE PARATHYROID GLANDS ALSO CONTROL BLOOD CALCIUM

The parathyroid glands (FIGURE 15.15) secrete a second, and perhaps more important, hormonal control on blood calcium, called parathyroid hormone (PTH). PTH removes calcium and phosphate from bones, stimulates uptake of calcium from the digestive tract, and prevents loss of calcium in the kidneys (where calcium is exchanged for phosphate) (FIGURE 15.16). This hormone is present throughout life and is the major force in maintaining blood calcium levels in adults.

The digestive tract cannot absorb calcium simply by interacting with PTH. PTH instead stimulates kidney cells to convert inactive vitamin D in the blood to its active form, which cells in the small intestine use to absorb calcium. Without vitamin D, no calcium can be absorbed. This interaction helps explain why we fortify calcium-rich milk with vitamin D. This supplies the extra vitamin D right when it is needed, helping us absorb more calcium from the milk.

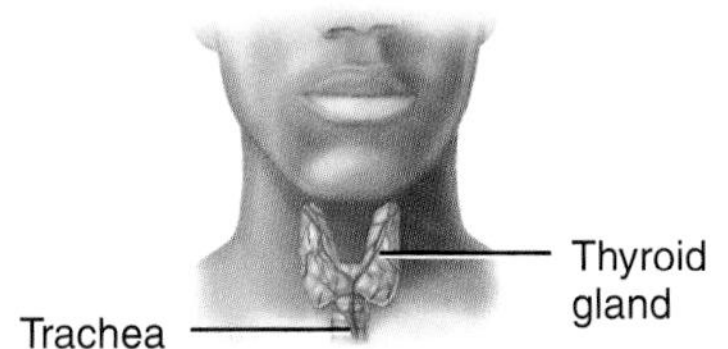

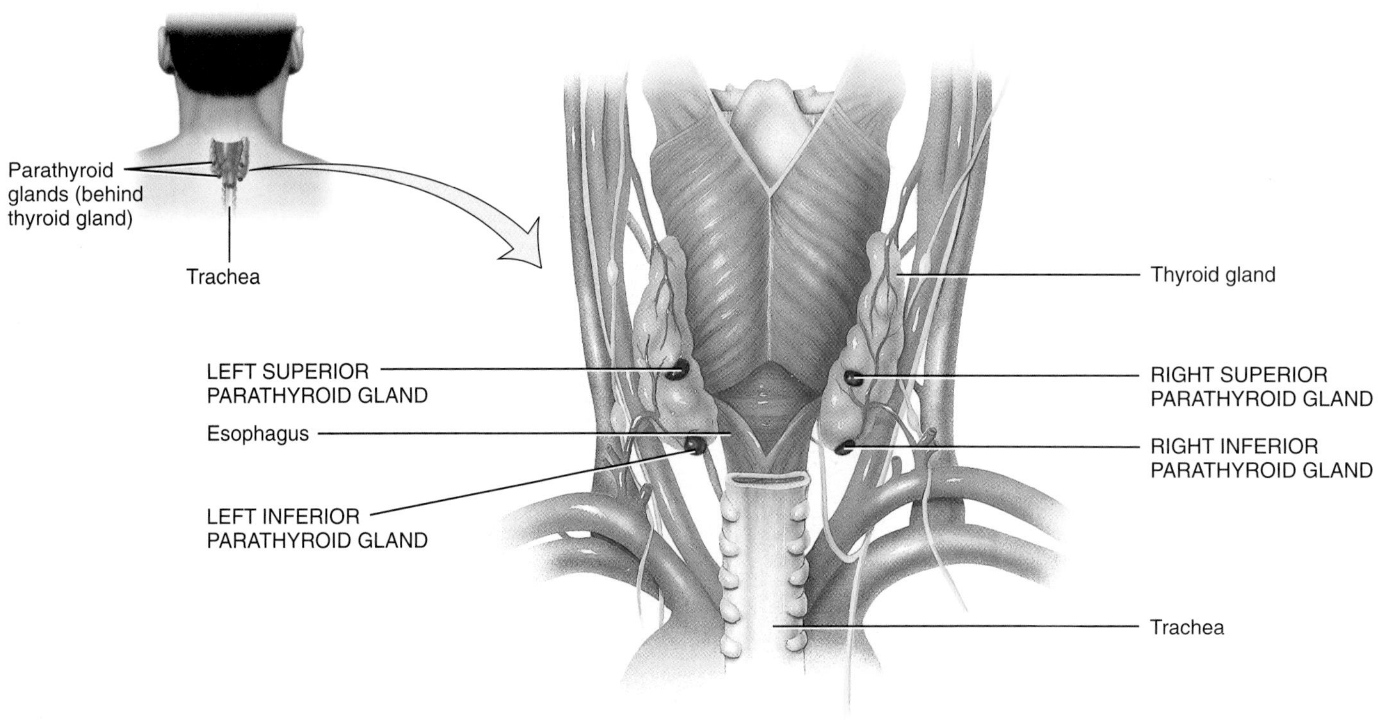

Thyroid and parathyroid glands FIGURE 15.15

Controlling calcium levels in the blood FIGURE 15.16

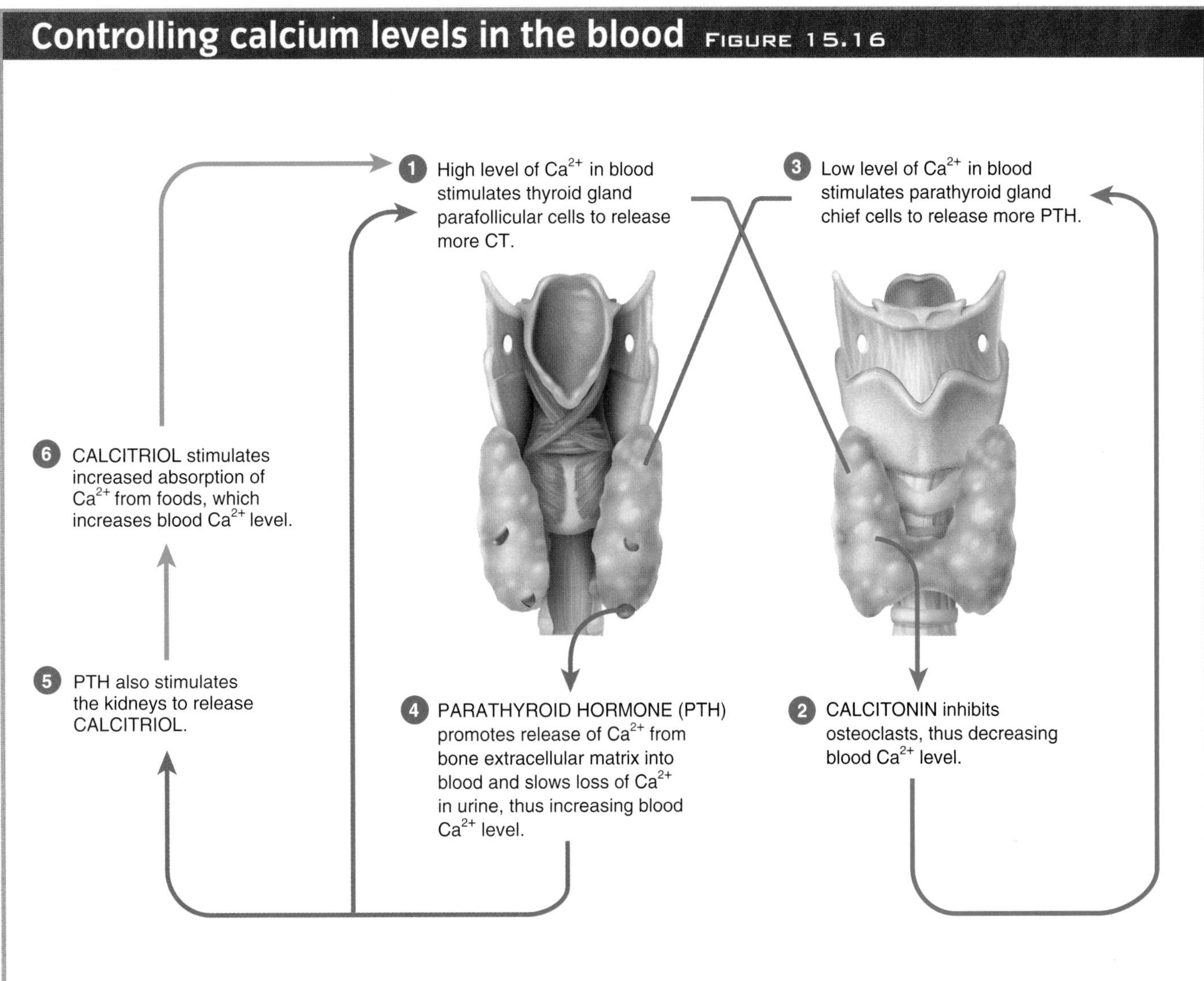
Process Diagram
1 High level of Ca^{2+} in blood stimulates thyroid gland parafollicular cells to release more CT.
3 Low level of Ca^{2+} in blood stimulates parathyroid gland chief cells to release more PTH.
6 CALCITRIOL stimulates increased absorption of Ca^{2+} from foods, which increases blood Ca^{2+} level.
5 PTH also stimulates the kidneys to release CALCITRIOL.
4 PARATHYROID HORMONE (PTH) promotes release of Ca^{2+} from bone extracellular matrix into blood and slows loss of Ca^{2+} in urine, thus increasing blood Ca^{2+} level.
2 CALCITONIN inhibits osteoclasts, thus decreasing blood Ca^{2+} level.

As with calcitonin, the trigger for PTH secretion is the blood calcium level. When blood calcium is low, PTH is produced. When blood calcium is high, PTH is inhibited.

Like the other hormones we have studied, parathyroid hormone is susceptible to hyposecretion and hypersecretion. When too little parathyroid hormone is in the blood (hypoparathyroidism), blood calcium drops precipitously. This in turn causes nerves to depolarize and muscle cells to begin contracting, causing twitches, spasms, and tetany (continuous contraction). With elevated PTH, blood calcium rises, and the bones are robbed of calcium, making them soft and prone to damage. High blood calcium leads to the formation of kidney stones. Less obviously, it also causes personality changes and fatigue.

The thymus is important during the early years

The thymus is a minor endocrine gland in adults, but an important one during infancy and childhood. This gland, located in the anterior **mediastinum**, secretes two hormones important in lymphatic cell

I WONDER . . .

Can I figure out my own basal metabolic rate?

Basal metabolic rate (BMR) is the rate of energy usage that your body needs to stay alive, without accomplishing anything more. It's the kilocalories you expend sleeping, fasting, and, to be technically accurate, probably not dreaming. The energy output in the basal state is used mainly in the liver, brain, skeletal muscle, other organs, heart, and kidneys, more or less in that order. In daily life, we seldom operate at BMR; this rate is a baseline to which we add the energy spent on activities like thinking, moving, and digesting.

The ultimate control on BMR is the hypothalamus, which regulates the autonomic nervous system. The autonomic nervous system controls smooth muscle and cardiac muscle, both of which are active in the basal state. The hypothalamus also regulates the thyroid gland, another key element in the control of metabolism. The higher the rate of thyroid secretion, the higher the BMR.

Many factors temporarily or permanently increase the rate of metabolism:

- stress hormones,
- fever,
- growing (children and pregnant women have higher metabolic rates),
- body type (being tall and thin increases metabolic rate),
- a hot or cold environment.

Other factors reduce energy usage. Older people tend to have more fat, which burns less energy than lean muscle. And fasting or malnutrition causes the body to compensate by slowing metabolism to save energy.

Because aerobic respiration supplies the metabolic energy expended during the basal state, the most accurate way to determine BMR is to measure oxygen consumption and carbon dioxide production. The Internet offers BMR calculators based on height, weight, age, and gender. These calculators are not accurate because body composition and genetics also affect basal metabolism.

Most people who are curious about their BMR probably want to design a diet for losing weight. But paradoxically, cutting calories causes the body to use energy more efficiently, which undermines the effectiveness of dieting for weight loss. Because lean muscle burns more energy than fat, weight training and other activities that build muscle may be a better way to shed pounds.

Calorie calculators try to estimate your real-world calorie expenditures, so they may be more helpful than BMR calculators. Instead of just height, weight, gender, and age, these calculators factor in your activity level. Because activity level strongly affects total energy expenditure, this gives a better picture of your true caloric needs. And because many people find it easier to increase activity than to reduce food intake, activity may be more relevant to the campaign to lose weight.

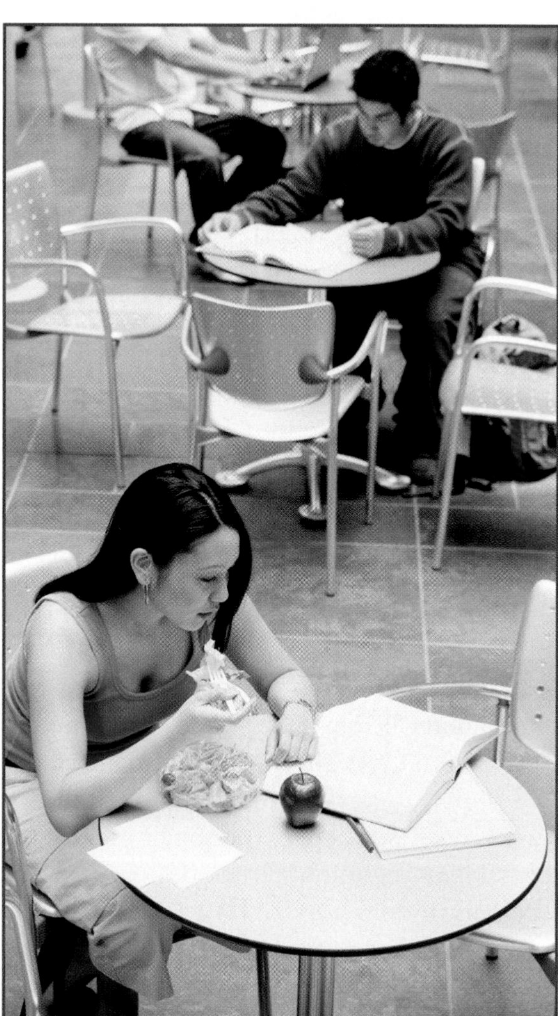

www.wiley.com/college/ireland

maturation, thymosin and thymopoietin. As we age, the thymus convolutes, becoming smaller, more wrinkled, and less functional. By adulthood, the thymus can be removed with no noticeable change in health. By age 60, the thymus functions at a mere 10 percent of its original rate. As mentioned earlier, this decline in thymic function compromises the overall functioning of your immune system.

The pineal gland is an internal clock The pineal gland is also more active in childhood and infancy. This small "brain pea" lies in the roof of the third ventricle and secretes the hormone melatonin. Pineal-gland cells are indirectly sensitive to light as they react to nerve impulses carried on the adjacent optic nerve. This odd interaction causes some to believe the pineal gland is the remnant of a third eye! Regardless, the pineal gland seems to be involved in sleep patterns and **circadian rhythms**.

Circadian rhythm
A daily predictable physiologic cycle based on a 24-hour day.

The pineal gland times its secretion of melatonin by monitoring the optic nerve. Melatonin is secreted only at night, when the optic nerve is quiet. During childhood, melatonin production is tremendously high, but the level drops to a low that usually correlates with the onset of puberty. Is this the elusive trigger that initiates puberty? We cannot be sure, but evidence seems to indicate some involvement.

Melatonin may also induce deeper sleep, as it is produced in infants and children while they sleep. The saying "sleeps like a baby" may have a physiological basis! Because of its apparent role in deep sleep, over-the-counter melatonin (see FIGURE 15.17) is marketed as a sleep aid. We really don't understand the mechanism of this hormone. Does melatonin somehow reduce the levels of FSH or LH? Does it help you sleep as advertised? These are interesting questions, but they require more research.

Other organs have endocrine functions Although they are not specifically endocrine organs, the kidney, heart, and several digestive organs also secrete hormones. The **kidney** secretes erythropoietin, calcitriol, and renin. As you learned in Chapter 11, erythropoietin stimulates the production of red blood cells, thereby increasing blood volume and blood pressure. Calcitriol is involved in calcium ion homeostasis by stimulating the absorption of calcium and phosphate along the digestive tract. Like erythropoietin, renin is involved in blood pressure and blood volume. When renal blood flow declines, the kidney cells secrete renin. This hormone begins an enzymatic chain that ends with the secretion of aldosterone, an increase in thirst, and water retention. Blood volume increases, renal blood flow increases, and the stimulus for renin secretion is removed.

The heart also secretes hormones in response to changes in blood volume. Specialized cells in the atria secrete atrial natriuretic peptide (ANP) when they are stretched by an increase in atrial blood volume. ANP functions opposite of renin: water is lost to the urine, thirst is suppressed, and blood volume decreases. The brain also produces a similar hormone with complementary functions—brain natriuretic peptide (BNP).

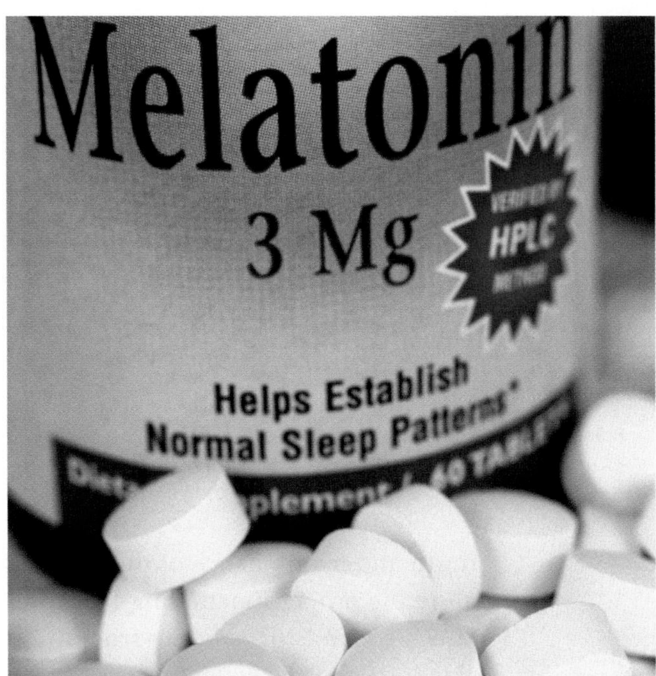

Synthetic melatonin FIGURE 15.17

The intestines produce hormones that coordinate the activity of other digestive organs, including gastrin, cholecystokinin, and secretin. Gastrin stimulates stomach secretions. Cholecystokinin causes the release of bile. Secretin initiates the exocrine functions of the pancreas, causing the pancreas to secrete digestive enzymes into the duodenum.

There is another type of hormone that can be produced by just about any cell of the body. This is called a local hormone, or paracrine. Paracrines are chemical compounds produced by one cell and released into the local environment, which affect only surrounding cells. Histamines and prostaglandins are examples of paracrine secretions. Both of these secretions cause localized inflammation and fluid leakage, but they become inactive just a short distance from their point of secretion.

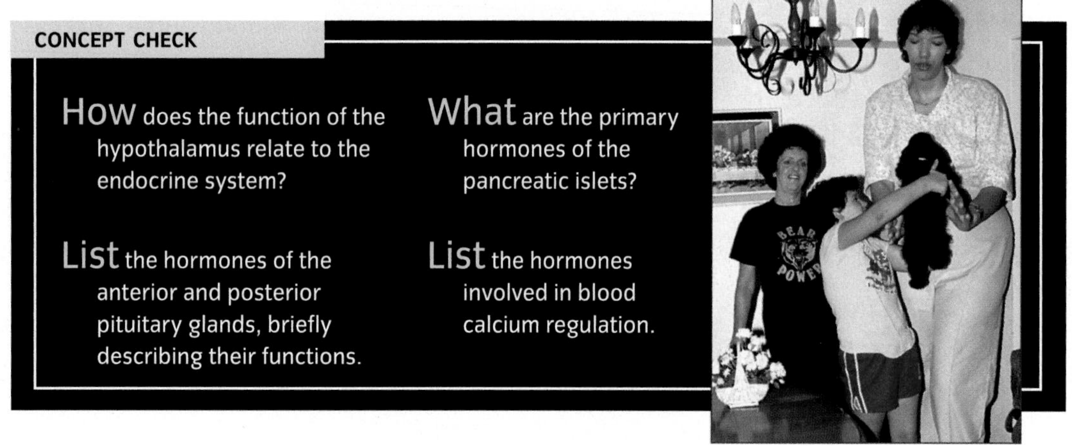

CONCEPT CHECK

How does the function of the hypothalamus relate to the endocrine system?

List the hormones of the anterior and posterior pituitary glands, briefly describing their functions.

What are the primary hormones of the pancreatic islets?

List the hormones involved in blood calcium regulation.

Development Takes Us from Infancy to Adulthood

LEARNING OBJECTIVES

Describe the stages of life.

Relate the events of development to the activity of the endocrine system.

Growth and maintenance of the body require the proper functioning of the endocrine glands. Hormones from these glands direct our sequential growth and maturation from the neonatal period and infancy through childhood, adolescence, and adulthood. Even **senescence**, or aging, has hormonal controls.

THE NEWBORN BABY IS A DEPENDENT BEING

The baby's first month is a time of dependency. The body systems are functioning on their own, but some are not yet fully functional. Have you ever held a newborn? Then you probably noticed that the head bobbed and jerked around if you did not support it. Maybe the hands and feet waved at random. Although the brain is formed, many connections remain to be developed or are not yet functioning well. Neurons will continue to be added for a few more years, and connections will be formed and re-formed throughout a lifetime. Mem-

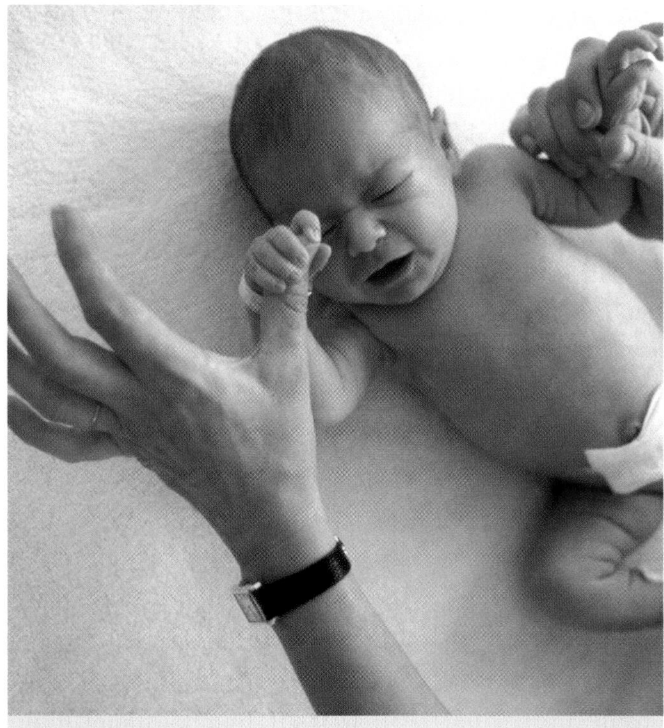

Neonate FIGURE 15.18

> **Neonate**
> The newborn child, from immmediately after birth to approximately one month of age.

ories cannot yet be formed, hearing is less acute than it will become, and muscular control is primitive. **Neonates** (FIGURE 15.18) spend most of their first month suckling, sleeping, urinating, and defecating. Even the digestive system is immature, not able to handle solid foods. The proportions of the head, limbs, and torso are much different from those of an adult, with the head as long as the torso. Rapid growth of torso and limbs will occur in good time.

Infancy is a time of rapid development From month 2 through month 15, body systems mature, control improves, and body proportions begin to shift (FIGURE 15.19). By the end of infancy, the head is one-third of the body length, and the limbs are lengthening. The amount of muscle tissue increases rapidly, and the brain grows quickly. Half of all **postnatal** brain development occurs in this period. The cerebral cortex

> **Postnatal**
> After birth.

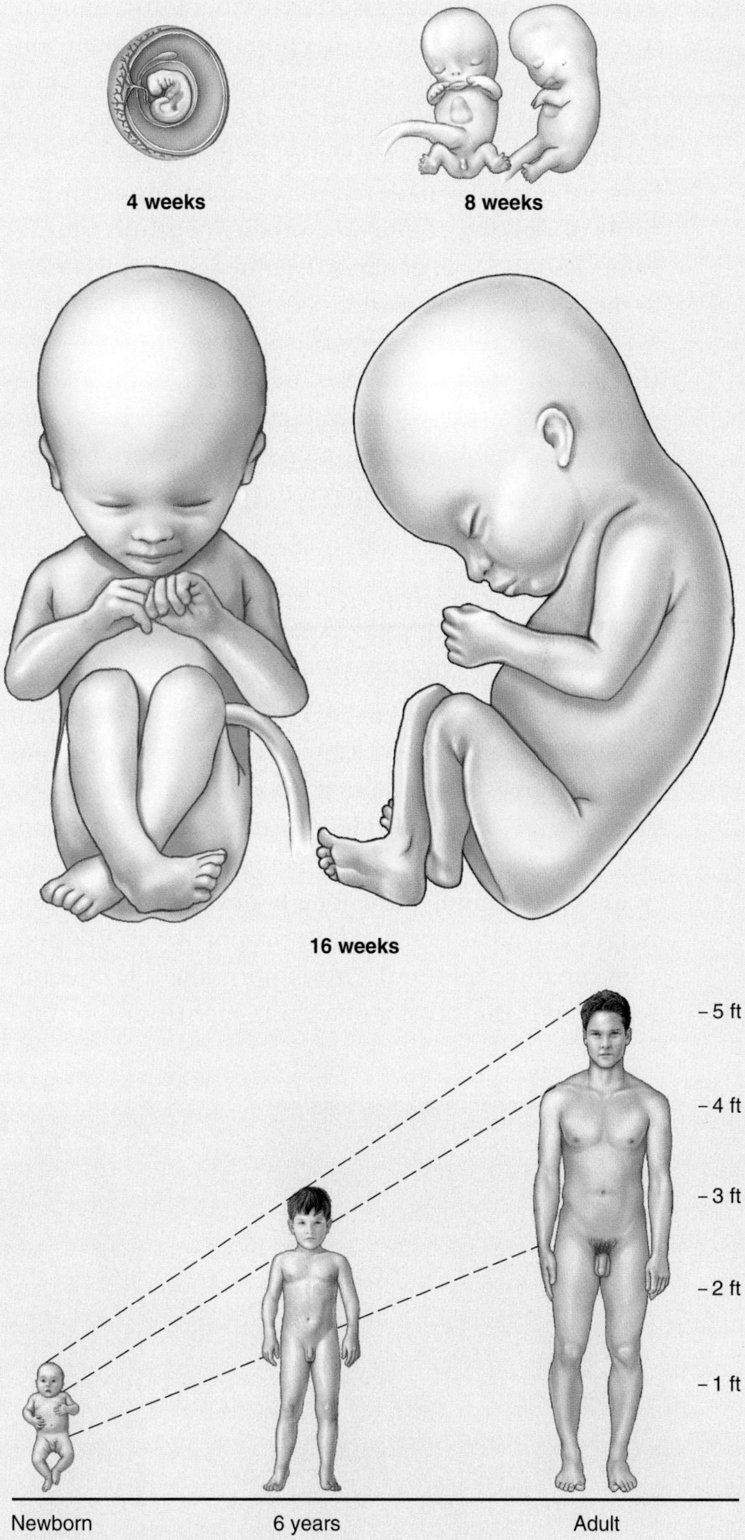

4 weeks

8 weeks

16 weeks

5 ft
4 ft
3 ft
2 ft
1 ft

Newborn
6 years
Adult

Development chart showing body proportions
FIGURE 15.19

expands, adding areas associated with motor functioning, speech, and sensory perception. The skeleton continues to harden, and ossification of the skull is almost complete. Teeth erupt, and solid foods can be eaten. Coordination rapidly improves, so that by 14 months, most infants have mastered the complex muscular patterns of walking. Their personality continues to develop, and they can generate laughter among observers as they explore their world.

The immune system is notably slower to mature. Many vaccinations are ineffective in infants because they cannot yet manufacture antibodies to the antigens. Vaccination regimes start with small doses, and boosters are administered to continually challenge the developing system.

Childhood is a time of almost steady growth

Childhood is much longer than infancy. From age 2 until approximately age 12, the body and all systems grow. The brain will reach 95 percent of its size when puberty ends childhood. Coordination improves, muscles and bones continue to grow, muscular strength increases, and weight is added. By the end of childhood, the average human (male or female) weighs 45.4 kilograms (100 pounds). The long bones lengthen, adding height as well as reach. Adult proportion is attained, with the torso and limbs now approximately equal in length and the head a mere one-eighth of the total body length. All systems are functioning, except the reproductive system.

The regular, sequential changes everyone experiences during their childhood years are governed by hormones. Through the end of childhood, the endocrine system has been directing the patterned growth of bones, muscles, and nervous tissue. The endocrine system has also been working to maintain a healthy metabolic rate, monitor sleep patterns, maintain ion and water balance, and regulate blood levels of calcium and glucose. When puberty begins, the endocrine system will also stimulate the appearance of secondary sexual characteristics and the production of eggs or sperm.

Puberty brings the ability to reproduce

From the neonatal stage through childhood, all of the endocrine glands have been functioning. By the end of childhood, the thymus is slowing its production of thymosin, and the pineal gland is shutting down melatonin production. Between ages 12 and 17, a final growth spurt occurs. Human growth hormone surges through the system, causing a rapid and obvious increase in size of the muscular and skeletal systems. This increase can be so rapid that adolescents often suffer a temporary loss of coordination. The internal organs also grow; the lungs, stomach, and kidneys double in size, and the brain increases by approximately 5 percent.

During puberty, which occurs near the time of this growth spurt, the reproductive organs begin to function. This is directed by the production of gonadotropin releasing hormone (GnRH) from the hypothalamus. GnRH causes the release of FSH and LH by the anterior pituitary gland. In females, the ovaries respond by maturing egg follicles (recall that FSH stands for follicle-stimulating hormone) and producing estrogen. Secondary sex characteristics appear as estrogen levels increase. Males begin to produce sperm and testosterone due to FSH and LH, respectively. Puberty is defined as the onset of the menstrual cycle in females (menarche), and the appearance of nocturnal emissions in males.

Adulthood and aging After puberty is reached, adulthood begins. This is a long stage, lasting from approximately age 15 to 18 until death (FIGURE 15.20). The average life expectancy of Americans at birth is now 77.6 years. Women live about five years longer than men. During this period, adults first see an improvement in overall health but then start to lose some of their physical prowess. The timing of aging varies with individuals, but the pattern is similar. Adults are at their physiological peak in their early 20s. Assuming they lead a healthy lifestyle and avoid serious illness, the peak performance of the 20s can be maintained for almost 20 years. By age 40, however, symptoms of aging begin to appear (TABLE 15.4 on p. 504). Even the most athletic and well-trained adult notices a slight loss in athletic performance by the late 40s. As we move past 50, predictable age-related changes arrive. Eventually eyesight weakens, hearing becomes less acute, muscles lose strength, and bones

FIGURE 15.20

Dating, mating, and marriage are some of the most exciting events of adulthood. Although the body has completed most of its growth and development, the hormones still have many jobs.

The effects of aging on the organ systems TABLE 15.4

System	*Changes associated with aging*
Integumentary system	1. The epidermis thins and weakens. 2. The immune cells of the skin diminish. 3. Vitamin D production decreases by up to 75%. 4. Melanocyte production decreases and the skin becomes paler. Blood supply to the skin is reduced, and sweat glands become less active. 5. Hair production slows, and hairs become thinner and less colorful. 6. The dermis weakens and wrinkles develop. 7. Secondary sex characteristics diminish, and fat deposition becomes similar in males and females.
Lymphatic system 	1. The entire immune system becomes less effective. 2. T cells become less responsive and their numbers drop. 3. B cell populations become less responsive. 4. Cancer increases, as does the susceptibility to viruses.
Skeletal system 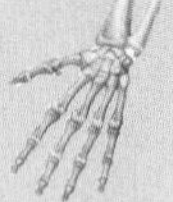	1. The bones become thinner and weaker. 2. Epiphyses, vertebrae, and the jaw lose mass, resulting in shorter stature and tooth loss. 3. Bones become fragile and limbs are more susceptible to breaking from simple actions like standing or walking.
Muscular system 	1. Skeletal muscle fibers become smaller in diameter, and therefore lose strength and endurance. 2. Skeletal muscle becomes less elastic and less flexible. 3. Exercise becomes difficult as fatigue comes more rapidly. 4. Recovery from muscular injuries slows.
Nervous system 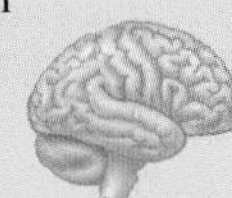	1. The brain is reduced in size and weight as the cerebral cortex shrinks. 2. The number of neurons decreases. 3. Blood flow to the brain declines. 4. Synaptic connections in the brain decrease, and neurotransmitter production declines. 5. Abnormal deposits and tangles may appear in neurons.
Cardiovascular system 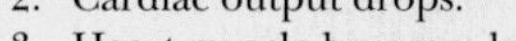	1. Blood hematocrit decreases; embolism and venous pooling are more likely. 2. Cardiac output drops. 3. Heart muscle becomes less elastic and responsive to bodily demands. 4. Scar tissue may build up in the heart. 5. Arterial walls lose elasticity. 6. Plaques and calcium deposits in vessels become more common.
Respiratory system 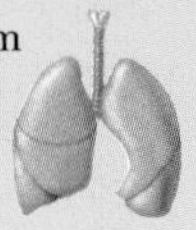	1. Elastic tissue in the system deteriorates. 2. Chest movements become more difficult as joints become less flexible. 3. Respiratory membrane is lost due to lifelong abrasions.
Digestive system 	1. The digestive epithelium is less able to regenerate, becoming more susceptible to disease and tearing. 2. Smooth muscle tone throughout the tract decreases, slowing the clearance of material. 3. Cumulative damage becomes apparent as areas that were slightly compromised early in life are now overwhelmed by lifelong damages.
Urinary system 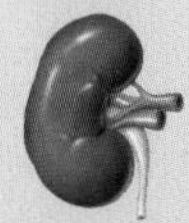	1. The number of functional nephrons declines. 2. Glomerular filtration decreases. 3. ADH sensitivity diminishes, increasing chance of dehydration. 4. Control over the external urinary sphincter is compromised, leading to incontinence.
Endocrine systems 	1. Reproductive hormone production declines. 2. The thymus shrinks, dramatically decreasing production of thymosin.
Reproductive systems 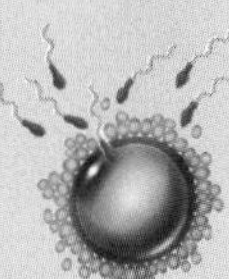	1. Menopause causes the cessation of the female reproductive cycle; hormone levels drop, follicles no longer respond to FSH, and unpleasant symptoms may occur. 2. The male climacteric occurs, reducing circulating testosterone levels.

often become brittle. A good diet and a regular exercise routine help slow the process but do not stop it.

Today three lines of thought are most often offered to explain aging: limits on cellular division, accumulated cellular damage, or the demise of organ systems.

Normal (noncancerous) cells can only go through a certain number of **cell divisions**, or a predetermined number of mitotic rounds. On the ends of each chromosome are strands of DNA that do not code for proteins, called telomeres. With each mitotic division, the telomeres get shorter, so it appears that telomere length may regulate the number of possible generations. Most tissue cells in the laboratory live for only 50 to 80 generations before they die out or become **senescent**. Cancer cells are an exception, as they can produce thousands of generations. Cancer cells produce the enzyme telomerase to rebuild shortened telomeres, which may explain their immortality.

Senescent Aging or growing old.

Free radicals Highly reactive ions, such as oxygen.

Cellular damage refers to damage to DNA. We have repair mechanisms to fix strands of DNA that have suffered various degrees of damage. But our metabolism produces **free radicals** and other noxious compounds that damage DNA, and some lifestyles cause further damage. If the damage overwhelms the repair mechanisms, the cell will die, and this is, according to one theory, a cause of aging. In the laboratory, putting animals on a restricted calorie diet prolongs life. This may be due to a slightly lower body temperature that develops as a result of less energy usage. Lower temperatures slow enzymatic reactions, thereby slowing the internal damage caused by the by-products of these reactions. Few people are willing to go on such restricted diets, but many are willing to swallow antioxidants in the hope that they will reduce free radicals and DNA damage, and thereby extend life. But some large studies have found no such benefit from the most popular antioxidant vitamins, including C and E.

The last theory on aging reminds us that "a chain is only as strong as its weakest link." If one small change occurs in an **organ system**, the repercussions may kill the organism. Many diseases provide evidence for this theory. If cutaneous cells that normally exclude pathogens are weakened by poor diet, viral infection, or physical damage, bacteria may find an "open door" into the body and cause septicemia. Without medical attention, homeostasis is compromised and, if the bacterial growth remains unchecked, death follows. Also, what would happen if the parathyroid glands stopped producing PTH, as a result of a viral infection or surgical removal of the gland? Without medication, blood calcium would plummet, causing uncontrollable muscle contractions. The diaphragm would cease to function and breathing would stop, causing oxygen starvation of the brain. Soon all organs would be affected; homeostasis would be lost, and the person would perish.

Luckily, just as death is inevitable, so is birth. Despite the daunting odds against everything necessary for pregnancy, fertilization, and growth falling into place, the miracle of life continues. The human body is awe-inspiring, functioning with precision yet tolerance; not merely maintaining homeostasis in the face of incredible odds, but in fact thriving.

CONCEPT CHECK
Describe the functioning of a typical neonate.
What event marks the end of childhood?
Describe the changing body proportions of an individual from neonate through puberty.
What hormones are directly involved in the onset of puberty?
Briefly describe three possible mechanisms of human aging.

CHAPTER SUMMARY

1 Hormones Are Chemical Messengers

The endocrine system is responsible for maintaining growth and development. It includes the hypothalamus, pituitary gland, thyroid, parathyroid glands, thymus, pancreas, adrenal glands, and pineal gland. Endocrine products, or hormones, communicate with distant target cells.

Hormones are either lipid soluble or water soluble. They bind to receptors and alter the functioning of the target cell. Steroid hormones reach receptors within the cell, while nonsteroid hormones bind to a membrane receptors. Nonsteroid hormones activate a second messenger inside the cell. They generally act faster than steroid hormones, because nonsteroid hormones alter proteins already present in the cell.

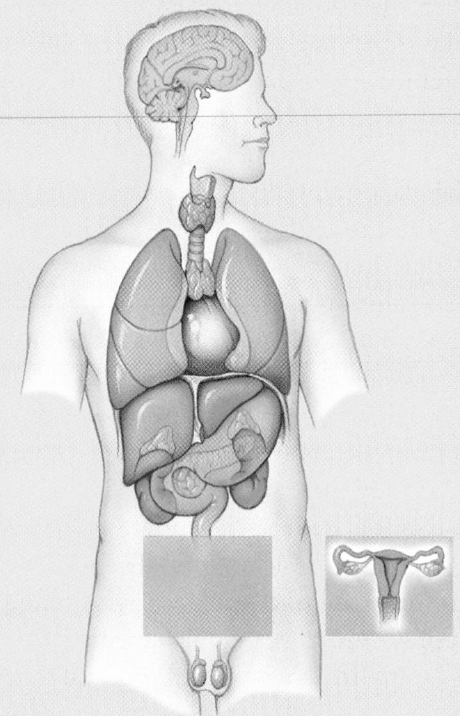

2 The Endocrine Glands Secrete Directly into the Bloodstream

The hypothalamus secretes factors that control the pituitary gland, which in turn secretes nine hormones: oxytocin, ADH, hGH, PRL, FSH, LH, MSH, ACTH, and TSH. Most of these hormones are controlled by negative feedback systems. ACTH stimulates the production of steroid hormones from the adrenal glands. TSH causes activation of the thyroid gland. FSH and LH affect the reproductive organs. ADH causes the kidneys to retain water. Oxytocin promotes smooth muscle contractions in the pregnant uterus and in the mammary glands after the baby's birth. MSH increases the production of melatonin. PRL promotes milk production in females; its role in males is not known.

3 Development Takes Us from Infancy to Adulthood

The endocrine system controls the long-term changes observed during growth and development in the neonate, infant, child, adolescent, and adult. Puberty marks the shift from child to adult. Aging is the progressive slowing of body functions, and while the timing of events differs with individuals, certain events are expected. Visual acuity diminishes, muscle strength is lost, skin thins, reproductive organs slow or stop functioning, bones lose density, and nervous system functioning is decreased.

KEY TERMS

- **autoimmune** p. 494
- **basal metabolic rate** p. 493
- **circadian rhythm** p. 499
- **cortex** p. 490
- **cortisol-releasing hormones** p. 490
- **cyclic AMP (cAMP)** p. 480
- **endorphins** p. 483
- **enkephalins** p. 483
- **free radicals** p. 505
- **gonadotropins** p. 486
- **kinase** p. 480
- **medulla** p. 490
- **mineralocorticoids** p. 486
- **neonate** p. 501
- **postnatal** p. 501
- **senescent** p. 505
- **somatostatin** p. 481

CRITICAL THINKING QUESTIONS

1. How does the hypothalamus govern the pituitary gland? Compare the route taken by hypothalamic releasing factors to the route taken by hormones that stimulate the anterior pituitary gland. Which is more direct? Why?
2. Compare the pathologies of Type I diabetes and Type II diabetes. What is similar in these two disease states?
3. Many hormones are associated with fluid balance. List those covered in the chapter, and describe each of their functions. How do they interact? Try to figure out which ones act together and which inhibit one another.
4. Which hormones are involved in puberty? GnRH is released from the hypothalamus, stimulating the release of the two gonadotropic hormones from the anterior pituitary gland. Which hormones are these? What other hormones are secreted in response to the action of these pituitary hormones? Do any other hormones arise during puberty?
5. Describe the aging process. What mechanisms could control it, and how could they be slowed or stopped?

SELF TEST

1. The functions of the endocrine system include
   a. cellular communication.
   b. precise timing of development.
   c. maintaining fluid balance.
   d. All of the above.

2. The gland indicated by the letter A is the
   a. adrenal glands.
   b. pancreas.
   c. pineal gland.
   d. pituitary gland.

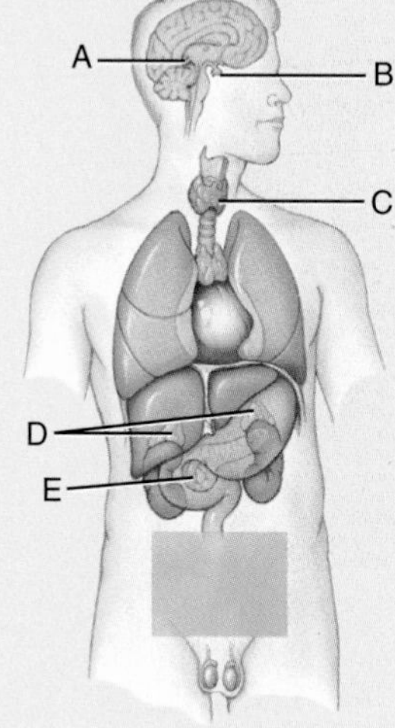

3. In the above figure, the endocrine gland that secretes both insulin and glucagon is labeled as
   a. A.
   b. B.
   c. C.
   d. D.
   e. E.

4. Of the glands indicated in the figure above, which serves as the "master gland"?
   a. A.
   b. B.
   c. C.
   d. D.
   e. E.

5. The figure at right demonstrates the action of
   a. steroid hormones.
   b. nonsteroid hormones.
   c. ADH.
   d. thyroid hormones.

6. True or False? Nonsteroid hormones act far more quickly than steroid hormones.

7. The portion of the pituitary gland indicated by the letter A secretes
   a. human growth hormone.
   b. hypothalamic releasing factors.
   c. oxytocin.
   d. ACTH.

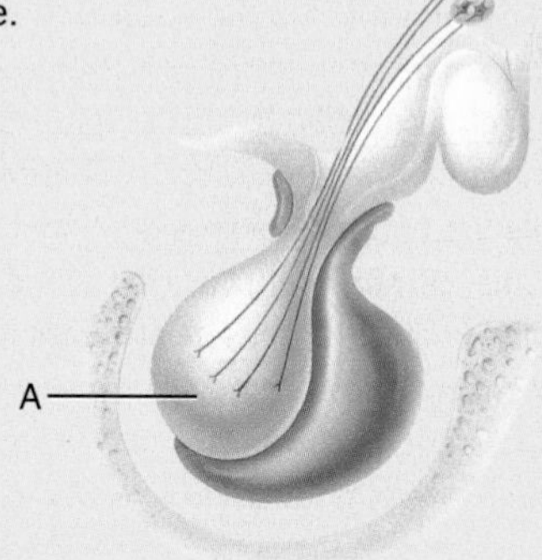

8. The hormone from the anterior pituitary gland that stimulates the adrenal glands to produce glucocorticoids is
   a. FSH.
   b. TSH.
   c. PRL.
   d. ACTH.

9. The individual shown in the photo below suffers from
   a. overproduction of one specific pituitary hormone.
   b. underproduction of the adrenal hormones.
   c. overproduction of thyroid hormones.
   d. underproduction of gonadotropic hormones.

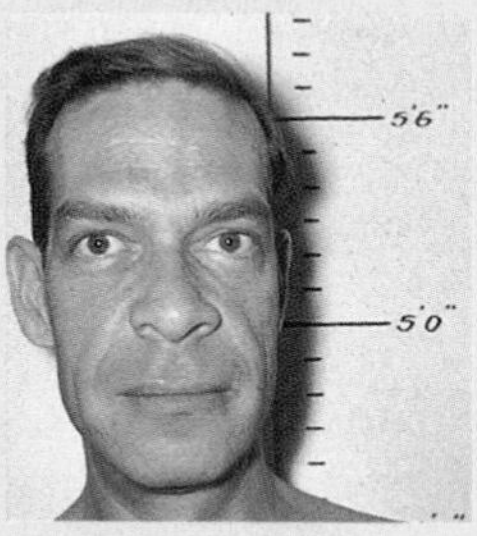

10. The hormones produced from this organ include
    a. glucocorticoids.
    b. glucagon.
    c. human growth hormone.
    d. endorphins.

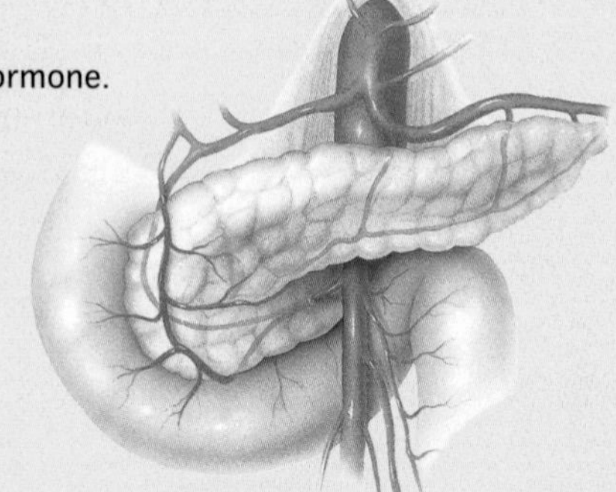

11. The vitally essential layer of the adrenal gland is the

a. cortex.
b. medulla.
c. capsule.
d. The entire gland is essential.

12. An example of an autoimmune disease of the endocrine glands is

a. Cushing's disease.
b. Addison's disease.
c. diabetes insipidus.
d. goiter.

13. The two endocrine glands shown in the illustration below both regulate

a. blood glucose levels.
b. metabolic rate.
c. production of estrogen and testosterone.
d. blood calcium levels.

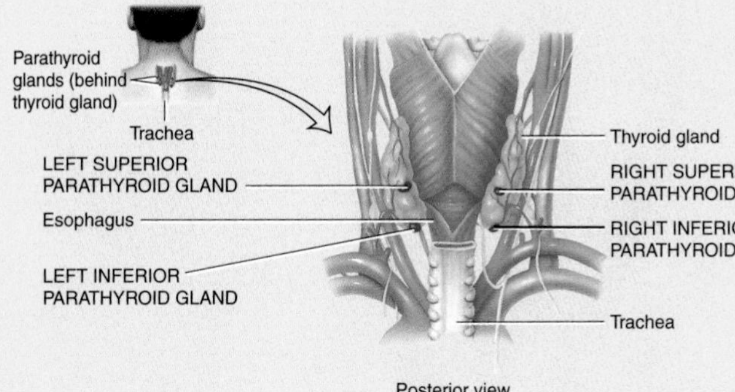

14. The endocrine gland that serves as an internal biological clock is the

a. pineal gland.
b. pituitary gland.
c. thymus.
d. thyroid.

15. The organ that serves as an endocrine gland and secretes erythropoietin, calcitriol, and renin is the

a. stomach.
b. pancreas.
c. brain.
d. kidney.

16. The hormone secreted by the atria of the heart is involved in

a. production of red blood cells.
b. digestion.
c. maintaining blood volume.
d. absorbing calcium.

17. The correct term for the stage of life of this human is

a. fetus.
b. neonate.
c. infant.
d. adolescent.

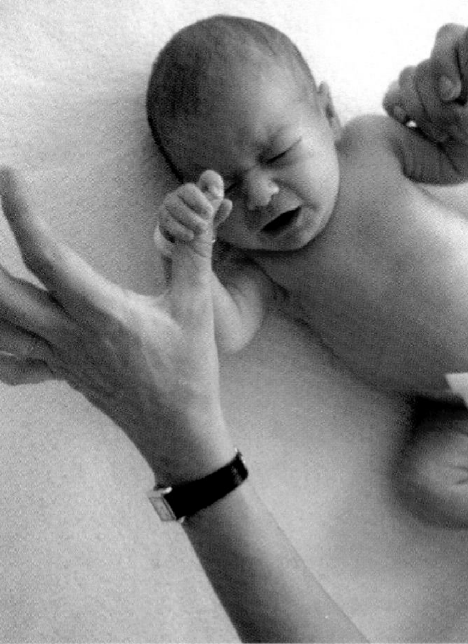

18. The brain grows most quickly during

a. the neonatal period.
b. adulthood.
c. infancy.
d. senescence.

19. The two hormones that are responsible for the onset of puberty are

a. FSH and ADH.
b. FSH and LH.
c. LH and ACTH.
d. melatonin and oxytocin.

20. Low supply of this element causes formation of a goiter:

a. calcium.
b. iron.
c. iodine.
d. potassium.

The Reproductive Systems: Maintaining the Species

16

"Birds do it. Bees do it. Even educated fleas do it. Let's do it. Let's fall in love."

Songwriter Cole Porter got it right decades ago, when he wrote this about sex. Okay, we'll admit that what he called "love" is actually "sexual reproduction," but you get the idea. There is nothing new about sex, which plants have been using to ensure reproductive success for hundreds of millions of years. The need to join gametes from two individuals traces back that long in life forms from fungi to flowering plants, from birds to bees and of course to humans.

Sexual reproduction has evolutionary benefits: it speeds up the formation of new genotypes (genetic configurations) that can be tested against the environment. It also dilutes harmful genes.

The urge to engage in sex is one of the strongest human desires, ranking second only to eating and breathing. Many biologists believe this urge originates in evolution through natural selection: without sex, we do not leave descendants. The genes of people who have sex and reproduce are found in the next generation, and to the extent that reproduction is a genetic urge, the mechanism is self-perpetuating.

Since reproduction is so critical to survival, the urge needs to be managed; many of the most common and critical human customs concern reproduction: marriage, childbirth, and family ties. In this chapter, we look at the physiology and anatomy of reproduction, and include some scientifically based suggestions for keeping the urge to reproduce in a healthy framework.

CHAPTER OUTLINE

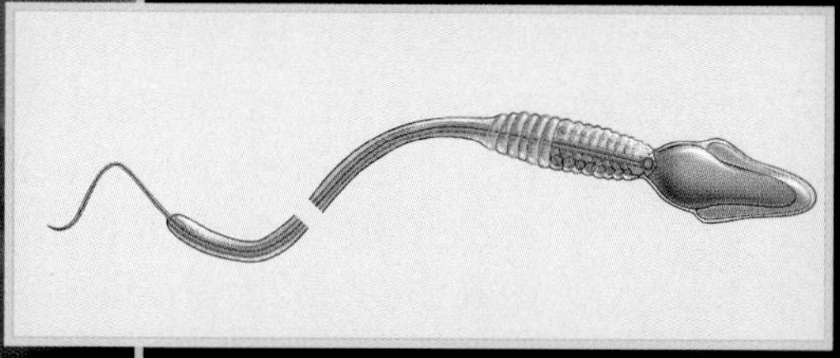

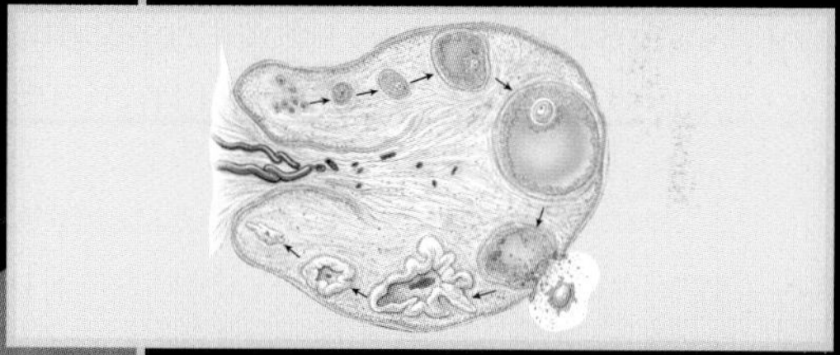

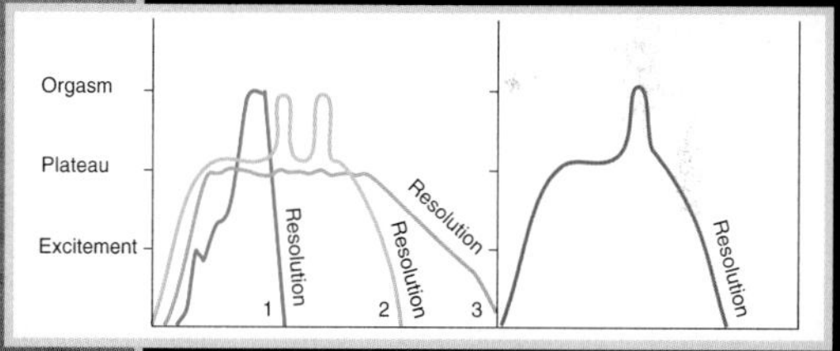

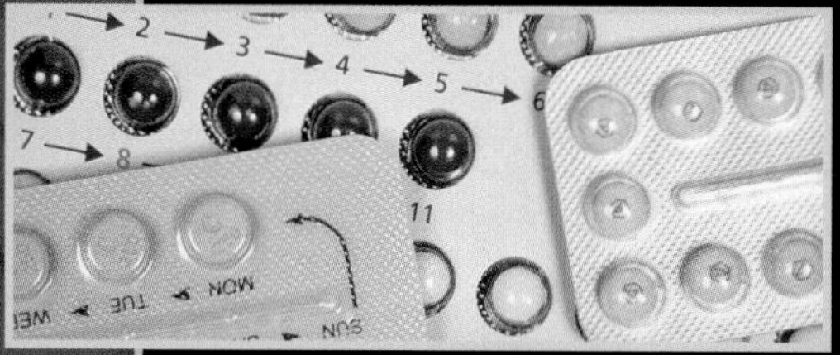

Survival of the Species Depends on Reproduction and Gamete Formation

LEARNING OBJECTIVES

Explain the functions of the reproductive system.

Place sexual reproduction in the context of the theory of evolution.

Gender is an obvious structural and functional difference between people. We are either male or female. We all know that the female produces eggs, and her anatomy is set up to house and nourish the developing baby. And we know that the male produces sperm, and his anatomy is designed to deliver that sperm to the egg. Because we rely on sexual reproduction, having two genders is necessary to perpetuate the species (FIGURE 16.1).

Aside from the obvious anatomical differences, are there any homeostatic differences between men and women? Are we so different as to verify the flippant pronouncement "men are from Mars, women are from Venus"? Are we worlds apart just because of a difference in one chromosome? To answer these questions, we will start by looking at reproduction in general, and then at male and female anatomy. We will explore hormonal differences, and finally, armed with this knowledge, we will explore birth control methods that help us to control when we reproduce.

The main purpose of the reproductive system is to produce **haploid** gametes (egg and sperm) and unite them to form a new individual. Sexual reproduction involves choosing a mate based on **phenotype** and mixing and shuffling genes from the two to form a new individual. This mixes and blends the **alleles** in the gene pool, creating new genetic combinations.

These new combinations are essential to the survival of the species. The genetic variation in populations of sexually reproducing organisms is the basis for adaptation of organisms to their environment. Given enough variation, some individuals will always be better suited to the environment than others so that they can survive and pass on their genes. These "more fit" individuals will produce more offspring, thereby increasing

Haploid
Having half the number of chromosomes of normal body cells, found in eggs and sperm.

Phenotype
An organism's observable characteristics.

Alleles
Genes found on the same spot on the same chromosome in different individuals, coding for subtle variations of the same protein.

Man and woman FIGURE 16.1

Sexual reproduction requires two sexes. Humans are **dioecious**, meaning the male and female reproductive organs are carried on different individuals. In contrast, earthworms and many plants are **monoecious**; one organism carries both male and female reproductive organs.

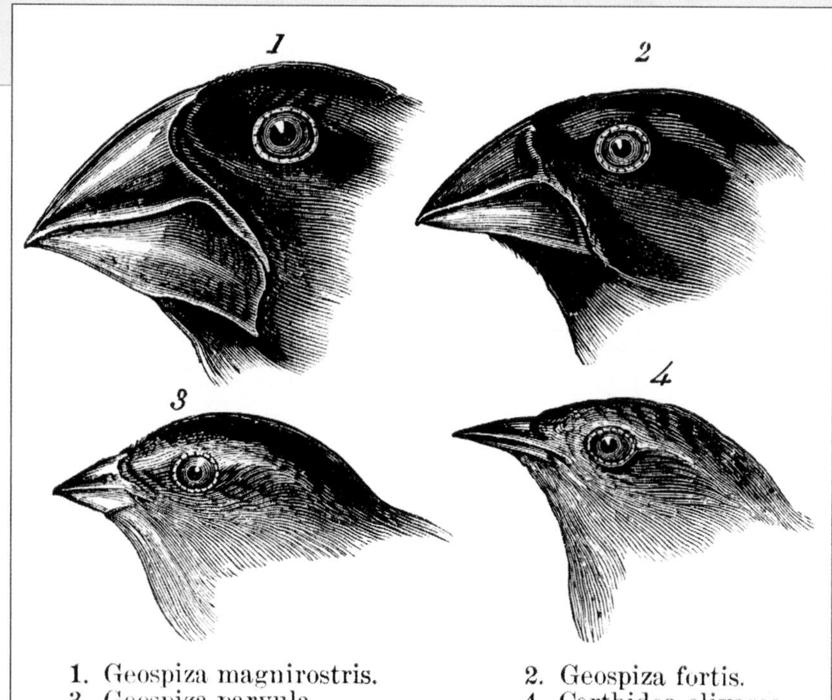

Natural selection: Darwin's finches

FIGURE 16.2

In observing the finches living on the Galapagos Islands, Charles Darwin noticed subtle variations in beak shape. He then observed that these variations correlated to the type of food available on each island. His conclusion was that those birds better able to eat the available food dominated the bird population of each island. We now understand that these successful birds were able to add their alleles to the population at a faster rate than those birds with less successful beak shapes.

the percentage of their alleles in the gene pool. This line of reasoning is the underpinning for Charles Darwin's theory of evolution through natural selection. The fittest organisms survive and pass their genes to the next generation (FIGURE 16.2).

In his classic discussion, Darwin noted that the finches on the Galapagos Islands had beaks specifically shaped to assist in eating the available food of that island. Some islands had large nuts and berries; those finches developed stronger, larger beaks. Other islands had grasses and thinner seeds; the finches on those islands developed delicate beaks able to pick the seeds from the grasses. In a recent press report, it has been shown that these finches' beaks are still evolving. Just two decades after a competing finch with a large heavy beak arrived on one of the Galapagos Islands, the native finch evolved a smaller, thinner beak to take advantage of a food source unavailable to the newcomer. As these species compete for food, they apparently can and do undergo descent with modification, or evolution.

Passing on your genes requires you to form haploid gametes. Gamete is a general term for the reproductive cells that will form a new individual, the egg and sperm. These are produced via **meiosis**, a specialized type of cell division that ensures the equal and orderly division of chromosomes (FIGURE 16.3, p. 514). In order to form gametes properly, the normally **diploid** chromosome number must be cut in half, with the resulting gametes having exactly half the usual complement of alleles. This way, when two haploid gametes unite to form a **zygote**, the original diploid number is restored. The division must be accomplished so that each gamete has a predictable and reliable half of the chromosomes. Rather than randomly splitting the chromosome, **homologous** chromosomes come together and are then separated, one to each new gamete.

Diploid
Having the total number of chromosomes of the body cells, twice that of the gametes.

Homologous
Similar in structure, function, or sequence of genetic information.

In the male, meiosis occurs exactly as depicted in FIGURE 16.3, and four sperm are produced from

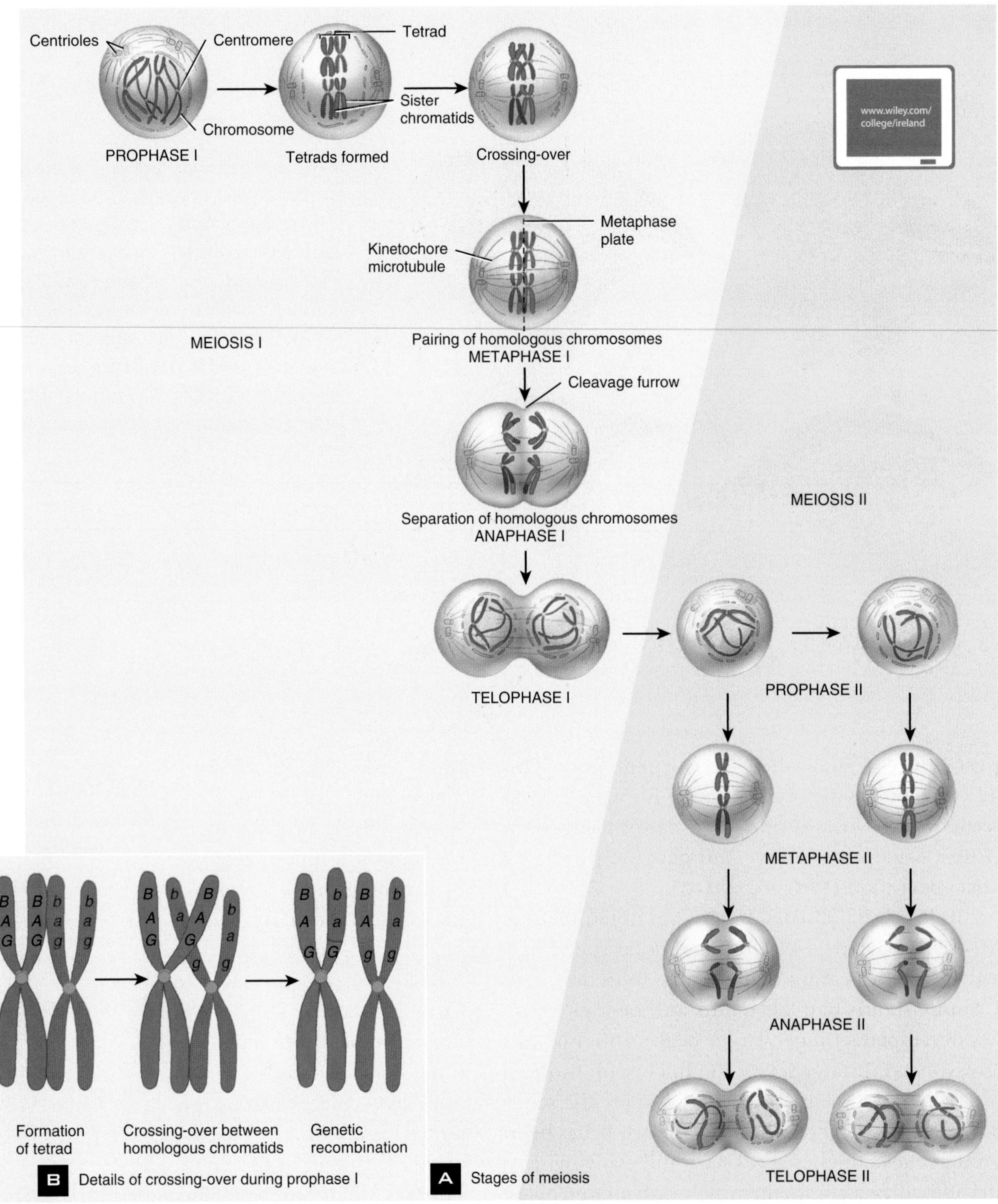

B Details of crossing-over during prophase I

A Stages of meiosis

Meiosis FIGURE 16.3

Meiosis is the orderly distribution of genetic material to newly formed haploid gametes. It includes steps very similar to those of mitosis, the main difference being the formation of tetrads in prophase I. These tetrads are pairs of homologous chromosomes that remain close to one another until they are pulled apart in anaphase I. Crossing over offers even more genetic variation, as the ends of these chromosomes are close enough to swap material. Telophase I then forms two "cells" that enclose doubled copies of half the chromosomes of the original diploid cell. The newly formed cell then immediately goes into prophase II, metaphase II, anaphase II and telophase II. These phases operate exactly the same way as those in mitosis, resulting this time in four haploid cells.

two divisions of a primary **spermatocyte**. Females produce only one egg from each round of meiosis, investing almost all of the cytoplasm and organelles in one gamete. The extra genetic material that is split out at anaphase I and anaphase II is ejected from the developing egg with very little associated cytoplasm. These tiny capsules of DNA are called **polar bodies**. They are not viable, and are quickly degraded in the female system.

Forming gametes is only one function of the reproductive system. The male and female gametes must be united in a protected environment, and the resulting embryo needs to be nourished and protected as it develops. In addition, the reproductive system must trigger puberty, maintain reproductive ability, stimulate secondary sex characteristics, and produce hormones involved in sexual maturation and general homeostasis.

Both the male and female reproductive systems are composed of gonads, ducts, and accessory glands. **Gonads** are the organs that produce gametes. **Ducts** transport the gametes and any fertilized egg that is present. **Accessory glands** facilitate gamete production and survival. Although all three components are found in both men and women, their structures and functions differ with gender, so we'll study each gender separately.

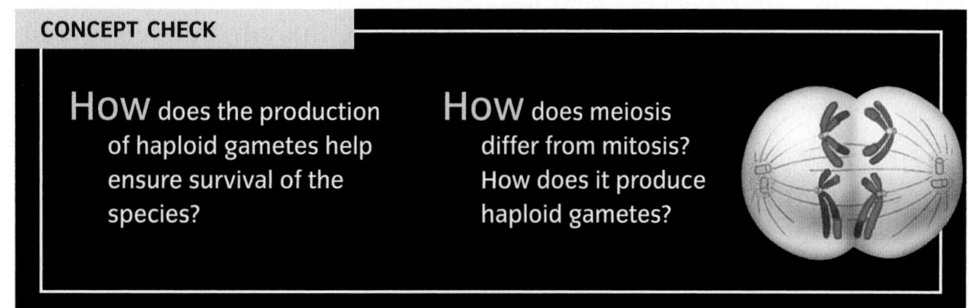

CONCEPT CHECK
How does the production of haploid gametes help ensure survival of the species?
How does meiosis differ from mitosis? How does it produce haploid gametes?

Structures of the Male Reproductive System Produce and Store Sperm

LEARNING OBJECTIVES

Trace the pathway of sperm through the male.

Describe sperm production.

Outline hormonal controls in the male.

The function of the male reproductive system is to produce sperm and deliver it to the female reproductive system. This requires a gonad to produce sperm, some tubes to carry the sperm, and three types of accessory glands to produce fluid to sustain the sperm (FIGURE 16.4, p. 516).

STRUCTURES OF THE MALE REPRODUCTIVE SYSTEM

The male reproductive system is essentially one long tube, with sperm generated in the gonads at one end, matured along the route, and released from the body at the other. Accessory glands add secretions to nourish and carry the sperm before it is released.

Sperm is produced in the testes. These paired organs are suspended in the **scrotal sac**, where their internal temperature can be regulated with ease. Viable sperm can only be produced at temperatures 2 to 3 degrees C below normal body temperature. The **cremaster** and **dartos** muscles of the scrotal sac move the testes to regulate their temperature. These muscles contract when the temperature drops. This elevates the testes, bringing them closer to the body and maintaining the required temperature by allowing the testes to absorb heat from the body. When the temperature within the testes rises, the muscles relax and the testes move away from the body, reducing their internal temperature.

The male reproductive organs usually begin development seven weeks after conception, forming from the embryonic mesonephros duct. By seven months after conception, the testes migrate from their position in the abdominal cavity to the scrotal sac through the inguinal canal, dragging their associated vessels, nerves, lymph, and reproductive cords with them (FIGURE 16.5). Their path leaves a weak spot in the abdominal wall, which can lead to a hernia later in life. A hernia is a rupture of the abdominal wall accompanied by the protrusion of internal organs, usually the small intestine. Hernias often require surgery to reposition the protruding organs and close the hole.

This "descending" of the testes is vital to reproductive health. Recall that production of viable sperm requires a temperature 3°C below body temperature. If the testes do not descend, the seminiferous tubules of the testes will be too warm for sperm creation. Additionally, when the testes remain in the body cavity, they are far more prone to testicular cancer.

In 3 percent of full-term male births and a full 30 percent of premature male births, the testes have yet to descend. The medical term for this condition is **cryptorchidism**, literally "hidden orchid." The male will be sterile if both testes remain in the body cavity, a condi-

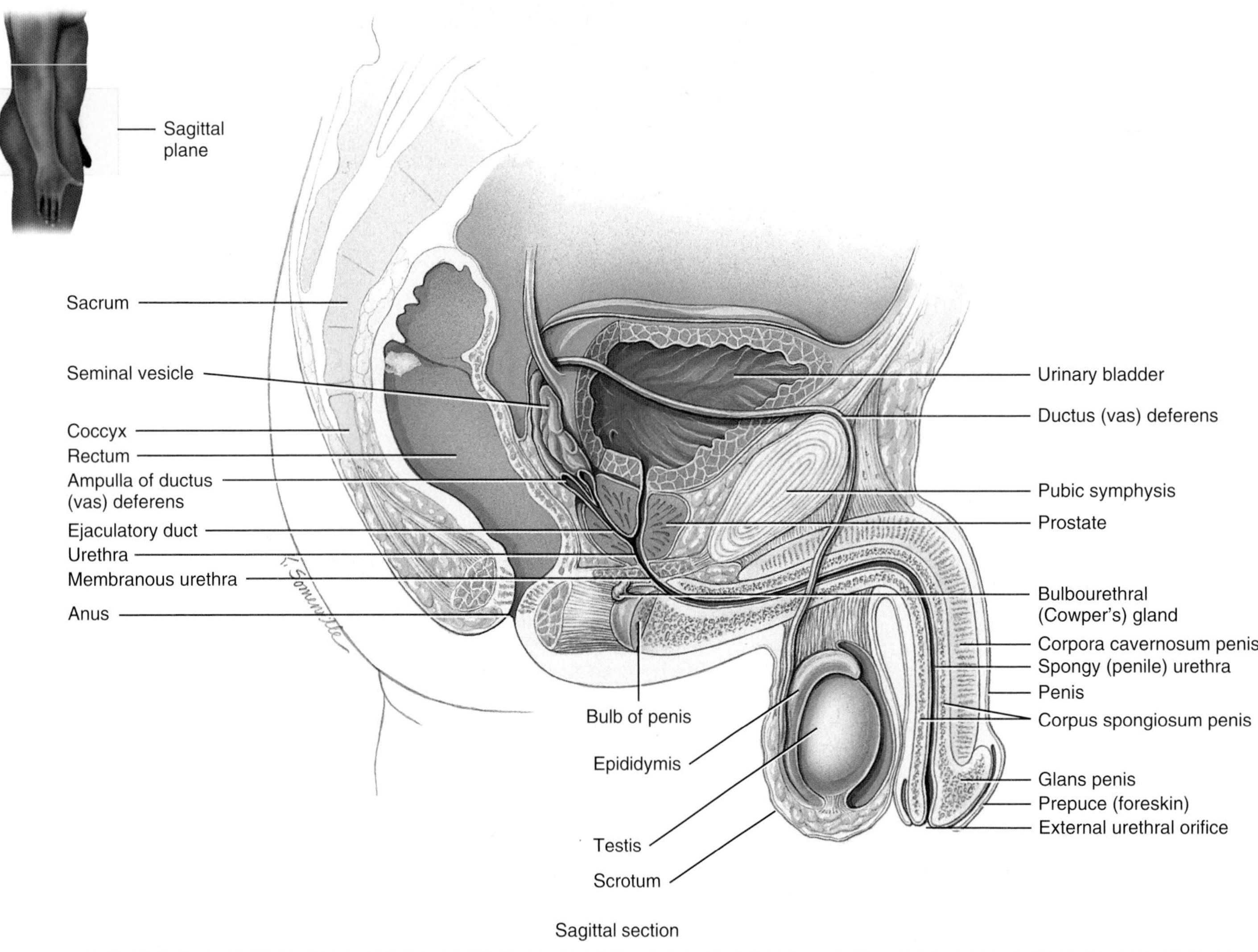

The male reproductive system FIGURE 16.4

tion called **bilateral cryptorchidism**. Luckily, among approximately 80 percent of cryptorchid males, the testes naturally descend within the first year. If they do not descend by 18 months, surgery is needed.

In normal development, each testis carries out **spermatogenesis** independently within the individual pouches of the scrotal sac. The testes are actually a densely packed mass of **seminiferous tubules**, which are contained in 200 to 300 lobules within each testis. An individual lobule holds up to three tubules, providing a large number of seminiferous tubules per testis (FIGURE 16.6, p. 518).

Spermatogenesis
The formation of spermatids (immature sperm cells).

Within the seminiferous tubules are two types of cell: **spermatogenic** cells and **Sertoli** cells. At puberty, the spermatogenic cells are stimulated to begin producing sperm. They first divide into spermatogonia. Spermatogonia in the walls of the seminiferous tubules divide, forming **primary spermatocytes**. As these cells continue to divide, they are pushed farther from the wall of the tubule into the lumen, where they become secondary spermatocytes and then **spermatids**.

During this stage of development, the cells become progressively less like the cells of the male body and more like separate entities. Eventually they become so different that these spermatids need protection from the immune system, which would otherwise destroy them as foreign cells. The Sertoli cells extend from the basement membrane of the seminiferous tubule all the way to the lumen. Their job is to isolate the developing sperm from the male blood supply, as protection against immune attack. The only cells of the seminiferous tubule

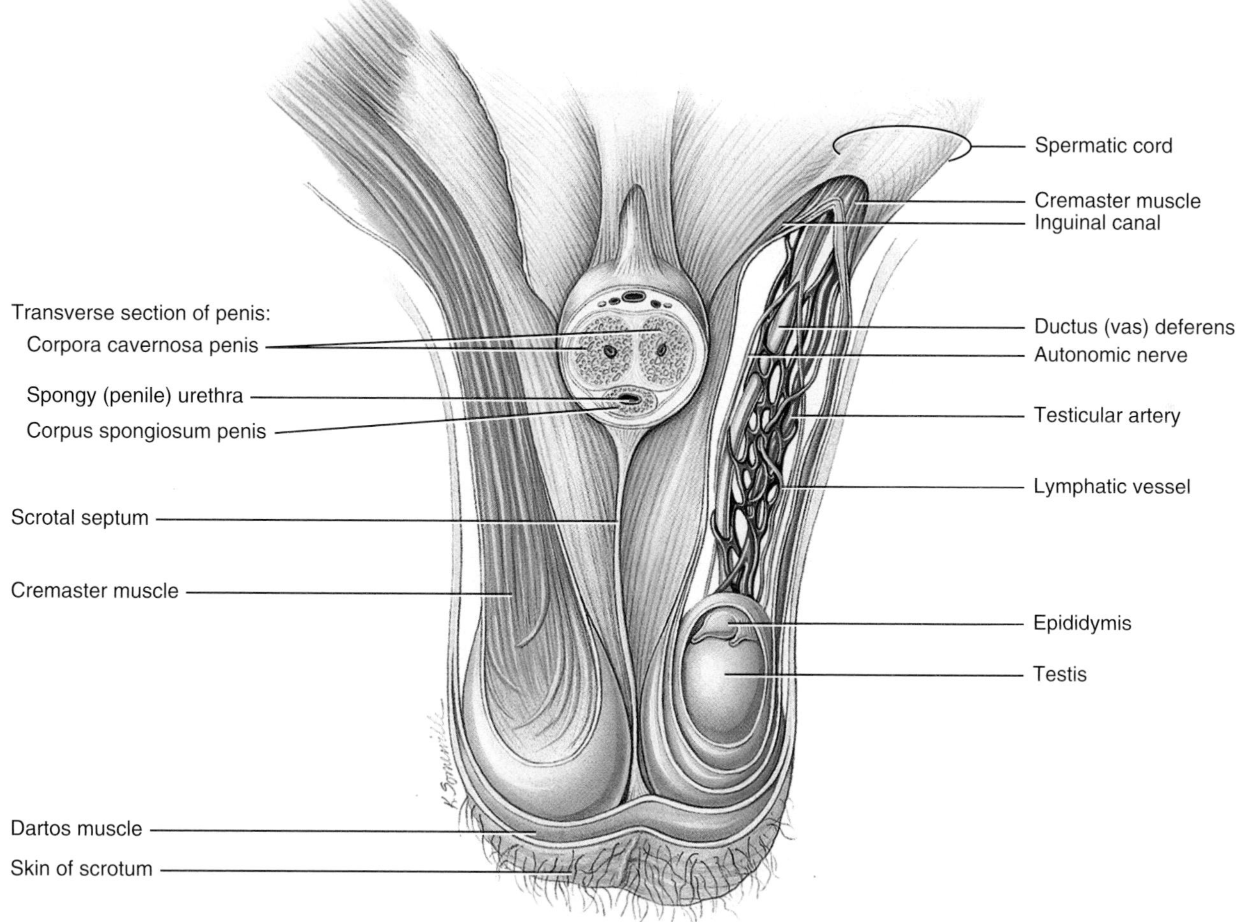

The scrotum, the supporting structure for the testes FIGURE 16.5

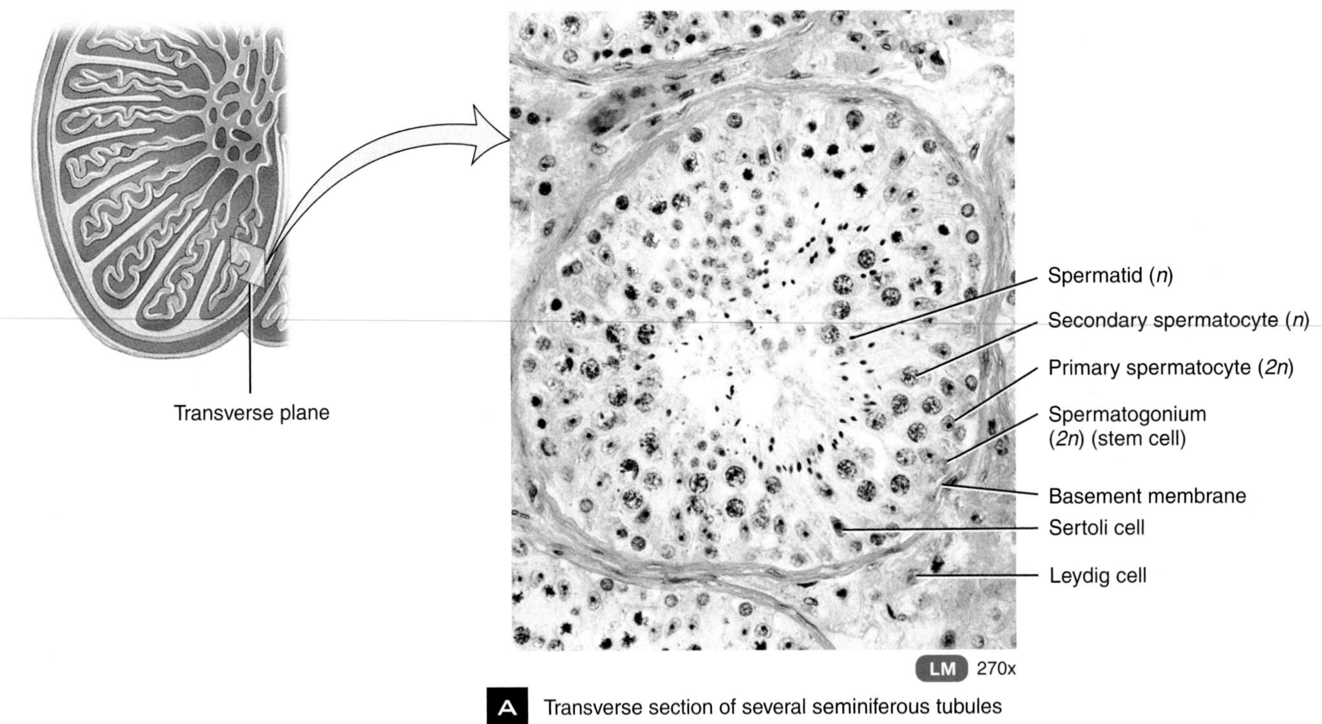

A Transverse section of several seminiferous tubules

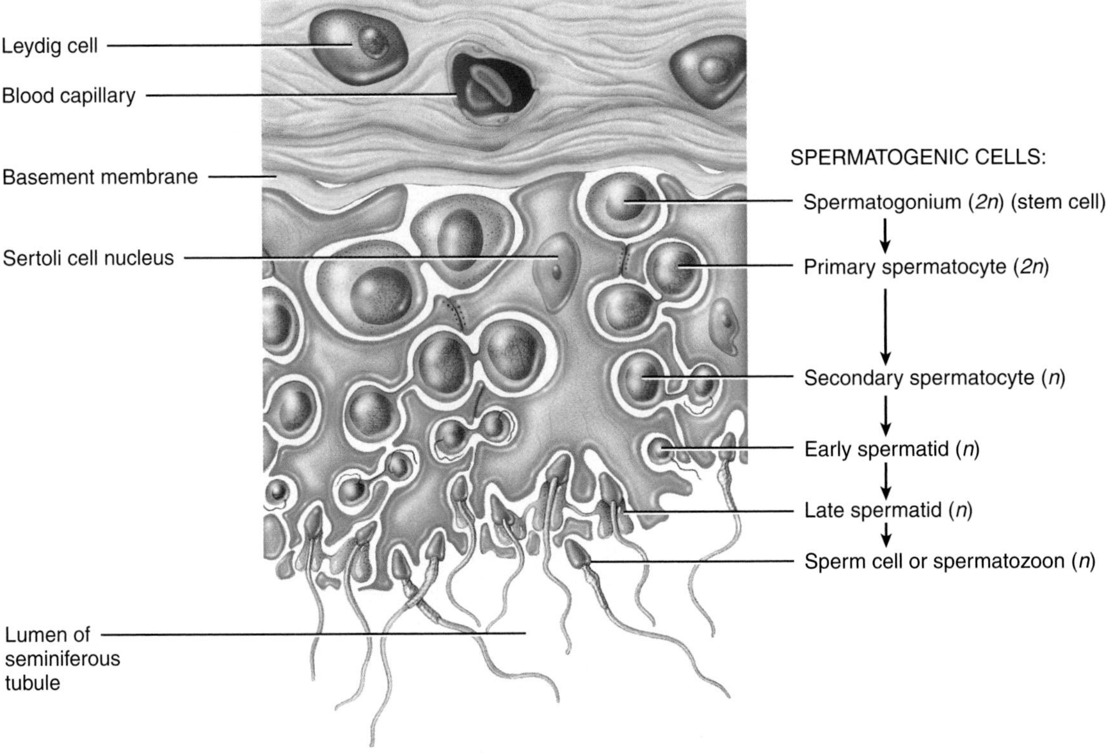

B Transverse section of a portion of a seminiferous tubule

Testis histology FIGURE 16.6

that do not require protection are the spermatogonia. These cells are identical to those of the male body, so they are in no danger of attack.

Beyond protecting the developing sperm, the Sertoli cells also assist in their survival. They provide nourishment for the developing sperm, assist in the final maturation of sperm by removing excess cytoplasm, control the release of sperm into the seminiferous tubule lumen, and mediate the effects of the hormones testosterone and inhibin.

One final type of cell in the testes occurs outside the seminiferous tubules, between them in the lobules. The **Leydig** cells, or interstitial endocrinocytes, produce the hormone **testosterone**. Testosterone stimulates spermatogonia to produce sperm; stimulates bone growth; increases hair production on the arms, legs, underarms, chest, groin, and face; stimulates cartilage growth of the larynx (thereby lowering the voice), and increases libido. In short, testosterone from the Leydig cells turns the adolescent male into a fully reproductive man.

SPERMATOGENESIS IS THE PROCESS OF SPERM FORMATION

The process of making and maturing a spermatozoon takes 65 to 75 days. It begins with the spermatogonia, which are **stem cells**. When the spermatogonia divide, one cell remains in contact with the basement membrane as a spermatogonium and the other moves toward the lumen to begin the process of spermatogenesis. This second cell moves into a Sertoli cell and transforms into a primary spermatocyte. Both primary spermatocytes and spermatogonia are diploid cells. Once safely protected by the Sertoli cells, meiosis can occur. At the end of meiosis I, two secondary spermatocytes are formed. Each one has 23 chromosomes, but each chromosome is doubled. This results in a haploid number of alleles but a diploid number of actual chromosomes. Rather than 46 different chromosomes, there are two identical copies of 23 chromosomes held together by a centromere.

Stem cell
A less differentiated cell that can give rise to a specialized cell.

As meiosis proceeds, each secondary spermatocyte divides further to produce two haploid spermatids. This yields a total of four haploid cells carrying the DNA of a sperm, but without the characteristic shape of the sperm cell. During the process of **spermiogenesis**, these spermatids are slowly ejected from the Sertoli cell as they mature. When the sperm are free of the seminiferous tubule, they are called spermatozoa and are fully formed, if not yet **capacitated** (FIGURE 16.7).

Capacitated
Activated, i.e., capable of fertilizing an ovum.

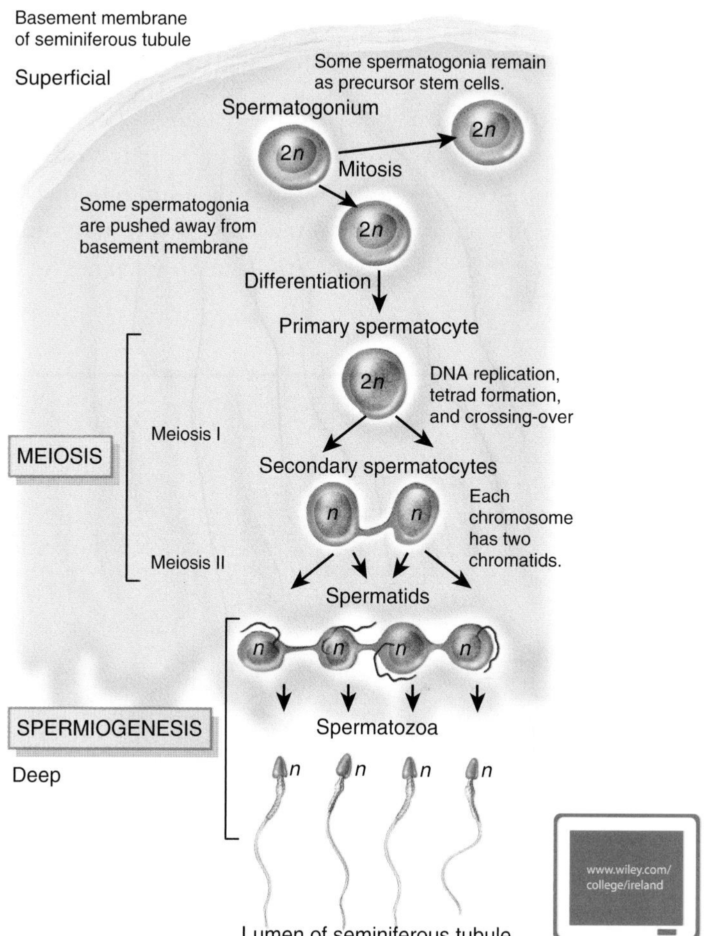

Sperm formation FIGURE 16.7

Spermatogonia undergo mitosis and produce two cells: one cell that migrates into the center of the seminiferous tubule becoming a primary spermatocyte, and a second one that remains on the periphery. The primary spermatocyte then undergoes meiosis I, producing two secondary spermatocytes. These spermatocytes go on to complete meiosis, producing a total of four spermatids. Once these spermatids undergo spermiogenesis, four functional sperm are produced.

A normal male produces about 300 million sperm (FIGURE 16.8) per day from puberty until death. Sperm exist to reach and penetrate an egg, and each part of the sperm has a role in meeting this goal. The head of the sperm includes the **acrosome** and the nucleus. The acrosome is a vesicle on the point of the sperm head that contains digestive enzymes. These will be useful when the sperm encounters the egg, as it will digest the **oocyte** membrane, allowing the nucleus of the sperm to penetrate. The midpiece of the sperm contains many mitochondria that produce the ATP needed to reach the egg. The tail of the sperm consists of one long flagellum. The sperm is the only human body cell with a flagellum and is one of few human cells that must propel itself from its place of origin in order to perform its function.

Oocyte
Egg; the female gamete.

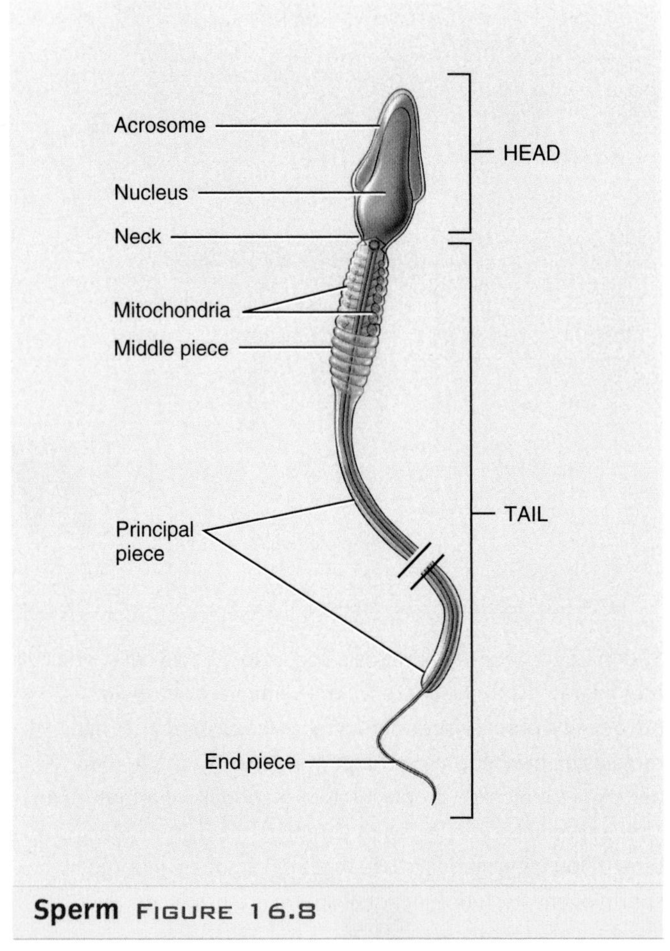

Sperm FIGURE 16.8

THE EPIDIDYMIS STORES AND MATURES NEWLY DEVELOPED SPERM

Once sperm is produced in the seminiferous tubules, it must be transported from the male to the female. This requires a series of ducts through which the sperm will pass. **Semen**, the fluid containing sperm and other ions, is formed as the sperm traverses these ducts.

The Sertoli cells create a fluid that fills the seminiferous tubule lumen and pushes the developing spermatozoa along. The spermatozoa leave the seminiferous tubules in the lobules of the testes and travel through the straight tubules to the rete testes before leaving this organ. This network of testicular tubules ends at the **epididymis**.

The function of the epididymis is similar to that of the aging cellar at a winery. It serves as a storage area and final maturation center for the spermatozoa, just as the casks at a winery serve as a suitable environment for the young wine to age and mature before being sold. Spermatozoa reach final form in the epididymis, losing that last bit of excess cytoplasm while gaining mobility and the ability to fertilize an ovum. This process takes about 10 to 14 days. Spermatozoa can remain **quiescent** in the epididymis for approximately one month. If an ejaculation event occurs, the walls of the epididymis aid in propelling the sperm forward. A small peristaltic wave is generated in the smooth muscle of the wall, helping to push the spermatozoa into the next tube in the system, the **ductus deferens** (vas deferens).

Quiescent
Resting, quiet, inactive.

The ductus deferens transports and stores sperm

The ductus deferens runs approximately 50 centimeters from the scrotal sac through the **inguinal** canal, looping over the **ureter** and posterior to the urinary bladder. It arises from the tail of the epididymis, where the epididymal tube expands in diameter. As the ductus deferens enters the abdominal wall, it sits at the anterior of the **spermatic cord**, readily acces-

Spermatic cord
The artery, vein, nerve, lymphatics, and ductus deferens that lead from the abdominal cavity to the testes.

sible immediately beneath the skin of the scrotum. The length of the ductus deferens is used to store sperm for as long as several months. If there is no ejaculation during that time, the sperm are broken down and reabsorbed. The placement of the vas deferens and its function as a sperm transport vessel make it a prime candidate for sterilization surgery. You may have heard of a vasectomy; the term literally means "cutting the vas deferens." (This procedure is explained in the section on birth control near the end of this chapter; see p. 541.)

Unlike the epididymis, the walls of the ductus deferens have a thick layer of smooth muscle to propel the sperm toward the urethra during ejaculation. The terminal end of the ductus deferens swells slightly as it runs down the back of the urinary bladder. This swelling is called the **ampulla** (FIGURE 16.9, p. 522).

Seminal vesicles help nourish the sperm

After the ampulla, glands add fluid to the developing semen as it moves through the male system during ejaculation. The paired **seminal vesicles**, located on the posterior base of the urinary bladder, are the first of these glands. They secrete an alkaline, fructose-rich fluid that serves two purposes. The high pH helps to neutralize the potentially lethal, acidic environment of the male urethra and the female reproductive tract. The fructose serves as an energy source for the sperm as they become motile. **Prostaglandins** are also released by the seminal vesicles. Prostaglandins have many physiological effects. They open airways, stimulate the sensation of pain, reduce stomach acid production, and cause local irritation. Prostaglandins also seem to stimulate sperm motility. A final important component of seminal vesicle fluid is a clotting factor, which may be responsible for the coagulation of semen after ejaculation. In all, the seminal vesicles secrete approximately 60 percent of the total ejaculate volume.

The ejaculatory duct runs through the prostate gland

After the seminal vesicles, sperm travels along the short ejaculatory duct to the prostatic urethra. The union of the ampulla of the ductus deferens and the duct from the seminal vesicle marks the beginning of the ejaculatory duct. Semen does not reach this duct except during ejaculation. During an ejaculation, sperm and semen are forcefully pushed into the prostatic urethra, causing the prostate gland to add secretions.

The prostate gland is a golf-ball-sized gland lying immediately at the base of the bladder. It completely surrounds the uppermost portion of the urethra, secreting a milky fluid into the passing semen. This fluid includes citric acid for ATP production, proteolytic enzymes to break up the clot formed by the seminal vesicle secretions, and acid phosphatase, whose function is unclear. Another 25 percent of the semen volume comes from this gland.

Physiology of the male orgasm

Directed by the sympathetic nervous system, the male orgasm propels sperm from the epididymis through the ductus deferens, the ejaculatory duct and the urethra, releasing it from the male body. As sperm enter the ejaculatory duct from the ductus deferens, the prostate and bulbourethral glands add their secretions, creating semen. Rhythmic, reflexive contractions of pelvic muscles cause the semen to be released from the penis in short bursts. During this reflex, the sphincter at the base of the urinary bladder closes, preventing urine from entering the urethra and sperm from entering the urinary bladder.

The total ejaculate released during orgasm is usually 2.5 to 5 ml. On average there are between 50 and 150 million sperm per ml, for a total of over 350 million sperm per ejaculate. If the sperm count drops below 20 million per ml, the male is said to be infertile. This number is usually too low to fertilize an egg because so few of the ejaculated sperm reach the ovum.

Often, a slight emission precedes ejaculation, as a peristaltic wave passes through the ejaculatory duct, ductus deferens, seminal vesicles, and prostate. The emission cleanses the urethra, removing potentially harmful crystals that might impair sperm function. While most ejaculations occur during the stimulation of sex, men can also experience "nocturnal emissions," or ejaculations during sleep. These are normal, and may or may not be associated with sexually arousing dreams. We will return to orgasm later in this chapter.

The urethra travels the length of the penis

Once through the prostatic urethra, the semen travels the rest of the urethra. Immediately upon leaving

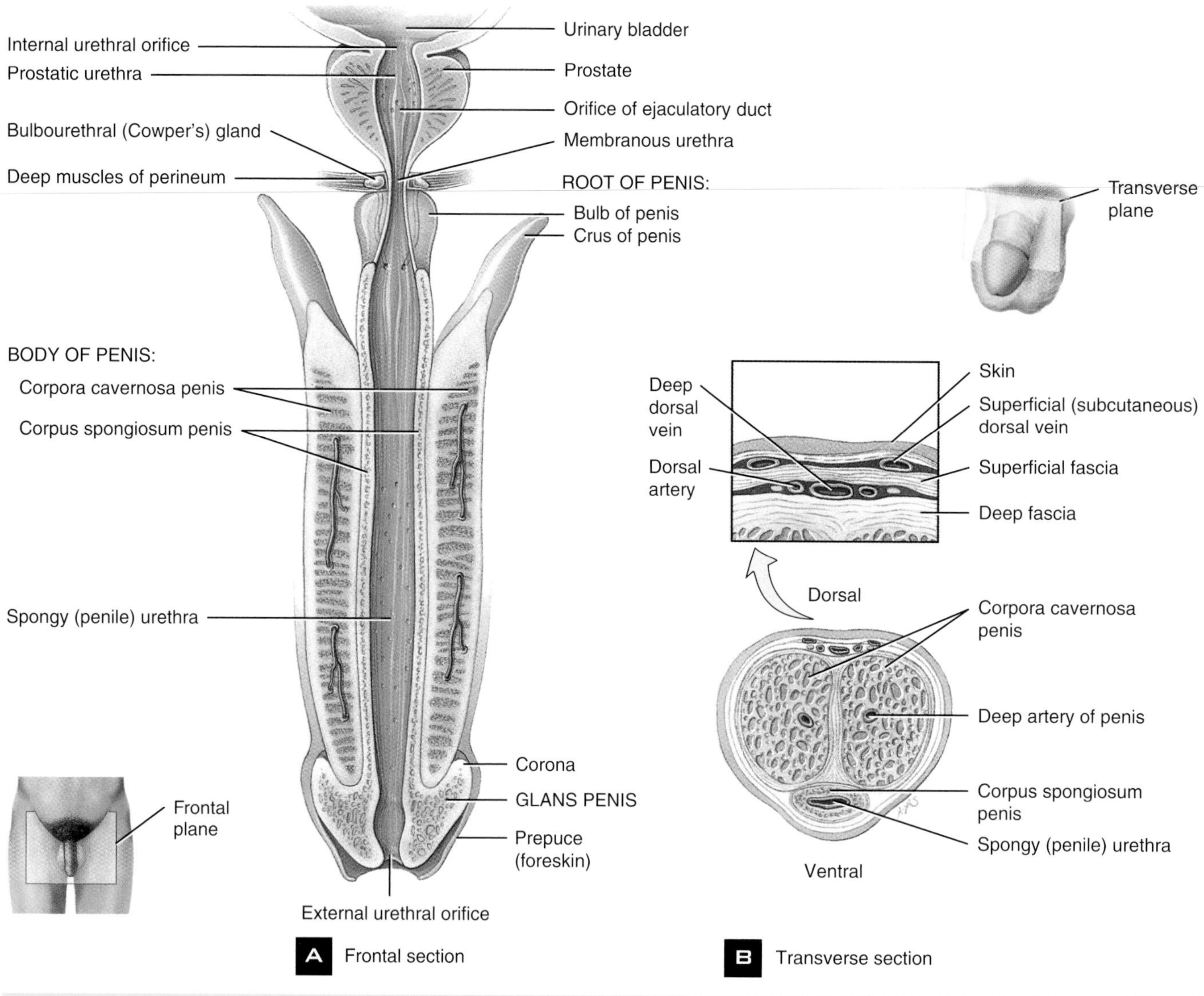

The prostatic urethra and penis FIGURE 16.9

Urogenital
Concerning both the urinary system and the reproductive system.

the prostate, the urethra dives through the **urogenital** diaphragm (see FIGURE 16.9). It then continues the length of the penis, through the spongy (or penile) urethra, and to the external urethral orifice.

Where the urethra enters the spongy tissue of the penis, a final set of glands adds fluid to the ejaculate. The **bulbourethral** glands lie on either side of the urethra, at the bulb, or base, of the penis. The glands are obviously named for their location. They secrete an alkaline, mucous fluid into the urethra prior to sperm arrival that protects the sperm from the normally acidic urethra. The mucus of these glands lubricates the tip of the penis and the urethral lining, preventing damage to the sperm as they travel the final portion of the male reproductive system. Refer to FIGURE 16.9 for the locations of these structures.

The penis itself is a passageway for both urine and semen. The **root** of the penis lies at the base of the prostate gland. The body of the penis contains three cylinders of erectile tissue. The two **corpora cavernosa** lie on the dorsolateral surfaces, and the single **corpus spongiosum** encircles the urethra and expands at the tip of the penis to form the **glans**. These three tissues contain numerous blood sinuses.

During arousal, the arteries that feed these tissues dilate under the influence of cyclic guanine monophosphate **(cGMP)**. cGMP allows more blood to enter the erectile tissue, filling the sinuses and compressing the veins. This combination results in an **erection**, an enlarged and stiffened penis. This process is reversed by constriction of the arteries, in turn lessening the pressure on the veins and allowing blood to drain from the tissues. Viagra® inhibits enzymes that naturally break down cGMP, thereby prolonging erections. cGMP is active in other processes in the body as well, for example, in processing visual and olfactory information, memory, and learning. The cGMP inhibitors in Viagra are extremely specific; otherwise the drug would have negative side effects on memory and vision.

cGMP
Cyclic guanine monophosphate, an energy molecule.

When boys are born, the tip of the penis is covered with a protective layer of skin, called the foreskin. Removal of this tissue does not seem to alter male functioning, nor does it have any clearly demonstrated positive physiological effects, except that removal has been shown to reduce the rate of infection by the AIDS virus. Regardless, male circumcision continues to be practiced in the United States and other countries around the world. The procedure involves the rapid removal of the entire foreskin of the penis, usually within the first few days of life.

The history of circumcision begins in East Africa long before biblical accounts. The practice was used to purify men and reduce sexuality and sexual pleasure. Jews, and later Muslims, adopted the practice, and it continues to this day. In the first century, Christians strongly opposed circumcision. Not until 1870 did the medical practice of circumcision begin in the United States. Once medical professionals embraced the procedure, it became safer and more accepted. Although medical circumcisions were performed without question, no scientific studies of its benefits or contraindications were undertaken. In 1949, Dr. Douglas Garnier wrote an article denouncing circumcision, explaining that there seemed to be no medical reason for the operation and that in fact it was causing unnecessary deaths. As a result, the number of circumcisions dropped dramatically in England, but the United States was slower in questioning the procedure. In 1971, another article stated that there was no medical reason for circumcision, despite common opinion that a circumcised penis is somehow "cleaner" or less likely to become infected. Since that time, the practice has slowly declined in the United States. Neonatal circumcision was performed in only 60 percent of male births in 1996 and declined further to 55 percent in 2001.

In 2006, however, a number of studies showed that the foreskin may transmit AIDS during sex. A study in Uganda, for example, found a 30 percent reduction in disease transmission among circumcised men. Researchers reported that the reduction may be due to the fact that HIV binds strongly to the foreskin. Similarly, a 2004 study from India found that circumcised men were six times less likely to acquire HIV during sex.

HORMONAL CONTROL OF THE MALE

Although it is true that males produce sperm endlessly from puberty until death, male hormones exert control over the rate of sperm production and the secretion of testosterone, which controls male secondary sex characteristics. The pituitary gland lies deep in the brain, protected by the sphenoid and attached to the brain by the hypothalamus. The anterior pituitary gland secretes **luteinizing hormone** and **follicle-stimulating hormone**, which are instrumental in governing the male reproductive system. The secretion of these hormones is governed by gonadotropin releasing factors produced by the hypothalamus.

The names of these two hormones reflect their roles in the female, not male, system. When reproduction was originally studied, it was assumed that only females exhibited hormonal controls. Therefore, these anterior pituitary hormones were first isolated and their functions were identified in females. Follicle-stimulating hormone (FSH) stimulates immature oocyte (egg) follicles in the female ovary. Luteinizing hormone stimulates the production of a yellow body; lutein roughly translates to yellow. After an oocyte is ovulated, a yellow body remains on the ovary, hence the name of the hormone responsible for ovulation. It came as a bit of a shock when scientists later discovered that the male pituitary secretes the same hormones, with subtly different effects.

In the male, luteinizing hormone (LH) stimulates the Leydig cells, causing the release of testosterone. For this reason, it is also called interstitial cell stimulating hormone (ICSH). The production of testosterone is governed by a typical negative feedback loop. As more testosterone is produced, its levels increase, inhibiting production of LH at the pituitary gland. In this way, the hormones testosterone and LH balance one another.

The functions of testosterone include stimulation of male patterns of development in utero, enlargement of male sex organs during puberty, development of male secondary sex characteristics, development of sexual function, and stimulation of **anabolism**.

Anabolism The building up of larger molecules from smaller ones (contrast to catabolism).

Secondary male sex characteristics are those associated with puberty: growth of skeleton and musculature; appearance of body and facial hair; cartilaginous growth of the ears, nose, and larynx; thickening of the skin; and increased oil secretion in the skin.

Some tissues of the male convert testosterone to **dihydrotestosterone**, or DHT. You may have heard this compound being blamed for male pattern baldness on Web sites or television infomercials, which make it sound as if everybody's hair will fall out if DHT concentration exceeds a certain level. In truth, male pattern baldness is due to varying sensitivity of hair follicles to circulating DHT. Hair follicles can produce the enzyme 5-alpha reductase, which converts circulating testosterone into dihydrotestosterone (DHT). This raises the concentration of DHT around these follicles, and DHT gets picked up by follicles that have many DHT receptors. A DHT-activated hair follicle has a shorter growing stage and a longer "resting" phase, and will produce a wispy hair. DHT also constricts blood vessels to the follicle, starving the hair of nutrition. All these factors eventually cause the hair to fall out, and any replacement hair will be thin and slow-growing. This extreme reaction to DHT is genetic, as are the patterns and numbers of susceptible hair follicles on the head. Because factors that increase the likelihood of developing male pattern baldness are carried on the X (female) chromosome, it is considered a sex-related trait (FIGURE 16.10).

FSH is secreted by the anterior pituitary gland in both sexes. In the male, where oocyte follicles are absent, FSH indirectly stimulates spermatogenesis. FSH and testosterone together cause the Sertoli cells to secrete **androgen-binding protein** (ABP). ABP moves to the interstitial spaces of the testes, binding available testosterone and maintaining it in high concentration near the seminiferous tubules. Testosterone stimulates the final production of spermatids. When the Sertoli cells are functioning to capacity to protect developing sperm, they secrete **inhibin**. This hormone inhibits FSH production from the anterior pituitary, slowing sperm production. In essence, the Sertoli cells are claiming that they are "full" and cannot protect any more developing sperm. In typical negative feedback, if sperm production slows too much, the process reverses. The Sertoli cells no longer release inhibin, the anterior pituitary increases production of FSH, and sperm production rises.

Male pattern baldness is hereditary
FIGURE 16.10

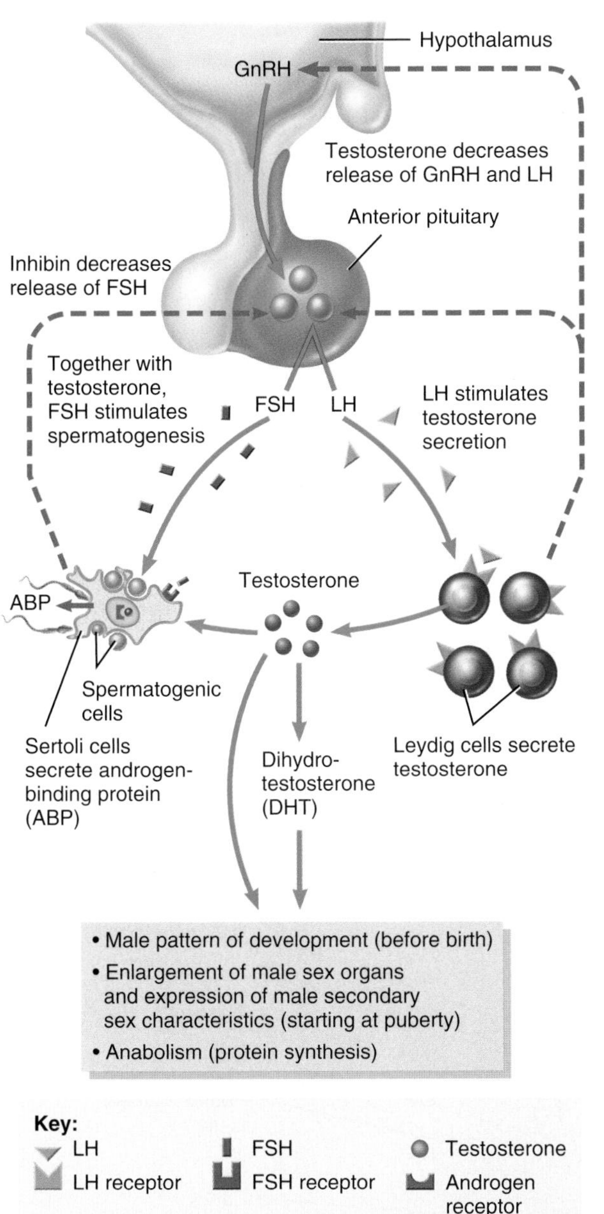

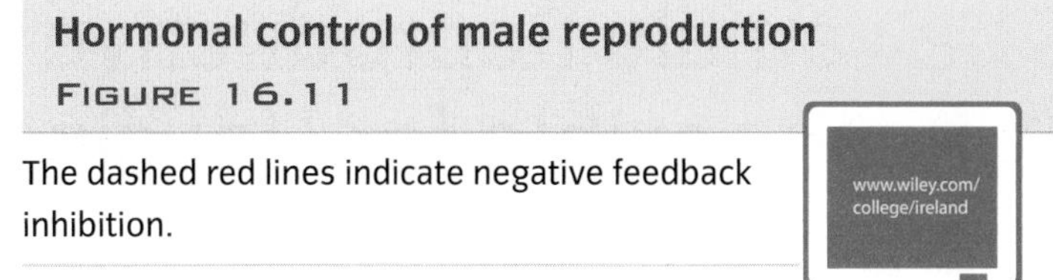

Hormonal control of male reproduction

FIGURE 16.11

The dashed red lines indicate negative feedback inhibition.

Testosterone itself also operates under negative feedback. If blood testosterone rises too high, it prevents the release of **GnRH** (gonadotropin-releasing hormone) from the hypothalamus. When released, GnRH goes directly to the anterior pituitary and stimulates release of LH. Recall that LH then increases secretion of testosterone by Leydig cells. If GnRH is blocked, LH is not released and the testosterone level will decline (FIGURE 16.11).

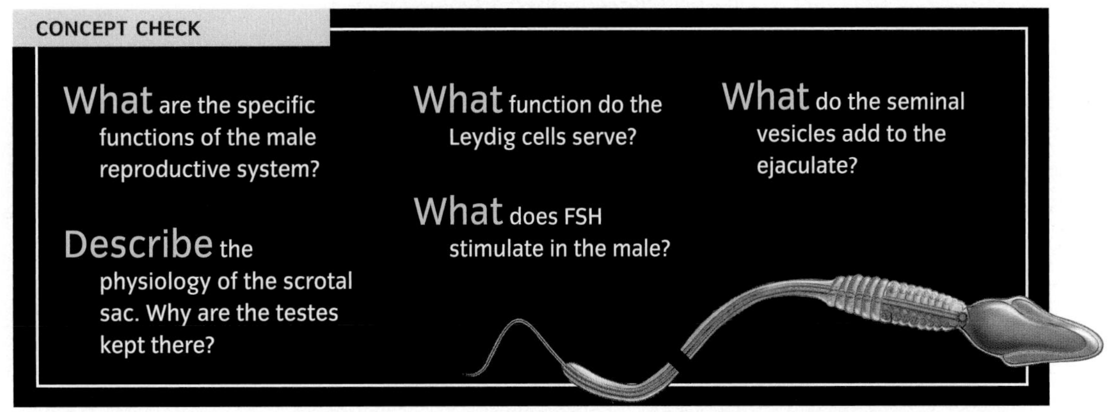

CONCEPT CHECK

What are the specific functions of the male reproductive system?

Describe the physiology of the scrotal sac. Why are the testes kept there?

What function do the Leydig cells serve?

What does FSH stimulate in the male?

What do the seminal vesicles add to the ejaculate?

The Female Reproductive System Is Responsible for Housing and Nourishing the Developing Baby

LEARNING OBJECTIVES

List the functions of each female reproductive organ.

Explain oogenesis.

Describe the female hormonal cycles.

If the purpose of the male reproductive system is to deliver sperm, one purpose of the female reproductive system must be to receive sperm. But the female system must also produce eggs (or **ova**) for fertilization, provide an area for the fertilized egg to develop into a fully developed fetus, and give birth. Like the male reproductive system, the female system also produces hormones that cause sexual maturity and stimulate the development of secondary sex characteristics.

The organs of the female reproductive system include the paired **ovaries**, the **fallopian** or **uterine tubes** leading from the ovaries to the uterus, the **uterus** itself, and the **vagina** (FIGURE 16.12). Accessory or-

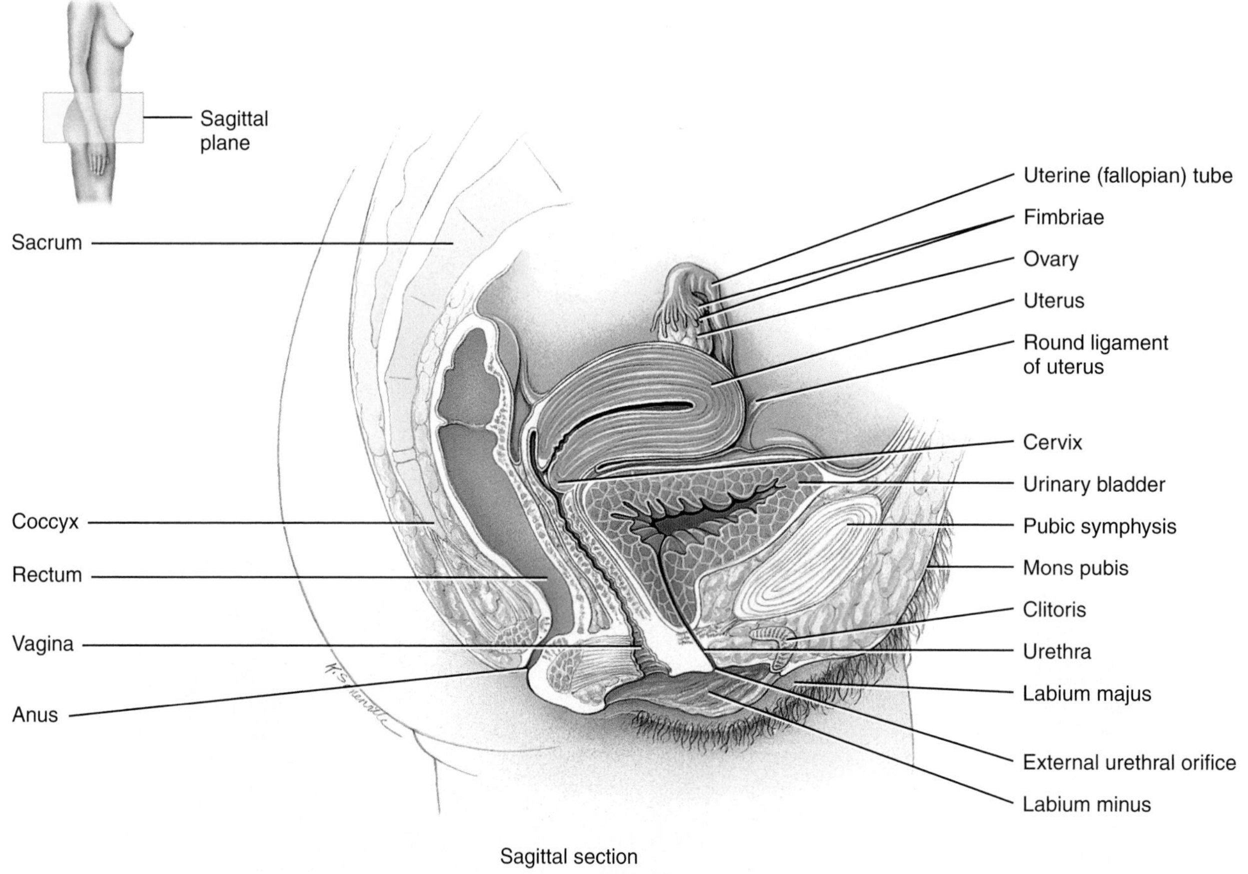

The female reproductive system FIGURE 16.12

gans of the female system are fewer than the male, represented mainly by the **mammary** glands and the external female genitalia.

While the anatomy of the female reproductive system is simpler than that of the male, the hormonal control of the female system is far more complex. This is because two interacting hormonal cycles occur simultaneously in the female. The anterior pituitary gland secretes FHS and LH, affecting the ovary, and the ovary then responds with the hormones **estrogen** and **progesterone** that affect the uterus. Ovarian hormones can inhibit the anterior pituitary gland, providing feedback control.

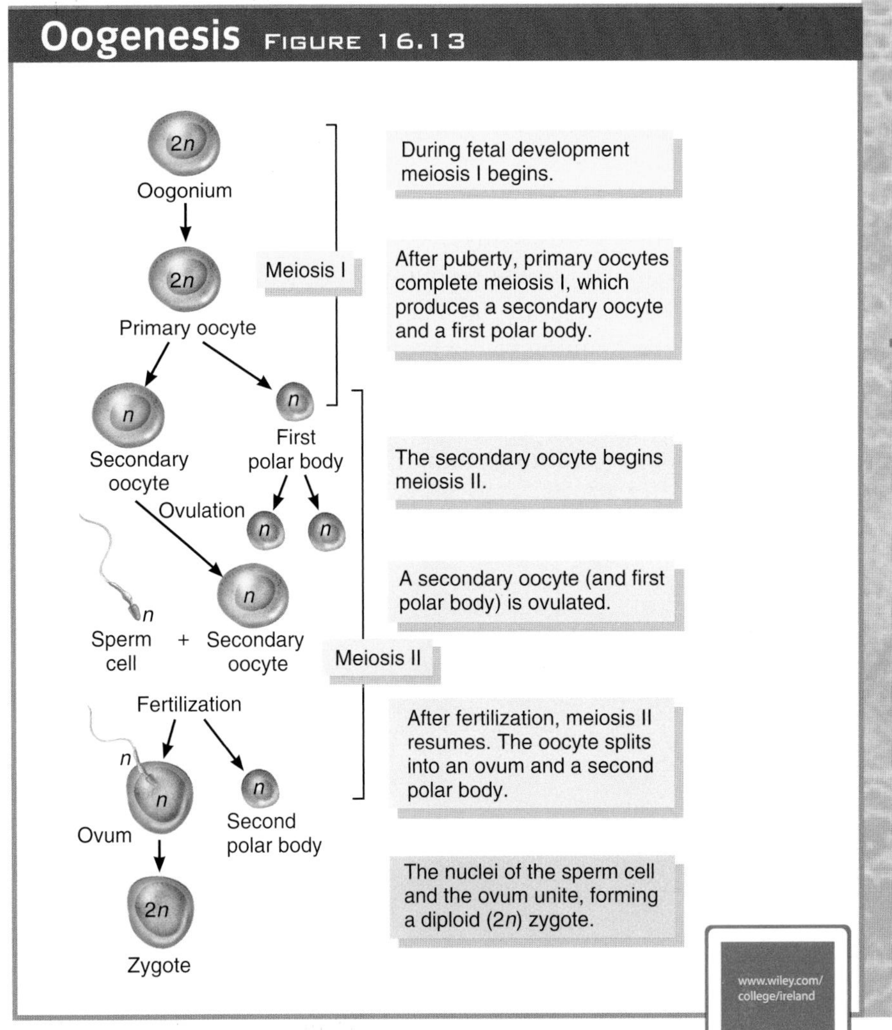

Oogenesis FIGURE 16.13
2n
Oogonium
During fetal development meiosis I begins.
Meiosis I
2n
Primary oocyte
After puberty, primary oocytes complete meiosis I, which produces a secondary oocyte and a first polar body.
n
Secondary oocyte
n
First polar body
The secondary oocyte begins meiosis II.
Ovulation
n
n
A secondary oocyte (and first polar body) is ovulated.
n
Sperm cell
+
Secondary oocyte
Meiosis II
Fertilization
After fertilization, meiosis II resumes. The oocyte splits into an ovum and a second polar body.
Ovum
Second polar body
The nuclei of the sperm cell and the ovum unite, forming a diploid (2n) zygote.
2n
Zygote
www.wiley.com/college/ireland
Process Diagram

OVARIES ARE RESPONSIBLE FOR OOGENESIS—EGG FORMATION

The ovaries are small, almond-shaped organs that lie in the pelvic cavity. They arise from the same embryonic tissue as the testes, making these organs homologous. Similar to the testes, the ovaries produce both gametes (ova) and hormones (**estrogens** and **progesterone**). **Oogenesis** occurs via meiosis but, unlike spermatogenesis, produces only one viable ovum per meiotic event (FIGURE 16.13).

Also unlike the production of sperm, oogenesis begins before the female is born, so that at birth the ovaries already contain all of the ova she will produce in her life (FIGURE 16.14, p. 528). The ova are found in the ovarian germinal epithelium, surrounded by the ovarian cortex. They wait there, suspended in early meiosis, until they receive hormonal signals to continue development. At birth, each ovary may contain from 200,000 to 2 million such cells. These **primary oocytes** undergo **atresia**, so that by puberty approximately 40,000 remain. Only 400 or so of these will actually mature to the point of ovulation during a woman's reproductive lifetime.

> **Atresia**
> Reabsorption of immature ova prior to birth.

Each primary oocyte sits in the center of a group of **follicular cells**, which are stimulated to develop alongside the oocyte. A **primary follicle** has one to seven layers of follicular cells surrounding the oocyte. These follicular cells produce the **zona pellucida**, a clear gel-like layer that surrounds the maturing oocyte. The innermost layer of follicular cells becomes attached to the zona pellucida, resembling a circular crown. These cells become the **corona radiata** of the oocyte.

Hormones released by the anterior pituitary gland affect these follicle cells, stimulating their maturation into a secondary follicle, and finally a mature, blister-like **graafian follicle**. The graafian follicle bursts during ovulation, releasing the secondary oocyte, along with its associated zona pellucida and corona radiata, into the pelvic cavity. Only if sperm are present and fertilization occurs will the secondary oocyte complete meiosis II to form an ovum. The ovulated egg itself is short-lived, remaining viable for about 24 hours. Therefore, either the immature egg is fertilized by the sperm within 24 hours, resulting in a **zygote**, or it degenerates and passes from the female body with the next menses.

THE UTERINE TUBES (FALLOPIAN TUBES) CONDUCT THE OVA

Once the oocyte is ovulated, it must be swept into the uterine tubes because the ovary has no physical contact with the uterine tubes. The open ends of the uterine tubes are expanded into a funnel-shaped **infundibulum** that ends in finger-like **fimbriae**. These tubes are extremely close, but not physically connected, to the ovaries. The small gap between the two is open to the entire abdominopelvic cavity. The fimbriae must collect the ovulated oocyte and sweep it into the infundibulum. Successful pregnancy can occur only in the uterus, so the fimbriae must get the newly ovulated egg heading in the right direction. This is done by rhythmic swaying of the fimbriae in response to the hormonal controls of

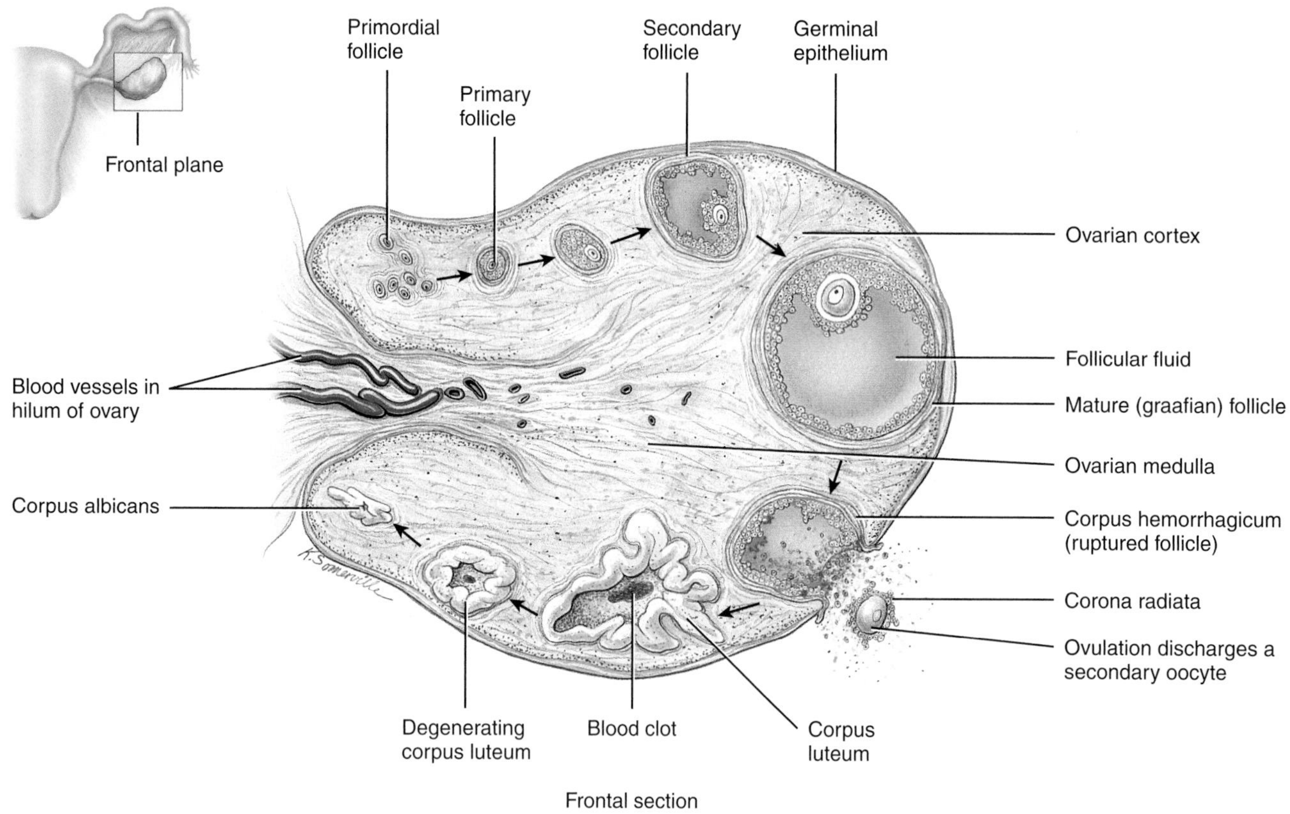

Histology of the ovary FIGURE 16.14

The follicles on the ovary are shown here in clockwise order, with the least mature primordial follicles in the upper left of the diagram. This arrangement of follicles maturing clockwise from left to right around the surface of the ovary is NOT how follicles appear in living ovaries! Follicles at various stages of maturity are randomly spread all over the ovarian germinal epithelium.

ovulation. The ends of these tubes fill with blood, distend, and sway, creating small currents in the abdominopelvic fluid, in turn drawing the newly ovulated oocyte into the uterine tubes. Once collected in the uterine tube, ciliated epithelia lining the tube help wash the oocyte (or developing zygote if fertilization occurs) into the uterus. Smoking can inhibit the movement of the cilia of the uterine tube; this is one reason women who smoke have difficulty conceiving.

Because the oocyte is only viable for a short while, fertilization must occur within 24 hours of ovulation. Usually the egg can travel only the upper one-third of the uterine tubes during this time, meaning that if fertilization does occur, it will happen there. Sperm introduced into the female system travel up through the uterus and into the uterine tubes, while from the other direction, the oocyte is collected and swept into the uterine tube. The oocyte takes six to seven days to reach the uterus itself, during which time it begins to degenerate unless fertilized.

The uterus is the site of development The uterus is the womb where fetal development occurs. This organ has an outer covering, the **perimetrium**, a middle layer of smooth muscle, the **myometrium**, and an inner **endometrium** (FIGURE 16.15). The endometrial lining thickens and sheds every 28 days or so in response to hormone levels, resulting in the menstrual flow. **Implantation** of the embryo occurs in the endometrial lining, which is built up every month in anticipation of receiving an embryo. If there is no successful fertilization, the endometrial lining is shed, resulting in most of the menstrual flow.

> **Implantation**
> Anchoring and settling of the embryo into the endometrial wall, starting placental formation.

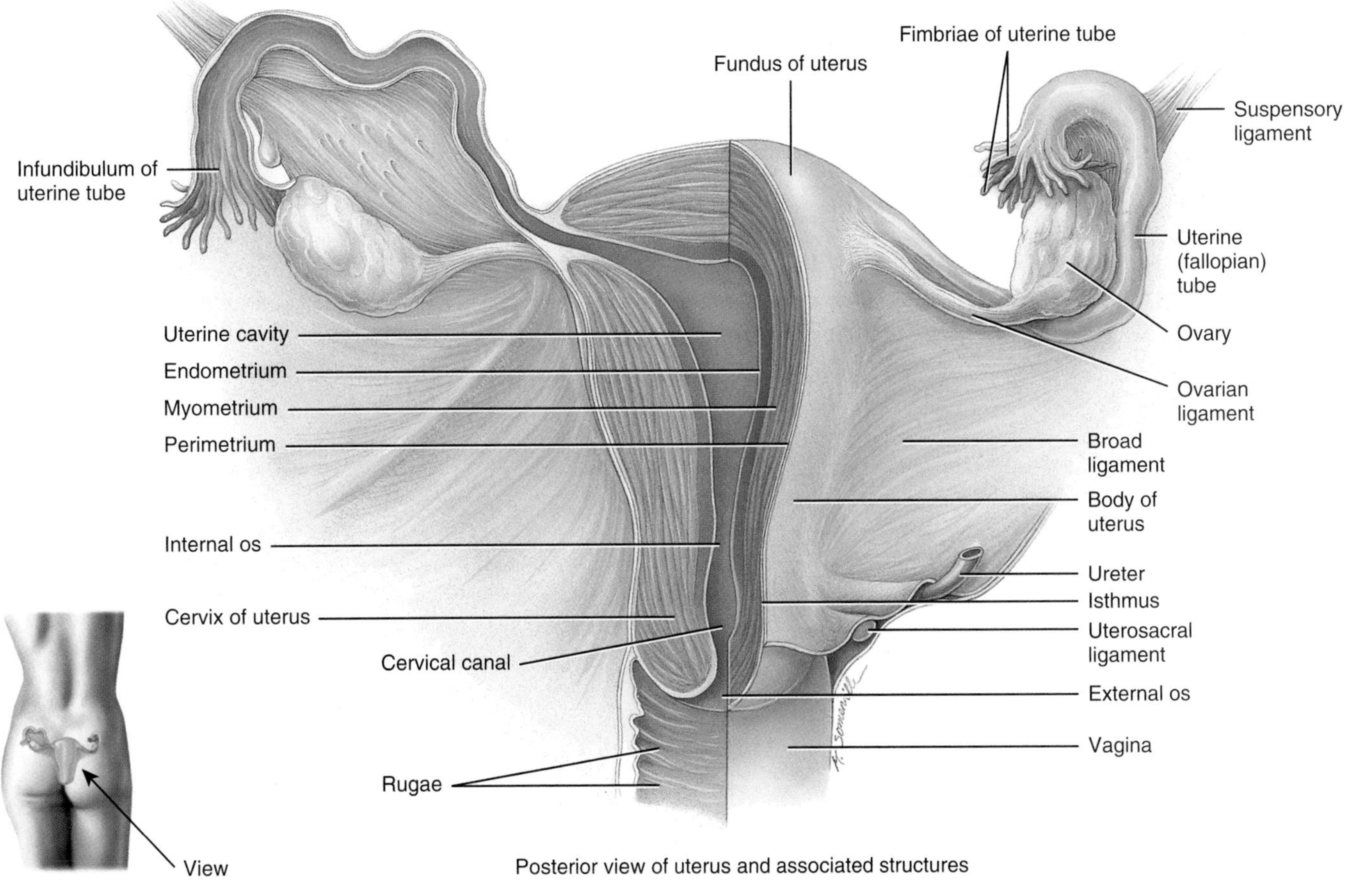

The anatomy of the uterus FIGURE 16.15

The cells that line the **cervix** produce a mucus that aids fertilization. During ovulation, the cervical mucus is thin and watery, allowing sperm to enter the uterus. The mucus also becomes more alkaline, improving sperm survival in the usually hostile acidic environment of the vagina. When no egg is present, the cervical mucus is thick and inhospitable to sperm, forming a cervical mucus plug.

Cervix
Base of the uterus.

Pregnancy is a phenomenally intricate process. Fertilization must occur within a specified window of time, and implantation must then precisely follow. To implant, the developing embryo must land on receptive endometrial tissue and then digest its way into the tissue and start to form the placental tissues.

In healthy females, endometrial tissue occurs only within the walls of the uterus. But in **endometriosis**, it also appears in the uterine tubes, on the external upper surface of the uterus, and even on the external surfaces of the urinary bladder and other pelvic organs. This causes trouble during menstruation when the endometrial lining is shed, since the tissue is trapped inside the abdominal cavity. This misplaced tissue can also cause abdominal cramps or pain as it grows.

Because the uterine tubes do not touch the ovaries, each ovulated egg floats in the abdominal cavity, hopefully swept into the uterine tubes by the fimbriae. Fertilization can occur outside the uterine tubes if sperm are present in the abdominopelvic cavity when the ovum is released. If endometrial tissue is present, this developing embryo can implant on the superior surface of the uterus or bladder. Equally alarming, the embryo could be swept into the uterine tubes, and implant on endometrial tissue on the walls of the tube. **Ectopic pregnancies** occur whenever implantation occurs outside the uterus (FIGURE 16.16). In all cases, the embryo will not survive. If the implantation occurs in the uterine tubes, the life of the mother is also in jeopardy. The tubes cannot expand to accommodate the developing embryo. As the embryo grows and the tube is stretched, the mother will feel pain, and if she does not get medical assistance, the tube will rupture, causing internal bleeding and perhaps death.

Some women past reproductive age develop uterine health problems, such as excessive bleeding related to the uterus, or uterine cancer. One of the options they are given is to undergo a **hysterectomy**. The suffix "ectomy" means to excise or remove a gland or organ. Hysterectomy means to remove the "hyster,"

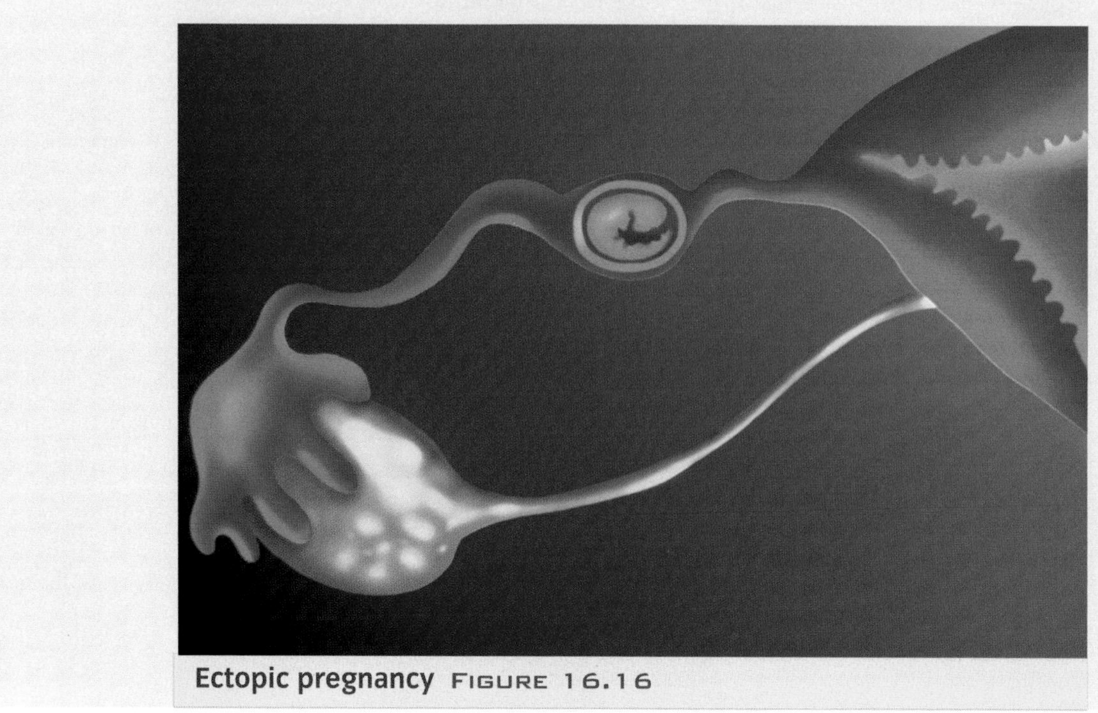

Ectopic pregnancy FIGURE 16.16

which derives from the Greek for "womb." What other words are rooted in "hyster"? Hysteria. Histrionics. All of these describe irrational behavior. Amazingly, it was once thought that the uterus was the root of this type of behavior, as it seemed that women suffered from more psychological disturbances than men. "Hyster" is still used to refer to the womb in medical terminology, even though the womb, or uterus, is not related to hysteria.

A hysterectomy, the removal of the uterus, is performed when uterine or ovarian cancer is detected, or as an emergency surgery to stop uterine hemorrhage. An elective hysterectomy can be used to alleviate difficult menstrual cycles. Severe cramping, bleeding, or other menstrual discomfort are eliminated with removal of the uterus. Fibroids, or benign tumors of the uterus, can also cause severe discomfort and excessive bleeding each month. If fibroids become troublesome, a hysterectomy is often recommended. Other reasons for electing a hysterectomy include endometriosis and uterine **prolapse,** which sometimes occurs in older women, usually after they have had children. The entire uterus drops slightly in the pelvic cavity, as the vaginal supporting ligaments sag. The bladder and rectum may be drawn down, causing discomfort and even displacement of these organs.

Prolapse
The dropping, sliding, or falling of an organ from its original position.

Uterine and ovarian cancers are common pathologies that often lead to the recommendation of a hysterectomy. In these cases, both the uterus and the ovaries are removed. The hormones produced by the ovaries may stimulate cancerous growth, so it is wise to remove them in either of these cancers, even if ovaries are healthy. If the patient suffers from endometriosis, the same principle holds. The ovaries are removed along with the uterus to prevent the misplaced endometrial tissue from responding to estrogens and progesterone. After the ovaries are removed, hormone replacement therapy is usually recommended to prevent postmenopausal symptoms such as night sweats, mood swings, and loss of bone density.

The vagina connects the uterus with the external environment The vagina serves as the receptacle for the penis during intercourse, an outlet for monthly menstrual flow, and the birth canal through which the developed fetus leaves the uterus. This 10-centimeter long muscular tube is lined with a mucous membrane. Because this tube must expand with the passage of the fetus, the walls feature transverse folds. The cells have a large store of glycogen, which breaks down to produce acids that retard microbial growth. Unfortunately, these acids are inhospitable to sperm as well and will kill them unless buffered. The aforementioned changes in cervical mucus during ovulation, together with the seminal vesicle fluids added to the semen, help the sperm to survive and reach the egg.

The vulva The external genitalia of the female are collectively called the **vulva** (FIGURE 16.17). The most sensitive area of the female external genitalia is the **clitoris**. This is a small tuft of erectile tissue homologous to the glans penis in males. It is extremely sensitive and plays a role in sexual stimulation.

Female external genitalia FIGURE 16.17

The mammary glands The mammary glands are modified sweat glands located above the **pectoralis major** muscles (FIGURE 16.18). These glands are supported by the Cooper's ligaments and are protected by a layer of adipose tissue. They are composed of **lactiferous** ducts, connected to lactiferous sinuses. Milk is produced in the lobules of the gland, stored in the lactiferous sinuses, and passed out of the breast via the lactiferous ducts. Naturally, this function is necessary only after childbirth. The breasts swell during the last weeks of pregnancy in response to the hormone **prolactin** (PRL) made by the anterior pituitary gland. Once milk is formed, it is released from the gland in response to **oxytocin**.

> **Lactiferous**
> Milk-producing.

Oxytocin is released from the posterior pituitary gland when an infant suckles, in the "let-down" reflex. This response can also occur when the mother hears her baby cry, or even thinks about nursing her baby.

HORMONAL CONTROL OF THE FEMALE REPRODUCTIVE SYSTEM

The female reproductive cycle is a study in feedback controls. Two separate cycles are occurring at once in the nonpregnant female: the **ovarian cycle** and the **uterine cycle**. Each affects the other, and together they cause the cyclic menstrual flow of the postpubescent female.

The ovarian cycle is a programmed series of events that occur in the ovary as eggs mature and ovulate, governed by hormones from the anterior pituitary gland. Hormones released from the ovary, in turn, affect the endometrium of the uterus. Ovarian hormones are the cause of the uterine cycle, which in turn is responsible for the appearance of the **menstrual flow**. The term *female reproductive cycle* usually includes both the ovarian and uterine cycles, as well as the hormones that regulate them and the associated cyclic changes in the breasts and cervix (FIGURE 16.19).

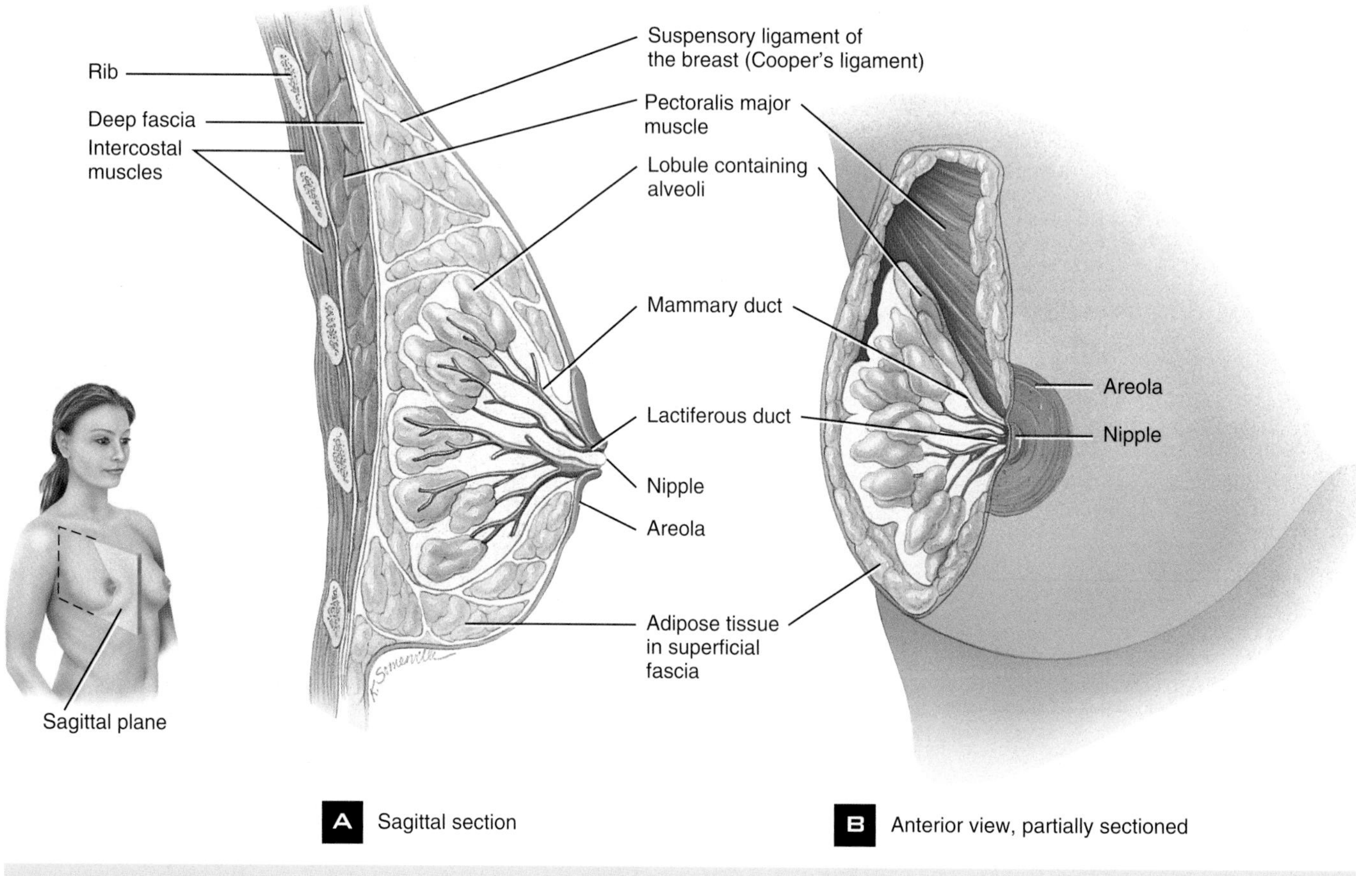

Mammary glands FIGURE 16.18

The female reproductive cycle is ultimately regulated by **GnRH** (gonadotropin-releasing hormone) from the hypothalamus. Through its effects, FSH and LH are produced in the anterior pituitary. Follicle-stimulating hormone (FSH) stimulates follicle cell growth in the ovaries, maturing the follicles and associated ova, hence the name. Luteinizing hormone (LH) causes the most mature follicle to burst (ovulate), leaving a yellow body of spent follicular cells (**corpus luteum**) on the ovary.

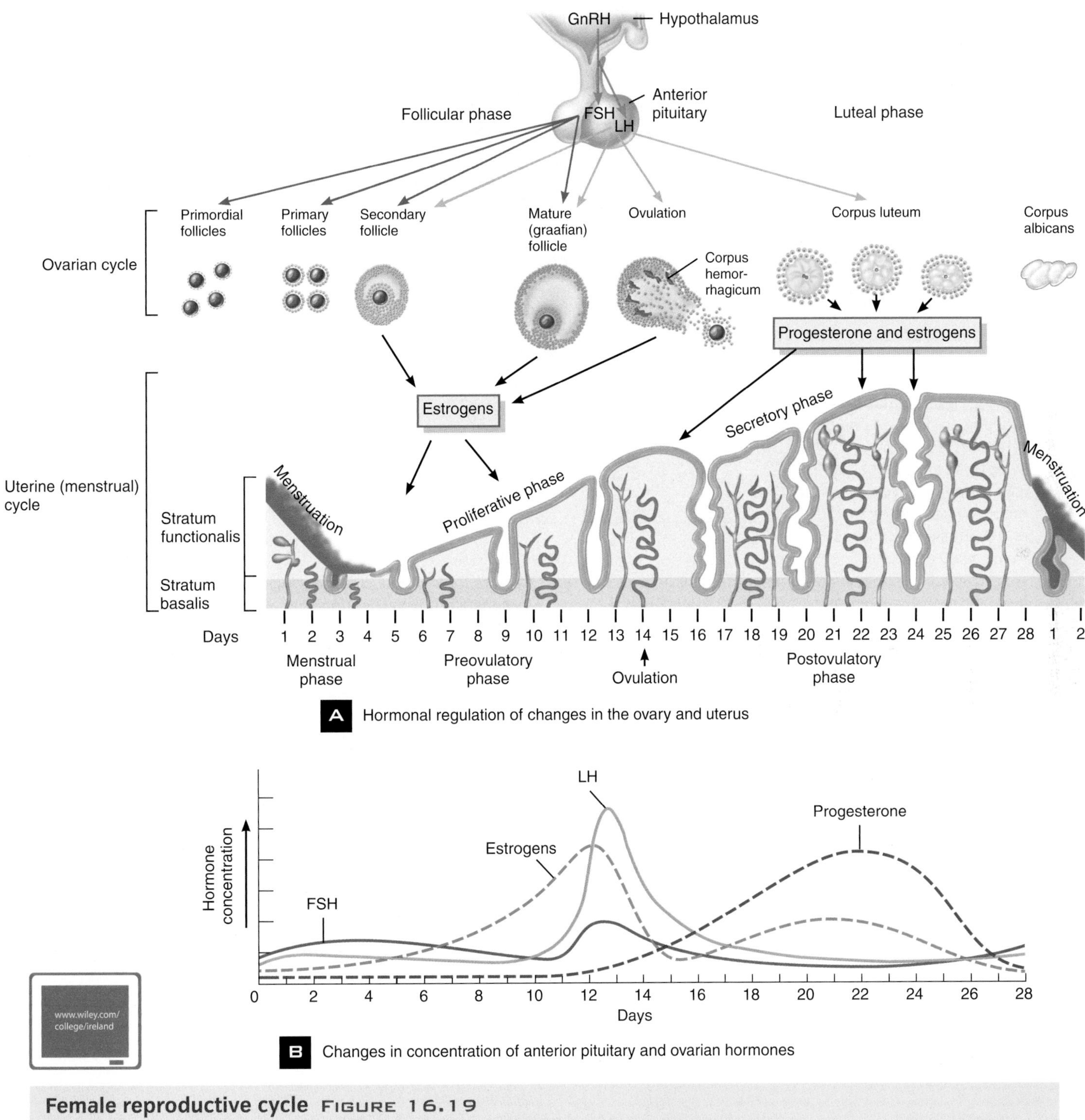

Female reproductive cycle FIGURE 16.19

I WONDER . . .

Can PMS really cause mood swings and emotional outbursts?

What is the truth to the scare stories? Can PMS cause mood swings and emotional outbursts? Premenstrual syndrome (PMS) is a cyclical disorder of severe physical and emotional distress that appears during the post-ovulatory phase of the female reproductive cycle and disappears when menstruation begins.

A severe form of PMS called premenstrual dysphoric disorder (PMDD), describes as many as 150 physical and emotional symptoms that are linked to the menstrual cycle. Common symptoms include nausea and acne. Breast tenderness and swelling are linked to fluid retention. But many symptoms are psychological, including severe mood swings, anxiety, and depression. While as many as 80 percent of American women may have some of these symptoms during their reproductive years, PMDD itself affects only 8 to 20 percent.

Women with severe cases of PMDD often have high blood levels of two stress hormones, cortisol and norepinephrine. That's significant because scientists link many of the symptoms of PMDD to the interaction of hormones and the brain. Just as the brain (acting through its control of the hypothalamus gland) can regulate hormones, hormones can affect the brain, as we see in the way that estrogen and testosterone can stimulate sexual arousal.

In PMDD, one example of how hormones and neural tissue interact concerns progesterone. Progesterone level varies with the menstrual cycle. Progesterone can be metabolized into a related compound called allopregnanolone, which has been linked to such PMDD symptoms as depression and anxiety. In studies using mice, whose menstrual cycle is similar to humans, scientists have found that hormones related to progesterone change a specific type of receptor in the brain, called GABA A receptor. These changes occur in the hippocampus, which is intimately involved in memory formation. Changes in the receptors affect the behavior of the hippocampal neurons, which affects the mouse's level of anxiety and therefore alters memory.

While any woman of reproductive age can suffer PMDD, the syndrome seems to be worse in women with a genetic predisposition or a family history of PMDD and extreme physiological reactions to normal hormone levels. Other risk factors for serious symptoms include high stress, multiple pregnancies, tubal ligation, use of oral contraceptives, excessive change in weight, lack of exercise, and poor diet. Tobacco, alcohol, and caffeine all aggravate PMDD symptoms.

Medicine, including anti-anxiety medications, can be prescribed for severe symptoms, yet many women can moderate their symptoms through behavior or diet. Some studies show a benefit from calcium supplements, and it's well established that stress reduction techniques such as exercise, yoga, and breathing exercises can help. If you or someone you love suffers from PMDD, it is important to be supportive and to seek medical assistance. A healthcare professional can work out a personal plan including medical intervention and behavioral changes that may reduce the intensity of this common syndrome.

The maturing follicle cells secrete estrogen into the bloodstream. Estrogen stimulates the development of the female secondary sex characteristics, including adipose deposition in the breasts, hips, and abdomen, and the development of groin and axillary hair. Estrogen also increases protein buildup, working in harmony with human growth hormone to increase body mass. In addition, estrogen lowers blood cholesterol. This hormone has been inplicated in PMS, the mood swings associated with the days immediately prior to beginning a new mentrual cycle. Investigate the truth of these accusations in the I Wonder . . . box. In the blood, estrogen serves as a feedback mechanism inhibiting the production of GnRH, FSH, and LH. As the estrogen level increases, GnRH, FSH, and LH levels all drop. Inhibin is also secreted by the cells of the growing follicle as well as the corpus luteum. Inhibin prevents secretion of FSH and LH, adding another level of feedback to the system.

Once the corpus luteum has been formed, it begins to secrete **progesterone**, which stimulates the growth of, and glandular secretion in, the endometrium. As the uterine lining thickens, the uterine glands begin to function. The corpus luteum also secretes small quantities of **relaxin**, a hormone that quiets smooth muscle. It is thought that relaxin aids in implantation. Perhaps implantation occurs more successfully in a quiescent uterus. Production of relaxin increases dramatically if implantation occurs, as the placenta begins secreting large quantities. A less irritable uterus provides a better environment for the developing embryo and permits placental development.

FEMALE REPRODUCTIVE CYCLE OVERVIEW

The physiological changes in the ovaries and uterus, and the hormonal changes during the female reproductive cycle, are part of an integrated system (see FIGURE 16.19). The uterine cycle is the regular growth and loss of the endometrial lining. At the beginning of the cycle, the month-old lining is shed. This usually takes from three to seven days to complete, allowing the female to know precisely when her "period," or menstrual flow, began. The low levels of all female hormones in the blood impair blood flow to the **stratum functionalis** of the endometrium, causing the lining to slough off. The volume of a typical menstrual flow is approximately 50 to 150 ml, made up of tissue fluid, mucus, blood, and epithelial cells.

> **Stratum functionalis**
> Outer layer of endometrium that grows and sheds in response to hormone levels in the blood.

The next 6 to 13 days make up the **preovulatory phase**. The variable length accounts for the individual differences in menstrual cycles. FSH secretion increases, stimulating follicles in the ovary, and causing maturation of approximately 20 follicles. By day 6, one follicle in one ovary has grown faster than the others, becoming the dominant follicle. This follicle secretes estrogen and inhibin, preventing further release of FSH and therefore quieting the development of the remaining follicles in both ovaries.

The dominant follicle will enlarge until it appears as a swollen area on the surface of the ovary. This Graafian follicle increases estrogen production under the influence of LH from the anterior pituitary. This stage of ovarian activity is called the **follicular phase** owing to the involvement of the follicle cells.

An increased estrogen level in the blood repairs the blood vessels damaged during the previous menstrual flow and stimulates mitosis of the endometrial cells. Glands develop in the stratum functionalis of the endometrium, but they do not yet function. Because the endometrium is growing (proliferating), this is the **proliferative phase**.

Increasing levels of estrogen stimulate increased production of GnRH, which in turn stimulates surges in LH. The Graafian follicle reacts to this LH spike by popping, extruding fluid and the ovum into the abdominopelvic cavity. This violent, often painful action is **ovulation**. A slight temperature increase indicates that ovulation has taken place. This normal response to trauma is the basis of some natural birth control methods, such as the sympto-thermal method, that involve charting body temperature every morning (see p. 546). A slight spike in recorded temperature indicates ovulation, when an ovum is released and made available for fertilization.

After ovulation, the follicle cells are dormant and the corpus luteum cells begin to function. This **postovulatory phase** has the most uniform duration, taking 14 days in almost every woman. The corpus luteum formed during ovulation will survive for exactly 14 days. If no fertilization occurs, the corpus luteum degenerates into the **corpus albicans**. During the lifespan of the corpus luteum, the progesterone level increases. As it degenerates, progesterone declines.

In the uterus, the endometrial lining is maintained by progesterone. The endometrial glands begin to function, and the lining is prepared for a possible implantation. This phase is often called the **secretory phase** in reference to these glandular activities. Assuming there is no implantation and no pregnancy, progesterone, estrogen, and inhibin levels all drop by the end of the postovulatory phase. As the progesterone levels decline, the endometrial lining loosens. With such low hormone levels in the blood, the endometrial lining cannot be maintained and is lost from the underlying tissues, and menstruation begins again.

Correct functioning of the female reproductive cycle depends on many variables. Lifestyle has a profound effect, as can be seen in postpubescent elite female athletes. True, girls who participate in sports are healthier, get better grades, and are less likely to suffer depression or use illegal substances. But intense involvement in sports can be risky. The **female athlete triad** is a condition in which health deteriorates due to overemphasis on sports (FIGURE 16.20). Three related problems can arise: disordered eating, **amenorrhea** (lack of a menstrual cycle), and osteoporosis.

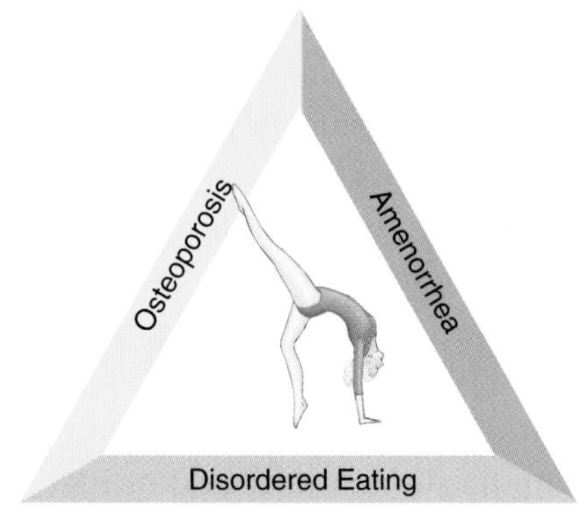

Female athlete triad FIGURE 16.20

This syndrome is more common in women who are perfectionists, highly competitive, and have low self-esteem.

Often coaches or others involved in girls' sports inadvertently feed into this triad by emphasizing intense training and success at all costs. Many female athletes are told to focus on their diet and weight, but if this focus is mainly on avoiding weight gain rather than quality of nutrition, it can contribute to eating disorders (see p. 436 in Chapter 13). Continued intense exercise and caloric restrictions can also interfere with a girl's reproductive cycle. It takes a fair amount of energy to sustain reproductive ability, and low caloric intake and increased muscular activity may make the necessary energy simply unavailable. Estrogen production slows, causing irregular menstrual cycles or ending them entirely, contributing to postmenopausal symptoms. A declining estrogen level reduces bone density, which is especially troublesome in teenagers, when the skeleton reaches its densest condition, forming a strong foundation for adult life. Some teenage female athletes can have a bone density typical of a 60-year-old woman, and training can lead to stress fractures and broken bones.

CONCEPT CHECK

Compare oogenesis to spermatogenesis. What are the main differences?

Describe the connection between the uterine tubes and the ovaries.

Describe the structure of a mammary gland, including the function of the Cooper's ligaments.

How, specifically, does the ovarian cycle direct the uterine cycle?

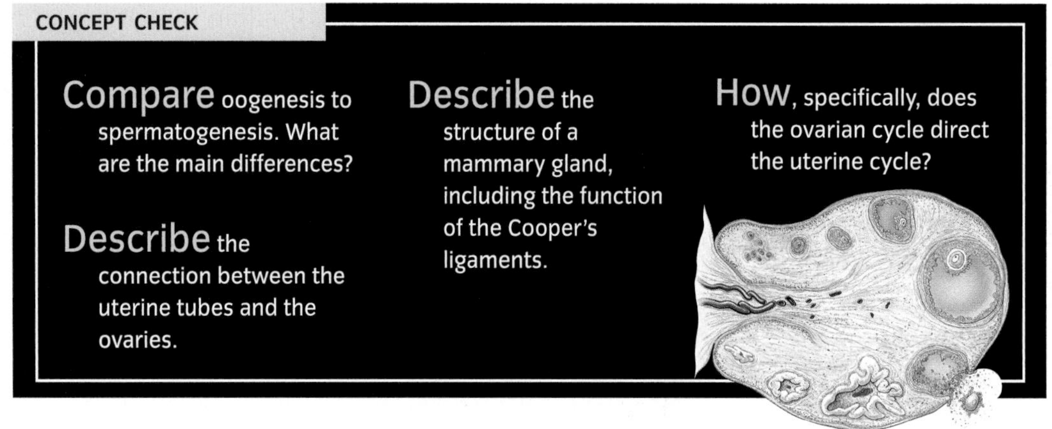

The Orgasm Is a Moment of Emotional and Physiological Epiphany

LEARNING OBJECTIVES

Understand the physiological role of orgasm in males and females.

Describe the four phases of the human sexual response.

Male and female sexual responses share some similarities (FIGURE 16.21). In both sexes, blood flow to the genitals is altered, gland secretion increases, and orgasm results in rhythmic contractions of pelvic muscles. In the mid-1950s, sex researchers Masters and Johnson began research on the human sexual response that spawned the study of human sexuality. They identified four phases of the human sexual response, which appear in both males and females: arousal, plateau, orgasm, and resolution.

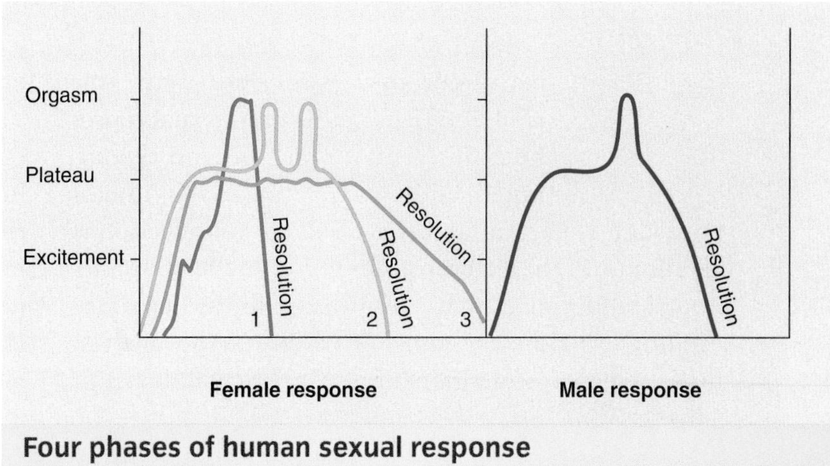

Four phases of human sexual response
FIGURE 16.21

During **arousal**, or excitement, blood flow to the penis or clitoris is altered, glands begin to secrete lubricating fluids, and heart rate and blood pressure increase. Arousal is governed by the parasympathetic nervous system. This phase is highly responsive to sensory stimulation, such as touching of the genitals, breasts, lips, or earlobes. Other sensory stimulation, including visual, auditory, or even olfactory stimuli, can increase or dampen the arousal.

As excitement builds, **plateau** is reached. This can last from a few seconds to many minutes. During this phase many females, and some males, experience a rashlike flush to the skin of the upper neck and face.

Orgasm, a series of wavelike muscular contractions, and an intense pleasurable sensation, marks the end of the plateau. Orgasm and resolution are controlled by the sympathetic nervous system. In the male, orgasm accompanies ejaculation. In the female, receiving the ejaculate does not provide much stimulation. Simultaneous orgasm is not automatic, nor should it be expected. Once males reach orgasm, they experience a refractory period of a few minutes to a few hours. During this time, a second ejaculation is physiologically impossible. Females do not require a refractory period and can experience two or more orgasms in rapid succession.

The last phase, **resolution**, begins with a sense of intense relaxation. Heart rate, blood pressure, and blood flow all return to prearousal levels. Resolution time is variable, taking longer to arrive when no orgasm occurred.

CONCEPT CHECK

How are the male and female orgasms similar? How are they different?

What are the four phases of the human sexual response?

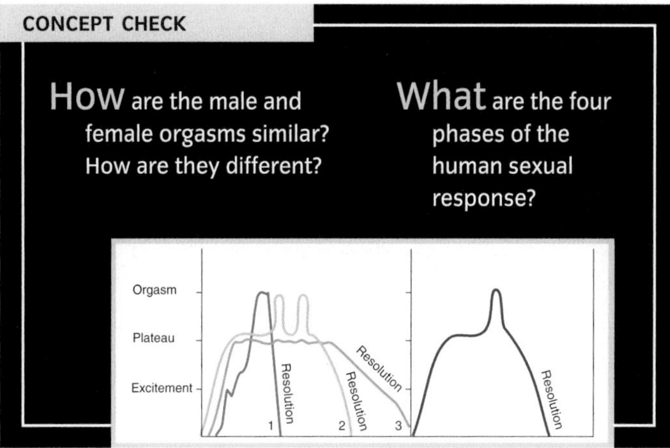

Sexually Transmitted Diseases Can Be a Side Effect of Sexual Contact

Learning Objectives

Define STD.

List the main categories of STD.

Understand the treatments for the most common STDs.

Sexual reproduction is critical to the survival of the species. During sexual reproduction, genes are mixed and recombined, adding variability to the human population. This variability is what natural selection is all about—slight differences in **genotype** lead to phenotypic differences that may give one individual more evolutionary fitness. Populations evolve as the environment selects for or against traits, as you will learn in Chapter 19.

Genotype
"Type o' genes"; the entire group of genes for an organism.

For all of its benefits to the population, sexual reproduction carries a real danger to individuals: sexually transmitted diseases (STDs) (Table 16.1). These range in severity from a mild discomfort that can be cured with antibiotics to severe, recurring infections to deadly diseases. The Health, Wellness, and Disease box discusses prevention of STDs. STDs may be caused by bacteria, viruses, fungi, insects, or protists. Bacterial STDs include **gonorrhea, syphilis**, and **chlamydia**. Viruses that can be transmitted through sex include **HIV** (see Chapter 10), **herpes simplex virus (HSV) 1 and 2**, **human papillomavirus (HPV)**, and **hepatitis B**. HSV 1 causes cold sores in the mouth, and less frequently genital lesions. HSV 2 causes most cases of genital herpes. Human papillomavirus, or

STDs Table 16.1

Common name	*Scientific name*	*Classification*	*Symptoms*	*Treatment*
Chlamydia	*Chlamydia trachomatis*	Bacterial (can only reproduce inside body cells)	Usually asymptomatic, may cause urethritis in males. Leads to pelvic inflammatory disease in females	Antibiotics
Gonorrhea or "the clap"	*Neisseria gonorrhoeae*	Bacterial	Urethritis with excess pus discharge, may be asymptomatic in females, leading to sterility	Antibiotics
Syphilis	*Treponema pallidum*	Bacterial (spiral bacterium)	Primary stage results in a painless open sore or chancre; secondary stage is a rash, fever, and joint pain; tertiary stage results when organs begin to degenerate	Antibiotics in primary or secondary stage
Genital herpes	Type II herpes simplex virus (HSV)	Virus	Painful blisters on the external genitals of males and females, with possible internal blistering in females	Incurable, but outbreaks can be controlled with anti-inflammatory drugs
Genital warts	Human papillomavirus (HPV)	Virus	Cauliflower growths on the external genital area, and internal growths in females; can also appear on or around the anus	Incurable; warts can be removed cryogenically
Trichomoniasis	*Trichomonas vaginalis*	Protozoan	Foul-smelling discharge and itching in females	Prescription drug metronidazole

STDs: The science of prevention

Sexually transmitted diseases have probably been around since the dawn of sex in animals. After all, sex is based on intimate contact between delicate tissues, and that allows pathogens to move directly from the blood or bodily fluids of one individual to another. STDs are—or should be—a constant concern among people with multiple sex partners.

Rates of STDs have shown no clear trend in the United States in recent years. Between 1996 and 2004, figures for syphilis went down, those for chlamydia went up, and the numbers for gonorrhea stabilized. HIV rates peaked in 1993, then declined. But since 2001, the rate for HIV has slowly gone back up, perhaps owing to fading memories of the severity of the disease, or perhaps to an overabundant faith in the drugs that treat HIV. (While effective, they are expensive, do not cure the disease, and often cause serious side effects.) One-third of all new cases of STD each year are attributable to human papillomavirus, which accounts for the current increases in STD rates.

The most effective way to prevent STDs is to abstain from sexual penetration of body orifices. But a number of other techniques can be used by those who prefer to be sexually active:

- Having sex only with people who have tested negative for STDs.
- Remaining in mutually monogamous relationships with a person who started the relationship free of STDs.
- Using barrier techniques, primarily condoms, that prevent pathogen transmission when used properly.
- For sexually active people with multiple partners, getting vaccinated. (Vaccines are only available against certain STDs.)

From a public health standpoint, STDs require a combination of prevention and treatment. For curable diseases, including many bacterial infections, treatment of sexually active patients can prevent them from spreading diseases. In this sense, treatment translates into prevention. For decades, public health officials have engaged in "partner tracing"—tracking down and testing sex partners of people with STDs, especially syphilis and gonorrhea, and then treating if needed. In all states, medical personnel are required to report cases of syphilis, gonorrhea, chlamydia, and AIDS, so that public health agencies can trace partners.

Due to the shame, confusion, and secrecy that commonly surround sexuality, finding and treating partners can be difficult. This was especially true during the early stages of the AIDS epidemic. In that period, AIDS patients aroused fear and suffered discrimination, and partner tracing was seen, rightly or wrongly, as one more way to discriminate. Now that antiviral therapy can reduce the viral load and therefore the rate of HIV transmission, there is more emphasis on tracing the source of infections. The acceptance of partner tracing has grown as discrimination against AIDS patients has declined; it also reflects a belated recognition that it's still better to prevent AIDS than to treat it.

Even during an epidemic, diseases are often considered a personal matter. But when treatment is imperfect or unavailable (as it is in AIDS, HSV, and avian influenza, for example), prevention becomes the first line of defense. For the good of the community, personal actions become a public affair.

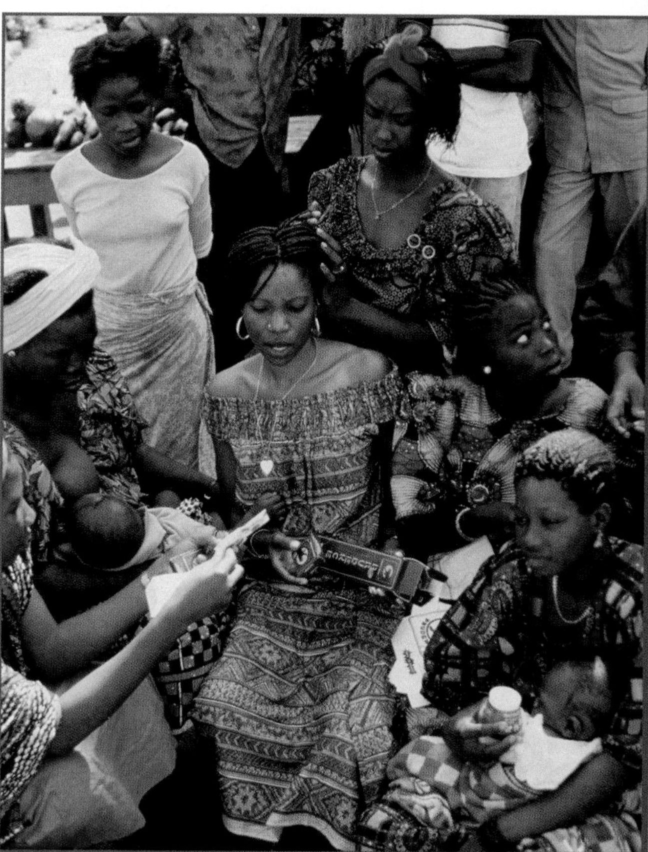

HPV, is a group of viruses that may be sexually transmitted. These viruses cause genital warts and in some cases can lead to cancer of the cervix, anus, penis, or vulva. Currently, the CDC estimates that over 20 million people are infected with HPV. This means that at least 50 percent of those sexually active get HPV at some point. An amazing 80 percent of women have contracted HPV by age 50. Because of the prevalence and seriousness of this virus, scientists have been working on a vaccine for a few years. The great news is that a promising vaccine against the most common strain, HPV-16, has been created and was recently deemed safe and effective by the U.S. Food and Drug Administration. Preliminary results are promising, with 100 percent protection against the most virulent forms of HPV. **Yeast infections** are caused by a fungus. Pubic **lice** are insects that burrow into the skin, and **vaginitis** is usually caused by a parasitic protozoan. Some have suggested that one alternative to sexual reproduction and its risk of STD, is the cloning of humans. See the Ethics and Issues box on pages 546–547 for a discussion of this topic.

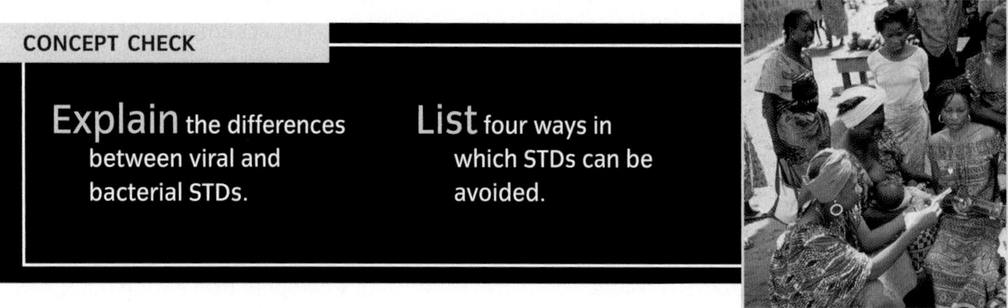

CONCEPT CHECK

Explain the differences between viral and bacterial STDs.

List four ways in which STDs can be avoided.

There Are Many Birth Control Choices, None of Them Perfect

LEARNING OBJECTIVES

Discuss the different types of birth control.

Understand the benefits and risks of each form of birth control.

While the biological function of the reproductive system is to propagate the species, pregnancy is not always the desired outcome of sexual activity. Preventing pregnancy is important to many couples, and there are now many good options that can fit just about anyone's lifestyle. Of course, the only absolute method of birth control is **abstinence**. If no sperm enters the female, pregnancy is impossible. Other birth control methods rely on **behavior modification, surgery, hormones, barriers**, or **spermicides**. Each form of birth control has advantages and disadvantages (TABLE 16.2), and choosing the optimum method can be confusing. The choice should be made after studying information on each form and considering the risks. It is also helpful to discuss the various methods with your partner. A birth control method that does not complement your lifestyle is likely to be less effective than one you can follow without changing your routine.

BIRTH CONTROL CAN BE HANDLED SURGICALLY

Surgical sterilization can prevent gametes from meeting (FIGURE 16.22). In either gender, the tube through which sperm travels to reach the egg can be blocked, preventing fertilization.

Failure rates of birth control methods

TABLE 16.2

Method	Failure Rates* Perfect Use†	Typical Use
None	85%	85%
Complete abstinence	0%	0%
Surgical sterilization		
Vasectomy	0.10%	0.15%
Tubal ligation	0.5%	0.5%
Hormonal methods		
Oral contraceptives	0.1%	3%‡
Depo-provera	0.05%	0.05%
Intrauterine device		
Copper T 380A	0.6%	0.8%
Spermicides	6%	26%‡
Barrier methods		
Male condom	3%	14%‡
Female condom	5%	21%‡
Diaphragm	6%	20%‡
Periodic abstinence		
Rhythm	9%	25%‡
Sympto-thermal	2%	20%‡

* Defined as percentage of women having an unintended pregnancy during the first year of use.

† Failure rate when the method is used correctly and consistently.

‡ Includes couples who forgot to use the method.

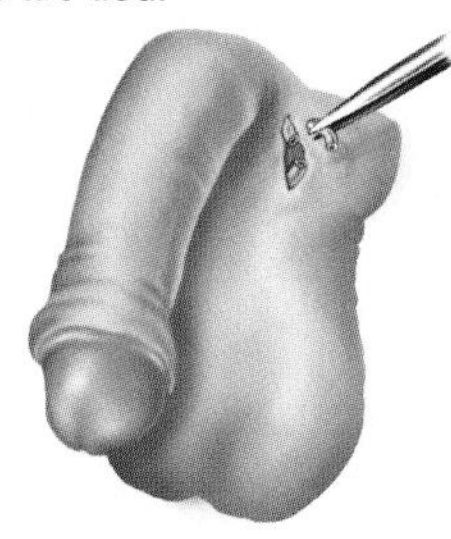
In a vasectomy, a small section of the vas deferens is removed and the ends are tied.

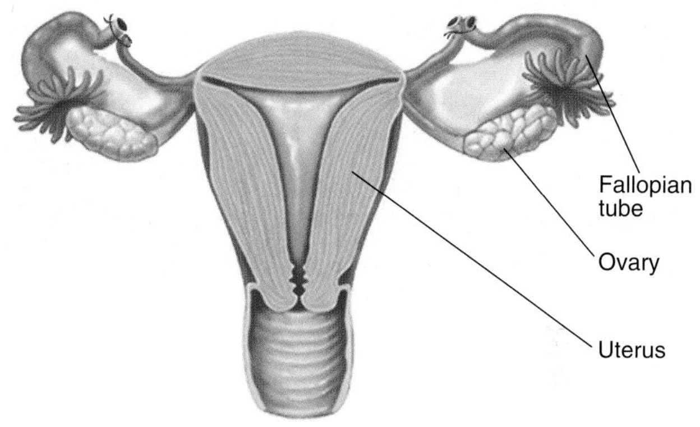
In a tubal ligation, the uterine tubes are cut and tied.
Fallopian tube
Ovary
Uterus

Surgical sterilization FIGURE 16.22

Male sterilization is an easy outpatient surgery that uses no scalpels and only two small punctures at the posterior base of the scrotal sac near the body. The spermatic cord is located; the ductus deferens isolated, pulled out slightly, and closed either by looping and **ligating**, by cutting and sealing, or by clamping. The skin of the scrotal sac is closed without stitches, and that's that. A local anesthetic prevents pain during the puncturing, and the patient may feel a slight pulling as the ductus deferens is located and pulled through the skin. Testosterone levels are not affected, so sexual desire does not change. Because the sperm contribute very little to the total volume of semen, a vasectomy is virtually undetectable in sexual performance.

Ligating
Tying off a tube to close it.

After vasectomy, sperm in the seminiferous tubules cannot pass the ductus deferens to reach the seminal vesicles. During ejaculation, sperm is forced from the epididymus to the blockage in the ductus deferens, and stops. The muscular contractions of the orgasm continue to push through the male system, causing the release of fluid from the seminal vesicles, the prostate, and the bulbourethral glands. Since there may be sperm in the ductus deferens above the vasectomy, sterility may be delayed for six weeks while any remaining sperm leave the system. After that, the male should be 100 percent sterile. This procedure costs between $500 and $1,000 and is covered by most insurance companies. As with any medical procedure, complications can arise, but the procedure is less risky than sterilization for females.

The female equivalent of a vasectomy is a **tubal ligation** which blocks the uterine tubes to prevent both the egg from reaching the uterus and the sperm from passing through the uterine tubes to an awaiting ovulated egg. Tubal ligation requires a brief stay in the hospital. The woman is anesthetized, her abdomen is

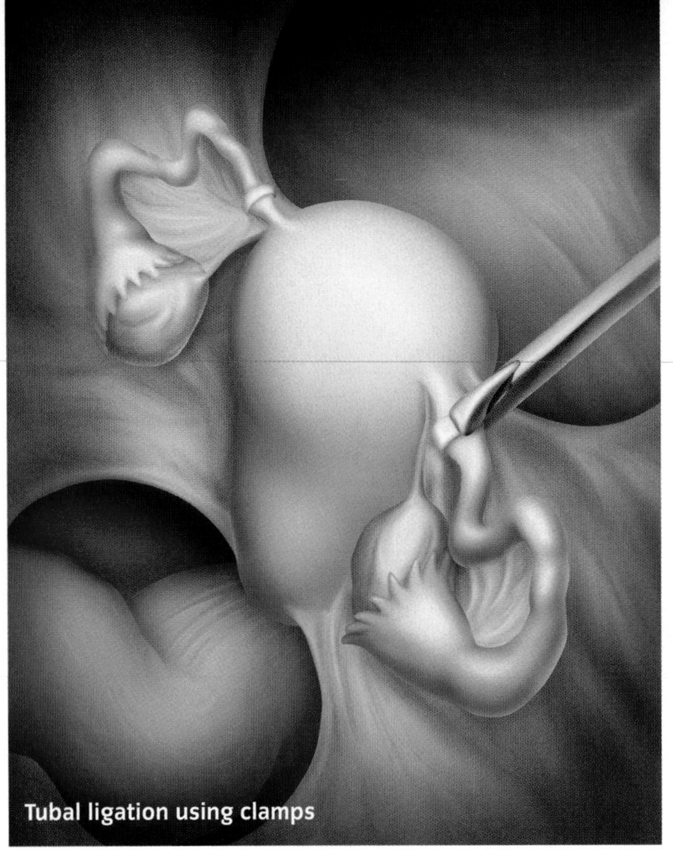
Tubal ligation using clamps

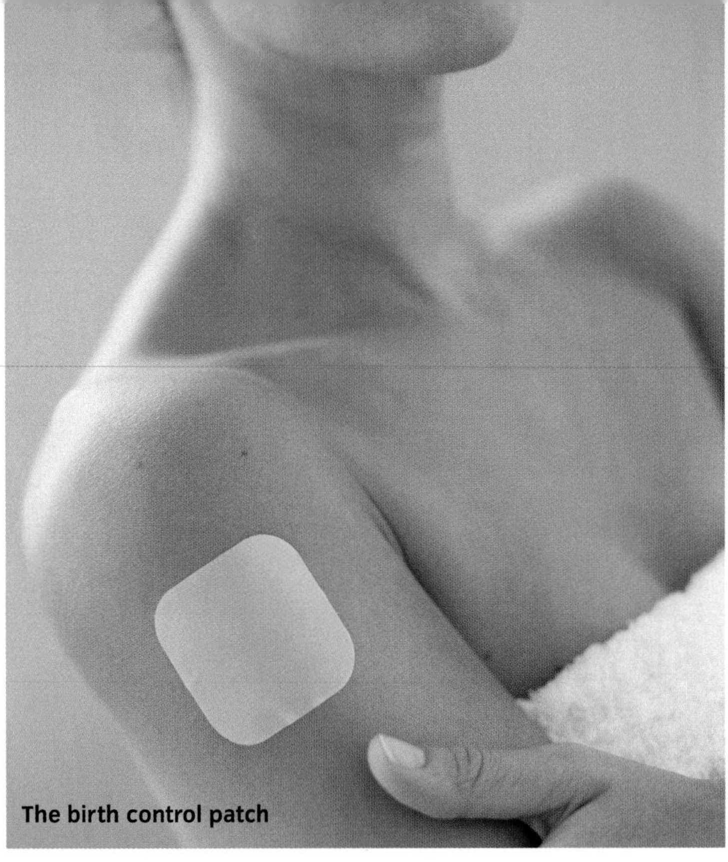
The birth control patch

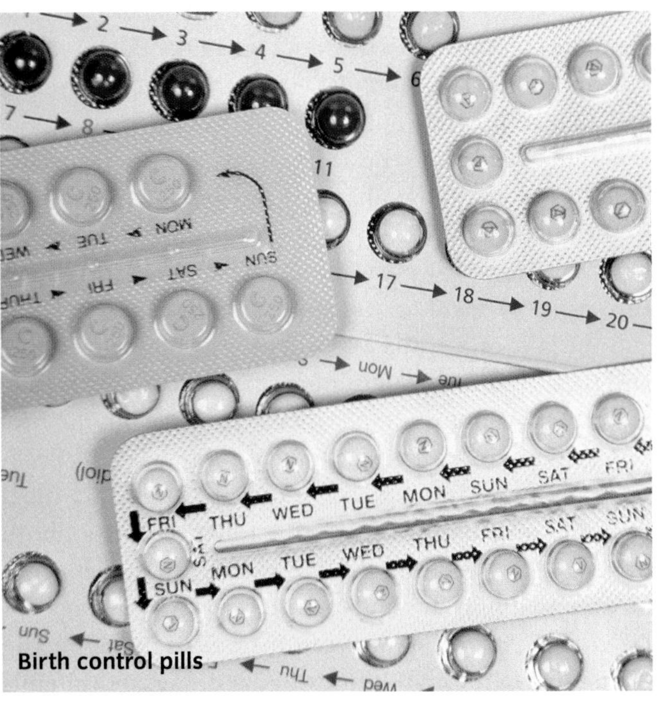

Birth control pills

distended with CO_2 gas to separate her organs, and the two uterine tubes are located via **laparoscopy**. The tubes can then be cut and tied similar to the vasectomy, sealed via electrocautery, or closed with titanium clamps. Alternatively, small coils called microinserts can be placed in the uterine tubes. These coils are brought in through the cervix and placed in the first third of the uterine tube, where they irritate the tube and cause scar tissue to form. After approximately three months, the scar tissue will block the uterine tube, preventing the passage of egg or sperm. While not technically a tubal ligation, these microinserts produce the same result.

Laparoscopy
Noninvasive surgery using fiber optic cables, remote control, and tiny surgical tools.

HORMONAL METHODS OF BIRTH CONTROL ARE ANOTHER OPTION

While surgery is permanent, hormonal methods (see photos) are temporary, delivered in pill or patch form. The birth control pill is an **oral contraceptive**, a com-

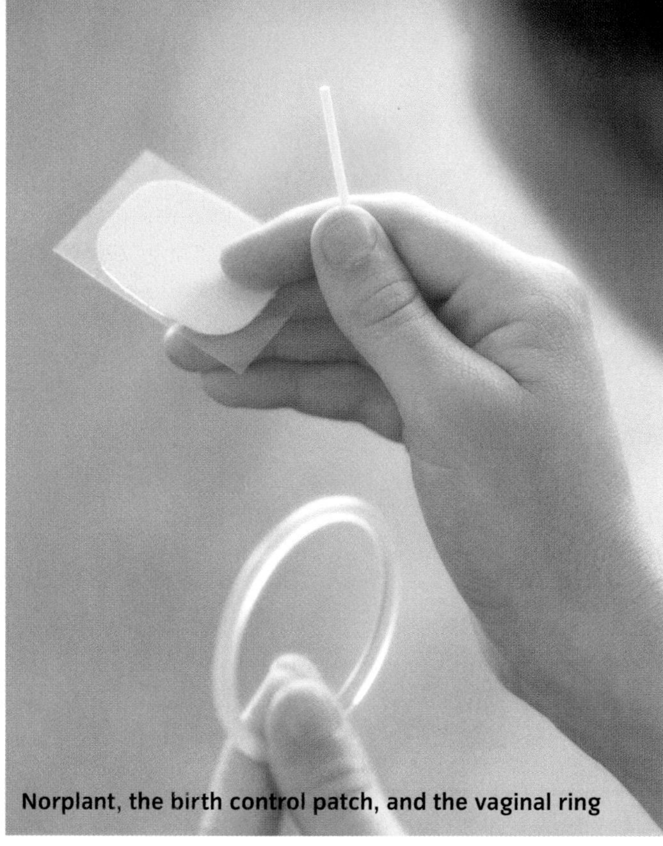
Norplant, the birth control patch, and the vaginal ring

bination of synthetic estrogens and progestins that alters the natural hormonal rhythms of the female. The birth control patch is a similar mixture of hormones absorbed through the skin rather than through the digestive membranes. In both cases, keeping estrogen and progestin levels high inhibits the secretion of FSH and LH from the anterior pituitary gland. Without FSH, the follicles in the ovaries do not mature, and no eggs are ready to ovulate. The hormone levels created by birth control pills almost guarantee that natural production of estrogen remains low, LH is not produced, and ovulation will not occur. Even birth control pills that maintain a very low estrogen level to alleviate side effects do not cut out the hormone entirely. Some birth control pills also alter mucus production of the cervix, creating an environment inhospitable to sperm. Taken correctly, the pill is close to 100 percent effective. But missing one dose can cause a dip in the artificial hormone levels, allowing natural rhythms to resume. In artificially regulating the menstrual cycle, the pill also provides beneficial side effects such as scant periods, a regulated and predictable menstrual cycle, and protection against endometriosis, breast cancer, and ovarian cancer. Because some women prefer not to have a menstrual period at all, there is now a form of birth control pill that provides three months of continuous hormonal control, rather than the usual three weeks of control and one week of placebo pills. This new form permits menstruation only four times a year. There has been little research to date on the side effects of this dosage of hormones. As with all medications, there are risks associated with taking oral contraceptives. Women with blood clotting disorders, frequent migraine headaches, blood vessel weaknesses, high blood pressure, or liver disease are advised not to take the pill. Also, women who smoke are at much greater risk of heart attack or stroke when taking birth control pills than nonsmokers.

Norplant, Depo-provera, and the **vaginal ring** are alternative forms of hormonal contraception. Norplant is a series of six hormone "sticks" surgically implanted under the skin of the upper arm. These sticks slowly leak progestins into the female system for five years, inhibiting ovulation and causing thickening of the cervical mucus. If the Norplant sticks are removed, fertility is restored. Depo-provera is an intramuscular injection of progestin given every three months. The initial months using Depo-provera can be difficult, as the body adjusts to the changes initiated by the progestins. Some women experience weight gain, PMS-type symptoms, fluid shifts, and inconsistent spotting and cramping. The vaginal ring is worn in the vagina for three weeks. It slowly releases estrogen and progestins in levels similar to the oral contraceptives discussed above. Removing the vaginal ring every fourth week allows the slight increase in endometrium to be shed, similar to a normal menstrual flow.

Emergency contraception, sometimes referred to as the "**morning after**" pill, prevents implantation of the fertilized ovum or causes an already implanted embryo to be lost as the endometrial lining weakens. The term *morning after* is misleading, for this form of birth control may be carried out within three days to seven weeks of unprotected sex. Emergency contraception can only be obtained with a prescription and may cause serious cramping and discomfort. This contraceptive method works similarly to the pill in that it requires altering the hormonal environment of the female. Two types of emergency contraception are available currently. Preven® is the brand name for a series of four pills, two to be taken within 72 hours of unprotected sex and two more to be taken 12 hours later. These pills cause the lining of the uterus to become inhospitable to implantation. The other form of emergency contraception is the drug mifepristone, or RU 486. It works by decreasing the uterine cells' sensitivity to progesterone. This in turn causes the uterine lining to be shed, just as it is at the end of a normal uterine cycle. Mifepristone essentially causes a chemical abortion of an implanted embryo.

Elective abortion, or more commonly simply "abortion," is the termination of a pregnancy. While early-stage pregnancy can be terminated using mifepristone, abortions are performed in medical offices, hospitals, or clinics. Elective abortions are performed only in the first trimester of pregnancy and can take one of several forms. The uterus can be scraped clean, removing the endometrial lining as well as the implanted embryo, the contents of the uterus can be suctioned out, or a strong saline solution can be injected into the womb causing loss of the endometrial lining. Abortions are performed for many reasons, including a pregnancy resulting from rape or incest, a pregnancy that endangers the life of the mother, or life-threatening malformations of the fetus. Because the procedure removes a potentially viable fetus, there is much controversy surrounding abortion. Currently, most states in the United States allow elective abortion, but the issue does arise in courts periodically and the ethical dilemma remains—life of the fetus versus the reproductive life of the mother.

Elective abortion
Removal of the developing embryo initiated by personal choice.

THE INTRAUTERINE DEVICE PROVIDES AN OBSTRUCTION TO CONCEPTION

The **intrauterine device (IUD)** is a foreign object that floats in the uterus and periodically hits the endometrial lining, preventing implantation. Most IUDs are made of plastic or copper. They can be almost any shape from a squiggly "s" to a number 7 to a capital T. Each IUD has a string that hangs out of the cervix in order to allow removal. The most common IUD is the Copper T 380 A. This small copper wire is placed in the uterus. It may cause cramping and bleeding upon implant, but these symptoms usually subside. The IUD can remain in the uterus for up to 10 years. IUDs that carry hormones further prevent implantation, but they must be replaced every five years.

IUDs lost popularity after the Dalkon Shield episode in the 1970s. This IUD was made of plastic, and looked like a bug with a rounded appearance and five leg-like structures extending from each side. Unlike other IUDs marketed with a single filament string extending from the cervix, this one had a larger, braided string. This large device was implicated in 12 deaths due to complications and infection allegedly introduced with the IUD. The thought was that the complicated string may have been a poor design, allowing bacteria to enter the braids and then enter the uterus. Test results did not confirm this theory, however. Despite no proof that the Dalkon Shield was responsible, plaintiffs

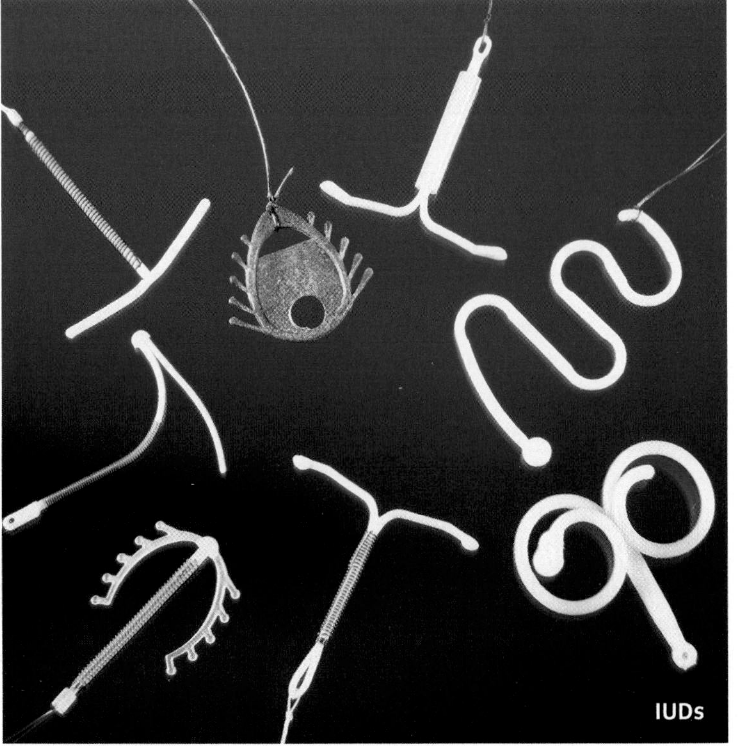
IUDs

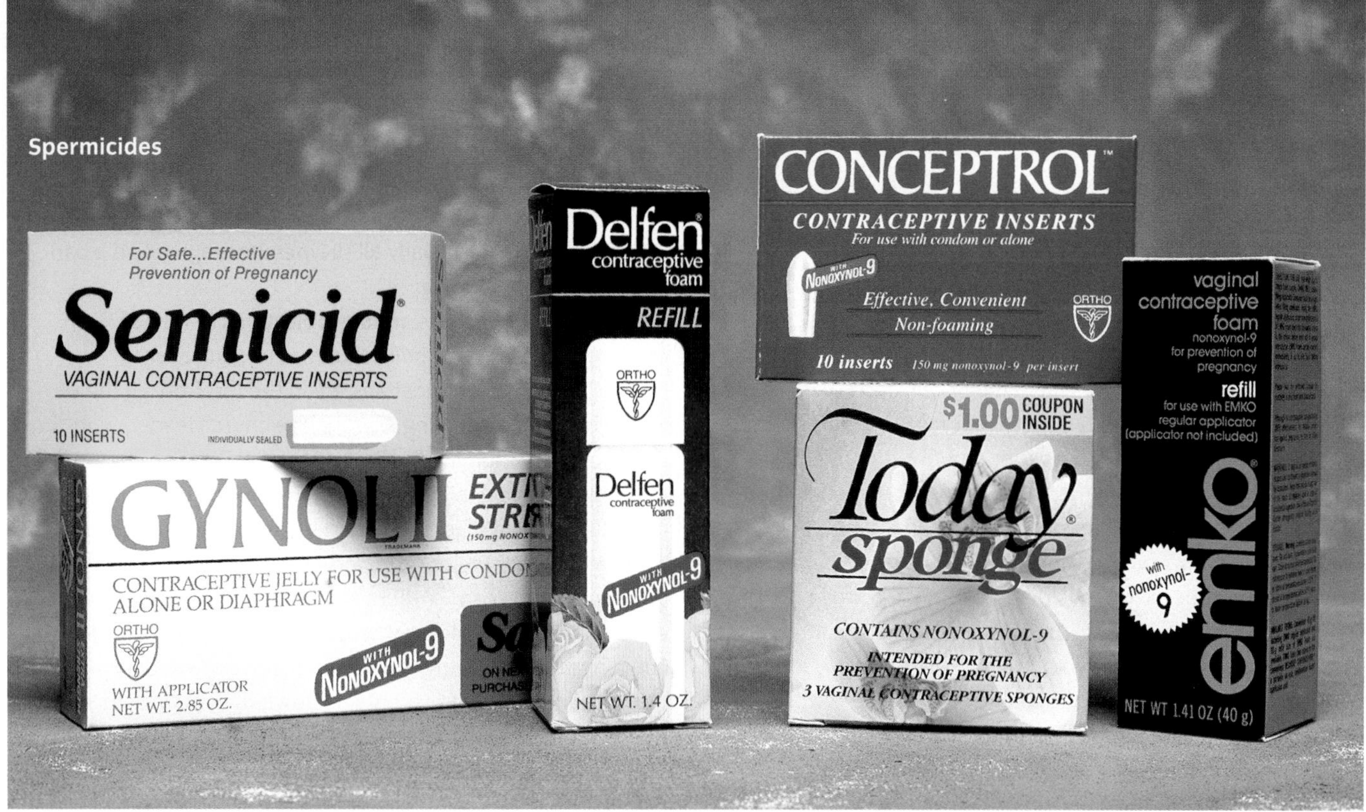

won a lawsuit, the Shield was pulled from the market, and many people erroneously still think all IUDs are dangerous.

SPERMICIDES KILL SPERM

Spermicides are creams and jellies that contain **nonoxynol 9**, a compound that kills sperm by disrupting the cell membrane. Recent evidence shows that nonoxynol 9 causes shedding of epithelial cells in alarmingly large sheets immediately after being introduced to the vagina. This loss of protective epithelium from the vagina or anal canal could allow entry of sexually transmitted diseases, trading one sexual problem for another. Spermicides are more effective when used in conjunction with a barrier method.

BARRIER METHODS BLOCK THE ENTRY OF SPERM; SOME PROTECT AGAINST STDS

Barrier methods of birth control establish a physical obstacle between sperm and egg. The **condom** is a barrier worn on the penis, while the **female condom, cervical cap**, and **diaphragm** are barriers worn in the vaginal area. Latex condoms are also effective against most STDs. Natural condoms, made of lambskin, do not block STDs, but do provide a barrier against sperm. The pores in these condoms are too large to block bacteria or viruses.

The diaphragm is a rubber disc held in the vagina by a flexible ring that blocks sperm but does not protect the vagina against STDs. A cervical cap is a smaller version of the diaphragm that is placed over the cervix. To be effective, both devices must be fitted by a physician. The female condom is a hybrid of diaphragm and condom, composed of two flexible rings connected by a latex sheath. The upper ring functions as a diaphragm, while the lower ring holds the latex sheath against the walls of the vagina, providing protection from disease along the entire tract. Combining a spermicide with a barrier method provides much greater protection against both STDs and pregnancy.

Ethics and Issues

Should we clone humans?

Should we do things just because we **can** do them? The question arises because scientists are almost able to clone human beings— to grow exact genetic copies of adults. In reproductive cloning, scientists usually take a cell from an adult, transfer its nucleus into an egg cell, and create an embryo that grows into an adult. Except for DNA found in the organelles, the clone is an identical twin of the original adult, only considerably younger. A second experimental procedure, called therapeutic cloning, takes embryonic stem cells from human embryos. These stem cells may develop into specialized cells that could, theoretically, be used to treat deadly diseases. Therapeutic cloning is controversial in its own right because it destroys embryos, but the research is proceeding under strict limitations, as we'll see later in this book.

The questions about reproductive cloning date to 1997, when a Scottish sheep named Dolly was cloned. Dolly was created after 276 failed attempts, proving that the process was highly experimental. After the headlines faded, she died early, a sign of major health problems, perhaps related to the "old" DNA in her cells.

So far as we know, nobody has cloned a human, which is widely considered risky, ethically objectionable, and illegal. The United Nations failed to ban both therapeutic and reproductive human cloning in 2005, but according to the *Washington Post*, "Virtually all UN members agree" on a ban for reproductive cloning.

Many religions have serious objections to human cloning. The Roman Catholic church is unequivocal: "Catholics and other Christians are in the forefront of the effort to ban human cloning," wrote one church leader, who cited scripture, human dignity, and medical reasons for avoiding cloning. Many other religious groups have also released statements. Some support a ban on the production of cloned human embryos, whether for reproduction or therapeutic purposes. Others see reproductive cloning as ethically wrong, but favor therapeutic cloning because it could alleviate suffering. A few see science as a manifestation of God's will for us to obtain knowledge, but others say that since a clone would have no parents, many of society's laws would be difficult to apply.

The U.S. Constitution separates church and state, which means the above arguments have no legal validity in the United States. However, reproductive human cloning raises other concerns such as:

- Causing physical or psychological harm to the cloned child.
- Treating children as objects for social purposes, as status enhancers, rather than as individuals with their own desires and needs.

NATURAL FAMILY PLANNING IS ANOTHER VIABLE TECHNIQUE

The female reproductive cycles provide clues about the timing of ovulation. If the female knows the exact timing of ovulation, she can avoid pregnancy by preventing the introduction of sperm into her reproductive tract during that time. Due to the timing of egg movement, the window of fertility is a six-day period beginning five days prior to ovulation and ending the day of ovulation. Test kits are available to help predict the timing of ovulation. Self-monitoring, such as charting daily morning temperature or observing changes in cervical mucus, also give a fairly accurate picture. By recording temperature or mucus condition on a calendar for a few months, the general ovulatory pattern becomes clear. This method, with temperature charts and precise information on when

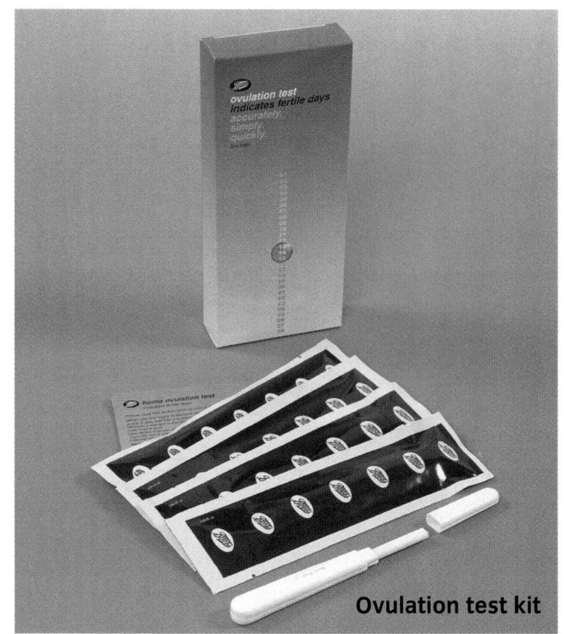

Ovulation test kit

- Producing social harm by having parents choose the child they want rather than "shooting craps" by randomly mixing their DNA. To take just one scenario, say a wife was cloned. What would happen when the clone passed through puberty and the husband found himself living with a carbon-copy of the woman he fell in love with, but who happened to be his daughter?
- Misappropriating medical resources to cloning rather than funding research with wider benefits. Reproductive cloning could benefit infertile couples who wanted to raise a child with one parent's genetics. It might also help people with an urge to reproduce their own incomparable genetics—in other words, people with serious tendency toward narcissism. But money spent on cloning might be money not spent, say, on curing hypertension or cancer.

What do you think? Once it becomes possible to clone humans, should we do it? If so, should we clone scientists or politicians? Average people or the rich and powerful? Would it be ethical to clone a human to produce perfectly matching organs for transplant? So many questions . . . so few answers. This technology is yet an infant, offering many opportunities, and perhaps an equal number of risks, for the next generation.

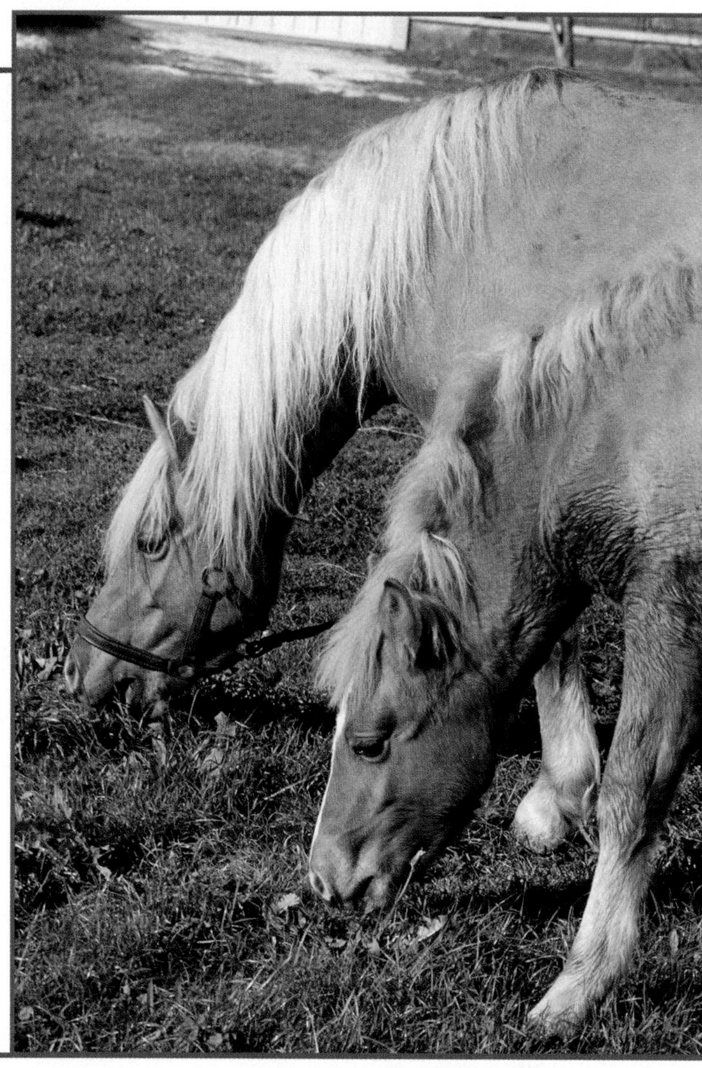

ovulation occurs is referred to as the sympto-thermal method of birth control. The rhythm method of birth control follows a similar practice but does not include temperature as a cue to ovulation. Couples who follow the rhythm method rely on consistency in the female's menstrual cycle. Based on history, ovulation is predicted. Practicing abstinence during her six-day window of fertility greatly reduces the chance of pregnancy. The more accurate her observations, the less likely is pregnancy. Another behavioral method of birth control is the withdrawal method. In this method, the penis is removed from the vagina prior to ejaculation. This method is very risky because some fluids are released prior to the ejaculation. These fluids may contain sperm, which could fertilize any available egg. Of all methods, withdrawal is the least reliable, resulting in pregnancy far more often than other methods.

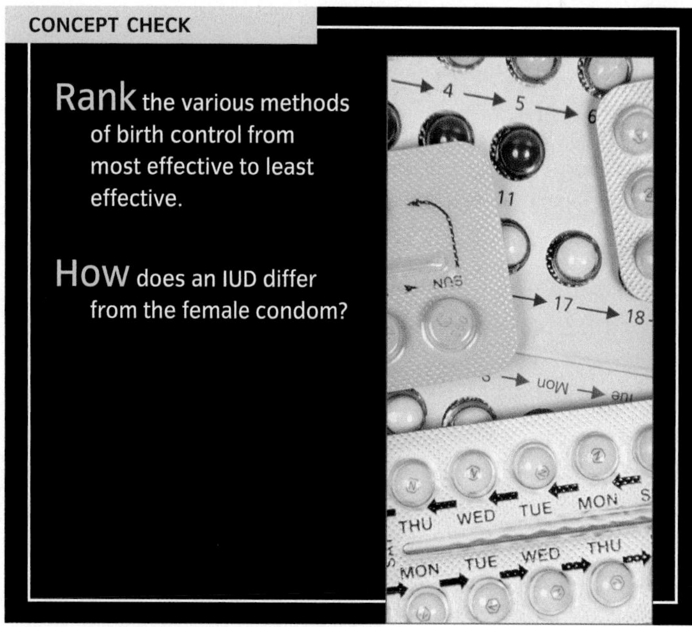

CONCEPT CHECK

Rank the various methods of birth control from most effective to least effective.

How does an IUD differ from the female condom?

CHAPTER SUMMARY

1 Survival of the Species Depends on Reproduction and Gamete Formation

Reproduction is among the most basic human urges, since it is essential to the survival of the species. The reproductive system produces gametes, provides a suitable place for the union of egg and sperm, nourishes the developing fetus, and produces the sexual characteristics associated with being male or female. Gametes are produced via meiosis, resulting in haploid cells. Males produce four sperm from each primary spermatocyte, while females produce one egg and three polar bodies.

2 Structures of the Male Reproductive System Produce and Store Sperm

The male system begins with the testes, the organs that produce sperm. The sperm travel down the epididymis and through the inguinal canal in the ductus deferens. The seminal vesicles add fluid, and then the sperm and developing semen travel through the prostate at the base of the urinary bladder. The semen leaves the male via the penile urethra. The Cowper's glands lubricate the tip of the penis. Hormones control the activity of the male reproductive system. FSH and LH are released from the anterior pituitary. FSH stimulates sperm production, while LH stimulates the interstitial cells, which produce testosterone, the hormone that creates male secondary sex characteristics. The male orgasm, directed by the sympathetic nervous system, causes the release of sperm from the male body.

3 The Female Reproductive System Is Responsible for Housing and Nourishing the Developing Baby

The female reproductive system is composed of the ovaries, the uterine tubes, the uterus and the vagina, and accessory organs, including the mammary glands. The ovaries produce eggs, estrogen, and progesterone. Estrogen creates the secondary sexual characteristics. The uterus houses the developing fetus, and the endometrial lining is shed once a month during the menstrual flow. Like the male reproductive system, the female reproductive system is controlled by hormones. The anterior pituitary secretes FSH, which stimulates the development of eggs. Developing eggs release estrogen, causing the lining of the uterus to build up. When estrogen levels get high, FSH is inhibited and LH is secreted by the anterior pituitary. LH causes ovulation, and the cells that surrounded the developing egg begin secreting progesterone, which causes the uterine lining to begin functioning, and secreting nutritive fluids. If there is no fertilization, the ovary stops producing progesterone, the blood levels of all female hormones decline, and the uterine membrane is shed.

4 The Orgasm Is a Moment of Emotional and Physiological Epiphany

Human sexual response has four phases: arousal, plateau, orgasm, and resolution. While the specifics are different in men and women, many similar physiological changes occur in both genders. Women, but not men, are able to have multiple orgasms.

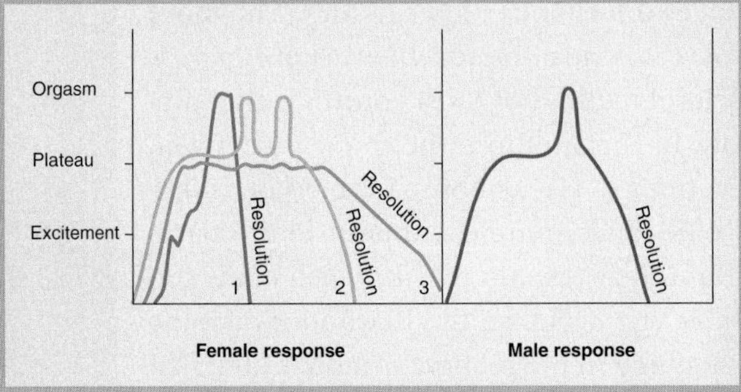

5 Sexually Transmitted Diseases Can Be a Side Effect of Sexual Contact

Human sexuality involves close physical contact, and that becomes an effective route for infection by pathogens, including bacteria, viruses, and parasites. To protect yourself, know your partner, avoid unprotected sex, and think carefully about your sexual practices. Sex is intimate, both physically and emotionally.

6 There Are Many Birth Control Choices, None of Them Perfect

Birth control is the prevention of conception or implantation. The types of birth control include abstinence, surgical procedures, hormonal controls, barrier methods, chemical methods such as spermicidal creams and jellies, and natural family planning.

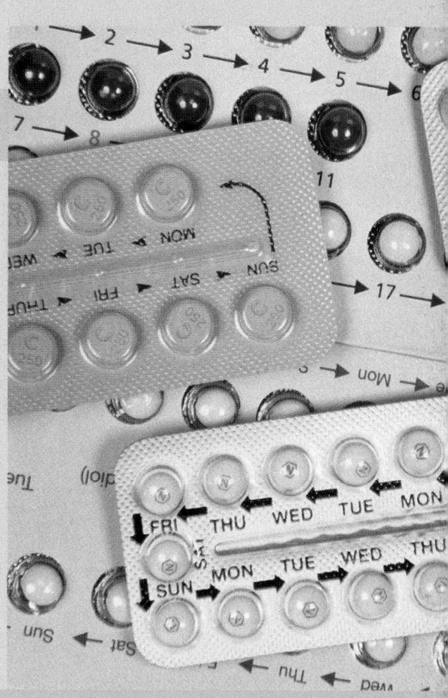

KEY TERMS

- **alleles** p. 512
- **anabolism** p. 523
- **atresia** p. 527
- **capacitated** p. 519
- **cervix** p. 530
- **cGMP** p. 523
- **diploid** p. 513
- **elective abortion** p. 544
- **genotype** p. 538
- **haploid** p. 512
- **homologous** p. 513
- **implantation** p. 529
- **lactiferous** p. 532
- **laparoscopy** p. 542
- **ligating** p. 541
- **oocyte** p. 520
- **phenotype** p. 512
- **prolapse** p. 531
- **quiescent** p. 520
- **spermatic cord** p. 520
- **spermatogenesis** p. 517
- **stem cell** p. 519
- **stratum functionalis** p. 535
- **urogenital** p. 522

CRITICAL THINKING QUESTIONS

1. FSH is secreted by the anterior pituitary in both males and females. What is the function of this hormone in males? How does that compare to its function in females? What are the similarities in the functioning of FSH in the two genders?

2. The male and female reproductive systems have many analogous structures. List the function of the male organs given below, then identify a female organ with similar function. Explain where the female organ is found, and describe the similarities between the two organs.
   a. Testes
   b. Ductus deferens
   c. Penis

3. Birth control pills maintain a high blood level of estrogen and progesterone. Study FIGURE 16.19 and explain how the pill prevents pregnancy. What is happening in the ovary when the blood level of estrogen is high? How is the uterus responding? How does this prevent pregnancy?

4. Look back at the anatomy of the female reproductive system in FIGURE 16.15. Note specifically the junction of the uterine tubes and the ovaries. Toxic shock syndrome (TSS) is caused by excessive growth of bacteria in the uterus, such as when tampons block the vaginal flow for an excessive time. Why does this cause serious concern? Where might the toxin made by the bacteria wind up if the infection is not treated properly? Can males get a form of TSS? Why or why not?

5. List five types of birth control. Explain how each method prevents pregnancy, and discuss its effectiveness. What is the most reliable method of birth control? What is the least reliable method? Which of these methods also prevent the spread of sexually transmitted diseases?

SELF TEST

1. What type of cell is produced from meiosis?
   a. Diploid body cell
   b. Haploid body cell
   c. Diploid gamete
   d. Haploid gamete

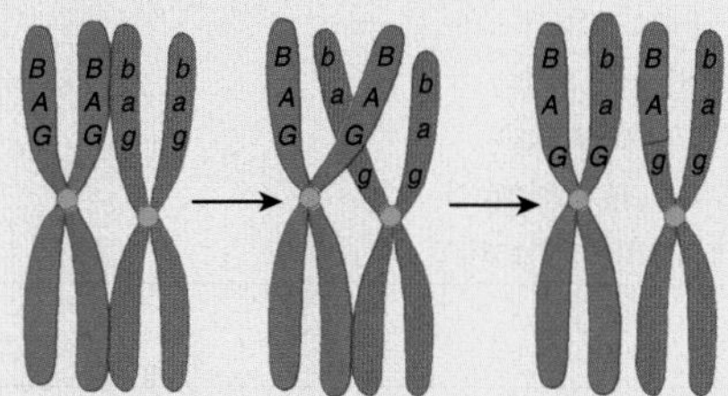

2. What is the significance of the crossing over shown above?
   a. It increases genetic variation.
   b. It reduces birth defects.
   c. It ensures reproductive success.
   d. It is necessary to the orderly separation of chromosomes.

3. True or False? In both males and females, four gametes are produced from each primary cell.

4. The function of the structure labeled A in the figure below is
   a. sperm production.
   b. sperm maturation.
   c. temperature regulation of sperm.
   d. sperm transport.

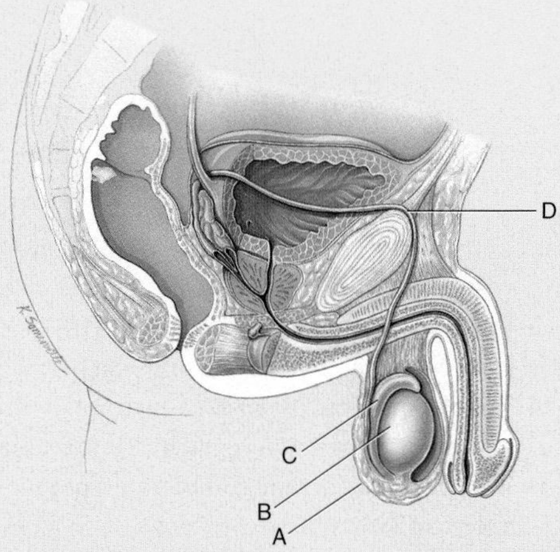

5. In the above figure, the epididymis is labeled
   a. A.
   b. B.
   c. C.
   d. D.

6. The function of the structure shown in the figure below is
   a. spermatid production.
   b. testosterone production.
   c. inhibin production.
   d. Both a and b are correct.

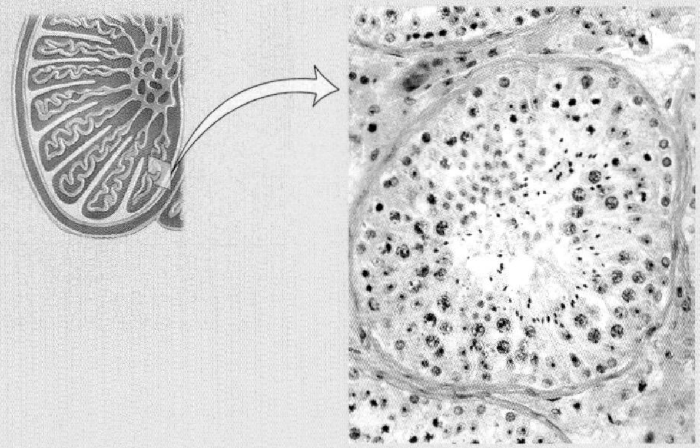

7. The function of the Sertoli cells is to
   a. produce testosterone.
   b. protect developing spermatids.
   c. promote development of secondary male sex characteristics.
   d. undergo meiosis to produce sperm.

8. The part of a mature sperm that includes many mitchondria, needed to produce energy for sperm propulsion through the female system, is the
   a. acrosome.
   b. head.
   c. midpiece.
   d. flagellum.

9. The correct sequence of glands that add fluid to semen during an ejaculation is
   a. bulbourethral glands → seminal vesicles → prostate gland.
   b. prostate gland → bulbourethral gland → seminal vesicles.
   c. seminal vesicles → bulbourethral gland → prostate gland.
   d. seminal vesicles → prostate gland → bulbourethral gland.

10. The gland in the male reproductive system that contributes most of the fluid of the semen, and buffers the potentially lethal acidic environment of the vagina is the
    a. seminal vesicles.
    b. prostate gland.
    c. bulbourethral glands.
    d. corpora spongiosum.

11. The function of FSH in the male is to

a. stimulate production of testosterone.
b. stimulate production of sperm.
c. inhibit release of testosterone from the testes.
d. FSH has no function in the male, only in the female.

12. The organ responsible for producing estrogen is labeled

a. A.
b. B.
c. C.
d. D.

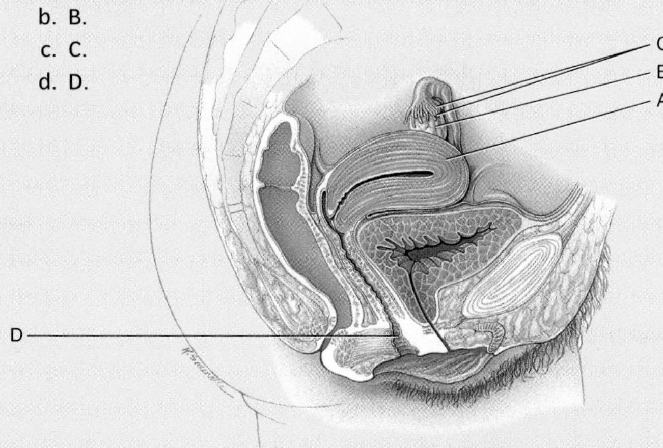

13. The function of the organ labeled C is to

a. produce estrogen.
b. sweep the ovulated egg toward the uterus.
c. nourish a developing embryo.
d. provide a passageway for delivery of sperm to the egg.

14. In the female, LH is directly responsible for

a. ovulation.
b. maturation of follicles.
c. build-up of the uterine lining.
d. menstruation.

15. The structure indicated by the letter A on the image below produces

a. mature eggs.
b. immature eggs.
c. estrogen only.
d. progesterone and estrogens.

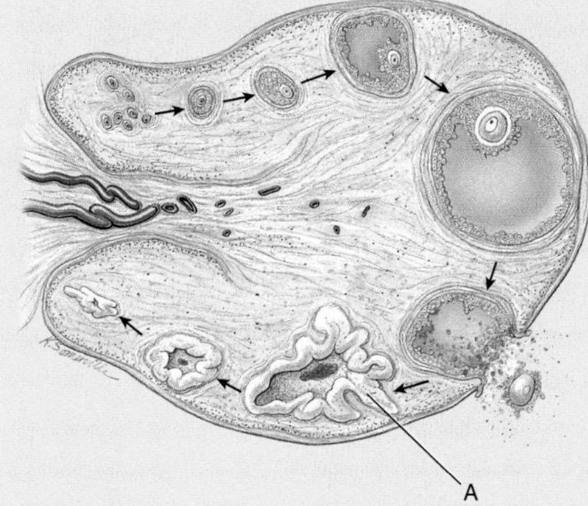

16. The layer of the uterus that repeatedly thickens and sheds under hormonal control is the

a. endometrium.
b. perimetrium.
c. myometrium.

17. The mammary glands release milk (the "let down" reflex) in response to the hormone

a. prolactin.
b. oxytocin.
c. estrogen.
d. progesterone.

18. The hormone responsible for proliferation of the uterine lining comes from

a. the hypothalamus.
b. the anterior pituitary gland.
c. secondary and mature follicles.
d. the corpus luteum.

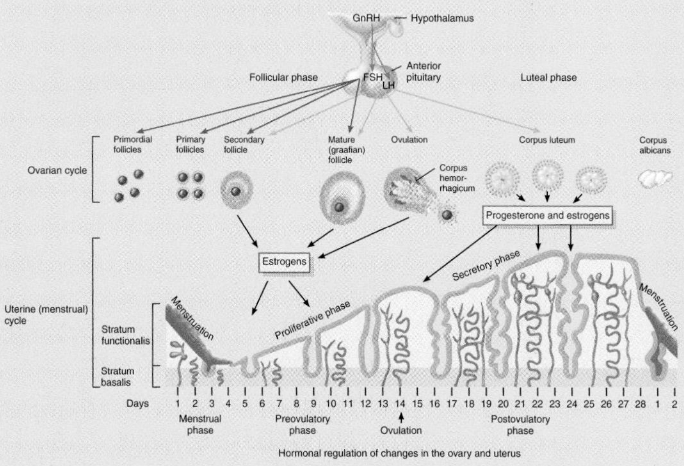

19. The birth control method that is also effective against STDs is

a. spermicidal creams and jellies.
b. the diaphragm.
c. the condom (either male or female).
d. a vasectomy or tubal ligation.

20. The most effective method of birth control, other than abstinence, is

a. hormonal methods such as the pill or Depo-provera injections.
b. surgical methods including vasectomy and tubal ligation.
c. natural family planning using temperature charts and observation of cervical mucus.
d. barrier methods combined with spermicidal creams and jellies.

Pregnancy: Development from Embryo to Newborn

17

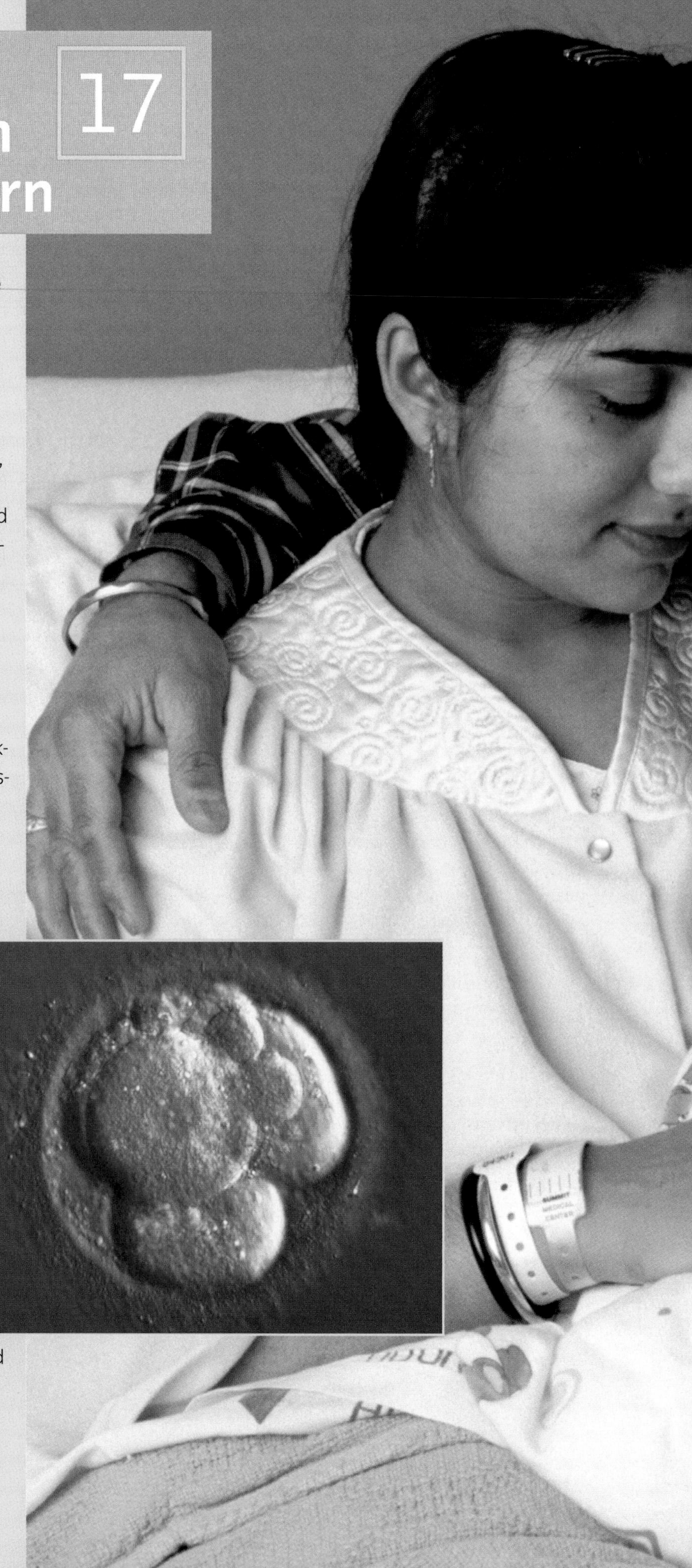

Since 1978, human reproduction has never been the same. Until then, every baby could be traced back to the introduction of male sperm into a woman's reproductive tract, usually through sex. And then came Louise Brown, a girl born in England in 1978 through in vitro ("in glass") fertilization, or IVF.

Louise's parents had tried to conceive for nine years, but her mother's uterine tubes were blocked. Her eggs could not meet her husband's sperm. Tubal ligation could not have made her more sterile. But Mrs. Brown was producing eggs, and her husband's sperm looked healthy. Her doctor proposed to withdraw an egg with laparoscopic surgery, expose it to sperm in a lab dish, and return the embryo into her uterus. This embryo implanted itself in the mother's uterus and developed into Louise Brown, the world's first "test tube baby."

IVF, like so many medical advances, rested on a background of animal research: in 1891, rabbit embryos transferred to another animal developed normally. In the 1960s, human eggs were first fertilized in the laboratory. Then in 1978, along came Louise Brown, the first of perhaps 1 million IVF babies. As IVF has grown to become a standard option for many infertile couples, steady improvements have emerged from our ever-better understanding of the biology of human reproduction and development. We now understand that during the nine months of pregnancy, an intimate dance of chemical messengers directs the development of a fertilized egg into a human being. We'll return to IVF later on, after we cover the overall choreography of development.

Pregnancy and birth create a bond among females of different cultures. Females carry the responsibilities of pregnancy: giving of their own body in order to provide nutrition and protection for a new life. What changes occur with pregnancy that allow an embryo and fetus to develop within the mother's uterus? How does that new life develop and grow properly? What regulates the timing of each stage of development? What can go wrong?

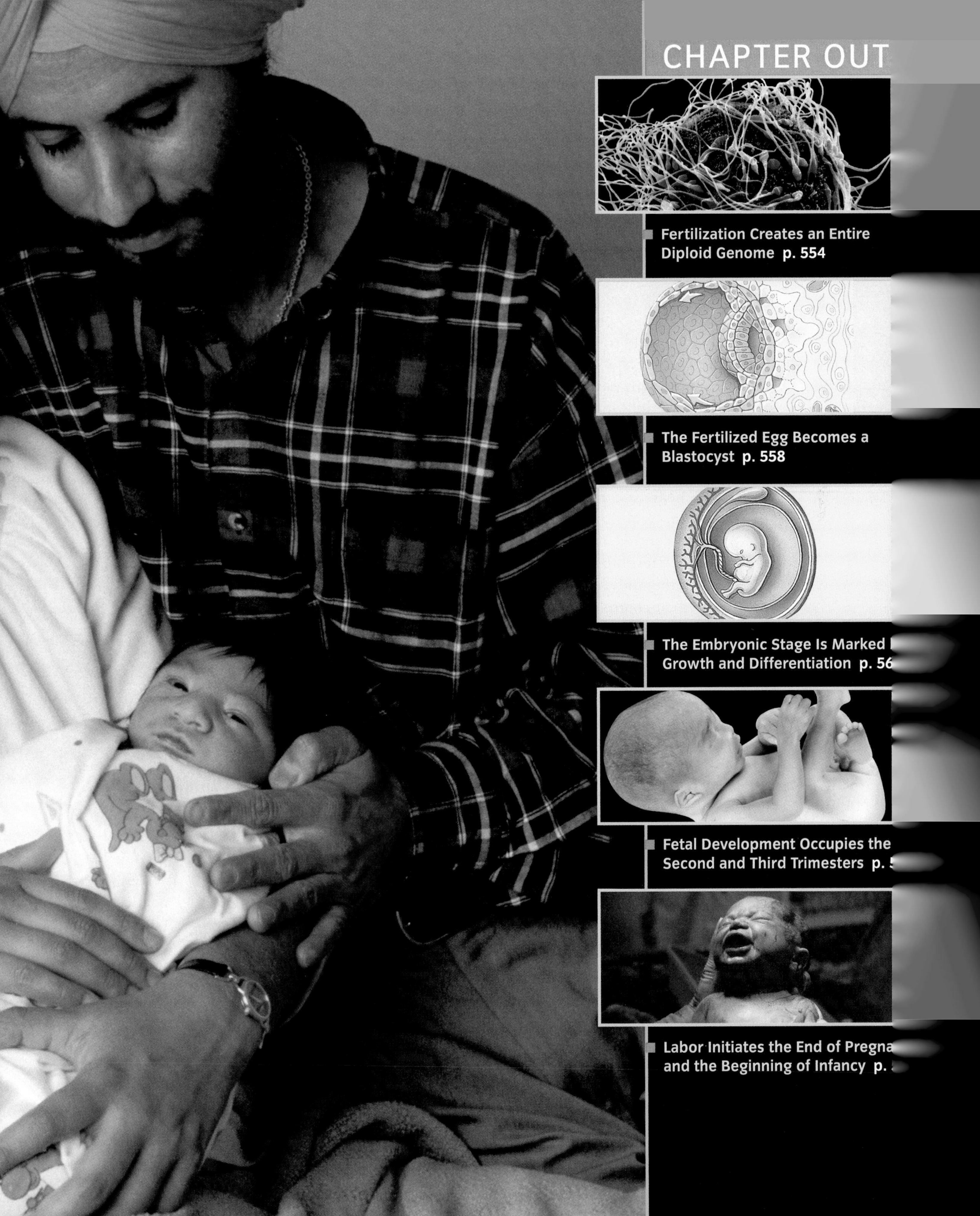
CHAPTER OUT
Fertilization Creates an Entire Diploid Genome p. 554
The Fertilized Egg Becomes a Blastocyst p. 558
The Embryonic Stage Is Marked
Growth and Differentiation p. 56
Fetal Development Occupies the
Second and Third Trimesters p.
Labor Initiates the End of Pregna
and the Beginning of Infancy p.

Fertilization Creates an Entire Diploid Genome

LEARNING OBJECTIVES

Define developmental biology.

Briefly explain the events of fertilization.

Compare a zygote, morula, and blastocyst.

Maintaining a healthy pregnancy and delivering a child able to survive and grow takes an amazing amount of energy. Inside the embryo and later the fetus, development needs precise control, as one cell divides over and over again to form the billions of different cells in the body of the infant. The timing of those divisions, the **differentiation** of the new cells, and the completion of major events, such as the development of a heart, a central nervous system, and limbs, are all tightly controlled.

Differentiation The process by which cells become specialized.

Embryonic Pertaining to the period from fertilization through the eighth week of development.

Fetal Pertaining to the period from week eight through birth.

Senescence Reaching old age, growing old.

Developmental biology is the study of these stages and their controls. Human development has two distinct prenatal periods: the **embryonic** period and the **fetal** period. The postnatal period is divided into further developmental stages, beginning with the **neonate** (the period immediately after birth), **infancy**, and then through the sequence of **childhood, adolescence**, **adulthood**, **senescence**, and finally death. Again each stage is carefully controlled, and the events need precise timing.

Many people consider that a new life begins in the female reproductive tract, as an ovulated ovum is fertilized. However, this new life will not be able to survive on its own for many months, raising considerable controversy. Does the new life start as soon as fertilization occurs, or only after that life can survive outside the mother? This is a question that individuals must answer for themselves.

Eggs are released inside the female when a spike of luteinizing hormone (LH) triggers ovulation of the most advanced ovarian follicle. After ovulation, the ovum drifts in the female abdominal cavity. LH also causes the fimbriae of the uterine tubes to swell with blood and to sway, creating fluid currents in the abdominal cavity. The waving fimbriae sweep both fluid and the ovum into the uterine tubes, where it may come in contact with sperm introduced into the female tract hours or even days earlier.

After ejaculation, semen does not remain liquid in the female tract, but rather it thickens in the acidic environment of the vagina. This causes the sperm to group together, possibly helping to protect those on the inside of the group from the inhospitable chemical environment of the female tract. This thickening dissolves after a few minutes, and the sperm travel en masse up the vagina, into the uterus and up the uterine tubes. It is important to note that sperm do not demonstrate a will—they do not "search for" the egg—they just whip their tails and move against the slight downward current created by the fimbriae. This movement carries sperm from the vagina into the cervical canal. From there, the muscular contractions of the female tract and the continued whipping of the sperm flagellum propel the sperm forward.

The traveling sperm undergo **capacitation**—changes that make them able to fertilize the egg:

- The flagellum moves faster.
- The membrane of the sperm head changes so that it can fuse with an ovum.

- The acrosomal enzymes are primed to digest the protective layers surrounding the egg, allowing the male DNA to enter the egg.

The process of capacitation takes approximately seven hours. Any sperm that reach the ovum before completing capacitation cannot fertilize the egg. However, the sperm get help: secretions in the female reproductive tract facilitate fertilization by degrading the sperm's outer surface, removing proteins and other membrane compounds from the head of the sperm.

The ovulated egg is surrounded by the **corona radiata** and the **zona pellucida**. The corona radiata (literally, circular crown) is composed of cells from the ovarian follicle, still clinging to the ovum. The zona pellucida is a clear-looking layer (*pellucida* means allowing light through) between the corona radiata and the ovum membrane. The zona pellucida has species-specific receptors for the sperm, explaining why only human sperm can fertilize human eggs. When a sperm binds to its receptor in the zona pellucida, the acrosome activates, releasing its load of digestive enzymes. These enzymes eat away the zona pellucida in front of the sperm while the flagellum continues pushing it forward.

One sperm works its way through the zona pellucida and fuses with the oocyte membrane during the process of **syngamy**. It's not entirely clear why only one sperm can enter the oocyte cytoplasm, when many sperm are bound to the zona pellucida receptors and beginning to digest their way through it (FIGURE 17.1). But within seconds after syngamy, the oocyte membrane depolarizes, blocking the entry of all other sperm. This immediate reaction is called a **fast block** to **polyspermy**. A parallel mechanism that takes a bit longer deactivates the sperm receptors in the zona pellucida and causes the zona pellucida to harden. This **slow block** to polyspermy prevents further interaction between the now fertilized egg and other sperm.

Polyspermy
Many sperm entering one ovum.

After syngamy, the oocyte finally completes meiosis, creating a mature ovum and a small polar body of "excess" DNA that will degenerate outside the ovum.

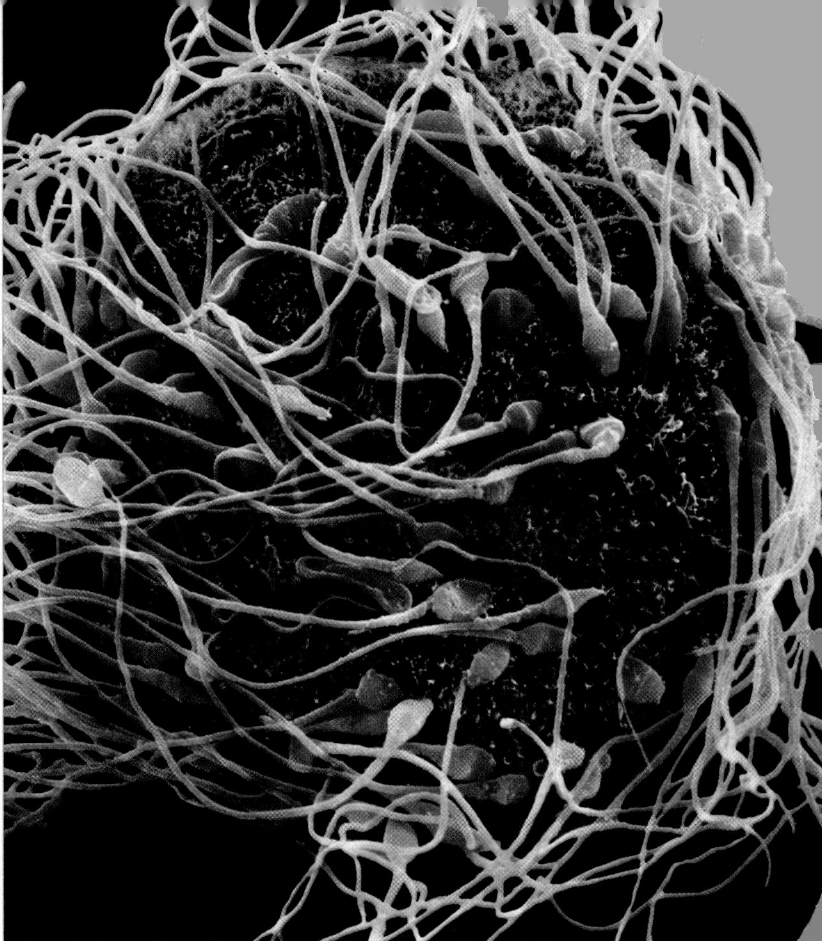

Egg surrounded by sperm FIGURE 17.1

The male DNA develops into a **male pronucleus** upon entering the ovum, and the mature ovum DNA simultaneously forms a **female pronucleus**. These two pronuclei will fuse, creating the diploid chromosome complement of the new life. At this point, fertilization is complete. All of the genetic instructions for the new individual are in place. Once they are activated in the proper order, life can begin.

This entire process usually happens in the upper third of the uterine tubes. The ovum is only **viable** for 24 to 48 hours after ovulation. At the normal traveling speed of the ovum, it barely reaches the halfway point of the uterine tubes in 48 hours. Sperm introduced to the female tract can survive for upwards of five days, so it may already be present in the upper reaches of the uterine tubes. Fertilization occurs where the living ovum contacts the sperm.

Viable
Capable of living.

I WONDER . . .

What causes twins, and how do they contribute to the study of genetics and human development?

Two babies at once can be a mixed blessing. Double the joy, double the diapers, double the college costs! We are all familiar with at least one set of twins. Some twins are difficult to tell apart, while others are no closer in appearance than any other pair of siblings.

Twins are either fraternal or identical. Fraternal twins grow from two separate fertilization events. If two eggs are ovulated in one ovarian cycle, two viable eggs may be present when sperm is introduced. If this happens, both may be fertilized. Two separate eggs and two different sperm form two unique zygotes that are, genetically, merely siblings. The only difference between these siblings and perhaps you and your sister or brother is that these siblings develop in the uterus at the same time. Two separate pregnancies will develop in the same nine-month time frame.

Identical twins, in contrast, arise when one zygote splits into two separate balls of cells as it rolls toward the uterus. Splitting does not harm the future embryos because each cell in the zygote can continue to develop on its own before reaching blastocyst stage. Human eggs undergo what is called radial cleavage, where every cell division from zygote to blastocyst forms two identically sized cells. These cells are totipotent, meaning that they retain the ability to eventually become any cell in the body. Splitting the dividing cells at this point does not harm the developing human; indeed, it is a natural form of cloning. Because these embryos trace their genetic lineage to one fertilization, both balls of cells carry exactly the same DNA, which is reproduced as the embryo implants and grows into a fetus and then a baby. Subtle changes in appearance and personality result from the regulation and timing of gene expression in each baby, but identical twins can be hard to distinguish, even sometimes for their parents.

Hollywood is fond of plots revolving around twins with psychic powers of communication. Although these wonders do not show up in laboratory studies, many people remain fascinated by the idea that people who shared the same womb and continue to share the same genes might have special powers of communication. It is true that both types of twins can have an unusually close bond, but this may be simply because they are always at the same stage of development.

Interestingly, when twins are raised separately, these bonds disappear. Studies of identical twins have been instrumental in resolving a thorny scientific dispute: the relative importance of genetics versus environment and upbringing in forming an adult. For example, in a 2006 study, the age at which Alzheimer's disease developed was much closer in identical twins than in fraternal twins. The researchers took that as a sign of a strong genetic component in susceptibility to this common brain disease, which kills neurons and eventually destroys memory and behavior.

A different study looked at identical twins who were infected with HIV at birth and found that their immune systems had reacted differently to the virus. By showing that people with identical genes had different immune systems, the study suggested that developing an AIDS vaccine could be unexpectedly difficult. So even as Hollywood continues its fascination with identical twins, so does science.

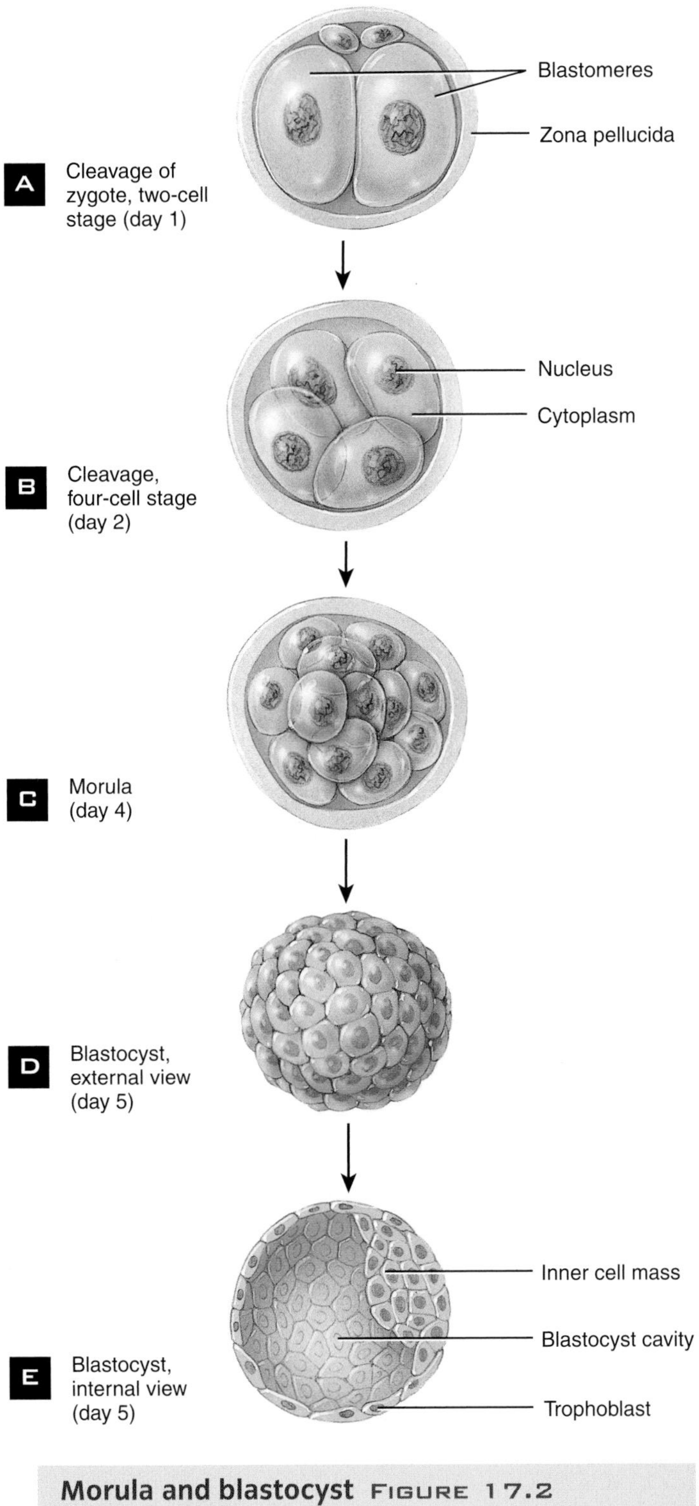

Morula and blastocyst FIGURE 17.2

The fertilized egg, now called a **zygote**, continues to float down the uterine tubes, carried on small fluid currents created by the cilia of the tubes. As it drifts toward the uterus, the zygote undergoes **cleavage**. All animals go through the same basic developmental stages at this point: first forming a **morula** (FIGURE 17.2). The morula is a solid mass of cells, each the same size and each with the capacity to develop into any of the myriad types of cell necessary to form a complete human being. The blastocyst follows the morula stage and is composed of even smaller individual cells. As the cells continue to divide, they push outward, forming a hollow ball. The **blastocyst** is the stage of development where cells begin to specialize, as you will learn in the next section. It is at this point that identical (monozygotic) twins can be formed (for more on twins, see the I Wonder . . . feature).

Cleavage
Repeated cell divisions with little time between rounds to enlarge the resulting daughter cells.

CONCEPT CHECK

List the stages of life in order from birth to death, and give one characteristic of each.

How is capacitation important to fertilization?

What is the role of the zona pellucida in the fertilization process?

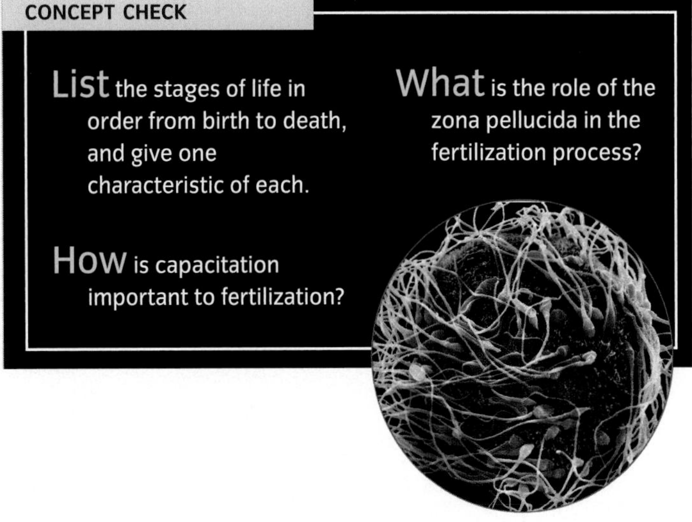

The Fertilized Egg Becomes a Blastocyst

Learning Objectives

Describe the changes that occur in the zygote.

Briefly explain implantation.

During the blastocyst stage of development, some cells of this hollow ball form an inner cell mass, and others remain as the outer surface. The placement of these cells determines what they will eventually become: embryo or supporting structures. Because most of the cells will form supporting structures rather than components of the actual embryo, this stage of development is sometimes referred to as the **pre-embryonic phase**. As the blastocyst forms, the cytoplasmic levels of RNA increase in each of its cells. The RNA level reflects the rate of protein synthesis. Originally, maternal mRNA was abundant in the egg. During fertilization and cleavage, ribosomal RNA increases along with overall translation activity. During the midblastula phase, the translation rate of the maternal mRNA is extremely high, but the sperm's mRNA is largely silent. This means that the proteins being created during blastula formation are mostly maternal in origin. As the embryo passes through the blastula stage, maternal mRNA translation is **downregulated** in favor of embryonic mRNA transcription and translation. The genes inherited from the mother and the father are expressed equally after this point.

Downregulated A slowed-down cellular function.

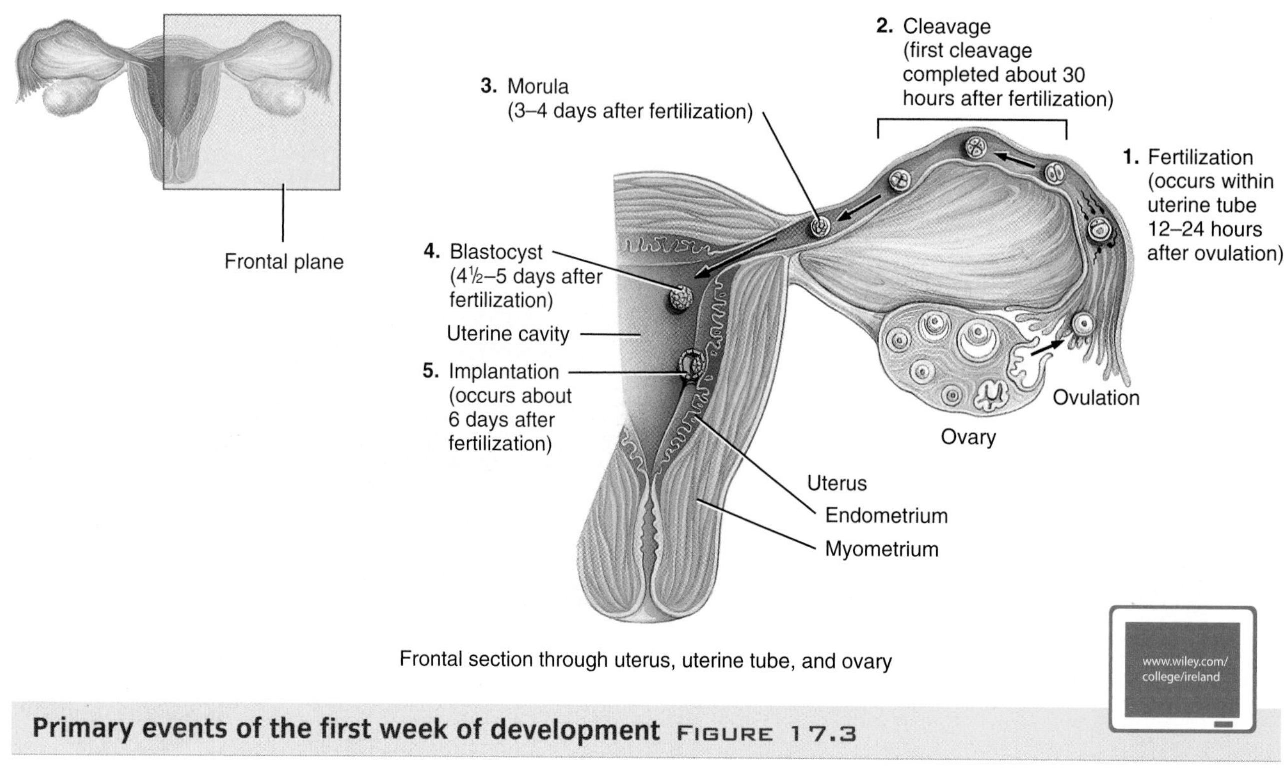

Primary events of the first week of development FIGURE 17.3

With the blastocyst formed, the embryonic cells begin to differentiate. Some will become nutritive layers, forming the **placenta;** others will become protective layers; still others will form the embryo itself.

The most remarkable cells in the blastocyst are a group of **pluripotent** cells at the center. Each of these embryonic stem cells has the potential to become any adult cell type, and these cells are the focus of current stem cell research.

An outer layer of cells, the **trophoblast**, surrounds the inner cell mass. This layer will contact the endometrial lining around day seven and begin to release enzymes to digest its way into the nutritive stratum functionalis of the endometrium. The trophoblast releases the hormone human chorionic gonadotropin (hCG), which is useful in detecting pregnancy, and is described later in this chapter.

The blastocyst travels the length of the uterine tube and reaches the uterus in about four to five days. (For an illustration of the events of the first week of development, see FIGURE 17.3.) Once it drops from the uterine tube, the blastocyst floats freely in the uterine cavity for a day or two. The uterus is now in the midst of the secretory phase, with the thick, spongy endometrium producing nutritive fluids. **Implantation** (FIGURE 17.4) occurs as the pre-embryo settles into the endometrium, and attaches to the uterine wall. A day later, the blastocyst attaches much more firmly to the endometrial wall. The cells of the trophoblast digest the endometrium, burrowing into the spongy tissue and leaving no trace on the surface. When implanted in the endometrial lining, the inner cell mass of the pre-embryo faces the **myometrium** of the uterus. Attachment causes endometrial glands near the blastocyst to enlarge and increase secretions. New blood vessels form to deliver more blood to the implanted blastocyst.

As the trophoblast enzymes digest the endometrium, the blastocyst trophoblast develops into the **chorion**. This tissue, one of the fetal membranes, forms the exchange membrane between fetal and maternal blood. As the embryo develops, this layer surrounds the new life. Eventually, the chorion becomes the main embryonic contribution to the placenta.

Two problems can arise related to the placement of the blastocyst and uterus. The blastocyst settles

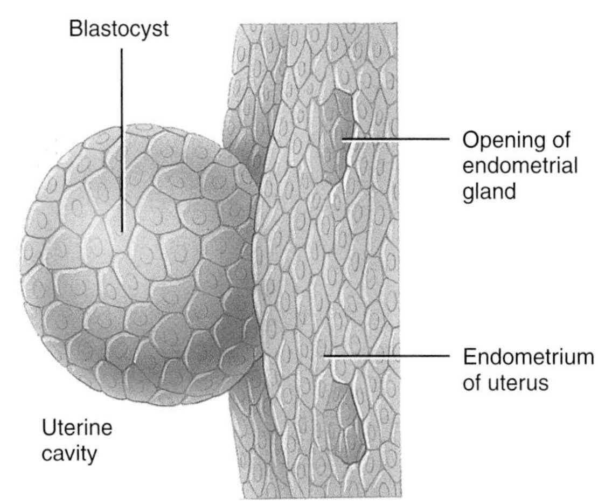

A External view of blastocyst, about 6 days after fertilization

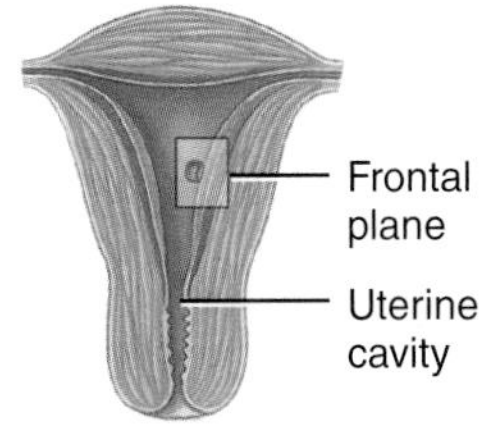

Frontal section through uterus

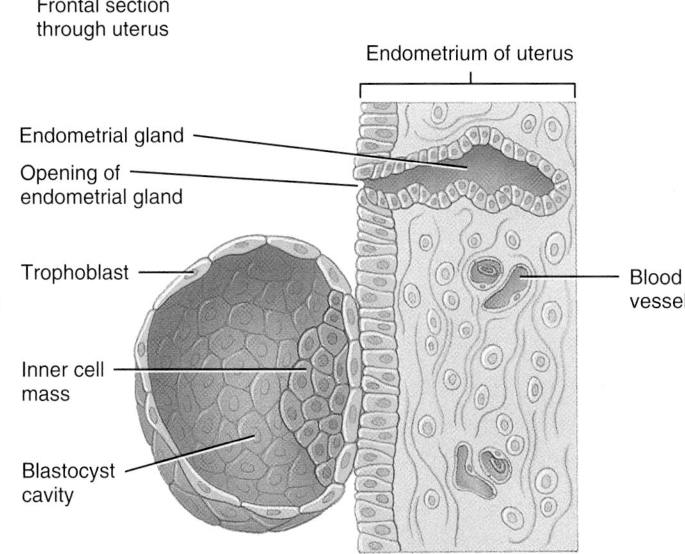

B Frontal section through endometrium of uterus and blastocyst, about 6 days after fertilization

Implantation FIGURE 17.4

Principal events of the second week of development

FIGURE 17.5

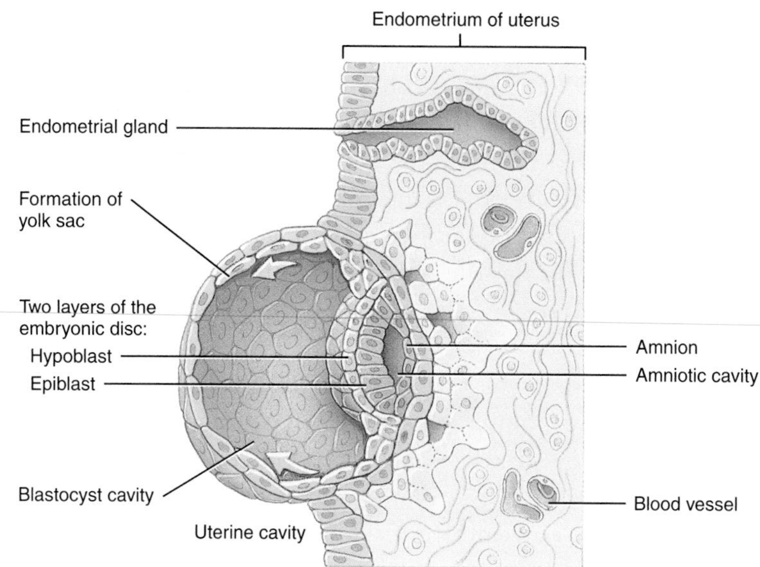

A Frontal section through endometrium of uterus showing blastocyst, about 8 days after fertilization

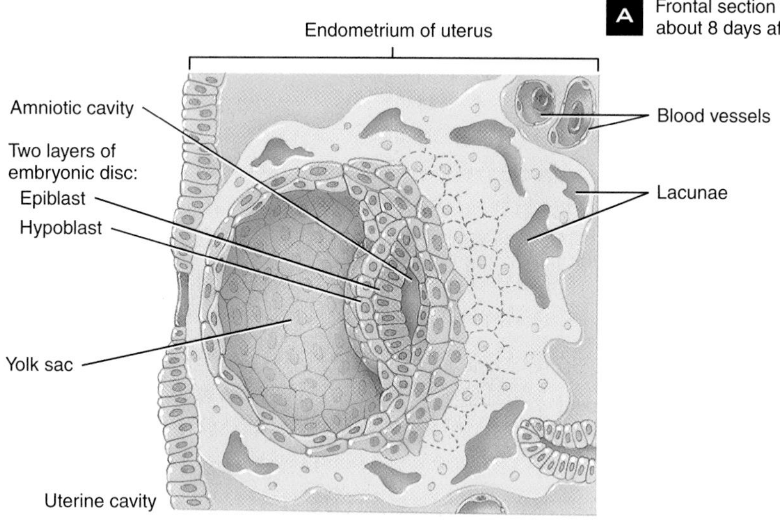

B Frontal section through endometrium of uterus showing blastocyst, about 9 days after fertilization

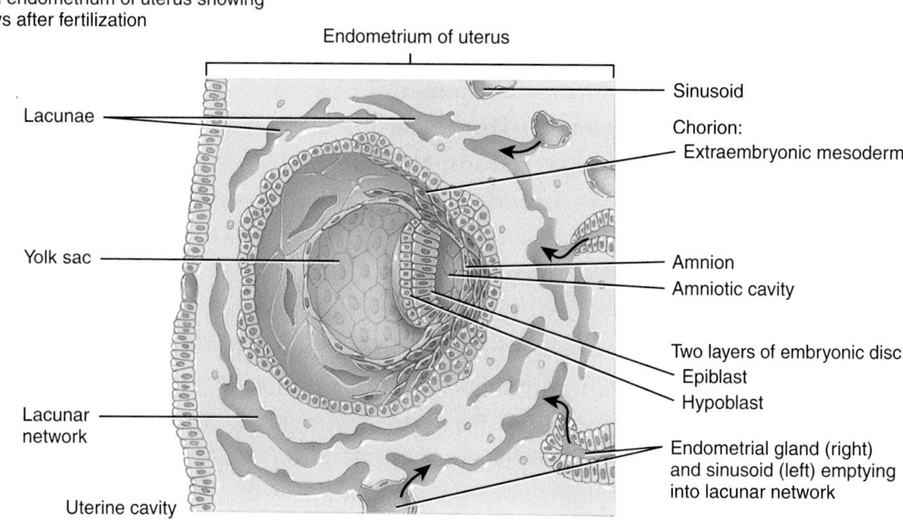

C Frontal section through endometrium of uterus showing blastocyst, about 12 days after fertilization

more or less at random in the uterus, usually in the upper back or the body of the uterus. But if it settles lower in the uterus, a life-threatening condition called **placenta previa** ("placenta first") may develop. The placenta grows near or over the cervical opening of the uterus, blocking the passage of the fetus during birth. This condition can cause maternal hemorrhaging before or during labor. A second, less troublesome condition concerns the orientation of the uterus, which is usually tipped toward the front, lying over the urinary bladder. Approximately one quarter of women have a **retroverted** or **tipped uterus**, with the uterus lying against the rectum. A tipped uterus may cause some pain during intercourse or menstruation but seems to have no effect on fertility or pregnancy. During implantation, the tipped position of the uterus does not seem to have any effect on the placement of the embryo. Regardless of its original position, the uterus will expand normally into the pelvic cavity as the pregnancy proceeds.

Amniotic cavity The fluid-filled cavity that bathes the developing embryo and fetus.

During implantation and development, the inner mass is developing. These cells first divide into two layers, called jointly the embryonic disc. Inside this disc, the **amniotic cavity** forms. Another saclike structure develops in the inner cell mass around day 8. The yolk sac is formed in humans, but it is noticeably low in the nutritive yolk that would be found, for example, in a chicken egg, because humans get their nutrients from the placenta rather than a yolk. Yet the yolk sac remains, providing some nutrition to the embryo during the second and third week. (See FIGURE 17.5 for a summary of events during the second week of development.) By the fourth week of development, cells from the yolk sac are migrating to the embryo and helping to form the respiratory and digestive systems. After nine weeks, the human yolk sac ceases all biological activities.

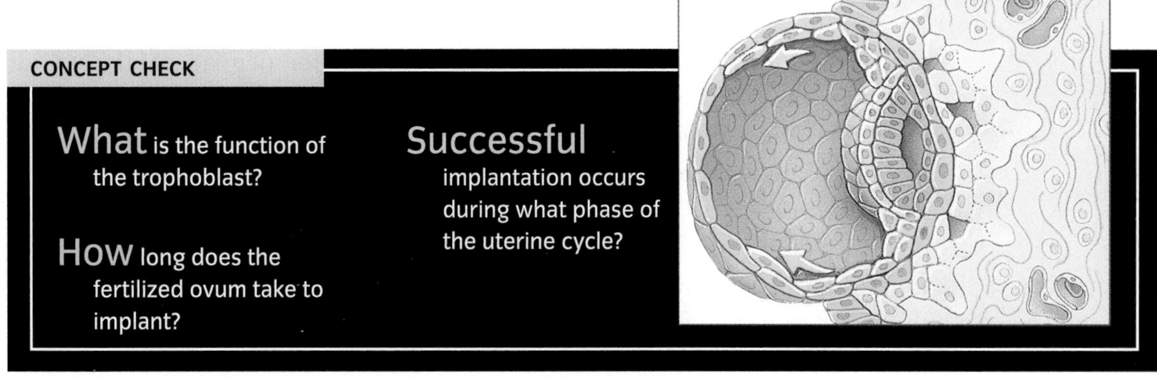

CONCEPT CHECK

What is the function of the trophoblast?

How long does the fertilized ovum take to implant?

Successful implantation occurs during what phase of the uterine cycle?

The Embryonic Stage Is Marked by Growth and Differentiation

LEARNING OBJECTIVES

Outline the function of the three germ layers in the embryo.

Explain the physiology of the extraembryonic membranes.

Describe the formation and functioning of the placenta.

By day 14, the inner cell mass begins to split, forming the **embryonic disc** that we have already mentioned. This disc includes two cell types: the **endoderm** and the **ectoderm**. Once the amniotic cavity and these two cell layers appear, the developing organism has passed into the **embryonic stage**.

All of the changes to date have occurred within the two weeks before the expected menstrual period. The mother-to-be is often unaware that she has conceived at this point, as her cycle is not yet visibly disrupted. She may not be taking precautions in her diet or changing her activity level to benefit the developing embryo. It is estimated that upwards of half of all conceptions do not result in successful implantation and pregnancy due to the myriad hazards of the pre-embryonic stage, which include congenital defects in the zygote, mistakes in genetic control during the intricate processes of the pre-embryo phase, or even subtle environmental disturbances in the uterus and endometrium caused by ingested toxins or maternal lack of essential vitamins, minerals, or macronutrients.

Endoderm
The innermost embryonic cell layer.

Ectoderm
The outer cell layer in the embryo.

Given how much must go right in order to create a viable embryo, it is a miracle that so many successful pregnancies occur. Some couples in fact, do find it impossible to get past the pre-embryo stage. Perhaps ovulation does not occur properly, so eggs are not released, or the uterine tubes are scarred, obstructing the fertilized egg. Or they have difficulty introducing sperm properly. Infertility can also arise from hormonal imbalances, anatomical malformations, defective eggs or sperm, or congenital defects of the reproductive system. Modern medical technologies can assist in conception in cases where the anatomy of the two people is intact, but the physiology is not functioning properly. One fertility procedure is IVF. Technically called in-vitro fertilization, pre-embryo transfer, this process was devised in 1978 by Dr. Robert Edwards (a physiologist at Cambridge University) and Dr. Patrick Steptoe (a gynecologist at Oldham General Hospital) in England.

In IVF, eggs are removed from the maternal ovaries, examined for health, and mixed with sperm from the paternal donor **in vitro**, or in laboratory glassware, as opposed to the usual "in vivo" fertilization process. Often the mother is given a two-week course of fertility drugs to ensure the availability of a large number of harvestable

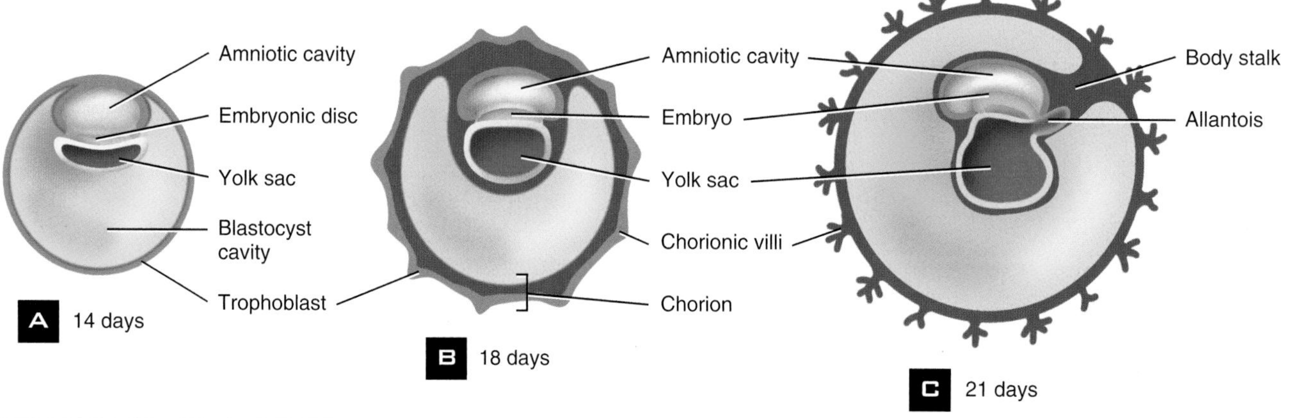

Embryonic development FIGURE 17.6

Morphogenesis Formation of organs and tissues during development.

Therapeutic abortion Removal of the developing embryo for medical reasons, recommended by a healthcare professional.

mature eggs. After the eggs and sperm are mixed, the physician will maintain the resulting pre-embryos in the laboratory in the correct physiological environment. After two days, the usual time needed for the pre-embryo to enter the uterus, the physician places two to four healthy pre-embryos in the mother's uterus, where they settle and implant as usual. If all goes well, the pre-embryos will implant, form a placenta, and at least one will develop normally. Success rates are increasing as technologies advance. The hormones used to stimulate follicle development are more potent, allowing more eggs to be harvested. Maintaining the endometrial lining is now easier with a better hormone mixture, and even the tools for harvesting and growing the embryos are advancing.

This procedure makes many twins and even triplets. Twins occur in the general populace approximately once in every 80 births, and triplets are even more rare, occurring on the order of 1 in 10,000. By comparison, the rate of twinning in successful multiple embryo IVF transplants is 25 percent, and the rate of triplets is 2 to 3 percent.

Embryonic development proceeds from weeks 3 through 8. During this time, the embryo undergoes rapid growth, differentiation, and **morphogenesis**. An almost incomprehensible array of biochemical changes occur within the next six weeks, from weeks 9 through 15. Everything must be timed exactly, or disaster can result. The rate of natural miscarriage during this period is thought to be nearly 20 percent. **Elective abortions**, removal of the developing embryo for personal reasons, are only performed in this early period of development, up to week 15 from the last menstrual flow. (See Chapter 16 for a more complete discussion of elective abortions.) **Therapeutic abortions** are usually performed during this period of development as well, although they can be performed later in the pregnancy.

THE THREE GERM LAYERS MAKE THE FOUR TISSUE TYPES

Mesoderm Middle layer of embryonic cells.

The development of a third cell type—the **mesoderm**—is the first landmark reached by the embryo. The three germ layers—endoderm, ectoderm, and mesoderm—eventually develop into the four tissue types of the body: epithelium, connective tissue, muscular tissue, and nervous tissue (FIGURE 17.6). Ectoderm is on the outer surface of the embryo, in contact with the

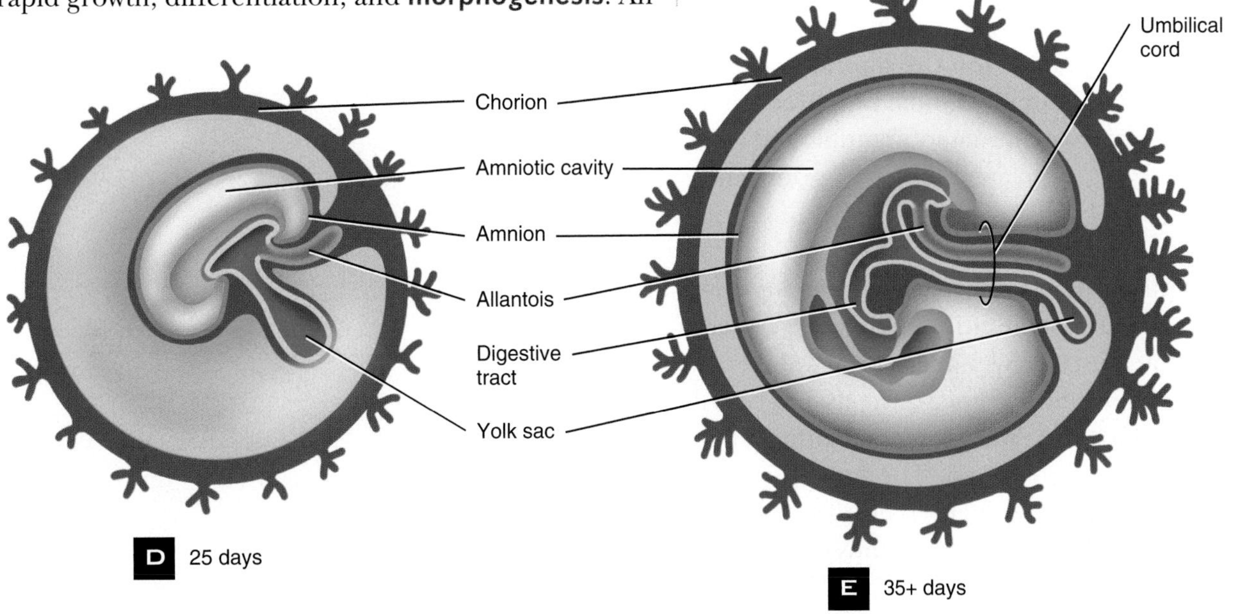

amniotic fluid. From this layer emerges the epidermis, the entire nervous system, portions of the eyes and teeth, the posterior pituitary gland, the adrenal medulla, and the epithelial linings of the digestive tract. Endoderm, the innermost layer, produces the alveoli, liver, most endocrine glands (pancreas, thyroid, parathyroids, anterior pituitary gland, and thymus), tonsils, and portions of the inner ear. The mesoderm develops between the endoderm and the ectoderm. It is responsible for the dermis, all connective tissues including the skeletal system, muscles, blood, kidneys, testes or ovaries, and the reproductive ducts, as well as the lymphatic vessels.

THE EXTRAEMBRYONIC MEMBRANES DEVELOP INTO ESSENTIAL CARRIERS OF NUTRIENTS

During the embryonic stage, as all this activity is taking place inside the embryonic disc, the **extraembryonic** membranes are also developing. The four extraembryonic membranes are the **amnion**, the **allantois**, the **yolk sac**, and the **chorion**. Each has a vital supporting function to the embryo and fetus.

Extraembryonic
Outside of the cells of an embryo.

Interstitial fluid
The clear, ion-rich fluid found between the cells in a tissue.

The amnion, closest to the embryo, lines the amniotic cavity, providing a diffusion area for the amniotic fluid. Amniotic fluid is derived from maternal **interstitial fluid** and is cleansed of embryonic and fetal waste products by diffusion across the amniotic membrane.

The amniotic fluid protects the developing embryo and fetus from outside injury, allows free fetal movements so the muscular system can grow symmetrically, maintains a constant temperature, and permits proper lung development. When the fetal kidneys begin to function, urine is formed. The fetus releases urine into the amniotic fluid where it diffuses across to the maternal bloodstream. The compounds of the fetal urine are then removed from the maternal bloodstream by her kidneys. The amniotic fluid also provides a protective cushion for the embryo and fetus. The fluid is noncompressible, so it transmits blows and shocks that could harm the fetus throughout the volume of the amnion. The amniotic fluid is mostly water, and water has a high latent heat—a large amount of energy is needed to change its temperature. Consequently, the embryo's temperature remains stable. The embryo and fetus can move within the amniotic fluid, stretching and pushing against the uterine walls; this activity helps develop muscle mass in the limbs. Finally, fetal development cannot occur in a dry environment. The amnion keeps the developing cells of all organs, including the lungs, from drying out.

As the embryo and fetus develop, cells are lost to the amniotic fluid. **Amniocentesis**, the collection of amniotic fluid for analysis, is shown in FIGURE 17.7 and described in more detail on page 576.

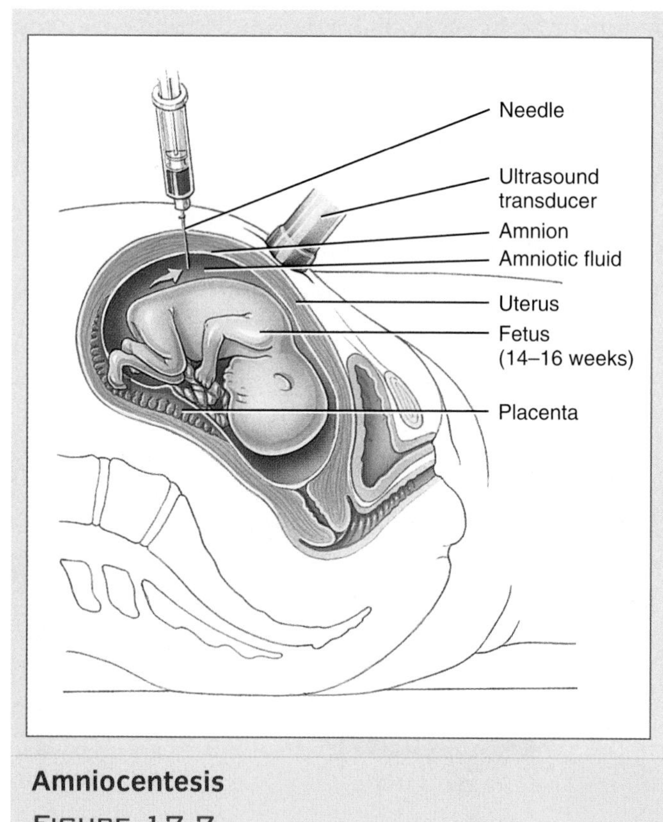

Amniocentesis
FIGURE 17.7

Umbilical cord The flexible cord that connects the fetal circulatory system with the placenta.

The allantois provides the starting material for the blood vessels of the **umbilical cord**. These vessels will transport fetal blood to and from the placental exchange surfaces where oxygen, nutrients, and waste materials are diffused. Once the vessels form, the allantois degenerates.

The yolk sac forms from the endoderm. In other animals, the yolk sac nourishes the developing embryo, but in humans the placenta plays this role. Our yolk sac eventually becomes part of the digestive and respiratory tracts. This membrane also produces fetal blood cells until the bone marrow can take over. The yolk sac may also be involved in gamete production; preliminary research indicates that **germ cells** are first produced here. These primordial germ cells migrate to the gonads where they differentiate into primary oocytes or spermatogonia.

Germ cell A cell destined to become an egg or sperm.

The chorion, the outermost layer of the extraembryonic membranes, develops from the trophoblast and makes up the exchange portions of the placenta. It is also responsible for producing **human chorionic gonadotropin (hCG)**, a hormone that maintains pregnancy until the placenta is fully functional, by preventing degeneration of the corpus luteum. With hCG present, the corpus luteum will continue to produce progesterone and other hormones, which maintain the uterine lining rather than permitting it to slough off, as usually occurs in the uterine cycle. hCG is the hormone detected in early pregnancy tests. These tests usually turn color when a particular subunit of the hormone hCG reacts with the test substances. These tests can boast 99 percent accuracy because if hCG is detected, it must be coming from an implanted embryo whose chorion is producing hormones to maintain the pregnancy. A positive test is an accurate indication of an implantation and a developing chorion. If no hCG is detected, the levels may be too low for the test to recognize. Therefore, a negative test does not guarantee that no developing chorion is present.

THE PLACENTA IS ESSENTIAL BUT DISPOSABLE

The placenta is unique: this organ is necessary for fetal development but is disposable. The placenta develops as the embryonic **chorionic villi** (FIGURE 17.8, p. 566) extend into the endometrium. The chorion develops fingerlike extensions that protrude into the thickened endometrial lining. Together these two tissues form diffusion surfaces, with only one layer of cells separating fetal blood from maternal blood. Just like their parent material (the trophoblast), the chorionic cells contain digestive enzymes that eat into the endometrium, damaging it and causing the maternal blood to pool. Chorionic villi, loaded with fetal capillaries, extend into these pools, allowing diffusion across their thin membranes.

Wastes circulating in the fetus leave via the fetal blood moving through the placenta, where the wastes diffuse to the mother's capillaries. This is carried out by simple diffusion because the concentrations of waste materials are lower in maternal blood than in fetal blood. The fetal urinary system begins functioning early in the tenth week. Fetal urine is then the main source for replenishing the amniotic fluid, supporting and cushioning the fetus with large volumes of fluid. The volume of amniotic fluid rises throughout the pregnancy, peaking around week 33 at about 750 ml. This volume is regulated by absorption into the maternal bloodstream and by fetal "respirations" where the fetus swallows small volumes of amniotic fluid continually. Nutrients from the mother enter the fetal circulation by diffusing down their concentration gradient just as fetal wastes leave by diffusing down theirs.

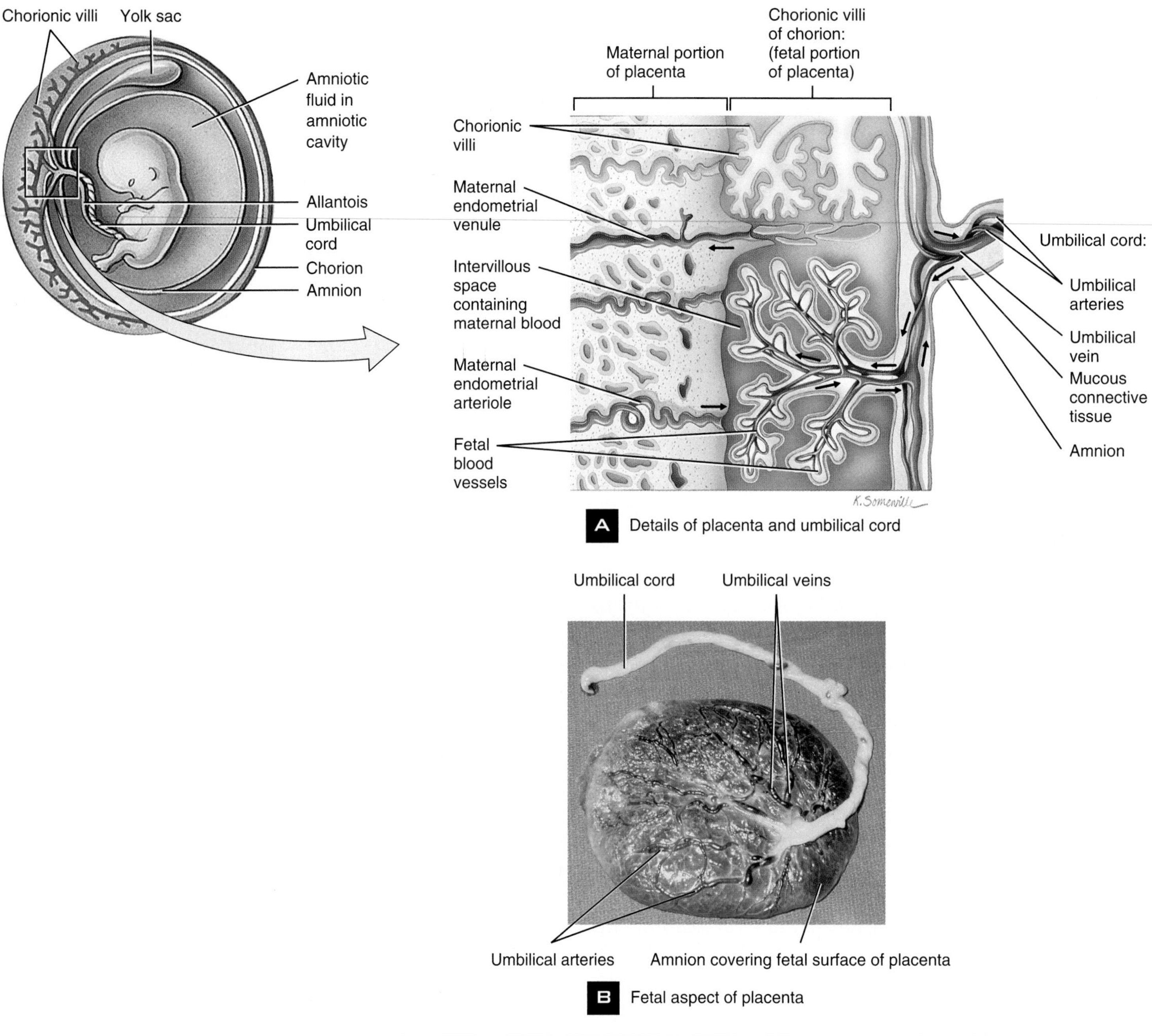

A Details of placenta and umbilical cord

B Fetal aspect of placenta

Chorionic villi FIGURE 17.8

Oxygen carried by the mother's hemoglobin is literally stolen from the maternal hemoglobin by fetal hemoglobin, which has a higher affinity for the gas.

All these activities require a fetal heart to pump blood to and from the placenta. By the fifth week, the embryonic heart is strong enough to take advantage of the two umbilical arteries created by the yolk sac. Fetal blood leaves the heart and moves through the umbilical arteries to the chorionic villi, where its associated gases and waste products diffuse with maternal blood (FIGURE 17.9). This chemically cleansed blood is then collected in placental veins and returned to the fetus through the single umbilical vein. The umbilical vein travels to the fetal liver where it is dropped into the hepatic vein. All that remains of these vessels after birth is the **round ligament**, which marks the path of the umbilical vein from the navel to the liver and is visible in your belly button. The umbilical arteries dissolve soon after birth, adding to the hepatic capillary system.

The placenta works like a large diffusion filter, allowing the exchange of nutrients, gases, and antibodies between mother and fetus, but the placenta is not a perfect filter. HIV, alcohol, cocaine, and other damaging substances can cross the placenta. Even prescription drugs, if introduced to the embryo during critical stages, can cause extreme damage (see the Health, Wellness, and Disease box on p. 572).

In addition to providing nutrition and oxygen, the placenta also produces a range of hormones responsible for maintaining pregnancy. Early in the pregnancy, the placenta secretes hCG, which in turn stimulates the corpus luteum to remain viable. Progesterone and estrogen from the corpus luteum prevent the loss of the endometrium while implantation occurs. Eventually, the placenta will secrete these hormones on its own in far larger quantities. The main effects of these hormones are to increase the size and strength of the uterine muscle, prevent loss of the endometrium, inhibit uterine contractions during pregnancy, and create a thick mucus plug at the cervix, which helps prevent uterine infections.

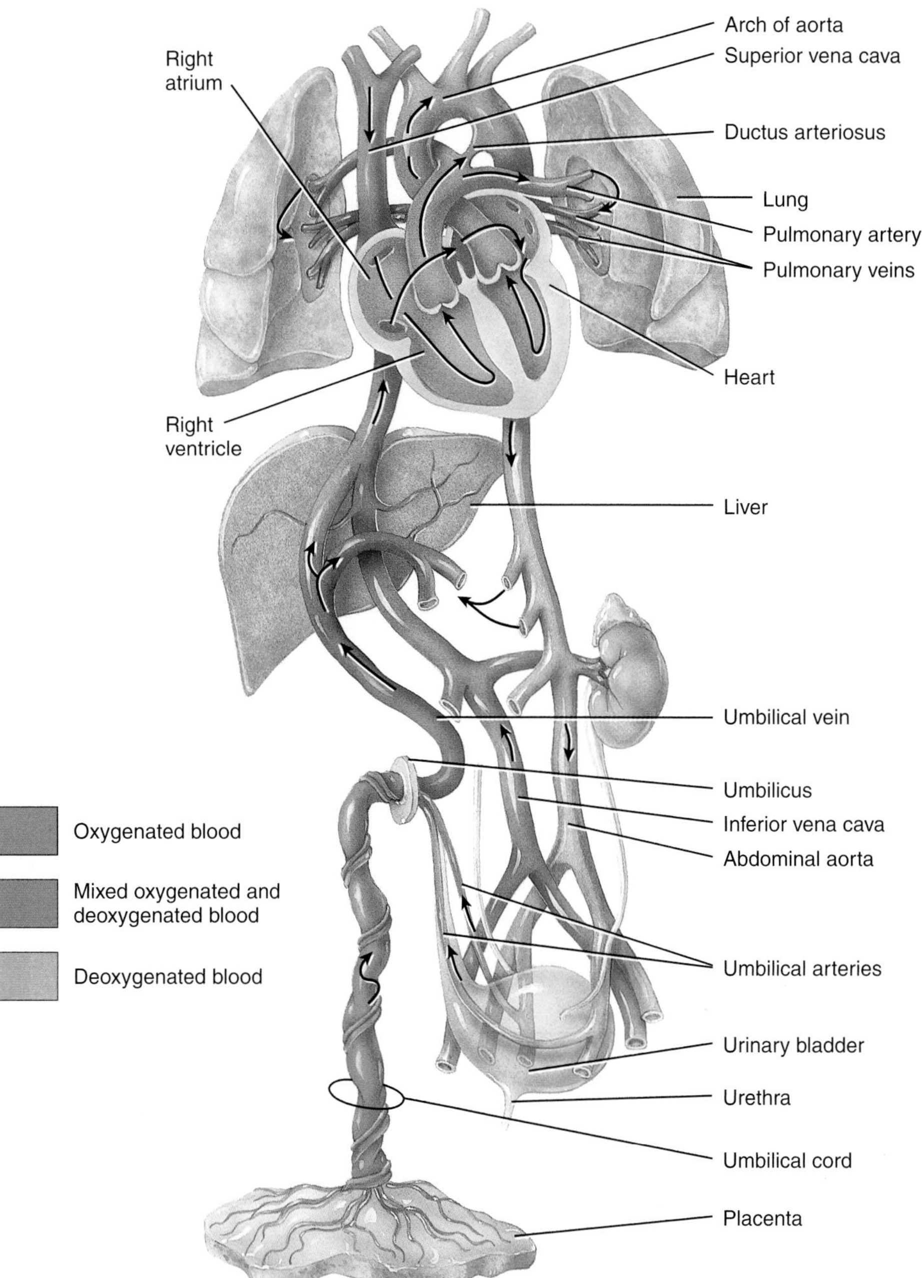

Placental–fetal circulation FIGURE 17.9

This image shows fetal circulation, beginning at the placenta and traveling via the umbilical artery through the fetal liver and on to the fetal heart. Note the hole between the right and left atria and the ductus arteriosus that both permit blood to move from the pulmonary circuit to the fetal body, circumventing the not-yet-functional fetal lungs. Deoxygenated fetal blood and wastes are returned to the placenta via the umbilical veins.

From weeks 5 through 8, the embryo becomes increasingly human in appearance. The tail that appeared in the first month regresses, the head enlarges, limb buds that appeared in the first month are forming structures that look very much like arms and legs, hands and feet, and the gonads are formed. The nose is flat, the eyes are widely spaced and open, but the face is obviously human. By week 8, all the major organs and organ systems are present, though not fully functional. Gender differentiation occurs at approximately 7 weeks (FIGURE 17.10). Prior to the seventh week, male and female development is exactly the same, with two distinct sets of reproductive tubes and no differentiation between male and female. In the seventh week, if a Y chromosome is present, the male tubes will be stimulated, testes will develop, and release of testosterone from the new testes will cause male sexual characteristics to form. If no Y chromosome is present, the organs destined to become the testes degenerate and the organs primed to develop into the ovaries will mature instead.

By the end of the embryonic period, the newly forming individual is approximately 2.5 centimeters (1 inch) long, with a recognizably human form. All the internal and external structures are present at the end of this phase, and the placenta is mature and functioning (FIGURE 17.11).

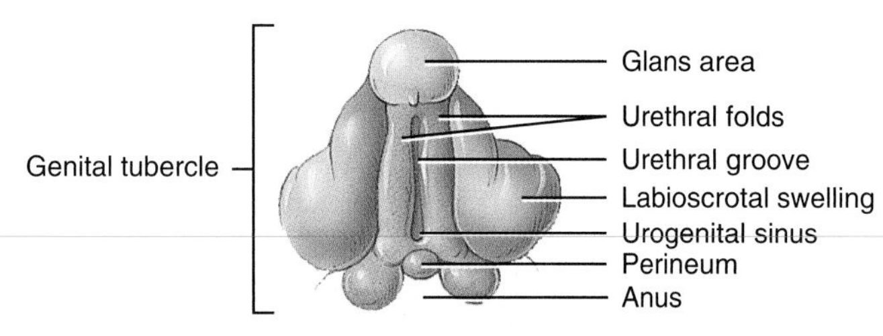

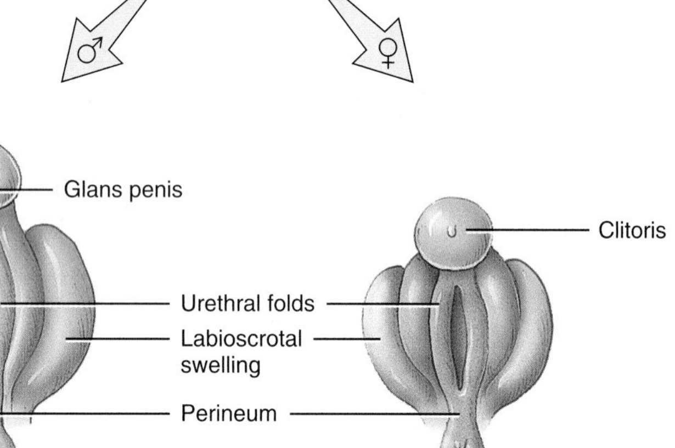

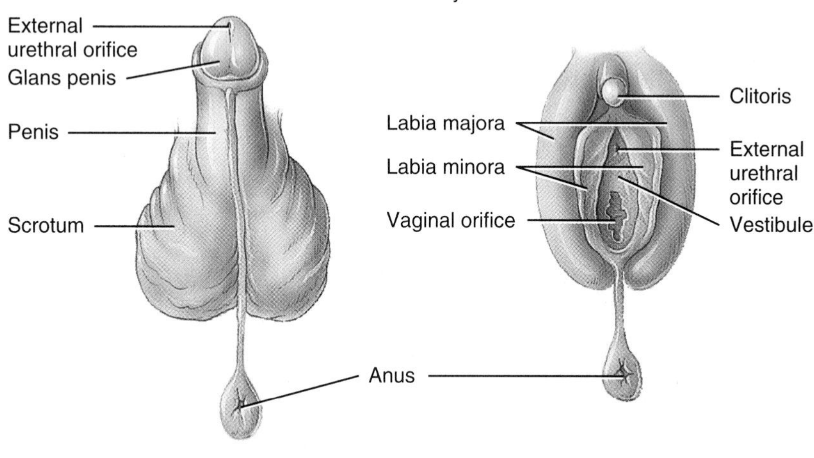

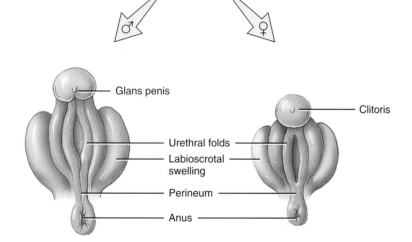

External genital development FIGURE 17.10

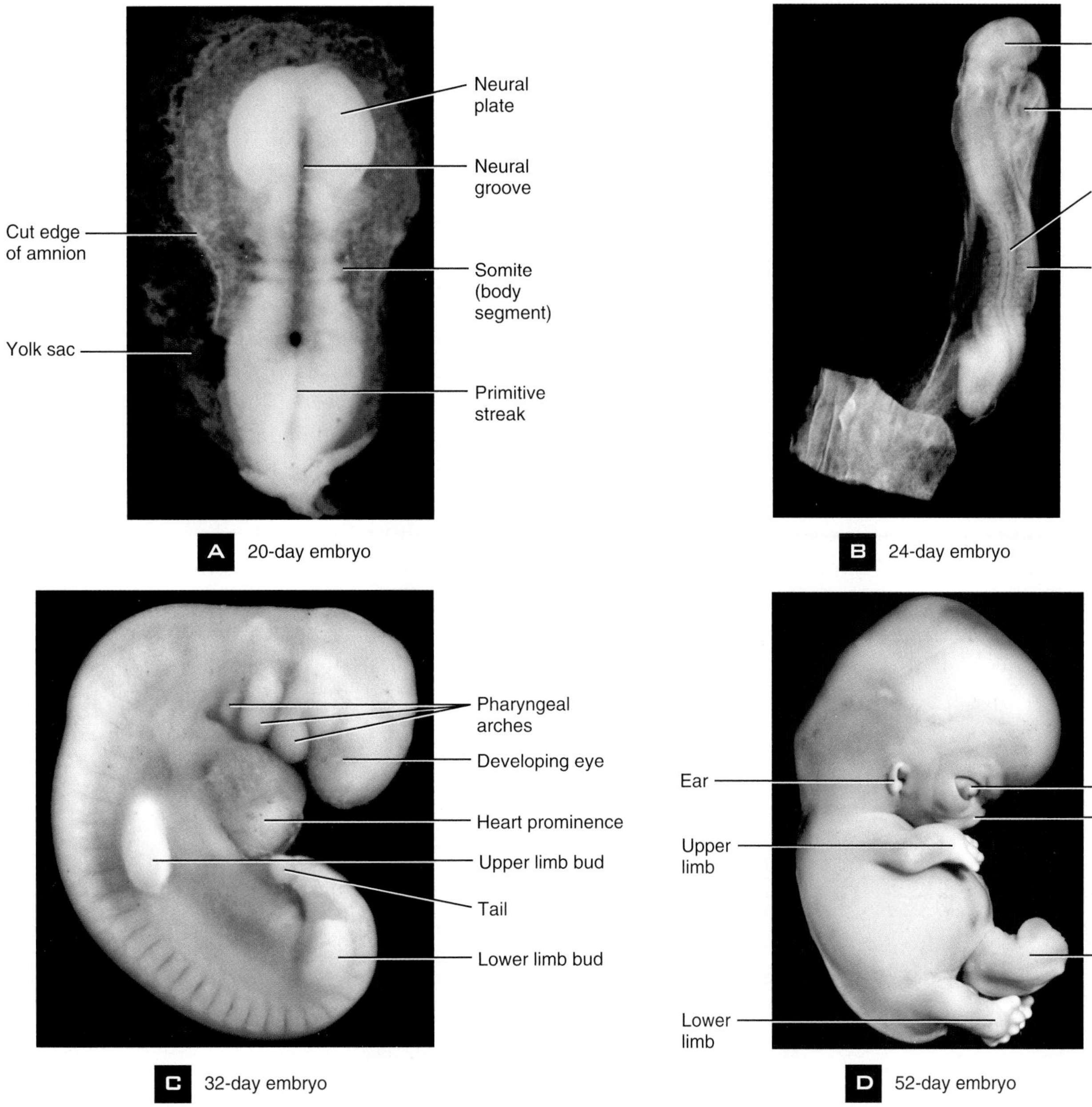

Neural plate
Neural groove
Cut edge of amnion
Somite (body segment)
Yolk sac
Primitive streak

A 20-day embryo

Developing brain
Heart prominence
Developing spinal cord
Somite

B 24-day embryo

Pharyngeal arches
Developing eye
Heart prominence
Upper limb bud
Tail
Lower limb bud

C 32-day embryo

Ear
Upper limb
Lower limb
Eye
Nose
Umbilical cord

D 52-day embryo

www.wiley.com/college/ireland

Summary of developmental events of the embryonic period FIGURE 17.11

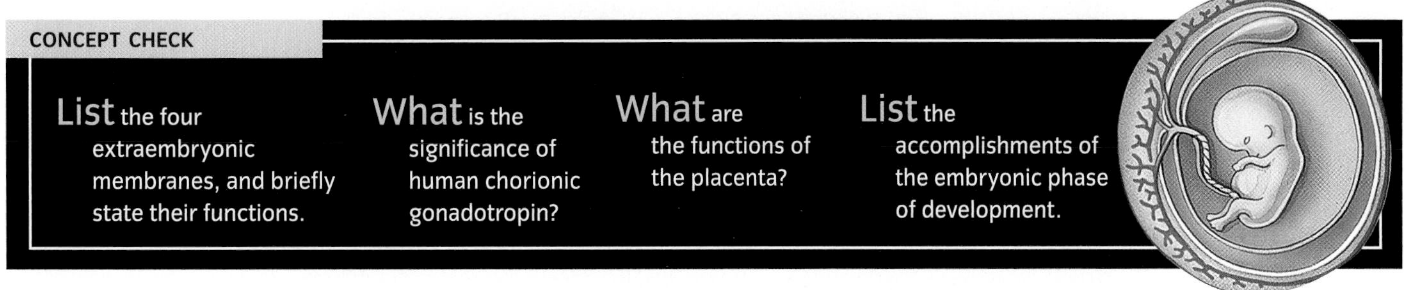

CONCEPT CHECK
List the four extraembryonic membranes, and briefly state their functions.
What is the significance of human chorionic gonadotropin?
What are the functions of the placenta?
List the accomplishments of the embryonic phase of development.

Fetal Development Occupies the Second and Third Trimesters

LEARNING OBJECTIVES

Describe the main events of the second and third trimesters of development.

Differentiate chorionic villus sampling from amniocentesis.

Understand the developmental changes that precede birth.

Fetal development, which begins at week 9 after conception, is a stage of rapid organ growth and maturation. The fetus begins this stage approximately 25 millimeters long, weighing about 1 gram. Within seven months, the fetus will grow to an average of 50 centimeters (20 inches) and weigh 3.75 kilograms (8 pounds). The pregnant woman's body will begin to show signs of the expanding uterus during the fetal stage (FIGURE 17.12). The fetus itself also begins to show signs of growth, becoming much more cramped in the confining uterus (FIGURE 17.13).

Fetal development is usually divided into **trimesters**. The first trimester includes all embryonic development and the first month of fetal development. By the end of the third month, the cartilage skeleton is starting to ossify, the kidneys and liver are functioning, teeth have formed, and external genitalia are clearly male or female.

Trimester One of three, three-month periods during pregnancy.

The **second trimester** includes months 4, 5, and 6. Month 4 sees continued rapid changes, as the face begins to resemble its final form. Blood cells are pro-

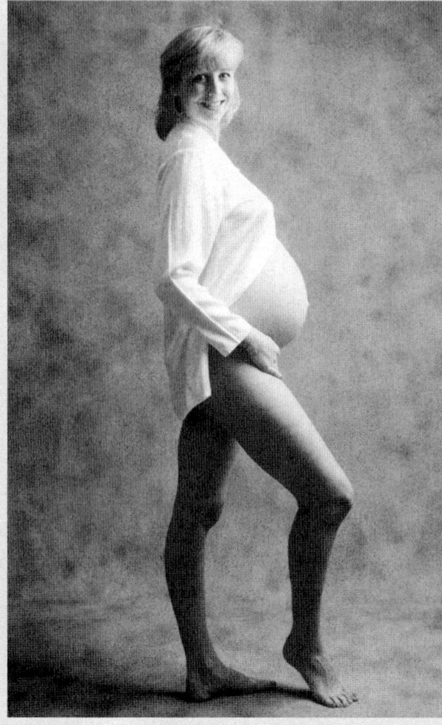

Pregnant woman in various months FIGURE 17.12

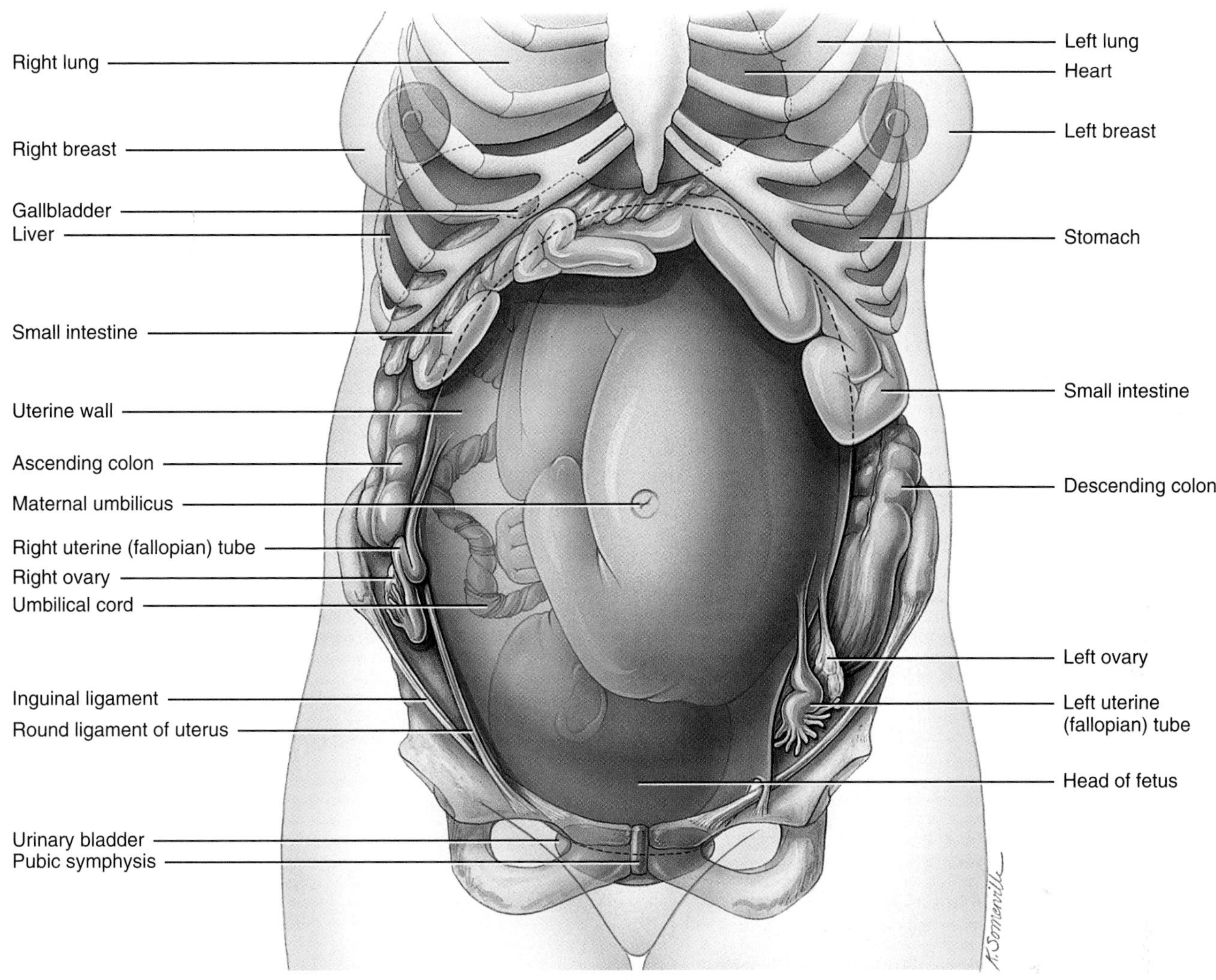

Anterior view of position of organs at end of full-term pregnancy

The fetus in utero FIGURE 17.13

Have you ever heard that you can predict the gender of an unborn fetus by how high or how low the mother is carrying the baby? Not true. Equally false is the attempt to determine the gender by watching the swing of a penny on a string held above the mother's belly. While on average a female fetus's heart does beat faster than a male's heart, it is not true that a fetal girl's heart rate is always faster than 140 while a fetal boy's rate is always below 140. The most reliable noninvasive way to predict the gender of the baby uses an ultrasound device, which can usually provide visualization of the developing male sex organs.

duced by the liver and bone marrow, and ovarian follicles are forming in the female ovaries. The fetus has grown from 25 to 153 millicentimeters in two months and has gained approximately 165 grams. The nervous and muscular systems have developed enough that by the fifth month movements may begin. The mother may feel this **quickening** as fluttering or "butterflies" in her abdomen. At this point, you can hear the fetal heartbeat through a stethoscope placed on the distended abdomen. The fetal skin is covered in soft hair called **lanugo**. By the end of the sixth month, the fetus weighs approximately 450 grams. With excellent and immediate medical care, it could survive outside the womb (we'll discuss the issue of prematurity later on). The lungs begin secreting **surfactant**, allowing the lungs to inflate and deflate without the alveolar walls sticking together.

Surfactant
Detergentlike compound that prevents alveolar membranes from sticking together.

Why are drug regulators so cautious about the effects of drugs on the unborn?

As we've seen, development in the womb is a precisely timed dance of hormones, structures, and genes. Normally, hormonal levels rise and fall just at the right time to activate the appropriate developmental processes.

But chemicals in the mother's blood can "cut the music" on this delicate dance. In the late 1950s, a pill called thalidomide was prescribed in Europe to treat morning sickness. Even one dose stunted limb growth, and babies were born without hands or feet, or with flipper-shaped limbs. The drug also caused severe defects in hearts, kidneys, genitals, digestive tracts, and nervous systems. Later, scientists learned that thalidomide blocks the formation of blood vessels, which is essential to fetal development. The U.S. Food and Drug Administration prevented thousands of deformities in American babies by refusing an early approval for the drug.

The devastation of thalidomide shows the danger of taking drugs during pregnancy. Some drugs cause babies to be born addicted, while others harm fetal development. Mothers who drink a lot of alcohol can give birth to babies with fetal alcohol syndrome, which includes growth problems or neurodevelopmental disorders, including mental retardation. Even moderate drinking can cause drunkenness in the developing baby, because the young liver is slow to decompose alcohol. Children of women who drink during pregnancy may also have trouble paying attention in school, making good decisions, or coordinating muscular activity.

Other substances cause other problems. Smoking during pregnancy increases the risk of premature delivery and doubles the chance of a low-birthweight baby. The federal Centers for Disease Control and Prevention estimate that almost 3 percent of pregnant women use illicit drugs such as marijuana, cocaine, heroin, and amphetamines. Some of these drugs can addict the unborn along with the mother. Cocaine babies may have abnormally small brains and can be jittery and irritable. Women who inject drugs expose themselves and their developing fetuses to HIV and hepatitis C. Heroin may cause miscarriage, poor fetal growth, premature delivery, low birthweight, long-term disability, and a tenfold increase in sudden infant death syndrome (SIDS).

Many psychoactive drugs, including alcohol, cocaine, painkillers, and antianxiety medicines, cause programmed cell death in the brain. For ethical reasons, the hypothesis cannot be studied in humans, but some researchers believe that even a single dose may cause widespread cell death. Here's one thing we know for sure: If you are pregnant, or could become pregnant, it's smart to eat right, get enough rest, and consult a physician before taking any drug.

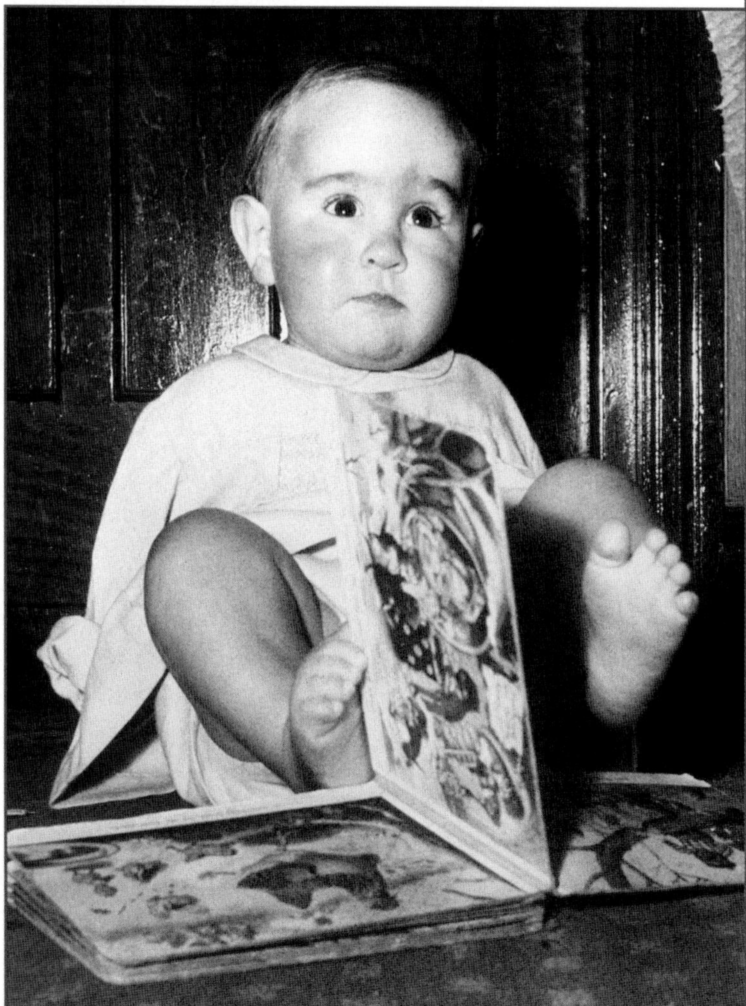

A thalidomide baby turning pages with his feet.

The **third trimester** is characterized by continued rapid growth and maturation. The eyes open and close, the sucking response develops (many fetuses begin to suck their thumbs or other fingers), and loud noises initiate a startle reaction. The fetus often moves regularly, perhaps looking for a comfortable position in the cramped womb. The lanugo is lost, and a layer of protective fat begins to develop. In males, the testes descend into the scrotal sac. (See FIGURE 17.14 for a summary of the events of fetal development.)

By the end of week 38, the fetus is prepared for life outside the uterus. It has maneuvered to a head-downward position, and "**dropped**." This means that the fetus is now resting on the cervix rather than filling the center of the uterus. While the maternal stomach and lungs now have a bit more room, the fetus puts more pressure on the bladder and rectum, stimulating more frequent voiding. Nevertheless, the mother often feels some relief during the last week or two of pregnancy. TABLE 17.1 on pages 574–575 gives a summary of the important milestones of the development process.

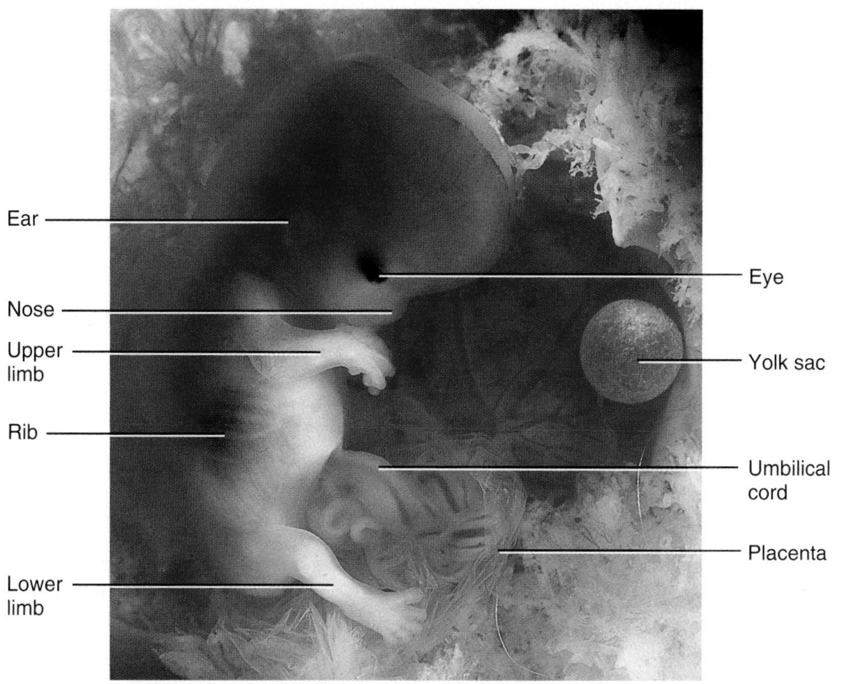

A Ten-week fetus

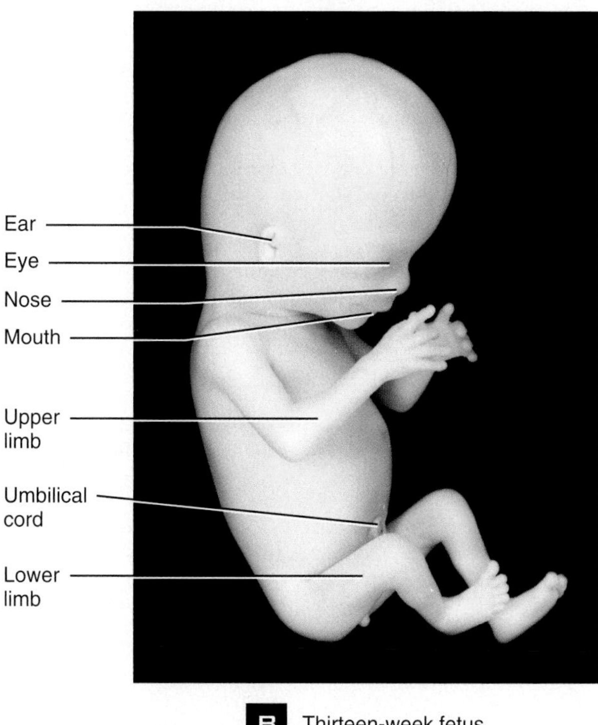

B Thirteen-week fetus

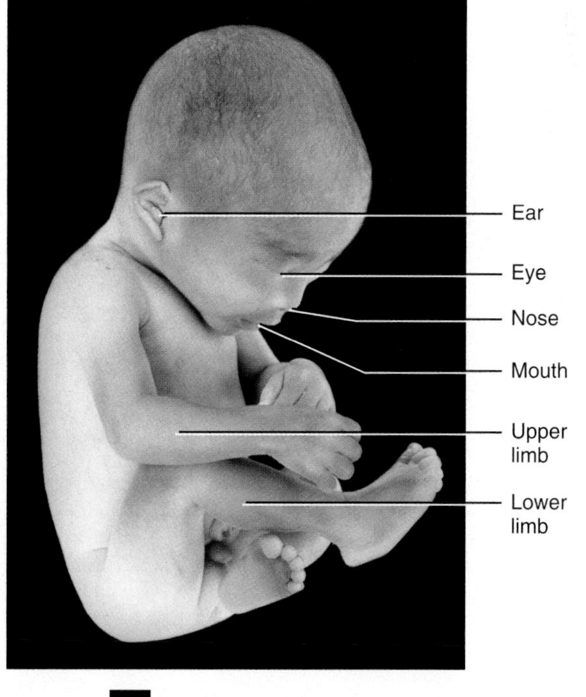

C Twenty-six-week fetus

Summary of fetal development FIGURE 17.14

Developmental summary TABLE 17.1

Time	*Approximate size and weight*	*Representative changes*
Embryonic period		
1–4 weeks	0.6 cm (3/16 in.)	Primary germ layers and notochord develop. Neurulation occurs. Primary brain vesicles, somites, and intraembryonic body cavity develop. Blood vessel formation begins and blood forms in yolk sac, allantois, and chorion. Heart forms and begins to beat. Chorionic villi develop and placental formation begins. The embryo folds. The primitive gut, pharyngeal arches, and limb buds develop. Eyes and ears begin to develop, tail forms, and body systems begin to form.
5–8 weeks	3 cm (1.25 in.) 1 g (1/30 oz)	Primary brain vesicles develop into secondary brain vesicles. Limbs become distinct and digits appear. Heart becomes four-chambered. Eyes are far apart and eyelids are fused. Nose develops and is flat. Face is more humanlike. Ossification begins. Blood cells start to form in liver. External genitals begin to differentiate. Tail disappears. Major blood vessels form. Many internal organs continue to develop.
Fetal period		
9–12 weeks	7.5 cm (3 in.) 30 g (1 oz)	Head constitutes about half the length of the fetal body, and fetal length nearly doubles. Brain continues to enlarge. Face is broad, with eyes fully developed, closed, and widely separated. Nose develops a bridge. External ears develop and are low set. Ossification continues. Upper limbs almost reach final relative length but lower limbs are not quite as well developed. Heartbeat can be detected. Gender is distinguishable from external genitals. Urine secreted by fetus is added to amniotic fluid. Red bone marrow, thymus, and spleen participate in blood cell formation. Fetus begins to move, but its movements cannot yet be felt by the mother. Body systems continue to develop.
13–16 weeks	18 cm (6.5–7 in.) 100 g (4 oz)	Head is relatively smaller than rest of body. Eyes move medially to their final positions, and ears move to their final positions on the sides of the head. Lower limbs lengthen. Fetus appears more humanlike. Rapid development of body systems occurs.
17–20 weeks	25–30 cm (10–12 in.) 200–450 g (0.5–1 lb)	Head is more proportionate to rest of body. Eyebrows and head hair are visible. Growth slows but lower limbs continue to lengthen. Vernix caseosa (fatty secretions of oil glands and dead epithelial cells) and lanugo (delicate fetal hair) cover fetus. Brown fat forms and is the site of heat production. Fetal movements are commonly felt by mother (quickening).
21–25 weeks	27–35 cm (11–14 in.) 550–800 g (1.25–1.5 Ib)	Head becomes even more proportionate to rest of body. Weight gain is substantial, and skin is pink and wrinkled. By 24 weeks, alveolar cells begin to produce surfactant.
26–29 weeks	32–42 cm (13–17 in.) 1110–1350 g (2.5–3 lb)	Head and body are more proportionate and eyes are open. Toenails are visible. Body fat is 3.5% of total body mass and additional subcutaneous fat smoothes out some wrinkles. Testes begin to descend toward scrotum at 28 to 32 weeks. Red bone marrow is major site of blood cell production. Many fetuses born prematurely during this period survive if given intensive care because lungs can provide adequate ventilation and central nervous system is developed enough to control breathing and body temperature.
30–34 weeks	41–45 cm (16.5–18 in.) 2000–2300 g (4.5–5 lb)	Skin is pink and smooth. Fetus assumes upside down position. Pupillary reflex is present by 30 weeks. Body fat is 8% of total body mass. Fetuses 33 weeks and older usually survive if born prematurely.
35–38 weeks	50 cm (20 in.) 3200–3400 g (7–7.5 lb)	By 38 weeks circumference of fetal abdomen is greater than that of head. Skin is usually bluish-pink, and growth slows as birth approaches. Body fat is 16% of total body mass. Testes are usually in scrotum in full-term male infants. Even after birth, an infant is not completely developed; an additional year is required, especially for complete development of the nervous system.

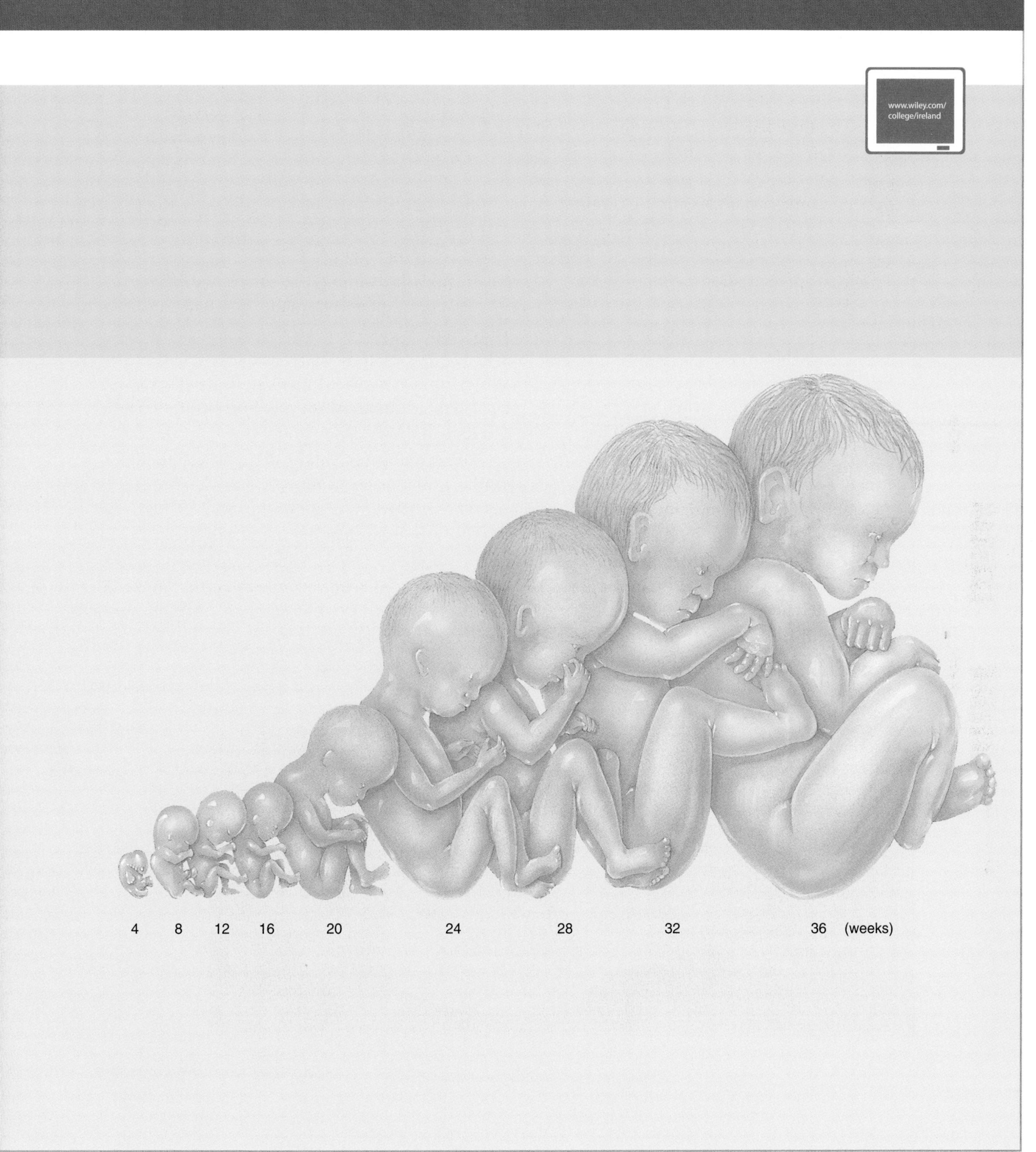
www.wiley.com/
college/ireland
4
8
12
16
20
24
28
32
36 (weeks)

Throughout the pregnancy, medical professionals and the prospective parents want assurance that the future child is developing correctly. Technology provides many ways to "see" inside the uterus and gauge the health or even the gender of the baby-to-be. The **obstetric** visit at 18 to 20 weeks in the United States routinely includes an **ultrasound examination**. This exam can visualize many things. A close approximation of gestational age can be determined by measuring and comparing the size of various fetal body parts; the heart can be seen beating, and internal organs and the skeleton can be viewed. Often even the gender of the developing fetus can be determined if it is in the right position. Birth defects such as spina bifida are also visible.

Obstetrics The medical field devoted to prenatal and maternal care.

Ultrasound examination Bouncing ultrasonic waves through the maternal skin into the uterus to observe the reflected patterns.

Information on genetic health is available through amniocentesis or chorionic villus sampling. Amniocentesis is performed at 15 to 18 weeks to determine gender and the condition of the chromosomes. Using ultrasound, the physician guides a needle into the amniotic fluid, being careful not to touch the fetus with the sharp end. A sample of amniotic fluid along with cells shed from the baby's skin is withdrawn and analyzed (see FIGURE 17.7 on page 564). The DNA within these cells is isolated and a **karyotype** is created. Abnormalities such as trisomy 21 (Down syndrome) or Klinefelter's syndrome (XXY chromosomes, causing a phenotypically male individual with enlarged breasts and female fat deposits) can be seen immediately. Chorionic villus sampling is used to detect genetic anomalies earlier in the pregnancy. In this test a small bit of the chorion is removed, usually between weeks 10 and 12—early enough to allow for an abortion if a serious defect is detected.

Karyotype A micrograph of the chromosomes, arranged to show chromosome pairs.

A newcomer to prenatal analysis, called 4D ultrasound, is becoming available. This is computer-enhanced ultrasound that produces a clear, lifelike view of the fetus. Movement can be seen as if the fetus was outside the womb, and facial features are much clearer. A few intriguing studies have shown that the father of the infant bonds much more strongly when he can clearly see the face of his baby-to-be.

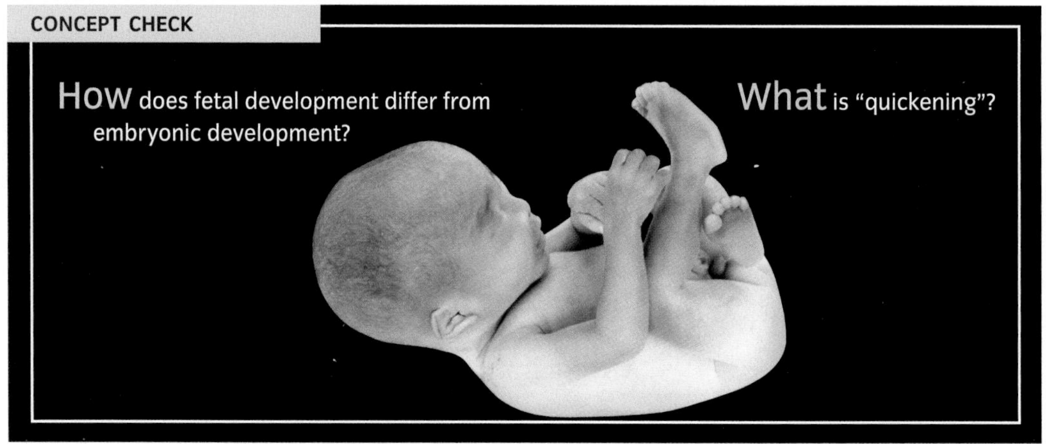

CONCEPT CHECK
How does fetal development differ from embryonic development?
What is "quickening"?

Labor Initiates the End of Pregnancy and the Beginning of Infancy

LEARNING OBJECTIVES

Describe the stages of labor.

Explain the functioning of the mammary glands after birth.

Labor and **birth** mark the end of prenatal life and of total reliance on the mother. The fetus goes from an aqueous environment with total life support to a dry environment where all life functions must come from within. Organ systems switch from standby status, or limited functioning, to the full-speed-ahead status they will occupy for the rest of the life span. The lungs, for example, must start exchanging gases before the umbilical cord is severed. The digestive system must start to work after the first suckle of milk. The heart must be able to pump blood against pressure. The skin must protect the body from damage, and fat layers must help maintain internal temperature.

But first, the fetus must be expelled from the uterus. Labor begins with hormonal triggers that are thought to originate in the fetal pituitary gland. The fetal **anterior pituitary gland** secretes **ACTH**, which triggers the fetal adrenal glands to secrete hormones that affect the placenta. The placenta increases production of estrogen and decreases production of progesterone. Estrogen increases **oxytocin** receptors on the placenta and increases placental **prostaglandin** production. This combination of factors makes the uterus much more sensitive to oxytocin levels. Maternal oxytocin then initiates rhythmic contractions in the uterus.

In a rare example of positive feedback in a healthy human (FIGURE 17.15), contractions of the uterus stimulate oxytocin production. More oxytocin means more and harder contractions, which in turn means more oxytocin. The contractions become stronger, harder, and closer together. Most first births take 24 hours from initial contractions to delivery. Subsequent births generally move much faster.

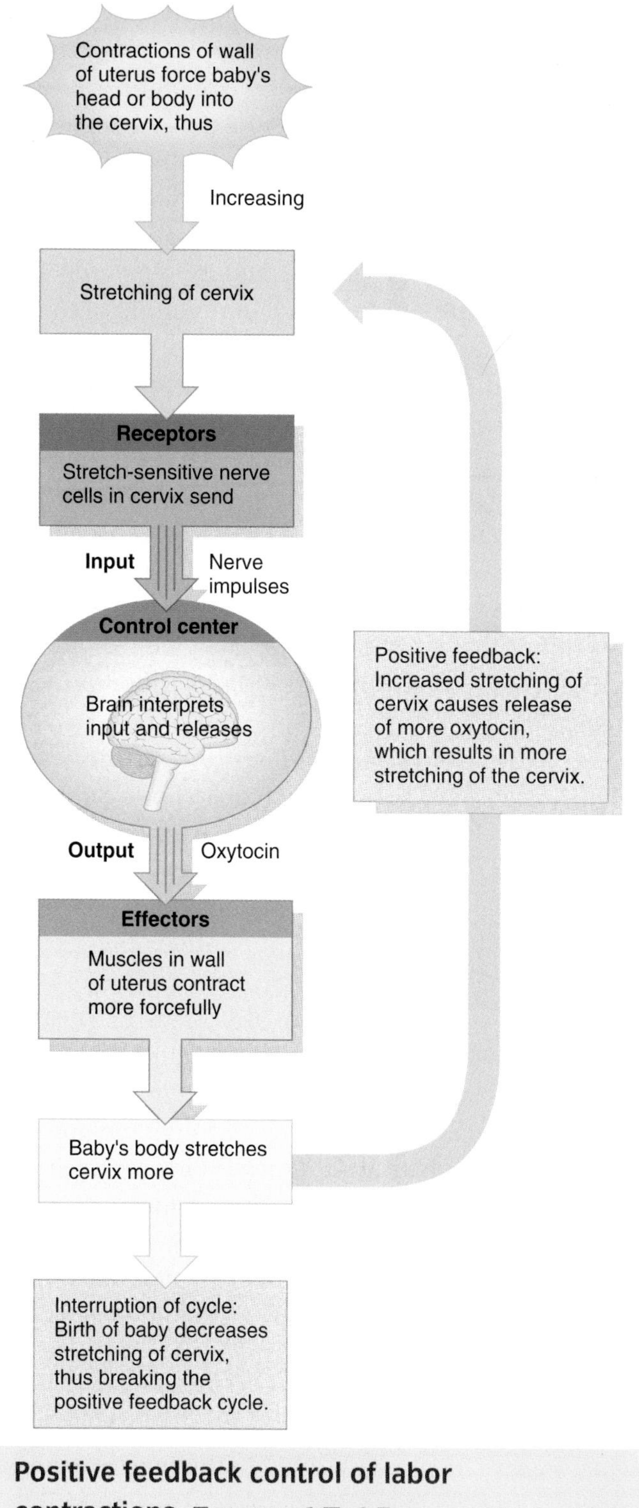

Positive feedback control of labor contractions FIGURE 17.15

DELIVERY HAS THREE STAGES

> **Dilation**
> The act of expanding or being expanded.

The first of the three stages of delivery (FIGURE 17.16) is **dilation**. The fetal head is pressing on the cervix. This pressure, combined with uterine contractions, stretches the cervical opening, which increases with each uterine contraction, going from slightly less than 2 centimeters to over 10 centimeters. As the opening enlarges, the mucus plug that was created by placental hormones drops out. The thin amnion is all that remains between the fetus and the external environment. This fragile membrane ruptures under increasing pressure, releasing a rush of amniotic fluid (in the vernacular, this is called the "water breaking"). After the amniotic fluid is lost, labor begins in earnest. Because the fetus is now subjected to the external environment without any protective fluid surrounding it, it is imperative that the baby be born within 24 hours. If true labor does not begin within that time, labor will be induced (artificially started) using injections of labor-inducing hormones.

The second stage of delivery, **expulsion**, is relatively short, usually lasting less than an hour. Expulsion is the time from full cervical dilation to delivery. Uterine contractions gain strength, and the mother experiences an overpowering desire to assist in the birth by pushing with voluntary muscles. With all this additional pushing, the baby moves through the cervix and out the vagina. Once the head **crowns**, or pushes through the opening of the vagina, the baby is on its way. A birthing attendant will help the baby breathe by suctioning mucus from the mouth and nose, even before the birth is completed. Once the baby's head is clear, the body slips out surprisingly quickly. The umbilical cord is clamped and then cut. The baby is now on its own, with no support from the maternal organs.

> **Expulsion**
> The act of forcing out.

The final stage of labor is the **afterbirth** or **placental stage**. The placenta is still in the uterus. As mentioned, the placenta is a disposable organ, and with the

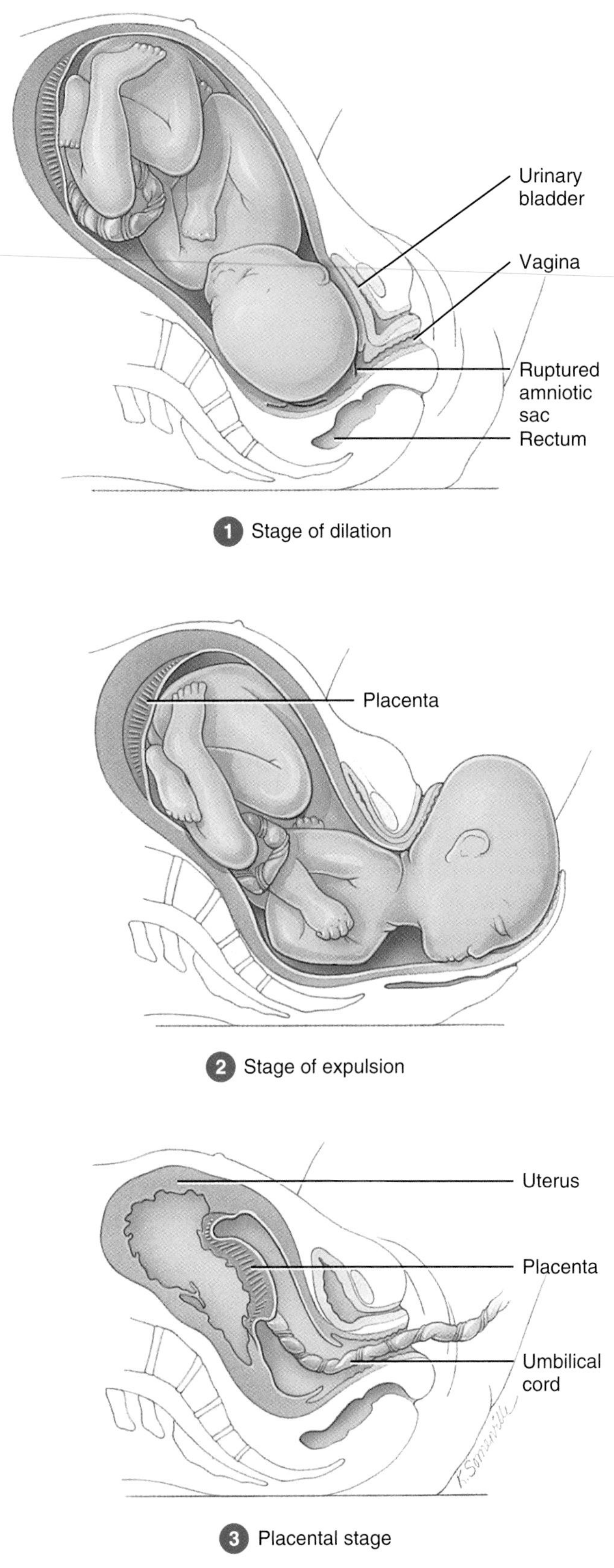

Stages of labor FIGURE 17.16

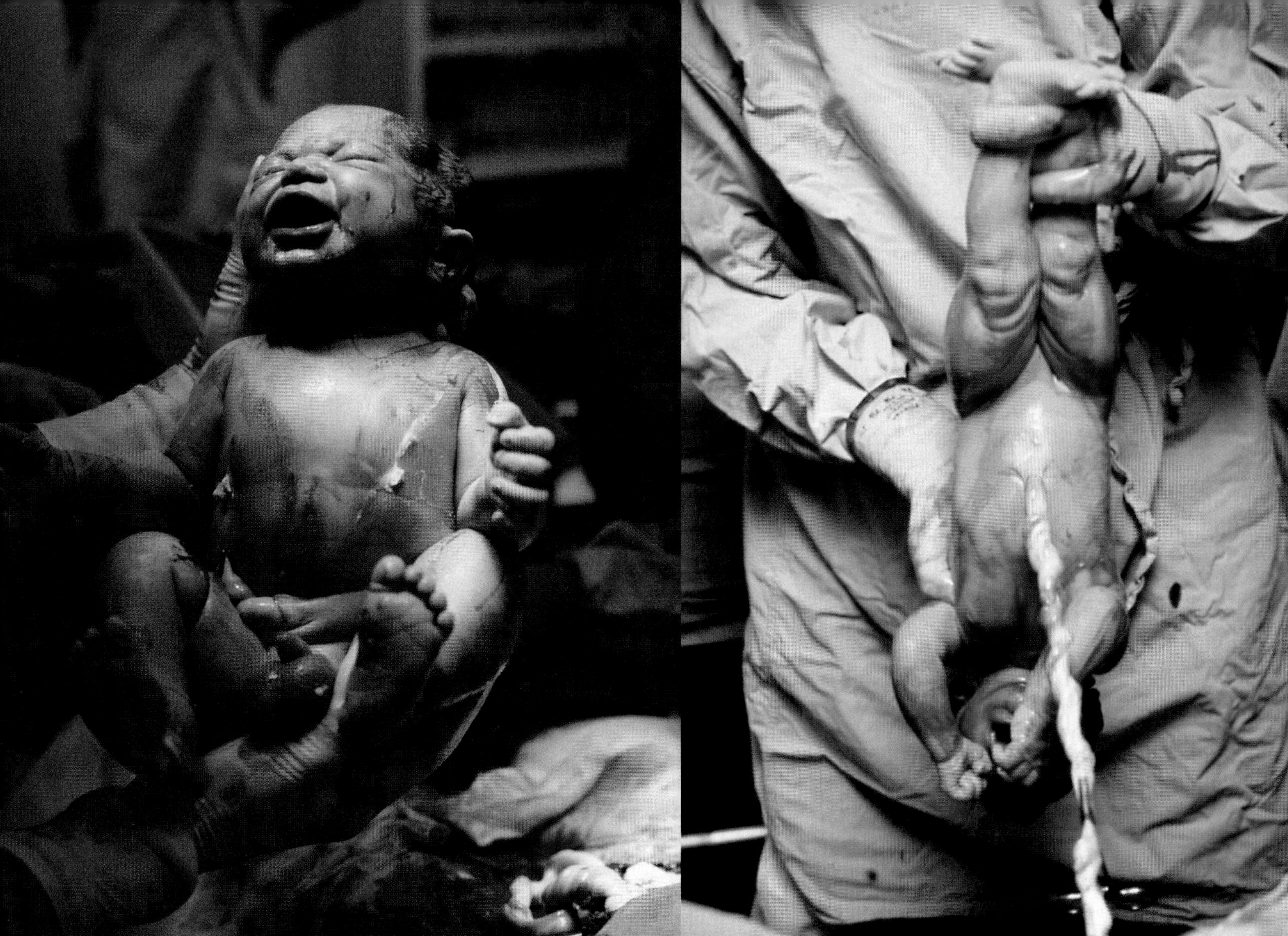

birth of the fetus, its utility is now over. Strong uterine contractions continue, and they tear the placenta from the walls of the now shrinking uterus.

The placenta and its attached umbilical cord are expelled through the birth canal and checked by medical personnel. The arrangement of the placental vessels may suggest the presence of congenital defects. The maternal and fetal surfaces of the placenta also show whether the entire organ has been expelled. Any pieces of placenta left in the uterus could become infected, causing life-threatening maternal septicemia if not removed. Some parents are having parts of the placenta frozen as a potential source of stem cells that may be useful in treating a serious illness in their child years afterward.

FETAL DEVELOPMENT CAN HAVE MANY COMPLICATIONS

Labor and delivery usually follow the same general pattern, but complications can require medical intervention. Some babies leave the sheltering environment of the uterus too soon. A **premature baby**, or a preemie, is defined as one born before 37 weeks of gestation (a full-term baby spends 37 to 42 weeks in utero). Preemies are usually quite small, and their organ systems are immature. They require specialized care in neonatal intensive care units until their organs have matured. The duration of care and the severity of the situation depend on the degree of prematurity. Typical complications include respiratory distress syndrome,

Ethics and Issues

Prematurity: How young is too young?

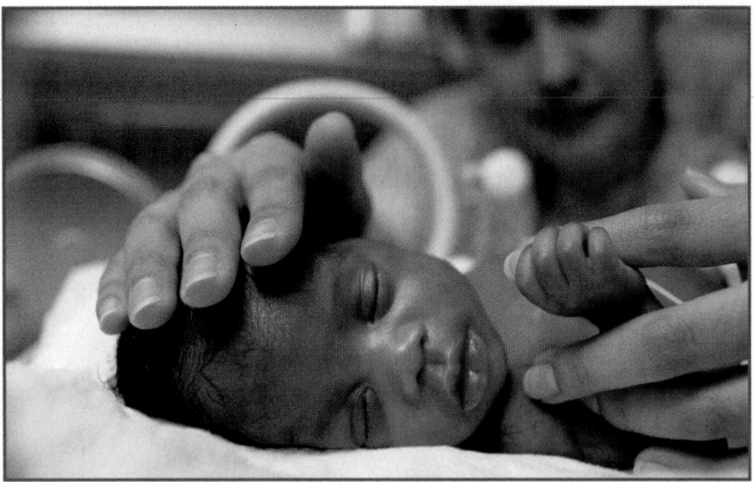

For a human fetus, the uterus is the optimum environment. Physically sheltered, warmed by the mother, protected by her skin and antibodies, and nourished by organic compounds in her blood, the womb is the perfect "home." But the fetus can be forcibly evicted, causing premature birth.

A variety of conditions can cause prematurity: Stress to the mother or fetus can give rise to corticotropin-releasing hormone, which may trigger other hormones that cause premature uterine contractions and delivery. Stress can result from illness, drug-taking, or a combination of factors.

A week or two of prematurity is nothing serious. But beyond that, critical considerations arise. According to the March of Dimes, 50 percent of neurological disability in children is related to being born too soon. These disabilities include cerebral palsy, mental retardation, and problems with learning, vision, and hearing. Another common consequence is chronic lung disease.

The medical specialty of neonatology has arisen to care for premature babies who are usually sheltered in neonatal intensive care units (NICUs). The field has made major strides in the past few decades, but as younger infants are routinely saved in NICUs, new problems arise. One recent concern is the effect of isolation, used to prevent infection and help the young lungs do their job. Many preemies spend weeks isolated in incubators. Studies of NICUs show that "procedural touch" used to sustain the baby can disturb the child physically and psychologically. After birth, everybody recognizes that children need comforting touch and soft voices. Why should this be any different in neonatal intensive care?

Studies using comforting, calming touch in neonatal intensive care have produced conflicting results. Many found better physiological signs, such as higher blood oxygen saturation, after comforting touch. In some cases, but not all, stroking and massage produce long-term benefits in infant health.

But is there a point, as medicine gains the ability to save ever-younger preemies, where rehabilitation no longer makes sense? Could it be less cruel to let nature take its course? The question is painful to ask and futile. Doctors and parents will always try to heal the patient—or child—in front of them and then deal with the long-term consequences as they arise.

The only good solution to this vexing question, of course, is prevention: to ensure that all babies live full term (38–40 weeks) in their ideal home. Alleviating the conditions that can cause prematurity is a key goal of prenatal medical care. Antibiotics can help reduce vaginal infections that are associated with increased risk of prematurity. Maternal treatment with the hormone betamethasone may help mature the fetal lungs and reduce lung damage in those fetuses that are born early.

The rate of prematurity is rising in the United States due to many factors, including IVF, increased age of pregnant women, poverty, poor health or eating habits during pregnancy, and increased environmental toxins. Neonatologists can now save infants weighing less than a kilogram. But as we marvel at these achievements, it's critical to remember the big picture. Ask any neonatologist, and you'll get the same answer: The best place for a developing baby is the womb.

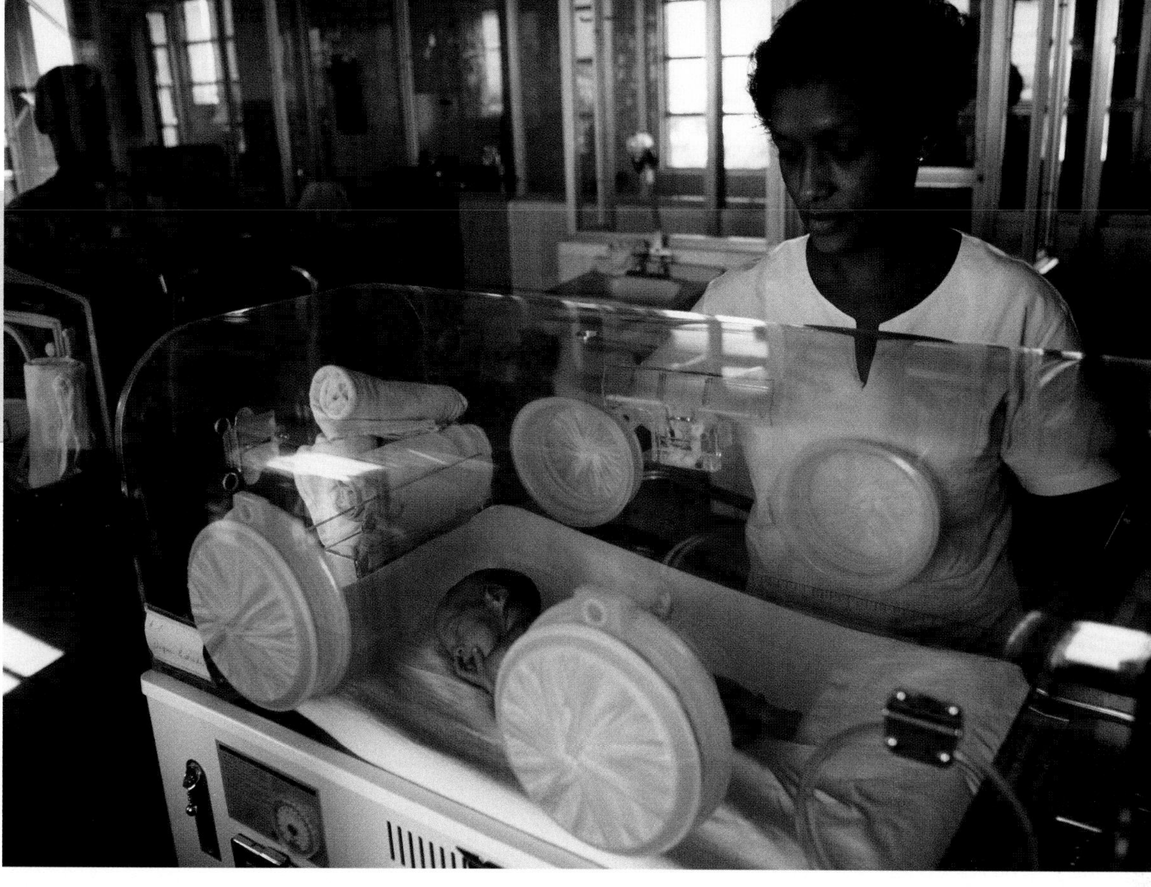

A "baby box" functioning as an external womb in the NICU (neonatal intensive care unit)

FIGURE 17.17

Technology can save fetuses born prematurely but often at great cost to the individual. The best place for development is always the womb, and doctors do everything they can to prevent premature delivery.

occasional cessation of breathing, inability to suck or swallow due to an immature nervous system, nutrient intolerance due to immature gastrointestinal lining, and improper blood filtration due to immature kidneys.

A premature baby looks different from a full-term baby (FIGURE 17.17). Typically, premature babies weigh less than 2.5 kilograms (5.5 pounds) and have thin, shiny skin through which veins are visible, wrinkled features, fine baby hair all over the body, and weak irregular breathing. Babies with these symptoms remain in the hospital under constant surveillance until they stabilize. Although premature babies grow and mature at their own rate, often they are not ready to go home with their parents until they reach at least 2.25 kilograms (5 pounds) in weight. By that time, they are usually feeding on their own and their breathing irregularities have improved.

The number of infants born before 37 weeks reached 12.3 percent of births in the United States in 2003. Reasons for this increase remain unclear, but according to the March of Dimes, many medical professionals believe industrial chemicals, pesticides, poor standards of living, and air pollutants are to blame. Others suggest that as women become pregnant later in life, premature births are more likely. In some cases, medical conditions such as placenta previa may require an early delivery, but in general, full term is desirable. Infants who are only moderately premature (34 to 36 weeks) have a threefold greater infant mortality rate, have higher medical costs, and spend more time in neonatal intensive care units. They also return to the hospital more frequently than full-term babies. Among highly preterm babies (less than 32 weeks gestation), the risk of death and long-term disabilities soars. These disabilities may include mental retardation, cerebral palsy, lung and gastrointestinal problems, and vision and hearing loss. (See the Ethics and Issues box for more on the complications of premature births.)

Sometimes it seems as though the baby just does not want to leave the womb. If there is no sign of labor after 42 weeks of development, the obstetrician will often induce labor. Although we still do not completely understand the hormonal controls on labor, we do know that increased levels of oxytocin initiate uterine contractions. To induce labor, the mother usually gets intravenous injections of a synthetic form of oxytocin called pitocin. Commonly called a "pit drip," pitocin pushes the uterus into strong contractions, beginning dilation and labor almost immediately.

The timing of labor seems to be a source of complication in the whole delivery process. Another complication associated with birth concerns the baby's position in the womb. A "breech baby" has the buttocks or feet below the head, which normally pushes the cervix open. Medical personnel can try to turn the baby using internal and external manipulations, or perform a cesarean delivery. Internal manipulation, using giant tongs called forceps, is an option for the mother who does not want a cesarean delivery. This can harm the baby, so forceps are used only after careful consideration. External manipulation involves putting pressure on the fetus through the maternal abdomen to try to shift the fetal position. Often fetuses in the breech position can be "turned" using gentle pressure from outside the uterus. The fetus will move its limbs, pressing back against the external pressure. If applied correctly, the fetus may turn itself in response to the gentle pressures from the physician.

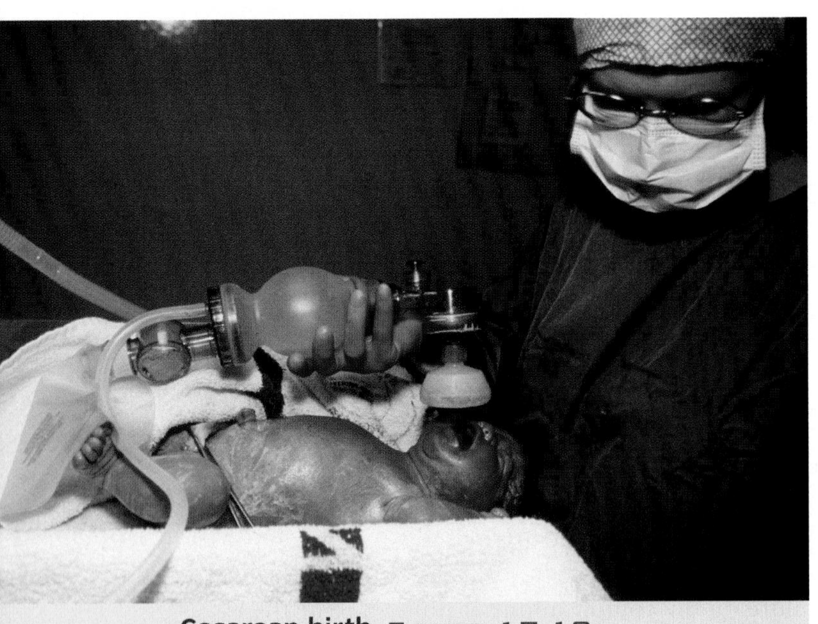

Cesarean birth FIGURE 17.18

Here doctors have performed a cesarean section, removing a baby from the uterus through the abdominal wall rather than through the birth canal. This is usually done when the baby cannot fit past the pelvic bones, or when the baby or mother is in mortal danger due to some difficulty with the natural birthing process.

A cesarean delivery (FIGURE 17.18) may be used when the baby cannot be delivered naturally—if, for example, the baby's head is too large to fit through the mother's pelvis, and for emergency deliveries, when the baby is in distress owing to lack of oxygen. A surgeon opens the maternal skin and uterus, lifts the baby out, examines the uterus, and removes all afterbirth. The oral cavity of the fetus is suctioned out to remove mucus and amniotic fluid that is normally removed while the baby is squeezed through the birth canal.

Another type of difficulty, called *failure to thrive*, can begin shortly after birth. Some infants and children fail to gain weight like others of their age. Because so many factors can affect growth and development, quick diagnosis is critical. At every doctor's visit, the infant is weighed and measured, and these numbers are plotted on a chart and compared to national standards. If there is cause for concern, medical, economic, social, and psychological factors should be investigated. Medical causes of failure to thrive include chromosomal defects, endocrine abnormalities, anemia, or malformed gastrointestinal organs. Economic and social causes are similar to those that are linked to high rates of prematurity and include poverty, parental neglect, poor eating habits, or exposure to toxic environments. Psychological factors include emotional deprivation and parental abuse.

THE MAMMARY GLAND PROVIDES MILK WHEN NEEDED

The neonate must obtain nutrients via the digestive system immediately after birth because the placenta is no longer supplying nutrients. The female does continue to nourish the infant, but now in the form of milk produced by the mammary glands. Each breast contains approximately 20 milk-producing lobules. These lobules are inactive until pregnancy, when they grow in size and number. The lobules end in ducts that drain to the nipple. No milk is produced until after birth, when prolactin is secreted by the anterior pituitary gland, causing the enlarged mammary glands to secrete milk (FIGURE 17.19).

The first substance produced by the mammary gland appears for two or three days after delivery. This watery fluid, called **colostrum**, is rich in proteins and antibodies. Actual milk production requires the infant to suckle the breast, which stimulates the areola and starts the release of oxytocin from the hypothalamus. Oxytocin stimulates the "let-down response," causing contractions of the larger **lactiferous ducts**.

Breast feeding is a personal choice, with products available now that can replace natural milk if the mother so chooses. However, there are benefits to breast feeding, as noted even on the labels of most commercial infant formulas. Breast feeding promotes bonding between mother and infant. Antibodies and lymphocytes present in breast milk, but not in baby formula, help protect infants during the first months of life from numerous diseases. There are some indications that the health benefits last well into their adult years.

According to the U.S. Food and Drug Administration, "Breast-fed infants have lower rates of hospital admissions, ear infections, diarrhea, rashes, allergies, and other medical problems than bottle-fed babies. . . . Breast-fed babies are protected, in varying degrees, from a number of illnesses, including pneumonia, botulism, bronchitis, staphylococcal infections, influenza, ear infections, and German measles. Furthermore, mothers produce antibodies to whatever disease is present in their environment, making their milk custom-designed to fight the diseases their babies are exposed to as well."

Within weeks of birth, the parents and relatives almost always start a guessing game: "Who does the baby resemble?" "Grandpa Frank?" "Grandma Miriam?" Even before birth, genetics play a role in the child's development, and genetics will be a key factor in health and disease for the new individual's entire life. Genetics governs a lot more than facial appearance: Genes play a role in physical and intellectual prowess, in susceptibility to disease, and perhaps even in personality. Why do babies resemble their parents or more distant relatives? How are physical and mental traits inherited? How are genetic problems inherited, and why do they express themselves in some children but not in others? To answer these questions, we turn next to genetics, inheritance, and DNA.

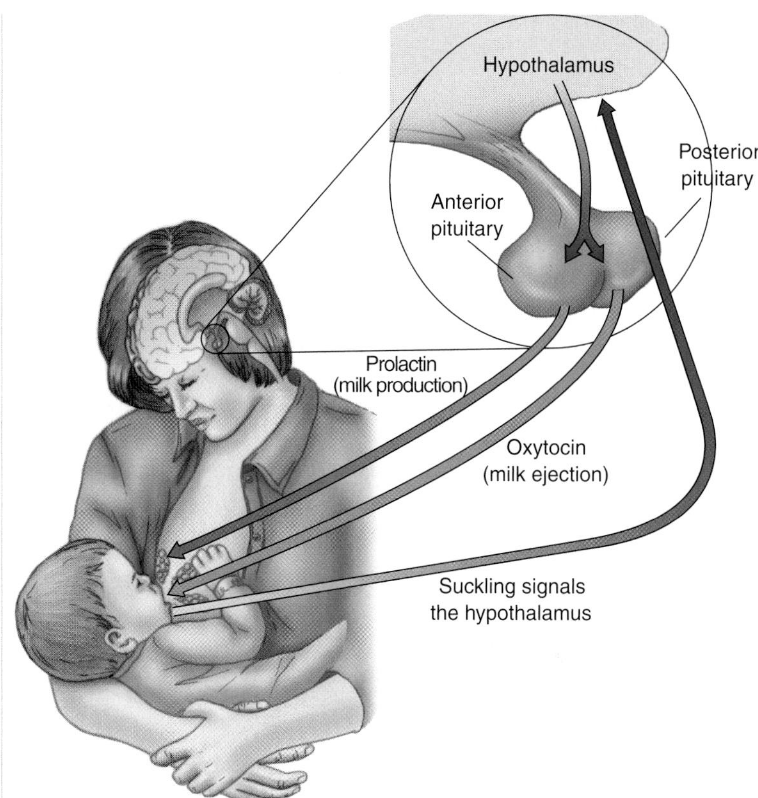

Hormonal controls on the mammary gland
FIGURE 17.19

CONCEPT CHECK

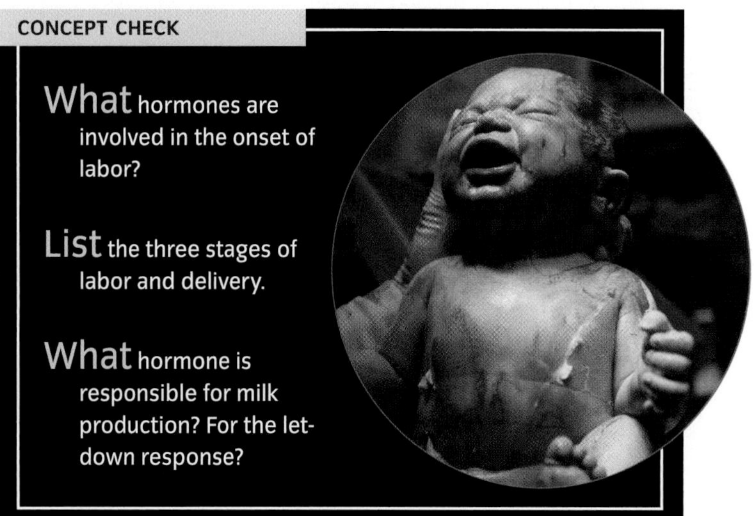

What hormones are involved in the onset of labor?

List the three stages of labor and delivery.

What hormone is responsible for milk production? For the let-down response?

CHAPTER SUMMARY

1 Fertilization Creates an Entire Diploid Genome

New life begins as sperm DNA fuses with egg DNA. The sperm travel up the female tract, propelled by muscular contractions of the female organs as well as the sperm's flagella. During the trip the sperm become capacitated, activating the acrosomal enzymes. Although many sperm may bind to the zona pellucida receptors of an egg, only one can enter the ovum cytoplasm and create a zygote. The zygote travels down the rest of the uterine tube, undergoing cleavage to form a morula and then a blastocyst.

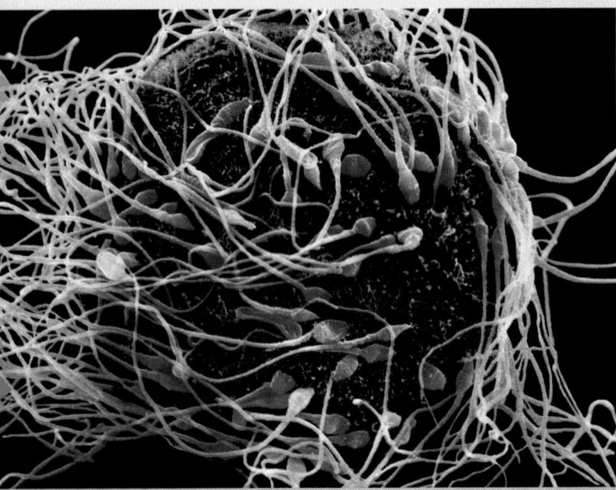

2 The Fertilized Egg Becomes a Blastocyst

By day 7, the blastocyst has entered the uterus and settled into the endometrium. In the next 24 hours, the blastocyst implants more permanently, as the trophoblast digests the endometrial tissue. The uterine lining thickens in the area where implantation occurs, and the chorion is developed. The beginnings of the embryo and supporting structures are created as the inner cell mass divides.

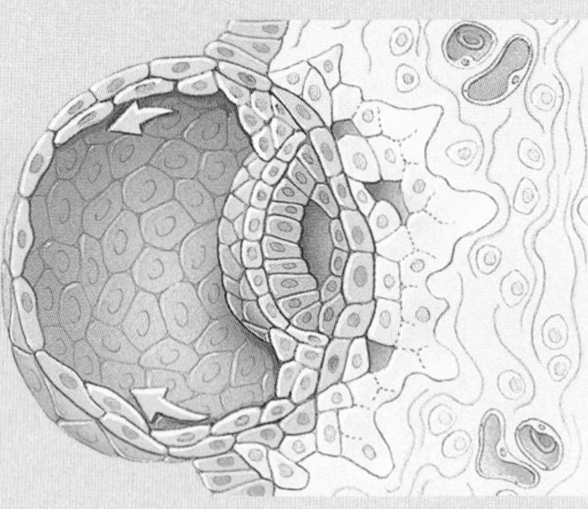

3 The Embryonic Stage Is Marked by Growth and Differentiation

The embryonic disc and the amnion form from the inner cell mass. Soon, the yolk sac, amnion, chorion, allantois, and the embryo differentiate. These tissues will become the extraembryonic membranes as well as the embryo itself. The umbilical cord develops, suspending the developing embryo in the amniotic fluid while still maintaining contact with the blood exchange areas of the placenta. At the end of the embryonic stage, all major organs are in place and the embryo looks distinctly human.

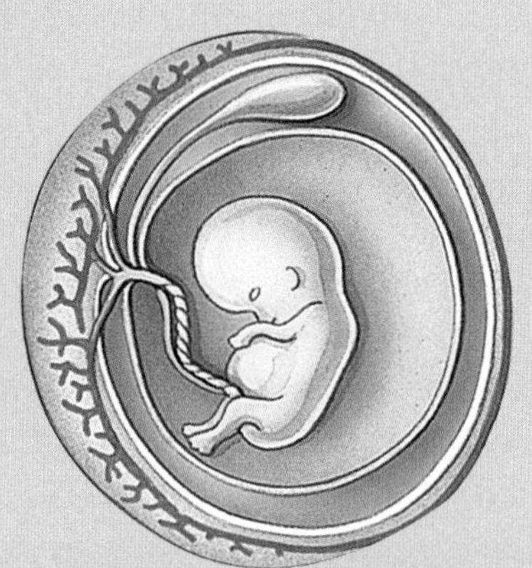

4 Fetal Development Occupies the Second and Third Trimesters

Fetal development starts at the end of the first trimester, as the organs enlarge and begin functioning. The fetus will grow to an average of 50 centimeters (20 inches) and gain approximately 3.75 kilograms (8 pounds). A huge amount of metabolic activity accompanies this growth. With difficulty, the fetus could live outside the womb by the end of the second trimester, but premature delivery is definitely to be avoided if possible. By the seventh month, surfactant is produced by the lungs, making life outside the womb easier.

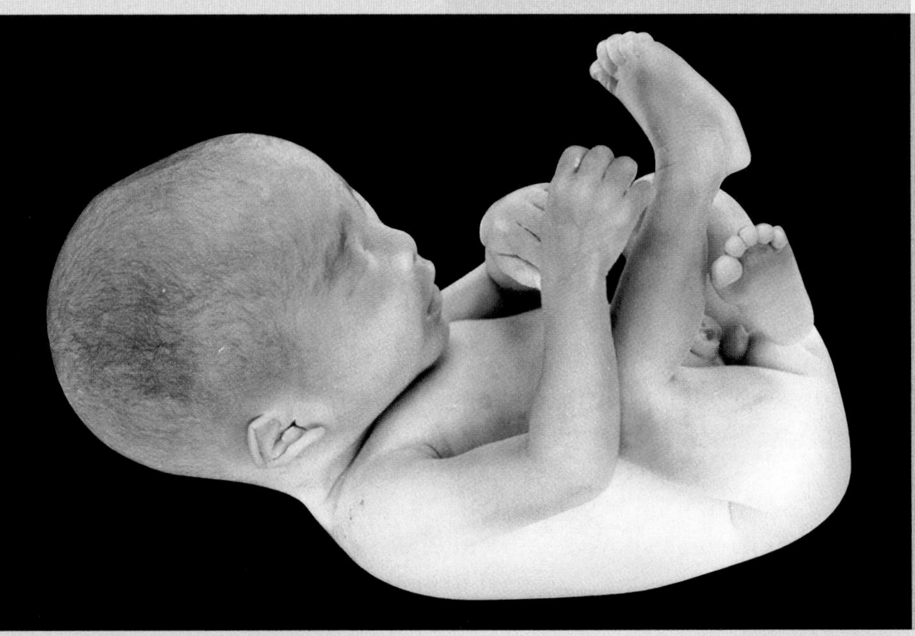

5 Labor Initiates the End of Pregnancy and the Beginning of Infancy

Labor starts at the end of pregnancy. It is believed that the fetal anterior pituitary gland initiates the positive feedback mechanism that leads to birth. The uterus becomes more susceptible to oxytocin, and as oxytocin levels increase in the maternal blood, the uterus begins to contract. Harder contractions stimulate production of more oxytocin, until the fetus is expelled, and the uterus shrinks to almost normal size. Labor includes dilating the cervix, expelling the fetus, and passing the placenta.

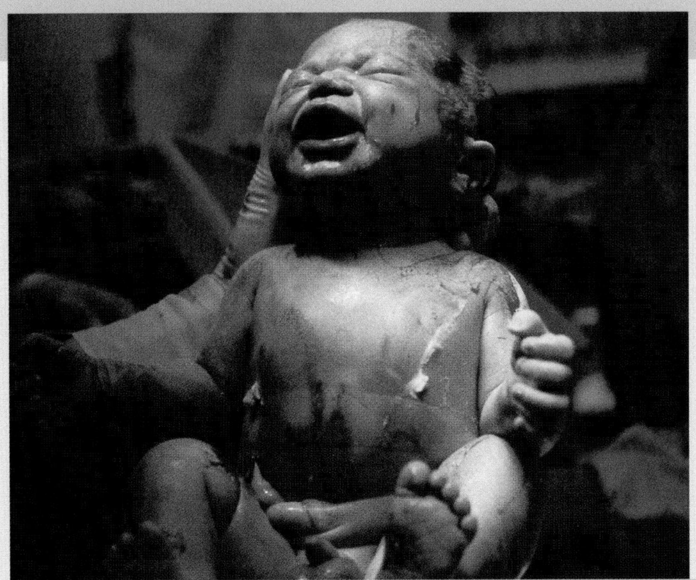

KEY TERMS

- **amniotic cavity** p. 561
- **cleavage** p. 557
- **differentiation** p. 554
- **dilation** p. 578
- **downregulated** p. 558
- **ectoderm** p. 562
- **embryonic** p. 554
- **endoderm** p. 562
- **expulsion** p. 578
- **extraembryonic** p. 564
- **fetal** p. 554
- **germ cell** p. 565
- **interstitial fluid** p. 564
- **karyotype** p. 576
- **mesoderm** p. 563
- **morphogenesis** p. 563
- **obstetrics** p. 576
- **polyspermy** p. 555
- **senescence** p. 554
- **surfactant** p. 571
- **therapeutic abortion** p. 563
- **trimester** p. 570
- **ultrasound examination** p. 576
- **umbilical cord** p. 565
- **viable** p. 555

CRITICAL THINKING QUESTIONS

1. The entire process of development in the womb can be confusing without a clear time line of activities. Return to the discussion of fertilization and embryonic development, and create such a time line. Indicate the order of events, beginning with the fusion of the male and female pronuclei and ending at the end of week 8.

2. Compare amniocentesis, chorionic villus sampling, and ultrasound. What are the strengths and weaknesses of each? Why is ultrasound routine, but not amniocentesis or chorionic villi sampling?

3. Look at the images of the pregnant woman in FIGURE 17.12. In the early months of pregnancy, most women experience the need to frequently urinate. They can eat only small meals by the eighth month, and the need for frequent urination returns in the ninth month. Breathing is also hindered in the seventh and eighth months, but may return close to normal in the ninth month. Sketch the approximate size of the uterus at each trimester. What organs are pushed out of place at each stage? Why the change in the last month?

4. Trace the hormonal controls on milk production and milk letdown. Many nursing women experience "accidental" milk flow when they hear their baby cry. Why?

5. At birth, the fetus transitions from an aqueous life protected in the womb to an arid, unprotected life in the atmosphere. What cardiovascular and respiratory changes must occur for the baby to survive this transition? Review fetal circulation in Chapter 9 and include the proper terminology for these changes. What other systems must now function to protect the infant?

SELF TEST

1. The two periods of development studied by developmental biologists are
   a. infancy and adulthood.
   b. fetal and senescent periods.
   c. embryonic and fetal periods.
   d. adolescence and adulthood.

2. Capacitation includes all of the following EXCEPT
   a. faster-moving flagella.
   b. changes to the sperm head membrane.
   c. acrosomal enzyme priming.
   d. biochemical changes in the corona radiata.

3. The very next step in fertilization following the one shown in this image will be
   a. syngamy.
   b. polyspermy.
   c. implantation.
   d. morula formation.

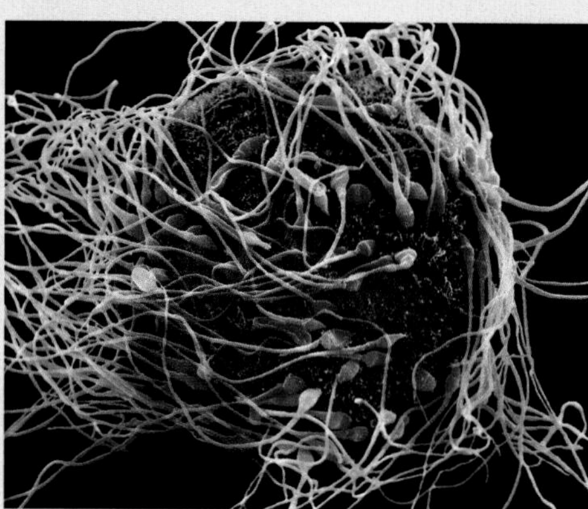

4. The formation of a ___________ involves cleavage.
   a. morula
   b. blastocyst
   c. pre-embryo
   d. All of the above are correct.

5. The correct term for the settling of the pre-embryo into the uterine lining is
   a. capacitation.
   b. implantation.
   c. fertilization.
   d. trophoblastation.

6. In the figure below, the cell layers destined to form the embryo are referred to as
   a. ectoderm only.
   b. mesoderm and endoderm.
   c. endoderm only.
   d. ectoderm, endoderm, and mesoderm.

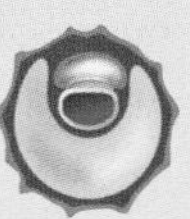
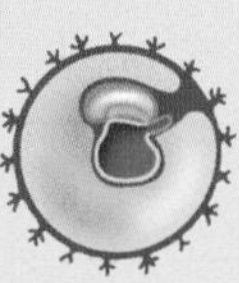
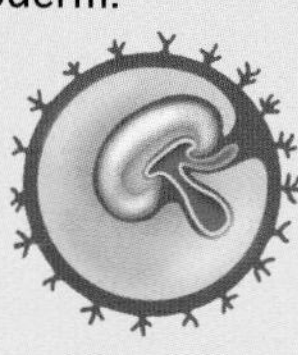
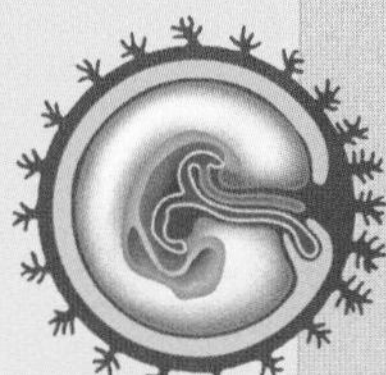

7. The extraembryonic membrane that develops into a protective fluid-producing membrane is the
   a. amnion.
   b. allantois.
   c. yolk sac.
   d. chorion.

8. The extraembryonic membrane that develops into the placenta is the
   a. amnion.
   b. allantois.
   c. yolk sac.
   d. chorion.

9. The test for hCG will give you an accurate reading only if you
   a. are pregnant.
   b. are female.
   c. have a corpus luteum still producing hormones.
   d. All of the above are true.

10. In the figure below, the structures labeled A is responsible for
    a. producing amniotic fluid.
    b. manufacturing fetal red blood cells.
    c. digesting maternal endometrium.
    d. producing hCG.

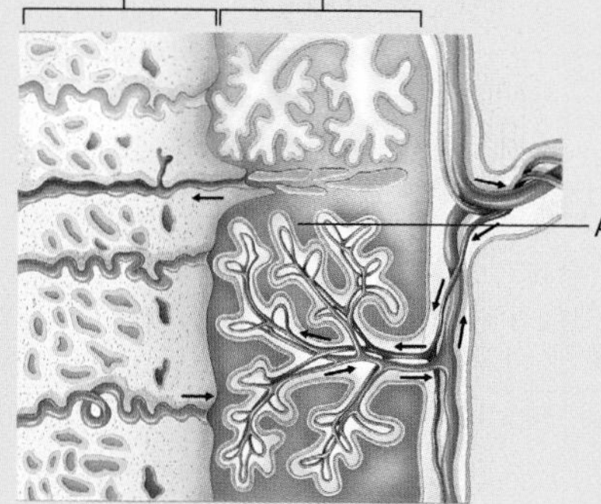

Details of placenta and umbilical cord

11. The umbilical vein shown in this figure eventually becomes the

a. foramen ovale.
b. hepatic portal system.
c. placenta.
d. round ligament.

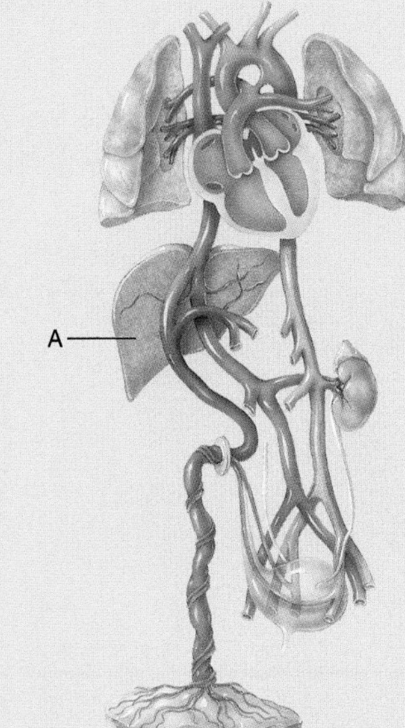

12. On the above figure, the structure labeled A is the

a. fetal liver.
b. umbilical vein.
c. umbilical artery.
d. fetal hepatic portal system.

13. True or False? The sucking response develops in fetuses immediately prior to birth.

14. The trimester characterized by rapid changes, including the formation of blood cells, human facial features, and ovarian follicles, and development of the nervous and muscular systems enough to stimulate "quickening," is the

a. first trimester.
b. second trimester.
c. third trimester.
d. fourth trimester.

15. The type of feedback seen in this example is ___________ feedback.

a. positive
b. negative
c. hormonal
d. unnatural

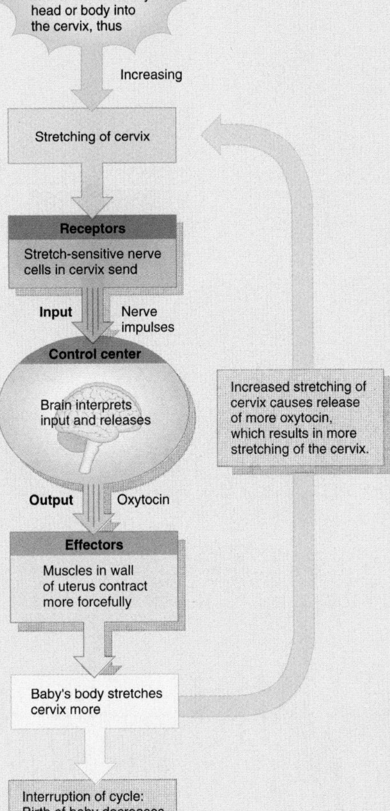

16. In the figure accompanying question 15, the hormone that initiates labor is

a. hGH.
b. hCG.
c. oxytocin.
d. ACTH.

17. The correct order of the stages of delivery is

a. dilation → contraction → expulsion.
b. dilation → expulsion → afterbirth.
c. afterbirth → expulsion → dilation.
d. contraction → expulsion → dilation.

18. Premature infants are those born before they reach ___________ weeks in utero.

a. 30
b. 34
c. 37
d. 40

19. The least invasive of the following means of assisting with labor and delivery is

a. cesarean section.
b. forceps delivery.
c. pitocin drips.
d. external manipulation.

20. The hormone that produces milk as shown below is

a. hGH.
b. prolactin.
c. oxytocin.
d. Both a and b are needed to produce milk.

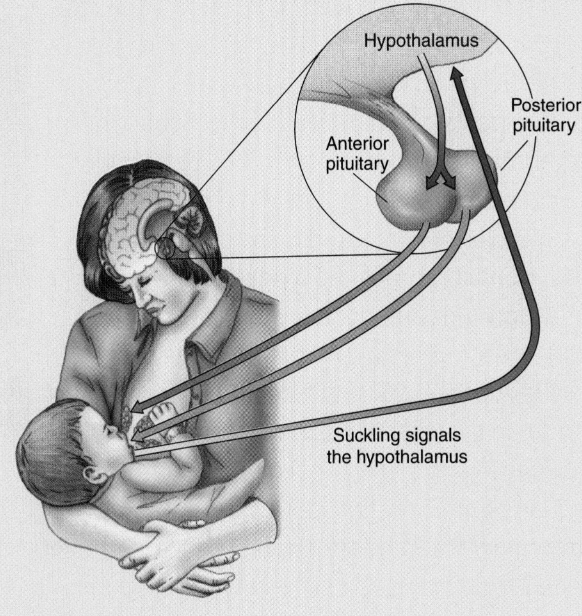

Inheritance, Genetics, and Molecular Biology

18

Dog breeding is one of the oldest uses of genetics. Ever since the domestic dog *(Canis familaris)* evolved from the gray wolf *(Canis lupus)*, humans have been changing the genetics of our best friend. The original shift from wild to domesticated probably came when friendly wolves hung around campfires, scrounging for scraps. Perhaps the proto-dogs made themselves useful by chasing away other scavengers, but it's likely that the wolf's social nature and its "psychology" (for example, its submission to the leader of the pack) predisposed the dog to be our friend and servant.

All this is speculation, but what happened next is clear. Humans began selecting and interbreeding the dogs they liked. Some became hunting dogs (labs or pointers). Others were chosen as sled dogs (huskies), fighters (pit bulls), or animal herders (border collies). Dogs bred for other purposes were put to use guiding the blind, rescuing people from mountains or inside buildings, tracking escaped prisoners, or sniffing for drugs or explosives.

How can one species take so many forms or have so many different behaviors? Long before the role of DNA in inheritance was discovered, humans were selectively breeding dogs, cattle, and corn; all descendants of "wild creatures" that our ancestors genetically altered for their own betterment. Genetic technology, this time of the modern variety, continues to play a role in the dog's life. In 2004, scientists completed a genetic analysis of the relationships among many of the 85+ domestic dog breeds. The huge variation in the size (think Pekinese and Great Dane) or behavior (Doberman pinscher and Labrador retriever) of dogs suggests to scientists that the dog could be not only man's but also a geneticist's best friend. In the study of how breeding affects genetics, dogs might well play the starring role.

CHAPTER OUT

Plant and Animal Traits Are Inherited in Specific Patterns

LEARNING OBJECTIVES

Define phenotype and genotype.

List factors that affect phenotype.

Explain the origin of the 23 pairs of chromosomes in our body cells.

Outline Mendel's basic experimental plan.

Describe Mendel's two laws of inheritance.

"What will my baby look like? Will it be intelligent? Short? Athletic? Oh my gosh, I hope it doesn't have my ridiculous ears or my skillet-flat feet!" These common questions reflect the fact that most of us are subtly aware that traits, appearances, and even intellect can be attributed to our genes. Genes, made of DNA, code for proteins and are found in the nucleus of almost every cell in your body. DNA is composed of a four-base "alphabet," where three bases (one codon) read as one "word." Your chromosomes are essentially strings of millions of these words. Your individual **DNA sequence** codes for the specific arrangement of amino acids in each of the millions of proteins in your body.

The "genetic alphabet" of DNA may contain only four letters, but it is phenomenally sophisticated. The 3 billion-plus individual microscopic base pairs in the nucleus of the human cell spell out everything you need to become a human. Furthermore, this DNA exists in trillions of cells, and it can be copied thousands of times with little or no appreciable error. Finally, the molecule is so durable that DNA found in fossils tens of thousands of years old can sometimes be analyzed!

DNA sequence The sequence of bases (adenine, cytosine, thymine, and guanine) on a chromosome.

Phenotype The appearance of an individual, as a direct result of the alleles being expressed.

We all have the same basic arrangement of genes in our chromosomes, despite individual differences in **phenotype**. These phenotypic differences emerge from subtle differences in **genotype**, as well as environmental factors. These factors, including the quality of our food, the type of shelter we live in, and even our financial "health," all play a role in our physical health and appearance.

Genotype The alleles carried on the chromosomes.

Genetic factors are important in determining our individuality. Chromosome 11, for example, carries the same basic information in all of us. It contains genes that code for some blood proteins, insulin, and the milk-digesting enzyme lactate dehydrogenase, as well as other proteins and regulating factors. The specifics of the information on your chromosome 11 are different from those found on the same chromosome in either of your parents. Think of building a planned neighborhood. Each house could have the same general blueprint and floor plan, but still look a bit different. Maybe the front doors are all centered, with different windows. The walls could be built the same, but garnished with different siding and paint. On first glance, the neighborhood might look diverse, but with some study, you would notice that important similarities among houses exist. The same could be said about human beings. Hair can change. Skin color can change. Facial proportions can change. But deep within the cells we are almost exactly the same.

Those similarities and differences both emerge from our genes. Genes are located on chromosomes (FIGURE 18.1). Humans have 23 pairs of chromosomes, for a total of 46 individual units of DNA. Twenty-three chromosomes came from the egg, and the matching 23 were delivered via the sperm during fertilization. This means the egg and sperm cannot have the usual

Process Diagram

Meiosis FIGURE 18.1

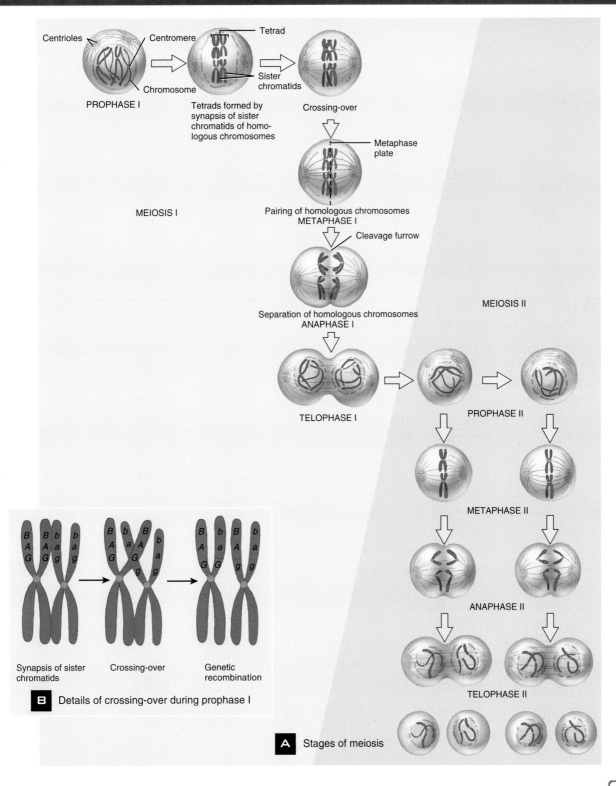

B Details of crossing-over during prophase I

A Stages of meiosis

Meiosis, the production of egg and sperm, was introduced in Chapter 16. Here we see an overview of the process, from prophase I through telophase II. Recall that meiosis I separates homologous chromosomes, breaking apart tetrads, while meiosis II produces four haploid gametes.

Diploid
Having two copies of every chromosome (the usual situation in body cells).

Haploid
Having one copy of every chromosome (the number in our eggs and sperm).

diploid chromosome complement (23 pairs). They must be **haploid**, carrying only 23 individual chromosomes (see FIGURE 18.1).

GREGOR MENDEL EXPLAINED PATTERNS OF INHERITANCE

The patterns of trait inheritance were manipulated long before chromosomes were even discovered. For thousands of years, herders and breeders of animals have known they could develop better animals through selective breeding. In the plant kingdom, farming apparently arose as early farmers learned they could improve on food crops by wise choice of the parent plants.

Farmers and herders brought wild plants and animals into domestication and greatly improved their yields, but they had no scientific understanding of the mechanisms of that improvement. Scientists and farmers alike tried to explain inheritance. Only in the 19th century did a monk from Central Europe provide a plausible—and accurate—theory.

Gregor Mendel devoted years to studying the inheritance of traits in many plants, including garden peas. The garden pea is an easy-to-grow plant with specific and definable traits, and it produces a simple flower that naturally self-pollinates. Left alone, pea plants will produce mature pollen (the male gamete) on the anthers before the flower opens. As in all flowers, when the pollen falls on the female reproductive parts of the plant, the stigma, pollen tubes grow through the female stigma into the ovary. Once this pollen contacts the eggs, fertilization occurs and seeds develop.

Mendel realized he could control this process. He cut into a closed flower and used a small paintbrush to gently remove mature pollen from the anthers. He then transferred that pollen to a different plant by cutting the second flower and painting the pollen on its female part, the stigma. To prevent self-fertilization, Mendel removed the anthers from this second flower.

Mendel was not alone in his quest to understand inheritance, but what set him apart was a combination of extensive research and studiously recorded, precise field notes. He performed his experiments in stages over seven years. He started by identifying traits in the pea plant that existed in only two forms and did not blend. For example, he noticed that pea flowers were purple or white, but never lavender, and the seeds were either yellow or green. The dried seeds were either smooth and round or wrinkled. In total, Mendel identified the seven non-blending traits listed in TABLE 18.1.

The seven traits Mendel used to study genetic inheritance TABLE 18.1

TRAIT	DOMINANT	×	RECESSIVE
Flower color	Purple	×	White
Seed color	Yellow	×	Green
Seed shape	Round	×	Wrinkled
Pod color	Green	×	Yellow
Pod shape	Round	×	Constricted
Flower and pod position	Axial (along stem)	×	Terminal (at top of stem)
Plant height	Tall	×	Dwarf

Self-pollinating Transferring the pollen of a flower directly to the stigma of the same flower.

Cross-pollinating Fertilizing the ovum of a flower with pollen from a different plant.

Using these traits, Mendel began his unparalleled experiments, **self-pollinating** and **cross-pollinating** his plants, then recording the phenotype for each trait in the offspring. In each test, Mendel observed hundreds of plants. As he followed these traits, he observed some surprising results, results that to this day accurately predict the outcome of genetic crosses (FIGURE 18.2).

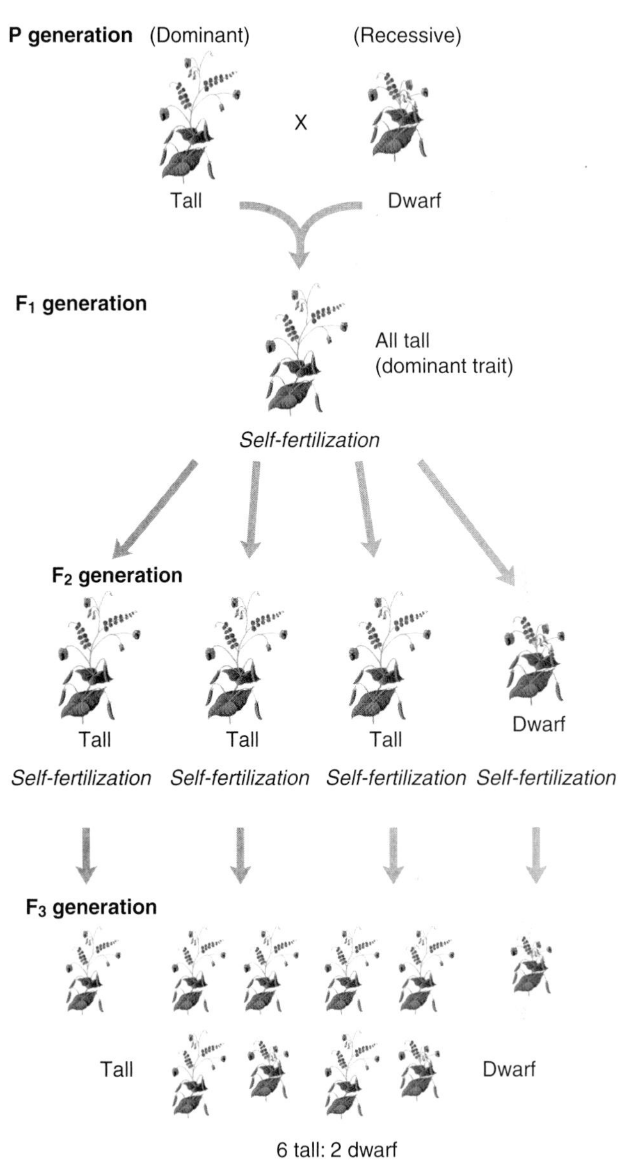

Mendel's experimental setup FIGURE 18.2

In the first generation, Mendel cross-pollinated two true breeding parents demonstrating opposite traits. In the example, he bred a tall plant with a short plant. He then recorded the phenotypes of the first generation from this cross. He permitted the first generation to self-pollinate, and again recorded the phenotypic results for every plant in the second generation. These data were what Mendel used to predict the phenotypic ratios resulting from cross-pollination.

HERITABLE UNITS RANDOMLY SEPARATE DURING GAMETE FORMATION

Mendel discovered that phenotype is inherited and that the proportion of each trait in the next generation is fixed. If he began by crossing **true-breeding** parents, all the offspring in the first generation (called F_1) had only one of the parental traits. For example, in crossing a purple-flowered plant with a white-flowered plant, the first generation was 100 percent purple. It appeared that purple was **dominant** over white.

True-breeding Plants that reliably produce the same phenotypic traits in their descendants with every self-pollination.

Dominant An allele of a gene that determines phenotype even if only one such allele is present.

Recessive An allele of a gene that determines phenotype only when two like alleles are present.

What had happened to the **recessive** white color? When Mendel self-pollinated the F_1 plants, the white flowers miraculously reappeared in the second generation, F_2. Oddly, flower color always had the same ratio: roughly one white-flowered plant for every three purple-flowered plants.

In seeking an explanation, Mendel decided there must be some "heritable unit," which we now understand as the various alleles or forms of genes. He hypothesized that these "heritable units" must exist in pairs in the parent and that these pairs separate as pollen and egg are formed. Each gamete would carry only one of the parent's "heritable units." Therefore, one of these "heritable units" from each parental plant is transferred to each offspring. Mendel called this the **Law of Segregation** and defined it as the random separation of parental "heritable units" during gamete formation.

MENDEL ALSO FORMULATED THE LAW OF INDEPENDENT ASSORTMENT

As Mendel's experiments got more sophisticated, he tracked several traits at once through dihybrid crosses (two trait crosses), and again he saw a pattern. There seemed to be no connection between the expression of one trait and the expression of the other. The expression of each trait was independent from the expression of another. In other words, even if a pea plant's flower color was dominant, he could not predict if its seed color would also be dominant. Mendel's **Law of Independent Assortment** states that each trait is carried in the egg and pollen as a separate entity, with no effect on any other trait (FIGURE 18.3).

Law of independent assortment: dihybrid cross FIGURE 18.3

The square in Part A shows the expected phenotypes of offspring when two traits are simultaneously observed in the F_1 generation. W indicates the dominant allele for smooth texture, and G indicates the dominant allele for yellow color. The lowercase letters indicate wrinkled texture (w) and green color (g). The law of independent assortment states that when a plant dominant for both traits is cross-pollinated with a plant recessive for both traits, the second, self-fertilized generation will show a predictable 9:3:3:1 ratio of dominant and recessive traits.

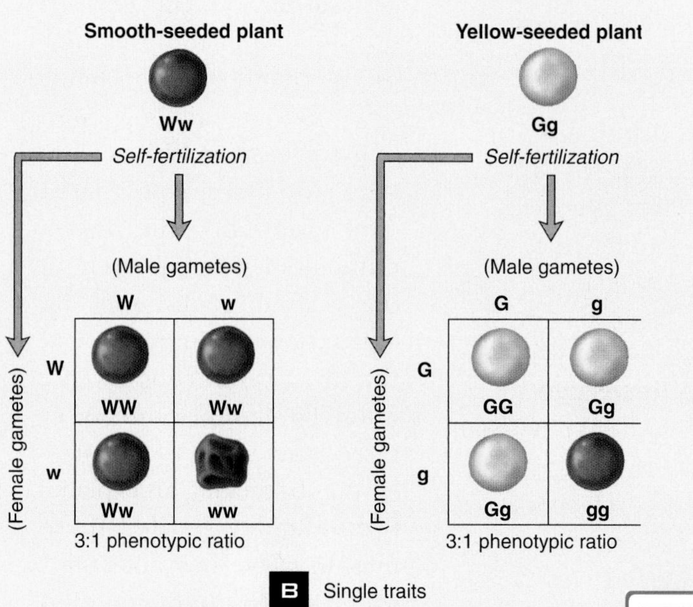

www.wiley.com/college/ireland

CONCEPT CHECK

Which term describes your appearance, phenotype or genotype?

How many chromosomes are carried in the egg? in the sperm?

Define the Law of Segregation.

Explain why peas were good test plants for Mendel's experiments.

How does the Law of Independent Assortment explain the seemingly random inheritance of two traits?

Modern Genetics Uncovers a More Complicated Picture

LEARNING OBJECTIVES

Explain the interaction of dominant and recessive alleles.

Understand that alleles can also be multifactorial, or codominant.

Analyze a Punnett square.

Mendel's experiments provided a great starting point for the science of genetics, although their significance was not recognized for almost 40 years. To understand inheritance as we now know it, we need more terms than dominant, recessive, phenotype, and genotype. When a gene can have alternate forms, such as white or purple color, these forms are called **alleles**. Each **somatic** cell contains two copies of every gene, one obtained from each parent. When the two alleles are identical, the genotype is **homozygous** for that trait. A homozygous gene is usually denoted by two identical letters, such as AA, or aa. Homozygous individuals can be **homozygous dominant**, meaning both alleles code for the dominant trait (AA), or **homozygous recessive** (aa). If one allele codes for the dominant trait and the other codes for a recessive trait, the genotype is **heterozygous**. Heterozygotes are usually indicated with a capital and a lowercase letter (Aa).

Somatic
Related to the body, in contrast to the gametes.

Only homozygous recessive individuals express a recessive phenotype. If one allele is dominant, the dominant phenotype must be expressed. This means that if your appearance includes a recessive trait, all of your gametes carry only the recessive allele. You are homozygous recessive for that trait. If that trait is dominant in your phenotype, you could be homozygous dominant or heterozygous, and it is hard to predict which allele any one of your gametes will carry.

We all have traits that define our phenotype: hair that is brown, black, blonde, or red; eyes that are blue, green, brown, or hazel; hair that is straight or curly; skin that is dark or light. But it turns out that human inheritance is more complicated than that of Mendel's pea plants. Very few of our phenotypic traits demonstrate simple dominant–recessive interactions

Dominant/recessive traits in humans TABLE 18.2	
Cleft in chin	No cleft dominant, cleft recessive
Hairline	Widow's peak dominant, straight hairline recessive
Eyebrow size	Broad dominant, slender recessive
Eyebrow shape	Separated dominant, joined recessive
Eyelash length	Long dominant, short recessive
Dimples	Dimples dominant, no dimples recessive
Earlobes	Free lobe dominant, attached recessive
Eye shape	Almond dominant, round recessive
Freckles	Freckles dominant, no freckles recessive
Tongue rolling	Roller dominant, nonroller recessive
Finger mid-digital hair	Hair dominant, no hair recessive
Hitchhiker's thumb	Straight thumb dominant, hitchhiker's thumb recessive
Interlaced fingers	Left thumb over right dominant, right over left recessive
Hair on back of hand	Hair dominant, no hair recessive

Fitness
Ability to produce living offspring and pass on DNA.

(TABLE 18.2). We can predict the possibility of passing these simple traits to our offspring just as Mendel did with his peas. While these traits are not critical to our overall **fitness**, they do demonstrate that a few human genes follow the same rules as Mendel's pea plants.

CODOMINANCE COMPLICATES THE PICTURE

Many traits in humans, including hair color, eye color, and facial structure, exhibit **incomplete dominance** or **codominance** rather than the complete dominance that Mendel found. Incomplete dominance tends to produce different phenotypes based on the combination of alleles present in heterozygotes. Codominance occurs when the effect of both alleles appears in the heterozygote.

How can this happen? Many human traits are **polygenic**, meaning the phenotype results from the interaction of many genes, not the expression of just one. Furthermore, many of our traits are **multifactorial traits**, meaning polygenic traits that are also influenced by environment. These traits express a continuum of phenotypes, usually producing a bell-shaped curve on a plot of their distribution in the population. Body type, muscular development, fat deposition, and height are all multifactorial traits.

Blood type is an excellent example of codominance (FIGURE 18.4). There are three alleles for blood type: the A allele, the B allele, and the O allele. The A allele codes for a modification of the original precursor erythrocyte surface protein randomly designated "A." Similarly, the B allele codes for modifications that produce the marker protein B. The O allele codes for no modified marker protein, effectively a null allele. If one of your alleles is A, and the other is A or O, you have type A blood. Similarly if you have two B alleles, or a B and an O, your blood type is B. If one allele is A and the other B, however, you have type AB blood. If you are homozygous O, you have type O blood. In each case, both alleles are expressed in the phenotype, which is the meaning of codominant. The alleles do not blend to form an entirely new AO marker protein, nor do they form an AB protein that is different from the individual A or B modified markers. Instead each allele codes for a separate protein, which is translated and added to the membrane of the red blood cells. Therefore, in type AB

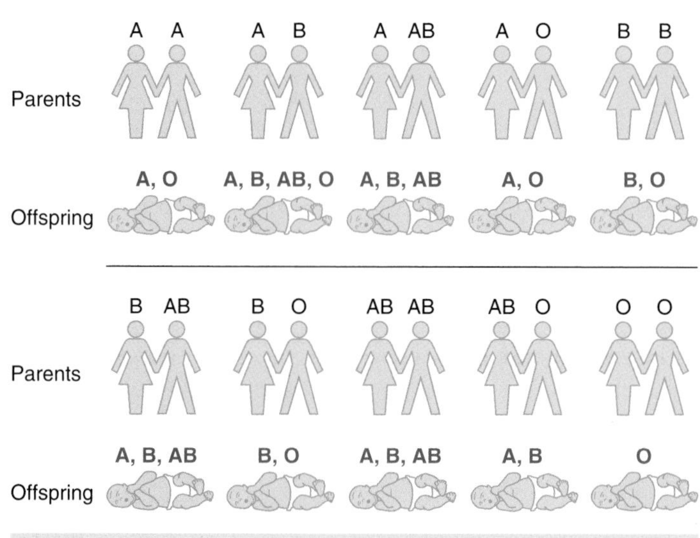

Blood type inheritance FIGURE 18.4

blood, the erythrocytes show both the A and the B protein, and with the genotype AO you will find both an A and an O marker on the red blood cells.

Incomplete dominance governs the human voice pitch, eye color, and hair curliness. The lowest and highest pitches in male voices occur in men who are homozygous dominant (AA) or homozygous recessive (aa) for the trait that determines pitch. All intermediate-range (baritone) voices are heterozygous (Aa). We see the same blending of traits in eye color. If one parent has green eyes and the other has brown, there is a good chance the children will express a blended, dark blue eye color. Recently, scientists have discovered that eye color is determined via an interaction of at least three different genes, each affecting the phenotype of the other. Although this trait requires more than one gene, the interactions between them can be understood in light of incomplete dominance. In Caucasians, hair can be straight (H′H′), wavy (HH′), or curly (HH). Wavy hair is an intermediate phenotype, indicating incomplete dominance of the curly trait (FIGURE 18.5).

PUNNETT SQUARES SHOW THE POSSIBILITIES

The Punnett square, a tool used to determine probability of genotypic combinations in offspring that we introduced in FIGURE 18.3 and shown below in FIGURE 18.5, works much like the multiplication tables you may remember from grade school. The alleles carried by one parent for the gene in question are listed across

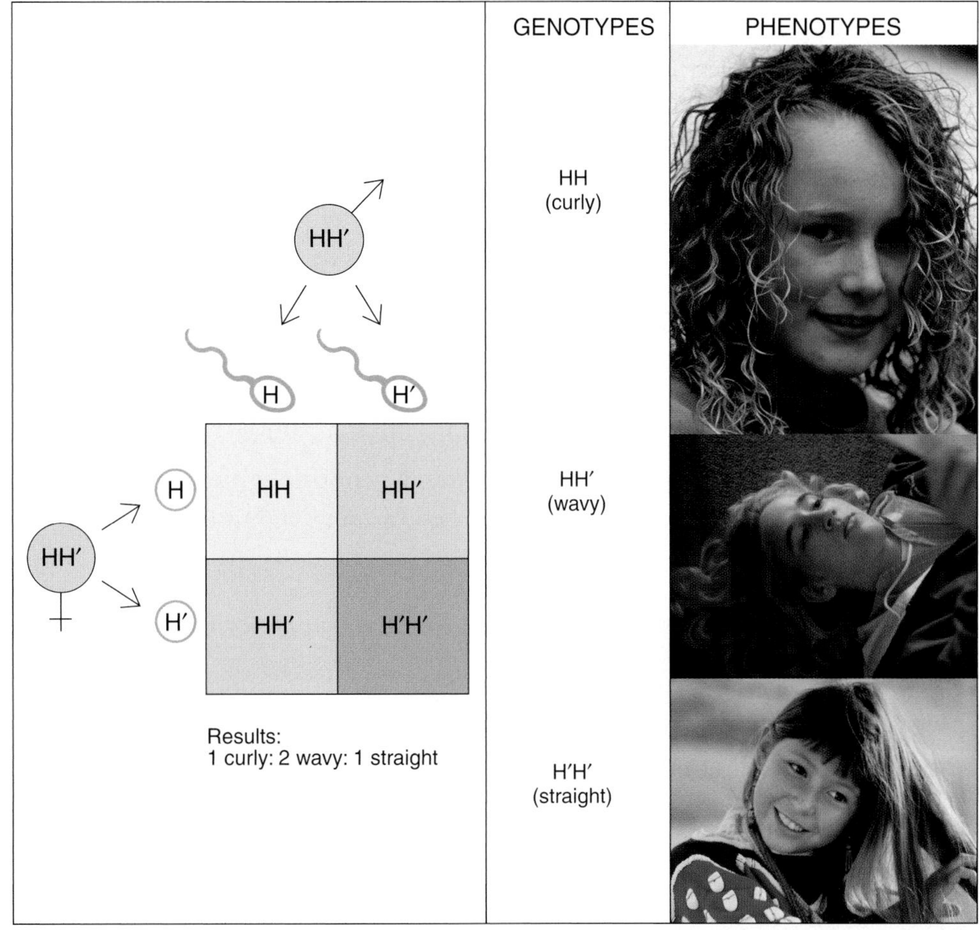

Hair patterns FIGURE 18.5

Note that the uppercase and lowercase conventions are not used here because one trait is not dominant over the other.

the top, representing that parent's potential gametes. The left side lists the other parent's alleles. In the center boxes of the table, the allele at the top and the one to the left are "multiplied" or combined, resulting in one possible allelic combination from these two parents.

Further complicating the inheritance pattern of humans, the sex chromosomes (X and Y) carry different traits. There are more alleles on the X chromosome than on the Y, so males (XY) have but one copy of the alleles found only on the X chromosome. Color blindness is one such trait. The gene for color discrimination is on the X chromosome but not the Y. If a female XX carries the gene for color blindness on only one of her two X chromosomes, she will not express the defect, but half of her eggs will carry the defective gene. Because the fertilizing sperm carries a Y chromosome, it cannot provide a correct copy of the gene, resulting in a colorblind male child. Despite these differences, inheritance patterns for these so-called sex-linked traits, which we will cover later in this chapter, can be predicted using a simple Punnett square (FIGURE 18.6).

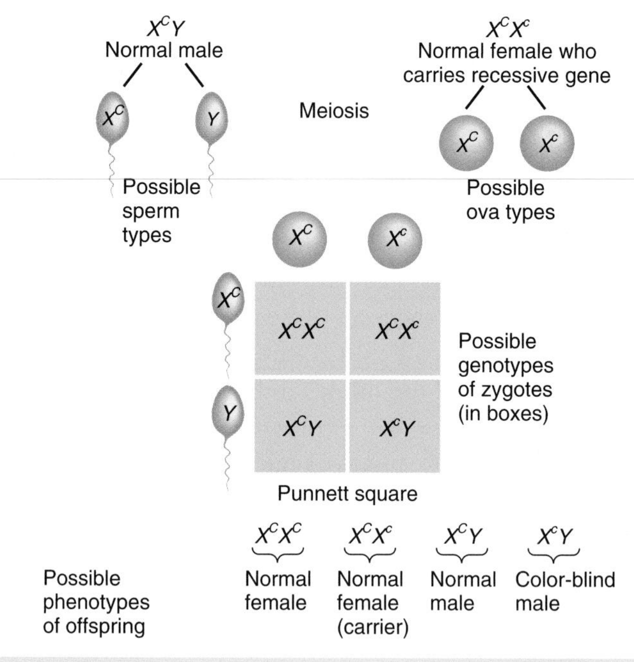

Punnet square for the inheritance of red–green color blindness FIGURE 18.6

Looking back at FIGURE 18.3, notice that Punnett squares predict the phenotypic ratios that Mendel observed in his pea plant experiments. Crossing a homozygous dominant individual and a homozygous recessive individual yields 100 percent heterozygous offspring, regardless of the trait. All of the offspring will express the dominant trait. Self-pollinating these heterozygotes yields three phenotypically dominant offspring and one phenotypically recessive individual (who has a homozygous recessive genotype). The same Punnett square can be used to represent flower color in peas or attached earlobes in humans. It is amazing that Mendel accurately explained this using his "heritable unit" without any knowledge of genes or chromosomes. Even with inheritance patterns of codominance or incomplete dominance, Punnett squares predict the proportions of potential genotypes of the offspring. The phenotypic expression of those genes may not yield the typical 3:1 or 9:3:3:1 ratios expected by Mendel, but the genotypes ratios remain the same.

CONCEPT CHECK

What is meant by "homozygous dominant"? "heterozygous"?

Define multifactorial traits.

What can be learned from a Punnett square?

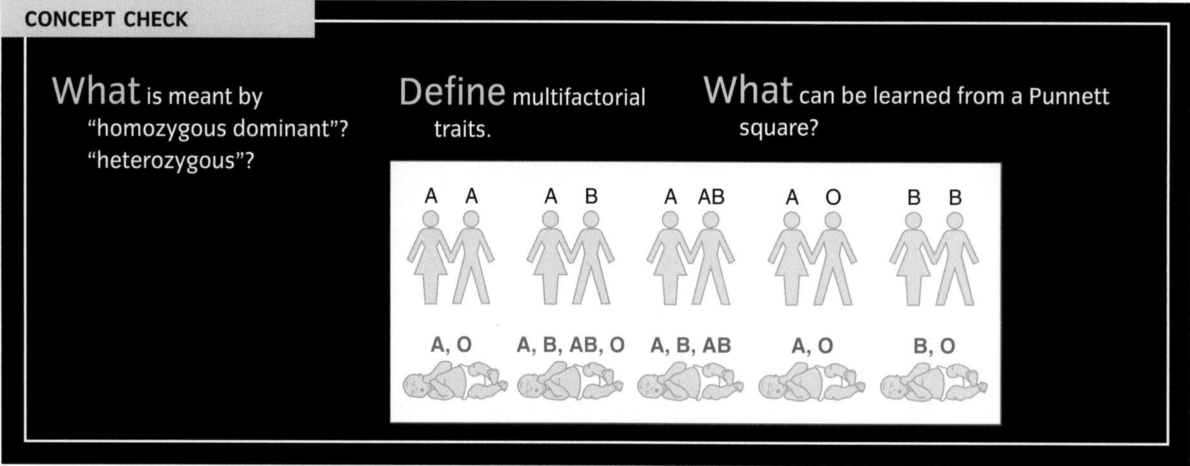

The Central Dogma of Genetics Is Still Accurate

Learning Objectives

Summarize the steps in transcription and translation.

Understand the significance of Beadle and Tatum's experiment.

How do we know that alleles are the heritable units of Mendel's observations? Although this seems obvious now, considerable time and several breakthroughs were required to identify the "heritable unit," find out where it existed in the cell, and determine how it worked. In 1941, two scientists demonstrated that DNA was the chemical in Mendel's "heritable unit." George Beadle and Edward Tatum, using cultures of the fungus *Neurospora*, showed that one sequence of DNA coded for one protein. This **one gene, one enzyme** idea marked the beginning of our understanding of how DNA produces proteins. Before this, it was thought that proteins might contain the unit of heredity because they occur in such enormous variety. Early scientists thought that since 20 amino acids make up the myriad proteins in the body, but only four nucleotides comprise DNA, surely the amino acids were the key to inheritance. This line of thought suggested that proteins were the basis of heredity.

Although proteins seemed a logical candidate as the genetic material, many scientists had begun to question this theory. Beadle and Tatum began looking for a way to conclusively identify the heritable unit. They understood that X rays caused mutations that could prevent proper functioning of some pathways in organisms. They reasoned that if they could "knock out" and then restore a function, they could learn what molecule was carrying the information that the radiation destroyed.

Beadle and Tatum first irradiated colonies of *Neurospora* to create mutant strains, and then they mated them with normal colonies (Figure 18.7, p. 600). They spread the resulting spores on **minimal media**, which lacks amino acids and vitamins, and moved copies of the resulting mold colonies onto various **supplemented media** containing certain needed amino acids. Colonies that could not grow on the minimal media, but did flourish on one of the supplemented media plates, demonstrated that the mutated molecule merely prevented the ability of the mold colony to metabolize one amino acid. By learning what was necessary for growth in minimal media, Beadle and Tatum demonstrated that knocking out one gene inhibits the function of one protein. This was good evidence that DNA controls protein production.

Minimal media Growth media consisting of only the essential requirements for survival.

Supplemented media Growth media with added nutrients and growth factors.

TRANSCRIPTION AND TRANSLATION CONVERT DNA INTO PROTEIN

The next step was to determine how DNA controls the production of proteins. This mechanism has two steps: **transcription** and **translation**. As you learned in Chapter 3, transcription and translation are the processes that convert the information carried on DNA into proteins for the cell. Transcription is copying information from one medium to another using the same language or alphabet. As you hear a lecture and take notes, you transcribe the information you hear into written form. Translation is converting information from one language to

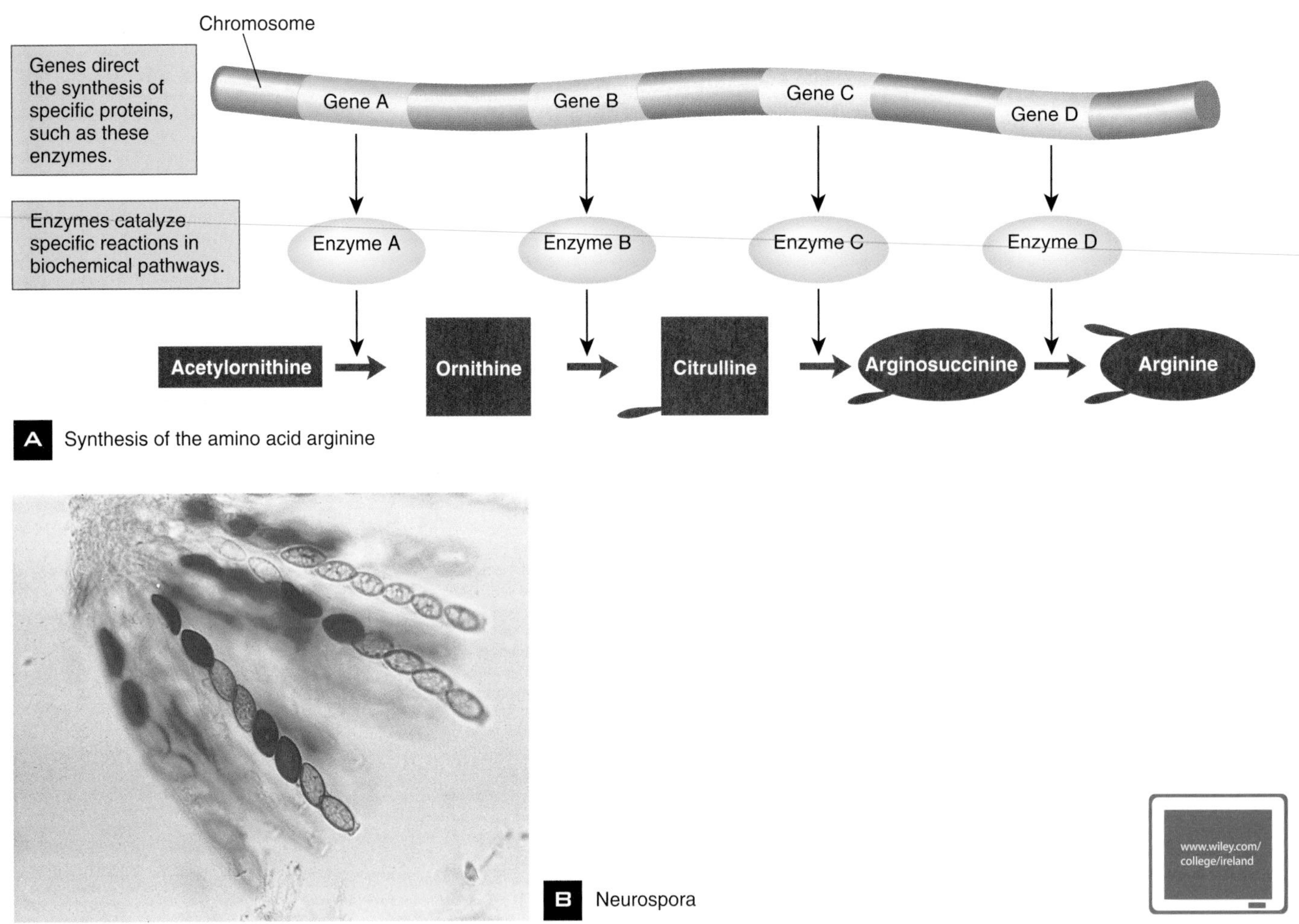

A Synthesis of the amino acid arginine

B Neurospora

The one-gene-one-protein theory FIGURE 18.7

another. If English is not your native tongue, you may be translating the words on the page into a more familiar language as you read them. In the formation of proteins, the meanings of transcription and translation have similar meanings.

Transcription is a change in medium

Recall that the information for new proteins is encoded in DNA, which is stored in your cell nuclei but the machinery for making proteins resides in the cytoplasm. Transcription is the copying of a sequence of nucleotide bases in DNA to **messenger RNA (mRNA)**. Unlike DNA, mRNA can leave the nucleus and carry information from the DNA to the cell's protein-producing machinery.

Recall from Chapter 2 that there are structural differences between DNA and RNA. RNA is a single-stranded molecule, composed of individual nitrogenous bases arranged along a sugar phosphate backbone. Although this backbone is similar to that in DNA, the sugar in RNA is ribose, not the deoxyribose of DNA. The nitrogenous base thymine is replaced by **uracil** during RNA synthesis. The usual base-pairing rule of DNA (A to T and C to G) is altered in RNA because of this substitution. Here the bases pair up A to U and C to G.

Translation makes the proteins

After the DNA code is transcribed to mRNA, it must be converted (translated) from nucleic acid "language" to amino acid "language." This occurs at the ribosomes, using **transfer RNA (tRNA)** to match up one base pair to the mRNA. Messenger RNA is "decoded" by tRNA three bases at a time. These three bases on mRNA are called a **codon**. The matching three bases on the tRNA molecule are the **anticodon**. When codon and anticodon meet at the

The codon table TABLE 18.3

Second Nucleotide in Codon

First nucleotide in codon (5′end)	U	C	A	G	
U	UUU Phe F *Phenylalanine*	UCU Ser S *Serine*	UAU Tyr Y *Tyrosine*	UGU Cys C *Cysteine*	U
U	UUC Phe F *Phenylalanine*	UCC Ser S *Serine*	UAU Tyr Y *Tyrosine*	UGC Cys C *Cysteine*	C
U	UUA Leu L *Leucine*	UCA Ser S *Serine*	UAA Stop codon	UGA Stop codon	A
U	UUG Leu L *Leucine*	UCG Ser S *Serine*	UAG Stop codon	UGG Trp W *Tryptophan*	G
C	CUU Leu L *Leucine*	CCU Pro P *Proline*	CAU His H *Histidine*	CGU Arg R *Arginine*	U
C	CUC Leu L *Leucine*	CCC Pro P *Proline*	CAC His H *Histidine*	CGC Arg R *Arginine*	C
C	CUA Leu L *Leucine*	CCA Pro P *Proline*	CAA Gin Q *Glutamine*	CGA Arg R *Arginine*	A
C	CUG Leu L *Leucine*	CCG Pro P *Prohne*	CAG Gin Q *Glutamine*	CGG Arg R *Arginine*	G
A	AUU Ile I *Isoleucine*	ACU Thr T *Threonine*	AAU Asn N *Asparagine*	AGU Ser S *Serine*	U
A	AUC Ile I *Isoleucine*	ACC Thr T *Threonine*	AAC Asn N *Asparagine*	AGC′ Ser S *Serine*	C
A	AUA Ile I *Isoleucine*	ACA Thr T *Threonine*	AAA Lys K *Lysine*	AGA Arg R *Arginine*	A
A	AUG Met M *Methionine* Start codon	ACG Thr T *Threonine*	AAG Lys K *Lysine*	AGG Arg R *Arginine*	G
G	GUU Val V *Valine*	GCU Ala A *Alanine*	GAU Asp D *Aspartic acid*	GGU Gly G *Glycine*	U
G	GUC Val V *Valine*	GCC Ala A *Alanine*	GAC Asp D *Aspartic acid*	GGC Gly G *Glycine*	C
G	GUA Val V *Valine*	GCA Ala A *Alanine*	GAA Glu E *Glutamic acid*	GGA Gly G *Glycine*	A
G	GUG Val V *Valine*	GCG Ala A *Alanine*	GAG Glu E *Glutamic acid*	GGG Gly G *Glycine*	G

In this table, the first nucleotide in the codon identifies the row of the amino acid. The second nucleotide in the codon identifies the column of the amino acid. This process identifies a box of four amino acids. The third nucleotide in the codon specifies the amino acid. Each triplet (e.g., UUU) is a codon. The table also shows the amino acids encoded by the codons and the three-letter and single-letter abbreviations for each amino acid. In many cases, as shown in the table, codons that encode the same amino acid differ in the third nucleotide only (e.g., CCU, CCC, CCA, and CCG all encode proline). UAA, UAG, and UGA are stop codons. AUG encodes methionine and is also a start codon.

ribosome, the amino acid carried by the tRNA is incorporated into the growing polypeptide chain. Each codon indicates one of the 20 amino acids.

The first and second base in the codon are vital, as they determine the specific amino acid added to the growing protein chain. The third base carries information, but it may "wobble," or change, without affecting the amino acid sequence of the protein. For example, from the codon chart shown in TABLE 18.3, you can see that CUU, CUA, CUG, and CUC all code for the amino acid leucine. Another example concerns the codons that begin with AC. No matter which base occupies position three, the amino acid specified is threonine. These RNA codons are termed redundant because several codons can indicate the same amino acid. All RNA codons are specific because each codon codes for only one amino acid.

Biologists call the mechanism of transcription and translation the "Central Dogma of Biology" because it has relevance to all aspects of their science.

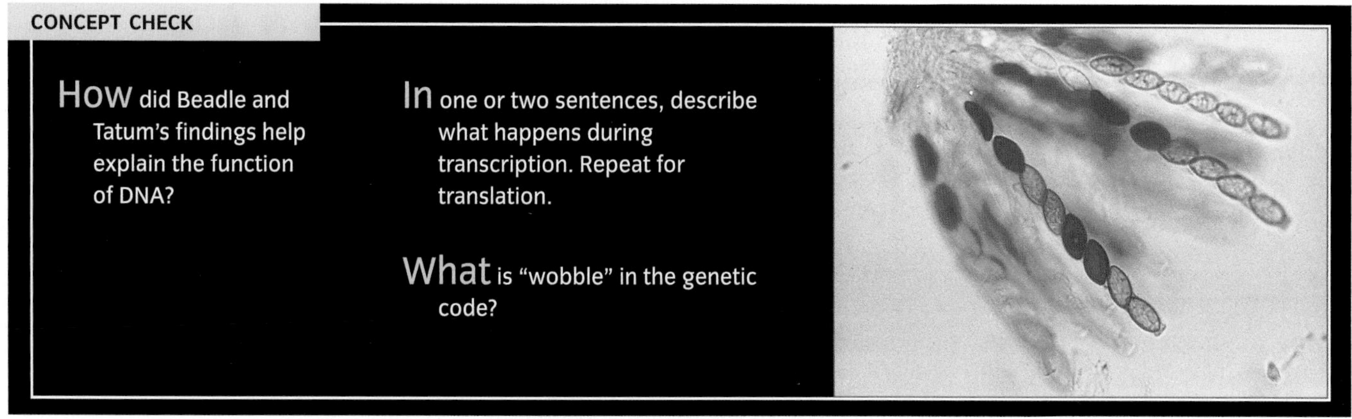

CONCEPT CHECK
How did Beadle and Tatum's findings help explain the function of DNA?
In one or two sentences, describe what happens during transcription. Repeat for translation.
What is "wobble" in the genetic code?

Genetic Counseling Puts Genetic Theory to Practical Use

LEARNING OBJECTIVES

Understand the information in a pedigree chart.

Define sex-linked traits.

Describe chromosomal disorders.

Relate population genetics to evolution.

Couples often request genetic counseling before they choose to conceive. Genetic counseling is the practice of predicting the potential combinations of alleles two individuals may produce. If there is a family history of congenital disease, or if the potential parents feel they are at risk of carrying a detrimental recessive allele, genetic counseling can help alleviate their fears. Knowing the probability of having a child with a genetic anomaly can help couples decide whether to conceive.

PEDIGREE CHARTS TRACE TRAITS THROUGH FAMILIES

Pedigree charts are symbolic representations of genetic transmission of phenotypic traits through families. Using a pedigree chart (FIGURE 18.8A), researchers can trace the pathway of a disease through families, and characteristics of its transmission can be deduced. If, for example, the disease is **autosomal** dominant, anyone with alleles Aa or AA will be afflicted (FIGURE 18.8B). If the disease shows up sporadically or appears in a child of two **asymptomatic** parents, the disease is probably autosomal recessive, and both parents were heterozygous carriers for the dysfunctional allele (FIGURE 18.8C).

Autosomal Any chromosome other than the sex chromosomes, X and Y.

Asymptomatic Without symptoms.

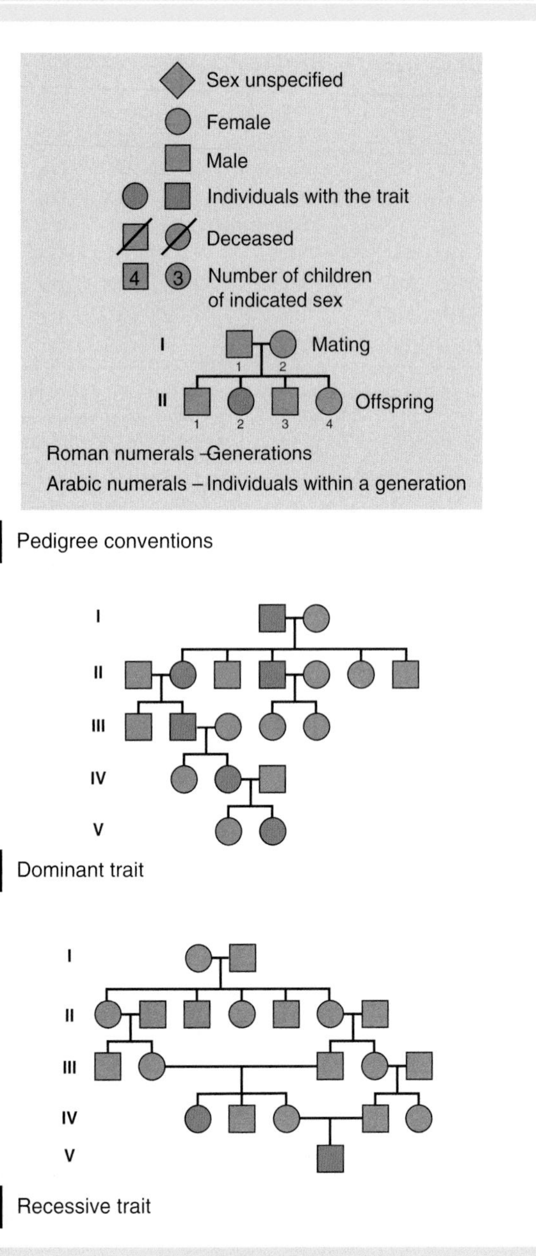

Pedigrees of autosomal dominant and autosomal recessive diseases FIGURE 18.8

SOME TRAITS ARE SEX-LINKED

Humans have one pair of chromosomes, called the sex chromosomes, that do not match in terms of size or content. The sex chromosomes include the large X chromosome and the smaller Y. These two determine gender. If a Y chromosome is present during development (XY), the fetus will become a male. If there is no Y present (XX), the fetus becomes female.

Females have two copies of every gene on the X chromosome, just like they do on their autosomal genes. Since only one copy of each allele is needed during normal growth and development, one X chromosome is randomly shut down. This shutdown occurs during development, leaving one condensed X chromosome as a **Barr body** within the nucleus (FIGURE 18.9). All the cloned progeny of this cell get the same functional X and the same Barr body. Thus, human females consist of patches of genetically distinct tissues, based on which X is inactivated. This is called mosaicism. In humans this mosaic pattern is difficult to discern, but through genetic testing, patches of tissue can be identified as clones of the same original cell. During development, the alleles on the active X chromosome are expressed and those on the inactivated X are repressed. Differences in the genes of the two X chromosomes are markers for these cloned cell populations. In other organisms, this mosaic patchiness is more easily discerned. (For more on this topic, see the I Wonder . . . box on p. 604.)

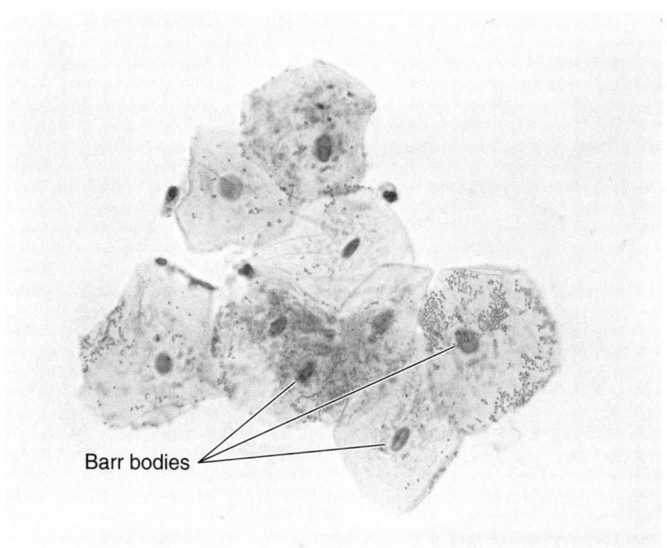

Nucleus with Barr bodies
FIGURE 18.9

The Y chromosome includes few functional genes, with the most recent count coming to just 78 genes. It had previously been assumed that there were no genes of consequence on the Y chromosome, but as the number of genes identified increases this seems illogical. Scientists are just beginning to understand the significance of the Y chromosome genes to the male. Only one, the ***SRY*** gene, codes for male anatomical traits. The remaining Y chromosome genes are "housekeeping" genes, genes that are active in most body cells and do not confer male characteristics. None of these genes have specific homologous counterparts on the X chromosome. This is a potential problem during nuclear division, as the Y chromosome does not condense and pair up with the X chromosome in the same fashion as autosomal chromosomes. As mentioned, these two sex chromosomes are not homologous, and therefore they show little pairing and crossing over during cell division. Instead, the Y chromosome includes a series of **palindromes** that allow it to fold back on itself during cell division. With limited ability to cross over during meiosis or to silence dysfunctional genes on either the X or the Y during development, mutations are more often retained and expressed in the developing male. In females, two copies of X chromosome genes double the chance of a functional allele. The male, however, has only one X. The alleles on that single chromosome must be used even if they are slightly defective.

> **Palindrome**
> A group of nucleotides with the same sequence when read in either direction (CGTTGC).

Genes carried on one sex chromosome with no counterpart on the other sex chromosome code for **sex-linked traits**. Because there are so many more functional genes on the X chromosome than the Y, these are the genes usually referred to when discussing sex-linked traits. Characteristics carried on the

I WONDER . . .

Why are tortoise shell cats always female?

Female animals receive one X chromosome from each parent, while males get an X from the mother and a Y from the father. This causes some traits to be "sex-linked," or present in only one gender. The Y chromosome carries less than 100 genes that govern the development of male anatomy and physiology. The X chromosome carries many more genes, some of which are reproductive and others that are essential to normal body functioning. The female embryo has two X chromosomes, but one randomly shuts down in each embryonic cell early in development. The same X chromosome is active in all daughter cells of any particular embryonic cell, creating a situation where female body cells have two distinct genetic lineages.

Biologists call such a person a mosaic, and the different expression rates of alleles on the X chromosomes can have visible and invisible consequences. In cats, X-pattern inheritance is visible. Fur color is partly encoded on the X chromosome, as shown in the diagram. Hair follicles may have an X chromosome that codes for either orange or black fur, which explains the speckled appearance of a tortoiseshell. Tortoise shell fur cannot appear in males because all of their X chromosomes came from the mother's X chromosome, so they are all identical. There is no shutting down of one X chromosome, as there is only one per cell.

Like cats, female humans are mosaics, and this affects traits coded by the X chromosome. If females have a defect on only one of their two X chromosomes, the functional copy of the gene may be able to compensate for the defective X. The varying proportion of each particular X chromosome in the adult tissues of the female may explain why the onset and intensity of X-linked genetic diseases vary more in women than in men.

Because males have only one type of X chromosome, any defective gene on it can cause disease. Scientists say this explains why males have higher rates of X chromosome diseases, such as Duchenne muscular dystrophy and hemophilia. But, paradoxically, having only one type of X chromosome may have accelerated evolution among males. X-linked diseases can be powerful enough to kill males in utero, or before reproductive age, which helps remove the defective genes from the population. In evolutionary terms, a weakness for the individual becomes a strength for the group.

Mosaicism is a complicated story, but the tortoise shell cat with her pretty, varied fur says it all. "All X chromosomes are not created equal!"

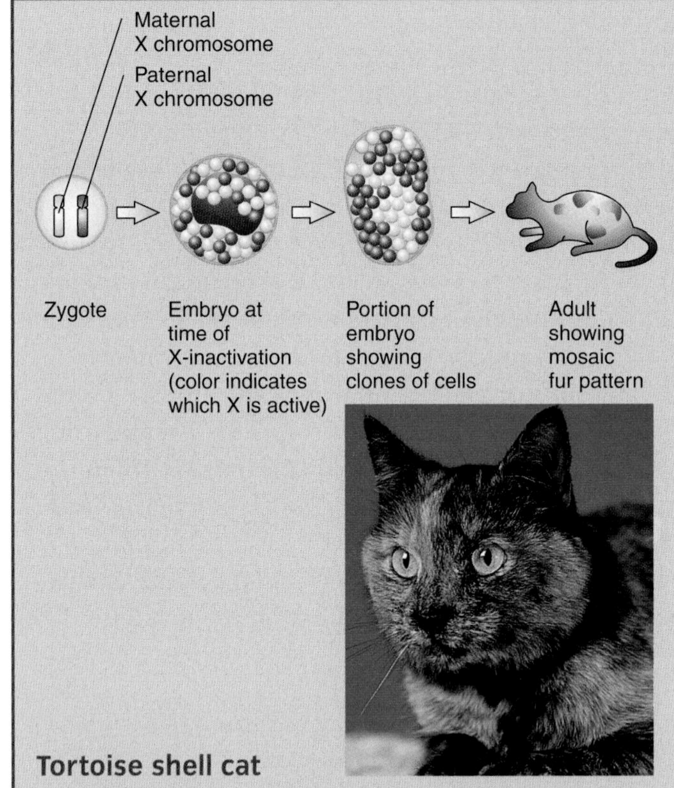

Tortoise shell cat

In a tortoise shell cat, the X chromosome may encode for fur that is orange or black. Because one X chromosome is inactivated in each embryonic cell, the female body contains two genetic clones: one expresses black fur, the other orange. These clones show up as patches of different colors on the cat.

X chromosome include color blindness and hemophilia. (FIGURE 18.10).

Male pattern baldness (see FIGURE 16.10 on p. 524) is not a sex-linked trait, which should immediately let you know that the gene causing the trait is not found on the X chromosome. Instead, the gene for this type of hair loss is carried on an autosomal chromosome. It is a polygenic trait, however, with one factor contributing to its phenotypic expression located on the X chromosome. Showing this trait is far easier, therefore, if the individual is male. **Sex-influenced** traits are carried on autosomal chromosomes but are more common in one sex than the other. Hormonal differences between the two genders are most often responsible for the altered expression of these genes.

Expression of the autosomal allele for male pattern baldness follows a simple dominant/recessive relationship. It is, however, controlled by hormonal levels.

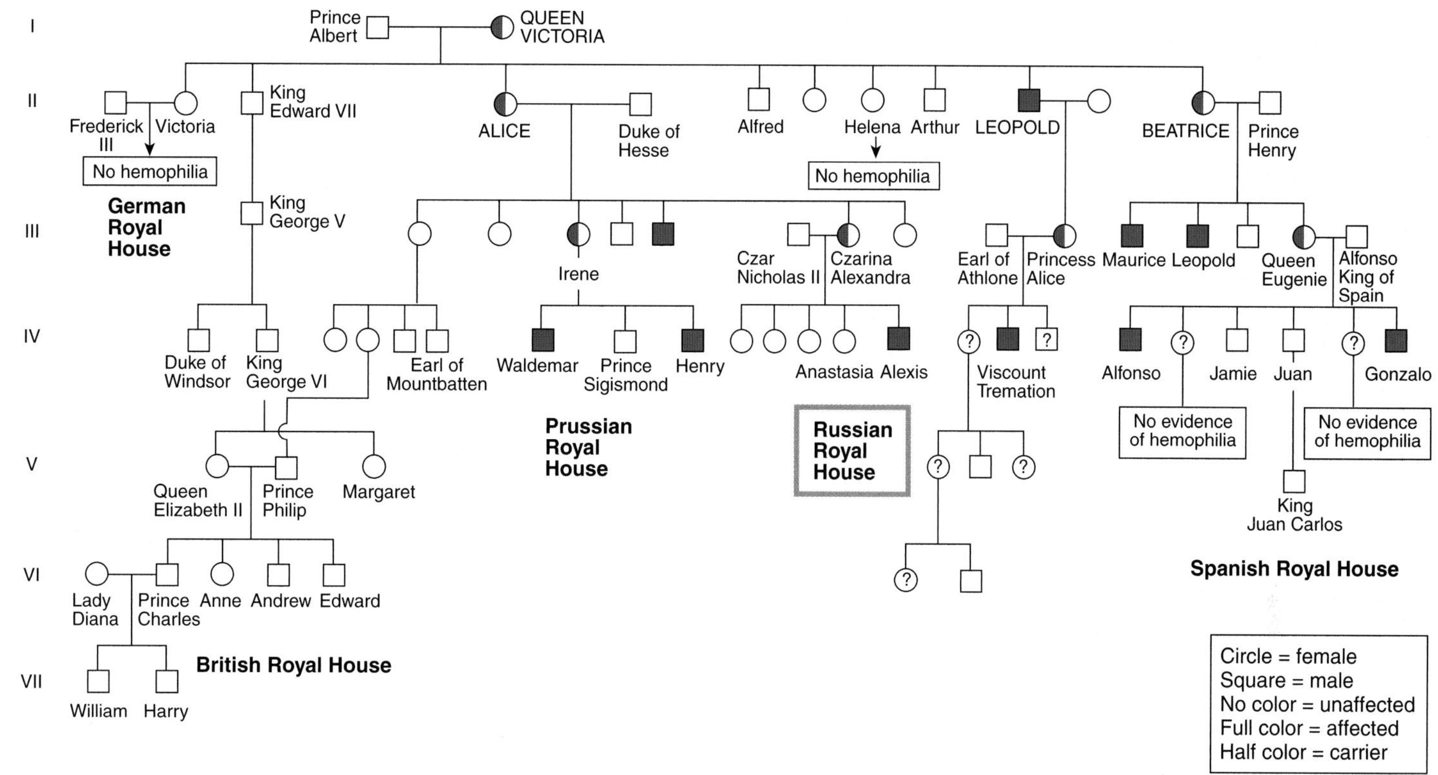

Pedigree of hemophilia in the royal family of Queen Victoria FIGURE 18.10

If testosterone is present in the body along with the dominant form of the allele, the dominant allele will be expressed and pattern baldness will result. If no testosterone is present, or if testosterone is present only in very low quantities as is typically true for females, pattern baldness may be present, but it will not be expressed (TABLE 18.4).

Many of the genetic variations we see in the human species are due to mutations that have been perpetuated in small, often isolated populations. Human populations, just like any animal population, will undergo more rapid **evolution** when they are reproductively isolated. Language and ethnicity can isolate human populations, even those that live in close physical proximity. Geographic structures such as mountains, deep valleys, or broad deserts, can also isolate human populations.

Evolution
Descent with modification.

Pattern baldness genotypes and phenotypes
TABLE 18.4

Genotype	Gender	Phenotype
BB	Male	Pattern baldness
BB	Female	Thinning hair late in life
Bb	Male	Pattern baldness
Bb	Female	Normal hair retention
bb	Male	Not bald
bb	Female	Not bald

(B = dominant bald allele)

Despite isolation, humans exhibit a range of expected phenotypes. For example, natural skin tones range from extremely pale tan to very dark brown. People are NOT blue, right? Wrong! Amazingly, a group of people in the Troublesome Creek area of Kentucky occasionally produce a blue child (FIGURE 18.11). These children are normal in all other respects. They are born the color of a bruised plum, and often they retain that color into adulthood. Some appear blue only when angry or cold.

What is going on? The people of this area are all descendants of Martin Fugate, a French orphan, and his red-headed American wife. Against all odds, both Martin and his wife carried a recessive gene coding for a nonfunctional enzyme in the blood. Methemoglobin is a blue oxygen-carrying compound, a variant of normal hemoglobin, that is usually present in small amounts in blood. Normally the enzyme diaphorase converts methemoglobin to functional hemoglobin, restoring blood's red color. These people in Kentucky cannot convert methemoglobin to hemoglobin, resulting in that blue coloration to the skin and mucous membranes. Geographic barriers have isolated the population with the "blue gene" to the Troublesome and Ball Creek valleys.

Fortunately for these blue people, the body has other compounds that will convert blue methemoglobin to red hemoglobin. All that is really needed is an electron donor. One of these is a compound you may be familiar with; methylene blue, the common laboratory blue stain, has been safely used to restore the usual red color when used medicinally. Happily, modern medicine can correct the blue color, allowing these people to lead normal lives without the stigma of blue skin, but the allele remains in the population. This strange phenomenon clearly demonstrates that inheritance and evolution follow the same rules in humans as in other animals. Genes carry traits, and traits can be either lost or enhanced in populations.

"Blue" phenotype from Troublesome Creek, Kentucky FIGURE 18.11

GENETIC COUNSELING CAN HELP AVOID CHROMOSOMAL DISORDERS

Genes and chromosomes can be damaged during cell division. Errors can occur in an entire chromosome, part of a chromosome, or a single gene. Gross errors, called chromosomal disorders, include variations in chromosome structure or number. These disorders include Down syndrome and fragile X mental retardation. According to the March of Dimes, chromosomal disorders affect about 7.5 percent of fertilizations, but many cause extreme deformities. Because many defective embryos abort spontaneously, only about 0.6 percent of live births show genetic defects due to chromosomal disorders. Just as it can be heartbreaking to lose a baby during pregnancy, it can be difficult to raise a child born with a **congenital** defect. Most babies born with congenital defects have trouble caused by a single gene with recessive or dominant allele. Therefore, the chances of conceiving a child with a congenital defect caused by recessive or dominant alleles can be predicted with the Punnett square.

Other genetic disorders are caused by a series of alleles spread over several genes, which, if present in one individual, lead to the expression of a genetic defect. These defects depend on the balance of interaction between several genes and the environment. These **multifactorial disorders** include cleft lip and

Congenital
A condition that is present at birth, due to genetic or environmental factors; usually detrimental.

Multifactorial disorder
Genetic disorder due to a combination of genetic and environmental factors.

palate, rheumatoid arthritis, epilepsy, and bipolar disorder. Simple traits such as skin color, hair color, and weight are also multifactorial traits. As you know from observing these traits in your own family, the inheritance of multifactorial traits is most apparent in the immediate generation. As individuals become farther removed from the affected individual (the carrier), the trait disappears. As an example, your hair color is probably closer to that of your parents than to your great-grandparents.

Sometimes unpredictable genetic disorders, caused by mutations or improper meiotic divisions, appear in families. Mutations occur with amazing frequency, at an estimated rate of about 1 misplaced base per 50 million nucleotides. That works out to 120 mutations per new cell. Although several enzymes "patrol" your DNA looking to repair these errors, and natural selection is constantly trying to delete defective genes from new generations, genes are not perfect. Things can go wrong. However, with our increasing knowledge of genetics, potential parents have tools at their disposal that take some of the guesswork out of producing healthy children. One option is the time-tested "let's fall in love, get married, and take our chances" approach. But some couples are more interested in taking control of their genetics. For these people, genetic counseling is a great choice. Table 18.5 lists some of the many genetic disorders that are discussed in genetic counceling.

Certain religious or ethnic groups have a higher proportion of detrimental recessive alleles than others, because their populations intermarry more than other groups. Ashkenazi (north European) Jews and French Canadians, for instance, have a higher likelihood of carrying the recessive allele for Tay-Sachs disease. Tay-Sachs is a fatal disease caused by a dysfunctional lysosomal enzyme in the brain. Normally, neurons create fatty substances that are easily removed from the brain by lysosomes. In Tay-Sachs, the allele that codes for the enzyme which breaks down these fatty substances is defective, so the fats build up. The affected homozygous recessive individual develops normally until age 4, but brain function then deteriorates rapidly. The gene for this defective lysozomal enzyme is

Genetic disorders, their symptoms, and their predominant carriers Table 18.5

Disorder	*Type*	*Symptoms*	*Carriers/Type of disease*
Huntington's disease	Chromosome abnormality	Affects the brain, causing poor memory, lack of coordination, mood swings, lack of fine motor control	Autosomal dominant disease
Turner syndrome	Chromosome abnormality	Short stature, improperly developed ovaries, stocky appearance, webbed neck, low hairline	Missing or incomplete X chromosome (XO female)
Klinefelter syndrome	Chromosome abnormality	After puberty, males develop breast tissue, have less muscle mass, and little facial hair	XXY males
Cri-du-chat syndrome	Chromosome abnormality	Distinctive cry due to abnormal larynx development, small birth weight, microcephaly, heart defects, facial deformities	Deletion in chromosome 5
Phenylketonuria	Single-gene disorder	Severe brain damage, epilepsy, eczema, microcephaly, and a musty body odor	Autosomal recessive
Severe combined immunodeficiency disorder (SCID)	Single-gene disorder	High rate of infections soon after birth, including pneumonia and meningitis	X-linked recessive trait
Sickle cell disease	Single-gene disorder	Loss of function in organs where oxygen delivery is compromised, shortened life span	Autosomal recessive trait
Cystic fibrosis	Single-gene disorder	Coughing, wheezing, respiratory illnesses, salty-tasting skin, weight loss	Defective gene on chromosome 7; autosomal recessive
Marfan syndrome	Single-gene disorder	Connective tissue disorder causing excessive growth with little strength, long fingers, toes, and shins, weak heart valves	Autosomal dominant disease; defective gene on chromosome 15

Ethics and Issues

Ethics of prenatal genetic testing

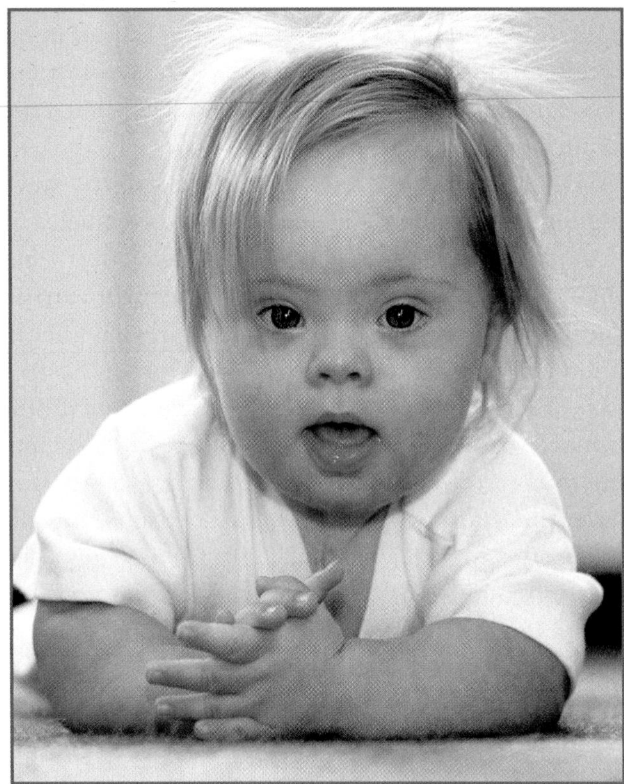

Rapid advances in genetics have raised the promise—or the peril—of studying the genetics of children yet to be born. Already, prenatal ultrasound can reveal the sex of a fetus, and some parents in the many cultures that favor male babies have responded by aborting female fetuses. This practice has increased the ratio of male to female children in China, and perhaps elsewhere.

As you remember from Chapter 17, prenatal genetic testing can take two forms: testing the developing baby with chorionic villus sampling or amniocentesis, and testing the genes of potential parents. Each process raises ethical questions.

Prenatal genetic sampling is done primarily to detect chromosomal abnormalities, such as Down syndrome, and focuses on women aged 35 or over who are more likely to have children with these abnormalities. These genetic problems cannot be corrected, and the parents must either abort such a "defective" fetus or understand and accept the challenges of raising such a child. If a problem is discovered, at the very least the test results can alert the parents to their future child's special needs. Any benefits of these tests must be weighed against the chance that the invasive sampling itself will harm the fetus. In the future, as the knowledge of genetics increases, we may see sampling of the embryo itself, in the hopes of intercepting genetic diseases even earlier.

A more complicated set of ethical questions arises when parents want to analyze their own genes before conception. In a few cases, the need for such analysis is clear and convincing. If a genetic disease like the deadly nerve disorder Huntington's disease, runs in the family, parents might want assurance that they will not pass the gene to their children. If testing reveals a high probability of their passing on this disease, the would-be parents may want to avoid pregnancy.

But genetic situations can present confusing ethical decisions, especially now that scientists are detecting the genetic components of dozens or even hundreds of diseases and conditions. Many of these genes do not amount to a death sentence. Would a genetic predisposition for cancer matter if the gene raised the child's risk of cancer by 10 percent? What if it doubled the risk of cancer? A gene for high cholesterol might, statistically, shorten the life span. But statistics apply to groups, not individuals. How would you respond to the idea that your child might face extreme cholesterol levels?

Prenatal genetic analysis also raises social and economic questions: Testing and genetic counseling are expensive, and yet the uncertainties are likely to outnumber the certainties for years to come. How much are you willing to spend to get a glimpse of your potential child's genetic future?

Genetic testing has aroused opposition among disability rights groups, who tend to see it as eugenics: an effort to "improve" the human **genome** by pruning out genes—and people—with "defective" genes.

Genome
Total genetic content of an organism.

The picture is complicated now, and the only thing we can say for sure is that better knowledge of genetics will make the issue of prenatal testing even more complex. A strong basic understanding of human genetics will help prepare you to answer the difficult questions you may confront during your reproductive years.

recessive, so phenotypically normal heterozygous carriers are not aware that they are carrying this potentially lethal mutation. Sadly, up to 1 in 25 Ashkenazi Jews are thought to be carriers. When marrying within the faith, Ashkenazi Jews often request a compatibility score from a genetic counseling service, which will indicate their probability that their child would have Tay-Sachs. As of May 2005, 646 marriages had been found to be incompatible, and over 150,000 people had been tested in just one testing facility. Each child from those 646 discouraged marriages would have had a 25 percent chance of suffering from the disease, and a 75 percent chance of carrying it. This type of testing has caused a dramatic reduction in deaths due to Tay-Sachs.

Tay-Sachs is a relatively simple disease to explain. If both parents are carriers, each child has a 25 percent chance of expressing the disease. But genetic counseling regarding other traits and disorders can be a lot more complicated, and it can force couples to make difficult decisions. If there is a family history of congenital disease, or if the potential parents feel they are at risk of carrying a detrimental recessive allele, genetic counseling can help them understand the situation or alleviate their fears (see the Ethics and Issues box).

CONCEPT CHECK

What can be learned from a family pedigree?

Which chromosome carries the genes for sex-linked traits?

How do the people in Troublesome Creek, Kentucky, help demonstrate the theory of evolution?

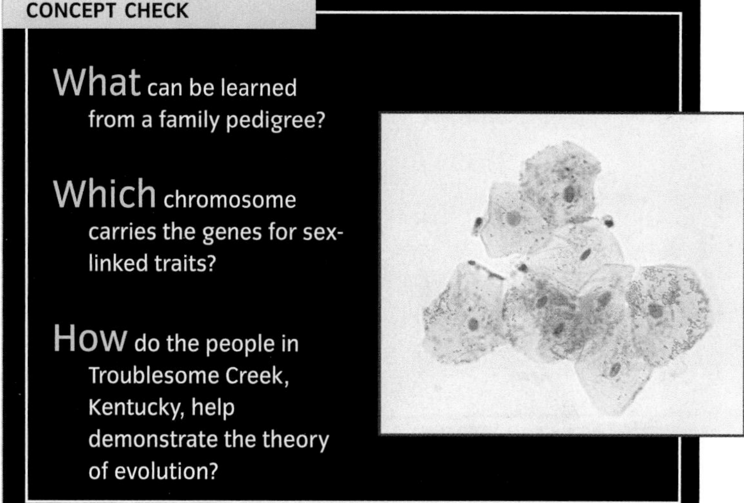

Biotechnology and DNA: The Facts, the Pros, and the Cons

LEARNING OBJECTIVES

List four biotechnologies used in modern research.

Understand how DNA is used in the courtroom.

Define genetic modification and transgenic organism.

Briefly describe the outcome of the Human Genome Project.

DNA is DNA (FIGURE 18.12, p. 610). When you work with it in the laboratory, it makes little difference where it came from, as all DNA is composed of the same four nucleotides, held in the same basic arrangement. What makes each organism, and each individual, unique is the sequence of nucleotides attached to the sugar-phosphate backbone.

To read the "language" of genetics, we must identify the sequence of bases encoded in DNA. The techniques used to isolate DNA and identify the base sequence include **nucleic acid hybridization, gel electrophoresis, PCR**, and **RFLP analysis**.

To isolate DNA from an organism, we pop the cell membrane and remove the nucleus. Through **centrifugation**, we separate the denser nucleus from the lighter organelles and cytoplasm, thus concentrating the DNA. With specific buffers and chemicals, we remove pure DNA from the nucleus.

Centrifugation Rapid spinning of a sample to separate components by density.

One of the simple techniques used to obtain a pure sample of DNA is to spin the impure DNA sample on a **cesium chloride gradient** (FIGURE 18.13). Pure DNA will form a band where its density matches that of the surrounding cesium chloride. The band can be visualized using the proper DNA staining techniques. Once located, the pure DNA can easily be removed from the cesium chloride gradient with a syringe.

DNA FIGURE 18.12

Francis Crick (shown at left in the photo) and James Watson discovered the double-helix structure of DNA in 1953.

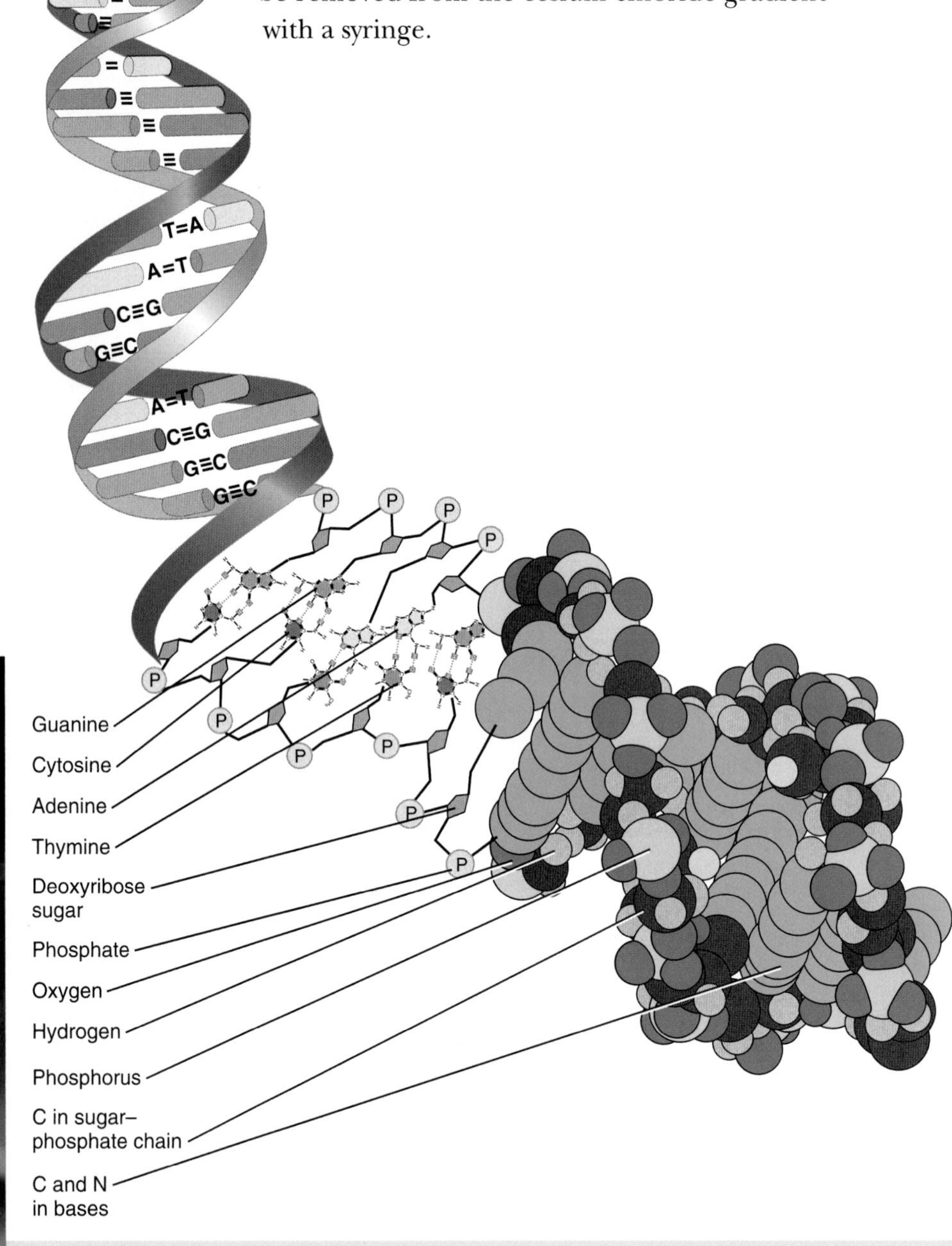

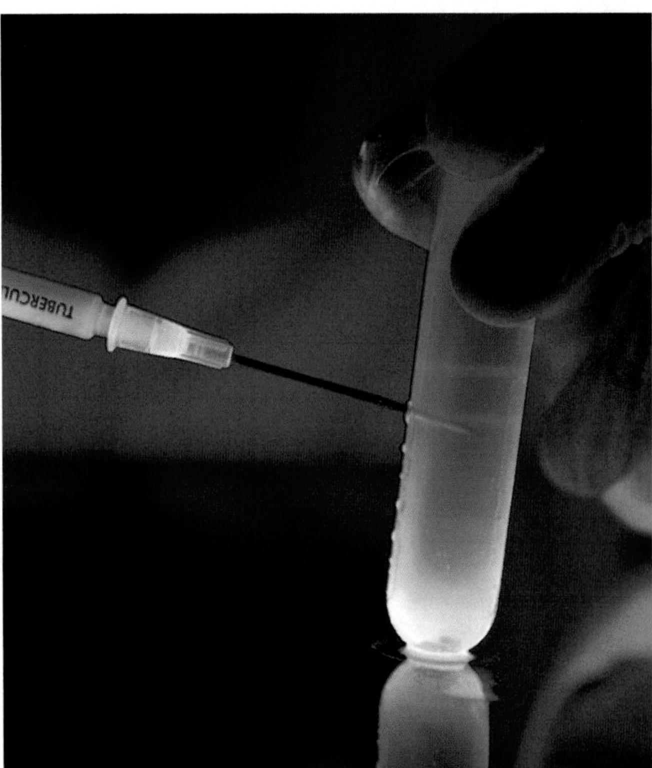

Cesium gradient FIGURE 18.13

When centrifuged at high speeds, cesium chloride solutions form a density gradient from most dense at the bottom of the tube to least dense at the top. The density of pure DNA lies near the middle of this gradient.

PURIFIED DNA CAN BE USED IN LABORATORY PROCEDURES

Once purified DNA is available, it is easy to work with in the lab. DNA behaves predictably. It is double-stranded, and the bases always pair up A to T, and C to G. If conditions favor the **dissociation** of the strands, they will fall apart (FIGURE 18.14). Increased heat is one factor that causes dissociation. To **re-anneal**, or reseal, the two complementary strands of DNA, we return the sample to body temperature. This is of no major consequence if we merely split and then recombine the same pieces of DNA. But if we add DNA from a different source or small pieces of RNA to the mix, things get more interesting. RNA will bind to the dissociated, single-stranded DNA where the bases match. We can then use specific kinds of RNA to locate genes for specific proteins on the DNA by matching the proper RNA from protein translation to the exact segment of DNA that created it.

How is this possible? Return to the central dogma and play the mental game of working backwards from the protein to the DNA sequence that codes for it. We must first choose a protein and determine the amino acid sequence. Insulin was the first human protein sequenced. Because of its relatively simple construction and the ease with which insulin could be harvested, it became the first protein used in this "identify the DNA sequence that codes for the protein" technology. It is a relatively small protein, composed of two polypeptide chains. The first chain has only 21 amino acids and the second has 30.

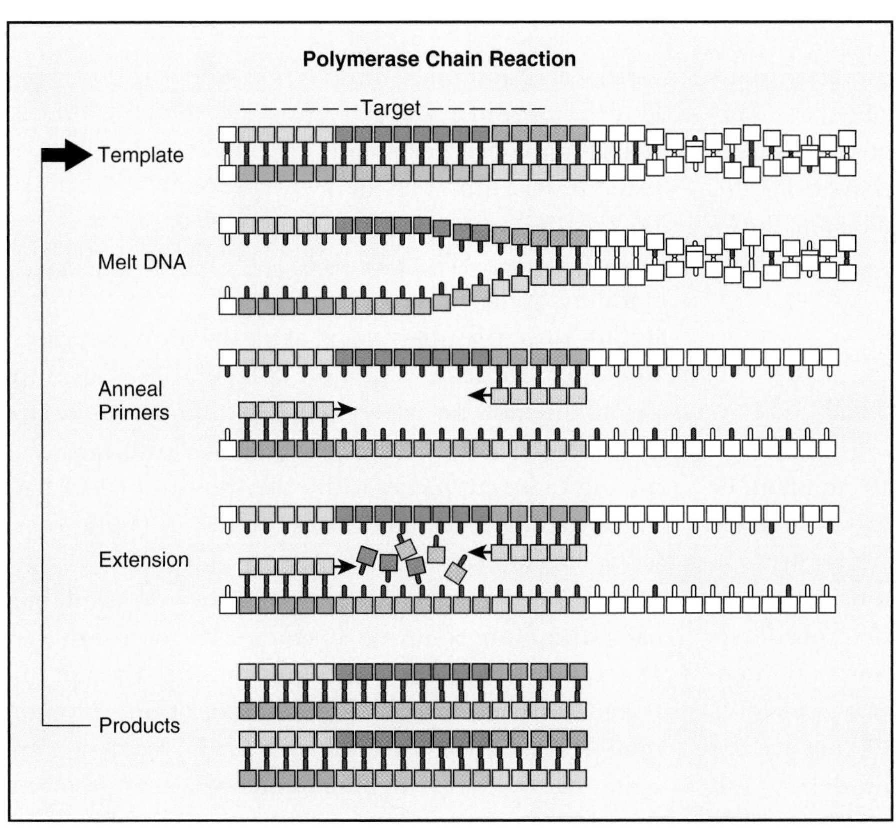

Dissociated DNA and probe annealing FIGURE 18.14

Template DNA is split apart (dissociated) using heat. The single strands of DNA are then exposed to RNA probes that will bind, or anneal, to specific matching areas of the DNA. Transcriptional enzymes will elongate the bound RNA, creating a new strand of DNA. Often the nucleotides provided in the elongation step are radiolabeled to allow experimental tracing of the newly created pieces of DNA.

To sequence a protein, we chemically disassemble it and identify each amino acid as it comes off. Because each amino acid is coded for by a specific three-base codon, we can recreate the tRNA molecules that created the original protein. Working backwards using the codon table (see TABLE 18.3 on p. 601), the amino acid sequence gives us the sequence of tRNA molecules that created the protein. This is the reverse of translation. Going back another step, we take the tRNA anticodon sequence and rebuild the necessary mRNA codon sequence. We can then build this mRNA sequence in the laboratory, which we often "label" with a radioactive compound. Radioactive phosphorus (P-32), being relatively easy to work with, is often used to mark the entire RNA backbone at the phosphate groups. P-32 is easy to trace during the experiment. If you add a laboratory-synthesized piece of P-32-labeled mRNA to dissociated DNA, the radio-labeled mRNA will bind to its complementary spot on the DNA. Once you locate the radioactive

Process Diagram

Insulin mapping FIGURE 18.15

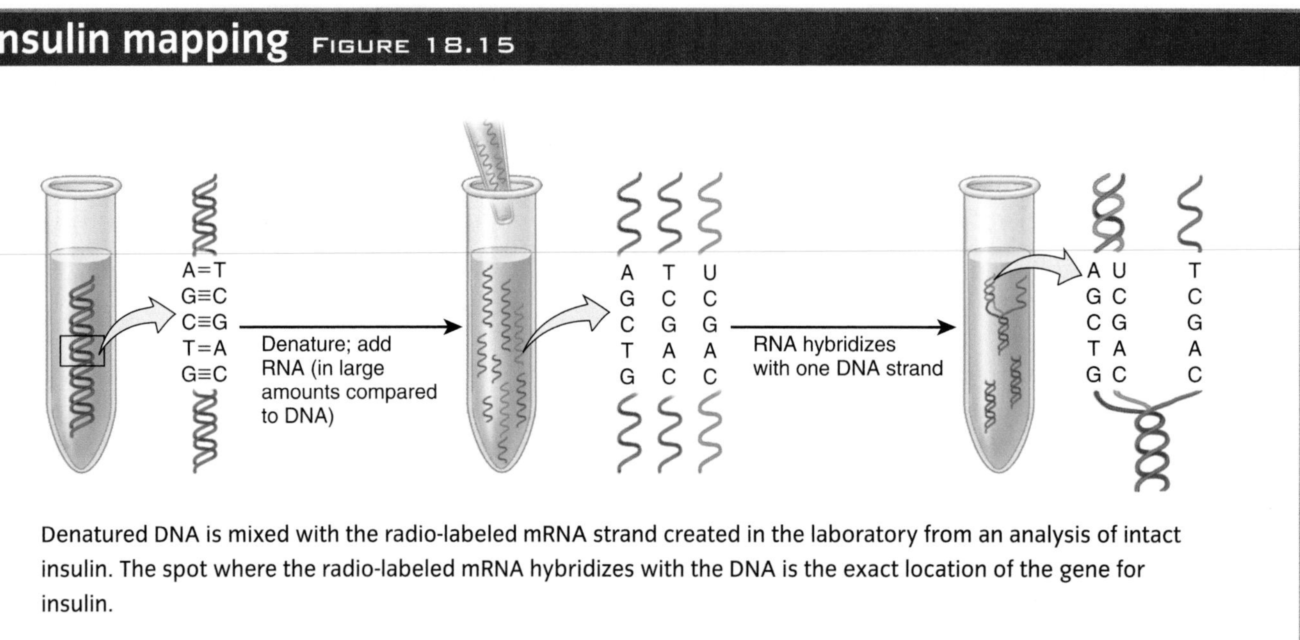

Denatured DNA is mixed with the radio-labeled mRNA strand created in the laboratory from an analysis of intact insulin. The spot where the radio-labeled mRNA hybridizes with the DNA is the exact location of the gene for insulin.

label, you have identified the exact spot on the chromosome where DNA and mRNA are complementary (FIGURE 18.15). And that spot is the location of the gene for your original protein. Scientists now have done this for many of our proteins, giving us a good map of the location of many of our genes.

RESTRICTION ENZYMES ARE THE "SCISSORS" OF BIOTECHNOLOGY

Once we locate a gene, we often want to isolate it, or cut it away from the adjacent DNA. In 1970, two scientists simultaneously discovered that bacteria carry enzymes that can cut DNA at specific palindromes. These enzymes are "restricted" to acting only at a specific sequence of DNA bases. These so-called restriction enzymes act as a kind of immune defense for the bacteria. Because they cut only sequences of DNA that are not found in the bacterial chromosome, they destroy foreign DNA, such as that from an invading virus. In 1970, these two scientists, Howard Temin and David Baltimore, purified a second type of nucleotide-altering enzyme: **reverse transcriptase**, an enzyme with **polymerization** properties opposite those normally found in eukaryotic cells. Reverse transcriptase is able to produce DNA from RNA templates, reversing the usual transcription sequence of creating mRNA from a DNA template. Their understanding of this enzyme, combined with the discovery of restriction enzymes, opened a whole new area of molecular biology, supplying the tools for precise DNA manipulation. By cutting DNA between known sequences of bases, and mixing these cut pieces with other pieces of DNA having matching ends, new genes can be **spliced** into existing chromosomes.

Polymerization The chemical bonding of monomers to form a larger molecule.

Spliced Two pieces of DNA artificially joined together to form new genetic combinations.

Many restriction enzymes are now available, each targeting a unique, specific recognition site (TABLE 18.6). These recognition sites, though different for each enzyme, all have one thing in common—they are DNA palindromes. Some restriction enzymes cut DNA, leaving **blunt ends**—meaning that the two backbones are cut parallel to each other at the same base pair. Others leave **sticky ends**, with the two pieces of DNA cut off-center in the palindrome. This produces two pieces of unequal length, with a 2- to 3-base, single-stranded tail sticking off the end. These two pieces of DNA are unstable and will immediately re-anneal unless prevented from doing so, thus their nickname, "sticky ends" (FIGURE 18.16).

Restriction enzymes, bacterial origins, and cutting sites (HindIII, Bam HI, ecoR1, Pst 2, etc.)
TABLE 18.6

Enzyme	Bacterial origin	Recognition sequence	Cutting site	
EcoRl	*Escherichia coli*	5′GAATTC 3′CTTAAG	5′—G 3′—CTTAA	AATTC—3′ G—5′
BamHl	*Bacillus amyloliquefaciens*	5′GGATCC 3′CCTAGG	5′—G 3′—CCTAG	GATCC—3′ G—5′
Hindlll	*Haemophilus influenzae*	5′AAGCTT 3′TTCGAA	5′—A 3′—TTCGA	AGCTT—3′ A—5′
MstII	*Microcoleus species*	5′CCTNAGG 3′GGANTCC		
Taql	*Thermus aquaticus*	5′TCGA 3′AGCT	5′—T 3′—AGC	CGA—3′ T—5′
Notl	*Nocardia otitidis*	5′GCGGCCGC 3′CGCCGGCG		
Hinfl	*Haemophilus influenzae*	5′GANTC 3′CTNAG		
Alul*	*Arthrobacter luteus*	5′AGCT 3′TCGA	5′—AG 3′—TC	CT—3′ GA—5′

* = blunt ends

Source: Wikipedia, *The Free Encyclopedia*, http://en.wikipedia.org/wiki/Restriction_enzymes#Types_of_restriction_enzymes

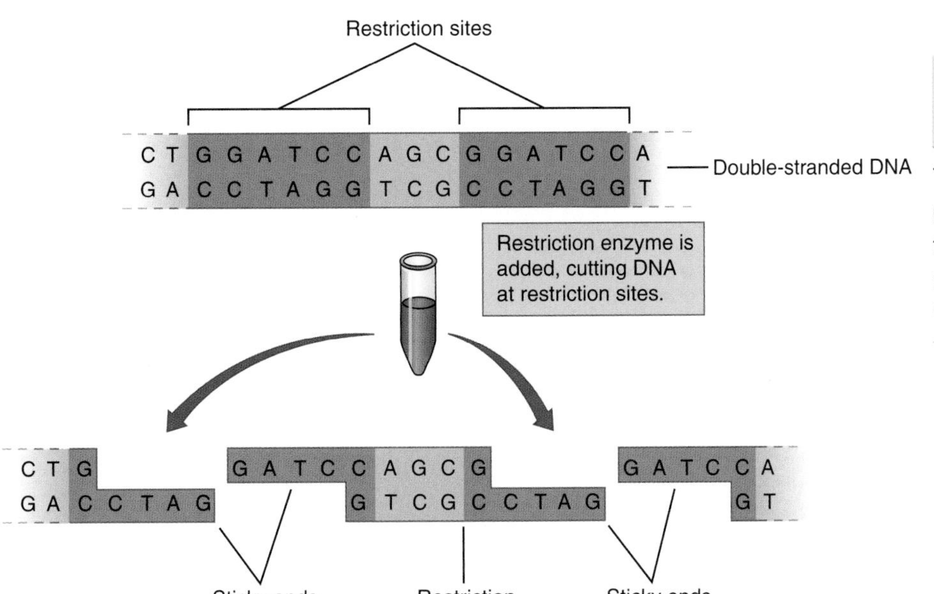

Restriction enzymes and sticky ends FIGURE 18.16

This restriction enzyme recognizes the palindrome GGATCC and it cuts between the first and second guanine bases. The resulting ends of the DNA are unequal, or "sticky."

Just two years after the Temin–Baltimore discovery, Paul Berg at Stanford University used restriction enzymes to create the first **recombinant DNA** molecule. Berg first purified the DNA he was interested in, then cut it with a restriction enzyme to open a slot for the new gene. Next he went to a different source of DNA and cut out the gene he wished to **transpose**, or move, using the same restriction enzyme. This en-

Recombinant DNA
The product of splicing genes.

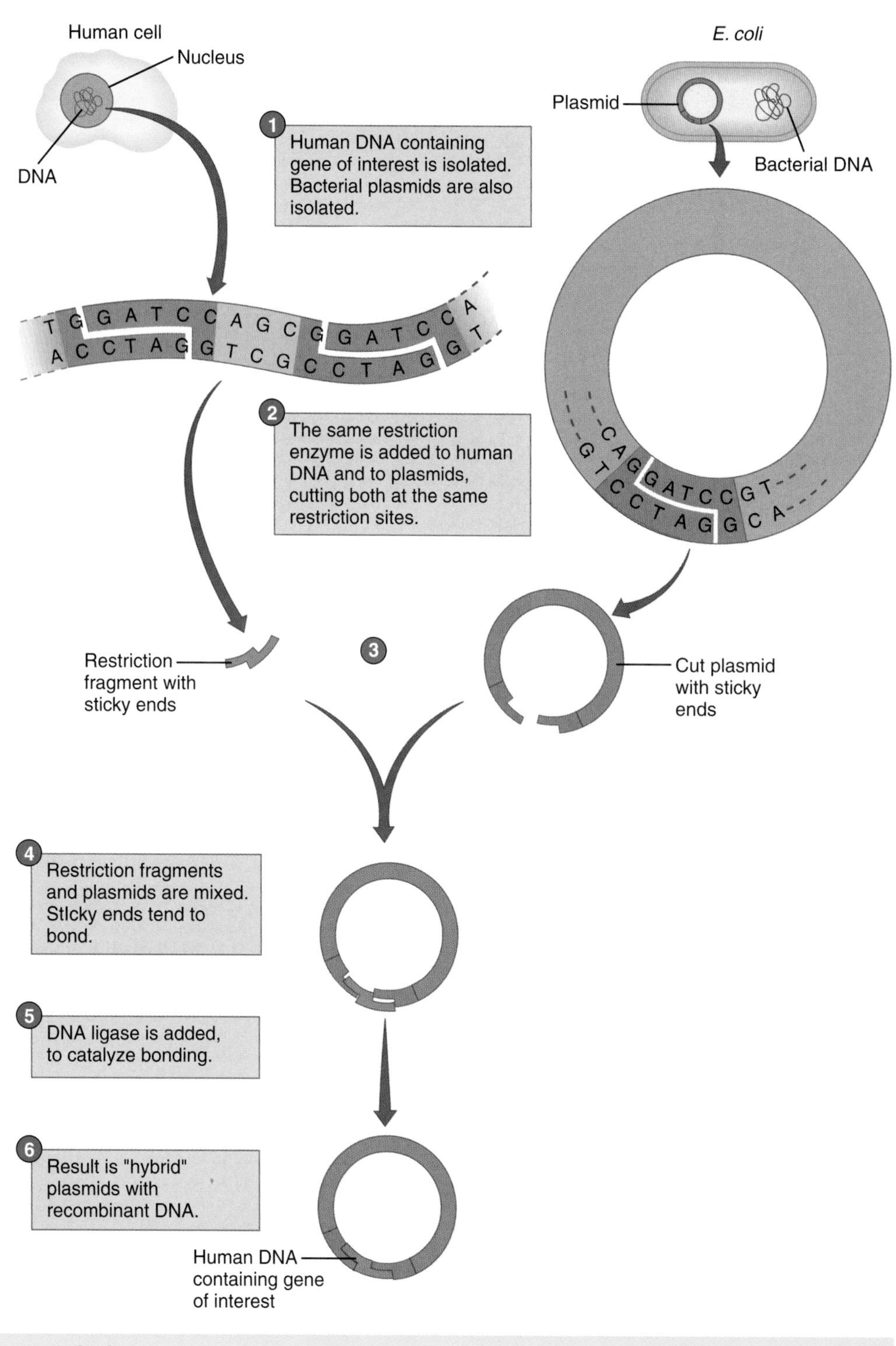

Human cell
Nucleus
DNA
E. coli
Plasmid
Bacterial DNA
1 Human DNA containing gene of interest is isolated. Bacterial plasmids are also isolated.
2 The same restriction enzyme is added to human DNA and to plasmids, cutting both at the same restriction sites.
3
Restriction fragment with sticky ends
Cut plasmid with sticky ends
4 Restriction fragments and plasmids are mixed. StIcky ends tend to bond.
5 DNA ligase is added, to catalyze bonding.
6 Result is "hybrid" plasmids with recombinant DNA.
Human DNA containing gene of interest

Restriction enzyme protocol FIGURE 18.17

sured that both types of DNA had matching sticky ends. Mixing the two pieces of DNA and adding **ligase** (an enzyme that seals and repairs DNA by reforming broken linkages between the phosphate groups and the sugars in the backbones), he inserted the new gene into the DNA. This was an early use of biotechnology, and the first **transgenic** organism.

Transgenic
An organism with a gene or group of genes in its genome that was transferred from another species or breed.

Plasmids
Circular pieces of double-stranded DNA outside the nucleus or the main DNA of the cell.

Recombinant DNA technology is now common and is often used to insert human genes into bacterial **plasmids** (FIGURE 18.17). If inserted properly, the modified bacteria will produce the human protein in large quantities, making it available for those people who cannot make the protein themselves. Following restriction enzyme protocol, the insulin gene was removed from human DNA, altered slightly to allow for production in the bacterium, and added to bacterial DNA. Bacteria carrying the human insulin gene now mass produce the protein, which is harvested for injection by diabetics. Using this technique, scientists have also engineered bacteria to produce vaccines and manufacture ethanol and **citric acid** (a natural preservative); they have even engineered bacteria that can clean up the environment by metabolizing toxic waste or petroleum spills.

TRANSGENICS AND CLONES ARE PART OF OUR BRAVE NEW WORLD

The first artificial transfer of genes occurred in bacteria, but by carefully timing the introduction of the DNA, genetic engineering of plants and animals is also possible. One way to alter plant DNA is to use a **vector**, such as bacteria that naturally infect the plant, to carry the new gene into the cells. The bacteria are first infected with a plasmid carrying the gene to be inserted. Embryonic plant cells are grown with the transgenic bacteria, and occasionally some plant cells pick up the plasmid. Alternatively, embryonic plant cells can be shocked with high voltage in the presence of the plasmid. As the cells respond to the electric shock, the plasmid is incidentally incorporated. Another route is to affix the DNA to a microscopic metal sphere and literally shoot it into embryonic plant cells, using a sterile gun in the laboratory (FIGURE 18.18, p. 616). Viruses can also be used to alter host cell DNA. The virus itself must be transformed, removing the pathogenic genetic material from within the viral coat and replacing it with the gene of interest. When the transformed virus infects the plant, it injects the gene of interest into the plant cells, rather than the pathogenic viral genes. Hopefully, the new gene will incorporate into the plant DNA just as the original viral genetic material would have.

All these methods have been used to create transgenic plants, with Monsanto Corporation, Syngenta, and Pioneer Hi-Bred International taking the lead in the production of transgenic corn, rice, cotton, and soybeans. These crops are catching on quickly around the world; about 100 million hectares were planted with transgenic crops in 2005. In corn and cotton, genetic engineers have added bacterial genes that impart insect resistance. Corn and soybeans have received a gene that makes the plants resistant to glyphosate, a popular and relatively nontoxic herbicide. There may be unintended consequences of these manipulations, which is why some scientists as well as many other people are uncomfortable about the increasing popularity of transgenic crops.

Genetic engineering can also be used to alter the nutritional properties of a food. Scientists have added a gene for beta carotene, a yellow nutrient that is lacking in the diet of many people in Third World countries, to rice. This "golden rice" was engineered to help alleviate malnutrition by supplying a raw material for vitamin A. Researchers are working to introduce genes that direct edible plants to produce vaccine proteins. Eating the raw plant would then provide both plant nutrients and a dose of vaccine. It sounds good,

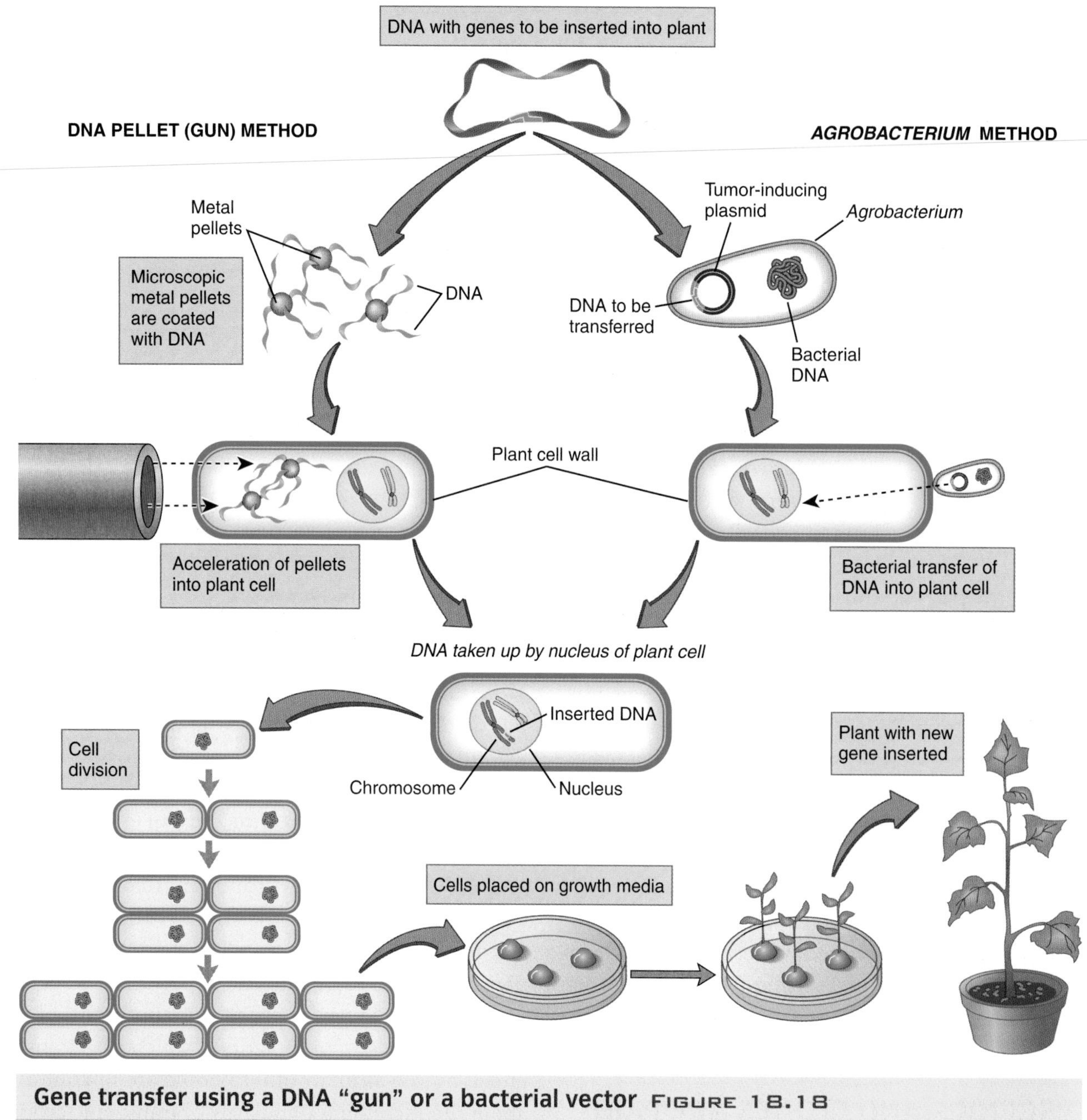

Gene transfer using a DNA "gun" or a bacterial vector FIGURE 18.18

but producing vaccines and other custom-made proteins also raises major public health concerns. If food plants start making vaccines, medicines, or industrial products, can we absolutely prevent food contamination with medicines or industrial chemicals? (For more on this topic, see the Health, Wellness, and Disease box.)

Transgenic animals are more difficult to produce, as animal cells do not take up genes as readily as bacterial or plant cells. Viral vectors are the most promising route of introducing foreign genes into animal cells to date. Once a transgenic animal cell is created, it must be cloned to produce a line of identical animals, and cloning animals is more difficult than cloning plants or bacteria.

Usually, transgenic animal cells are created by inserting a gene into a fertilized egg. If the gene is taken up, it will appear in every cell and hopefully will

Are genetically modified foods safe for the environment? Are they healthy to eat?

Did you enjoy the transgenic food you ate today? If you ate a food containing corn or soybeans, you probably ate genetically modified food, unless the food was all organic, which by definition does not contain genetically engineered ingredients. In the United States, most corn and soybeans are transgenic (genetically modified or GM), to make them easier to grow.

Is this safe? Is this wise? For a decade, these questions have swirled around GM seeds, which were introduced in 1996. Advocates say that moving genes from other organisms can improve the resistance of plants to insects and disease, raise yields, allow cropping on marginal land, increase farmer profits, and improve nutrition. Good monitoring and a careful understanding of the genes that are moved will control risks, advocates say. And they stress that farmers have been altering genes in their crops since the dawn of agriculture.

Critics dispute many of these claims. To date, yield improvements have been slight at best. No transgenics have yet been approved for salty or droughted soils. And golden rice, with its high level of vitamin A, has not yet begun to prevent blindness caused by shortages of vitamin A in developing countries, despite the considerable publicity it attracted years ago. Critics also say that transgenic crops are a threat to the environment. They maintain that herbicide-resistant genes have already been carried by pollen into several weeds in Canada and the United States. If true, this proves that early fears about breeding "super-weeds" with genetic resistance to herbicides were on the mark.

Transgenics could also pose threats to human health. For example, several years ago, a seed company transferred a gene from the Brazil nut into soybean, in an effort to make the protein more complete. Before the seed reached the market, the company asked a university lab to test the new seed, and the lab warned that the crop would trigger food allergies among many people who are allergic to nuts. The seed company halted the research project. Depending on your perspective, the incident proved either that the safety system works or that transgenic food can be dangerous. Nobody has yet observed a wave of disease due to transgenic crops, but critics say research into the human health effects of GM crops is too scarce to make us confident of its safety.

The soaring popularity of organic food indicates that a substantial percentage of the U.S. population is leery of GM food. But the rapid acceptance of GM crops by farmers around the world indicates that farmers are happy to avoid spraying insecticide and would rather buy a seed that "makes its own pesticide."

The debate over GM food has many facets. For example, many people express more concern about cloning animals than plants. In any case, great power brings great responsibility. Can we wisely handle the responsibility that comes with deliberately moving genes from one species to another? Does the need to feed a soaring population without harming more of the Earth mandate that we adopt every promising farm technology? How do we balance threats to human health with threats to the environment?

be expressed as intended. Although this sounds simple, it is not. Often transgenic animals are sterile, requiring that they be cloned to reproduce, introducing another level of technological difficulty. In 1998, researchers at the University of Hawaii, Manoa, cloned 50 transgenic mice from adult cells after years of failures, giving the mice a gene for a green fluorescent protein as a marker that they were, in fact, cloned. All 50 mice glowed green under UV light.

The Hawaii method, as it is called, was a simplification and blending of two methods already in use and sparked renewed interest in animal transgenics for pharmaceutical uses as well as livestock improvement. Using mice, the research team first identified a donor cell as the source of the adult DNA. The donor cell was found in the follicle near the developing egg. Dr. Ryuzo Yanagimachi removed these follicle cells from the ovary and harvested their nuclei. Nuclei are removed from cells using a tiny needle and some suction. This nucleus, like all nuclei in the body, contains all the DNA of the mouse, although not all of it is expressed. The scientists injected this nucleus into the nearby unfertilized egg. The egg, now holding all the DNA of the mother, was effectively fertilized, meaning that it did not hold the usual haploid complement of chromosomes found in eggs, but a complete diploid set of chromosomes. The only difference between this artificially fertilized egg and one that was fertilized via introducing sperm was that this egg held exactly the same DNA as the female who produced it. There was no sexual reproduction, no mixing of genetic material, not even any crossing over during gamete formation! This artificially fertilized egg then divided just like one fertilized normally. The resulting mouse pup was a genetic clone of the donor.

This simple procedure has been a boon to genetic technology and has been used to create larger cloned animals. Where can this technology lead? The guar is an extremely rare animal living in India. Scientists have attempted to clone a guar into a cow egg using older fusion methods. Should they again have the opportunity to harvest cells from this endangered animal, the Hawaii method may provide a larger percentage of living embryos, perhaps leading to a successful cloning. With only 90 of these organisms left in the wild, genetic engineering might be their only chance.

GENE THERAPY CAN CORRECT DEFECTS AND TREAT DISEASE

As genetic engineers work on cloned and transgenic animals, health researchers are also considering a less drastic step—**gene therapy** for humans. In transgenic animals, the entire animal gets a new gene, which was first inserted into the fertilized egg. In gene therapy, genes are inserted into cells to correct defects or treat disease. Defective or inactive genes are supplemented with active, functional copies of those genes in the adult human. Many difficulties could arise from this seemingly simple idea. An astronomical number of cells might need to express the gene. How could we get the gene inserted properly and ensure that it is working correctly in all of those cells?

One answer is to use viruses as vectors for gene insertion. As discussed previously, viruses normally inject their pathogenic DNA into the host's chromosomes, either directly or through reverse transcription of the viral RNA in retroviruses (review the discussion of viral life cycles and retroviruses in Chapter 10 if necessary). Removing the pathogenic viral genes and inserting transgenes take advantage of this viral mechanism. The viruses become tiny gene therapy injectors, delivering their modified and now helpful genetic contents to cells. Of course, these viral particles cannot reproduce in the cells, limiting the number of cells that can be "infected." Bits of tissue requiring the gene therapy may be removed from the patient, exposed to the virus, and then replaced, to ensure that the virus reaches the target cells. Alternatively, the viral vectors can be injected directly into the target tissue.

Even after succeeding in the daunting task of getting the modified gene to the necessary cells, another issue remains. Unless the gene was inserted into the patient's germ cells (gamete-producing cells), the children of these genetically altered adults would likely have the same genetic defect and the disease, so they might also need gene therapy. On the positive side, gene therapy was successfully used to treat a four-year-old child who was unable to produce an enzyme that caused a severe immunodeficiency. Following closely on that success, in 2003 more than 600 clinical trials using gene therapy were underway in the United States for such diseases as severe combined deficiency disease (SCID) and cystic fibrosis. Thus far, it seems that gene therapy is a successful treatment for SCID, but successful treatment of other diseases remains elusive. It's fair to say that gene therapy has not met its early promise, and research is needed to understand why not.

DNA TECHNOLOGIES CAN BE USED TO IDENTIFY INDIVIDUALS

Turning DNA technology to societal needs rather than food or medicine, scientists have perfected ways to purify DNA from crime scenes, separate it into small pieces, and compare it to other DNA. The challenge in this research is to compare DNA from various sources in order to identify similarities. Many crime scene samples provide precious little DNA for examination. The DNA in that sample must be amplified to provide enough for the investigators to analyze. In 1983, in what some have called the greatest single achievement in modern molecular biology, Kary Mullis developed **polymerase chain reaction**, or PCR. Mullis worked for a biotech company and was probably not thinking about *CSI*, but his clever invention has brought the power of biotechnology to crime investigations.

PCR is a series of reactions that amplifies DNA using the same enzymes that cells use to synthesize DNA. You add a small amount of the sample DNA to a test tube, along with the four DNA nucleotides as building blocks: DNA **polymerase** (the enzyme that adds nucleotides during DNA duplication), **RNA or DNA primer**, and the appropriate buffers. The tube is set in a **thermo-cycler**, which raises and lowers the temperature in a precise sequence. Rising temperature dissociates the double-stranded DNA into two single strands. As the temperature falls, the primers anneal to each strand and DNA polymerase binds to the primer/DNA combination, creating a new complementary strand on each strand of the original DNA (FIGURE 18.19). The new complementary strands are made from the A, T, C, and G bases present in the tube. As the temperature again rises, the newly created DNA dissociates, and the process repeats itself.

RNA or DNA primer A short segment of RNA or DNA binding to the original DNA strand, initiating DNA replication.

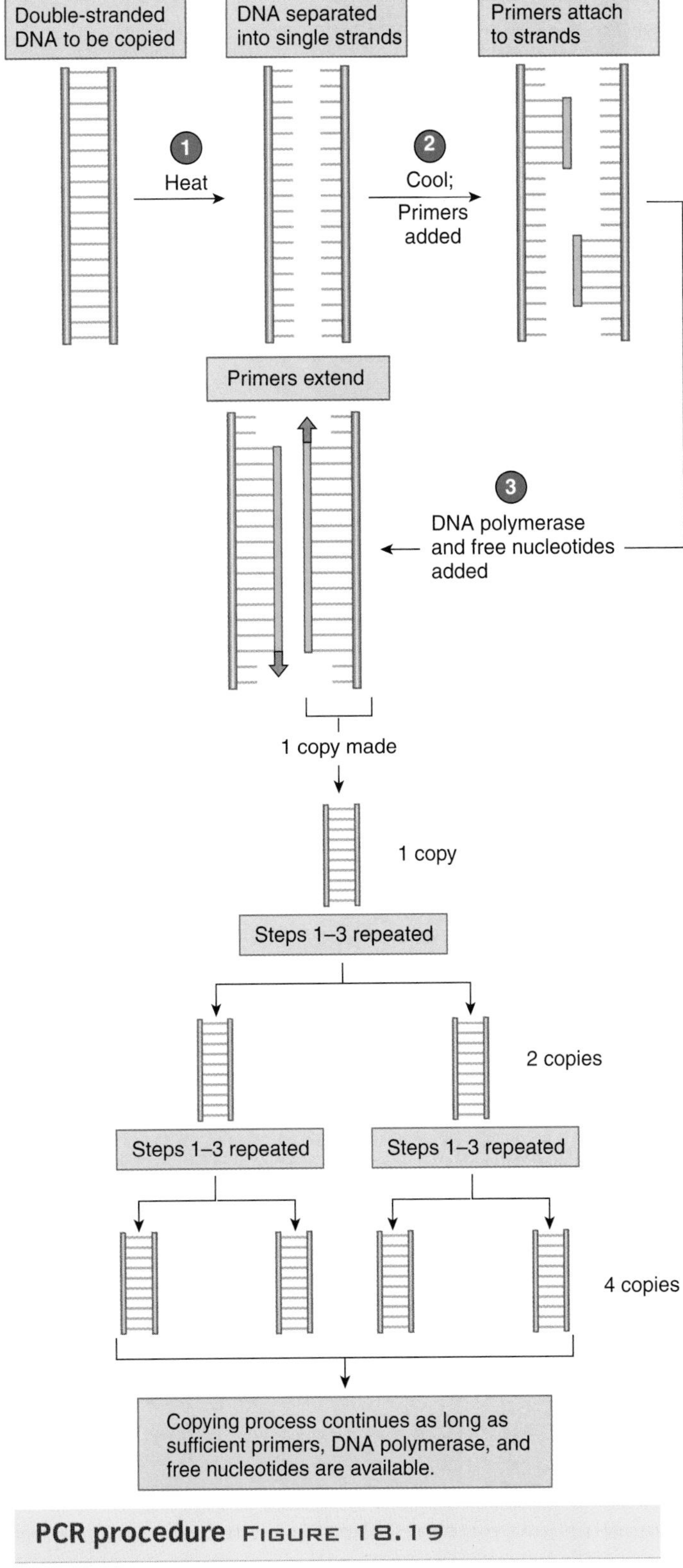

PCR procedure FIGURE 18.19

After two to three hours, a huge multiplication is possible, since the sample doubles with each thermal cycle. Because DNA nucleotides pair with high fidelity, the resulting DNA is almost 100 percent identical to the original sample. This simple technique amplifies a small amount of DNA, providing enough sample to begin to analyze the base-pair sequence and match it with the DNA of suspects.

A large sample allows the next step in crime scene analysis: differentiating the DNA of several individuals. Much of the DNA in the human genome does not code for proteins, and because mutations and alterations in these regions do not affect function, they are highly variable. Each individual has a different sequence of DNA bases in these regions. Recall that restriction enzymes cut DNA only where they find their specific restriction site. In the variable regions of chromosomes, the locations of the restriction sites along the chromosomes change from one person to the next. Subjecting DNA to a variety of restriction enzymes should cut the DNA strands in different places. Since we each have a unique series of nucleotides in our variable regions, each of us will generate different lengths of DNA from samples cut with the same restriction enzymes. The resulting restriction digest is then analyzed to compare the lengths of the restriction fragments.

GEL ELECTROPHORESIS SORTS OUT DNA FRAGMENTS

Separating these myriad cut pieces of DNA based on length allows us to view them as an organized group as we compare a suspect's DNA to samples from the crime scene. One easy way to view these fragment lengths is to spread them out based on size. Change-counting machines use this principle. It is hard to count the nickels, dimes, and quarters in one mixed pile of coins, but if you separate them by size and line them up in rows, it's suddenly quite easy. Gel electrophoresis does the same thing with pieces of DNA. The mass of fragmented DNA is loaded into the top well of an **agarose** gel. The gel is floating in a salt buffer, and an electrical current is passed through it. DNA, being slightly negative, is pulled through the gel toward the positive pole by the current.

Agarose
A gel-like compound obtained from agar that provides a flexible, yet solid, medium for separation of DNA fragments.

Imagine racing alone through a crowded room, dodging chairs to reach the front of the room. If you then linked arms with four other people, the spaces between chairs would seem much smaller, and you would

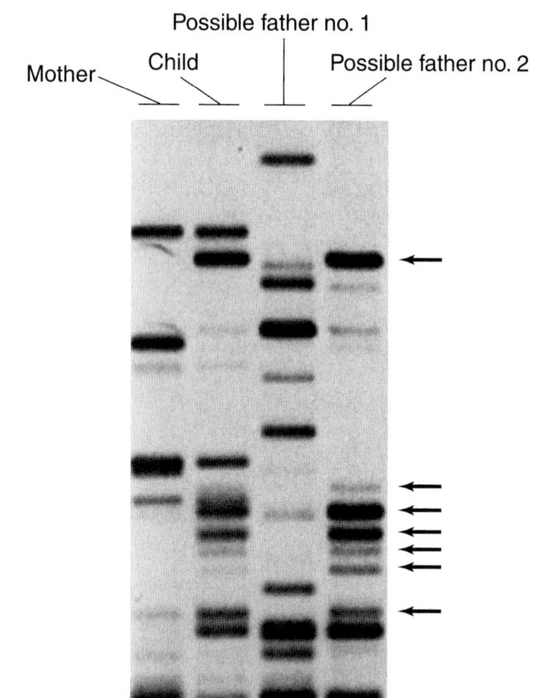

DNA fingerprints of a mother, her child, and two possible fathers FIGURE 18.20

DNA fingerprinting is also used to establish paternity. Each of the men whose DNA was tested claimed to be the child's father. Arrows pointing to the DNA of possible father #2 indicate bands that are the same as those of the child's. There are no comparable bands in the DNA of father #1, indicating that #2 is the biological father.

reach the front much later. This principle holds for DNA moving through a gel, when a current is applied to the gel, larger fragments move more slowly, and the separation that results can be used to distinguish the various-sized DNA fragments created by the restriction enzymes (FIGURE 18.20). The name of this technique, restriction fragment-length polymorphism (RFLP), means just that. Restriction enzymes create DNA fragments that are of different lengths in different individuals. Using gel electrophoresis, the pattern of those fragment-length polymorphisms becomes visible.

This series of bands is your personal **DNA fingerprint**. Approximately one in one billion people will match your DNA fingerprint, unless you have an identical twin. Taking a sample of your DNA, exposing it to the same restriction enzymes, and running it on a gel next to the crime-scene sample will allow comparison of your DNA to that from the scene. If the banding pattern of each sample is the same, this is a sure sign that you were near the scene at some point, and you had best start working on an alibi! (See FIGURE 18.21.)

DNA fingerprinting can also be used to exonerate the innocent. In more than 100 capital cases in the past decade or so, "criminals" have been sprung from prison on the basis of DNA evidence. Many of these men had served more than a decade in prison for horrific crimes that they had not committed. Though more expensive and time-consuming than traditional methods, DNA fingerprinting can also be used to establish paternity in rigorously contested cases. In this case, the infant's DNA fingerprint must show a high degree of similarity to both the maternal and real paternal fingerprints, demonstrating banding patterns that can be matched with either one parent or the other. The trouble with using DNA fingerprinting to identify paternity is that the baby's DNA will be a combination of maternal and paternal DNA, showing new banding patterns unique to the new individual. It is easier and more reliable at this point to match the infant's protein profiles with those of the father—for example, those proteins that appear on the red blood cells.

Perhaps the most famous use of DNA fingerprinting in the courtroom in recent history is the O. J. Simpson trial. At this trial, DNA evidence was ruled inconclusive because of questions concerning the quality and purity of the sample collection, and the validity of the testing. This was certainly not the first time this technology had hit the courtroom floors, however.

The entire field of genetic research remained out of

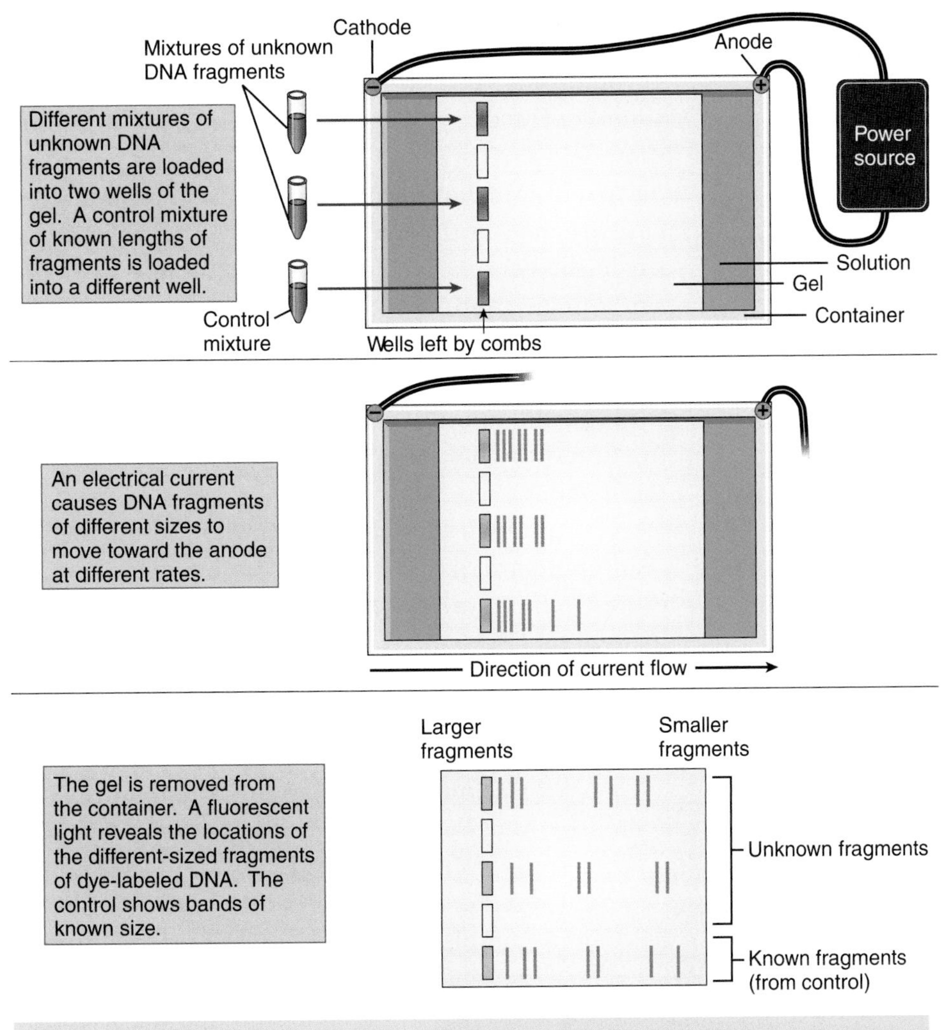

DNA fingerprinting FIGURE 18.21

the courtrooms for many years after being introduced to the laboratory. Not until 1985 did genetic fingerprinting appear as legal evidence in court cases. The identification of large stretches of repeating patterns of DNA in the human genome had literally just occurred. Alec Jeffries and his laboratory associates at the University at Leicester, UK, had no sooner discovered that these repeating patterns of DNA differed in length from one person to the next, than the information was used in court. The relationship of a woman and a child needed to be established for an immigration case. Through RFLP analysis the child was shown to be closely related to the woman. Using this evidence, the courts allowed the child to immigrate to the UK to be with her relative. As a matter of fact, Jeffries coined the term *genetic fingerprinting* in his paper describing RFLP analysis as a sort of tongue-in-cheek joke.

Following closely on the heels of this case was the first murder case ever solved with genetic fingerprinting. In November 1987, Colin Pitchfork was convicted of the murders of two teenage girls in Narborough, Leicester. While Colin was the first murderer to be convicted based on DNA technology, Richard Buckland was the first person to be exonerated of murder using DNA evidence. He was suspect #1, but his DNA banding patterns did not match those at the crime scene at all.

THE HUMAN GENOME PROJECT MAPPED HUMAN GENETICS

Beyond catching crooks and identifying fathers, genetic technology is also used for more basic purposes. One of the most interesting projects in modern genetics was the sequencing of the entire human genome.

The Human Genome project, begun in 1990 and essentially finished in 2003, was a massive research undertaking. This project required numerous new technologies, as the speed of sequencing increased by several orders of magnitude during the process. After a flurry of invention, fast, simple DNA technologies were introduced and used to expedite the mapping of the human genome. By 2002, **DNA sequencing** and transgenic bacteria production were commonplace enough to be available in many American high school biology labs.

DNA sequencing Determining the sequence of A, C, T, and G on a gene or chromosome.

The goals of the Human Genome Project included:

- identifying all the genes in human DNA,
- determining the sequences of the 3 billion + nitrogenous base pairs in human DNA,
- storing this information in databases,
- improving tools for data analysis,
- transferring related technologies to the private sector,
- addressing the ethical, legal, and social issues that would arise from this knowledge.

In completing the map of the human genome, scientists were able to locate the precise chromosome, and even the location on that chromosome, of the genes responsible for many congenital diseases (FIGURE 18.22). Duchenne's muscular dystrophy, Marfan's syndrome, and Alzheimer's disease are among the many diseases we now can identify in the genome. In addition, we can now compare the human genome to that of other organisms, giving us a better look at evolutionary relationships. Scientists can trace the history of particular genes through the animal or plant kingdoms, hypothesizing about the meaning of conserved or radically altered genes.

Although the science behind this information is accepted and part of mainstream biology classes, it is quite new (TABLE 18.7). Mendel's laws of heredity were first presented in 1865, Beadle and Tatum uncovered gene function in 1941, and in 1953, Watson and Crick proposed the double helix structure of DNA. April 25, 2002, was designated the first official **National DNA Day** to commemorate 50 years of DNA research, rather arbitrarily beginning with Watson and Crick's model of the double helix and ending with the completion of the sequence of the human genome. Although not on most calendars, this day is commemorated in the scientific community, and perhaps in your biology

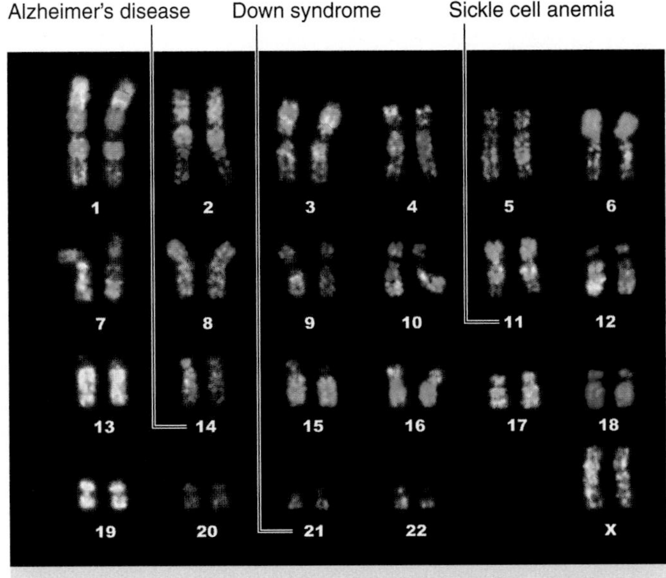

Human karyotype showing selected genetic disorders FIGURE 18.22

Based on the information gleaned from the Human Genome Project, many of the genes responsible for congenital diseases have been precisely located. This figure shows the chromosomal locations of just a few of these genes.

class, as a day to reflect on all that we have learned in such a short period.

With rapid knowledge comes the need for ethical debate. What do we do with this information? Should we sequence the genotypes of every individual soon after birth? Should we make the **genetic fingerprint** of each individual as accessible as his or her dermal fingerprint is today? These questions are currently being debated in both the scientific and public communities.

DNA discovery: selected events TABLE 18.7

1865	Mendel's laws of heredity presented.
1868	Miescher isolated "Nuclein," a compound that includes nucleic acid, from pus cells.
1905–1908	Bateson and Punnett showed that genes modify action of other genes.
1911	Morgan showed genes to be units of inheritance.
1926	Morgan published Theory of the Gene.
1939	Belozersky began work showing DNA and RNA are always present in cells.
1941	Beadle and Tatum discovered gene function.
1944	McClintock found genes can be transposed from one chromosome position to another.
1953	Watson and Crick proposed double-stranded, helical, complementary, antiparallel model for DNA.
1966	Nirenberg, Mathaei, and Ochoa demonstrated that sequence of three nucleotides (a codon) determines each of 20 amino acids.
1973	First human gene-mapping conference held.
1990	Human Genome Project launched.
2000	Working draft of human genome sequence completed.
2003	Sequencing of the human genome completed.
2003	Celebrated 50 years of DNA's double helix.
2004	Family Tree DNA's 1st International Conference on Genetic Genealogy.
2005	The Genographic Project announced by National Geographic, IBM, et al.

Adapted from Access Excellence at the National Health Museum, http://www.accessexcellence.org/RC/AB/BC/Search-for-DNA.html

CONCEPT CHECK

What are restriction enzymes?

Describe two roadblocks to gene therapy.

How is RFLP analysis used in criminal cases?

What happens to fragments of DNA as they are pulled through an agarose gel by an electrical current?

Armed with our understanding of genes and gene manipulations, we have the potential to greatly affect evolution. Is that a wise thing? Explain your answer.

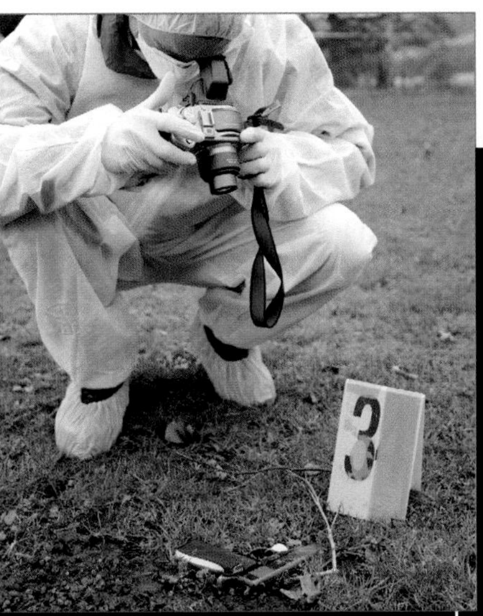

CHAPTER SUMMARY

1 Plant and Animal Traits Are Inherited in Specific Patterns

Discussing inheritance and genetics requires a set of terms that precisely define genetic characteristics. Genotype is used when discussing the actual genes present on the chromosomes, while phenotype is used when discussing the outward appearance resulting from the presence of those genes. On the one hand, eggs and sperm carry only half the chromosome number of the entire human, a condition referred to as haploid. Mature human cells, on the other hand, have a full set of chromosomes, referred to as diploid.

Working with seven traits in pea plants, Gregor Mendel explained the predictable relationship between dominant and recessive traits. He also described the laws of segregation and of independent assortment.

2 Modern Genetics Uncovers a More Complicated Picture

Alternate forms of a gene are called alleles. During meiosis, these alleles separate so each gamete gets one allele instead of the normal pair. At fertilization, maternal and paternal alleles are joined to form the diploid content of the new individual. If both alleles are the same, the individual is homozygous. If the two alleles are different, the individual is heterozygous. The vast majority of human traits are codominant, incompletely dominant, polygenic, or multifactorial.

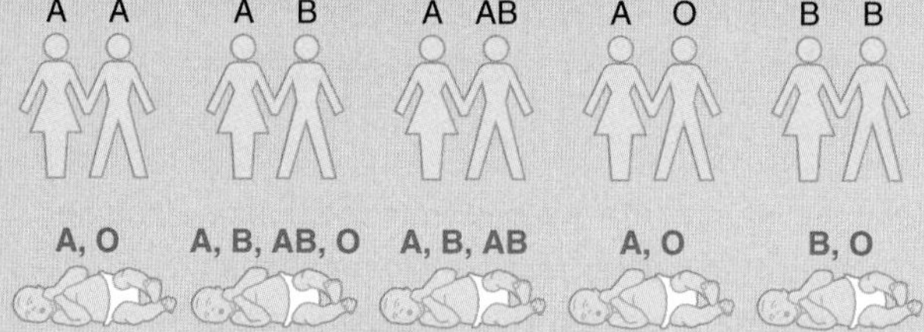

3 The Central Dogma of Genetics Is Still Accurate

One gene, one protein – DNA codes for protein formation. That DNA is the molecule of inheritance was proven by Beadle and Tatum in their benchmark experiment with *Neurospora*. By knocking out genes with radiation, and then replacing the products of those nonfunctional genes, these two men demonstrated that DNA and not protein was the molecule of inheritance. Transcription "reads" nuclear DNA to an mRNA molecule that can leave the nucleus. Translation of that message in the cytoplasm produces a protein.

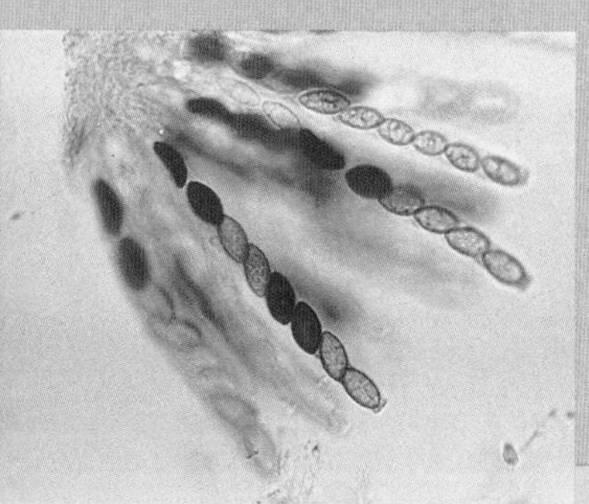

4 Genetic Counseling Puts Genetic Theory to Practical Use

Sex-linked traits are carried on either the X or the Y chromosome. Because the X chromosome is much larger and seems to include more functional genes, most sex-linked traits are carried on it. Sex-influenced traits are those that are not present on the X or Y chromosome, but are more prevalent in one gender. Often the hormones estrogen or testosterone aggravate or promote these traits. Chromosomal disorders can occur during gamete formation. These include gross chromosomal alterations (gain or loss of entire chromosomes), loss or gain of portions of chromosomes or even smaller alterations in individual genes. Most infants born with congenital defects suffer from individual gene troubles.

Testing is available for prospective parents concerned that they may produce a child with genetic defects. Human genes evolve, just like those of other organisms.

5 Biotechnology and DNA: The Facts, the Pros, and the Cons

Understanding DNA has opened up a whole new field: molecular biology, the study and manipulation of DNA. Molecular biology uses nucleic acid hybridization, gel electrophoresis, PCR, and RFLP analysis. Physicians can use these techniques to identify the risks of certain cancers.

DNA technology has left the classroom and research labs and has become a household word. If your DNA fingerprint (banding pattern) matches a sample found at a crime scene, you had better get a good lawyer. . . As we become more proficient at DNA splicing and creating transgenic clones, DNA technologies will play an increasingly greater role in our daily lives, but we should be prepared for mistakes along the way.

KEY TERMS

- **agarose** p. 620
- **asymptomatic** p. 602
- **autosomal** p. 602
- **centrifugation** p. 609
- **congenital** p. 606
- **cross-pollinating** p. 593
- **diploid** p. 592
- **DNA sequence** p. 590
- **DNA sequencing** p. 622
- **dominant** p. 593
- **evolution** p. 605
- **fitness** p. 596
- **genome** p. 608
- **genotype** p. 590
- **haploid** p. 592
- **minimal media** p. 599
- **multifactorial disorder** p. 606
- **palindrome** p. 603
- **phenotype** p. 590
- **plasmid** p. 615
- **polymerization** p. 612
- **recessive** p. 593
- **recombinant DNA** p. 614
- **RNA or DNA primer** p. 619
- **self-pollinating** p. 593
- **somatic** p. 595
- **spliced** p. 612
- **supplemented media** p. 599
- **transgenic** p. 615
- **true-breeding** p. 593

CRITICAL THINKING QUESTIONS

1. Multifactorial traits are influenced by genetics and the environment. These traits, such as height and weight, are expressed in a range of phenotypes in the population. This complication leads to the long-running "nature versus nurture" argument. How could you determine how much the environment affects a particular genetic trait? Design an experiment that would, at least theoretically, shed light on this age-old debate.

2. Transcription and translation are precisely controlled. There is almost no error in the base pairing of nucleotides, ensuring the DNA code is transcribed reliably. List the steps in transcription and translation in order, and indicate which can introduce mutations.

3. PCR, DNA transcription, and simple inheritance are all based on the integrity of the DNA base-pairing rules. What are these rules? What enzymes ensure that these rules are not violated? How does this relate to cancer, as described in Chapter 3?

4. Restriction enzymes occur naturally in bacteria. What is their function in those cells? Eukaryotic cells do not have these enzymes. What organelles and functions normally found in eukaryotic cells are these prokaryotic enzymes replacing? What specific function of the restriction enzymes do we exploit while creating transgenic organisms?

5. What is a transgenic organism? In your own words, explain the process of creating such an organism. If such an organism is created, we may wish to clone it to get many individuals with the new gene. Is this the same as asexual reproduction? Why or why not?

SELF TEST

1. The term that describes the appearance of an organism is
   a. genotype.
   b. phenotype.
   c. DNA sequence.

2. The structure(s) indicated as A on this diagram is (are)
   a. centrioles.
   b. chromosomes.
   c. centromeres.
   d. a tetrad.

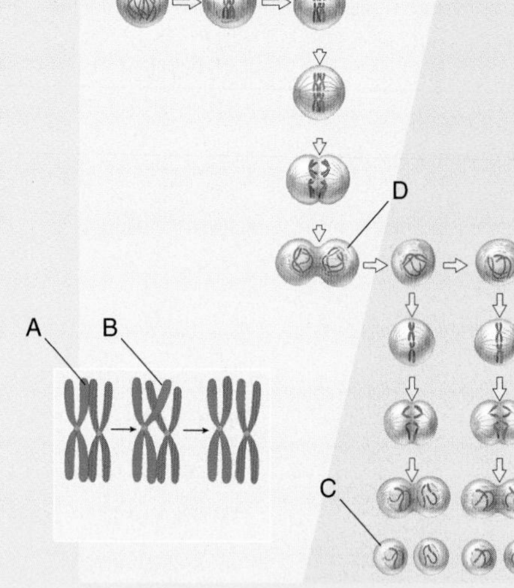

3. The structure indicated as B on the figure is important because it
   a. increases chromosome number in the gametes.
   b. increases the genetic combinations of the resulting gametes.
   c. organizes the chromosomes before splitting them apart.
   d. Two of these answers are correct.

4. In the above figure, the structure indicated by the letter D marks the end of what phase?
   a. Metaphase I
   b. Anaphase I
   c. Telophase I
   d. Prophase I

5. Gregor Mendel is known as the father of genetics because
   a. he was the only person looking at genetic variation in the 1800s.
   b. his research was thorough and included quantifiable data.
   c. he used pea plants when others were using cows and corn.
   d. he studied both the F1 and the F2 generations.

6. In Mendel's experiments, he produced the F_2 generation through
   a. self-fertilization.
   b. natural fertilization.
   c. cross-fertilization.
   d. random matings.

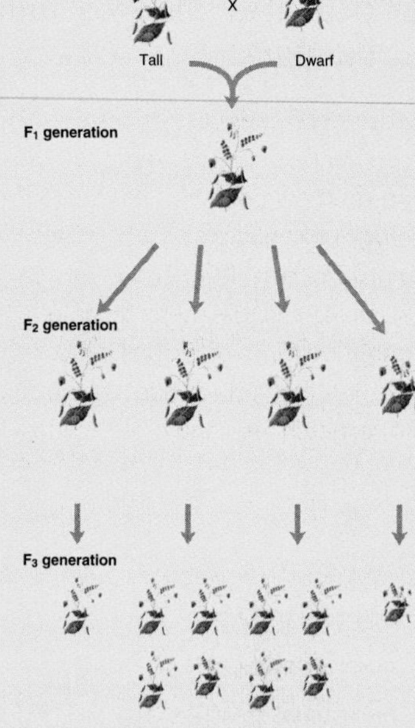

7. In the F_2 generation of his pea plants, Mendel consistently observed a phenotypic ratio of (See the above diagram for assistance in answering this question.)
   a. 1:2:1.
   b. 3:1.
   c. 9:3:3:1.
   d. 1:1.

8. True or False? The law of independent assortment states that the heritable units of parent plants are randomly separated during gamete formation.

9. As shown in this Punnett square, what is the probability of an individual F_2 pea plant showing both dominant traits in its phenotype?
   a. 13/16
   b. 9/16
   c. 4/16
   d. 1/16

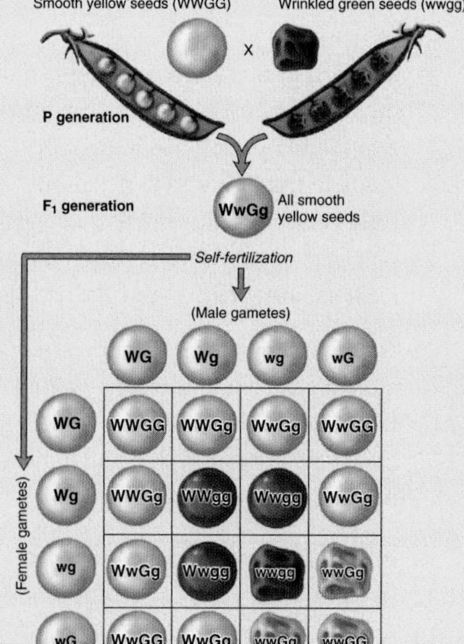

10. An individual who expresses the dominant phenotype for an allele must be

a. heterozygous.
b. homozygous dominant.
c. heterozygous dominant.
d. either heterozygous or homozygous dominant.

11. Most human traits show a __________ pattern of inheritance.

a. dominant / recessive
b. co-dominant
c. polygenic
d. multifactorial

12. Beadle and Tatum's experiment can be summarized by the saying

a. "You are what you eat."
b. "DNA is the molecule of inheritance."
c. "Proteins are the molecules of inheritance."
d. "X-rays cause knock-outs."

13. The central dogma of biology states that the process of __________ is relevant to all of biology.

a. meiosis
b. polymerase chain reaction
c. transcription / translation
d. evolution

14. In this pedigree, the trait being characterized appears to be a __________ trait.

a. dominant
b. recessive
c. sex-linked
d. environmental

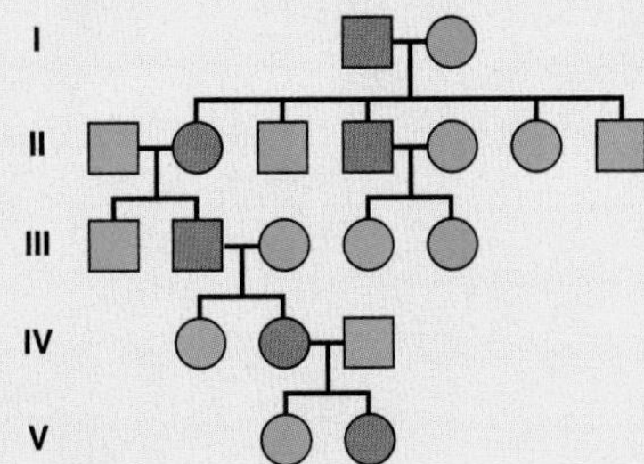
I
II
III
IV
V

15. One good reason for undergoing genetic counseling is to determine

a. emotional compatibility.
b. your likelihood of contracting a genetic disease.
c. the number of mutations in your DNA.
d. your likelihood of passing on a deleterious gene to your offspring.

16. The process of __________ pulls DNA through a semisolid matrix, separating out pieces of DNA by size.

a. nucleic acid hybridization
b. gel electrophoresis
c. polymerase chain reaction (PCR)
d. restriction fragment-length polymorphism (RFLP)

17. __________ is used to produce and compare DNA fingerprints.

a. Nucleic acid hybridization
b. Gel electrophoresis
c. Polymerase chain reaction (PCR)
d. Restriction fragment-length polymorphism (RFLP)

18. The process shown in this figure can be used to

a. prepare samples of DNA for fingerprinting comparisons.
b. amplify the amount of DNA in a sample.
c. isolate purified DNA.
d. create transgenic organisms.

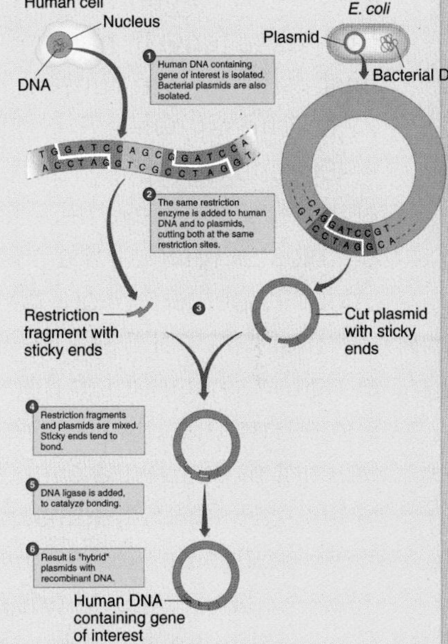
Human cell
Nucleus
DNA
E. coli
Plasmid
Bacterial DNA
Human DNA containing gene of interest is isolated. Bacterial plasmids are also isolated.
The same restriction enzyme is added to human DNA and to plasmids, cutting both at the same restriction sites.
Restriction fragment with sticky ends
Cut plasmid with sticky ends
Restriction fragments and plasmids are mixed. Sticky ends tend to bond.
DNA ligase is added, to catalyze bonding.
Result is "hybrid" plasmids with recombinant DNA.
Human DNA containing gene of interest

19. The Human Genome Project was undertaken to

a. identify all the genes in human DNA.
b. improve tools for both DNA and data analysis.
c. address the legal, ethical, and social issues surrounding DNA research.
d. All of the above were stated goals of the project.

20. This DNA gel indicates the DNA fingerprints of a mother, her child and two possible fathers. What can be determined from this?

a. The mother has a congenital disease.
b. The baby's father is one of the two men tested.
c. The baby has a recessive phenotype.
d. No useful information can be obtained from this technique.

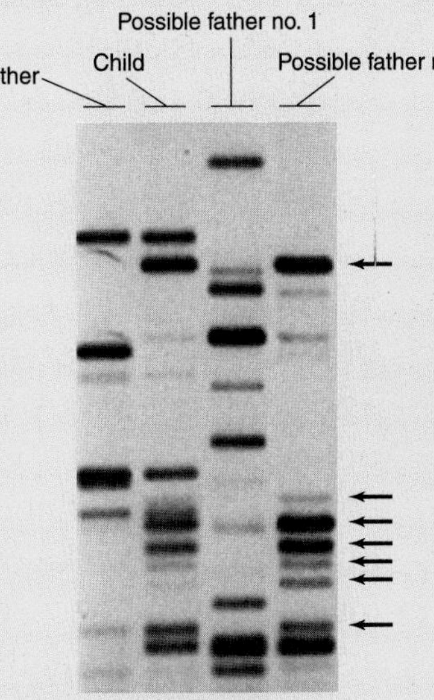
Possible father no. 1
Mother
Child
Possible father n

Evolution

19

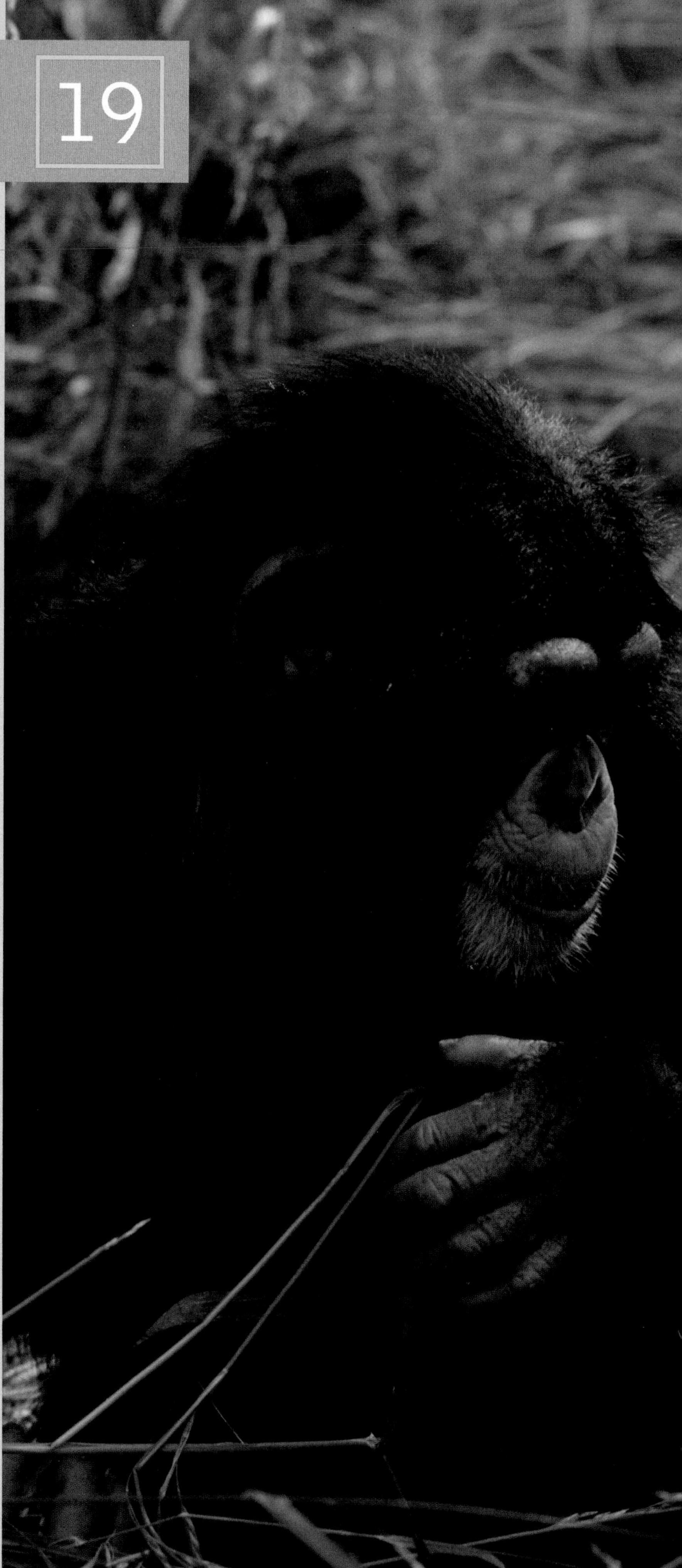

DNA is the most fascinating four-letter language in history, as it is the basis for our genetic code and that of all of living things. To date, thousands of genes have been identified in organisms ranging from bacteria to trees to fungi to humans. According to the best estimate, we humans have just over 22,000 genes, while *E. coli* bacteria have about 4,400, and a simple virus may have just six or eight. You may have heard that humans and our closest relative, the chimpanzee, share about 99 percent of our genes.

That may not be much of a surprise, considering the similarities in body type. But here's something that may surprise you. About 60 percent of our genes overlap closely with those in fungi: yes, mushrooms, mildew, and mold. How can this be? The simplest answer is that nature is lazy: once it finds a solution to a particular problem, it tends to reuse it. By "solution to a problem," we mean one of the metabolic processes, such as making proteins or oxidizing sugars to continue life, a situation that commonly confronts organisms. Once an early form of life solves a problem, the genes that underlie that solution are passed on to descendant organisms.

We also see this "laziness" in DNA itself. All living organisms, and most viruses, house their genetic code in DNA. Once DNA evolved, there was no need for a better system to encode the information that an organism needs, so the DNA system was passed down again and again. Indeed, DNA is so important that early organisms evolved a way to "proofread" it and correct mistakes after cell division. As with other biological solutions, this proofreading mechanism was passed down and is probably active in your cells at this very moment.

CHAPTER OUT

The Theory of Evolution through Natural Selection Is the Foundation of Biology

LEARNING OBJECTIVES

Appreciate the scientific nature of the theory of evolution.

Understand the history of our current "evolution in the classroom" debate.

Briefly describe natural selection.

List the five criteria that would allow a population's gene pool to remain unchanging.

Give two examples of evolution in action.

Evolution. Even the word can cause an argument. What is evolution? Why does the theory of evolution hold such emotional sway over us, while few nonscientists give a second thought to cell theory or the atomic theory? Although the theory of gravity is much less understood than the theory of evolution, we don't jump out a skyscraper window and claim "gravity is just a theory."

Many who loudly criticize the teaching of evolution have serious misconceptions about what it really means. The theory of evolution, as outlined by Charles Darwin in 1859 and refined by thousands of scientists in the intervening century and a half, is an explanation for the appearance, relationships, and distribution of the myriad life forms on Earth. Darwin studied life—in the barnyard and the backyard, on islands and volcanoes, in rain forests and deserts—for decades. He attributed the differences in the life forms he observed to natural selection, the advantage that one phenotype may have over another in any given situation. Interestingly, Darwin was not the only scientist to put forth this notion. Alfred Russell Wallace developed similar views on the origin of species. In 1858, Wallace wrote Darwin outlining a theory of evolution—almost exactly the same theory that Darwin himself had spent decades developing. Wallace and Darwin both presented their ideas and published important works (including Darwin's book *On the Origin of Species*) in 1858 and 1859, respectively.

We now define evolution more precisely as any change in the frequency of **alleles** in a population. In every population, genes are encoded in several forms. The frequency of these different forms, the allele frequencies, can be calculated. If those frequencies change, evolution has occurred. The most common force that changes these allele frequencies is natural selection. Natural selection occurs because an organism's environment may favor one phenotype over another at a given time, so individuals with the "right" phenotype have a greater chance of reaching reproductive age and passing on their "better" genes to the next generation. The layperson's understanding of natural selection is embodied in the phrase "survival of the fittest." Those organisms with the best possible combination of genes for their particular situation, those fit to survive and reproduce, do so, and their offspring do likewise.

Allele A particular sequence of DNA bases, one of two or more different forms of a single gene.

Creationism Belief in a literal interpretation of the Biblical story of the creation of the universe, the Earth, and life.

Intelligent design The hypothesis that complex biological creatures were designed by intelligent beings rather than simply evolving through natural selection processes.

The theory of evolution says nothing of a planned universe, an intelligent designer, or any supernatural or external guiding power of any sort. (Natural selection may properly be thought of as an internal response to external and impartial forces acting on the individual.) Evolution's two strongest antagonistic schools of thought, **creationism** and **intelligent design** or ID, both require the presence of a higher power investing energy in the life forms on Earth.

Although some tenets of the theory of natural selection are difficult to test experimentally, leaving questions that scientists have yet to answer, neither of the alternative suggestions is based on the masses of scientific evidence that uphold the theory of Darwin and Wallace. Both alternatives include nonnatural (supernatural) intervention and therefore cannot be investigated with the scientific method. In fact, they make no testable predictions whatsoever.

Recall that scientific hypotheses must be testable and falsifiable. The overall theory of natural selection is testable despite difficulties in directly testing some of its specifics. Experiments can be designed to show natural selection in action. We see examples all around us: in the rapid change of the HIV virus that can make it resist drugs after repeated exposure, or the appearance of antibiotic-resistant bacteria, or even the changes in prey species that allow them to avoid being eaten by predators.

The principles of creationism and intelligent design are neither testable nor falsifiable. While the statement "God created heavens and Earth" may for argument's sake be labeled a hypothesis, it is not one that can be tested, and therefore it is not a *scientific* hypothesis. For this reason, it is incorrect and inappropriate to include such theological or philosophical principles in a science curriculum, except as examples of untestable, and therefore nonscientific, hypotheses.

Nevertheless, as recently as February 2005, members of the Kansas Board of Education seriously considered adding intelligent design to their high school science teaching curriculum. In April 2005, a public forum to hear constituents' opinions was held, followed by a formal legal hearing to determine the legalities of teaching ID and evolution. Elections in 2006 produced a moderate majority on the state board, which moved to eliminate intelligent design and return to a more mainstream approach to teaching evolution, based on science. In Kansas and elsewhere, there is evidence that, to avoid controversy, many teachers simply shy away from teaching much about evolution. **Table 19.1** includes some of the more recent proceedings surrounding this issue.

Evolution, school, and politics: a contentious history TABLE 19.1

State	*Year*	*Ruling*
Tennessee	1925	The Scopes trial, in which the state ban on teaching evolution in classrooms (the Butler law) was challenged. Scopes was found guilty of willfully violating the Butler law, but the verdict was overturned on a technicality in a higher court. The Butler law remained in effect.
Kentucky	1976; revised in 1990	Public school teachers may teach "the theory of creation as presented in the Bible."
Alabama	1996	Textbooks must carry a disclaimer stating that evolution is controversial.
Georgia	January 2002	A federal judge mandated a sticker be put in high school textbooks stating that evolution is "a theory, not a fact," and "should be approached with an open mind, studied carefully, and critically considered." The school board is appealing the decision.
Pennsylvania	2005	A federal judge ruled that Pennsylvania school districts cannot teach ID in science classes, stating that "teaching Intelligent Design would violate the Constitutional separation of church and state." His exact words include: "We have concluded that it is not [science], and moreover that ID cannot uncouple itself from its creationist, and thus religious, antecedents."
Alabama	February 2005	Evolution disclaimer removed from books, but remains in standards urging students to "wrestle with the unanswered questions and unresolved problems still faced by evolutionary theory."
Arkansas	2005	ACLU and school officials in Beebe, Arkansas, argued over the legality of stickers placed in high school textbooks questioning the theory of evolution, saying that evolution alone is "not adequate to explain the origins of life."
Kansas	April 2005	State Board of Education required evolution to be taught as a critical thinking exercise rather than a scientific theory, with ID included as a viable alternative.
Maryland	2005	State Board of Education approved a biology text that emphasized the importance of the theory of evolution in biology.
Missouri	2005	State Board of Education required biology texts used in high school to include at least one or two chapters that critically examined the theory of evolution.
Ohio	2005	State Board of Education voted in favor of a curriculum emphasizing the "debate over evolution."

You may decide, for personal reasons, that creationism makes more sense to you as an explanation for the natural world, but to be a responsible citizen and take an active part in this debate, you must have a clear understanding of the theory of evolution. Evolution is so central to a biological view of the natural world that it is often called the unifying theme of biology. "Nothing in biology makes sense except in the light of evolution," said a highly respected Russian American biologist almost a century ago, and it's even more true today, as yet more evidence for evolution accumulates. Let's take a look at the theory of evolution and specifically at how it relates to the study of human biology.

Evolution equals changes in the gene pool

Charles Darwin's evidence for his theory of evolution through natural selection is presented in the I Wonder . . . box: How did Darwin figure out his theory of evolution? on page 634. Curiously, Darwin seldom used the word "evolution" in his epochal book, *On the Origin of Species* (1859). Instead, he preferred "descent with modification," considering it a better description of his ideas. Darwin proposed that natural selection caused the modifications he and other scientists documented. His definition of natural selection included four general statements:

1. All organisms produce more offspring than can survive and reproduce in subsequent generations.
2. Organisms show differences that can be inherited.
3. Variations among organisms can increase or decrease each individual's ability to reproduce.
4. Variations that increase the likelihood of successful reproduction will be passed on to future generations.

Darwin recognized that the excessive number of offspring in natural populations caused competition for resources like food and shelter, and that individuals with more ability to acquire these resources would survive and reproduce, so traits that helped the parents survive would be passed to the next generation.

After many years, and despite our understanding of the molecular processes of evolution, no one has yet found a basic flaw with Darwin's notion of descent with modification. The current picture of evolution is as an unpredictable and natural process of descent over time through genetic modifications. The important phrases in this understanding are: descent over time, genetic modification, and unpredictable and natural.

Evolution takes time. Individuals do not evolve; populations evolve. Although Darwin envisioned gradual and subtle changes that would build up over many generations, we now know that changes can also be rapid. But in either case, evolution alters the frequency of alleles in a population. These allelic differences show up as phenotypic differences in individuals and may eventually cause enough divergence to create new species over long periods of time. Evolution, therefore, is a change in allele frequency in a population over time. Small adaptive changes in allele frequency in a population's gene pool are called **microevolution**. The term **macroevolution** is used when a new species is created through these changes in allele frequencies in a population's gene pool, leading to more dramatic changes over longer periods, such as the transformation of a fish into a tetrapod.

Microevolution Evolution occurring through a series of small genetic changes, typically referring to changes within populations.

Macroevolution Evolution over long periods of time, resulting in vastly different organisms, typically referring to changes leading directly to new species.

THE HARDY-WEINBERG EQUATION SPECIFIES HOW ALLELES CHANGE

Many factors can contribute to changes in allele frequencies. Independently, the population biologists Godfray Hardy and Wilhelm Weinberg described a list of characteristics in a population that would prevent changes in the alleles and their frequencies in the gene pool over time. For no evolution to occur, a population must meet these requirements: (1) The population must be extremely large, in fact, effectively infinite, to eliminate the possibility of random **genetic drift**. (2) The individuals must reproduce sexually and mate ran-

Genetic drift
Random differences in the frequency of an allele within a small or isolated population due to chance events.

Selection pressure
Any external forces that cause differences in the fitness of individuals having particular alleles.

Gene flow
Gain or loss of alleles in the gene pool of a population as individuals enter or leave by migration (as opposed to by birth and death).

domly within the population, meaning that the only criterion for mate selection is gender. (3) No random mutations can occur, a condition that does not occur in the natural world. (4) There is no **selection pressure** on the population. (5) There is no **gene flow** in or out of the population.

These criteria are useful for describing an ideal or a benchmark model population, even though such a population does not exist in the real world. We know the frequencies of alleles do change in natural populations, and therefore evolution is a continuous and ever present process. However, by looking at this list, we can see why evolutionary changes are occurring in that population.

Hardy and Weinberg were not content to generate a list of characteristics for genetic stability; they also saw a need for a mathematical model to predict allele frequencies. The Hardy-Weinberg equilibrium equation is a mathematical representation of the expected genotypic frequencies in a non-evolving population. This equation allows us to compare frequencies of genotypes from one generation to the next, looking for differences between the ideal H-W model population and the actual population.

The frequencies of two alleles of one gene are designated with the variables p (dominant allele) and q (recessive allele). If there are only two alleles to choose from, the total of the frequencies of p and q must add up to 1. Mathematically, $p + q = 1$. For example, if 32 percent of the alleles in a population code for a recessive trait such as attached ear lobes, the other 68 percent must code for the dominant trait (unattached ear lobes).

As mentioned, those five requirements for a nonevolving population do not occur in the real world. But the value of the Hardy-Weinberg equation lies in its ability to compare allele frequencies between

$$p^2 + 2pq + q^2 = 1$$

Hardy-Weinberg equation FIGURE 19.1

This formula is used to determine the extent of allelic frequency changes that are occurring in a natural population. p = the dominant allele, and q = the recessive allele. Some deviation from predicted frequencies is expected, as the conditions for a non-evolving population are impossible to obtain here on Earth.

the model population (which is not evolving) and the natural population (which is most likely evolving) over time. The equation also lets you predict the number of individuals in the **model** population that carry a trait (FIGURE 19.1). Mathematically, the equation for this prediction is: $p^2 + 2pq + q^2 = 1$. We use the equation to calculate the frequency of homozygous dominant (p^2), heterozygous (pq), and homozygous recessive individuals (q^2). We can compare this calculated frequency for the model population to the natural population's predicted allele frequencies (based on the expected allele frequencies with and without evolutionary pressures). A difference between calculated frequencies and observed frequencies indicates that evolution is occurring in the natural population, and that at least one of the Hardy-Weinberg criteria is not satisfied.

Let's look at sickle cell anemia, a recessive trait. Homozygous recessive individuals may die young due to their fatal anemia, but they and, more importantly, the heterozygous carriers of the allele, are less affected by malaria. In a population where 9 percent are homozygous recessive (have sickle cell anemia), the Hardy-Weinberg equation can calculate the expected frequency of carriers. Nine percent (0.09) is the value for q^2. Taking the square root finds $q = 0.3$. Since $p + q = 1$, then $p = 1 - q$, or $1 - 0.3$, or 0.7. Knowing the frequency of each allele, you can easily calculate the frequency of heterozygous individuals. $2pq = 2(0.7)(0.3) = 0.42$. Forty-two percent of the population are expected to be sickle cell anemia carriers and therefore partly protected from malaria. This is unlikely to be the case in a real population, even in a nonmalarial area, owing to other selection pressures.

I WONDER . . .

How did Darwin figure out his theory of evolution?

The theory of evolution through natural selection is rightly considered a crowning achievement of science. But how did an amateur naturalist named Charles Darwin figure it out? In the early 1800s, evolution was in the scientific air, in the sense that somebody needed to explain the many life forms that biologists were bringing back from the far corners of the Earth, and various individuals had speculated about the mechanism of evolution.

In that era, science was less compartmentalized than today, and Darwin was able to draw on many sources. He began his masterwork, *On the Origin of Species* (1859) with a discussion of plant and animal breeding, observing that our crop animals and plants sometimes produce offspring with what we now call new phenotypes. By analogy, individuals in natural species could vary.

In 1831, Darwin signed on as naturalist on a round-the-world voyage on the good ship *Beagle*. From geology, Darwin gained clues to the nature of time and gradual change. In 1832, the *Beagle* stopped in Cape Verde, near West Africa, and Darwin noticed a band of sea shells and corals in rocks 30 feet above the sea. Logically, the shells must have been deposited below water, so why were they above water now? Darwin recalled geologist Charles Lyell, who argued that Earth was continuously changing. Darwin realized he was looking at evidence for Lyell's idea. Time matters. Solid ground changes over long spans of time; these rocks had once been submerged!.

In 1833, Darwin gawked at dinosaur fossils in Argentina, and wrote: "It is impossible to reflect on the changed state of the American continent without the deepest astonishment. Formerly it must have swarmed with great monsters: now we find mere pygmies . . . What has exterminated so many species?" Life changes. Death plays a part.

In the Galapagos Islands, off the coast of Ecuador, Darwin observed a series of related birds, now called Darwin's finches, that had adapted to many different ecological niches. As he wrote in his diary, "a most singular group of finches, related to each other in the structure of their beaks, short tails, form of body, and plumage: there are thirteen species . . . all . . . peculiar to this archipelago." Species occur in some kind of related groups.

By the time Darwin returned home in 1836, all the ingredients of his theory were in place: variation, time, competition, change, death, related groups of species.

Darwin had intended to become a country pastor, and he was a religious man. But he wrote in 1837 that intense exposure to nature changed him: "The old argument of design in nature, which formerly seemed to me so conclusive, fails, now that the law of natural selection has been discovered. . . . Everything in nature is the result of fixed laws."

But Darwin was in no rush to publish: He worked privately on his theory until 1858, when he received a letter from Alfred Russel Wallace, a little-known biological collector. On his own, during an eight-year expedition through the Malay Archipelago (now Malaysia and Indonesia), Wallace had come up with a strikingly similar theory.

Darwin immediately saw that Wallace's ideas closely paralleled his own unpublished theory. In 1859, both scientists published their parallel papers, describing their common theory that species vary because some are better suited to survival than others. But Darwin had conceived it first, and thus it is correct to give him primary credit for discovering evolution through natural selection. Wallace agreed, and the two biologists became lifelong friends after Wallace returned from the Malay Peninsula.

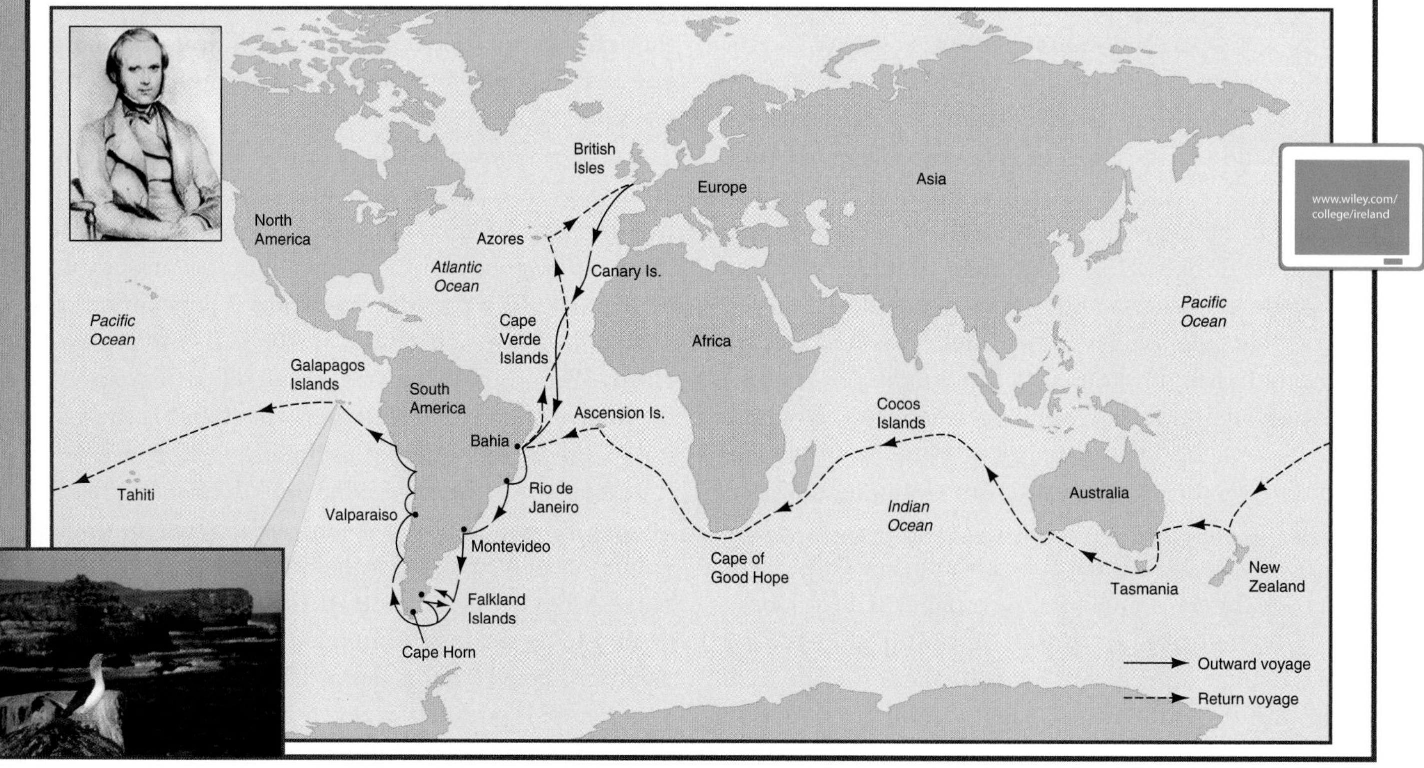

FIGURE 19.2

The descent with modification of our present day wolves is depicted here. As the environmental conditions of the Earth changed, different characteristics became helpful in survival. These changes in allele frequency led to great changes in wolf form and function over long periods of time.

Despite the fact that the Hardy-Weinberg equation can help us analyze the course of evolution, keep in mind that evolution is neither linear nor directed. One of the largest misconceptions about evolution is that it is progressive or has an end goal; that it is aiming toward a perfect life form. Evolution is a natural process, and it has no more purpose than gravity. However, evolution can, but does not always, maximize the **fitness** of a population. Allele frequency changes that persist in a population allow the population to exploit the available resources more effectively than other organisms. Sometimes these changes form a new species, and the old one dies off, in a linear alteration. Or similar modifications may lead to the formation of two or more **divergent** species (FIGURE 19.2).

Fitness The relative ability of an individual to produce viable (living) offspring that survive to reproduce.

Divergent Separating from a common point; growing farther apart.

Allele changes can also lead to neutral modifications because some mutations or mistakes in copying DNA during mitosis have little or no effect on phenotype. These neutral modifications are neither beneficial nor detrimental. But as environmental conditions change, their significance may change as well. The fitness of any trait is affected by chance events, natural selection, man-made alterations in the environment, and natural changes to the environment.

CONCEPT CHECK

How do evolution and intelligent design differ?

What does it mean for a statement to be testable?

What are the five criteria for genetic stability in the Hardy-Weinberg equilibrium?

Define evolutionary fitness.

Compare microevolution to macroevolution.

Evolution Is Backed by Abundant Evidence

LEARNING OBJECTIVES

List and describe the five types of evidence for evolution.

Give an example of evolutionary change, supported by evidence.

What evidence do we have to support or refute the theory of evolution?

New species don't appear in our back yards very often, at least in the human time frame. No one person has witnessed macroevolution, although thousands of studies have detected it over long periods of time.

A classic example of how changing selection pressure affects fitness appeared in the peppered moth populations (*Biston betularia*) of Britain. Before industrialization, the peppered moths included a small number of moths with black wings and a higher percentage with light wings. Peppered moths rest with their wings open. The lighter wings camouflaged the moths on gray (and lichen-covered) tree bark, but the darker moths made an easy target for birds. As dark soot from factory pollution built up on the trees, selection pressure shifted to favor the darker moths, and the lighter ones were disproportionately eaten. The frequency of the dark allele in the moth population increased as dark phenotype moths were eaten. Then, as air pollution laws controlled pollution, tree bark once again became lighter. The dark peppered moth phenotype changed again with lighter colored moths once again predominating.

A similar phenomenon appeared among moths in Michigan, again tracking the rise and fall of air pollution. In 1895, approximately 98 percent of the peppered moths around Manchester were dark colored, again attributable to the increased air pollution from factories. As clean air legislation forced these factories to curb their emissions, the moth population underwent a dramatic shift in phenotype. In 1959, the percentage of dark moths in Michigan had declined by an astounding 90 percent, and another drop of 6 percent was recorded by 2001.

Another compelling example of evolution is the current rise of antibiotic-resistant bacteria (see Chapter 10). Antibiotics place enormous selective pressure on bacterial populations. If one bacterium gains a plasmid (an extra bit of DNA-carrying functional genes) that confers resistance to that antibiotic, the bacterial cell may survive and thrive. Or a single bacterium may possess a small phenotypic difference that allows it to survive the antibiotic. In either case, this single bacterium can reproduce where others cannot, producing a new colony of antibiotic-resistant bacteria.

Since we cannot observe macroevolution directly, how do we determine the scientific validity of the theory of evolution? The main lines of evidence are the fossil record, biogeography, comparative anatomy, comparative embryology, and comparative biochemistry.

FOSSILS ARE THE OLDEST EVIDENCE FOR EVOLUTION

Fossils are evidence of past life that includes teeth, bones, seeds, shells, and other hard parts of organisms. A second category of fossils shows evidence of softer tissues. Imprints of leaves, for example, show us the structure of early plants; scientists have even analyzed fossilized dinosaur footprints and feces for hints about dinosaur behavior.

Fossils form when organisms die and are covered with sediment or volcanic ash. The soft tissues usually decompose, but the hard tissues are slowly transformed to minerals. Water percolating through the overlying sediment brings in ions that start a chemical reaction in the organic material, creating a permanent stony material. As more sediment is deposited, heat and pressure build up, speeding fossilization.

The fossil record gives an incomplete but intriguing look at past life. Soft tissue usually does not fossilize, which means that whole phyla of plants and animals have left no fossil record. We have no fossil record of jellyfish, for example, but we believe they must have been present from a very early time.

Strata Layers.

Fossils can be dated by looking at the **strata** where they reside. Sedimentary rock is deposited in layers, one on top of the previous one, creating a repeated layer cake effect (FIGURE 19.3). You can see millions of years of sedimentary rock at the Grand Canyon. Although the strata form horizontally, geological forces can cause uplifting or faulting that tilts or disrupts the sedimentary layers. If you know when a particular stratum was deposited, you can tell the age of the fossils it contains. These layers can be

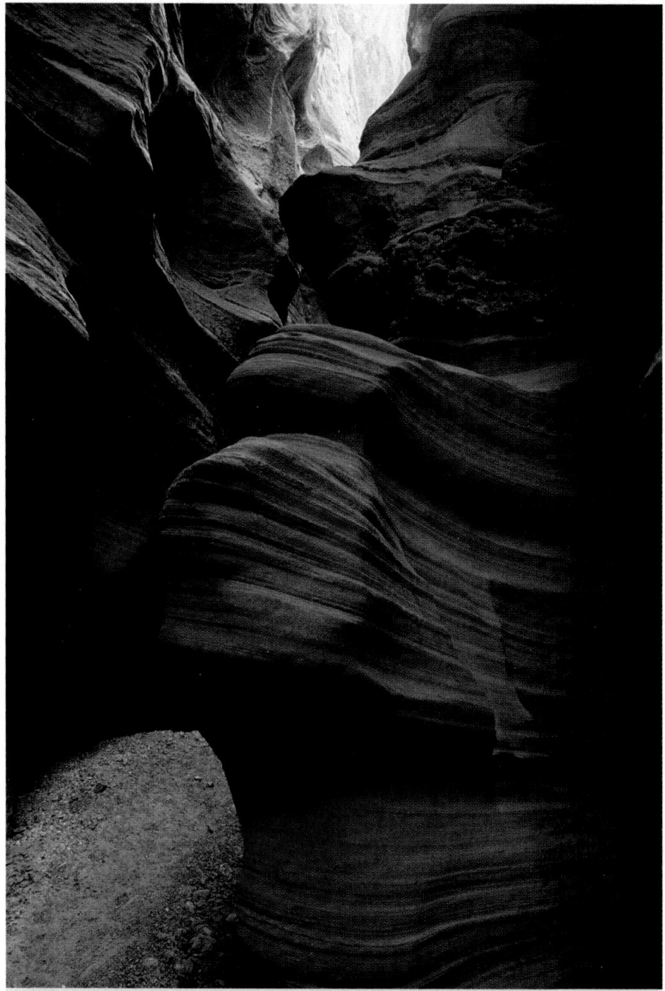

Stratification of sediment and fossil formation
FIGURE 19.3

Fossils can be dated by observing the layer of rock in which they lie. As more rock is layered on top of the dead organism, the remains become trapped in sediment. Under the proper conditions, fossil imprints of the organism are left behind.

Radiometric dating FIGURE 19.4

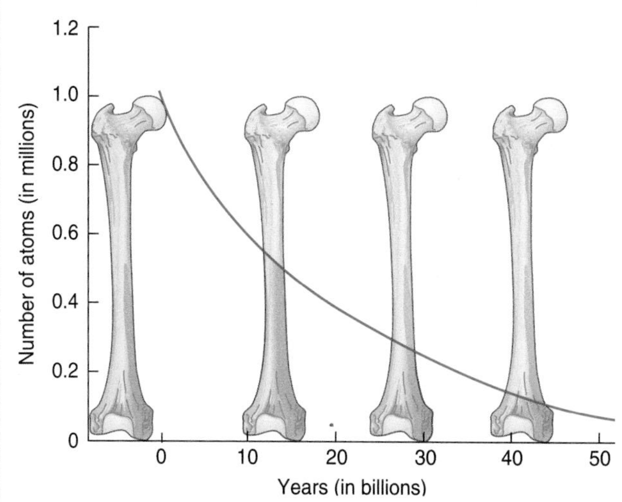

When an organism dies, the radioactive carbon in its skeleton is fixed. Over the years, as the bone ages, the carbon 14 decays. By the end of one half-life, the amount of radioactivity within the femur is half of that in a newly formed bone. After a second half-life, the remaining radioactivity is only one-fourth of that in new bone. With each passing half-life, the amount of radioactive carbon 14 is cut in half. Knowing the exact half-life and the original concentration of the radioactive isotope, we can determine the age of the bone.

Process Diagram

dated using simple logic as well as the comparison to layers in other areas. For example, it is clear that older layers of sediment are beneath newer layers. It is also logical to assume that fossils found within a particular layer are the same age as that layer. By accepting these two assumptions, fossils can be arranged in order from older to younger. This type of comparison is referred to as relative or stratigraphic dating.

A second, more accurate way to date layers uses radiometric dating (FIGURE 19.4), which takes advantage of the fact that radioactive elements decay at well-understood rates. During decay, these elements release radioactive particles and move down an isotopic "decay chain," eventually forming stable atoms. The amount of time it takes for 50 percent of a particular radioisotope to convert to another isotope, which may or may not be stable, is called its half-life. During the

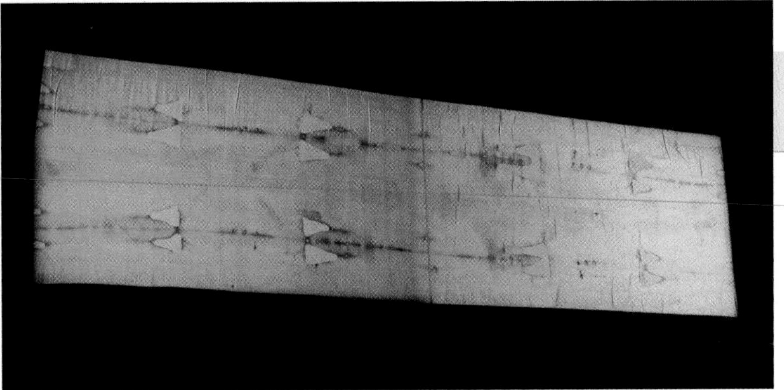

Carbon dating: Was this the Shroud of Turin?
FIGURE 19.5

This artifact was originally thought by some to be the shroud used to cover the body of Christ after his crucifixion. Carbon dating of a small section of the material, with its apparent bloodstain, was performed to determine the age of the shroud. This technique indicated that the shroud was not old enough to have been used during Christ's time.

second half-life, half of the remaining sample decays, leaving one-fourth of the original sample. As this process continues, the amount of the original sample decreases by geometric progression.

To use radioisotope dating, we must know the original composition of the rock or material in question. In other words, if we know the original amount of radioactive isotope present in the sediment layer, we can compare that original level with the present level of radioactivity. The change in radioactivity is due to the slow decay of the radioisotope to its stable form. The half-life of the radioisotope is used to calculate the age of the sediment fairly accurately. For example, radioactive uranium is soluble in water, but stable thorium is not, so a rock that precipitated at the bottom of the sea would contain uranium, not thorium. After precipitation, the uranium 234 would start to decay, forming thorium 230. Because uranium 234 has a half-life of about 245,000 years, we can calculate, based on the proportions of these two isotopes, how long ago the thorium began to accumulate.

For fossils formed within the last 50,000 years or so, carbon 14 (^{14}C) is the isotope of choice for dating. ^{14}C decays into ^{12}C with a half-life of 5,730 years. Because ^{14}C continually forms in the atmosphere, its concentration in the atmosphere has been relatively stable over time. Living organisms take up both isotopes of carbon from the atmosphere, in the proportions found in the atmosphere. After death, no further carbon is taken up by the organism, and its store of ^{14}C decays, so the proportions of ^{14}C and ^{12}C show the sample's age. This type of radiodating is used to determine the age of archaeological finds, such as the Shroud of Turin (FIGURE 19.5).

PLATE TECTONICS: SHIFTING CONTINENTS SHOW EVOLUTION

Another key to interpreting the fossil record is **plate tectonics**. The crust of the Earth is not a solid sheet, but rather a patchwork of huge fragments. Each fragment, or plate, may consist of ocean floor, continental land, or a combination of the two. These plates form the Earth's crust, which floats on the relatively fluid **asthenosphere**. **Convection currents** in the magma of the asthenosphere rise and transfer heat from the core of the Earth. When they reach the crust, they may flow sideways, pulling and pushing the plates from their original positions.

Plate tectonics Geological theory that explains the gradual movement of the plates of Earth's crust.

Asthenosphere Zone of deformable rock immediately below Earth's hardened crust.

Convection currents Transfer of heat by the mass movement of heated fluid to cooler areas.

Tectonic plates spread above these convection currents. Plates slide past one another where the currents run laterally beneath the crust. If two plates butt against each other, one is usually forced under the other. This is the force that raised the giant Himalayan Mountains.

Tectonic movements have shaped the Earth's surface (FIGURE 19.6). **Convergent zones** form mountains and oceanic trenches (and often nearby strings of volcanoes), **divergent zones** form oceanic ridges and continental rifts, and **transverse zones** are extremely unstable. The Pacific Ocean holds the most

Plate tectonics and continental drift FIGURE 19.6

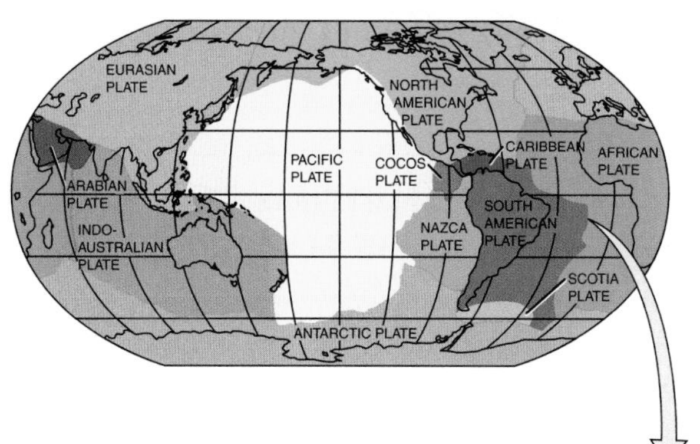

Continental drift and the plate tectonics theory are based on the movement of magma under the Earth's crust. 1 Magma rises to the surface. 2 Cooling magma spreads out under the plates of the divergent zone, traveling away from the upwelling area and eventually sinking as it cools. 3 The magma moving under the surface of the crust drags the crust along, spreading the plates at divergent zones. 4 Where magma sinks, plates either butt up against one another to form mountains, or one sinks beneath the other to form trenches. The deepest portion of the ocean is just such a trench.

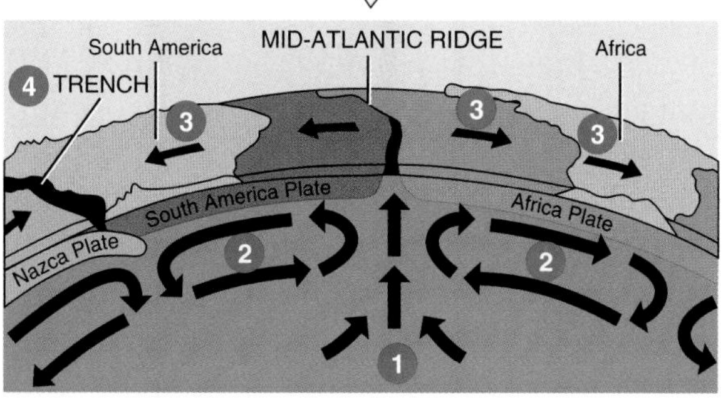

Process Diagram

Convergent zones Areas where two or more plates are moving toward one another, creating mountains or oceanic trenches.

Divergent zones Areas where two or more plates are separating from one another, leaving rifts.

Transverse zones Areas where two or more plates are sliding past one another.

active plates on the planet. Some move relative to nearby plates at a blistering speed of up to 15 centimeters (about 6 inches) a year. Surrounding the Pacific Ocean, the "Ring of Fire" has more seismic and volcanic activity than any other place on the planet. The ring is located at the intersection of tectonic plates encircling the Pacific.

This movement of plates causes the phenomenon called continental drift. Geologists believe that about 200 million years ago, all the continents were connected in one large landmass called Pangaea, which broke into smaller continents (FIGURE 19.7). These pieces, today's continents, drifted away from Pangaea to today's position, carrying both fossil-bearing rock and current

240 MILLION YEARS AGO
PANGAEA

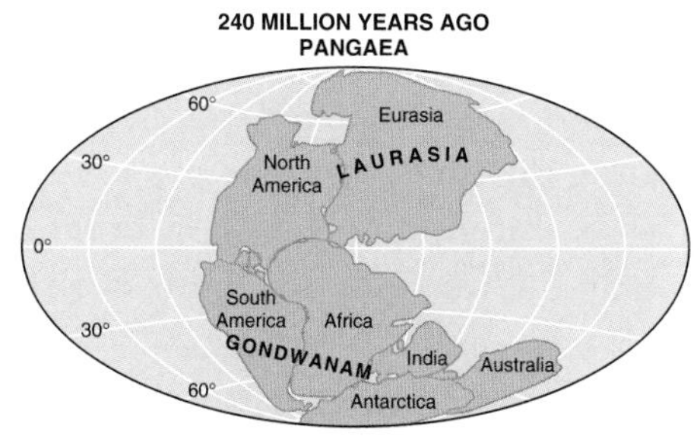

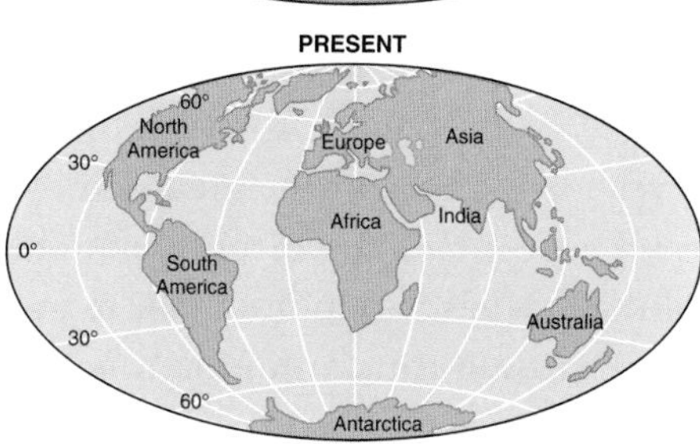

Movement of continental plates over time
FIGURE 19.7

life forms. This movement separated the fossil record. Imagine dividing a completed jigsaw puzzle into seven pieces; to see the whole picture, you would have to fit those pieces back together.

In 1917, German meteorologist Alfred Wegener noticed identical fossils on either side of the Atlantic Ocean. Wegener recognized that most of the early life forms living on the land could not have crossed the Atlantic, so he proposed that the land must have been one large piece. In other words, he hypothesized that Pangaea must have existed. Not until the 1960s was his proposal finally taken seriously, after scientists mapping the ocean floor discovered that the new-formed crust at the midoceanic ridges could provide a mechanism for continental drift. In interpreting the fossil record, scientists must take into account where a particular landmass was located when the fossils were deposited. Therefore, the fossil record not only supports the theory of plate tectonics, but it also records the action of evolution.

When Pangaea split and formed early supercontinents, land animals and plants became isolated. As these life forms reproduced, natural selection and random genetic drift occurred. The climate on a particular piece of land also slowly changed as it moved to a new latitude, altering the fitness of traits and leading to natural selection of different alleles in different climates. Populations on the various continents evolved along similar but separate paths.

Isolation also contributed to the evolution of new species, as you can see from the **extant** animals of Australia. Seventy percent of the world's marsupials are found in Australia, including the well-known kangaroos and koalas. Marsupials are mammals whose young complete their development outside the female's body, in her pouch rather than inside the uterus. Despite their unusual reproductive behavior, Australian marsupials use similar resources, have the same general body appearance, and act much like placental mammals. This type of evolution, where adaptations in marsupial lineages parallel those in placental mammals, is an example of **convergent evolution**. For example, the marsupial wolf and the placental wolf are thought to have independently evolved their current dog-like appearance. And both marsupial moles and placental moles have short, strong forelimbs for digging, and sleek tubular bodies.

Extant
Living; opposite of extinct.

COMPARATIVE ANATOMY IS A TRADITIONAL BASIS FOR OBSERVING EVOLUTIONARY RELATIONSHIPS

Comparative anatomy is the study of structural similarities and differences in body forms. Comparative anatomy remains a key piece of evidence for the evolutionary relationships of organisms and their organs. Evolutionarily speaking, organs can be **homologous** and perhaps **vestigial**, or they can be **analogous**. Anatomical structures that were found in different living populations and on a common fossil ancestor are homologous structures (FIGURE 19.8). Homologous structures have a similar structure but perhaps different functions. When different populations or categories of organisms share homologous anatomical features, they presumably also share a common ancestor. A good example appears in the forelimbs of virtually every vertebrate. Even though some vertebrates use their front limbs for flying, others for swimming, swinging through trees, running, or paddling a canoe, all vertebrate forelimbs have the same basic structure. This observation indicates that the common ancestor of vertebrates is more recent than that of, say, crustaceans and invertebrates. The degree of anatomical similarity suggests the degree of relationship between two organisms.

Homologous
Structures with a common ancestral origin, but not necessarily the same current function.

Vestigial
A persistent but currently unused structure.

Analogous
Structures with the same function, but without a common ancestral origin.

Some structures that serve the same function in different animal groups actually arose independently. Analogous structures share a common function but not

Forelimbs of vertebrates FIGURE 19.8

The striking similarity between the front limbs of these animals indicates a relatively recent common ancestor. Each limb has one bone in the proximal section, two bones in the distal section, a series of short bones, and then another series of long bones making up the digits. These are all homologous structures.

a common ancestry. Bird wings and insect wings are analogous (FIGURE 19.9). They are both aerodynamic structures used to fly, but they are not structurally similar. The gills of a fish and the lungs of a frog are analogous—they both provide respiratory surfaces, but they did not arise from the same ancestral organ. Many analogous structures resulted from convergent evolution, showing that there is more than one way to engineer flight, to take one example.

One last evolutionary term for organs is vestigial. You have the remnants of a tail. It is small, hidden under skin, and tucked under your sacral bone, but it is there. This tail is no longer functional but is obviously homologous to tails in other primates. The appendix is another vestigial organ, as are the muscles that move the ears. We humans cannot move our ears individually or with any precision (even though we can use large muscles to "wiggle our ears"). Other vertebrates, such as cats or horses, can pivot their ears precisely toward the source of a sound. Knowledge of the origins of vestigial structures can also help scientists deduce evolutionary relationships.

Bird wing and insect wing comparison FIGURE 19.9

Despite the fact that bird wings and insect wings are both used for flight, they have little structural similarity. The bird wing is composed of bones and muscle surrounded by skin and feathers, while the insect wing has no internal muscle or bone and is essentially a membranous fold held over a rigid protein framework.

EMBRYOLOGY PROVIDES MORE CLUES TO EVOLUTION

Embryology, the study of development, provides other clues about evolutionary relationships. Some stages of the developing human embryo are strikingly similar to the embryos in other organisms. Scientists used to say, "ontogeny recapitulates phylogeny." (Ontogeny is development from embryo to adult; to recapitulate is to repeat concisely; phylogeny is the study of the development and history of a species.) So this mouthful of jargon implied that the development from embryo to adult echoes the development and history of that species. This seems to be true with a cursory look at vertebrate embryonic development, but we know this statement is an oversimplification. Human embryos do not pass from a single-cell stage, through a fish-like vertebral stage on their way to becoming an adult (FIGURE 19.10). Our embryo does, however, follow the same pathway as other vertebrate embryos. It is uncanny how human embryos resemble reptile or even fish embryos, and therefore it is easy to see how early embryologists could have interpreted superficial appearances as an inference of phylogenetic relationships.

The phylogenetic tree (FIGURE 19.11) visually represents the genetic and evolutionary relationships among organisms. When comparing two organisms on the phylogenetic tree, long branches indicate

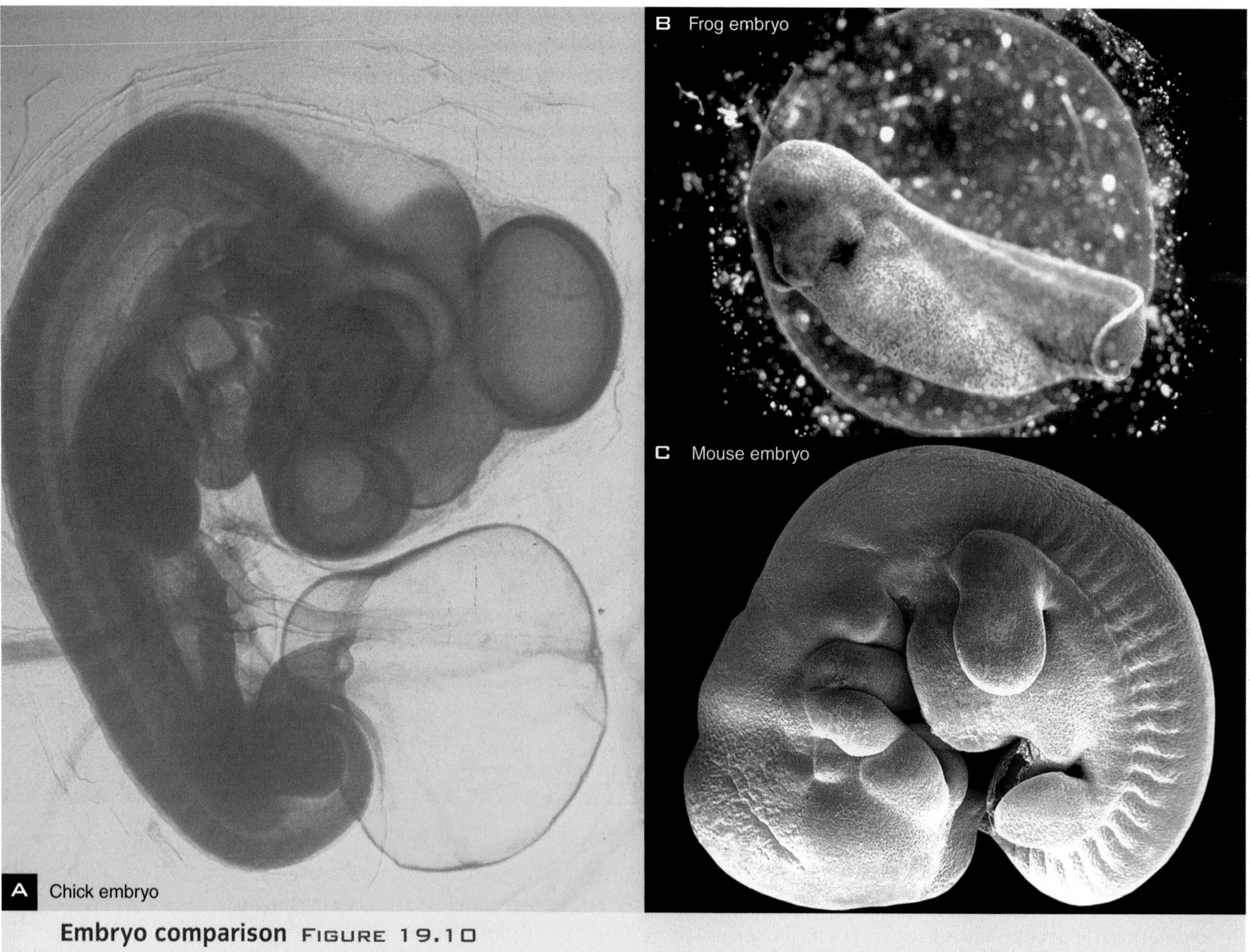

Embryo comparison FIGURE 19.10

less similarity, while shorter branches indicate closer relationships and/or recent speciation (and therefore less time for divergence).

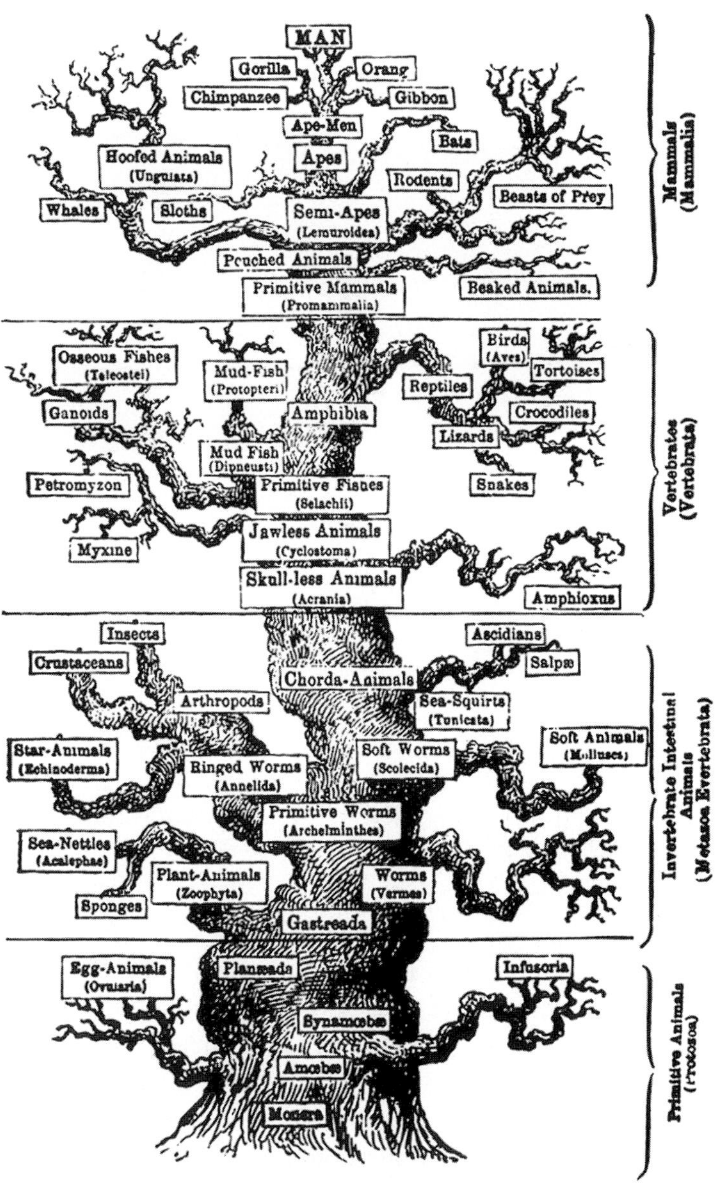

Phylogenetic tree FIGURE 19.11

This phylogenetic tree indicates the relationships of extant animals, as we now understand them. Each branch in the tree is a splitting point, where life forms diverged enough to create a new taxonomic group.

BIOCHEMISTRY: EVOLUTION IS IN THE GENES

Recent technical breakthroughs have provided even more support for the theory of evolution (FIGURE 19.12). The structures of both proteins and DNA are biochemical evidence for evolutionary relationships. Closely related species have nearly identical DNA sequences; as the relationships become more distant, we see fewer matching sequences. As species develop separately, mutations build up in the DNA. The longer the two species have been separated, the more mutations will have occurred and the more differences there will be in the DNA.

For example, humans apparently diverged from chimpanzees about 5 million years ago. During that time, our DNA mutated and accumulated differences. These differences were recently cataloged by Cornell University scientists, who were looking for the genes

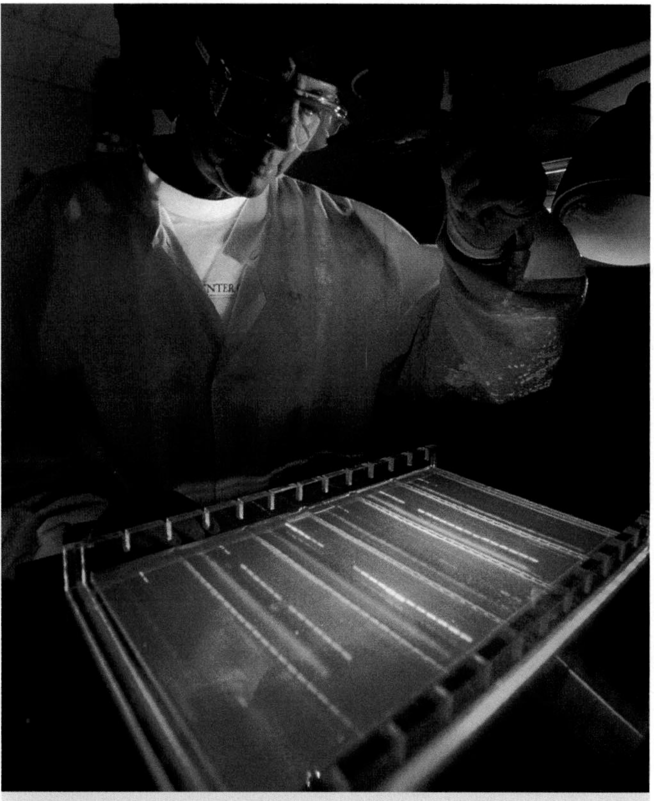

Sequencing gel of DNA bands FIGURE 19.12

This gel shows the bands produced as a sample of DNA is "run." Each band represents a different size piece of DNA.

that had changed the most when humans and chimpanzees evolved into their present forms. Surprisingly, they discovered the largest differences in genes on the X chromosome and genes associated with the immune system. These areas showed a much faster rate of mutation and therefore much more evolution than, for example, genes that code for proteins expressed in the brain. This was surprising, as the biggest evolutionary difference between humans and chimpanzees seems to be the composition or functioning of our brains.

Protein sequences also show the distance between species. Hemoglobin is a ubiquitous protein that differs only slightly among vertebrate organisms. As with DNA sequences, the degree of variability between hemoglobin sequences suggests the length of reproductive isolation between two species. Our hemoglobin and that of a common grass frog differ by 67 of the 147 amino acids. In dogs, the same protein differs from ours by 32 amino acids, and in macaque monkeys our hemoglobin matches in all but 8 amino acids.

CONCEPT CHECK

How did continental drift affect populations on the continents?

What is the difference between homologous limbs and analogous limbs?

How do the protein sequences on hemoglobin show evolutionary relationships?

Natural Selection Has Far-Reaching Effects on Populations

LEARNING OBJECTIVES

Define fitness in evolutionary terms.

Explain how the bottleneck effect, the founder effect, and adaptive radiation affect allele frequency.

One of the Hardy-Weinberg criteria for a nonevolving population is the lack of natural selection. Natural selection refers to many forces acting on species, such as the need to react to climate, the formation of new mutations, and inter- and intraspecific competition for limited resources. The result of natural selection is successful reproduction of only the best-adapted organisms. This selective pressure is the backbone of Darwin's theory of descent with modification and is ever-present in nature.

The raw materials for natural selection are the random mutations that occur in DNA and the different genetic combinations resulting from sexual reproduction. Mutations occur in nonessential, even unused portions of the DNA over time, as well as in the genes that determine the phenotype. These altered alleles can persist for generations, with little or no detrimental effect. An accumulation of these random mutations over millions of years may be enough to produce new species, assuming selective pressures change to benefit individuals with the mutations.

Many people summarize natural selection as "survival of the fittest." Fitness is the ability of an organism to survive and successfully reproduce, not to run 1,000 meters. The key is to leave more copies of your genes in the next generation. A woman who dies at age 25 but leaves six surviving children is biologically more "fit" than a woman who runs marathons and lives to be 98 but only has two children.

When the environment changes, for example, and **successional changes** occur in the ecosystem, different pressures are put on the resident life forms. These new pressures may require a different foraging strategy, faster reproduction, or perhaps a faster running speed. Mutations may produce phenotypic variations that are beneficial in a changing environment, conferring an advantage to those organisms carrying the mutation. If these organisms reproduce, the mutations may pass to future generations, eventually becoming more common in the population. For example, as wolves prey on deer, the average speed of the deer population increases. The fastest individuals can escape the wolves, while the slower ones get eaten. Those that outrun the wolves breed and pass on the alleles for larger muscles, faster muscle contraction, or more efficient joints, which produces faster offspring. Natural selection causes individuals with the combination of traits most suited to the environment to reproduce and leave a larger proportion of their offspring in the next generation. Assuming this also holds true for humans, the question in human biology becomes, do civilizations rise and fall due to environmental changes? See the Health, Wellness, and Disease box on page 646 for more on this topic.

Successional changes Predictable changes in the dominant species found in a particular area as the area reaches ecological stability.

POPULATIONS LOSE ALLELES

Stable populations can be devastated by natural events, such as tsunami or fire. These catastrophes upset the balance of the ecosystem and promote evolution without regard to fitness. In other words, those individuals in the path of the disaster die, regardless of their genetic makeup. When a large portion of any population is suddenly removed, the frequency of alleles in the remaining population may not be representative of the original population. This is the **bottleneck effect**.

Among humans, we witness the bottleneck effect after ecological disasters. The tsunami of December 26, 2004, killed more than 175,000 villagers in Southeast Asia without regard to age, gender, or health. Few individuals from the original populations were left to repopulate their villages. If there is little immigration, the alleles among the remaining individuals will be all that are available for the next generations. If these alleles occur in a different proportion than what was found in the original population, a bottleneck has occurred and the gene pool is different than it would have been without the tsunami (FIGURE 19.13).

The best modern example of an ongoing genetic bottleneck is the cheetah. These animals used to live on vast tracts from the Middle East to India. Due to loss of habitat and increasing humans hunting of the cheetah and its prey, their numbers have plummeted. Currently, there are approximately 17,000 cheetahs left in the wild. These are isolated in small areas of Africa and Iran, with little mixing between populations. Many surviving cheetah carry a harmful allele that decreases fertility. There may not be enough genetic diversity in

Bottleneck effect FIGURE 19.13

Health, Wellness, and Disease

Does environmental change cause civilizations to disappear?

We have talked about human health throughout this book. Now let's turn our attention to the health of the societies on which depend our collective health, education, support, defense, and culture. History is written by the winners, so we often lose track of civilizations that retreated or failed after a period of economic and cultural glory. Historically, the many examples of declined civilizations include the Maya of Central America and the Norse colony, which occupied Greenland from 1000 to 1400.

Jared Diamond, a professor at the University of California at Los Angeles, looked at the relationship between environmental damage and the crash of civilizations, in his book *Collapse*. Why have societies in Japan, the New Guinea highlands, and Switzerland flourished for many centuries, while others declined after a period of prosperity?

Diamond outlined five factors that influence the long-term sustainability of a society:

- Environmental damage: Overgrazing, overpopulation, overfishing, and clear-cutting forests can all lead to degradation of land and water, reducing the ability of land and sea to support a population. Long-term damage to soils is occurring in many agricultural regions today; when topsoil washes away or turns saline due to overirrigation, agricultural productivity declines. More people have to work on farms, reducing the surplus labor available for skilled trades, administration, culture, and the military.
- Climate change: Long-term shifts in temperature and rainfall can have disastrous impacts on civilizations that may, for example, have grown dependent on steady rainfall. A drought can cause starvation.
- Hostile neighbors: Military attacks by neighbors often spell doom for various societies. However, Diamond sees many of these conquests as ultimately rooted in environmental destruction.
- Friendly trade partners: Trade partners can become allies in time of need. Our current reliance for oil on unstable countries in the Middle East shows, Diamond argues, how the fate of trade partners can affect both sides of the partnership.
- Society's response to threats: Highland New Guinea and Japan have both developed management plans to conserve forests and end rapid deforestation, helping to ensure their continued cultural survival.

Other factors also play a role in the destruction of civilizations. For example, occupants of the Western Hemisphere lacked a properly challenged immune system to combat many of the diseases carried by the colonizing Europeans 500 years ago. The result was a largely inadvertent genocide as smallpox, measles, and other pathogens swept through Native Americans. And of course the availability of advanced weapons plays a role in who survives to write the history books. But in a world that grows ever-more crowded, Diamond's broad picture of environmental survival or decline helps provide a framework for intelligent action to make sure our society survives as well.

www.wiley.com/college/ireland

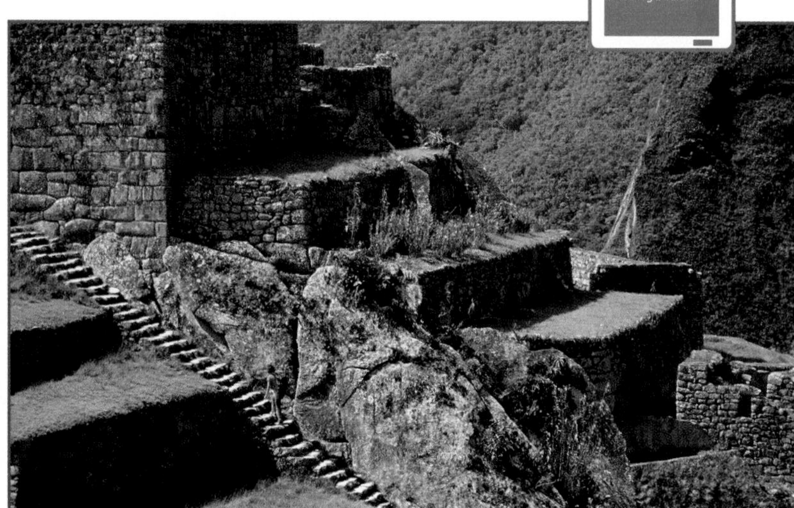
Inca ruins in Machu Picchu, Peru

the remaining population to overcome the increased frequency of this detrimental allele. If so, the species is doomed, even if further habitat destruction and hunting are reversed. North Pacific humpback whales may fall into this category as well. By 1966, whaling had decreased the number of these whales to less than 1,000. By 2004, the population had rebounded to 6,000 to 10,000. It is not clear if the remaining gene pool is diverse enough to avoid any detrimental effects of the 1966 bottleneck. For example, as the population increases, a new viral attack may threaten its numbers. If there is not enough genetic diversity in the recovering population, a viral infection may wipe out the entire species.

A similar evolutionary process, called the **founder effect**, occurs when a small group of individuals splinters off to form a new population elsewhere. The main Hawaiian Islands are a great place to witness the founder effect. The islands are home to many species found nowhere else, called **endemic** species. The ancestors of these species arrived in very small numbers and then expanded to fill all available spaces. The founding populations likely had a different allele frequency than their continental source populations, so those organisms did not carry a full range of genetic possibilities to their new environment. In adapting to their new island habitats, these founding individuals underwent natural selection to evolve into the present island species.

Gene flow can also create new allele frequencies and sometimes even new species. Gene flow mixes genes from different populations when individuals migrate between populations. When individuals leave one population (**emigration**) and join another (**immigration**), they are subtracting alleles from one gene pool and adding to the next. Gene flow may affect allele frequencies by delivering new genetic combinations or removing deleterious ones. The allele frequency of people in the United States has been dramatically altered by gene flow. As a simple demonstration, the Native American population has a high percentage of type B blood. Had they been the only founder population in the United States, type B blood would be common, but the most common blood type in the United States is type O, closely followed by type A. Gene flow when individuals immigrated from Europe and Africa has changed these allelic frequencies.

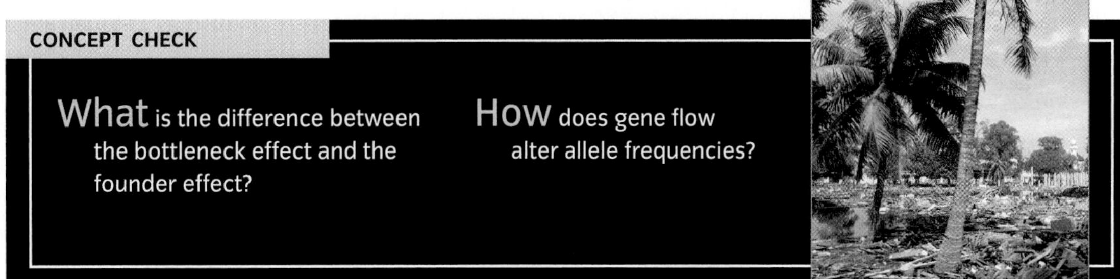

CONCEPT CHECK

What is the difference between the bottleneck effect and the founder effect?

How does gene flow alter allele frequencies?

Mass Extinctions Can Be Followed by Regrowth

LEARNING OBJECTIVES

Identify the two largest extinction events.

Explain the importance of the amniotic egg.

When looking at species development over time, we see periods of rapid speciation alternating with major die-offs, called **extinction events**. Extinction occurs when a species is completely removed from Earth, because all of the individuals died instead of adapting (see the Ethics and Issues box on p. 648). In the last 530 million years, the fossil record shows five mass extinctions, when at least 50 percent of the species disappeared. During the largest, the Permian–Triassic extinction, 70 percent of the land animals and 90 percent of ocean life went extinct. The Cretaceous period ended with the best-known extinction event, which extinguished the dinosaurs. Scientists believe that some or all of these mass extinctions were caused by a massive impact of an asteroid or comet, but mammoth volcanic eruptions and other factors may also be responsible.

The opposite of a mass extinction is **adaptive radiation**. Adaptive radiation refers to changes in organisms resulting from new resource combinations,

Ethics and Issues

Why should endangered species matter to me?

About 20 years ago, biologists began to realize that they would start to run out of things to study due to the accelerating wave of extinctions shaking the planet. Extinctions occur for many reasons; overhunting, destruction of habitat by fire, construction or ecological change, and invasion of exotic species can all play a role.

But what's the big deal? Some extinction is natural, after all. Why is it important to prevent endangered species from going extinct? The answers range from scientific to economic to spiritual:

- Organisms can be useful. A species of plant called the rosy periwinkle was the source of a key drug that defeats one type of leukemia. Scientists are actively looking in many unusual ecosystems for useful chemicals that organisms have evolved for specific reasons. Many antibiotics, for example, were derived from fungi that evolved these compounds for protection against bacteria.
- Life is unique. As far as we know, this is the only planet with life. If we respect life, we should respect its myriad forms as well: the whales, swans, lobsters, and even the endangered fish and mussels in our streams.
- Life has scientific value. To understand the wonders of evolution, we need to study the results of evolution.

- Life is a web. Organisms in the wild have complex interactions that we are only beginning to understand. Extinguishing one organism can have cascading effects throughout an ecosystem.

It's hard to know exactly how far along we are in the current wave of extinction because biologists are not even sure how many species inhabit Earth. Today, about 1.9 million species have been described, but estimates of the total number are many multiples of that. The World Conservation Union says that 748 species are already extinct, and another 16,119 are threatened with extinction. These threatened organisms include one in three amphibians, one in four coniferous trees and mammals, and one bird in eight. The group also says, "56% of the 252 endemic freshwater Mediterranean fish are threatened with extinction."

Perhaps the worst part of species loss is this: Most of the organisms going extinct today are things we have not yet even identified. Their utility and their beauty will go completely unrecognized. And while evolution may eventually restore biodiversity to its current levels, that will take millions of years. In biodiversity, as in so many things, a gram of prevention is worth a kilo of cure!

Amniotic Related to the membranous fluid-filled sac that protects the developing mammalian embryo.

new ecological **niches**, becoming available to the population. Often through history, conditions were favorable for the sudden growth and expansion of animal species, usually due to the opening of new habitat. When amphibian vertebrates developed the **amniotic** egg during the Carboniferous period (about 320 million years ago), they became able to reproduce without returning to the water. This seemingly small development opened vast new habitats for vertebrates by allowing them to occupy the centers of the continents, away from aquatic breeding grounds. On the phylogenetic tree, periods of adaptive radiation show up as the sudden appearance of many branches. These blooms of life are often followed by die-offs of less adapted forms.

Adaptive radiation continues to create new species in many parts of the world (FIGURE 19.14). For example, in Hawaii, a full 30 percent of the organisms on the near-shore reefs are endemic. The ances-

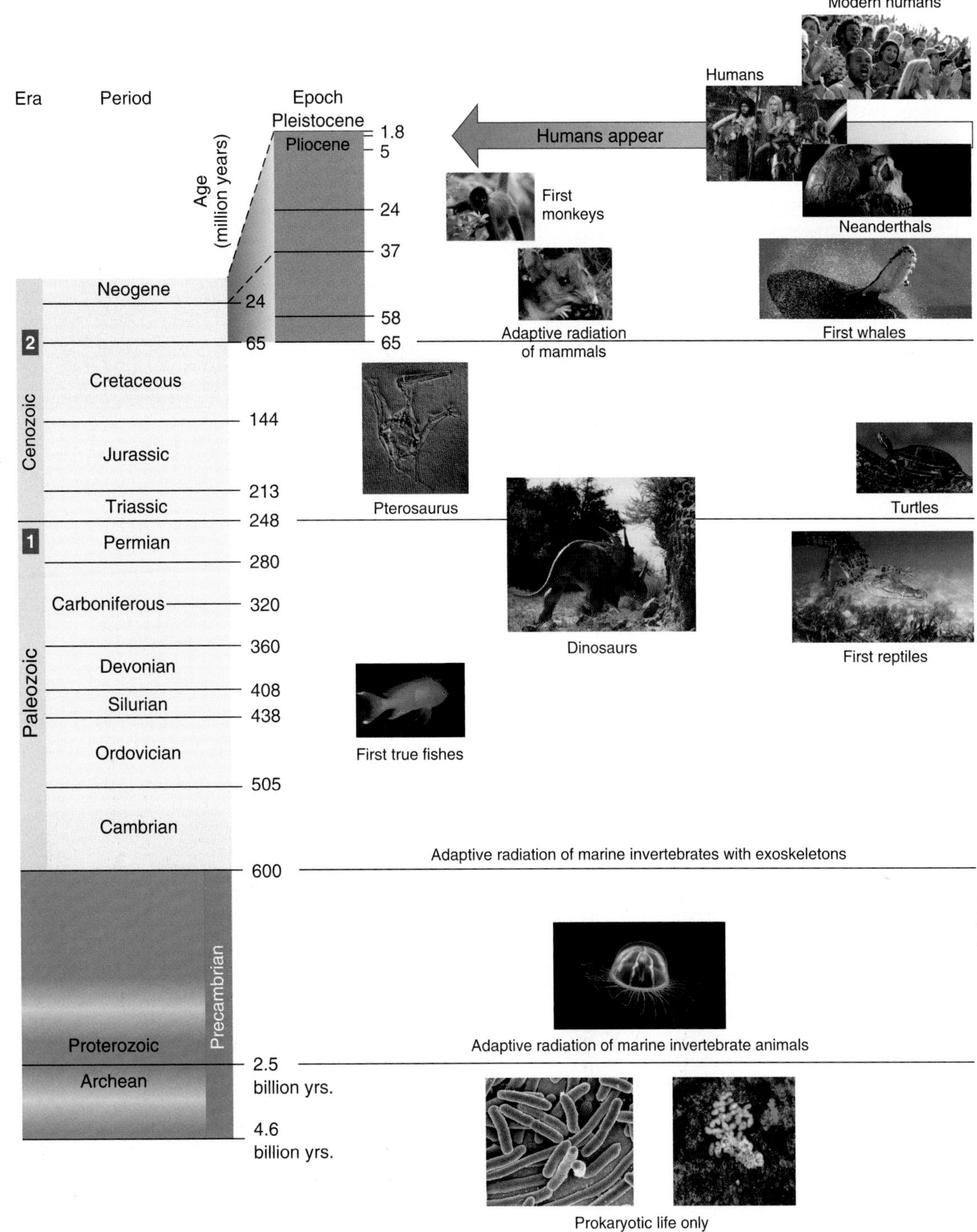

Adaptive radiation and the phylogenetic tree FIGURE 19.14

This figure demonstrates the changing forms of life through time. Originally, life on Earth consisted of bacteria, and then marine invertebrates, such as the sea jellies and sea worms. Over time, many more marine organisms appeared in the primordial sea. By the Silurian period, fishes had appeared, as had insects and terrestrial vertebrates. As we approach modern times, the changes in life forms escalate, and in all the plants and animals we now see appeared quite rapidly.

tors of these fishes and other creatures reached the less populated Hawaiian reefs as larvae floating on ocean currents. Because few organisms were competing for the resources of Hawaii's reef, these fishes, crabs, coral, and the like, began to occupy different habitats and adopt different diets, which caused speciation. Adaptive radiation alters population allele frequency in response to different environmental conditions and therefore different selective pressures.

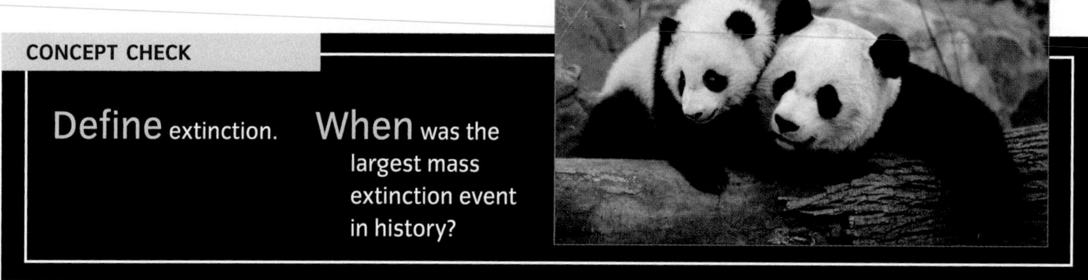

CONCEPT CHECK

Define extinction.

When was the largest mass extinction event in history?

Amazingly, a Planet Forms and Life Begins

LEARNING OBJECTIVES

Describe conditions on the primitive Earth.

Explain the appearance of organic molecules.

Relate the presence of oxygen to evolving life forms.

Press "rewind" to about 4.6 billion years ago. We observe a planetary nebula, a whirling disk of hot gas and dust. Over time, these materials cool and coalesce, forming the sun at the center and a series of planets orbiting it. Young Earth was an inhospitable place—semimolten rock, with a liquid core of nickel, iron, and other metals (FIGURE 19.15). We have now looked at several processes that have caused life to evolve, survive, or go extinct (FIGURE 19.16). But how did those processes play a role in the formation and development of life on Earth?

Volcanoes cracked the thin skin of the Earth, venting toxic gases into the atmosphere. The atmosphere was composed mainly of carbon dioxide, water

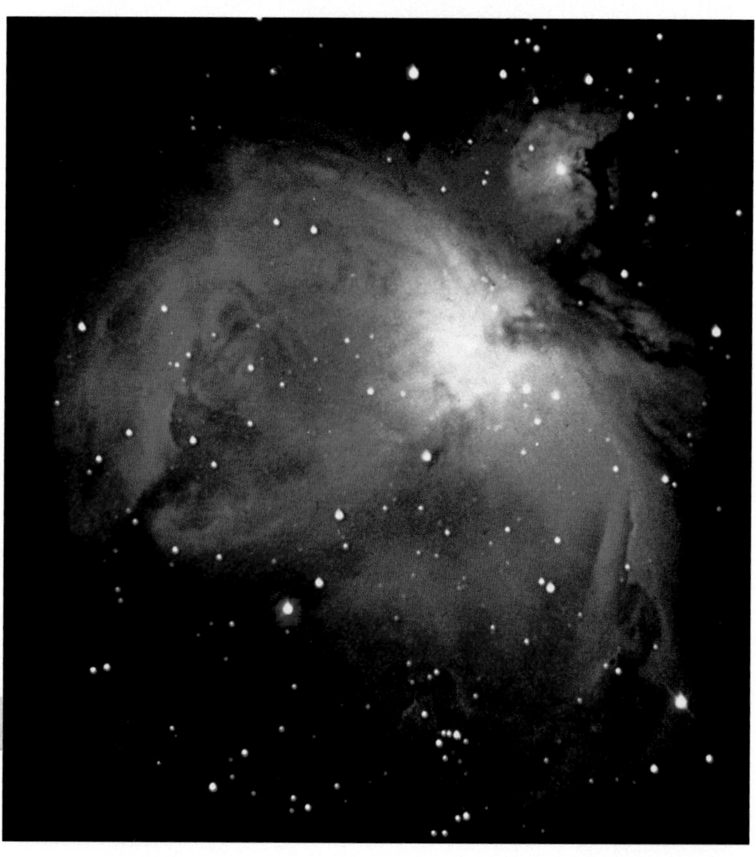

Early Earth FIGURE 19.15

Time line of major dates in formation of Earth FIGURE 19.16

A summary of the major biological and geological events in Earth's history. The time line isn't drawn to scale: If 0.1 inch represented 1 million years, the time line would be almost 40 feet high! (mya = millions of years ago)

Era	*Period*	*Epoch*	*MYA*	*Major biological and geological events*
Cenozoic	Quaternary	Pleistocene	1.8	End of last ice age. (10,000 years ago) Modern humans (Cro-Magnons) appear. (40,000 years ago) Neanderthals appear. (230,000 years ago) *Homo sapiens* (early human beings). (500,000 years ago)
	Tertiary	Pliocene	53	*Homo erectus* (tools; fire). (1.8 mya) Human forms (genus *Homo*) appear. (2 mya) *Australopithecus anamensis* (first human ancestor to be fully bipedal). (4 mya)
		Miocene	26	Hominins (humanlike forms) appear. (6-7 mya)
		Oligocene	38	*Aegyptopithecus* (and ancestor of early hominids). (27 mya) *Parapithecus* (ancestor of Old World monkeys and hominids). (30 mya)
		Eocene	54	Anthropoid primates appear. (50 mya)
		Paleocene	65	Primates appear. Mammals become abundant.
Mesozoic	Cretaceous		145	**Major extinction (dinosaurs, many other animals, many plants, and most marine genera disappear).** (65 mya) Flowering plants appear. (140 mya)
	Jurassic		210	*Archaeopteryx* (oldest fossil bird). (145 mya) Large dinosaurs dominate Earth. Birds appear. Mammals appear. (200 mya)
	Triassic		250	**Major extinction (most marine species and some terrestrial animal species disappear).** (210 mya) Pangaea (supercontinent) forms. (240 mya) Dinosaurs appear.
Paleozoic	Permian		285	**Major extinction (most marine species disappear).** (250 mya)
	Carboniferous		360	Reptiles appear. Conifers appear. Coal deposits form. Horsetails, ferns, and seed-bearing plants become abundant.
	Devonian		417	**Major extinction (most marine invertebrates and many fishes disappear).** (360 mya) Amphibians appear. Bony fishes appear. Insects appear. Jawed fishes appear. (410 mya)
	Silurian		443	Land plants appear.
	Ordovician		490	**Major extinction (almost all corals and fish and many other species disappear).** (443 mya)
	Cambrian		544	Vertebrates appear. (500 mya) Chordates appear. (520 mya) "Cambrian explosion" of living forms, including origin of main invertebrate phyla.
Proterozoic			2600	Multicellular animals appear. (600 mya) Fungi appear. Multicellular organism appear. (1200 mya) Oxygenation of atmosphere and oceans.
Archean			4600	Stromalites form. First living cells appear. (3500 mya) First rocks form. (3800 mya) Origin of Earth. (4600 mya)

vapor, hydrogen, nitrogen, methane, and ammonia. The raw surface of the Earth was exposed and too hot for liquid water. Today's protective ozone layer had not yet developed, so killing levels of ultraviolet light reached the ground. Lightning storms were common and fierce, and a rain of asteroids during the "heavy bombardment phase" pummeled the surface with an intensity you can appreciate by looking at how those same asteroids scarred the face of our moon.

As the Earth continued to cool, water vapor condensed in the polar regions, forming rain that helped to cool the Earth through evaporation. The planet eventually radiated enough heat to space so liquid water could fill surface depressions, forming oceans that were warm, shallow, and not very salty. The salt came later, from minerals and ions that dissolved from the rock on the ocean floor.

LIFE ON EARTH BEGINS WITH ORGANIC COMPOUNDS

As near as we can tell, life arose in this hot, steamy environment 3.8 billion years ago, less than a billion years after the planet formed. This life consisted of single-celled organisms that were rather similar to certain groups of modern bacteria. Where did these organisms come from? Many scientists believe that life formed from organic compounds, which in turn formed from atmospheric gases interacting with the intense heat, ultraviolet light, and lightning of the primitive Earth. In the laboratory, scientists have recreated these conditions and formed some of the molecules that must have been present.

We may never know the exact origin of life, as the evidence is dimmed by the mists of time. Even today, scientists argue fiercely over the first evidence for life, which appears in rocks that have been torturously deformed by heat and pressure for 3.8 billion years.

Scientists do agree that complex organic compounds must have been present at the dawn of life. Some scientists believe that organic compounds reached Earth on comets. Amino acids, simple sugars, and small fatty acids formed and dissolved in the seas, creating a warm, nutrient-rich, oxygen-deficient (because of an oxygen-poor atmosphere) soup. The absence of oxygen may actually have helped because this highly reactive element would quickly have oxidized these complex molecules.

ORGANIC COMPOUNDS BECOME LIFE

What caused the giant jump from small nutrient molecules to a living cell (FIGURE 19.17)? One major possibility is that life arose in pools of geothermally heated water at the spreading centers of the mid-oceanic ridges, where the oceans would have protected the young organisms from asteroids and ultraviolet light. The molecules floating in the seas somehow came together to form self-replicating molecules. The only self-replicating molecules we know are DNA and RNA. Both are highly organized, making a spontaneous appearance highly unlikely. But of the two, the single-stranded RNA is less complex, and therefore is the focus of speculation about the formation of life. One hypothesis is that RNA assembled on a slick mud flat near the sea. With mud as substrate, RNA formed, replicated, and perhaps directed the formation of DNA.

RNA is a plausible candidate for the nucleic acid of early life. Scientists think it could have originally directed the formation of DNA, as RNA primers are required for most transcription events. Once created, the DNA could have directed the synthesis of additional RNA and proteins, including enzymes. At some point, RNA probably shifted to its many current functions: directing protein formation, regulating gene expression, transferring molecules, and monitoring timing of events within cells. Logically, the formation of DNA must have come after RNA because DNA is double-stranded, more complex, and more stable. Although a few viruses store their genetic instructions on RNA, most life forms use DNA as their self-replicating molecule, likely due to its stability. The base-paired structure of DNA makes mutations less likely and, easier to identify and correct. At any rate, once self-replicating molecules appeared, independent life was probably not far behind.

Once DNA and other macromolecules were available, what else was needed to form a cell? First, a lipid/protein bilayer was needed to encase this new, self-replicating DNA. The process by which this might

Process Diagram

Creation of biological molecules from early Earth's atmosphere

FIGURE 19.17

1. A sample of nutrient- laden water representative of the ancient seas is heated to boiling.
2. The water evaporates, carrying some of the smaller compounds.
3. As water vapor collects in the chamber, electrodes deliver a spark similar to a natural bolt of lightning. The collected gases in this chamber simulate the ancient atmosphere, and the electrodes provide the necessary spark of energy to form molecules.
4. Chilling the gases from the simulated atmosphere causes condensation and precipitation of any molecules formed during the "lightning storm."
5. As the atmospheric gases condense and collect, the fluid is analyzed for the presence of macromolecules. In this closed system, fluid can be passed repeatedly through these chambers, reconstructing the possible events in the early atmosphere.

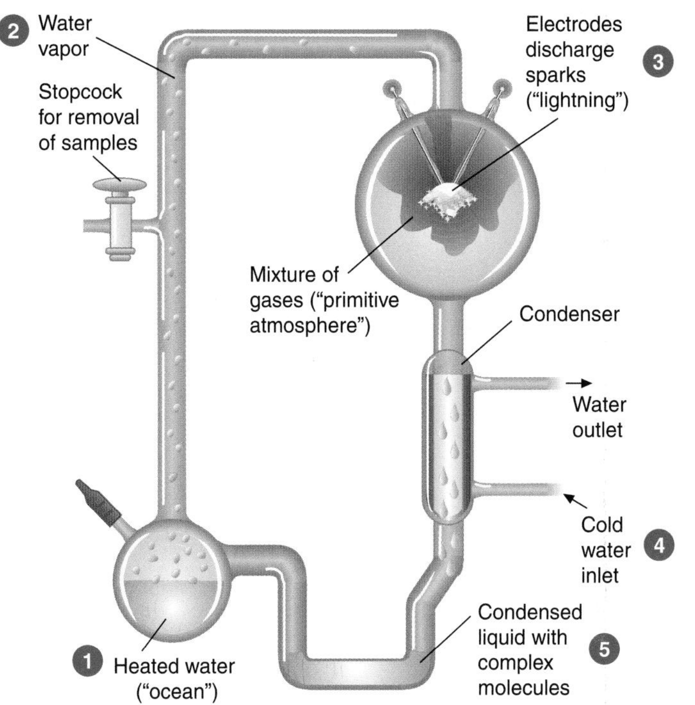

have happened is hard to envision, but in the lab, tiny spheres called micelles can spontaneously assemble under certain conditions. These hollow, water-filled, phospholipid spheres could have been produced spontaneously in the primordial oceans. Assuming the forming spheres captured some DNA, a primitive prokaryotic cell would be the result.

Importantly, these first cells had to be **anaerobic**. They used the nutrients in their environment and harvested energy from the ambient heat and atmospheric compounds. No molecular oxygen was available at that time to participate in their metabolism (FIGURE 19.18).

Anaerobic
Living and metabolizing in the absence of free oxygen in the environment.

As you've noticed, we have spoken tentatively about exactly how life began because we have no good record of that epochal event. Each of the events we

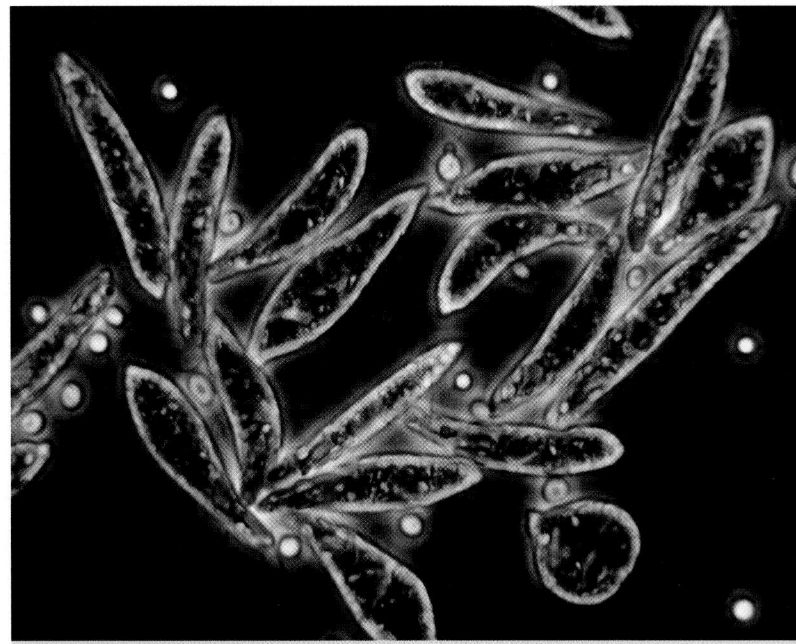

Prokaryote containing DNA
FIGURE 19.18

Geological time FIGURE 19.19

The one-year geological calendar gives a sense of the relative length of the geological time divisions.

Period	*Epoch*	*MYA*	*Major biological and geological events*	*One-year geological calendar*
Quaternary	Pleistocene	1.8	End of last ice age. Modern humans appear Neanderthals appear.	December 31 (11:59 PM) December 31 (11:55 PM) December 31 (11:34 PM) December 31 (11:02 PM)
Tertiary	Pliocene	53	Human forms (genus *Homo*) appear.	December 31 (late evening) December 31 (early evening) December 31 (midafternoon)
	Miocene	26	Hominins (humanlike forms) appear.	December 31 (mid-day)
	Oligocene	38	*Aegyptopithecus* (and ancestor of early hominids) appear.	December 30 (early morning) December 29 (early afternoon)
	Eocene	54	Anthropoid primates appear. (50 mya)	December 27
	Paleocene	65	Primates appear. Mammals become abundant.	
Cretaceous		145	**Major extinction (dinosaurs, many other animals, many plants, and most marine genera disappear).**	December 26 December 20
Jurassic		210	Large dinosaurs dominate Earth. Birds appear. Mammals appear.	December 19 December 15
Triassic		250	**Major extinction (most marine species and some terrestrial animal species disappear).** Pangaea forms. Dinosaurs appear.	December 14 December 12
Permian		285	**Major extinction (most marine species disappear).**	December 11
Carboniferous		360	Reptiles appear. Conifers appear.	
Devonian		417	**Major extinction (most marine invertebrates and many fishes disappear).** Amphibians appear. Bony fishes appear. Insects appear. Jawed fishes appear.	December 2 November 28
Silurian		443	Land plants appear.	
Ordovician		490	**Major extinction (almost all corals and fish and many other species disappear).**	November 26
Cambrian		544	Vertebrates appear. Chordates appear. "Cambrian explosion" of living forms, including origin of main invertebrate phyla.	November 21 November 20
		2600	Multicellular animals appear. Fungi appear. Multicellular organisms appear. Oxygenation of atmosphere and oceans.	November 13 September 27 July 18
		4600	First living cells appear. First rocks form. Origin of Earth.	March 29 March 4 January 1

have discussed may be extremely unlikely, but consider: We are discussing events on an entire planet, over hundreds of millions of years. And so what is extremely unlikely to occur at any one place at any one time suddenly becomes likely to occur, somewhere, sometime. On our Earth, a once-in-a-million-years event has occurred approximately 4,000 times! The same argument holds true for many of the major events of evolution; sure, it's unlikely that an eye will develop overnight. But the ability to detect light is found in primitive organisms, and eventually, over uncountable trillions of mutations, this helpful ability evolved into the eyes on your face. Trial and error is a great process for perfecting mechanisms; no engineer would be without it.

THE APPEARANCE OF OXYGEN CHANGED EVERYTHING

These early years occurred in an atmosphere with little free oxygen. Then **photosynthesis** began, perhaps after a series of mutations led to the formation of **chlorophyll**. The organisms that developed this photo-pigment suddenly became able to use the carbon dioxide, water, and light energy to make simple sugars. The present-day cyanobacteria (blue-green bacteria) closely resemble the first photosynthetic organisms.

Photosynthesis
The metabolic process of creating glucose using light energy, pigments such as chlorophyll, water, and carbon dioxide; the process of carbon fixing observed in green plants, algae, and certain bacteria.

Chlorophyll
A green pigment that absorbs light energy and creates simple sugars.

Recall that oxygen is a by-product of photosynthesis. As photosynthetic organisms increased in numbers, oxygen began to accumulate in the atmosphere and to react with simple compounds there, such as ammonia, hydrogen, methane, and water. These oxidation reactions removed compounds from the atmosphere and put selective pressure on organisms that could not photosynthesize.

Oxygen harms anaerobic cells by reacting with the nutrients these cells need, so anaerobic populations declined, creating niches for organisms that **respire** aerobically. Aerobic metabolism developed after anaerobic metabolism and photosynthesis. As the atmospheric oxygen level rose, resources shifted, new opportunities arose, and life evolved to take advantage of new conditions. Oxygen in the atmosphere changed the whole course of evolution.

Respire
To acquire oxygen for use in the metabolism of glucose or other nutrients.

THE ARRIVAL OF THE MULTICELLULAR ORGANISMS

Life was "content" to live as single organisms for several billion years. The first photosynthetic organisms appeared about 3.8 billion years ago, and the first eukaryotic organisms appeared 2 billion years ago. Only after another billion years did multicellular organisms develop. Since about 1 billion years ago, land plants developed, dinosaurs appeared, mammals appeared, and dinosaurs dramatically disappeared, allowing mammals to occupy the various habitats on Earth. Primitive humans finally showed up on Earth less than 5 million years ago, an eyeblink in the history of life (FIGURE 19.19).

CONCEPT CHECK

Which organic molecule probably appeared first, DNA or RNA?

How did photosynthesis alter the path of life on Earth?

How did oxygen change the atmosphere and the course of evolution?

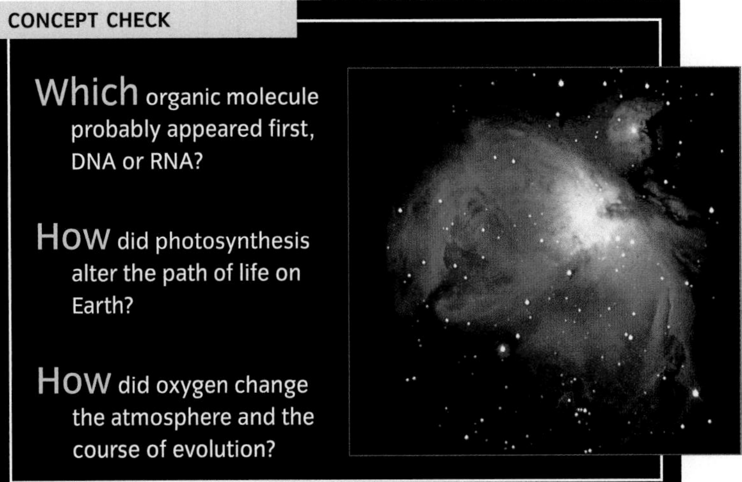

The Human Family Tree Is a Confusing One

Learning Objectives

Understand the origins of modern humans.

Describe the characteristics of primates.

Differentiate *Homo habilis, Homo erectus, Homo neanderthalensis,* and *Homo sapiens*.

Appreciate the variety in modern humans.

Discuss the evolutionary forces currently affecting the human population.

In Chapter 1, we learned the taxonomic classification of humans: We belong to the class Mammalia, which also includes whales, dogs, squirrels, and bears. We are further separated into the order Primates, along with lemurs, monkeys, and apes. Primates share a common ancestor that lived about 60 million years ago. The order is characterized by five-digit hands with an opposable thumb, fingernails and toenails rather than claws, and stereoscopic vision with forward-facing eyes. All of these shared characteristics were adaptations to life in the trees. Our opposable thumb was a great evolutionary advance, allowing us to grasp firmly yet with precise control.

Twenty-five million years ago, the ancestor of apes and humans diverged from the ancestors of old-world monkeys (FIGURE 19.20). Apes and humans are larger, with larger brains and smaller tails than monkeys. Our tails are so small, in fact, that they are not visible outside the body. Apes and humans are further distinguished by their complex social interactions. Biochemical comparisons show that gibbons diverged first, followed by orangutans, gorillas, chimpanzees, and humans. To be clear, we did not develop from a

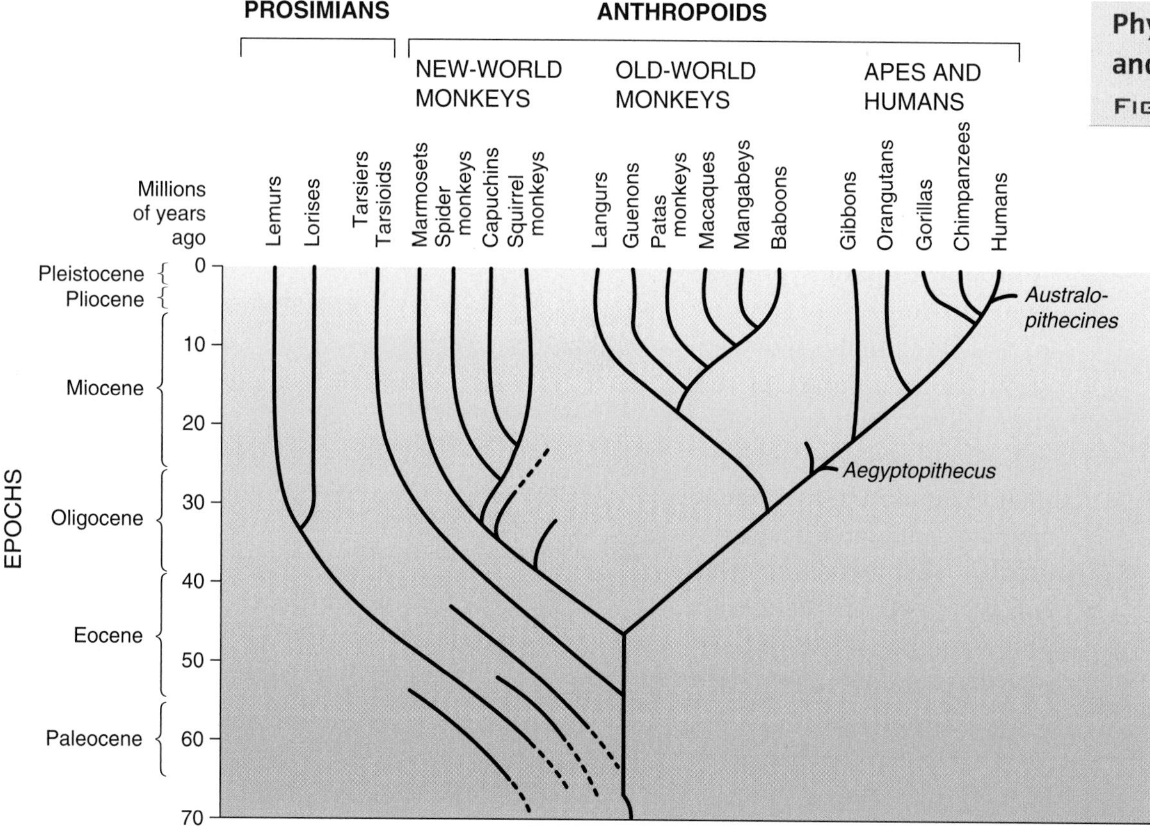

Phylogeny of apes and humans
FIGURE 19.20

Hominid evolution
FIGURE 19.21

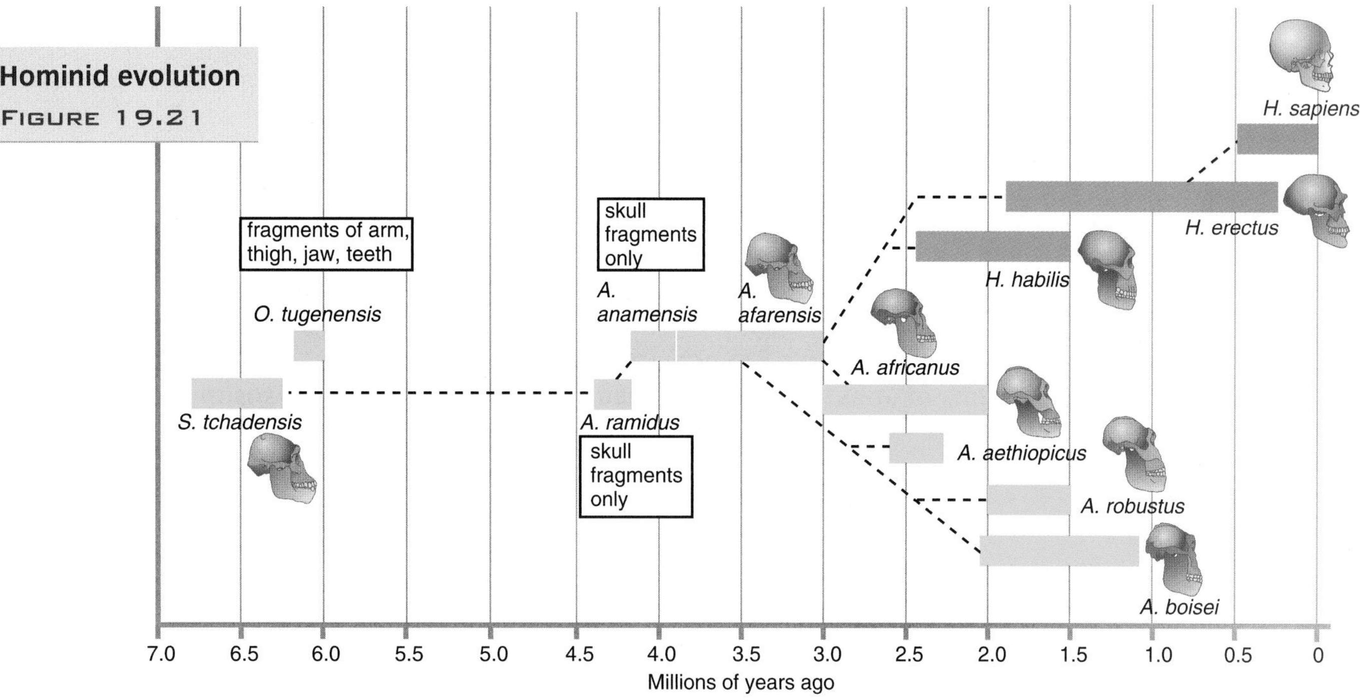

chimpanzee, but rather chimpanzees and humans diverged most recently from a common ancestor that probably have looked something like a chimpanzee.

Continuing with the human taxonomic classification, we belong to the genus *Homo*, with the species epithet *sapiens*. The fossil record contains many other *Homo* species, each suited to different environments. Although scientists are still debating the specifics of human evolution, most agree on the basic pathway: that humans evolved in Africa when a primate began to walk upright as its usual form of locomotion (FIGURE 19.21).

THE HUMAN ANCESTORS ARE DEAD TWIGS ON THE FAMILY TREE

Australopithecus was the first member of the family Hominidae. This organism walked upright, and its cranium was slightly larger than that of previous, nonhuman primates. Interestingly, the first hominid was an omnivore and relatively small in stature. A second Australopithecine, *A. afarensis* (FIGURE 19.22), was slightly larger and, based on dentition, ate like a modern vegetarian. These organisms showed social behaviors and **sexual dimorphism** similar to the apes.

Sexual dimorphism Morphological differences between the two genders.

About 3 million years ago, *Homo habilis* appeared to share the planet with *A. afarensis*. This organism had a larger brain, new types of teeth allowing it to eat a more varied diet, and perhaps the ability to make and use tools. *Homo habilis* literally means "handy man," and many of these fossils are surrounded by stones that could be primitive tools.

A. afarensis FIGURE 19.22

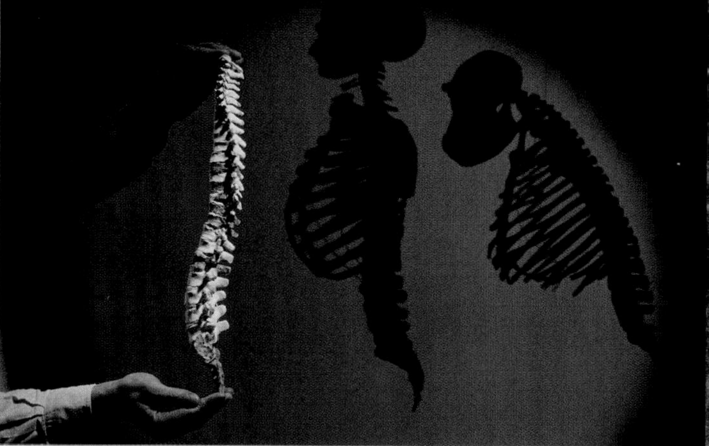

A comparison of the skeletons of apes and *Homo erectus* FIGURE 19.23

Neanderthals and *Homo sapiens* in *Clan of the Cave Bear* FIGURE 19.24

And 1.8 million years ago, another speciation event produced *Homo erectus* and *H. ergaster.* These lighter, more graceful organisms can be classified as humans, for they had subtle differences in cranial capacity, stature, and gait (FIGURE 19.23). Originally, these two were classified together as *H. erectus. H. ergaster* was distinguished in 1994, when scientists discovered that their skulls were different. *H. ergaster* has a high skull bone, thin cranial bones, a slim brow ridge, and a generally lighter skeleton than *H. erectus.* Both had a swift gait, long muscled limbs, narrow hips, and body proportions like those of modern tropical humans. Sexual dimorphism was effectively lost in this group, indicating that both males and females probably participated in the same societal activities. Infant development was extended, allowing a longer family period for passing on learned traits and culture. These primates continued to make hunting tools and eating equipment.

Although scientists are not clear on the exact date, it appears that *Homo erectus* and *H. ergaster* migrated out of Africa approximately 1 million years ago, and began to populate other continents. *H. erectus* may have left Africa to avoid environmental changes during an ice age. They remained a part of the biota of Java as recently as 500,000 years ago, making them contemporaries of modern *Homo sapiens.*

We have all heard of Neanderthals. Some scholars believe these hominids evolved as a separate species from *H. erectus.* Others think *H. erectus* first evolved into a form that was very close to modern humans, which then gave rise to both modern humans and Neanderthals. Neanderthal fossils are anywhere from 200,000 to 30,000 years old. They show characteristics of our morphology, along with those of *H. erectus. H. neanderthalensis* probably represents an evolutionary dead end; however there is still debate as to its exact position in human evolution. Are Neanderthals and modern humans related closely enough to be subspecies of *Homo sapiens*? In 1964, this was the accepted wisdom, based on anatomical similarities. Apparently, the two existed on Earth at the same time, as indicated by fossil sites in Israel where geologic strata indicate that *H. sapiens* lived at that location before *H. neanderthalensis. H. sapiens* could not have lived there before *H. neanderthalensis* if the Neanderthals had died out before *Homo sapiens* arrived. We know that *H. sapiens* did not die out; therefore, the two must have been around simultaneously. The fiction series *Clan of the Cave Bear* by Jean M. Auel, is based on the fossil findings of Neanderthal and modern man (FIGURE 19.24). The books and the movie depict many interactions between these two species. The author vividly portrays the imagined similarities and differences of these two hominids.

Neanderthals certainly had many social customs that would be familiar to us, as we can tell from their burial rituals. Both Neanderthals and modern humans had a well-developed communication system and lived in places we would recognize as human dwellings.

HOMO SAPIENS MAKES THE SCENE AND STARTS TO CHANGE EVERYTHING

It is difficult to pinpoint the exact beginning of *Homo sapiens.* Some scientists believe that all modern humans came from one small population in Africa that splintered, migrated, and populated the globe. This splintering must have happened approximately 140,000 to

100,000 years ago. Wherever *H. sapiens* appeared, they replaced all other hominids. We cannot be certain why, as the fossil record gives no indication of violence between species of hominids, nor does it provide evidence of disease. Did *H. sapiens* really fight and kill Neanderthals? Did Neanderthals fall victim to viruses that did not harm *H. sapiens*? Did Neanderthals breed with *H. sapiens*, eventually losing their characteristics as their genes were diluted in the larger *H. sapiens* gene pool? The questions are tantalizing, but we may never know their answers.

FIGURE 19.25

This group of ethnically diverse school children exemplifies the many different phenotypes now found in the human population.

HUMAN POPULATION DIFFERENCES AND ETHNICITY ARE TANGLED CONCEPTS

The bottom line on the evolution of humans is that we are all one species. Do we look different? A bit (FIGURE 19.25). Humans have subtle physical differences that are heritable and that are usually associated with one group of people.

For almost all of our history, human populations were isolated by geographic barriers like forests, deserts, oceans, rivers, and mountains. During this isolation, natural selection, gene flow, bottleneck effects, founder effects, and subsequent natural, and, perhaps, sexual, selection favored different genetic traits in the various populations. These differences formed what we used to call racial differences, such as skin color, hair color, hair texture, eye shape, and body stature.

But these subtle differences can be overblown and used as a tool of oppression more than a tool of understanding. As a concept, the scientific validity of human races is questionable at best. Genetically, we now know that people can have more genetic differences with their nearest neighbors than with people living on other continents.

Some of these traits developed as selective advantages in local environments. Dark skin offers better protection against UV light. Facial features, hair texture, and even blood types may have developed in response to environmental pressures. It is important to remember that phenotypic differences are nothing more than products of that slight genetic variation on the overall successful gene plan of the human species. Now that geographical barriers have been lifted, *Homo sapiens* may become even more uniform in appearance. We can jump on a jet and reach another continent in no time flat. This raises the opportunity for jet-speed gene flow.

The last step in understanding our role in the environment is to study the environment itself. How do we interact with other organisms? What is our role in the biosphere? Our final chapter will place humans in the ecosystem by looking at the science of ecology.

CONCEPT CHECK

What was the first hominid?

How do *Homo erectus* and *H. habilis* differ?

Trace the descent of man, from the first hominid to *Homo sapiens*.

Discuss the causes of variation in the human population.

What evolutionary factors are currently at work in the human population? (Refer to the Hardy-Weinberg principles to formulate your answer.)

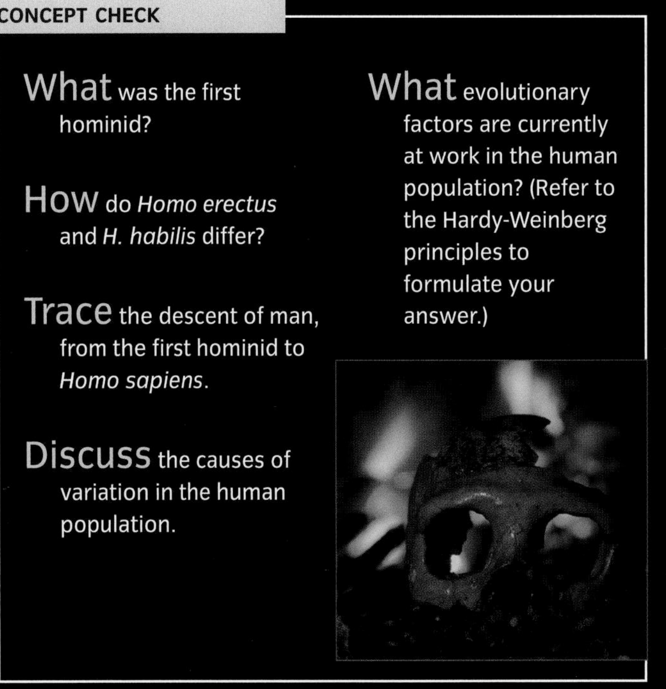

CHAPTER SUMMARY

1 The Theory of Evolution through Natural Selection Is the Foundation of Biology

Charles Darwin proposed the theory of evolution in 1859 to explain the diversity of life on Earth. Evolution is a change in allele frequencies in a population over time. Creationism and intelligent design are nonscientific explanations for life. Darwin's theory is based on natural selection and can be quantified with the Hardy-Weinberg equation.

2 Evolution Is Backed by Abundant Evidence

Evidence for evolution can be found in many areas. Evolutionary evidence appears as changes accumulate over time in the fossil record. Organisms living in similar ranges of climate and ecological conditions develop similar structures and functions for coping with their environment. Comparative anatomy shows homologous and vestigial structures that are evidence of evolutionary relationships. Developmental similarities also indicate common ancestors. Sequencing the DNA or the protein structure of organisms provides yet another indication of ancestral relationships.

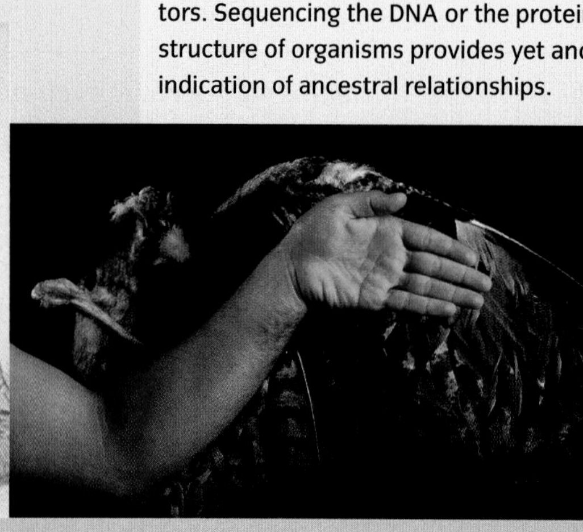

3 Natural Selection Has Far-Reaching Effects on Populations

Natural selection can alter the allele frequency of a population by reducing the number of less fit or unfit individuals. Allele frequencies are also altered via the bottleneck effect, the founder effect, and gene flow.

4 Mass Extinctions Can Be Followed by Regrowth

In recent history, there have been five major extinction events. The Permian–Triassic extinction event destroyed approximately 70 percent of land animals and 90 percent of ocean life on the planet. The dinosaurs became extinct at the end of the Cretaceous period. Asteroid impacts, volcanism, and other geological forces may have caused these events. We continue to lose species at an alarming rate, forcing us to look for effective conservation methods. Counteracting extinction is adaptive radiation, where a new niche is opened and a previously unselected phenotype imparts an advantage to those organisms that carry it.

5 Amazingly, a Planet Forms and Life Begins

The solar system originated as a hot swirling cloud of dust and gas. Hundreds of millions of years after Earth came together and cooled, it became hospitable to life. The earliest life forms were anaerobic bacteria. Some became photosynthetic and released compounds, including oxygen, as a by-product of metabolism. Multicellular plants and animals followed as the atmosphere became more conducive to life as we know it. Humans are newcomers, having evolved less than 5 million years ago.

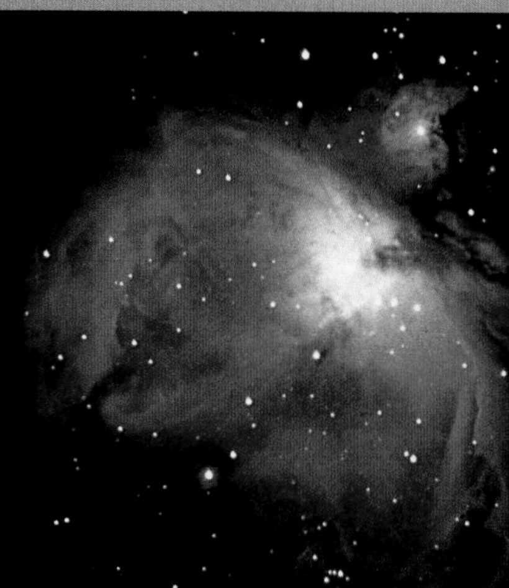

6 The Human Family Tree Is a Confusing One

Humans are lighter, more graceful primates that evolved from *Australopithecus* and *Homo habilis*, *H. ergaster*, and *H. erectus*; 200,000 to 30,000 years ago, Neanderthals appeared. These most recent ancestral species were so close to modern humans that their classification has long been debated. Neanderthals are now considered a subspecies of modern man, not our ancestor. The fossil record shows many varied appearances of humanoids over the last few million years. All modern humans are of one species and have extremely similar genetics.

KEY TERMS

- **allele** p. 630
- **amniotic** p. 648
- **anaerobic** p. 653
- **analogous** p. 640
- **asthenosphere** p. 638
- **chlorophyll** p. 655
- **convection currents** p. 638
- **convergent zones** p. 639
- **creationism** p. 630
- **divergent** p. 635
- **divergent zones** p. 639
- **extant** p. 640
- **fitness** p. 635
- **gene flow** p. 633
- **genetic drift** p. 633
- **homologous** p. 640
- **intelligent design** p. 630
- **macroevolution** p. 632
- **microevolution** p. 632
- **photosynthesis** p. 655
- **plate tectonics** p. 638
- **respire** p. 655
- **selection pressure** p. 633
- **sexual dimorphism** p. 657
- **strata** p. 637
- **successional changes** p. 645
- **transverse zones** p. 639
- **vestigial** p. 640

CRITICAL THINKING QUESTIONS

1. Charles Darwin published his theory of descent with modification in 1859, when the Anglican Church was a dominant force in British society. Reflect on what scientific research must have been like at that time, and explain why Darwin's ideas were seen as dangerous. Why do his ideas still cause concern to some individuals and groups today?

2. Intelligent design is based on the belief that the origin and change of species is directed by some super-intelligent force or being. This theory states that the complex forms of life we see on Earth now could not have arisen simply by chance, but rather that they are the result of an intelligent designer directing the path of change. Why is intelligent design not considered a scientific theory? What rules of science does it violate or ignore?

3. Evidence, observation, and testing are the cornerstones of scientific investigations. List three types of evidence for the theory of evolution. Tell how each supports the theory, and give specific examples if possible.

4. The Hardy-Weinberg equation is directly related to the information expressed in a Punnett square (see Chapter 18). Prepare a typical heterozygous-cross Punnett square. In each square, indicate the predicted genotype, as well as the correct symbol (p or q) from H-W equilibrium. In other words, locate the square that represents p^2, the $2pq$ squares, and the q^2 square. If 36 percent of the population is phenotypically recessive (q^2 is $= 0.36$), what percentage of heterozygous individuals are expected? (*Hint*: Solve for q and then calculate $p + q = 1$. Use the resulting values in the H-W equilibrium equation: $p^2 + 2pq + q^2 = 1$.)

5. Humans have a great effect on the evolution of other organisms. What activities do we engage in that directly affect that evolution? How do humans affect our own evolution?

SELF TEST

1. The idea that life on Earth has been controlled and guided by an extraterrestrial force or higher power is referred to as
   a. evolution.
   b. creationism.
   c. intelligent design.
   d. Both b and c are correct.

2. Which of the following is NOT a statement concerning natural selection?
   a. Organisms produce more offspring than will survive.
   b. Organisms show variation that can be inherited.
   c. Variations can alter the individual's ability to reproduce.
   d. Variations that decrease fitness will be passed on to subsequent generations.

3. The alteration of a population over time to produce a subtle change in that population's phenotype is called
   a. microevolution.
   b. macroevolution.
   c. genetic flow.
   d. genetic drift.

4. The premise of this equation is that

$$p^2 + 2pq + q^2 = 1$$

   a. evolution is always occurring.
   b. evolution can be quantified by comparing ideal to observed conditions.
   c. allele frequencies in populations never change.
   d. *p* and *q* measure population fitness.

5. In the above equation, solving for pq will give the percentage of ____________ individuals in the population.
   a. homozygous recessive
   b. heterozygous
   c. homozygous dominant
   d. evolved

6. All of the following are requirements for a population to be in Hardy-Weinberg equilibrium EXCEPT:
   a. The population must be infinitely large.
   b. There can be no selection pressure on the population.
   c. Mate choice must be selective and based on phenotype.
   d. Gene flow must be prevented.

7. The newest line of evidence for evolution is
   a. fossil dating.
   b. comparative anatomy.
   c. biogeography.
   d. comparative biochemistry.

8. Using this graph, predict how many half-lives have passed for a bone that contains only 1/4 of the original radioactivity.
   a. 1
   b. 2
   c. 3
   d. 4

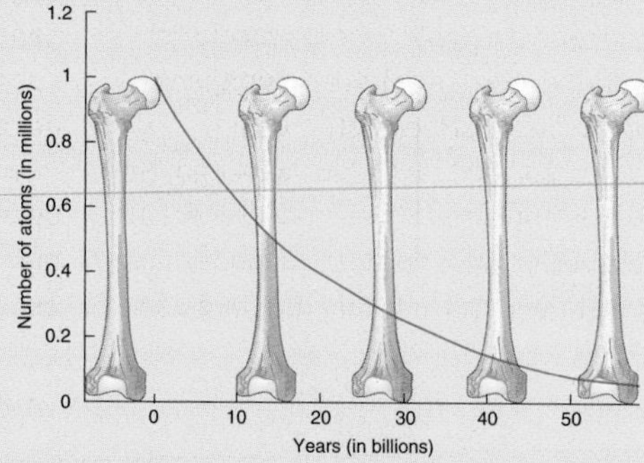

9. The label A on the image below indicates the
   a. trench.
   b. convection currents.
   c. divergent zone.
   d. convergent zone.

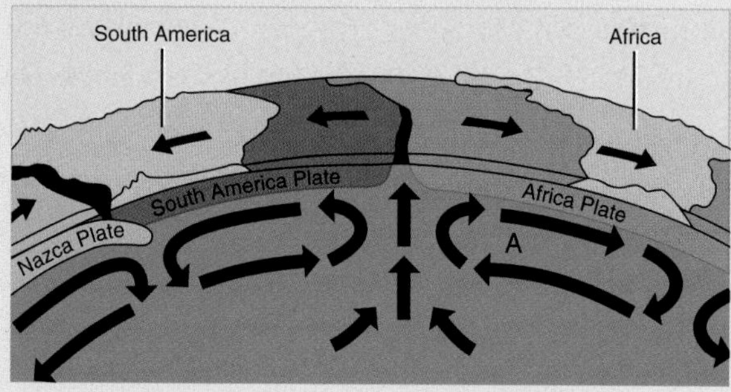

10. The fact that marsupials on Australia and placental mammals on the North American continent share common solutions to life's problems is an example of
    a. homologous evolution.
    b. analogous evolution.
    c. divergent evolution.
    d. convergent evolution.

11. The image below represents
    a. homologous structures.
    b. analogous structures.
    c. vestigial structures.
    d. ontogeny recapitulating phylogeny.

12. What evolutionary process is depicted in the photo below?
   a. Genetic flow
   b. Genetic drift
   c. Bottleneck effect
   d. Founder effect

13. The population of Troublesome Creek (Chapter 18, p. 605–606) suffers from a rare congenital disease obtained through
   a. genetic flow.
   b. genetic drift.
   c. bottleneck effect.
   d. founder effect.

14. True or False? The opposite of extinction is adaptive radiation.

15. The device shown below was used to simulate
   a. the reactions occurring on the surface of the early Earth.
   b. the effects of oxygen on the early atmosphere.
   c. the formation of complex molecules and eventually life on Earth.
   d. the formation of the seas.

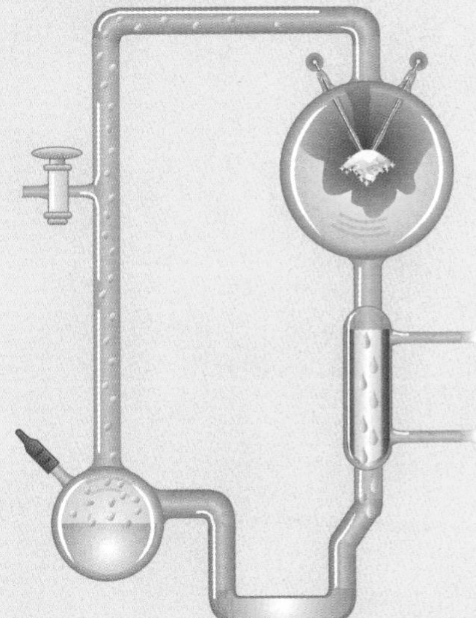

16. Which process most likely developed first?
   a. Photosynthesis
   b. Aerobic respiration
   c. Anaerobic respiration
   d. Phytochemical respiration.

17. Which of the following periods did NOT include a major extinction event?
   a. Quaternary period
   b. Devonian period
   c. Permian period
   d. Cretaceous period

18. The first member of the family Hominidae that is identified from more than a skull fragment is
   a. *H. sapiens*.
   b. *H. habilis*.
   c. *A. afarensis*.
   d. *S. tchadensis*.

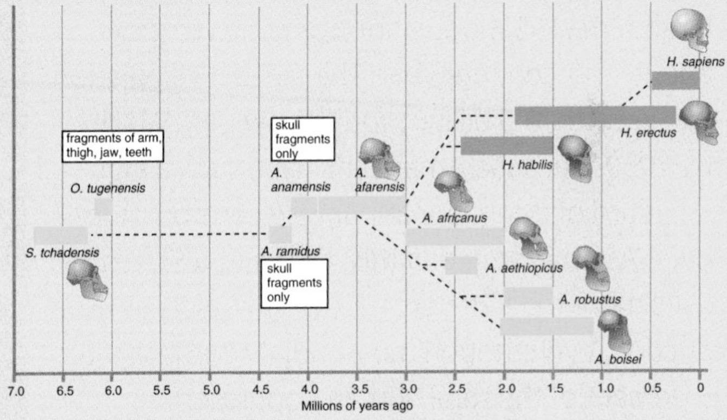

19. As shown on the above figure, the species of man that has the longest survivorship thus far is
   a. *H. sapiens*.
   b. *H. erectus*.
   c. *H. habilis*.
   d. This information is not given on the figure.

20. In this image, you can see ___________ different species of human.
   a. 1
   b. 2
   c. 4
   d. 7

Ecology and Societal Issues

20

How many people can Earth support? Perhaps no question is more important, and yet it is devilishly difficult to answer. What is meant by support? As a sprawl of dense cities surrounded by factory farms, with every hectare put "to use"? Or as a planet where some nature survives to provide spiritual comfort to its people, where the plants and animals that evolved along with us still live alongside us?

These questions are pressing. Even as the population of Japan and Western Europe has stabilized, the U.S. population continues to grow about 1 percent per year. The current population of the Earth, 6.5 billion, could grow past 9 billion by 2050. Because Americans have such a high standard of living, our impact exceeds our numbers. As an example of this, the United States uses more than 25 percent of the world's oil production, even though our population is less than 5 percent of the total human population.

But this impact is not caused by people alone: the best way to view the human impact is with this simple equation: population $\times$ technology $=$ impact. Hopefully, the advancing science of environmental economics will provide a better idea of how we can live on the Earth without destroying it. This chapter examines environmental and ecological science from a particular point of view: Our decisions should be based on how our actions will affect the seventh generation. The long-term goal is to practice sustainable development: in other words, producing the goods we need, while making sure that our grandchildren can do the same.

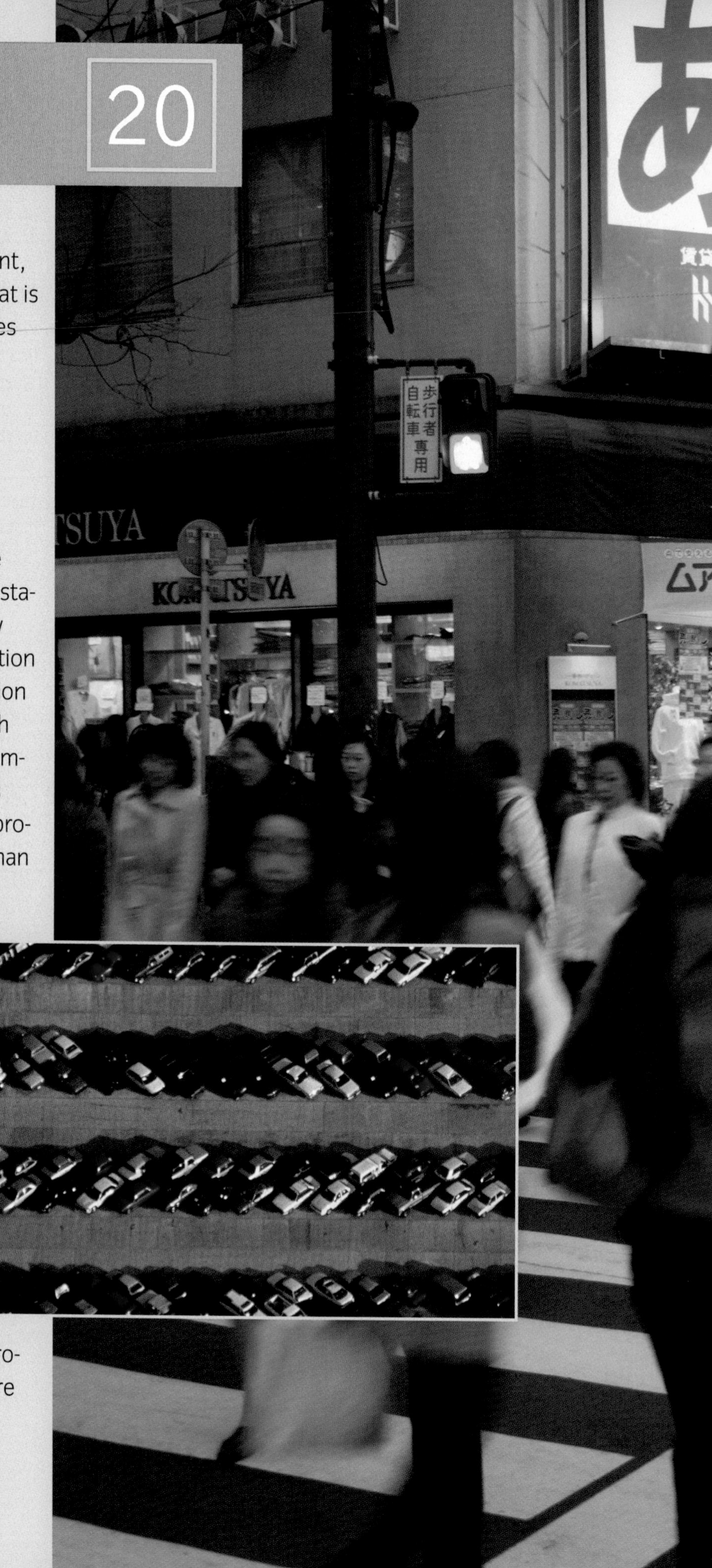

CHAPTER OUT

Introduction

Throughout this book, you have been studying human life, examining the inner workings of an individual. One thing that should be abundantly clear by now is that in biology, nothing happens in isolation. Every muscle contraction, every chemical reaction, every breath you take, affects your entire body. Homeostasis is a constant balancing act between the inevitable environmental consequences of vital activities and the need to remain in the optimal range of conditions needed for those activities. We have looked at a closed system, the human organism, to appreciate this interplay.

But now we need to step back a bit. The intricate interplay of energy and molecules within your body has parallels in our exterior environment. Water moves through the environment in a predictable pattern, much as it moves through your body. Energy is harvested in the environment and used to create and power organisms just as your body harvests and uses energy to create proteins and power activities. Does the entire North American continent, or even the entire Earth, need to maintain a similar homeostatic balance (FIGURE 20.1)? How do we as humans fit into the larger world picture?

Humans and the globe FIGURE 20.1

Human activity leaves visible footprints on the globe. The Great Wall of China has stood for centuries and is one of the few man-made structures on Earth visible from outer space. It is time we seriously consider the impact of our actions.

Ecosystems Define Plants and Animals Living Together

LEARNING OBJECTIVES

Define ecosystem, population, and community.

List and describe the major biomes.

Understand the changes communities go through as they mature.

Compare primary and secondary succession.

The land, water, and air of Earth, with life in all its varied forms, comprise our **biosphere**. Within this biosphere are smaller interrelated units called **ecosystems**. The field of **ecology** attempts to interpret and explain the interactions between the biotic (living) and abiotic (nonliving) components of ecosystems. The teeming, diverse life forms that exist all around us are part of our ecosystem. We interact with these organisms and the physical environment, sharing the resources and hazards of the area.

There are many ecosystems on Earth, each of which interacts with the others in an intricate web of dependency. The interaction between plants and animals in a defined area provides the basis of that ecosystem. The area can be as large as an ocean or as small as a park in the center of a city. The key to an ecosystem is that it is in balance and that the organisms in the ecosystem work together as a functional unit. In ecosystems, this type of interdependency is exemplified by a nonlinear relationship, such as most predator–prey interactions (FIGURE 20.2). When prey populations increase, predators have more food. This in turn allows the predator population to increase. But more predators mean less prey. Eventually, the prey population decreases, and the predator population follows. The balance between these two populations keeps the ecosystem functioning, so the habitat is not destroyed through the depletion of ecosystem resources.

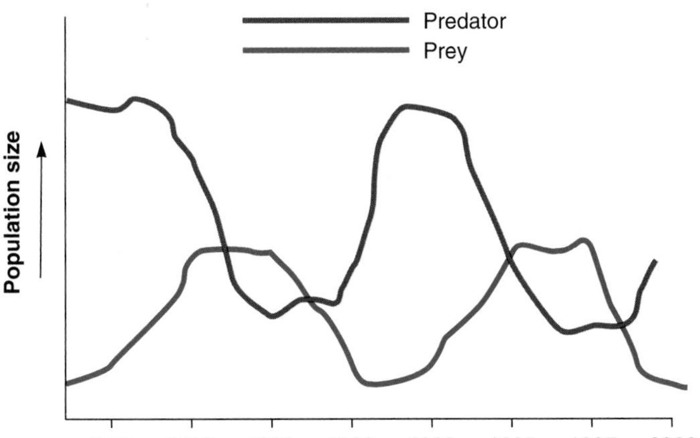

Typical predator–prey interactions

FIGURE 20.2

This graph shows the relationship between a predator species and its prey over a 20-year span. Note that when the prey population increases, the predator population soon follows. When the predator population gets too high, the prey population plummets, which in turn causes the predator population to drop.

COMMUNITIES ARE GROUPS OF POPULATIONS INTERACTING WITH ONE ANOTHER

Within ecosystems are **communities**—groups of organisms interacting with one another, living in the same area, and surviving under the same physical conditions. In New York City, for example, there are as many as 59 distinct communities. One such community includes the plants, animals, and people that live and interact in midtown Manhattan, bordering Central Park. The community of this area consists of the grasses and plants of Central Park, along with the people who live and work around the park. Their pets, pests, and indigenous animals are also part of this community, including dogs, cats, various insects, birds, and rodents. Each of these organisms interacts with the others, living in close proximity under the same or extremely similar conditions.

BIOMES ARE REGIONAL COMMUNITIES

Biome
A regional community characterized by its dominant plant life and climate.

In an effort to describe large-scale ecological situations, ecologists have defined nine **biomes**: ice (permafrost), tundra, taiga (coniferous forest), temperate forest, tropical rain forest, grassland, desert, marine, and freshwater. The characteristics of each biome are listed in TABLE 20.1.

POPULATIONS CAN INTERBREED

Communities are made up of different **populations**. A population includes all the members of one species living in the same area. All members of a population can interbreed and produce living offspring. In the example just presented, the people living and working in midtown Manhattan are a single population within the community. When communities are discussed, only the populations living together are considered. If the physical environment is included, the discussion re-

Biomes of the world TABLE 20.1

	Boreal forest/ Alpine	*Temperate forest*	*Tropical forest*	*Tundra*	*Grasslands*
Location	Northern hemisphere between latitudes 50° and 60° N	Eastern North America, northeastern Asia, Western and Central Europe	Near the equator between latitudes 23.5° N and 23.5° S	55° to 70° N	Middle latitudes, in the interiors of continents
Temperature	Very low	−30° to 30°C	Varies little between 20° and 25°C	Ice covered; 56°C	−40° to 21°C
Annual precipitation	400 to 1000 mm	750 to 1500 mm	May exceed 2000 mm	150 to 250 mm (usually snow)	250 to 1500 mm
Soil type	Deficient in nutrients, thin and acidic	Fertile and enriched with decaying litter	Deficient in nutrients and acidic	Permafrost	Thin and dry, rich
Dominant flora	Evergreen conifers such as jack pine, balsam fir, and black spruce	Broad-leaved species such as oak, hickory, beech, hemlock, maple, elm, and willow	Trees reach 25 to 35 m while plants include orchids, bromeliads, vines, ferns, mosses, and palms	Shrubs, sedges, mosses, lichens, and grasses, flowers	Buffalo grass, sunflower, crazy weed, asters, blazing stars, coneflowers, goldenrods, clover, and wild indigos
Dominant fauna	Woodpeckers, hawks, woodland caribou, bears, weasels, lynxes, foxes, wolves, deer, hares, chipmunks, and shrews	Squirrels, rabbits, skunks, birds, deer, mountain lions, bobcats, timber wolves, and foxes	Birds, bats, small mammals, and insects	Caribou, musk ox, polar bear, shrews, hares, rodents, wolves, foxes, bears, and deer	Coyotes, eagles, bobcats, the gray wolf, wild turkey, flycatcher, Canada geese, crickets, dung beetle, bison, and prairie chicken

turns to ecosystems (FIGURE 20.3). It is difficult to talk only of communities because the physical environment plays such a large role in determining which populations are able to survive.

ECOLOGICAL SUCCESSION CAN BE PREDICTED

One amazing thing about communities is their fluidity. We interact with other populations in our community daily; therefore, we are not often aware of subtle changes. Natural communities undergo constant change, with the **dominant population** shifting with conditions. This sequential change in species dominance is called **succession**. We could observe succession in a lawn in a humid region if we suddenly quit caring for it. The "weeds" we constantly fight in a manicured lawn would outcompete the grass for the sunlight, take over the yard, and choke out slower growing plants. Insects that pollinate the weeds would move in, altering the dominant insect species. The weeds would slowly be replaced by shrubs or trees, which are slower growing but able to reach above the weeds and catch more sunlight. Birds and other insect predators would move in to the shrubs. The shrubs may eventually be outcompeted by trees. Larger mammals that can live beneath the trees would infiltrate the area. Slow-growing hardwood trees would finally appear, turning your yard into a forest.

Dominant population The population with the largest number of individuals in an area.

Chaparral	*Desert*	*Savanna*
West coast of the United States, the west coast of South America, Cape Town area of South Africa, western tip of Australia, and coastal areas of Mediterranean	Hot and dry deserts are near the Tropic of Cancer or the Tropic of Capricorn, cold deserts in polar regions	Wide band on either side of the equator on the edges of tropical forests
10°–40°C	20 to 49°C; 2 to 26°C	Averages 21°C
350 to 600 mm annually	150 to 270 mm	100 mm in dry season, 600 mm in wet season
Rocky, sandy, gravelly, or heavy soils	Sand, exposed bedrock, thin deficient soil	Varies; rocky and sandy to thin to rich
Poison oak, scrub oak, yucca wiple, and other shrubs, trees, and cacti	Turpentine bush, prickly pears, brittle bush, sagebrush	Short twisted trees, grasses, plants specialized for nutrient storage
Coyotes, jack rabbits, mule deer, alligator lizards, horned toads, praying mantis, honeybees, and ladybugs	Small nocturnal carnivores, borrowers, mourning wheatears, horned vipers, antelope, ground squirrels, jackrabbits, and kangaroo rats	Lions, zebras, elephants, giraffes, herds of ungulates, capybara and marsh deer, birds of prey

Relationship between biosphere, ecosystem, community, population, and individual
FIGURE 20.3

Individual organisms interact in populations. All the reef populations taken together comprise the reef community. When you include the sandy ocean bottom and the water column along with the organisms, you are discussing the reef ecosystem. The entire marine ecosystem is a part of the biosphere.

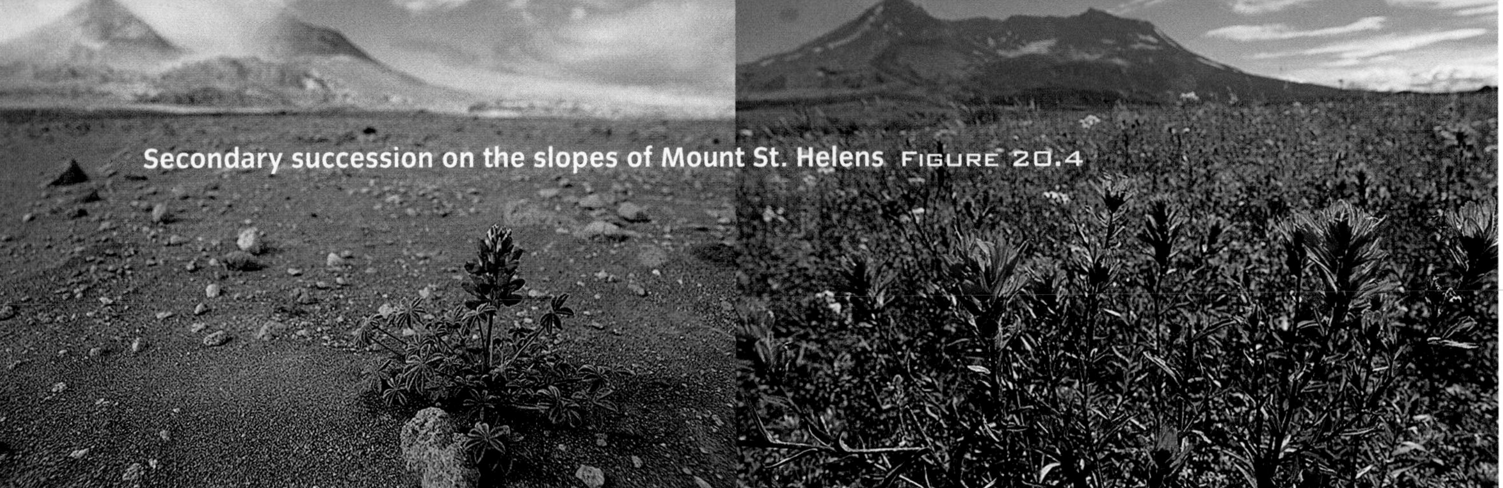
Secondary succession on the slopes of Mount St. Helens FIGURE 20.4

PRIMARY AND SECONDARY SUCCESSION

Pioneer species First plant species to colonize a newly established area.

The life forms that appear during succession are more or less predictable for each ecosystem or niche. When an area begins with bare rock or sand, we call the process **primary succession**. Primary succession may occur on fresh lava, beaches, river deltas, or areas recently gouged by glaciers. The **pioneer species** hold the newly formed soil and add to it as they drop organic material, allowing grasses and then larger plants to take over. As the dominant populations change, the process of succession occurs.

Secondary succession occurs when a disturbance has disrupted a stable ecosystem of plants and animals. Organisms associated with one of the earlier stages of primary succession again become dominant so the process of succession starts anew. A graphic example of secondary succession occurred after the Mount St. Helens volcano erupted. The stable community living along the slopes of the mountain was destroyed, causing the return of the pioneer species (FIGURE 20.4). Leaving your yard to its own devices would be a less dramatic example of secondary succession, as the plants slowly return to the original community that was there before your house was built.

When scientists first noticed this progression of communities, they supposed that there was a predictable and stable end to the succession. They looked for **climax communities** and predicted that they would be similar in similar locations. In the dry parts of the U.S. Great Plains, for example, the climax community is prairie. In the same latitudes along the Atlantic coast, however, the climax community is deciduous forest. Climax communities are less predictable than once assumed, because they reflect the interplay of many factors, including biota, soil, and weather, not just the vegetation. They are, however, stable communities that do not change appreciably in dominant species over many years.

Climax communities Relatively stable, mature communities that have reached equilibrium after passing through a series of established steps.

CONCEPT CHECK

How are communities, populations, ecosystems, and the biosphere related?

Briefly describe the community you live in. Now add to that and describe your ecosystem.

What are the characteristics used to define biomes?

Define how succession may change the dominant population.

Compare primary and secondary succession.

Organisms Have Specific Habitats and Niches

LEARNING OBJECTIVES

Define habitat.

Describe the niche of an organism.

Explain the outcome of niche competition.

Each of the organisms in a biome has a specific **habitat** and **niche**. Habitat is loosely defined as where the organism lives. White-tailed deer are found in deciduous forests in North America; adult green sea turtles are found in near-shore waters of the Central Pacific; tsetse flies live in the low-lying rain forest and savannah of Africa. Assuming the habitat is large enough, it is usually shared by many populations. Rabbits and field mice share grassy fields near forests. Polar bears and seals make the Arctic Ocean their habitat.

Habitat is limited by physical obstacles and competition for resources. Physical obstacles can be obvious structures such as mountain ranges, rivers, and deserts, or subtle variations such as salinity and density gradients in the open ocean, or sunlight availability in the forest. Habitat limits create a geographic range of population distributions. **Biogeographic ranges** are so precise they have been used to predict the location of populations along the entire ocean floor (FIGURE 20.5).

Biogeographic ranges The expected geographic range of an organism, based on its habitat requirements.

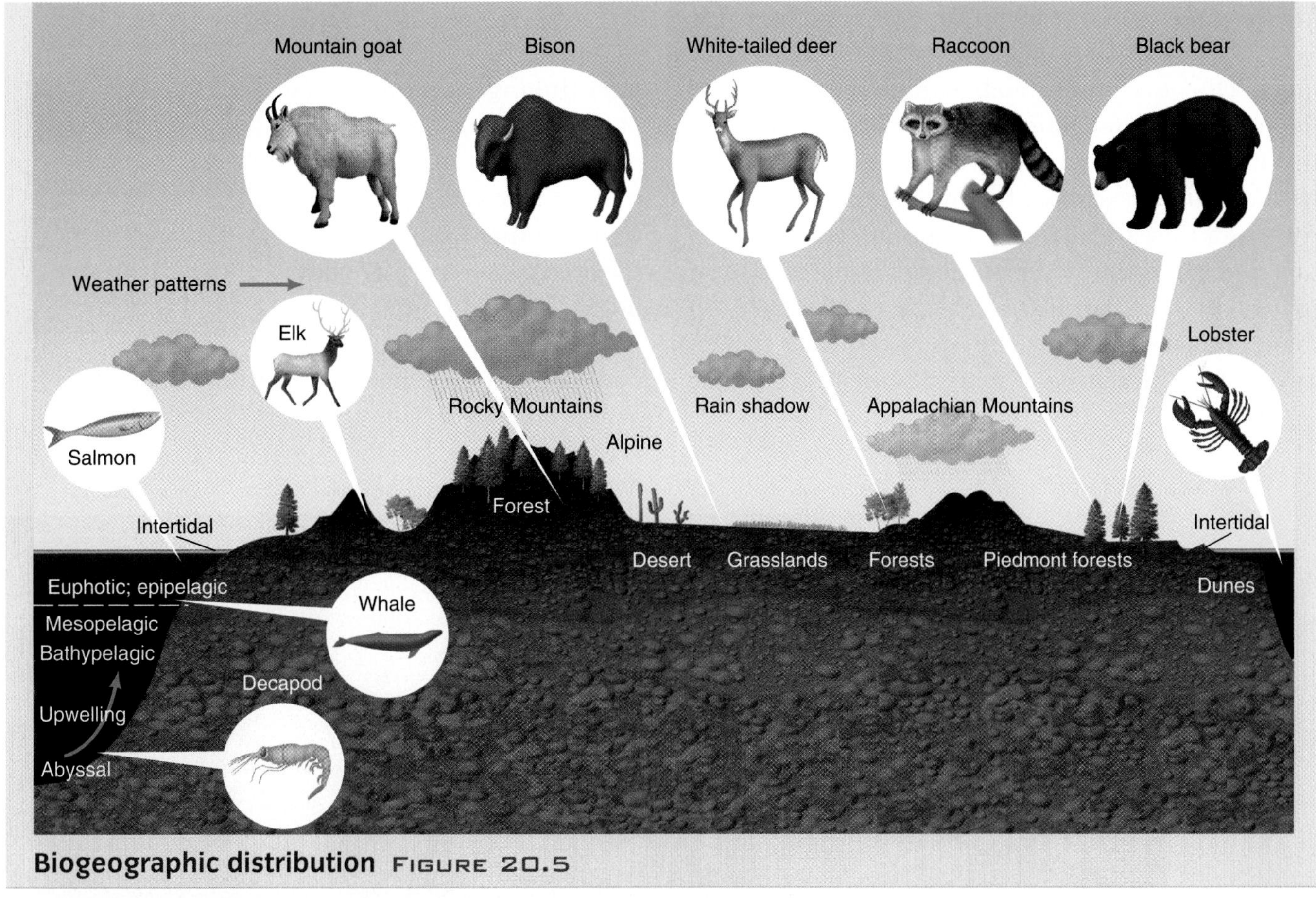

Biogeographic distribution FIGURE 20.5

An organism's location can be predicted based on its ecological requirements. These requirements define the biogeographic range of that organism.

By understanding the habitat needs of a bottom-dwelling organism, we can predict where it will be found. Knowing the habitat requirements of any organism allows us to predict its location.

Niche, on the other hand, defines the organism's "job," or role in the community. Everything from where they live to what they eat to what time of day they are active helps define a niche. If you are a typical college student, your habitat is your campus. Your niche includes your dorm room, your class schedule, your extracurricular activities, your dietary choices, your study habits, and even your wardrobe. No two organisms can occupy the same niche in the same habitat. Imagine how difficult your existence would be if another student was following your exact schedule, living in your room, and eating the same food at exactly the same time! One of you would have to alter your routine in order to coexist.

Often we describe the niche of an entire species rather than each individual. Individuals of a species utilize the same resources in the same fashion; therefore, we can speak of an entire species when we describe niche. Of course, individuals within species compete for resources, but a more global view would indicate that different species compete for niches, while individuals in that species share the resources of that niche. Although they are all using the resource at the same time and in the same fashion, enough resources remain to support the population. Different species, on the other hand, usually share resources in the habitat by altering the timing of activities, or by otherwise **partitioning** the resources (FIGURE 20.6). Perhaps one species of insect will eat only the flowers of a plant, while another will ingest only the leaves. One species of bird might nest in the lower branches of a tree, while a second species nests in holes in the trunk.

Resource partitioning FIGURE 20.6

The crab and the marine iguana both forage in the intertidal zone, but the flattened crab can obtain food nearer to the surf zone than can the larger iguana.

Partitioning
Dividing available resources into discrete parts to reduce competition.

CONCEPT CHECK

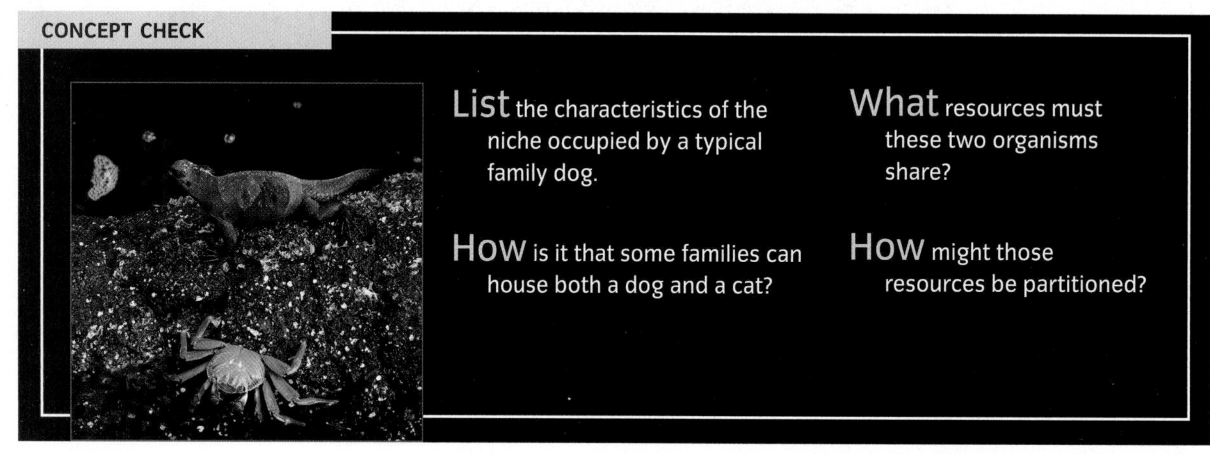

List the characteristics of the niche occupied by a typical family dog.

How is it that some families can house both a dog and a cat?

What resources must these two organisms share?

How might those resources be partitioned?

Carrying Capacity and Population Growth Are Regulated by the Environment

Learning Objectives

Relate carrying capacity to biotic potential.

Discuss different population growth patterns.

The sizes of populations continually change as they exploit available habitat. The **carrying capacity** of the ecosystem is the number of individuals in each population the area can support indefinitely without permanently reducing the productivity of the area. Carrying capacity represents a balance between resources and competition on one hand and population growth on the other (FIGURE 20.7). Carrying capacity varies with species, with ecological conditions, and with time. Your vegetable garden may be able to sustain only two rabbits, while at the same time supporting thousands of aphids. The carrying capacity for each population is different in that same small plot of land. In each case, the populations in the habitat will grow to the maximum number of individuals the resources can support without intrinsic damage. In natural ecosystems, populations often stabilize near their carrying capacity, but they do not remain static. Instead they tend to bounce up and down, around limits determined by the physical environment. One theory is that under steady environmental conditions, carrying capacity is determined by the limiting resource, often food.

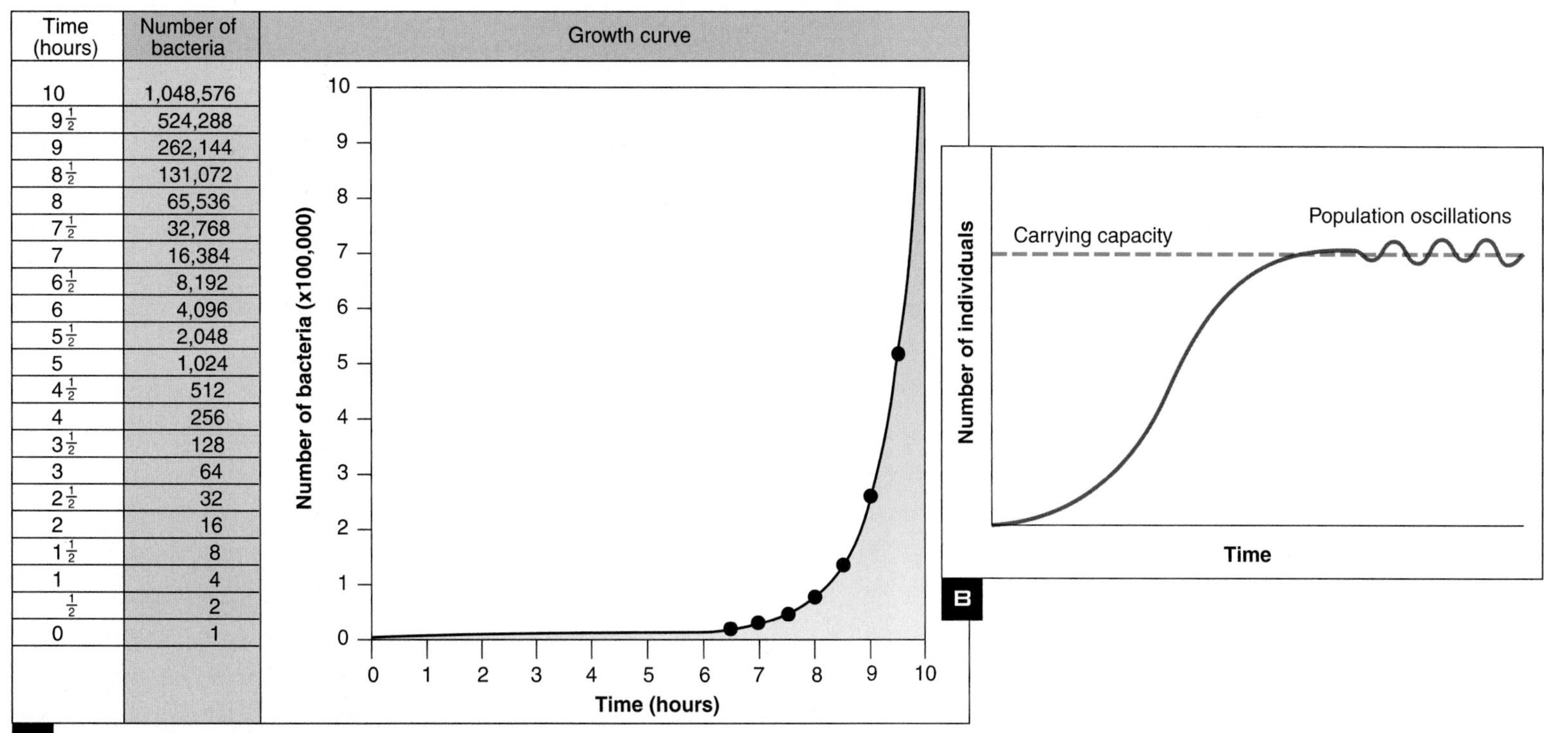

Time (hours)	Number of bacteria
10	1,048,576
$9\frac{1}{2}$	524,288
9	262,144
$8\frac{1}{2}$	131,072
8	65,536
$7\frac{1}{2}$	32,768
7	16,384
$6\frac{1}{2}$	8,192
6	4,096
$5\frac{1}{2}$	2,048
5	1,024
$4\frac{1}{2}$	512
4	256
$3\frac{1}{2}$	128
3	64
$2\frac{1}{2}$	32
2	16
$1\frac{1}{2}$	8
1	4
$\frac{1}{2}$	2
0	1

Bacterial population growth and human population growth FIGURE 20.7

A Bacterial populations tend to increase exponentially until they hit and exceed the carrying capacity. The population then crashes as resources are depleted. B The human population, on the other hand, tends to reproduce more slowly, which may allow it to fluctuate around the carrying capacity rather than simply overshoot it and die back.

BIOTIC POTENTIAL MEASURES THE MAXIMUM GROWTH RATE

The **biotic potential** of a population is its maximum growth rate under ideal conditions. Biotic potential in sexually reproducing populations depends on (1) the number of offspring produced per female, (2) the time to reproductive maturity, (3) the ratio of males to females, and (4) the number of reproductively active individuals. If the population is below the carrying capacity of the environment, most populations grow exponentially, creating a J-shaped curve (see FIGURE 20.7A). Under ideal conditions, **phytoplankton** grow rapidly enough to cause visible changes in ocean waters. A red tide is actually a population explosion, or **bloom**, of a dinoflagellate containing a droplet of reddish oil in its nearly transparent body. When the population grows unchecked, uncountable numbers of microscopic dinoflagellates seem to stain the ocean. Red tides are called "harmful algal blooms" because they produce neurotoxins that can harm the nervous systems of people who eat shellfish from affected waters (FIGURE 20.8).

Phytoplankton Microscopic plants and photosynthetic protists that float in the water column.

POPULATION GROWTH REFLECTS MULTIPLE FACTORS

Algal blooms are examples of exponential growth in areas where carrying capacity has not yet been reached. Many populations however, especially those composed of larger organisms, do not grow so fast. The population expands slowly, decreasing as it nears carrying capacity, forming an S-shaped growth curve (see FIGURE 20.7B).

Population growth curves vary. The steeper the growth curve, the faster that

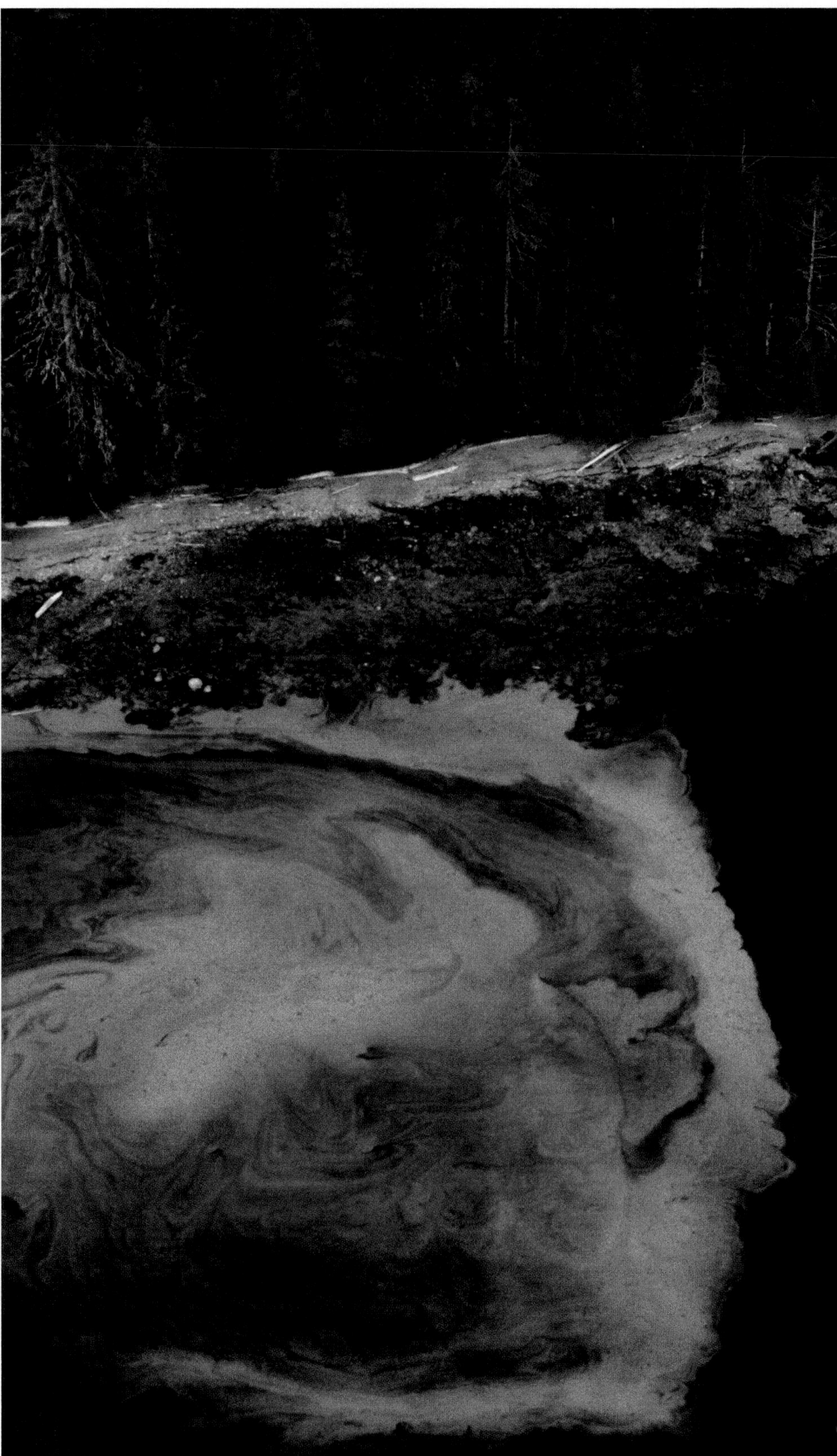

Red tide FIGURE 20.8

population doubles. Under ideal conditions, bacterial population growth curves are quite steep, while human population growth curves are much flatter. The average generation time for *E. coli* bacteria is a mere 20 minutes. Humans take a minimum of 12 to 14 years to reach sexual maturity, and many people do not reproduce for some years after that.

The two basic control patterns of population growth are extrinsic and intrinsic control. If the population is extrinsically controlled, organisms colonize new habitat, produce many offspring, invest little energy in each one, and usually widely overshoot the carrying capacity. Because economist Thomas Malthus first described this type of population regulation in his analysis of the human condition, these organisms are called Malthusian strategists.

In contrast, intrinsic control appears among organisms that follow the logistic strategy. These organisms grow and mature more slowly, live longer, produce fewer offspring, and invest more energy per offspring. Logistic strategists are usually large animals that prey on smaller ones, while Malthusian strategists tend to be producers or animals that eat plants. Malthusian strategists tend to be pioneer species that invade new ecosystems first and take over the resources for a short period. Typical Malthusian strategists include many of the grasses, insects, and those red tide algae we just discussed. See TABLE 20.2 for a comparison of these strategies.

Extrinsic and intrinsic growth patterns are related to a population's survivorship curve (FIGURE 20.9). Three basic age distributions show patterns of mortality in a population. Type I survivorship curves describe organisms that provide considerable parental care. Individuals tend to survive through young adulthood and die out at advanced ages. Type II populations have a constant death rate regardless of age. Type III populations produce many young but provide no more than a bit of parental care. Those few individuals that survive infancy are likely to live a long time. The green sea turtle falls into this category. Many eggs hatch, and most of the young return to the sea, but only a handful of the hatchlings survive to reproduce. Once a turtle reaches age 5, however, predation risks drop, and it will probably survive into old age.

Comparison of Malthusian and logistic strategists TABLE 20.2

Malthusian	*Logistic*
Low trophic levels	Higher trophic levels
Pioneer species in succession	Climax community species
Generalists or opportunists	Specialists
Rapid growth	Slow growth
Early maturation	Late maturation
Produce many offspring with little to no parental care	Produce few offspring but invest large amounts of parental care
Limited by extrinsic factors	Limited by intrinsic factors
Examples: insects, rodents, annual plants	Examples: elephants, whales, wolves, primates, deciduous trees

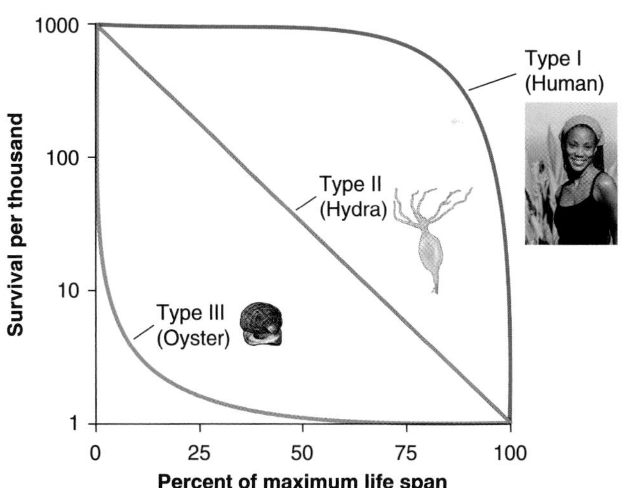

Survivorship curves FIGURE 20.9

A survivorship curve depicts the expected decrease in a population over time. There are three types of survivorship curve: I those populations that produce few young and invest a lot of energy in keeping them alive. Most die-off occurs in old age, resulting in a steep drop in numbers only after many years of life; II those populations that suffer a uniform death rate throughout life, regardless of age of the individual; and III those populations of organisms that tend to produce many young, most of which do not survive beyond the first few days or weeks, resulting in an early and steep drop in numbers.

Populations rarely reach their biotic potential because competition among individuals for finite resources impedes population growth. Environmental limits on growth include diseases, predation, environmental toxins, and both inter- and intra-population competition for food, shelter, and water. When environmental conditions deteriorate, the carrying capacity of the ecosystem is reduced, and populations decline.

Another way to look at the regulation of population is to classify the factors as density-dependent or density-independent. **Density-dependent factors** increase in intensity along with increasing population, including factors such as predation and competition for territory or mates. **Density-independent factors** affect the entire population or ecosystem, and would include factors such as tsunami, earthquake, and volcanism.

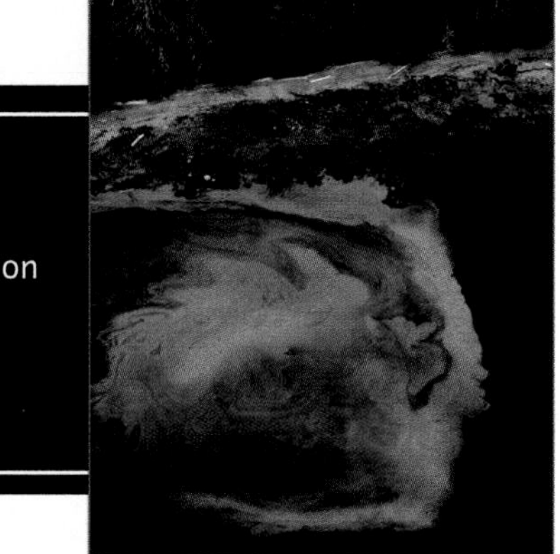

CONCEPT CHECK

Define carrying capacity.

List the four characteristics that limit an organism's biotic potential.

Give two examples of density-dependent population factors. Do these affect the human population?

Organisms Are Either Producers or Consumers

LEARNING OBJECTIVES

Characterize a producer.

List and describe the three types of consumer.

Organisms, and entire populations, can be classified as **producers** or **consumers**. Producers assemble usable food molecules through photosynthesis or (more rarely) chemosynthesis. Green plants and bacteria that live off chemicals emitted from oceanic thermal vents are examples of producers. Consumers cannot create food molecules, but must instead obtain them from other organisms. Animals, whether they eat plants or other animals, are consumers.

Producers are **autotrophic**, meaning they carry out photosynthesis or chemosynthesis and make food for themselves. They do not eat like humans, or even like mosquitoes or dung beetles. Almost the entire biosphere relies on producers to create organic fuel from the sun's energy. On land, green plants and **cyanobacteria** are the main producers. In freshwater and marine ecosystems, algae and phytoplankton fill this niche. A few of communities survive on chemical energy instead of solar energy. In these communities, found mainly at deep-sea vents, the primary producers are chemosynthetic. (FIGURE 20.10). Their role

Producers Organisms that create their own nutrients from inorganic substances; mainly green plants.

Consumers Organisms that must ingest nutrients because they cannot manufacture their own.

Cyanobacteria Blue-green, photosynthetic bacteria.

Fixing
Converting gaseous elements to organic compounds.

in **fixing** organic compounds remains the same, however.

Consumers are **heterotrophs**. They cannot manufacture organic fuel with solar power, but must instead ingest existing organic fuel. The four types of consumer are classified by food source:

- **Herbivores** eat green plants. They get their energy directly from the producer. Because they feed directly on autotrophs, herbivores are also called **primary consumers**. Herbivores include bison, humans who are strict vegetarians, fish that graze on vegetation, and fruit- and grain-eating birds such as parrots.
- **Carnivores** eat other animals, and meet their protein and caloric requirements through this "complete" nutrition source. Carnivores usually eat less often and/or require smaller portions than herbivores. But it does take more energy to be a carnivore, since herbivores do not have to waste energy chasing plants! Carnivores that feed on herbivores are called **secondary consumers.** If they feed on other carnivores, they may be **tertiary** or rarely, **quaternary, consumers**.
- **Omnivores** are animals that can eat either plants or animals. The benefit of being an omnivore is that food can be obtained much more efficiently from both plant and animal sources. The human is an omnivore that can eat such bizarre and diverse foods as artichokes and lobster, and obtain nutrition from each.

Chemotrophic communities are found at thermal vents in the deep ocean
FIGURE 20.10

- **Decomposers,** or **detritovores**, obtain their nutrients from **detritus**, returning most of the material to the soil. Decomposers don't get much respect, but bacteria, fungi, earthworms, and small soil organisms such as nematodes and isopods are essential to a healthy ecosystem. These organisms recycle dead plant and animal matter into nutrients that primary producers can use, ensuring that the limited resources of the ecosystem are available for reuse and that dead bodies do not pile up.

Detritus
Loose fragments of organic and inorganic matter obtained from decomposition and weathering.

CONCEPT CHECK

Compare a producer to a consumer.

Give examples of a producer, a primary consumer, a secondary consumer, and a decomposer.

What is the name for a species that eats both plants and animals?

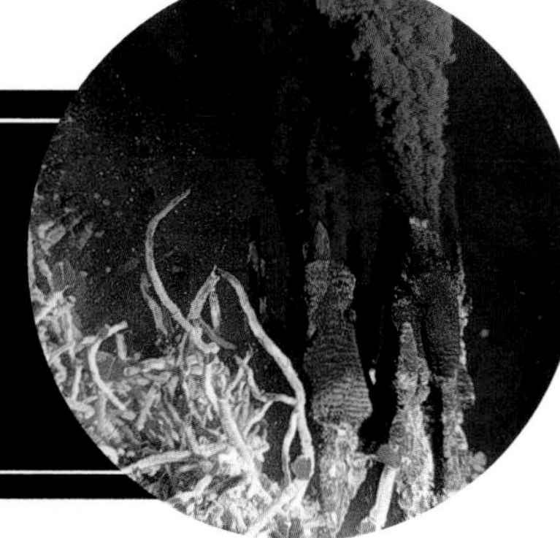

Energy Flows through an Ecosystem while Chemicals Cycle

Learning Objectives

Summarize photosynthesis.

Describe a typical food web.

Describe the water cycle.

Compare the water cycle to the nitrogen and phosphorus cycles.

Understand the importance of the carbon cycle.

Ecology is all about flow. When studying the interactions of the biotic and abiotic factors in the biosphere, we use the repeated appearance of two factors: energy and fundamental chemicals. **Energy** flows through ecosystems on a one-way trajectory, while many chemicals cycle repeatedly through the biosphere (FIGURE 20.11).

Energy is constantly supplied to the ecosystem by the sun. Producers pick up that energy and use it to convert chemicals to useful organic compounds, which often cycle repeatedly through the biosphere. As consumers eat producers, both the energy and chemicals are transferred to the next organism in line. Energy continues to move through the ecosystem until it is lost as heat to the atmosphere. Some of that heat is generated by metabolic activity. Ecosystems need a constant supply of energy to compensate for this heat loss. Chemicals, on the other hand, cycle through organisms and the abiotic portion of the biosphere. The original inorganic compounds used by producers are often made available to other organisms through decomposers and other natural activities. Some chemicals leave the biosphere as they become trapped in geologic sediments, but they may return to the biosphere later when resulting rock breaks down.

Energy Usable heat or power.

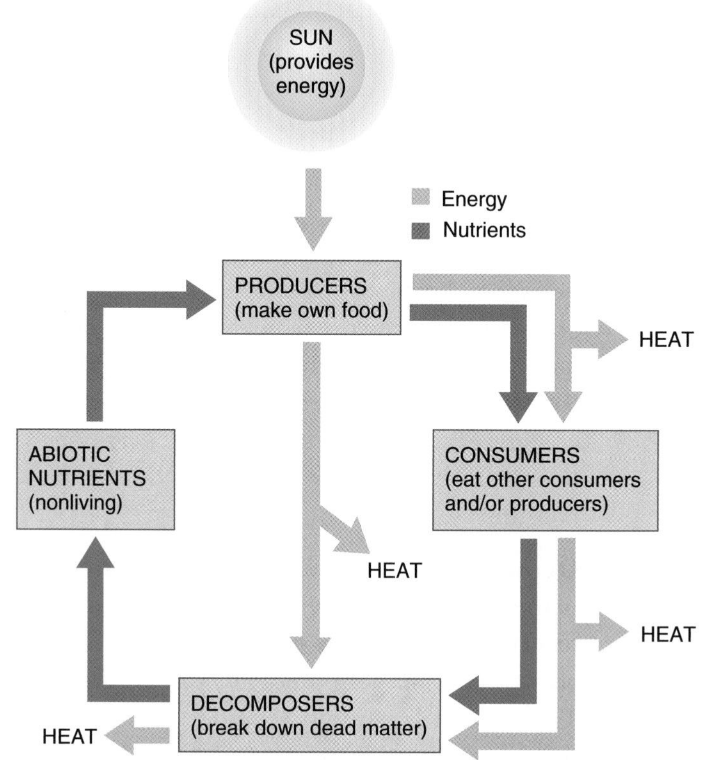

Energy flow and resource cycling

FIGURE 20.11

In this image, energy from the sun travels through the producers, consumers, and decomposers, escaping the system as heat at each step. In contrast, the nutrients cycle through the organisms and abiotic segments of the biosphere.

PHOTOSYNTHESIS CONVERTS SOLAR ENERGY INTO CARBOHYDRATE COMPOUNDS

Energy is harvested by green plants using a **photopigment** called **chlorophyll**, which is usually found in small green organelles called **chloroplasts** (FIGURE 20.12). Photosynthesis occurs in these chloroplasts in two stages—the **light reaction** and the **dark reaction**. In the light reaction, chlorophyll absorbs a photon of light and releases an excited (energy-carrying) electron. This excited electron is captured by a specialized protein and transferred through a series of compounds, releasing its energy in a slow, controlled fashion. The released energy is collected in ATP and another high-energy compound, **NADPH**. This phase of photosynthesis is called the light reaction because it begins when light is absorbed (FIGURE 20.13A, p. 680).

Photopigment
An organic compound that changes in response to light.

Chlorophyll
A blue-green photopigment found in plants and algae.

The electrons that popped off the chlorophyll molecule during the light reaction must be replaced for the pigment to continue absorbing light. To accomplish this, water is **hydrolyzed**. As the water molecule splits, it replaces the missing electron and releases an oxygen atom to the atmosphere. This oxygen is critical to animal life. Green plants supply the atmosphere with the oxygen needed to sustain the metabolic reactions of respiration in autotrophs and heterotrophs.

Hydrolyzed
A water molecule that has been split, releasing H^+ and OH^-.

The dark reaction, or **Calvin cycle**, of photosynthesis occurs in the chloroplasts without needing photons. During the Calvin cycle, energy stored in ATP and NADPH is used to convert carbon dioxide into glucose molecules (FIGURE 20.13B, p. 681). Because no photons are absorbed, the Calvin cycle can occur day or night. For environmental reasons, some plants run

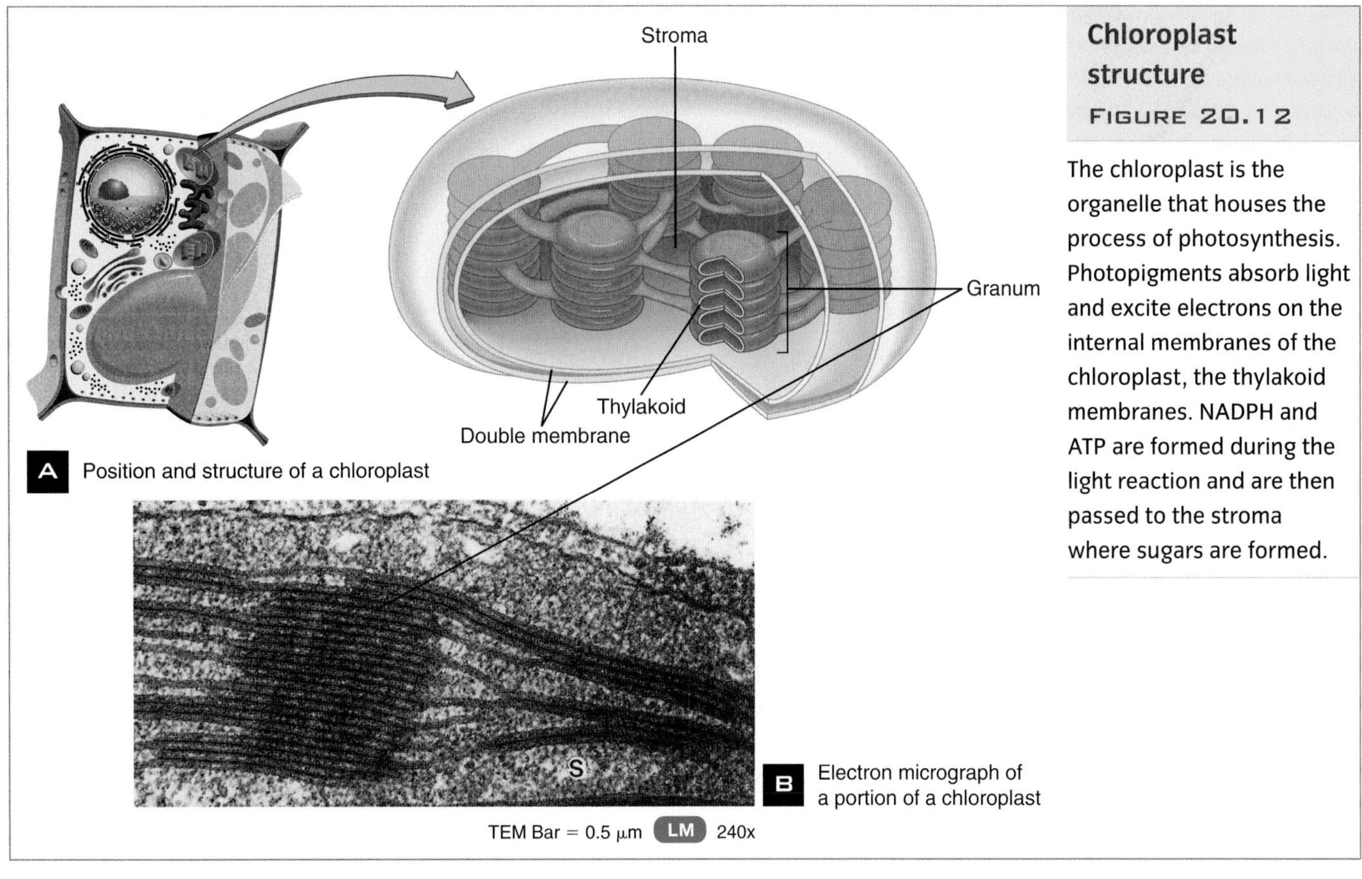

Chloroplast structure FIGURE 20.12

The chloroplast is the organelle that houses the process of photosynthesis. Photopigments absorb light and excite electrons on the internal membranes of the chloroplast, the thylakoid membranes. NADPH and ATP are formed during the light reaction and are then passed to the stroma where sugars are formed.

Process Diagram

The two stages of photosynthesis FIGURE 20.13

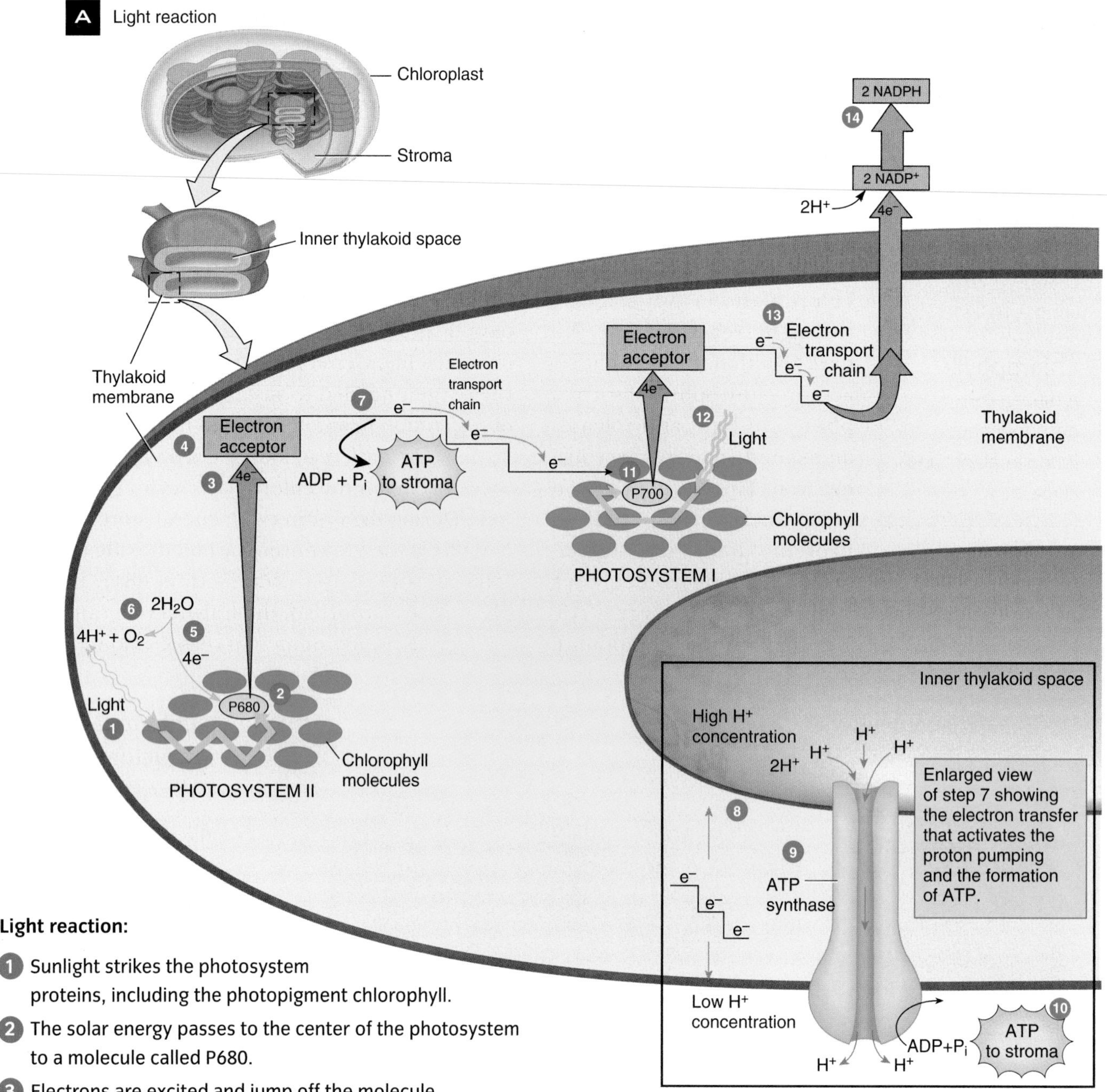

Light reaction:

1. Sunlight strikes the photosystem proteins, including the photopigment chlorophyll.
2. The solar energy passes to the center of the photosystem to a molecule called P680.
3. Electrons are excited and jump off the molecule.
4. These electrons are grabbed by a primary electron acceptor rather than falling back to P680. If they were to simply fall, red light would be emitted.
5. Electrons are stripped from a molecule of water to replace those that left P680.
6. The remnant of the water molecule, oxygen gas, is released as a by-product of photosynthesis.
7. The electrons from P680 are passed down an electron transport chain in the thylakoid membrane within the chloroplast. ATP is generated as they fall down this cascade.

8–10 The steps of ATP production. Hydrogen ions are used to drive the reaction. 8 As hydrogen ions travel through the membrane protein ATP synthase, a membrane protein, they cause ADP to be phosphorylated.

(continues on next page)

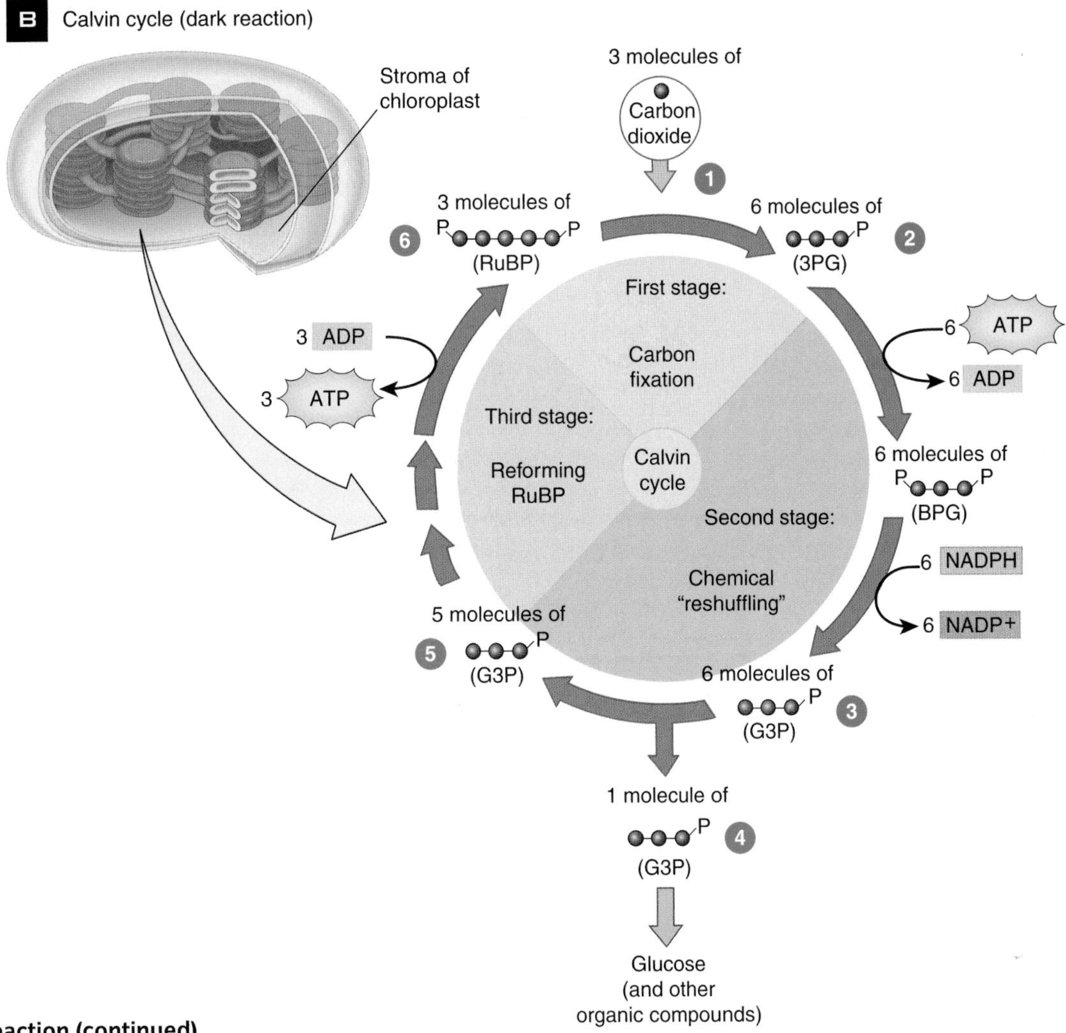

Light reaction (continued)

9 The electrons from P680 pass from the electron transport chain to the second photosystem, arriving finally at P700.

10 Another photon strikes the chloroplast, this time affecting the second photosystem.

11–14 The electrons that arrived at P700 are again energized to jump from the photopigment to a second electron acceptor and then through a transport chain, this time slowly releasing their energy to phosphorylate the high-energy compound NADPH.

Dark reaction:

The Calvin cycle is a series of carbon compounds undergoing "shape shifting" in order to fix the carbon in CO_2 into glucose. Six carbon dioxide molecules create one glucose molecule.

1 Carbon dioxide is attached to a 5-carbon compound through the action of the enzyme rubisco.

2 The unstable 6-carbon compound just formed immediately breaks into two 3-carbon compounds.

3 Energy is added in the form of ATP and NADPH, resulting in a slight shift in carbon atom arrangement.

4 A molecule of glucose is pushed out of the Calvin cycle for every three CO_2 molecules that enter.

5 The remaining molecules are reshuffled to generate the original 5-carbon compound that rubisco needs in order to fix CO_2.

6 The original 5-carbon compound is then regenerated and ready to serve as a substrate for rubisco.

Photosynthesis/respiration FIGURE 20.14

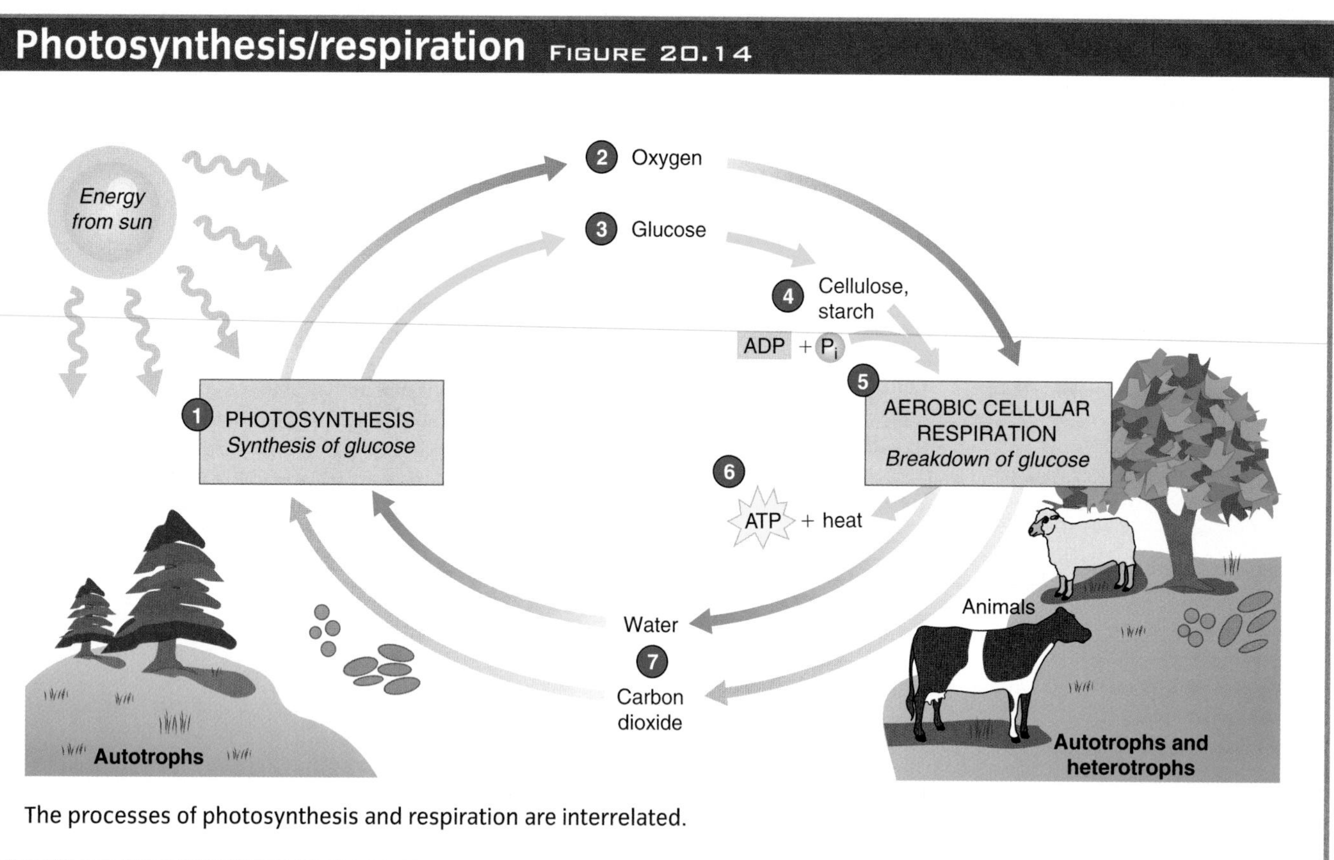

The processes of photosynthesis and respiration are interrelated.

the Calvin cycle mainly at night. Collecting CO_2 for the Calvin cycle during the heat of the day would cause their cells to dry out. The pineapple, for example, which grows in hot, dry climates, collects CO_2 much more efficiently than other plants and does so at night, storing it for use the next day to avoid opening its cells to the drying air during the day.

Once the photosynthetic cycle is completed, plants can use the energy in the newly created glucose for cellular respiration. Plants burn glucose and therefore respire just like animals. They need to produce structural and functional proteins, build the support and storage carbohydrates cellulose and starch, and create the lipids needed for survival. Excess glucose can be stored as starch. Each year, green plants produce an estimated 145 billion tons of carbohydrates, equal to about 23 tons per person.

Metabolize
All chemical processes in an organism, including both buildup and breakdown of organic compounds.

You might have noticed that photosynthesis is the reverse of respiration. Both plants and animals require oxygen and glucose, or another carbon source, for survival. As animals **metabolize**, they produce carbon dioxide and water vapor as waste products. Plants require carbon dioxide and water and release oxygen and glucose during photosynthesis. This balance between plants and animals is what drives energy flow through ecosystems (FIGURE 20.14).

FOOD CHAINS CAN FORM FOOD WEBS

Plants secure usable energy, and animals take advantage of that energy. The simplest depiction of this relationship is to isolate a simple **food chain** (FIGURE 20.15). A food chain begins when the producer obtains energy from the sun. The producer, for example, a corn plant, is eaten by a primary consumer. In this case, the herbivore might be a grasshopper. The primary consumer then becomes food for a secondary consumer. Toads, songbirds, and feral cats all eat grasshoppers. The secondary consumer is in turn captured and eaten by a tertiary consumer, such as a snake or redtailed hawk.

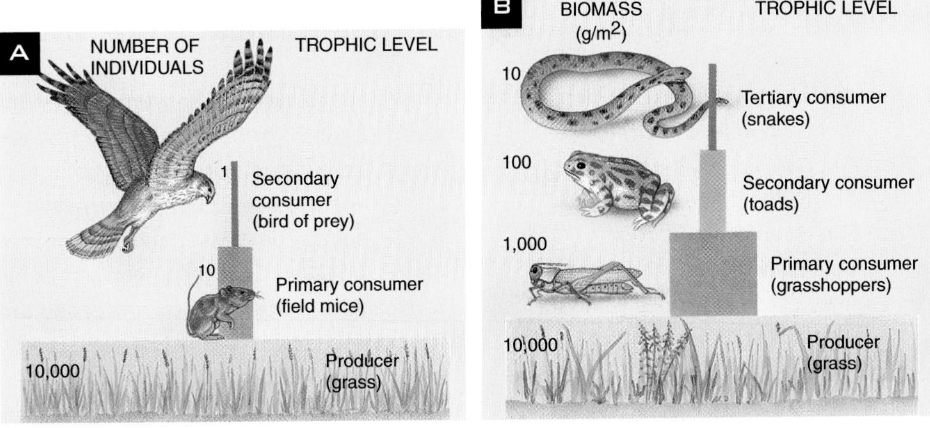

Food chains, trophic levels, and energy consumption FIGURE 20.15

At each level of the food chain, energy is transferred, but a great deal of energy is lost. In fact, only about 10 percent of the energy at one level of the food chain is transferred into the tissues of the consumers in the level above it. So only 10 percent of the energy stored in a plant becomes stored in the herbivore that eats that plant. And the carnivore that eats the herbivore gets 10 percent of the herbivore's energy, and winds up with only 1 percent of the energy that was stored in the plants that the herbivores ate. Some people argue that all this wasted energy could be put to better use; see the Health, Wellness, and Disease feature on page 684.

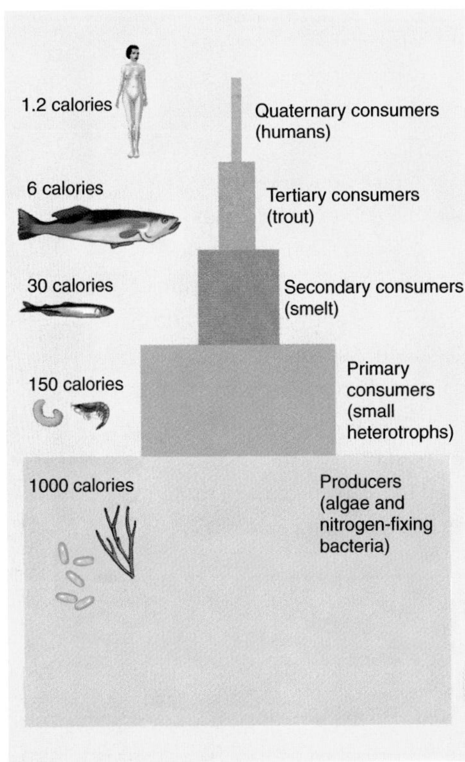

Ecological Pyramid: Trophic levels and energy loss FIGURE 20.16

Each level in the food chain includes less available energy. The width of the bars at each level indicates the energy available from consuming that level. By the fifth trophic level, so little energy is available that a sixth level is not practical.

TROPHIC LEVELS COMPRISE THE ECOLOGICAL PYRAMID

Owing to the major loss of energy at each level, food chains can have no more than five **trophic levels** including the producer. These are usually portrayed in an **ecological pyramid** (FIGURE 20.16). The size of each level indicates assumed energy, measured biomass, or the number of individuals living at each level.

> **Trophic levels**
> All the organisms that occupy the same energy tier in a community, such as primary producers, primary consumers, or secondary consumers.

Ecological pyramids give a strong indication of

Health, Wellness, and Disease

Could vegetarianism help feed the world?

With the global population surging toward 7 billion, we have to wonder: Can the world feed itself? In 2003, the World Food Program pegged the number of hungry people at 852 million. But some say the problem is based more on inequalities of food distribution than on problems with food production. People with enough money get to eat; those without money do not.

As we know, it takes up to 10 calories of grain to produce 1 calorie of meat. Could it help to eat more grain directly instead of feeding it to animals? Yes, say vegetarians and vegans, who argue that eating meat is wrong. (Vegetarians eat no meat, but may use leather or eat dairy products, while vegans consume no animal products in any portion of their lives.) From the standpoint of the food supply, would that make the world a better place?

Ecologically, eating vegetable matter puts a person lower on the food chain and therefore uses available energy more efficiently. Humans occupy the primary consumer level when we eat vegetables and fruits, and the secondary or even tertiary consumer level when eating meat. If people directly ate the quantity of primary producers needed to raise one cow to maturity, many more people could eat.

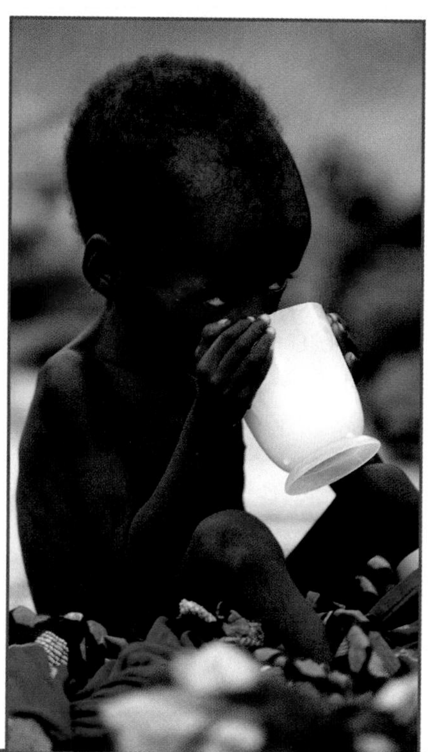

Although livestock consume 35 percent of the world's grain and 90 percent of the soybeans, the situation is complicated. For one thing, a large proportion of meat comes from animals that are raised partly or wholly on rangeland. The grasses and other primary producers on this land would not feed people at all because, unlike ruminants, we lack the bacteria needed to break down the cellulose in grasses. In this sense, cattle expand, not contract, our food supply.

In any case, the trend in world food consumption is veering away from vegetarianism: more livestock and poultry are being raised and eaten. Meat consumption doubled between 1977 and 2002 and is rising much faster in the newly developing world than in the developed world (although the developed world continues to eat the majority of meat). In many cultures, eating meat is a sign of wealth, and meat provides a much-needed source of protein and micronutrients.

Finally, if vegetarianism was able to solve the world's food crisis, it would already have done so. In 2002, 48 percent of human food calories came directly from grain. As a personal matter, choosing not to eat meat or consume animal products is perfectly valid. But in the larger scale, it is unlikely to solve the pressing need for more food.

ecosystem stability. The open ocean sometimes has an inverted ecological pyramid (see **Figure 20.30c** on p. 701), with producers comprising a smaller biomass at a given time than primary consumers. This happens in areas where the primary productivity is reduced, perhaps owing to extreme wind or temperature changes in the surface waters. The phytoplankton count drops, but the **zooplankton** count remains high. Either the phytoplankton will recover in numbers sufficient to sustain the zooplankton, or the zooplankton will thin out due to the food shortage.

Zooplankton
Microscopic and macroscopic animals that float in the water column and move at the mercy of the currents.

FOOD WEBS SHOW REAL-LIFE INTERACTIONS

To complicate matters, energy does not travel only in a straight line through an ecosystem. For example, many different herbivores consume the same primary producers. These herbivores could, in turn, become prey to any number of carnivores, or they could die of old age. When many interacting food chains are depicted together, we see a **food web** (FIGURE 20.17), which more accurately shows the movement of energy through an ecosystem.

Because many herbivores and carnivores can eat multiple food sources, **extinction** is usually rare in ecosystems where humans play no role (TABLE 20.3, p. 686). When one population in the food web declines, organisms often have other population ways to obtain energy, and although populations fluctuate in response to changes in the food web, the animals survive. Extinction occurs when a population is reduced below its ability to reproduce, often by some combination of predation, habitat destruction, and disease.

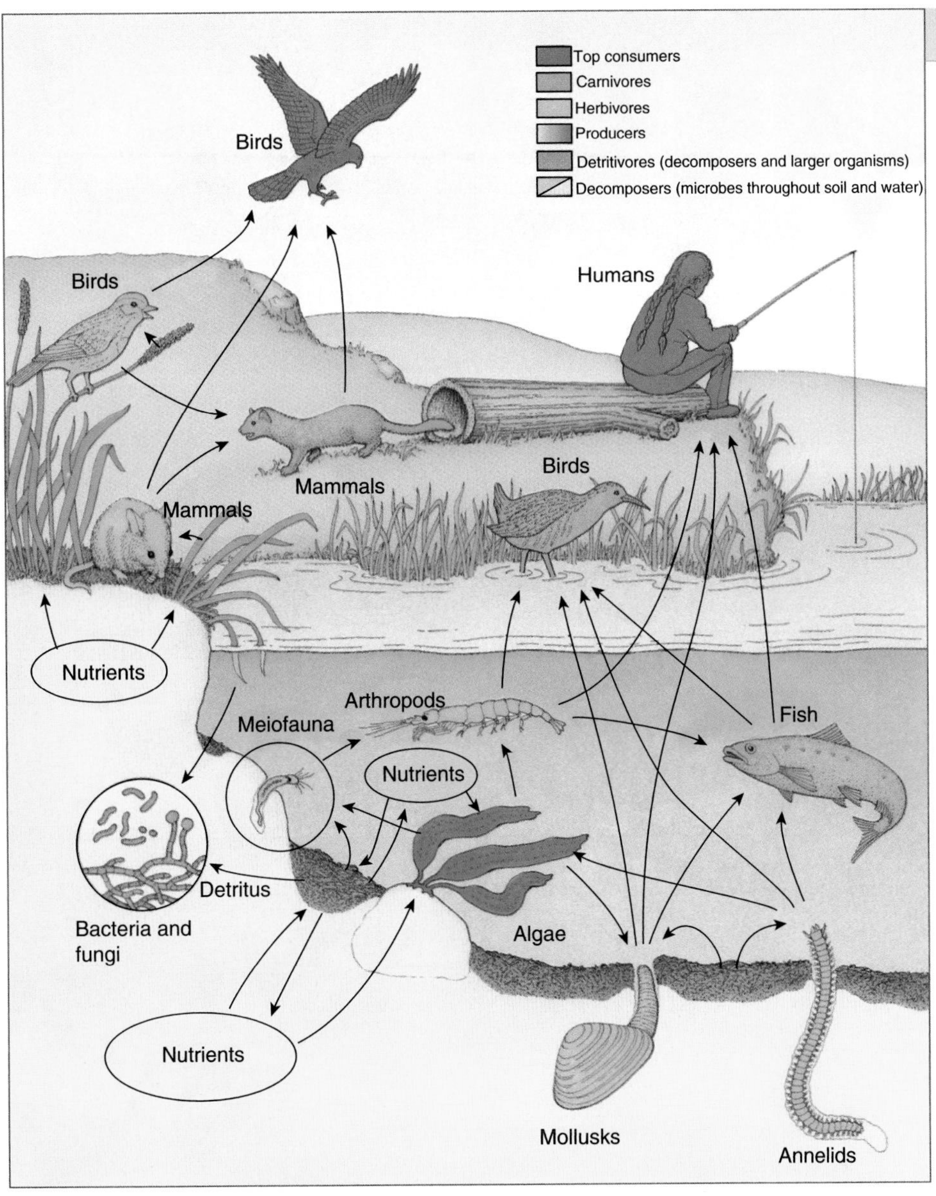

Food web FIGURE 20.17

The food web of a salt marsh includes an interwoven community of algae, mollusks, fishes, arthropods, birds, and man. There are many routes to each organism, indicating the many pathways energy can take through the salt marsh.

Representative endangered or extinct species TABLE 20.3

Atitlan giant pied-billed grebe - considered extinct in 1936, but observed in 1960 in Guatemala

Ivory-billed woodpecker - considered extinct in 1945, but confirmed sighting in 2005 in Arkansas

GLOBAL BIOGEOCHEMICAL CYCLES MOVE ATOMS ON THE PLANET

Energy may follow a one-way path through the ecosystem, but many essential elements are caught in continuous cycles. Recall from Chapter 1 which elements predominate in your body: carbon, nitrogen, hydrogen, and oxygen, with phosphorus also high on the list. Each of these elements is necessary for life of any sort, from producer to tertiary consumer to detritivore.

The cycles of elements through an ecosystem can be complicated and can take years, centuries, or much longer to complete. During the cycles, elements pass through both the biological and geological components of the biosphere. Each cycle usually includes a **reservoir**, an **exchange pool,** and the **biotic community**. The reservoir holds the chemical in a way that is inaccessible to the producer. Rocks, deep-sea sediments, and fossils are good examples of reservoirs. The exchange pool is the area where the chemical occurs in a form that is usable to the biological community. The atmosphere, soil, and water all serve as exchange pools. The biotic community holds a surprisingly large store of chemicals in each of the trophic levels. They are passed along the food chain, sometimes remaining in a particular trophic level for long periods (FIGURE 20.18).

Elements of the biogeochemical cycle FIGURE 20.18

The two main types of cycle are **gaseous** and **sedimentary**. As the names imply, chemicals that follow gaseous cycles are drawn from and returned to the atmosphere as a gas. Those in a sedimentary cycle are absorbed from the soil by plant roots and returned to the soil (usually very close to its release point) by decomposers.

Even though biogeochemical cycles appear isolated from one another, they do overlap. Nutrients and chemicals can and do flow from the aquatic environment to the terrestrial, and back from the terrestrial to the aquatic environment. One major way that phosphorus moves from the ocean into terrestrial ecosystems is through the droppings of sea-going birds. The birds eat marine fishes that have absorbed phosphorus at sea. This phosphorus passes through the bird and is deposited on land in bird **guano**.

Guano
Seabird or bat dung.

Substances introduced into one ecosystem can appear in another, as we see in the movement of toxic chemicals through the biosphere. Arctic carnivores carry the remains of agricultural pesticides in their tissues, even though no agricultural pesticides were ever used in their ecosystem. These persistent pesticides have drifted north over the decades and accumulated in the fat of polar bears and other carnivores.

THE HYDROLOGIC CYCLE RECYCLES WATER THROUGH THE ENVIRONMENT

Perhaps the best-studied of the chemical cycles is the **hydrologic cycle**, or water cycle. The reservoir for water on Earth is the world's oceans, where nearly 98 percent of all water is found (FIGURE 20.19, p. 688). The remaining 2.15 percent is freshwater. Glaciers and ice caps contain 1.7 percent of the Earth's water, almost all of the freshwater! Groundwater is the next largest pool, holding 0.4 percent of total water. Surface water, such as rivers and lakes, contain a mere 0.04 percent of total water, and the atmosphere holds only 0.01 percent. As you can see, the water available for life, human and otherwise, is an almost vanishingly small percentage of the total water. Wasting this water by leaving it running while brushing your teeth, or watering your lawn after a rain, is not only unmindful, but environmentally risky. (For more on this topic, see the Ethics and Issues box on p. 688.)

How bad is the freshwater crisis?

Aside from fresh air, nothing is more important than fresh water—for drinking, cleaning, and growing food. Yet fresh, clean water is scarce in many regions. An estimated 5 million people die due to shortage or lack of freshwater each year, mostly due to water-borne diseases.

A rising world population is placing more pressure on freshwater supplies. As a result of groundwater pumping, the giant Ogallala aquifer under the North American Great Plains has lost 6 percent of its capacity since 1940, with the worst declines in Texas and Kansas. California farmers are selling water rights to cities, which allows the cities to expand but can reduce farm output. In 2003, the federal government directed California to stop taking so much water from the Colorado River to give a larger share to the booming populations of Arizona and Nevada. But no relief is in sight for the once-rich wetlands near the Sea of Cortez in Mexico, at the Colorado's mouth. Almost no fresh water reaches these wetlands, an apparent violation of a 1944 treaty.

The freshwater shortage is more extreme in the long band of land reaching across North Africa, through the Middle East and Arabian Peninsula, into Pakistan and India, and ending in northern China. Despite conservation measures, Israel and adjacent countries use every bit of available freshwater. Egypt, where the population is increasing by more than 1 million per year, gets 97 percent of its water from the Nile River. The river's source is in Ethiopia, Uganda, and Sudan. Egypt has already threatened war if its water supply is endangered.

The Aral Sea in Central Asia is polluted with pesticides and has shrunk so much that fishing boats are grounded miles from the sea. Groundwater under the Punjab, a major grain-producing region of India, is falling by 1 meter per year. In North China, the water table is also dropping as ever-deeper wells pump water for a huge population enjoying a historic industrial expansion. Farmers are left with less water, and the grain harvest has been falling for years.

Future trends don't look promising. With many major rivers already tapped out, desalinizing ocean water is still expensive and energy-intensive. Conservation may be the best solution. In California, urban water districts are funding water-saving technology on farms; in return they receive the water that is saved. Raising the price of fresh water is a free-market approach to saving water. But in poor countries, high prices can force residents to choose between food and water. A water privatization scheme in Bolivia was dropped after major riots. In theory, selling water would restrain demand and raise money for piping improvements, but average Bolivians could not afford the water.

Water conservation is difficult, but living without water is impossible. If we are not careful, the twenty-first century could see water wars, environmental damage to wetlands, and a decline in living standards caused by poor sanitation and an impaired food supply.

Percentages of water on Earth FIGURE 20.19

Oceans	97.85%	Rivers and lakes	0.04%
Glaciers	1.7 %	Atmosphere	0.01%
Groundwater	0.4 %		

Water cycle Figure 20.20

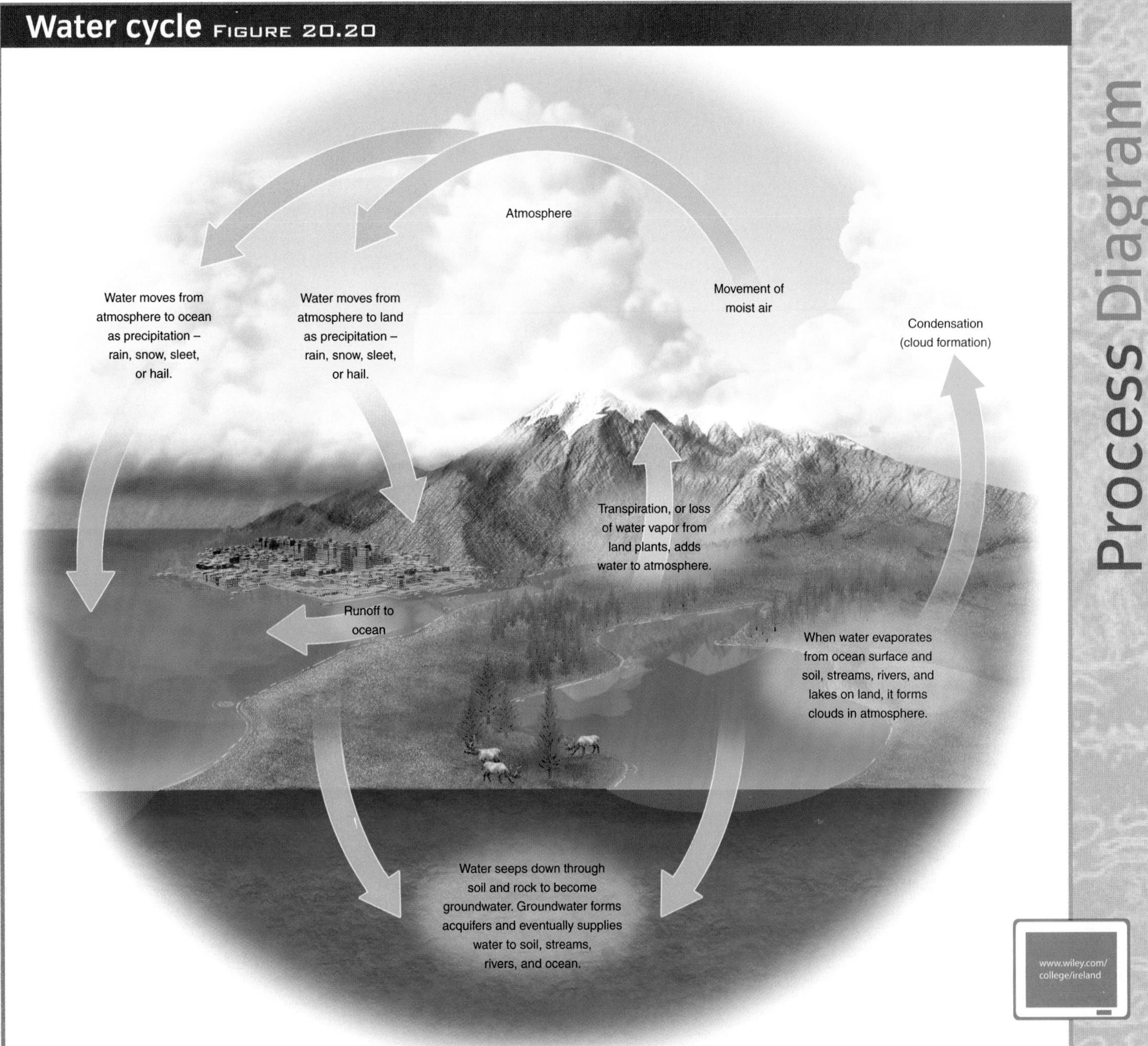

Atmosphere
Water moves from atmosphere to ocean as precipitation – rain, snow, sleet, or hail.
Water moves from atmosphere to land as precipitation – rain, snow, sleet, or hail.
Movement of moist air
Condensation (cloud formation)
Transpiration, or loss of water vapor from land plants, adds water to atmosphere.
Runoff to ocean
When water evaporates from ocean surface and soil, streams, rivers, and lakes on land, it forms clouds in atmosphere.
Water seeps down through soil and rock to become groundwater. Groundwater forms acquifers and eventually supplies water to soil, streams, rivers, and ocean.
www.wiley.com/college/ireland
Process Diagram

Water **evaporates** from the oceans and from freshwater and land. As temperatures increase, evaporation takes place more quickly. The water rises into the atmosphere as water vapor and condenses into clouds as the air and water vapor cool. The vapor continues to condense, eventually coalescing into drops that fall back to the land or ocean as **precipitation**.

Evaporates Changes from a liquid to a vapor through the addition of energy.

Because land is above sea level, and water naturally seeks the lowest level, surface water eventually returns to the sea. Precipitation over land can either run off into a river or percolate into **groundwater,** which flows underground at various depths. Groundwater saturates the sediment to a constant level called the **water table,** which is where our water wells must reach. A large body of groundwater is called an **aquifer**. Rainfall or snow melt can recharge an aquifer (Figure 20.20).

Phosphorus cycle FIGURE 20.21

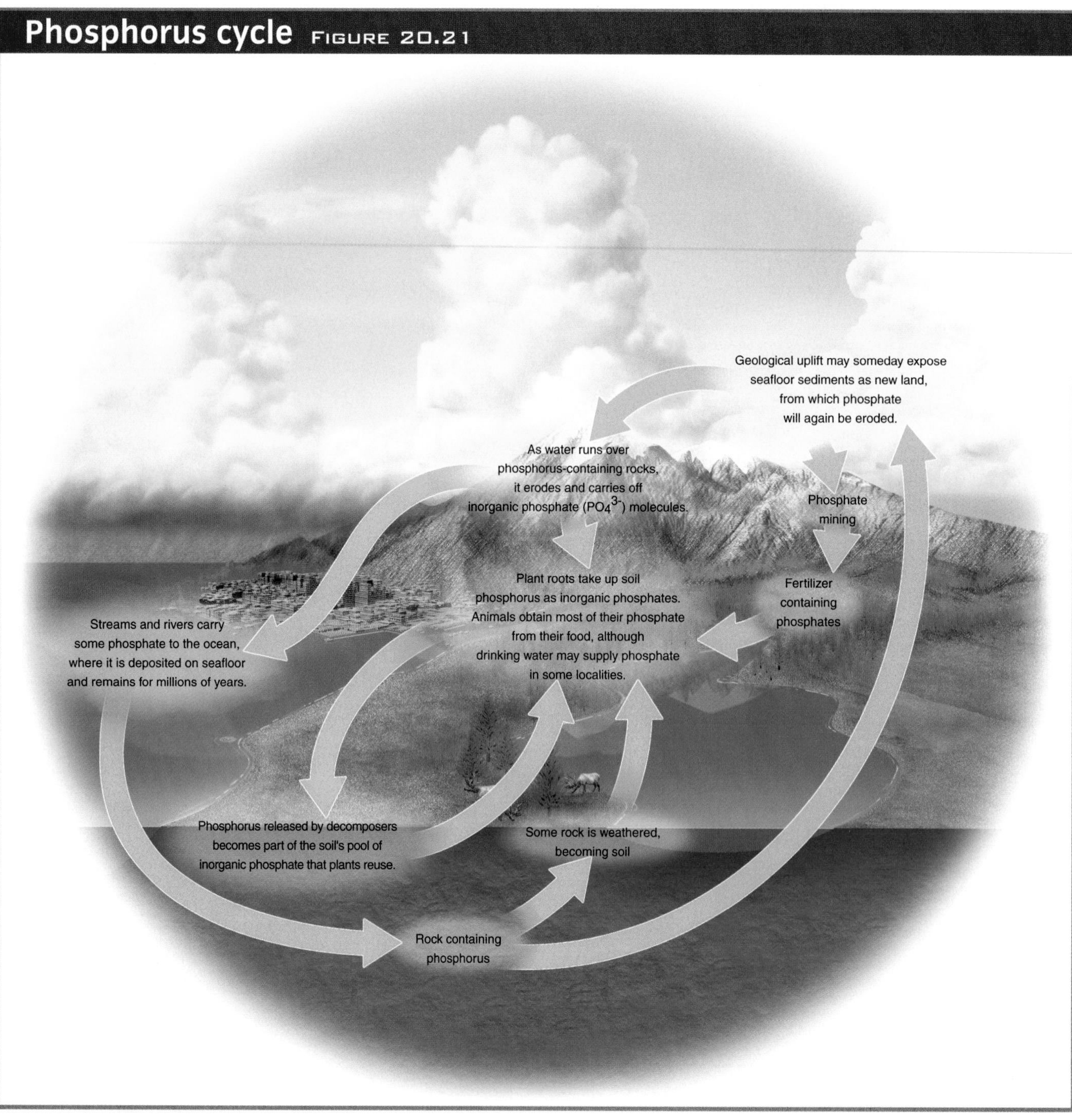

Geological uplift may someday expose seafloor sediments as new land, from which phosphate will again be eroded.
As water runs over phosphorus-containing rocks, it erodes and carries off inorganic phosphate (PO_4^{3-}) molecules.
Phosphate mining
Plant roots take up soil phosphorus as inorganic phosphates. Animals obtain most of their phosphate from their food, although drinking water may supply phosphate in some localities.
Fertilizer containing phosphates
Streams and rivers carry some phosphate to the ocean, where it is deposited on seafloor and remains for millions of years.
Phosphorus released by decomposers becomes part of the soil's pool of inorganic phosphate that plants reuse.
Some rock is weathered, becoming soil
Rock containing phosphorus

Populated areas often get their drinking water from aquifers. As the human population increases, many aquifers are being drained faster than their recharge rate. This is called **groundwater mining**, and it is lowering the water table and will eventually result in a loss of water to these areas. The problem is particularly acute in many parts of China, India, and in the North American Great Plains, where the Ogallala Aquifer has been drained by years of pumping for irrigation.

THE PHOSPHORUS CYCLE IS A SEDIMENTARY CYCLE

While the water cycle includes a large atmospheric portion, the phosphorus cycle (FIGURE 20.21) is mainly a sedimentary cycle. When some rocks weather, they release phosphate ions into the soil. Plants take up these phosphates through their roots and use them to create phospholipid bilayers, ATP, and DNA or RNA nucleotides. As the plants are eaten, the phosphates move into higher trophic levels. In animal tissue, phosphorus is incorporated into teeth, shell, and bones as well as ATP, cell membranes, and nucleotides. When organisms decay, the phosphorus returns to the soil. Phosphate may run off the land in rivers, and be either absorbed by phytoplankton and seaweeds or lost into sediment. Only when there is an upwelling of bottom sediment will this phosphorus return to the biosphere.

The availability of phosphorus is often the limiting factor in the growth of algae, which can have profound effects on the health of the entire community. In 1998, thirteen Florida residents suddenly became ill with skin lesions, nausea, diarrhea, and neurological problems. Their symptoms resembled those in a nearby fishing community, and among scientists and tourists in Maryland and North Carolina, which were blamed on a toxic single-celled algal bloom. How could people in both Florida and Maryland become ill from a microscopic alga normally found in the water? In the waters off Florida, phosphates and nitrates are exceptionally high. In Florida, 400 million gallons of treated sewage is injected underground daily through 120 wells. The effluent from these wells seeps into the nearby ocean, carrying a high concentration of phosphorus, nitrogen, and other nutrients. Low levels of these nutrients in the oceanic environment makes them an extrinsic control on population growth. As more nutrients become available, populations expand. One species that quickly takes advantage of this increased resource is *Pfiesteria piscicida*, an alga that produces a toxin responsible for the symptoms listed above.

Eutrophying
Encouraging blooms of plants and algae that eventually deplete the resources of a body of water, leading to the destruction of that ecosystem.

The hazards of nutrient dumping have been recognized since the mid-1970s. Laundry detergents used to include phosphates as a cleaning aid, but the excess phosphates in wastewater were **eutrophying** aquatic environments. Although phosphates are banned from detergents in many places, fertilizers are loaded with phosphates and other nutrients. Runoff from farms, golf courses, and lawns that use fertilizer includes high levels of nutrients, creating problems for aquatic ecosystems. Sewage effluent and farm manure runoff continue to cause trouble. Diligent controls on water treatment plants and runoff can reduce nutrient pollution. As with many environmental concerns, greater public awareness will go a long way toward alleviating the problems.

NITROGEN CYCLES BETWEEN THE SOIL AND THE ATMOSPHERE

The majority of the world's nitrogen is in the atmosphere, where nitrogen comprises 78 percent of air's volume. Despite this large reservoir, plants are often starved for nitrogen because they cannot use the molecular nitrogen in the atmosphere. Plants can only use nitrogen that has been "fixed" in the soil. Fixing reduces nitrogen molecules to **ammonia** (NH_3) or **nitrates** (NO_3^-). Nitrogen fixing is a three-step process. First, one set of bacteria converts atmospheric nitrogen to ammonia; then a second set converts that ammonia in the soil to **nitrites** (NO_2^-). Nitrate-producing bacteria in the soil and plant **root nodules** then convert nitrite to **nitrate**. Once in plant tissues, nitrates are converted to ammonium ions (NH_4^+), which are found in amino acids and nucleic acids. (See FIGURE 20.22 on p. 692.)

Root nodules
Swellings on the root hairs of legumes and other plants containing nitrogen-fixing bacteria.

Nitrogen cycle Figure 20.22

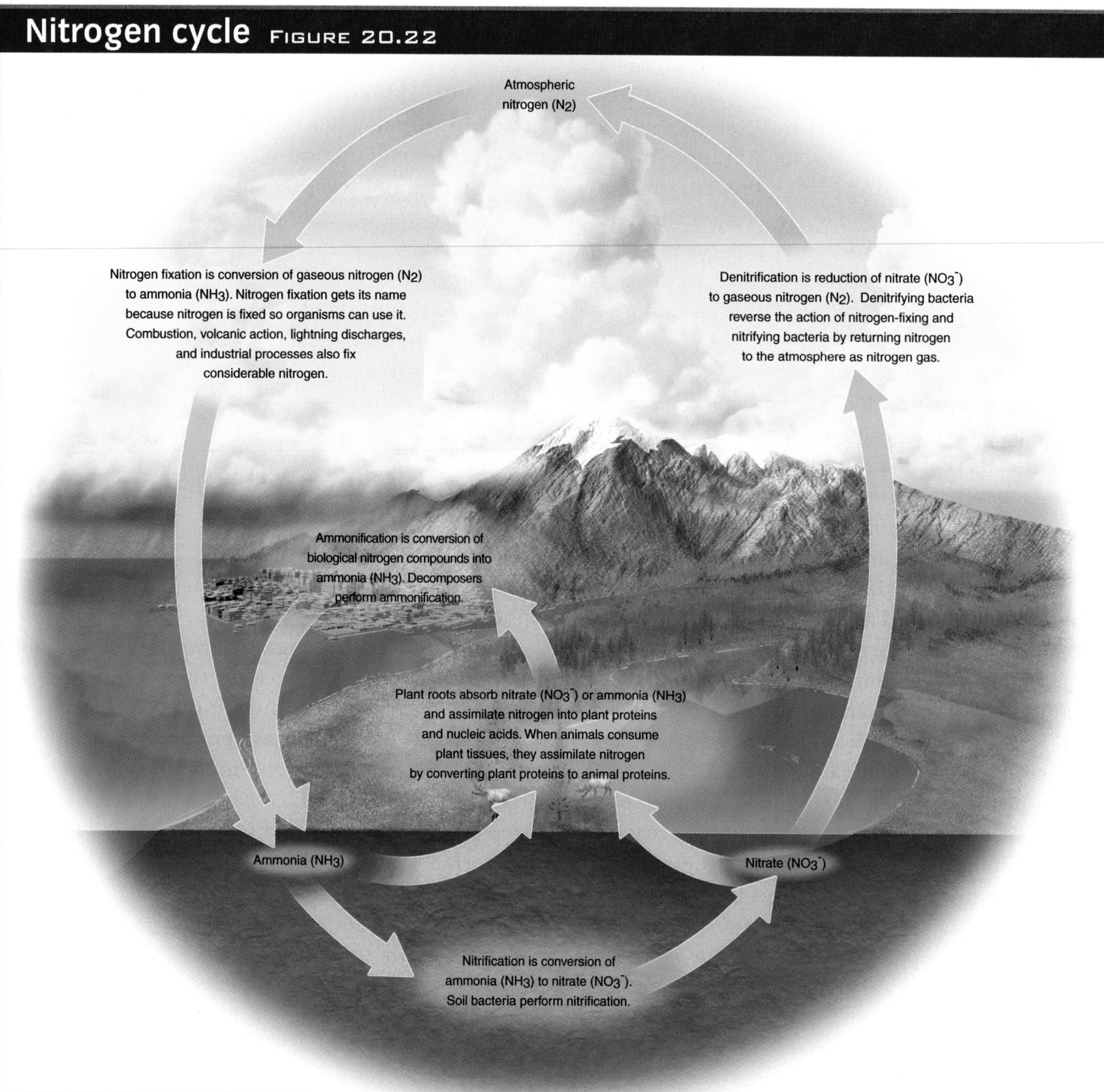

Atmospheric nitrogen (N_2)
Nitrogen fixation is conversion of gaseous nitrogen (N_2) to ammonia (NH_3). Nitrogen fixation gets its name because nitrogen is fixed so organisms can use it. Combustion, volcanic action, lightning discharges, and industrial processes also fix considerable nitrogen.
Denitrification is reduction of nitrate (NO_3^-) to gaseous nitrogen (N_2). Denitrifying bacteria reverse the action of nitrogen-fixing and nitrifying bacteria by returning nitrogen to the atmosphere as nitrogen gas.
Ammonification is conversion of biological nitrogen compounds into ammonia (NH_3). Decomposers perform ammonification.
Plant roots absorb nitrate (NO_3^-) or ammonia (NH_3) and assimilate nitrogen into plant proteins and nucleic acids. When animals consume plant tissues, they assimilate nitrogen by converting plant proteins to animal proteins.
Ammonia (NH_3)
Nitrate (NO_3^-)
Nitrification is conversion of ammonia (NH_3) to nitrate (NO_3^-). Soil bacteria perform nitrification.

Nitrification occurs when soil bacteria convert ammonia to nitrite and then nitrate in the soil. **Denitrification** is the reverse process, whereby nitrates are converted to nitrous oxide and nitrogen gas, which reenter the atmosphere. Before humans started manufacturing fertilizer, denitrification and nitrogen fixation were balanced at the ecosystem level. Now excess nitrates are being introduced via fertilizer runoff, adding to our water pollution troubles.

CARBON IS FOUND ALMOST EVERYWHERE

Carbon also cycles through the biosphere. Large carbon reserves are found in many places: oceans, plants, animals, soil, and geologic formations. Moving through these reservoirs, carbon follows either a **short-term** or a **long-term** cycle (Figure 20.23).

The short-term cycle involves the interactions between the oceans and the biosphere, and the land plants and animals and the atmosphere. This carbon is

Carbon cycle FIGURE 20.23

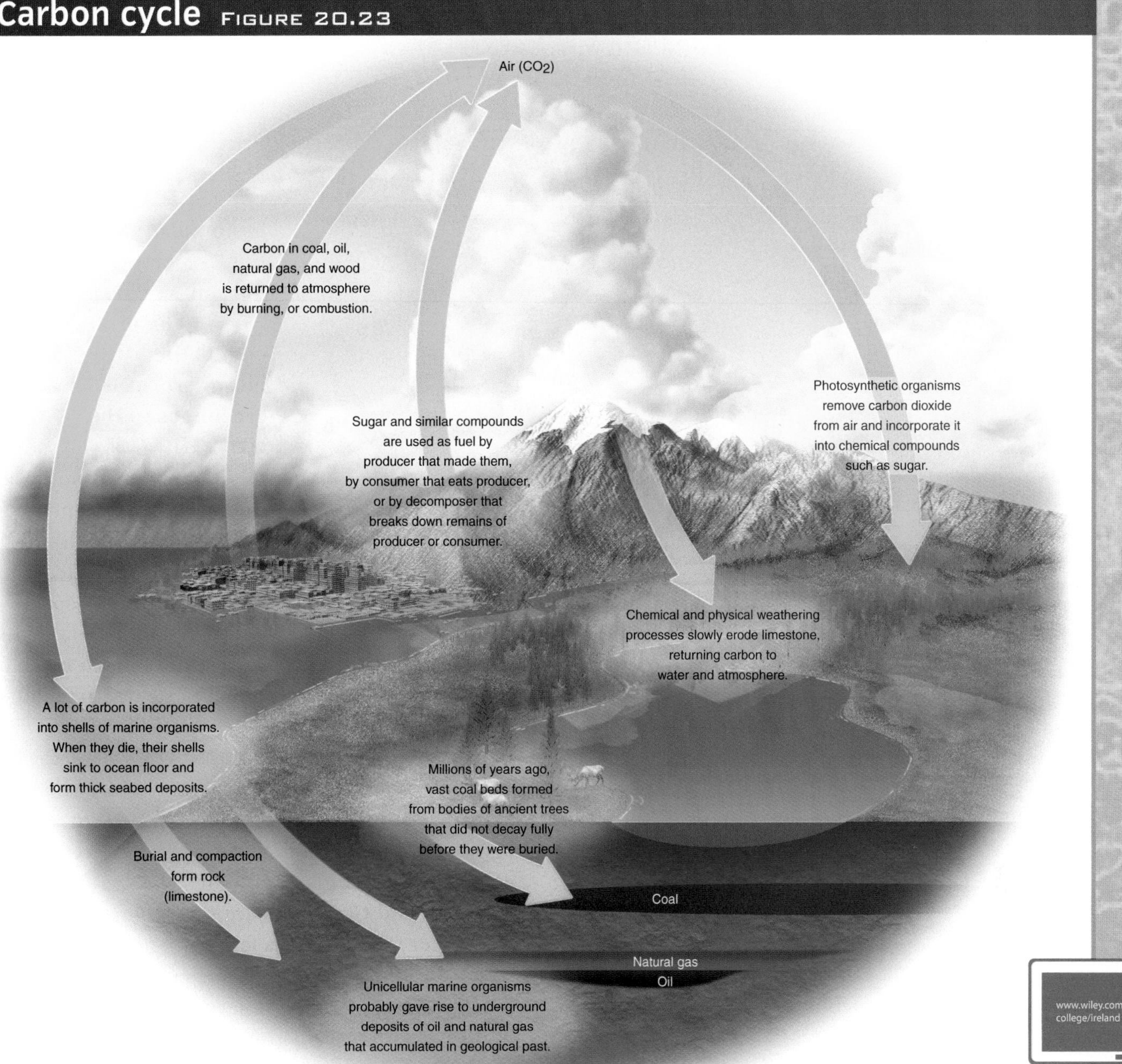

tied directly to the activities of living organisms. As you know, carbon is taken up during photosynthesis and released during cellular respiration. The rate of removal from the atmosphere by terrestrial plants is about equal to the rate of return through cellular respiration. The same opposing processes of photosynthesis and cellular respiration also work in the ocean. However, the carbon in the ocean must first diffuse from the atmosphere into the water. Once it enters the water column, carbon is removed via photosynthesis by primary producers, and released again by cellular respiration in the bodies of producers and consumers. A small amount of carbon is lost to ocean-floor sediments, returning to the water column only during upwelling. The amount of carbon in the water column is relatively constant, maintained by constant diffusion with the atmosphere. If aquatic carbon levels increase, more is released into the air.

Air (CO_2)
Carbon in coal, oil, natural gas, and wood is returned to atmosphere by burning, or combustion.
Sugar and similar compounds are used as fuel by producer that made them, by consumer that eats producer, or by decomposer that breaks down remains of producer or consumer.
Photosynthetic organisms remove carbon dioxide from air and incorporate it into chemical compounds such as sugar.
Chemical and physical weathering processes slowly erode limestone, returning carbon to water and atmosphere.
A lot of carbon is incorporated into shells of marine organisms. When they die, their shells sink to ocean floor and form thick seabed deposits.
Millions of years ago, vast coal beds formed from bodies of ancient trees that did not decay fully before they were buried.
Burial and compaction form rock (limestone).
Coal
Natural gas
Oil
Unicellular marine organisms probably gave rise to underground deposits of oil and natural gas that accumulated in geological past.
Process Diagram
www.wiley.com/college/ireland

Carbon also cycles through the sediment, entering the soil community as dead organisms and animal waste. The living biota contains a staggering 800 billion tons of organic carbon, with an additional 1,000

Mean Annual Air-Sea Flux for 2000 (NCEP II Wind, 2,800K, Γ=.27)

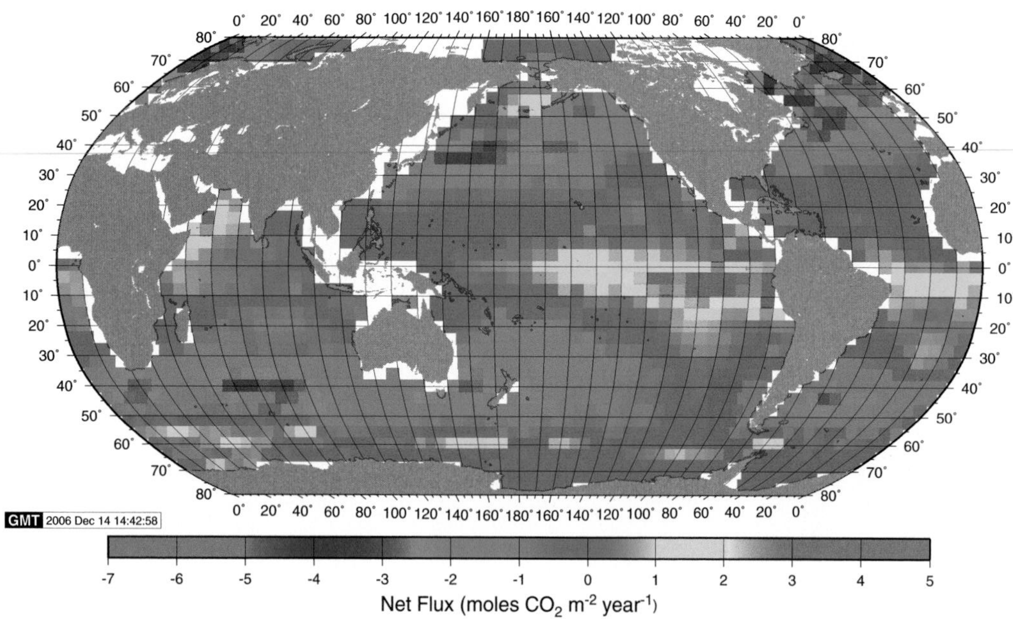

Satellite view of carbon sinks FIGURE 20.24

Carbon is held in the biosphere in organisms as well as in the sediment and atmosphere. Those areas of large carbon stores are called carbon sinks. This satellite image indicates the areas where carbon is concentrated on the Earth.

Fossil fuels Energy source derived from organic matter stored in hydrocarbon deposits.

to 3,000 billion tons held in dead or decaying matter in the soil (FIGURE 20.24). These either undergo immediate decay, releasing carbon into the atmosphere, or become **fossil fuels**.

Fossil fuels are part of the carbon cycle

Fossil fuels are the basis of the long-term organic carbon cycle. Under the proper geochemical conditions, decaying organic matter is converted to coal, oil, or natural gas. These fossil fuels are collected and burned for transportation and heat. As humans continually remove and oxidize carbon from the fossil fuel reservoir, the amount of carbon in each pool is shifting. In the last 20 years, there has been a substantial increase in the level of carbon dioxide in the atmosphere, equivalent to approximately 42 billion metric tons of carbon. Not only are we adding carbon to the atmosphere, but we are reducing the carbon store in plants as we burn rain forests. This reduction in plant material in turn reduces the amount of photosynthesis, exacerbating the shifting carbon levels. With fewer photosynthetic organisms, less carbon is being removed from the atmosphere.

The problem with accumulating carbon in the atmosphere centers on the **greenhouse effect** (FIGURE 20.25). Carbon dioxide, methane, and other gases in the atmosphere capture heat radiating from Earth's surface. Normally, a great deal of this heat escapes into space. But more greenhouse gases in the atmosphere mean more heat remains near Earth, raising the average temperature. In the past century, near-Earth temperatures have risen 0.6°C. This may not sound like much, but most scientists agree: the dangers of global warming are real. Sea levels are rising as glaciers melt. The ice caps on the Antarctic and Greenland are showing signs of instability, further raising the sea level.

Scientists who study ancient climates have seen rapid changes just over a decade or two, proving that climate is not a steady-state affair, but a dynamic phenomenon that can change quickly. After many years of discussion and research, scientists are almost unanimous in their assessment. The climate is warming, and almost certainly the result of the increase in human activity. Global warming is likely to make hurricanes and droughts more intense. We could see more wildfires and more deaths due to heat waves. Diseases are moving into new areas. Malaria, for example, is moving into the highlands, which are now warm enough for malarial mosquitoes. Changes in temperature and rainfall could devastate farmlands, but perhaps open other areas to the plow.

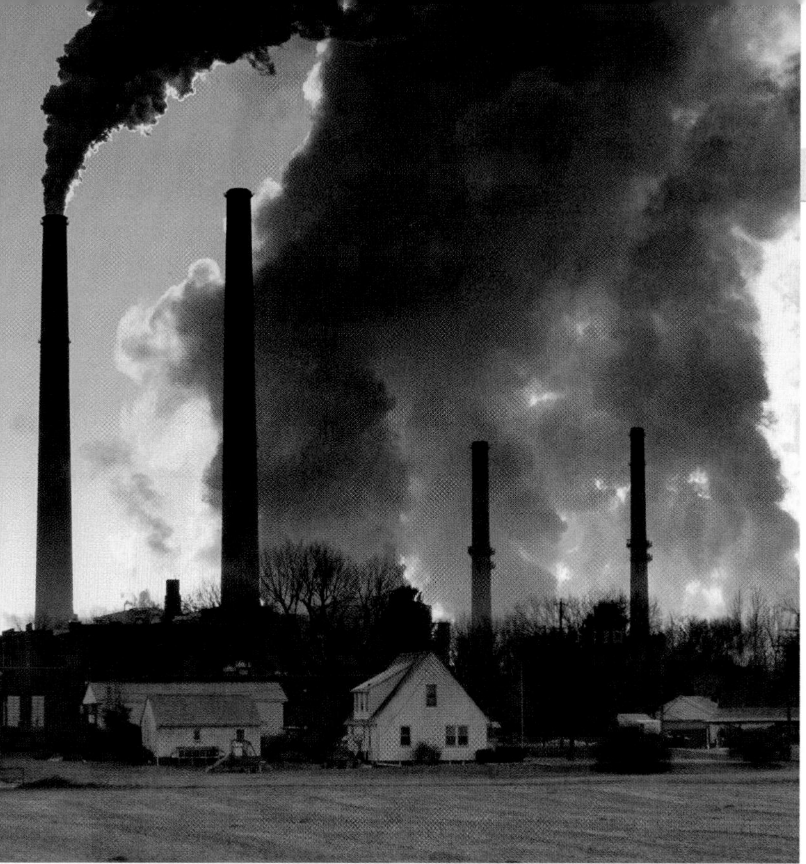

Greenhouse effect FIGURE 20.25

Gases emitted from industrial stacks includes carbon dioxide, methane, and other heat-capturing compounds. The heat radiated back from the Earth is caught in these gases, remaining close to the Earth's surface, just as the panes of glass trap the heat in a greenhouse. This in turn elevates the Earth's temperature.

CONCEPT CHECK

Describe the movement of energy through ecosystems.

Sketch out an example of a food chain and label the trophic levels.

Why are there usually no more than five trophic levels in a food chain or web?

Describe a typical biogeochemical cycle.

Compare the phosphorus cycle with the nitrogen cycle. Are the reservoirs the same in both cases?

How is the carbon cycle related to the greenhouse effect?

Humans Have a Tremendous Impact on the Environment

LEARNING OBJECTIVES

Summarize the effects of humans on the biosphere.

Understand the origins of smog and acid rain.

Relate eutrophication to water pollution.

Describe the process of biomagnification.

List three human activities that decrease biodiversity.

Although it seems harsh to view human beings as a plague or a weed on the Earth, there is an element of truth to that description. Humans do not interact with the environment like other animals. When a large number of humans populate an area, they alter the landscape to suit their needs. Rather than die out due to lack of resources, the population continues to increase and cities spring up where once plants and animals lived. They use up local resources and then take from surrounding areas. Carrying capacity seems to have no meaning to humans. Even in less-industrialized countries, humans are altering the environment to suit their own needs. And as the human population grows and industrialization increases, many observers of the global environment are pessimistic that we will be able to solve our problems before they overwhelm us.

AGRICULTURAL PRACTICES AND "CIVILIZED" USE OF OUR RESOURCES

Monoculture
The practice of planting a single species over large tracts of land.

Humans living in one place tend to drastically alter the vegetation, often planting only one or two species of food crops. A drive through the Midwestern Corn Belt will demonstrate this. **Monoculture** agriculture abounds—there is literally nothing but corn for miles and miles!

What does this do to the ecosystem? Originally, the Midwest was short-grass and tall-grass prairie, with hundreds of species of grasses and wildflowers. Insect populations were diverse, occupying myriad niches in the prairie. Larger animals were also represented by a good number of species, including buffalo. When plant diversity is decreased, the ecosystem's ability to support diverse animal life decreases as well. With the same primary producer on acre after acre, ecosystem diversity declines, and so does the resilience of the ecosystem. That one crop year after year pulls a specific set of nutrients from the soil, whereas many different plants remove and replace different nutrients, allowing the soil to maintain its diversity and health. Also, diseases or insects that attack the dominant species could wipe out the entire crop, further reducing the diversity of the area.

Of course, hunger is a basic human drive, one that we must constantly work to satisfy, and monoculture does provide vast quantities of food. Perhaps we can find a smarter, more environmentally sound way to produce food to meet our growing needs? Currently, some agricultural research is focusing on the quest to produce quality food with minimal environmental devastation; this is called sustainable agriculture.

WATER AND AIR POLLUTION ARE HUMAN HEALTH ISSUES

Other ecosystem damage comes from widespread pollution of water, soil, and air. Air pollution includes anything suspended in the atmosphere that decreases the quality of life for those organisms breathing it. Polluted air causes problems in the respiratory tracts of organisms, either by adding particles that clog or damage respiratory membranes, or by creating compounds that otherwise harm the body's tissues. Water pollution deprives us of water's usual benefits. Adding too many chemicals to water prevents us from using that water for industrial or personal needs. Soil can also be contaminated with persistent toxic compounds. These get into the soil through human activities and include salts, pesticides, radioactive materials, or biological factors such as pathogenic bacteria or viruses. Finally, simple garbage can also be an environmental problem (see the I Wonder . . . box).

In many parts of the globe, legislatures have attempted to curb the destruction of the environment by passing laws limiting air and water pollution. For example, a 1987 international forum called the Montreal Protocol began the phaseout of CFCs (**chlorofluorocarbons**). CFCs were used in air conditioners and refrigerators until scientists discovered that they deplete the **ozone layer** after being released into the atmosphere (FIGURE 20.26).

Chlorofluorocarbons
Compounds made of hydrogen, carbon, fluorine, and chlorine, once used as refrigerants.

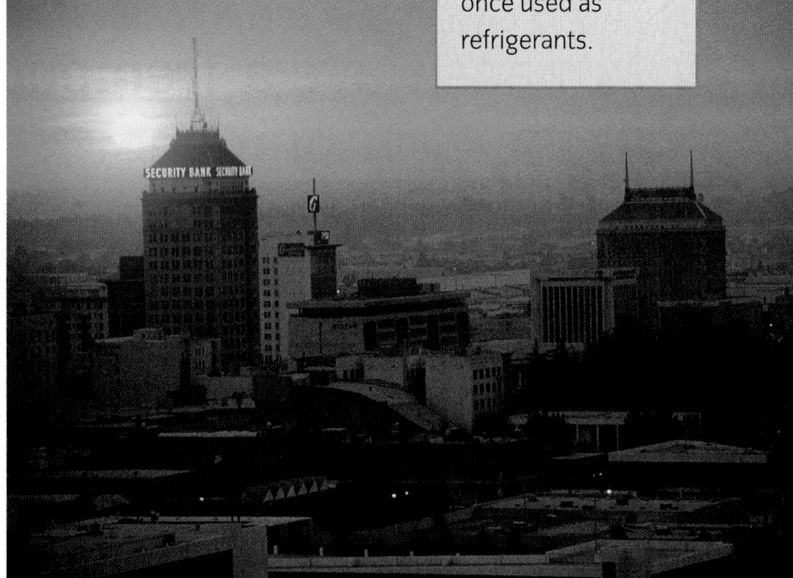

Ozone layer FIGURE 20.26

Ozone, or O_3, provides a protective layer of UV absorption in our upper atmosphere. Damage to this layer causes thinning of the O_3, which in turn allows more dangerous ultraviolet (UV) rays to reach the Earth.

I WONDER . . .

Where does all the garbage go?

For centuries, Americans have found place after place to put a garbage dump. But a growing population and continued economic prosperity have translated into an ever-growing stream of products—and garbage. In 2003, the daily per capita solid waste production in the United States reached 4.5 pounds, up from 2.7 pounds in 1960.

The bury-and-forget approach to garbage came under fire in the 1970s, when plumes of groundwater pollution were detected streaming away from garbage dumps. In 1978, the massive chemical contamination from Love Canal in New York State awakened people to the fact that the dumps used to dispose of household and toxic wastes eventually leak into the groundwater, which provides much of our drinking water. Unfortunately, the containment systems that the Environmental Protection Agency required in 1991 to protect groundwater will also "ultimately fail," as the agency has acknowledged, thereby only postponing instead of preventing pollution. At the same time, municipalities have had more difficulty finding landfill sites, and the cost of disposal is starting to soar.

These factors have created a fertile ground for recycling, especially as energy prices soared after the two oil crises of the 1970s. Why throw out an aluminum can when it takes so much electrical energy to refine the aluminum to replace it? In the 1980s, recycling started to play a major role in reducing the amount of garbage. By 2003, recycling rates ranged from 22 percent for glass bottles to 93 percent for auto batteries. Beyond reducing the need for landfills, recycling can reduce the impact of mining or logging to produce raw material, and also cut fuel use and air pollution.

Recycling can take many forms:

- Composting converts organic material, such as food waste or lawn trimmings, into a soil amendment. Composting reduces waste volume and retains precious nutrients. A second form of composting can be used to dispose of leftovers from municipal sewage treatment: The solid matter remaining can be tested for contaminants and composted into an excellent, odorless fertilizer for farm fields.
- Curbside recycling programs gather plastic, metal, glass, newspaper, and cardboard for reuse.
- Construction recycling handles massive amounts of material. Road builders commonly reuse asphalt and concrete in new roads. In addition, millions of tons of coal ash produced by electric generators can be made into building material, such as concrete blocks.

Other countries have their own approaches to recycling. In Europe, small appliance manufacturers are required to take their product back at the end of its lifetime, which creates a strong incentive to manufacture products that are easy to recycle. In Japan, a conservation ethic combined with much higher population density, has led to extremely high recycling rates.

Here in the United States, there is much more that can be done. In 2003, we generated 236 million tons of municipal solid waste and recycled only 72 million tons of that. Promoting markets for recycled materials and buying recycled material can both help raise the incentive to recycle. Another way to reduce the garbage glut is to urge manufacturers to reduce their packaging and to make products that can be repaired and will last longer before needing disposal. So could stopping to think before buying that next trinket. Do you really need it?

www.wiley.com/college/ireland

Ozone (O_3) in the upper atmosphere blocks harmful ultraviolet radiation that can cause cancer and other problems in the biosphere. CFCs released near the Earth's surface rise to the **troposphere** where the chlorine destroys ozone. In the early 1990s, a giant hole in the ozone over Antarctica raised the prospect of widespread biological damage through UV radiation. Since the CFC phaseout began, this hole has begun to recede. CFCs are stable and will remain in the atmosphere for a long period, but the episode does show that global action can slow or reverse a clear environmental threat.

Troposphere
The portion of the atmosphere immediately above the Earth's surface.

Air pollution includes **smog** and **acid rain**. Smog is a general term for nitrogen oxides and hydrocarbons that, in sunlight, turn a brown or gray and form smog. Smog contains **PAN (peroxyacetyl nitrates)** and ozone, both of which irritate mucous membranes in the eyes and respiratory tract (FIGURE 20.27). Ozone is helpful in the upper atmosphere, but closer to the ground it can make serious diseases like asthma and emphysema even worse. Regulations on automotive and industrial emissions have reduced smog in some countries, but the problem is by no means solved. TABLE 20.4 ranks the most polluted major American cities, using suspended particulate matter as the indicator of pollution level.

Acid rain is caused by the release of compounds that can convert to acids in the atmosphere (FIGURE 20.28). Nitrogen oxides from auto emissions are easily converted to nitric acid, for example. The largest source of acid rain is **sulfur oxides** (SO_2 and SO_3) released from burning fossil fuels. These oxides combine with water va-

Smog caused by a temperature inversion FIGURE 20.27

We all know that warm air rises, but occasionally cold air can form a "lid," trapping warm air next to the Earth. If this happens over an industrialized area, smog produced in that area will not move. Smog ratings are provided in cities where this is a common occurrence, such as Los Angeles, California.

Metropolitan areas most polluted by year-round particle pollution

TABLE 20.4

Rank	Metropolitan areas	Rank	Metropolitan areas
1	Los Angeles–Long Beach-Riverside, CA	15	Canton–Massillon, OH
2	Visalia–Porterville, CA	16	Charleston, WV
3	Bakersfield, CA	17	Modesto, CA
4	Fresno–Madera, CA	18	New York, NY–Newark, NJ–Bridgeport, CT
5	Pittsburgh–New Castle, PA	18	Merced, CA
6	Detroit–Warren–Flint, MI	20	St. Louis–St. Charles, MO–Farmington, IL
7	Atlanta–Sandy Springs–Gainesville, GA	21	Washington, DC–Baltimore, MD–Northern Virginia, VA
8	Cleveland–Akron–Elyria, OH	22	Louisville, KY–Elizabethtown–Scottsburg, IN
9	Hanford–Corcoran, CA	22	Huntington, WV–Ashland, OH
9	Birmingham–Hoover–Cullman, AL	24	York–Hanover–Gettysburg, PA
11	Cincinnati, OH–Middletown, KY–Wilmington, IN	24	Lancaster, PA
12	Knoxville–Sevierville–La Follette, TN	24	Columbus–Marion–Chillicothe, OH
13	Weirton, WV–Steubenville, OH		
14	Chicago, IL–Naperville, IN–Michigan City, WI		

Source: http://www.lungusa.org/site/pp.asp?c=dvLUK900E&b=50752

Scrubbers Equipment in a smokestack that removes impurities from the escaping gas.

por and form sulfuric acid. The acid falls back to Earth as acid rain, damaging biotic and abiotic structures alike. Regulations in some places require industries to reduce sulfur dioxide emissions. **Scrubbers** reduce sulfur emissions from coal-fired electric generators, for example, and automobile manufacturers must meet minimum standards for tailpipe emissions.

The atmosphere is not the only resource that can be damaged by human activity. Water pollution is a serious threat. As mentioned, there is precious little available freshwater. We water crops, cool factory machinery and electric generators, and flush away feces and urine with freshwater, and all of these actions pollute the water. Water can be polluted by organic or inorganic nutrients, as well as toxic chemicals. Organic nutrients include compounds from sewage treatment plants, paper mills, and food-processing factories. Inorganic nutrients usually come from fertilizer runoff.

Acid rain FIGURE 20.28

Acid rain can cause widespread death of vegetation, and can lower the pH of lakes and ponds to the point where nothing can survive. The acidity of the water can also erode both synthetic and natural stone structures with amazing speed. These raindrops have a pH near 2, as shown by the red color of the indicator dye.

One consequence of water pollution is **eutrophication** (FIGURE 20.29). When nutrient levels skyrocket in a shallow body of freshwater, plant life multiplies rapidly. Eventually, the excess mass of producers will die, and the decomposers will work on the biomass. The death and decomposition of these plants reduces the water's oxygen concentration, killing fishes and invertebrates. Eutrophication is a natural process that converts ponds to wetlands and dry land, but human activity greatly enhances its rate.

Groundwater can be polluted as chemicals percolate from the surface into the groundwater. This is doubly troublesome, as groundwater serves as drinking water sources for the majority of humans. The slow **turnover** of groundwater means those pollutants will remain in that aquifer for many years.

Toxic pollutants can be particularly dangerous. **Biomagnification**, or the concentrating of toxins as they move up the food chain, can lead to serious health concerns (FIGURE 20.30). One dramatic example occurred in the 1960s, where 111 people died in Minimata, Japan, after eating fish containing high levels of mercury. How did this happen? Fish naturally carry low levels of mercury in their tissues, due to a small amount of mercury vapor in the atmosphere that becomes dissolved in water. Bacteria convert dissolved mercury to methyl mercury, which is more toxic. Fish concentrate methyl mercury in their tissues as water passes over

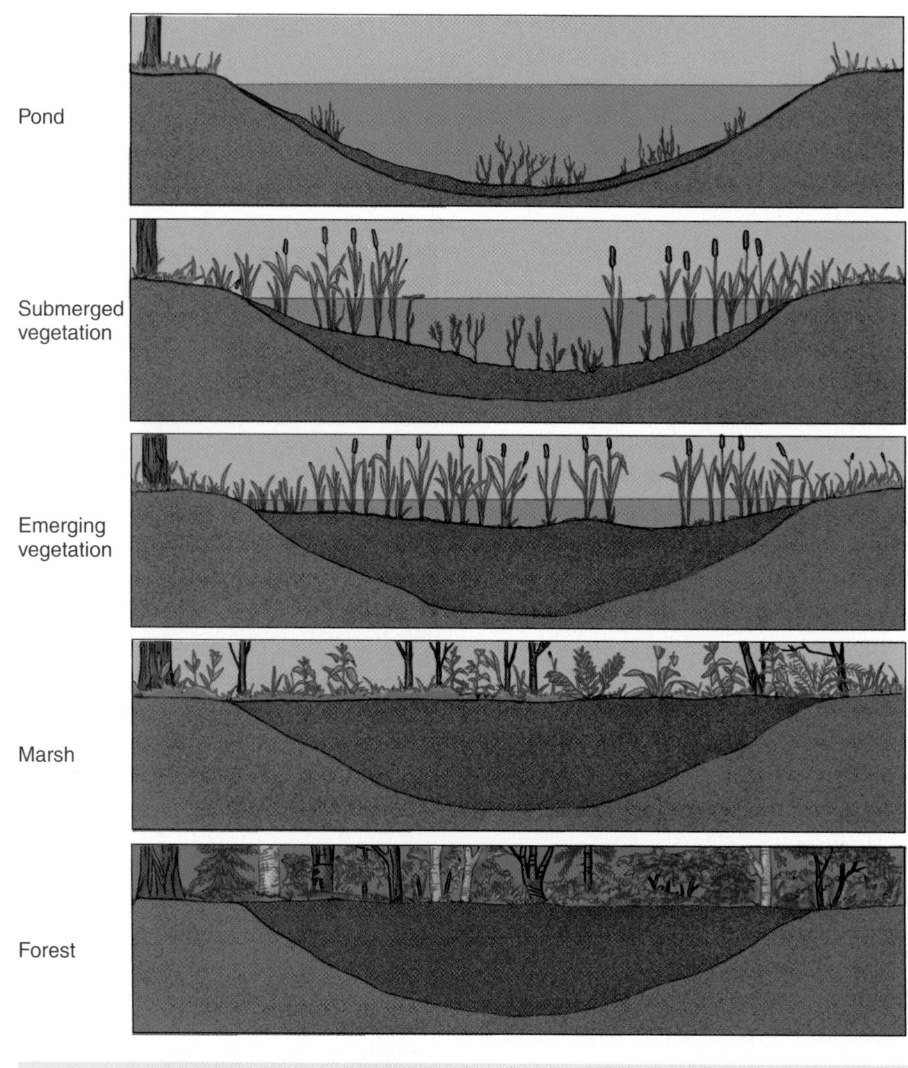

Eutrophication FIGURE 20.29

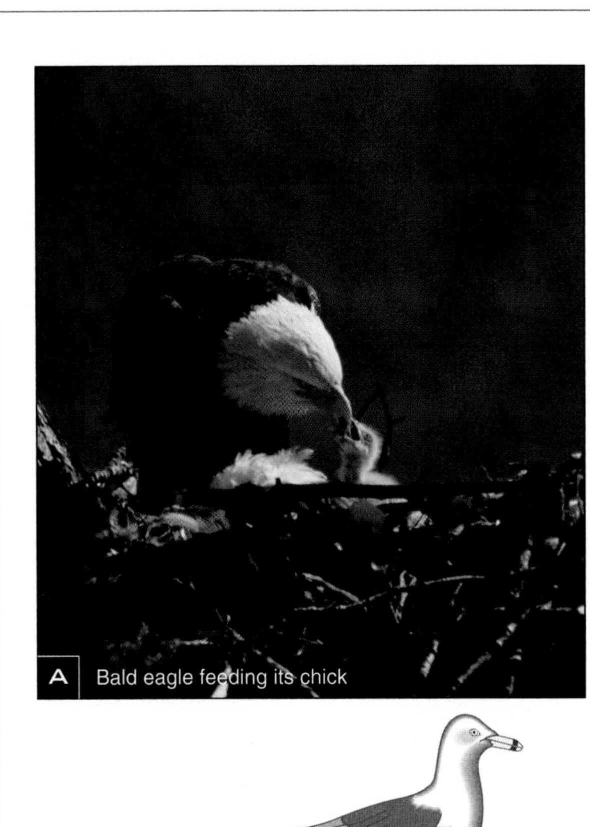

A Bald eagle feeding its chick

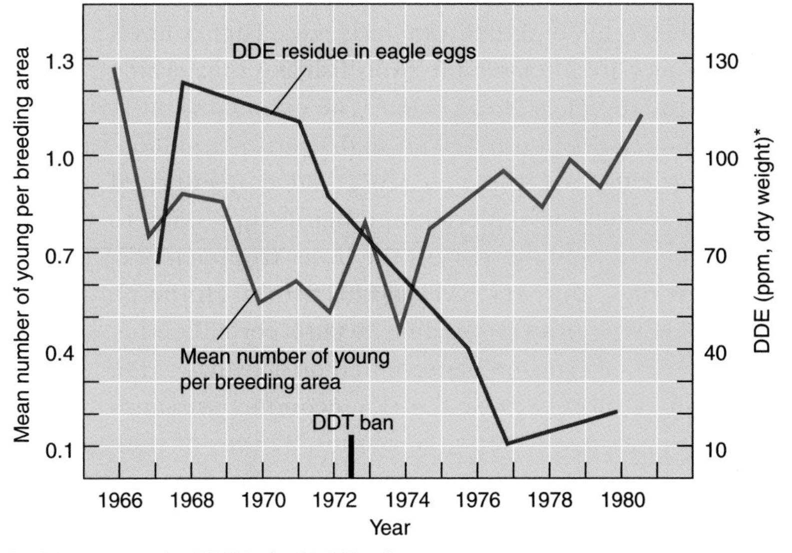

*DDT is converted to DDE in the birds' bodies.

B Effect of DDT ban on bald eagle offspring

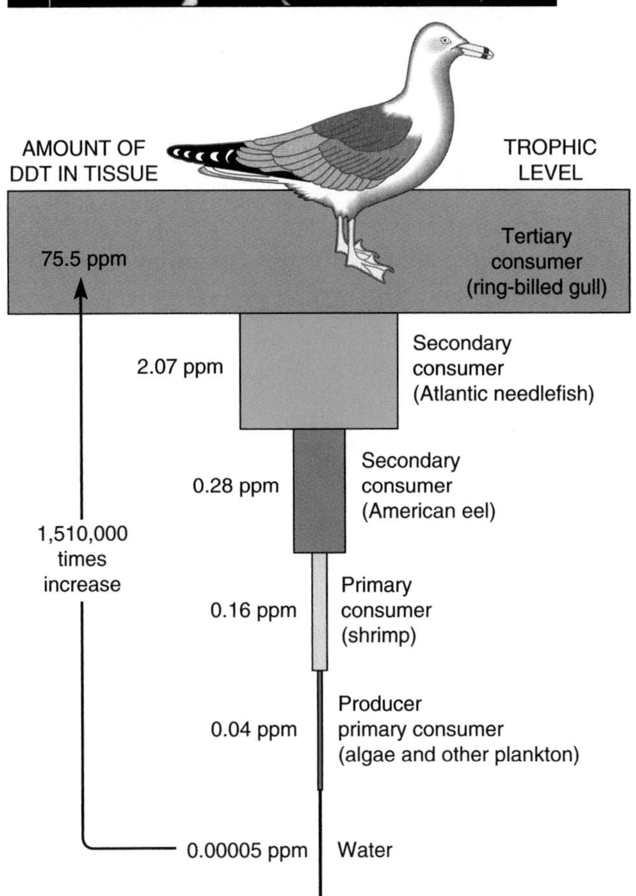

C Biomagnification of DDT in a salt marsh

Biomagnification FIGURE 20.30

One of the most well-understood examples of biomagnification is the story of DDT and the bald eagle. DDT was used to control insects and was sprayed with abandon during the summer to kill mosquitoes. Through biomagnification, the low level of sprayed DDT was magnified many times through the food chain. In the top predators, the bald eagles, DDT levels were alarmingly high. The worst effect of high DDT in eagles and other birds was the inability to produce a thick, healthy eggshell. Fertilized eggs broke in the nest, and the bird populations plunged. After the ban on DDT in 1972, the eagle populations have slowly recovered.

their gills. The absorbed methyl mercury binds to proteins. When larger fish eat the smaller fish, they ingest methyl mercury and their own level rises. Higher levels of mercury are often seen in swordfish and tuna, which are at the top of the food chain. The usual concentration of open-ocean fish is 0.001 to 0.5 parts per million. In Minimata, the top predators had concentrated alarming amounts of mercury in their tissues owing to small increases in mercury at the lower trophic levels. The average mercury level of the fish in Minimata ranged from 9 to an astounding 40 parts per million!

This discussion has centered on activities of the U.S. population because the U.S. economy uses and, some say, abuses so many resources: 22 tons of fuel, metals, minerals, food and forest products per person, per year. All of these materials are used to construct cities, suburbs, and roads, to drive long distances in our 215 million cars and trucks; we also allow soil erosion in our one-crop farming system, and divert waterways to suit our needs. Taking all this into account, some scientists have estimated that the typical American consumes 88 tons of resources per year.

Americans are not alone in their use of resources. For example, in 2001, Asia consumed 2,370,000 thousand metric tons of fossil fuel. During the same year, the Middle East consumed 560,000 thousand metric tons, Europe consumed 3,038,000 thousand metric tons, and North America consumed 2,157,000 thousand metric tons. Another resource that the world uses at a tremendous rate is paper and paper products. In 2004, 103,861,000 metric tons of paper were used in Asia; 7,260,000 tons in the Middle East and North Africa, 94,191,000 tons in Europe, and 101,058,000 tons by the North American population.

Around the world, people are moving into urban areas that are rapidly expanding to take over the countryside, while some rural areas are being depopulated or running deeper into poverty. In the United States, **urban sprawl** is eating up valuable farmland as we turn rural areas into housing divisions and bedroom communities. Eventually, as the population rises, we could regret the day we decided that farmland was ideal for growing subdivisions.

Biodiversity is a measure of species richness, measured at any one of three levels. **Genetic diversity** is the variation in individuals in a population. Genetic diversity promotes reproductive success and is the raw material for adaptation. **Species diversity** is the number of species alive today. Taxonomists believe there are at least 30 million species on Earth, but some estimates run up to 80 million. **Community diversity** measures the diverse forms of life living together in a community.

Human activity tends to decrease biodiversity, as we alter landscapes and force animals and plants to move or die out. Removing habitat through farming, construction, mining, or recreation can cause extinction. When we pollute the environment, overfish, or overhunt animals, we cause extinction. We can even cause extinction by bringing new species to an environment. These organisms may have no natural predators in the new ecosystem and can thus outcompete native organisms. Introduced species often take over large parts of ecosystems, preying on **endemic** and **indigenous** species or otherwise destroying the natural balance of the ecosystem.

Endemic
Unique to a region.

Indigenous
Native to a region.

Hawaii and other remote islands are currently fighting the introduction of new species, which have an especially easy time colonizing remote ecosystems. One good example comes from the introduction of the brown tree snake (*Boiga irregularis*) to Guam between 1945 and 1952. The snake probably arrived as a stowaway on a ship. With no natural predators and an abundance of easy prey, the brown tree snake population boomed, eating birds, lizards, bird and reptile eggs, even pets. The brown tree snake populations grew so large that villages would suffer power outages as the weight of snakes basking on power lines caused them to snap. Eventually, the snake exterminated most of the small vertebrate populations of the forests in Guam, including birds and mammals. To this day the island is eerily quiet due to the lack of birds. Control efforts are underway, as are strict monitoring programs on neighboring islands, where this snake could cause similar havoc. In Hawaii, the brown tree snake would cause an ecological disaster similar to the one it caused in Guam. The Hawaiian Islands are already fighting ecological invaders like the little fire ants, coqui frogs, and miconia plants. Each of these invaders is capable of harming the Islands' precious biodiversity by outcompeting endemic species and pushing them toward extinction.

Although oceanic islands are an extreme example, species extinctions are worldwide problems. In the last century, 10 percent of the 297 known mussel and clam species and 40 of about 950 freshwater fishes in North America have disappeared. Plants and animals of the rain forests, as well as coral reefs, are also threatened. While biologists cannot predict exactly how accelerating global warming and global growth will affect extinctions, most expect the problem to intensify.

Due mainly to human activity, the current extinction rate is estimated to be 100 to 1,000 times higher than in recent history. Dr. Donald Levin, a researcher at the University of Texas at Austin, claims that one additional species becomes extinct every 20 minutes. He predicts that within 200 to 300 years Earth will lose at least half of all animal and plant species. We could be in the early stages of a sixth major extinction, since the current rate is higher than that at any time except during the five mass extinctions in geological history. This is reason for concern.

The ecological future is not pretty. An increasing population needs more food. We abuse our resources and produce more waste. Some steps toward **sustainability** include better waste treatment, land and water conservation, planned community growth, reduced population growth, smarter use of energy, and more recycling. These steps may seem small, but they can move us in the right direction. The primary reason for hope is this: Human beings are phenomenally creative and inventive!

Sustainability The wise exploitation of resources and energy, to ensure resources for future generations.

LIFE ON EARTH GOES ON

Throughout this book, we have studied the processes and concepts of biology through the human organism. We began with what it means to be alive, and we gave a short explanation of the classification of humans. From there, we progressed through biochemistry, cells, tissues, and organ systems. We discussed the notion of homeostasis and described how major body processes continuously attempt to restore balance. Later, we explored how the human animal fits into the history of life and into the environment.

As your understanding of life processes increased, we added ecological and evolutionary ideas. Social consciousness was an underlying theme throughout the chapters, as ethical dilemmas regarding the effects and consequences of science and technology abound in all cultures. Hopefully, as you complete this book, you are feeling better prepared to make sound political and social decisions based on facts. Your decisions and opinions are firmly grounded in understanding and should not be swayed by partial or misleading arguments, whether they come from friends, family, politicians, or the media.

CONCEPT CHECK

What are the environmental effects of our modern agricultural practices?

What is the primary cause of acid rain?

Explain how the brown tree snake affected biodiversity in Guam.

What can be done to prevent the loss of biodiversity on a global scale? On a local scale? In your own yard?

CHAPTER SUMMARY

1 Ecosystems Define Plants and Animals Living Together

Ecology is the study of organisms and their interactions with one another and the environment. Individuals of the same species make up a population. The many populations in an area define a community. Ecosystems are composed of all the communities in a larger geographical area. The entire Earth is referred to as the biosphere when discussed in ecological terms. Ecosystems can be defined by the dominant organisms and are grouped in nine categories of biomes. When a biome is new, organisms fill the available niches in predictable patterns. Primary succession occurs on newly created land, while secondary succession takes place after a devastating fire, or another disaster.

2 Organisms Have Specific Habitats and Niches

Organisms live in a particular habitat and occupy a specific niche. No two organisms can occupy the same niche at the same time. Because of these specific requirements, an organism's biogeographic range can be predicted and mapped. When organisms compete for resources, they can partition the resource, so more organisms can use it without direct competition.

3 Carrying Capacity and Population Growth Are Regulated by the Environment

Mortality rates indicate the health of a population. When carrying capacity is reached, organisms cannot obtain needed resources and the population declines. As the population dips below carrying capacity, organisms find resources more easily, and the population again increases. This oscillation is natural.

Growth curves of biotic potential can start steeply and then level off, or start out flat and then increase with age. Malthusian strategists have many offspring and tend to overshoot carrying capacity. Mortality curves can show three trends. Type I organisms have few offspring but provide extensive care for them. Type II organisms are apt to perish at any point in their life cycle. Type III organisms show a high infant mortality rate, with a subsequent leveling of mortality in adulthood.

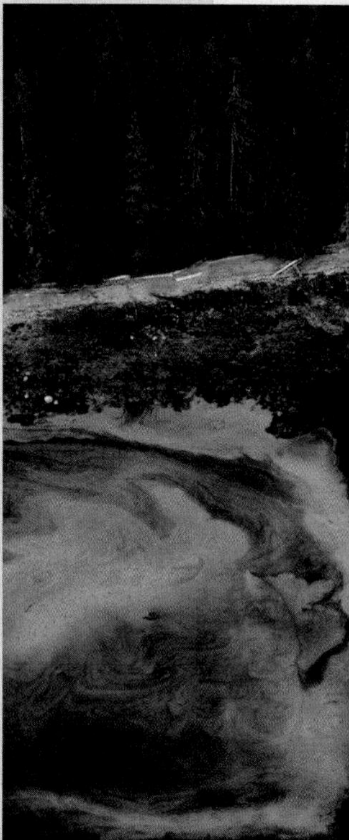

4 Organisms Are Either Producers or Consumers

Photosynthesis is the production of sugars from energy, water, and carbon dioxide. The opposite process is respiration, which releases energy, carbon dioxide, and water by burning glucose and other sugars. Primary producers photosynthesize or (much more rarely) chemosynthesize (use chemical energy) macromolecules. Both producers and consumers respire, releasing and using the energy stored in glucose to do work for the cell. These processes are related in a food chain or food web. At each step up in a food chain, 90 percent of the energy is lost. Most food chains can sustain a maximum of four or five trophic levels.

5 Energy Flows through an Ecosystem while Chemicals Cycle

Energy flows through an ecosystem, while nutrients usually cycle within it. Energy is constantly lost as heat and it is gained from sunlight. Nutrients cycle through a gaseous or a sedimentary cycle. Carbon, water, nitrogen, and phosphorus are the main nutrients that ecologists study. Phosphorus is mainly stored in the ground, while nitrogen and water have a strong atmospheric component. The reality of global warming, caused largely by carbon dioxide accumulation in the atmosphere, has focused special attention on the carbon cycle.

6 Humans Have a Tremendous Impact on the Environment

Humans upset the natural balance of ecosystems every day. We pollute the waters, add particle pollution to the air, and speed the process of eutrophication in freshwater lakes and ponds. We destroy biodiversity and speed the loss of endemic and indigenous species. These challenges will not go away with wishful thinking alone.

KEY TERMS

- **biogeographic ranges** p. 671
- **biome** p. 668
- **chlorofluorocarbons** p. 696
- **chlorophyll** p. 679
- **climax communities** p. 670
- **consumers** p. 676
- **cyanobacteria** p. 676
- **detritus** p. 677
- **dominant population** p. 669
- **endemic** p. 702
- **energy** p. 678
- **eutrophying** p. 691
- **evaporates** p. 689
- **fixing** p. 677
- **fossil fuels** p. 694
- **guano** p. 687
- **hydrolyzed** p. 679
- **indigenous** p. 702
- **metabolize** p. 682
- **monoculture** p. 696
- **partitioning** p. 672
- **photopigment** p. 678
- **phytoplankton** p. 674
- **pioneer species** p. 670
- **producers** p. 676
- **root nodule** p. 691
- **scrubbers** p. 699
- **sustainability** p. 703
- **trophic levels** p. 683
- **troposphere** p. 698
- **zooplankton** p. 684

CRITICAL THINKING QUESTIONS

1. Describe your population, your community, and your ecosystem. What is your habitat? Your niche? How and with whom do you partition the resources you need?

2. Relate the survivorship curves in FIGURE 20.9 (see p. 675) to the concepts of density-dependent and density-independent factors. Which type of factor contributes to the mortality of each type of organism?

3. Diagram the process of photosynthesis. Using the products shown in your diagram, continue the flow to show how respiration is related to photosynthesis.

4. Some people believe that being a vegan is the most responsible way to get nutrients while still maintaining the balance of the ecosystem. Explain this rationale, giving solid information that supports or discredits this claim.

5. List three ways in which humans can negatively impact the environment. What can you do in your life to leave behind a healthier and more productive ecosystem?

SELF TEST

1. The proper order of terms, from most inclusive (largest grouping) to least inclusive (smallest grouping) is:
   a. biosphere → ecosystem → community → population → individual.
   b. ecosystem → population → community → biosphere → individual.
   c. population → community → biosphere → ecosystem → individual.
   d. community → population → ecosystem → biosphere → individual.

2. The graph below shows that
   a. prey populations are uncontrolled and grow exponentially.
   b. predator populations are always smaller than prey populations.
   c. predator population trends mirror prey population trends, with some delay.
   d. there is no relationship between predator and prey population sizes.

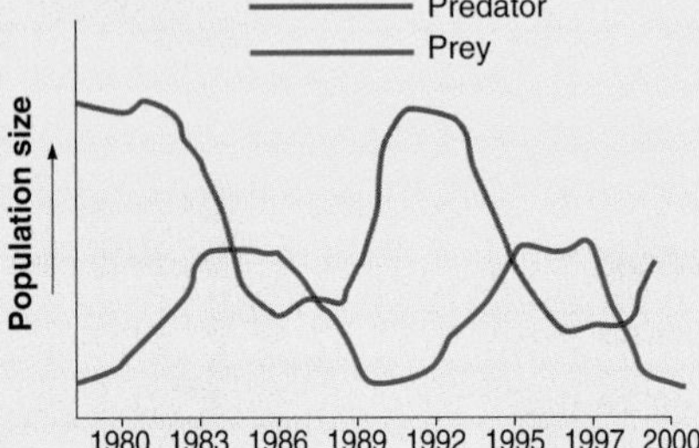

3. Which of the following would NOT be included in a farm-town community?
   a. People
   b. Farm animals
   c. Farm crops
   d. Soil type

4. The biome pictured here is the
   a. temperate grasslands.
   b. tundra.
   c. tropical rainforest.
   d. chaparral.

5. The biome with the least amount of yearly rainfall is the
   a. savanna.
   b. chaparral.
   c. desert.
   d. tundra.

6. A clear-cut portion of the rainforest that is abandoned and allowed to return to the natural state is an example of
   a. primary succession.
   b. climax community.
   c. secondary succession.
   d. partitioned succession.

7. True or False? The plant life seen in the image below is a pioneer species.

8. The ___________ of the raccoon includes temperate forests, suburban communities, and streams. It also includes fresh water and trees available for nesting.
   a. habitat
   b. niche
   c. biogeographical range

9. The two organisms in this image are
   a. sharing one niche.
   b. actively competing, in which case one will be the winner, and the other will have to find a new home.
   c. involved in competitive selection.
   d. partitioning the resource to avoid direct competition.

10. Populations that reproduce quickly, produce many offspring but offer little or no parental care, and overshoot the carrying capacity of the environment are called
    a. Malthusian strategists.
    b. logistic strategists.
    c. type I survivorship curve organisms.
    d. type II survivorship curve organisms.

11. The organisms that are responsible for fixing the organic compounds used by the rest of the food chain are referred to as
    a. consumers.
    b. cyanobacteria.
    c. producers.
    d. heterotrophs.

12. The final link (highest-ranking energy utilizer) in the energy web of any biome is occupied by
   a. producers.
   b. tertiary and, rarely, quaternary consumers.
   c. detritivores.
   d. omnivores.

13. The reaction in the figure below is occurring
   a. both during the day and at night.
   b. in the mitochondrion.
   c. on the choroplast inner membrane, the thylakoid.
   d. in animal cells only.

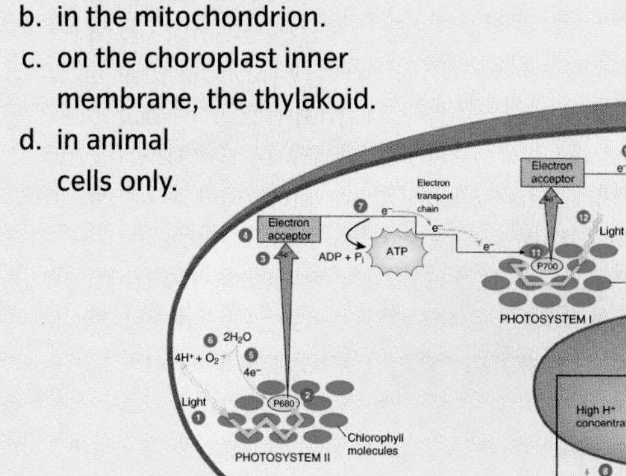

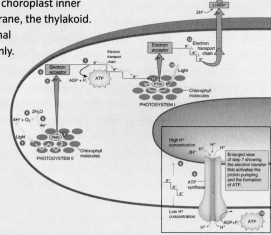

14. Photosynthesis is biochemically related to cellular respiration because
   a. they both require oxygen in order to begin.
   b. they both occur in the chloroplast.
   c. photosynthesis produces the materials needed for respiration and vice versa.
   d. both require ATP and produce sugars.

15. The diagram below indicates that if the fish were removed from the food web,
   a. annelids and man would necessarily die out.
   b. mollusk and arthropod populations would increase in size.
   c. algae would die out.
   d. humans would move to a more reliable food source.

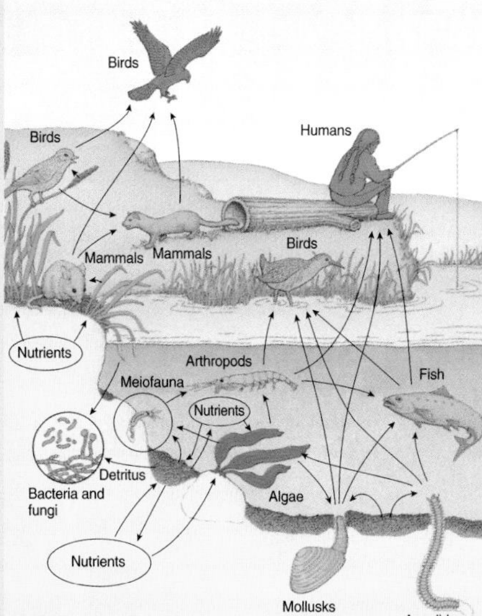

16. The phosphorus cycle is an example of a/an ___________ cycle.
   a. gaseous
   b. sedimentary
   c. open
   d. exchange pool

17. Water is found in greatest abundance in the
   a. atmosphere.
   b. glaciers.
   c. rivers and groundwater.
   d. oceans.

18. The diagram below illustrates the natural process of
   a. ecology.
   b. eutrophication.
   c. climax community development.
   d. deforestation.

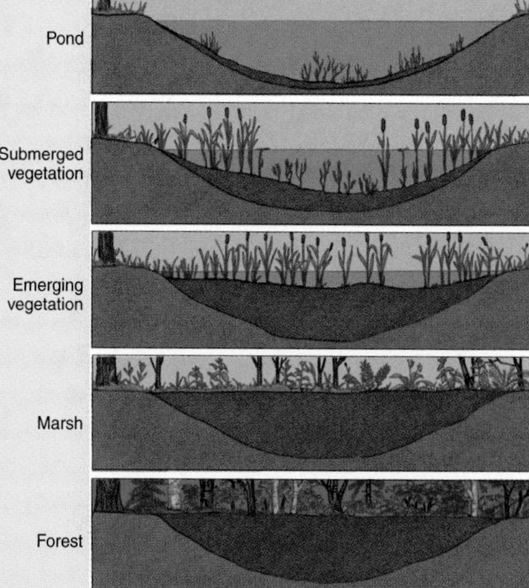

19. When a chemical compound is found in trace quantities in the environment but becomes more and more concentrated in the flesh of the animals living there, we say ___________ is occurring.
   a. evolution
   b. succession
   c. biomagnification
   d. bioluminescence

20. Which of the following activities is most in agreement with the practices of sustainability?
   a. running factories 24 hours a day, 7 days a week to produce more goods
   b. monocultural farming practices
   c. diet practices of a strict vegan-type vegetarian
   d. reducing population growth by closely controlling reproductive behavior

APPENDIX A *Periodic Table*

The periodic table lists the known **chemical elements,** the basic units of matter. The elements in the table are arranged left-to-right in rows in order of their **atomic number,** the number of protons in the nucleus. Each horizontal row, numbered from 1 to 7, is a **period.** All elements in a given period have the same number of electron shells as their period number. For example, an atom of hydrogen or helium each has one electron shell, while an atom of potassium or calcium each has four electron shells. The elements in each column, or **group,** share chemical properties. For example, the elements in column IA are very chemically reactive, whereas the elements in column VIIIA have full electron shells and thus are chemically inert.

Scientists now recognize 113 different elements; 92 occur naturally on Earth, and the rest are produced from the natural elements using particle accelerators or nuclear reactors. Elements are designated by **chemical symbols,** which are the first one or two letters of the element's name in English, Latin, or another language.

Twenty-six of the 92 naturally occurring elements normally are present in your body. Of these, just four elements—oxygen (O), carbon (C), hydrogen (H), and nitrogen (N) (coded blue)—constitute about 96% of the body's mass. Eight others—calcium (Ca), phosphorus (P), potassium (K), sulfur (S), sodium (Na), chlorine (Cl), magnesium (Mg), and iron (Fe) (coded pink)—contribute 3.8% of the body's mass. An additional 14 elements, called **trace elements** because they are present in tiny amounts, account for the remaining 0.2% of the body's mass. The trace elements are aluminum, boron, chromium, cobalt, copper, fluorine, iodine, manganese, molybdenum, selenium, silicon, tin, vanadium, and zinc (coded yellow).

23 ← Atomic number
V ← Chemical symbol
50.942 ← Atomic mass (weight)

Percentage of body mass
- 96% (4 elements)
- 3.8% (8 elements)
- 0.2% (14 elements)

	IA	IIA	IIIB	IVB	VB	VIB	VIIB	VIIIB			IB	IIB	IIIA	IVA	VA	VIA	VIIA	VIIIA
1	1 Hydrogen H 1.0079																	2 Helium He 4.003
2	3 Lithium Li 6.941	4 Beryllium Be 9.012											5 Boron B 10.811	6 Carbon C 12.011	7 Nitrogen N 14.007	8 Oxygen O 15.999	9 Fluorine F 18.998	10 Neon Ne 20.180
3	11 Sodium Na 22.989	12 Magnesium Mg 24.305											13 Aluminum Al 26.9815	14 Silicon Si 28.086	15 Phosphorus P 30.974	16 Sulfur S 32.066	17 Chlorine Cl 35.453	18 Argon Ar 39.948
4	19 Potassium K 39.098	20 Calcium Ca 40.08	21 Scandium Sc 44.956	22 Titanium Ti 47.87	23 Vanadium V 50.942	24 Chromium Cr 51.996	25 Manganese Mn 54.938	26 Iron Fe 55.845	27 Cobalt Co 58.933	28 Nickel Ni 58.69	29 Copper Cu 63.546	30 Zinc Zn 65.38	31 Gallium Ga 69.723	32 Germanium Ge 72.59	33 Arsenic As 74.992	34 Selenium Se 78.96	35 Bromine Br 79.904	36 Krypton Kr 83.80
5	37 Rubidium Rb 85.468	38 Strontium Sr 87.62	39 Yttrium Y 88.905	40 Zirconium Zr 91.22	41 Niobium Nb 92.906	42 Molybdenum Mo 95.94	43 Technetium Tc (99)	44 Ruthenium Ru 101.07	45 Rhodium Rh 102.905	46 Palladium Pd 106.42	47 Silver Ag 107.868	48 Cadmium Cd 112.40	49 Indium In 114.82	50 Tin Sn 118.69	51 Antimony Sb 121.75	52 Tellurium Te 127.60	53 Iodine I 126.904	54 Xenon Xe 131.30
6	55 Cesium Cs 132.905	56 Barium Ba 137.33		72 Hafnium Hf 178.49	73 Tantalum Ta 180.948	74 Tungsten W 183.85	75 Rhenium Re 186.2	76 Osmium Os 190.2	77 Iridium Ir 192.22	78 Platinum Pt 195.08	79 Gold Au 196.967	80 Mercury Hg 200.59	81 Thallium Tl 204.38	82 Lead Pb 207.19	83 Bismuth Bi 208.980	84 Polonium Po (209)	85 Astatine At (210)	86 Radon Rn (222)
7	87 Francium Fr (223)	88 Radium Ra (226)		104 Rutherfordium Rf (261)	105 Dubnium Db (262)	106 Seaborgium Sg (263)	107 Bohrium Bh (264)	108 Hassium Hs (269)	109 Meitnerium Mt (268)	110 Uun (281)	111 Uuu (272)	112 Uub (277)		114 Uuq (289)				

57–71, Lanthanides

57	58	59	60	61	62	63	64	65	66	67	68	69	70	71
Lanthanum	Cerium	Praseodymium	Neodymium	Promethium	Samarium	Europium	Gadolinium	Terbium	Dysprosium	Holmium	Erbium	Thulium	Ytterbium	Lutetium
La	Ce	Pr	Nd	Pm	Sm	Eu	Gd	Tb	Dy	Ho	Er	Tm	Yb	Lu
138.91	140.12	140.907	144.24	144.913	150.35	151.96	157.25	158.925	162.50	164.930	167.26	168.934	173.04	174.97

89–103, Actinides

89	90	91	92	93	94	95	96	97	98	99	100	101	102	103
Actinium	Thorium	Protactinium	Uranium	Neptunium	Plutonium	Americium	Curium	Berkelium	Californium	Einsteinium	Fermium	Mendelevium	Nobelium	Lawrencium
Ac	Th	Pa	U	Np	Pu	Am	Cm	Bk	Cf	Es	Fm	Md	No	Lr
(227)	232.038	(231)	238.03	(237)	244.064	(243)	(247)	(247)	242.058	(254)	257.095	258.10	259.10	260.105

APPENDIX B *Measurements*

U.S. Customary System

Parameter	Unit	Relation to Other U.S. Units	SI (Metric) Equivalent
Length	inch	1/12 foot	2.54 centimeters
	foot	12 inches	0.305 meter
	yard	36 inches	9.144 meters
	mile	5,280 feet	1.609 kilometers
Mass	grain	1/1000 pound	64.799 milligrams
	dram	1/16 ounce	1.772 grams
	ounce	16 drams	28.350 grams
	pound	16 ounces	453.6 grams
	ton	2,000 pounds	907.18 kilograms
Volume (Liquid)	ounce	1/16 pint	29.574 milliliters
	pint	16 ounces	0.473 liter
	quart	2 pints	0.946 liter
	gallon	4 quarts	3.785 liters
Volume (Dry)	pint	1/2 quart	0.551 liter
	quart	2 pints	1.101 liters
	peck	8 quarts	8.810 liters
	bushel	4 pecks	35.239 liters

International System (SI)

Base Units

Unit	Quantity	Symbol
meter	length	M
kilogram	mass	Kg
second	time	S
liter	volume	L
mole	amount of matter	Mol

Prefixes

Prefix	Multiplier	Symbol
tera-	10^{12} = 1,000,000,000,000	T
giga-	10^{9} = 1,000,000,000	G
mega-	10^{6} = 1,000,000	M
kilo-	10^{3} = 1,000	k
hecto-	10^{2} = 100	h
deca-	10^{1} = 10	da
deci-	10^{-1} = 0.1	d
centi-	10^{-2} = 0.01	c
milli-	10^{-3} = 0.001	m
micro-	10^{-6} = 0.000,001	μ
nano-	10^{-9} = 0.000,000,001	n
pico-	10^{-12} = 0.000,000,000,001	p

Temperature Conversion

Fahrenheit (F) To Celsius (C)

$$°C = (°F - 32) \div 1.8$$

Celsius (C) To Fahrenheit (F)

$$°F = (°C \times 1.8) + 32$$

U.S To SI (Metric) Conversion

When you know	Multiply by	To find
inches	2.54	centimeters
feet	30.48	centimeters
yards	0.91	meters
miles	1.61	kilometers
ounces	28.35	grams
pounds	0.45	kilograms
tons	0.91	metric tons
fluid ounces	29.57	milliliters
pints	0.47	liters
quarts	0.95	liters
gallons	3.79	liters

SI (Metric) To U.S. Conversion

When you know	Multiply by	To find
millimeters	0.04	inches
centimeters	0.39	inches
meters	3.28	feet
kilometers	0.62	miles
liters	1.06	quarts
cubic meters	35.32	cubic feet
grams	0.035	ounces
kilograms	2.21	pounds

CHAPTER 1

1b. Low degree of organization **2d.** Cell **3b.** B **4c.** The cell **5b.** negative feedback systems

6b.

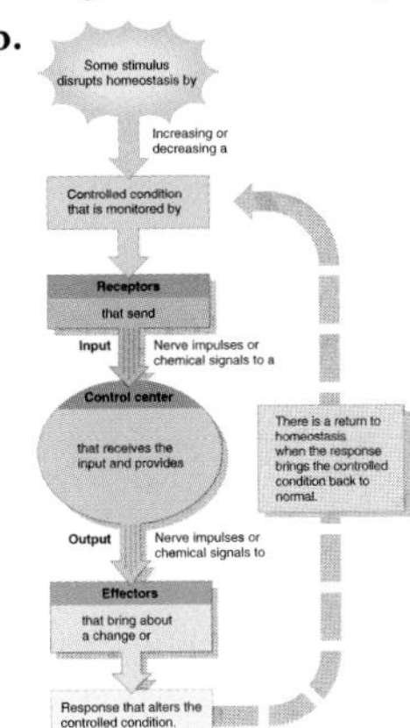

7b. Effector **8a.** Negative feedback **9e.** chemical level **10a.** Organism level **11d.** D **12b.** species **12s.** true **14.** true **15c.** inductive reasoning **16b.** observing **17c.** read and evaluate every article that you can find on the subject

CHAPTER 2

1e. both c and d **2c.** electron **3a.** positive **4c.** A **5b.** 6 **6c.** ionic **7c.** polar covalent bond **8d.** all of the above utilize positive / negative attraction **9b.** hydrophilic **10c.** it has a high specific heat **11a.** its polarity **12d.** 10,000 units **13a.** carbohydrate **14c.** polysaccharide **15b.** saturated **16c.** phospholipids **17a.** first level, the amino acid sequence **18d.** all of the above **19c.** Thymine **20d.** the last phosphate bond

CHAPTER 3

1b. cells cannot arise from preexisting cells **2b.** a structure within the cytosol that performs at least one vital cellular function **3d.** All of the above. **4.** true **5d.** sodium potassium ATPase **6c.** C **7d.** allowing cellular interaction with the aqueous environment of the body **8d.** D **9b.** expanding as water moves into the cell **10c.** exocytosis

11.

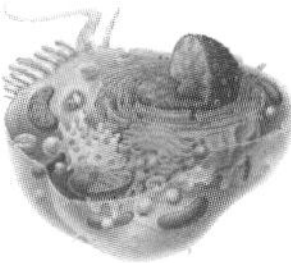

12d. digesting worn-out organelles **13e.** all of the above are correct **14a.** mitochondrion **15c.** cilia **16b.** Golgi complex **17.** true **18c.** local hormones **19c.** anaphase **20e.** interphase

CHAPTER 4

1c. areolar tissue **2a.** epithelial tissue **3d.** epithelial and some types of connective tissue **4c.** simple epithelium **5b.** a protective membrane **6a.** squamous epithelial cell **7c.** matrix **8d.** dense regular connective tissue **9b.** hyaline cartilage **10c.** adipose **11d.** smooth muscle **12b.** cardiac muscle **13b.** dendrites **14d.** two of the above are correct **15c.** tissue, organ, organ system, organism **16a.** superior **17c.** proximal **18d.** thoracic cavity **19c.** peritoneum **20a.** A

CHAPTER 5

1e. movement **2c.** the ribs **3a.** long **4e.** E **5.** true **6c.** osteoblasts **7b.** Haversian systems **8b.** epiphysis **9b.** osteoclast **10a.** fracture hematoma **11d.** sphenoid **12d.** all of the above **13c.** cervical vertebrae **14c.** vitamin D surplus **15a.** clavicle **16c.** metacarpal **17a.** male **18c.** patella **19a.** synarthrotic joint **20c.** flexion and extension

CHAPTER 6

1e. all the above are functions of the muscular system **2c.** the radius **3b.** agonist **4a.** epimysium **5d.** carry the impulse to contract quickly through the entire cell **6a.** sarcomere **7c.** C **8e.** both a and b are correct. **9b.** neurotransmitter being dumped into the neuromuscular synapse **10c.** troponin **11.** false **12c.** powerful contractions can only be generated in a very narrow range of sarcomere lengths **13c.** contraction **14a.** aerobic pathways **15a.** fast glycolytic fiber **16d.** flexion **17a.** rectus abdominus **18d.** triceps brachii **19b.** plantar flexion **20c.** shoulder

CHAPTER 7

1c. the neuron **2a.** afferent division of the PNS **3.** true **4d.** oligodendrocyte **5b.** sending and receiving motor information **6c.** voltage gated channel **7a.** −70 mV **8c.** sodium **9b.** absolute refractory period **10e.** both a and c are correct **11c.** oligodendrocyte **12.** false **13c.** arachnoid **14b.** CSF formation **15a.** brainstem **16b.** proprioception **17d.** limbic system **18a.** a highway for information traveling up and down the cord **19d.** sensory receptor, sensory neuron, spinal cord, motor neuron, effector organ **20a.** sympathetic division **21b.** increased respiratory and heart rate

CHAPTER 8

1d. proprioception **2e.** C and d are correct **3.** false **4c.** taste bud **5b.** block the passageway from mouth to nose **6a.** collect and transmit sound **7c.** dynamic equilibrium **8b.** B **9c.** F **10a.** hearing **11d.** all of the above **12b.** fibrous tunic **13b.** collect tears **14e.** retina **15b.** regulate the amount of light entering the eye **16c.** C **17a.** in front of the retina **18b.** a convex lens **19.** false **20d.** ganglionic neurons, bipolar neurons, rods and cones, back of eye **21e.** both a and b are correct

CHAPTER 9

1d. all of the above are stressors **2c.** antibodies and immune cells **3a.** the alarm phase **4c.** dermis **5c.** produce oil **6c.** D **7a.** A **8d.** E **9a.** A **10c.** produce dark pigments to absorb light **11a.** Meisner's corpuscles **12.** false **13d.** an eccrine sweat gland **14a.** hair follicles and nails **15b.** serous membrane **16b.** complement **17c.** pyrogens **18a.** inflammation **19d.** all of the above **20d.** interferon

CHAPTER 10

1. false **2c.** defending against bacterial invasion via fever **3d.** spleen **4c.** clean body fluids passing through the organ **5b.** thymus **6a.** both red and white blood cells **7d.** All of the above **8.** true **9c.** specific receptor on lymphatic cell **10b.** cloned B memory cells **11d.** helper T cell **12a.** the CD4 protein complex of the T cell **13c.** agglutinate the pathogen

14b. polyclonal antibodies 15b. cytotoxic T cell 16. true 17c. passive natural immunity 18d. increase in number and specificity in the blood 19d. streptococcus 20b. lysogenic phase 21b. helper T cell

CHAPTER 11

1b. heart, arteries, capillaries, veins, heart **2c.** D **3b.** C **4e.** both G and H **5.** true **6b.** ventricular systole **7d.** SA node **8b.** slow the impulse to contract and pass it to the AV bundle and on to the ventricles **9a.** 1 **10c.** vein **11c.** pulmonary arteries carry oxygen-poor blood **12d.** atherosclerosis **13d.** all of the above are true **14c.** erythrocyte **15.** false **16c.** basophil **17a.** release **18d.** AB **19c.** fibrinogen **20a.** anemia

CHAPTER 12

1a. warm incoming air **2b.** oropharynx **3c.** C **4d.** D **5c.** trachea, bronchi, bronchioles, respiratory bronchioles **6b.** larynx **7a.** carina **8d.** all of the above are true **9d.** cardiac notch of the left lung **10b.** produce surfactant **11a.** diffusion of gases into and out of the blood **12c.** relaxes, increasing **13.** false **14d.** inspiratory reserve volume **15a.** internal respiration **16b.** the partial pressure of carbon dioxide is lower in the blood **17c.** iron portion of the hemoglobin molecule **18c.** high, high **19d.** sinusitis **20.** true

CHAPTER 13

1c. vitamins **2d.** cellular respiration **3d.** all of the above describe the first reaction shown **4b.** B **5c.** tips on healthy eating based on your gender, age, and activity level **6.** false **7a.** serosa, muscularis, submucosa, mucosa **8b.** creating the peristaltic wave **9c.** premolars **10c.** liver **11b.** B **12c.** chemical digestion of proteins **13c.** pancreas **14c.** a spiral bacterium **15a.** cephalic phase **16d.** all of the above are true of this organ. **17b.** small intestine, increase surface area **18c.** hepatitis C **19d.** severely undereating **20a.** in the colon

CHAPTER 14

1a. production of urine **2d.** storage of produced urine **3d.** D **4c.** collecting ducts **5a.** renal artery, arcuate artery, afferent arteriole, efferent arteriole, peritubular capillaries, renal vein **6c.** reabsorb necessary nutrients **7d.** the glomerulus **8d.** PCT, capsule, and DCT **9b.** B and C **10a.** PCT **11c.** loop of Henle **12b.** overly concentrated urine is produced **13.** false **14b.** ADH is absent **15b.** ANP **16.** true **17b.** bicarbonate ions **18c.** viral infection **19b.** a possible UTI **20b.** lower levels

CHAPTER 15

1d. all of the above **2c.** pineal gland **3e.** E **4b.** B **5a.** steroid hormones **6.** true **7c.** oxytocin **8d.** ACTH **9a.** overproduction of one specific pituitary hormone **10b.** glucagons **11a.** cortex **12b.** Addison's disease **13d.** blood calcium levels **14a.** pineal gland **15d.** kidney **16c.** maintaining blood volume **17b.** neonate **18c.** infancy **19b.** FSH and LH **20c.** iodine

CHAPTER 16

1d. haploid gamete **2a.** it increases genetic variation **3.** false **4c.** temperature regulation of sperm **5c.** C **6d.** both a and b are correct **7b.** protect developing spermatids **8c.** midpiece **9d.** seminal vesicles, prostate gland, bulbourethral gland **10a.** seminal vesicles **11b.** stimulate production of sperm **12b.** B **13b.** sweep the ovulated egg toward the uterus **14a.** ovulation **15d.** progesterone and estrogens **16a.** endometrium **17b.** oxytocin **18c.** secondary and mature follicles **19c.** the condom (either male or female) **20b.** surgical methods including vasectomy and tubal ligation

CHAPTER 17

1c. embryonic and fetal periods **2d.** biochemical changes in the corona radiata **3a.** syngamy **4d.** all of the above are correct **5b.** implantation **6d.** ectoderm, endoderm, and mesoderm **7a.** amnion **8d.** chorion **9a.** are pregnant **10c.** digesting maternal endometrium **11d.** round ligament **12a.** fetal liver **13.** false **14b.** second trimester **15a.** positive **16c.** oxytocin **17a.** dilation, contraction, expulsion **18c.** 37 **19d.** external manipulation **20b.** prolactin

CHAPTER 18

1b. phenotype **2d.** a tetrad **3d.** two of these answers are correct **4c.** telophase I **5b.** his research was thorough and included quantifiable data **6a.** self-fertilization **7b.** 3:1 **8.** true **9b.** 9/16 **10d.** either heterozygous or homozygous dominant **11d.** multifactorial **12b.** "DNA is the molecule of inheritance" **13c.** transcription / translation **14a.** dominant **15d.** your likelihood of passing on a deleterious gene to your offspring **16b.** gel electrophoresis **17d.** restriction fragment length polymorphism (RFLP) **18d.** create transgenic organisms **19d.** all of the above were stated goals of the project **20b.** the baby's father is one of the two men tested

CHAPTER 19

1d. both b and c are correct **2d.** variations that decrease fitness will be passed on to subsequent generations **3a.** microevolution **4b.** evolution can be quantified by comparing ideal to observed conditions **5b.** heterozygous **6c.** mate choice must be selective and based on phenotype **7d.** comparative biochemistry **8b.** 2 **9b.** convection currents **10d.** convergent evolution **11a.** homologous structures **12c.** bottleneck effect **13d.** founder effect **14.** true **15c.** the formation of complex molecules and eventually life on Earth **16d.** photosynthesis and anaerobic respiration probably developed simultaneously **17a.** Quaternary period **18c.** *A. afarensis* **19b.** *H. erectus* **20a.** 1

CHAPTER 20

1a. biosphere, ecosystem, community, population, individual **2c.** predator population trends mirror prey population trends, with some delay **3d.** soil type **4a.** temperate grasslands **5c.** desert **6c.** secondary succession **7.** true **8a.** habitat **9d.** partitioning the resource to avoid direct competition **10a.** Malthusian strategists **11c.** producers **12b.** tertiary and, rarely, quaternary consumers **13c.** on the chloroplast inner membrane, the thylakoid **14c.** photosynthesis produces the materials needed for respiration and vice versa. **15b.** mollusk and arthropod populations would increase in size **16b.** sedimentary **17d.** oceans **18b.** eutrophication **19c.** biomagnification **20d.** reducing population growth by closely controlling reproductive behavior

Abdominal cavity Cavity that contains stomach, small intestine, spleen, liver, gallbladder, and most of large intestine.

Abdominopelvic (ab-dom'-i-nō-PEL-vik) **cavity** Cavity that contains abdominal and pelvic cavities.

Absolute refractory period The period of time immediately after an action potential when the neuron is physically incapable of beginning a second action potential while membrane channels are reset to their original position.

Acetylcholine (as'-ē-til-KŌ-lēn) **(ACh)** Common neurotransmitter used to signal muscle contraction.

Acetylcholinesterase An enzyme found in neuromuscular junctions and in neuron synapses that quickly breaks down acetylcholine, preventing continuous stimulation of the postsynaptic cell.

Acid rain Acidic precipitation caused by the sulfur- and nitrogen-oxide pollution that combine with water in the atmosphere.

Acidosis Acidic condition in the blood.

Acquired immunodeficiency syndrome (AIDS) A fatal disease caused by the human immunodeficiency virus (HIV). Characterized by a positive HIV-antibody test, low helper T cell count, and certain indicator diseases. Other symptoms include fever or night sweats, coughing, sore throat, fatigue, body aches, weight loss, and enlarged lymph nodes.

Acromegaly (ak-rō-MEG-al-ē) The secretion of excess growth hormone after the closure of the epiphyseal plates in the long bones.

Acrosome (AK-rō-sōm) A vesicle on the point of the sperm head that contains digestive enzymes.

Actin (AK-tin) Protein that functions in muscle contraction; see myosin.

Active transport Movement of a molecule or ion through the cell membrane, against the concentration gradient.

Acute (a-KYOOT *or* a-KUTE) Having rapid onset, severe symptoms, and a short course; not chronic.

Adaptive radiation Creation of several species from one ancestor that reaches habitat with empty niches.

Addison's disease The hyposecretion of glucocorticoids and aldosterone, usually due to autoimmune destruction of the adrenal cortex.

Adenosine diphosphate (a-DEN-ō-sēn dī-FOS-fāt) **(ADP)** The molecule formed when ATP releases one phosphate group.

Adenosine triphosphate (ATP) The primary form of energy that can be used to perform cellular functions.

Adhesive Having the ability to stick to other surfaces.

Adipocytes Specialized cells (fat cells) that store large quantities of lipid in a vacuole.

Adrenocorticotropic (ad-rē'-nō-kor-ti-kō-TRŌP-ik) **hormone (ACTH)** A hormone produced by the anterior pituitary that influences the production and secretion of certain hormones of the adrenal cortex.

Aerobic Requiring oxygen to metabolize.

Aerobic pathway Metabolic pathway that requires oxygen to burn glucose completely.

Afferent (AF-er-ent) Sensory impulses moving toward the brain.

Afterbirth (placental stage) Stage of delivery when the placenta is released and expelled.

Agarose A gel-like compound obtained from agar that provides a flexible, yet solid, medium for separation of DNA fragments.

Agglutinate (a-GLOO-ti-nāte) To clump with other cells due to the adhesion of surface proteins.

Agglutinin (a-GLOO-tin-in) Agent that causes cells to clump together or agglutinate.

Agonist (AG-ō-nist) The muscle in an antagonistic pair that shortens during a specific movement; prime mover.

Aldosterone (al-DOS-ter-ōn) Hormone that affects water balance by regulating sodium and potassium excretion.

Allantois (a-LAN-tō-is) An outpouching of the yolk sac that is an early site for blood formation and development of the urinary bladder.

Alleles (a-LEELZ *or* a-LĒLZ) Genes found on the same spot on the same chromosome in different individuals, coding for subtle variations of the same protein.

Alveolar macrophages (MAK-rō-fāj-ez) **(dust cells)** In alveoli, immune cells that remove any inhaled particles that escaped the mucus and cilia of the conducting zone.

Alveolar sac A cluster of alveoli that share a common opening.

Alveolus (al-VE-ō-lus) A small hollow or cavity; an air sac in the lungs; milk-secreting portion of a mammary gland. *Plural* is alveoli (al-VE-ol-ī).

Amenorrhea (ā-men-ō-RE-a) Absence of menstruation.

Amino acids The building blocks of proteins.

Aminopeptidase Secreted from the edges of the intestinal cells, this enzyme digests proteins.

Amniocentesis The collection of a small amount of amniotic fluid for analysis for genetic defects.

Amnion (AM-nē-on) Extra-embryonic membrane that lines the amniotic cavity, providing a diffusion area for the amniotic fluid.

Amniotic cavity Fluid-filled cavity that bathes the developing embryo and fetus.

Amoeba A single-celled organism that moves using pseudopods (false feet formed by oozing a portion of the body forward).

Amphiarthrotic A joint that is partly movable.

Anabolic steroids Lipid-soluble cholesterol-based compounds that stimulate increased muscle development, among other effects.

Anabolism (a-NAB-ō-lizm) The building up of larger molecules from smaller ones (contrast to catabolism).

Anaerobic (an-ar-Ō-bik) Metabolism that occurs without oxygen present.

Anaerobic pathways Metabolic pathways that occur in the cytoplasm and burn glucose to lactic acid, releasing some energy.

Anastomoses (a-nas-tō-MŌ-sēz) Networks or connections between two or more vessels.

Anatomical position Human body arranged in standard position; used to describe location of parts.

Androgens (AN-drō-jenz) Masculinizing sex hormones produced by the testes in males and the adrenal cortex in both sexes; responsible for libido.

Anemia (a-NĒ-mē-a) Condition of the blood in which the number of functional red blood cells or their hemoglobin content is below normal.

Aneurysm (AN-ū-rizm) Usually fatal, it occurs when a blood vessel wall balloons under pressure, forming a weak spot that can be burst by the increased blood pressure generated with each heartbeat.

Angina pectoris (an-JĪ-na *or* AN-ji-na PEK-tō-ris) A pain in the chest related to reduced coronary circulation due to coronary artery disease (CAD) or spasms of vascular smooth muscle in coronary arteries.

Animalia The kingdom of life that includes animals.

Anorexia nervosa A disorder characterized by severely limiting caloric intake; symptoms include osteoporosis, brittle hair, intolerance of cold, and muscle wasting.

Antagonist (an-TAG-ō-nist) The muscle in an antagonistic pair that lengthens during a specific movement.

Antagonistic (an-tag-ō-NIST-ik) **pair** Two muscles or muscle groups that combine to control the movement of one joint; synergistic pair.

Antibiotics Drugs that interfere with cellular processes in bacterial cells.

Antibodies (AN-ti-bod'-ēz) Proteins produced by B lymphocytes and directed against specific pathogens or foreign tissue.

Anti-codon Three bases on tRNA, carrying the same information as a codon.

Antidiuretic (an'-ti-dī-ū-RET-ik) **hormone** Hormone that prevents water loss by altering the permeability of the distal convoluted tubule cells to water.

Apical membrane Membrane at the free end, or top, of the intestinal cells.

Apneustic (ap-NOO-stik) A part of the respiratory center in the pons that stimulates deep, gasping breathing.

Apocrine (AP-ō-krin) A cellular secretion that pinches off the upper portion of the cell with the secretion.

Apoptosis (ap-ō-TŌ-sis *or* ap'-ōp-TŌ-sis) Programmed cell death.

Appendicitis A blockage in the appendix that prevents normal flow through the large intestine, leading to a buildup of pressure, decreased blood flow, and inflammation.

Appendicular skeleton System of appendages: limbs, pelvic girdle, and shoulders.

Appositional growth Growth at the outer surface of bone.

Aqueous (AK-wē-us) Solution of material dissolved in water.

Aqueous humor (AK-wē-us HŪ-mer) The watery fluid, similar in composition to cerebrospinal fluid, that fills the anterior cavity of the eye.

Aquifer A large body of groundwater.

Arachnoid (a-RAK-noyd) The middle of the three meninges (coverings) of the brain and spinal cord.

Archaebacteria Single-celled organisms, considered the most ancient forms of life; the kingdom that includes them.

Artery (AR-ter-ē) A blood vessel that carries blood away from the heart.

Arterioles (ar-TE-rē-ōl) A small, almost microscopic, artery that delivers blood to a capillary.

Articulating cartilage (KAR-ti-lij) Hyaline cartilage that prevents bones from grinding against each other.

Arytenoid (ar'-i-TE-noyd) **cartilages** A pair of small, pyramidal cartilages of the larynx that move the vocal folds.

Association area Areas of the brain that integrate new information with previously stored information, associating new and old information.

Asthma (AZ-ma) A constrictive pulmonary disease that can be life-threatening.

Asymptomatic Without symptoms.

Atherosclerosis (ath'-er-ō-skle-RŌ-sis) (literally "hardened vessels") Disease of the blood vessels wherein plaques of fatty compounds are deposited in the artery lumen, slowing blood flow.

Atomic mass The mass of the atom; different isotopes have different atomic masses.

Atomic number The number of protons in the nucleus of an atom.

Atoms The smallest unit of an element that has the properties of that element.

Atresia (a-TRE-zē-a) Reabsorption of immature ova prior to birth.

Atria (Ā-trē-a) Small, thin-walled chambers sitting atop the thick-walled, muscular ventricles in heart.

Atrioventricular (AV) (ā'-trē-ō-ven-TRIK-ū-lar) **bundle** The part of the conduction system of the heart that begins at the atrioventricular (AV) node, passes through the interventricular wall, then extends a short distance down the interventricular wall before splitting into right and left bundle branches.

Atrioventricular (AV) **node** The part of the conduction system of the heart made up of a compact mass of conducting cells located in the wall between the two atria.

Attention Deficit Hyperactivity Disorder (ADHD) A disorder in which behaviors are uncontrolled, resulting in impaired learning.

Auditory canal The hole that leads from the pinna to the tympanic membrane through which sound waves pass.

Autoimmune An immune response launched against healthy tissues of the individual's body destroying normal organs.

Autonomic (aw'-tō-NOM-ik) **division** Division of the nervous system regulating functions such as blood vessel diameter and stomach activity.

Autosomal Any chromosome other than the sex chromosomes, X and Y.

Autotroph Organism that can make its own food, usually through photosynthesis.

Avascular (Ā-vas'-kū-lar) Without blood vessels.

Axial skeleton Bone structures parallel to the body's core; head, vertebrae.

Axillary nodes Lymph nodes located in the armpit.

Bacteriolytic agent Agent that lyses (destroys) bacteria.

Balloon angioplasty (an'-jē-ō-PLAS-tē) Medical procedure in which a balloon is inserted into an atherosclerotic artery. The tip inflated, flattening the plaque and flattening it, to improve blood flow.

Basal metabolic rate Rate of energy use when body is resting and fasting.

Basophil (BĀ-sō-fil) A white blood cell with a pale nucleus and large granules that stain blue-purple with basic dyes.

Biceps brachii (BRĀ-kē) The anterior muscle of the upper arm.

Bicuspid (bī-KUS-pid) The valve between the left atrium and left ventricle, composed of two opposing cusps or flaps of connective tissue.

Bile Compound formed by the liver as a byproduct of the breakdown of hemoglobin and cholesterol.

Biodiversity A measure of species richness in a location.

Biogeographic range The expected geographical range of an organism, based on its habitat and niche requirements.

Biomagnification The concentration of toxins as they move up the food chain.

Biome A regional community characterized by a dominant plant life and climate.

Biosphere Earth's land, water, and air, plus all life.

Biotic community All the organisms in a particular location or relationship.

Biotic potential The maximum growth rate of a population under ideal conditions.

Blastocyst (BLAS-tō-sist) The stage of development where cellular specialization begins.

Blastomere (BLAS-tō-mēr) Small cell created during the rapid cell division of cleavage.

Blastula (BLAS-tyū-la) An early stage in the development of a zygote.

Bleached Rhodopsin, that has decomposed and cannot recombine.

Blood–brain barrier A barrier consisting of specialized brain capillaries and astrocytes that prevents the passage of materials from the blood to the cerebrospinal fluid and brain.

Blood The fluid that circulates through the heart, arteries, capillaries, and veins and that constitutes the chief means of transport within the body.

Bolus (BŌ-lus) A round, soft mass of chewed food within the digestive tract.

Bottleneck effect Drastic reduction in species population; reduces diversity of species genes.

Brain The part of the central nervous system within the cranial cavity.

Broca's (BRŌ-kaz) **area** Motor area of the brain in the frontal lobe that translates thoughts into speech.

Bronchi (BRON-kē) Branches of the respiratory passageway including primary bronchi, and divisions of the primary bronchi that are distributed to the lobes of the lung. *Singular* is bronchus.

Bronchial tree The trachea, bronchi, and their branching structures up to and including the terminal bronchioles.

Bronchiole (BRONG-kē-ōl) Smaller division of a tertiary bronchus which gives rise to terminal bronchioles and then respiratory bronchioles that deliver air to the alveolar sacs.

Bronchitis An inflammation of the mucous membrane lining the bronchi.

Bronchodilator Inhalant that relaxes the smooth muscle of the bronchi, opens the constricted tubes, and helps clear unwanted mucus.

Bronchopulmonary (brong'-kō-PUL-mō-ner-ē) **segment** One of the smaller divisions of a lobe of a lung supplied by its own branches of a bronchus.

Brush border Entire surface of a cell covered with microvilli.

Buffer A compound that absorbs hydrogen ions or hydroxide ions, stabilizing pH.

Bulbourethral (bul'-bō-ū-RE-thral) **gland** One of a pair of glands inferior to the prostate that secretes alkaline fluid into urethra.

Bulimia (boo-LIM-ē-a *or* boo-LĒ-mē-a) **nervosa** A disorder characterized by overeating at least twice a week, followed by purging by self-induced vomiting, strict dieting or fasting, vigorous exercise, or use of laxatives or diuretics. Also called binge purge syndrome.

Bursa (BUR-sa) Fluid-filled sac between the bones or tendons of a joint and the skin, positioned to reduce friction.

Bypass surgery Heart surgery that bypasses clogged arteries of the heart. These bypasses are looped over the damaged coronary artery and sewn in place so blood can flow around the damage and continue to nourish the heart tissue.

Callus Thickened formation on bone in response to wear.

Calorie The amount of heat stored in food, equal to the amount of heat it takes to raise the temperature of 1 kilogram of water 1 degree Celsius.

Canaliculi (kan'-a-LIK-ū-lī) Canals that connect cells in ossified bone. *Singular* is canaliculus.

Capacitated Activated, i.e., capable of fertilizing an ovum.

Capacitation (ka'-pas-i-TĀ-shun) Changes that make sperm able to fertilize an egg.

Capillaries (KAP-i-lar'-ēz) Very small diffusion vessels located between an arteriole and venule.

Capillary bed Interwoven mat of capillaries threading through a tissue.

Carbaminohemoglobin Carbon dioxide bound to the protein portion of hemoglobin.

Carbohydrates The most efficient source of energy for humans; molecules composed of carbon, hydrogen, and oxygen in a 1:2:1 ratio.

Carbon monoxide (CO) A molecule composed of one atom of carbon and one atom of oxygen, covalently bound.

Carbonic anhydrase Enzyme that allows red blood cells to remove most of the carbon dioxide from the blood.

Carboxypeptidase Pancreatic enzyme that digests proteins.

Cardiac sinus Large vein on the dorsal surface of the right atrium that collects blood from the cardiac veins and returns it to the chambers of the heart.

Cardiovascular system The system that consists of the heart, veins, blood vessels, and blood. It transports blood, carrying nutrients, wastes, and dissolved gases to and from the tissues.

Carina Extremely sensitive area where the trachea divides into the left and right primary bronchi, at the lower base of the trachea.

Carnivore (secondary consumer) Animal that eats other animals.

Carotene A yellow-orange pigment.

Carrying capacity The number of individuals in each population an area can support in a sustainable manner.

Cartilage (KAR-ti-lij) A type of connective tissue consisting of chondrocytes in lacunae embedded in a dense network of collagen and elastic fibers.

Casts Small structures formed by mineral or fat deposits on the walls of the renal tubules.

Catabolism (ka-TAB-ō-lizm) Metabolic activity that breaks down tissue.

CD4 Recognition elements in major histocompatibility complex (MHC) class II immune responses; identifies certain T cells.

Cell division Process by which a cell reproduces, includes nuclear division (mitosis) and cytoplasmic division (cytokinesis).

Cell theory Overall understanding of the role of cells in biology.

Cell The smallest unit of life, contained in a membrane or cell wall.

Cellulite Adipose tissue dimpled by differential expansion of connective and lipid components.

Cellulose Insoluble carbohydrate that provides structure to plant cells.

Central nervous system (CNS) That portion of the nervous system that consists of the brain and spinal cord.

Central vacuole Container inside plant cells that maintains turgor.

Centrifugation Rapid spinning of a sample to separate fragments by density.

Cephalic (se-FAL-ik) **phase** In digestion, the initial phase consisting of reflexes initiated by the senses.

Cerebral edema Fluid accumulation in the brain or cerebral area.

Cerebrospinal (se-rē'-brō-SPĪ-nal) **fluid (CSF)** A liquid similar to blood, but with less dissolved material and no blood cells, that maintains uniform pressure within the brain and spinal cord.

Cerebrum (SER-e-brum *or* se-RĒ-brum) The two hemispheres of the forebrain, making up the largest part of the brain.

Cervical nodes Lymph nodes located in the neck.

Cervix (SER-viks) Base of the uterus.

cGMP Cyclic guanine monophosphate, an energy molecule.

Chemical digestion Breaking down food using enzymes to alter the chemical structure of the food.

Chemically regulated Membrane channels that open or close in response to a specific chemical, such as sodium.

Chemiosmosis (ke'-mē-oz-MŌ-sis) The diffusion of hydrogen ions across a membrane, generating ATP as they move from high concentration to low.

Chemoreceptor (kē'-mō-rē-SEP-tor) Sensory receptor that detects the presence of a specific chemical.

Chloride shift An exchange reaction that requires no ATP because it merely switches the positions of the anions.

Chlorofluorocarbons (CFC's) (klor-ō-FLOR-ō-kar-bunz) Compounds once used as refrigerants that destroy stratospheric ozone.

Chlorophyll (KLOR-a-fil) A green photosynthetic pigment that absorbs light energy.

Chloroplast Green organelle in plants that contains chlorophyll.

Cholecystokinin (CCK) (kō'-lē-sis-TO-kīn-in) Hormone that inhibits stomach emptying.

Cholesterol A class of steroid found in animals; aids in membrane fluidity.

Chondroblasts Immature cartilage cells, not yet completely surrounded by the cartilage matrix.

Chondrocyte (KON-drō-sīt) Cartilaginous cell that secretes a gel-like matrix that eventually surrounds and imprisons it.

Chordae tendineae (KOR-dē TEN-di-nē-ē) (literally chords of tendons) The "heart strings" that anchor the cusps of the valves to the papillary muscles.

Chorion (KŌ-rē-on) Tissue that forms the exchange membrane between fetal and maternal blood.

Chorionic villi (kō-rē-ON-ik VIL-lī) Fingerlike extensions of the chorion that protrude into endometrial lining.

Choroid (KŌ-royd) One of the vascular layers of the eyeball that carries the blood supply and the melanin of the inner eye.

Chromatin (KRŌ-ma-tin) Threadlike material that packages DNA.

Chromosome (KRŌ-mō-sōm) Genetic material, consisting of multiple genes strung end to end.

Chronic (KRON-ik) Long-term or frequently recurring; applied to a disease.

Chronic obstructive pulmonary disease (COPD) Emphysema or chronic bronchitis, disease that severely obstructs airflow.

Chylomicrons (kē'-lō-MĪ-krōnz) Small lipoproteins carrying ingested fat from the intestinal mucosa to the liver.

Chyme (KĪM) The thick, partially digested fluid in the stomach and small intestine.

Chymotrypsin Pancreatic enzyme that digests proteins.

Cilium (SIL-ē-um) Hairlike appendages of cell, used to move extracellular fluid. *Plural* is cilia.

Circadian rhythm A daily predictable physiologic cycle based on a 24-hour day.

Cirrhosis (si-RŌ-sis) Scar tissue build-up in the liver generally caused by alcohol consumption, chronic hepatitis infection, autoimmune diseases that attack the liver, or congenital defects.

Class A taxonomic subcategory of phyla.

Class II MHC (major histocompatibility complex) Recognition proteins present on the membranes of antigen-presenting cells and lymphocytes.

Cleavage Repeated cell divisions with little time between rounds to enlarge the resulting daughter cells.

Climax community Relatively stable, mature community that has reached equilibrium after passing through a series of stages.

Clitoris (KLI-to-ris) An erectile organ of the female external genitalia that is homologous to the penis.

Codominant Neither form of a gene will overshadow the other when both forms are present, the individual will express both equally.

Codon Three bases on mRNA, corresponding to one amino acid.

Cohesive Able to stick to itself.

Collagen (KOL-a-jen) Group of tough molecules, often found in connective tissue.

Colon The portion of the large intestine consisting of ascending, transverse, sigmoid, and descending portions.

Colony-stimulating factors Blood-borne compounds that cause cells in the bone marrow to produce new blood cells.

Colostrum (kō-LOS-trum) The first substance produced by the mammary gland, a watery fluid rich in proteins and antibodies.

Columnar epithelium Tissue composed of cylindrical epithelial cells.

Community diversity A measure of the diverse forms of life in a community.

Community Group of interacting organisms.

Complement system A series of plasma proteins that, when activated, associate in a specific order to destroy pathogenic bacteria.

Compound Molecule composed of at least two elements.

Conducting zone Portion of the respiratory tract that conducts air to the respiratory membrane.

Conduction deafness Deafness resulting from poor conduction of sound to the inner ear.

Congenital (kon-JEN-i-tal) Present at birth.

Congestive heart failure A condition in which the heart weakens to the point that it cannot push the blood through the circulatory system. Blood builds up in the lungs, causing difficulty breathing.

Connective tissue Stretchy, strong tissue that connects body structures, providing support.

Constipation Difficult or infrequent defecation, leading to dry, potentially painful fecal evacuation.

Constrictive In the respiratory system, indicating that the airways have been narrowed in some way.

Consumer heterotroph Organism that must ingest nutrients because it cannot manufacture its own.

Contraceptive Chemical, anatomical or physical modification that prevents pregnancy.

Convergent evolution Evolution of similar structures in unrelated organisms.

Cornea (KOR-nē-a) The nonvascular, transparent fibrous coat on front of the eye.

Corona radiata The inner layer of granulosa cells around a secondary oocyte.

Coronary arteries Arteries that supply oxygen and nutrients to cardiac muscle.

Coronary sinus A wide venous channel on the back of the heart that collects the blood from the coronary circulation and returns it to the right atrium.

Corpus albicans (KOR-pus AL-bi-kanz) A white fibrous patch in the ovary that forms after the corpus luteum regresses.

Corpus luteum (LOO-tē-um) Spent follicular cells on the ovary.

Cortex (KOR-teks) Thin outer layer of any organ.

Cortisol A suppresses immune system, raises blood pressure, and raises blood glucose.

Cortisol-releasing hormone A compound secreted by the hypothalamus into the portal system leading to the pituitary gland that causes the release of ACTH, a pituitary hormone.

Countercurrent multiplication (CCM) Mechanism that increases the diffusion rate by flowing solutions in opposite directions on either side of a diffusion membrane.

Covalent bond Relatively weak bond between atoms, made by sharing electrons.

Cranial cavity Cavity that contains the brain.

Cranial nerves Twelve pairs of nerves that leave the brain and supply sensory and motor neurons to the head, neck, part of the trunk, and viscera of the thorax and abdomen. Each is designated by a Roman numeral and a name.

Creatine phosphate Compound that stores energy during anaerobic metabolism in muscle cell.

Cribriform plate A fragile, porous area of the ethmoid bone at the superior portion of the nasal cavity.

Cricoid cartilage The only complete ring of cartilage in the respiratory system, it is narrow in front but thick in the back of the larynx.

Cristae Folds of a membrane inside mitochondria.

Cross-pollinate Fertilize the ovum of a flower with pollen from a different plant.

Cuboidal epithelium Tissue composed of cube-shaped epithelial cells.

Cupula (KŪ-pū-la) A mass of gelatinous material covering the hair cells of a crista; a sensory receptor in the ampulla of a semicircular canal stimulated when the head moves.

Cushing's syndrome Condition caused by a hypersecretion of adrenal cortex hormones, characterized by spindly legs, "moon face," "buffalo hump," pendulous abdomen, flushed facial skin, poor wound healing, hyperglycemia, osteoporosis, hypertension, and susceptibility to disease.

Cutaneous Of or pertaining to the skin.

Cyanobacteria (SĪ-nō-bak-ter-ē-a) Blue-green, photosynthetic bacteria.

Cyclic AMP (cAMP) A form of adenosine monophosphate in which the phosphate appears in ring formation carrying little energy (not enough to harness for metabolic processes).

Cystic fibrosis (CF) Congenital disease that causing thick mucus in the lungs.

Cytokines (sī'-tō-kine) Chemical signals released by immune cells during the immune response.

Cytology The study of cells. A cytologist is a scientist who studies cells.

Cytoskeleton The internal framework of a cell.

Cytotoxic T cells (T_C) Subset of T lymphocytes responsible for killing virally infected cells.

Dalton A unit of mass, equal to the mass of one proton.

Dalton's law Law stating that gases move independently down their pressure gradient, toward lower pressure.

Decomposer (detritivore) Organism that feeds upon dead organisms and returns nutrients to the soil.

Deductive reasoning Moving from the general hypothesis to a specific situation.

Defecation (def-e-KĀ-shun) The discharge of feces from the rectum.

Deglutition (dē-gloo-TISH-un) The act of swallowing.

Dentin Bony tissue that lies below the enamel, inside the tooth.

Dentrification The conversion of nitrates into nitrogen gas.

Depolarizing Altering the neuron transmembrane potential so a weaker stimulus can begin an action potential.

Dermis (DER-mis) The underlying, vascularized, connective layer of the skin.

Dialysis (dī-AL-i-sis) The removal of waste products from blood by diffusion through a selectively permeable membrane.

Diapedesis A process by which macrophages escape the bloodstream by squeezing between cells of the vessel wall.

Diaphysis (dī-AF-i-sis) Shaft of a long bone.

Diarrhea (dī-a-RE-a) Frequent defecation of liquid feces caused by irritation of the colon.

Diarthrotic A joint that is fully movable.

Diastole (dī-AS-tō-lē) Relaxation of the heart.

Differentiation The process by which cells become specialized.

Diffusion Movement from a region of higher concentration to a region of lower concentration.

Dihydrotestosterone (DHT) Male sex hormone that works with testosterone to grow and develop male reproductive organs, secondary sex characteristics, and body.

Dilation The act of expanding or being expanded.

Dipeptidase Secreted from the edges of the intestinal cells, this enzyme digests proteins.

Diploid (DIP-loid) Having the total number of chromosomes of the body cells, twice that of the gametes.

Dissociation Separation of the strands of DNA.

Distal (DIS-tal) Farther from the attachment of a limb to the trunk; farther from the point of origin or attachment.

DNA fingerprint Process of identifying individuals based on their genetic sequences.

DNA sequence The sequence of bases (adenine, cytosine, thymine, and guanine) on a chromosome.

Dominant An allele of a gene that determines phenotype even if only one such allele is present.

Dominant population The population with the largest number of individuals in an area.

Dorsal root The sensory neurons of each spinal nerve that split off and enter the spinal cord from the posterior (dorsal) surface.

Downregulated A slowed-down cellular function.

Ductus deferens (vas deferens) The duct that carries sperm from the epididymis to the ejaculatory duct.

Duodenum (doo'-ō-DE-num *or* doo-OD-e-num) Region of the small intestine, extending about 25 centimeters from the pyloric sphincter.

Dura mater (DOO-ra MĀ-ter) The outermost of the three meninges (coverings) of the brain and spinal cord.

Eccrine (EK-r ē n) A secretion that does not include any portion of the secreting cell.

Ecology The study of the interactions between plants, animals, microbes, and ecosystems.

Ecosystem A subdivision of the biosphere.

Ectoderm The outer cell layer in the embryo.

Ectopic (ek-TOP-ik) **pregnancy** Embryo implanted outside the uterus.

Edema (e-DE-ma) Abnormal swelling in tissues.

Efferent (EF-er-ent) Motor impulses moving from the brain to the effect or organ.

Elastase Pancreatic enzyme that digests proteins.

Elastin Springy type of connective tissue.

Electrocardiogram (e-lek-trō-KAR-dē-ō-gram) A graphic representation of the electrical pattern during a heartbeat.

Electrolytes Compounds with ions that can conduct electricity.

Electron transport chain Step three in aerobic respiration, as electrons are passed in a series of chemical reactions, eventually producing ATP.

Electron The negative particle, found in orbitals surrounding the nucleus.

Element A substance made entirely of one type of atom; cannot be broken down via chemical processes.

Embolism (EM-bō-lism) A blood clot, bubble of air or fat from broken bones, mass of bacteria, or other debris or foreign material floating in the blood.

Embryonic Pertaining to the period from fertilization through the eighth week of development.

Emigration Departure from a location.

Emphysema (em-fi-SE-ma) A lung disorder in which alveolar walls disintegrate, producing abnormally large air spaces and loss of elasticity in the lungs.

Endemic Native to a region.

Endocardium (en-dō-KAR-dē-um) The inside lining of the heart wall, covers the valves and tendons that hold the valves open.

Endochondral (en'-dō-KON-dral) Within cartilage.

Endocytosis (en'-dō-sī-TŌ-sis) Movement of compounds into a cell.

Endoderm (EN-dō-derm) The innermost embryonic cell layer.

Endometriosis (en'-dō-ME-trē-ō'-sis) The growth of endometrial tissue outside the uterus.

Endometrium (en'-dō-ME-trē-um) The mucous membrane lining the uterus.

Endomysium (en'-dō-MĪZ-ē-um) The innermost connective tissue lining, on top of the muscle cell membrane.

Endoplasmic reticulum (en'-dō-PLAS-mik re-TIK-ū-lum) A type of organelle; *see* Rough endoplasmic reticulum or Smooth endoplasmic reticulum.

Endorphins and enkephalins Naturally occurring compounds that reduce the sensation of pain and produce a feeling of well-being.

Endothermic Organisms that maintain an internal temperature within a narrow range despite environmental conditions.

Energy Usable heat or power.

Eosinophil (ē-ō-SIN-ō-fil) A type of white blood cell characterized by granules that stain red or pink with acid dyes.

Epidemic Disease outbreak.

Epidermis (ep'-i-DERM-is) The outermost, nonvascular layer of the skin.

Epididymis (ep'-i-DID-i-mis) Storage area and final maturation center in the testes, for spermatozoa.

Epiglottis (ep'-i-GLOT-is) Large, leaf-shaped piece of cartilage lying over top of the larynx.

Epimysium (ep-i-MĪZ-ē-um) The outermost covering on a muscle, separating one muscle from the next.

Epinephrine (ep-ē-NEF-rin) A hormone released from the adrenal gland in response to stress.

Epiphyseal plate (ep-i-FIZ-ē-al) Area of cartilage where long bones grow during childhood and adolescence.

Epiphysis (e-PIF-i-sis) End of a bone.

Epithelial (ep-i-THE-lē-al) **tissue** Tissue that covers the body, lines all cavities, and composes the glands.

Erythrocytes Red blood cells.

Erythropoiesis (e-rith'- rō-POY-e-sis) The formation of red blood cells (*erythro* = red, *poiesis* = to form).

Esophagus (e-SOF-a-gus) The hollow muscular tube that connects the pharynx and the stomach.

Essential amino acids Eight amino acids that must be consumed by humans, since the body does not manufacture them.

Estrogens (ES-tro-jenz) Feminizing sex hormones produced by the ovaries; govern development of oocytes, maintenance of female reproductive structures, and secondary sex characteristics.

Eubacteria Single-celled organisms without nuclei; the kingdom that includes them.

Eukaryotic (YOO-kar-ē-ō'-tik) Cells that contain a distinct membrane-bound nucleus.

Eustachian (ū-STĀ-shun *or* ū-STĀ-kē-an) **tube** The tube that connects the middle ear with the nose and nasopharynx region of the throat.

Eutrophication A natural process that converts ponds to wetlands and dry land.

Evolution Descent with modification.

Exchange pool Area where a chemical resource is in a form that the biotic community can use.

Excitatory postsynaptic potentials (EPSPs) A stimulus that moves the postsynaptic neuron membrane potential closer to threshold, without causing an action potential.

Exhalation Decreasing lung volume, expelling air.

Exocrine (EK-sō-krin) **glands** Glands that secrete directly into ducts.

Exocytosis (ex'-ō-sī-TŌ-sis) Movement of compounds out of a cell.

Exophthalmos (ek'-sof-THAL-mas) Fluid buildup behind the eyes, may cause the eyes to "pop" from their sockets.

Exothermic (ex'-ō-THER-mik) Chemical reaction that releases energy.

Expiratory reserve volume (ERV) Additional volume of air that can be expelled from the lungs after a normal exhalation. 700 ml for females, 1000 ml for males.

Extension Condition of diarthrotic joint where the joint angle is maximal; contrast with Flexion.

External nares (NĀ-rez) The nostrils themselves, the paired openings into the nasal cavity.

External respiration The exchange of gases between the air in the alveoli and the blood in the respiratory capillaries.

External sphincter (SFINGK-ter) **muscle** Ring of voluntary skeletal muscle that closes the urethra.

Extinction Death of an entire species, often due to some combination of predation, habitat destruction, and disease.

Extraembryonic Outside of the cells of an embryo.

Extrinsic controls The hearbeat control used to modulate the intrinsic baseline rate to meet the body's immediate demands.

Facilitated diffusion Diffusion across a membrane, down the concentration gradient, without energy expenditure.

Fallopian (fal-LŌ-pē-an) **tube** (uterine tube) Duct that transports ova from the ovary to the uterus.

Family A taxonomic subcategory of order.

Fast block Depolarization of the oocyte membrane immediately after syngamy.

Fast-twitch fiber Myofibril that contracts quickly.

Feedback system System whose effects change its own rate.

Fenestrations Windows or openings between cells in the lining of the glomerulus.

Fetal Pertaining to the period from week eight through birth.

Fever State of hyperthermia, usually a sign of disease.

Fiber Undigestible carbohydrate fibers that pass through the digestive tract without releasing any stored energy.

Fibrin A thread-like protein formed by platelets during clot formation.

Fibrocartilage Cartilage with strengthening fibers in the matrix.

Filtration Process that removes some solids from a liquid.

Fitness Ability to produce living offspring and pass on DNA.

Flagellum (fla-JEL-um) Whiplike appendage to cell, used for movement, found on sperm. *Plural* is flagella (fla-JEL-a).

Flexion (FLEK-shun) State of a diarthrotic joint where the angle at the joint is minimal; contrast Extension.

Follicle (FOL-i-kul) A small cavity such as where hair originates.

Follicle stimulating hormone (FSH) A hormone that stimulates the growth and functioning of the ovaries and testes.

Follicular cells Cells stimulated to develop alongside the oocyte.

Food chain System of energy transfer that starts with green plants and moves upward through various trophic levels to top carnivores.

Fossa (FOS-a) A pit, groove, or depression.

Fossil fuels Energy source derived from organic matter and stored in hydrocarbon deposits.

Founder effect Genetic consequence of a few organisms that occupy a new habitat.

Fracture hematoma (hē'-ma-TŌ-ma) A bruise that develops over the site of a fractured bone.

Free radicals Highly reactive.

Functional group A subunit on an organic molecule that helps determine how it reacts with other chemicals.

Fundus (FUN-dus) The bottom portion of any hollow organ.

Fungi (*Singular:* **fungus**) Eukaryotic decay organism; kingdom that includes fungi.

Gallbladder A small pouch, inferior to the liver, that stores bile and empties through the cystic duct.

Ganglia (GANG-glē-a) A group of neuronal cell bodies lying outside the central nervous system.

Gangrene (Gang-GREN) Tissue death due to lack of blood flow.

Gap junction Gap between nearby cells; used for communication.

Gastric (GAS-trik) Related to the stomach.

Gastric juice Fluid produced in the stomach.

Gastric lipase Enzyme that digests short fatty acids, such as those found in milk.

Gastric phase In digestion, hormonal and neural pathways that cause an increase in gastric wave force and secretion from gastric pits.

Gated channels Membrane channels that open or close in response to a specific stimulus; are not open at all times.

Gene flow Gain or loss of alleles in the gene pool of a population as individuals enter or leave by migration (as opposed to birth and death).

General adaptation syndrome (GAS) The body's response to any stressor, in three stages: alarm, resistance, and exhaustion.

Genetic diversity The genetic variation among individuals in a population or species.

Genotype (JE-nō-tīp) The alleles carried on the chromosomes.

Genus A taxonomic subcategory of family.

Germ cell A cell destined to become an egg or sperm.

Gigantism A condition caused by hypersecretion of human growth hormone before closure of the epiphyseal plates.

Glucagon (GLOO-ka-gon) A hormone produced by the alpha cells of the pancreatic islets (islets of Langerhans) that increases blood glucose level.

Glucocorticoid (gloo'-kō-KOR-ti-koyd) Steroid hormones that maintain mineral balance, and control inflammation and stress.

Glycocalyx Outside layer of a cell, composed of glycolipids and glycoproteins.

Glycogen (GLĪ-kō-jen) A large polysaccharide easily broken down to release individual glucose molecules.

Glycolipid Lipid plus at least one carbohydrate group.

Glycolysis (GLĪ-kō-lis-is') The enzymatic breakdown of glucose to pyruvate, occurring within the cytoplasm.

Glycoprotein Proteins plus a carbohydrate.

Goiter (GOY-ter) An enlarged thyroid.

Golgi (GOL-jē) **complex** Organelle involved with processing proteins and fatty acids.

Gonad (GŌ-nad) A gland that produces gametes and hormones; ovary in female and testis in male.

Gonadotropin (Gō -nad-ō-TRŌ-pin) releasing hormone (GnRH) Hormone released from the hypothalamus that governs the female reproductive cycle.

Graded contraction A smooth transition from a small, weak contraction to a forceful contraction.

Graves disease The most common hyperthyroidism disease. It may be treated with surgical removal of part of the thyroid or the application of radioactive iodine to the thyroid.

Gray matter Neuron cell bodies and dendrites within the CNS.

Greenhouse effect Reflection of heat back to Earth by carbon dioxide and other compounds; primary cause of global warming.

Groin (GROYN) The depression between the thigh and the trunk.

Gustation The sense of taste.

Gyrus (JĪ-rus) (*plural* is gyri) Depressions separating individual sulci in brain.

Habitat Where an organism lives.

Haploid (HAP-loyd) Having half the number of chromosomes of normal body cells, found in eggs and sperm.

Haustra (HAWS-tra) Pouches created by strands of muscle in walls of large intestine that fill with undigested material.

Haversian system Concentric rings of matrix laid by osteocytes, formed around a central canal; osteon.

Hematocrit (hē-MAT-ō-krit) The percentage of blood made up of red blood cells.

Hematopoiesis (hem'-a-tō-poy-E-sis) Process that forms blood cells.

Hemispheric lateralization The isolation of a task to either the left or right hemisphere of the cerebrum.

Hemoglobin (hē'-mō-GLŌ-bin) (Hb) A substance in red blood cells consisting of the protein globin and the iron containing red pigment heme that transports most of the oxygen and some carbon dioxide in blood.

Hemolymph An oxygen-carrying fluid that circulates through the tissues of many invertebrates with open circulatory systems.

Hemolytic disease of the newborn A blood disease caused by the destruction of the infant's red blood cells by antibodies produced by the mother; usually due to Rh blood type incompatibility.

Hemothorax (hem'-ō-THŌ-raks) Blood in the pleural space.

Hepatitis Viral inflammation of the liver caused by ingested toxins or other materials.

Hepatocytes (he-PAT-ō-cyte) Liver cells (*hepato* = liver; *cyte* = cell).

Herbivore (primary consumer) Organism that eats green plants.

Heterotopic bone Bone that forms outside the usual areas for bone formation.

Heterotrophs Organisms that cannot manufacture their own organic compounds and must obtain them from the environment.

Heterozygous (het-er-ō-ZĪ-gus) Organism where one allele codes for the dominant trait and the other codes for a recessive trait (Aa).

Hilum (HĪ-lum) *or* **Hilus** Site of entry and exit for the nerves, blood vessels, and lymphatic vessels on most organs.

Histamine (HISS-ta-mēn) A compound involved in allergic reactions that causes capillary leakage and increased fluid movement to affected tissues.

HIV Human (human immunodeficiency virus) The retrovirus that causes the disease AIDS.

Homeostasis (hō'-mē-ō-STĀ-sis) Staying the same; the condition in which the body's internal environment remains relatively constant and within physiological limits.

Homologous Similar in structure, function, or genetic sequence.

Homozygous (HŌ-mō-zī-gus) Gene in which both alleles are identical.

Homozygous dominant Both alleles code for the dominant trait (AA).

Homozygous recessive Both alleles code for the recessive trait (aa).

Homunculus A proportional diagram of the structures of the human body as they are represented in the brain rather than the proportions in which they physically exist.

Hormones (HOR-mōn) Compounds secreted in one area of the body that are active in another, usually carried by the blood.

Host cell A cell that harbors a virus.

Human chorionic gonadtrophin (hCG) (kō-rē-ON-ik gō-nad-ō-TRŌ-pin) A hormone that maintains pregnancy until the placenta is fully functional, by preventing degeneration of the corpus luteum.

Human growth hormone (hGH) Hormone that stimulates the growth of muscle, cartilage, and bone, and causes many cells to speed up protein synthesis, cell division, and burning fats for energy.

Hydrogen bond Weak bond formed by electrical attraction between molecules.

Hydrolases Digestive enzymes that catalyze the breakdown of large polymers by inserting water molecules between monomers.

Hydrolytic enzymes Proteins that help decompose compounds by splitting bonds with water molecules.

Hydrophilic (hī-drō-FIL-ik) Having an affinity for water.

Hydrophobic (hī-drō-FOB-ik) Lacking affinity for water.

Hyperpolarizing Altering the neuron transmembrane potential so that a stronger stimulus is needed to begin an action potential.

Hypertension High blood pressure, defined as a diastolic number above 90.

Hypertonic (hī'-per-TON-ik) Solution that causes cells to shrink due to loss of water by osmosis.

Hypertrophy (hī-PER-trō-fē) Enlargement of an organ owing to enlarged cells rather than an increasing number of cells; growth of new myofibrils within the endomysium of individual muscle cells.

Hypodermis The layer of connective tissue that holds the skin to the deeper organs composed of areolar connective tissue, adipose tissue, a large blood supply, and many connective tissue fibers.

Hypothalamus (hī'-pō-THAL-a-mus) A portion of the diencephalon, lying beneath the thalamus and forming the floor and part of the wall of the third ventricle.

Hypothyroidism Condition that occurs when the thyroid secretes too little T3 and T4.

Congenital hypothyroidism Glandular defect that can lead to mental retardation and stunted bone growth.

Hypotonic (hī-pō-TON-ik) Solution that causes cells to swell and perhaps rupture due to gain of water by osmosis.

Hysterectomy (hiss-te-REK-tō-mē) The surgical removal of the uterus.

Ileum (IL-ē-um) The longest region of the small intestine, measuring approximately 3 meters.

Immigration Movement to a location.

Immune response The disease-fighting activity of an organism's immune system.

Immunize To stimulate resistance to a specific disease through exposure to a nonpathogenic form of the disease.

Implantation Anchoring and settling of the embryo into the endometrial wall, starting placental formation.

In vitro (VE-trō) Literally, in glass; outside the body and in an artificial environment such as a test tube.

Incisors Teeth that function as cutting tools.

Incomplete dominance Genetics that produces different phenotypes, based on the combination of alleles present in heterozygotes.

Incontinence (in-KON-ti-nens), **urinary** The inability to prevent urine leakage; types include stress, urge, and overflow.

Incus (IN-kus) The second of the auditory ossicles, joined to the malleus and the stapes.

Indigenous Found only in a particular region.

Inductive reasoning Creating a general statement from observations.

Inferior (in-FĒR-ē-or) Away from the head or toward the lower part of a structure.

Inflammation A localized method for increasing enzyme function, including swelling, redness, heat, and pain. The benefits include temporary tissue repair, blockage of continued pathogen entry, slowing of pathogen spreading, and quicker repair of the damaged tissue.

Inguinal (IN-gwi-nal) **nodes** Lymph nodes located in the groin; pertaining to the groin.

Inhalation The act of pulling air into the lungs.

Inhibin Hormone that inhibits FSH production from the anterior pituitary, slowing sperm production.

Inhibitory postsynaptic potentials (IPSPs) A stimulus that moves the postsynaptic neuron membrane potential farther from threshold, making it more difficult to begin an action potential.

Innate immunity Our inborn ability to defend against daily stresses and invasions of fungal, bacterial, or viral pathogens.

Inner ear The portion of the ear that lies completely within the temporal bone, from oval window to round window. This area is filled with fluid and supports the membranous labyrinth.

Insertion (of muscle) End of muscle that moves during contraction.

Inspiratory reserve volume (IRV) Additional volume of air that can be added to the lungs after a normal inspiration; 1,900 ml in females, 3300 ml in males.

Insulin (IN-suh-lin) A hormone produced by the beta cells of a pancreatic islet (islet of Langerhans) that decreases the blood glucose level.

Integral protein A protein that spans the plasma membrane.

Interferon A protein produced by virally infected cells that helps other cells respond to viral infection.

Intermediate filament Protein in cytoskeleton that protects cell from mechanical stresses.

Internal nares The twin openings at the back of the nasal passageway, leading to the upper throat.

Internal respiration The exchange of gases between the blood in the systemic capillaries and the body's cells.

Internal urethral sphincter (SFINGK-ter) Ring of involuntary smooth muscle that keeps the urethra closed.

Interneurons (in'-ter-NOO-ronz) Neurons whose axons extend only for a short distance and lie completely within the brain, spinal cord, or a ganglion; they connect one neuron to another.

Interosseus Between bones.

Interstices (in'-ter-STISH-es) The small fluid-filled spaces between cells.

Interstitial (in'-ter-STISH-al) **fluid** Fluid that fills the interstices.

Intestinal phase The final phase of gastric digestion.

Intramembranous Between membranes.

Intrauterine device (IUD) Birth-control device made of plastic or copper that floats in the uterus and periodically hits the endometrial lining, preventing implantation.

Intrinsic controls The heartbeat control maintained from within the heart that establishes the usual, day-in, day-out pace of heartbeats.

Intrinsic factor Hormone produced by the parietal cells of the gastric pits that facilitates absorption of vitamin B_{12}.

Intubation The insertion of a tube through the mouth or nose, through the larynx and into the trachea.

Invertebrate Organism without a vertebral column, such as an earthworm, crab, or starfish.

Ion (I-on) Charged atom.

Ionic bond Strong molecular bond, formed between atoms with opposite charges.

Iris The colored portion of the vascular tunic of the eyeball visible through the cornea, which contains circular and radial smooth muscle.

Ischemia (is-KE-mē-a) Lack of oxygen to a tissue because of constriction or blockage of the blood vessels.

Isotonic A solution with the same concentration as the cell cytoplasm.

Isotope Chemically identical forms of an atom with different numbers of neutrons.

Jejunum (je-JOO-num) The middle region of the small intestine, measuring approximately 2 meters.

Karyotype A micrograph of the chromosomal complement of the fetus, arranged to show chromosome pairs.

Keratin (KER-a-tin) A waterproof substance that accumulates in the epidermal cells as they move toward the skin surface.

Keratinized (KER-a-tin-īzd) Filled with keratin and therefore waxy.

Kidney (KID-nē) One of the paired organs in the lumbar region that regulates the composition, volume, and pressure of blood and produces urine.

Kinase A group of enzymes, all of which carry a phosphate from one compound to another.

Kingdom A high-level taxonomic classification.

Krebs (TCA) **cycle** Chemical cycle occurring in the mitochondria that slowly releases the energy stored in pyruvate, producing $FADH_2$ and NADH.

Lacrimal glands Secretory cells, located at the lateral upper portion of each orbit, that secrete tears into ducts that open onto the surface of the eye.

Lacrimal punctae Small holes in the corners of the eyelids that collect tears.

Lactase Enzyme that digests carbohydrates.

Lactiferous Milk-producing.

Lacuna (la-KOO-na) Hole in bone matrix that houses blood or nerve cell.

Lanugo (la-NOO-gō) Soft hair covering the fetal skin.

Laparoscopy Noninvasive surgery using fiber optic cables, remote control, and tiny surgical tools.

Laryngopharynx (la-rin'-gō-FAR-inks) The lowest level of the pharynx and the last part of the respiratory tract shared by the digestive and respiratory systems.

Larynx (LAR-inks) Voice box or Adam's apple.

Law of Independent Assortment Each trait is carried in the gametes as a separate entity, with no effect on any other trait.

Law of Segregation The separation of parental "heritable units" during gamete formation.

Leukemia (loo-KE-mē-a) A malignant disease of the blood-forming tissues characterized by either uncontrolled production and accumulation of immature leukocytes (acute) or an accumulation of mature leukocytes in the blood because they do not die at the end of their normal life span (chronic).

Leukocytes (LOO-kō-sītz) White blood cells.

Leutenizing hormone (LH) A hormone that stimulates the growth and functioning of the ovaries and testes.

Leydig (LĪ-dig) **cells** Cells in the testes that secrete testosterone.

Ligament Dense regular connective tissue connecting bone to bone.

Ligating Tying off a tube to close it.

Light reaction Reaction in which chlorophyll absorbs a photon of light and releases an electron.

Limbic system A part of the forebrain concerned with various aspects of emotion and behavior.

Lingual (LIN-gwal) Relating to speech or the tongue.

Lipase An enzyme that chemically digests lipids.

Lipid (LIP-id) Class of macronutrient made of long chains of carbon molecules, with many more carbon atoms and far fewer oxygen atoms than carbohydrates; fats.

Lipoprotein lipase Enzyme that breaks chylomicrons down to short-chain fatty acids and glycerol.

Liver Large organ under the diaphragm. It produces bile; detoxifies substances; stores glycogen, iron, and vitamins.

Lobules Structures in the liver composed of a hepatic portal vein, a hepatic artery, and a bile duct.

Lower esophageal sphincter Circular muscle located at the base of the esophagus.

Lower respiratory tract Respiratory organs within the thoracic cavity, including the bronchial tree and the lungs.

Lumen (LOO-men) The inner, hollow portion of a tubular structure; the center of the blood vessel.

Lungs Main organs of respiration that lie on either side of the heart in the thoracic cavity.

Lymph (LIMF) **nodes** Small, encapsulated glands that are located to filter large volumes of lymph.

Lymphatic (lim-FAT-ik) **system** The tissues, vessels, and organs that produce, transport, and store cells that fight infection.

Lymphocytes (LIM-fō-sītz) White blood cells that patrol the body, fight infection, and prevent disease.

Lysosome (LĪ-sō-zīm) Chemical package produced by the Golgi complex, contains hydrolytic enzymes.

Macerated Soaked until soft and separated into constituent parts.

Macronutrients Carbohydrates, lipids, and proteins.

Macrophage (MAK-rō-fāj) Large phagocytic cell that patrols tissue, ingesting foreign material and stimulating immune cells.

Macula (MAK-ū-la) A small, thickened region on the wall of the utricle and saccule that contains receptors for static equilibrium.

Macula lutea (MAK-ū-la LOO-tē-a) *Macula* = spot; *lutea* = yellow; yellow spot near center of retina.

Malleus (Mal-ē-us) The first of the auditory ossicles, attached to the tympanic membrane.

Maltase Enzyme that digests carbohydrates.

Mammary (MAM-ar-ē) **gland** Gland of the female that produces milk.

Mass The amount of "substance" in an object ("weight" is the mass under a particular amount of gravity).

Matrix The "ground substance" secreted by connective-tissue cells; determines the characteristics of the connective tissue.

Mechanical digestion The chopping, cutting, and tearing of food into smaller pieces.

Mechanically regulated Membrane channels that open or close in response to physical deformation of the channel.

Mechanoreceptor (me-KAN-ō-rē-sep-tor) Sensory receptor that detects mechanical deformation of the receptor or adjacent cells; detecting touch, pressure, vibration, proprioception, hearing, equilibrium, and blood pressure.

Mediastinum (mē'-dē-as-TĪ-num) The central portion of the thoracic cavity between the lungs, housing the heart, major blood vessels, and lymphatics.

Medulla (me-DOO-la) Inner portion of the organ.

Medulla oblongata (me-DOO-la ob'-long-GA-ta) Portion of the brainstem immediately adjacent to the spinal cord, associated with heart rate, breathing controls, and blood.

Megakaryocytes Large cells in the bone marrow that produce platelets.

Meissner corpuscles (MĪS-ner KOR-pusulz) Structures in the dermis that register light touch.

Melanin (MEL-a-nin) A brown pigment produced by melanocytes.

Melanocytes (MEL-a-nō-sīt') Cells that produce melanin.

Membrane Structure that delineates a component, such as a cell or organ.

Membrane potential The difference in electrical charge between two sides of a membrane.

Meninges (me-NIN-jēz) Three protective membranes covering the brain and spinal cord.

Meningitis Inflammation of the meninges.

Menisci (men-IS-kī) Fat pads within joints that cushion bones and assist in "fit.".

Menstruation (men'-stroo-Ā-shun) Periodic discharge of blood, tissue fluid, mucus, and epithelial cells; the menstrual cycle or menses.

Mesenteric (MEZ-en-ter'-ik) Pertaining to the membranous fold in the abdominal cavity attaching many of the abdominal organs to the body.

Mesenteries (MEZ-ēn-ter'-ez) Folds in the lining of the abdominal cavity that help to secure the digestive organs.

Mesoderm Middle layer of embryonic cells.

Messenger RNA (mRNA) RNA that takes information from DNA into the cytoplasm.

Metabolize To perform a process in an organism, including both breakdown and buildup of organic compounds.

Microfilament Protein in cytoskeleton, responsible for basic shape, cellular locomotion, muscle contractions, and movement during cell division.

Micronutrients Vitamins and minerals.

Microphage (MIK-rō-fāj) A small phagocyte mainly found in the nervous system.

Microtubule Long strings of coiled tubulin that serve as tracks for organelle movement.

Microvilli (mī'-krō-VIL-ē) Folded parts of the cell membrane that increase the cell's surface area.

Middle ear The portion of the ear from the tympanic membrane to the oval window, encased within the temporal bone and filled with air.

Migrating motility complexes Part of the peristaltic wave that moves the chyme along the small intestine.

Mineralocorticoids (min'-er-al-ō-KOR-ti-koyds) Steroid hormones involved in maintaining water and ion balance.

Mitochondrion (mī-tō-KON-drē-on) Organelle that processes energy; (*plural* is mitochondria).

Mitosis (mī-TŌ-sis) Division of a cell into two daughter cells.

Mitral Pertaining to the left ventricle of the heart.

Molars and premolars Teeth that function as grinding instruments.

Molecule Group of similar or dissimilar atoms bound together.

Monocyte (MON-ō-sit') The largest type of white blood cell, characterized by a granular cytoplasm.

Morphogenesis (MORF-ō-JEN-e-sis) Formation of organs and tissues during development.

Morula (MOR-ū-la) A solid mass of cells that can develop into any type of cell.

Motor neurons Neurons that conduct impulses from the brain.

Motor unit The group of muscle cells controlled by one motor neuron.

Mucosa (mū-KŌ-sa) A membrane that lines a body cavity that opens to the exterior, also called mucous membrane.

Mucosa-associated lymphoid tissue (MALT) Lymphoid tissue in the tonsils, small intestine and other regions in contact with the exterior.

Multifactorial disorder Genetic disorder due to a combination of genetic and environmental factors.

Multifactorial trait Polygenic trait that is also influenced by environment.

Mumps A common infection of the salivary glands; causes swelling of the glands, sore throat, tiredness, and fever.

Muscular tissue Dense tissue that provides movement and heat.

Muscularis mucosae (MUS-kū-la'-ris mū-KŌ-sē) A thin layer of smooth muscle fibers underlying the mucosa of the GI tract that gives the tract the ability to move substances lengthwise.

Myelin (MĪ-e-lin) White lipids and phospholipids wrapped around neural processes that aids impulse transmission.

Myeloid Pertaining to bone marrow.

Myocardial infarction (mī'-ō-KAR-dē-al in-FARK-shun) **(MI)** Large-scale death of heart tissue due to interrupted blood supply (heart attack).

Myofiber Muscle cell.

Myofibrils Bundles of the contractile proteins actin and myosin.

Myoglobin (mī-ō-GLŌB-in) Oxygen-carrying protein in muscle cells.

Myometrium (mī'-ō-MĒ-trē-um) The smooth muscle layer of the uterus.

Myosin (MĪ-ō-sin) Protein that functions in muscle contraction; see Actin.

MyPyramid Personalized dietary guidelines from the U.S. Department of Agriculture (www.mypyramid.gov).

Myxedema (mik-sē-DĒ-ma) Condition where thyroid works normally at birth but fails to secrete enough hormone in adult life, causing slow heart rate, low body temperature, dry hair and skin, muscular weakness, general tiredness, and a tendency to gain weight.

Nasopharynx (nā'-zō-FAR-inks) Upper throat, including the nasal openings and the soft palate.

Negative feedback System that tends to return to homeostasis.

Neonate (nē-ō-NĀT) The newborn child, from immediately after birth to approximately one month of age.

Nephron (NEF-ron) The filtering unit of the kidney.

Nerve deafness Condition where sound is either not detected or the nerve impulse is not transmitted to the brain.

Nerves A bundle of axons and/or dendrites covered with connective tissue found outside the central nervous system.

Nervous tissue Tissue that responds to the environment by detecting, processing, and coordinating information.

Neuroendocrine Cells that can both carry nerve impulses and produce hormones.

Neuroglia (noo-RŌG-lē-a) Supporting and protecting cells within the nervous system, providing nutrients, removing debris, and speeding impulse transmission.

Neuromuscular junction Junction between a nerve cell and the motor unit it controls.

Neuron (NOO-ron) A nerve cell that sends and receives electrical signals.

Neurotransmitter A chemical used to transmit a nervous impulse from one cell to the next.

Neutron The neutral particle in the atomic nucleus.

Neutrophil (NOO-trō-fil) A type of white blood cell characterized by granules that stain pale lilac with a combination of acidic and basic dyes.

Niche A specific part of a habitat that can be occupied by one type of organism.

Nitrification The formation of nitrates in the atmosphere.

Nitrogenous wastes Compounds containing nitrogen, such as urea, produced during protein metabolism.

Nociceptors (nō-sē-SEP-torz) Nonadapting pain receptors in the skin (*noci* = pain).

Nonpolar Molecule that is electrically balanced.

Norepinephrine (nor'-ep-ē-NEF-rin) **(NE)** A hormone secreted by the adrenal medulla that produces actions similar to those that result from sympathetic stimulation. Also called noradrenaline (nor-a-DREN-a-lin).

Nuclear envelope Membrane surrounding nucleus.

Nuclear pore Opening in nuclear envelope that allows material to enter and exit nucleus.

Nuclei Areas of concentrated neuronal cell bodies in the brain.

Nucleoli Dark regions of chromatin that produce ribosomal RNA and assemble ribosomes. *Singular* is nucleolus.

Nucleoplasm Fluid within the nucleus, containing the DNA.

Nucleus (NOO-klē-us) Compartment of a cell that contains genetic information.

Nucleus pulposus A soft, elastic substance in the center of intervertebral discs.

Nutrients Ingredients in food that are required by the body.

Obesity (o-BĒ-si-tē) Body weight more than 20% above a desirable standard due to excessive fat.

Obligate anaerobes Bacteria that require an oxygen-free environment.

Obstetrics (ob-STET-riks) The medical field devoted to prenatal and maternal care.

Obstructive In the respiratory system, blocking the normal flow of gases through the lungs.

Olfaction (ōl-FAK-shun) The sense of smell.

Olfactory bulb A mass of gray matter containing neurons that form synapses with neurons of the olfactory (I) nerve, lying below the frontal lobe of the cerebrum on either side of the ethmoid bone.

Oligodendrocyte (OL-i-gō-den'-drō-sīt) A neuroglial cell that supports neurons and produces a myelin sheath around axons of neurons of the central nervous system.

Omega-3 Fatty acid Alpha-linoleic acid; a fat with an omega functional group on the third carbon. Found in vegetable and fish oils.

Omnivore Animal that can eat either plants or animals.

Oocyte Egg; the female gamete.

Oogenesis (ō'-ō-JEN-e-sis) Formation and development of female gametes (oocytes).

Open system A system with a starting point and an ending point rather than a continuous circular flow.

Optic chiasma The physical crossing of the left and right optic nerves.

Oral contraceptive A combination of estrogens and progestins that alters the natural hormonal rhythms of the female to prevent ovulation.

Orbital Region where electrons are found around atomic nucleus.

Order A taxonomic subcategory of class.

Organ Group of tissues having one primary function.

Organ of Corti The organ responsible for transmitting sound waves to the brain via nerve impulses.

Organ system A group of organs that perform a broad biological function, such as respiration or reproduction.

Organelle Typically a membrane-bound structure suspended in the cytosol; hair-like projections from the cell may also be called organelles.

Organism One living individual.

Orgasm A series of wavelike muscular contractions, and an intense pleasurable sensation, during sex.

Origin (of muscle) End of muscle that remains stationary during contraction.

Oropharynx (or'-ō-FAR-inks) The area directly behind the tongue, is covered by the uvula when it hangs down.

Osmolarity Osmotic pressure of a solution.

Osmosis (oz-MŌ-sis) Movement of water across a membrane, driven by differences in concentration on each side of the membrane.

Ossify Process of forming hard bone.

Osteoblasts (OS-tē-ō-blasts') Immature bone cells not yet surrounded by bony matrix.

Osteoclast (OS-tē-ō-clast') Cell that breaks down bone by removing calcium.

Osteocytes Mature bone cells surrounded by bone matrix.

Osteoid Bone matrix before it is calcified.

Osteon (OS-tē-on) The basic unit of structure in adult compact bone, consisting of a central (haversian) canal with its concentrically arranged lamellae, lacunae, osteocytes, and canaliculi. Also called a haversian (ha-VER-shan) system.

Otitis media Inflammation of the middle ear that fills it with fluid, distending the eardrum.

Outer ear The portion of the ear that extends from the fleshy pinna (external ear cartilage) through the auditory canal of the ear to the eardrum (tympanic membrane).

Oval window The fibrous connective tissue covering on the opening into the inner ear; the stapes attaches to the oval window.

Ovarian (ō-VAR-ē-an) **cycle** A monthly series of events in the ovary associated with the maturation of a secondary oocyte.

Ovary (Ō-var-ē) Female gonad, produces oocytes and the estrogens, progesterone, inhibin, and relaxin hormones.

Ovulation (ov-ū-LĀ-shun) The rupture of a mature ovarian (Graafian) follicle with discharge of a secondary oocyte into the pelvic cavity.

Oxygen debt The amount of oxygen needed to convert the lactic acid produced by anaerobic respiration into pyruvic acid and burn it entirely to CO_2, H_2O, and energy.

Oxyhemoglobin (ok'-sē-HĒ-mō-glō-bin) Hemoglobin molecule with at least one oxygen molecule bound to the iron center.

Oxytocin (ok'-sē-TŌ-sin) Hormone that initiates labor and causes the cells of the mammary gland to contract during the "milk let-down" response.

P wave The deflection wave of an electrocardiogram that signifies atrial depolarization.

Pacinian (pa-SIN-ē-an) **corpuscles** Structures deep in the dermis, near the hypodermis, that register pressure.

Pancreas (PAN-krē-as) A soft, oblong organ lying along the greater curvature of the stomach.

Pancreatic (pan'-krē-AT-ik) **amylase** Enzyme that digests carbohydrates.

Pancreatic juice The fluid produced by the pancreas and released into the small intestine.

Pancreatic lipase The enzyme that removes two of the three fatty acids from ingested triglycerides.

Pandemic An epidemic in a wide geographic region.

Papilla (pa-PIL-a) Any small rounded projection extending above a surface.

Papillary muscles Tufts of muscle extending from the walls of the ventricles, anchoring the valves.

Paracrine Hormone that affects only local cell.

Parasympathetic (par'-a-sim-pa-THET-ik) **division** One of the two subdivisions of the autonomic nervous system, originating in the brain stem and the sacral portion of the spinal cord; primarily concerned with activities that conserve and restore body energy.

Parathyroid (par-a-THĪ-royd) **gland** One of usually four small endocrine glands embedded in the posterior surfaces of the lateral lobes of the thyroid gland.

Parathyroid hormone (PTH) A hormone secreted by the chief cells of the parathyroid glands that increases blood calcium level and decreases blood phosphate level.

Parietal (pa-RĪ-e-tal) Pertaining to the parietal lobe of the brain.

Parotid glands Salivary glands located below and in front of the ears.

Partial pressure The percentage of the total gas pressure exerted by a single gas in the mixture.

Partition Dividing available resources into discrete parts to reduce competition.

Pathogens (PATH-ō-jenz) Agents that produce disease.

Pectoral (PEK-tō-ral) **girdle** The bones that attach the arm to the axial skeleton; the shoulder bones.

Pedigree chart Representation of genetic transmission of traits through families.

Pelvic cavity Cavity that contains urinary bladder, internal organs of reproduction, and part of large intestine.

Penis (PĒ-nis) The organ of urination and copulation in males; used to deposit semen into the female vagina.

Pepsin Enzyme that digests proteins.

Pepsinogen An inactive precursor of the enzyme pepsin.

Peptide bond Bond between the carboxyl group of one amino acid and the amino group of the adjacent amino acid.

Perforins Molecules released by T cell that break through the plasma membrane of the infected cell.

Pericardium (per-i-KAR-dē-um) Membrane surrounding heart.

Perimetrium (per'-i-MĒ-trē-um) The outer covering of the uterus.

Perimysium (per-i-MĪZ-ē-um) An inner connective tissue covering and supporting a group of muscle cells.

Periodic table Table that organizes all atoms by structure.

Periosteum (per'-ē-OS-tē-um) Membrane that covers bone.

Peripheral nervous system (PNS) That portion of the nervous system that consists of the nerves and sensory organs.

Peripheral protein A protein that sits on the inside or the outside of the cell membrane.

Peristaltic (per'-i-STAL-tic) **wave** Rhythmic muscular contractions of a tube that force contents lengthwise.

Peritoneum (per-i-tō-NĒ-um) Membrane lining the abdominal cavity.

Peritubular capillaries (*peri* = around; *tubular* = nephron tubules) Capillaries that surround the nephron.

Peyer's (PĪ-erz) **patches** Clusters of lymph nodules that are most numerous in the ileum.

Phagocytes (FAG-ō-sītz) Cells that endocytose (engulf) pathogens.

Phagocytosis (fag'-ō-sī-TŌ-sis) Cell eating, or taking in large molecules and particles through vacuoles.

Pharynx (FAR-inks) Throat.

Phenotype (FĒ-nō-tīp) An organism's observable characteristics.

Phospholipids Compounds containing phosphoric acid and a fatty acid.

Photoreceptor Receptor that detects light in the retina.

Photosynthesis Process of producing carbohydrates with solar energy, chlorophyll, carbon dioxide, and water.

Phylum A taxonomic subcategory of kingdoms. *Plural* is phyla.

Pia mater (PĪ-a MĀ-ter *or* PĒ-a MA-ter) The innermost of the three meninges (coverings) of the brain and spinal cord.

Pinna (PIN-na) The projecting part of the external ear composed of elastic cartilage and covered by skin and shaped like a trumpet.

Pinocytosis (pin- ō-sī-TŌ-sis) Cell drinking, or taking in a small quantity of the extracellular fluid.

Pituitary (pi-TOO-i-tār-ē) **dwarfism** A condition caused by hyposecretion of human growth hormone during development.

Placenta (pla-SEN-ta) Structure that connects uterus to fetus, providing nourishment.

Placenta previa Condition in which the placenta grows near or over the cervical opening of the uterus, blocking the passage of the fetus during birth.

Plantae The kingdom that includes plants.

Plaque (PLAK) A combination of bacterial colonies, their wastes, leftover sugars from chewed up food, epithelial cells from the host, and saliva.

Plaques (PLAKS) Fatty deposits of cholesterol that form in the arteries.

Plasma (PLAZ-ma) The extracellular fluid in blood vessels; blood minus the formed elements.

Plasma The clear, yellowish fluid portion of blood.

Plasmid Circular piece of double-stranded DNA outside the nucleus or the main DNA of the cell.

Platelet plug The first step in formation of a clot; a fragile plug that slows blood flow in small wounds.

Pleura (PLOO-ra) The serous membrane that covers the lungs and lines the walls of the chest and the diaphragm. The visceral pleura lines the lungs themselves, while the parietal pleura adheres to the walls of the cavity.

Pleural cavity Small potential space between the visceral and parietal pleurae filled with serous fluid.

Pleurisy (PLOO-ra- sē) Inflammation of the pleura.

Pluripotent cells Cells with the potential to become any adult cell type.

Pneumonia (noo-MŌ-n ē-a) Buildup of fluid in the lung, often in response to bacterial or viral infection.

Pneumotaxic (noo-mō-TAK-sik) A part of the respiratory center in the pons that sends inhibitory impulses to the inspiratory area, preventing overinflation of the lungs.

Pneumothorax (noo'-mō-THŌ-raks) Air in the pleural space.

Polar covalent bond Covalent bond that is electrically unbalanced, for example, water.

Polygenic Trait coded on several genes.

Polymerase The enzyme that adds nucleotides during DNA duplication.

Polymerase chain reaction (PCR) A series of reactions that amplifies DNA using the same enzymes that cells use to synthesize DNA.

Polymer Long chain of repeating subunits.

Polyp Growth protruding from a mucous membrane.

Polyspermy Many sperm entering one ovum.

Pons (PONZ) The area superior to the medulla oblongata, involved in transfer of information and respiratory reflexes.

Population All representatives of a specific organism found in a defined area.

Portal systems Vascular systems that carry blood from arteries to veins to capillaries to veins, back to capillaries, then on to veins and the heart.

Postsynaptic (pōst-sin-AP-tik) **neuron** The neuron that begins after the synapse; its dendrites pick up diffusing neurotransmitters.

Precapillary sphincter (SFINGK-ter) A ring of smooth muscle cells at the origin of the capillaries to regulate blood flow into them.

Premature baby A baby born before 37 weeks of gestation.

Premenstrual dysphoric disorder (PMDD) Group of physiological and emotional symptoms linked to the menstrual cycle.

Presynaptic (prē-sin-AP-tik) **neuron** The neuron before the synapse, whose axon leads to the synapse.

Primary motor area A region of the cerebral cortex in the frontal lobe that controls specific muscles or groups of muscles.

Primary succession Colonization of life on bare rock or sand.

Prime mover The muscle in an antagonistic pair that shortens during a specific movement; agonist.

Producer autotroph Organism that creates its own nutrients from inorganic substances; mainly green plants.

Progesterone (prō-JES-te-rōn) A female sex hormone produced by the ovaries; helps prepare the uterus for implantation and mammary glands for milk secretion.

Prokaryotic A cell with no internal membrane-bound compartments, usually having only ribosomes and chromosomes as recognizable organelles.

Prolactin (PRL) Hormone that stimulates milk production in females.

Prolapse (PRŌ-laps) Movement of an organ from its original position, usually because of gravity or other pressure.

Proprioception (prō-prē-ō-SEP-shun) Stimuli from within the body giving information on body position and posture.

Prostaglandin (pros'-ta-GLAN-din) A membrane-associated lipid; released in small quantities and acts as a local hormone.

Protein A macronutrient consisting of carbon, hydrogen, oxygen, nitrogen, and sometimes sulfur and phosphorus; synthesized on ribosomes and made up of amino acids linked by peptide bonds.

Prothrombin (prō-THROM-bin) Blood-clotting factor synthesized by the liver, released to the blood, and converted to active thrombin during blood clotting.

Proton The positive particle in the atomic nucleus.

Proximal (PROK-si-mal) Nearer the attachment of a limb to the trunk; nearer to the point of origin or attachment.

Puberty (PŪ-ber-tē) The time of life when the secondary sex characteristics begin to appear and sexual reproduction becomes possible; usually between ages 10 and 17.

Pulmonary (PUL-mo-ner'-ē) **trunk** The vessels leaving the right side of the heart, going toward the lungs.

Pulmonary and aortic (ā-OR-tik) **valves** Valves between the ventricles of the heart and the great vessels. The pulmonary valve lies between the right ventricle and the pulmonary arch; the aortic valve lies between the left ventricle and the aorta.

Pulmonary circuit Blood flow from the heart to the lungs and back to the heart.

Pulmonary edema Fluid build-up in the lungs due to congestive heart failure.

Pulp Soft tissue near the tooth's nerve.

Pupil The hole in the center of the iris.

Purkinje (pur-KIN-jē) **fibers** Conduction myofibers that reach individual cells of the ventricles.

Pyloric (pī-LOR-ik) **sphincter** Located at the end of the stomach, it opens to allow chyme to enter the small intestine when it is chemically ready.

Pyrogens (PĪ-ro-jenz) Proteins that reset the body's thermostat to a higher temperature.

Pyruvate Three-carbon compounds that form in the cytoplasm, from the initial breakdown of glucose.

QRS complex The deflection waves of an electrocardiogram that represent beginning of ventricular depolarization.

Q-T interval The total time of ventricular contraction and relaxation.

Quadrant Four-part division of the body used to describe organ location.

Radiation The transfer of heat from a warm body to the surrounding atmosphere.

Radioactive decay Change of an atom into another element through division of the nucleus and the release of energy.

Radioisotope Isotope that decays spontaneously, releasing energy.

Recessive An allele of a gene that determines phenotype only when two like alleles are present.

Recombinant DNA The product of splicing genes.

Referred pain Pain the brain interprets as coming from an area other than its actual origin.

Reflex Fast response to a change (stimulus) in the internal or external environment.

Relative refractory period The period immediately after an action potential when the sodium channels are in their original position, but the transmembrane potential has not yet stabilized at resting levels.

Relaxin A female hormone produced by the ovaries and placenta that relaxes the smooth muscle and helps dilate the cervix to ease delivery.

Remission A decrease of disease symptoms leading to an apparent curing of the disease; indicates that the disease is still present, but dormant.

Renal (RĒ-nal) **pyramids** Cone-shaped structures formed from an accumulation of collecting ducts in the medulla of the kidney.

Renal pelvis A cavity in the center of the kidney that collects urine and passes it to the ureters.

Reservoir Location that holds a compound in a way that is inaccessible to the user.

Residual volume (RV) The amount of air that always remains in the lungs.

Respiratory membrane The thin, membranous part of the respiratory system, where gases are exchanged.

Respiratory system The system that brings oxygen to the blood and removes carbon dioxide from it.

Respiratory zone Portion of the respiratory tract where gas exchange occurs.

Reticular (re-TIK-ū-lar) **activating system (RAS)** A portion of the reticular formation that has many ascending connections with the cerebral cortex; produces generalized alertness or arousal from sleep when active.

Reticular formation A network of small groups of neuronal cell bodies beginning in the medulla oblongata and extending superiorly through the central part of the brain stem.

Retina (RET-i-na) The deep coat of the posterior portion of the eyeball consisting of nervous tissue that detect light and create nerve impulses.

Retroverted (tipped) uterus Condition in which the uterus lies against the rectum.

Retrovirus Virus that carries RNA as its genetic material.

Reverse transcriptase An enzyme that forms DNA from RNA.

Rhinoplasty Surgery on the nose done to treat a "deviated septum" or for cosmetic reasons.

Rhodopsin Visual pigment that responds to low levels of white light.

Ribosome (RĪ-bō-sōm) Organelle that synthesizes proteins.

Ribs Flattened bones that emerge from the cervical or thoracic spine to shape the thorax.

Rigor mortis Rigidity that occurs in muscles after death.

Rough endoplasmic reticulum (RER) Membrane that processes and sorts proteins synthesized by ribosomes.

Rugae (ROO-gē) Folds in the wall of the stomach that permit expansion.

Saccule (SAK-ūl) Small circular vesicle used to transport substances within a cell.

Salivary glands Glands in the oral cavity that secrete saliva to moisten the oral mucosa.

Sarcolemma (sar'-kō-LEM-ma) The cell membrane of a muscle fiber (cell), especially of a skeletal muscle fiber.

Sarcomere (SAR-kō-mēr) The contractile unit of a myofiber.

Schwann (SCHWON) **cell** A neuroglial cell of the peripheral nervous system that forms the myelin sheath wrapping around the axon in jelly-roll fashion.

Scientific method System of study that includes observation, hypothesis generation, testing, data collection, drawing conclusions, and communication of the results of the experiment.

Sclera (SKLE-ra) The white coat of fibrous tissue that forms the superficial protective covering over the eyeball.

Sebaceous (se-BĀ-shus) **glands** Oil glands found in the dermis of the skin, associated with hair follicles.

Secondary succession Ecological change in an ecosystem after a disturbance.

Secrete To move from the blood to the filtrate, using energy.

Secretin Hormone that decreases gastric secretions.

Secretory phase Phase of the menstrual cycle when the endometrial glands function.

Self-pollinate Transfer the pollen of a flower directly to the stigma of the same flower.

Semen The fluid containing sperm and other components, formed as the sperm moves through the reproductive system.

Seminal vesicles Glands on the posterior base of the urinary bladder that secrete an alkaline, fructose-rich fluid.

Semipermeable Describes a membrane that is permeable to some compounds, but not others.

Senescence Reaching old age, growing old.

Sensory neurons Neurons that carry sensory information.

Septal cells Cells found in the alveolar membrane that secrete surfactant.

Septicemia (sep-ti-SĒ-mē-a) A life-threatening condition in which the blood carries bacterial toxin.

Septum (SEP-tum) A wall dividing two cavities.

Sex-influenced trait Trait carried on autosomal chromosomes that is more common in one sex than another.

Sex-linked trait Trait coded by genes that are carried on the one sex chromosome with no counterpart on the other sex chromosome.

Simple epithelium Single layer of cells that often functions as a diffusion or absorption membrane.

Sinoatrial (si-nō-Ā-trē-al) **(SA) node** A small mass of heart cells located in the right atrium that spontaneously depolarize and generate the resting heartbeat. Also called the pacemaker.

Skeletal muscle Contractile tissue composed of protein filaments arranged to move the skeletal system.

Sleep apnea (*a-* = without, *pnea* = breath) The periodic cessation of breathing during sleep.

Sliding filament model Standard explanation of how a muscle cell creates contraction.

Slow block Deactivation of the sperm receptors in the zona pellucida, preventing interaction between a fertilized egg and a second sperm.

Slow-twitch fiber Myofibril that contracts relatively slowly.

Smog Nitrogen oxides and hydrocarbon pollution that turns brown or gray in sunlight.

Smooth endoplasmic reticulum (SER) Membrane that synthesizes fatty acids and steroid hormones.

Sodium potassium exchange pump (Na^+/K^+ ATPase) An active transport pump located in the plasma membrane that transports sodium ions out of the cell and potassium ions into the cell at the expense of cellular ATP.

Soft connective tissue Connective tissue with a matrix composed of a semifluid ground substance, fibroblasts, and white blood cells.

Solute Substance dissolved in the solvent.

Somatic (sō-MAT-ik) Related to the body, in contrast to the gametes.

Somatostatin A water-soluble hormone that prevents the secretion of growth hormone; literally to keep the body the same (*soma* = body).

Species diversity The variation in species in a particular location.

Specific gravity A ratio of the density of a substance to the density of pure water.

Specific immunity Immunity directed by white blood cells, antibodies, and macrophages that specifically target individual pathogens.

Spermatic cord The artery, vein, nerve, lymphatics, and ductus deferens that lead from the abdominal cavity to the testes.

Spermatogenesis (sper'-ma-tō-JEN-e-sis) The formation of spermatids (immature sperm cells).

Spermicide Birth-control measure that kills sperm inside female reproductive tract.

Spider veins Small visible yet flat veins on the surface of the body, made visible by trapped red blood cells.

Spinal (SPĪ-nal) **cord** The part of the central nervous system contained within the vertebral canal (spinal cavity).

Spinal cavity Cavity inside vertebral column; houses spinal cord.

Spinal nerves One of the 31 pairs of nerves that originate on the spinal cord from posterior and anterior roots.

Spliced Two pieces of DNA artificially joined together to form new genetic combinations.

Squamous (SKWĀ-mus) **cell** Flattened cell; squamous epithelium forms a diffusion membrane.

Stable angina Pain that develops in the heart only under specific and identifiable conditions, such as strenuous exercise or smoking.

Stapes The third of the auditory ossicles, transferring the movement of the tympanic membrane directly to the oval window and the fluid of the inner ear.

Starling's law When the ventricles are stretched by increased blood volume, they recoil with matching force. This increased blood flow to the heart, which occurs when we start hard physical work, causes the heart to respond with more forceful pumping.

Stem cells Undifferentiated cells that remain able to divide and specialize into functional cells.

Stent Medical instrument that is inserted into a weakened blood vessel for support.

Stereocilia (ste'-rē-ō-SIL-ē-a) Groups of extremely long, slender, nonmotile microvilli.

Stereoscopic Depth perception gained through use of both eyes.

Stomach A J-shaped organ that lies beneath the esophagus and is divided from the esophagus and the small intestine by two sphincter muscles.

Stratified epithelium Several layers of epithelial cells.

Stratum corneum The top layer of the epidermis that is composed of dead cells joined by strong cell-to-cell junctions.

Stratum functionalis Outer layer of endometrium that grows and sheds in response to hormone levels in the blood.

Striation A series of parallel lines.

Submandibular glands Salivary glands located under the tongue that produce thick, ropy saliva with a large concentration of mucus.

Submucosa Second layer of the GI tract, found under the mucosa, and including the glands, nerves, and blood supply for the GI tract.

Succession The sequential change in species in an ecosystem.

Sucrase Enzyme that digests carbohydrates.

Sulcus (SUL-kus) A shallow groove on the surface of the brain. *Plural* is *sulci* (SUL-sī).

Summation Buildup of contractions inside a myofiber.

Superior (soo-PĒR-ē-or) Toward the head or upper part of a structure.

Surfactant (sur-FAK-tant) Detergent-like fluid that moistens the alveoli and prevents the walls from sticking together during exhalation.

Suture (SOO-chur) Inflexible joint between two fixed bones.

Symbiotic Intimate co-existence of two organisms in a mutually beneficial relationship.

Sympathetic (sim'-pa-THET-ik) **division** One of two subdivisions of the autonomic nervous system, originating in the thoracic segment and the first two or three lumbar segments of the spinal cord; primarily concerned with processes involving the expenditure of energy.

Sympathetic chain A cluster of cell bodies of sympathetic neurons close to the body of a vertebra. These are found in the neck, thorax, and abdomen on both sides of the vertebral column and are connected in a chain on each side of the vertebral column.

Synapse (SIN-aps) Gap between neurons, across which a nerve impulse must flow via chemical signal.

Synarthrotic A joint that is not movable.

Synergistic (syn-er-JIS-tik) **pair** Two muscles or muscle groups that combine to control the movement of one joint; antagonistic pair.

Syngamy (SIN-ga-mē) Process in which one sperm penetrates the zona pellucida and fuses with the oocyte membrane.

Synovial (sī-NŌ-vē-al) **fluid** Fluid secreted by the inner membrane of a synovial joint, similar in viscosity to egg white.

Synovial joint A fully movable or diarthrotic joint in which a synovial (joint) cavity is present between the two articulating bones.

Systemic circuit Blood flow from the heart to the tissues of the body and back to the heart.

Systole (SIS-tō-lē) Contraction phase of heartbeat.

T tubules Areas where the sarcolemma forms a tube that crosses through the muscle cell, carrying contractile impulses to the opposite side of the cell.

T wave The deflection of an electrocardiogram that represents ventricular repolarization.

Tachycardia (tak'-i-KAR-dē-a) Resting heart rate above 100 beats per minute.

Target cell A cell whose activity is affected by a particular hormone.

Taxonomy The study of classification, based on structural similarities and common ancestry.

Tectoral membrane The structure within the organ of Corti that deforms with sound waves and generates nerve impulses .

Tendon Dense regular connective tissue connecting muscle and bone.

Terminal bulb The swollen terminal end of the axon that releases neurotransmitters into the synapse.

Testosterone (tes-TOS-te-rōn) A male sex hormone needed for development of sperm, male reproductive organs, secondary sex characteristics, and body.

Tetanus State of continuous contraction in a myofibril.

Thalamus (THAL-a-mus) A large, oval structure located on either side of the third ventricle of the brain; main relay center for sensory impulses heading to the cerebral cortex.

Therapeutic abortion Removal of the developing embryo for medical reasons.

Thoracic (thor-AS-ik) **cavity** Chest and its contents.

Threshold stimulus The minimum stimulation needed to cause a reaction.

Thymus (THĪ-mus) **gland** A bilobed organ, located in the upper thoracic cavity behind the sternum and between the lungs, in which T cells mature.

Thyroid (THĪ-royd) **cartilage** Shield-shaped cartilage that composes the front of the larynx.

Thyroid (THĪ-royd) **stimulating hormone (TSH)** Activates the thyroid to produce T3 (triiodothyronine) and T4 (thyroxin), which maintain basal metabolic rate.

Tidal volume (TV) The volume of air inhaled per minute during normal breathing, approximately 500 ml.

Tissue Group of cells with similar function.

Titer (TĪ-ter) Level of a compound or antibody in the blood.

Tonsils (TON-silz) A group of large lymphatic nodules embedded in the mucous membrane of the throat.

Trabeculae (tra-BEK-ū-lē) Struts that form in response to stress in spongy bone.

Trachea (TRĀ-kē-a) Windpipe; tubular air passageway extending from the larynx to the fifth thoracic vertebra.

Tracheotomy Insertion of a temporary breathing tube in order to prevent suffocation due to a crushed larynx.

Tracts Axons and/or dendrites with a common origin, destination, and function.

Transcription Process of copying information from DNA to RNA.

Transfer RNA (tRNA) RNA that "reads" mRNA at ribosome, one codon at a time.

Transgenic An organism with a gene or group of genes in its genome that was transferred from another species.

Translation Converting information from one language to another.

Transport protein Protein that assists in facilitated diffusion.

Treppe (trep') The increased strength of contraction after successive identical stimuli.

Tricuspid (trī-KUS-pid) The valve between the right atrium and right ventricle, composed of three points (cusps) of connective tissue.

Trimester One of three, three-month periods during pregnancy.

Trophic level All the organisms that occupy the same energy tier in a community, such as primary producers, primary consumers, and secondary consumers.

Trophoblast (TRŌF-ō-blast) The superficial covering of cells of the blastocyst.

Trypsin Pancreatic enzyme that digests proteins.

Tubal ligation Surgical procedure that blocks fallopian tubes, preventing union of sperm and egg.

Tuberculosis Disease caused by *Mycobacterium tuberculosis* infection.

Turgor Internal pressure in living cells.

Tympanic (tim-PAN-ik) **membrane** A thin, semitransparent partition of fibrous connective tissue between the auditory canal and the middle ear; eardrum.

Tympanic canal The lower compartment of the cochlea, continuous with the round window, where sound waves are released from the fluid of the inner ear.

Ulcers Open wounds that remain aggravated in the GI tract.

Ultrasound examination Bouncing sound waves through the maternal skin into the uterus to observe the reflected patterns.

Umbilical cord The flexible cord that connects the fetal circulatory system with the placenta.

Unstable angina Pain that develops in the heart seemingly randomly, with no connection to activity or situation.

Upper respiratory tract Respiratory organs in the face and neck.

Ureter (Ū-rē-ter) One of two tubes that connect the kidney with the urinary bladder.

Urethra (ū-RĒ-thra) The duct from the urinary bladder to the exterior of the body that conveys urine in females and urine and semen in males.

Urinalysis (ū-ri-NAL-i-sis) An analysis of the volume and physical, chemical, and microscopic properties of urine.

Urinary (Ū-ri-ner-ē) **bladder** A hollow, muscular organ situated in the pelvic cavity that stores urine until it is excreted through the urethra.

Urine The fluid produced by the kidneys that contains wastes and excess materials.

Urogenital (ū'-rō-JEN-i-tal) Concerning both the urinary and reproductive systems.

Uterus (Ū-te-rus) The hollow, muscular organ in females; site of menstruation, implantation, development of fetus, and labor.

Utricle (Ū-tri-kul) The larger of the two divisions of the membranous labyrinth located inside the vestibule of the inner ear, containing a receptor organ for static equilibrium.

Uvula (Ū-vū-la) The tab of soft tissue that hangs down in the back of the throat, visible as a pointed tab.

Vagina (va-JĪ-na) A tubular organ leading from the uterus to the vestibule.

Vagus nerve Cranial nerve X that innervates the muscles of the throat, and thoracic and abdominal organs.

Valence shell A group of electron orbitals around the nucleus.

Van der Waals force Weak interaction between resonating molecules.

Variable A factor that can be changed in an experiment to test whether and how it affects the outcome.

Varicose veins A medical condition in which superficial veins fill with blood, but do not empty, resulting in a distended, often painful swelling on the surface of the body.

Vegan A vegetarian who consumes only plant products, eating no animal products whatsoever.

Vein A blood vessel that carries blood from tissues back to the heart.

Vena cavae (VĒ-na CĀ-vē) The two large veins that open into the right atrium, returning to the heart all of the deoxygenated blood from the systemic circulation except that from the coronary circulation.

Ventral cavity Entire ventral aspect of torso; belly and chest.

Ventricle (VEN-tri-kul) A cavity in the brain filled with cerebrospinal fluid. An inferior chamber of the heart.

Venule (VEN-ūl) Small vein that drains blood from capillaries to larger veins.

Vertebrae Bony structures that comprise the vertebral column.

Vertebral (VER-te-bral) **body** The vertebra, exclusive of the vertebral arch.

Vestibular (ves-TIB-ū-lar) **canal** The uppermost compartment of the cochlea, continuous with the oval window, where sound waves travel on their way to the auditory nerves.

Vestibulocochlear (ves-tib'-ū-lō-KOK-lē-ar) **nerve** Cranial nerve VIII that carries impulses from the ear to the brain.

Villi (VIL-ī) Fingerlike digestive extensions from intestinal mucosal cells. *Singular* is *villus* (VIL-lus).

Visual acuity The resolving power of the eye.

Vital capacity (VC) The sum of inspiratory reserve volume, tidal volume, and expiratory reserve volume.

Vitreous (VIT-rē-us) **humor** A gel-like fluid that holds the third tunic, the retina, in place.

Vocal cords A pair of cartilaginous cords stretched across the laryngeal opening that produce the voice.

Voltage-regulated Membrane channels that open or close in response to changes in the transmembrane electrical charge (membrane potential), of the cell.

Water potential Osmotic pressure of resting cells in an isotonic solution; equals pressure from the environment plus the cell's solute concentration.

Yolk sac A structure that provides some nutrition to the embryo.

Zona pellucida (pe-LOO-si-da) A gel-like layer surrounding the maturing oocyte.

Zooplankton Microscopic and macroscopic animals that float in the water column and move at the mercy of the currents.

Zygote (ZĪ-got) The cell resulting from the union of male and female gametes; the fertilized ovum.

ART CREDITS

Chapter 1

Pages 2-3: Norbert Rosing/NG Image Collection; page 2 (inset) PhotoDisc/Getty Images; page 5: Dr. Gopal Murti/Getty Images; page 4: Janeart/Getty Images; page 7: Michael S. Quinton/NG Image Collection; page 8: Stacy Gold/NG Image Collection; page 10 (bottom left) Rubberball Productions/Getty Images; page 10 (bottom right) Courtesy Michael Ross, University of Florida; page 11 (top left) James L. Stanfield/NG Image Collection; page 11 (top right) Joel Sartore/NG Image Collection; page 11 (center left) William Albert Allard/NG Image Collection; page 11 (bottom right) Raymond Gehman/NG Image Collection; page 11 (bottom left) Randy Olson/NG Image Collection; page 12: Beverly Joubert/NG Image Collection; page 12: Norbert Rosing/NG Image Collection; page 12: Dr. Richard Kessel/Getty Images; page 12: NIAID/CDC/Photo Researchers; page 12: T. Stevens & P. McKinley, PNNL/Photo Researchers; page 12: Raymond Gehman/NG Image Collection; page 13: Jim & Jamie Dutcher/NG Image Collection; page 14: Medford Taylor/NG Image Collection; page 14: Tim Laman/NG Image Collection; page 14: Joel Sartore/NG Image Collection; page 14: Tim Laman/NG Image Collection; page 14: Kenneth Garrett/NG Image Collection; page 14: Kenneth Garrett/NG Image Collection; page 14: Mark Cosslett/NG Image Collection; page 17: Mariea Stenzel/NG Image Collection; page 18: Paul Nicklen/NG Image Collection; page 20: Karen Kasmauski/NG Image Collection; page 21 (top) Paula Bronstein/Getty Images; page 21 (bottom left) SuperStock Inc.

Chapter 2

Pages 26-27: Courtesy NASA; page 26 (inset) Sisse Brimberg/NG Image Collection; page 29: Michael Keller/Corbis; page 30: Kenneth Eward/Photo Researchers; page 32: Mark Harmel/Stone/Getty Images; page 36: Roy Toft/NG Image Collection; page 37: Wes C. Skiles/NG Image Collection; page 39: L.S. Stepanowicz/Visuals Unlimited; page 40: David Woodfall/Riser/Getty Images; page 45: David Mclain/NG Image Collection; page 44: Susan Van Etten/PhotoEdit; page 48 (right) Dr. Tim Evans/Photo Researchers; page 50: Dr. Dennis Kunkel/Visuals Unlimited; page 52: Doug Struthers/Stone/Getty Images

Chapter 3

Pages 60-61: Flip Nicklin/NG Image Collection; page 60 (inset) Carsten Peter/NG Image Collection; page 64 (left) Richard Nowitz/NG Image Collection; page 64 (right) Richard Nowitz/NG Image Collection; page 66: Andy Washnik; page 67 (bottom left) David Phillips/Photo Researchers, Inc.; page 67 (bottom center) David Phillips/Photo Researchers, Inc.; page 67 (bottom right) David Phillips/Photo Researchers, Inc.; page 69 (top left) © Dennis Kunkel/Phototake; page 69 (top right) Biophoto Associates/Photo Researchers, Inc.; page 69 (bottom) Steve Gschmeissner/Science Photo Library/Photo Researchers, Inc.; page 70: Ben Edwards/Stone/Getty Images; page 72: © Dr. David M. Phillips/Visuals Unlimited; page 73 (right) © Robert Bolender & Dr. Donald Fawcett/Visuals Unlimited; page 74: Biology Media/Photo Researchers, Inc.; page 75 (bottom right) Dr. Gopal Murti/Visuals Unlimited; page 76 (left) Dr. Kari Lounatmaa/Photo Researchers, Inc.; page 79 (center) Keith R. Porter/Photo Researchers, Inc.; page 81: Biophoto Associates/Photo Researchers, Inc.; page 83: Eric V. Grave/Photo Researchers, Inc.; page 85: Photodisc Blue/Getty Images; page 86: Courtesy Michael Ross, University of Florida; page 86: Courtesy Michael Ross, University of Florida; page 86: Courtesy Michael Ross, University of Florida; page 86: Courtesy Michael Ross, University of Florida; page 86: Courtesy Michael Ross, University of Florida; page 86: Courtesy Michael Ross, University of Florida

Chapter 4

Pages 92-93: ?Brand X/SuperStock; page 94: Eye of Science/Photo Researchers, Inc.; page 104: Associated Press; page 102: Mike Kemp/Rubberball/Getty Images; page 97 (top left) Courtesy Michael Ross, University of Florida; page 97 (center left) Courtesy Andrew J. Kuntzman; page 97 (bottom left) Courtesy Michael Ross, University of Florida; page 98 (top left) Courtesy Michael Ross, University of Florida; page 98 (center left) Courtesy Michael Ross, University of Florida; page 98 (bottom left) Courtesy Michael Ross, University of Florida; page 99 (top center) John Burbidge/Photo Researchers; page 99 (bottom center) Courtesy Michael Ross, University of Florida; page 100 (center) Courtesy Michael Ross, University of Florida; page 101 (bottom center) Courtesy Michael Ross, University of Florida; page 101 (top center) Courtesy Michael Ross, University of Florida; page 103: Science VU/Visuals Unlimited; page 107 (top center) Rubberball Productions/Getty Images; page 107 (bottom left) Rubberball Productions/Getty Images; page 107 (bottom right) Courtesy Michael Ross, University of Florida; page 109: Pablo Corral Vega/NG Image Collection; page 108: Courtesy Michael Ross, University of Florida

Chapter 5

Pages 118-19: Neil Borden/Photo Researchers; page 118 (inset) Patrick McFeeley/NG Image Collection; page 127 (top right) John Burbidge/Photo Researchers; page 120: S. Fraser/Photo Researchers; page 133: Dr. M.A. Ansary/Photo Researchers; page 137 (top) Larry Mulvehill/Photo Researchers; page 148: Science Photo Library/Photo Researchers; page 150: John Wilson White; page 144: Edward Kinsman/Photo Researchers; page 149 (bottom) CNRI/Phototake

Chapter 6

Pages 156-57: Jim Cummins/Taxi/Getty Images; page 158: George F. Mobley/NG Image Collection; page 159 (right) Suza Scalora/PhotoDisc/Getty Images; page 173: Joel Sartore/NG Image Collection; page 174: Photofest; page 177 (left) Biophoto Associates/Photo Researchers; page 177 (right) Bruce Dale/NG Image Collection; page 180: Jodi Cobb/NG Image Collection; page 181 (top) David Gifford/Photo Researchers; page 179: Firstlight/Getty Images; page 178: Photodisc Blue/Getty Images

Chapter 7

Pages 186-87: Dr. Scott T. Grafton/Visuals Unlimited; page 186 (inset) REZA/NG Image Collection; page 195 (bottom) Kent Wood/Photo Researchers; page 199: Dr. John Zajicek/Photo Researchers; page 201: Associated Press; page 203: Ralph Hutchings/Visuals Unlimited; page 207: Alfred Pasieka/Photo Researchers; page 210: NG Image Collection; page 216: Stephen Alvarez/NG Image Collection

Chapter 8

Pages 228-29: NASA/NG Image Collection; page 228 (inset) Tomasz Tomaszewski/NG Image Collection; page 232: Mark W. Moffett/NG Image Collection; page 234: Taylor Kennedy/NG Image Collection; page 235: Anatomical Travelogue/Photo Researchers; page 242: Antonia Reeve/Science Photo Library/Photo Researchers; page 423: Rohen JW, Yokochi C, Lutjen-Drecoll E. *Color Atlas of Anatomy.* 6ed. Stuttgart, New York: Schattauer 2006: 294; page 244: Courtesy Michael Ross, University of Florida; page 247: O. Louis Mazzatenta/NG Image Collection; page 248: Joe McNally/Sygma/NG Image Collection; page 249: Taylor Kennedy/NG Image Collection

Chapter 9

Pages 254-55: Associated Press; page 254 (inset) Associated Press; page 257: Stephanie Adams/AgeFotostock; page 260: Karen Kasmausk/NG Image Collection; page 263 (right) Courtesy Michael Ross, University of Florida; page 264 (top) Tomasz Tomaszewski/NG Image Collection; page 264 (bottom) Sylvain Grandadam/Photo Researchers; page 266 (left) CNRI/Photo Researchers, Inc.; page 266 (center) Dr. P. Marazzi/Photo Researchers, Inc.; page 266 (right) Dr. P. Marazzi/Photo Researchers, Inc.; page 270: Scot Soka/NG Image Collection; page 267: Carl Rohrig/NG Image Collection; page 268 (right) Randy Olson/NG Image Collection; page 277: Dr. Dennis Kunkel/Visuals Unlimited

Chapter 10

Pages 282-83: CORBIS; page 282 (inset) James Cavallini/Photo Researchers, Inc.; page 287 (bottom) M.I. Walker/Photo Researchers, Inc.; page 289: Carl Rohrig/NG Image Collection; page 291 (top) Martin Dohrn/Photo Researchers, Inc.; page 291 (bottom) James Cavallini/Photo Researchers, Inc.; page 290: Todd Gipstein/NG Image Collection; page 292 (top) Anatomical Travelogue/Photo Researchers, Inc.; page 292 (bottom) Chris Hondros/Getty Images News; page 294: David Scharf/Science Photo Library/Photo Researchers; page 304 (left) Richard Lord/The Image Works; page 304 (right) Stacy Gold/NG Image Collection; page 305: Custom Medical Stock Photo; page 306: Courtesy Heide Schulz, Max Planck Institute for Marine Mikrobiology, Bremen, Germany; page 307: Michael Abbey/Visuals Unlimited; page 307: Michael Abbey/Visuals Unlimited; page 307: Michael Abbey/Visuals Unlimited; page 307: Michael Abbey/Visuals Unlimited; page 307: Michael Abbey/Visuals Unlimited; page 309: David R. Frazier/Photo Researchers, Inc.; page 315 (bottom) Karen Kasmauski/NG Image Collection; page 316: Gideon Mendel/NG Image Collection; page 318: James P. Blair/NG Image Collection; page 319 (top) Stephen Alvarez/NG Image Collection

Chapter 11

Pages 324-25: Yann Arthus-Bertrand/NG Image Collection; page 333 (bottom) Courtesy Michael Ross, University of Florida; page 334: Antonia Reeve/Photo Researchers, Inc.; page 335: Antonia Reeve/Photo Researchers, Inc.; page 344 (bottom left) Vu/Cabisco/Visuals Unlimited; page 344 (bottom right) W. Ober/Visuals Unlimited; page 344 (top) Howard Sochurek/NG Image Collection; page 345 (bottom) ISM/Phototake; page 346: Science Photo Library/Photo Researchers; page 350 (bottom) Courtesy Michael Ross, University of Florida; page 352: Science Photo Library/Photo Researchers; page 353 (right) Courtesy Michael Ross, University of Florida; page 356 (bottom) Dennis Kunkel/Phototake; page 359: Biophoto Associates/Photo Researchers, Inc.; page 360: Lewin/Royal Free Hospital/Photo Researchers; page 338 (left) Forrest Anderson/?AP/Wide World Photos; page 338 (right) Associated Press; page 347: Michael Nichols/NG Image Collection

Chapter 12

Pages 366-67: David Allen Harvey/NG Image Collection; page 366 (inset) Pr. M. Brauner/Photo Researchers, Inc.; page 370: Brian Evans/Photo Researchers, Inc.; page 375: Anatomical Travelogue/Photo Researchers, Inc.; page 377: Mark Nielsen; page 378: Zephyr/Photo Researchers, Inc.; page 379 (top right) Biophoto Associates/Photo Researchers; page 390: Jodi Cobb/NG Image Collection; page 381: Biophoto Associates/Photo Researchers; page 386 (top) Joel Sartore/NG Image Collection; page 390: Illustration, Irving Geis. Image from the Irving Geis Collection/Howard Hughes Medical Institute. Rights owned by HHMI. Reproduction by permission only.; page 394: Dr. P. Marazzi/Photo Researchers, Inc.; page 396: Bluestone/Photo Researchers, Inc.; page 395: Biophoto Associates/Photo Researchers, Inc.; page 398: Dr. P. Marazzi/Photo Researchers, Inc.; page 397: Annabella Bluesky/Photo Researchers, Inc.; page 399: Simon Fraser/Royal Victoria Infirmary/Photo Researchers, Inc.

Chapter 13

Pages 404-405: Susie M. Elsing Food Photography/StockFood America; page 404 (inset) Stacy Gold/NG Image Collection; page : Innerspace Imaging/Photo Researchers, Inc.; page 406: Justin Guariglia/NG Image Collection; page 410: Pixtal/AgeFotostock; page 418: Charles Kogod/NG Image Collection; page 420: Science Photo Library/Photo Researchers, Inc.; page 422: David M. Martin, M.D./Photo Researchers, Inc.; page : From Johannes W. Rohen, Chihiro Yokochi and Elke L[&~rom~Å~normal~&]tjen-Drecoll, Color Atlas of Anatomy, Schattauer Publishing, Stuttgart, Germany. Reproduced iwth permission; page 424: Steve Gschmeissner/Photo Researchers, Inc.; page 426: Innerspace Imaging/Photo Researchers, Inc.; page 427 (left) Science Photo Library/Photo Researchers, Inc.; page 427 (right) George Wilder/Visuals Unlimited; page 428: Dennis Kunkel/Phototake, Inc.; page 429: David M. Martin, M.D./Photo Researchers, Inc.; page 430: Science Photo Library/Photo Researchers, Inc.; page 431: Du Cane Medical Imaging Ltd/Photo Researchers, Inc.; page 437 (bottom) Jodi Cobb/NG Image Collection; page 438: Jim Richardson/NG Image Collection; page 438 (inset) Eye of Science/Photo Researchers, Inc.

Chapter 14

Pages 444-45: Tim Laman/NG Image Collection; page 448: Mark Nielsen; page 469: Ian Hooten/Photo Researchers, Inc.; page 454: Karen Kasmauski/NG Image Collection; page 458: Stephen J. Krasemann/Photo Researchers, Inc.; page 459 (left) Du Cane Medical Imaging Ltd/Photo Researchers, Inc.; page 459 (right) CNRI/Photo Researchers, Inc.; page 464: Lea Paterson/Photo Researchers, Inc.; page 465: Steve Raymer/NG Image Collection; page 467: Jean Luc Morales/The Image Bank/Getty Images; page 471: AJPhoto/Photo Researchers, Inc.

Chapter 15

Pages 476-77: Donald Bowers/SuperStock; page 487 (top) Bettina Cirone/Photo Researchers, Inc.; page 487 (bottom) Lester Bergman/The Bergman Collection; page 488: BananaStock/AgeFotostock; page 489 (bottom right) Courtesy Michael Ross, University of Florida; page 495: Garo/Photo Researchers, Inc.; page 498: Blend Images/Getty Images; page 492 (bottom center) Biophoto Associates/Photo Researchers, Inc.; page 492 (bottom right) Bettmann/CORBIS; page 493 (top) Lester Bergman/The Bergman Collection; page 493 (bottom) Martin Rotker/Phototake; page 499: Cordelia Molloy/Photo Researchers, Inc.; page 501 (left) Petit Format/Photo Researchers, Inc.; page 502 (left) Justin Guariglia/NG Image Collection; page 502 (right) Richard Nowitz/NG Image Collection; page 503 (bottom left) Richard Nowitz/NG Image Collection; page 503 (bottom right) David Edwards/NG Image Collection; page 503 (top) Deborah Jaffe/Photodisc/AgeFotostock

Chapter 16

Pages 510-11: Joe McNally/Getty Images; page 510 (inset) Tim Laman/NG Image Collection; page 512: Maria Stenzel/NG Image Collection; page 518 (top) Courtesy Michael Ross, University of Florida; page 513: Library of Congress/NG Image Collection; page 524: Royalty-free/CORBIS; page 526: Mark Nielsen; page 530: Veronique Estiot/Photo Researchers, Inc.; page 534: PhotoDisc, Inc./Getty Images; page 547: Mauro Fermariello/Photo Researchers, Inc.; page 542 (bottom) Saturn Stills/Photo Researchers, Inc.; page 542 (top right) Gusto/Photo Researchers, Inc.; page 545 (bottom) Photodisc Blue/Getty Images; page 544: Saturn Stills/Photo Researchers, Inc.; page 545 (top) Aaron Haupt/Photo Researchers, Inc.; page 542 (top left) Brian Evans/Photo Researchers, Inc.; page 546: Mark Thomas/Photo Researchers, Inc.; page 539: Karen Kasmauski/NG Image Collection; page 543: Phanie/Photo Researchers, Inc.

Chapter 17

Pages 552-53: Jennifer Leigh Sauer/Getty Images; page 552 (inset) Pascal Goetgheluck/Photo Researchers, Inc.; page 555: David M. Phillips/Photo Researchers, Inc.; page 556: James A. Sugar/NG Image Collection; page 566 (bottom) Siu, Biomedical Comm./Custom Medical Stock Photo, Inc.; page 569 (top left) Photo provided courtesy of Kohei Shiota, Congenital Anomaly Research Center, Kyoto University, Graduate School of Medicine; page 569 (top right) Courtesy National Museum of Health and Medicine, Armed Forces Institute of Pathology; page 569 (bottom left) Courtesy National Museum of Health and Medicine, Armed Forces Institute of Pathology; page 569 (bottom right) Courtesy National Museum of Health and Medicine, Armed Forces Institute of Pathology; page 572: Bettmann/CORBIS; page 570 (left) Chris Lowe/PhototakeUSA.com; page 570 (center) Chris Lowe/PhototakeUSA.com; page 570: Chris Lowe/PhototakeUSA.com; page 573 (top) Photo by Lennart Nilsson/Albert Bonniers Fîrlag AB, A Child is Born, Dell Publishing Company. Reproduced with permission; page 573 (bottom left) Photo provided courtesy of Kohei Shiota, Congenital Anomaly Research Center, Kyoto University, Graduate School of Medicine; page 573 (bottom right) Photo provided courtesy of Kohei Shiota, Congenital Anomaly Research Center, Kyoto University, Graduate School of Medicine; page 580: Louie Psihoyos/Getty Images; page 579 (left) Stephen St. John/NG Image Collection; page 579 (right) Katrina Thomas/Photo Researchers, Inc.; page 581: Steve Raymer/NG Image Collection; page 582: Antonio Reeve/Science Photo Library/Photo Researchers, Inc.

Chapter 18

Pages 588-89: Zuma Press/Newscom; page 588 (inset) Alfred Pasieka/Photo Researchers, Inc.; page 597 (top right) Annie Griffiths Belt/NG Image Collection; page 597 (center right) Joel Sartore/NG Image Collection; page 597 (bottom right) Bruce Dale/NG Image Collection; page 600 (bottom) Science Source/Photo Researchers, Inc.; page 603: Eye of Science/Photo Researchers, Inc.; page 604: Grant Heilman Photography; page 606: Walt Spitzmiller; page 610 (top) Cold Springs Harbor Laboratory Archives/Photo Researchers, Inc.; page 610 (bottom) Ted Spiegel/Corbis; page 617: Shaun Best/Reuters/Corbis; page 620 (top) Michael Donne/Photo Researchers, Inc.; page 608: Lauren Shear/Photo Researchers, Inc.; page 620 (bottom) Courtesy Cellmark Diagnostics/Zeneca; page 623: L. Willatt/Photo Researchers, Inc.

Chapter 19

Pages 628-29: Karen Hunt/NG Image Collection; page 634 (top inset) SPL/Photo Researchers, Inc.; page 634 (bottom inset) Tim Laman/NG Image Collection; page 635: NG Image Collection; page 637 (left) Bill Hatcher/NG Image Collection; page 638: Victor R. Boswell/NG Image Collection; page 641 (top) O. Louis Mazzatenta/NG Image Collection; page 641 (bottom left) Ben Hall/The Image Bank/Getty Images; page 641 (bottom right) Robert Sisson/NG Image Collection; page 642 (left) Dr. John Cunningham/Visuals Unlimited; page 642 (top right) Perennou Nuridsany/Photo Researchers, Inc.; page 642 (bottom right) Dr. Fred Hossler/Visuals Unlimited; page 643 (right) Jim Richardson/NG Image Collection; page 648: Nick Nichols/NG Image Collection; page 645: Dimas Ardian/Getty Images; page 646: John Burcham/NG Image Collection; page 643: Bill Crutsinger/NG Image Collection; page 643: Tim Laman/NG Image Collection; page 643: Jonathan Blair/NG Image Collection; page 643: Stuart Westmorland/The Image Bank/Getty Images; page 643: First Light/Getty Images; page 643: Jonathan Blair/Corbis; page 643: © Gary Meszaros/Visuals Unlimited; page 643: Stuart Westmorland/Stone/Getty Images; page 643: Roy Toft/NG Image Collection; page 643: Kenneth Garrett/NG Image Collection; page 643 (top right) Ryan McVay/Stone/Getty Images; page 650 (bottom) NASA; page 653 (bottom) Eric V. Grave/Photo Researchers, Inc.; page 657 (bottom) Kenneth Garrett/NG Image Collection; page 658 (left) Kenneth Garrett/NG Image Collection; page 658 (right) The Kobal Collection, Ltd.; page 659 (top) Richard Nowitz/NG Image Collection

Chapter 20

Pages 664-65: Justin Guariglia/NG Image Collection; page 664 (inset) Jim Richardson/NG Image Collection; page 666: Raymond Gehman/NG Image Collection; page 669 (bottom right) Tim Laman/NG Image Collection; page 670 (left) Gary Braasch; page 670 (right) Gary Braasch; page 668 (left) Maria Stenzel/NG Image Collection; page 668 (center) James P. Blair/NG Image Collection; page 668 (center) Frans Lanting/Minden Pictures, Inc.; page 668 (center) Paul Nicklen/NG Image Collection; page 668 (right) Richard Olsenius/NG Image Collection; page 669 (top left) John Cunningham/Visuals Unlimited; page 669 (top center) Tom Bean/NG Image Collection; page 669 (top right) Jason Edwards/NG Image Collection; page 672: Michael Lustbader/Photo Researchers, Inc.; page 674: Bill Curtsinger/NG Image Collection; page 677: Courtesy Dr. Verena Tunnicliffe, University of Victoria; page 679 (bottom) M. Gillot/from Electron Microsopy 2e by John J. Bozzola and Lonnie D. Russell, Jones and Bartlett Publishers, Inc.; page 684: Daniel Berehulak/Getty Images; page 687: Leonardo Papini/SambaPhoto/Getty Images; page 688 (left) Ralph Lee Hopkins/NG Image Collection; page 688 (right) L. Lefkowitz/Taxi/Getty Images; page 694 (right) Peter Essick/NG Image Collection; page 694 (left) Courtesy Taro Takahashi, Lamont-Doherty Earth Observatory, Columbia University; page 696: Sarah Leen/NG Image Collection; page 697: David Grossman/Photo Researchers, Inc.; page 698: Gabriel Bouys/AFP/Getty Images; page 699: Robert Sisson/NG Image Collection; page 701 (top left) Roy Toft/National Geographic Society

ILLUSTRATION CREDITS

Chapter 1

Figure 1.1: Keith Kasnot. Figure 1.2: Jared Schneidman Design/Imagineering. Figure 1.4: Imagineering. Figure 1.5: Imagineering. Figure 1.6: Imagineering. Figure 1.8: Imagineering. Figure 1.9: from *Visualizing Environmental Science* by Linda R. Berg and Mary Catherine Hager, John Wiley & Sons, Copyright 2007. Figure 1.10: Imagineering.

Chapter 2

Figure 2.1: Imagineering. Figure 2.2B: Imagineering. Figures 2.4-2.6: Imagineering. Figure 2.10: Adapted from *Biology: Understanding Life* by Sandra and Brian Alters, John Wiley & Sons, Copyright 2005. Figures 2.12-2.14: Imagineering. Figures 2.16-2.18: Imagineering. Figure 2.20: Imagineering. Figure 2.22: Imagineering. Figures 2.24-2.25: Imagineering.

Chapter 3

Figures 3.1-3.2: Tomo Narashima. Figure 3.4: Imagineering. Figure 3.5: from *Biology: Understanding Life* by Sandra and Brian Alters, John Wiley & Sons Copyright 2006. Figure 3.8: from *Biology: Understanding Life* by Sandra and Brian Alters, John Wiley & Sons Copyright 2006. Figure 3.10: Tomo Narashima. Figures 3.11-3.12: Imagineering. Figure 3.13: Tomo Narashima. Figures 3.14-3.16: Imagineering. Figure 3.18: Imagineering. Table 3.1: Imagineering. Figure 3.21: Tomo Narashima. Figure 3.22: Imagineering.

Chapter 4

Figures 4.2-4.9: Imagineering. Figure 4.10: Kevin Somerville. Figure 4.11: Imagineering. Figure 4.12: Molly Borman. Figures 4.13-4.14: Imagineering.

Chapter 5

Figure 5.2: Leonard Dank/Imagineering. Figure 5.3: Imagineering. Figures 5.4-5.5: Kevin Somerville. Figure 5.6a, c: Kevin Somerville/Imagineering. Figure 5.7: Leonard Dank/Imagineering. Figures 5.8-5.9: Imagineering. Figure 5.11: Imagineering. Figure 5.12: Leonard Dank/Imagineering. Figures 5.13-5.30: Imagineering. Figure 5.32: Leonard Dank/Imagineering.

Chapter 6

Figures 6.1-6.3: Leonard Dank. Figure 6.4: Kevin Somerville. Figure 6.5: Kevin Somerville/Imagineering. Figure 6.6-6.7: Imagineering. Figure 6.8: Kevin Somerville. Figures 6.9-6.11: Imagineering. Figures 6.12-6.13: Jared Schneidman Design. Figure 6.16: Imagineering. Figure 6.17: Imagineering. Figure 6.18: Jared Schneidman Design.

Chapter 7

Figure 7.1: Kevin Somerville. Figure 7.2: Kevin Somerville/Imagineering. Figure 7.3a: Kevin Somerville. Table 7.1: Jared Schneidman Design/Imagineering. Table 7.2: Kevin Somerville/Imagineering. Figures 7.4-7.5: Imagineering. Figures: 7.7-7.10: Imagineering. Figure 7.13: Imagineering. Figures 7.14-7.15: Kevin Somerville/Imagineering. Figure 7.17-7.18: Kevin Somerville/Imagineering. Figure 7.19: Kevin Somerville/Imagineering. Figure 7.20a: Kevin Somerville. Figure 7.21: Kevin Somerville. Figure 7.22: Kevin Somerville/Imagineering. Table 7.5: Kevin Somerville/Imagineering. Figure 7.23: Steve Oh/Imagineering. Figure 7.24: Imagineering.

Chapter 8

Figure 8.1: Molly Borman. Figure 8.2: Molly Borman/Imagineering. Figure 8.4: Tomo Narashima. Figure 8.6: Tomo Narashima/Imagineering. Figure 8.7: Tomo Narashima/Imagineering/Sharon Ellis. Figure: 8.8: Tomo Narashima/Imagineering. Figure 8.9: Sharon Ellis/Imagineering. Table 8.1: Kevin Somerville. Figure 8.11: Imagineering. Figure 8.13: Sharon Ellis/Imagineering. Figure 8.14: Lynn O'Kelley/Imagineering.

Chapter 9

Figure 9.2: Imagineering. Figure 9.3: Jared Schneidman Design. Figure 9.5: Kevin Somerville. Figures 9.6-9.7: Kevin Somerville. Figures 9.9-9.10: Kevin Somerville/Imagineering. Figure 9.11: Kevin Somerville. Figure 9.12: Imagineering. 9.13: Imagineering. Figure 9.14: from *Cell and Molecular Biology: Concepts and Experiments*, fourth edition by Gerald Karp, John Wiley & Sons, Copyright 2005. Figure 9.15: Molly Borman.

Chapter 10

Figure 10.1: Molly Borman/Imagineering. Figures 10.2-10.3: Sharon Ellis/Imagineering. Figure 10.4a: Imagineering. Figure 10.5: Molly Borman. Figures 10.12-10.14: Imagineering. Figure 10.15: from *Microbiology: Principles and Explorations*, sixth edition by Jacquelyn Black, John Wiley & Sons, Copyright 2005. Figure 10.16: from *Microbiology: Principles and Explorations*, sixth edition by Jacquelyn Black, John Wiley & Sons, Copyright 2005. Figure 10.17: from *Microbiology: Principles and Explorations*, sixth edition by Jacquelyn Black, John Wiley & Sons, Copyright 2005. Figure 10.18: from *Microbiology: Principles and Explorations*, sixth edition by Jacquelyn Black, John Wiley & Sons, Copyright 2005. Figure 10.19: Imagineering. Figure 10.20: from *Microbiology: Principles and Explorations*, sixth edition by Jacquelyn Black, John Wiley & Sons, Copyright 2005. Figure 10.25: : from *Microbiology: Principles and Explorations*, sixth edition by Jacquelyn Black, Wiley & Sons, Copyright 2005. Figure 10.26: from *Microbiology: Principles and Explorations*, sixth edition by Jacquelyn Black, John Wiley & Sons, Copyright 2005. Figure 10.27: from *Microbiology: Principles and Explorations*, sixth edition by Jacquelyn Black, John Wiley & Sons, Copyright 2005. Figure 10.28: Nadine Sokol/Imagineering. Figures 10.29-10.30: Imagineering.

Chapter 11

Figure 11.1: Imagineering. Figure 11.2: Kevin Somerville/Imagineering. Figure 11.3: Kevin Somerville. Figure 11.4: Kevin Somerville/Imagineering. Figure 11.5: Kevin Somerville. Figure 11.6: Kevin Somerville/Imagineering. Figure 11.7: Imagineering. Figure 11.9: Kevin Somerville/Imagineering. Figure 11.12: Kevin Somerville. Figure 11.13: Imagineering and

Jared Schneidman Design. Figure 11.14: Imagineering. Figure 11.15: Kevin Somerville/Imagineering. Figures 11.16-17: Kevin Somerville. Figure 11.20: Hilda Muinos/Imagineering. Figures 11.12, 11.23, Table 11.1: Imagineering. Figure 11.26: Jared Schneidman Design, Figure 11.28: Jean Jackson. Figure 11.29 Nadine Sokol. Figure 11.30: Imagineering.

Chapter 12

Figure 12.1: Molly Borman. Figure 12.2: Adapted from Kevin Somerville/Imagineering. Figures 12.4, 12.5: Molly Borman/Imagineering. Figure 12.6: Imagineering. Figure 12.7: Molly Borman. Figure 12.9: Molly Borman/Imagineering. Figure 12.10: Imagineering. Figure 12.12: Imagineering. Figure 12.13, 12.14: Kevin Somerville/Imagineering. Figure 12.15: Jared Schneidman Design. Figure 12.16: Imagineering. Figure 12.17: Jared Schneidman Design. Figure 12.18: Imagineering. Figures 12.20, 12.22, 12.24-25: Jared Schneidman Design. Figure 12.26: Kevin Somerville.

Chapter 13

Figure 13.1: Imagineering. Figure 13.2: from *Nutrition: Everyday Choices* by Mary Grosvenor and Lori Smolin, John Wiley & Sons, Copyright 2006. Figure 13.3: from *Nutrition: Science and Applications* by Lori Smolin and Mary Grosvenor, John Wiley & Sons, Copyright 2008. Figure 13.4: Steve Oh. Figure 13.5: Kevin Somerville. Figure 13.6: Nadine Sokol. Figure 13.8: Nadine Sokol. Figure 13.10: Nadine Sokol. Figure 13.12: Steve Oh. Figure 13.14: Imagineering. Figure 13.22: from *Nutrition: Everyday Choices*, by Mary Grosvenor and Lori Smolin, John Wiley & Sons Copyright 2006.

Chapter 14

Figure 14.1: Kevin Somerville. Figure 14.2: Kevin Somerville/Imagineering. Figure 14.3: Steve Oh. Figure 14.4: Steve Oh/Imagineering. Figures 14.5-14.7: Imagineering. Figures 14.8-14.11: Jared Schneidman Design. Figure 14.15: Kevin Somerville/Imagineering. Figure 14.16: Jared Schnedman Design. Figure 14.17: Imagineering. Figures14.18-14.20: Jared Schneidman Design.

Chapter 15

Figure 15.1: Steve Oh/Imagineering. Figure 15.2: Imagineering. Figure 15.3: Imagineering. Figure 15.4: Lynn O'Kelley/Imagineering. Figure 15.5: Nadine Sokol. Figures 15.8-15.9: Molly Borman/Imagineering. Figure 15.10: Imagineering. Figure 15.15: Molly Borman/Imagineering. Figure 15.16: Imagineering. Figure 15.19: Imagineering.

Chapter 16

Figure 16.3: Imagineering. Figures 16.4-16.5: Kevin Somerville/Imagineering. Figures 16.6-16.9: Imagineering. Figure 16.11: Imagineering. Figure 16.12: Kevin Somerville/Imagineering. Figure 16.13: Imagineering. Figures 16.14-16.15: Kevin Somerville/Imagineering. Figure 16.17: Kevin Somerville. Figure 16.18: Kevin Somerville/Imagineering. Figure 16.19: Imagineering. Figure 16.20: from *Nutrition: Everyday Choices* by Mary Grosvenor and Lori Smolin, John Wiley & Sons Copyright 2006. Figure 16.21: *Biology: Understanding Life* by Sandra and Brian Alters, John Wiley & Sons, Copyright 2006. Figure 16.22: *Biology: Understanding Life* by Sandra and Brian Alters, John Wiley & Sons Copyright 2006.Unnumbered figure Laparoscopy: Imagineering.

Chapter 17

Figure 17.2: Kevin Somerville. Figure 17.3: Kevin Somerville/Imagineering. Figures 17.4-17.5: Kevin Somerville. Figure 17.6: Imagineering. Figures 17.7-17.8: Kevin Somerville. Figure 17.9: from *Biology: Understanding Life* by Sandra and Brian Alters, John Wiley & Sons, Copyright 2006. Figure 17.10, 17.3: Kevin Somerville. Figure 17.15: Jared Schneidman Design. Figure 17.16: Kevin Somerville. Figure 17.9: *Biology: Understanding Life* by Sandra and Brian Alters, John Wiley & Sons, Copyright 2006.

Chapter 18

Figure 18.1: Imagineering. Figure 18.2: from *Biology: Understanding Life* by Sandra and Brian Alters, John Wiley & Sons, Copyright 2006. Figure 18.3: from *Biology: Understanding Life* by Sandra and Brian Alters, John Wiley & Sons, Copyright 2005. Figure 18.4: Jared Schneidman Design. Figure 18.5: Imagineering. Figure 18.6: Jared Schneidman Design. Figure 18.7: from *Biology: Understanding Life* by Sandra and Brian Alters, John Wiley & Sons, Copyright 2005. Figure 18.8: from *Principles of Genetics*, Fourth edition by Peter Snustad and Michael Simmons, John Wiley & Sons, Copyright 2006. Figure 18.10: from *Biology: Understanding Life* by Sandra and Brian Alters, John Wiley & Sons, Copyright 2006. Figure 18.13: Imagineering. Figure 18.14: Imagineering. Figure 18.16: from *Biology: Understanding Life* by Sandra and Brian Alters, John Wiley & Sons, Copyright 2006. Figure 18.17: from *Biology: Understanding Life* by Sandra and Brian Alters, John Wiley & Sons, Copyright 2006. Figure 18.18: from *Biology: Understanding Life* by Sandra and Brian Alters, John Wiley & Sons, Copyright 2006. Figure 18.19: from *Biology: Understanding Life*, by Sandra and Brian Alters, John Wiley & Sons, Copyright 2006. Figure 18.21: from *Biology: Understanding Life*, by Sandra and Brian Alters, John Wiley & Sons, Copyright 2006.

Chapter 19

Figure 19.2: National Geographic Society. Figure 19.4: Imagineering. Figures 19.6-19.7: from *Biology: Understanding Life* by Sandra and Brian Alters, John Wiley & Sons, Copyright 2006. Figure 19.11: from *Biology: Understanding Life*, by Sandra and Brian Alters, John Wiley & Sons, Copyright 2006. Figure 19.14: Adapted from *Biology: Understanding Life*, by Sandra and Brian Alters, John Wiley & Sons, Copyright 2006. Figure 19.17: from *Biology: Understanding Life* by Sandra and Brian Alters, John Wiley & Sons, Copyright 2006. Figure 19.20: from *Biology: Understanding Life* by Sandra and Brian Alters, John Wiley & Sons, Copyright 2005. Figure 19.21: from *Biology: Understanding Life* by Sandra and Brian Alters, John Wiley & Sons, Copyright 2006.

Chapter 20

Figure 20.2: *Biology: Understanding Life* by Sandra and Brian Alters, John Wiley & Sons, Copyright 2006. Figure 20.5: Botkin and Keller, Environmental Science, Copyright 2006. Figures 20.7, 20.9, 20.11-17 *Biology: Understanding Life* by Sandra and Brian Alters, John Wiley & Sons, Copyright 2006. Table 20.3: Raven and

Berg, Environment, Copyright 2006. Figure 20.20 from *Visualizing Environmental Science* by Linda R. Berg and Mary Catherine Hager, John Wiley & Sons, Copyright 2007. Figure 20.21 from *Visualizing Environmental Science* by Linda R. Berg and Mary Catherine Hager John Wiley & Sons, Copyright 2007. Figure 20.22 *Visualizing Environmental Science* by Linda R. Berg and Mary Catherine Hager, John Wiley & Sons, Copyright 2007. Figure 20.23: *Visualizing Environmental Science* by Linda R. Berg and Mary Catherine Hager, John Wiley & Sons, Copyright 2007. Table 20.4: Adapted from State of the Air 2004: Metropolitan areas most polluted by year-round particle pollution, The American Lung Association, © 2007, www.lungusa.org. 20.29: *Biology: Understanding Life* by Sandra and Brian Alters, John Wiley & Sons, Copyright 2006. 20.30: *Environment* by Peter H. Raven and Linda R. Berg, John Wiley & Sons, Copyright 2006.

INDEX

In this index, page ranges in boldface type indicate a major discussion of the topic. A page number followed by an "f" indicates a figure, and one followed by a "t" indicates a table.

N

O

P

Q

R